4th Symposium on Prospects in the Physics of Discrete Symmetries

(DISCRETE2014)

AA003038

Journal of Physics: Conference Series Volume 631

London, United Kingdom
2-6 December 2014

Part 1 of 2

Editors:

Antonio Di Demenico
Vasiliki A. Mitsou
Nick E. Mavronatos
Dimitri Skliros

ISBN: 978-1-5108-0983-3
ISSN: 1742-6588

Printed from e-media with permission by:

Curran Associates, Inc.
57 Morehouse Lane
Red Hook, NY 12571

Some format issues inherent in the e-media version may also appear in this print version.

Printed by Curran Associates, Inc. (2015)

Institute of Physics
Dirac House, Temple Back
Bristol BS1 6BE UK

Phone: 44 1 17 929 7481
Fax: 44 1 17 920 0979

techtracking@iop.org

Additional copies of this publication are available from:

Curran Associates, Inc.
57 Morehouse Lane
Red Hook, NY 12571 USA
Phone: 845-758-0400
Fax: 845-758-2634
Email: curran@proceedings.com
Web: www.proceedings.com

Table of contents

Volume 631

4th Symposium on Prospects in the Physics of Discrete Symmetries (DISCRETE2014)
2–6 December 2014, London, UK

Accepted papers received: 24 June 2015
Published online: 30 July 2015

PARALLEL SESSIONS

Discrete symmetries (T, C, P), flavour, accidental symmetries

Higgs physics at LHC, discrete symmetries at LHC, new facilities

Experiments – Discrete Symmetries

Strongly Coupled gauge Theories

Discrete Symmetries in Strings and in GUT theories

Early universe Physics: Inflation, Lepto(Baryo)genesis

Journal of Physics: Conference Series **631** (2015) 011001 doi:10.1088/1742-6596/631/1/011001

Preface

The DISCRETE 2014: Fourth Symposium in the Physics of Discrete Symmetries took place at King's College London, Strand Campus, London WC2R 2LS, from Tuesday, December 2 2014 till Saturday, December 6 2014. This is the fourth Edition of the DISCRETE conference series, which is a biannual event, having been held previously in Valencia (Discrete'08), Rome (Discrete2010) and Lisbon (Discrete2012).

The topics covered at the DISCRETE series of conferences are: T, C, P, CP symmetries; accidental symmetries (B, L conservation); CPT symmetry, decoherence and entangled states, Lorentz symmetry breaking (phenomenology and current bounds); neutrino mass and mixing; implications for cosmology and astroparticle physics, dark matter searches; experimental prospects at LHC, new facilities. In DISCRETE 2014 we have also introduced two new topics: cosmological aspects of non-commutative space-times as well as PT symmetric Hamiltonians (non-Hermitian but with real eigenvalues), a topic that has wide applications in particle physics and beyond.

The conference was opened by the King's College London Vice Principal on Research and Innovation, Mr Chris Mottershead, followed by a welcome address by the Chair of DISCRETE 2014 (Professor Nick E. Mavromatos). After these introductory talks, the scientific programme of the DISCRETE 2014 symposium started. Following the tradition of DISCRETE series of conferences, the talks (138 in total) were divided into plenary-review talks (25), invited research talks (50) and shorter presentations (63) - selected by the conveners of each session in consultation with the organisers - from the submitted abstracts. We have been fortunate to have very high-quality, thought stimulating and interesting talks at all levels, which, together with the discussions among the participants, made the conference quite enjoyable. There were 152 registered participants for the event.

The first scientific talk of the conference was the plenary talk by Clerk Maxwell Professor of Physics at King's College London, Professor John Ellis, FRS, CBE, who lectured on the status of the Physics landscape after the Higgs discovery, putting emphasis on Discrete Symmetry aspects; the talk was followed by another plenary talk by Professor Sir Tajinder (Jim) Virdee, summarising the current status of the searches for new Physics in colliders, in particular LHC, and future perspectives; then we have had a special talk by Professor Subir Sarkar on the status of precision cosmological measurements using the cosmic microwave background and their implication for particle physics and inflationary models.

This year at King's College London we also celebrate the 150th anniversary of the publication of Maxwell's equations, while James Clerk Maxwell was the Professor of Natural Philosophy here. Moreover it is the 60th Anniversary of the discovery of Yang-Mills Gauge theories, the 50th anniversary of the discovery of CP Violation and the 50th anniversary of the proposal of Bell's Inequalities in Quantum Physics. Hence DISCRETE 2014 has been combined with the celebration of the above anniversaries.

Journal of Physics: Conference Series **631** (2015) 011001 doi:10.1088/1742-6596/631/1/011001

In particular, we have had a whole session with invited talks on Yang-Mills theories, starting with a historical lecture by Professor Sir Tom Kibble on the history of spontaneous symmetry breaking for non-Abelian gauge theories, where, as one of the proponents of the mechanism, he gave, apart from the underlying science, his own account of events and facts that lead to the discovery, thus constituting an important contribution to the history of Science that will be quite useful for future generations of physicists. This work has been published as part of the DICE2014 proceedings (T W B Kibble 2015 *J. Phys.: Conf. Ser.* **626** 012001) at http://iopscience.iop.org/1742-6596/626/1/012001. The special session on Yang-Mills theory covered the state of the art of the field, in particular its non-perturbative aspects relevant to strongly coupled gauge theories and their future perspectives, including also experimental talks on searches for Physics beyond the Standard Model at the CERN LHC and in future colliders, as well as on the status of quark-gluon plasma and in general of the heavy-ion programme at the CERN LHC. There were also talks on experimental and theoretical advances in Higgs physics at LHC and beyond, with emphasis on discrete symmetry aspects.

The celebration of 60 Years of CP Violation started with a talk by Prof. Cecilia Jarlskog, on the history of CP Violation, where Prof. Jarlskog, as a leading expert in the field but also a former member of the Nobel Prize Committee, presented quite useful details on the road to the discovery and its aftermath. As dedicated to one of the main themes of the DISCRETE conference series, the session encompassed several topics on CP, and T Violation by leading experts on the field, with reports on the latest theoretical and experimental results, including prospects for the observation of CP Violation in the leptonic sector of the Standard Model and beyond. In a more speculative part of the conference, we also had quite interesting talks on the phenomenology and current constraints of Lorentz Invariance and/or CPT Violation. We also have had several important talks on discrete symmetries, including flavour, in supersymmetric extensions of the Standard Model and Grand Unified Theories. There was a whole parallel session dedicated to man-made antimatter factories, and tests of CPT and in general comparison of basic properties of matter with antimatter.

The session on foundations of quantum theory celebrating 50 years of Bell's inequalities included important talks summarising the state of the art of tests of Bell's inequalities using particle physics systems, such as neutral Kaons, as well as talks dedicated to experimental falsification of theoretical models involving wave-function collapse and tests of the Pauli principle via the VIP2 experiment. There was also a theoretical talk discussing compatibility of discrete cellular automata with quantum mechanics.

Several other parallel sessions of the conference were dedicated to neutrino physics (ranging from models of neutrino masses and mixing to neutrino experiments and neutrino astrophysics), astroparticle physics experiments, including very high energy cosmic probes, as well as dark matter searches, direct and indirect. As we have mentioned above, we also hosted for the first time a session on PT symmetric Hamiltonians, including experimental and theoretical talks on this relatively novel but quite rich field of research.

As the year 2015 is also the 100th anniversary of the theory of General Relativity, we linked this event to our conference by having important lectures on the status of Supergravity theories, a local version of supersymmetry which incorporates gravity, as well as string theory, with emphasis on discrete symmetries. In the parallel sessions we also have had highlight talks on experimental searches for supersymmetry at LHC and future colliders. Moreover, there were plenary, invited and contributed talks dedicated to cosmology (including models in non-commutative space times), inflation, novel theories of gravity beyond Einstein's theory, TeV scale Baryogenesis, Leptogenesis and the role of neutrinos in the early universe.

On the plenary session of the last day of the conference there was also a highlight talk on experimental searches for magnetic monopoles and other highly ionising particles at the CERN LHC by Professor James Pinfold, the spokesperson of the MoEDAL, an LHC experiment dedicated to such searches, which has just commenced operation and in which King's College London particle theory and cosmology group plays a leading role. The conference ended with a plenary talk by Professor Jose Bernabeu, chair of the steering committee of the DISCRETE series, on the status and future prospects of the discrete symmetries C,P and T, and CPT in particle physics.

DISCRETE 2014 was also combined with two cultural events, providing a nice mixture of art with science. The first event was the association of the conference with the Art and Science Project of the celebrated Venetian-Glass artist Ms Laura Pesce, whose kind contribution was the production of small hand-made glass artworks for the participants, reflecting the main theme of the conference, discrete symmetries, in particular time-reversal, CP violation and time irreversibility, and their connection to the Universe. Each participant got a personalized item; the artworks were all different, and expressed the influences of the artist by the science of the topics discussed in the conference. The second event was an evening Piano recital, held in the Great Hall of King's College London, in which a soprano (Ms Katerina Fürstová) accompanied by a pianist (who is also a particle theorist and one of our invited speakers, Professor Massimo Blasone) sang songs based on musical works of Mozart, Schubert and Chopin, as well as some Neapolitan Folk Songs. Last but not least the Conference Dinner was held on the evening of Friday 5th December, in the headquarters of the Royal Society of London. The after-dinner speech was given by King's College London Clerk Maxwell Professor, Professor John Ellis, FRS, CBE.

The organisers wish to thank the conveners, speakers and participants of DISCRETE 2014 for providing an intellectually challenging and thought stimulating atmosphere through the scientific presentations and discussions during this exciting meeting, thus contributing to its success. The meeting kept the tradition of the DISCRETE series of conferences in providing a research output of very high standards. As editors, we are indebted to all the contributors of this volume for their prompt and diligent submission of very high quality papers, covering the state of the art of the respective fields.

London, June 2015

The Editors,

Antonio Di Domenico (U. Roma-Sapienza and INFN),
Nick E. Mavromatos (Chair, King's College London),
Vasiliki A. Mitsou (IFIC-Valencia),
Dimitri Skliros (King's College London).

Journal of Physics: Conference Series **631** (2015) 011001 doi:10.1088/1742-6596/631/1/011001

Conference photographs

Conference Structure and Sponsors

Local Organising Committee

Jean Alexandre
Nick E. Mavromatos – *Chair*
Sarben Sarkar – *Co-chair*
Dimitri Skliros

Organisation Support Team

Krzysztof Bozek
Marco de Cesare
Nicholas Houston
Julio Leite
Tevong You

Conference Secretariat

Julia Kilpatrick

e-mail: `Julia.Kilpatrick@kcl.ac.uk`
web page: `https://indico.cern.ch/e/discrete2014`

Session Conveners

B.S. Acharya (KCL London & ICTP)
J. Alexandre (King's College London)
C.M. Bender (U. Washington St. Louis)
J. Bernabeu (U. Valencia)
A. Di Domenico (U. Roma Sapienza)
A. Faraggi (U. Liverpool)
V.A. Mitsou (IFIC Valencia)
J. Papavassiliou (U. Valencia)
A. Pilaftsis (U. Manchester)
M.N. Rebelo (IST Lisbon)

International Advisory Committee

Francisco del Aguila (Granada)
Luis Alvarez-Gaume (CERN)
Jose Bernabeu (Valencia)
Gustavo C. Branco (Lisbon)
Andrzej J. Buras (Munich)
Antonio Di Domenico (Rome)
Jonathan R. Ellis (CERN & KCL)
Domenec Espriu (Barcelona)
David G. Hitlin (Caltech)
Cecilia Jarlskog (Lund)
Antonio Masiero (Padua)
Nikolaos E. Mavromatos (KCL)
Vasiliki A. Mitsou (Valencia)
Hitoshi Murayama (Berkeley)
Keith Olive (Minnesota)
Antonio Pich (Valencia)
Apostolos Pilaftsis (Manchester)
Mariano Quiros (Barcelona)
Subir Sarkar (Oxford)
Fabio Zwirner (Padua)

DISCRETE Symposia Steering Committee

J. Bernabeu (Valencia) – *Chair*
G. Branco (IST Lisbon)
A. Di Domenico (U. Roma Sapienza)
G. Isidori (LN Frascati)
N.E. Mavromatos (King's College London)
V.A. Mitsou (IFIC Valencia)
M.N. Rebelo (IST Lisbon)

Plenary Speakers

B.S. Acharya (KCL London and ICTP)
C.M. Bender (U. Washington St. Louis)
J. Bernabeu (U. Valencia)
A.J. Bevan (Queen Mary U. of London)
G. Branco (IST Lisbon)
J.P. Derendinger (U. Bern)
A. Di Domenico (U. Roma Sapienza)
J.F.R. Ellis (KCL and CERN)
D. Evans (U. Birmingham)
A. Faraggi (U. Liverpool)
N. Harnew (U. Oxford)
C. Jarlskog (U. Lund)
T. Kibble (Imperial College London)
S.F. King (U. Southampton)
M. Krawczyk (U. Warsaw)
S. Liberati (SISSA)
R. Mohapatra (U. Maryland)
J. Papavassiliou (U. Valencia)
S. Pascoli (U. Durham)
A. Pilaftsis (U. Manchester)
J.A. Pinfold (U. Alberta)
M. Sakellariadou (King's College London)
Subir Sarkar (U. Oxford & Niels Bohr Institute)
M. Trodden (U. Pennsylvania)
T. Virdee (Imperial College London)

Invited Research Talks

A.C. Aguilar (U. Campinas)
A. Aranda (U. Colima)
P. Athanasopoulos (U. Liverpool)
W.A. Bertsche (U. Manchester & Daresbury Lab.)
D. Binosi (ECT Trento & Fond. Bruno Kessler)
M. Blanke (CERN & Karlsruher Inst. Tech.)
M. Blasone (U. Salerno)
F. Botella (U. Valencia & IFIC)
V. Branchina (U. Catania)
J. Casaus (CIEMAT, Madrid)
M.-C. Chen (U. California Irvine)
C. Curceanu (LNF-INFN)
E. Daw (U. Sheffield)
J.S. Diaz (Karlsruher Institut für Technologie)
G. Eigen (U. Bergen)
H.T. Elze (U. Pisa)
I. Florakis (CERN)
S. Groot Nibbelink (U. Munich)
I. Garzia (U. Ferrara)
F. Hahn (CERN)
B. Hiesmayr (U. Vienna)
G. Honecker (Johannes Gutenberg Univ. Mainz)
V.V. Konotop (U. Lisbon)
S. Krippendorf (U. Oxford)
K. Langfeld (U. Plymouth)
G. Leontaris (U. Ioannina)
B. Lucini (U. Swansea)
C. Luhn (U. Siegen)
P. Mannheim (U. Connecticut)
J. Marton (Austrian Academy of Sciences)
J. Matias (U. Autonoma Barcelona)
V. Mehta (U. Heidelberg)
P. Millington (Technische Univ. München)
V.A. Mitsou (IFIC Valencia)
M. Mondragon (UNAM, Mexico)
L. Nellen (UNAM, Mexico)
S. Olejnik (Slovak Academy of Sciences)
O. Philipsen (Goethe-University Frankfurt)
A. Rajantie (Imperial College London)
M.N. Rebelo (IST Lisbon)
N. Rius (IFIC, Valencia)
J. Rodriguez Quintero (U. Huelva)
J. Rubio (EPFL)
H. Sanchis Alepuz (Justus-Liebig Un. Giessen)
F. Siringo (U. Catania & INFN)
H. Sonmez (U. Liverpool)
D. Teresi (U. Manchester)
G. Vitiello (U. Salerno)
E. Widmann (Austrian Academy of Sciences)

Participants

Name	Institute
ABEL, Steven Adam	University of Durham, Durham, United Kingdom
AGUILAR, Arlene Cristina	University of Campinas - Unicamp Campinas, Brazil
ALEXANDRE, Jean	King's College London, London, United Kingdom
ARANDA, Alfredo	Universidad de Colima, Colima, Mexico
ATHANASOPOULOS, Panos	University of Liverpool, Liverpool, United Kingdom
BARNBY, Lee	University of Birmingham, Birmingham, United Kingdom
BENAKLI, Karim	LPTHE, Paris, France
BENDER, Carl	Washington University in St. Louis, St. Louis, United States
BERNABEU, Jose	Instituto de Fisica Corpuscular, Valencia, Spain
BERTSCHE, William Alan	University of Manchester, Geneva, Switzerland
BEVAN, Adrian	Queen Mary University of London, London, United Kingdom
BINOSI, Daniele	ECT* Fondazione Bruno Kessler, Trento, Italy
BJÖRKEROTH, Fredrik	University of Southampton, Southampton, United Kingdom
BLANKE, Monika	CERN & KIT, Karlsruhe, Germany
BLASONE, Massimo	Università di Salerno, Salerno, Italy
BLOISE, Caterina	Istituto Nazionale Fisica Nucleare, Frascati, Italy
BOTELLA OLCINA, Francisco Jose	Instituto de Fisica Corpuscular, Valencia, Spain
BOZEK, Krzysztof	King's College London, London, United Kingdom
BRANCHINA, Vincenzo	University of Catania, Catania, Italy
BRANCO, Gustavo C.	CFTP / IST/ U. Lisboa, Lisboa, Portugal

cont'ed on next page

Journal of Physics: Conference Series **631** (2015) 011001 doi:10.1088/1742-6596/631/1/011001

cont'ed from previous page

Name	**Institute**
BRISTER, James	King's College London, London, United Kingdom
CASAUS, Jorge	CIEMAT, Madrid, Spain
CERDENO, David G.	University of Durham, Durham, United Kingdom
CHALA, Mikael	DESY, Hamburg, Germany
CHEN, Mu-Chun	University of California, Irvine, United States
CLOUGH, Katy	King's College London, London, United Kingdom
CRISPIM ROMAO, Miguel	University of Southampton, Southampton, United Kingdom
CRISTINA, Volpe	Astroparticule et Cosmologie (APC), Paris, France
CURCEANU, Catalina Oana	LNF-INFN Frascati, Roma, Italy
D'AGOSTINI, Giulio	University La Sapienza & INFN Roma 1, Roma, Italy
D'OLIVO, Juan Carlos	Depto. de Física Teórica, Universidad de Valencia, Valencia, Spain
DAW, Ed	University of Sheffield, Sheffield, United Kingdom
DAY, Melanie	IceCube, Madison, WI, United States
DE ANDA, Francisco J.	Universidad de Guadalajara, Tepatitlán, Mexico
DE CESARE, Marco	King's College London, London, United Kingdom
DE MEDEIROS VARZIELAS, Ivo	University of Southampton, Southampton, United Kingdom
DERENDINGER, Jean-Pierre	Albert Einstein Center, Bern University, Bern, Switzerland
DEV, Bhupal	University of Manchester, Manchester, United Kingdom
DI DOMENICO, Antonio	Sapienza University of Rome & INFN sez., Roma, Italy
DIAZ, Jorge S.	Karlsruhe Institute of Technology, Karlsruhe, Germany
DIMOU, Maria	Southampton University, Southampton, United Kingdom
DUERR, Michael	MPIK Heidelberg, Heidelberg, Germany
EFRATI, Aielet	WIS, Rehovot, Israel
EIGEN, Gerald	University of Bergen, Bergen, Norway
ELLIS, Jonathan	CERN and King's College London, Geneva, Switzerland
ELZE, Hans-Thomas	Università di Pisa, Pisa, Italy
EVANS, David	University of Birmingham, Birmingham, United Kingdom
FARAGGI, Alon	University of Liverpool, Liverpool, United Kingdom
FARAKOS, Fotis	Masaryk University of Brno, Brno, Czech Republic
FIGUEIREDO, António	CFTP / IST / U. Lisbon, Lisbon, Portugal
FLECHL, Martin	Austrian Academy of Sciences, Vienna, Austria
FLORAKIS, Ioannis	CERN, Geneva, Switzerland
FRANKE, Michael	Technical University Munich, Munich, Germany
FUENTES-MARTIN, Javier	Instituto de Física Corpuscular, Valencia, Spain
GAJOS, Aleksander	Jagiellonian University, Krakow, Poland
GARGIULO, Maria Vittoria	Università degli Studi di Salerno, Salerno, Italy
GARZIA, Isabella	INFN-Ferrara, Ferrara, Italy
GHEZELBASH, Masoud	University of Saskatchewan, Saskatoon, Canada
GIRARDI, Ivan	SISSA, Trieste, Italy
GOMEZ, Melina	Universidad de las Americas Puebla, Cholula, Puebla, Mexico
GRAZIANI, Enrico	Roma Tre Università Degli Studi, Roma, Italy
GROOT NIBBELINK, Stefan	Munich University, Munich, Germany
HAGEDORN, Claudia	Excellence Cluster 'Universe', TU Munich, Munich, Germany
HAHN, Ferdinand	CERN, Geneva, Switzerland
HARNEW, Neville	University of Oxford, Oxford, United Kingdom
HEAL, John	King's College London, London, United Kingdom
HEWITT, Mike	Canterbury Christ Church University, Canterbury, United Kingdom
HICKERSON, Kevin	University of California Los Angeles, Los Angeles, United States
HIESMAYR, Beatrix	University of Vienna, Vienna, Austria
HONECKER, Gabriele	Johannes Gutenberg University Mainz, Mainz, Germany
HORI, Masaki	Max-Planck Institute of Quantum Optics, Garching, Germany

cont'ed on next page

Journal of Physics: Conference Series **631** (2015) 011001 doi:10.1088/1742-6596/631/1/011001

cont'ed from previous page

Name	**Institute**
HOUSTON, Nick	KCL, London, United Kingdom
JARLSKOG, Cecilia	Lund University, Lund, Sweden
KAMINSKA, Daria	Institute of Physics, Jagiellonian University, Krakow, Poland
KIMURA, Mitsuhiro	Universitaet Bern, Bern, Switzerland
KING, Stephen	Department of Physics (SHEP), Southampton, United Kingdom
KONOTOP, Vladimir	Universidade de Lisboa, Lisboa, Portugal
KRAWCZYK, Maria	University of Warsaw, Warsaw, Poland
KRIPPENDORF, Sven	University of Oxford, Oxford, United Kingdom
LAMBIASE, Gaetano	Università di Salerno, Salerno, Italy
LANCASTER, Mark	University College London, London, United Kingdom
LANGFELD, Kurt	Plymouth University, Plymouth, United Kingdom
LATHAM, Thomas	University of Warwick, Coventry, United Kingdom
LAVEDER, MARCO	University of Padua and INFN, Padua, Italy
LEITE, Julio	KCL, London, United Kingdom
LENTI, Massimo	INFN Sezione di Firenze, Firenze, Italy
LEONARDO, Nuno	Purdue University, IN, United States
LEONTARIS, Georgios	University of Ioannina, Ioannina, Greece
LIBERATI, Stefano	SISSA, Trieste, Italy
LONGHIN, Andrea	Istituto Nazionale Fisica Nucleare, Frascati, Roma, Italy
LUCINI, Biagio	Swansea University, Swansea, United Kingdom
LUHN, Christoph	University of Siegen, Siegen, Germany
MALEK, Fairouz	LPSC-Grenoble, CNRS-IN2P3, Grenoble, France
MANNHEIM, Philip	University of Connecticut, Storrs, United States
MARRONE, Antonio	University of Bari, Bari, Italy
MARTON, Johann	Austrian Academy of Sciences, Vienna, Austria
MASSRI, Karim	University of Birmingham, Birmingham, United Kingdom
MATIAS, Joaquim	Universitat Autonoma Barcelona, Barcelona, Spain
MAVROMATOS, Nikos	King's College London, University of London, United Kingdom
MEADOWCROFT, Andrew	University of Southampton, Southampton, United Kingdom
MEHTA, Viraf	Universität Heidelberg, Heidelberg, Germany
MILLINGTON, Peter	Technische Universität München (TUM), München, Germany
MITSOU, Vasiliki	IFIC Valencia, Valencia, Spain
MOHAPATRA, Rabindra	University of Maryland, College Park, United States
MONDRAGON, Myriam	Instituto de Fisica, UNAM, Mexico City, Mexico
MOULIN, Emmanuel	CEA Saclay, Gif-sur-Yvette, France
NEBOT GOMEZ, Miguel	CFTP - Lisbon, Lisbon, Portugal
NEDER, Thomas	University of Southampton, Southampton, United Kingdom
NELLEN, Lukas	Universidad Nacional Autonoma, Mexico City, Mexico
OLEJNIK, Stefan	Institute of Physics, Slovak Acad. Sci., Bratislava, Slovakia
PALOMARES-RUIZ, Sergio	IFIC-Valencia, Valencia, Spain
PAPAVASSILIOU, Joannis	University of Valencia, Valencia, Spain
PETROV, Yevgeniy	University of British Columbia, Vacnouver, Canada
PHILIPSEN, Owe	Goethe-University Frankfurt, Frankfurt, Germany
PILAFTSIS, Apostolos	University of Manchester, Manchester, United Kingdom
PILLONI, Alessandro	Sapienza U., Rome, Italy
PINFOLD, James	University of Alberta, Edmonton, Canada
PONGKITIVANICHKUL, Chakrit	King's College London, London, United Kingdom
POTTER, Franklin	Sciencegems.com, Huntington Beach, United States
RAJANTIE, Arttu	Imperial College / King's College London, United Kingdom
RAMIREZ, Monica F.	Universidad de Guadalajara, Guadalajara, Mexico
RAMMENSEE, Michael Christian	CERN, Geneva, Switzerland

cont'ed on next page

cont'ed from previous page

Name	**Institute**
REBELO, Margarida Nesbitt	CFTP/IST/UL, Lisboa, Portugal
RIUS, Nuria	Valencia University, Valencia, Spain
RODRIGUEZ QUINTERO, José	University of Huelva, Huelva, Spain
RUBIO, Javier	EPFL, Lausanne, Switzerland
SAKELLARIADOU, Mairi	King's College London, London, United Kingdom
SANCHIS ALEPUZ, Hèlios	Justus-Liebig University Giessen, Giessen, Germany
SANTELJ, Luka	High Energy Accelerator Research Organization (KEK), Tsukuba, Japan
SARKAR, Subir	University of Oxford, UK & Niels Bohr Institute Copenhagen, Denmark
SARKAR, Sarben	King's College London, London, United Kingdom
SHARMA, Sukhdev	HNB Garhwal Central University, Srinagar, Uttarakhand, India
SIMOES, Catarina	IFPA, University of Liège, Liège, Belgium
SIRINGO, Fabio	Università di Catania, Catania, Italy
SKLIROS, Dimitris	King's College London, London, United Kingdom
SONMEZ, Hasan	University of Liverpool, Liverpool, United Kingdom
SOOMRO, Fatima	Ecole Polytechnique Federale de Lausanne, Lausanne, Switzerland
SUKHORUCHKIN, Sergey	PNPI, Gatchina, Russian Federation
TALBERT, Jim	University of Oxford, Oxford, United Kingdom
TAMSETT, Matthew	University of Sussex, Falmer, United Kingdom
TASSIELLI, Giovanni F.	INFN Lecce / Università del Salento, Lecce, Italy
TERESI, Daniele	University of Manchester, Manchester, United Kingdom
VICINO, Denise	INFN, University of Padova, Padova, Italy
VITIELLO, Giuseppe	University of Salerno, Salerno, Italy
WALTER, Manuel	Physik Institut, University of Zurich, Zurich, Switzerland
WIDMANN, Eberhard	Austrian Academy of Sciences, Vienna, Austria
YAMAOKA, Jared	PNNL, Richland, United States
YOU, Tevong	King's College London, London, United Kingdom

Sponsors

London Centre for Terauniverse Studies (LCTS), using funding from the European Research Council via the Advanced Investigator Grant 267352 (PI: John Ellis)
King's College London
London Institute for Field Theory And Particle Physics (LIFTAPP)

London Centre for Terauniverse Studies (LCTS) AdV 267352

erc

KING'S College LONDON

LIFTaPP London Institute for Field Theory and Particle Physics

Journal of Physics: Conference Series **631** (2015) 011001 doi:10.1088/1742-6596/631/1/011001

Conference Poster

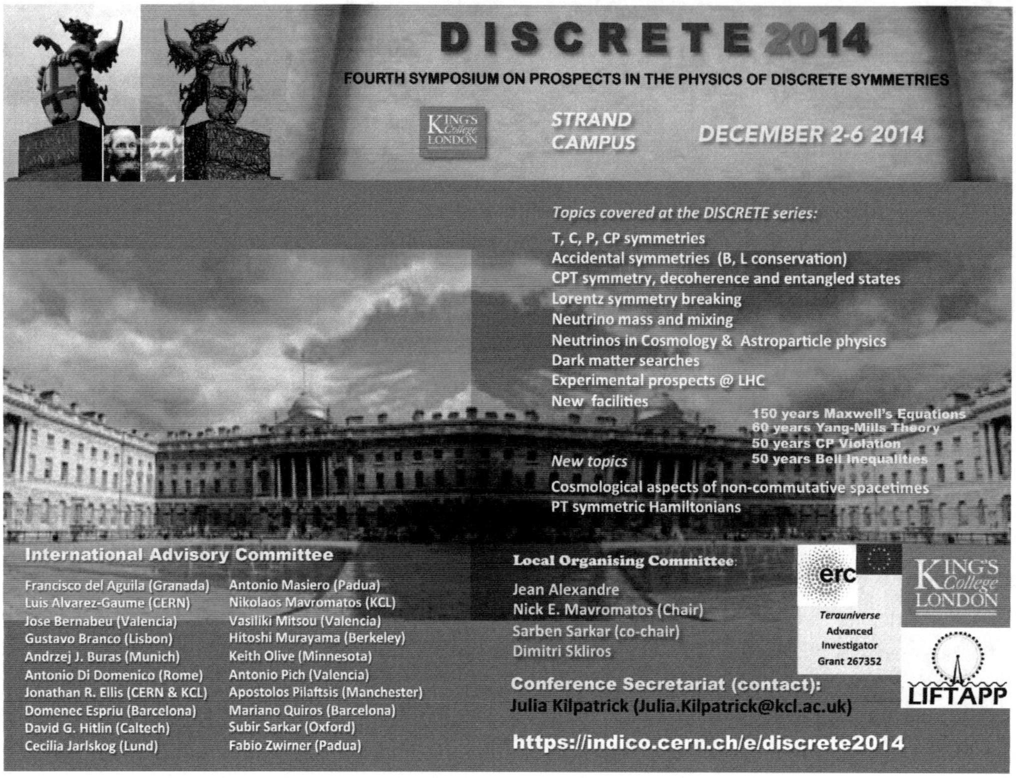

Journal of Physics: Conference Series **631** (2015) 011002 doi:10.1088/1742-6596/631/1/011002

Peer review statement

All papers published in this volume of *Journal of Physics: Conference Series* have been peer reviewed through processes administered by the proceedings Editors. Reviews were conducted by expert referees to the professional and scientific standards expected of a proceedings journal published by IOP Publishing.

4th Symposium on Prospects in the Physics of Discrete Symmetries (DISCRETE2014) IOP Publishing
Journal of Physics: Conference Series **631** (2015) 012001 doi:10.1088/1742-6596/631/1/012001

Discrete Glimpses

of the Physics Landscape after the Higgs Discovery

John Ellis

Physics Department, King's College London, Strand, London WC2R 2LS, U.K; Theory Division, Physics Department, CERN, CH 1211 Geneva 23, Switzerland

E-mail: John.Ellis@cern.ch

Abstract. What is the Higgs boson telling us? What else is there? How do we find it? This talk discusses these current topics in particle physics in the wake of the Higgs discovery, with particular emphasis on the discrete symmetries CP and R-parity, not forgetting flavour physics and dark matter, and finishing with some remarks about possible future colliders.

1. The Story so far

Run 1 of the LHC has brought the apotheosis of the Standard Model (SM). The SM has predicted successfully many cross sections for particle and jet production measured at the LHC [1], as seen in Fig. 1. These successes include QCD jet production cross sections, which agree with SM predictions over large ranges in energy and many orders of magnitude, measurements of single and multiple $W^\pm$ and Z^0 production, as well as multiple measurements of top quark production, both pairwise and singly and in association with vector bosons. Moreover, Run 1 of the LHC has also brought the discovery by CMS and ATLAS of a (the?) Higgs boson [2], whose production has by now been observed in three different production channels, as also seen in Fig. 1, again with cross sections in agreement with the SM predictions. A major theme of this talk is what we already know about this newly-discovered particle.

Meanwhile, over there in flavour space, the Cabibbo-Kobayashi-Maskawa (CKM) description of flavour mixing and CP violation is also a pillar of the SM, but here the picture is more complex. It is in general very successful, as seen in the left panel of Fig. 2 [3], but there are a number of anomalies. On the one hand, the SM predicted successfully the branching ratio for the rare decay $B_s \to \mu^+\mu^-$:

$$BR(B_s \to \mu^+\mu^-) \; = \; 2.8^{+0.7}_{-0.6} \times 10^{-9} \, , \tag{1}$$

measured by the CMS and LHCb Collaborations [4]: see Fig. 3. However, the joint CMS and LHCb analysis [4] also has an suggestion of a $B_d \to \mu^+\mu^-$ signal that is considerably larger than the SM prediction (which is an ironclad prediction also of models with minimal flavour violation (MFV), including many SUSY scenarios):

$$BR(B_d \to \mu^+\mu^-) \; = \; 3.9^{+1.6}_{-1.4} \times 10^{-10} \, , \tag{2}$$

Journal of Physics: Conference Series **631** (2015) 012001 doi:10.1088/1742-6596/631/1/012001

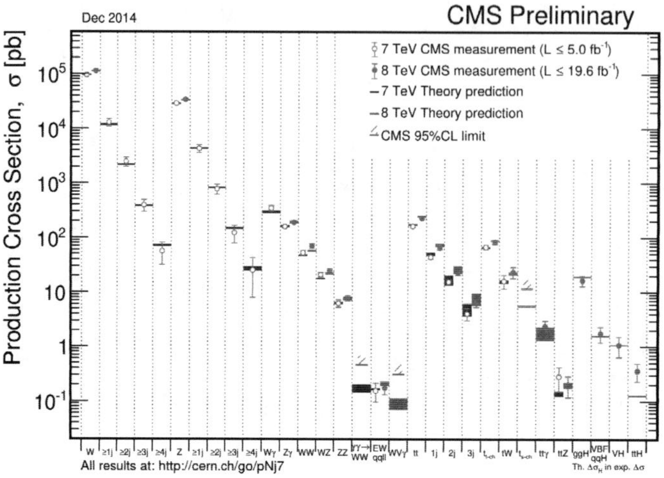

Figure 1. *A compilation of cross sections at the LHC measured by the CMS Collaboration [1].*

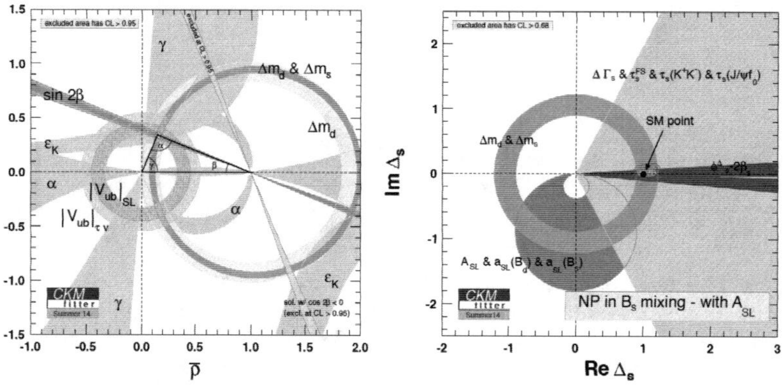

Figure 2. *Left panel: Flavour and CP violation measurements generally agree well with the CKM paradigm. Right panel: Experimental constraint on a possible non-Standard Model contribution to* B_s *mixing [3].*

as also seen in Fig. 3. This will be something to watch during Run 2: Parhaps MFV is not the whole story? *Wait and see!*

There is scope elsewhere for deviations from CKM predictions: for example, the data allow an important contribution to the mixing amplitude for B_s mesons from physics beyond the SM (BSM), as seen in the right panel of Fig. 2 [3]. Moreover, there are several indications of anomalies in flavour physics: for example, the branching ratio for $B^\pm \to \tau^\pm \nu$ decay differs from the SM prediction by $\sim 2\sigma$, and there are issues with $e - \mu$ universality in semileptonic B decays [5]. The most significant anomaly appears in the P_5' angular distribution for $B^0 \to K^{*0}\mu^+\mu^-$ [6], though the non-perturbative corrections need to be understood better. It has recently been suggested that these may be related to the intriguing excess in $H \to \mu\tau$ decay reported by the CMS Collaboration [7], as discussed later. Also worth noting are discrepancies in the determinations of the V_{ub} CKM matrix element, and there is still an anomaly in the

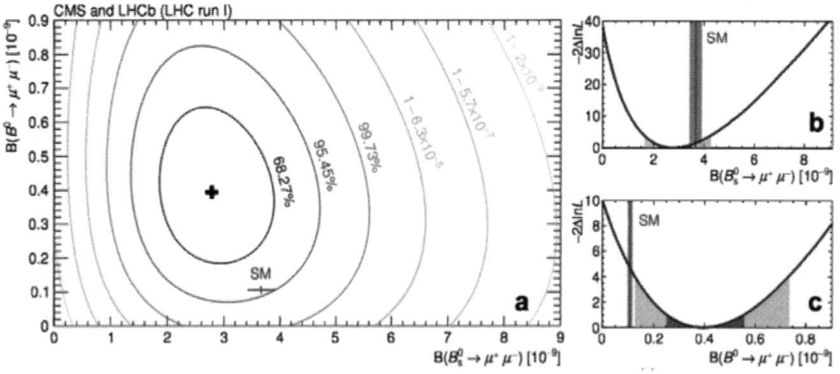

Figure 3. *Panel* **a**: *measurements by the CMS and LHCb Collaborations [4] of* $B_{s,d} \to \mu^+\mu^-$ *decays. Panel* **b**: *The CMS and LHCb Collaborations see a clear signal for* $B_s \to \mu^+\mu^-$ *decay. Panel* **c**: *They also see a possible hint of* $B_d \to \mu^+\mu^-$ *decay.*

diimuon asymmetry at the Tevatron [8]. On the other hand, some anomalies do seem to be going away, such as the forward-backward asymmetry in $t\bar{t}$ production, which now agrees with higher-order QCD calculations [9], as does the $t\bar{t}$ rapidity asymmetry measured at the LHC. However, the discrete charm of flavour physics leaves many issues to be addressed during LHC Run 2 and at SuperKEK-B.

One of the principal focuses during Run 2 of the LHCwill be the more detailed study of the Higgs boson and probes whether its properties deviate from SM predictions, e.g., in the flavour sector. As I discuss later, the measurement of the Higgs mass has produced new reasons to expect BSM physics, and the search for BSM physics will start anew at Run 2, with its greatly increased centre-of-mass energy and increased integrated luminosity. My personal favourite candidate for BSM physics is supersymmetry (SUSY), and I also discuss later in this talk how SUSY models are constrained by flavour physics, as well as by the observations to date of the Higgs boson and searches for BSM physics with Run-1 data. This talk concludes with some remarks about searches for particle dark matter at the LHC and elsewhere, and some advertisements for possible future colliders.

2. Higgs Physics

2.1. Mass Measurements

The mass of the Higgs boson can be measured most accurately in the $\gamma\gamma$ and $ZZ^* \to 2\ell^+2\ell^-$ final states, and ATLAS and CMS both report accurate measurements in both these final states. ATLAS measures [10]

$$\text{ATLAS combined}: m_H \;=\; 125.36 \pm 0.37 \pm 0.18 \text{ GeV} \;=\; 125.36 \pm 0.41 \text{ GeV}, \qquad (3)$$

and CMS measures [11]

$$\text{CMS combined}: m_H \;=\; 125.03^{+0.26}_{-0.27}\,{}^{+0.13}_{-0.15} \text{ GeV} \;=\; 125.03 \pm 0.30 \text{ GeV}. \qquad (4)$$

Some interest has been generated by the differences in the masses measured in these channels, but these have opposite signs in the two experiments:

$$\begin{aligned} \text{ATLAS}: \Delta m_H &= 1.47 \pm 0.67 \pm 0.18 \text{ GeV}, \\ \text{CMS}: \Delta m_H &= -0.9 \pm 0.4 \pm 0.2^{+0.34}_{-0.35} \text{ GeV}, \end{aligned} \qquad (5)$$

so are presumably statistical and/or systematic artefacts. Combining naively the ATLAS and CMS measurements yields

$$m_H = 125.15 \pm 0.24 \text{ GeV}. \quad (6)$$

In addition to being a fundamental measurement in its own right, and casting light on the possible validity of various BSM models, the precise value of m_H is also important for the stability of the electroweak vacuum in the Standard Model, as discussed later.

2.2. The Higgs Spin and Parity

The fact that the Higgs boson has been observed to decay into $\gamma\gamma$ excludes spin 1, and spin 0 is expected, but spins 2 and higher are also possible in principle. There have been many probes of the Higgs spin [12, 13, 14], including its production and decay rates [15], the kinematics of its production in association with massive vector bosons [16], as well as angular distributions in its decays into W^+W^-, ZZ and $\gamma\gamma$ [17]. The results of many of these tests are shown in Fig. 4. By now there is overwhelming evidence that the Higgs boson has spin 0 and that its couplings to W^+W^- and ZZ are predominantly CP-even.

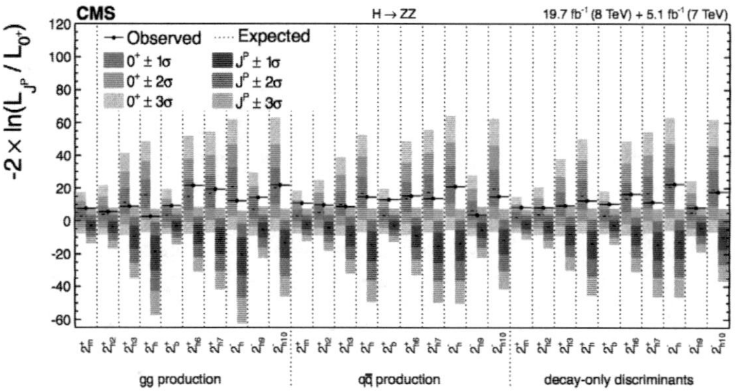

Figure 4. *Tests of spin-parity hypotheses for the Higgs boson. All favour strongly the 0^+ assignment expected in the SM [13].*

On the other hand, it is possible that there may be an admixture of CP-odd couplings, and their fraction may depend on the particle whose coupling to the Higgs boson are being probed. In particular, the leading CP-odd H coupling to fermions would have the *same* (zero) dimension as the leading CP-even coupling, whereas the leading CP-odd H coupling to massive vector bosons would have *higher* dimension than the leading CP-even coupling, so it may be more suppressed. Various ways to probe CP violation in the $H\tau^+\tau^-$ couplings have been proposed [18], and it is also possible to probe CP violation in the $Ht\bar{t}$ couplings [19]. These would affect the total cross sections for associated $Ht\bar{t}$, Ht and $H\bar{t}$ production, as seen in the left panel of Fig. 5 as functions of $\zeta_t \equiv$ arctan(CP-odd coupling/CP-even coupling). If $\zeta_t \neq 0$, a CP-violating transverse polarization asymmetry is in principle observable in Ht and $H\bar{t}$ production, as shown in the right panel of Fig. 5.

2.3. Higgs Couplings

As seen in Fig. 6, the strengths of the Higgs signals measured by ATLAS and CMS individual channels are generally compatible with the SM predictions within the statistical fluctuations [20, 21, 11], which are inevitably large at this stage. Combining their measurements

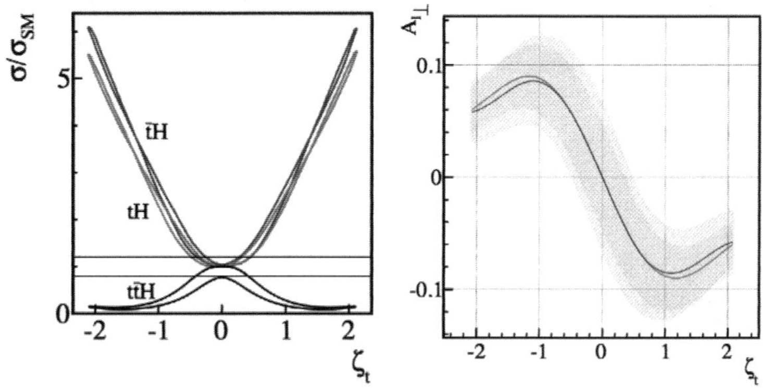

Figure 5. *Left panel: The effects of a CP-violating coupling on the $Ht\bar{t}$, Ht and $H\bar{t}$ production cross sections. Right panel: A CP-violating transverse polarization asymmetry observable in Ht and $H\bar{t}$ production [19].*

in the $\gamma\gamma$, ZZ^*, WW^*, $b\bar{b}$ and $\tau^+\tau^-$ channels, ATLAS and CMS report the following overall signal strengths:

$$\begin{aligned} \text{ATLAS} : \mu \; &= \; 1.30 \pm 0.12 \pm 0.10 \pm 0.09 \, , \\ \text{CMS} : \mu \; &= \; 1.00 \pm 0.09 \; ^{+0.08}_{-0.07} \pm 0.07 \, . \end{aligned} \tag{7}$$

These averages are again quite compatible with each other and with the SM, and measurements at the Tevatron are also compatible with SM predictions for the Higgs boson [22].

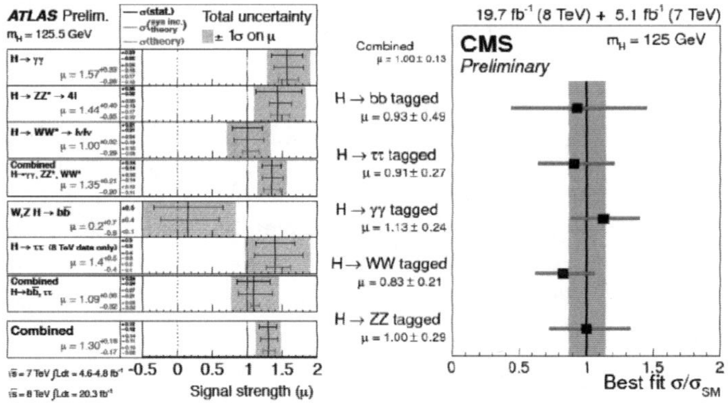

Figure 6. *The Higgs signal strengths μ, normalised to unity for the SM, as measured by ATLAS [20] (left panel) and CMS [11] (right panel).*

One distinctive feature of the Higgs couplings to other particles in the SM is that they should be related to their masses: linearly for fermions, quadratically for bosons, and be proportional to the Higgs vev $v = 246$ GeV. These predictions are verified indirectly by the measurements in Fig. 6, but one may also test these predictions directly, as seen in Fig. 7. This shows the result of a global fit to the data parametrising the Higgs couplings as [23]

$$\lambda_f \; = \; \sqrt{2}\left(\frac{m_f}{M}\right)^{(1+\epsilon)}, \; g_V \; = \; 2\left(\frac{M_V^{2(1+\epsilon)}}{M^{(1+\epsilon)}}\right). \tag{8}$$

As seen in the left panel of Fig. 7, the data yielded

$$\epsilon \; = \; -0.022^{+0.020}_{-0.043}, \; M \; = \; 244^{+20}_{-10}\ \mathrm{GeV}, \tag{9}$$

quite compatible with the SM predictions $\epsilon = 0$, $M = 246$ GeV. Similar results have also been found recently in an analysis by the CMS Collaboration [21]. It seems that Higgs couplings have a very similar flavour structure to particle masses.

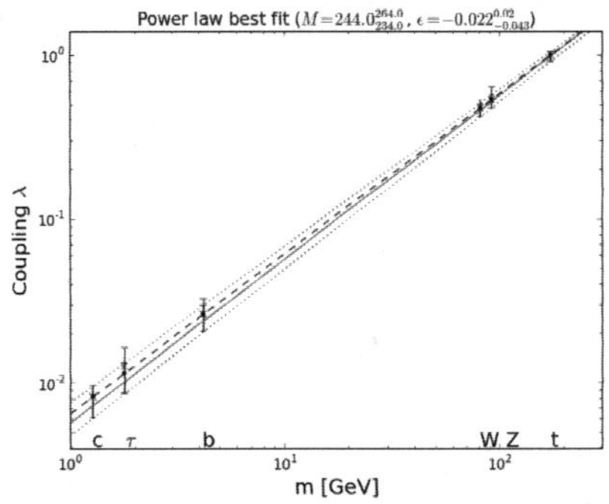

Figure 7. *A global fit to the H couplings of the form (8) (central values as dashed and* $\pm 1\sigma$ *values as dotted lines), which is very compatible with the expected linear mass dependence for fermions and quadratic mass dependence for bosons (solid red line) [23].*

A related aspect of the SM is the expectation is that flavour should be conserved to a very good approximation in Higgs couplings to fermions. This is indeed consistent with the upper limits on low-energy effective flavour-changing interactions, but these would allow also lepton-flavour-violating Higgs couplings much larger than the SM predictions, so looking for such interactions is a possible window on BSM physics. We estimated on the basis of low-energy data that the branching ratios for $H \to \mu\tau$ and $H \to e\tau$ decays could each be as large as $\mathcal{O}(10)\%$, i.e., as large as BR($H \to \tau\tau$, whereas the branching ratio for $H \to \mu e$ could only be $\lesssim 10^{-5}$ [24]. The CMS Collaboration has recently found [7]

$$\mathrm{BR}(H \to \mu\tau) \; = \; 0.89^{+0.40}_{-0.37}\,\%\,, \tag{10}$$

corresponding to a background-only p-value of 0.007, a $\sim$2.46σ effect. LHC Higgs measurements are therefore already testing SM flavour physics predictions more stringently than previous low-energy experiments, and we are on tenterhooks to see corresponding results from ATLAS and from Run 2 of the LHC!

With the discovery of the Higgs boson at the LHC, one of the key questions was whether it is elementary or composite. It used to be thought that a composite Higgs boson would normally have a mass comparable to the scale of compositeness, but this can be reduced if it is a pseudo-Nambu-Goldstone boson whose mass is protected by some approximate symmetry, possibly compatible with the measured Higgs mass $\sim$ 125 GeV. The compositeness possibility may be

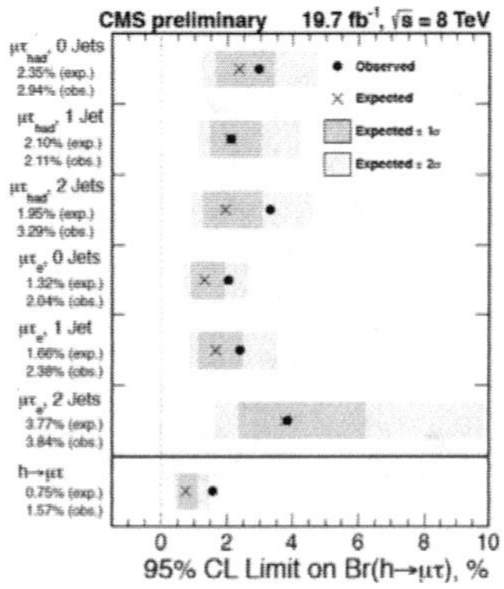

Figure 8. *Results from the CMS search for $H \to \mu\tau$ decay [7].*

probed using a phenomenological Lagrangian $\mathcal{L}$ with free parameters to describe H interactions, that may be constrained using H production and decay data. In view of the consistency of the Standard Model relation $\rho \equiv m_W/m_Z \cos\theta_W = 1$ with data, one may assume a custodial symmetry in this phenomenological Lagrangian: SU(2)×SU(2) → SU(2). In this case, one may parametrize as follows the leading-order terms in $\mathcal{L}$:

$$\begin{aligned}
\mathcal{L} &= \frac{v^2}{4}\mathrm{Tr}D_\mu\Sigma^\dagger D^\mu\Sigma\left(1 + 2a\frac{H}{v} + b\frac{H^2}{v^2} + \ldots\right) \\
&- \bar{\psi}_L^i\Sigma\left(1 + c\frac{H}{v} + \ldots\right) \\
&+ \frac{1}{2}(\partial_\mu H)^2 + \frac{1}{2}m_H^2 H^2 + d_3\frac{1}{6}\left(\frac{3m_H^2}{v}\right)H^3 + d_4\frac{1}{24}\left(\frac{3m_H^2}{v}\right)H^4 + \ldots\,,
\end{aligned} \tag{11}$$

where

$$\Sigma \equiv \exp\left(i\frac{\sigma^a\pi^a}{v}\right). \tag{12}$$

The free coefficients a, b, c, d_3 and d_4 are all normalized such that they are unity in the SM, whereas composite models may give observable deviations from these values.

Fig. 9 shows the result of one analysis [23], that looked for possible rescalings of the H couplings to bosons by a factor a and to fermions by a factor c [1]. Clearly there is no sign of a significant deviation from the SM prediction $a = c = 1$, and the specific composite models shown as yellow lines in Fig. 9 are excluded unless (in some cases) their predictions can be adjusted to be very similar those of the SM.

3. The SM as an Effective Field Theory

In view of the continuing successes of the SM, now also in the Higgs sector, a common approach is to regard it as an effective field theory valid at low energies $\lesssim 1$ TeV. The effects of higher-scale physics may then be expressed, as a first approximation, via higher-dimensional operators constructed out of SM fields, whose coefficients may be constrained by precision electroweak data, Higgs data and triple-gauge couplings (TGCs). Table 1 lists the operators entering electroweak

[1] For a similar recent result from the CMS Collaboration, see [11]. The ATLAS and CMS collaborations follow the Higgs Cross Section Working group in defining the quantities $\kappa_V \equiv a$ and $\kappa_f \equiv c$ [25].

Journal of Physics: Conference Series **631** (2015) 012001 doi:10.1088/1742-6596/631/1/012001

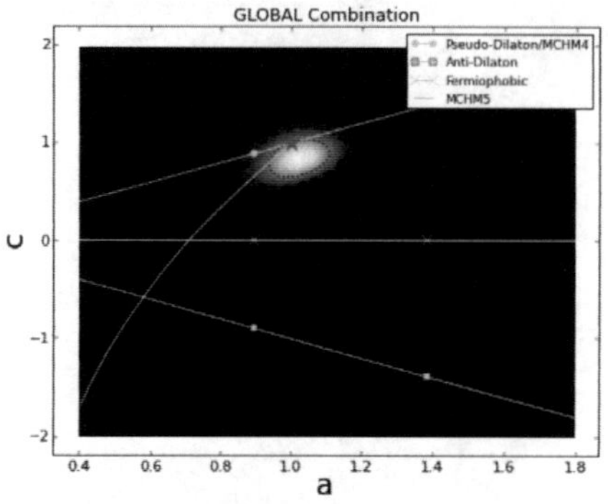

Figure 9. *A global fit to bosonic and fermionic H couplings rescaled by factors a and c, respectively, indicating possible predictions of some composite models. The SM prediction $a = c = 1$ is shown as the green star [23].*

precision tests (EWPTs) at LEP, together with 95% CL bounds on their individual coefficients when they are switched on one at a time, and also when marginalised in a simultaneous global fit [26]. For the first four coefficients we list the constraints from the leptonic observables alone, while the constraints on the remaining coefficients also include hadronic observables.

Operator	Coefficient	LEP Constraints Individual	LEP Constraints Marginalized
$\mathcal{O}_W = \frac{ig}{2}\left(H^\dagger\sigma^a \leftrightarrow D^\mu H\right) D^\nu W^a_{\mu\nu}$ $\mathcal{O}_B = \frac{ig'}{2}\left(H^\dagger \leftrightarrow D^\mu H\right)\partial^\nu B_{\mu\nu}$	$\frac{m_W^2}{\Lambda^2}(c_W + c_B)$	$(-0.00055, 0.0005)$	$(-0.0033, 0.0018)$
$\mathcal{O}_T = \frac{1}{2}\left(H^\dagger \leftrightarrow D_\mu H\right)^2$	$\frac{v^2}{\Lambda^2}c_T$	$(0, 0.001)$	$(-0.0043, 0.0033)$
$\mathcal{O}_{LL}^{(3)\,l} = (\bar{L}_L\sigma^a\gamma^\mu L_L)\,(\bar{L}_L\sigma^a\gamma_\mu L_L)$	$\frac{v^2}{\Lambda^2}c_{LL}^{(3)l}$	$(0, 0.001)$	$(-0.0013, 0.00075)$
$\mathcal{O}_R^e = (iH^\dagger \leftrightarrow D_\mu H)(\bar{e}_R\gamma^\mu e_R)$	$\frac{v^2}{\Lambda^2}c_R^e$	$(-0.0015, 0.0005)$	$(-0.0018, 0.00025)$
$\mathcal{O}_R^u = (iH^\dagger \leftrightarrow D_\mu H)(\bar{u}_R\gamma^\mu u_R)$	$\frac{v^2}{\Lambda^2}c_R^u$	$(-0.0035, 0.005)$	$(-0.011, 0.011)$
$\mathcal{O}_R^d = (iH^\dagger \leftrightarrow D_\mu H)(\bar{d}_R\gamma^\mu d_R)$	$\frac{v^2}{\Lambda^2}c_R^d$	$(-0.0075, 0.0035)$	$(-0.042, 0.0044)$
$\mathcal{O}_L^{(3)\,q} = (iH^\dagger\sigma^a \leftrightarrow D_\mu H)(\bar{Q}_L\sigma^a\gamma^\mu Q_L)$	$\frac{v^2}{\Lambda^2}c_L^{(3)q}$	$(-0.0005, 0.001)$	$(-0.0044, 0.0044)$
$\mathcal{O}_L^q = (iH^\dagger \leftrightarrow D_\mu H)(\bar{Q}_L\gamma^\mu Q_L)$	$\frac{v^2}{\Lambda^2}c_L^q$	$(-0.0015, 0.003)$	$(-0.0019, 0.0069)$

Table 1. *Operators and coefficients contributing to LEP electroweak precision tests (EWPTs), includeng 95% CL bounds when individual operators are switched on, and also when they are marginalized in a simultaneous global fit [26].*

Our results for fits to the coefficients $\bar{c}_{LL}^{(3)l}, \bar{c}_T, \bar{c}_W + \bar{c}_B$ and $\bar{c}_R^e$ that affect the leptonic observables $\{\Gamma_Z, \sigma^0_{\rm had}, R^0_e, R^0_\mu, R^0_\tau, A^{0,e}_{\rm FB}, m_W\}$ are shown in the left panel of Fig. 10. The upper (green) bars show the ranges for each of the coefficients varied individually, assuming that the other coefficients vanish, and the lower (red) bars show the ranges for a global fit in which all the coefficients are varied simultaneously. The legend at the top of the left panel of Fig. 10 translates the ranges of the coefficients into ranges of sensitivity to a large mass scale Λ. We see that all the sensitivities are in the multi-TeV range. The right panel of Fig. 10 shows the effect of including the hadronic observables $\{R^0_b, R^0_c, A^{0,b}_{\rm FB}, A^{0,c}_{\rm FB}, A_b, A_c\}$, and the operator coefficients that contribute directly to them, namely $\bar{c}_L^q, \bar{c}_L^{(3)q}, \bar{c}_R^u$ and $\bar{c}_R^d$.

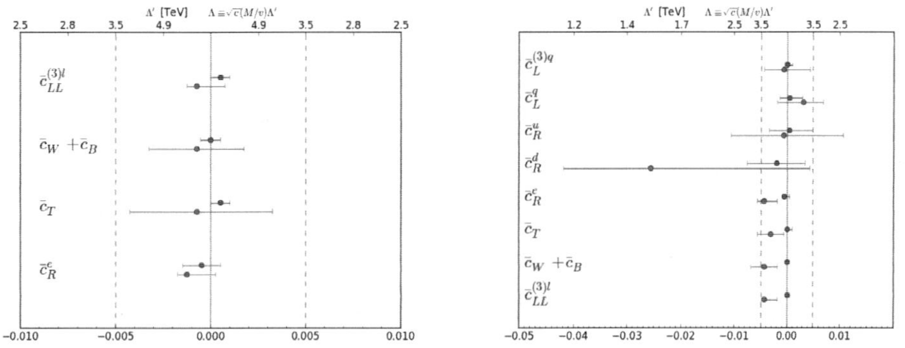

Figure 10. *The 95% CL ranges found in analyses of the leptonic observables (left panel) and including also the hadronic observables (right panel). In each case, the upper (green) bars denote single-coefficient fits, and the lower (red) bars denote multi-coefficient fits. The upper-axis should be read* $\times \frac{m_W}{v} \sim 1/3$ *for* $\bar{c}_W + \bar{c}_B$*. [26]*

The operators with the largest effects on Higgs physics and TGCs are listed in Table 2 [2], together with 95% CL bounds obtained when individual operators are switched on one at a time, and also when they are marginalized in a simultaneous global fit [26]. Important information is provided by kinematic distributions [27], as well as by total rates, as seen in Fig. 11.

Operator	Coefficient	LHC Constraints Individual	Marginalized
$\mathcal{O}_W = \frac{ig}{2}\left(H^\dagger \sigma^a \overleftrightarrow{D^\mu} H\right) D^\nu W^a_{\mu\nu}$ $\mathcal{O}_B = \frac{ig'}{2}\left(H^\dagger \overleftrightarrow{D^\mu} H\right) \partial^\nu B_{\mu\nu}$	$\frac{m_W^2}{\Lambda^2}(c_W - c_B)$	$(-0.022, 0.004)$	$(-0.035, 0.005)$
$\mathcal{O}_{HW} = ig(D^\mu H)^\dagger \sigma^a (D^\nu H) W^a_{\mu\nu}$	$\frac{m_W^2}{\Lambda^2} c_{HW}$	$(-0.042, 0.008)$	$(-0.035, 0.015)$
$\mathcal{O}_{HB} = ig'(D^\mu H)^\dagger (D^\nu H) B_{\mu\nu}$	$\frac{m_W^2}{\Lambda^2} c_{HB}$	$(-0.053, 0.044)$	$(-0.045, 0.075)$
$\mathcal{O}_{3W} = \frac{1}{3!} g \epsilon_{abc} W^{a\,\nu}_\mu W^b_{\nu\rho} W^{c\,\rho\mu}$	$\frac{m_W^2}{\Lambda^2} c_{3W}$	$(-0.083, 0.045)$	$(-0.083, 0.045)$
$\mathcal{O}_g = g_s^2 \lvert H \rvert^2 G^A_{\mu\nu} G^{A\mu\nu}$	$\frac{m_W^2}{\Lambda^2} c_g$	$(0, 3.0) \times 10^{-5}$	$(-3.2, 1.1) \times 10^{-4}$
$\mathcal{O}_\gamma = g'^2 \lvert H \rvert^2 B_{\mu\nu} B^{\mu\nu}$	$\frac{m_W^2}{\Lambda^2} c_\gamma$	$(-4.0, 2.3) \times 10^{-4}$	$(-11, 2.2) \times 10^{-4}$

Table 2. *List of operators entering in LHC Higgs and TGC physics, together with 95% CL bounds when individual coefficients are switched on one at a time, and marginalized in a simultaneous fit [26].*

The results of a global fit to the Higgs data (including associated production kinematics) and LHC TGC measurements are summarised in Fig. 12 [26]. The individual 95% CL constraints obtained by switching one operator on at a time are shown as green bars. The other lines are the marginalised 95% ranges obtained using just the LHC signal-strength data in combination with the kinematic distributions for associated $H + V$ production measured by ATLAS and D0 (blue bars), in combination with the LHC TGC data (red lines), and in combination with both

[2] EWPTs constrain the operators in Table 1 so strongly that they are not important for Higgs physics and TGCs.

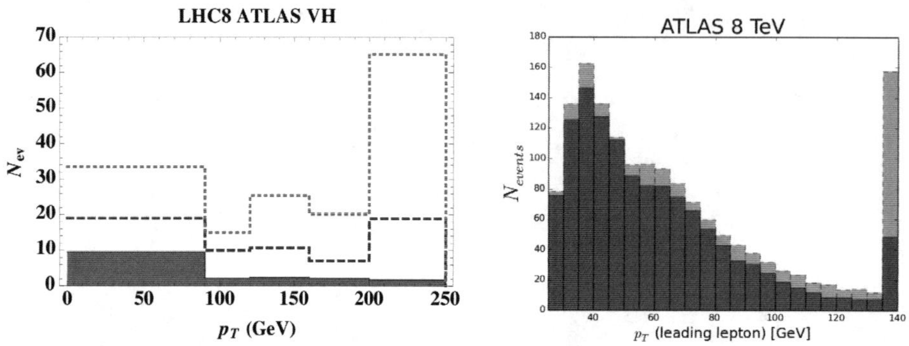

Figure 11. *Left panel: Simulation of the p_T^V distribution in $(V \to 2\ell) + (H \to \bar{b}b)$ events at the LHC after implementing ATLAS cuts. The solid distribution is the SM expectation, and the red-dotted and blue-dashed lines correspond to the distributions with $\bar{c}_W$ =0.1 and 0.05, respectively [27]. Right panel: The same-flavour p_T distribution of the leading lepton after the TGC analysis cuts for ATLAS at 8 TeV. The Standard Model distribution is shown in blue with solid lines, and the effect of $\bar{c}_{HW} = 0.1$ is superimposed in green with dashed lines. In both cases the last (overflow) bin provides significant extra information compared to the overall normalisation [26].*

the associated production and TGC data (black bars). We see that the LHC TGC constraints are the most important for $\bar{c}_W$ and $\bar{c}_{3W}$, whereas the Higgs constraints are more important for $\bar{c}_{HW}$, $\bar{c}_{HB}$ and $\bar{c}_g$. Numerical results for the 95% CL ranges for these coefficients are shown alongside the operator definitions in Table 2.

4. The SM is not enough!

Albert Michelson infamously asserted in 1894 that *"The more important fundamental laws and facts of physical science have all been discovered"*, just before the discoveries of radioactivity and the electron. Likewise, Lord Kelvin asserted in 1900 that *"There is nothing new to be discovered in physics now, all that remains is more and more precise measurement"*, just before Einstein postulated the photon and proposed special relativity. Even after the discovery of a (the?) Higgs boson, there are many reasons to expect physics beyond the SM, as I now discuss.

Inspired by James Bond [28], here are 007 of them. 1) The measured values of m_t and m_H suggest that the electroweak vacuum is *probably* unstable, unless some BSM physics intervenes. 2) The SM cannot provide the dark matter required by astrophysics and cosmology. 3) Additional CP violation beyond the CKM model is required to explain the origin of the matter in the Universe. 4) The small sizes of the neutrino masses seem to require BSM physics. 5) New physics at the TeV scale is needed for the hierarchy of mass scales to seem more natural. 6) BSM physics is required for cosmological inflation, notably because in the SM the effective Higgs potential would seem to become negative at high scales. 7) A consistent quantum theory of gravity would certainly require going (far) beyond the SM. Some of these issues are discussed in the following.

4.1. The Instability of the Electroweak Vacuum

The effective electroweak potential of the SM resembles a Mexican hat, invariant under the SM SU(2)×U(1) symmetry. The origin is unstable, and is surrounded by a valley where $\langle H \rangle \equiv v = 246$ GeV, the electroweak vacuum. At larger Higgs field values, the brim of the

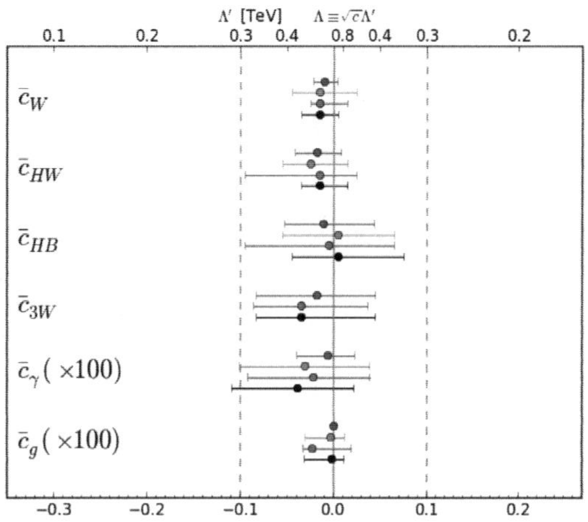

Figure 12. *The 95% CL constraints obtained for single-coefficient fits (green bars), and the marginalised 95% ranges for the LHC signal-strength data combined with the kinematic distributions for associated $H+V$ production measured by ATLAS and D0 (blue bars), combined with the LHC TGC data (red lines), and the global combination with both the associated production and TGC data (black bars). Note that $\bar{c}_{\gamma,g}$ are shown $\times 100$ for which the upper axis should therefore be read $\times 10$ [26].*

hat rises, at least for a while. However, calculations in the SM show that, for the measured values of m_t and m_H renormalization of the Higgs self-coupling by the top quark overwhelms that by the Higgs itself, turning the hat brim down at large field values. Thus, in the SM the present electroweak vacuum is unstable, in principle, with quantum tunnelling though the brim generating collapse into an anti-de-Sitter 'Big Crunch'.

According to the SM calculations [29] shown in the left panel of Fig. 13, the brim turns down at a Higgs scale Λ:

$$\log_{10}\left(\frac{\Lambda}{\text{GeV}}\right) = 11.3+1.0\left(\frac{m_H}{\text{GeV}} - 125.66\right) - 1.2\left(\frac{m_t}{\text{GeV}} - 173.10\right) + 0.4\left(\frac{\alpha_s(M_Z) - 0.1184}{0.0007}\right). \tag{13}$$

Substituting the world average values of m_t, m_H and $\alpha_s(M_Z)$, this formula yields

$$\Lambda = 10^{10.5\pm1.1}\ \text{GeV}. \tag{14}$$

This calculation is very sensitive to m_t, as seen in the right panel of Fig. 13. The D0 Collaboration has recently reported a new, higher, value of m_t [30], which would tend to destabilise further the electroweak vacuum, whereas the CMS Collaboration has reported lower values of m_t from new analyses [31] that would tend to make the vacuum more stable. A more accurate experimental measurement of m_t would help us understand the fate of the Universe within the SM, but this experimental effort must be matched by improved understanding of the relationship between the parameter m_t in the SM Lagrangian and the effective mass parameter appearing in the Monte Carlos used by experiments [32].

According to these calculations, the lifetime of the electroweak vacuum would (probably) be much longer than the age of the Universe, but this does not mean that one can simply ignore

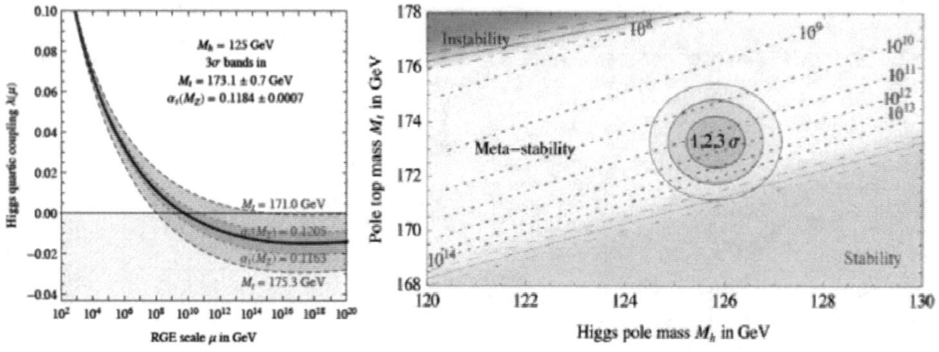

Figure 13. *Left panel: Within the SM, normalisation by the top quark appears to drive the Higgs self-coupling* $\lambda < 0$. *Right panel: The regions of vacuum stability, metastability and instability in the* (m_H, m_t) *plane. Both panels are from [29].*

the problem. We think that the Universe once had a very high energy density, e.g., during an inflationary epoch [33], and quantum and thermal fluctuations at that time would have tended to populate the anti-de-Sitter 'Big Crunch' region of the effective potential [34]. Of course, we might have struck lucky, living in a non-anti-de-Sitter region and thereby surviving - though this would require anthropic special pleading [3]. The problem could be avoided with suitable higher-dimensional terms in the effective potential [36], or other examples of new physics beyond the SM. One example is supersymmetry [37], which would have prevented the turn-down of the brim of the Mexican hat.

4.2. Supersymmetry

There are many longstanding reasons for loving supersymmetry (SUSY), such as alleviating the fine-tuning aspect of the hierarchy problem, providing a natural candidate for the cold dark matter, facilitating grand unification and playing an essential rôle in string theory. Run 1 of the LHC has added three more reasons to this list, namely the mass of the Higgs boson, which is within the range predicted by supersymmetry [38, 39], the fact that the couplings resemble those of the SM Higgs boson, as expected in simple realisations of the minimal supersymmetric extension of the SM (MSSM) [40], and the stabilisation of the electroweak vacuum, as mentioned at the end of the previous Section. How could Nature resist SUSY's manifold charms?

However, despite our ardent love for SUSY, so far she has kept out of sight. Direct searches for SUSY at the LHC have drawn blanks so far. This is also the case for searches for the scattering of dark matter particles, indirect searches in flavour physics, etc.. How can we interpret these searches, and where may SUSY be hiding? Unfortunately, we know that SUSY must be a broken symmetry, but we do not know how: there are no signposts in superspace! It is often assumed that there is a residual discrete R-symmetry that guarantees the stability of the lightest supersymmetric particle (LSP), which is the dark matter candidate mentioned above. Beyond that, it is often assumed (without much theoretical justification) that the SUSY-breaking sparticle masses are universal at some high renormalisation scale, usually the GUT scale. The simplest model in which all the SUSY-breaking contributions m_0 to the squark, slepton and Higgs masses are equal at the GUT scale, and the SU(3), SU(2) and U(1) gauging masses $m_{1/2}$ are also universal, is called the constrained MSSM (CMSSM). One may also consider models in which the SUSY-breaking contributions to the masses of the two Higgs doublets of the MSSM

[3] It has also been pointed out recently [35] that small black holes would catalyse rapid vacuum decay.

are equal but different from those of the squarks and leptons (the NUHM1), or where they are also different from each other (the NUHM2).

The results of global fits to the CMSSM, NUHM1 and NUHM2, combining all experimental and phenomenological constraints, and requiring that the relic supersymmetric particle density be within the cosmological range, are shown projected on the $(m_0, m_{1/2})$ plane in the left panel of Fig. 14 [41, 42]. In these models the 95% CL lower limits on the squark and gluino masses are ~ 1.5 GeV, as seen in Fig. 15. The right panel of Fig. 14 displays the corresponding $(m_{\tilde{q}}, m_{\tilde{g}})$ plane, showing prospective exclusion and discovery reaches of the LHC in future runs with 300 and 3000/fb of luminosity at high energy [43]. Superposed on this plane are the 68 and 95% CL contours found in the global fit to the CMSSM. As already seen in the left panel of Fig. 14, there are two distinct regions, the lower-mass one being favoured by the disagreement between experiment and the SM prediction of $g_\mu - 2$. We see that future runs of the LHC could detect squarks and gluinos if Nature is described by supersymmetry with parameters in this lower-mass region of the CMSSM.

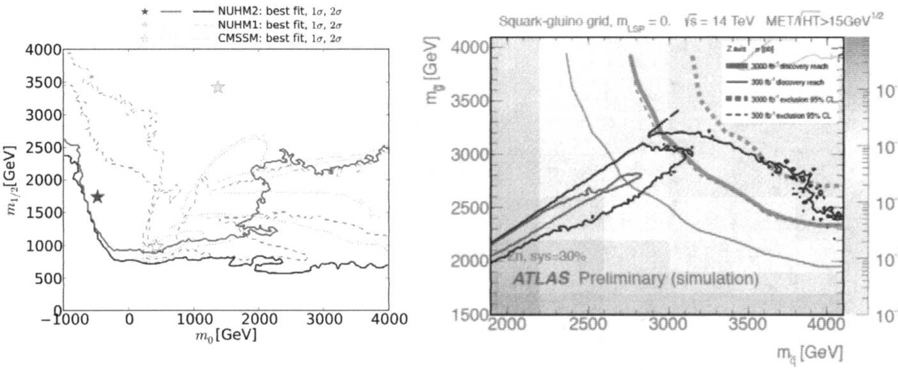

Figure 14. *Left panel: The 68% CL (red) and 95% CL contours (blue) in the $(m_0, m_{1/2})$ panes for the CMSSM (dotted lines), NUHM1 (dashed lines) and NUHM2 (solid lines) [42]. Right panel: The reach of ATLAS in the $(m_{\tilde{q}}, m_{\tilde{g}})$ plane for exclusion and discovery with 300 and 3000/fb of integrated LHC luminosity at high energy [43], compared with the 68% CL (red) and 95% regions in the CMSSM.*

One may, instead, consider the phenomenological MSSM (pMSSM) in which no universality is assumed. In this case, the the lower limits on the gluino and squark masses are reduced, compared with the CMSSM, NUHM1 and NUHM2, as seen in Fig. 15, enhancing the prospects for discovering SUSY in LHC Run 2 [44]. The left panel of Fig. 15 displays the one-dimensional profile likelihood function for the mass of the gluino in the pMSSM, and the right panel is the corresponding plot for the first- and second-generation squarks.

The pMSSM offers anew the possibility that supersymmetry could explain the discrepancy between the SM calculation of $g_\mu - 2$ and the experimental measurement. The LHC constraints on the CMSSM, NUHM1 and NUHM2 all predict values of the $g_\mu - 2$ that are very similar to the unsuccessful SM prediction, as seen in the left panel of Fig. 16, whereas the experimental measurement pMSSM could be accommodated within the pMSSM [44]. Fortunately, there are plans for two new experiments to measure $g_\mu - 2$ [45], and other low-energy e^+e^- experiments will help refine the SM predictions, clarifying the discrepancy between the SM and experiment. Turning to $B_s \to \mu^+\mu^-$, as seen in the right panel of Fig. 16, all of the CMSSM, NUHM1, NUHM2 and pMSSM predict branching ratios very similar to the SM.

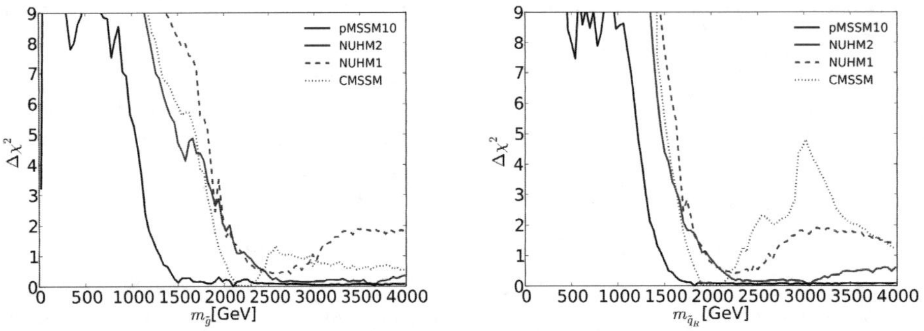

Figure 15. *The one-dimensional profile likelihood functions for the gluino mass (left panel) and the first- and second-generation squark masses (right panel). In each panel the solid black line is for the pMSSM, the solid blue line for the NUHM2, the dashed blue line for the NUHM1 and the dotted blue line for the CMSSM [44].*

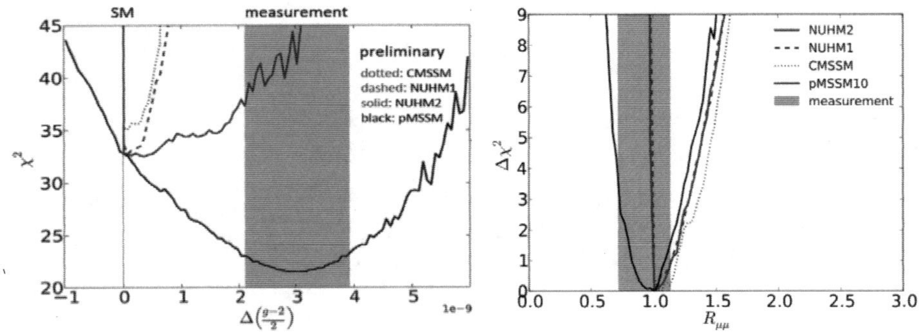

Figure 16. *The one-dimensional χ^2 likelihood function for $g_\mu - 2$ (left panel) and $B_s \to \mu^+\mu^-$ (right panel) in the CMSSM, NUHM1, NUHM2 and pMSSM [44]. The vertical shaded bands represent the 68% CL ranges of $g_\mu - 2$ and $B_s \to \mu^+\mu^-$.*

5. CP-Violating MSSM Scenarios

In the CMSSM and related models, one can introduce 6 CP-violating phases, even if one assumes minimal flavour violation: there are 3 phases in the gaugino masses $M_{1,2,3}$, and 3 more in the third-generation trilinear SUSY-breaking couplings $A_{t,b,\tau}$ [46]. There are 4 important constraints on these phases coming from upper limits on the electric dipole moments (EDMs) of the neutron, thorium monoxide, thallium and mercury. These 4 constraints in the 6-dimensional CP-violating parameter space leave a 2-dimensional blind subspace where combinations of the phases may be large, as seen in the case of the NUHM2 in Fig. 17 [47]. In the left panel we see the correlation these constraints impose between the phases of M_3 and A_t, and in the right panel the correlation between the phases of M_3 and A_b. In both cases we see diagonal features including populations of points with large phases: the EDM constraints do not require all the CP-violating phases to be small simultaneously.

We have explored what values of the CP-violating asymmetry in $b \to s\gamma$ decays, A_{CP}, are compatible with the EDM constraints. We find that values of $A_{CP} \lesssim 2\%$ can be found in the NUHM2 for values of the $b \to s\gamma$ branching ratio lying within the experimentally-allowed range,

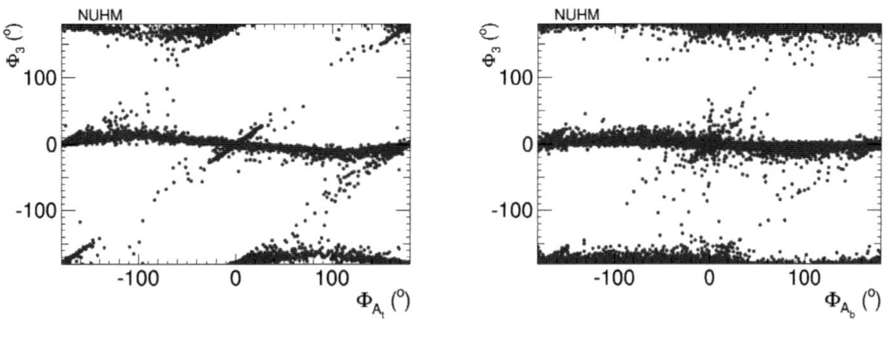

Figure 17. *Correlations of the phase of M_3 with that of A_t (left panel) and with that of A_b (right panel) imposed by the EDM constraints in the NUHM2 scenario [47].*

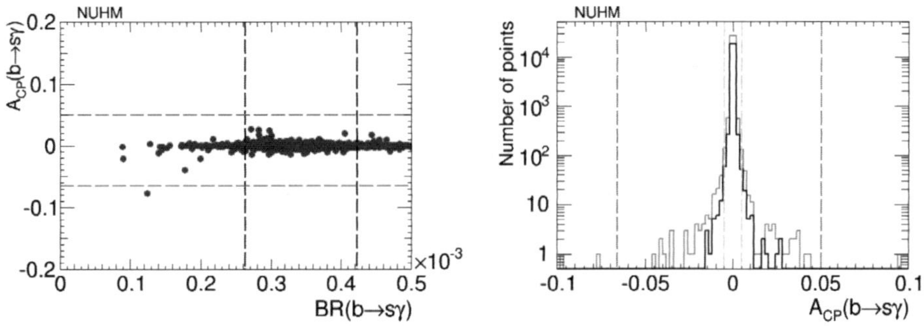

Figure 18. *Left panel: Scatter plot of the branching ratio for $b \to s\gamma$ decay versus its CP-violating asymmetry, A_{CP}, in the NUHM2 scenario. Right panel: Histogram of A_{CP} in the NUHM2, imposing only the Higgs mass and EDM cuts (grey: full sample, black: points satisfying the EDM constraints). The vertical red dashed lines represent the present experimental limits, and the vertical green dashed lines represent the prospective future improvement in the sensitivity to A_{CP} by a factor of ten [47].*

as shown in the left panel of Fig. 18, and the right panel of this Figure displays a histogram of the NUHM2 results (grey: full sample, black: points satisfying the EDM constraints). The vertical red dashed lines are the present experimental constraints on A_{CP} [48]. According to a study of the prospective Belle II sensitivity [49], it should be possible to improve the present experimental sensitivity by a factor of ten, as shown by the vertical green dashed lines, and we see that there are some CP-violating NUHM2 models that could be explored with such an improvement. We conclude that the EDM constraints allow an observable value of A_{CP} within the NUHM2, and the same is true in the pMSSM.

6. Dark Matter Searches

As already commented, if a supersymmetric model conserves R-parity it provides a natural candidate for a cold dark matter particle. The same is true in some other TeV-scale extensions of the SM, such as some extra-dimensional models with K-parity and little Higgs models with

T-parity. The subject of searches for TeV-scale dark matter particles is vast, and there is no time or space here to discuss it in detail. However, I take the opportunity to display in Fig. 19 a recent comparison between the current reaches of the LHC via monojet searches and direct astrophysical searches for the scattering of generic TeV-scale dark matter particles, for the cases of spin-dependent (axial) couplings (left panel) and spin-independent (vector) couplings (right panel) [50]. We see that in the former case the LHC monojet searches generally have greater sensitivity than the direct searches, except for dark matter particle masses $\gtrsim$ 1 TeV where the LHC runs out of phase space. On the other hand, direct searches for spin-independent interactions are stronger than the LHC searches for masses $\gtrsim$ 4 GeV. Supersymmetry models generally favour a relatively large mass for the dark matter particle.

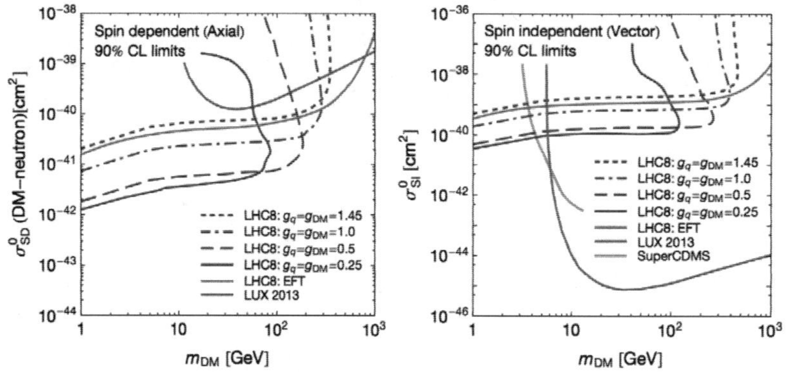

Figure 19. *A comparison of the current 90% CL direct search limits from LUX and SuperCDMS (red and orange lines, respectively), the monojet limits in simple models (blue lines) and the limits in an effective field theory framework (green line) in the cross section vs m_{DM} plane used by the direct detection community. The left and right panels show, respectively, the limits on the spin-dependent and spin-independent cross sections appropriate for axial- vector and vector mediators [50].*

7. Possible Future Colliders

In conclusion, here are a few words about possible future colliders. There has long been interest in building a high-energy linear e^+e^- collider such as the ILC ($E_{CM} \lesssim 1$ TeV) or CLIC ($E_{CM} \lesssim 3$ TeV), and there is now increasing interest in Europe and China in a possible large circular tunnel that could accommodate either an e^+e^- collider with $E_{CM} \lesssim 350$ GeV and/or a *pp* collider with $E_{CM} \lesssim 100$ TeV [51]. A circular e^+e^- collider could also provide unparalleled accuracy in measuring the properties of the Z and Higgs bosons, in particular, as seen in Fig. 20 [52]. In principle, these could be able to distinguish between the predictions of the SM and various fits in the CMSSM, NUHM1 and NUHM2, as shown. However, this will require considerable efforts to reduce correspondingly the present theoretical uncertainties, shown by the shaded green bars.

A future high-energy *pp* collider would also offer the possibility of producing very large numbers of Higgs bosons, as seen in the left panel of Fig. 21 [53], and studies are underway to estimate better the accuracies experiments could give in measuring Higgs couplings, including in particular the elusive triple-Higgs coupling, which will be very difficult to measure at the LHC or in any but a very high-energy e^+e^- collider. A high-energy *pp* collider would also offer unique possibilities to discover and/or measure the properties of supersymmetric particles. Even

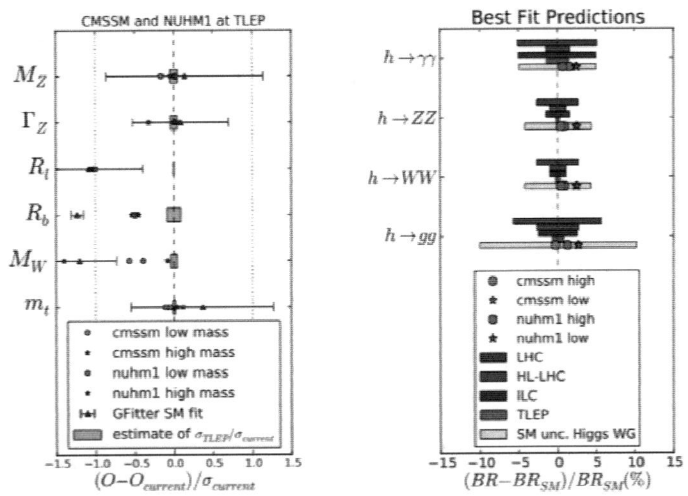

Figure 20. *Comparison of the present precisions in measurements of various Z properties (left panel) and Higgs couplings (right panel), together with the prospective precisions of possible measurements at future colliders, the current theoretical uncertainties, and the deviations from the SM predictions found at the best-fit points in various SUSY models. From [52].*

the lightest of these could weigh several TeV, as seen in the right panel of Fig. 21 [54]. In the example shown, only a *pp* collider with $E_{CM} \sim 100$ TeV would be capable of exploring the full range of particle masses compatible with SUSY providing dark matter weighing $\lesssim 3$ TeV (solid and upper dashed blue lines), and the green lines show that for all this range calculations of the Higgs mass are compatible with the experimental value (represented by the yellow band), considering the theoretical uncertainties represented by the green lines.

The physics cases for future large circular colliders are still to be developed. Certainly there will be a strong bedrock of high-precision Higgs and other SM measurements to test possible scenarios for physics beyond the SM. As for direct searches for new physics, the search for dark matter particles looks to provide the strongest case, but this still requires further study. Needless to say, the physics landscape will look completely different when/if future runs of the LHC at higher energy and luminosity provide some evidence for new physics beyond the SM. Some of my supersymmetric friends are disappointed that SUSY has not shown up yet, but the LHC saga has only just begun. It took 48 years for the Higgs boson to be discovered [55] (the record time-lag so far), but four-dimensional supersymmetric were first written down only 41 years ago [56], so let us be patient for a while longer!

Acknowledgments

I am grateful to Nick Mavromatos and Sarben Sarkar for their kind invitation to speak at this fun conference. The work was supported in part by the London Centre for Terauniverse Studies (LCTS), using funding from the European Research Council via the Advanced Investigator Grant 267352 and from the UK STFC via the research grant ST/J002798/1.

References

[1] CMS Collaboration, `https://twiki.cern.ch/twiki/bin/view/CMSPublic/PhysicsResultsSMP`.
[2] G. Aad *et al.* [ATLAS Collaboration], Phys. Lett. B **716** (2012) 1 [arXiv:1207.7214 [hep-ex]]; S. Chatrchyan *et al.* [CMS Collaboration], Phys. Lett. B **716** (2012) 30 [arXiv:1207.7235 [hep-ex]].

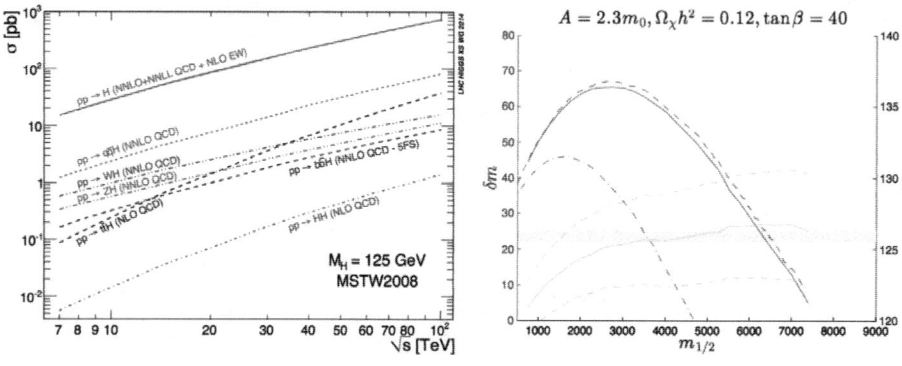

Figure 21. *Left panel: Cross sections for various Higgs production processes at pp colliders as functions of the centre-of-mass energy [53]. Right panel: One of the possibilities for a relatively heavy supersymmetric dark matter particle weighing* $\sim 0.4 m_{1/2} \lesssim 3$ *TeV. The vertical axis is the mass difference between the dark matter particle and the next-to-lightest supersymmetric particle, in this case the lighter stop squark. The solid and upper dashed blue lines correspond to the current central and* $+1\sigma$ *values of the dark matter density, the horizontal yellow band represents the experimental value of the Higgs mass, and the green lines represent the central value and* $\pm 1\sigma$ *uncertainties in theoretical calculations of the Higgs mass [54].*

[3] J. Charles, O. Deschamps, S. Descotes-Genon, H. Lacker, A. Menzel, S. Monteil, V. Niess and J. Ocariz *et al.*, arXiv:1501.05013 [hep-ph].
[4] V. Khachatryan *et al.* [CMS and LHCb Collaborations], arXiv:1411.4413 [hep-ex] and references therein.
[5] R. Aaij *et al.* [LHCb Collaboration], Phys. Rev. Lett. **113** (2014) 151601 [arXiv:1406.6482 [hep-ex]].
[6] R. Aaij *et al.* [LHCb Collaboration], Phys. Rev. Lett. **111** (2013) 19, 191801 [arXiv:1308.1707 [hep-ex]].
[7] CMS Collaboration, `https://cds.cern.ch/record/1740976?ln=en`.
[8] V. M. Abazov *et al.* [D0 Collaboration], Phys. Rev. Lett. **105** (2010) 081801 [arXiv:1007.0395 [hep-ex]].
[9] T. Aaltonen *et al.* [CDF Collaboration], Phys. Rev. D **87** (2013) 9, 092002 [arXiv:1211.1003 [hep-ex]].
[10] G. Aad *et al.* [ATLAS Collaboration], Phys. Rev. D **90** (2014) 052004 [arXiv:1406.3827 [hep-ex]].
[11] V. Khachatryan *et al.* [CMS Collaboration], arXiv:1412.8662 [hep-ex].
[12] G. Aad *et al.* [ATLAS Collaboration], Phys. Lett. B **726** (2013) 120 [arXiv:1307.1432 [hep-ex]];
[13] V. Khachatryan *et al.* [CMS Collaboration], arXiv:1411.3441 [hep-ex].
[14] T. A. Aaltonen *et al.* [CDF Collaboration], arXiv:1501.04875 [hep-ex].
[15] J. Ellis, V. Sanz and T. You, Phys. Lett. B **726** (2013) 244 [arXiv:1211.3068 [hep-ph]].
[16] J. Ellis, D. S. Hwang, V. Sanz and T. You, JHEP **1211** (2012) 134 [arXiv:1208.6002 [hep-ph]].
[17] J. Ellis, R. Fok, D. S. Hwang, V. Sanz and T. You, Eur. Phys. J. C **73** (2013) 2488 [arXiv:1210.5229 [hep-ph]], and references therein.
[18] A. Askew, P. Jaiswal, T. Okui, H. B. Prosper and N. Sato, arXiv:1501.03156 [hep-ph], and references therein.
[19] J. Ellis, D. S. Hwang, K. Sakurai and M. Takeuchi, JHEP **1404** (2014) 004 [arXiv:1312.5736 [hep-ph]].
[20] ATLAS Collaboration, `https://twiki.cern.ch/twiki/bin/view/AtlasPublic/HiggsPublicResults`.
[21] S. Chatrchyan *et al.* [CMS Collaboration], JHEP **1306** (2013) 081 [arXiv:1303.4571 [hep-ex]].
[22] CDF and D0 Collaborations, `http://tevnphwg.fnal.gov`.
[23] J. Ellis and T. You, JHEP **1306** (2013) 103 [arXiv:1303.3879 [hep-ph]]. References to the original literature can be found here.
[24] G. Blankenburg, J. Ellis and G. Isidori, Phys. Lett. B **712** (2012) 386 [arXiv:1202.5704 [hep-ph]].
[25] A. David *et al.* [LHC Higgs Cross Section Working Group Collaboration], arXiv:1209.0040 [hep-ph].
[26] J. Ellis, V. Sanz and T. You, arXiv:1410.7703 [hep-ph]. References to the original literature can be found here.
[27] J. Ellis, V. Sanz and T. You, JHEP **1407** (2014) 036 [arXiv:1404.3667 [hep-ph]].
[28] J. Bond *et al.*, `http://www.imdb.com/title/tt0143145/`.
[29] D. Buttazzo, G. Degrassi, P. P. Giardino, G. F. Giudice, F. Sala, A. Salvio and A. Strumia, JHEP **1312**

(2013) 089 [arXiv:1307.3536].
[30] A. Jung, on behalf of the D0 Collaboration, `https://indico.cern.ch/event/279518/session/27/contribution/36/material/slides/0.pdf`.
[31] CMS Collaboration, `http://cds.cern.ch/record/1951019/files/TOP-14-015-pas.pdf`.
[32] S. Moch, S. Weinzierl, S. Alekhin, J. Blumlein, L. de la Cruz, S. Dittmaier, M. Dowling and J. Erler *et al.*, arXiv:1405.4781 [hep-ph].
[33] P. A. R. Ade *et al.* [Planck Collaboration], Astron. Astrophys. **571** (2014) A22 [arXiv:1303.5082 [astro-ph.CO]]; P. A. R. Ade *et al.* [BICEP2 Collaboration], Phys. Rev. Lett. **112** (2014) 241101 [arXiv:1403.3985 [astro-ph.CO]].
[34] M. Fairbairn and R. Hogan, Phys. Rev. Lett. **112** (2014) 201801 [arXiv:1403.6786 [hep-ph]]; A. Hook, J. Kearney, B. Shakya and K. M. Zurek, arXiv:1404.5953 [hep-ph].
[35] P. Burda, R. Gregory and I. Moss, arXiv:1501.04937 [hep-th].
[36] V. Branchina, E. Messina and M. Sher, arXiv:1408.5302 [hep-ph].
[37] J. R. Ellis and D. Ross, Phys. Lett. B **506** (2001) 331 [hep-ph/0012067].
[38] J. R. Ellis, G. Ridolfi and F. Zwirner, Phys. Lett. B **257** (1991) 83; H. E. Haber and R. Hempfling, Phys. Rev. Lett. **66** (1991) 1815; Y. Okada, M. Yamaguchi and T. Yanagida, Prog. Theor. Phys. **85** (1991) 1.
[39] T. Hahn, S. Heinemeyer, W. Hollik, H. Rzehak and G. Weiglein, Phys. Rev. Lett. **112** (2014) 14, 141801 [arXiv:1312.4937 [hep-ph]].
[40] J. R. Ellis, S. Heinemeyer, K. A. Olive and G. Weiglein, Phys. Lett. B **515** (2001) 348 [hep-ph/0105061].
[41] O. Buchmueller, R. Cavanaugh, A. De Roeck, M. J. Dolan, J. R. Ellis, H. Flacher, S. Heinemeyer and G. Isidori *et al.*, Eur. Phys. J. C **74** (2014) 6, 2922 [arXiv:1312.5250 [hep-ph]].
[42] O. Buchmueller, R. Cavanaugh, M. Citron, A. De Roeck, M. J. Dolan, J. R. Ellis, H. Flaecher and S. Heinemeyer *et al.*, arXiv:1408.4060 [hep-ph].
[43] ATLAS Collaboration, arXiv:1307.7292 [hep-ex].
[44] O. Buchmueller, R. Cavanaugh, M. Citron, A. De Roeck, M. J. Dolan, J. R. Ellis, H. Flaecher and S. Heinemeyer *et al.*, *The pMSSM after LHC Run 1*, in preparation.
[45] FNAL g-2 Collaboration, `http://muon-g-2.fnal.gov`; H. Iinuma (for the J-PARC New g-2/EDM experiment Collaboration), `http://iopscience.iop.org/1742-6596/295/1/012032/pdf/1742-6596_295_1_012032.pdf`.
[46] J. R. Ellis, J. S. Lee and A. Pilaftsis, Phys. Rev. D **76** (2007) 115011 [arXiv:0708.2079 [hep-ph]]; JHEP **1010** (2010) 049 [arXiv:1006.3087 [hep-ph]].
[47] A. Arbey, J. Ellis, R. M. Godbole and F. Mahmoudi, arXiv:1410.4824 [hep-ph].
[48] K. A. Olive *et al.* [Particle Data Group Collaboration], Chin. Phys. C **38** (2014) 090001.
[49] T. Aushev, W. Bartel, A. Bondar, J. Brodzicka, T. E. Browder, P. Chang, Y. Chao and K. F. Chen *et al.*, arXiv:1002.5012 [hep-ex].
[50] O. Buchmueller, M. J. Dolan, S. A. Malik and C. McCabe, JHEP **1501** (2015) 037 [arXiv:1407.8257 [hep-ph]]; S. Malik, C. McCabe, H. Araujo, A. Belyaev, C. Boehm, J. Brooke, O. Buchmueller and G. Davies *et al.*, arXiv:1409.4075 [hep-ex].
[51] `https://espace2013.cern.ch/fcc/Pages/default.aspx`.
[52] M. Bicer *et al.* [TLEP Design Study Working Group Collaboration], JHEP **1401** (2014) 164 [arXiv:1308.6176 [hep-ex]].
[53] LHC Higgs Cross-Section Working Group, as reported in M. Klute, `http://indico.cern.ch/event/300048/session/13/contribution/60/material/slides/0.pdf`.
[54] J. Ellis, K. A. Olive and J. Zheng, Eur. Phys. J. C **74** (2014) 2947 [arXiv:1404.5571 [hep-ph]].
[55] `http://www.economist.com/blogs/graphicdetail/2012/07/daily-chart-1`.
[56] J. Wess and B. Zumino, Phys. Lett. B **49** (1974) 52.

$\mathcal{PT}$-symmetric quantum theory

Carl M. Bender

Physics Department, Washington University, St. Louis, MO 63130, USA

E-mail: cmb@wustl.edu

Abstract. The average quantum physicist on the street would say that a quantum-mechanical Hamiltonian must be Dirac Hermitian (invariant under combined matrix transposition and complex conjugation) in order to guarantee that the energy eigenvalues are real and that time evolution is unitary. However, the Hamiltonian $H = p^2 + ix^3$, which is obviously not Dirac Hermitian, has a positive real discrete spectrum and generates unitary time evolution, and thus it defines a fully consistent and physical quantum theory. Evidently, the axiom of Dirac Hermiticity is too restrictive. While $H = p^2 + ix^3$ is not Dirac Hermitian, it is $\mathcal{PT}$ symmetric; that is, invariant under combined parity $\mathcal{P}$ (space reflection) and time reversal $\mathcal{T}$. The quantum mechanics defined by a $\mathcal{PT}$-symmetric Hamiltonian is a complex generalization of ordinary quantum mechanics. When quantum mechanics is extended into the complex domain, new kinds of theories having strange and remarkable properties emerge. In the past few years, some of these properties have been verified in laboratory experiments. A particularly interesting $\mathcal{PT}$-symmetric Hamiltonian is $H = p^2 - x^4$, which contains an upside-down potential. This potential is discussed in detail, and it is explained in intuitive as well as in rigorous terms why the energy levels of this potential are real, positive, and discrete. Applications of $\mathcal{PT}$-symmetry in quantum field theory are also discussed.

1. Introduction

Physical systems whether they are classical or quantum-mechanical, are described by a quantity called the *Hamiltonian.* In conventional quantum physics two kinds of Hamiltonians are used, (i) Hermitian Hamiltonians, which govern the behavior of isolated systems, and (ii) non-Hermitian Hamiltonians, which govern the behavior of systems in contact with the environment. Hermitian Hamiltonians describe idealized systems in equilibrium whose total energy and probability are conserved; the energy levels of such systems are real. Non-Hermitian Hamiltonians describe the phenomenological behavior of experimental systems involved in scattering or decay processes. Such systems receive energy from and/or deposit energy into their environment, so they are typically not in equilibrium, the total energy and probability are not conserved, and the energy levels are complex numbers; the imaginary part of an energy level is associated with the lifetime of the physical state or the width of a scattering resonance.

This talk introduces and summarizes new kinds of Hamiltonians called *$\mathcal{PT}$-symmetric* Hamiltonians, which are intermediate between Hermitian and non-Hermitian Hamiltonians. Like non-Hermitian systems, $\mathcal{PT}$-symmetric systems are not isolated, but their contact with the environment is highly constrained so that gain from the environment and loss to the environment is exactly balanced. As a consequence, even though they are not isolated, $\mathcal{PT}$-symmetric systems behave like Hermitian systems in that they are in equilibrium and their energy levels are real.

The special kinds of systems discussed here are called $\mathcal{PT}$ symmetric because they are invariant under the combined operation of space reflection (designated parity $\mathcal{P}$) and reversing the arrow of time (designated $\mathcal{T}$). $\mathcal{PT}$-symmetric quantum systems were discovered in 1998 (in a highly theoretical and abstract context) by Prof. Bender [1], who proposed the idea of extending real (Hermitian) quantum theory into the complex (non-Hermitian) domain. (By analogy, in mathematics it is extremely useful to extend the real numbers to the complex numbers; doing so helps to elucidate the properties of the real number system.) Since its inception sixteen years ago there have been dozens of experimental confirmations published in *Nature*, *Science*, and *Physical Review Letters* in diverse research areas such as optics [2, 3, 4, 5, 6, 7, 8, 9, 10, 11, 12, 13, 14], NMR [15], superconductivity [16, 17], microwave cavities [18, 19], lasers [20, 21], atomic diffusion [22], electronic circuits [23, 24], chaos and noise [25, 26], mechanical systems [27], and graphene [28, 29]. Some particularly interesting recent experimental work involves the study of $\mathcal{PT}$-symmetric whispering-gallery microcavities [30, 31, 32]. New experiments involving Bose-Einstein condensates are in the planning stage [33, 34, 35] and new $\mathcal{PT}$-symmetric metamaterials are being developed [36, 37, 38, 39]. As an indication that the study of $\mathcal{PT}$ symmetry is now part of the scientific and technological mainstream, it is noteworthy that in August 2012 the U.S. Air Force Office of Scientific Research issued a Broad Agency Announcement (BAA) calling for scientific proposals, including research and development of $\mathcal{PT}$-symmetric optical systems and synthetic $\mathcal{PT}$ metamaterials [40]. (A \$5 million experimental MURI grant was awarded in June 2013.) This BAA is a significant recognition of $\mathcal{PT}$ symmetry, which was originated by Prof. Bender just 14 years earlier.

Over the past sixteen years, the research area of $\mathcal{PT}$ symmetry has become highly interdisciplinary with an explosion of theoretical and experimental research and new papers appearing on the arXiv daily. There are now nearly two thousand published papers on the subject of $\mathcal{PT}$ symmetry. This explosive growth is reflected in the fact that there have now been more than twenty international conferences entirely devoted to $\mathcal{PT}$-symmetric quantum theory.

2. Elementary $\mathcal{PT}$-symmetric quantum-mechanical systems

An example of a class of non-Hermitian $\mathcal{PT}$-symmetric Hamiltonians having entirely real, positive, and discrete spectra is

$$H = p^2 + x^2(ix)^\varepsilon \quad (\varepsilon \text{ real}) \tag{1}$$

(see Fig. 1). The spectrum of H for $\varepsilon > 0$ is real because its $\mathcal{PT}$ symmetry is *unbroken* (that is, the eigenstates of H are also eigenstates of $\mathcal{PT}$). When $\varepsilon < 0$, the $\mathcal{PT}$ symmetry of H is broken and the spectrum is not real. Conventional Hermitian quantum mechanics (here, the harmonic oscillator) sits at the transition between the regions of unbroken and broken $\mathcal{PT}$ symmetry. Ref. [1] proposed that the physical requirement of $\mathcal{PT}$ symmetry (that is, spacetime reflection symmetry) could be used in place of the mathematical condition of Hermiticity. Because $\mathcal{PT}$ symmetry is a weaker requirement than Hermiticity, Hamiltonians that previously would have been rejected as unphysical can now be considered as potentially valid descriptions of physical processes.

A reader unfamiliar with $\mathcal{PT}$-symmetric quantum mechanics will regard Fig. 1 as astonishing because it shows (contrary to what one would naively expect) that the eigenvalues of $H = p^2 + ix^3$ (pure imaginary potential) and $H = p^2 - x^4$ (upside-down potential) are *real*, *positive*, and *discrete*. An intuitive explanation of why the eigenvalues of the upside-down $-x^4$ potential are real, positive, and discrete was given by Bender along with Berry and Ahmed [41] and an exact calculation can be found in Ref. [42]. Dorey, Dunning, and Tateo proved *rigorously* that for all $\varepsilon \geq 0$ the eigenvalues of $H = p^2 + x^2(ix)^\varepsilon$ are real [43, 44]. Their proof establishes a beautiful mathematical correspondence between ordinary-differential-equation eigenvalue problems and integrable models, which is known as the *ODE/IM correspondence* [45].

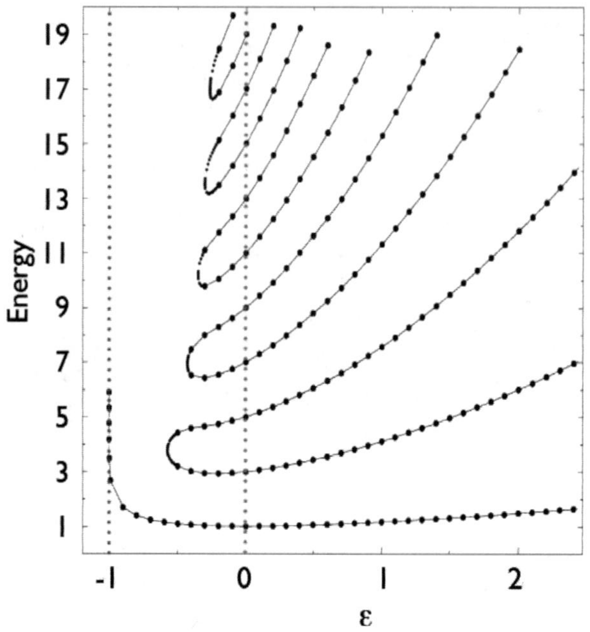

Figure 1. Energy levels of the parametric family of Hamiltonians $H = p^2+x^2(ix)^\varepsilon$ in (1). When $\varepsilon \geq 0$, the eigenvalues are all real and positive, and they increase with increasing ε. When ε decreases below 0, the eigenvalues disappear into the complex plane as complex conjugate pairs. Eventually, only one real eigenvalue remains when ε is less than about -0.57, and as ε approaches -1 from above, this eigenvalue becomes infinite.

3. The $\mathcal{PT}$ phase transition

A crucial aspect of the energy levels in Fig. 1 is that they exhibit a transition, known as the *$\mathcal{PT}$ phase transition*, in which they become pairwise degenerate at exceptional points and move off into the complex plane. This transition is a generic characteristic of $\mathcal{PT}$-symmetric systems but cannot occur in conventional Hermitian quantum systems because such systems can never have complex energies. This phase transition is the signature and characteristic behavior that was observed in many of the experiments mentioned above.

4. Extending physical theories into the complex plane tames instabilities

The discovery of complex numbers was a major development in mathematics because it allowed one to solve equations such as $1 = -x^4$ that were thought to have no solution. Analogously, extending physics into the complex plane also yields new insights. For example, the conventional belief is that one should reject theories based on apparently unstable upside-down potentials such as $V(x) = -x^4$ because a classical particle on the real axis subject to this potential will zoom off to $\pm\infty$ (unless it is precariously balanced at rest at $x = 0$). As a consequence, one would believe that the spectrum of the Hamiltonian for the quantum system is not bounded below and that there is no ground state. This belief is wrong!

How does an upside-down quartic potential confine classical and quantum particles? Let us calculate the time for a classical particle of energy E in an $-x^4$ potential to travel to infinity. The time of flight T from $x = 0$ to $x = \infty$ for a particle described by the Hamiltonian $H = p^2 - x^4$ is

$$T = \int_{t=0}^{T} dt = \int_{x=0}^{\infty} \frac{dx}{\dot{x}} = \int_{x=0}^{\infty} \frac{dx}{2\sqrt{E + x^4}}.$$

This time is *finite* because the integral exists. After the particle reaches infinity in finite time, where does it go next? To answer this question we examine the particle's motion in the complex-x plane. Figure 2 shows four trajectories in the upper-half x plane; the trajectories are all *closed*, *periodic*, and *stable* orbits. (One trajectory represents oscillatory motion between the turning points at $x = e^{i\pi/4}$ and $x = e^{3i\pi/4}$.) We emphasize that all trajectories of a particle of positive energy E in a $-x^4$ potential are *stable* $\mathcal{PT}$-symmetric (left-right symmetric) orbits encircling pairs of turning points, which are the solutions to the equation

$$E = -x^4.$$

The unstable motion on the real-x axis becomes infinitesimally rare (with probabilistic measure zero). The classical particles spend almost all their time near the origin $x = 0$ because this is where the particles are moving slowest, so $x = 0$ is *NOT* a point of instability. As the trajectories approach the real axis, the classical particle does not simply disappear at $x = \pm\infty$. Rather, as the particle reaches $\pm\infty$, it instantly reappears at $\mp\infty$. Thus, the classical particle *periodically* completes the transit from $\pm\infty$ to $\mp\infty$. (This periodic motion is equivalent to a flux of particles on the real axis with a source at $\pm\infty$ and a sink at $\mp\infty$.) A three-dimensional plot of the absolute value of the complex classical probability is shown in Fig. 3.

If we quantize the theory, we find that an upside-down potential has discrete quantum bound states. Indeed, the $-x^4$ potential has bound states that are *localized* at $x = 0$. This counterintuitive, but rigorous fact contradicts the naive view that the origin is unstable. To see why the quantum energies are discrete and not continuous, recall the Bohr-Sommerfeld concept of quantization. A quantum particle is a wave, and if the classical motion is periodic, the wave must interfere with itself constructively. The condition for constructive interference is

$$\oint dx \sqrt{E - V(x)} = (n + 1/2)\pi \quad (n = 0,\, 1,\, 2,\, \ldots),$$

which is the complex version of the WKB quantization condition. Thus, complex classical mechanics provides the insight to explain how complex quantum-mechanical systems work and why complex systems can be stable when their real counterparts appear to be unstable.

5. Differences between $\mathcal{PT}$-symmetric and conventional quantum theory

$\mathcal{PT}$-symmetric quantum theory is not in conflict with conventional quantum theory; rather, it is generalization of it. An important early discovery regarding $\mathcal{PT}$-symmetric quantum theory is that there is a (positive-metric) Hilbert space if the $\mathcal{PT}$ symmetry of the quantum theory is unbroken (which means that the energy spectrum is real and bounded below) [46]. This *PRL* showed that when the $\mathcal{PT}$ symmetry is unbroken, the theory possesses a hidden symmetry represented by a linear operator $\mathcal{C}$, which commutes with the Hamiltonian. $\mathcal{C}$ represents a new *observable quantum number* mathematically related to particle-antiparticle symmetry. The Hilbert-space metric is the $\mathcal{CPT}$ inner product, and with respect to this new inner product (rather than the conventional Dirac inner product) the non-Hermitian $\mathcal{PT}$-symmetric Hamiltonian generates unitary time evolution. Thus, *a $\mathcal{PT}$-symmetric theory is consistent with the fundamental axioms of quantum mechanics.* These results were extended from $\mathcal{PT}$-symmetric quantum mechanics to $\mathcal{PT}$-symmetric (spacetime symmetric) quantum field theory in a subsequent *PRL* [47] and in a *PRD* paper [48]. (For recent work on the $\mathcal{C}$ operator, see Refs. [49, 50, 51].)

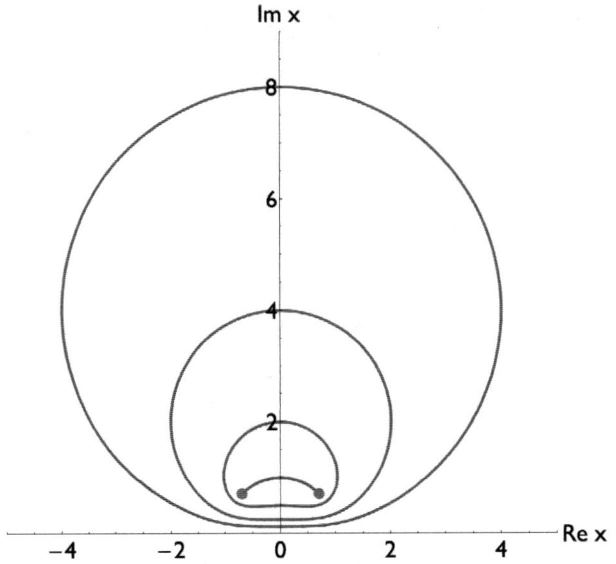

Figure 2. Closed periodic trajectories for $H = p^2 - x^4$ in the upper-half complex-x plane. The energy is $E = 1$. Turning points located at $x = e^{i\pi/4}$ and $x = e^{3i\pi/4}$ are indicated by dots. (The two other turning points in the lower-half plane are not shown.) Initial conditions are $x_0 = i,\ i/2,\ i/4,\ i/8$. A key feature of $\mathcal{PT}$-symmetric systems is that while they appear open from the narrow perspective of the real axis, they become closed when extended to the complex plane.

To see that $\mathcal{PT}$ symmetry is a generalization of standard Hermiticity, we take the limit $\varepsilon \to 0$ in the Hamiltonian $H = p^2 + x^2(ix)^\varepsilon$. In this limit the $\mathcal{CPT}$ norm reduces to the Hermitian conjugate norm of conventional quantum theory. The connection between the $\mathcal{C}$ operator in $\mathcal{PT}$-symmetric quantum field theory and the charge-conjugation operator in conventional quantum field theory is not yet understood. They are both Lorentz scalars [52], but in particle physics parity $\mathcal{P}$ and charge conjugation $\mathcal{C}$ commute; in $\mathcal{PT}$-symmetric quantum theory they do not. Thus, $\mathcal{PT}$-symmetric "particles" (states of positive $\mathcal{C}$) and "antiparticles" (states of negative $\mathcal{C}$) may have different masses. It is conceivable that this fact might account for the so-far-unexplained baryon-antibaryon asymmetry of the universe.

$\mathcal{PT}$-symmetric quantum systems have characteristically different behaviors from ordinary quantum systems. In addition to the $\mathcal{PT}$ phase transition, $\mathcal{PT}$-symmetric systems can evolve *faster* than conventional Hermitian quantum-mechanical systems, as claimed in Ref. [53]. The predicted fast time evolution has been observed in NMR experiments [15]. Also, DeKieviet, an experimentalist at the University of Heidelberg (probably the world's expert on atomic-beam spin-echo experiments) observed this predicted fast time evolution in experiments involving He^3 beams. He presented his work at an ECT conference in 2012 [54]. Another characteristic difference: Hermitian periodic potentials, such as $V(x) = \sin x$, exhibit energy bands and gaps. At one edge of each band of allowed energies the wave function is bosonic (2π periodic); at the other edge the wave function is fermionic (4π periodic). However, while $\mathcal{PT}$-symmetric periodic potentials such as $V(x) = i \sin x$ also have real energy bands and gaps, there are half as many gaps [55]. At the band edges the wave function is always bosonic and never fermionic. It would

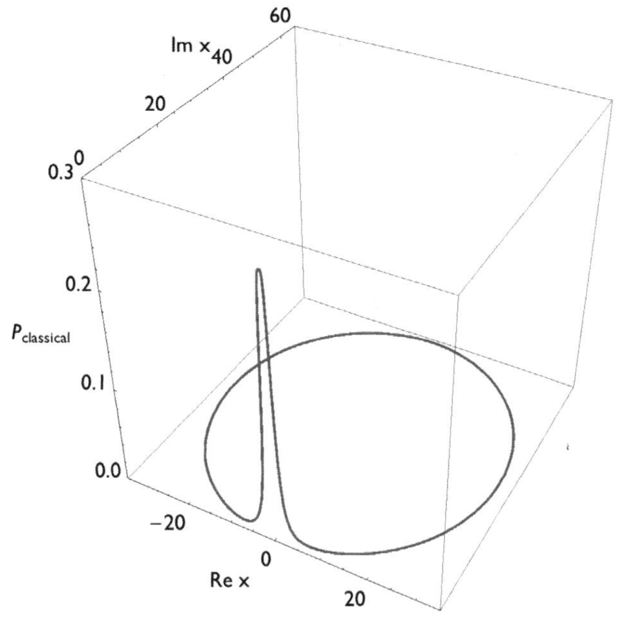

Figure 3. Three-dimensional plot of the probability (inverse of the absolute value of $\dot{x}$) for a contour beginning at $x_0 = i/64$ in the complex plane. The classical energy is $E = 1$. The classical particle executes a rapid loop in the complex plane but goes slowly near the origin.

be a difficult experiment (possibly involving neutron scattering), but a laboratory observation of a material having this kind of band structure would be clear evidence of a fundamental physical system described by a $\mathcal{PT}$-symmetric Hamiltonian; such a material could be regarded as a complex $\mathcal{PT}$-symmetric crystal.

6. Illustrative 2×2 matrix example of a $\mathcal{PT}$-symmetric Hamiltonian

Let us consider the elementary 2×2 Hamiltonian matrix

$$H = \begin{pmatrix} re^{i\theta} & s \\ s & re^{-i\theta} \end{pmatrix},$$

where the three parameters r, s, and θ are real. (This matrix Hamiltonian was discussed in Ref. [56].) The Hamiltonian H is, of course, not Hermitian, but it is easy to see that it is $\mathcal{PT}$ symmetric, where we define the parity operator as

$$\mathcal{P} = \begin{pmatrix} 0 & 1 \\ 1 & 0 \end{pmatrix}$$

and we define the operator $\mathcal{T}$ to perform complex conjugation.

As a first step in analyzing the Hamiltonian H, we calculate its two eigenvalues:

$$E_\pm = r\cos\theta \pm (s^2 - r^2\sin^2\theta)^{1/2}.$$

There are clearly two parametric regions to consider, one for which the square root is real and the other for which it is imaginary. When $s^2 < r^2\sin^2\theta$, the energy eigenvalues form a complex conjugate pair. This is the region of broken $\mathcal{PT}$ symmetry. On the other hand, when $s^2 \geq r^2 \sin^2\theta$, the eigenvalues are real. This is the region of unbroken $\mathcal{PT}$ symmetry. In the unbroken region the simultaneous eigenstates of the operators H and $\mathcal{PT}$ are

$$|E_+\rangle = \frac{1}{\sqrt{2\cos\alpha}}\begin{pmatrix} e^{i\alpha/2} \\ e^{-i\alpha/2} \end{pmatrix} \quad \text{and} \quad |E_-\rangle = \frac{i}{\sqrt{2\cos\alpha}}\begin{pmatrix} e^{-i\alpha/2} \\ -e^{i\alpha/2} \end{pmatrix},$$

where

$$\sin\alpha = \frac{r}{s}\sin\theta.$$

The $\mathcal{PT}$ inner product (which is *not* the correct inner product) gives

$$(E_\pm, E_\pm) = \pm 1 \quad \text{and} \quad (E_\pm, E_\mp) = 0,$$

where $(u, v) = (\mathcal{PT}u)\cdot v$. With respect to the $\mathcal{PT}$ inner product, the vector space spanned by the energy eigenstates has a metric of signature $(+,-)$. If the condition $s^2 > r^2\sin^2\theta$ for an unbroken $\mathcal{PT}$ symmetry is violated, the states are no longer eigenstates of $\mathcal{PT}$ because α becomes imaginary. When $\mathcal{PT}$ symmetry is broken, the $\mathcal{PT}$ norm of the energy eigenstate vanishes.

Next, we construct the operator $\mathcal{C}$:

$$\mathcal{C} = \frac{1}{\cos\alpha}\begin{pmatrix} i\sin\alpha & 1 \\ 1 & -i\sin\alpha \end{pmatrix}.$$

Note that $\mathcal{C}$ is distinct from H and $\mathcal{P}$ and it has the key property that

$$\mathcal{C}|E_\pm\rangle = \pm|E_\pm\rangle.$$

The operator $\mathcal{C}$ commutes with H and satisfies $\mathcal{C}^2 = 1$. The eigenvalues of $\mathcal{C}$ are precisely the signs of the $\mathcal{PT}$ norms of the corresponding eigenstates. Using the operator $\mathcal{C}$ we construct the new inner product structure $\langle u|v\rangle = (\mathcal{CPT}u)\cdot v$. This inner product is positive definite because $\langle E_\pm|E_\pm\rangle = 1$. Thus, the two-dimensional Hilbert space spanned by $|E_\pm\rangle$, with inner product $\langle\cdot|\cdot\rangle$, has signature $(+,+)$.

Finally, we show that the $\mathcal{CPT}$ norm of any vector is positive. For the arbitrary vector $\psi = \binom{a}{b}$, where a and b are any complex numbers, we see that

$$\begin{aligned} \mathcal{T}\psi &= \begin{pmatrix} a^* \\ b^* \end{pmatrix}, \\ \mathcal{PT}\psi &= \begin{pmatrix} b^* \\ a^* \end{pmatrix}, \\ \mathcal{CPT}\psi &= \frac{1}{\cos\alpha}\begin{pmatrix} a^* + ib^*\sin\alpha \\ b^* - ia^*\sin\alpha \end{pmatrix}. \end{aligned}$$

Thus,

$$\langle\psi|\psi\rangle = (\mathcal{CPT}\psi)\cdot\psi = \frac{1}{\cos\alpha}[a^*a + b^*b + i(b^*b - a^*a)\sin\alpha].$$

Now let $a = x + iy$ and $b = u + iv$, where x, y, u, and v are real. Then

$$\langle\psi|\psi\rangle = \frac{1}{\cos\alpha}\left(x^2 + v^2 + 2xv\sin\alpha + y^2 + u^2 - 2yu\sin\alpha\right),$$

which is explicitly positive and vanishes only if $x = y = u = v = 0$.

Since $\langle u|$ denotes the $\mathcal{CPT}$-conjugate of $|u\rangle$, the completeness condition reads

$$|E_+\rangle\langle E_+| + |E_-\rangle\langle E_-| = \begin{pmatrix} 1 & 0 \\ 0 & 1 \end{pmatrix}.$$

Furthermore, using the $\mathcal{CPT}$ conjugate $\langle E_\pm|$, we can represent $\mathcal{C}$ as

$$\mathcal{C} = |E_+\rangle\langle E_+| - |E_-\rangle\langle E_-|.$$

In the limit $\theta \to 0$, the matrix Hamiltonian H for this two-state system becomes Hermitian and $\mathcal{C}$ reduces to the parity operator $\mathcal{P}$. Thus, $\mathcal{CPT}$ invariance reduces to the standard condition of Hermiticity for a symmetric matrix; namely, $H = H^*$.

7. $\mathcal{PT}$ symmetry clarifies questions in conventional quantum mechanics and quantum field theory

$\mathcal{PT}$ symmetry has explained some long-standing mysteries regarding non-Hermitian Hamiltonians that were used in the past. For example, in 1959 Wu showed that the ground state of a Bose system of hard spheres is described by a non-Hermitian Hamiltonian [57]. Wu found that the ground-state energy is real and conjectured that all energy levels were real. In 1992 Hollowood found that while the Hamiltonian for a complex Toda lattice is non-Hermitian, the energy levels are real [58]. Non-Hermitian Hamiltonians of the form $H = p^2 + ix^3$ and cubic quantum field theories having imaginary self-interaction terms also arise in studies of the Lee-Yang edge singularity [59, 60, 61, 62] and in various Reggeon field-theory models [63, 64, 65]. In all these cases a non-Hermitian Hamiltonian having a real spectrum was confusing at the time; the explanation is now clear: In each case the non-Hermitian Hamiltonian is $\mathcal{PT}$-symmetric.

8. $\mathcal{PT}$ symmetry tames instabilities in quantum field theory

When a quantum field theory has *ghosts* (negative-norm states) that arise when renormalizing the field theory, it is because the Hamiltonian has become non-Hermitian as a consequence of renormalization. Often, as in the case of the Lee Model, this Hamiltonian is $\mathcal{PT}$-symmetric, and if it is quantized using the $\mathcal{CPT}$ metric appropriate for the Hamiltonian instead of the standard Dirac inner product, these "ghost states" can be re-interpreted as conventional physical states having positive norm. A case in point is the Lee model, which was proposed by T. D. Lee in 1954 as a quantum field theory in which the renormalization program could be carried out exactly and in closed form [66]. One year later, Källén and Pauli found that upon renormalization the model develops a new negative-norm state that makes the S matrix nonunitary [67]. The problem of how to treat this ghost state was studied by well-known physicists, including Pauli, Heisenberg, and Wick, but it remained unsolved for 50 years. The prevailing belief was that the Lee model was fundamentally flawed [68, 69]. However, in 2005 this 50-year-old problem solved [70]. This paper showed that the renormalized Lee-model Hamiltonian is $\mathcal{PT}$-symmetric. The $\mathcal{C}$ operator was calculated exactly, the "ghost" state was shown to be a physically acceptable state, and that the S matrix was demonstrated to be unitary. Smilga *et al.* [71] and Curtright *et al.* [72, 73] followed this approach and showed that other quantum theories that were thought to contain ghost states actually have physically acceptable positive-norm states when the theory is quantized according to the techniques of $\mathcal{PT}$-symmetric quantum theory.

9. $\mathcal{PT}$ symmetry tames instabilities in the Pais-Uhlenbeck model

In 2008 it was shown that $\mathcal{PT}$ symmetry also eliminates the ghost states in the Pais-Uhlenbeck Model, which is a model higher-derivative quantum field theory. The solution to this 50-year-old problem was given in Refs. [74, 75]. Again, the Hamiltonian for this model is not Hermitian but it is $\mathcal{PT}$-symmetric, and the long-standing belief that this model possesses states of negative norm is false. These papers show that there may not be a problem with quantizing higher-derivative field theories. Indeed, the "ghost" in Pauli-Villars regularization may not be a ghost at all, but rather an indication that the wrong Hilbert-space metric is being used.

10. $\mathcal{PT}$ symmetry tames instabilities in the double-scaling limit

The conventional double-scaling limit of an O(N)-symmetric quartic quantum field theory is inconsistent because the critical coupling constant is negative. Thus, the partition function integral for the critical theory does not exist. Bender, Moshe, and Sarkar showed that $\mathcal{PT}$-symmetric quantum theory resolves this problem. In a first paper they demonstrated that a zero-dimensional O(N)-symmetric quantum field theory avoids this difficulty if the original quartic theory is replaced by its $\mathcal{PT}$-symmetric analog [76]. A second paper on one-dimensional O(N)-symmetric quartic quantum field theory [O(N)-symmetric quantum mechanics] showed that the $\mathcal{PT}$-symmetric formulation of the Schrödinger equation provides a consistent double-scaling limit [77].

11. Stable $\mathcal{PT}$-symmetric timelike logarithmic Liouville quantum field theory

Complex Liouville quantum field theory has a self-interaction term of the form $e^{i\phi}$ and is a topic of great interest at both the experimental and the theoretical level [78, 79, 80, 81, 82, 83]. A quantum field theory having such a self-interaction term is $\mathcal{PT}$ symmetric. However, it is a remarkable fact that the seemingly unstable $\mathcal{PT}$-symmetric self-interaction term of the form $i\phi e^{i\phi}$ also gives a theory with a stable vacuum. In fact, this interaction leads to an infinite number of quantum field theories each of which has a stable vacuum [84].

12. Possible applications of $\mathcal{PT}$-symmetry in the Higgs sector of the SM

There are a number of interesting possible applications of $\mathcal{PT}$-symmetric quantum theory to the Higgs sector of the Standard Model. In the conventional formulation the Higgs field has a quartic ϕ^4 interaction term. However, in four dimensions a ϕ^4 quantum field theory is not asymptotically free and the theory is trivial (noninteracting). In contrast, a $\mathcal{PT}$-symmetric $-\phi^4$ theory is asymptotically free and nontrivial. Furthermore, the vacuum expectation value $\langle\phi\rangle$ is nonzero, not because of spontaneous symmetry breaking, but simply because the boundary conditions on the partition functional integral automatically imply that $\langle\phi\rangle \neq 0$. Of course, in the past a $-\phi^4$ model for the Higgs particle was not considered because it was assumed that the energy spectrum would be unbounded below. In fact, this is not a problem; a $-\phi^4$ theory is stable and has a positive spectrum.

There are further problems with the Higgs model. If one renormalizes the model, one generates a new interaction term of the form $-\phi^4 \log\phi$ in addition to the original ϕ^4 interaction term, which suggests that the Higgs vacuum becomes unstable [85]. One possible interpretation of this result is to concede that the vacuum really is unstable but that the lifetime of the vacuum is sufficiently long that one need not be concerned. We believe that this is a rather unappealing interpretation. However, as we have repeatedly pointed out, by extending this quartic interation into the complex domain and treating the interaction term by using the methods of $\mathcal{PT}$-symmetric quantum theory, we believe that the vacuum actually becomes stable.

References

[1] Bender C M and Boettcher S 1998 Real spectra in non-Hermitian Hamiltonians having $\mathcal{PT}$ symmetry *Physical Review Letters* **80** 5243

[2] Guo A, Salamo G J, Duchesne D, Morandotti F, Volatier-Ravat M, Aimez V, Siviloglou G A and Christodoulides D N 2009 Observation of $\mathcal{PT}$-symmetry breaking in complex optical potentials *Phys. Rev. Lett.* **103** 093902

[3] Longhi S 2009 Bloch oscillations in complex crystals with $\mathcal{PT}$ symmetry *Phys. Rev. Lett.* **103** 123601
[4] Rüter C E, Makris, El-Ganainy R, Christodoulides D N, Segev M and Kip D 2010 Observation of $\mathcal{PT}$ symmetry in optics *Nat. Phys.* **6** 192
[5] Longhi S 2010 Optical realization of relativistic non-Hermitian quantum mechanics *Phys. Rev. Lett.* **105** 013903
[6] Lin Z, Ramezani H, Eichelkraut T, Kottos T, Cao H and Christodoulides D N 2011 Unidirectional invisibility induced by $\mathcal{PT}$-symmetric periodic structures *Phys. Rev. Lett.* **106** 213901
[7] Feng L, Ayache M, Huang J, Xu Y-L, Lu M-H, Chen Y-F, Fainman Y and Scherer A 2011 Nonreciprocal light propagation in a silicon photonic circuit *Science* **333** 729
[8] Ramezani H, Christodoulides D N, Kovanis V, Vitebskiy I and Kottos T 2012 $\mathcal{PT}$-symmetric Talbot effects *Phys. Rev. Lett.* **109** 033902
[9] Regensberger A, Bersch C, Miri M-A, Onishchukov G and Christodoulides D N 2012 Parity-time synthetic photonic lattices *Nature* **488** 167
[10] Razzari L and Morandotti R 2012 Gain and loss mixed in the same cauldron *Nature* **488** 163
[11] Liang G Q and Chong Y D 2013 Optical resonator analog of a two-dimensional topological insulator *Phys. Rev. Lett.* **110** 203904
[12] Luo X, Huang J, Zhong H, Qin X, Xie Q, Kivshar Y S and C Lee C 2013 Pseudo-parity-time symmetry in optical systems *Phys. Rev. Lett.* **110** 243902
[13] Regensberger A, Miri M-A, Bersch C, Nager J, Onishchukov G, Christodoulides D N and Peschel U 2013 Observation of defect states in $\mathcal{PT}$-symmetric optical lattices *Phys. Rev. Lett.* **110** 223902
[14] Hang C, Huang G and Konotop V V 2013 $\mathcal{PT}$ symmetry with a system of three-level atoms *Phys. Rev. Lett.* **110**, 083604
[15] Zheng C, Hao L and Long G L 2013 Observation of a fast evolution in a parity-time-symmetric system *Phil. Trans. R. Soc.* A **371** 20120053
[16] Rubinstein J, Sternberg P and Ma Q 2007 Bifurcation siagram and pattern formation of phase slip centers in superconducting wires driven by electric currents *Phys. Rev. Lett.* **99** 167003
[17] Chtchelkatchev N, Golubov A, Baturina T and Vinokur V 2012 Stimulation of the fluctuation superconductivity by $\mathcal{PT}$ symmetry *Phys. Rev. Lett.* **109** 150405
[18] Dietz B, Harney H L, Kirillov O N, Miski-Oglu M, Richter A and Schäfer F 2011 Exceptional points in a microwave billiard with time-reversal invariance violation *Phys. Rev. Lett.* **106** 150403
[19] Bittner S, Dietz B, Günther U, Harney H L, Miski-Oglu M, Richter A and Schäfer F 2012 $\mathcal{PT}$ symmetry and spontaneous symmetry breaking in microwave billiards *Phys. Rev. Lett.* **108** 024101
[20] Chong Y D, Ge L and Stone A D 2011 $\mathcal{PT}$-symmetry breaking and laser-absorber modes in optical scattering systems *Phys. Rev. Lett.* **106** 093902
[21] Liertzer M, Ge L, Cerjan A, Stone A D, Tureci H and Rotter S 2012 Pump induced exceptional points in lasers *Phys. Rev. Lett.* **108** 173901
[22] Zhao K F, Schaden M and Wu Z 2010 Enhanced magnetic resonance signal of spin-polarized Rb atoms near surfaces of coated cells *Phys. Rev.* A **81** 042903
[23] Schindler J, Li A, Zheng M C, Ellis F M and Kottos T 2011 Experimental study of active LRC circuits with $\mathcal{PT}$ symmetries *Phys. Rev.* A **84** 040101(R)
[24] Bender N, Factor S, Bodyfelt J D, Ramezani H, Christodoulides D N, Ellis F M and Kottos T 2013 Observation of asymmetric transport in structures with active nonlinearities *Phys. Rev. Lett.* **110** 234101
[25] Schomerus H 2010 Quantum noise and self-sustained radiation of $\mathcal{PT}$-symmetric systems *Phys. Rev. Lett.* **104** 233601
[26] West C T, Kottos T and Prosen T 2010 $\mathcal{PT}$-symmetric wave chaos *Phys. Rev. Lett.* **104** 054102
[27] Bender C M, Berntson B, Parker D and Samuel E 2013 Observation of $\mathcal{PT}$ phase transition in a simple mechanical system *Am. J. Phys.* **81** 173
[28] Fagotti M, Bonatti C, Logoteta D, Marconcini P and Macucci M 2011 Armchair graphene nanoribbons: $\mathcal{PT}$-symmetry breaking and exceptional points without dissipation *Phys. Rev.* B **83** 241406(R)
[29] Szameit A, Rechtsman M C, Bahat-Treidel O and Segev M 2012 $\mathcal{PT}$-symmetry in honeycomb photonic lattices *Phys. Rev.* A **84** 021806
[30] Bender C M, Gianfreda M, Peng B, Özdemir S K and Yang L 2013 Twofold transition in $\mathcal{PT}$-symmetric coupled oscillators *Phys. Rev.* A **88** 062111
[31] Peng B, Özdemir S K, Lei F, Monifi F, Gianfreda M, Long G L, Fan S, Nori F, Bender C M and Yang L 2014 Parity-time-symmetric whispering-gallery microcavities *Nature Physics* **10** 394
[32] Peng B, Özdemir S K, Rotter S, Yilmaz H, Liertzer M, Monifi F, Bender C M, Nori F and Yang L 2014 Loss-induced suppression and revival of lasing *Science* **346**, 328
[33] Cartarius H and Wunner G 2012 Model of a $\mathcal{PT}$-symmetric Bose-Einstein condensate in a δ-function double-well potential *Phys. Rev.* A **86** 013612

[34] Kreibich M, Main J, Cartarius H and Wunner G (2013) Hermitian four-well potential as a realization of a $\mathcal{PT}$-symmetric system *Phys. Rev.* A **87** 051601
[35] Kartashov Y V, Konotop V V and Abdullaev F Kh 2013 Gap solitons in a spin-orbit-coupled Bose-Einstein condensate *Phys. Rev. Lett.* **111** 060402
[36] Castaldi G, Savoia S, Galdi V, Alu A and Engheta N 2013 $\mathcal{PT}$ metamaterials via complex coordinate transformation optics *Phys. Rev. Lett.* **110** 173901
[37] Lazarides N and Tsironis G P 2013 Gain-driven discrete breathers in $\mathcal{PT}$-symmetric nonlinear metamaterials *Phys. Rev. Lett.* **110** 053901
[38] Yin X and Zhang X 2013 Unidirectional light propagation at exceptional points *Nature Materials* **12** 175
[39] Feng L, Xu Y-L, Fegadolli W S, Lu M-H, Oliveira J E B, Almeida V R, Chen Y-F and Scherer A 2013 Experimental demonstration of a unidirectional reflectionless parity-time metamaterial at optical frequencies *Nature Materials* **12** 108
[40] See http://www.wpafb.af.mil/news/story.asp?id=123352368
[41] Ahmed Z, Bender C M and Berry M V 2005 Reflectionless potentials and $\mathcal{PT}$ symmetry *J. Phys. A: Math. Gen.* **38** L627
[42] Bender C M, Brody D C, Chen J-H, Jones H F, Milton K A and Ogilvie M C 2006 Equivalence of a complex $\mathcal{PT}$-symmetric quartic Hamiltonian and a Hermitian quartic Hamiltonian with an anomaly *Phys. Rev.* D **74** 025016
[43] Dorey P E, Dunning C and Tateo R 2001 Supersymmetry and the spontaneous breakdown of $\mathcal{PT}$ symmetry *J. Phys. A: Math. Gen.* **34** L391
[44] Dorey P E, Dunning C and Tateo R 2001 Spectral equivalences, Bethe ansatz equations, and reality properties in $\mathcal{PT}$-symmetric quantum mechanics *J. Phys. A: Math. Gen.* **34** 5679
[45] Dorey P E, Dunning C and Tateo R 2007 The ODE/IM correspondence *J. Phys. A: Math. Gen.* **40** R205
[46] Bender C M, Brody D C and Jones H F 2002 Complex extension of quantum mechanics *Phys. Rev. Lett.* **89** 270401
[47] Bender C M, Brody D C and Jones H F 2004 Scalar quantum field theory with complex cubic interaction *Phys. Rev. Lett.* **93** 251601
[48] Bender C M, Brody D C and Jones H F 2004 Extension of $\mathcal{PT}$-symmetric quantum mechanics to quantum field theory with cubic interaction *Phys. Rev.* D **70** 025001
[49] Bender C M and Klevansky S P 2009 Nonunique $\mathcal{C}$ operator in $\mathcal{PT}$ quantum mechanics *Phys. Lett.* A **373** 2670
[50] Bender C M and Kuzhel S 2012 Unbounded $\mathcal{C}$ symmetries and their nonuniqueness *J. Phys. A: Math. Theor.* **45** 444005
[51] Bender C M and Gianfreda M 2013 Nonuniqueness of the $\mathcal{C}$ operator in $\mathcal{PT}$-symmetric quantum mechanics *J. Phys. A: Math. Theor.* **46** 275306
[52] Bender C M, Brandt S F, Chen J-H and Wang Q 2005 The $\mathcal{C}$ operator in $\mathcal{PT}$-symmetric quantum field theory transforms as a Lorentz scalar *Phys. Rev.* D **71** 065010
[53] Bender C M, Brody D C, Jones H F and Meister B K 2007 Faster than Hermitian quantum mechanics *Phys. Rev. Lett.* **98** 040403
[54] ECT* conference, "Many-body Open Quantum Systems: From Atomic Nuclei to Quantum Optics," http://www.ectstar.eu/node/44
[55] Bender C M, Dunne G V and Meisinger P N 1999 Complex periodic potentials with real band structure *Phys. Lett.* A **252** 272
[56] Bender C M, Brody D C and Jones H F 2003 Must a Hamiltonian be Hermitian? *Am. J. Phys.* **71**, 1095
[57] Wu T T 1959 Ground state of a Bose system of hard spheres *Phys. Rev.* **115** 1390
[58] Hollowood T 1992 Solitons in affine Toda field theories 1992 *Nucl. Phys.* B **384** 523
[59] Fisher M E 1978 Yang-Lee edge singularity and ϕ^3 field theory *Phys. Rev. Lett.* **40** 1610
[60] Cardy J L 1985 Conformal invariance and the Yang-Lee edge singularity in two dimensions *Phys. Rev. Lett.* **54** 1345
[61] Cardy J L and Mussardo G 1989 S-matrix of the Yang-Lee edge singularity in two dimensions *Phys. Lett.* B **225** 275
[62] Zamolodchikov A B 1991 Two-point correlation function in scaling Lee-Yang model *Nucl. Phys.* B **348** 619
[63] Brower R, Furman M and Moshe M 1978 Critical exponents for the Reggeon quantum spin model *Phys. Lett.* B **76** 213
[64] Harms B, Jones S and Tan C-I 1980 Complex energy spectra in Reggeon quantum mechanics with quartic interactions *Nucl. Phys.* **171** 392
[65] Harms B, Jones S and Tan C-I 1980 New structure in the energy spectrum of Reggeon quantum mechanics with quartic couplings *Phys. Lett.* B **91** 291
[66] Lee T D 1954 Some special examples in renormalizable field theory *Phys. Rev.* **95** 1329

[67] Källén G and Pauli W 1955 On the mathematical structure of TD Lee's model of a renormalizable field theory *Mat.-Fys. Medd.* **30**, No. 7
[68] Schweber S S 1961 *An Introduction to Relativistic Quantum Field Theory* (Row, Peterson and Co., Evanston) Chap. 12
[69] Barton G 1963 *Introduction to Advanced Quantum Field Theory* (John Wiley & Sons, New York)
[70] Bender C M, Brandt S F, Chen J-H and Wang Q 2005 Ghost busting: $\mathcal{PT}$-symmetric interpretation of the Lee model *Phys. Rev.* D **71** 025014
[71] Ivanov E A and Smilga A V 2007 "Cryptoreality of nonanticommutative Hamiltonians *JHEP* **07** 036
[72] Curtright T, Ivanov E, Mezincescu L and Townsend P K 2007 Planar super-Landau models revisited *JHEP* **04** 020
[73] Curtright T and Veitia A 2007 Quasi-Hermitian quantum mechanics in phase space *J. Math. Phys.* **48** 102112
[74] Bender C M and Mannheim P D 2008 No-ghost theorem for the fourth-order derivative Pais-Uhlenbeck oscillator model *Phys. Rev. Lett.* **100** 110402
[75] Bender C M and Mannheim P D 2008 Exactly solvable $\mathcal{PT}$-symmetric Hamiltonian having no Hermitian counterpart, *Phys. Rev.* D **78** 025022
[76] Bender C M, Moshe M and Sarkar S 2013 $\mathcal{PT}$-symmetric interpretation of double-scaling *J. Phys. A: Math. Theor.* **46** 102002
[77] Bender C M and Sarkar S 2013 Double-scaling limit of the O(N)-symmetric anharmonic oscillator" J. Phys. A: Math. Theor. **46** 442001
[78] Berry M V and O'Dell D H J 1998 Diffraction by volume gratings with imaginary potentials *J. Phys. A: Math. Gen.* **31** 2093 (Some experimental results have been obtained by M. Zeilinger (private communication, M. Berry).
[79] Affleck T, Hofstetter W, Nelson D R and Schollwock U 2004 Non-Hermitian Luttinger liquids and flux line pinning in planar superconductors *J. Stat. Mech.* **0410** P003
[80] Schomerus V 2003 Rolling tachyons from Liouville theory *JHEP* **0311** 043
[81] Fredenhagen S and Schomerus V 2005 Boundary Liouville theory at $c = 1$ *JHEP* **0505** 025
[82] Strominger A and Takayanagi T 2003 Correlators in timelike bulk Liouville theory *Adv. Theor. Math. Phys.* **7** 369
[83] Gutperle M and Strominger A 2003 Timelike boundary Liouville theory *Phys. Rev.* D **67** 126002
[84] Bender C M, Hook D W, Mavromatos N and Sarkar S 2004 Infinite class of $\mathcal{PT}$-symmetric theories from one timelike Liouville Lagrangian *Phys. Rev. Lett.* **113** 231605
[85] Sher M 1989 Electroweak Higgs potentials and vacuum stability *Phys. Repts.* **179** 273

Experimental prospects for C, P, T, CP, and CPT tests

Adrian Bevan
Queen Mary University of London, Mile End Road, London, E1 4NS, UK
E-mail: a.j.bevan@qmul.ac.uk

Abstract. Discrete symmetry violation in the weak interaction is central to the Standard Model of particle physics, however the origin of these violations is not well understood. Nor are we able to provide a satisfactory explanation of the Universal dominance of matter over antimatter, an issue related to CP violation. As a result study of discrete symmetry violation remains a topic of broad interest. These proceedings discuss experimental prospects of studying C, P, T, CP and CPT symmetries in a number of contexts, including the use of triple product asymmetries and entangled neutral meson systems.

1. Introduction

The discrete symmetries parity (P), charge conjugation (C), time reversal which is also commonly referred to as motion reversal (T), and their combinations CP and CPT have been instrumental in probing a number of fundamental aspects of nature in the context of particle physics. The strong and electromagnetic forces conserve these symmetries. The weak interaction violates P, C, CP, and T, however the combination CPT is found to be conserved by experimental observations made so far. CP has been found to be violated in weak interactions involving quark transitions, and the quest for the complementary effect with neutrinos has started via combinations of data from all possible sources; several new experiments have been proposed that may constrain or measure the CP phase for the neutrino sector. Several tests of CP and CPT have been performed using charged leptons, in particular τ decay studies of decays to $K_S\pi^{\pm}\nu$ and the mass difference between τ^+ and τ^-, respectively [1]. The underlying reasons behind the observed pattern of behaviour is not established, however the observed phenomena have been parameterised within the Standard Model of particle physics (SM).

A common feature to testing symmetry violation is to identify a process ψ and its symmetry conjugate ψ'. If a symmetry is conserved then the asymmetry given by

$$A = \frac{\mathcal{P}(\psi') - \mathcal{P}(\psi)}{\mathcal{P}(\psi') + \mathcal{P}(\psi)} \quad (1)$$

is zero as the probability for the process and its conjugate to occur are equal. If a measurement is performed where it is found that $A \neq 0$, then the symmetry is violated. We can construct asymmetries using both particle transitions and ensembles of particle transitions as reference processes. The key is to identify reference states where it is possible to physically define ψ' for the symmetry of interest.

The remainder of these proceedings discuss the use of triple product asymmetries to probe P, C and CP violation followed by the use of entangled neutral meson systems to test CP, T and CPT symmetries, before summarising.

2. Triple product asymmetries

Triple product asymmetries have been used to probe for CP violation in a number of different systems, for example see [2, 3] and references therein. One can use the momenta (or spins) of final state particles in four body decays of some mother particle M to a state $abcd$ to compute a scalar triple product $\vec{p}_c \cdot (\vec{p}_a \times \vec{p}_b)$. The asymmetry of Eq. (1) can be computed either integrating over positive (+) and negative (−) values of the triple product, or as a function of ϕ, the angle between the decay planes defined by ab and cd. The scalar triple product is even under C and odd under P and CP transformations. It is also odd under T, however it is not possible to identify physical pairings of ψ and ψ' for $M \to abcd$. Some of the literature chooses to invoke CPT in order to trivially relate CP violation to a T violation and call these CP violating triple product asymmetry measurements T violating. With such a mind set one ignores the full set of tests that can be performed and chooses the limiting case where all CP violation signals might equally be called T violation. For such an approach there is simply no point in making the distinction between triple product CP tests and other CP violation manifestations, hence no point in discussing T violation. For some time a more general approach has been advocated in the literature, and in recent years experiment has gone beyond the limiting case of CPT conservation for a number of measurements. The invocation of CPT conservation is no longer reasonable for a general discussion of discrete symmetry violations. The latter part of these proceedings discuss tests of CP, T and CPT symmetries, and these proceedings regard CP violating triple products to be just that. It is worth noting that for entangled systems one could define ψ and ψ' under a T or CPT transformation and hence study triple products using decays of $M\overline{M} \to (abcd)(efgh)$ as well as mixing CP filters with the four body decays in analogy with the methodology outlined in Section 3. We do not discuss this possibility further here as there have not been any attempts to perform such a measurement thus far, however this is a further reason to ensure clarity of the language used when talking about symmetry violations[1]. Figure 1 illustrates the decay $M \to abcd$ and the corresponding P, C and CP transformed processes. The T transformed process $\overline{abcd} \to \overline{M}$ never happens and is not shown.

By considering the possible conjugate pairings illustrated in the figure one can construct six distinct asymmetries from triple products. These are [3]

$$A_P = \frac{\Gamma_+ - \Gamma_-}{\Gamma_+ + \Gamma_-}, \qquad \overline{A}_P = \frac{\overline{\Gamma}_+ - \overline{\Gamma}_-}{\overline{\Gamma}_+ + \overline{\Gamma}_-}. \tag{2}$$

$$A_C = \frac{\overline{\Gamma}_- - \Gamma_-}{\overline{\Gamma}_- + \Gamma_-}, \qquad \overline{A}_C = \frac{\overline{\Gamma}_+ - \Gamma_+}{\overline{\Gamma}_+ + \Gamma_+}. \tag{3}$$

$$A_{CP} = \frac{\overline{\Gamma}_+ - \Gamma_-}{\overline{\Gamma}_+ + \Gamma_-}, \qquad \overline{A}_{CP} = \frac{\overline{\Gamma}_- - \Gamma_+}{\overline{\Gamma}_- + \Gamma_+}, \tag{4}$$

where the quantities $\Gamma_\pm$ and $\overline{\Gamma}_\pm$ denote the rate for particle (anti-particle) decays with the sign of the triple product denoted by the subscript. The subscript on the asymmetry A denotes the symmetry used to construct the conjugate pair. A further six asymmetries can be constructed by considering the two other symmetries acting on A and $\overline{A}$ in turn. These are given by [3]

$$a_C^P = \frac{1}{2}\left(A_P - \overline{A}_P\right),$$

[1] Several authors have pointed out that the usual definitions of discrete operators need to be modified in the context of CPT violation. For example see Ref. [4].

Journal of Physics: Conference Series **631** (2015) 012003 doi:10.1088/1742-6596/631/1/012003

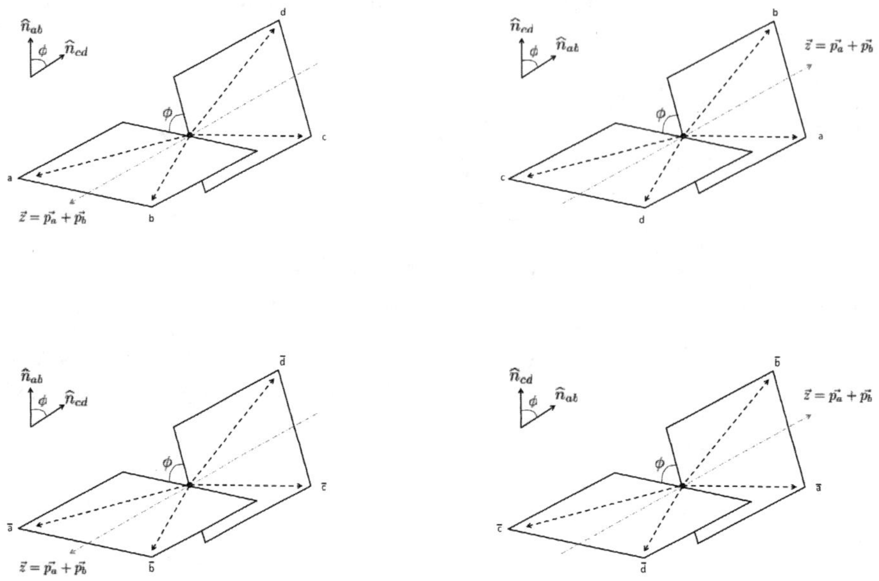

Figure 1. The decay (top-left) $M \to abcd$ and the symmetry transformed decays under (top-right) P, (bottom-left) C, and (bottom-right) CP.

$$
\begin{aligned}
a_{CP}^{P} &= \frac{1}{2}\left(A_P + \overline{A}_P\right), \\
a_{P}^{C} &= \frac{1}{2}(A_C - \overline{A}_C), \\
a_{CP}^{C} &= \frac{1}{2}(A_C + \overline{A}_C), \\
a_{P}^{CP} &= \frac{1}{2}(A_{CP} - \overline{A}_{CP}), \\
a_{C}^{CP} &= \frac{1}{2}(A_{CP} + \overline{A}_{CP}).
\end{aligned} \tag{5}
$$

Here these secondary asymmetries are denoted by a lower case a, where the superscript corresponds to the first symmetry used for conjugate pairing and the subscript corresponds to the second one. One can determine which symmetry is under scrutiny by multiplying the superscripts and subscripts together. For example a_C^P is a test of CP, a_{CP}^P a test of $CP^2 = C$, and so on. In general these asymmetries receive contributions from all interactions, and the interest is in isolating effects that are dominated by weak interactions (i.e. theoretically clean), or where that may not be possible, to isolate effects that can either be understood in the longer term, or signify a non-trivial weak interaction effect even in the presence of pollution from strong force induced final state interactions (i.e. soft QCD or re-scattering). A concrete example of this issue is discussed below in the context of the measurement of α_b from Λ_b decays. Hence interpretation of these asymmetries depends on the decay under study (and hence the model required to interpret data).

There are three categories of event that can be studied; for the first two all twelve quantities are non-trivial and distinct. For the third category the non-trivial asymmetries are degenerate. These categories are defined by: (i) $M \neq \overline{M}$ and $abcd \neq \overline{abcd}$, (ii) $M \neq \overline{M}$ and $abcd = \overline{abcd}$, and (iii) $M = \overline{M}$ and $abcd = \overline{abcd}$. The asymmetry relevant for type (iii) decays is

$$A_{P,CP} = \frac{\langle\Gamma\rangle_+ - \langle\Gamma\rangle_-}{\langle\Gamma\rangle_+ + \langle\Gamma\rangle_-}, \tag{6}$$

which can be constructed by considering P and CP using the rates measured.

2.1. Meson decays

Triple product asymmetry measurements have been made in K, D, and B systems. Until recently only three of these were reported by experiments; A_P, $\overline{A}_P$, and a^P_{CP}. The rare kaon decays $K_{L,S} \to \pi^+\pi^-e^+e^-$ are type (iii) processes and they have been studied experimentally by KTeV (K_L only) and NA48 (both decays). The PDG reports an asymmetry of $13.7 \pm 0.5\%$ for $K_L \to \pi^+\pi^-e^+e^-$ [5]. The results for the K_L decay are in excellent agreement with theoretical prediction; a 14% CP asymmetry is observed [6] in this decay. The corresponding asymmetry found for $K_S \to \pi^+\pi^-e^+e^-$ is consistent with zero. It was pointed out some time ago that a^P_C could also be measured [7] while studying triple product asymmetries relevant for charm studies. The full set of asymmetries has been measured by *BABAR* for $D^\pm_{(s)}$ decays into $K_S K^\pm\pi^+\pi^-$ and for D^0 decays into $K^+K^-\pi^+\pi^-$ [8]. LHCb have recently measured the three asymmetries A_P, $\overline{A}_P$ and a^P_C for these modes [9].

The presence of soft QCD and re-scattering gives rise to the question, what is one testing with these decays. It is well known that some asymmetries can be manifestly non-zero driven solely by strong phase differences and hence do not provide clear insight into weak dynamics. The converse is also true for other asymmetries. If one considers a simple model of two interfering amplitudes it is straightforward to show that half of the asymmetries defined above fall into the former category and half into the latter one. Hence six of these quantities are interesting to measure from the perspective of the weak interaction even in the limit that we are unable to control hadronic contributions.

2.2. Other possible measurements

It is possible to perform these measurements using baryon decays. One possible advantage of such measurements is that the energy regime is increased relative to meson decays and so it may be easier to compute the expected contributions from hadronic uncertainties to estimate the underlying weak effects. This is analogous to the effort underway for determining α from Λ_b decays at the LHC [10, 11] as a follow up to the corresponding study in Λ decays performed by Cronin and Overseth [12]. Similar measurements are proposed for Λ_c decays [13, 14, 15]. It is also possible to study top quark and tau lepton systems as outlined in [3].

Since the advent of the LHC it has been possible to entertain the prospects of analysing data involving Z or H boson decay, or associated production of pairs of bosons (ZH, WH, HH, etc.) that subsequently decay into a four particle final state. The SM predictions for C and CP effects are clear for all of these transitions - the associated asymmetries should be zero as non-zero violations arise from W boson interaction with up and down type quark fields. It is possible to estimate the precision with which the LHC can measure asymmetries for a number of modes based on the available Run 1 publications, and expectations through to the high luminosity LHC phase expected to start in the a decade from now. One of the most straightforward measurements to make is that of a Z or H boson decay to four leptons; $\ell^+\ell^-\ell'^+\ell'^-$. The final states where $\ell \neq \ell'$ have the largest rate and no ambiguity in terms of combinatorics of the signal reconstruction. The LHC has observed both $Z^0 \to \mu^+\mu^-e^+e^-$ [16]

and $H \to \mu^+\mu^-e^+e^-$ [17, 18]. Expectations for the precision attainable for the triple product asymmetry measurement in these processes are given in Table 1. Percent-level precision could be achieved at the end of the high luminosity LHC (HL-LHC) assuming that trigger rates can be maintained and systematic uncertainties can be controlled in reconstructing these final states. Final states where $\ell = \ell'$ have rates a factor of two smaller than those for $\ell \neq \ell'$, and may also suffer from a small dilution coming from ambiguities in reconstruction of the final states. Hence one can expect that the 4ℓ modes, where $\ell = e, \mu, \tau$, have precisions at least $\sqrt{2}$ larger than the estimates shown in the table.

Table 1. Estimated precisions on the asymmetry of Eq. (6) for Z^0 and H decays to $\mu^+\mu^-e^+e^-$ for ATLAS and CMS at the LHC and HL-LHC. These estimates are based on the published run 1 yields, assuming that the asymmetry is zero.

Data sample	$Z^0 \to \mu^+\mu^-e^+e^-$	$H \to \mu^+\mu^-e^+e^-$
Run 1 ($\sim 25\,\text{fb}^{-1}$)	0.12	0.38
Run 2 ($\sim 125\,\text{fb}^{-1}$)	0.04	0.11
Run 3 ($\sim 300\,\text{fb}^{-1}$)	0.03	0.07
HL-LHC ($\sim 3000\,\text{fb}^{-1}$)	0.01	0.02

Another potential channel of interest is the associated production of a H boson in conjunction with a W or Z. The use of triple product asymmetries to probe NP is WH decays has been discussed in Ref. [19]. The latest experimental searches from the LHC are reported in [20, 21]. It might be possible to find evidence for these decays during Run 2. If that were to happen one could start to measure triple product asymmetries for combinations of H boson couplings with a W or Z. The proposed ILC is expected to accumulate large samples of ZH bosons, and assuming the yields outlined in [22] we would expect a statistical precision of 2.1 (3.7)% and 2.5 (4.7)% to be attainable on the asymmetry of Eq. (6) using inclusive (exclusive) Z decays to lepton pairs at 250 and 500 GeV, respectively. The proposed Chinese Higgs Factory (CEPC) aims to integrate a sample of about 1 million ZH decays [23]. If built the proposed Future Circular e^+e^- Collider (FCC-ee) would accumulate similar statistics. Such a sample size could be used to archive a precision about a factor of five times better than that of the ILC.

3. Entangled states

John Bell was the first person to understand the concept of quantum entanglement [24, 25] and in doing so solved an old problem in quantum mechanics. This concept has been tested extensively in a number of systems, and is used in a practical sense today in the context of quantum information and to under-pin efforts to construct quantum computers. This concept is also central to the study of discrete symmetry violation tests in neutral mesons performed at $K(\phi)$ and B meson factories.

Pairs of entangled neutral mesons $M = K$, D, $B_{s,d}$ can be prepared in decays of ϕ, $\psi(3770)$, $\Upsilon(4S)$ mesons. If we assume the fundamentals of quantum mechanics such as locality and coherence of the entangled wave function Ψ, then we may write

$$\Psi = \frac{1}{\sqrt{2}}\left(M_1^0\overline{M}_2^0 - \overline{M}_1^0 M_2^0\right), \quad (7)$$

$$= \frac{1}{\sqrt{2}}\left(M_{1,+}M_{2,-} - M_{1,-}M_{2,+}\right). \quad (8)$$

Here the subscript 1 and 2 refers to the first and second meson, respectively and M^0 is a particle, while $\overline{M}^0$ is the anti-particle for the flavour filter basis set $\{M^0, \overline{M}^0\}$. The corresponding CP

filter basis set is $\{M_+, M_-\}$, where the subscript indicates the CP eigenvalue of the filter. One can probe CP, T, and CPT using pairs of entangled mesons by considering the 15 non-trivial possible ways of constructing asymmetries. These pairings are obtained by considering all possible combinations of decay for the first and second meson illustrated in Table 2.

Table 2. Possible combinations of decay for an entangled wave function. The first meson decay type is indicated in the first row, and that of the second meson is given by the first column.

	M^0	$\overline{M}^0$	M_+	M_-
M^0	$(+1, +1)$	$(-1, +1)$	$(CP_{even}, +1)$	$(CP_{odd}, +1)$
$\overline{M}^0$	$(+1, -1)$	$(-1, -1)$	$(CP_{even}, -1)$	$(CP_{odd}, -1)$
M_+	$(+1, CP_{even})$	$(-1, CP_{even})$	(CP_{even}, CP_{even})	(CP_{odd}, CP_{even})
M_-	$(+1, CP_{odd})$	$(-1, CP_{odd})$	(CP_{even}, CP_{odd})	(CP_{odd}, CP_{odd})

Traditionally experiments have studied entangled states in terms of flavour filter basis pairs only. This permits two tests to be performed: the comparison of $M^0 \to \overline{M}^0$ with the CP and T conjugate process $\overline{M}^0 \to M^0$; the comparison of $M^0 \to M^0$ with the CP and CPT conjugate process $\overline{M}^0 \to \overline{M}^0$. These two tests are typically characterised by reconstructing same sign and opposite sign di-lepton final states in order to measure $A_{CP,T}$ and $A_{CP,CPT}$, respectively [26, 27, 28, 29, 30, 31, 32]. Some measurements have also been performed using hadronically tagged flavour filters to complement those of semi-leptonic decays. The study of hadronically tagged decays is motivated by the fact that the semi-leptonic decay results are dominated by systematic uncertainties at the B Factories. In particular measurements of $A_{CP,T}$ at the B Factories mirror efforts by CPLEAR to measure T violation via the so-called Kabir asymmetry method [33, 34]. The Tevatron and LHC experiments measure semi-leptonic asymmetries to perform these tests, see for example Ref. [35]. It is interesting to note that these measurements serve as simultaneous tests of two symmetries at a time [c.f. the type (iii) triple product asymmetry above] and are constructed using only the flavour basis pairing. Experimental results from the B Factories on $A_{CP,T}$ and $A_{CP,CPT}$ measurements can be found in Chapter 17.5 of [1] and references therein.

Twelve asymmetries can be constructed as outlined by Bernabeu et al. in Refs. [36, 37, 38, 39, 40]. The full set of reference and conjugate pairs used to test CP, T and CPT are listed in Table 3. These are each tests of a single symmetry by virtue of invoking both flavour and CP filter basis pairs reconstructed from the decays of the two mesons. Figure 2 illustrates this for the reference process $\overline{B}^0 \to B_-$ under the symmetry T. The conjugate pairing to this under T interchanges the initial and final states. The $\overline{B}^0$ meson in the reference process is tagged by the decay of the other B in the event being filtered by an $\ell^+ X$ decay and the B_- filter is identified through the channel $J/\psi K_S$. The T conjugate of this scenario requires that the B_- be tagged by the decay of the other B in the event filtered via the channel $J/\psi K_L$, and the $\overline{B}^0$ is identified by the flavour filter channel $\ell^- X$. It is not sufficient in this case to just consider the interchange of initial and final states, one also has to consider how to filter on the flavour and CP components in order to correctly construct the desired asymmetry test.

The *BABAR* experiment has performed the measurements proposed by Bernabeu et al., the results of which can be found in [40]. *BABAR* find T violation occurs with a significance, assuming Gaussian errors, of 14σ. This balances the level of CP violation observed, and CPT is found to be conserved. The asymmetries are measured as a function of the time difference $\Delta t = t_2 - t_1$, which is positive by definition and not to be confused with the signed difference in proper time

Table 3. The twelve pairings of reference and symmetry conjugated transitions used to study CP, T and CPT following the approach described in the text.

Symmetry	Reference transition	Conjugate transition
CP	$\overline{M}^0 \to M_-$	$M^0 \to M_-$
	$M_+ \to M^0$	$M_+ \to \overline{M}^0$
	$\overline{M}^0 \to M_+$	$M^0 \to M_+$
	$M_- \to M^0$	$M_- \to \overline{M}^0$
T	$\overline{M}^0 \to M_-$	$M_- \to \overline{M}^0$
	$M_+ \to M^0$	$M^0 \to M_+$
	$\overline{M}^0 \to M_+$	$M_+ \to \overline{M}^0$
	$M_- \to M^0$	$M^0 \to M_-$
CPT	$\overline{M}^0 \to M_-$	$M_- \to M^0$
	$M_+ \to M^0$	$\overline{M}^0 \to M_+$
	$M^0 \to M_-$	$M_- \to \overline{M}^0$
	$M_+ \to \overline{M}^0$	$M^0 \to M_+$

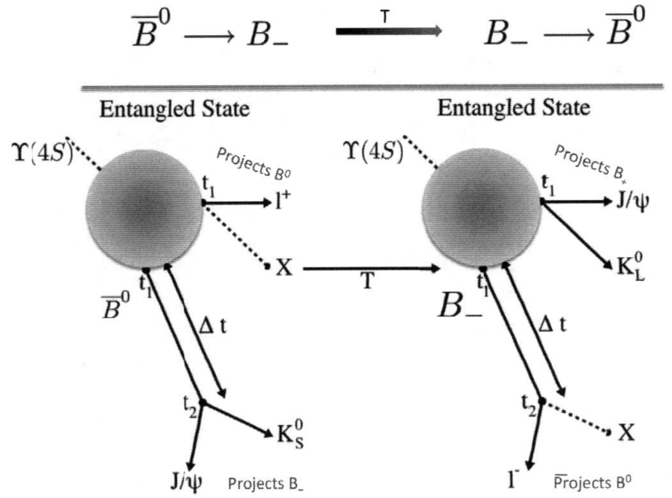

Figure 2. Illustration of the conjugate pairings used to test T for a reference process of $\overline{B}^0 \to B_-$ and the T conjugate $B_- \to \overline{B}^0$.

used in usual time-dependent CP analysis approach. The time-dependent rates are given by

$$g^{\pm}_{\alpha,\beta}(\Delta t) \propto e^{-\Gamma\Delta t}\left[1 + C^{\pm}_{\alpha,\beta}\cos(\Delta m\Delta t) + S^{\pm}_{\alpha,\beta}\sin(\Delta m\Delta t)\right], \tag{9}$$

Journal of Physics: Conference Series **631** (2015) 012003 doi:10.1088/1742-6596/631/1/012003

where Γ is the width of the B, Δm is the mixing frequency between B^0 and $\overline{B}^0$ mesons, α and β are the flavour and CP filters and $\pm$ indicates if the flavour filter occurs before the CP filter $(+)$ or not $(-)$. One can relate the S and C parameters to the real and imaginary parts of $\lambda = (q/p)(\overline{A}/A)$, where q and p are mixing parameters and $\overline{A}/A$ is the ratio of amplitudes for antiparticle and particle decays to CP conjugate final states. This means that these measurements can be used to constrain the phase in the CKM matrix. For example the T asymmetry can be written as

$$A_T \simeq \frac{\Delta C^{\pm}}{2}\cos(\Delta m \Delta t) + \frac{\Delta S^{\pm}}{2}\sin(\Delta m \Delta t), \quad (10)$$

where in the SM one would expect $\Delta C^{\pm} = 0$ and $\Delta S^{\pm} = \mp \sin 2\beta$ for the CP filters $J/\psi K_{S,L}$.

It has been noted that one has to take care in defining the operators used to test symmetries when CPT is relaxed. A potential consequence of this is a second interpretation of the *BABAR* results; it is possible that those results are a sign of CPT violation masquerading as fake T violation [4]. The *BABAR* interpretation of the data is the more conservative approach, and it will be very interesting to see the results of complementary measurements made by Belle II in the fullness of time in case this might shed light on the interpretational ambiguity.

It has been noted that one can extend the set of CP filter basis pairs from $\{J/\psi K_S, J/\psi K_L\}$ to other final states in order to broaden the possible measurements that can be made [41, 42]. In addition to studying $c\overline{c}s$ CP filter transitions where the reconstructed kaon is either used to select a K_S or K_S filter state. One can use decays into VV final states to filter out CP even an odd amplitudes via a time-dependent angular analysis as noted in [41]. These measurements are yet to be performed, however it is expected that Belle II would accumulate sufficient statistics in $50\,\text{ab}^{-1}$ to observe T violation using CP basis pairs from $(\eta^{(\prime)}, \omega)K_{S,L}$, ϕK^* and D^*D^* decays.

The remaining asymmetry that can be constructed with entangled states is done so by considering only the CP filter basis pairing. This enables one to compare $M_+ \to M_-$ with $M_- \to M_+$ and in doing so perform a test of T and CPT; the rate difference between this pair of processes is $A_{T,CPT}$. Experimental prospects for such a measurement are bleak at the current time. *BABAR* and Belle should have of the order of 1 event, and Belle II is expected to accumulate of the order of 20 events for such a measurement. A third generation B factory may be able to start measuring $A_{T,CPT}$.

As remarked in Section 2 it is possible to also test C and P using entangled meson systems if one studies decays into double-tagged pairs of four body final states, or other combinations of final state using flavour or CP filter decays. Hence it is possible to probe the full set of weak interaction behaviour under C, P, CP, T and CPT using the same sample of data to fully over-constrain the consistency of results with expectations of the weak interaction.

4. Summary

The weak interaction is well known to violate P, C, T and CP. These effects are accommodated in the SM Lagrangian, however a deep understanding of the origin of these phenomena eludes us. The potential to learn something new requires a next generation of innovative and exhaustive measurements in the hope that some may go beyond our current expectations and hint at a deeper truth. The combined symmetry of CPT has important ramifications, and as a consequence we require CPT to be conserved if Lorentz symmetry is respected in nature. While one expects to find that this symmetry is conserved, at some point we believe that CPT violation may be manifest (for example in theories of quantum gravity). These measurements are important to make as the implications, should we find CPT violation, would be far reaching for modern physics. The use of triple product asymmetries in testing symmetry violation for different final states may yield a deeper insight into the weak interaction. Likewise the use of entangled meson systems at experiments will continue to enhance our understanding of violations of these discrete symmetries. By systematically probing a broad range of observables spanning quarks, leptons and bosons one hopes that the data may point to a more complete understanding of nature.

References

[1] A. J. Bevan *et al.* [BaBar and Belle Collaborations], Eur. Phys. J. C **74** (2014) 11, 3026 [arXiv:1406.6311 [hep-ex]].
[2] M. Gronau and J. L. Rosner, Phys. Rev. D **84** (2011) 096013 [arXiv:1107.1232 [hep-ph]].
[3] A. J. Bevan, arXiv:1408.3813 [hep-ph].
[4] M. Fidecaro, H. -J. Gerber and T. Ruf, arXiv:1312.3770 [hep-ph].
[5] K. A. Olive *et al.* [Particle Data Group Collaboration], Chin. Phys. C **38** (2014) 090001.
[6] P. Heiliger and L. M. Sehgal, Phys. Rev. D **48** (1993) 4146 [Phys. Rev. D **60** (1999) 079902].
[7] X. W. Kang and H. B. Li, Phys. Lett. B **684** (2010) 137 [arXiv:0912.3068 [hep-ph]].
[8] M. Martinelli, contribution to the 8th international workshop on the CKM Unitarity Triangle, Vienna (2014).
[9] M. Martinelli, arXiv:1411.4140 [hep-ex].
[10] R. Aaij *et al.* [LHCb Collaboration], Phys. Lett. B **724** (2013) 27 [arXiv:1302.5578 [hep-ex]].
[11] G. Aad *et al.* [ATLAS Collaboration], Phys. Rev. D **89** (2014) 9, 092009 [arXiv:1404.1071 [hep-ex]].
[12] J. W. Cronin and O. E. Overseth, Phys. Rev. **129** (1963) 1795.
[13] W. Bensalem, A. Datta and D. London, Phys. Lett. B **538** (2002) 309 [hep-ph/0205009].
[14] W. Bensalem, A. Datta and D. London, Phys. Rev. D **66** (2002) 094004 [hep-ph/0208054].
[15] X. W. Kang, H. B. Li, G. R. Lu and A. Datta, Int. J. Mod. Phys. A **26** (2011) 2523 [arXiv:1003.5494 [hep-ph]].
[16] G. Aad *et al.* [ATLAS Collaboration], Phys. Rev. Lett. **112** (2014) 23, 231806 [arXiv:1403.5657 [hep-ex]].
[17] G. Aad *et al.* [ATLAS Collaboration], Phys. Lett. B **716** (2012) 1 [arXiv:1207.7214 [hep-ex]].
[18] S. Chatrchyan *et al.* [CMS Collaboration], Phys. Lett. B **716** (2012) 30 [arXiv:1207.7235 [hep-ex]].
[19] C. Delaunay, G. Perez, H. de Sandes and W. Skiba, Phys. Rev. D **89** (2014) 3, 035004 [arXiv:1308.4930 [hep-ph]].
[20] S. Chatrchyan *et al.* [CMS Collaboration], Phys. Rev. D **89** (2014) 1, 012003 [arXiv:1310.3687 [hep-ex]].
[21] G. Aad *et al.* [ATLAS Collaboration], JHEP **1501** (2015) 069 [arXiv:1409.6212 [hep-ex]].
[22] T. Tanabe, contribution to the workshop on the Interplay between Particle and Astroparticle Physics, London (2014).
[23] M. Ahmad *et al.* [CEPC-SPPC Study Group], IHEP-CEPC-DR-2015-01.
[24] A. Einstein, B. Podolsky, N. Rosen Phys. Rev. **47**, 777-780 (1935).
[25] J. S. Bell, Physics **1**, 195 (1964).
[26] B. Aubert *et al.* [BaBar Collaboration], Phys. Rev. Lett. **88** (2002) 231801 [hep-ex/0202041].
[27] B. Aubert *et al.* [BaBar Collaboration], Phys. Rev. Lett. **92** (2004) 181801 [hep-ex/0311037].
[28] B. Aubert *et al.* [BaBar Collaboration], Phys. Rev. D **70** (2004) 012007 [hep-ex/0403002].
[29] E. Nakano *et al.* [Belle Collaboration], Phys. Rev. D **73** (2006) 112002 [hep-ex/0505017].
[30] B. Aubert *et al.* [BaBar Collaboration], Phys. Rev. Lett. **96** (2006) 251802 [hep-ex/0603053].
[31] B. Aubert *et al.* [*BABAR* Collaboration], Phys. Rev. Lett. **100**, 131802 (2008) [arXiv:0711.2713 [hep-ex]].
[32] T. Higuchi *et al.* [Belle Collaboration], Phys. Rev. D **85**, 071105 (2012) [arXiv:1203.0930 [hep-ex]].
[33] P. K. Kabir, Phys. Rev. D **2**, 540 (1970).
[34] A. Angelopoulos *et al.* [CPLEAR Collaboration], Phys. Lett. B **444**, 43-51 (1998).
[35] A. Lenz and U. Nierste, arXiv:1102.4274 [hep-ph].
[36] M. C. Banuls and J. Bernabeu, Phys. Lett. B **464**, 117 (1999) [hep-ph/9908353].
[37] M. C. Banuls and J. Bernabeu, Nucl. Phys. B **590**, 19 (2000) [hep-ph/0005323].
[38] E. Alvarez and A. Szynkman, Mod. Phys. Lett. A **23**, 2085 (2008) [hep-ph/0611370].
[39] J. Bernabeu, F. Martinez-Vidal and P. Villanueva-Perez, JHEP **1208**, 064 (2012) [arXiv:1203.0171 [hep-ph]].
[40] J. P. Lees *et al.* [*BABAR* Collaboration], Phys. Rev. Lett. **109**, 211801 (2012) [arXiv:1207.5832 [hep-ex]].
[41] A. Bevan, PoS EPS **-HEP2013** (2013) 001 [arXiv:1311.0920 [hep-ph]].
[42] R. Dadisman, S. Gardner and X. Yan, arXiv:1409.6801 [hep-ph].

4th Symposium on Prospects in the Physics of Discrete Symmetries (DISCRETE2014) IOP Publishing
Journal of Physics: Conference Series **631** (2015) 012004 doi:10.1088/1742-6596/631/1/012004

Tests of discrete symmetries in K systems

Antonio Di Domenico

Dipartimento di Fisica, Sapienza Università di Roma,
and I.N.F.N. Sezione di Roma, P.le A. Moro, 2, I-00185 Rome, Italy

E-mail: `antonio.didomenico@roma1.infn.it`

Abstract. The status of present experiments and future projects with kaons is reviewed, focusing on prospects for discrete symmetries tests.

1. Introduction

In 2014 there are several anniversaries to be celebrated, as reminded by Prof. Mavromatos in the opening of the DISCRETE 2014 symposium: the 150th anniversary of the publication of Maxwell's equations, the 60th anniversary of the discovery of Yang-Mills Gauge theories, the 50th anniversary of the discovery of CP Violation, and the 50th anniversary of the proposal of Bell's Inequalities in Quantum Physics.

Among them there is also the 70th anniversary of the first publication reporting an indication of a new charged particle with mass ~ 0.5 GeV/c^2 in cosmic rays (Leprince-Ringuet, L'héritier [1]), i.e. the first experimental evidence for a K^+ meson, remarkably even before the discovery of the pion. Since then, K mesons, or kaons, turned out to be one of the most interesting and promising system in particle physics, especially in the study of discrete symmetries. The discovery in 1957 of the $\tau - \theta$ anomaly due to parity non-conservation in weak decays [2], the discovery in 1964 of CP violation in the $\mathrm{K}^0 - \bar{\mathrm{K}}^0$ mixing [3], and of direct CP violation almost 40 years later [4, 5], are bright examples of the outcomes of these studies.

After 70 years from the first observation one might wonder whether investigations on kaons have exhausted the information that such relatively simple but rich system can provide. The present paper, trying to convince the reader that this is not the case, focuses on the most recent progresses in the field, updating a previous report on the status of experiments and future projects [6]; for general and historical reviews the reader is referred to several excellent papers and books in the literature [7].

2. Kaon experiments at hadron machines

2.1. The intensity frontier for New Physics searches

Precise measurements in the Flavor sector can probe extremely high energy scales, in a complementary way with respect to the LHC high energy frontier. In fact, the present LHC searches at high transverse momenta exclude wide regions of the parameter space for New Physics (NP), thus increasing the importance of the study of rare processes sensitive to very high energy. This is true, in particular, for the $\mathrm{K} \to \pi\nu\bar{\nu}$ decays, whose SM predictions are known with very high precision [8]. The branching ratio expectation of the CP-violating $\mathrm{K_L} \to \pi^0\nu\bar{\nu}$ channel is: $\mathrm{BR}(\mathrm{K_L} \to \pi^0\nu\bar{\nu}) = (0.243 \pm 0.039 \pm 0.006) \times 10^{-10}$, while for the charged mode is:

$BR(K^+ \to \pi^+ \nu\bar{\nu}) = (0.781 \pm 0.075 \pm 0.029) \times 10^{-10}$. In both cases, the first error summarizes the parametric uncertainties, while the second one the remaining theoretical error. The extraction of the hadronic part of the amplitude is obtained from the very well measured $K_{\ell 3}$ decay. The main uncertainties are due to the knowledge of the CKM matrix elements.

Processes mediated by FCNC are suppressed in the SM by the GIM mechanisms [9]. Further suppression can originate from hierarchies in CKM matrix or helicity. In some models of NP large enhancement with respect to SM rates are possible. In this sense the $K \to \pi\nu\bar{\nu}$ modes play a key role in seeking NP beyond the SM. This is clear from Fig.1, which reproduces the BR values predicted by the SM and by some models of NP [10]. Apart from establishing a direct signal of NP, the correlation of the BR of the two modes can be exploited to probe the flavour structure of NP theories and, therefore, to distinguish among different classes of NP scenarios. The exclusion regions given by the Grossman-Nir consistency condition [11], limiting the BR ratio of the two decay modes, and by the only measurement of the charged mode [12] are also reported in Fig.1.

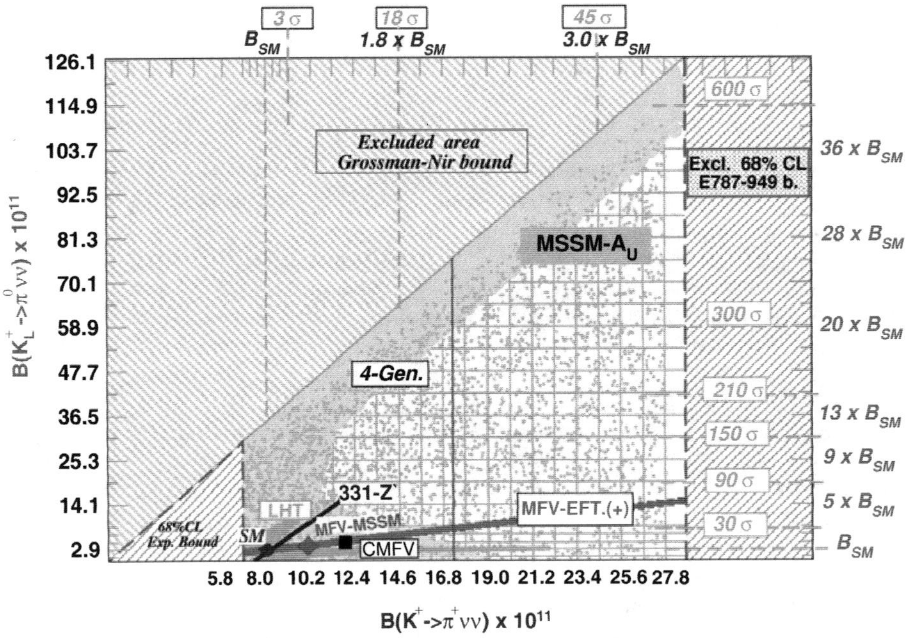

Figure 1. Correlation between the branching ratios of the charged and neutral $K \to \pi\nu\bar{\nu}$ decays showing the predictions of SM and some NP models [10].

The measurement $BR(K^+ \to \pi^+ \nu\bar{\nu}) = (1.73 \pm 1.15) \times 10^{-10}$ is based on 7 events observed by E959 and E787 experiments at BNL [12], and is compatible with the SM prediction. The CP-violating neutral $K_L \to \pi^0 \nu\bar{\nu}$ decay has not been observed yet. An upper limit on the BR was set by the E391a collaboration at KEK [13]: $BR(K_L \to \pi^0 \nu\bar{\nu}) < 260 \times 10^{-10}$ (90% C.L.).

2.2. NA62 experiment at CERN

The main goal of the NA62 experiment at CERN [14] is to collect about 100 $K^+ \to \pi^+ \nu\bar{\nu}$ events with 10% of background in two years of data taking. This implies to collect more than 10^{13} K decays with a background rejection factor of at least 10^{12}, assuming a signal acceptance of 10%.

The experimental method relies on exploiting a decay-in-flight technique with an intense charged kaon beam at high energy. This requires a beam of unseparated charged hadrons (with 6% K^+), a long decay region with an extended detector and the event-by-event measurement of kaon momenta. Positive charged kaons will be used in order to get, at the same proton energy and flux, higher kaon fluxes and lower pion background.

The incoming kaon is measured by the Gigatracker system in the beam. The charged decay particle is measured by the straw-chamber spectrometer and is identified by the Ring Imaging Cerenkov (RICH) detector and the muon-veto sampling calorimeter. The LKr calorimeter, originally built for the NA48 experiment, is used as a veto for forward photons. Photons at large angles are intercepted by a series of 12 ring-shaped veto counters constructed using lead-glass blocks from the OPAL electromagnetic barrel calorimeter.

The NA62 analysis strategy is based on the accurate kinematic reconstruction of all the particles detected in the event to disentangle the signal from the huge amount of background processes, a precise timing to associate correctly the π^+ with the parent K^+, a system of efficient vetoes to reject events with γ and μ in the final state, effective particle identification systems to identify K^+ among other particles of the intense hadron beam and to distinguish π^+ from μ^+ and e^+ in the final states.

The construction and installation of the experimental apparatus have been completed in May 2014, as shown in the pictures of Fig.2 (see Ref.[15] for more details). The first physics run has successfully started at the end of 2014 at a lower intensity beam, while completing the commissioning of hardware and readout, with the aim of reaching the SM sensitivity. The nominal intensity beam is foreseen for the full physics run in 2015-2017.

2.3. KOTO experiment at J-PARC

The KOTO experiment [16] at the high-intensity proton accelerator facility J-PARC (Japan Proton Accelerator Research Complex) [17] aims to observe for the first time the rare decay $K_L \to \pi^0 \nu\bar{\nu}$, which is rather challenging because both initial state and final decay products are completely neutral.

A decay volume for K_L is surrounded by particle detectors. The signature of a $K_L \to \pi^0 \nu\bar{\nu}$ decay is that there are two photons from a π^0 decay and no other visible particles in the final state. An electromagnetic calorimeter is placed downstream of the decay volume to detect the two photons. All the K_L decay modes except $K_L \to \pi^0 \nu\bar{\nu}$ and $K_L \to \gamma\gamma$ have at least two charged particles, or two or more extra photons in the final state. These decays can be rejected by detecting additional particles with the surrounding detectors. The $K_L \to \gamma\gamma$ decays can be rejected by requiring a finite transverse momentum for the two photon system. In the case of $K_L \to \pi^0 \nu\bar{\nu}$ decay, the two photon system has a finite transverse momentum, because the undetected two neutrinos take some momentum away. The decay vertex is calculated and the $K_L \to \pi^0 \nu\bar{\nu}$ decay reconstructed from two photons in the calorimeter, with the assumption that the two photons come from a π^0 decay on the line of flight.

The new neutral beam-line has been built at the Hadron Experimental Hall of J-PARC. The KOTO detector, consisting mainly of the CsI crystal calorimeter, the charged-particle veto counters (CV) in front of it, and the main-barrel photon veto counters (MB) surrounding the decay volume in the vacuum vessel, have been built and installed by May 2013.

After the major Earthquake in Japan on 11 March 2011 (no damages to the beam line and the KOTO detector), the KOTO first physics run started on May 17 2013 with a beam power of 24 kW (10% of design intensity). Unfortunately, the data taking was terminated after only 100 hours due to a radiation accident in J-PARC Hadron hall (a gold target partially melted and radioactive material diffused in the atmosphere).

The analysis of the 100 hours run data [18], after determining all the selection criteria and

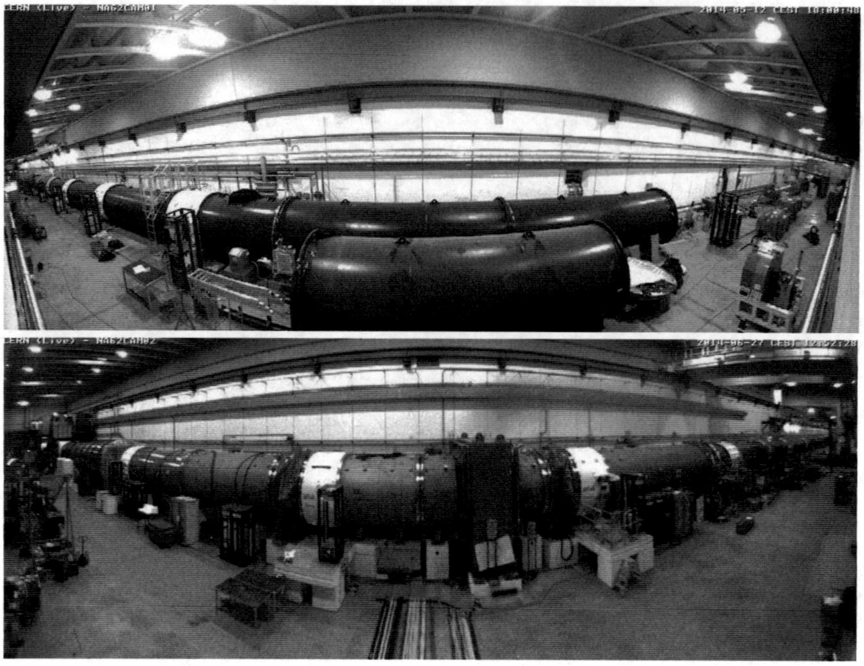

Figure 2. Live webcam pictures taken before and after the completion of the installation of the decay beam pipe of the NA62 experiment at CERN.

estimating background levels, showed one observed event in the signal region, as shown in Fig.3, which was statistically consistent with the number of expected background events. Based on the number of simultaneously measured $K_L \to 2\pi^0$ events, the evaluated single event sensitivity (S.E.S) for the $K_L \to \pi^0 \nu\bar{\nu}$ decay results to be 1.29×10^{-8}, while the S.E.S of the E391a experiment was 1.11×10^{-8}. Although data taking time was only 100 hours, about the same sensitivity as E391a, which took data for 1100 hours, has been achieved.

The next data taking is planned to start in 2015 improving the experimental sensitivity by preparing new detectors to reduce the background events and improving the analysis methods to remove hadron interaction events.

2.4. TREK experiment at J-PARC

In the $K^+ \to \pi^0 \mu^+ \nu$ decay, the transverse muon polarization p_t (the perpendicular component of the muon spin vector relative to the decay plane determined by the momentum vectors of the muon and the pion in the K^+ rest frame) is a T -odd quantity. The SM prediction of p_t is almost vanishing ($\sim 10^{-7}$) and a non-zero value for such quantity at a level above 10^{-4} would be an indication of time-reversal violation in NP, still allowed by the current limit ($p_t < 5 \times 10^{-3}$) set by the KEK E246 experiment [19].

A successor to E246 is another new kaon experiment at J-PARC, E06 TREK [20], aiming at a p_t sensitivity of 10^{-4}. This experiment is conducted in conjunction with the E36 one aiming to measure the $R_K = \Gamma(K^\pm \to e\nu_e)/\Gamma(K^\pm \to \mu\nu_\mu)$ ratio with a relative uncertainty of $\sim 2.5 \times 10^{-3}$, which is about half of the current world record. The physics run of E36 is expected in 2015.

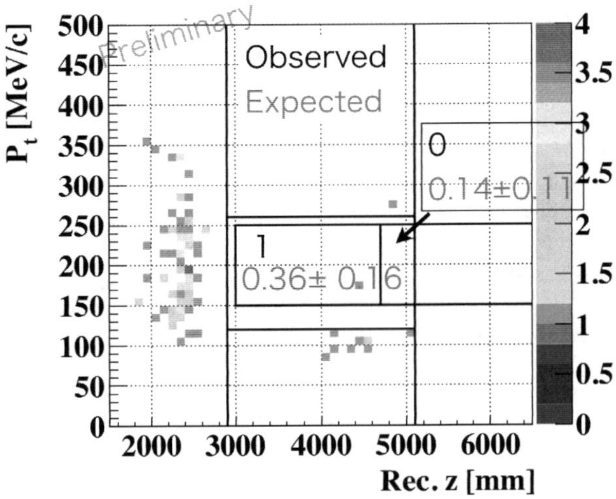

Figure 3. Scatter plot in the Rec. P_T vs Rec. z after imposing all selection criteria. The inner box is the signal region.

2.5. Searches for rare kaon decays at LHCb

It's worth mentioning that searches for rare kaon decays are also possible at LHC, at least for the cleanest decay channels with a closed kinematics and a controllable background. This holds for the $K_S \to \mu^+\mu^-$ decay, which is a Flavour Changing Neutral Current (FCNC) transition that has not yet been observed. This decay is suppressed in the SM, with an expected branching fraction [21] $BR(K_S \to \mu^+\mu^-) = (5.0 \pm 1.5) \times 10^{-12}$, while the current experimental upper limit is 3.2×10^{-7} at 90% confidence level (C.L.) [22]. Although the dimuon decay of the K_L meson is known to be $BR(K_L \to \mu^+\mu^-) = (6.84 \pm 0.11) \times 10^{-9}$ [23], in agreement with the SM, effects of new particles can still be observed in $K_S \to \mu^+\mu^-$ decays.

The LHCb detector, described in detail in Ref.[24], is a single-arm forward spectrometer covering the pseudorapidity range $2 < \eta < 5$. A search for the decay $K_L \to \mu^+\mu^-$ is performed, based on a data sample of 1.0 fb^{-1} of pp collisions at $\sqrt{s} = 7$ TeV collected by the LHCb experiment at the Large Hadron Collider. The observed number of candidates is consistent with the background only hypothesis, yielding an upper limit of $BR(K_S \to \mu^+\mu^-) < 11(9) \times 10^{-9}$ at 95 (90)% confidence level [25]. This limit is a factor of thirty below the previous measurement.

In the future, probing the interesting region of $BR(K_S \to \mu^+\mu^-) < 10^{-10}$ will require dedicated trigger studies and improvements [26]. The LHCb collaboration is also considering to study other rare kaon decays as $K_S \to e^+e^-\mu^+\mu^-$, $K_S \to e^+e^-e^+e^-$, and $K_S \to \pi^0\mu^+\mu^-$, thanks to the copious production of kaons at LHC and the excellent LHCb detector reconstruction capabilities, even though the presence of a huge background to fight, and trigger limitations still remain the major concerns for these kind of studies.

2.6. Other projects

The KLOD R&D program for a K_L measurement at Protvino should also be mentioned, as well as the OKA project, in the same laboratory, using a separated charged kaon beam.
Unfortunately the ORKA experiment [27], and more broadly kaon physics research, have been terminated at Fermilab for the foreseeable future.

Journal of Physics: Conference Series **631** (2015) 012004 doi:10.1088/1742-6596/631/1/012004

3. Kaon experiments at e^+e^- collider

3.1. The KLOE-2 experiment at DAΦNE

DAΦNE, the Frascati ϕ-factory is an e^+e^- collider working at a center of mass energy of $\sqrt{s} \sim 1020$ MeV, corresponding to the peak of the ϕ resonance. The ϕ-meson production cross section is $\sim 3\mu$b, and its decay into $K^0\bar{K}^0$ has a branching fraction of 34%, yielding $\sim 10^6$ $K^0\bar{K}^0$ pairs per pb^{-1} of integrated luminosity.

The KLOE detector is a 4π detector setup, which is able to measure both charged and neutral particles. It consists of a large volume drift chamber [28], which provides excellent momentum and vertex reconstruction for charged particles, and a barrel shaped electromagnetic calorimeter with two end-caps [29], made from lead and scintillating fibers, which surrounds the drift chamber. The energy deposits of charged and neutral particles in the calorimeter are measured with very good time resolution, which allows for the identification of charged particles based on their time of flight. Drift chamber and calorimeter are enclosed in a superconducting solenoid, providing an axial 0.52 T magnetic field.

The KLOE experiment at DAΦNE completed its first data taking campaign in March 2006 with a total integrated luminosity of ~ 2.5 fb^{-1}, corresponding to a production of $\sim 7.5 \times 10^9$ ϕ-mesons and $\sim 2.5 \times 10^9$ $K^0\bar{K}^0$ pairs.

After an experimental test [30], DAΦNE has been upgraded implementing an innovative collision scheme based on a *crab-waist* configuration, providing an improvement in the peak luminosity of a factor ~ 3.

The KLOE-2 experiment aims to continue and extend the physics program of its predecessor by collecting $\mathcal{O}(10\ fb^{-1})$ of data at the upgraded DAΦNE with an improved KLOE detector. The KLOE-2 physics program has been described in detail in Ref.[31], and among the main issues includes neutral kaon interferometry and tests of discrete symmetries and quantum mechanics. Improvements of about one order of magnitude in almost all present limits on CPT violation and decoherence parameters are expected [32, 33, 34, 31], thanks to the increased luminosity and the better quality of reconstructed data.

The upgrade of the KLOE detector consists of the addition of (i) an inner tracker [35, 36] based on cylindrical GEM technology for the improvement of tracking and decay vertex resolution close to the interaction point (IP), (ii) a $e^{\pm}$ tagging system [37, 38, 39] for the $\gamma\gamma$ physics, and (iii) two calorimeters [40, 41, 42] in the final focusing region to improve acceptance and efficiency for photons coming from the IP and neutral kaon decays inside the detector volume.

The installation of the new KLOE-2 sub-detectors has been completed in 2013, and a large consolidation and maintenance program of the DAΦNE complex has been carried out. During the commissioning phase, it emerged that the advantages of the new collision scheme appear more difficult to get, due to the strong perturbation of the KLOE solenoid on the machine optics – absent in the test experiment described in Ref.[30] – and to the increased machine background. A new data taking campaign has started in November 2014 with the aim of collecting ~ 1 fb^{-1} of integrated luminosity by mid 2015, and to assess the real DAFNE/KLOE-2 capabilities of integrating the required luminosity in a running period of 2-3 years.

In the following only the most recent published result will be briefly reported, while a novel method to directly test discrete symmetries with entangled neutral kaon pairs at KLOE-2 will be discussed in some details. For a review of previous KLOE results about tests of discrete symmetries and quantum mechanics the reader is referred to the existing literature[32, 33, 34, 31].

3.1.1. CPT and Lorentz symmetry test CPT invariance holds for any realistic Lorentz-invariant quantum field theory. However a very general theoretical possibility for CPT violation is based on spontaneous breaking of Lorentz symmetry, as developed by Kostelecký [43, 44, 45], which appears to be compatible with the basic tenets of quantum field theory and retains the property of gauge invariance and renormalizability (Standard Model Extensions - SME). In SME for neutral kaons, CPT violation manifests to lowest order only in the mixing parameter δ, (e.g. vanishes at first order in the decay amplitudes), and exhibits a dependence on the 4-momentum of the kaon:

$$\delta \approx i \sin\phi_{SW} e^{i\phi_{SW}} \gamma_K (\Delta a_0 - \vec{\beta_K} \cdot \Delta\vec{a})/\Delta m \tag{1}$$

where γ_K and $\vec{\beta_K}$ are the kaon boost factor and velocity in the observer frame, ϕ_{SW} is the so called *superweak* phase, $\Delta m = m_L - m_S$, and Δa_μ are four CPT- and Lorentz-violating coefficients for the two valence quarks in the kaon.

Using 1.7 fb^{-1} of data collected at KLOE and studying the interference pattern of the entangled neutral kaon pairs in the $\phi \to \mathrm{K}^0\bar{\mathrm{K}}^0 \to \pi^+\pi^-\pi^+\pi^-$ final state, as a function of sidereal time and particle direction in celestial coordinates, the following results have been obtained [46]:

$$\begin{aligned}
\Delta a_0 &= (-6.0 \pm 7.7_{stat} \pm 3.1_{syst}) \times 10^{-18}\ \mathrm{GeV} \\
\Delta a_x &= (\ \ 0.9 \pm 1.5_{stat} \pm 0.6_{syst}) \times 10^{-18}\ \mathrm{GeV} \\
\Delta a_y &= (-2.0 \pm 1.5_{stat} \pm 0.5_{syst}) \times 10^{-18}\ \mathrm{GeV} \\
\Delta a_z &= (\ \ 3.1 \pm 1.7_{stat} \pm 0.5_{syst}) \times 10^{-18}\ \mathrm{GeV}
\end{aligned}$$

These results [46] constitute the most sensitive measurements in the quark sector of SME, and can be compared to similar results obtained in the B and D meson systems, where an accuracy of $\mathcal{O}(10^{-13}\mathrm{GeV})$ has been reached [47].

3.1.2. Direct tests of discrete symmetries with kaons In general any *theoretical* connection among the parameters describing T, CP or CPT violation, as given by the CPT theorem, does not imply a corresponding connection among the *experimental* results of the T, CP or CPT tests. A *direct test* means a test whose outcome is independent from the result of any other discrete symmetry test, as discussed in detail in Refs.[48, 49, 50, 51, 52].

In order to implement a direct test of the T and CPT symmetries, which requires in both cases the inversion of *in* and *out* states, it has been suggested to exploit the Einstein-Podolsky-Rosen (EPR) entanglement of neutral mesons produced at a ϕ-factory (or B-factory[1]) [49, 50, 51, 52]. In fact in this case the initial state of the neutral kaon pair produced in $\phi \to \mathrm{K}^0\bar{\mathrm{K}}^0$ decay can be rewritten in terms of any pair of orthogonal states $|\mathrm{K}_+\rangle$ and $|\mathrm{K}_-\rangle$:

$$|i\rangle = \frac{1}{\sqrt{2}}\{|\mathrm{K}^0\rangle|\bar{\mathrm{K}}^0\rangle - |\bar{\mathrm{K}}^0\rangle|\mathrm{K}^0\rangle\} = \frac{1}{\sqrt{2}}\{|\mathrm{K}_+\rangle|\mathrm{K}_-\rangle - |\mathrm{K}_-\rangle|\mathrm{K}_+\rangle\}\ . \tag{2}$$

Here one can consider the states $|\mathrm{K}_+\rangle$, $|\mathrm{K}_-\rangle$ defined as follows: $|\mathrm{K}_+\rangle$ is the state filtered by the decay into $\pi\pi$ ($\pi^+\pi^+$ or $\pi^0\pi^0$), a pure $CP = +1$ state; analogously $|\mathrm{K}_-\rangle$ is the state filtered by the decay into $3\pi^0$, a pure $CP = -1$ state. Their orthogonal states correspond to the states which cannot decay into $\pi\pi$ or $3\pi^0$, defined, respectively, as

$$\begin{aligned}
|\widetilde{\mathrm{K}}_-\rangle &\propto [|\mathrm{K_L}\rangle - \eta_{\pi\pi}|\mathrm{K_S}\rangle] \\
|\widetilde{\mathrm{K}}_+\rangle &\propto [|\mathrm{K_S}\rangle - \eta_{3\pi^0}|\mathrm{K_L}\rangle]\ ,
\end{aligned} \tag{3}$$

with $\eta_{\pi\pi} = \langle\pi\pi|T|\mathrm{K_L}\rangle/\langle\pi\pi|T|\mathrm{K_S}\rangle$ and $\eta_{3\pi^0} = \langle 3\pi^0|T|\mathrm{K_S}\rangle/\langle 3\pi^0|T|\mathrm{K_L}\rangle$. With these definitions of states, it can be shown that the condition of orthogonality $\langle\mathrm{K}_-|\mathrm{K}_+\rangle = 0$, (i.e. $|\mathrm{K}_+\rangle \equiv |\widetilde{\mathrm{K}}_+\rangle$ and $|\mathrm{K}_-\rangle \equiv |\widetilde{\mathrm{K}}_-\rangle$) corresponds to assume negligible direct CP (or CPT) violation contributions, assumption quite well satisfied for neutral kaons (see detailed discussion in Appendix A of Ref.

[1] It's worth mentioning that in the neutral B meson system a direct T test has been already accomplished [53], exploiting a similar methodology as discussed in the following.

[52]). The validity of the $\Delta S = \Delta Q$ rule is also assumed, so that the two flavor orthogonal eigenstates $|\mathrm{K}^0\rangle$ and $|\bar{\mathrm{K}}^0\rangle$ are identified by the charge of the lepton in semileptonic decays, i.e. a $|\mathrm{K}^0\rangle$ can decay into $\pi^-\ell^+\nu$ and not into $\pi^+\ell^-\bar{\nu}$, and vice-versa for a $|\bar{\mathrm{K}}^0\rangle$.

Thus, exploiting the perfect anticorrelation of the states implied by Eq. (2), it is possible to have a "flavor-tag"or a "CP-tag", i.e. to infer the flavor (K^0 or $\bar{\mathrm{K}}^0$) or the CP (K_+ or K_-) state of the still alive kaon by observing a specific flavor decay ($\pi^+\ell^-\nu$ or $\pi^-\ell^+\bar{\nu}$) or CP decay ($\pi\pi$ or $\pi^0\pi^0\pi^0$) of the other (and first decaying) kaon in the pair.

In this way one can experimentally access – for instance – the transition $\mathrm{K}^0 \to \mathrm{K}_+$, taken as reference, and the $\mathrm{K}_+ \to \mathrm{K}^0$, $\bar{\mathrm{K}}^0 \to \mathrm{K}_+$ and $\mathrm{K}_+ \to \bar{\mathrm{K}}^0$ transitions, i.e. the T, CP and CPT conjugated transitions of the reference one, respectively.

For the direct T symmetry test one can define the following ratios of probabilities:

$$\begin{aligned}
R_1(\Delta t) &= P\left[\mathrm{K}^0(0) \to \mathrm{K}_+(\Delta t)\right] / P\left[\mathrm{K}_+(0) \to \mathrm{K}^0(\Delta t)\right] \\
R_2(\Delta t) &= P\left[\mathrm{K}^0(0) \to \mathrm{K}_-(\Delta t)\right] / P\left[\mathrm{K}_-(0) \to \mathrm{K}^0(\Delta t)\right] \\
R_3(\Delta t) &= P\left[\bar{\mathrm{K}}^0(0) \to \mathrm{K}_+(\Delta t)\right] / P\left[\mathrm{K}_+(0) \to \bar{\mathrm{K}}^0(\Delta t)\right] \\
R_4(\Delta t) &= P\left[\bar{\mathrm{K}}^0(0) \to \mathrm{K}_-(\Delta t)\right] / P\left[\mathrm{K}_-(0) \to \bar{\mathrm{K}}^0(\Delta t)\right] . \qquad (4)
\end{aligned}$$

The measurement of any deviation from the prediction $R_i(\Delta t) = 1$ imposed by T invariance is a signal of T violation.

At a ϕ-factory the corresponding observable quantities are two ratios, $R_2^{\mathrm{exp}}(\Delta t)$ and $R_4^{\mathrm{exp}}(\Delta t)$ of double decay rates $I(f_1, f_2; \Delta t)$ into decay products f_1 and f_2 as a function of the difference of kaon decay times Δt [52, 32], with f_1 occurring before f_2 decay for $\Delta t > 0$, and viceversa for $\Delta t < 0$. For $\Delta t > 0$ one has:

$$R_2^{\mathrm{exp}}(\Delta t) \equiv \frac{I(\ell^-, 3\pi^0; \Delta t)}{I(\pi\pi, \ell^+; \Delta t)} = R_2(\Delta t) \times D \qquad (5)$$

$$R_4^{\mathrm{exp}}(\Delta t) \equiv \frac{I(\ell^+, 3\pi^0; \Delta t)}{I(\pi\pi, \ell^-; \Delta t)} = R_4(\Delta t) \times D \qquad (6)$$

while for $\Delta t < 0$:

$$R_2^{\mathrm{exp}}(\Delta t) = \frac{D}{R_3(|\Delta t|)} \qquad (7)$$

$$R_4^{\mathrm{exp}}(\Delta t) = \frac{D}{R_1(|\Delta t|)} . \qquad (8)$$

Here the normalization constant D, assuming no CPT violation in semileptonic decays, is $D = \{\mathrm{BR}\left(\mathrm{K_L} \to 3\pi^0\right) \cdot \Gamma_L\} / \{\mathrm{BR}\left(\mathrm{K_S} \to \pi\pi\right) \cdot \Gamma_S\}$.

For the direct CPT symmetry test one can define the following ratios of probabilities, similarly as for the T test:

$$\begin{aligned}
R_{1,CPT}(\Delta t) &= P\left[\mathrm{K}^0(0) \to \mathrm{K}_+(\Delta t)\right] / P\left[\mathrm{K}_+(0) \to \bar{\mathrm{K}}^0(\Delta t)\right] \\
R_{2,CPT}(\Delta t) &= P\left[\mathrm{K}^0(0) \to \mathrm{K}_-(\Delta t)\right] / P\left[\mathrm{K}_-(0) \to \bar{\mathrm{K}}^0(\Delta t)\right] \\
R_{3,CPT}(\Delta t) &= P\left[\bar{\mathrm{K}}^0(0) \to \mathrm{K}_+(\Delta t)\right] / P\left[\mathrm{K}_+(0) \to \mathrm{K}^0(\Delta t)\right] \\
R_{4,CPT}(\Delta t) &= P\left[\bar{\mathrm{K}}^0(0) \to \mathrm{K}_-(\Delta t)\right] / P\left[\mathrm{K}_-(0) \to \mathrm{K}^0(\Delta t)\right] . \qquad (9)
\end{aligned}$$

The measurement of any deviation from the prediction $R_{i,CPT}(\Delta t) = 1$ imposed by CPT invariance is a signal of CPT violation.

At a ϕ-factory the corresponding observable quantities are, for $\Delta t > 0$:

$$R^{\rm exp}_{2,CPT}(\Delta t) \equiv \frac{I(\ell^-, 3\pi^0; \Delta t)}{I(\pi\pi, \ell^-; \Delta t)} = R_{2,CPT}(\Delta t) \times D_{CPT} \tag{10}$$

$$R^{\rm exp}_{4,CPT}(\Delta t) \equiv \frac{I(\ell^+, 3\pi^0; \Delta t)}{I(\pi\pi, \ell^+; \Delta t)} = R_{4,CPT}(\Delta t) \times D_{CPT} \tag{11}$$

while for $\Delta t < 0$:

$$R^{\rm exp}_{2,CPT}(\Delta t) = \frac{D_{CPT}}{R_{1,CPT}(|\Delta t|)} \tag{12}$$

$$R^{\rm exp}_{4,CPT}(\Delta t) = \frac{D_{CPT}}{R_{3,CPT}(|\Delta t|)} \ . \tag{13}$$

Here the normalization constant D_{CPT} is

$$D_{CPT} = \{\mathrm{BR}\left(\mathrm{K_L} \to 3\pi^0\right) \cdot \Gamma_L\}/\{\mathrm{BR}\left(\mathrm{K_S} \to \pi\pi\right) \cdot \Gamma_S\}$$

without any assumption on CPT violation in semileptonic decays.

The KLOE-2 experiment at DAΦNE with an integrated luminosity of $\mathcal{O}(10\ \mathrm{fb}^{-1})$ [31] could make statistically significant T and CPT tests, measuring the ratios $R^{\rm exp}_2(\Delta t)$, $R^{\rm exp}_4(\Delta t)$, $R^{\rm exp}_{2,CPT}(\Delta t)$ and $R^{\rm exp}_{4,CPT}(\Delta t)$ integrated in the statistically most populated Δt region, $0 \leq \Delta t \leq 300\ \tau_S$ [52]. In this region these ratios are expected to be constant and a precise knowledge of the normalization constants D and D_{CPT} is needed in order to not weaken the significance of the test. A precise reconstruction of $\mathrm{K_S} \to \pi\ell\nu$ and $\mathrm{K_L} \to 3\pi^0$ decays is needed, and specific studies are in progress at KLOE-2 [54, 55].

4. Conclusions

The new generation of kaon experiments at hadron machines (NA62 and KOTO) are operational, and very soon will allow us to definitely enter a new precision era in probing the SM, and in searching for New Physics in a complementary way to LHC.

Some specific issues in the field, e.g. the transverse muon polarization in K^+ decays or the detection of very rare kaon decays, will be addressed by dedicated experiments or by general purpose experiments, as LHCb.

The neutral kaon system still constitutes an excellent laboratory for the study of discrete symmetries, and a ϕ-factory represents a unique opportunity to push forward these studies. CPT symmetry and quantum mechanics tests will be one of the key issues at KLOE-2, and their precision will be further improved, while new ideas - e.g. the direct T and CPT symmetry tests - will be implemented.

Acknowledgements

The author would like to warmly thank Prof. N. Mavromatos, and all the organizing committee for the invitation to the DISCRETE 2014 symposium, and the very pleasant stay in London.

References

[1] L. Leprince-Ringuet and M. L'héritier, Comptes Rendus Acad. Sciences de Paris, séance du 13 Dec. 1944, p. 618.
[2] M. Baldo-Ceolin et al., Nuovo Cim. **6** (1957) 84.
[3] J. H. Christenson et al., Phys. Rev. Lett. **13** (1964) 138.
[4] E. Abouzaid et al., KTeV collaboration, Phys. Rev. D **83** (2011) 092001.
[5] J. R. Batley, NA48 collaboration, et al., Phys. Lett. B **544** (2002) 97.

[6] A. Di Domenico, J.Phys. Conf. Ser. 447 (2013) 012008.
[7] see e.g. G. C. Branco, L. Lavoura, and J. P. Silva, *CP violation*, Oxford Univ. Press, 1999;
M. Sozzi, *Discrete symmetries and CP violation*, Oxford Univ. Press, 2008;
T. K. Komatsubara, Progr. Part. Nucl. Phys. **67** (2012) 995;
M. Sozzi, J. Phys. Conf. Ser. 335 (2011) 012002.
[8] J. Brod, M. Gorbahn and E. Stamou, Phys. Rev. D **83** (2011) 034030.
[9] S.L. Glashow, J. Iliopoulos, L. Maiani, Phys. Rev. D **2** (1970) 1285.
[10] C. Smith, arXiv:1012.3698v1[hep-ph] (2010).
[11] Y. Grossman and Y. Nir, Phys. Lett. B **398** (1997) 163
[12] A.V. Artamonov et al., Phys. Rev. D **79** (2009) 092004.
[13] J. K. Ahn et al., Phys. Rev. D **81** (2010) 072004.
[14] G. Anelli et al., CERN-SPSC-2005-013, CERN-SPSC-P-326 (2005);
NA62 Collaboration, CERN-SPSC-2007-035, SPSC-M- 760 (2007).
[15] F. Hahn, these proceedings.
[16] H. Watanabe, for the J-PARC E14 KOTO collaboration, PoS ICHEP 2010 (2010) 274;
see also http://koto.kek.jp/.
[17] http://j-parc.jp/.
[18] Koji Shiomi for the KOTO collaboration, arXiv:1411.4250v1 [hep-ex] 16 Nov 2014.
[19] M. Abe et al, E246 Collaboration, Phys. Rev. Lett. **93** (2004) 131601.
[20] J. Imazato, PoS KAON09 (2009) 007;
see also http://trek.kek.jp/.
[21] G. Ecker and A. Pich, Nucl. Phys. B **366** (1991) 189;
G. Isidori and R. Unterdorfer, JHEP 01 (2004) 009.
[22] S. Gjesdal et al., Phys. Lett. B **44** (1973) 217.
[23] K.A. Olive et al. (Particle Data Group), Chin. Phys. C, **38**, 090001 (2014).
[24] A. Augusto Alves, Jr. et al., LHCb collaboration, JINST 3 (2008)S08005
[25] R. Aaij et al., LHCb Collaboration, JHEP 01 (2013) 090.
[26] F. Soomro, LHCb Collaboration, these proceedings.
[27] E.T. Worcester for the ORKA Collaboration, Nucl. Phys. B (Proc. Suppl.) **233** (2012) 285.
[28] M. Adinolfi et al., KLOE collaboration, Nucl. Instr. and Meth. A **488** (2002) 51.
[29] M. Adinolfi et al., KLOE collaboration, Nucl. Instr. and Meth. A **482** (2002) 364.
[30] M. Zobov et al., Phys. Rev. Lett. **104**, 174801 (2010).
[31] G. Amelino-Camelia et al., Eur. Phys. J. C **68**, 619 (2010).
[32] A. Di Domenico (Editor) , Handbook on Neutral Kaon Interferometry at a ϕ-factory, Frascati Phys. Ser. **43** (2007).
[33] A. Di Domenico and the KLOE collaboration, J. Phys. Conf. Ser. **171**, 012008 (2009).
[34] A. Di Domenico and the KLOE collaboration, Found. Phys. **40**, 852 (2010).
[35] A. Balla et al., JINST 9, C01014 (2014).
[36] A. Di Cicco, G. Morello, Acta Phys. Pol. B **46** 73 (2015).
[37] D. Babusci et al., Nucl.Instrum.Meth. **A617** 81 (2010).
[38] F. Archilli et al., Nucl.Instrum.Meth. **A617** 266 (2010).
[39] D. Babusci et al, Acta Phys. Pol. B **46** 81 (2015).
[40] F. Happacher et al., Nucl. Phys. Proc. Suppl. **197**, 215 (2009).
[41] M. Cordelli et al., Nucl. Instr. & Meth. **A 617**, 105 (2010).
[42] F. Happacher, M. Martini, Acta Phys. Pol. B **46** 87 (2015).
[43] V. A. Kostelecký, Phys. Rev. Lett. **80**, 1818 (1998).
[44] V. A. Kostelecký, Phys. Rev. D **61**, 016002 (1999).
[45] V. A. Kostelecký, Phys. Rev. D **64**, 076001 (2001).
[46] D. Babusci et al. (KLOE-2 Collaboration), Phys. Lett. B **730** (2014) 89
[47] V.A. Kostelecky, N. Russell, Rev. Mod. Phys. **83** (2011) 11
[48] L. Wolfenstein, Int. J. Mod. Phys. E **8**, 501 (1999).
[49] M. C. Banuls and J. Bernabeu, Phys. Lett. B **464**, 117 (1999); Nucl. Phys. B **590**, 19 (2000).
[50] J. Bernabeu, J. Phys. Conf. Ser. **335** 012011 (2011).
[51] J. Bernabeu, F. Martínez-Vidal, P. Villanueva-Pérez, JHEP **08**, 064(2012).
[52] J. Bernabeu, A. Di Domenico, P. Villanueva-Perez, Nucl. Phys. B. **868** (2013) 102.
[53] J. P. Lees et al., BABAR Collaboration, Phys. Rev. Lett. **109** 211801 (2012).
[54] D. Kaminska, Acta Phys. Pol. B **46** 19 (2015), and these proceedings.
[55] A. Gajos, Acta Phys. Pol. B **46** 13 (2015), and these proceedings.

Discrete Symmetries and Models of Flavour Mixing

Stephen F. King

Physics and Astronomy, University of Southampton, Southampton SO17 1BJ, UK

E-mail: king@soton.ac.uk

Abstract. In this talk we shall give an overview of the role of discrete symmetries, including both CP and family symmetry, in constructing unified models of quark and lepton (including especially neutrino) masses and mixing. Various different approaches to model building will be described, denoted as direct, semi-direct and indirect, and the pros and cons of each approach discussed. Particular examples based on $\Delta(6n^2)$ will be discussed and an A to Z of Flavour with Pati-Salam will be presented.

1. Introduction

We begin with a brief reminder about neutrino mass and mixing. The reactor angle was unmeasured before 2012 but is now measured to incredible accuracy: $\theta_{13} \approx 8.5° \pm 0.2°$ [1]. The other angles are determined from global fits [2, 3, 4] to be: $\theta_{12} \approx 34° \pm 1°$ and $\theta_{23} \approx 45° \pm 3°$, and first hints of the CP-violating (CPV) phase $\delta \sim -\pi/2 \pm \pi/3$ have been reported, however with a large error. The meaning of the angles is given in Fig.1. Two possible mass squared orderings are possible as explained in Fig.2. The above quoted angles are extracted from the global fits which are displayed in Fig.3 for the normal ordering case.

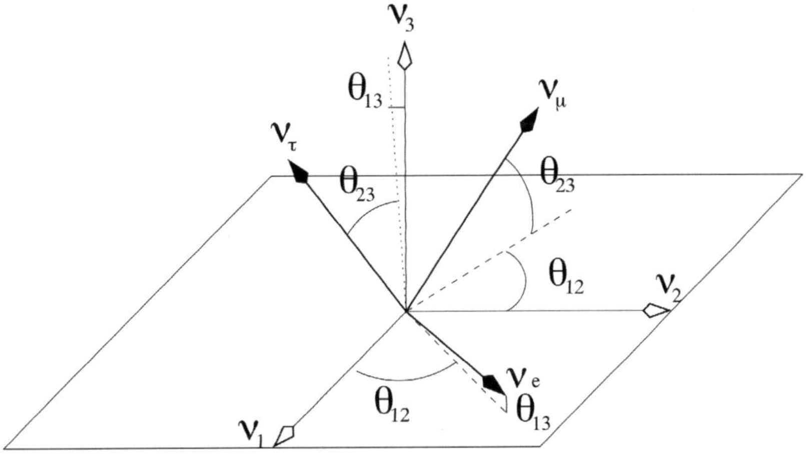

Figure 1. Neutrino mixing angles (assuming zero CP violation) may be represented as Euler angles relating the charged lepton mass basis states $(\nu_e, \nu_\mu, \nu_\tau)$ to the mass eigenstate basis states (ν_1, ν_2, ν_3).

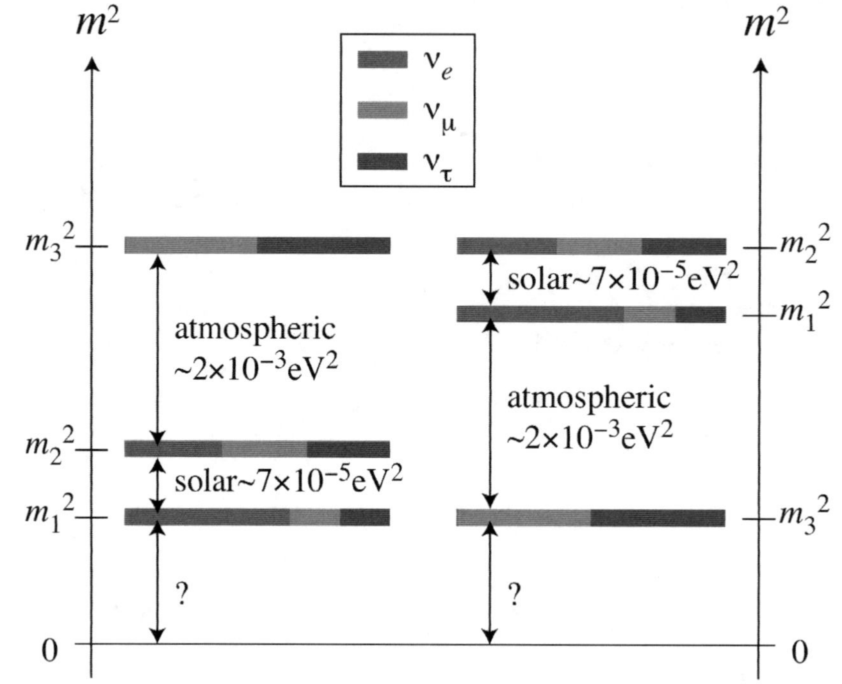

Figure 2. The probability that a particular neutrino mass state ν_i with mass m_i contains a particular charged lepton mass basis state $(\nu_e, \nu_\mu, \nu_\tau)$ is represented by colours. The left and right panels of the figure are referred to as normal or inverted mass squared ordering, respectively. The value of the lightest neutrino mass is presently unknown.

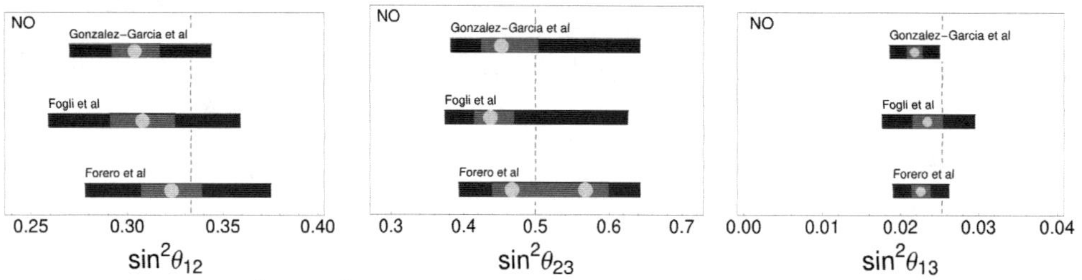

Figure 3. Global fits [2, 3, 4] to the lepton mixing angles for the case of normal neutrino mass squared ordering. The green dots are the best fit points, the red (blue) areas indicate the one (three) sigma ranges. The dashed lines indicate tri-bimaximal-Cabibbo (TBC) mixing [5], namely the values: $s_{12}^2 = 1/3$, $s_{23}^2 = 1/2$ $s_{13}^2 = \theta_C^2/2$. The Fogli et al fits do not include the latest Daya Bay results, whereas the other two do. This is a modified version of a figure provided by Stefano Morisi.

The measurement of the reactor angle has had a major impact on models of neutrino mass and mixing as reviewed in [6, 7] (for earlier reviews with extensive references with see e.g. [8, 9, 10]). This talk will attempt to summarise the most recent model building directions that survive or have emerged following the reactor angle measurement (see [6, 7] for comprehensive references).

2. The Standard Model Puzzles

Even though the Standard Model (SM) is essentially complete, following the Higgs boson discovery, we are far from satisfied since it offers no solutions to the cosmological puzzles of matter-antimatter asymetry, dark matter and dark energy. It therefore cannot be the final answer. In addition it leaves three unresolved puzzles in its wake:

The origin of mass - the origin of the Higgs vacuum expectation value, its stability under radiative corrections, and the solution to the hierarchy problem (most urgent problem of LHC).

The quest for unification - the question of whether the three known forces of the standard model may be related into a grand unified theory, and whether such a theory could also include a unification with gravity.

The problem of flavour - the problem of the undetermined fermion masses and mixing angles (including neutrino masses and lepton mixing angles) together with the CP violating phases, in conjunction with the observed smallness of flavour changing neutral currents and very small strong CP violation. In particular the extreme smallness of neutrino mass *for all three families* may offer a clue as to what lies beyond.

The differences between quark and lepton mixing may also offer clues concerning the flavour problem. Certainly the flavour problem has now become much richer, following the discovery of neutrino mass and mixing, so we shall discuss more about this.

Before proceeding to discuss the flavour problem in more detail, we disgress slightly to discuss an alternative point of view that frequently is voiced and that came up again at this Workshop. Namely, all that is required for neutrino masses is to add two or three right-handed neutrinos with zero Majorana mass due to a conserved $B-L$, and Yukawa couplings for all neutrino families of about 10^{-11} and that no new physics beyond this is required. However this conservative point of view involves a new mystery, namely why the third family Yukawa couplings are of order unity for the top quark, and not very small for the bottom quark and τ lepton, but are of order 10^{-11} for the third family of neutrinos. The see-saw mechanism [11], i.e. large third family neutrino Yukawa couplings, with physical neutrino masses suppressed by heavy right-handed Majorana masses with $B - L$ broken at a high scale, provides an elegant solution to this mystery, and opens the door to leptogenesis. However, the see-saw mechanism by itself does not account for the observed large lepton mixing, so we need to go further.

It has been one of the long standing goals of theories of particle physics beyond the Standard Model to predict quark and lepton masses and mixings. With the discovery of neutrino mass and mixing, this quest has received a massive impetus. Indeed, perhaps the greatest advance in particle physics over the past decade has been the discovery of neutrino mass and mixing involving large mixing. The largeness of the lepton mixing angles contrasts with the smallness of the quark mixing angles, and this observation, together with the smallness of neutrino masses, provides new and tantalising clues in the search for the origin of quark and lepton flavour. For example, it is amusing to note that the smallest lepton mixing may be related to the largest quark mixing, $|U_{e3}| \approx \theta_C/\sqrt{2}$ where θ_C is the Cabibbo angle. The quest to understand the origin of the three families of quarks and leptons and their pattern of masses and mixing parameters is called the flavour puzzle, and motivates the introduction of family symmetry. In particular, as we shall see, lepton mixing provides a motivation for discrete family symmetry, which will form the central part of this review. As we shall also see, such theories demand a high precision knowledge of the lepton mixing angles, beyond that currently achieved.

The SM does not address the following questions which loosely define the flavour problem:

- Why are there three families of quarks and leptons?
- Why are all charged fermion masses so hierarchical with down-type quark masses being of the same order as charged lepton masses, and up-type quark masses are much more hierarchical?

Journal of Physics: Conference Series **631** (2015) 012005 doi:10.1088/1742-6596/631/1/012005

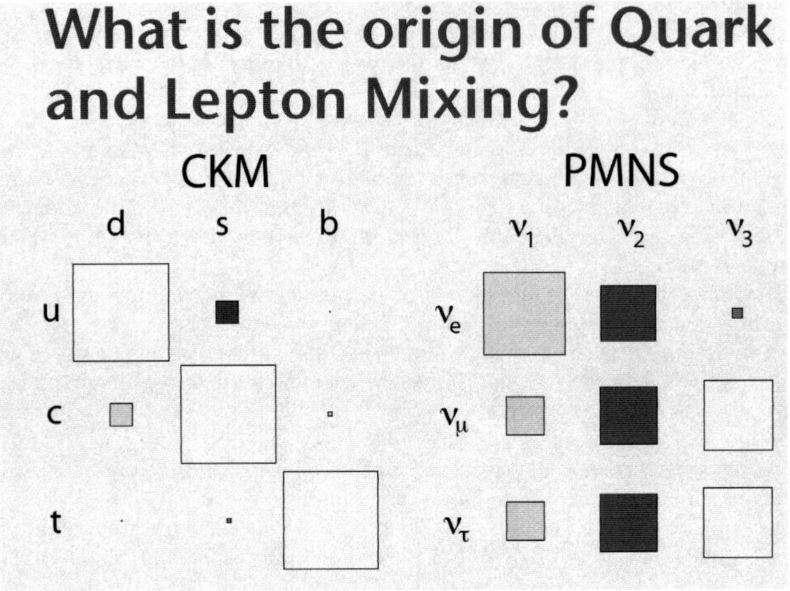

Figure 4. The sides of the squares represent the magnitude of the CKM and PMNS elements.

- Why are at least two neutrino masses not very hierarchical?
- What is the origin of the neutrino mass?
- Why are neutrino masses so tiny compared to charged fermion masses?
- What is the origin of fermion mixing (both CKM and PMNS matrices)?
- Why are CKM mixing angles smaller than PMNS mixing angles apart from the Cabibbo angle which is of the same order as the reactor angle?
- What is the origin of CP violation in the quark (and lepton) sectors?

For completeness, we display the CKM and the PMNS mixing matrices are both given by:

$$\begin{pmatrix} c_{12}c_{13} & s_{12}c_{13} & s_{13}e^{-i\delta} \\ -s_{12}c_{23} - c_{12}s_{13}s_{23}e^{i\delta} & c_{12}c_{23} - s_{12}s_{13}s_{23}e^{i\delta} & c_{13}s_{23} \\ s_{12}s_{23} - c_{12}s_{13}c_{23}e^{i\delta} & -c_{12}s_{23} - s_{12}s_{13}c_{23}e^{i\delta} & c_{13}c_{23} \end{pmatrix} \quad (1)$$

where δ is the CP violating phase in each sector (quark and lepton) and $s_{13} = \sin\theta_{13}$, etc. with (very) different angles for quarks and leptons. In the case of Majorana neutrinos, the PMNS matrix also involves the phase matrix: $\mathrm{diag}(1, e^{i\frac{\alpha_{21}}{2}}, e^{i\frac{\alpha_{31}}{2}})$ which post-multiplies the above matrix. The different shapes of the quark and lepton mixing matrices is schematically illustrated in Fig.4. The large lepton mixing must arise in conjunction with the mechanism responsible for the smallness of neutrino mass, which however is unknown.

3. Neutrino Mass Models

3.1. The open questions from neutrino physics

Despite the great progress coming from neutrino oscillation experiments there are still some outstanding questions. Are the lepton mixing angles consistent with TBC mixing? If not then

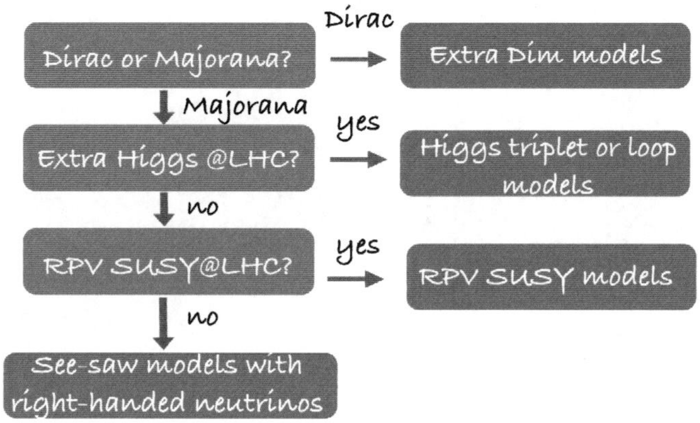

Figure 5. Roadmap of neutrino mass models.

is the atmospheric angle in the first or second octant? What is the CP violating phase δ? Is the current hint $\delta \sim -\pi/2$ going to hold up? Maybe there is no CP violation in the lepton sector? Are neutrino mass squared ordered normally or inverted? [1] What is the lightest neutrino mass? Are neutrino masses Majorana or Dirac in nature? Many neutrino experiments are underway or in the planning stages to address these questions such as T2K, NOνA, Daya Bay, JUNO, RENO, KATRIN, LBNF and many neutrinoless double beta decay experiments running and planned.

3.2. Road Map of Neutrino Mass Models

Everyone can invent her or his personal roadmap of neutrino mass models, one example being that shown in Fig.5. The blue boxes contain experimental questions and the red boxes possible theoretical consequences. Since no new physics on the right-hand side has yet emerged from the LHC, in this talk we shall focus on the see-saw mechanism [11] with right-handed neutrinos.

3.3. See-saw mechanism with right-handed neutrinos

Both the number of species and the mass spectrum of right-handed (or sterile) neutrinos is completely unknown [12]. As shown in Fig.6 the mass spectrum can cover the whole range with different physical consequences as indicated. It is one of the goals of neutrino physics to determine this spectrum.

3.4. See-saw mechanism and sequential dominance

Suppose that the Standard Model is supplemented by just three right-handed neutrinos with masses in the classic see-saw range TeV to M_{GUT} in Fig.6. Then the light neutrino mass matrix

[1] It is common but incorrect to refer to this question as the "neutrino mass hierarchy" since the "ordering" question is separate from whether neutrinos are hierarchical in nature or approximately degenerate, which is to do with the lightest neutrino mass.

Journal of Physics: Conference Series **631** (2015) 012005 doi:10.1088/1742-6596/631/1/012005

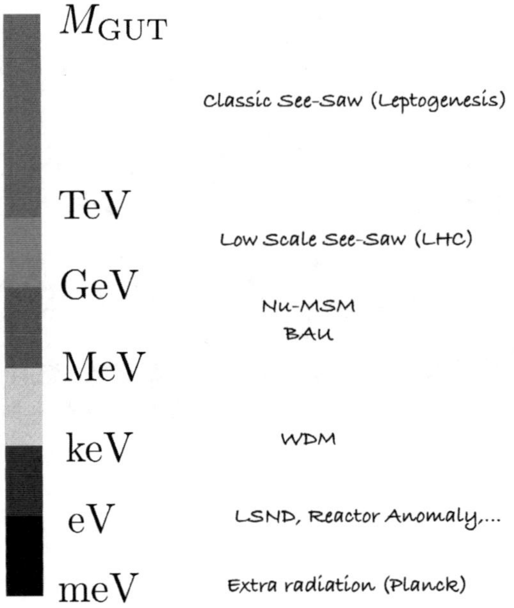

Figure 6. Possible mass spectrum of right-handed (or sterile) neutrinos corresponding to the physical implications as shown.

emerges from the see-saw formula [11],

$$m^{\nu} = -m^D M_R^{-1} (m^D)^T, \tag{2}$$

where m^{ν} is the the light effective left-handed Majorana neutrino mass matrix (i.e. the physical neutrino mass matrix), m^D is the Dirac mass matrix in LR convention and M_R is the (heavy) Majorana mass matrix.

In this case it is possible to implement the see-saw mechanism using sequential dominance (SD) mechanism [13], in the basis, $M_R = \mathrm{diag}(M_{\rm atm}, M_{\rm sol}, M_{\rm dec})$ where the Dirac mass matrix is constructed from three columns $m^D = (m^D_{\rm atm}, m^D_{\rm sol}, m^D_{\rm dec})$,

$$\frac{m^D_{\rm atm}(m^D_{\rm atm})^T}{M_{\rm atm}} > \frac{m^D_{\rm sol}(m^D_{\rm sol})^T}{M_{\rm sol}} \gg \frac{m^D_{\rm dec}(m^D_{\rm dec})^T}{M_{\rm dec}}. \tag{3}$$

SD immediately predicts a normal neutrino mass hierarchy $m_3 > m_2 \gg m_1$ since the lightest physical neutrino mass m_1 is much smaller than the others since the corresponding right-handed neutrino $\nu_R^{\rm dec}$ is approximately decoupled from the see-saw mechanism. The SM with three such right-handed neutrinos is depicted in Fig.7.

The observed pattern of lepton mixing angles can be understood in the above SD framework as follows. In the diagonal charged lepton and right-handed neutrino mass basis, if the dominant "atmospheric" right-handed neutrino has couplings $(m^D_{\rm atm})^T = (0, a_1, a_2)$ to $(\nu_e, \nu_\mu, \nu_\tau)$, then this implies $\tan\theta_{23} \sim a_1/a_2$ [13] and a bound $\theta_{13} \leq m_2/m_3$ [14]. The subdominant "solar" right-handed neutrino couplings $(m^D_{\rm sol})^T = (b_1, b_2, b_3)$ to $(\nu_e, \nu_\mu, \nu_\tau)$ further yield $\tan\theta_{12} \sim \sqrt{2}b_1/(b_2 - b_3)$ [13, 14]. The lepton mixing angles are of course insensitive to the "decoupled" right-handed neutrino couplings $(m^D_{\rm dec})^T$.

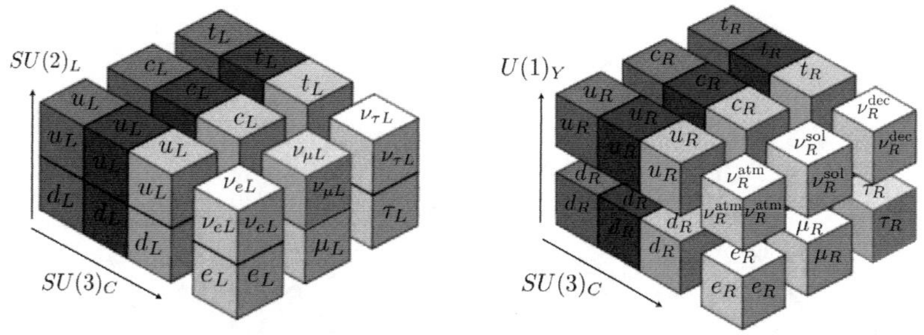

Figure 7. The Standard Model with three right-handed neutrinos defined as $(\nu_R^{\rm atm}, \nu_R^{\rm sol}, \nu_R^{\rm dec})$ which in sequential dominance are mainly responsible for the m_3, m_2, m_1 physical neutrino masses, respectively.

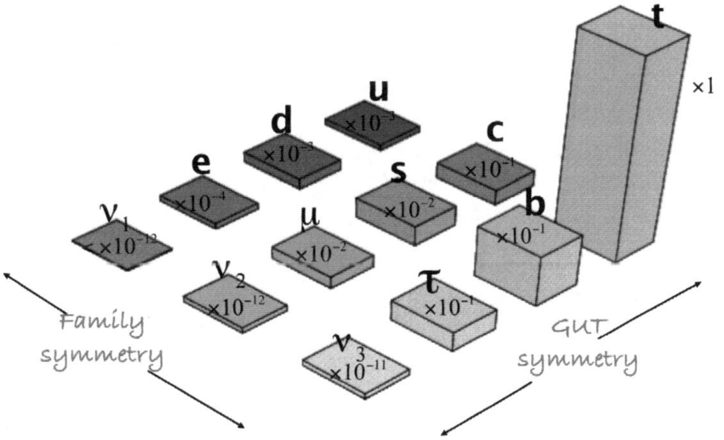

Figure 8. Quark and lepton masses lego plot (true heights need to be scaled by the factors shown).

4. Towards a theory of flavour

4.1. GUTs and Family Symmetry

If we wish to understand the pattern of quark and lepton masses and mixing angles (and CP violation) then a fruitful approach is to introduce GUT and family symmetry acting in the directions shown in Fig.8. Some popular family symmetries which admit triplet representations are shown in Fig.9. Some possible candidate unified gauge groups are shown in Fig. 10.

4.2. Klein symmetry

The starting point for family symmetry models is to consider the Klein symmetry of the neutrino mass matrix. First consider the phase symmetry of the diagonal charged lepton mass matrix

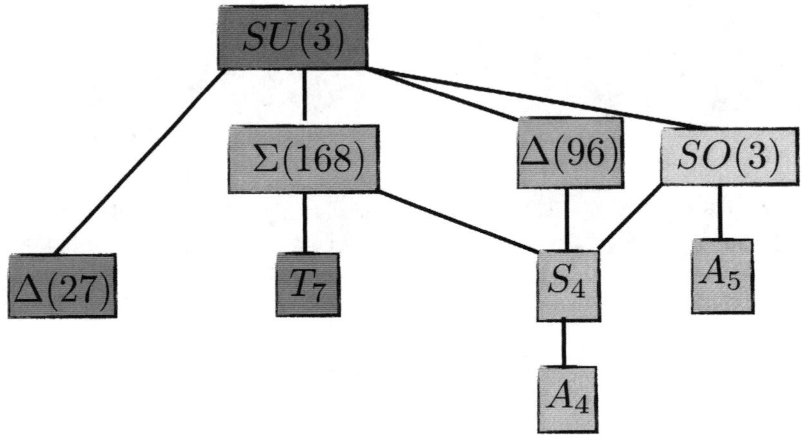

Figure 9. Some popular family symmetries and their relationships.

M_e,

$$T^\dagger(M_e^\dagger M_e)T = M_e^\dagger M_e \tag{4}$$

where $T = \text{diag}(1, \omega, \omega^2)$ and $\omega = e^{2\pi/n}$. For example for $n = 3$ clearly T generates the group Z_3^T. In any case, the Klein symmetry of the neutrino mass matrix is given by,

$$m^\nu = S^T m^\nu S, \quad m^\nu = U^T m^\nu U \tag{5}$$

where [15]

$$S = U_{\text{PMNS}}^* \, \text{diag}(+1, -1, -1) \, U_{\text{PMNS}}^T \tag{6}$$
$$U = U_{\text{PMNS}}^* \, \text{diag}(-1, +1, -1) \, U_{\text{PMNS}}^T \tag{7}$$
$$SU = U_{\text{PMNS}}^* \, \text{diag}(-1, -1, +1) \, U_{\text{PMNS}}^T \tag{8}$$

and

$$\mathcal{K} = \{1, S, U, SU\} \tag{9}$$

is called the Klein symmetry $Z_2^S \times Z_2^U$.

4.3. Direct models and simple lepton mixing patterns

The idea of direct models is that the three generators S, T, U introduced above are embedded into a discrete family symmetry G which is broken by new Higgs fields called "flavons" of two types: ϕ^l whose VEVs preserve T and ϕ^ν whose VEVs preserve S, U. These flavons are segregated such that ϕ^l only appears in the charged lepton sector and ϕ^ν only appears in the neutrino sector as depicted in Fig.11, thereby enforcing the symmetries of the mass matrices. Note that the full Klein symmetry $Z_2^S \times Z_2^U$ of the neutrino mass matrix is enforced by symmetry in the direct approach. The Direct approach typically leads to simple patterns of mixing, as discussed below.

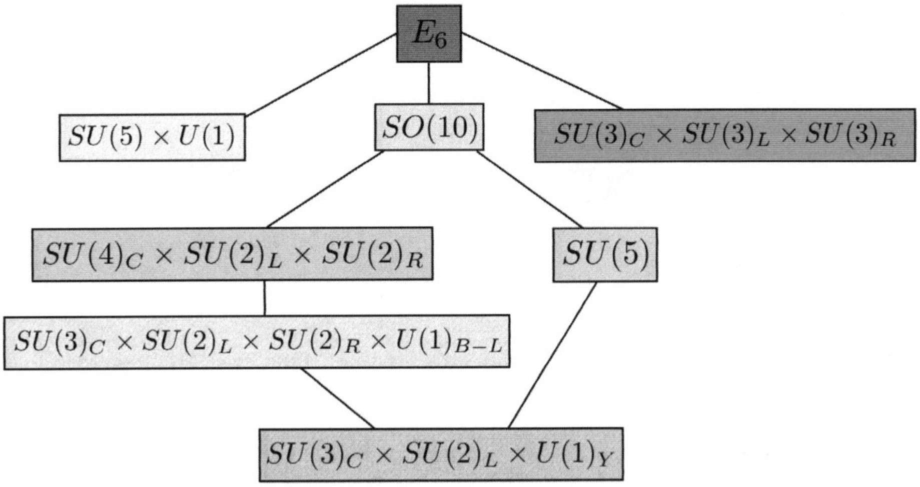

Figure 10. Some possible candidate unified gauge groups.

An early suggested pattern of lepton mixing is known as bimaximal (BM) mixing with $s_{13}^2 = 0$ and $s_{12}^2 = s_{23}^2 = 1/2$ which could originate from the discrete group S_4, where the T generator is preserved in the charged lepton sector, and the S, U generators are preserved in the neutrino sector (identified with the Klein symmetry) as shown in Fig.11. It has a maximal solar mixing angle [16], and is given by a matrix of the form

$$U_{\rm BM} = \begin{pmatrix} \frac{1}{\sqrt{2}} & \frac{1}{\sqrt{2}} & 0 \\ -\frac{1}{2} & \frac{1}{2} & \frac{1}{\sqrt{2}} \\ \frac{1}{2} & -\frac{1}{2} & \frac{1}{\sqrt{2}} \end{pmatrix}. \tag{10}$$

A second pattern of lepton mixing which came to dominate the model building community until the measurement of the reactor angle is the tribimaximal (TB) mixing matrix [17]. This has been associated with models based on the flavour symmetries A_4 and S_4. Like BM mixing it predicts $s_{13}^2 = 0$ and $s_{23}^2 = 1/2$ but differs in that it predicts a solar mixing angle given by $s_{12} = 1/\sqrt{3}$, i.e. $\theta_{12} \approx 35.3°$. The mixing matrix is given explicitly by

$$U_{\rm TB} = \begin{pmatrix} \sqrt{\frac{2}{3}} & \frac{1}{\sqrt{3}} & 0 \\ -\frac{1}{\sqrt{6}} & \frac{1}{\sqrt{3}} & \frac{1}{\sqrt{2}} \\ \frac{1}{\sqrt{6}} & -\frac{1}{\sqrt{3}} & \frac{1}{\sqrt{2}} \end{pmatrix}. \tag{11}$$

Another pattern of lepton mixing which was viable until the reactor angle measurement associates the golden ratio $\varphi = \frac{1+\sqrt{5}}{2}$ with the solar mixing angle. The original golden ratio

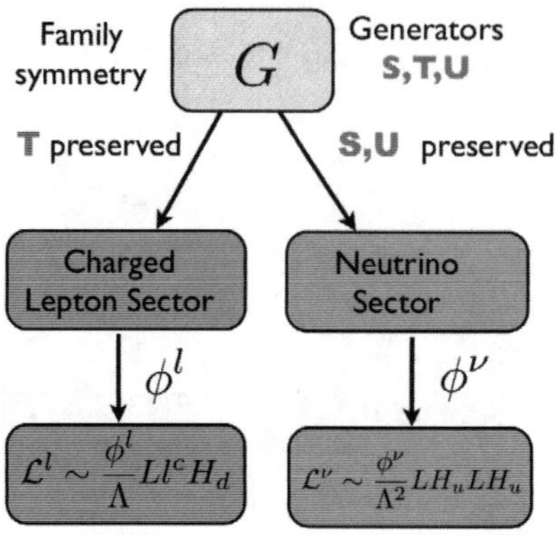

Figure 11. The direct approach to models of lepton mixing.

(GR) mixing pattern is related to the flavour symmetry A_5 [18]. As above, it predicts $s^2_{13} = 0$ and $s^2_{23} = 1/2$ but differs by having a solar mixing angle given by $t^\nu_{12} = 1/\varphi$, i.e. $\theta^\nu_{12} \approx 31.7^\circ$, resulting in the mixing matrix

$$U_{\mathrm{GR}} = \begin{pmatrix} \frac{\varphi}{\sqrt{2+\varphi}} & \frac{1}{\sqrt{2+\varphi}} & 0 \\ -\frac{1}{\sqrt{4+2\varphi}} & \frac{\varphi}{\sqrt{4+2\varphi}} & \frac{1}{\sqrt{2}} \\ \frac{1}{\sqrt{4+2\varphi}} & -\frac{\varphi}{\sqrt{4+2\varphi}} & \frac{1}{\sqrt{2}} \end{pmatrix}. \tag{12}$$

Following the measurement of the reactor angle, all of the above simple patterns are excluded and it has emerged that the only viable direct models are those based on $\Delta(6N^2)$ which leads to more complicated patterns of lepton mixing consistent with data [19, 20, 21]. Unfortunately large N is required in order to achieve the desired reactor angle. Moreover such models generally predict the CP phase $\delta = 0, \pi$ resulting in the atmospheric sum rule [20],

$$\theta_{23} = 45^o \mp \theta_{13}/\sqrt{2}. \tag{13}$$

which follows since the PMNS matrix has the TM2 form shown later in Eq.17.

4.4. Spontaneous CP violation

The direct approach can be generalised to the case of a conserved CP (see [22] and references therein) which is spontaneously broken as shown in Fig.12. This approach has been studied for $\Delta(6N^2)$ in [23]. However since we already know that $\delta = 0, \pi$ in this case it only fixes the Majorana phases.

The generalised CP approach has also been used in the semi-direct approach (defined below) where the phase δ is undetermined without CP. Here the results are more interesting since for the smaller groups like A_4 and S_4 one generally predicts a discrete choice including $\delta = \pm\pi/2$ [24]. However for larger groups in the series $\Delta(6N^2)$ and $\Delta(3N^2)$, broken in a semi-direct way, the discrete predictions for δ proliferate [25].

5. Semi-direct approach and mixing sum rules

Taking a less constrained approach to model building one may suppose that we start from only smaller discrete family groups such as S_4, which leads to either TB or BM mixing at leading

Journal of Physics: Conference Series **631** (2015) 012005 doi:10.1088/1742-6596/631/1/012005

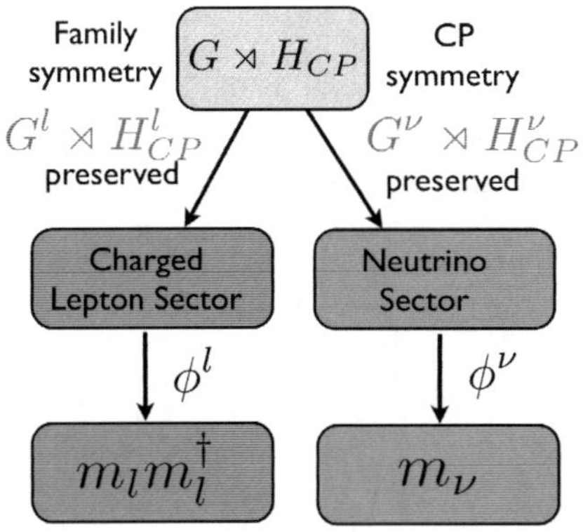

Figure 12. The direct approach including CP.

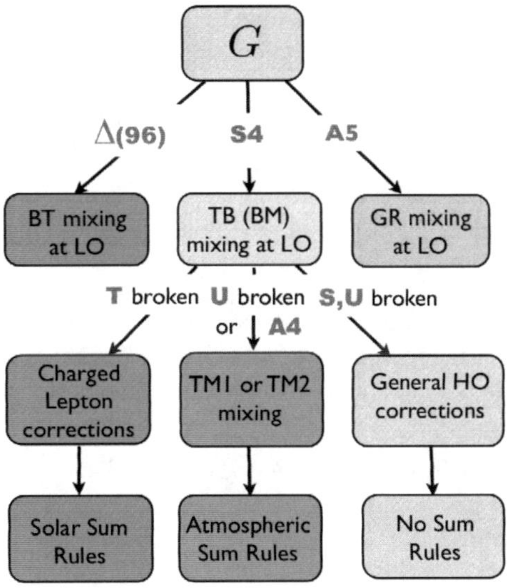

Figure 13. The semi-direct approach to models of lepton mixing.

order, or A_5 which leads to GR mixing at leading order as shown in Fig.13. Then we suppose that at higher order, one or more of the generators S, T, U is broken, which is necessary in this approach since the resulting BM, TB and GR mixing patterns discussed in Eqs.10,11,12 are excluded. There are two interesting possibilities depicted in Fig.13 as follows:

(i) If the U generator is broken then this leads to either TM1 or TM2 mixing depending on whether SU or S is preserved, leading to atmospheric sum rules, as discussed below.

(ii) If the T generator protecting the charged lepton mass matrix is broken, then we can expect charged lepton corrections leading to the solar sum rules also as discussed below.

The semi-direct approach was first used in [26, 27] for A_4 where there is no U generator to start with and also S_4 which is broken to A_4 at higher order [27]. It was subsequently generalised to von Dyck groups in [28]. In all cases the reactor angle is not predicted but described by a free parameter. This is a retreat from the original goal of predicting lepton mixing angles using symmetry.

5.1. TB deviation parameters and atmospheric sum rules

After the measurement of the reactor angle, TB mixing is excluded. However, TB mixing still remains a reasonable approximation to lepton mixing for the solar and atmospheric angles. It therefore makes sense to expand the angles about their TB values [29, 30]:

$$\sin\theta_{12} = \frac{1}{\sqrt{3}}(1+s), \tag{14}$$

$$\sin\theta_{23} = \frac{1}{\sqrt{2}}(1+a), \tag{15}$$

$$\sin\theta_{13} = \frac{r}{\sqrt{2}}, \tag{16}$$

where s, a, and r are the (s)olar, (a)tmospheric and (r)eactor deviation parameters such that TB mixing [17] is recovered for $s = a = r = 0$. For example, TBC mixing [5] corresponds to $s = a = 0$ and $r = \theta_C$, where θ_C is the Cabibbo angle, which is consistent with data at three sigma as shown in Fig.3.

Tri-maximal mixing is a variation which preserves either the first or the second column of the TB mixing mixing matrix in Eq.11, leading to two versions called TM1 or TM2,

$$U_{\mathrm{TM1}} = \begin{pmatrix} \frac{2}{\sqrt{3}} & - & - \\ -\frac{1}{\sqrt{6}} & - & - \\ \frac{1}{\sqrt{6}} & - & - \end{pmatrix}, U_{\mathrm{TM2}} = \begin{pmatrix} - & \frac{1}{\sqrt{3}} & - \\ - & \frac{1}{\sqrt{3}} & - \\ - & -\frac{1}{\sqrt{3}} & - \end{pmatrix}, \tag{17}$$

where TM1 preserves the product SU of S_4, while TM2 preserves the S generator of S_4. The dashes indicate that the other elements are undetermined. However these are fixed once the reactor angle is specified (it is a free parameter here). These imply the relations

$$\mathrm{TM1}: \quad |U_{e1}| = \sqrt{\frac{2}{3}} \text{ and } |U_{\mu 1}| = |U_{\tau 1}| = \frac{1}{\sqrt{6}}\,; \tag{18}$$

$$\mathrm{TM2}: \quad |U_{e2}| = |U_{\mu 2}| = |U_{\tau 2}| = \frac{1}{\sqrt{3}}\,. \tag{19}$$

The *atmospheric mixing sum rule*

$$a = \lambda r\cos\delta + \mathcal{O}(a^2, r^2), \quad \text{with } s = \mathcal{O}(a^2, r^2), \tag{20}$$

was first derived in [29] by expanding the PMNS matrix to first order in r, s, a. It also follows from a first order expansion of Eqs.18,19, where $\lambda = 1$ for TM1 and $\lambda = -1/2$ for TM2. The study of correlations of this type, and their application to the discrimination between underlying models, has been shown to be a realistic aim for a next-generation superbeam experiment [31], see for example Fig.14.

5.2. Charged lepton corrections and solar sum rules

Now suppose that neutrino mixing U^{ν}_{TB} obeys TB exactly so that the PMNS matrix is given by $U = U^e U^{\nu}_{\mathrm{TB}}$ where U^{ν}_{TB} is equated to Eq.11 while U^e encodes some unknown charged lepton corrections which must be small since U is not far from TB mixing. Charged lepton corrections involve some violation of the T generator.

The *solar mixing sum rule* [32, 33, 34] then follows from the assumption that $\theta^e_{23} = \theta^e_{13} = 0$,

$$\theta_{12} \approx 35.26^o + \theta_{13}\cos\delta \quad \text{or} \quad \cos\delta \approx \frac{\theta_{12} - 35.26^o}{\theta_{13}} \tag{21}$$

where $35.26^o = \sin^{-1}\frac{1}{\sqrt{3}}$, which can be recast as [29],

$$s = r\cos\delta + \mathcal{O}(a^2, r^2). \tag{22}$$

Recently it has been realised that, assuming only $\theta^e_{13} = 0$, but allowing $\theta^e_{23} \neq 0$, the following exact result can be simply obtained [35]:

$$\frac{|U_{\tau 1}|}{|U_{\tau 2}|} = \frac{1}{\sqrt{2}}, \tag{23}$$

which readily leads to an exact prediction for $\cos\delta$ in terms of the other physical lepton angles,

$$\cos\delta = \frac{t_{23}s^2_{12} + s^2_{13}c^2_{12}/t_{23} - \frac{1}{3}(t_{23} + s^2_{13}/t_{23})}{\sin 2\theta_{12}s_{13}}, \tag{24}$$

as displayed in Fig.15. For the case of TB neutrino mixing the error in $\cos\delta$ incurred by using the leading order sum rule is $\Delta(\cos\delta) \sim 0.08$ for normal ordering and $\Delta(\cos\delta) \sim 0.01$ for inverted ordering [35]. These errors are unimportant for current levels of accuracy as shown in Fig.15 but they may (hopefully) become significant in the future.

The leading order sum rule offers a simple way to understand the results in Fig.15. For example from Fig.15 it seems that TB neutrino mixing predicts $\cos\delta \approx 0$ if $\theta_{12} \approx 35.26^o$, which is obvious from Eq.21 [2]. This can also be understood from Eq.24 where we see that for $s^2_{12} = 1/3$ the leading terms $t_{23}s^2_{12}$ and $\frac{1}{3}t_{23}$ in the numerator cancel, leaving $\cos\delta = s_{13}/(2\sqrt{2}t_{23}) \approx 0.05$ which is consistent with the numerical estimates of the error given above for a range of θ_{12}.

Solar sum rules can also be obtained for different types of neutrino mixing such as $U^\nu_{\rm BM}$ (which is almost excluded) and $U^\nu_{\rm GR}$ (which gives similar results to the case $U^\nu_{\rm TB}$ considered here). The general formula given in [35] is,

$$\cos\delta = \frac{t_{23}s^2_{12} + s^2_{13}c^2_{12}/t_{23} - s^{\nu 2}_{12}(t_{23} + s^2_{13}/t_{23})}{\sin 2\theta_{12}s_{13}}, \tag{25}$$

where $s^{\nu 2}_{12} = \frac{1}{3}, \frac{1}{2}$ for $U^\nu_{\rm TB,BM}$ and so on. The prospects for studying *solar sum rules* at JUNO and LBNF is discussed in [35]. A slightly more lengthy but equivalent formula to Eq.24 had been previously derived [36] by an alternative method involving an auxiliary phase ϕ without using the elegant result Eq.23,

$$\cos\delta = \frac{t_{23}}{\sin 2\theta_{12}s_{13}}\left[\cos 2\theta^\nu_{12} + (s^2_{12} - c^{\nu 2}_{12})(1 - \cot^2\theta_{23}s^2_{13})\right]. \tag{26}$$

We prefer the simpler form in Eq.25 which involves θ^ν_{12} in only one place since it exhibits the cancellation between the terms $t_{23}s^2_{12}$ and $s^{\nu 2}_{12}t_{23}$ when $s^2_{12} = s^{\nu 2}_{12}$ responsible for the prediction $\cos\delta \approx 0$ in this case

Finally we give a word of caution that when comparing leading order sum rules to the exact results the ratio $(\cos\delta)_{\rm exact}/(\cos\delta)_{\rm LO}$ used in [37] will lead to misleading results when $(\cos\delta)_{\rm LO} \approx 0$. In general it is safer to compare them using the experimentally relevant quantity $\Delta(\cos\delta) = (\cos\delta)_{\rm exact} - (\cos\delta)_{\rm LO}$ defined in [35].

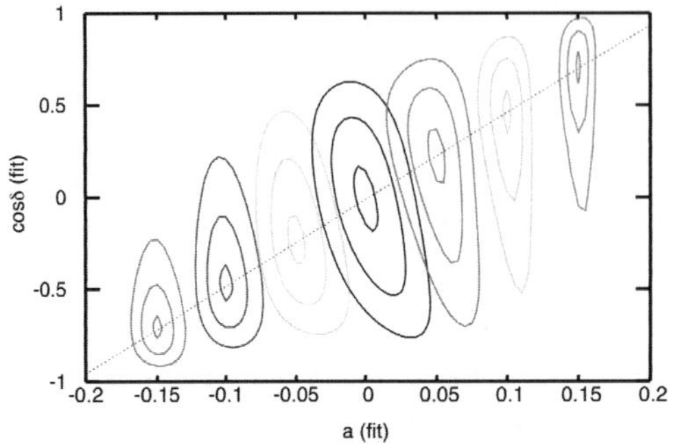

Figure 14. Expectation for the determination of the TM1 atmospheric mixing sum rule $a \approx r\cos\delta$ at one, three and five sigma for a low energy neutrino factory with a magnetised iron detector (for more details see [31]).

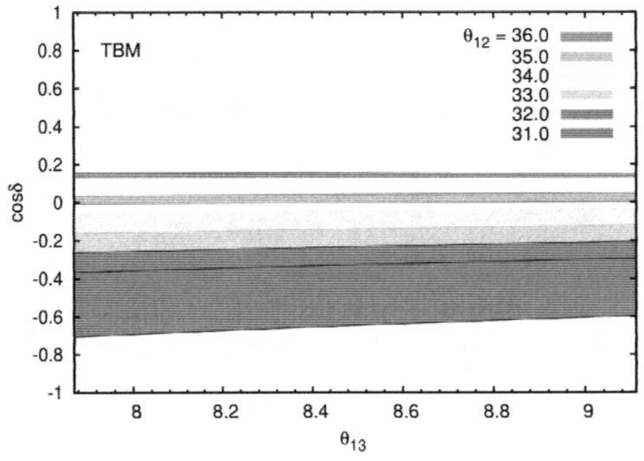

Figure 15. Solar sum rules prediction for $\cos\delta$ using the exact result in Eq.24 for TB neutrino mixing (for more details see [35]).

6. Indirect models

The final logical possibility is that the family symmetry is completely broken as shown in Fig.16. In this approach, we allow the flavons ϕ^l and ϕ^ν to have not only symmetry preserving vacuum alignments, but also new alignments which are orthogonal to them and break the symmetry.

For example in A_4, we have the symmetry preserving alignments,

$$\begin{pmatrix} v \\ 0 \\ 0 \end{pmatrix}, \begin{pmatrix} 0 \\ v \\ 0 \end{pmatrix}, \begin{pmatrix} 0 \\ 0 \\ v \end{pmatrix}, \begin{pmatrix} \pm v \\ \pm v \\ \pm v \end{pmatrix}, \tag{27}$$

where the first three alignments may be used to make the charged lepton mass matrix diagonal. In the neutrino sector we consider new orthogonal alignments such as,

$$\begin{pmatrix} 0 \\ a \\ a \end{pmatrix} \perp \begin{pmatrix} v \\ v \\ -v \end{pmatrix}, \begin{pmatrix} v \\ 0 \\ 0 \end{pmatrix}, \tag{28}$$

[2] Note that $\cos\delta \sim 0$ is consistent with the current hint $\delta \sim -\pi/2$.

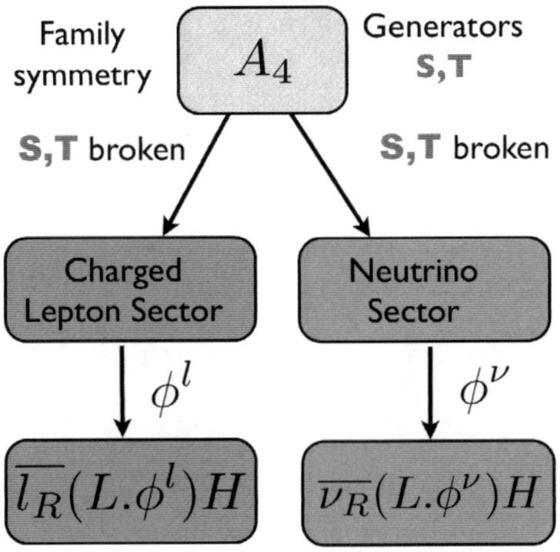

Figure 16. The indirect approach to models of lepton mixing.

as well as ones which are orthogonal to them, and so on. Such alignments completely break the A_4 symmetry. The following symmetry breaking alignment may be obtained using orthogonality arguments

$$\begin{pmatrix} 2 \\ -1 \\ 1 \end{pmatrix} \perp \begin{pmatrix} 1 \\ 1 \\ -1 \end{pmatrix}, \begin{pmatrix} 0 \\ 1 \\ 1 \end{pmatrix}. \tag{29}$$

Using such "orthogonal" alignments various types of constrained sequential dominance (CSD) have been proposed, with the dominant right-handed "atmospheric" couplings $(0, a, a)$ to $(\nu_e, \nu_\mu, \nu_\tau)$, provided by the alignment in Eq.28, and hence an approximate maximal atmospheric angle $\tan\theta_{23} \sim a_1/a_2 \sim 1$. Further orthogonal alignments are responsible for the subdominant "solar" right-handed neutrino couplings to $(\nu_e, \nu_\mu, \nu_\tau)$ as follows:

- CSD(1): $(b, b, -b) \to \theta_{13} = 0$ [32].
- CSD(2): $(b, 2b, 0) \to \theta_{13} \sim 7^\circ$ [38].
- CSD(3): $(b, 3b, b) \to \theta_{13} \sim 8.5^\circ$ [39].
- CSD(4): $(b, 4b, 2b) \to \theta_{13} \sim 9^\circ$ [39].

These examples are chosen to maintain an approximate trimaximal solar angle $\tan\theta_{12} \sim \sqrt{2}b_1/(b_2 - b_3) \sim 1/\sqrt{2}$ but have different reactor angles depending on the phase $\arg(b/a)$, leading to a prediction for δ in each case.

The CSD(n) alignment $(1, n, (n-2))^T$ is orthogonal to the alignment in Eq. 29,

$$\begin{pmatrix} 1 \\ n \\ (n-2) \end{pmatrix} \perp \begin{pmatrix} 2 \\ -1 \\ 1 \end{pmatrix}, \tag{30}$$

where the orthogonality in Eq. 30 is maintained for any value of n (not necessarily integer). To pin down the value of n and show that it is a particular integer requires a further orthogonality condition as discussed in [41].

In a CSD(n) framework [41], after the see-saw mechanism, the neutrino mass matrix simplifies in the two right-handed neutrino case to

$$m^{\nu}_{(n)} = m_a \begin{pmatrix} 0 & 0 & 0 \\ 0 & 1 & 1 \\ 0 & 1 & 1 \end{pmatrix} + m_b e^{i\eta} \begin{pmatrix} 1 & n & n-2 \\ n & n^2 & n(n-2) \\ n-2 & n(n-2) & (n-2)^2 \end{pmatrix}, \tag{31}$$

where η is a physically important phase. This case immediately predicts the lightest physical neutrino mass to be zero, $m_1 = 0$. It also immediately predicts TM1 mixing since,

$$m^{\nu}_{(n)} \begin{pmatrix} 2 \\ -1 \\ 1 \end{pmatrix} = \begin{pmatrix} 0 \\ 0 \\ 0 \end{pmatrix}. \tag{32}$$

In other words the column vector $(2,-1,1)^T$ is an eigenvector of $m^{\nu}_{(n)}$ with a zero eigenvalue, i.e. it is the first column of the PMNS mixing matrix, corresponding to $m_1 = 0$, which means TM1 mixing.

For a given choice of alignment n, there are three real input parameters m_a, m_b and η from which two light physical neutrino masses m_2, m_3, three lepton mixing angles, the CP-violating phase δ_{CP} and two Majorana phases are derived; a total of nine physical parameters from three input parameters, i.e. six predictions for each value of n. As the Majorana phases are not known and δ_{CP} is only tentatively constrained by experiment, this leaves five presently measured observables, namely the two neutrino mass squared differences and the three lepton mixing angles, from only three input parameters. The resulting predictions [41] are shown in Fig.17.

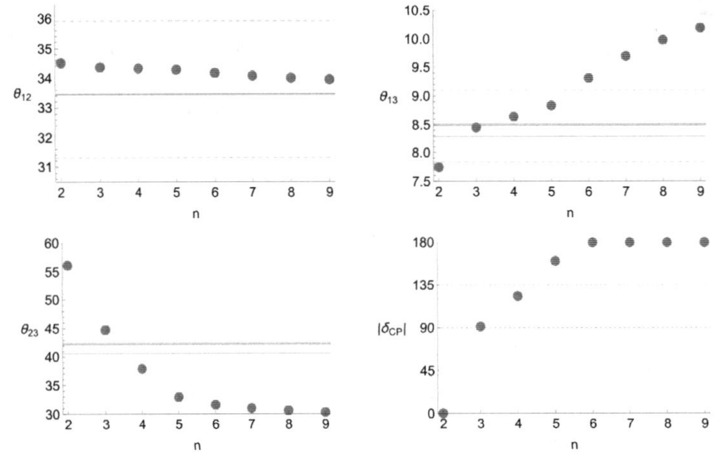

Figure 17. Best-fit PMNS mixing angles and CP-violating phase with respect to n, for the two right-handed neutrino CSD(n) model, after performing a global fit to the three real input parameters m_a, m_b and η. We emphasise that $|\delta_{\rm CP}|$ is a genuine prediction here since have not used the one sigma hint from experiment as an input constraint. It is striking that both CSD(3) and CSD(4) both yield predictions within the preferred range $|\delta_{\rm CP}| \sim 90° \pm 45°$ but may be distinguished by their differing predictions for the atmospheric angle $\theta_{23} \approx 45°$ and $\theta_{23} \approx 38°$, respectively.

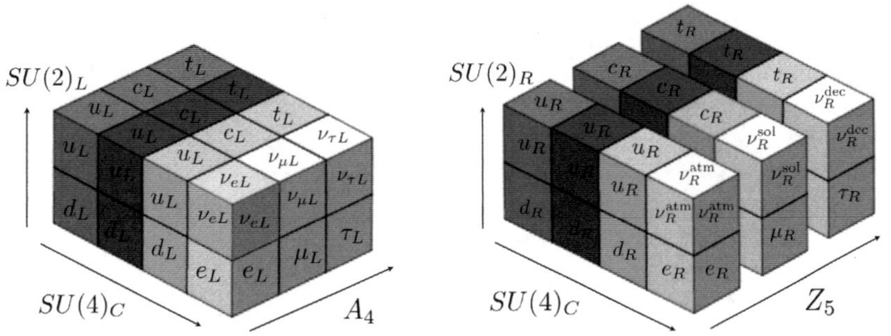

Figure 18. A to Z of flavour with Pati-Salam, where $A \equiv A_4$ and $Z \equiv Z_5$. The left-handed families form a triplet of A_4 and are doublets of $SU(2)_L$. The right-handed families are distinguished by Z_5 and are doublets of $SU(2)_R$. The $SU(4)_C$ unifies the quarks and leptons with leptons as the fourth colour, depicted here as white.

6.1. A to Z of flavour with Pati-Salam

As an example of an "indirect" model, an "A to Z of flavour with Pati-Salam" based on the Pati-Salam gauge group has been proposed [42] as sketched in Fig.18. The Pati-Salam symmetry leads to $Y^u = Y^\nu$, with the first column proportional to the orthogonal alignment $(0, e, e)$ the second column proportional to the CSD(4) orthogonal alignment, $(a, 4a, 2a)$ and the third column is proportional to the symmetry preserving alignment $(0, 0, c)$, where $e \ll a \ll c$ gives the hierarchy $m_u \ll m_c \ll m_t$. This structure predicts a Cabibbo angle $\theta_C \approx 1/4$ in the diagonal $Y^d \sim Y^e$ basis enforced by the first three alignments in Eq.27. It also predicts a normal neutrino mass hierarchy with $\theta_{13} \approx 9°$, $\theta_{23} \approx 45°$ and $\delta \approx 260°$ [42].

The model is based on the Pati-Salam (PS) gauge group, with $A_4 \times Z_5$ family symmetry,

$$SU(4)_C \times SU(2)_L \times SU(2)_R \times A_4 \times Z_5. \tag{33}$$

The quarks and leptons are unified in the PS representations as follows,

$$\begin{aligned} F_i &= (4, 2, 1)_i = \begin{pmatrix} u & u & u & \nu \\ d & d & d & e \end{pmatrix}_i \to (Q_i, L_i), \\ F_i^c &= (\bar{4}, 1, 2)_i = \begin{pmatrix} u^c & u^c & u^c & \nu^c \\ d^c & d^c & d^c & e^c \end{pmatrix}_i \to (u_i^c, d_i^c, \nu_i^c, e_i^c), \end{aligned} \tag{34}$$

where the SM multiplets $Q_i, L_i, u_i^c, d_i^c, \nu_i^c, e_i^c$ resulting from PS breaking are also shown and the subscript i $(= 1, 2, 3)$ denotes the family index. The left-handed quarks and leptons form an A_4 triplet F, while the three (CP conjugated) right-handed fields F_i^c are A_4 singlets, distinguished by Z_5 charges $\alpha, \alpha^3, 1$, for $i = 1, 2, 3$, respectively. Clearly the Pati-Salam model cannot be embedded into an $SO(10)$ Grand Unified Theory (GUT) since different components of the 16-dimensional representation of $SO(10)$ would have to transform differently under $A_4 \times Z_5$, which is impossible. On the other hand, the PS gauge group and A_4 could emerge directly from string theory.

The Pati-Salam gauge group is broken at the GUT scale to the SM,

$$SU(4)_C \times SU(2)_L \times SU(2)_R \to SU(3)_C \times SU(2)_L \times U(1)_Y, \tag{35}$$

by PS Higgs, H^c and $\overline{H^c}$,

$$\begin{aligned} H^c &= (\bar{4}, 1, 2) = (u_H^c, d_H^c, \nu_H^c, e_H^c), \\ \overline{H^c} &= (4, 1, 2) = (\bar{u}_H^c, \bar{d}_H^c, \bar{\nu}_H^c, \bar{e}_H^c). \end{aligned} \tag{36}$$

These acquire vacuum expectation values (VEVs) in the "right-handed neutrino" directions, with equal VEVs close to the GUT scale 2×10^{16} GeV,

$$\langle H^c \rangle = \langle \nu^c_H \rangle = \langle \overline{H^c} \rangle = \langle \overline{\nu}^c_H \rangle \sim 2 \times 10^{16} \text{ GeV}, \tag{37}$$

so as to maintain supersymmetric gauge coupling unification.

Our starting point is to assume that the high energy theory, above the PS breaking scale, conserves CP. Under a CP transformation, the A_4 singlet fields ξ, Σ_u, Σ_d transform into their complex conjugates,

$$\xi \to \xi^*, \quad \Sigma_u \to \Sigma_u^*, \quad \Sigma_d \to \Sigma_d^*, \tag{38}$$

where the complex conjugate fields transform in the complex conjugate representations under $A_4 \times Z_5$. For example if $\xi \sim \alpha^4$, under Z_5, then $\xi^* \sim \alpha$. Similarly if $\Sigma_u \sim 1'$, $\Sigma_d \sim 1''$, under A_4, then $\Sigma_u^* \sim 1''$, $\Sigma_d^* \sim 1'$. On the other hand, in a particular basis, for A_4 triplets $\phi \sim (\phi_1, \phi_2, \phi_3)$, a consistent definition of CP symmetry requires the second and third triplet components to swap under CP,

$$\phi \to (\phi_1^*, \phi_3^*, \phi_2^*). \tag{39}$$

With the above definition of CP, all coupling constants g and explicit masses m are real due to CP conservation and the only source of phases can be the VEVs of fields which break $A_4 \times Z_5$. In the model of interest, all the physically interesting CP phases will arise from Z_5 breaking.

Let us now consider the A_4 triplet fields ϕ which also carry Z_5 charges. In the full model there are four such triplet fields, or "flavons", denoted as ϕ^u_1, ϕ^u_2, ϕ^d_1, ϕ^d_2. The idea is that ϕ^u_i are responsible for up-type quark flavour, while ϕ^d_i are responsible for down-type quark flavour.

The structure of the Yukawa matrices depends on the so-called CSD(4) vacuum alignments of these flavons, with the overall phases quantised due to Z_5,

$$\langle \phi^u_1 \rangle = \frac{V^u_1}{\sqrt{2}} e^{im\pi/5} \begin{pmatrix} 0 \\ 1 \\ 1 \end{pmatrix}, \qquad \langle \phi^u_2 \rangle = \frac{V^u_2}{\sqrt{21}} e^{im\pi/5} \begin{pmatrix} 1 \\ 4 \\ 2 \end{pmatrix}, \tag{40}$$

and

$$\langle \phi^d_1 \rangle = V^d_1 e^{in\pi/5} \begin{pmatrix} 1 \\ 0 \\ 0 \end{pmatrix}, \qquad \langle \phi^d_2 \rangle = V^d_2 e^{in\pi/5} \begin{pmatrix} 0 \\ 1 \\ 0 \end{pmatrix}. \tag{41}$$

We note here that the vacuum alignments in Eq.41 and the first alignment in Eq.40 are fairly "standard" alignments that are encountered in tri-bimaximal mixing models, while the second alignment in Eq.40 is obtained using orthogonality arguments, as discussed above.

The model will involve Higgs bi-doublets of two kinds, h_u which lead to up-type quark and neutrino Yukawa couplings and h_d which lead to down-type quark and charged lepton Yukawa couplings. In addition a Higgs bidoublet h_3, which is also an A_4 triplet, is used to give the third family Yukawa couplings.

After the PS and A_4 breaking, most of these Higgs bi-doublets will get high scale masses and will not appear in the low energy spectrum. In fact only two light Higgs doublets will survive down to the TeV scale, namely H_u and H_d. The basic idea is that the light Higgs doublet H_u with hypercharge $Y = +1/2$, which couples to up-type quarks and neutrinos, is a linear combination of components of the Higgs bi-doublets of the kind h_u and h_3, while the light Higgs doublet H_d with hypercharge $Y = -1/2$, which couples to down-type quarks and charged leptons, is a linear combination of components of Higgs bi-doublets of the kind h_d and h_3,

$$h_u, h_3 \to H_u, \qquad h_d, h_3 \to H_d. \tag{42}$$

The renormalisable Yukawa operators, which respect PS and A_4 symmetries, have the following form, leading to the third family Yukawa couplings shown, using Eqs.34,42,

$$F.h_3F_3^c \to Q_3H_uu_3^c + Q_3H_dd_3^c + L_3H_u\nu_3^c + L_3H_de_3^c, \tag{43}$$

where we have used Eqs.34,42. The non-renormalisable operators, which respect PS and A_4 symmetries, have the following form,

$$F.\phi_i^u h_u F_i^c \;\to\; Q.\langle\phi_i^u\rangle H_u u_i^c + L.\langle\phi_i^u\rangle H_u\nu_i^c, \tag{44}$$
$$F.\phi_i^d h_d F_i^c \;\to\; Q.\langle\phi_i^d\rangle H_d d_i^c + L.\langle\phi_i^d\rangle H_d e_i^c, \tag{45}$$

where $i = 1$ gives the first column of each Yukawa matrix, while $i = 2$ gives the second column and we have used Eqs.34,42. Thus the third family masses are naturally larger since they correspond to renormalisable operators, while the hierarchy between first and second families arises from a hierarchy of flavon VEVs.

Inserting the vacuum alignments in Eqs.40 and 41 into Eqs.44 and 45, together with the renormalisable third family couplings in Eq.43, gives the Yukawa matrices of the form,

$$Y^u = Y^\nu = \begin{pmatrix} 0 & a & 0 \\ e & 4a & 0 \\ e & 2a & c \end{pmatrix}, \quad Y^d \sim Y^e \sim \begin{pmatrix} y_d^0 & 0 & 0 \\ 0 & y_s^0 & 0 \\ 0 & 0 & y_b^0 \end{pmatrix}. \tag{46}$$

The PS unification predicts the equality of Yukawa matrices $Y^u = Y^\nu$ and $Y^d \sim Y^e$, while the A_4 vacuum alignment predicts the structure of each Yukawa matrix, essentially identifying the first two columns with the vacuum alignments in Eqs.40 and 41. With a diagonal right-handed Majorana mass matrix, Y^ν leads to a successful prediction of the PMNS mixing parameters. Also the Cabibbo angle is given by $\theta_C \approx 1/4$ [40]. Thus Eq.46 is a good starting point for a theory of quark and lepton masses and mixing, although the other quark mixing angles and the quark CP phase are approximately zero. However the above discussion ignores the effect of Clebsch factors which will alter the relationship between elements of Y^d and Y^e, which also include off-diagonal elements responsible for small quark mixing angles in the full model discussed in [42].

7. Conclusion

In conclusion, although the reactor angle has been accurately measured, which rules out simple patterns of lepton mixing such as BM, TB and GR, it is still possible to have simple patterns of lepton mixing with the first or second column of the TB matrix preserved, namely TM1 or TM2, with atmospheric sum rules. It is also possible to maintain BM, TB and GR mixing for *neutrinos* with the reactor angle is due to charged lepton corrections, leading to solar sum rules.

Although adding right-handed neutrinos is a very simple and minimal thing to do, the number of right-handed (sterile) neutrinos is undetermined by anomaly cancellation, and their mass spectrum is completely unknown. The classic see-saw mechanism would correspond to having three right-handed neutrinos with masses in the range TeV-M_{GUT}. Sequential dominance provides an elegant and natural way to understand neutrino mixing angles, with the dominant right-handed neutrino couplings providing the atmospheric mixing angle, the sub-dominant solar right-handed neutrino couplings providing the solar mixing angle and the decoupled right-handed neutrino couplings being irrelevant.

Turning to theories of flavour, a very promising approach is the combination of GUT and family symmetry. The large lepton mixing angles suggest some sort of discrete family symmetry at work, although not in the most simple direct way imagined before the reactor angle was measured. However the direct symmetry approach in which the symmetries of the mass matrices are directly embedded into the family symmetry, drives us to family symmetry groups in the $\Delta(6N^2)$ series with large N values necessary in order to explain the reactor angle.

A retreat from this is to suppose that only part of the symmetries of the mass matrices can be found in the family symmetry group, which is called the semi-direct approach. This allows smaller family groups such as S_4, A_4, A_5 whose generators S, T, U may only partly survive. For example if T is broken but the Klein symmetry S, U survives in the neutrino sector this would correspond to BM, TB or GR neutrino mixing but with charged lepton corrections, leading to solar sum rules. If T is preserved but U is broken then this corresponds to TM1 or TM2 mixing with atmospheric sum rules.

Alternatively, it is possible to have a small family symmetry such as A_4 where the family symmetry is completely broken, as in the indirect approach. In this case new vacuum alignments are possible which can be used to give interesting Yukawa couplings corresponding to different types of constrained sequential dominance. Although not enforced by symmetry, the indirect approach leads to highly predictive models which can be tested experimentally.

Acknowledgments
SK acknowledges support from the European Union FP7 ITN-INVISIBLES (Marie Curie Actions, PITN- GA-2011- 289442) and thanks the organisers for a very stimulating and enjoyable workshop.

References

[1] Talk by Soo-Bong Kim (RENO), these proceedings; talk by Jen-Chieh Peng (Daya Bay), these proceedings.
[2] M. C. Gonzalez-Garcia, M. Maltoni and T. Schwetz, arXiv:1409.5439.
[3] F. Capozzi, G. L. Fogli, E. Lisi, A. Marrone, D. Montanino and A. Palazzo, Phys. Rev. D **89** (2014) 093018 [arXiv:1312.2878].
[4] D. V. Forero, M. Tortola and J. W. F. Valle, arXiv:1405.7540.
[5] S. F. King, Phys. Lett. B **718** (2012) 136 [arXiv:1205.0506 [hep-ph]].
[6] S. F. King and C. Luhn, Rept. Prog. Phys. **76** (2013) 056201 [arXiv:1301.1340].
[7] S. F. King, A. Merle, S. Morisi, Y. Shimizu and M. Tanimoto, New J. Phys. **16** (2014) 045018 [arXiv:1402.4271].
[8] S. F. King, Rept. Prog. Phys. **67** (2004) 107 [hep-ph/0310204].
[9] G. Altarelli and F. Feruglio, Rev. Mod. Phys. **82**, 2701 (2010) [arXiv:1002.0211 [hep-ph]].
[10] H. Ishimori, T. Kobayashi, H. Ohki, Y. Shimizu, H. Okada and M. Tanimoto, Prog. Theor. Phys. Suppl. **183**, 1 (2010) [arXiv:1003.3552 [hep-th]]; Lect. Notes Phys. **858**, 1 (2012); Fortsch. Phys. **61**, 441 (2013).
[11] P. Minkowski, Phys. Lett. B **67** (1977) 421; M. Gell-Mann, P. Ramond and R. Slansky in Sanibel Talk, CALT-68-709, Feb 1979, and in *Supergravity*, North Holland, Amsterdam (1979); T. Yanagida in *Proc. of the Workshop on Unified Theory and Baryon Number of the Universe*, KEK, Japan (1979); S.L.Glashow, Cargese Lectures (1979); R. N. Mohapatra and G. Senjanovic, Phys. Rev. Lett. **44** (1980) 912.
[12] M. Drewes, arXiv:1502.06891 [hep-ph].
[13] S. F. King, Phys. Lett. B **439** (1998) 350 [hep-ph/9806440]; S. F. King, Nucl. Phys. B **562** (1999) 57 [hep-ph/9904210]; S. F. King, Nucl. Phys. B **576** (2000) 85 [hep-ph/9912492];
[14] S. F. King, JHEP **0209** (2002) 011 [hep-ph/0204360]; S. F. King, Phys. Rev. D **67** (2003) 113010 [hep-ph/0211228].
[15] S. F. King and C. Luhn, JHEP **0910** (2009) 093 [arXiv:0908.1897].
[16] M. Fukugita, M. Tanimoto and T. Yanagida, Phys. Rev. D **57** (1998) 4429 [hep-ph/9709388]; V. D. Barger, S. Pakvasa, T. J. Weiler and K. Whisnant, Phys. Lett. B **437** (1998) 107 [hep-ph/9806387]; S. Davidson and S. F. King, Phys. Lett. B **445** (1998) 191 [hep-ph/9808296].
[17] P. F. Harrison, D. H. Perkins and W. G. Scott, Phys. Lett. B **530** (2002) 167 [hep-ph/0202074].
[18] A. Datta, F. S. Ling and P. Ramond, Nucl. Phys. B **671** (2003) 383 [hep-ph/0306002]; L. L. Everett and A. J. Stuart, Phys. Rev. D **79** (2009) 085005 [arXiv:0812.1057]; F. Feruglio and A. Paris, JHEP **1103** (2011) 101 [arXiv:1101.0393].
[19] M. Holthausen, K. S. Lim and M. Lindner, Phys. Lett. B **721** (2013) 61 [arXiv:1212.2411 [hep-ph]].
[20] S. F. King, T. Neder and A. J. Stuart, Phys. Lett. B **726** (2013) 312 [arXiv:1305.3200 [hep-ph]].
[21] R. M. Fonseca and W. Grimus, JHEP **1409** (2014) 033 [arXiv:1405.3678 [hep-ph]].

[22] M. Holthausen, M. Lindner and M. A. Schmidt, JHEP **1304**, 122 (2013) [arXiv:1211.6953 [hep-ph]].
[23] S. F. King and T. Neder, Phys. Lett. B **736** (2014) 308 [arXiv:1403.1758 [hep-ph]].
[24] F. Feruglio, C. Hagedorn and R. Ziegler, JHEP **1307**, 027 (2013) [arXiv:1211.5560 [hep-ph]]; F. Feruglio, C. Hagedorn and R. Ziegler, Eur. Phys. J. C **74**, 2753 (2014) [arXiv:1303.7178 [hep-ph]]; G. -J. Ding, S. F. King, C. Luhn and A. J. Stuart, JHEP **1305**, 084 (2013) [arXiv:1303.6180 [hep-ph]]; G. -J. Ding, S. F. King and A. J. Stuart, JHEP **1312** (2013) 006 [arXiv:1307.4212].
[25] C. Hagedorn, A. Meroni and E. Molinaro, arXiv:1408.7118 [hep-ph]; G. J. Ding, S. F. King and T. Neder, JHEP **1412** (2014) 007 [arXiv:1409.8005 [hep-ph]].
[26] Y. Shimizu, M. Tanimoto and A. Watanabe, Prog. Theor. Phys. **126** (2011) 81 [arXiv:1105.2929].
[27] S. F. King and C. Luhn, JHEP **1109** (2011) 042 [arXiv:1107.5332].
[28] D. Hernandez and A. Y. Smirnov, Phys. Rev. D **86** (2012) 053014 [arXiv:1204.0445]; D. Hernandez and A. Y. Smirnov, Phys. Rev. D **87** (2013) 5, 053005 [arXiv:1212.2149].
[29] S. F. King, Phys. Lett. B **659** (2008) 244 [arXiv:0710.0530 [hep-ph]].
[30] S. Pakvasa, W. Rodejohann and T. J. Weiler, Phys. Rev. Lett. **100** (2008) 111801 [arXiv:0711.0052 [hep-ph]].
[31] P. Ballett, S. F. King, C. Luhn, S. Pascoli and M. A. Schmidt, Phys. Rev. D **89** (2014) 016016 [arXiv:1308.4314].
[32] S. F. King, JHEP **0508** (2005) 105 [hep-ph/0506297].
[33] I. Masina, Phys. Lett. B **633** (2006) 134 [hep-ph/0508031].
[34] S. Antusch and S. F. King, Phys. Lett. B **631** (2005) 42 [hep-ph/0508044]; S. Antusch, P. Huber, S. F. King and T. Schwetz, JHEP **0704** (2007) 060 [hep-ph/0702286].
[35] P. Ballett, S. F. King, C. Luhn, S. Pascoli and M. A. Schmidt, arXiv:1410.7573 [hep-ph].
[36] D. Marzocca, S. T. Petcov, A. Romanino and M. C. Sevilla, JHEP **1305** (2013) 073 [arXiv:1302.0423]; S. T. Petcov, arXiv:1405.6006.
[37] I. Girardi, S. T. Petcov and A. V. Titov, arXiv:1410.8056 [hep-ph].
[38] S. Antusch, S. F. King, C. Luhn and M. Spinrath, Nucl. Phys. B **856** (2012) 328 [arXiv:1108.4278].
[39] S. F. King, JHEP **1307** (2013) 137 [arXiv:1304.6264]; S. F. King, Phys. Lett. B **724** (2013) 92 [arXiv:1305.4846].
[40] S. F. King, JHEP **1401** (2014) 119 [arXiv:1311.3295 [hep-ph]];
[41] F. Bjrkeroth and S. F. King, arXiv:1412.6996 [hep-ph].
[42] S. F. King, JHEP **1408** (2014) 130 [arXiv:1406.7005 [hep-ph]]; S. F. King, JHEP **1401** (2014) 119 [arXiv:1311.3295 [hep-ph]].

4th Symposium on Prospects in the Physics of Discrete Symmetries (DISCRETE2014) IOP Publishing
Journal of Physics: Conference Series **631** (2015) 012006 doi:10.1088/1742-6596/631/1/012006

Unraveling the organization of the QCD tapestry

J. Papavassiliou

Department of Theoretical Physics and IFIC, University of Valencia and CSIC, E-46100, Valencia, Spain

E-mail: Joannis.Papavassiliou@uv.es

Abstract. I review some key aspects of the ongoing progress in our understanding of the infrared dynamics of the QCD Green's functions, derived from the close synergy between Schwinger-Dyson equations and lattice simulations. Particular attention is dedicated to the elaborate nonperturbative mechanisms that endow the fundamental degrees of freedom (quarks and gluons) with dynamical masses. In addition, the recently established connection between the effective interaction obtained from the gauge sector of the theory and that needed for the veracious description of the ground-state properties of hadrons is briefly presented.

1. Introduction

Sixty years after their invention [1], Yang-Mills theories occupy the center stage of elementary particle physics, providing a fundamental description for an impressive array of physical phenomena. The remarkable property of asymptotic freedom makes these theories particularly attractive, and, in a sense, self-contained. However, while perturbation theory works well in their ultraviolet regime, the infrared dynamics represent a challenge of notorious complexity. In fact, in the case of QCD [2], the most characteristic phenomena, such as confinement, mass generation, and bound-state formation, are purely nonperturbative.

In this presentation we will focus on the paradigm-shifting picture that emerges for some of the aforementioned phenomena from the systematic study of Green's functions within a global framework, where the analysis of the Schwinger-Dyson equations (SDEs) is complemented and refined by inputs obtained from large-volume lattice simulations.

The present study is restricted to the special case of the Landau gauge, where the vast majority of recent lattice simulations have been performed. The SDE part of the problem is addressed within the formalism obtained from the fusion between the pinch technique (PT) [3, 4, 5, 6, 7] and the background field method (BFM) [8], denoted simply as "PT-BFM" (for a representative sample of different approaches, see [9, 10, 11, 12, 13, 14, 15, 16, 17, 18, 19, 20]).

This presentation is organized as follows. In section 2 we review the PT-BFM formalism, and highlight its most salient features. Then, in section 3 we summarize the subtle filed-theoretic mechanism that leads to the gauge-invariant generation of a dynamical gluon mass. In section 4 we present a brief overview of the impressive coincidence achieved between SDE calculations and lattice simulations of the ghost propagator, while in section 5 we address some of the intricate issues surrounding the dynamical generation of the constituent quark masses. Then, in section 6 we review a recent *ab-initio* derivation of the interaction kernel that has been extensively used

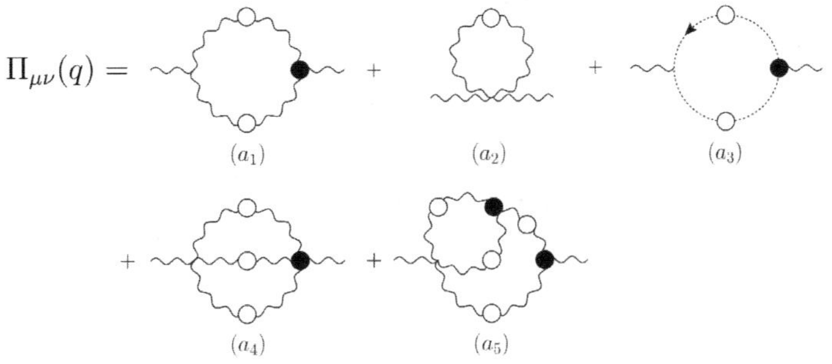

Figure 1. The conventional SDE of the standard gluon propagator (QQ). Black blobs represent fully dressed one-particle irreducible vertices, whereas the white ones denote fully dressed propagators.

in the contemporary description of hadron phenomenology. Finally, in section 7 we summarize our conclusions.

2. The PT-BFM formalism: a powerful framework for studying SDEs

In what follows we will work exclusively in the Landau gauge, where the gluon propagator assumes the totally transverse form

$$i\Delta_{\mu\nu}(q) = -iP_{\mu\nu}(q)\Delta(q^2); \qquad P_{\mu\nu}(q) = g_{\mu\nu} - q_\mu q_\nu/q^2\,. \tag{1}$$

The scalar form factor $\Delta(q^2)$ is related to the all-order gluon self-energy $\Pi_{\mu\nu}(q)$. Specifically, as a consequence of the BRST symmetry, $\Pi_{\mu\nu}(q)$ is transverse, both perturbatively and nonperturbatively; one has then

$$q^\mu \Pi_{\mu\nu}(q) = 0; \qquad \Pi_{\mu\nu}(q) = \Pi(q^2)P_{\mu\nu}(q), \tag{2}$$

and $\Delta^{-1}(q^2) = q^2 + \Pi(q^2)$. In addition, for later convenience, we introduce the *inverse* of the gluon dressing function, to be denoted by $J(q^2)$, namely

$$\Delta^{-1}(q^2) = q^2 J(q^2). \tag{3}$$

The nonperturbative dynamics of the gluon propagator are governed by the corresponding SDE. In particular, within the conventional formulation [21, 22], the gluon self-energy $\Pi_{\mu\nu}(q)$ is given by the fully dressed diagrams shown in Fig. 1. One longstanding difficulty with this equation is that it cannot be truncated in any obvious way without compromising the validity of Eq. (2). This happens because the conventional fully dressed vertices appearing in the diagrams of Fig. 1 satisfy complicated Slavnov-Taylor identities (STIs), and it is only after the inclusion of all diagrams that Eq. (2) may be enforced.

Instead, the formulation of this SDE in the context of the PT-BFM formalism furnishes considerable advantages, because it allows for a systematic truncation that respects manifestly, and at every step, the crucial identity of Eq. (2). The deeper reason why such a truncation becomes possible may be traced back to the fact that the PT-BFM Green's functions satisfy (by construction) Abelian-like Ward identities (WIs) instead of STIs. In particular, if we employ the BFM language and distinguish the gluon fields into background (B) and quantum (Q)

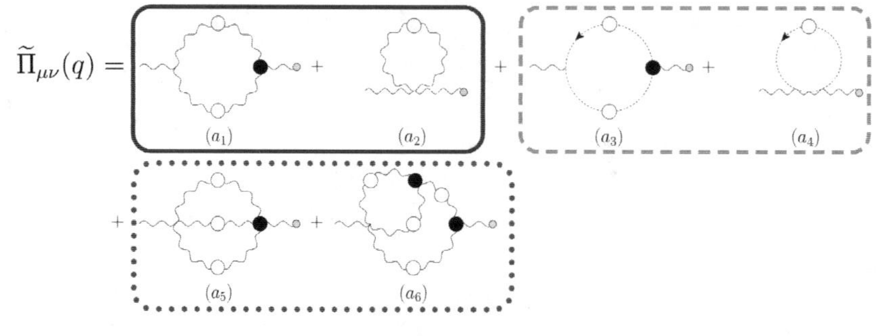

Figure 2. The SDE obeyed by the QB gluon propagator. Black blobs represents fully dressed 1-PI vertices; the small gray circles appearing on the external legs (entering from the right, only!) are used to indicate background gluons. The diagrams contained in each box form individually transverse subsets.

ones, vertices such as BQQ, $B\bar{c}c$, and $BQQQ$, to be denoted by $\widetilde{\Gamma}_{\alpha\mu\nu}(q,r,p)$, $\widetilde{\Gamma}_{\alpha}(q,r,-p)$, and $\widetilde{\Gamma}^{mnrs}_{\mu\nu\rho\sigma}(q,r,p,t)$, respectively, satisfy the simple WIs [23]

$$q^{\alpha}\widetilde{\Gamma}_{\alpha\mu\nu}(q,r,p) = i\Delta^{-1}_{\mu\nu}(r) - i\Delta^{-1}_{\mu\nu}(p), \tag{4}$$

$$q^{\alpha}\widetilde{\Gamma}_{\alpha}(q,r,-p) = D^{-1}(q+r) - D^{-1}(r), \tag{5}$$

and

$$\begin{aligned} q^{\mu}\widetilde{\Gamma}^{mnrs}_{\mu\nu\rho\sigma}(q,r,p,t) &= f^{mse}f^{ern}\Gamma_{\nu\rho\sigma}(r,p,q+t) + f^{mne}f^{esr}\Gamma_{\rho\sigma\nu}(p,t,q+r) \\ &+ f^{mre}f^{ens}\Gamma_{\sigma\nu\rho}(t,r,q+p). \end{aligned} \tag{6}$$

These particular vertices (instead of the conventional ones) appear in the SDE of Fig. 2, which controls the self-energy $\widetilde{\Pi}_{\mu\nu}(q)$ of the the mixed background-quantum gluon propagator (BQ), denoted by $\widetilde{\Delta}(q^2)$. Consequently, it is relatively straightforward to prove the block-wise transversality of $\widetilde{\Pi}_{\mu\nu}(q)$, namely [23]

$$q^{\mu}[(a_1)+(a_2)]_{\mu\nu} = 0; \qquad q^{\mu}[(a_3)+(a_4)]_{\mu\nu} = 0; \qquad q^{\mu}[(a_5)+(a_6)]_{\mu\nu} = 0. \tag{7}$$

From Eq. (2) it is clear that the SDE of $\widetilde{\Delta}(q^2)$ contains inside its defining diagrams the full propagator $\Delta(q^2)$; therefore, in that sense, it cannot be considered as a dynamical equation for neither $\widetilde{\Delta}(q^2)$ nor $\Delta(q^2)$. At this point a crucial identity relating $\Delta(q^2)$ and $\widetilde{\Delta}(q^2)$ [24, 25] enters into the game. Specifically, one has that

$$\Delta(q^2) = [1+G(q^2)]\widetilde{\Delta}(q^2), \tag{8}$$

where $G(q^2)$ is the $g_{\mu\nu}$ co-factor of a certain two-point function that originates from the ghost sector of the theory.

The fundamental observation put forth in a series of works [23, 26, 27] is that one may use the SDE for $\widetilde{\Delta}(q^2)$, take advantage of its improved truncation properties, and then convert it to an equivalent equation for $\Delta(q^2)$ (the propagator simulated on the lattice) by means of the special relation given in Eq. (8). Thus, the PT-BFM version of the SDE for the conventional gluon propagator reads

$$\Delta^{-1}(q^2)P_{\mu\nu}(q) = \frac{q^2 P_{\mu\nu}(q) + i\sum_{i=1}^{6}(a_i)_{\mu\nu}}{1+G(q^2)}. \tag{9}$$

3. Dynamical gluon mass generation

The possibility that Yang-Mills theories generate dynamically a gluon mass was first proposed and explored in the early eighties [3, 28, 29], and has attracted particular attention in recent years, mainly due an accumulation of hard evidence obtained from large-volume lattice simulations, both in SU(3) [30, 31, 32, 33] and in SU(2) [34, 35, 36, 37] (for a different lattice approach, see [38]).

What these simulations reveal is that the Landau gauge gluon propagator saturates in the deep infrared at a fixed nonvanishing value, as shown on the left panel of Fig. 3, which is the smoking gun signal for gluon mass generation. A conclusive demonstration of how this may happen at the level of the gluon SDE within the PT-BFM framework has been given in [39].

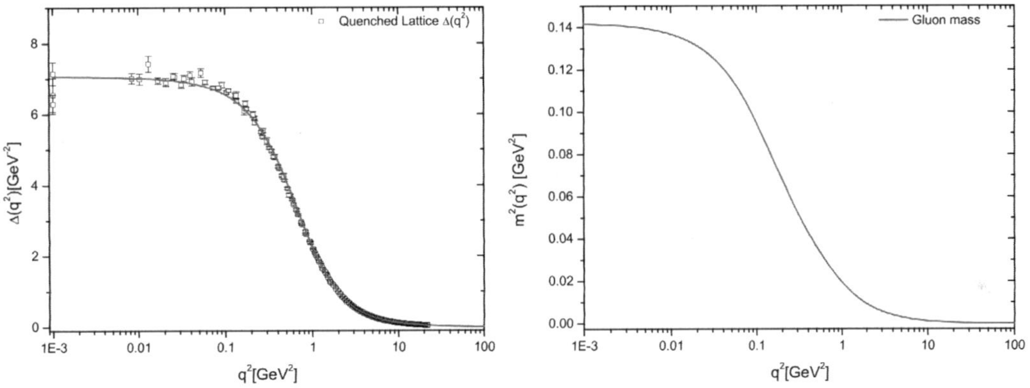

Figure 3. The quenched lattice gluon propagator $\Delta(q^2)$ (left panel), and the corresponding gluon mass (right panel).

From the kinematic point of view we will describe the transition from a massless to a massive gluon propagator by implementing the replacement (in Euclidean space)

$$\Delta^{-1}(q^2) = q^2 J(q^2) \quad \longrightarrow \quad \Delta^{-1}(q^2) = q^2 J(q^2) + m^2(q^2), \tag{10}$$

where $m^2(q^2)$ is the (momentum-dependent) dynamically generated mass, with the property that $m^2(0) > 0$; evidently, $\Delta^{-1}(0) = m^2(0)$.

The actual field-theoretic mechanism that permits the generation of the term $m^2(q^2)$ can be traced back to the seminal work of Schwinger [40, 41], which may be summarized in the statement that the vacuum polarization of a gauge boson that is massless at the level of the original Lagrangian may develop a massless pole, whose residue will be eventually identified with $m^2(0)$. In the case of Yang-Mills theories, the origin of the aforementioned poles is due to purely nonperturbative dynamics: for sufficiently strong binding, the mass of certain (colored) bound states may be reduced to zero [42, 43, 44, 45, 46]. As has been shown in [47] through a

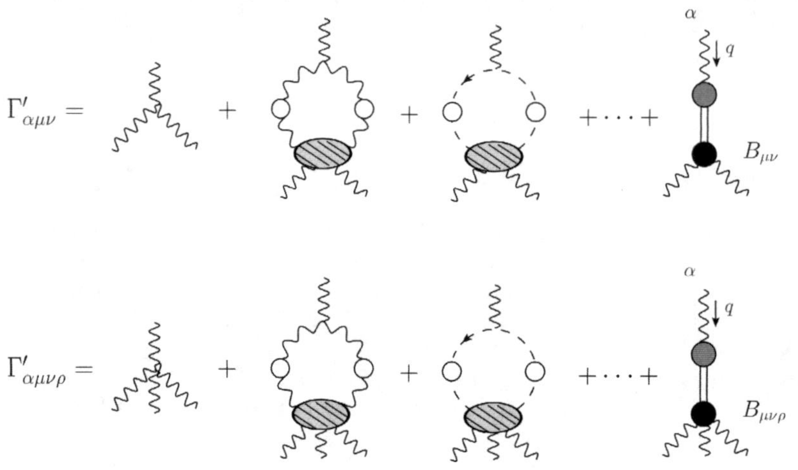

Figure 4. The SDEs for the full three and four gluon vertices in the presence of their corresponding pole parts.

detailed study of the Bethe-Salpeter equation that controls the formation of such a composite massless pole, this dynamical scenario is indeed realized.

In addition to triggering the Schwinger mechanism, these bound-state poles act as composite, longitudinally coupled Nambu-Goldstone bosons, maintaining gauge invariance; notice, however, that they differ from ordinary Nambu-Goldstone bosons as far as their origin is concerned, since they are not associated with the spontaneous breaking of any continuous symmetry. In particular, gauge invariance requires that the replacement described schematically in Eq. (10) be accompanied by a simultaneous replacement of all relevant vertices by

$$\widetilde{\Gamma} \quad \longrightarrow \quad \widetilde{\Gamma}' = \widetilde{\Gamma} + \widetilde{V}, \tag{11}$$

where $\widetilde{V}$ contains the aforementioned massless poles (see Fig. 4), and guarantees that the new vertex $\widetilde{\Gamma}'$ satisfies the same WIs as $\widetilde{\Gamma}$, but now replacing the gluon propagators appearing on their rhs by massive ones. For example, in order to maintain Eq. (4) intact, the corresponding $\widetilde{V}$ must satisfy the WI

$$q^{\alpha}\widetilde{V}_{\alpha\mu\nu}(q,r,p) = m^2(r^2)P_{\mu\nu}(r) - m^2(p^2)P_{\mu\nu}(p); \tag{12}$$

similarly, when contracted with respect to the other two momenta, $\widetilde{V}_{\alpha\mu\nu}(q,r,p)$ satisfies the STIs necessary to enforce the validity of the corresponding STIs satisfied by $\widetilde{\Gamma}_{\alpha\mu\nu}(q,r,p)$, but now with massive instead of massless gluon propagators on the rhs.

The fact that the massless poles must be longitudinally coupled is reflected at the level of $\widetilde{\Gamma}'$ through the condition

$$P^{\alpha'\alpha}(q)P^{\mu'\mu}(r)P^{\nu'\nu}(p)\widetilde{V}_{\alpha'\mu'\nu'}(q,r,p) = 0, \tag{13}$$

and a exactly analogous relation for the four-gluon $\widetilde{V}_{\alpha\mu\nu\rho}(q,r,p,t)$. These latter conditions guarantee that the massless poles decouple completely from the on-shell S-matrix or other physical observables. Moreover, when Eq. (13) is combined with Eq. (12) and the additional STIs not reported here, one obtains the exact closed expression for $\widetilde{V}_{\alpha'\mu'\nu'}(q,r,p)$ [48].

Journal of Physics: Conference Series **631** (2015) 012006 doi:10.1088/1742-6596/631/1/012006

The inclusion of the aforementioned vertices into the gluon SDE leads to its separation into two integral equations of the generic form

$$\begin{aligned} J_m(q^2) &= 1 + \int_k \mathcal{K}_1(q^2, m^2, \Delta_m), \\ m^2(q^2) &= \int_k \mathcal{K}_2(q^2, m^2, \Delta_m), \end{aligned} \tag{14}$$

such that $q^2\mathcal{K}_1(q^2, m^2, \Delta_m) \to 0$, as $q^2 \to 0$, whereas $\mathcal{K}_2(q^2, m^2, \Delta_m) \neq 0$ in the same limit, precisely because it includes the $1/q^2$ terms contained inside the $\widetilde{V}$ terms. This set of equations bares a close analogy to those studied in the case of the more familiar phenomenon of quark mass generation; in particular, $m^2(q^2)$ plays the role of the quark mass function (denoted by $B(p^2)$ in Eq. (22)).

The explicit closed form of the integral equation that determines $m^2(q^2)$, together with a variety of related theoretical considerations, have been presented in a series of works [49, 50, 51]. The upshot is that one obtains a dynamical gluon mass of the form shown on the right panel of Fig. 3, whose functional dependence may be fitted by

$$m^2(q^2) = \frac{m_0^2}{1 + (q^2/\mathcal{M}^2)^{1+p}}, \tag{15}$$

with $m_0 = 375\,\text{MeV}$, $p = 0.11$ and $\mathcal{M} = 431\,\text{MeV}$.

The nontrivial momentum dependence of the gluon mass is mainly responsible for the fact that, contrary to a propagator with a constant mass, the gluon propagator of Fig. 3 displays an inflection point. It turns out that the presence of such a feature is a sufficient condition for the spectral density of the gluon propagator, ρ, to be non-positive definite.

Specifically, the Källén-Lehman representation of the gluon propagator reads

$$\Delta(q^2) = \int_0^\infty d\sigma \frac{\rho(\sigma)}{q^2 + \sigma}, \tag{16}$$

and if $\Delta(q^2)$ has an inflection point at $q_\star^2$, then its second derivative vanishes at that point (see Fig. 5), namely [52]

$$\Delta''(q_\star^2) = 2\int_0^\infty d\sigma \frac{\rho(\sigma)}{(q_\star^2 + \sigma)^3} = 0\,. \tag{17}$$

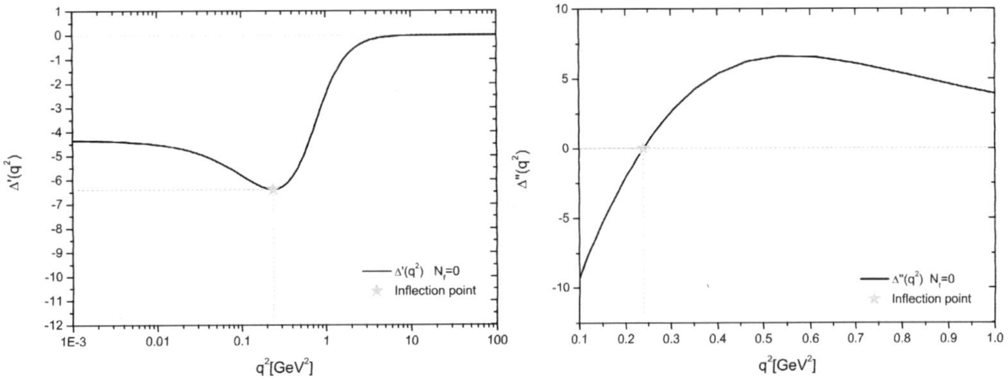

Figure 5. The first and second derivatives of the gluon propagator.

Given that $q_\star^2 > 0$, then $\rho(\sigma)$ is forced to reverse sign at least once. The non-positivity of $\rho(\sigma)$, in turn, may be interpreted as a signal of confinement (see [53], and references therein), since the breeching of the axiom of reflection positivity excludes the gluon from the Hilbert space of observables states (for related works, see [54, 55, 56, 57, 58, 59, 60]).

As can be seen in Fig. 5, the first derivative of $\Delta(q^2)$ displays a minimum at $q_\star^2 = 0.238\,\text{GeV}^2$, and, consequently, the second derivative vanishes at that same point.

4. Ghost sector

The nonperturbative behavior of the ghost propagator, $D(p^2)$, has been the focal point of intensive studies, because in the past it had been linked to a certain confinement scenario [9], whose realization required that the corresponding ghost dressing function, $F(p^2) = p^2 D(p^2)$, should diverge at $p^2 = 0$. However, a plethora of high quality lattice simulations (for SU(3) [31, 32, 30, 33] and for SU(2) [34, 35, 36, 37]) have conclusively established that $F(p^2)$ is finite at the origin. In fact, the finiteness of $F(p^2)$ is tightly interwoven with the massiveness of the gluon, because it is precisely the presence of the gluon mass that tames the logarithm associated with $F(p^2)$, and prevents it from diverging in the infrared [39].

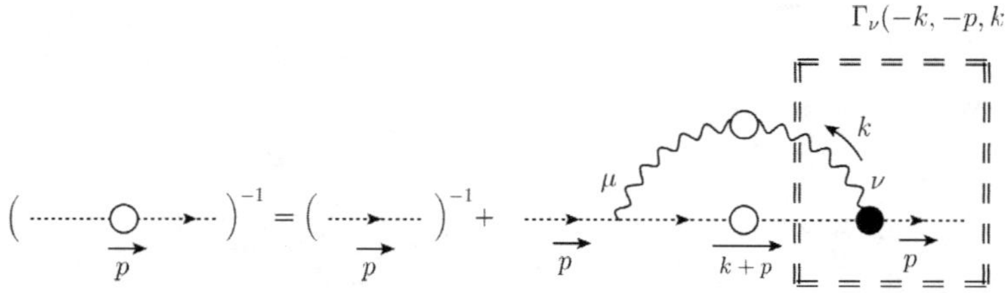

Figure 6. The SDE for the ghost propagator given by Eq. (18). The white blobs represent the fully dressed gluon and ghost propagators, while the black blob denotes the dressed ghost-gluon vertex.

Even though the finiteness of $F(p^2)$ may be explained qualitatively in the above relatively simple terms, a more strenuous effort is required in order to obtain from the corresponding SDE the full curve of $F(p^2)$ observed on the lattice.

The SDE for the ghost propagator is diagrammatically represented in the Fig. 6, and assumes the form

$$iD^{-1}(p^2) = ip^2 - g^2 C_{\rm A} \int_k \Gamma_\mu^{[0]}(k,-k-p,p)\Delta^{\mu\nu}(k)\Gamma_\nu(-k,-p,k+p)D(k+p)\,. \tag{18}$$

In the above equation C_A represents the Casimir eigenvalue of the adjoint representation (N for $SU(N)$), $d = 4-\epsilon$ is the space-time dimension, and we have introduced the integral measure

$$\int_k \equiv \frac{\mu^\epsilon}{(2\pi)^d}\int \mathrm{d}^d k, \tag{19}$$

with μ the 't Hooft mass. The vertex $\Gamma_\nu(-k,-p,k+p)$ is the fully dressed ghost-gluon vertex, whose tensorial decomposition is given by ($r = k+p$)

$$\Gamma_\nu(-k,-p,r) = A(-k,-p,r)\,p_\nu + B(-k,-p,r)\,k_\nu\,; \tag{20}$$

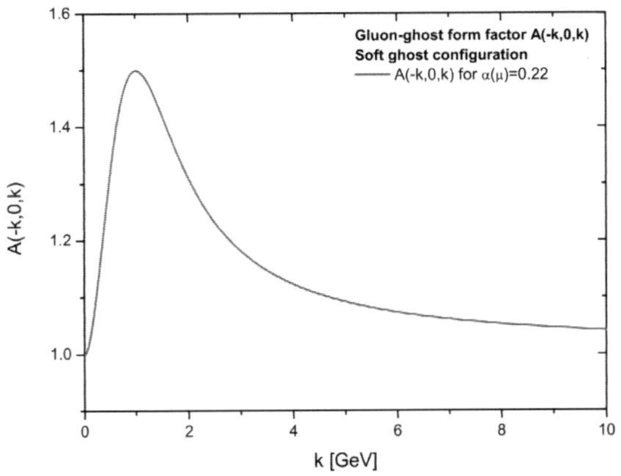

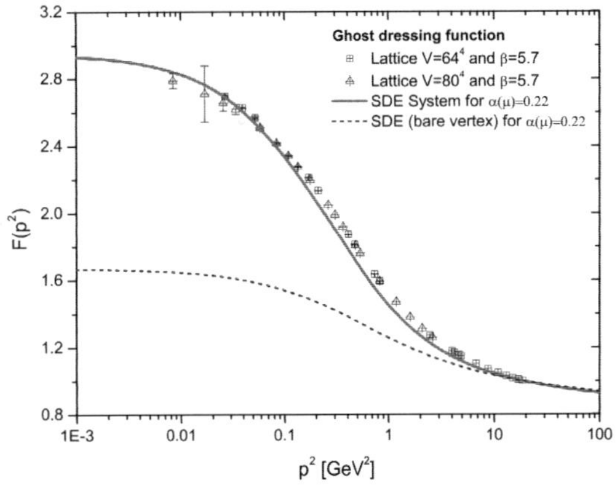

Figure 7. *Left panel*: The numerical result for the form factor $A(-k,0,k)$ in the soft ghost configuration (red continuous line) [64]. *Right panel*: The $F(p^2)$ obtained from the ghost SDE with a bare ghost-gluon vertex (blue dashed curve) and the corresponding solution when the $A(-k,0,k)$ is used (red continuous line) , using in both cases $\mu = 4.3$ GeV and $\alpha_s(\mu) = 0.22$. The lattice data are from Ref. [31].

at tree-level, the two form factors assume the values $A^{[0]}(-k,-p,r) = 1$ and $B^{[0]}(-k,-p,r) = 0$, reproducing the standard bare vertex $\Gamma_\nu^{[0]} = p_\nu$. Clearly, due to the full transversality of $\Delta_{\mu\nu}(k)$, any reference to the form factor B disappears from the ghost SDE of Eq. (18), and one obtains.

$$F^{-1}(p^2) = 1 + ig^2 C_{\rm A} \int_k \left[1 - \frac{(k\cdot p)^2}{k^2p^2}\right] A(-k,-p,k+p)\Delta(k)D(k+p)\,. \tag{21}$$

The dependence of this equation on both $\Delta(k)$ and $A(-k,-p,k+p)$ is just another example of the perennial difficulty inherent to the SDE framework, namely the fact that, in principle, all Green's functions of the theory are coupled to each other. In the case at hand, $\Delta(k)$ is known from the lattice (Fig. 3), and there is no need to resort to its own SDE. However, A is more problematic, since there are no lattice data for the kinematic configuration relevant to this problem (for SU(2), see [61, 62, 63]). The obvious easy way out, namely the use of the tree-level value for A is quantitatively insufficient: one obtains the result shown on the right panel of Fig. 7 (blue dashed curve), which, even though qualitatively correct, it is clearly considerably suppressed compared to the lattice data.

To ameliorate this situation, an approximate version of the SDE equation that controls the evolution of A has been considered in [64], in the special kinematic limit $p \to 0$ (for a related study, see [65]); the result for $A(-k,0,k)$ is shown in the left panel of Fig. 7.

Even though the increase with respect to the tree level value is not too dramatic, the peak of $A(-k,0,k)$ at around 1 GeV turns out to be just right: when coupled with Eq. (21), a considerable enhancement of the resulting $F(p^2)$ is produced, due to the nonlinear nature of this system of integral equations. As shown in the right panel of Fig. 7, one may actually reproduce the lattice data very accurately, using the correct value for the strong coupling α_s [66, 67].

5. Quark sector

It is well known that the nonperturbative mechanism that endows quarks with their constituent masses accounts for about 98% of the visible matter in the Universe. In fact, it has often been denominated as the most efficient mass generating mechanism known, practically generating masses from nothing (see [53] and references therein). This fundamental phenomenon (and the related dynamical breaking of the chiral symmetry) is encoded in the SDE of the quark propagator, know also as quark "gap equation".

The standard decomposition of the quark propagator, $S(p)$, is given by [21]

$$S^{-1}(p) = A(p^2)\not{p} - B(p^2)\mathbb{I} = A(p^2)[\not{p} - \mathcal{M}(p^2)\mathbb{I}]\,, \tag{22}$$

where $\mathbb{I}$ is the identity matrix, and $\mathcal{M}(p^2)$ is the dynamical quark mass; its generation is inextricably connected with the dynamical breaking of the chiral symmetry.

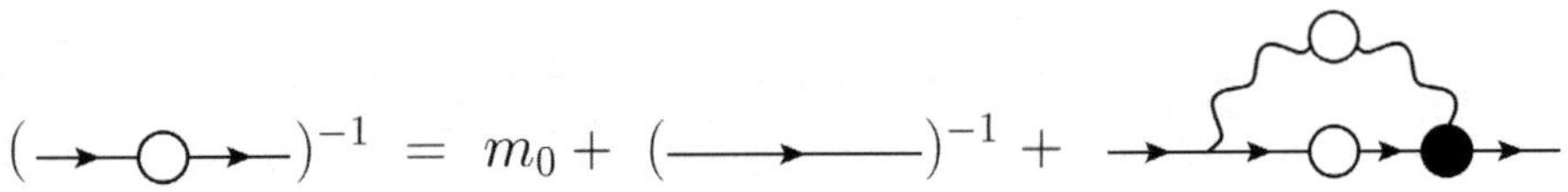

Figure 8. The SDE (gap equation) for the quark propagator, given by Eq. (23). The white blobs represent the fully dressed gluon and quark propagators, while the black blob denotes the dressed quark-gluon vertex.

The SDE for the quark propagator is shown diagrammatically in Fig. 8, and is given by

$$S^{-1}(p) = \not{p} - m_0 - C_{\rm F} g^2 \int_k \Gamma_\mu^{[0]} S(k)\Gamma_\nu(-p,k,q)\Delta^{\mu\nu}(q)\,, \tag{23}$$

where $q \equiv p - k$, $\Gamma_\nu(-p,k,q)$ is the fully dressed quark-gluon vertex, $\Gamma_\mu^{[0]}$ is the tree-level value, and $C_{\rm F}$ is the Casimir eigenvalue in the fundamental representation. In this talk we will only consider the case of vanishing current quark mass, $m_0 = 0$, *i.e.*, the chiral symmetry is kept intact at the Lagrangian level. One of the characteristic features of this integral equation is that, in order to give rise to nontrivial solutions, the support of its kernel in a special momentum regime must overcome a critical amount.

As happens in the case of the ghost equation, the nonperturbative vertex constitutes a crucial ingredient for overcoming the aforementioned critical amount and arriving to a physically acceptable solution. The situation, however, is much more complicated now, given that the tensorial decomposition of $\Gamma_\nu(-p,k,q)$ consists of twelve independent structures [72] instead of only two. Projecting them out of the SDE that governs $\Gamma_\nu(-p,k,q)$ represents a major technical challenge, let alone solving the resulting intricate system of coupled integral equations. Lattice simulations, on the other had, have obtained some of the corresponding form factors for special kinematic configurations [68, 69, 70], but their relevance to the problem at hand is relatively limited.

The alternative that has been extensively adopted is to employ gauge-technique inspired Ansätze (see, *e.g.*, [71] and references therein) for $\Gamma_\nu(-p,k,q)$, namely construct it in such a way that it satisfies the STI imposed by the BRST symmetry. In particular, $\Gamma_\mu(p_1,p_2,p_3)$ obeys the fundamental STI [2]

$$p_3^\mu \Gamma_\mu(p_1,p_2,p_3) = F(p_3)[S^{-1}(-p_1)H(p_1,p_2,p_3) - \overline{H}(p_2,p_1,p_3)S^{-1}(p_2)]\,, \tag{24}$$

where the fermion-ghost scattering kernel $H(p_1,p_2,p_3)$ is defined diagrammatically in Fig. 9, and has the tensorial decomposition

$$H(p_1,p_2,p_3) = X_0\mathbb{I} + X_1\not{p}_1 + X_2\not{p}_2 + X_3\tilde{\sigma}_{\mu\nu}p_1^\mu p_2^\nu, \tag{25}$$

$H(p_1, p_2, p_3) \; = \;$ 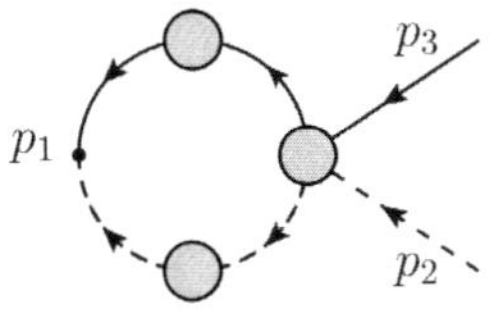

Figure 9. Diagrammatic representation of the ghost-gluon kernel H.

with $\tilde{\sigma}_{\mu\nu} \equiv \frac{1}{2}[\gamma_\mu, \gamma_\nu]$; $\overline{H}(p_2, p_1, p_3)$ is the "conjugated" version of $H(p_1, p_2, p_3)$ [72].

Note that when both F and H are set to unity (*i.e.*, the ghost sector is switched off), Eq. (24) reduces to the text-book WI relating the photon-electron vertex with the electron propagator in QED. The the so-called Ball-Chiu (BC) vertex [73],

$$\Gamma^\mu_{\rm BC}(p_1, p_2, p_3) = \frac{A(p_1) + A(p_2)}{2}\gamma^\mu + \frac{(p_1 - p_2)^\mu}{p_1^2 - p_2^2}\left\{[A(p_1) - A(p_2)]\frac{\not{p}_1 - \not{p}_2}{2} + [B(p_1) - B(p_2)]\right\}. \quad (26)$$

satisfies precisely this particular WI, instead of the full STI of Eq. (24), and has served as the starting point of numerous studies.

Specifically, use of Eq. (26) into Eq. (23) leads to a coupled system for $A(p^2)$ and $B(p^2)$

$$\begin{aligned} A(p^2) &= 1 + C_F g^2 \int_k \frac{\mathcal{K}(p-k)}{A^2(k^2)k^2 + B^2(k^2)} \mathcal{K}_A^{\rm BC}(k,p)\,, \\ B(p^2) &= C_F g^2 \int_k \frac{\mathcal{K}(p-k)}{A^2(k^2)k^2 + B^2(k^2)} \mathcal{K}_B^{\rm BC}(k,p)\,, \end{aligned} \quad (27)$$

where the kernel $\mathcal{K}_B^{\rm BC}(k,p)$ originates from Eq. (26) (for its explicit form, see, *e.g.*, [74]), while the kernel $\mathcal{K}(q)$ encompasses all remaining contributions.

If the ghost contributions to Γ_μ are switched off, $\mathcal{K}(q)$ consists simply of $\Delta(q)$; in this case, however, the resulting combined kernel (right panel of Fig. 10) fails to furnish a nontrivial solution (left panel of Fig. 10).

Given this unsatisfactory situation, a next possible improvement consists in restoring the omitted ghost-related contributions to Γ_μ, thus achieving the gradual "non-Abelianization" of the Ball-Chiu vertex. In particular, one may include the contributions of $F(q)$ and H, as dictated by the STI of Eq. (24). Evidently, since $F(q)$ is very well known (see previous section), the main challenge is to obtain an approximation for H, or at least, for some of its form factors. In addition, renormalization introduces (effectively) into $\mathcal{K}(q)$ an additional multiplicative factor of $F(q)$, an approximation which guarantees that the anomalous dimension of the resulting dynamical quark mass is consistent with the known one-loop expression (see also [75]). On the right panel of Fig. 10 one may follow the gradual built-up of $\mathcal{K}(q)$ and on the left the form of the quark mass $\mathcal{M}(p^2)$ obtained at each step. Evidently, when $\mathcal{K}(q) = F^2(q)\Delta(q)$, a considerable amount of $\mathcal{M}(p^2)$ is obtained from Eq. (27); however, one deviates by a factor of about 30% from the phenomenologically accepted value.

It turns out that the missing amount of dynamical quark mass may be obtained by supplying to $\mathcal{K}(q)$ some of the strength coming from H. In particular, the form factor X_0 was computed within the one-loop dressed approximation, $X_0^{[1]}$, for the symmetric kinematic configuration $(-q/2, -q/2, q)$ [74]. The result is shown in Fig. 11; note that the deviation from unity (peaked around 600 MeV) is modest, but, due to the nonlinearity of the gap equation, it gives rise to a considerable amount of additional quark mass, as can be appreciated in Fig. 10.

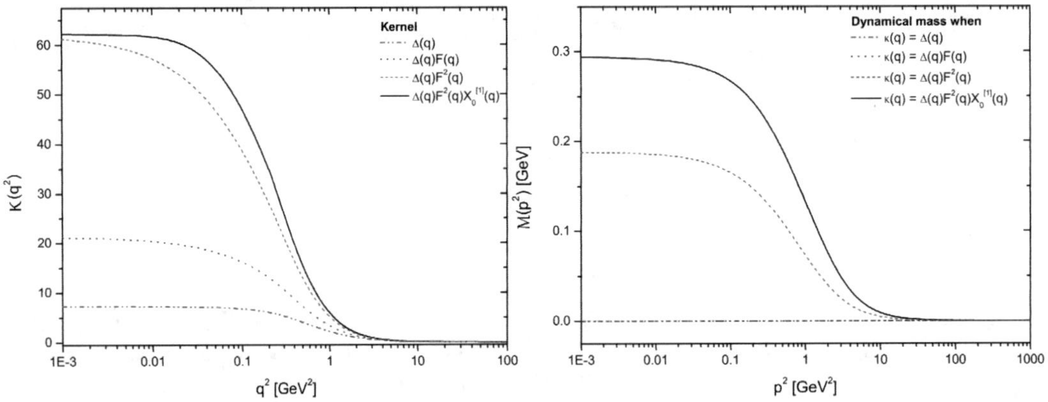

Figure 10. *Left panel:* The individual contribution of the ingredients composing $\mathcal{K}(q)$. The green dotted-dashed line represents the case where $\mathcal{K}(q) = \Delta(q)$, the blue dotted line the case where $\mathcal{K}(q) = \Delta(q)F(q)$, in the red dashed curve $\mathcal{K}(q) = \Delta(q)F^2(q)$ and, finally the black continuous line represents the case where $\mathcal{K}(q)$ assumes the full form used, namely $\mathcal{K}(q) = \Delta(q)F^2(q)X_0^{[1]}(q)$. *Right panel:* The corresponding dynamical quark mass generated when we use in Eq. (27) the different forms of $\mathcal{K}(q)$ shown in the left panel.

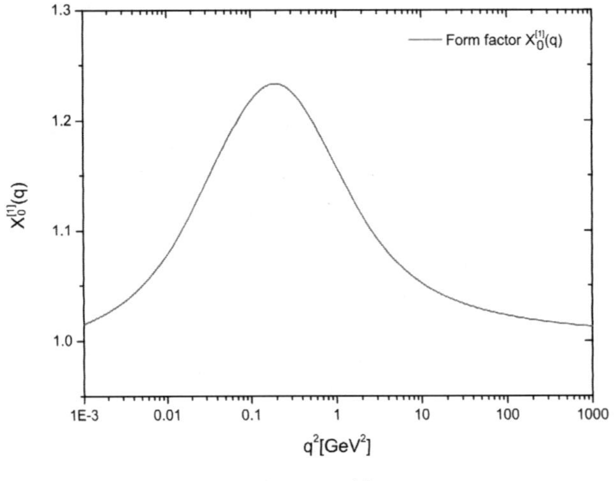

Figure 11. Numerical result for the form factor $X_0^{[1]}(q)$, obtained within the one-loop dressed approximation, using fully dressed propagators and bare vertices [74].

6. Hadron observables from continuum QCD

One of the longstanding challenges of QCD is to furnish quantitatively accurate *ab initio* predictions for the observable properties of hadrons [76, 77, 78, 79]. In a recent work [80], considerable progress has been made in this direction; in what follows we will highlight the main aspects of these new developments.

Let us consider the kernel appearing in a typical Bethe-Salpeter equation [81], shown in Fig. 12, which is contained in the blue box. This kernel receives a "universal" (process-

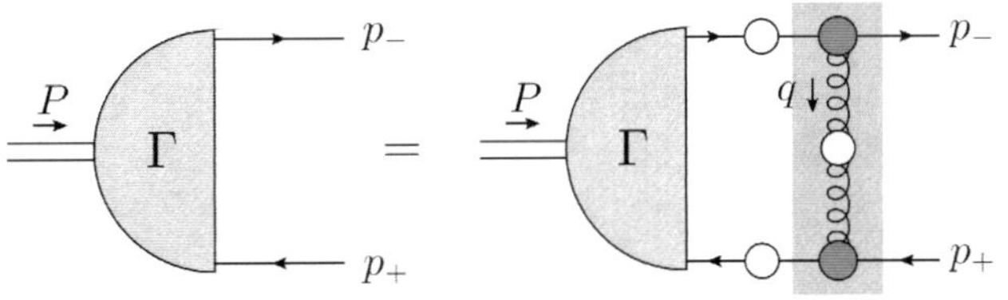

Figure 12. A typical Bethe-Salpeter equation. The blue box contains the interaction kernel, whose form is crucial for the type of solutions (spectrum) obtained.

independent) contribution, whose origin is the pure gauge sector of the theory. In that sense, this contribution constitutes the common ingredient of any such kernel, regardless of the nature of the particles between which it is embedded; most notably, it does not depend on the valence-quark content of the Bethe-Salpeter equation.

The systematic diagrammatic identification of the precise pieces that constitute this particular quantity may be carried out following the PT. As has been explained in detail in the literature, the upshot of this construction is the rearrangement of a physical amplitude into sub-amplitudes with very special properties [7]. In addition, it is well-known that the result of this particular rearrangement coincides with the BB gluon propagator defined in the BFM (see [6] and references therein).

Specifically, the standard $\Delta(q^2)$ introduced in Eq. (1) and the corresponding quantity of the BB gluon propagator, denoted by $\widehat{\Delta}(q^2)$, are related by the exact relation

$$\Delta(q^2) = \widehat{\Delta}(q^2)[1 + G(q^2)]^2\,, \tag{28}$$

which is completely analogous to Eq. (8).

At the one-loop level, and keeping only ultraviolet logarithms, one has [39]

$$\begin{aligned} 1 + G(q^2) &= 1 + \left(\frac{9}{4}\right)\frac{\alpha_s C_A}{12\pi}\ln\left(\frac{q^2}{\mu^2}\right), \\ \Delta^{-1}(q^2) &= q^2\left[1 + \left(\frac{13}{2}\right)\frac{\alpha_s C_A}{12\pi}\ln\left(\frac{q^2}{\mu^2}\right)\right]^{-1}, \end{aligned} \tag{29}$$

and thus

$$\widehat{\Delta}^{-1}(q^2) = q^2\left[1 + b\alpha_s\,\ln\left(\frac{q^2}{\mu^2}\right)\right]. \tag{30}$$

where $b = 11C_A/12\pi$ is the first coefficient of the Yang-Mills β function, as it should [8].

Due to the special WIs satisfied by the PT-BFM Green's functions, the (dimensionful) universal combination

$$\widehat{d}(q^2) = \alpha_s\widehat{\Delta}(q^2) = \frac{\alpha_s\Delta(q^2)}{[1 + G(q^2)]^2} \tag{31}$$

is renormalization-group invariant. In particular, the renormalization constants of the gauge-coupling and of $\widehat{\Delta}^{-1}(p^2)$, defined as

$$\begin{aligned} g(\mu^2) &= Z_g^{-1}(\mu^2)g_0, \\ \widehat{\Delta}(p^2,\mu^2) &= Z_{\widehat{A}}^{-1}(\mu^2)\Delta_0(p^2), \end{aligned} \tag{32}$$

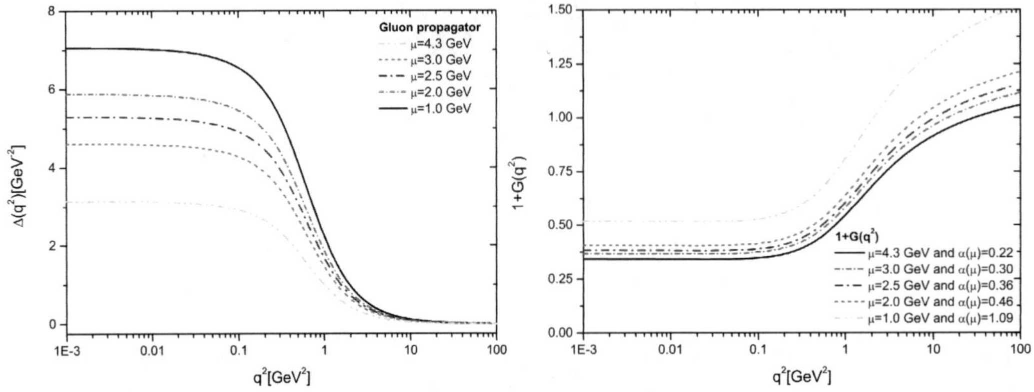

Figure 13. *Left panel:*: The gluon propagator for different values of the renormalization point μ. *Right panel*: The auxiliary function $1 + G(q^2)$ for the same set of values of μ.

where the "0" subscript indicates bare quantities, satisfy the QED-like relation

$$Z_g = \widehat{Z}_A^{-1/2}. \tag{33}$$

Therefore,

$$\widehat{d}_0(q^2) = \alpha_{s0}\widehat{\Delta}_0(q^2) = \alpha_s(\mu^2)\widehat{\Delta}(q^2,\mu^2) = \widehat{d}(q^2) \tag{34}$$

maintains the same form before and after renormalization, *i.e.*, it forms a μ-independent quantity.

This last property has been explicitly verified to an impressive accuracy, following the procedure summarized in Fig. 13 and Fig. 14. The gluon propagator $\Delta(q^2)$, obtained from the lattice data of [31], is multiplicatively renormalized (in the momentum-subtraction scheme) at five different values of μ. The dynamical equation furnishing $1 + G(q^2)$ (not reported here, see [82, 83]) is subsequently solved, for the same values of μ. When the resulting curves (given in the two panels of Fig. 13) are combined according to Eq. (31), it turns out that a unique curve emerges, provided that the values of each $\alpha_s(\mu^2)$ are those shown in the insert of Fig. 14.

From the dimensionful quantity $\widehat{d}(q^2)$ one may extract a dimensionless effective charge, to be denoted by $\mathcal{I}(q^2)$, by simply pulling out a factor of q^2, namely

$$\mathcal{I}(q^2) = q^2\widehat{d}(q^2)\,. \tag{35}$$

This SDE-derived quantity allows for a direct comparison with an analogous quantity commonly employed by practitioners for describing the (momentum-dependent) interaction between quarks, obtained through a well-defined truncation of the equations in the matter sector that are relevant to bound-state properties. In particular, one uses

$$\mathcal{I}(q^2) = k^2\mathcal{G}(q^2); \quad \mathcal{G}(q^2) = \frac{8\pi^2}{\omega^4}De^{-q^2/\omega^2} + \frac{8\pi^2\gamma_m(1-e^{-q^2/4m_t^2})}{q^2\ln[\tau + (1+q^2/\Lambda_{\rm QCD}^2)^2]}, \tag{36}$$

where $\gamma_m = 12/(33 - 2N_f)$ [typically, $N_f = 4$], $\Lambda_{\rm QCD} = 0.234$ GeV; $\tau = e^2 - 1$, $m_t = 0.5$ GeV. A large body of observable properties of ground-state vector- and isospin-nonzero pseudoscalar mesons (and even numerous properties of the nucleon and Δ resonance) are practically insensitive

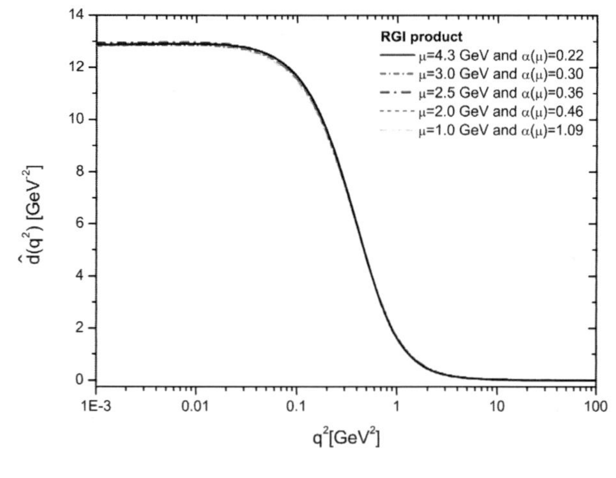

Figure 14. The RG-invariant product $\widehat{d}(q^2)$ put together according Eq. (31), using the results of Fig. 13: one practically obtains one single curve for all different values of the renormalization point μ

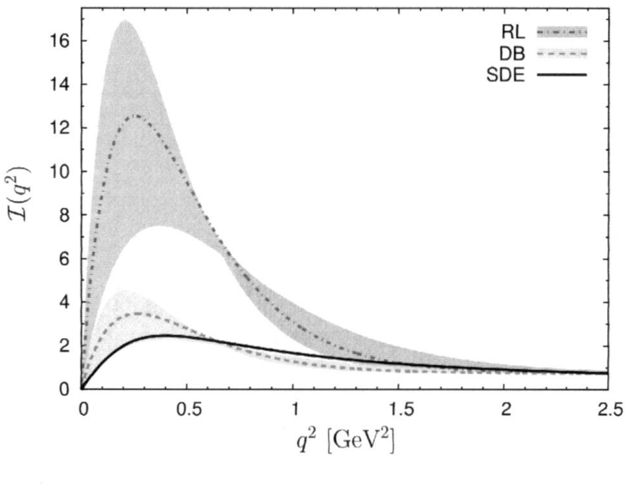

Figure 15. Comparison between the SDE-derived effective interaction obtained from the gauge-sector [Eq. (35), solid, black] with those obtained using Eq. (36): *pale-red band*: RL truncation, *pale-green band*: DB truncation. The bands denote the domain of constant ground-state physics $0.45 < \omega < 0.6$, with the curve within each band obtained with $\omega = 0.5\,\text{GeV}$.

to variations of $\omega \in [0.4, 0.6]$ GeV, so long as $\varsigma^3 = D\omega$ =const. The value of ς is typically chosen to obtain the measured value of f_π; however its value depend on the truncation scheme used when deriving (36). The usual rainbow-ladder (RL) truncation [84] yields $\varsigma_{\text{RL}} = 0.87$ GeV, while the improved truncation scheme (DB) of [76] gives $\varsigma_{\text{DB}} = 0.55$ GeV.

As can be seen in Fig. 15, the direct comparison between the $\mathcal{I}(q^2)$ defined through Eq. (35) and Eq. (36) reveals an excellent agreement. This seems to suggest that a QCD-derived description of measurable hadron properties is well within our reach.

7. Conclusions

In this presentation we have reviewed a large body of developments related to the study of the off-shell Green's functions of pure Yang-Mills theories and QCD, using SDEs and the lattice simulations in the Landau gauge. The general picture that surfaces may be summarized as follows:

i) Gluons just as quarks acquire a dynamical mass, which is controlled by the corresponding "gap equation".

ii) Ghosts remain massless, and have a finite dressing function that can be accurately obtained from the corresponding SDE.

iii) The quality of the ingredients entering into the quark gap equation is steadily improving, giving rise to phenomenologically successful quark masses.

iv) A significant step has been taken for bridging the gap between nonperturbative continuum QCD predictions and hadron phenomenology.

Even though a lot of work remains to be done in all directions, Richard Feynman's famous quote from the first of his seven "Messenger Lectures" at Cornell University in 1964 comes to mind: *Nature uses only the longest threads to weave her patterns, so that every small piece of the fabric reveals the organization of the entire tapestry*". We seem to have reached a similar point with QCD: even though the Green's functions that have been studied are but a small piece of the QCD fabric, unequivocal signs of an underlying organization begin to emerge.

Acknowledgments

This research is supported by the Spanish MEYC under grant FPA2011-23596 and the Generalitat Valenciana under grant PrometeoII/2014/066. I would like to thank the organizers of the "Discrete 2014" for their kind invitation and hospitality, and all participants for providing a most stimulating and pleasant atmosphere.

References

[1] C. N. Yang and R. L. Mills, Phys. Rev. **96**, 191 (1954).
[2] W. J. Marciano and H. Pagels, Phys. Rept. **36**, 137 (1978).
[3] J. M. Cornwall, Phys. Rev. D **26**, 1453 (1982).
[4] J. M. Cornwall and J. Papavassiliou, Phys. Rev. D **40**, 3474 (1989).
[5] A. Pilaftsis, Nucl. Phys. B **487**, 467 (1997).
[6] D. Binosi and J. Papavassiliou, Phys. Rev. D **66**, 111901 (2002).
[7] D. Binosi and J. Papavassiliou, Phys. Rept. **479**, 1 (2009).
[8] L. F. Abbott, Nucl. Phys. B **185**, 189 (1981).
[9] R. Alkofer and L. von Smekal, Phys. Rept. **353**, 281 (2001).
[10] A. C. Aguilar and A. A. Natale, JHEP **0408**, 057 (2004).
[11] C. S. Fischer, J. Phys. G **32**, R253 (2006).
[12] J. Braun, H. Gies and J. M. Pawlowski, Phys. Lett. B **684**, 262 (2010).
[13] P. Boucaud, J. P. Leroy, A. L. Yaouanc, J. Micheli, O. Pene and J. Rodriguez-Quintero, JHEP **0806**, 012 (2008).
[14] P. Boucaud, J. P. Leroy, A. Le Yaouanc, J. Micheli, O. Pene and J. Rodriguez-Quintero, JHEP **0806**, 099 (2008).
[15] D. Dudal, J. A. Gracey, S. P. Sorella, N. Vandersickel and H. Verschelde, Phys. Rev. D **78**, 065047 (2008).
[16] C. S. Fischer, A. Maas and J. M. Pawlowski, Annals Phys. **324**, 2408 (2009).
[17] M. R. Pennington and D. J. Wilson, Phys. Rev. D **84**, 119901 (2011).
[18] J. Serreau and M. Tissier, Phys. Lett. B **712**, 97 (2012).
[19] M. Tissier and N. Wschebor, Phys. Rev. D **84**, 045018 (2011).
[20] F. Siringo, Phys. Rev. D **90**, no. 9, 094021 (2014).
[21] C. D. Roberts and A. G. Williams, Prog. Part. Nucl. Phys. **33**, 477 (1994).
[22] P. Maris and C. D. Roberts, Int. J. Mod. Phys. E **12**, 297 (2003).

[23] A. C. Aguilar and J. Papavassiliou, JHEP **0612**, 012 (2006).
[24] P. A. Grassi, T. Hurth and M. Steinhauser, Annals Phys. **288**, 197 (2001).
[25] D. Binosi and J. Papavassiliou, Phys. Rev. D **66**, 025024 (2002).
[26] D. Binosi and J. Papavassiliou, Phys. Rev. D **77**, 061702 (2008).
[27] D. Binosi and J. Papavassiliou, JHEP **0811**, 063 (2008).
[28] C. W. Bernard, Nucl. Phys. B **219**, 341 (1983).
[29] J. F. Donoghue, Phys. Rev. D **29**, 2559 (1984).
[30] I. L. Bogolubsky, E. M. Ilgenfritz, M. Muller-Preussker and A. Sternbeck, Phys. Lett. B **676**, 69 (2009).
[31] I. L. Bogolubsky, E. M. Ilgenfritz, M. Muller-Preussker and A. Sternbeck, PoS LAT **2007**, 290 (2007).
[32] P. O. Bowman, U. M. Heller, D. B. Leinweber, M. B. Parappilly, A. Sternbeck, L. von Smekal, A. G. Williams and J. b. Zhang, Phys. Rev. D **76**, 094505 (2007).
[33] O. Oliveira and P. J. Silva, PoS LAT **2009**, 226 (2009).
[34] A. Cucchieri and T. Mendes, PoS LAT **2007**, 297 (2007).
[35] A. Cucchieri and T. Mendes, Phys. Rev. Lett. **100**, 241601 (2008).
[36] A. Cucchieri and T. Mendes, Phys. Rev. D **81**, 016005 (2010).
[37] A. Cucchieri and T. Mendes, PoS QCD **-TNT09**, 026 (2009).
[38] O. Philipsen, Nucl. Phys. B **628**, 167 (2002).
[39] A. C. Aguilar, D. Binosi and J. Papavassiliou, Phys. Rev. D **78**, 025010 (2008).
[40] J. S. Schwinger, Phys. Rev. **125**, 397 (1962).
[41] J. S. Schwinger, Phys. Rev. **128**, 2425 (1962).
[42] R. Jackiw and K. Johnson, Phys. Rev. D **8**, 2386 (1973).
[43] R. Jackiw, In *Erice 1973, Proceedings, Laws Of Hadronic Matter*, New York 1975, 225-251 and M I T Cambridge - COO-3069-190 (73,REC.AUG 74) 23p
[44] J. M. Cornwall and R. E. Norton, Phys. Rev. D **8**, 3338 (1973).
[45] E. Eichten and F. Feinberg, Phys. Rev. D **10**, 3254 (1974).
[46] E. C. Poggio, E. Tomboulis and S.-H. H. Tye, Phys. Rev. D **11**, 2839 (1975).
[47] A. C. Aguilar, D. Ibanez, V. Mathieu and J. Papavassiliou, Phys. Rev. D **85**, 014018 (2012).
[48] D. Ibanez and J. Papavassiliou, Phys. Rev. D **87**, no. 3, 034008 (2013)
[49] A. C. Aguilar and J. Papavassiliou, Phys. Rev. D **81**, 034003 (2010).
[50] D. Binosi, D. Ibanez and J. Papavassiliou, Phys. Rev. D **86**, 085033 (2012).
[51] A. C. Aguilar, D. Binosi and J. Papavassiliou, Phys. Rev. D **89**, no. 8, 085032 (2014).
[52] This demonstration is taken from Peter C. Tandy's talk "Some Chapters from the Do-It-Yourself Hadron Theory Manual".
[53] I. C. Cloet and C. D. Roberts, Prog. Part. Nucl. Phys. **77**, 1 (2014).
[54] L. Del Debbio, M. Faber, J. Greensite and S. Olejnik, Phys. Rev. D **55** (1997) 2298.
[55] A. P. Szczepaniak and E. S. Swanson, Phys. Rev. D **65**, 025012 (2002).
[56] K. Langfeld, H. Reinhardt and J. Gattnar, Nucl. Phys. B **621**, 131 (2002).
[57] A. P. Szczepaniak, Phys. Rev. D **69**, 074031 (2004).
[58] J. Greensite, Prog. Part. Nucl. Phys. **51**, 1 (2003).
[59] J. Gattnar, K. Langfeld and H. Reinhardt, Phys. Rev. Lett. **93**, 061601 (2004).
[60] J. Greensite, H. Matevosyan, S. Olejnik, M. Quandt, H. Reinhardt and A. P. Szczepaniak, Phys. Rev. D **83**, 114509 (2011).
[61] A. Cucchieri, T. Mendes and A. Mihara, JHEP **0412**, 012 (2004).
[62] A. Cucchieri, A. Maas and T. Mendes, Phys. Rev. D **74**, 014503 (2006).
[63] A. Cucchieri, A. Maas and T. Mendes, Phys. Rev. D **77**, 094510 (2008).
[64] A. C. Aguilar, D. Ibanez and J. Papavassiliou, Phys. Rev. D **87**, no. 11, 114020 (2013).
[65] D. Dudal, O. Oliveira and J. Rodriguez-Quintero, Phys. Rev. D **86**, 105005 (2012).
[66] P. Boucaud, F. de Soto, J. P. Leroy, A. Le Yaouanc, J. Micheli, H. Moutarde, O. Pene and J. Rodriguez-Quintero, Phys. Rev. D **74**, 034505 (2006).
[67] P. Boucaud, F. De Soto, J. P. Leroy, A. Le Yaouanc, J. Micheli, O. Pene and J. Rodriguez-Quintero, Phys. Rev. D **79**, 014508 (2009).
[68] J. Skullerud and A. Kizilersu, JHEP **0209**, 013 (2002).
[69] J. I. Skullerud, P. O. Bowman, A. Kizilersu, D. B. Leinweber and A. G. Williams, JHEP **0304**, 047 (2003).
[70] A. Kizilersu, D. B. Leinweber, J. I. Skullerud and A. G. Williams, Eur. Phys. J. C **50**, 871 (2007).
[71] A. Kizilersu and M. R. Pennington, Phys. Rev. D **79**, 125020 (2009).
[72] A. I. Davydychev, P. Osland and L. Saks, Phys. Rev. D **63**, 014022 (2001).
[73] J. S. Ball and T. W. Chiu, Phys. Rev. D **22**, 2542 (1980).
[74] A. C. Aguilar and J. Papavassiliou, Phys. Rev. D **83**, 014013 (2011).
[75] C. S. Fischer and R. Alkofer, Phys. Rev. D **67**, 094020 (2003).

[76] L. Chang and C. D. Roberts, Phys. Rev. Lett. **103**, 081601 (2009).
[77] L. Chang, Y. X. Liu and C. D. Roberts, Phys. Rev. Lett. **106**, 072001 (2011).
[78] S. x. Qin, L. Chang, Y. x. Liu, C. D. Roberts and D. J. Wilson, Phys. Rev. C **85**, 035202 (2012).
[79] L. Chang and C. D. Roberts, Phys. Rev. C **85**, 052201 (2012).
[80] D. Binosi, L. Chang, J. Papavassiliou and C. D. Roberts, Phys. Lett. B **742**, 183 (2015).
[81] G. Eichmann, arXiv:0909.0703 [hep-ph].
[82] A. C. Aguilar, D. Binosi and J. Papavassiliou, JHEP **0911**, 066 (2009).
[83] A. C. Aguilar, D. Binosi, J. Papavassiliou and J. Rodriguez-Quintero, Phys. Rev. D **80**, 085018 (2009).
[84] P. Maris and P. C. Tandy, Phys. Rev. C **60**, 055214 (1999).

TeV Scale Lepton Number Violation and Baryogenesis

P. S. Bhupal Dev[1], Chang-Hun Lee[2], R. N. Mohapatra[2]

[1] Consortium for Fundamental Physics, School of Physics and Astronomy, University of Manchester, Manchester M13 9PL, United Kingdom.
[2] Maryland Center for Fundamental Physics and Department of Physics, University of Maryland, College Park, Maryland 20742, USA.

Abstract. Contrary to the common lore based on naive dimensional analysis, the seesaw scale for neutrino masses can be *naturally* in the TeV range, with small parameters coming from radiative corrections. We present one such class of type-I seesaw models, based on the left-right gauge group $SU(2)_L \times SU(2)_R \times U(1)_{B-L}$ realized at the TeV scale, which fits the observed neutrino oscillation parameters as well as other low energy constraints. We discuss how the small parameters of this scenario can arise naturally from one loop effects. The neutrino fits in this model use quasi-degenerate heavy Majorana neutrinos, as also required to explain the matter-antimatter asymmetry in our Universe via resonant leptogenesis mechanism. We discuss the constraints implied by the dynamics of this mechanism on the mass of the right-handed gauge boson in this class of models with enhanced neutrino Yukawa couplings compared to the canonical seesaw model and find a lower bound of $m_{W_R} \geq 9.9$ TeV for successful leptogenesis assuming maximal *CP* asymmetry for each flavor. We also present a model with explicit neutrino mass fit, where the lower bound goes up to 13.1 TeV due to less than maximal primordial *CP* asymmetry predicted by the model.

1. Introduction

The seesaw mechanism [1] seems to provide a simple way to understand the smallness of neutrino masses. The simplest among the various seesaw mechanisms is the type-I seesaw, which requires the introduction of Standard Model (SM) gauge-singlet right-handed (RH) neutrinos (N's). The Majorana mass M_N of these RH neutrinos explicitly breaks the global $B - L$ symmetry of the SM by two units and gives rise to a light neutrino mass matrix of the form

$$M_\nu \simeq -M_D M_N^{-1} M_D^{\mathsf{T}} , \tag{1}$$

where M_D is the Dirac mass generated due to the neutrino Yukawa couplings Y_ν after the electroweak symmetry breaking. If one assumes maximal allowed values for the third-generation Dirac mass $M_{D,33}$, as is implied by certain Grand Unified Theories such as $SO(10)$, then the seesaw scale is pushed up to $\sim 10^{14}$ GeV making it virtually inaccessible to both collider experiments at the energy frontier and other low energy searches at the intensity frontier. Such naive ("dimensional counting") considerations however can be misleading in the context of specific ultraviolet (UV) complete seesaw models where there may be dynamical suppressions of neutrino Dirac masses. One class of such models that are based on the SM gauge group generally go by the name of radiative seesaw models; see e.g. [2]. Within the type-I seesaw framework, there exists another possibility to suppress the light neutrino masses by assigning specific textures to the Dirac and Majorana mass matrices in the seesaw formula [3, 4]. The stability of these textures can in principle be guaranteed by enforcing some symmetries in the lepton sector.

Extending the SM gauge group to the Left-Right (L-R) Symmetric gauge group $SU(2)_L \times SU(2)_R \times U(1)_{B-L}$ provides a simple UV-complete seesaw model [5]. Here, the RH neutrinos are a necessary part of the model and do not have to be added 'by hand' just to implement the seesaw mechanism. An important point is that the RH neutrinos acquire a Majorana mass as soon as the $SU(2)_R$ symmetry is broken at a scale v_R. This is quite analogous to the way the charged fermions get masses in the SM by the Higgs mechanism when the $SU(2)_L$ gauge symmetry is broken at a scale v. The Higgs field that gives mass to the RH neutrinos becomes the analogue of the 125 GeV Higgs boson discovered at the LHC.

In this proceedings, we focus on a new class of L-R seesaw models [6], where in addition to the RH gauge bosons as well as other Higgs fields, light (TeV scale) RH neutrinos with sizable mixing with the left-handed (LH) neutrinos may be accessible at colliders and other low energy experiments (for a review, see e.g. [7]). With the LHC ramping up to a higher center of mass energy $\sqrt{s} = 13$ TeV and later to 14 TeV, our model as well as other low scale seesaw models, should provide motivation for testing the origin of neutrino mass in the coming years. For the collider phenomenology of TeV-scale LRSM, see e.g. [8, 9, 10], and for the current experimental status, see [11]. Such considerations should also be of interest [12] for setting physics goals for future higher energy colliders such as the $\sqrt{s} = 80$ or 100 TeV VLHC [13]. It is also important to mention here that the new class of L-R seesaw models with relatively large left-right neutrino mixing can be distinguished from the minimal LRSM using different kinematic variables [10, 14].

An attractive consequence of the seesaw mechanism is the possibility that it could solve another important puzzle of cosmology, i.e. the origin of matter-antimatter asymmetry in our Universe, via a mechanism called leptogenesis [15]. At the heart of this mechanism is the out-of-equilibrium decay of the RH Majorana neutrinos via the decay modes $N \to L\Phi, L^c\Phi^c$ (where the superscript c denotes the *CP*-conjugate), which violate L, C and CP, thereby dynamically generating an asymmetry in the lepton sector. This primordial lepton asymmetry undergoes thermodynamic evolution as the Universe expands and finally gets converted to a baryon asymmetry via equilibrated electroweak sphaleron interactions [16]. The attractive aspect of this mechanism is that the *same* Yukawa couplings Y_ν that give rise to neutrino masses via the seesaw mechanism in Eq. (1) are also responsible for the origin of matter, thus implying an intimate connection between two seemingly disparate pieces of evidence for beyond the SM physics. This beautiful idea can be tested, provided the seesaw scale, and hence, the scale at which leptogenesis takes place, is accessible to current and near future laboratory experiments. Low-scale leptogenesis is possible when at least two RH Majorana neutrinos have a small mass difference comparable to their decay widths, thus resonantly enhancing the ε-type *CP* asymmetry [17, 18] due to RH neutrino self-energy effects [19].

An important question to explore is whether the leptogenesis mechanism works in a TeV-scale realization of L-R seesaw, or in other words, whether it is possible to falsify the leptogenesis mechanism if a RH gauge boson is discovered at the LHC. This problem was analyzed in detail in Ref. [20] and it was pointed out that the additional dilution and washout effects on the lepton asymmetry due to W_R-mediated $\Delta L = 1$ scattering processes become important for a TeV-scale L-R seesaw. Since in generic versions of the seesaw model, sub-eV neutrino masses would require the Dirac Yukawa couplings to be $Y_\nu \lesssim 10^{-11/2}$, the washout effects would lead to an efficiency factor $\kappa \sim \frac{Y_\nu^2 M_{W_R}^4}{g_R^4 M_N^4}$ (g_R being the $SU(2)_R$ gauge coupling, which is assumed to be equal to the $SU(2)_L$ gauge coupling in the minimal scenario) which is too small for a low-scale leptogenesis to work. Through a detailed analysis of the relevant Boltzmann equations, it was concluded in [20] that even for maximal *CP*-asymmetry $\varepsilon \sim \mathcal{O}(1)$, the observed value of the baryon asymmetry can be explained by leptogenesis in these L-R models, only if $m_{W_R} \geq 18$ TeV. Turning this argument around, if a positive signal for W_R is observed at the LHC, this will falsify leptogenesis as a mechanism for understanding the origin of matter in L-R seesaw framework. Since this is such an important issue, we have reanalyzed this [21] to examine the robustness of the lower bound on m_{W_R} and to investigate if there exists *any* allowed parameter space with successful leptogenesis for smaller values of m_{W_R}. We find that for the generic class of L-R seesaw models with large light-heavy neutrino mixing, successful leptogenesis requires $m_{W_R} \geq 9.9$ TeV for the maximal *CP* asymmetry. For illustration, we will present an explicit neutrino mass fit which satisfies the observed baryon asymmetry. The main new results presented here should supersede those reported in [21].

This proceedings is organized as follows: in Section 2, we review the basic features of the generic L-R seesaw models as well as our new L-R seesaw model with special Dirac and Majorana textures resulting in light neutrinos via type-I seesaw with large light-heavy neutrino mixing generated in a natural manner. In Section 3, we present an alternative neutrino mass fit relying on cancellations in the seesaw formula. In Section 4, we discuss resonant leptogenesis in this model and the impact of a TeV-scale W_R on the lepton asymmetry. Our conclusions are given in Section 5.

2. A TeV-scale left-right seesaw model

In the minimal L-R symmetric model (LRSM), the fermions are assigned to the gauge group $SU(2)_L \times SU(2)_R \times U(1)_{B-L}$ as follows: denoting $Q_i \equiv (u,d)_i^\mathsf{T}$ and $\psi_i \equiv (\nu_l, l)_i^\mathsf{T}$ as the quark and lepton doublets of the ith generation respectively, $Q_{L,i}$ and $\psi_{L,i}$ (also denoted simply by L_i) are assigned to doublets under the $SU(2)_L$ group, while $Q_{R,i}$ and $\psi_{R,i}$ (also denoted by R_i) as the doublets under the $SU(2)_R$ group. The Higgs sector of the model can consist of one or several bidoublets ϕ_a and triplets $\Delta_{R,b}$:

$$\phi \equiv \begin{pmatrix} \phi_1^0 & \phi_2^+ \\ \phi_1^- & \phi_2^0 \end{pmatrix}, \qquad \Delta_R \equiv \begin{pmatrix} \Delta_R^+/\sqrt{2} & \Delta_R^{++} \\ \Delta_R^0 & -\Delta_R^+/\sqrt{2} \end{pmatrix}. \tag{2}$$

The gauge symmetry $SU(2)_R \times U(1)_{B-L}$ is broken by the vacuum expectation value (VEV) $\langle \Delta_R^0 \rangle = v_R$ to the group $U(1)_Y$ of the SM. The LH counterpart (Δ_L) to Δ_R is not considered here, assuming that parity and $SU(2)_R$ gauge symmetry scales are decoupled. In this case, the Δ_L fields become heavy when the discrete parity symmetry is broken, and disappear from the low energy theory [22]. The VEV of the ϕ field given by $\langle \phi \rangle = \mathrm{diag}(\kappa, \kappa')$ breaks the SM gauge group to $U(1)_{\rm em}$. The fermion masses are obtained from the following Yukawa Lagrangian:

$$\begin{aligned} -\mathcal{L}_Y &= h_{ij}^{q,a} \bar{Q}_{Li} \phi_a Q_{R,j} + \tilde{h}_{ij}^{q,a} \bar{Q}_{L,i} \tilde{\phi}_a Q_{R,j} + h_{ij}^{l,a} \bar{L}_i \phi_a R_j \\ &+ \tilde{h}_{ij}^{l,a} \bar{L}_i \tilde{\phi}_a R_j + f_{ij} (R_i^\mathsf{T} C i \tau_2 \Delta_R R_j + L_i^\mathsf{T} C i \tau_2 \Delta_L L_j) + \text{H.c.}, \end{aligned} \tag{3}$$

where $\tilde{\phi} = \tau_2 \phi^* \tau_2$ (τ_2 being the second Pauli matrix). After electroweak symmetry breaking, the Dirac fermion masses are given by the generic formula $M_f = h^f \kappa + \tilde{h}^f \kappa'$ for up-type fermions, and for down-type quarks and charged leptons, it is the same formula with $\kappa \leftrightarrow \kappa'$. The Yukawa Lagrangian (3) leads to the Dirac neutrino mass matrix $M_D = h^l \kappa + \tilde{h}^l \kappa'$ and the Majorana mass matrix for the heavy RH neutrinos $M_N = f v_R$ which go into the seesaw formula (1) for calculating the neutrino masses and the heavy-light neutrino mixing.

Since the Higgs sector relates the neutrino Yukawa couplings with the charged-lepton ones in the LRSM , it is interesting to see how a TeV-scale L-R seesaw can be realized in Nature. There are essentially three ways to do so: (i) by choosing one set of Yukawa couplings to be $\lesssim 10^{-11/2}$ for a particular VEV assignment for the bidoublet Higgs fields; (ii) by choosing particular symmetry-motivated textures for the Yukawa couplings which ensure that the leading order seesaw contribution to neutrino masses vanish; and (iii) by choosing large Yukawa couplings as allowed by specific textures and invoking cancellations in the seesaw formula (1). The strategy (i) with small Yukawa couplings does not lead to any new interesting phenomenology over the minimal LRSM. Therefore, we will focus on strategies (ii) and (iii) in the following sections, and in particular, use strategy (iii) for our numerical analysis for the calculation of lepton asymmetry. A similar analysis can be performed following strategy (ii) as well. For the phenomenological implications of this new class of L-R seesaw models, see e.g. [6, 10].

2.1. A symmetry-protected L-R seesaw with large light-heavy neutrino mixing

As mentioned above, the basic strategy for understanding small neutrino masses with a low-scale type-I seesaw is to have the appropriate textures for M_D and M_N in Eq. (1). There are several examples of this type [3, 4] in the context of the minimal seesaw extensions of the SM. Here, we discuss the embedding of one such texture (from Ref. [3]) in the L-R model using an appropriate family symmetry [6]. The symmetry must not only guarantee the special leptonic textures but also must be free of light scalar bosons which can result if the effect of the discrete symmetry is to automatically lead to a $U(1)$ symmetry. Moreover in LRSM, the charged-lepton mass matrix M_l is related to M_D which puts additional constraints on phenomenological viability of the model. Therefore, we find it remarkable that the model presented below remains a viable TeV-scale L-R type-I seesaw model for neutrinos, and as a result, has interesting phenomenological implications beyond the minimal LRSM.

The Dirac and Majorana mass matrices M_D and M_N considered here have the following form:

$$M_D = \begin{pmatrix} m_1 & \delta_1 & \epsilon_1 \\ m_2 & \delta_2 & \epsilon_2 \\ m_3 & \delta_3 & \epsilon_3 \end{pmatrix}, \qquad M_N = \begin{pmatrix} \delta M & M_1 & 0 \\ M_1 & 0 & 0 \\ 0 & 0 & M_2 \end{pmatrix}, \tag{4}$$

with $\epsilon_i, \delta_i \ll m_i$ and $\delta M \ll M_i$. In the limit of $\epsilon_i, \delta_i, \delta M \to 0$, the neutrino masses vanish, although the heavy-light mixing parameters given by $\xi_{ij} = m_i/M_j$ can be quite large. The light neutrino masses given by the seesaw formula (1) become proportional to either ϵ_i, δ_i or their products. If the smallness of δ_i and ϵ_i can be guaranteed by some symmetry, then we have a "natural" TeV-scale seesaw model with tiny neutrino masses. As we show below, the mass textures (4) can be successfully embedded into the L-R framework, while satisfying all current experimental constraints. Note that the choice of texture for the RH neutrino mass matrix in Eq. (4) is necessary to keep the model compatible with strong bounds from tree-level $\mu^- \to e^- e^+ e^-$ decay via TeV-scale Δ_R exchange [21].

In order to obtain the special Dirac and Majorana textures given in Eq. (4), we supplement the L-R gauge group $SU(2)_L \times SU(2)_R \times U(1)_{B-L}$ with a global discrete symmetry $D \equiv Z_4 \times Z_4 \times Z_4$ [6]. For the Higgs sector, we choose three bi-doublets ($\phi_{1,2,3}$) with $B - L = 0$ and two RH triplets ($\Delta_{R,1}, \Delta_{R,2}$) with $B - L = 2$. The fermion and Higgs multiplets are assigned the D quantum numbers as shown in Table 1. The leptonic part of the Yukawa Lagrangian (3) invariant under the D-symmetry is given by

$$-\mathcal{L}_{l,Y} = h_{i1}\bar{L}_i\tilde{\phi}_1 R_1 + h_{i2}\bar{L}_i\phi_2 R_2 + h_{i3}\bar{L}_i\phi_3 R_3 + f_{12}R_1R_2\Delta_{R,1} + f_{33}R_3R_3\Delta_{R,2} + \text{H.c.} \tag{5}$$

In the symmetry limit, the VEVs of $\phi_{2,3}$ have the form $\langle\phi_a\rangle = \text{diag}(0, \kappa'_a)$. This is because the terms of the form $\text{Tr}(\tilde{\phi}_a\phi_b^\dagger)$ which would change the ϕ VEV to the form $\text{diag}(\kappa, \kappa')$ are forbidden by the D-symmetry from appearing in the scalar potential. Thus in the symmetry limit, the M_D elements in one column are big and non-zero, and the charged lepton mass matrix has one eigenvalue zero which can be identified as the electron flavor. The RH neutrino Majorana mass

Field	$Z_4 \times Z_4 \times Z_4$ transformation
$L_{1,2,3}$	$(1,\ 1,\ 1)$
R_1	$(-i,\ 1,\ 1)$
R_2	$(1,\ -i,\ 1)$
R_3	$(1,\ 1,\ -i)$
ϕ_1	$(-i,\ 1,\ 1)$
ϕ_2	$(1,\ i,\ 1)$
ϕ_3	$(1,\ 1,\ i)$
$\Delta_{R,1}$	$(i,\ i,\ 1)$
$\Delta_{R,2}$	$(1,\ 1,\ -1)$

Table 1: The discrete symmetry assignments for the fermion and Higgs fields in our L-R model that lead naturally to the special Dirac and Majorana textures given in Eq. (4) [6].

matrix has the form in Eq. (4) with $\delta M = 0$. To make the model realistic, we add small soft symmetry-breaking terms of the form

$$\delta V(\phi) = \sum_{a,b} \mu_{ab}^2 \mathrm{Tr}(\tilde{\phi}_a \phi_b^\dagger) + \mathrm{H.c.} \tag{6}$$

to the scalar potential. This will induce the ϕ_a VEVs of the form $\langle \phi_a \rangle = \mathrm{diag}(\delta\kappa_a, \kappa'_a)$, where

$$\delta\kappa_a \propto \frac{\sum_b \mu_{ab}^2 \kappa'_b}{\sum_{a,b} \lambda'_{ab} \kappa'_a \kappa'_b + \sum_a \lambda_a v_{R,a}^2}, \tag{7}$$

with λ, λ' respectively being the scalar self-couplings of the bidoublet and triplet fields in the scalar potential. Choosing appropriately small μ_{ab}^2, we can get very small $\delta\kappa_a$ as required to satisfy the neutrino oscillation data.

2.2. Naturalness of small $\delta\kappa$

It is possible to generate the small parameters $\delta\kappa_a$ naturally through loop effects involving the $W_L - W_R$ mixing. To explain this, first we note that if in a L-R model, $\delta\kappa$'s are set to zero in a natural manner, it will lead to zero $W_L - W_R$ mixing. By the same token, when $W_L - W_R$ mixing is nonzero, it will induce a nonzero $\delta\kappa$. Thus there is an intimate connection between $\delta\kappa$ and $W_L - W_R$ mixing. In our model discussed above, $\delta\kappa = 0$ is guaranteed by $(Z_4)^3$ discrete symmetry. However, if bi-doublets are used to generate quark masses, it is not possible to set either κ or κ' to zero, since in that case the symmetry that stabilizes the vanishing of κ or κ' makes it impossible to get realistic quark masses and mixing. Once in the quark sector both κ and κ' are set to nonzero values, they will induce tree level $W_L - W_R$ mixing and hence arbitrary $\delta\kappa \neq 0$ at the tree level in the leptonic sector, thus invalidating the naturalness of the whole scenario. There is however an alternative way to get quark flavor pattern right without introducing bidoublet Higgs fields and thereby avoiding nonzero $W_L - W_R$ mixing at the tree level. If we introduce $SU(2)_{L,R}$ singlet vector-like up ($\mathcal{U}$) and down ($\mathcal{D}$) quarks, and only LH and RH doublet Higgs fields $\chi_{L,R}$ with nonzero VEVs which give mass to the quarks via a quark seesaw mechanism as in [23], it keeps the $\delta\kappa$'s naturally small. To briefly introduce the quark seesaw, we write the Yukawa interactions responsible for fermion masses in this model as

$$-\mathcal{L}_Y = Y_u \bar{Q}_L \tilde{\chi}_L \mathcal{U}_R + Y_d \bar{Q}_L \chi_L \mathcal{D}_R + (L \leftrightarrow R) + \bar{\mathcal{U}}_L M_{\mathcal{U}} \mathcal{U}_R + \bar{\mathcal{D}}_L M_{\mathcal{D}} \mathcal{D}_R + \mathrm{H.c.}. \tag{8}$$

where $\tilde{\chi}_{L,R} = i\tau_2\chi^*_{L,R}$. This leads to generic seesaw type quark mass relations, e.g.

$$M_u \simeq -\frac{Y_u^2 v_L v_R}{2M_{\mathcal{U}}}. \quad (9)$$

The leptonic sector bidoublets do not couple to the quark sector due to the $(Z_4)^3$ discrete symmetry under which the quark sector fields are singlets. As a result, the $W_L - W_R$ mixing is zero also in the quark sector, and hence, $\delta\kappa$ is not induced at the tree level. However, $W_L - W_R$ mixing is induced at the one loop level via the exchange of top and bottom quarks in the quark sector (see Figure 1), which in turn induces leptonic $\delta\kappa$ at the two loop level from the gauge interaction $g_L g_R \phi_1^{0\dagger}\phi_2^0 W_L^+ W_R^-$ in the notation of Eq. (2). This makes $\delta\kappa$ finite and small, thus keeping the model technically natural.

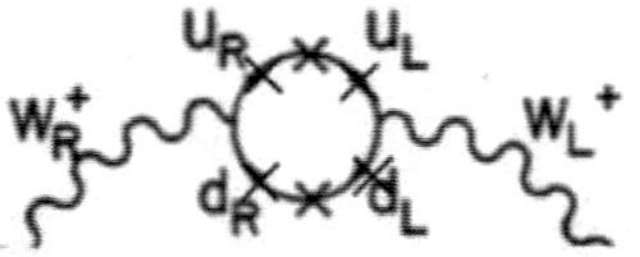

Figure 1: One loop graph for $W_L - W_R$ mixing in the quark seesaw model. The crosses in the middle of the two quark propagators are for the vectorlike quark mass insertions, while the other four crosses denote the mixing between the vectorlike quarks and SM quarks.

An important point to notice is that $\delta\kappa_a$'s are responsible for the electron mass as well as neutrino masses via type-I seesaw. Thus getting a fit to the observed neutrino masses and mixing while at the same time keeping electron mass at its desired value is a nontrivial task since in the lepton sector, the model has only 14 free parameters (9 Yukawa couplings, 3 Majorana mass parameters and two VEVs) to start with. One can always go to a basis where M_D takes an upper-triangular form, in which case the number of free parameters reduces to 11. Out of this 11 model parameters, we must not only get fits for the three charged lepton masses, the two neutrino mass-squared differences and three mixing angles (total of 8 outputs), but must also satisfy the unitarity constraints on the new light neutrino mixing matrix as well as constraints from rare lepton decays which involve only the RH mixing matrix for charged leptons. It was explicitly shown in [6] that this model does indeed provide a fit to all observables while satisfying all the necessary constraints.

3. An alternative neutrino mass fit

In this section, we discuss neutrino fit in a model with M_D and M_N texture as given in Eq. (4) without worrying about its origin from discrete symmetries and use this fit in the subsequent section for the discussion of leptogenesis. We consider a minimal scenario with only one bidoublet ϕ and a single right-Higgs triplet Δ_R to illustrate the fit. Clearly, in this case we will have to adjust parameters to get the desired fermion mass textures.

In this class of L-R models, we can always choose a basis of the LH sector prior to $SU(2)_L \times U(1)_Y$ breaking such that the Dirac mass matrix M_D can be written in an upper-triangular form without affecting the RH neutrino texture. In this basis, the mass matrices obtained from the Yukawa Lagrangian in Eq. (3) are given by

$$M_l = \begin{pmatrix} h_{11}\kappa' + \tilde{h}_{11}\kappa^* & h_{12}\kappa' + \tilde{h}_{12}\kappa^* & h_{13}\kappa' + \tilde{h}_{13}\kappa^* \\ \frac{|\kappa'|^2-|\kappa|^2}{\kappa'^*} h_{21} & h_{22}\kappa' + \tilde{h}_{22}\kappa^* & h_{23}\kappa' + \tilde{h}_{23}\kappa^* \\ \frac{|\kappa'|^2-|\kappa|^2}{\kappa'^*} h_{31} & \frac{|\kappa'|^2-|\kappa|^2}{\kappa'^*} h_{32} & h_{33}\kappa' + \tilde{h}_{33}\kappa^* \end{pmatrix}, \tag{10}$$

$$M_D = \begin{pmatrix} h_{11}\kappa + \tilde{h}_{11}\kappa'^* & h_{12}\kappa + \tilde{h}_{12}\kappa'^* & h_{13}\kappa + \tilde{h}_{13}\kappa'^* \\ 0 & h_{22}\kappa + \tilde{h}_{22}\kappa'^* & h_{23}\kappa + \tilde{h}_{23}\kappa'^* \\ 0 & 0 & h_{33}\kappa + \tilde{h}_{33}\kappa'^* \end{pmatrix}, \tag{11}$$

$$M_N = \begin{pmatrix} \delta M & f_{12}v_{R,1} & 0 \\ f_{12}v_{R,1} & 0 & 0 \\ 0 & 0 & 2f_{33}v_{R,2} \end{pmatrix}, \tag{12}$$

Note that since parity is a broken symmetry in this low energy version, we need not have the Yukawa couplings h_{ij} to satisfy hermiticity. We choose the parameters of the model, i.e. h_{ij}'s and κ, κ' such that all the experimental constraints in the lepton sector are satisfied. As an example fit, we consider the mass matrices of the following form:

$$M_l = \begin{pmatrix} 0.00120 & -0.0507 & -1.41 \\ 0 & 0.0929 & -0.657 \\ 0 & 0 & -0.861 \end{pmatrix} \text{GeV}, \tag{13}$$

$$M_D = \begin{pmatrix} 0.0676\, e^{i0.557\pi} & -1.06\times 10^{-7}\, e^{-i0.557\pi} & -1.06\times 10^{-4} \\ 0 & 1.95\times 10^{-7}\, e^{-i0.557\pi} & -4.78\times 10^{-5} \\ 0 & 0 & -6.38\times 10^{-5} \end{pmatrix} \text{GeV}, \tag{14}$$

$$M_N = \begin{pmatrix} 2.59\times 10^{-6} & 585 & 0 \\ 585 & 0 & 0 \\ 0 & 0 & -585 \end{pmatrix} \text{GeV}, \tag{15}$$

In addition, we choose the VEVs $\kappa = 6.46$ GeV and $\kappa' = 173.88\, e^{i0.700\pi}$ GeV, which satisfy $|\kappa^2| + |\kappa'^2| = v^2$ with $v = 174$ GeV. These are all the input parameters for the model. The M_l is diagonalized by a bi-unitary transformation: $\widehat{M_l} = (V_l^L)^\dagger M_l V_l^R$, where

$$V_l^L = i\begin{pmatrix} 0.426 & 0.436 & 0.793 \\ 0.232 & -0.900 & 0.369 \\ -0.875 & -0.0271 & 0.484 \end{pmatrix}, \quad V_l^R = i\begin{pmatrix} 1.00 & 0.00495 & 0.000535 \\ 0.00495 & -1.00 & -0.00332 \\ 0.000519 & 0.00332 & -1.00 \end{pmatrix}. \tag{16}$$

The light neutrino mass matrix M_ν is diagonalized by a unitary transformation: $\widehat{M_\nu} = V_\nu^{\mathrm{T}} M_\nu V_\nu$, where

$$V_\nu = \begin{pmatrix} 0.338i & -0.0149 & 0.941 \\ 0.742i & 0.619 & -0.257 \\ -0.579i & 0.785 & 0.221 \end{pmatrix}. \tag{17}$$

The Pontecorvo-Maki-Nakagawa-Sakata (PMNS) mixing matrix is then given by

$$V_{\mathrm{PMNS}} = (V_l^L)^\dagger V_\nu = \begin{pmatrix} 0.823 & 0.549 & 0.148 \\ -0.505 & 0.585 & 0.635 \\ 0.262 & -0.597 & 0.758 \end{pmatrix}\begin{pmatrix} 1 & 0 & 0 \\ 0 & i & 0 \\ 0 & 0 & -i \end{pmatrix}, \tag{18}$$

The resulting neutrino masses and mixing angles as well as the charged-lepton masses are given in Table 2.

Parameter	Value
m_e	0.511 MeV
m_μ	106 MeV
m_τ	1.78 GeV
m_{ν_1}	6.49×10^{-3} eV
m_{ν_2}	1.08×10^{-2} eV
m_{ν_3}	5.02×10^{-2} eV
θ_{12}	33.7°
θ_{23}	40.0°
θ_{13}	8.51°
m_{N_1}	585 GeV
m_{N_2}	585 GeV
m_{N_3}	585 GeV

Table 2: Masses and mixing angles calculated from the mass matrices given by Eqs. (13)-(15). The mass splitting between the RH neutrinos is of order of $\delta M \sim 10^{-6}$ GeV.

Since $M_l = h\kappa' + \tilde{h}\kappa^*$ and $M_D = h\kappa + \tilde{h}\kappa'^*$, the Yukawa couplings h, $\tilde{h}$ are easily calculated using VEV's chosen as above, and they are found to be

$$h = \begin{pmatrix} -(0.15+1.98i)\times10^{-5} & (1.72+2.36i)\times10^{-4} & (4.77+6.57i)\times10^{-3} \\ 0 & -(3.14+4.33i)\times10^{-4} & (2.22+3.06i)\times10^{-3} \\ 0 & 0 & (2.91+4.01i)\times10^{-3} \end{pmatrix}, \quad (19)$$

$$\tilde{h} = \begin{pmatrix} -(2.69+2.81i)\times10^{-4} & 1.08\times10^{-5} & 3.02\times10^{-4} \\ 0 & -1.99\times10^{-5} & 1.41\times10^{-4} \\ 0 & 0 & 1.84\times10^{-4} \end{pmatrix}. \quad (20)$$

The leptonic CP asymmetry generated by the RH neutrino decays is governed by these Yukawa couplings. If we assume that the two SM doublets in the LR bi-doublet i.e. ϕ_1 and ϕ_2 are approximate mass eigenstates with $m_{\phi_1} > m_N > m_{\phi_2}$, the dominant contribution to the CP asymmetry comes solely from the self-energy correction by the ϕ_2-loop to the decay process $N \to \phi_2 L_l$. Therefore, only the Yukawa coupling matrix $\tilde{h}$ determines the flavor effect in RH neutrino interactions relevant to resonant leptogenesis.

The RH neutrino mixing effect at resonance can be captured by the one-loop resummed Yukawa couplings [18]. Explicitly, in the heavy-neutrino mass eigenbasis, we obtain for the resummed Yukawa couplings $\widehat{\mathbf{h}}_{l\alpha}$ and their CP-conjugates $\widehat{\mathbf{h}}^c_{l\alpha}$ the following values:

$$\widehat{\mathbf{h}} = \begin{pmatrix} (-2.52+1.03i)\times10^{-4} & (2.39+1.00i)\times10^{-4} & (-0.03+3.46i)\times10^{-4} \\ -(2.82+5.07i)\times10^{-5} & (4.37-3.68i)\times10^{-5} & (-0.08+1.49i)\times10^{-4} \\ -(3.63+4.69i)\times10^{-5} & (3.77-4.89i)\times10^{-5} & (0.00+1.84i)\times10^{-4} \end{pmatrix}, \quad (21)$$

$$\widehat{\mathbf{h}}^c = \begin{pmatrix} -(1.59+1.26i)\times10^{-4} & (1.40-1.25i)\times10^{-4} & (1.02-7.03i)\times10^{-5} \\ (2.71+4.90i)\times10^{-5} & (-1.57+3.67i)\times10^{-5} & (0.02-1.39i)\times10^{-4} \\ (3.62+4.70i)\times10^{-5} & (-3.78+4.88i)\times10^{-5} & (0.00+1.84i)\times10^{-4} \end{pmatrix}. \quad (22)$$

The heavy neutrino decay rates for the Yukawa-mediated two-body processes $N_\alpha \to L_l\phi,\ L_l^c\phi^c$ are calculated with these resummed Yukawa couplings:

$$\Gamma(N_\alpha \to L_l\phi) = \frac{m_{N_\alpha}}{16\pi}\widehat{\mathbf{h}}_{l\alpha}\widehat{\mathbf{h}}^*_{l\alpha}, \qquad \Gamma(N_\alpha \to L_l^c\phi^c) = \frac{m_{N_\alpha}}{16\pi}\widehat{\mathbf{h}}^c_{l\alpha}\widehat{\mathbf{h}}^{c*}_{l\alpha}. \quad (23)$$

which are crucial parameters for the calculation of lepton asymmetry, as we will see in the next section.

4. Resonant leptogenesis with TeV-scale W_R

As eluded to above, for low scale seesaw models, a simple way to generate enough lepton asymmetry is to use resonant leptogenesis [17, 18]. It requires at least two RH neutrinos to be nearly degenerate, which is guaranteed in our model by the choice of the texture in the RH neutrino mass matrix given by Eq. (12). Thus, in addition to guaranteeing suppression of lepton flavor violating processes such as $\mu \to 3e$, this choice of RH neutrino texture also provides a necessary condition for resonant leptogenesis. As far as the Dirac neutrino mass matrix M_D is concerned, we choose the pattern given in Eq. (11) so that it fits neutrino oscillation data. Since our focus is on the feasibility of low scale leptogenesis in a realistic model for neutrinos, we do not stress here on the naturalness of the texture, although a similar exercise can be performed with the fit presented in Section 2.1. The details of leptogenesis in this class of models was discussed in [21] and we do not repeat the basic equations here. However, we have extended the work of [21] and also present here some revised numerical results due to a computational error in the code used there.

The basic steps in our analysis are as follows: first we calculate the flavored CP asymmetry

$$\varepsilon_{l\alpha} \;=\; \frac{1}{\Gamma_{N_\alpha}}[\Gamma(N_\alpha \to L_l\phi) - \Gamma(N_\alpha \to L_l^c\phi^c)]\,, \tag{24}$$

where the partial decay widths are given by Eq. (23) and the total decay width is given by

$$\Gamma_{N_\alpha} \;=\; \sum_l [\Gamma(N_\alpha \to L_l\phi) + \Gamma(N_\alpha \to L_l^c\phi^c)] + 2\,\Gamma(N_\alpha \to l_R q_R \bar{q}_R') \,. \tag{25}$$

Note that in LRSM, there is an additional contribution to the total decay width due to the three-body decay mediated by W_R:

$$\Gamma(N_\alpha \to l_R q_R \bar{q}_R') \;=\; \Gamma(N_\alpha \to \bar{l}_R \bar{q}_R q_R') \;=\; \frac{3g_R^4}{2^9\pi^3 m_{N_\alpha}^3}\int_0^{m_{N_\alpha}^2} ds \frac{m_{N_\alpha}^6 - 3m_{N_\alpha}^2 s^2 + 2s^3}{(s-M_{W_R}^2)^2 + M_{W_R}^2\Gamma_{W_R}^2}\,, \tag{26}$$

where $\Gamma_{W_R} \simeq (g_R^2/4\pi)M_{W_R}$ is the total decay width of W_R, assuming that all three heavy neutrinos are lighter than W_R. The CP asymmetry (24) in our case is determined by the RH neutrino texture, specifically, the magnitudes of $M_{1,2}$ and δM in Eq. (4) as well as the magnitudes of the Yukawa couplings, and these parameters are constrained by the neutrino fit.

The second step in our calculation is to calculate the thermodynamic evolution of the normalized heavy neutrino and lepton doublet number densities η_α^N and $\eta_l^{\Delta L}$ respectively, where $\eta^X \equiv n^X/n^\gamma$ and $n^\gamma = 2m_{N_1}^3\zeta(3)/(\pi^2 z^3)$ is the photon number density, $\zeta(x)$ being the Riemann zeta function. For simplicity and for a fair comparison with the results of [20], here we use the flavor-diagonal Boltzmann equations [24][1]:

$$\frac{d\eta_\alpha^N}{dz} \;=\; -\left(\frac{\eta_\alpha^N}{\eta_{\rm eq}^N} - 1\right)(D_\alpha + S_\alpha)\,, \tag{27}$$

$$\frac{d\eta_l^{\Delta L}}{dz} \;=\; \sum_\alpha \varepsilon_{l\alpha}\left(\frac{\eta_\alpha^N}{\eta_{\rm eq}^N} - 1\right)\widetilde{D}_\alpha - \frac{2}{3}\eta_l^{\Delta L} W_l\,, \tag{28}$$

where $z = m_{N_1}/T$ is a dimensionless variable (T being the temperature of the Universe) and $\eta_{\rm eq}^N \equiv n_{\rm eq}^N/n^\gamma = z^2K_2(z)/2\zeta(3)$ is the heavy neutrino equilibrium number density, $K_n(x)$ being

[1] Including flavor off-diagonal effects could lead to additional enhancement of the final lepton asymmetry [25].

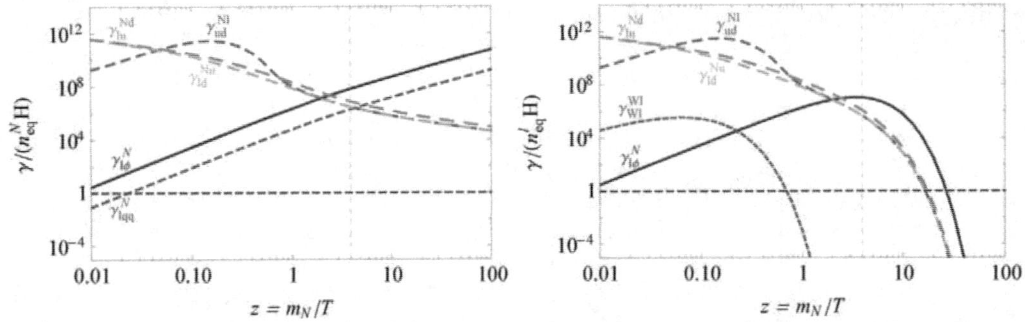

Figure 2: Evolution of $\gamma/(n_{\rm eq}^N H)$ [left panel] and $\gamma/(n_{\rm eq}^l H)$ [right panel]. For the latter, we have also included the scattering rate of $W_R^+ l_R^- \to W_R^- l_R^+$ with $m_{\Delta^{++}} = 10$ TeV and $m_{W_R} = 13.1$ TeV.

the n-th order modified Bessel function of the second kind. The various decay (D_α, $\widetilde{D}_\alpha$), scattering (S_α) and washout (W_l) rates appearing in Eqs. (27) and (28) are given by

$$\widetilde{D}_\alpha = \frac{z}{n^\gamma H_N} \sum_k \widetilde{\gamma}^D_{k\alpha}, \tag{29}$$

$$D_\alpha = \frac{z}{n^\gamma H_N} \sum_k \gamma^D_{k\alpha}, \tag{30}$$

$$S_\alpha = \frac{z}{n^\gamma H_N} \sum_k (\gamma^{S_L}_{k\alpha} + \gamma^{S_R}_{k\alpha}), \tag{31}$$

$$W_l = \frac{z}{n^\gamma H_N} \left[\sum_\alpha \left(B_{l\alpha} \sum_k \gamma^D_{k\alpha} + \widetilde{\gamma}^{S_L}_{l\alpha} + \widetilde{\gamma}^{S_R}_{l\alpha} \right) + \sum_k \left(\gamma^{(\Delta L=2)}_{lk} + \gamma^{(\Delta L=0)}_{lk} \right) \right]$$
$$\equiv \frac{1}{2\zeta(3)} z^3 K_1(z) K^{\rm eff}_l(z)\,, \tag{32}$$

where $H_N \equiv H(z=1) \simeq 17 m^2_{N_1}/M_{\rm Pl}$ is the Hubble parameter at $z = 1$, assuming only SM degrees of freedom in the thermal bath, $M_{\rm Pl} = 1.2 \times 10^{19}$ GeV is the Planck mass and $B_{l\alpha}$ is the branching fraction of the RH neutrino decays relevant for the generation of *CP* asymmetry:

$$B_{l\alpha} = \frac{1}{\Gamma_{N_\alpha}} [\Gamma(N_\alpha \to L_l \phi) + \Gamma(N_\alpha \to L_l^c \phi^c)]\,. \tag{33}$$

The various γ's appearing in Eqs. (29)-(32) represent the reaction rates that involve the RH neutrino decays and inverse decays as well as other $2 \leftrightarrow 2$ scattering processes in the model (see Appendix A). Only the two-body decays of the RH neutrinos involving complex Yukawa couplings are responsible for building up the asymmetry, whereas all other processes lead to washout effects.

We present in Figure 2 the evolution of the various collision rates as given in Appendix A. The left panel of the Figure is relevant to the first Boltzmann equation (27), while the second one is relevant to (28). The vertical dashed line shows the critical value $z = z_c$ beyond which the sphaleron processes become ineffective. The horizontal dashed line is shown for easy comparison with the Hubble rate. Here we have used $m_{W_R} = 13.1$ TeV which is the lowest value of m_{W_R} we found in this model with an *explicit* neutrino fit that satisfies the leptogenesis constraints.

Recently, it was pointed out [28] that the scattering process $W_R^+ l_R^- \to W_R^- l_R^+$ mediated by the doubly-charged component of the RH triplet Δ_R^{++} might spoil leptogenesis, if m_{W_R} and $m_{\Delta_R^{++}}$

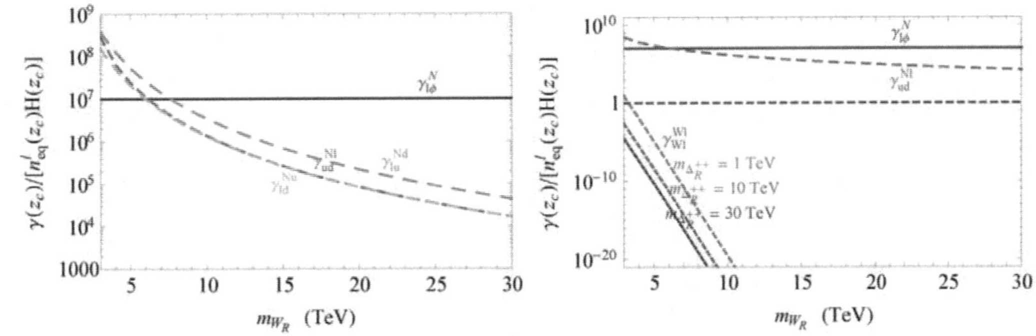

Figure 3: Various reaction rates $\gamma(z_c)/[n_l^{\rm eq}(z_c)H(z_c)]$ as a function of m_{W_R} for $m_N = 585$ GeV. The green lines in the right panel correspond to the process $W_R^+ l_R^- \to W_R^- l_R^+$ for various Δ_R^{++} masses.

are relatively small. We have explicitly calculated the washout effect due to this process for various values of m_{W_R} and $m_{\Delta^{++}}$. As shown in Figure 3 and also in Figure 2 (right panel), this effect is negligible compared with the decay rate and other scattering rates when m_{W_R} and $m_{\Delta^{++}}$ are in the regime of interest.

In the strong washout regime, the flavored lepton asymmetry is approximately given by [26, 21]

$$\eta^{\Delta L}(z)_{l\alpha} \simeq \frac{3}{2zK_l^{\rm eff}(z)}\varepsilon_{l\alpha}\frac{\widetilde{D}_\alpha(z)}{D_\alpha(z)+S_\alpha(z)}, \tag{34}$$

where $K_l^{\rm eff}$ is the effective washout parameter as defined in Eq. (32) and $\widetilde{D}/(D+S)$ is the effective dilution factor. For the numerical fit given in Section 3, we obtain

$$\eta_{l\alpha}^{\Delta L} = \begin{pmatrix} 5.16\times10^{-9} & 5.13\times10^{-9} & 1.32\times10^{-8} \\ 6.44\times10^{-11} & 4.89\times10^{-10} & 6.48\times10^{-10} \\ 0 & 0 & 0 \end{pmatrix}, \tag{35}$$

and the total lepton asymmetry is found to be $\eta^{\Delta L} = \sum_{l\alpha}\eta_{l\alpha}^{\Delta L} = 2.47\times10^{-8}$. This is well consistent with the observed value of the baryon asymmetry $\eta^{\Delta B} = (6.105^{+0.086}_{-0.081})\times10^{-10}$ [27], after taking into account the sphaleron conversion rate and the entropy dilution factors.

4.1. Lower bound of m_{W_R}

Here we present our procedure to derive a lower bound of m_{W_R} which is compatible with successful leptogenesis in a LRSM with all RH neutrinos being quasi-degenerate in mass. The total lepton asymmetry is given by Eq. (34), which can be rewritten using Eq. (32) as

$$\eta^{\Delta L}(z) = \frac{3}{4\zeta(3)}z^2K_1(z)\sum_{l,\alpha}\frac{1}{W_l(z)}\varepsilon_{l\alpha}\frac{\widetilde{D}_\alpha(z)}{D_\alpha(z)+S_\alpha(z)}. \tag{36}$$

For the purpose of deriving a lower bound on m_{W_R}, we assume that the dominant washout effect comes from the W_R-mediated scattering processes, i.e., $S_\alpha \approx \sum_k S_{R\alpha} = 3S_{R\alpha}$ where $S_{R\alpha} \equiv z\widetilde{\gamma}_{k\alpha}^{S_R}/(n^\gamma H_N)$. This follows from the fact that gauge interactions are flavor-blind, and hence, $\widetilde{\gamma}_{k\alpha}^{S_R}$ has no dependence on lepton flavors. Similarly, when all the RH neutrino masses are quasi-degenerate, we can write $S_R \equiv \sum_\alpha S_{R\alpha} \approx 3S_{R\alpha}$. We also assume that $D = \sum_\alpha D_\alpha \approx 3D_\alpha$ for simplicity. Note that S_R is not summed over lepton flavors while D is. We further define $\varepsilon^Y_{l\alpha}$ and $B^Y_{l\alpha}$ as the *CP* asymmetry and branching ratio, respectively, without the 3-body decay width included in the denominator of Eqs. (24) and (33). For simplicity, we further assume that the branching ratios of the decay process $N_\alpha \to L_l\phi$ are the same for all the lepton flavors. With all the above-mentioned reasonable assumptions, we can approximate the total lepton asymmetry in Eq. (36) at the critical temperature as

$$|\eta^{\Delta L}(z_c)| \approx \frac{3}{4\zeta(3)} z_c^2 K_1(z_c) \sum_l \frac{1}{\sum_\alpha [B_{l\alpha} D_\alpha(z_c) + S_{R\alpha}(z_c)]} \left| \sum_\alpha \varepsilon_{l\alpha} \frac{\widetilde{D}_\alpha(z_c)}{D_\alpha(z_c) + 3S_{R\alpha}(z_c)} \right| \quad (37)$$

$$\approx \frac{9}{4\zeta(3)} \frac{z_c^2 K_1(z_c)}{S_R(z_c)} \frac{r_s r_d^2}{(3+r_s r_d)(3+r_s)} \varepsilon^Y_{\rm tot} \quad (38)$$

where the ratios $r_s \equiv D(z_c)/S_R(z_c)$ and $r_d \equiv \widetilde{D}(z_c)/D(z_c)$ parametrize the relative washout strengths of the 2-body decay, W_R-mediated decay and W_R-mediated scattering processes. The values of r_s and r_d depend on m_N, m_{W_R}, and the Yukawa coupling h. Assuming a specific value of h and calculating the 2-body decay width simply as $\Gamma(N_\alpha \to L_l\phi) = \Gamma(N_\alpha \to L_l^c\phi^c) = h^2 m_N/16\pi$, we can evaluate these parameters as functions of m_N and m_{W_R}.

Figure 4 shows the contour plots of $|\eta^{\Delta L}(z_c)| = 2.47 \times 10^{-8}$ for two different Yukawa couplings $h = 10^{-3.8}$ and $h = 10^{-3.5}$. The red curves correspond to $\varepsilon^Y_{\rm tot} = 1$ which is the total *CP* asymmetry in the example fit given in Section 3. Any region outside the red curve is incompatible with successful leptogenesis under the assumptions we introduced to obtain it, i.e. $m_{N1} \approx m_{N2} \approx m_{N3}$, $B^Y_{1\alpha} \approx B^Y_{2\alpha} \approx B^Y_{3\alpha} \approx 1/3$, $\varepsilon^Y_{\rm tot} = 1$ for the two specific values of h. For $h = 10^{-3.8}$, we find that the lowest value of m_{W_R} allowed for $\varepsilon^Y_{\rm tot} = 1$ is 13 TeV at around $m_N =$ 580 TeV. Note that the example fit we have presented in the previous section has $m_{W_R} = 13.1$ TeV, $m_N = 585$ GeV, and the resummed Yukawa coulings of order around $\sim 10^{-3.8}$. Thus, our example fit, shown as a green dot in Figure 4, is very close to the minimum value of m_{W_R} obtained here in a simplified manner, which justifies the validity of this approach. In other words, we have derived the lower bound in two different ways: (i) in Section 3 by carefully scanning the parameter space in order to find an explicit fit with the lowest possible m_{W_R} compatible with leptogenesis, and (ii) using the approximate expression in Eq. (38). The remarkable agreement between these two approaches shows that the simplified expression in Eq. (38) is very effective in finding the region in parameter space compatible with leptogenesis and predicting the lower bound of m_{W_R} as well as the position in the parameter space where it exists. If we use the same expression and take the maximal *CP* asymmetry allowed in principle, i.e. $\varepsilon^Y_{\rm tot} \equiv \sum_{l,\alpha} \varepsilon^Y_{l\alpha} = 3$, then we obtain the blue curves Figure 4. With this maximal *CP* asymmetry, we have found that the Yukawa coupling $h = 10^{-3.5}$ gives the lower bound of $m_{W_R} = 9.9$ TeV at $m_N = 630$ GeV.

The Yukawa couplings cannot be increased arbitrarily without spoiling the lepton asymmetry, since not only the source term due to the two-body decay of the RH neutrinos, but also the washout effects due to inverse decay and $\Delta L = 2$ scattering increase with the Yukawa couplings. Similarly, for very small values of the Yukawa couplings, the branching fraction of the two-body decay mode becomes comparable or smaller than the three-body decay mode due to W_R interactions, and therefore, the washout effect again increases. Thus, successful leptogenesis works only in a range of the Yukawa coupling parameter space. This is shown in the three-dimensional plot given in Figure 5, where we see that leptogenesis constraints in our model require the Yukawa coupling to be $10^{-5.6} \leq h \leq 10^{-3.2}$ for $m_{W_R} \lesssim 30$ TeV. The robustness of the lower bound on m_{W_R} obtained in Figure 4 can also be verified from Figure 5.

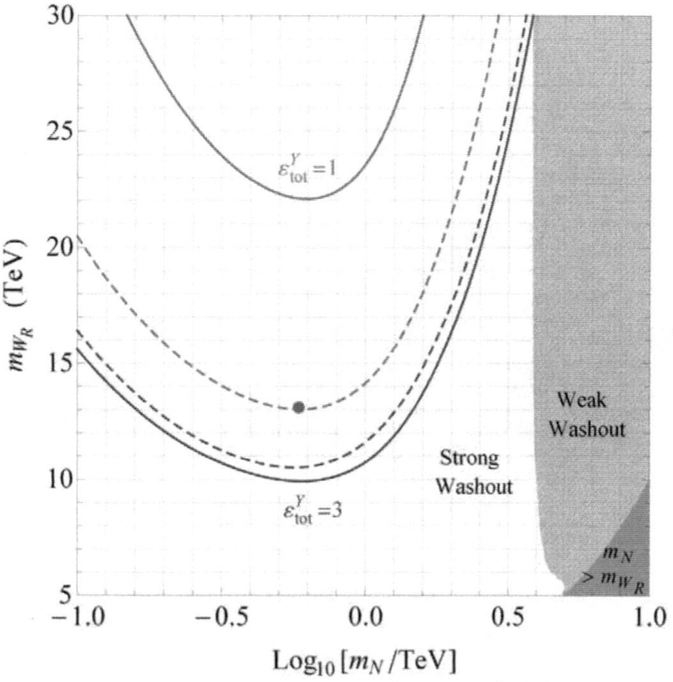

Figure 4: Contour plots of $|\eta^{\Delta L}(z_c)| = 2.47 \times 10^{-8}$ for $h = 10^{-3.8}$ (dashed lines) and $h = 10^{-3.5}$ (solid lines) with $\varepsilon^Y_{\mathrm{tot}} = 1$ (red lines) and $\varepsilon^Y_{\mathrm{tot}} = 3$ (blue lines). The green dot corresponds to the example fit value presented in Section 3.

5. Summary

In this proceedings, we address two issues related to seesaw models for neutrino masses. The first one deals with whether the TeV scale can be naturally in the TeV range without fine-tuning of parameters. We present a natural TeV scale left-right model which achieves this goal and therefore provides a counterexample to the common lore that either the seesaw scale must be superheavy or the active-sterile neutrino mixing must be tiny in a UV-complete seesaw model. The second issue addresses the important question: Whether in such low scale models, one can have successful leptogenesis, and if so, what constraints are implied by this on the mass of the RH gauge boson. In an explicit TeV-scale LR model, with an explicit fermion mass fit, we find the lower bound to be 13.1 TeV and for generic models in this class of L-R seesaw with enhanced neutrino Yukawa couplings compared to the canonical seesaw case and with maximal possible *CP* asymmetry for each flavor, this bounds becomes 9.9 TeV.

Acknowledgement

The work of P. S. B. D. is supported by the Lancaster-Manchester-Sheffield Consortium for Fundamental Physics under STFC grant ST/L000520/1. The work of C. H. L. and R. N. M. is supported by the NSF grant No. PHY-1315155.

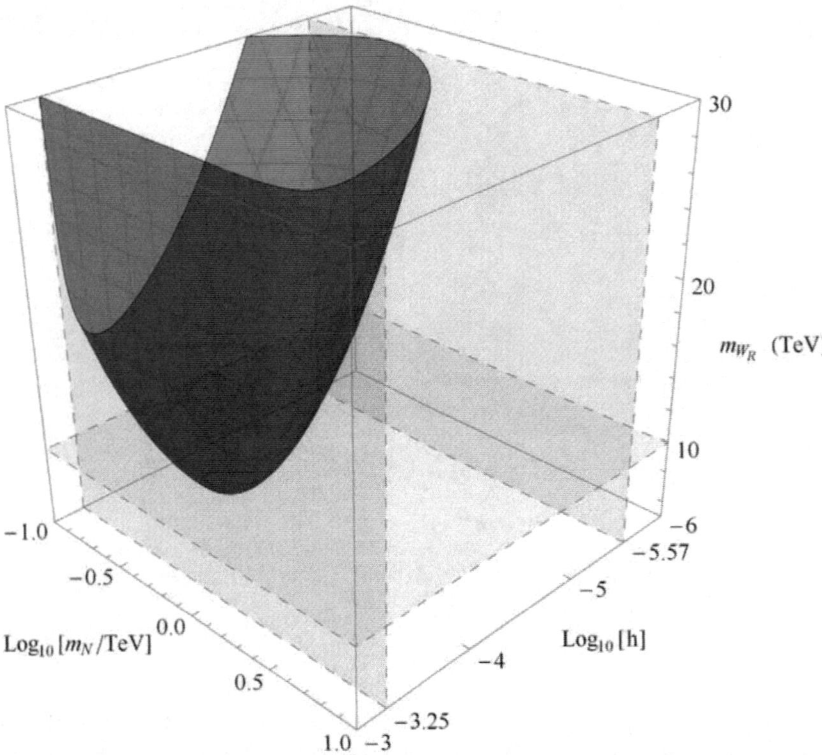

Figure 5: The allowed region of parameter space (shaded region) yielding successful leptogenesis in our L-R seesaw model. The vertical gray surfaces show the bound on Yukawa couplings, while the horizontal pink surface shows the lower bound on m_{W_R}.

Appendix A. Collision Rates

The various collision rates for decay and scattering processes in Eqs. (27) and (28) are given below:

$$\gamma^D_{l\alpha} = \gamma^{N_\alpha}_{L_l\phi_l} + \gamma^{N_\alpha}_{l_R q\bar{q}'}, \tag{A.1}$$

$$\widetilde{\gamma}^D_{l\alpha} = \gamma^{N_\alpha}_{L_l\phi_l}, \tag{A.2}$$

$$\gamma^{S_L}_{l\alpha} = \gamma^{N_\alpha L_l}_{Qu^c} + \gamma^{N_\alpha u^c}_{L_l Q^c} + \gamma^{N_\alpha Q}_{L_l u} + \gamma^{N_\alpha L_l}_{\phi^\dagger V_\mu} + \gamma^{N_\alpha V_\mu}_{L_l\phi} + \gamma^{N_\alpha\phi^\dagger}_{L_l V_\mu}, \tag{A.3}$$

$$\widetilde{\gamma}^{S_L}_{l\alpha} = \frac{\eta^N_\alpha}{\eta^N_{\text{eq}}}\gamma^{N_\alpha L_l}_{Qu^c} + \gamma^{N_\alpha u^c}_{L_l Q^c} + \gamma^{N_\alpha Q}_{L_l u} + \frac{\eta^N_\alpha}{\eta^N_{\text{eq}}}\gamma^{N_\alpha L_l}_{\phi^\dagger V_\mu} + \gamma^{N_\alpha V_\mu}_{L_l\phi} + \gamma^{N_\alpha\phi^\dagger}_{L_l V_\mu}, \tag{A.4}$$

$$\gamma^{S_R}_{l\alpha} = \gamma^{N_\alpha l_R}_{\bar{u}_R d_R} + \gamma^{N_\alpha \bar{u}_R}_{l_R \bar{d}_R} + \gamma^{N_\alpha d_R}_{l_R u_R}, \tag{A.5}$$

$$\widetilde{\gamma}^{S_R}_{l\alpha} = \frac{\eta^N_\alpha}{\eta^N_{\text{eq}}}\gamma^{N_\alpha l_R}_{\bar{u}_R d_R} + \gamma^{N_\alpha \bar{u}_R}_{l_R \bar{d}_R} + \gamma^{N_\alpha d_R}_{l_R u_R}, \tag{A.6}$$

$$\gamma^{(\Delta L=2)}_{lk} = \gamma'^{L_l\phi_l}_{L^c_k\phi^\dagger_k} + \gamma^{L_l L_k}_{\phi^\dagger_l\phi^\dagger_k}, \tag{A.7}$$

$$\gamma^{(\Delta L=0)}_{lk} = \gamma'^{L_l\phi_l}_{L_k\phi_k} + \gamma^{L_l\phi^\dagger_l}_{L_k\phi^\dagger_k} + \gamma^{L_l L^c_k}_{\phi_l\phi^\dagger_k}. \tag{A.8}$$

The scattering terms involving two heavy neutrinos in the initial state, e.g. induced by a t-channel W_R or e_R, and by an s-channel Z_R, are not included here since their rates are doubly Boltzmann-suppressed and numerically much smaller than the scattering rates given above [20, 29]. The decay rates are explicitly given by

$$\gamma^{N_\alpha}_{L_l\phi} = \frac{m^3_{N_\alpha}}{\pi^2 z} K_1(z) \left[\Gamma(N_\alpha \to L_l\phi) + \Gamma(N_\alpha \to L^c_l\phi^\dagger)\right], \tag{A.9}$$

$$\gamma^{N_\alpha}_{l_R q\bar{q}'} = \frac{m^3_{N_\alpha}}{\pi^2 z} K_1(z) \left[\Gamma(N_\alpha \to l_R q_R \bar{q}'_R) + \Gamma(N_\alpha \to \bar{l}_R \bar{q}_R q'_R)\right] . \tag{A.10}$$

The various collision terms for the $2 \leftrightarrow 2$ scattering processes $XY \leftrightarrow AB$ are defined as

$$\gamma^{XY}_{AB} = \frac{m^4_{N_1}}{64\pi^4 z} \int_{x_{\rm thr}}^{\infty} dx \sqrt{x} K_1(z\sqrt{x}) \hat{\sigma}^{XY}_{AB}(x), \tag{A.11}$$

where $x = s/m^2_{N_1}$ with the kinematic threshold value $x_{\rm thr} = \max[(m_X+m_Y)^2, (m_A+m_B)^2]/m^2_{N_1}$, and $\hat{\sigma}^{XY}_{AB}(x)$ are the relevant reduced cross sections, whose explicit expressions can be found in [21]. The reduced cross section of the additional process $W^-_R l^+_R \to W^+_R l^-_R$ mediated by Δ^{++}_R is given by

$$\hat{\sigma}^{W_R l_R}_{W_R l_R}(s) = \frac{g^4_R m^2_{N_1}}{8\pi s} \left[\log\left(\frac{(s - m^2_{W_R})^2 + m^2_{\Delta^{++}_R} s}{m^2_{\Delta^{++}_R} s} \right) + \frac{m^2_{\Delta^{++}_R} s}{(s - m^2_{W_R})^2 + m^2_{\Delta^{++}_R} s} - 1 \right] . \tag{A.12}$$

References

[1] P. Minkowski, Phys. Lett. B **67**, 421 (1977); R. N. Mohapatra and G. Senjanović, Phys. Rev. Lett. **44**, 912 (1980); T. Yanagida, Conf. Proc. **C7902131**, 95 (1979); M. Gell-Mann, P. Ramond and R. Slansky, Conf. Proc. **C790927**, 315 (1979); S. L. Glashow, NATO Sci. Ser. B **59**, 687 (1980).
[2] A. Pilaftsis, Z. Phys. **C55**, 275 (1992); E. Ma, Phys. Rev. Lett. **81**, 1171 (1998); E. Ma, Phys. Rev. D **73**, 077301 (2006); P. S. B. Dev and A. Pilaftsis, Phys. Rev. **D86**, 113001 (2012); S. S. Law and K. L. McDonald, Int. J. Mod. Phys. **A29**, 1450064 (2014); A. J. R. Figueiredo, Eur. Phys. J. C **75**, 99 (2015).
[3] J. Kersten and A. Yu. Smirnov, Phys. Rev. **D76**, 073005 (2007).
[4] W. Buchmüller, C. Greub, and P. Minkowski, Phys. Lett. **B267**, 395 (1991); J. Gluza, Acta Phys. Polon. B **33**, 1735 (2002); A. Pilaftsis, Phys. Rev. Lett. **95**, 081602 (2005); Z.-z. Xing, Prog. Theor. Phys. Suppl. **180**, 112 (2009); X.-G. He, S. Oh, J. Tandean and C.-C. Wen, Phys. Rev. D **80**, 073012 (2009); A. Ibarra, E. Molinaro and S. T. Petcov, JHEP **1009**, 108 (2010); R. Adhikari and A. Raychaudhuri, Phys. Rev. D **84**, 033002 (2011); M. Mitra, G. Senjanović and F. Vissani, Nucl. Phys. B **856**, 26 (2012).
[5] J.C. Pati and A. Salam, Phys. Rev. D **10**, 275 (1974); R. N. Mohapatra and J. C. Pati, Phys. Rev. D **11** 2558 (1975); G. Senjanović and R. N. Mohapatra, Phys. Rev. D **12** 1502 (1975).
[6] C. H. Lee, P. S. B. Dev and R. N. Mohapatra, Phys. Rev. D **88**, 093010 (2013).
[7] F. F. Deppisch, P. S. B. Dev and A. Pilaftsis, arXiv:1502.06541 [hep-ph].
[8] W.-Y. Keung and G. Senjanović, Phys. Rev. Lett. **50**, 1427 (1983).
[9] A. Ferrari *et al.*, Phys. Rev. D **62**, 013001 (2000); M. Nemevsek, F. Nesti, G. Senjanovic and Y. Zhang, Phys. Rev. D **83**, 115014 (2011); J. Chakrabortty, J. Gluza, R. Sevillano and R. Szafron, JHEP **1207**, 038 (2012); S. P. Das, F. F. Deppisch, O. Kittel and J. W. F. Valle, Phys. Rev. D **86**, 055006 (2012); J. A. Aguilar-Saavedra and F. R. Joaquim, Phys. Rev. D **86**, 073005 (2012).
[10] C.-Y. Chen, P. S. B. Dev and R. N. Mohapatra, Phys. Rev. D 88, 033014 (2013); P. S. B. Dev and R. N. Mohapatra, arXiv:1308.2151 [hep-ph].
[11] V. Khachatryan *et al.* [CMS Collaboration], Eur. Phys. J. C **74**, 3149 (2014).
[12] T. G. Rizzo, Phys. Rev. D **89**, 095022 (2014).
[13] W. Barletta *et al.*, Nucl. Instrum. Meth. A **764**, 352 (2014).
[14] T. Han, I. Lewis, R. Ruiz and Z. g. Si, Phys. Rev. D **87**, 035011 (2013) [Erratum-ibid. D **87**, 039906 (2013)].
[15] M. Fukugita and T. Yanagida, Phys. Lett. B **174**, 45 (1986).
[16] V. A. Kuzmin, V. A. Rubakov and M. E. Shaposhnikov, Phys. Lett. **B155**, 36 (1985).
[17] A. Pilaftsis, Phys. Rev. **D 56**, 543 (1997).
[18] A. Pilaftsis and T. E. J. Underwood, Nucl. Phys. **B 692**, 303 (2004).
[19] M. Flanz, E. A. Paschos and U. Sarkar, Phys. Lett. **B345**, 248 (1995); L. Covi, E. Roulet and F. Vissani, Phys. Lett. **B384**, 169 (1996).

[20] J.-M. Frere, T. Hambye and G. Vertongen, JHEP **0901**, 051 (2009).
[21] P. S. B. Dev, C. H. Lee and R. N. Mohapatra, Phys. Rev. D **90**, 095012 (2014).
[22] D. Chang, R. N. Mohapatra and M. Parida, Phys. Rev. Lett. **52**, 1072 (1984).
[23] Z. G. Berezhiani, Phys. Lett. B **129**, 99 (1983); D. Chang and R. N. Mohapatra, Phys. Rev. Lett. **58**, 1600 (1987); S. Rajpoot, Mod. Phys. Lett. A **2**, 307 (1987); A. Davidson and K. C. Wali, Phys. Rev. Lett. **59**, 393 (1987); K. S. Babu and R. N. Mohapatra, Phys. Rev. Lett. **62**, 1079 (1989) and Phys. Rev. D **41**, 1286 (1990).
[24] F. F. Deppisch and A. Pilaftsis, Phys. Rev. D **83**, 076007 (2011).
[25] P. S. B. Dev, P. Millington, A. Pilaftsis and D. Teresi, Nucl. Phys. B **886**, 569 (2014); arXiv:1409.8263 [hep-ph]; arXiv:1502.07987 [hep-ph].
[26] W. Buchmuller, P. Di Bari and M. Plumacher, Annals Phys. **315**, 305 (2005).
[27] P. A. R. Ade *et al.* [Planck Collaboration], arXiv:1502.01589 [astro-ph.CO].
[28] M. Dhuria, C. Hati, R. Rangarajan and U. Sarkar, arXiv:1502.01695 [hep-ph].
[29] S. Blanchet, Z. Chacko, S. S. Granor and R. N. Mohapatra, Phys. Rev. D **82**, 076008 (2010); S. Blanchet, P. S. B. Dev and R. N. Mohapatra, Phys. Rev. D **82**, 115025 (2010).

Symmetry Improved 2PI Effective Action and the Infrared Divergences of the Standard Model

Apostolos Pilaftsis, Daniele Teresi

Consortium for Fundamental Physics, School of Physics and Astronomy,
University of Manchester, Manchester M13 9PL, United Kingdom.

E-mail: `Apostolos.Pilaftsis@manchester.ac.uk`, `Daniele.Teresi@manchester.ac.uk`

Abstract. Resummations of infinite sets of higher-order perturbative contributions are often needed both in thermal field theory and at zero temperature. For instance, the behaviour of the Standard Model (SM) effective potential extrapolated to very high energies is known to be extremely sensitive to higher-order effects. The 2PI effective action provides a systematic approach to consistently perform such resummations. However, one of its major limitations was that its loopwise expansion introduces residual violations of possible global symmetries, thus giving rise to massive Goldstone bosons in the spontaneously broken phase of the theory. We review the recently developed symmetry-improved 2PI formalism for consistently encoding global symmetries in the 2PI approach, and discuss its satisfactory field-theoretical properties. We then apply the formalism to study the infrared divergences of the SM effective potential due to Goldstone bosons, which may affect the stability analyses of the SM. We present quantitative comparisons, for the scalar sector of the SM, with the approximate partial resummation procedure recently developed to address this problem, and show the quantitative discrepancy of the latter with the more complete 2PI approach, thus motivating further studies in this direction.

1. Introduction

In thermal field theory, finite-order perturbative expansions break down at high temperatures and one needs to resum higher-order contributions to deal with this problem. On the other hand, also at zero temperature there are situations where higher-order effects may potentially play an important role, even in a small-coupling regime. For instance, it has recently become well known that the behaviour of the Standard Model (SM) effective potential, extrapolated to very high energies, is extremely sensitive to the physics at the electroweak scale [1, 2, 3]. Thus, a formalism to incorporate and resum higher-order effects in a consistent manner is highly desirable for of both thermal and non-thermal applications.

A natural framework to address such problems is the formalism introduced by Cornwall, Jackiw and Tomboulis (CJT) [4]. Its simplest version is known as the Two-Particle-Irreducible (2PI) effective action. This is an effective action expressed not only in terms of the field, but also in terms of dressed propagators. When one considers a given truncation to the 2PI effective action, at any given order of the loopwise expansion, the 2PI effective action resums, automatically, an infinite set of higher-order diagrams induced by partially resummed propagators, without the danger of over-counting graphs.

So far, the 2PI formalism has been mainly used in thermal contexts, both at equilibrium [5, 6,

7, 8, 9] and out of equilibrium [10, 11, 12, 13]. One of the major limitations for its application to zero-temperature problems like, for example, the study of the SM effective potential, has been the well-known difficulties of the CJT formalism to incorporate symmetries in a satisfactory way. In particular, global symmetries are not exactly maintained at a given loop order of the 2PI expansion, since they get distorted by higher-order effects, and this results in violation of the Goldstone theorem [14, 15] by higher-order terms, giving rise to a massive Goldstone boson in the Spontaneous Symmetry Breaking (SSB) phase of the theory [16, 17, 18, 19]. In the past, several studies were presented in the literature, attempting to provide a satisfactory solution to this problem (see references in [20]).

We have recently developed a new *symmetry-improved CJT formalism* [20] that addresses this long-standing problem and allows one to consistently use the 2PI effective action to study theories with global symmetries. In the symmetry-improved 2PI formalism, the Ward Identities (WIs) associated with global symmetries provide additional constraints for the extremal solutions of the fields and propagators. This formalism has a number of satisfactory field-theoretical properties, and actually provides an improvement over the standard CJT formulation, in the sense that the behaviour expected for the full theory is recovered already at a low orders of approximation. The symmetry-improved formalism has been also used to study the chiral phase transition [21], and it has recently been extended to higher nPI effective actions [22], confirming its consistency also when one goes beyond the 2PI approach.

It has been recently pointed out [23] that the SM effective potential suffers from infrared (IR) divergences due to the Goldstone bosons of the electroweak gauge group, which appear at three-loop order. They appear at two loops for the derivative of the potential, i.e. for the minimization condition that fixes the value of the Higgs vacuum expectation value (VEV). This problem needs to be addressed for a two-fold reason. From the conceptual point of view, the effective potential should be well-defined for all values of the Higgs background field ϕ. On the other hand, from the quantitative point of view, the appearance of these IR divergences formally lowers the order of the involved three- and higher-loop contributions, thus breaking down perturbation theory and potentially giving a significant contribution to the threshold corrections to the VEV. Because of the extreme sensitivity of the effective potential extrapolated at high energies to the matching conditions at the electroweak scale, this could result on quantitative sizable effects on the stability analyses of the SM potential. These issues have been recently addressed by means of an approximate partial resummation procedure, developed in [24, 25]. In addition to confirming the disappearance of the IR divergences within our approach, we will show that the more complete 2PI analysis, based on our symmetry-improved formalism, gives quantitatively different results, at least for the scalar sector of the SM. This suggests that even for the full SM there might be quantitative discrepancies between the approximate resummation scheme of [24, 25] and the 2PI approach, thus motivating further studies in this direction [26].

The layout of the paper is as follows. In Sections 2 and 3 we review the standard and symmetry-improved 2PI formalisms, respectively. In Section 4 we consider the first non-trivial truncation of the effective action, known as the Hartree-Fock (HF) approximation, for a scalar $\mathbb{O}(2)$ model, and demonstrate some of the satisfactory field-theoretical properties of the formalism. In Section 5 we go beyond this approximation and show that the symmetry-improved approach describes correctly the thresholds of the Higgs and Goldstone self-energies, in agreement with unitarity requirements. Moreover, we demonstrate explicitly how the formalism implicitly resums sets of processes at arbitrarily high order. Finally, in Section 6 we use the symmetry-improved 2PI effective action to study the issue of IR divergences in a global $\mathbb{SU}(2) \times \mathbb{U}(1)$ scalar model. We draw our conclusions in Section 7.

2. CJT Effective Action

In this section we briefly review the CJT formalisms. For concreteness, we consider a $\mathbb{O}(N)$ scalar model and show that, as opposed to the 1PI formalism, loopwise truncations of the CJT generating effective action lead to violation of the Goldstone theorem through higher-order terms.

The model under consideration here is the $\mathbb{O}(N)$ scalar model described by the Lagrangian

$$\mathcal{L}[\phi] \;=\; \frac{1}{2}\,(\partial_\mu\phi^i)\,(\partial^\mu\phi^i) \,+\, \frac{m^2}{2}\,(\phi^i)^2 \,-\, \frac{\lambda}{4}\,(\phi^i)^2\,(\phi^j)^2\,, \tag{2.1}$$

where $\phi^i = \big(\phi^1\,,\,\phi^2\,,\,\cdots\,,\,\phi^N\big)$ represents the $\mathbb{O}(N)$ scalar multiplet and summation over the repeated indices $i, j = 1, 2, \ldots, N$ is implied. At zero temperature, $T = 0$, the model has a SSB phase for $m^2 > 0$, with breaking pattern: $\mathbb{O}(N) \to \mathbb{O}(N-1)$. As a consequence of the Goldstone theorem [14, 15], there are $N-1$ Goldstone bosons and one Higgs particle H. For the simple case $N = 2$, which we are considering here, the field components $\phi^{1,2}$ may be decomposed as

$$\phi^H \;\equiv\; \phi^1 \;=\; \langle\widehat{\phi}^1\rangle \,+\, H\,, \qquad \phi^G \;\equiv\; \phi^2 \;=\; G\,, \tag{2.2}$$

where we have denoted field operators with a caret. Here, G is the Goldstone field, which is massless at the minimum of the potential, whereas H is the Higgs boson, which is in general massive.

The CJT effective action is the generalization of the 1PI one, where in addition to the local source J_x, multi-local sources are introduced. In its simplest version, the 2PI formalism, we consider one local and one bi-local source, i.e. J_x and K_{xy}. Therefore, the connected generating functional $W[J, K]$ is given by

$$W[J,K] \;=\; -i\,\ln\int\mathcal{D}\phi^i\,\exp\left[i\bigg(S[\phi] \,+\, J^i_x\,\phi^i_x \,+\, \frac{1}{2}\,K^{ij}_{xy}\,\phi^i_x\,\phi^j_y\bigg)\right], \tag{2.3}$$

where $S[\phi] = \int_x \mathcal{L}[\phi]$ is the classical action. Here and in the following, repeated spacetime coordinates will denote integration with respect to these coordinates. The background fields ϕ^i_x and the connected propagator Δ^{ij}_{xy} are obtained by single and double functional differentiation of $W[J, K]$ with respect to the source J^i_x:

$$\frac{\delta W[J,K]}{\delta J^i_x} \;\equiv\; \phi^i_x\,, \qquad -i\,\frac{\delta W[J,K]}{\delta J^i_x\,\delta J^j_y} \;=\; \langle\widehat{\phi}^i_x\widehat{\phi}^j_y\rangle - \langle\widehat{\phi}^i_x\rangle\langle\widehat{\phi}^j_y\rangle \;\equiv\; i\Delta^{ij}_{xy}\,. \tag{2.4}$$

We also have

$$\frac{\delta W[J,K]}{\delta K^{ij}_{xy}} \;=\; \frac{1}{2}\left(i\Delta^{ij}_{xy} \,+\, \phi^i_x\,\phi^j_y\right). \tag{2.5}$$

The 2PI effective action $\Gamma[\phi, \Delta]$ is obtained as the double Legendre transform of $W[J, K]$ with respect to J and K:

$$\Gamma[\phi,\Delta] \;=\; W[J,K] \,-\, J^i_x\,\phi^i_x \,-\, \frac{1}{2}\,K^{ij}_{xy}\left(i\Delta^{ij}_{xy} \,+\, \phi^i_x\phi^j_y\right). \tag{2.6}$$

This has the explicit form [4]:

$$\Gamma[\phi,\Delta] \;=\; S[\phi] \,-\, \frac{i}{2}\,\mathrm{Tr}\left(\ln\Delta\right) \,+\, \frac{i}{2}\,\mathrm{Tr}\left(\Delta^{(0)\,-1}\,\Delta\right) \,-\, i\Gamma^{(\geq 2)}\,, \tag{2.7}$$

where $\Delta^{(0)\,-1,ij}_{xy} = \delta^2 S[\phi]/(\delta\phi^i_x\,\delta\phi^j_y)$ is the inverse tree-level propagator matrix and $\Gamma^{(\geq 2)}$ is the diagrammatic series of all two- and higher-loop 2PI vacuum diagrams in which the dressed

Figure 1. Infinite set of perturbation-theory diagrams implicitly resummed, in the 2PI formalism, by the self-energy on the LHS. Thick lines represent the full propagator, thin lines the tree-level one.

propagator matrix Δ is used, instead of the tree-level one. The equations of motions (EoMs) are obtained as

$$\frac{\delta\Gamma[\phi,\Delta]}{\delta\phi^i_x} = -J^i_x - K^{ij}_{xy}\,\phi^j_y\,, \qquad \frac{\delta\Gamma[\phi,\Delta]}{\delta\Delta^{ij}_{xy}} = -\frac{i}{2}\,K^{ij}_{xy}\,. \tag{2.8}$$

In the limit of vanishing external sources, the physical solution is obtained by extremizing the 2PI effective action $\Gamma[\phi,\Delta]$. In particular, the EoM for the propagator has the form

$$\Delta^{-1} = \Delta^{(0)\,-1}[\phi] + \Pi[\phi,\Delta]\,, \tag{2.9}$$

where $\Pi[\phi,\Delta]$ is the 1PI self-energy, in which the propagator lines are given by the dressed propagator matrix Δ. When a given truncation of $\Gamma[\phi,\Delta]$ is explicitly considered, this EoM implicitly resums an infinite set of perturbation-theory Feynman graphs, as shown in Fig. 1.

By considering the $\mathbb{O}(N)$ invariance of the 2PI effective action $\Gamma[\phi,\Delta]$ we may find a WI valid in the 2PI formalism at the extremal point of the 2PI effective action [20]:

$$v\int_x \frac{\delta^2\Gamma[\phi,\Delta]}{\delta G_x\delta G_y} + 2\,\frac{\delta^2\Gamma[\phi,\Delta]}{\delta G_y\,\delta\Delta^{GH}_{xz}}\left(\Delta^{HH}_{xz} - \Delta^{GG}_{xz}\right) = 0\,. \tag{2.10}$$

When $\mathbb{O}(N)$-symmetric truncations of the 2PI effective action are considered, the approximate solutions ϕ^i_x and Δ^{ij}_{xy} satisfy (2.10), which is the WI of the 2PI formalism, but not, in general, the Goldstone theorem, which is a WI of the 1PI formalism. Unlike the latter, the WI (2.10) does not protect the masslessness of the Goldstone boson. In fact, as it has explicitly been shown in [17, 18, 19], the Goldstone boson G is not massless in the HF approximation, manifesting itself as a pole at $k^2 \neq 0$ in the Goldstone-boson propagator $\Delta^{GG}(k)$. We stress here that the exact solutions obtained from the *complete* 2PI effective action satisfy, instead, the Goldstone theorem, since the complete 1PI and 2PI effective actions are fully equivalent at their extremal points [4].

3. Symmetry-improved 2PI Effective Action

In this section we present the symmetry-improved formalism for the 2PI effective action $\Gamma[\phi,\Delta]$, which respects the Goldstone theorem for any $\mathbb{O}(N)$-symmetric truncation of $\Gamma[\phi,\Delta]$.

From the discussion in the previous section, it is clear that when a given truncation of the 2PI effective action $\Gamma_{\rm tr}[\phi,\Delta]$ is considered, the solution obtained extremizing $\Gamma_{\rm tr}[\phi,\Delta]$ with respect to ϕ and Δ fails to satisfy the Goldstone theorem

$$v\int_x \Delta^{-1,GG}_{xy} = v\,\Delta^{-1,GG}(k)\Big|_{k=0} = 0\,. \tag{3.1}$$

The symmetry-improved EoMs are obtained by imposing the 1PI WI (3.1) directly as a constraint for the Goldstone propagator. This can be achieved, in a variational formulation, by performing a constrained extremization of the 2PI effective action, i.e. by finding the extremum of

$$\widetilde{\Gamma}[\phi,\Delta,\ell] = \Gamma_{\rm tr}[\phi,\Delta] - \ell^0_y\,\phi^H_x\,\Delta^{-1,GG}_{xy}\,, \tag{3.2}$$

Journal of Physics: Conference Series **631** (2015) 012008 doi:10.1088/1742-6596/631/1/012008

where ℓ_x^0 is a Lagrange-multiplier field. In [20] it is shown that this procedure results in replacing the 2PI EoM with the 1PI one

$$\frac{\delta\Gamma_{\rm tr}[\phi,\Delta]}{\delta\phi_x^H} = 0 \quad \longrightarrow \quad \frac{\delta\Gamma_{\rm 1PI}[\phi]}{\delta\phi_x^H} = 0\,. \tag{3.3}$$

The latter is explicitly given by the 1PI WI

$$\frac{\delta\Gamma_{\rm 1PI}[\phi]}{\delta\phi_x^H} = \phi_x^H\,\Delta_{xy}^{-1,GG}[\phi]\,. \tag{3.4}$$

Therefore, assuming homogeneity, the symmetry-improved EoMs to solve in the SSB phase are

$$\frac{\delta\Gamma_{\rm tr}[v,\Delta]}{\delta\Delta^{ij}(k)} = 0\,, \tag{3.5a}$$

$$v\,\Delta^{-1,GG}(k)\Big|_{k=0} = 0\,, \tag{3.5b}$$

namely the standard 2PI EoM for the propagators and the Goldstone theorem itself. In the symmetric phase of the theory we have $v = 0$ and only (3.5a) needs to be solved.

For homogeneous background field values ϕ away from the minimum v, i.e. for $\phi \neq v$, we may define a symmetry-improved effective potential $\widetilde{V}_{\rm eff}(\phi)$ by means of the 1PI WI (3.4), seen as a differential equation defining the improved potential. We thus have the definition [20]

$$-\,\frac{d\widetilde{V}_{\rm eff}(\phi)}{d\phi} \equiv \phi\,\Delta^{-1,GG}(k=0;\phi)\,, \tag{3.6}$$

where $\Delta^{-1,GG}(k = 0;\phi)$ is the Goldstone component of the solution of the EoM for the propagators (3.5a) for generic $\phi \neq v$. This definition is manifestly compatible with the EoMs at $\phi = v$ (3.5). The integral form of (3.6) is

$$\widetilde{V}_{\rm eff}(\phi) = -\int_0^\phi d\phi\,\phi\,\Delta^{-1,GG}(k=0;\phi) + \widetilde{V}_{\rm eff}(\phi=0)\,. \tag{3.7}$$

We show in [20] that the additive constant $\widetilde{V}_{\rm eff}(\phi = 0)$ can be chosen such that the formalism is thermodynamically consistent in the sense discussed first by Baym in [27]. However, for the purposes of this work, we do not need to determine it. We refer the interested reader to [20] for a more detailed discussion.

4. The Hartree-Fock Approximation

In this section we apply the symmetry-improved approach outlined in the previous section to the HF approximation of the 2PI effective action. We show that the predicted Goldstone boson is massless, as it should be, and that the phase-transition is second order already in this approximation. These predictions are in agreement with general field-theoretic properties that hold for the full effective action of the theory, thus showing that the symmetry-improved formalism gives actually *improved* predictions as compared to the standard truncated effective action.

The HF approximation consists in retaining only the *double-bubble* graphs (a)–(c) of Fig. 2. The unrenormalized 2PI effective action in this approximation is given by

$$\begin{aligned}\Gamma_{\rm HF}[v,\Delta^H,\Delta^G] = {} & \int_x \left(\frac{m^2}{2}v^2 - \frac{\lambda}{4}v^4\right) - \frac{i}{2}{\rm Tr}\left(\ln\Delta^H\right) - \frac{i}{2}{\rm Tr}\left(\ln\Delta^G\right) \\ & + \frac{i}{2}{\rm Tr}\left(\Delta^{(0)\,-1,H}\,\Delta^H\right) + \frac{i}{2}{\rm Tr}\left(\Delta^{(0)\,-1,G}\,\Delta^G\right) \\ & - i\,\frac{-6i\lambda}{8}\,i\Delta_{xx}^H\,i\Delta_{xx}^H - i\,\frac{-2i\lambda}{4}\,i\Delta_{xx}^H\,i\Delta_{xx}^G - i\,\frac{-6i\lambda}{8}\,i\Delta_{xx}^G\,i\Delta_{xx}^G\,,\end{aligned} \tag{4.1}$$

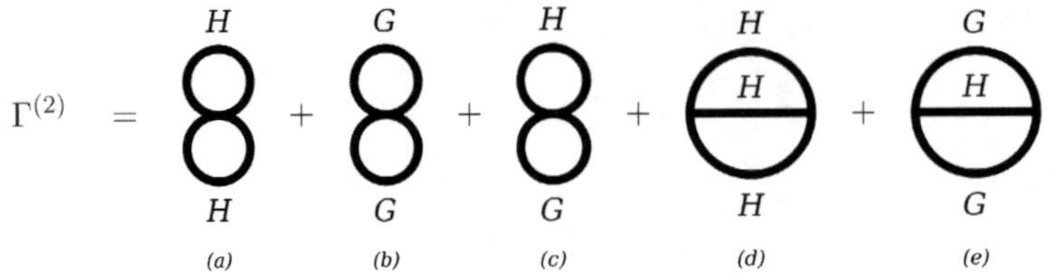

Figure 2. Unrenormalized two-loop contributions to $\Gamma[\phi, \Delta]$, with thick lines denoting dressed propagators. The HF approximation consists of the graphs (a), (b) and (c) and the sunset approximation includes the graphs (d) and (e).

In the above, we have simplified the notation for the diagonal propagators as $\Delta^H \equiv \Delta^{HH}$ and $\Delta^G \equiv \Delta^{GG}$.

4.1. Renormalization

It has been shown in [28, 29, 30] that the 2PI effective action is renormalizable with temperature-independent counterterms (CTs).

To renormalize the theory, we consider all parameters occurring in the CJT effective action (4.1) to be bare (denoted with the subscript B):

$$\phi^i_B = Z^{1/2}\,\phi^i\,, \qquad m^2_B = Z^{-1}\big(m^2 + \delta m^2\big)\,, \qquad \lambda_B = Z^{-2}\big(\lambda + \delta\lambda\big)\,, \qquad \Delta^{ij}_B = Z\,\Delta^{ij}\,. \tag{4.2}$$

In the 2PI formalism, there are two distinct 2-point operators appearing in the effective actions, corresponding to the two different derivatives

$$\frac{\delta^2\Gamma_{\rm tr}[\phi,\Delta]}{\delta\phi\,\delta\phi}\,, \qquad \frac{\delta\Gamma_{\rm tr}[\phi,\Delta]}{\delta\Delta}\,, \tag{4.3}$$

When a generic truncation of $\Gamma[\phi, \Delta]$ is considered, these two operators are independent (including their divergences), and need to be renormalized by two different CTs [30]. Analogously, in the truncated 2PI formalism we have the independent 4-point functions

$$\frac{\delta^4\Gamma_{\rm tr}[\phi,\Delta]}{\delta\phi\,\delta\phi\,\delta\phi\,\delta\phi}\,, \qquad \frac{\delta^3\Gamma_{\rm tr}[\phi,\Delta]}{\delta\phi\,\delta\phi\,\delta\Delta}\,, \qquad \frac{\delta^2\Gamma_{\rm tr}[\phi,\Delta]}{\delta\Delta\,\delta\Delta}\,. \tag{4.4}$$

Moreover, in the $\mathbb{O}(N)$ model these operators appear in two different $\mathbb{O}(N)$-invariant combinations, denoted by A and B. It is shown in [30] that only the standard perturbation-theory coupling-constant CT $\delta\lambda$ is required to renormalize the higher-loop graphs of the 2PI effective action, when going beyond the HF approximation. Hence, a finite number of CTs is needed to make the effective action finite, thus guaranteeing its renormalizability. With the

Journal of Physics: Conference Series **631** (2015) 012008 doi:10.1088/1742-6596/631/1/012008

above considerations, the effective action (4.1) reads:

$$
\begin{aligned}
\Gamma_{\rm HF}[v,\Delta^H,\Delta^G] \;=\; & \int_x \left(\frac{m^2+\delta m_0^2}{2}\,v^2 \;-\; \frac{\lambda+\delta\lambda_0}{4}\,v^4\right) \;-\; \frac{i}{2}\,{\rm Tr}\left(\ln\Delta^H\right) \;-\; \frac{i}{2}\,{\rm Tr}\left(\ln\Delta^G\right) \\
& -\; \frac{i}{2}\,{\rm Tr}\left\{\left[Z\,\partial^2 \;+\; \left(3\lambda+\delta\lambda_1^A+2\delta\lambda_1^B\right)v^2 - \left(m^2+\delta m_1^2\right)\right]\Delta^H\right\} \\
& -\; \frac{i}{2}\,{\rm Tr}\left\{\left[Z\,\partial^2 \;+\; \left(\lambda+\delta\lambda_1^A\right)v^2 - \left(m^2+\delta m_1^2\right)\right]\Delta^G\right\} \\
& -\; i\,\frac{-i\,(3\lambda+\delta\lambda_2^A+2\delta\lambda_2^B)}{4}\,i\Delta^H_{xx}\,i\Delta^H_{xx} \;-\; i\,\frac{-2i\,(\lambda+\delta\lambda_2^A)}{4}\,i\Delta^H_{xx}\,i\Delta^G_{xx} \\
& -\; i\,\frac{-i\,(3\lambda+\delta\lambda_2^A+2\delta\lambda_2^B)}{4}\,i\Delta^G_{xx}\,i\Delta^G_{xx}\,,
\end{aligned}
\tag{4.5}
$$

where we may set $Z = 1$ at this order of loop expansion.

By canceling separately divergences and subdivergences proportional to temperature-dependent terms, in [20] we find the following CTs in the HF approximation, in the $\overline{\rm MS}$ scheme of dimensional regularization $d = 4 - 2\epsilon$:

$$
\delta\lambda_2^A \;=\; \delta\lambda_1^A \;=\; \frac{2\lambda^2}{16\pi^2\epsilon}\,\frac{3-\dfrac{4\lambda}{16\pi^2\epsilon}}{1-\dfrac{6\lambda}{16\pi^2\epsilon}+\dfrac{8\lambda^2}{(16\pi^2\epsilon)^2}}\,, \tag{4.6a}
$$

$$
\delta\lambda_2^B \;=\; \delta\lambda_1^B \;=\; \frac{2\lambda^2}{16\pi^2\epsilon}\,\frac{1}{1-\dfrac{2\lambda}{16\pi^2\epsilon}}\,, \qquad \delta m_1^2 \;=\; \frac{4\lambda m^2}{16\pi^2\epsilon}\,\frac{1}{1-\dfrac{4\lambda}{16\pi^2\epsilon}}\,. \tag{4.6b}
$$

The above $T = 0$ CTs are sufficient to renormalize the EoMs for the propagators, also when thermal effects are considered.

4.2. Thermal Phase Transition

In the HF approximation, the self-energies are momentum independent. Therefore, we may parameterize the propagators as $\Delta^{H/G}(k) = (k^2 - M^2_{H/G} + i\varepsilon)^{-1}$, where the effective Higgs and Goldstone masses, M_H^2 and M_G^2, depend only on the temperature T.

In the symmetric phase of the theory the constraint (3.5b) is automatically satisfied, and the renormalized EoMs (3.5a) take on the form

$$
M_H^2 \;=\; -m^2 \;+\; 3\lambda\frac{M_H^2}{16\pi^2}\ln\frac{M_H^2}{2m^2} \;+\; \lambda\frac{M_G^2}{16\pi^2}\ln\frac{M_G^2}{2m^2} \;+\; 3\lambda\int_{\mathbf{k}}\frac{n(\omega_{\mathbf{k}}^H)}{\omega_{\mathbf{k}}^H} \;+\; \lambda\int_{\mathbf{k}}\frac{n(\omega_{\mathbf{k}}^G)}{\omega_{\mathbf{k}}^G}\,, \tag{4.7a}
$$

$$
M_G^2 \;=\; -m^2 \;+\; \lambda\frac{M_H^2}{16\pi^2}\ln\frac{M_H^2}{2m^2} \;+\; 3\lambda\frac{M_G^2}{16\pi^2}\ln\frac{M_G^2}{2m^2} \;+\; \lambda\int_{\mathbf{k}}\frac{n(\omega_{\mathbf{k}}^H)}{\omega_{\mathbf{k}}^H} \;+\; 3\lambda\int_{\mathbf{k}}\frac{n(\omega_{\mathbf{k}}^G)}{\omega_{\mathbf{k}}^G}\,, \tag{4.7b}
$$

where $\int_{\mathbf{k}} \equiv \int d^3\mathbf{k}/(2\pi)^3$, $\omega_{\mathbf{k}} = \sqrt{\mathbf{k}^2+M^2}$ is the on-shell energy of the (quasi)particle, and $n(\omega) = (e^{\omega/T}-1)^{-1}$ is the Bose–Einstein distribution function. In the symmetric phase we obtain, as expected, a single solution with $M_G^2 = M_H^2$.

In the HF approximation the constraint (3.5b) reads

$$
v\,M_G^2 \;=\; 0\,. \tag{4.8}
$$

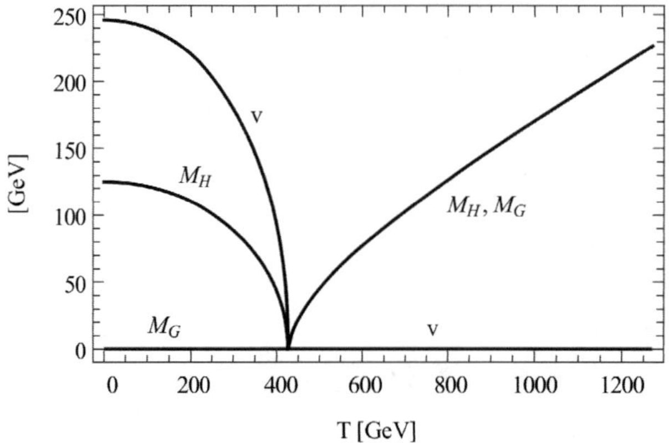

Figure 3. The values of M_H^2, M_G^2 and the VEV v, as functions of T, predicted in the HF approximation of the symmetry-improved 2PI formalism.

In the SSB phase of the theory this implies $M_G^2 = 0$, yielding the EoMs [20]

$$M_H^2 = 3\lambda v^2 - m^2 + 3\lambda \frac{M_H^2}{16\pi^2} \ln \frac{M_H^2}{2m^2} + 3\lambda \int_{\mathbf{k}} \frac{n(\omega_{\mathbf{k}}^H)}{\omega_{\mathbf{k}}^H} + \lambda \int_{\mathbf{k}} \frac{n(\omega_{\mathbf{k}}^G)}{\omega_{\mathbf{k}}^G}, \tag{4.9a}$$

$$0 = \lambda v^2 - m^2 + \lambda \frac{M_H^2}{16\pi^2} \ln \frac{M_H^2}{2m^2} + \lambda \int_{\mathbf{k}} \frac{n(\omega_{\mathbf{k}}^H)}{\omega_{\mathbf{k}}^H} + 3\lambda \int_{\mathbf{k}} \frac{n(\omega_{\mathbf{k}}^G)}{\omega_{\mathbf{k}}^G}, \tag{4.9b}$$

We point out that have chosen the $\overline{\text{MS}}$ mass scale μ such that the tree-level relations $M_H^2 = 2m^2$, $v^2 = m^2/\lambda$ are satisfied at $T = 0$. The mass-gap equations (4.9) are solved analytically to be

$$M_H^2 = 2m^2 - \frac{8\,\lambda T^2}{12}, \tag{4.10a}$$

$$M_G^2 = 0, \tag{4.10b}$$

$$v^2 = \frac{m^2}{\lambda} - \frac{M_H^2}{16\pi^2} \ln \frac{M_H^2}{2m^2} - \int_{\mathbf{k}} \frac{n(\omega_{\mathbf{k}}^H)}{\omega_{\mathbf{k}}^H} - \frac{3\,T^2}{12}. \tag{4.10c}$$

In Fig. 3, we exhibit the dependence of the squared thermal masses, M_H^2 and M_G^2, and the thermally-corrected VEV v, as functions of the temperature T. The parameters of the model are chosen such that $M_H = 125$ GeV and $v = 246$ GeV at $T = 0$. We observe that we predict a second-order phase transition at $T = T_c = \sqrt{3}v(T = 0)$, already in the HF approximation, in agreement with theoretical expectations to all orders. This is in sharp contrast with the incorrect first-order phase-transition predicted in the HF approximation by the previous approaches [31, 32, 33].

4.3. Symmetry-improved Effective Potential

Let us now calculate the effective potential at high temperatures in the HF approximation of the symmetry-improved formalism. Extending the renormalized EoMs (4.9) from $v \to \phi$, we obtain

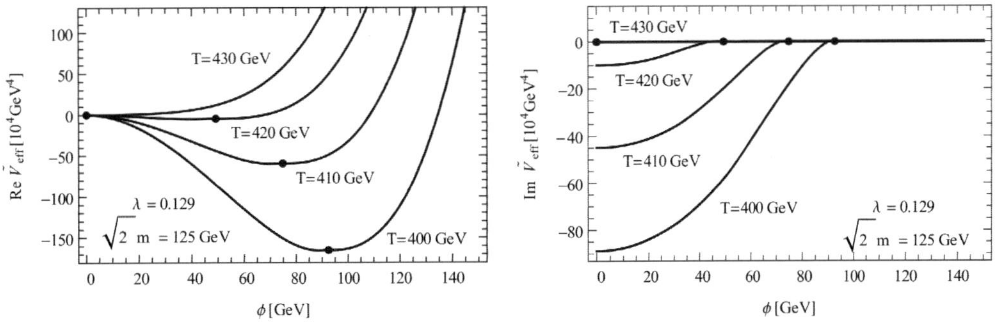

Figure 4. Symmetry-improved HF effective potential in the high-temperature approximation. The large dots denote the minimum solutions $\phi = v$, obtained in Section 4.2.

$$
\begin{aligned}
M_H^2(\phi) \; &= \; 3\lambda\phi^2 \, - \, m^2 \, + \, 3\lambda \, \frac{M_H^2(\phi)}{16\pi^2} \ln\left(\frac{M_H^2(\phi)}{2m^2}\right) \, + \, \lambda \, \frac{M_G^2(\phi)}{16\pi^2} \ln\left(\frac{M_G^2(\phi)}{2m^2}\right) \\
&\quad + \, 3\lambda \int_{\mathbf{k}} \frac{n[\omega_{\mathbf{k}}^H(\phi)]}{\omega_{\mathbf{k}}^H(\phi)} \, + \, \lambda \int_{\mathbf{k}} \frac{n[\omega_{\mathbf{k}}^G(\phi)]}{\omega_{\mathbf{k}}^G(\phi)} \; , \qquad (4.11a)\\
M_G^2(\phi) \; &= \; \lambda\phi^2 \, - \, m^2 \, + \, \lambda \, \frac{M_H^2(\phi)}{16\pi^2} \ln\left(\frac{M_H^2(\phi)}{2m^2}\right) \, + \, 3\lambda \, \frac{M_G^2(\phi)}{16\pi^2} \ln\left(\frac{M_G^2(\phi)}{2m^2}\right) \\
&\quad + \, \lambda \int_{\mathbf{k}} \frac{n[\omega_{\mathbf{k}}^H(\phi)]}{\omega_{\mathbf{k}}^H(\phi)} \, + \, 3\lambda \int_{\mathbf{k}} \frac{n[\omega_{\mathbf{k}}^G(\phi)]}{\omega_{\mathbf{k}}^G(\phi)} \; , \qquad (4.11b)
\end{aligned}
$$

$$
\frac{1}{\phi} \, \frac{d\widetilde{V}_{\text{eff}}(\phi)}{d\phi} \; = \; M_G^2(\phi) \; . \qquad (4.11c)
$$

The first two equations are the EoMs for the propagators, for general $\phi \neq v$. The latter equation (4.11c) results from the definition of the symmetry-improved effective potential (3.6), i.e. the 1PI WI for $\phi \neq v$. Figure 4 presents our numerical estimates for the high-temperature symmetry-improved effective potential, as functions of ϕ, for different temperatures T [20]. Again, a second-order phase transition is described, already in the HF approximation. The fact that $\widetilde{V}_{\text{eff}}(\phi)$ acquires an imaginary part when $\phi < v$ signals the instability of the homogeneous vacuum in this region [34].

5. Threshold Properties

In this section, we study the threshold properties of the Higgs and Goldston particles, by including the contributions from the sunset diagrams (d) and (e) in Fig. 2. In particular, we will show that the resummed Higgs- and Goldstone-boson propagators predicted within our symmetry-improved CJT formalism exhibit the correct threshold properties arising from on-shell Higgs and Goldstone particles in the loop. Therefore, we explicitly demonstrate that the symmetry-improved approach is consistent with the optical theorem and unitarity.

One of the common approaches in the literature, when studying the issue of symmetries in the CJT formalism, is to define an additional 2-point function as

$$
\Delta_{\text{ext}}^{-1} \; \equiv \; \frac{\delta^2 \Gamma_{\text{tr}}[\phi, \Delta(\phi)]}{\delta\phi \, \delta\phi} \; , \qquad (5.1)
$$

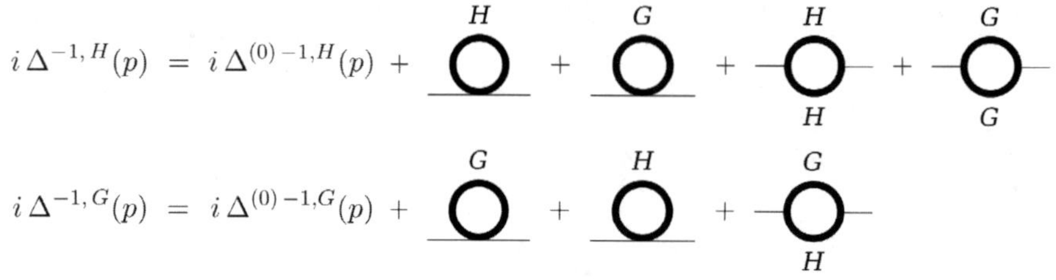

Figure 5. Higgs and Goldstone EoMs in the sunset approximation.

where $\Delta(\phi)$ is the solution of the standard 2PI EoM for the propagator. This 2-point function is sometimes referred to as the *external propagator*. It can be shown that this function satisfies the Goldstone theorem, since the second derivative (in the Goldstone direction) of any symmetric ϕ-only functional does so [20]. Thus, it is often claimed that this function should be considered as the true approximation to the propagator, in the CJT formalism, rather than Δ. However, the problem of this approach is that what appears in the diagrammatic series of $\Gamma[\phi, \Delta]$ is the propagator Δ, not the external propagator $\Delta_{\rm ext}$. The former does not satisfy the Goldstone theorem, in the standard CJT formalism, and therefore the thresholds of the particles are described incorrectly, since the Goldstone bosons propagating within quantum loops are massive. In this sense, this approach is not capable of describing the Goldstone bosons as consistently massless quantum-mechanically, i.e. within loops. From the above discussion, it is clear that studying the threshold properties described by the symmetry-improved 2PI effective action is an important step in establishing the consistency of the formalism and show further its advantages as compared to previous approaches.

As diagrammatically represented in Fig. 5, the symmetry-improved 2PI EoMs derived in the sunset approximation are given by

$$\begin{aligned}\Delta^{-1,H}(p) \;&=\; p^2 \;-\; (3\lambda + \delta\lambda_1^A + 2\delta\lambda_1^B)\,v^2 \;+\; m^2 \;+\; \delta m_1^2 \;-\; (3\lambda + \delta\lambda_2^A + 2\delta\lambda_2^B)\,\mathcal{T}_H \\ &\quad -\; (\lambda + \delta\lambda_2^A)\,\mathcal{T}_G \;+\; \frac{1}{i}\frac{(-6i\lambda v)^2}{2}\,\mathcal{I}_{HH}(p) \;+\; \frac{1}{i}\frac{(-2i\lambda v)^2}{2}\,\mathcal{I}_{GG}(p)\,, \qquad (5.2a)\end{aligned}$$

$$\begin{aligned}\Delta^{-1,G}(p) \;&=\; p^2 \;-\; (\lambda + \delta\lambda_1^A)\,v^2 \;+\; m^2 \;+\; \delta m_1^2 \\ &\quad -\; (\lambda + \delta\lambda_2^A)\,\mathcal{T}_H \;-\; (3\lambda + \delta\lambda_2^A + 2\delta\lambda_2^B)\,\mathcal{T}_G \;+\; \frac{1}{i}(-2i\lambda v)^2\,\mathcal{I}_{HG}(p)\,, \qquad (5.2b)\end{aligned}$$

$$v\,\Delta^{-1,G}(0) \;=\; 0\,. \qquad (5.2c)$$

Here, we have abbreviated the loop integrals as follows:

$$\mathcal{T}_a \;=\; \overline{\mu}^{2\epsilon}\int_k i\Delta^a(k)\,, \qquad\qquad \mathcal{I}_{ab}(p) \;=\; \overline{\mu}^{2\epsilon}\int_k i\Delta^a(k-p)\,i\Delta^b(k)\,, \qquad (5.3)$$

where $a, b = H, G$, $\ln\overline{\mu}^2 = \ln\mu^2 + \gamma - \ln(4\pi)$ and μ is the $\overline{\rm MS}$ renormalization scale. In (5.2) we have introduced the shorthand notation $\int_k \equiv \int d^4k/(2\pi)^4$. Details of the renormalization and the numerical approach to solve these self-consistent nonlinear equations are given in [20]. Here, we only outline the physical content of their solution.

It is useful to define the effective energy-dependent squared masses $\widehat{M}^2_{H/G}(s)$ as

$$\Delta^{-1,H/G}(s) \;=\; s \;-\; \widehat{M}^2_{H/G}(s)\,, \qquad (5.4)$$

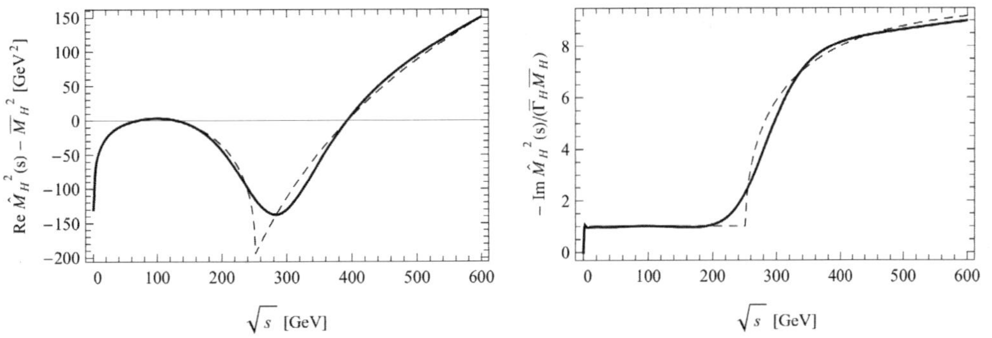

Figure 6. Numerical solutions for $\mathrm{Re}[\widehat{M}_H^2(s)] - \overline{M}_H^2$ (left frame) and $\mathrm{Im}[\widehat{M}_H^2(s)]/(\overline{\Gamma}_H\overline{M}_H)$ (right frame), where $\overline{M}_H$ and $\overline{\Gamma}_H$ are the Higgs-boson pole mass and width, respectively. The dashed lines are the one-loop results in the 1PI formalism.

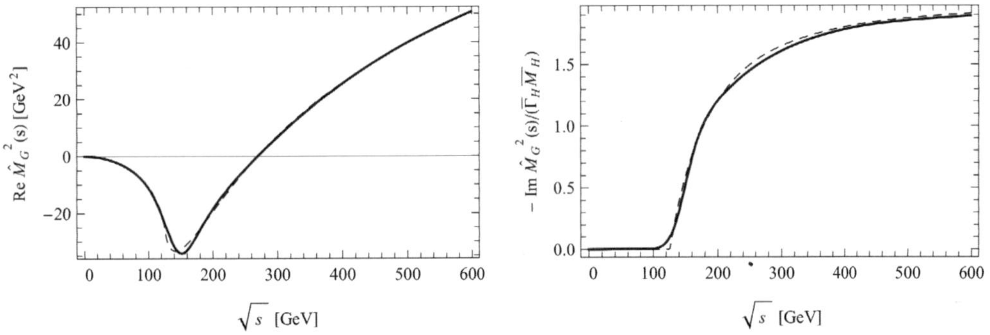

Figure 7. Numerical solutions for $\mathrm{Re}[\widehat{M}_G^2(s)]$ (left frame) and $\mathrm{Im}[\widehat{M}_G^2(s)]/(\overline{\Gamma}_H\overline{M}_H)$ (right frame).

where $s \equiv p^2$ is the Lorentz-invariant energy-squared parameter. In Fig. 6 and Fig. 7, we plot the dispersive (real) and absorptive (imaginary) Higgs- and Goldstone-boson mass squares, as functions of s. We see that there is a non-vanishing absorptive part $\mathrm{Im}\,\widehat{M}_H^2(s)$ that results from the on-shell decay of the Higgs particle into two Goldstone bosons, i.e. $H \to GG$. The threshold for this process is at $s = 0$, thus demonstrating that *the Goldstone bosons in the loop are consistently treated as massless within our symmetry-improved 2PI formalism.* This is, again, in sharp contrast with previous approaches in the literature, e.g. with the results found in [31] for the absorptive part of the external propagator, where the Goldstone boson exhibits a non-zero mass in the loop, as discussed above.

The dashed lines in Figs. 6 and 7 show the predictions obtained at one-loop level in the 1PI formalism. We point out that the kinematic opening of the thresholds is very sharp, as opposed to the smooth thresholds predicted by our symmetry-improved CJT formalism. As illustrated in Fig. 8 this is due to the fact that *the 2PI formalism automatically resums infinitely-many higher-order processes, without the need of explicitly considering them*, including also processes that take place below the 1PI threshold.

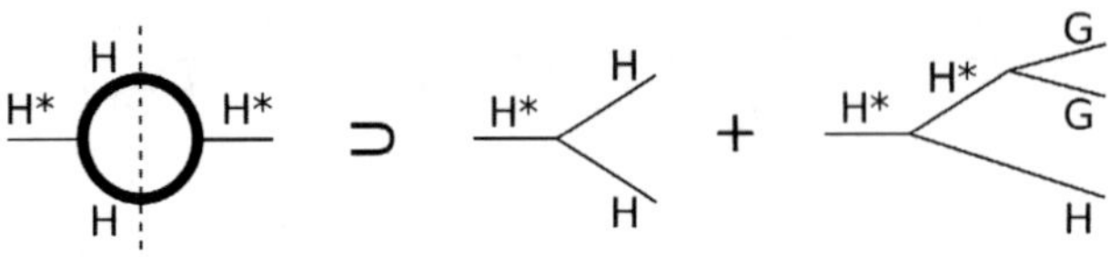

Figure 8. Some of the on-shell processes whose description is included in the absorptive part of the 1-loop CJT self-energy on the LHS.

6. The Infrared Divergences of the Standard-Model Effective Potential

In this section we describe the issue, recently pointed out in [23], of the IR divergences of the SM effective potential due to the Goldstone bosons of the electroweak group. Then, by considering the scalar sector of the SM, we show how the symmetry-improved 2PI effective action can be used to study this problem, and compare the results obtained in this framework with the existing approach [24, 25] in the literature. More technical details of the 2PI analysis will be given in a forthcoming publication [26].

The effective potential V_{eff} of the SM, calculated in perturbation theory, suffers from IR divergences due to the appearance of Goldstone bosons in ring diagrams, as shown in Figure 9. At 3 loops the divergence is logarithmic, but it becomes more and more severe with increasing loop order. Moreover, these divergences are more severe when one considers the derivative of the effective potential $\frac{dV_{\text{eff}}}{d\phi}$. For the latter, IR divergences start from 2-loop order (see Figure 9).

In perturbation theory these divergences appear when the tree-level Goldstone propagators become massless, i.e. at the tree-level minimum of the potential. This is instead finite, together with its derivatives, at the dressed minimum $\phi = v$. Nevertheless, this problem needs to be addressed, for the following reasons:

(i) The effective potential $V_{\text{eff}}(\phi)$ should be well-defined for all values of ϕ, not only at its minimum $\phi = v$. Among other things, the off-shell $\phi \neq v$ effective potential governs the dynamics of the background field in inflationary scenarios.

(ii) At the dressed minimum $\phi = v$, the dressed masses of the Goldstone bosons vanish, and this implies that the tree-level mass m_G^2 is formally of the same order as the 1-loop Goldstone self-energy $\Pi_G^{(1)}$. Thus, since starting from 3-loop order the IR divergences of $\frac{dV_{\text{eff}}}{d\phi}$ are of the form $1/(m_G^2)^n$, $n \geq 1$ (see Figure 9) all *these higher-loop contributions to* $\frac{dV_{\text{eff}}}{d\phi}$ *are formally at 2-loop order*. This means that perturbation theory breaks down and these diagrams can potentially have a significant impact on 2-loop results for $\frac{dV_{\text{eff}}}{d\phi}$, i.e. on the state-of-the-art threshold corrections to the VEV v. In view of the extreme sensitivity of the SM effective potential, extrapolated at very high energies, to the matching conditions at the electroweak scale [1, 2, 3], these issues can potentially affect the stability analyses of the SM.

Moreover, a seemingly unrelated problem is that the tree-level mass of the Goldstone boson can be negative at the dressed minimum $\phi = v$, thus generating an unphysical imaginary part for the SM effective potential at its minimum, which does not correspond to a true instability of the homogeneous vacuum. This fact should be contrasted with the discussion in Section 3, where we have shown that the symmetry-improved effective potential acquires an imaginary part only in the concave region corresponding to a physical instability. This suggests that a resummation of higher-loop diagrams is needed to address this conceptual issue as well.

Finally, we point out that the location of the IR divergence depends on the value of the gauge-fixing parameter ξ, but the divergence is nonetheless present in any renormalizable R_ξ gauge at the value of the field ϕ for which $m_G^2(\phi; \xi) = 0$.

Journal of Physics: Conference Series **631** (2015) 012008 doi:10.1088/1742-6596/631/1/012008

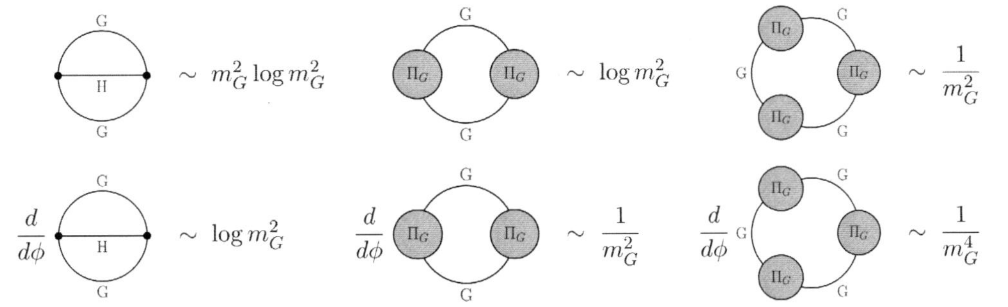

Figure 9. IR behaviour of Goldstone-boson ring diagrams contributing to the effective potential and its derivative. The IR divergences start at three loops for $V_{\rm eff}(\phi)$, already at two loops for its derivative $dV_{\rm eff}/d\phi$.

6.1. Approximate Partial Resummation

Let us outline the approximate partial resummation procedure presented in [24, 25] to address these IR issues. To facilitate an objective comparison with the 2PI approach discussed in this work, we will limit ourselves to the scalar sector of the SM, i.e. we will consider a global $\mathbb{SU}(2) \times \mathbb{U}(1)$ model with only the scalar Higgs doublet. Although this does not allow to draw quantitative conclusion for the complete SM case, this simplified model is sufficient to study this issue qualitatively and to compare the results obtained in the 2PI approach with the ones in [24, 25].

The partial resummation procedure consists in considering ring diagrams, as in Figure 9, with insertions of 1-loop Goldstone self-energies $\Pi_G(k)$. One approximates these 1-loop self-energies with their zero-momentum value $\Pi_G(0)$. With this important simplification, one can resum these diagrams by replacing the 1-loop Coleman-Weinberg contribution of the Goldstone bosons with

$$V^{(1)}_{\rm eff,\mathit{G}} = \frac{3\,m_G^4}{4\,(16\pi^2)}\left[\log\left(\frac{m_G^2}{\mu^2}\right)-\frac{3}{2}\right] \longrightarrow \frac{3\,(m_G^2+\Pi_G(0))^2}{4\,(16\pi^2)}\left[\log\left(\frac{m_G^2+\Pi_G(0)}{\mu^2}\right)-\frac{3}{2}\right]. \tag{6.1}$$

However, it can be shown that the derivative of this resummed term is still divergent. This problem can be solved by limiting $\Pi_G(0)$ to contain only the terms not proportional to m_G^2, i.e. by replacing

$$\Pi_G(0) \longrightarrow \Pi_g \equiv \Pi_G(0) - \frac{3\lambda}{(16\pi^2)}\,m_G^2\Big(\log(m_G^2/\mu^2)-1\Big)\,. \tag{6.2}$$

Notice that the subtracted term does not correspond to the contribution of a given diagram, but it is contained in the contribution of both the Goldstone tadpole integral and the Higgs-Goldstone sunset diagram. Finally, one needs to subtract from $V_{\rm eff}$ the diagrams that would be double-counted otherwise. In conclusion, one adds to the effective potential the term [24, 25]:

$$V^{(resum)}_{\rm eff,\mathit{G}} \equiv \frac{3\,(m_G^2+\Pi_g)^2}{4\,(16\pi^2)}\left[\log\left(\frac{m_G^2+\Pi_g}{\mu^2}\right)-\frac{3}{2}\right] - V^{(d.c.)}_{\rm eff,\mathit{G}}\,, \tag{6.3}$$

where $V^{(d.c.)}_{\rm eff,\mathit{G}}$ is the contribution of the double-counted diagrams, as discussed above.

As we show in Figure 10 for the scalar sector of the SM, this procedure effectively resums the IR divergences present in $dV_{\rm eff}/d\phi$ at two- and three-loop orders. We include only the 3-loop

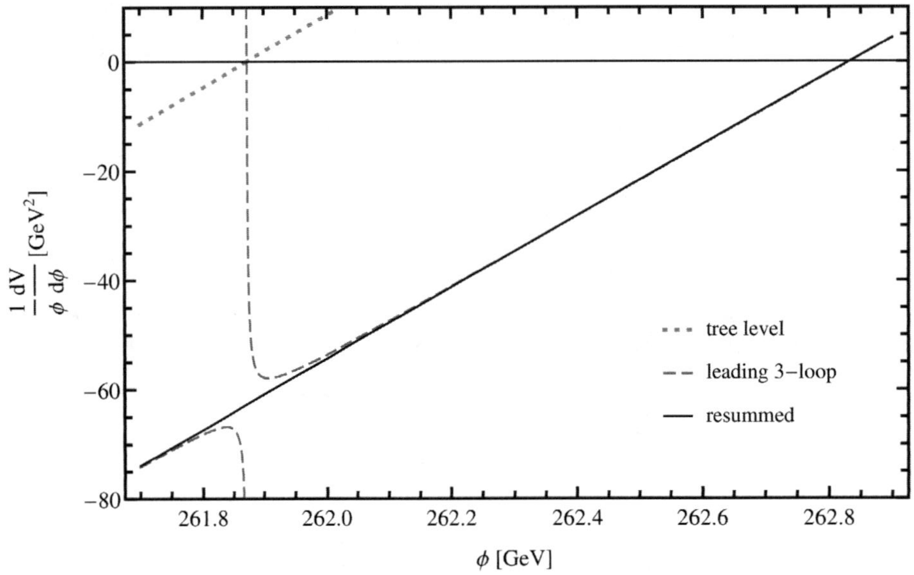

Figure 10. IR divergence of the derivative of the effective potential, as calculated in perturbation theory at 3-loop order, considering only the scalar sector of the SM. The gray dotted line is the tree-level contribution, the red dashed line the leading 3-loop one, whereas the black continuous line is the result of the approximate partial resummation procedure developed in [24, 25].

contributions coming from the Goldstone ring diagrams in Figure 9, since they are proportional to $1/m_G^2$ and so responsible for the IR divergence. Moreover, as we discussed above, their order gets formally lowered at the dressed minimum.

6.2. 2PI Approach to the Resummation of IR Divergences

We now study these issues by means of the symmetry-improved 2PI formalism. This provides a more complete resummation, as compared to the approach outlined above, for several reasons. First, it is a first-principle approach, and no ad-hoc subtraction is needed. Second, as we are going to show below, the 2PI approach takes into account more topologies and does not necessitate to neglect the momentum dependence of the self-energy insertions that are resummed. Moreover, as we showed in the previous section, the threshold properties are correctly described within the symmetry-improved formalism. In particular, the Goldstone propagator in the formalism is massless at the dressed minimum of the potential and nevertheless the effective potential will be shown to be free of IR pathologies.

We consider the global $\mathbb{SU}(2) \times \mathbb{U}(1)$ scalar model. By including, in the EoMs, the 1-loop (HF + sunset) CJT diagrams shown in the first line of Figure 11 we actually resum, automatically, a large class of diagrams, as depicted in Figure 12. Moreover, in order to be able to compare our results with the perturbative 2-loop calculation, we include the 2-loop CJT diagrams in the second line of Figure 11. However, to simplify the treatment, we approximate the propagators appearing in these diagrams as the tree-level ones $\Delta(\phi) \approx \Delta^{(0)}(\phi)$. In this way, we take into account the full contribution of 2-loop topologies and, in addition, resum diagrams as the ones shown in Figure 13. Therefore, a much larger class of diagrams is included, as compared to the

Journal of Physics: Conference Series **631** (2015) 012008 doi:10.1088/1742-6596/631/1/012008

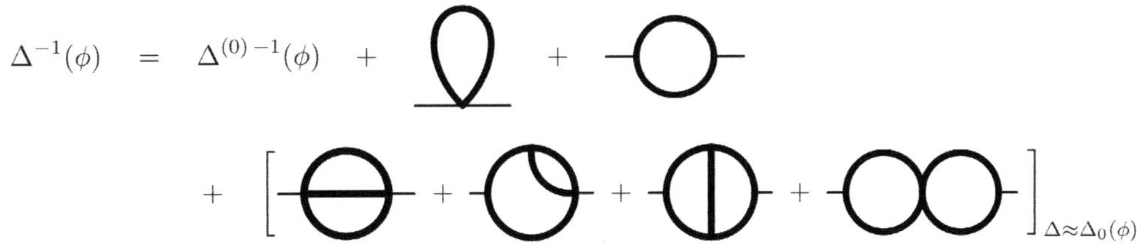

Figure 11. Diagrammatic representation of the approximation scheme for the EoMs in Section 6.

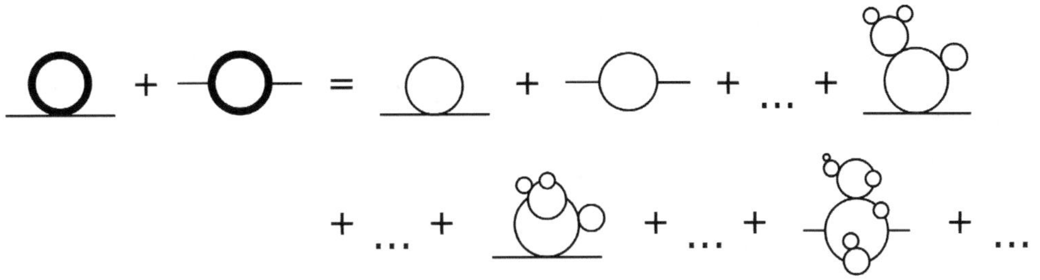

Figure 12. Some of the topologies implicitly resummed by the 1-loop 2PI self-energies in the first line of Figure 11.

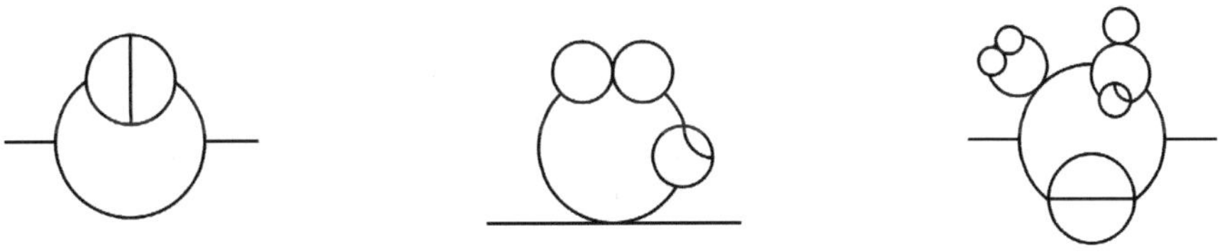

Figure 13. Some examples of the topologies resummed by including the 2-loop 2PI self-energies in the second line of Figure 11. Notice that the propagators belonging to 2-loop 2PI topologies do not get dressed, because of the approximation $\Delta(\phi) \approx \Delta^{(0)}(\phi)$ used, for these, in Figure 11.

method outlined in Section 6.1, and the momentum dependence of the resummed insertions is also retained.

In the symmetry-improved 2PI approach, the IR divergences are absent by construction: IR divergences can be present only when two or more Goldstone propagators carry the same momentum, as in the ring diagrams shown in Figure 9. In other words, the IR pathologies originate from chains of Goldstone lines with self-energy insertions between them. However, such topologies are necessarily *2-particle-reducible* and thus do not appear in the diagrammatic series of $\Gamma[\phi, \Delta]$. Therefore, the resummation of IR divergences is achieved automatically by the construction of the 2PI effective action.

The EoMs can be easily obtained by generalizing (5.2) and including the 2-loop self-energies, in which $\Delta(\phi) \approx \Delta^{(0)}(\phi)$. Their explicit form is given in Appendix A. The numerical solution, in the vicinity of the dressed minimum $\phi = v$, is plotted in Figure 14. The black dots represent the numerical solution obtained in our approach. It is apparent that the results from the partial

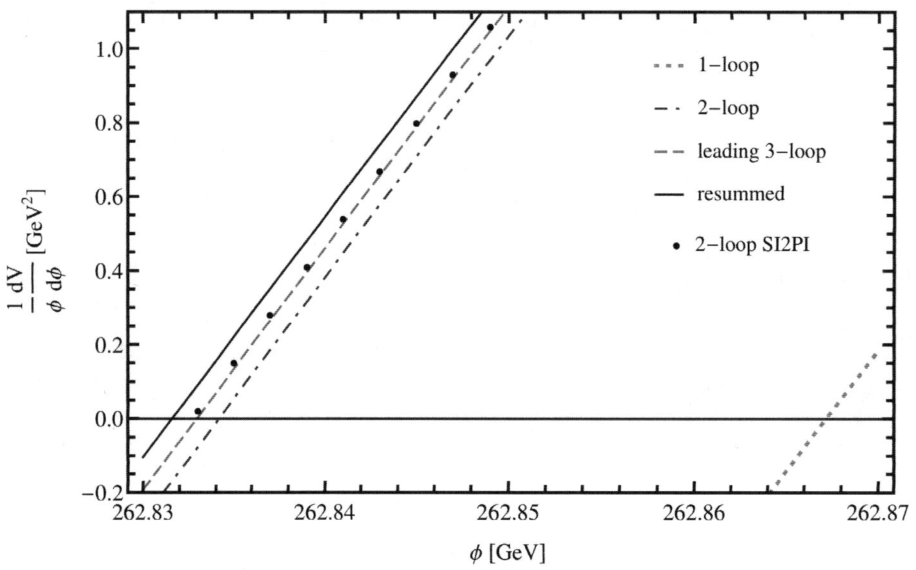

Figure 14. The derivative of the effective potential, near its dressed minimum. We show results from 1-loop (green dotted line), 2-loop (blue dash-dotted line) and leading 3-loop (red dashed line) perturbation theory. The black continuous line is the result of the approximate partial resummation procedure discussed in Section 6.1. The black dots are the results obtained from the symmetry-improved 2PI effective action.

resummation outlined in Section 6.1 do not reproduce the ones obtained in the more complete 2PI resummation. We have checked explicitly that, expanding the EoMs at 2-loop order, we reproduce numerically the results coming from 2-loop perturbation theory. Notice also that the 3-loop result is larger than what one would naively expect (roughly about $\lambda/16\pi^2$ times the 2-loop one), because of the breakdown of perturbation theory, as discussed at the beginning of this section. In Figure 14 the symmetry-improved 2PI solution appears to be similar in size to the leading 3-loop result, potentially indicating that the effect of the resummation is small near the dressed minimum. However, this is a coincidence and depends on the class of topologies of graphs considered here. A more detailed study will be given in [26].

7. Conclusions

The 2PI effective action provides a powerful theoretical tool to consistently resum infinite series of perturbation-theory diagrams of different topologies. However, its loopwise expansion introduces residual violations of possible global symmetries by higher-order terms. In the case of global symmetries, this leads to the appearance of massive Goldstone bosons in the spontaneously broken phase of the theory. In this work we have reviewed the symmetry-improved CJT formalism, developed in [20] for consistently encoding global symmetries in loopwise expansions of the 2PI effective action.

We have demonstrated, in a simple $\mathbb{O}(2)$ scalar model, the key field-theoretical properties of the formalism. In detail, we have shown that the Goldstone bosons are described as massless, also within quantum loops, thus providing a consistent description of the threshold properties of the Higgs and Goldstone particles. Moreover, the thermal phase transition is at second order already in the HF approximation. Thus, the behaviour expected for the full theory is recovered already in the first non-trivial approximation, contrary to other approaches in the literature.

Thanks to the satisfactory field-theoretical properties of the formalism, we have applied the symmetry-improved 2PI effective action to the study of the IR divergences of the SM effective potential due to the electroweak Goldstone bosons. By limiting ourselves to the scalar sector of the electroweak group, we have confirmed that these IR divergences are indeed an artifact of perturbation theory, and are absent in the symmetry-improved 2PI formalism, as it should be. We have compared quantitatively our results with the predictions given by the existing approach in the literature, consisting in an approximate partial resummation of Goldstone-boson ring diagrams. We have shown that the latter, whilst correctly resumming qualitatively the IR divergences, does not reproduce quantitatively, at least in the simplified scalar model considered here, the results of the more complete 2PI resummation, near the dressed minimum of the effective potential.

In view of the extreme sensitivity of the SM effective potential, extrapolated to very high energies, to the matching conditions at the electroweak scale, it is important to assess these issues in a more realistic way, going beyond the scalar sector of the SM. A detailed study of this matter, including the quantitatively most important contributions, will be given in a forthcoming publication.

Acknowledgments
The work of A.P. is supported by the Lancaster-Manchester-Sheffield Consortium for Fundamental Physics under STFC grant ST/L000520/1. The work of D.T. is supported by a fellowship of the EPS Faculty of the University of Manchester.

Appendix A. Equations of Motion for the $\mathbb{SU}(2) \times \mathbb{U}(1)$ Scalar Model

In this appendix we give the renormalized EoMs used in Section 6.2 to study the IR divergences of the effective potential. Since the scalar $\mathbb{SU}(2)_L \times \mathbb{U}(1)$ model automatically possesses the custodial symmetry $\mathbb{SU}(2)_L \times \mathbb{SU}(2)_R \sim \mathbb{O}(4)$, the HF and sunset contributions to the EoMs are easily inferred from the $\mathbb{O}(2)$ case (see [20]). We find, in Euclidean space:

$$\begin{aligned}\Delta^{-1,\,H}(p;\phi) \ &= \ p^2 \,+\, 3\lambda\phi^2 \,-\, m^2 \,+\, 3\lambda\,\mathcal{T}_H^{\rm ren} \,+\, 3\lambda\,\mathcal{T}_G^{\rm ren} \,-\, 18\lambda^2\phi^2\,\mathcal{I}_{HH}^{\rm ren}(p) \,-\, 6\lambda^2\phi^2\,\mathcal{I}_{GG}^{\rm ren}(p) \\ &\quad +\, \Pi_H^{\rm 2PI,(2)}(p;\phi)\,, \qquad\qquad \text{(A.1a)}\\ \Delta^{-1,\,G}(p;\phi) \ &= \ p^2 \,+\, \lambda\phi^2 \,-\, m^2 \,+\, \lambda\,\mathcal{T}_H^{\rm ren} \,+\, 5\lambda\,\mathcal{T}_G^{\rm ren} \,-\, 4\lambda^2\phi^2\,\mathcal{I}_{GH}^{\rm ren}(p) \,+\, \Pi_G^{\rm 2PI,(2)}(p;\phi)\,, \qquad \text{(A.1b)}\\ \frac{1}{\phi}\,\frac{d\widetilde{V}_{\rm eff}(\phi)}{d\phi} \ &= \ \Delta^{-1,\,G}(0;\phi)\,, \qquad\qquad \text{(A.1c)}\end{aligned}$$

where the $\overline{\rm MS}$ renormalized sunset and tadpole integrals are given by [20]

$$\begin{aligned}\mathcal{I}_{ab}^{\rm ren}(p) \ &\equiv \ \int_k \bigg(\Delta^a(k-p)\,\Delta^b(k) \ -\ \frac{1}{(k^2+\mu^2)^2}\bigg)\,, \qquad \text{(A.2)}\\ \mathcal{T}_a^{\rm ren} \ &\equiv \ \int_k \bigg[\Delta^a(k) \ -\ \frac{1}{k^2+\mu^2} \ -\ \frac{1}{(k^2+\mu^2)^2}\bigg(\mu^2 - M_a^2 + \frac{\nu_a\,\lambda^2\phi^2}{16\pi^2}B(p;\mu^2,\mu^2)\bigg)\bigg] \\ &\quad -\ \frac{\mu^2}{16\pi^2} \ +\ \frac{\nu_a\,\lambda^2\phi^2}{(16\pi^2)^2}\,\frac{\eta}{2}\,, \qquad \text{(A.3)}\end{aligned}$$

where $\nu_H = 24$, $\nu_G = 4$ and we have introduced

$$\frac{1}{16\pi^2} B(p;\mu^2,\mu^2) \equiv \mathcal{I}_{00}^{\rm ren}(p)\,, \qquad \text{with} \quad \Delta^0(k) \equiv \frac{1}{k^2+\mu^2}\,, \tag{A.4}$$

$$\eta = 1 - \frac{4i}{\sqrt{3}}\left(\mathrm{Li}_2\frac{1-i\sqrt{3}}{2} - \frac{\pi^2}{36}\right) \simeq -1.34391\,. \tag{A.5}$$

Here, we seize the opportunity to eliminate an error in the value of η reported in [20]. The terms in the second line of (A.3) are the $\overline{\rm MS}$ finite part of the CT integrals used to renormalize $\mathcal{T}_a$ by the method discussed in [20]. Their subtraction is needed to guarantee that the renormalization scheme used here matches the standard $\overline{\rm MS}$ one in perturbation theory. Finally, the 2-loop 2PI self-energies $\Pi_{H,G}^{\rm 2PI,(2)}(p;\phi)$ are calculated in perturbation theory by standard techniques [35, 36]. We approximate them by their zero-momentum value $\Pi_{H,G}^{\rm 2PI,(2)}(\phi)$, since the error introduced in this way is expected to be negligible for the purposes of this work. Adopting the compact notation used in [35], in terms of the functions defined there we find

$$\begin{aligned}(16\pi^2)^2\,\Pi_H^{\rm 2PI,(2)}(\phi) &= 54\lambda^3\phi^2\,\overline{\ln}^2 H + 36\lambda^3\phi^2\,\overline{\ln}\,H\,\overline{\ln}\,G + 30\lambda^3\phi^2\,\overline{\ln}^2 G \\ &- 6\lambda^2\,I(H,H,H) - 6\lambda^2\,I(H,G,G) \\ &- 216\lambda^3\phi^2\,I(H',H,H) - 72\lambda^3\phi^2\,I(H',G,G) - 24\lambda^3\phi^2\,I(G',G,H) \\ &- 648\lambda^4\phi^4\,I(H',H',H) - 144\lambda^4\phi^4\,I(H',G',G) - 24\lambda^4\phi^4\,I(G',G',H)\,,\end{aligned} \tag{A.6a}$$

$$\begin{aligned}(16\pi^2)^2\,\Pi_G^{\rm 2PI,(2)}(\phi) &= 8\lambda^3\phi^2\,B(G,H)^2 - 24\lambda^2 I(H,H,H) + 22\lambda^2 I(G,H,H) \\ &- 16\lambda^2 I(G,G,H) + 6\lambda^2 I(G,G,G)\,,\end{aligned} \tag{A.6b}$$

where all functions have to be evaluated at zero momentum.

References

[1] Bezrukov F, Kalmykov M Y, Kniehl B A and Shaposhnikov M 2012 *JHEP* **1210** 140 (*Preprint* arXiv:1205.2893 [hep-ph])
[2] Degrassi G, Di Vita S, Elias-Miro J, Espinosa J R, Giudice G F, Isidori G and Strumia A 2012 *JHEP* **1208** 098 (*Preprint* arXiv:1205.6497 [hep-ph])
[3] Buttazzo D, Degrassi G, Giardino P P, Giudice G F, Sala F, Salvio A and Strumia A 2013 *JHEP* **1312** 089 (*Preprint* arXiv:1307.3536 [hep-ph])
[4] Cornwall J M, Jackiw R and Tomboulis E 1974 *Phys. Rev.* D **10** 2428–2445
[5] Blaizot J P, Iancu E and Rebhan A 2001 *Phys. Rev.* D **63** 065003 (*Preprint* hep-ph/0005003)
[6] Berges J, Borsanyi S, Reinosa U and Serreau J 2005 *Phys. Rev.* D **71** 105004 (*Preprint* hep-ph/0409123)
[7] Markó G, Reinosa U and Szép Z 2013 *Phys. Rev.* D **87** 105001 (*Preprint* arXiv:1303.0230 [hep-ph])
[8] Alford M G, Mallavarapu S K, Schmitt A and Stetina S, 2014 *Phys. Rev.* D **89** 085005 (*Preprint* arXiv:1310.5953 [hep-ph])
[9] Fejos G 2014 *Phys. Rev.* D **90** 116001 (*Preprint* arXiv:1410.1337 [hep-ph])
[10] Berges J 2005 *AIP Conf. Proc.* **739** 3 (*Preprint* (hep-ph/0409233)
[11] Millington P and Pilaftsis A 2013 *Phys. Rev.* D **88** 085009 (*Preprint* arXiv:1211.3152 [hep-ph])
[12] Meistrenko A, Wesp C, van Hees H and Greiner C 2014 *J. Phys. Conf. Ser.* **503** 012003 (*Preprint* arXiv:1311.7444 [hep-ph])
[13] Dev P S B, Millington P, Pilaftsis A and Teresi D 2015 *Nucl. Phys.* B **891** 128 (*Preprint* arXiv:1410.6434 [hep-ph])
[14] Goldstone J 1961 *Nuovo Cim.* **19** 154–164
[15] Goldstone J, Salam A and Weinberg S 1962 *Phys. Rev.* **127** 965–970
[16] Baym G and Grinstein G 1977 *Phys. Rev.* D **15** 2897–2912
[17] Amelino-Camelia G 1997 *Phys. Lett.* B **407** 268–274
[18] Petropoulos N 1999 *J. Phys.* G **25** 2225–2241

[19] Lenaghan J T and Rischke D H 2000 *J. Phys.* G **26** 431–450
[20] Pilaftsis A and Teresi D 2013 *Nucl. Phys.* B **874** 594 (*Preprint* arXiv:1305.3221 [hep-ph])
[21] Mao H 2014 *Nucl. Phys.* A **925** 185 (*Preprint* arXiv:1305.4329 [hep-ph])
[22] Brown M J and Whittingham I B 2015 *Preprint* arXiv:1502.03640 [hep-th]
[23] Martin S P 2014 *Phys. Rev.* D **89** 013003 (*Preprint* arXiv:1310.7553 [hep-ph])
[24] Martin S P 2014 *Phys. Rev.* D **90** 016013 (*Preprint* arXiv:1406.2355 [hep-ph])
[25] Elias-Miro J, Espinosa J R and Konstandin T 2014 *JHEP* **1408** 034 (*Preprint* arXiv:1406.2652 [hep-ph])
[26] Pilaftsis A and Teresi D, in preparation
[27] Baym G 1962 *Phys. Rev.* **127** 1391
[28] van Hees H and Knoll J 2002 *Phys. Rev.* D **65** 025010
[29] Blaizot J P, Iancu E and Reinosa U 2004 *Nucl. Phys.* A **736** 149–200
[30] Berges J, Borsanyi S, Reinosa U and Serreau J 2005 *Annals Phys.* **320** 344–398
[31] van Hees H and Knoll J 2002 *Phys. Rev.* D **66** 025028
[32] Ivanov Y, Riek F and Knoll J 2005 *Phys. Rev.* D **71** 105016
[33] Ivanov Y, Riek F, van Hees H and Knoll J 2005 *Phys. Rev.* D **72** 036008
[34] Weinberg E J and Wu A-Q 1987 *Phys. Rev.* D **36** 2474
[35] Martin S P 2004 *Phys. Rev.* D **70** 016005 (*Preprint* hep-ph/0312092)
[36] Martin S P 2003 *Phys. Rev.* D **68** 075002 (*Preprint* hep-ph/0307101)

Journal of Physics: Conference Series **631** (2015) 012009 doi:10.1088/1742-6596/631/1/012009

On supergravity theories, after $\sim$ 40 years

Jean-Pierre Derendinger

Albert Einstein Center for Fundamental Physics, Institute for Theoretical Physics,
Bern University, Sidlerstrasse 5, 3012 Bern, Switzerland

E-mail: derendinger@itp.unibe.ch

Abstract. An introduction to and a partial review of supergravity theories is given, insisting on concepts and on some important technical aspects. Topics covered include elements of global supersymmetry, a derivation of the simplest $\mathcal{N} = 1$ supergravity theory, a discussion of $\mathcal{N} = 1$ matter–supergravity couplings, of the scalar sector and of some simple models. Space-time is four-dimensional.

1. Foreword

Almost forty years ago, in 1974 and soon after, most of the important properties of quantum field theories with linear supersymmetry[1] were displayed in a brilliant series of papers. These fundamental results include what is now called the Wess-Zumino model [1], super-Yang-Mills theory [2][2], exceptional renormalization properties [4], spontaneous supersymmetry breaking [5, 6], the current structure [7] and the development of superspace and superfield techniques [8, 9].

Two attractive complementary aspects of supersymmetry were recognized. Firstly, it is an extension compatible with quantum mechanics of the Poincaré algebra, the symmetry of relativistic field theory [10, 11, 4]. Secondly, it can be implemented in the Standard Model of particle interactions and be experimentally tested [12], leading to numerous models and analysis and, now, to LHC (preliminary and future) results.

Almost forty years ago, in 1976, the first supergravity theory, the gauge theory of supersymmetry, was invented by Ferrara, Freedman and van Nieuwenhuizen [13] and by Deser and Zumino [14], opening decades of developments which installed supergravity at the meeting point of two independent approaches to the unification program.

[1] In four dimensions.
[2] And later on [3].

On the low energy side, the success of the Standard Model emphasizes the enigma of the large hierarchical ratio M_P/M_W. Global supersymmetry helps with his capacity to forbid destabilizing quantum corrections [4, 15]. But it does not spontaneously break, as observations obviously require, and this is where supergravity helps, by proposing a source for supersymmetry breaking effects at low energies [16, 17, 18] and also in proposing a scheme to radiatively induce a small scale (M_W) from supersymmetry breaking at a much higher scale [19, 20, 21, 22, 23].

On the high energy scale, at present, the only concrete proposal for a coherent description of all particle interactions including gravitation involves models based on strings with scale close or related to the Planck scale. Coherence and stability of these models require some amount of supersymmetry. Supergravity appears then as the natural tool to describe the light sector of superstring theories, at energy scales where string excitations decouple and where supersymmetry breaking should be able to generate the so-called soft breaking terms needed in the supersymmetric Standard Model [16, 17, 18].

More pragmatically, the possible relevance of supersymmetry to particle physics, in relation with the weak interaction scale, is concrete and testable at LHC (and some other) experiments. This is maybe a strong enough motivation.

Over the years, supergravity theories have of course found many other more theoretical roles, in gauge/gravity dualities, to understand the structure of gravitational scattering amplitudes, in the description of superstring solutions and compactifications, in studies of black holes, solutions of gravitational theories with tensor fields, attractors, in sigma models living on particular geometries ...

This contribution tries to provide an introduction to (four-dimensional) supergravity and a detailed discussion of some aspects of simple theories and of their applications. It begins with basic aspects of global supersymmetry (section 2), with a detailed discussion of the role of auxiliary fields. The simplest $\mathcal{N} = 1$ supergravity and its Anti-de Sitter deformation are then derived (section 3). After a brief discussion of theories with extended supersymmetry, general properties of the $\mathcal{N} = 1$ supergravity–matter couplings are described (section 4), with particular attention to the scalar potential and the gravitino sector. The examples of simple no-scale or dilaton supergravities are the subject of section 5.

2. Elements of global supersymmetry

Relativistic quantum field theories and strong or electroweak interactions are invariant under global transformations of the Poincaré group, *i.e.* under Lorentz (proper and orthochronous) transformations and translations. In other words, there are Lorentz generators $M^{\mu\nu} = -M^{\nu\mu}$ and translation generators P^μ acting on coordinates and fields and leaving the dynamical equations unchanged. Translation generators are universal, $P_\mu = -i\partial_\mu$. Lorentz generators act on coordinates according to

$$\delta x^\mu = \frac{i}{2}\omega_{\rho\sigma} M^{\rho\sigma} x^\mu = -\omega^{\mu\nu} x_\nu, \qquad\qquad M^{\rho\sigma} = -i(x^\rho\partial^\sigma - x^\sigma\partial^\rho). \tag{2.1}$$

They act on fields with generators in representations depending on the spins of the fields. For a set of fields $\Phi(x)$,

$$\delta\Phi(x) = \frac{i}{2}\omega_{\rho\sigma}\Sigma^{\rho\sigma}\Phi(x) - \delta x^\mu \partial_\mu \Phi(x) = \frac{i}{2}\omega_{\rho\sigma} M^{\rho\sigma}\Phi(x) \tag{2.2}$$

and the information on spins is in the choice of linear operators $\Sigma^{\mu\nu}$. These variations represent the Poincaré Lie algebra

$$\begin{aligned} [M^{\mu\nu}, M^{\rho\sigma}] &= -i\left(\eta^{\mu\rho}M^{\nu\sigma} + \eta^{\nu\sigma}M^{\mu\rho} - \eta^{\mu\sigma}M^{\nu\rho} - \eta^{\nu\rho}M^{\mu\sigma}\right), \\ [P^\mu, M^{\nu\rho}] &= i\left(\eta^{\mu\nu}P^\rho - \eta^{\mu\rho}P^\nu\right), \\ [P^\mu, P^\nu] &= 0. \end{aligned} \tag{2.3}$$

2.1. The Poincaré superalgebra

Global supersymmetry is an extension compatible with quantum field theory of Poincaré symmetry adding spin 1/2 generators, the *supercharges*, with an algebra of anticommutators. The relevant algebra includes commutators

$$[M^{\mu\nu}, Q^i_\alpha] = -\frac{i}{4}([\sigma^\mu, \overline{\sigma}^\nu]Q^i)_\alpha, \qquad\qquad [P_\mu, Q^i_\alpha] = 0. \tag{2.4}$$

The index $i = 1, \ldots, \mathcal{N}$ labels the supercharges and the first relation indicates that the Q^i_α's have spin 1/2. The superalgebra is completed by the anticommutators[3]

$$\{Q^i_\alpha, \overline{Q}^j_{\dot\alpha}\} = -2i\,\delta^{ij}(\sigma^\mu)_{\alpha\dot\alpha}\partial_\mu = 2\,\delta^{ij}(\sigma^\mu)_{\alpha\dot\alpha}P_\mu. \tag{2.5}$$

The representations of $\mathcal{N}$–extended supersymmetry share several important properties:

- Firstly, the particle states in a supermultiplet have helicities extending from a maximal λ to $\lambda - \mathcal{N}/2$ (and the opposite helicities if $\lambda - \mathcal{N}/2 \neq -\lambda$). Since quantum field theory admits helicities $|\lambda| \leq 1$, it admits at most $\mathcal{N} = 4$ global supersymmetries. For supergravity, $|\lambda| \leq 2$ and then $\mathcal{N} \leq 8$.
- Secondly, all particle states have the same mass (which can be zero) and the numbers of fermionic and bosonic particle states are equal, $n_B = n_F$. Similarly, for representations in terms of fields (unconstrained by a dynamical field equation) bosonic and fermionic component fields come in equal numbers. The number of helicity zero states is then always even.

[3] We disregard the possibility of central changes for $\mathcal{N} > 1$.

- Thirdly, for a supermultiplet of fields, the allowed interactions are strongly constrained and related.
- Finally, the divergences of supersymmetric quantum field theories are much softer than in a generic case: quadratic divergences in the scalar sector are absent and, in particular, scalar and Yukawa (scalar–fermion) interactions are not renormalized.

Of course, these properties immediately indicate that supersymmetry cannot be an exact symmetry of Nature. Realistic theories with supersymmetry must include a mechanism of supersymmetry breaking, a condition which turns out to be a challenge to model builders.

It should also be mentioned that Poincaré supersymmetry is a limiting case of the supersymmetric extension of anti-de Sitter space-time symmetry $SO(2,3)$ (a contraction of the AdS superalgebra). But it is not compatible with the de Sitter $SO(1,4)$ algebra. This algebraic fact has important dynamical implications for supergravity theories, which are our main subject of interest here.

2.2. The simplest supermultiplet and auxiliary fields

The simplest representation of $\mathcal{N}=1$ supersymmetry is the chiral multiplet, which describes particle states with helicities $\pm 1/2, 0, 0$. It then includes a Weyl (or Majorana) spinor ψ and a complex scalar z. But it also introduces the concept of auxiliary fields of central importance in field representations of the supersymmetry algebra. The relevance of auxiliary fields, with algebraic, non-propagating field equations, is related to the requirement $n_B = n_F$, and the fact that counting degrees of freedom on-shell and off-shell gives different numbers.

To illustrate the use of auxiliary fields, consider the sum of the free massless Klein-Gordon and Dirac lagrangians for a complex scalar and a Weyl (or Majorana) spinor:

$$\mathcal{L}_0 = (\partial_\mu \overline{z})(\partial^\mu z) + \frac{i}{2}\psi\sigma^\mu\partial_\mu\overline{\psi} - \frac{i}{2}\partial_\mu\psi\sigma^\mu\overline{\psi}. \tag{2.6}$$

It is a supersymmetric theory: under the variations

$$\delta z = \sqrt{2}\,\epsilon\psi, \qquad \delta\psi_\alpha = -\sqrt{2}i\,\partial_\mu z(\sigma^\mu\overline{\epsilon})_\alpha, \tag{2.7}$$

the lagrangian changes by a derivative and the action is then invariant. There is however trouble in the algebra. Firstly,

$$[\delta_1, \delta_2]z = -2i(\epsilon_2\sigma^\mu\overline{\epsilon}_1 - \epsilon_1\sigma^\mu\overline{\epsilon}_2)\partial_\mu z, \tag{2.8}$$

which is a translation $\delta x^\mu = 2(\epsilon_2\sigma^\mu\overline{\epsilon}_1 - \epsilon_1\sigma^\mu\overline{\epsilon}_2) = 2\,\overline{\epsilon}_2\gamma^\mu\epsilon_1$ as required by the supersymmetry algebra (2.5). But

$$\begin{aligned}[\delta_1, \delta_2]\psi_\alpha &= -2i(\epsilon_2\sigma^\mu\overline{\epsilon}_1 - \epsilon_1\sigma^\mu\overline{\epsilon}_2)\partial_\mu\psi_\alpha \\ &\quad +2i(\partial_\mu\psi\sigma^\mu\overline{\epsilon}_2)\epsilon_{1\alpha} - 2i(\partial_\mu\psi\sigma^\mu\overline{\epsilon}_1)\epsilon_{2\alpha}.\end{aligned} \tag{2.9}$$

The first term is as expected, but the second only vanishes if the spinor solves the Dirac equation $\partial_\mu \psi \sigma^\mu = 0$ implied by the lagrangian. Hence, variations (2.7) only close the supersymmetry algebra for on-shell fields. This is certainly a problem if one wishes to construct more complicated, interacting lagrangians with nonlinear field equations. One must then simultaneously invent the lagrangian and the corresponding supersymmetry variations (which become nonlinear as well).

However, modify the variation of the spinor:

$$\delta\psi_\alpha = -\sqrt{2} f \epsilon_\alpha - \sqrt{2} i \partial_\mu z (\sigma^\mu \overline{\epsilon})_\alpha \tag{2.10}$$

where f is a complex scalar field. The new term adds

$$-\sqrt{2}\, \delta_1 f \, \epsilon_{2\alpha} + \sqrt{2}\, \delta_2 f \, \epsilon_{1\alpha}$$

to $[\delta_1, \delta_2]\psi_\alpha$ and choosing then

$$\delta f = -\sqrt{2} i \, (\partial_\mu \psi \sigma^\mu \overline{\epsilon}) \tag{2.11}$$

leads to the expected algebra

$$\begin{aligned} [\delta_1, \delta_2]\psi_\alpha &= -2i(\epsilon_2 \sigma^\mu \overline{\epsilon}_1 - \epsilon_1 \sigma^\mu \overline{\epsilon}_2)\, \partial_\mu \psi_\alpha, \\ [\delta_1, \delta_2] f &= -2i(\epsilon_2 \sigma^\mu \overline{\epsilon}_1 - \epsilon_1 \sigma^\mu \overline{\epsilon}_2)\, \partial_\mu f \end{aligned} \tag{2.12}$$

for all three fields z, ψ and f. The modification of the spinor variation also adds a new contribution to the variation of $\mathcal{L}_0$:

$$-f \delta \overline{f} - \overline{f} \delta f + \frac{i}{\sqrt{2}} \partial_\mu [f \, \epsilon \sigma^\mu \overline{\psi} - \overline{f} \, \psi \sigma^\mu \overline{\epsilon}].$$

This in turn imposes to modify the lagrangian to

$$\mathcal{L} = (\partial_\mu \overline{z})(\partial^\mu z) + \frac{i}{2} \psi \sigma^\mu \partial_\mu \overline{\psi} - \frac{i}{2} \partial_\mu \psi \sigma^\mu \overline{\psi} + \overline{f} f, \tag{2.13}$$

with field equations

$$\Box z = 0, \qquad \partial_\mu \psi \sigma^\mu = 0, \qquad f = 0 \tag{2.14}$$

and the scalar f is *auxiliary*: it describes $n_B = 2$ off-shell fields and $n_B = 0$ on-shell states. Since ψ includes $n_F = 4$ off-shell and $n_F = 2$ on-shell degrees of freedom while for z $n_B = 2$ on-shell and off-shell, the equality $n_B = n_F$ is verified on-shell *and* off-shell in the supermultiplet (z, ψ, f). On shell, $\delta f = 0$ and one returns to the original expressions (2.6) and (2.7).

In general, the equality of the number of bosonic and fermionic physical (on-shell) degrees of freedom is imposed by the supersymmetry algebra while a mismatch in the numbers of bosonic and fermionic off-shell fields suggests that adding auxiliary fields is necessary to obtain an off-shell representation, if possible at all.

The canonical dimensions (in energy unit) of z, ψ and f are respectively 1, 3/2 and 2 and the parameter ϵ has dimension $-1/2$. Hence f must transform in a field with dimension 5/2, which is then a derivative of ψ. This suggests a method to construct supersymmetric lagrangians: starting with an off-shell supermultiplet like (z, ψ, f), combine supermultiplets into a new supermultiplet (*tensor calculus*) and take its component with the highest dimension as a lagrangian term: it necessarily transforms as a derivative. The simplest example is

$$Z = z^2, \qquad \Psi = 2z\psi, \qquad F = 2fz + \psi\psi. \tag{2.15}$$

One easily verifies that (Z, Ψ, F) and (z, ψ, f) have identical transformations. Hence, since the variation of F is a derivative,

$$\begin{aligned} \mathcal{L}_m &= (\partial_\mu \overline{z})(\partial^\mu z) + \tfrac{i}{2}\psi\sigma^\mu\partial_\mu\overline{\psi} - \tfrac{i}{2}\partial_\mu\psi\sigma^\mu\overline{\psi} + \overline{f}f \\ &\quad -m[fz + \tfrac{1}{2}\psi\psi] - m[\overline{f}\overline{z} + \tfrac{1}{2}\overline{\psi\psi}] \end{aligned} \tag{2.16}$$

is supersymmetric. Eliminating f with its field equation $f = m\overline{z}$ leads to

$$\mathcal{L}_m = (\partial_\mu \overline{z})(\partial^\mu z) - m^2 z\overline{z} + \frac{i}{2}\psi\sigma^\mu\partial_\mu\overline{\psi} - \frac{i}{2}\partial_\mu\psi\sigma^\mu\overline{\psi} - \frac{m}{2}[\psi\psi + \overline{\psi\psi}], \tag{2.17}$$

with a common mass m for z and ψ, and to the supersymmetry variations

$$\delta z = \sqrt{2}\,\epsilon\psi, \qquad \delta\psi_\alpha = -\sqrt{2}\,m\overline{z}\,\epsilon_\alpha - \sqrt{2}i\,\partial_\mu z(\sigma^\mu\overline{\epsilon})_\alpha. \tag{2.18}$$

Contrary to variation (2.10), $\delta\psi_\alpha$ now depends on the lagrangian parameter m. Note that with Dirac equation $i\partial_\mu\psi\sigma^\mu = -m\overline{\psi}$, the on-shell variation of the auxiliary field is

$$\delta f = -\sqrt{2}i\,(\partial_\mu\psi\sigma^\mu\overline{\epsilon}) = m\sqrt{2}\,\overline{\epsilon\psi} = m\,\delta\overline{z}, \tag{2.19}$$

as indicated by $f = m\overline{z}$.

Similarly, a renormalizable interaction would follow from the observation that

$$\begin{aligned} Z &= \tfrac{m}{2}z^2 + \tfrac{\lambda}{3}z^3, \equiv W(z), \qquad & \Psi_\alpha &= (mz + \lambda z^2)\,\psi_\alpha, \\ F &= (mz + \lambda z^2)f + \tfrac{1}{2}(m + 2\lambda z)\,\psi\psi \end{aligned} \tag{2.20}$$

is a chiral multiplet. The holomorphic function W of z only is the *superpotential.* The supersymmetric lagrangian

$$\begin{aligned} \mathcal{L}_{m,\lambda} &= (\partial_\mu \overline{z})(\partial^\mu z) + \tfrac{i}{2}\psi\sigma^\mu\partial_\mu\overline{\psi} - \tfrac{i}{2}\partial_\mu\psi\sigma^\mu\overline{\psi} + \overline{f}f \\ &\quad -(mz + \lambda z^2)f - (m\overline{z} + \lambda\overline{z}^2)f - \tfrac{m}{2}[\psi\psi + \overline{\psi\psi}] - \lambda z\psi\psi - \overline{\lambda}\overline{z}\overline{\psi\psi} \\ &= (\partial_\mu \overline{z})(\partial^\mu z) - V(z, \overline{z}) \\ &\quad +\tfrac{i}{2}\psi\sigma^\mu\partial_\mu\overline{\psi} - \tfrac{i}{2}\partial_\mu\psi\sigma^\mu\overline{\psi} - \tfrac{m}{2}[\psi\psi + \overline{\psi\psi}] - \lambda z\psi\psi - \overline{\lambda}\overline{z}\overline{\psi\psi}, \end{aligned} \tag{2.21}$$

using $\overline{f} = mz + \lambda z^2$ in the second expression, includes the scalar potential

$$V(z, \overline{z}) = |f|^2 = |mz + \lambda z^2|^2 = \left|\frac{d}{dz}W(z)\right|^2. \tag{2.22}$$

Journal of Physics: Conference Series **631** (2015) 012009 doi:10.1088/1742-6596/631/1/012009

It is a renormalizable quantum field theory, and supersymmetry holds to all orders of perturbation theory. This result shows how supersymmetry relates all three scalar and Yukawa interactions and how the scalar potential is related to the spinor variations

$$\begin{aligned} \delta\psi_{i\alpha} &= -\sqrt{2}\,\mathcal{A}_i(\varphi_i)\epsilon_\alpha + \partial_\mu(\ldots) + \ldots \qquad \longleftrightarrow \qquad V = \sum_i |\mathcal{A}_i(\varphi_i)|^2, \\ \mathcal{A} &= f = \frac{d\overline{W}(\overline{z})}{d\overline{z}}. \end{aligned} \tag{2.23}$$

In the first line, the index i would label the various spinor and scalar fields φ_i in the theory. The second line refers to our example of a single chiral multiplet with superpotential W. This relation between the potential and spinor variations is a universal property, even in theories and supermultiplets for which auxiliary fields (and then the second line) do not exist [24]. Notice that the on-shell spinor variation is not linear in an interacting theory.

The generalization of the "square" (2.15) of a chiral multiplet is as follows. Consider a set of chiral multiplets (z^i, ψ^i, f^i) and an arbitrary superpotential function $W(z^i)$ of the scalar fields z^i. Then,

$$Z = W(z^i), \qquad \Psi_\alpha = \frac{\partial W}{\partial z^i}\psi^i_\alpha, \qquad F = \frac{\partial W}{\partial z^i} f^i + \frac{1}{2}\frac{\partial^2 W}{\partial z^i \partial z^j}\psi^i\psi^j \tag{2.24}$$

are the components of a chiral multiplet. Another important chiral multiplet is called *kinetic*. Its component fields are

$$Z = \overline{f}, \qquad \Psi_\alpha = i(\sigma^\mu \partial_\mu \overline{\psi})_\alpha, \qquad F = -\Box\overline{z}, \tag{2.25}$$

where (z, ψ, f) is a chiral multiplet. Multiplying then the kinetic multiplet with (z, ψ, f) using the tensor product rule (2.24) leads to the kinetic lagrangian (2.13):

$$zF + fZ + \psi\Psi = -z\Box\overline{z} + i\psi\sigma^\mu\partial_\mu\overline{\psi} + \overline{f}f = \mathcal{L} + \partial_\mu\Big[\frac{i}{2}\psi\sigma^\mu\overline{\psi} - z\partial^\mu\overline{z}\Big]. \tag{2.26}$$

The method of tensor calculus [1] can be systematically applied to construct lagrangians invariant under global supersymmetry. It has found a beautiful synthesis, at least for the case of $\mathcal{N} = 1$ supersymmetry (in four dimensions), in *superspace* and *superfield* techniques [8, 9], building on the idea that supersymmetry generators act like "square roots of translations", as suggested by the superalgebra (2.5). The operators Q_α are realized in terms of derivatives (like translations) acting in a superspace extended with fermionic, Grassmann (fictitious) coordinates. A tensor calculus also exists for conformal supersymmetry (gauge theories of the superconformal algebra). It probably offers the most efficient procedure to construct supergravity theories, with local supersymmetry.[4]

A similar discussion could be made for the supermultiplet with helicities $\pm 1, \pm 1/2$, which is realized by a gauge field A_μ and a Majorana spinor λ_α, the *gaugino*. Since the gaugino includes

[4] See section 3.5.

four off-shell fields while the gauge field has three, the off-shell supermultiplet includes one real scalar auxiliary field D. Again, the Yang-Mills and Dirac lagrangians, with their non-abelian covariantizations, provide a supersymmetric theory: the *super-Yang-Mills* (SYM) lagrangian is then simply

$$\mathcal{L}_{SYM} = -\frac{1}{4}F^A_{\mu\nu}F^{A\mu\nu} + \frac{i}{2}\lambda^A\sigma^\mu D_\mu\overline{\lambda}^A - \frac{i}{2}D_\mu\lambda^A\sigma^\mu\overline{\lambda}^a + \frac{1}{2}D^AD^A, \tag{2.27}$$

where D_μ and $F^a_{\mu\nu}$ are the usual covariant derivative and field-strength tensor of a non-abelian gauge theory[5], and $D^A = 0$ by its field equation. The SYM theory can be coupled to chiral multiplets in an anomaly-free representation of the gauge group, to give the supersymmetric extension of gauge theories. This is the framework of the *minimal supersymmetric Standard Model* (MSSM), and of its variations.

It is then not surprising that the supermultiplet with helicity states $\pm 2, \pm 3/2$ would lead to a field theory combining the Einstein-Hilbert lagrangian of general relativity and the Rarita-Schwinger lagrangian for the helicities $\pm 3/2$: this leads to $\mathcal{N} = 1$ supergravity.

2.3. *Breaking supersymmetry*

With respect to standard "bosonic" symmetries, spontaneous breaking of supersymmetry [5, 6] is peculiar and difficult to achieve. Breaking a local or global symmetry is usually "parameter-controlled", in the sense that the scalar potential which defines the ground state depends on parameters and the various possible phases correspond in general to sizeable domains in the parameter space of the theory. Selecting values of parameters selects the phase. The spontaneous breaking of supersymmetry is "algebra-controlled": the scalar potential is a sum of positive terms, each term proportional to the square of an auxiliary field, either f_i for chiral superfields or D^A for gauge multiplets. In the renormalizable theory,

$$V = \sum_i |f_i|^2 + \frac{1}{2}\sum_A D^AD^A. \tag{2.28}$$

By their algebraic field equations, the auxiliary fields are functions of the chiral scalars z^i and if equations

$$f_i(z_i) = D^A(z_i, \overline{z}_i) = 0 \tag{2.29}$$

have a solution, this solution is the true ground state of the theory and supersymmetry is not broken.[6] We then have an algebraic condition for supersymmetry breaking, that these equations cannot be solved.

Spontaneous supersymmetry breaking has two undesired consequences. Firstly, it generates a massless spin 1/2 Goldstone particle, the Goldstino. This can be seen for instance in the

[5] All fields are in the adjoint representation.
[6] The potential vanishes then at a supersymmetric minimum. But since general relativity is absent, the value of the potential at the ground state, sometimes called *vacuum energy*, does not have any physical significance.

variation of ψ_α, eq. (2.10). If f acquires a vacuum expectation value $\langle f\rangle$

$$\delta\psi_\alpha = \sqrt{2}\langle f\rangle\epsilon_\alpha + \ldots \tag{2.30}$$

and the inhomogeneous term is typical of Goldstone particles. Secondly, if spontaneous supersymmetry breaking is able to lift fermion–boson mass degeneracies, it in general moves the mass of some spin zero states below their fermionic partner, in contradiction with observations. These obstructions can be avoided if the spontaneously broken supersymmetry is local, *i.e.* in a theory with gauged supersymmetry. This is a first motivation, from a low-energy perspective, for supergravity, with the idea that the residual, effective effects of the breaking will produce the mass terms necessary in a realistic particle spectrum. Models realising this idea are actually easy to construct, and they are at the origin of the various supersymmetric extensions of the Standard Model under test at LHC experiments.

3. Supergravity

Supergravity is the theory of gauged supersymmetry. The spin 1/2 parameter ϵ_α is then local, $\epsilon_\alpha(x)$, and since translation generators P_μ appear in the supersymmetry algebra (2.5), translations are local as well. This calls for coordinate diffeomorphisms (general coordinate transformations, GCT) and then for general relativity and gravitation. The theory of gauged supersymmetry is then a field theory of gravitation.

A local symmetry requires a gauge field (a connection) to construct tensors and invariants involving derivatives, needed in lagrangians and dynamical field equations. Since supercharges Q_α and parameters ϵ_α are Lorentz spinors, the gauge field of supersymmetry is a vector-spinor field, $\psi_{\alpha\mu}$, the *gravitino*, and the physical (massless) states will have helicities $\pm 3/2$. To construct a supersymmetric lagrangian, we first need a kinetic lagrangian for the gravitino: the Rarita-Schwinger lagrangian. We may then add this term to the Einstein-Hilbert lagrangian for the metric tensor and maybe find supersymmetry variations leaving the action invariant and closing the supersymmetry algebra for solutions of the field equations, on-shell.

Or we may try to directly obtain an off-shell representation of supersymmetry including the metric tensor, the gravitino and, if needed, auxiliary fields. Let us count off-shell degrees of freedom:

- The metric tensor $g_{\mu\nu}$ has ten components, four can be removed by gauge transformations (local translations) to remain with six fields, 6_B.
- The gravitino is a Majorana vector–spinor with four local gauge supersymmetries. It includes then $4\times 4 - 4 = 12_F$ component fields.[7]
- With $6_B + 12_F$ propagating fields, $(6+n)_B + n_F$ non-propagating auxiliary fields are then needed to construct an off-shell representation.

[7] Alternatively, the supercharge Q_α includes four operators and one gauge field (3_F) is needed for each of them.

It turns out that several possibilities exist.[8] Minimal supergravities have 6_B auxiliary fields ($n = 0$). *Old minimal* supergravity [26, 27] has a complex scalar (2_B) and a vector field not associated with a gauge symmetry (4_B), *new minimal* supergravity [28] has an antisymmetric tensor $B_{\mu\nu}$ and a vector field A_μ with gauge symmetries

$$\delta B_{\mu\nu} = \partial_\mu \Lambda_\nu - \partial_\nu \Lambda_\mu \qquad (6-3)_B, \qquad\qquad \delta A_\mu = \partial_\mu \Lambda \qquad (4-1)_B. \tag{3.1}$$

And there are *non-minimal* versions with $n \neq 0$, including some versions which have supplementary propagating fields.[9] In pure supergravity, with $\pm 3/2$ and ± 2 physical states, the formulation does not matter much since auxiliary fields anyway vanish. They play however a role when coupling supergravity to other supermultiplets: they define classes of admissible interactions which depend directly on the choice of supergravity auxiliary fields [25].

3.1. Spinors and the vierbein

From here on, x^μ denotes coordinates of a space-time with metric tensor $g_{\mu\nu}(x)$ and line element $ds^2 = g_{\mu\nu}(x) dx^\mu dx^\nu$.

Spinor fields live in the flat minkowskian space tangent at each point x. This tangent space would have local coordinates $\zeta^a(x)$ with line element[10]

$$ds^2 = \eta_{ab}\, d\zeta^a d\zeta^b = \eta_{ab} (\partial_\mu \zeta^a)(\partial_\nu \zeta^b) dx^\mu dx^\nu. \tag{3.2}$$

Hence, the transition from the curved space-time to the flat tangent space is given by the sixteen fields $e^a_\mu(x) = \partial_\mu \zeta^a(x)$, in other words, we can define a *vierbein* e^a_μ and its inverse e^μ_a (since the metric has an inverse) such that

$$g_{\mu\nu} = \eta_{ab}\, e^a_\mu e^b_\nu, \qquad\qquad e^\mu_a e^a_\nu = \delta^\mu_\nu, \qquad e^a_\mu e^\mu_b = \delta^a_b. \tag{3.3}$$

At each point x, a Lorentz algebra acts in the tangent space. This local Lorentz symmetry allows to eliminate six components of the vierbein and since ds^2 in (3.2) is Lorentz invariant, the ten remaining components are the ten components of $g_{\mu\nu}$. It also acts on spinors:

$$\delta\psi(x) = \frac{1}{2} \omega_{ab}\, \sigma^{ab} \psi(x), \qquad\qquad \sigma^{ab} = \frac{1}{2}\gamma^{ab}, \qquad \gamma^{ab} = \frac{1}{2}[\gamma^a, \gamma^b]. \tag{3.4}$$

Covariant derivatives of spinors are provided by the local Lorentz gauge field, the *spin connection* ${\omega_\mu}^{ab} = -{\omega_\mu}^{ba}$:

$$D_\mu \psi = \partial_\mu \psi + \frac{1}{2} \omega_{\mu\, ab}\, \sigma^{ab} \psi \tag{3.5}$$

and the Dirac lagrangian takes then the form

$$e^{-1} \mathcal{L}_\psi = i\overline{\psi} \gamma^\mu D_\mu \psi \tag{3.6}$$

where $e = \sqrt{|\det g_{\mu\nu}|} = \det e^a_\mu$ and $\gamma^\mu = e^\mu_a \gamma^a$. It is invariant under GCT and local Lorentz.

[8] For a review and a comparison of different choices in the superconformal approach, see [25].
[9] "$16_B + 16_F$" supergravity [29, 30, 31] for instance is related to string theory compactifications [32], or to supercurrent structures [33, 34].
[10] $\eta_{ab} = \mathrm{diag}(1, -1, -1, -1)$ is the flat Minkowski metric.

3.2. The gravitino and the Rarita-Schwinger action

The Rarita-Schwinger action describes the propagation of a vector-spinor field $\psi_{\alpha\mu}$ in the background defined by the vierbein e^a_μ and the spin connection ${\omega_\mu}^{ab}$. Its form is dictated by invariance requirements and reduction to the relevant helicity components only. Under the Lorentz algebra, the field $\psi_{\alpha a} = e^\mu_a \psi_{\alpha\mu}$ (α is a spinor index) transforms in the reducible representation

$$\begin{array}{ccccccc} \text{spinor} & \otimes & \text{vector} & = & \text{gravitino} & \oplus & \text{spinor}\,, \\ [(2,1)\oplus(1,2)] & \otimes & (2,2) & = & [(3,2)\oplus(2,3)] & \oplus & [(1,2)\oplus(2,1)]\,. \end{array}$$

The second equation indicates the representations of the Lorentz algebra $SO(1,3) \sim Sl(2,\mathbb{C})$, the numbers are the dimensions of $Sl(2,\mathbb{R})$ representations. The spinor part of $\psi_{\alpha a}$ is $\gamma^a\psi_{\alpha a}$ and the gravitino part is then isolated by the condition

$$(\gamma^a\psi_a)_\alpha = (\gamma^\mu\psi_\mu)_\alpha = 0 \qquad \Longrightarrow \qquad \widetilde{\psi}_{\alpha a} = \psi_{\alpha a} - \frac{1}{4}(\gamma_a\gamma^b\psi_b)_\alpha. \tag{3.7}$$

An action for the gravitino should in principle include this projection condition in its field equations.

Consider then the following free lagrangian density, in Minkowski space (coordinates ζ^a and metric η_{ab}):

$$\mathcal{L}_0 = \frac{1}{2\kappa^2}\,\overline{\psi}_a\gamma^{abc}\partial_b\psi_c \tag{3.8}$$

where κ is a constant with dimension (mass)$^{-1}$ and $\gamma^{abc} = \gamma^{[a}\gamma^b\gamma^{c]} = \frac{1}{6}\gamma^a\gamma^b\gamma^c \pm 5$ terms. The gravitino ψ_a is Majorana and $\mathcal{L}_0$ is hermitian. It implies the field equation

$$\gamma^{abc}\partial_b\psi_c = 0. \tag{3.9}$$

Invariance under the gauge transformation $\delta\psi_a = \partial_a\lambda$, with an arbitrary Majorana spinor λ, can be used to impose the projection condition (3.7) by solving $\gamma^a\partial_a\lambda = -\gamma^a\psi_a$. This leaves a residual gauge symmetry $\delta\psi_a = \partial_a\tilde{\lambda}$ with $\tilde{\lambda}$ solution of the massless Dirac equation $\gamma^a\partial_a\tilde{\lambda} = 0$. In the gauge $\gamma^a\psi_a = 0$, the field equation reduces to

$$\gamma^a\partial_b\psi^b = \gamma^b\partial_b\psi^a \tag{3.10}$$

and multiplication by γ_a leads to

$$\gamma^a\psi_a = 0 \quad \text{(gauge choice)}, \qquad \partial_b\psi^b = 0, \qquad \gamma^b\partial_b\psi_a = 0 \qquad \text{(Dirac)}. \tag{3.11}$$

The Dirac equation indicates that the field is massless and the count of physical degrees of freedom is as follows. Starting with 16_F fields, the gauge choice and $\partial^a\psi_a = 0$ remove two spinors (8_F), the massless Dirac equation removes four of the 8_F remaining fields and finally the residual gauge symmetry eliminates one of the massless Dirac spinor (2_F) to leave only two degrees of freedom, which turn out to have helicities $\pm3/2$.[11]

[11] Plane waves $\epsilon_{\alpha a}(k)e^{-ikx}$ can be used to see this.

Coupling theory (3.8) to the background described by the vierbein e^a_ν leads to the Rarita-Schwinger lagrangian

$$e^{-1}\mathcal{L}_{RS} = \frac{1}{2\kappa^2}\,\overline{\psi}_\mu\gamma^{\mu\nu\rho}\widetilde{D}_\nu\psi_\rho, \tag{3.12}$$

where $\gamma^{\mu\nu\rho} = e^\mu_a e^\nu_b e^\rho_c \gamma^{abc}$. In principle, since the gravitino field $\psi_{\alpha\mu}$ is a space-time vector (index μ), its covariant derivative should be

$$D_\mu\psi_{\alpha\rho} = \partial_\mu\psi_{\alpha\rho} - \Gamma^\sigma_{\mu\rho}(g)\psi_{\alpha\sigma} + \frac{1}{2}\omega_\mu{}^{ab}(\sigma_{ab}\psi_\rho)_\alpha \tag{3.13}$$

with affine connection

$$\Gamma^\mu_{\nu\rho}(g) = \frac{1}{2}g^{\mu\sigma}[\partial_\nu g_{\rho\sigma} + \partial_\rho g_{\nu\sigma} - \partial_\sigma g_{\nu\rho}]. \tag{3.14}$$

But the antisymmetry of $\gamma^{\mu\nu\rho}$ removes the symmetric affine connection and

$$\widetilde{D}_\mu\psi_\nu = \partial_\mu\psi_\nu + \frac{1}{2}\omega_\mu{}^{ab}\,\sigma_{ab}\psi_\nu \tag{3.15}$$

appears in the Rarita-Schwinger lagrangian (3.12).

3.3. Simple $\mathcal{N} = 1$ supergravity

The need for spinor fields imposes a formulation of general relativity in terms of the vierbein. Since the Rarita-Schwinger also uses the spin connection, it is natural to use the *first order* (or Palatini) formalism in which e^a_μ and $\omega_\mu{}^{ab}$ are independent fields, their relation being a field equation. One then introduces the curvature tensor of the spin connection,

$$R_{\mu\nu}{}^{ab} = \partial_\mu\omega_\nu{}^{ab} - \partial_\nu\omega_\mu{}^{ab} + \omega_\mu{}^{ac}\omega_{\nu c}{}^b - \omega_\nu{}^{ac}\omega_{\mu c}{}^b = -R_{\nu\mu}{}^{ab} = -R_{\mu\nu}{}^{ba}, \tag{3.16}$$

the curvature scalar

$$R = R_{\mu\nu}{}^{ab}e^\mu_a e^\nu_b, \tag{3.17}$$

and the gravity lagrangian

$$\mathcal{L}_{grav.} = \frac{1}{2\kappa^2}\,eR. \tag{3.18}$$

All quantities are diffeomorphism and Lorentz tensors or scalars and the gravitational coupling constant is $\kappa = \sqrt{8\pi}M_P^{-1}$ in terms of the Planck scale $M_P \simeq 1.2\times 10^{19}$ GeV. Under a variation of the vierbein,

$$\delta e^\mu_a \frac{\delta}{\delta e^\mu_a}\mathcal{L}_{grav.} = \frac{1}{\kappa^2}e\left[R_{\mu\nu}{}^{ab}e^\nu_b - \frac{1}{2}e^a_\mu R\right]\delta e^\mu_a \tag{3.19}$$

leads to Einstein equation, after the elimination of the spin connection.

Since the action (3.18) is quadratic in the spin connection and linear in its first derivative, the Euler-Lagrange equation for $\omega_\mu{}^{ab}$ is algebraic only. To calculate this field equation, rewrite

$$\begin{aligned} eR &= e(e^\mu_a e^\nu_b - e^\mu_b e^\nu_a)\left(\partial_\mu\omega_\nu{}^{ab} + \omega_\mu{}^{ac}\omega_{\nu c}{}^b\right) \\ &= -\omega_\nu{}^{ab}\partial_\mu[e(e^\mu_a e^\nu_b - e^\mu_b e^\nu_a)] + e(e^\mu_a e^\nu_b - e^\mu_b e^\nu_a)\omega_\mu{}^{ac}\omega_{\nu c}{}^b + \text{derivative} \end{aligned} \tag{3.20}$$

and the field equation leads to

$$\begin{aligned}\omega_{\mu\,cd} &= -\tfrac{1}{2}(\partial_\mu e_{\nu c} - \partial_\nu e_{\mu c})e^\nu_d + \tfrac{1}{2}(\partial_\mu e_{\nu d} - \partial_\nu e_{\mu d})e^\nu_c - \tfrac{1}{2}e^\rho_c e^\nu_d(\partial_\rho e_{\nu a} - \partial_\nu e_{\rho a})e^a_\mu \\ &\equiv \omega_{\mu\,cd}(e).\end{aligned} \tag{3.21}$$

In terms of $g_{\mu\nu}$ and of the symmetric $\Gamma^\lambda_{\mu\nu}(g)$ (3.14), the Ricci tensor $R_{\mu\nu}$ corresponding to definitions (3.16) and (3.17) is

$$R_{\mu\nu} = \partial_\rho \Gamma^\rho_{\mu\nu}(g) - \partial_\nu \Gamma^\rho_{\rho\mu}(g) - \Gamma^\rho_{\sigma\mu}(g)\Gamma^\sigma_{\rho\nu}(g) + \Gamma^\rho_{\mu\nu}(g)\Gamma^\sigma_{\rho\sigma}(g) \;=\; R_{\nu\mu} \tag{3.22}$$

and $R = g^{\mu\nu}R_{\mu\nu}$.

Next, we combine the gravity and Rarita-Schwinger lagrangians:

$$\mathcal{L}_{ERS}(e^a_\mu, \psi_{\alpha\mu}, \omega_{\mu\,ab}) = \frac{1}{2\kappa^2}\, e\left(R + \overline{\psi}_\mu \gamma^{\mu\nu\rho}\tilde{D}_\nu \psi_\rho\right). \tag{3.23}$$

After elimination of the spin connection, using its algebraic field equation, we will obtain an interacting theory for the propagating vierbein and gravitino. Since the gravitino lagrangian also includes a term linear in ${\omega_\mu}^{ab}$ in the Lorentz covariant derivative $\widetilde{D}_\mu$, its field equation and its solution are modified. As a consequence, the spin connection acquires *contorsion*,

$${\omega_\mu}^{ab} = {\omega_\mu}^{ab}(e) + {\kappa_\mu}^{ab}, \tag{3.24}$$

and the contorsion tensor is quadratic in the gravitino field:

$$\kappa_{\mu\,ab} = -\frac{1}{4}\left[\overline{\psi}_\mu \gamma_a \psi_b - \overline{\psi}_\mu \gamma_b \psi_a + \overline{\psi}_a \gamma_\mu \psi_b\right] = -\kappa_{\mu\,ba}. \tag{3.25}$$

In the Rarita-Schwinger lagrangian,

$$\tilde{D}_\mu \psi_\nu - \tilde{D}_\nu \psi_\mu = \widehat{D}_\mu \psi_\nu - \widehat{D}_\nu \psi_\mu + 2S^\lambda_{\mu\nu}\psi_\lambda, \qquad \widehat{D}_\mu \psi_\nu = \partial_\mu \psi_\nu + \frac{1}{2}{\omega_\mu}^{ab}(e)\sigma_{ab}\psi_\nu, \tag{3.26}$$

with torsion tensor

$$S^\lambda_{\mu\nu} = -\frac{1}{4}\overline{\psi}_\mu \gamma^\lambda \psi_\nu \tag{3.27}$$

defined as the antisymmetric part of the affine connection $\Gamma^\mu_{\nu\rho} = \Gamma^\mu_{\nu\rho}(g) + S^\mu_{\nu\rho}$, $S^\mu_{\nu\rho} = -S^\mu_{\rho\nu}$.

Useful formulas are obtained by inserting the decomposition (3.24) into the gravity lagrangian, after some partial integrations:

$$\begin{aligned} eR &= eR\,(\omega(e)) - e[{\kappa^a}_{ac}{\kappa_b}^{bc} - \kappa_{abc}\kappa^{cba}] + \text{ derivative}, \\ eR\,(\omega(e)) &= e[{\omega^a}_{ac}(e){\omega_b}^{bc}(e) - \omega_{abc}(e)\omega^{cba}(e)] + \text{ derivative}, \end{aligned} \tag{3.28}$$

where $\omega_{abc}(e) = e^\mu_a\, \omega_{\mu\,bc}(e)$ and $\kappa_{abc} = e^\mu_a\, \kappa_{\mu\,bc}$ and the derivatives can be dropped in the lagrangian.

Inserting the spin connection (3.24) with contorsion (3.25) in $\mathcal{L}_{ERS}$ leads to the lagrangian density of $\mathcal{N}=1$ pure supergravity, as a function of e^a_μ and ψ_μ only (*second order formalism*):

$$\begin{aligned}\mathcal{L} &= \frac{1}{2\kappa^2}eR\left(\omega(e)\right)+\frac{1}{2\kappa^2}e\overline{\psi}_\mu\gamma^{\mu\nu\rho}\widehat{D}_\nu\psi_\rho \\ &+\frac{e}{32\kappa^2}\left[4(\overline{\psi}^\mu\gamma_\mu\psi_\rho)(\overline{\psi}^\nu\gamma_\nu\psi^\rho)-(\overline{\psi}_\mu\gamma_\nu\psi_\rho)(\overline{\psi}^\mu\gamma^\nu\psi^\rho)-2(\overline{\psi}_\mu\gamma_\nu\psi_\rho)(\overline{\psi}^\mu\gamma^\rho\psi^\nu)\right],\end{aligned}\tag{3.29}$$

with now $\widehat{D}_\nu\psi_\rho=\partial_\nu\psi_\rho+\frac{1}{2}\omega_{\nu\,ab}(e)\sigma^{ab}\psi_\rho$.

With some efforts,[12] one can show that $\mathcal{L}$ transforms with a derivative under the local supersymmetry variations

$$\begin{aligned}\delta e^a_\mu &= -\tfrac{1}{2}\bar\epsilon\gamma^a\psi_\mu\,, & \delta e^\mu_a &= \tfrac{1}{2}\bar\epsilon\gamma^\mu\psi_a\,,\\ \delta\psi_\mu &= D_\mu\epsilon = \partial_\mu\epsilon+\tfrac{1}{2}\omega_{\mu\,ab}\,\sigma^{ab}\epsilon\,, & \delta\overline{\psi}_\mu &= D_\mu\bar\epsilon = \partial_\mu\bar\epsilon-\tfrac{1}{2}\omega_{\mu\,ab}\,\bar\epsilon\sigma^{ab}\,,\end{aligned}\tag{3.30}$$

in the first-order formalism. Using eqs. (3.24) and (3.25), the gravitino supersymmetry variation acquires a fermionic nonlinear contribution through the contorsion tensor:

$$\delta\psi_\mu=\widehat{D}_\mu\epsilon-\frac{1}{8}\left(2\overline{\psi}_\mu\gamma_a\psi_b+\overline{\psi}_a\gamma_\mu\psi_b\right)\sigma^{ab}\epsilon\,,\qquad \widehat{D}_\mu\epsilon=\partial_\mu\epsilon+\frac{1}{2}\omega_{\mu\,ab}(e)\,\sigma^{ab}\epsilon. \tag{3.31}$$

The gravitino transforms as the gauge field of supersymmetry: the first term is the derivative of the transformation parameter ϵ.

One easily obtains the algebra

$$[\delta_1,\delta_2]\,e^a_\mu=\delta_1[-\frac{1}{2}\bar\epsilon_2\gamma^a\psi_\mu]-\delta_2[-\frac{1}{2}\bar\epsilon_1\gamma^a\psi_\mu]=-\frac{1}{2}\,D_\mu(\bar\epsilon_2\gamma^a\epsilon_1)\,, \tag{3.32}$$

the covariant derivative acting on the Lorentz vector $\bar\epsilon_2\gamma^a\epsilon_1$:

$$D_\mu(\bar\epsilon_2\gamma^a\epsilon_1)=\partial_\mu(\bar\epsilon_2\gamma^a\epsilon_1)+\omega_\mu{}^{ab}(\bar\epsilon_2\gamma_b\epsilon_1)\,.$$

The quantity $\xi^\mu=e^\mu_a(\bar\epsilon_2\gamma^a\epsilon_1)$ is then the parameter of the infinitesimal coordinate transformation predicted by the supersymmetry algebra. But as earlier mentioned, we do not expect with 6_B+12_F off-shell fields that the supersymmetry algebra closes without the field equations. Auxiliary fields would be needed and, in the case of $\mathcal{N}=1$ supergravity, exist and are not unique.

3.4. Anti-de Sitter supergravity: the cosmological constant

The natural background geometry of the supergravity lagrangian (3.29) is flat Minkowski space. Since the matter energy-momentum tensor is entirely generated by the gravitino, it vanishes in a Lorentz-invariant background.[13] But we know that Poincaré supersymmetry is a limit case

[12] Using properties of Majorana spinors, Fierz rearrangements and partial integrations.
[13] There could be gravitino condensates [35], recently reviewed in [36].

of the more general supersymmetry in anti-de Sitter (AdS) space-time. A supergravity theory with natural AdS background geometry should then exist.

We use the following standard definition of the cosmological constant Λ: it should contribute to Einstein equations as

$$\begin{aligned} R_{\mu\nu}^{ab} e_b^\nu - \tfrac{1}{2} e_\mu^a R &= -\Lambda e_\mu^a + \text{contributions from other fields}, \\ R &= 4\Lambda + \text{contributions from other fields}. \end{aligned} \tag{3.33}$$

A positive (negative) Λ leads to de Sitter (anti-de Sitter) space-time. The field equation (3.33) follows from the lagrangian density

$$\frac{e}{\kappa^2}\left[\frac{1}{2}R - \Lambda\right]. \tag{3.34}$$

The situation is somewhat similar to the introduction of mass in the chiral supermultiplet theory discussed in section 2.2. Keeping the supersymmetry variation δe_μ^a unchanged, the gravitino variation is modified to

$$\delta\psi_\mu = D_\mu\epsilon - \frac{1}{2}M\,\gamma_\mu\epsilon\,, \tag{3.35}$$

with a real number M.[14] The modified variation of the gravitino kinetic term requires the presence in $\mathcal{L}$ of a quadratic, mass-like term for the gravitino, and the variation of this term implies the existence of a negative cosmological constant proportional to M^2. The resulting lagrangian is then:[15]

$$\mathcal{L}_{AdS} = \frac{e}{\kappa^2}\left[\frac{1}{2}R + \frac{1}{2}\overline{\psi}_\mu\gamma^{\mu\nu\rho}\tilde{D}_\nu\psi_\rho + \frac{M}{2}\overline{\psi}_\mu\gamma^{\mu\nu}\psi_\nu + 3M^2\right]. \tag{3.36}$$

The gravitino mass-like term in the lagrangian density (3.36) does not mean that the gravitino is massive: the theory is supersymmetric, the graviton is massless, the gravitino must then be massless. Actually, the cosmological constant $\Lambda = -3M^2$ in Einstein equation (3.33) propagates graviton waves along light-like curves in the anti-de Sitter geometry. Similarly, the gravitino mass-like term is precisely the contribution required to propagate gravitino waves on light-like curves in this geometry. Again, we find that a positive cosmological constant is not compatible with supersymmetry, as the basic superalgebra already indicates.

3.5. The superconformal derivation, old minimal supergravity

In section 2.2, we have constructed the (globally) supersymmetric theory of a chiral multiplet (z, ψ, f), to illustrate the role of auxiliary fields. In this paragraph, we outline a similar approach, in the context of superconformal symmetry, to construct $\mathcal{N} = 1$ supergravity with the old minimal set of auxiliary fields. The reason to consider this construction here is that this

[14] Reality follows from the Majorana property of ψ_μ.
[15] In terms of $\omega_\mu{}^{ab} = \omega_\mu{}^{ab}(e) + \kappa_\mu{}^{ab}$.

procedure generalizes very well to theories describing generic $\mathcal{N} = 1$ supersymmetric gauge theories coupled to supergravity [37, 38], however at the price of a considerable increase in technical complexity.

The Poincaré and Anti-de Sitter $\mathcal{N} = 1$ superalgebras are subalgebras of the $\mathcal{N} = 1$ superconformal algebra $SU(2,2|1)$. Its bosonic sector $SU(2,2) \times U(1)_R$ includes the conformal algebra $SU(2,2) \sim SO(2,4)$ and $U(1)_R$ symmetry, and $SO(2,4) \supset SO(1,3) \times SO(1,1)$, where $SO(1,3)$ is Lorentz algebra and $SO(1,1)$ generates dilatation or scale or Weyl transformations. A supermultiplet of Poincaré supersymmetry is also a representation of the superconformal algebra once a $SO(1,1)$ Weyl weight w and a $U(1)_R$ chiral charge q have been assigned to all fields and with the appropriate symmetry and supersymmetry variations. There are restrictions on these quantum numbers. For instance $w = q$ for (the lowest component of) a chiral multiplet.[16]

To construct $\mathcal{N} = 1$ Poincaré supergravity, one uses a chiral supermultiplet S_0 with Weyl and $U(1)_R$ weights $w = q = 1$ and component fields z_0 $(w = q = 1)$, ψ_α $(w = 3/2,\ q = -1/2)$ and f_0 $(w = 2,\ q = -2)$.[17] Its conjugate antichiral $\overline{S}_0$, with weights $w = -q = 1$ has components $\overline{z}_0$ $(w = -q = 1)$, $\overline{\psi}_{\dot\alpha}$ $(w = 3/2,\ q = 1/2)$ and $\overline{f}_0$ $(w = 2,\ q = 2)$. The chiral kinetic multiplet of S_0, denoted by $T(S_0)$, analogous to expressions (2.25), has components

$$\begin{aligned} Z_0 &= \overline{f}_0 && (w = q = 2), \\ \Psi_{0\,\alpha} &= i(\sigma^\mu D^C_\mu \overline{\psi}_0)_\alpha && (w = 5/2,\ q = 1/2), \\ F_0 &= -\Box^C \overline{z}_0 && (w = 3,\ q = -1). \end{aligned} \tag{3.37}$$

The derivatives D^C_μ and $\Box^C$ are covariant under the full local superconformal algebra. Gauge fields are:

$$\begin{array}{llll} e^a_\mu & \text{(vierbein, translations)}, & {\omega_\mu}^{ab} & \text{(spin connection, Lorentz)}, \\ \psi_{\alpha\mu} & \text{(gravitino, supersymmetry)}, & & \\ f^a_\mu & \text{(conformal boosts)}, & \phi_{\alpha\mu} & \text{(special supersymmetry)}, \\ b_\mu & \text{(dilatation)}, & A_\mu & (U(1)_R). \end{array} \tag{3.38}$$

Constraints lead to algebraic expressions for ${\omega_\mu}^{ab}$ (as earlier), $\phi_{\alpha\mu}$ and f^a_μ, leaving bosonic gauge fields e^a_μ, b_μ, A_μ $(6 + 3 + 3 = 12_B)$ and the gravitino $\psi_{\alpha\mu}$ (12_F). The idea is then to write a superconformal lagrangian for the chiral multiplet S_0 and then to reduce the symmetry to local Poincaré symmetries, by applying appropriate gauge fixing conditions for conformal boosts,

[16] Two conventions, different by the normalization of $U(1)_R$, exist in the literature: either $w = q$ as used here, or $q = \frac{2}{3}w$.
[17] The weights of a supermultiplet are the weights of its "lowest" component, in our cse z_0. Apart from minor differences (metric sign, two-component spinors), we use the notation of ref. [39].

dilatation, $U(1)_R$ and special supersymmetry. These conditions assign values to

$$\begin{array}{ll} z_0: & \text{dilatation and } U(1)_R, \text{ (modulus and phase of } z_0), \\ \psi_0: & \text{special supersymmetry,} \\ b_\mu: & \text{conformal boosts.} \end{array} \tag{3.39}$$

We are then left with the propagating fields of Poincaré supergravity, e_μ^a and $\psi_{\alpha\mu}$ and the auxiliary field A_μ (the gauge field of gauge-fixed $U(1)_R$) and f_0 (in S_0).

In global supersymmetry, we obtain invariant lagrangians by combining supermultiplets into other supermultiplets (tensor calculus), or by multiplying superfields, and by selecting the highest-dimensional component which transforms with a derivative. In the superconformal case, there are two (related) possibilities to produce invariant action terms. Firstly, we can combine supermultiplets to obtain a real supermultiplet with weights $w = 2$, $q = 0$ and take the D–*density formula*. For instance, symbolically,

$$\Big[S_0\overline{S}_0\Big]_D.$$

Secondly, we can combine chiral multiplets into another chiral multiplet with weights $w = q = 3$ and use the F–*density formula*. In our case,

$$\Big[S_0\,T(S_0)\Big]_F \qquad \text{or} \qquad \Big[S_0^3\Big]_F.$$

Up to conventions (and partial integration), $[S_0\overline{S}_0]_D$ and $[S_0T(S_0)]_F$ are equivalent.

To obtain the $\mathcal{N} = 1$ Poincaré supergravity, start then with the superconformal lagrangian

$$\mathcal{L} = -\frac{3}{2}\Big[S_0\overline{S}_0\Big]_D + \lambda\Big[S_0^3\Big]_F \tag{3.40}$$

calculated using superconformal tensor calculus and the density formulas [39]. Applying the gauge fixing conditions

$$b_\mu = 0, \qquad\qquad \psi_0 = 0, \tag{3.41}$$

but retaining z_0 for a moment, the superconformal lagrangian reads

$$\begin{aligned} e^{-1}\mathcal{L} \;=\;& \tfrac{1}{2}z_0\overline{z}_0\Big[R\left(\omega(e)\right) + \overline{\psi}_\mu\gamma^{\mu\nu\rho}\widehat{D}_\nu\psi_\rho\Big] \\ & -\tfrac{3}{2}\Big[2(D_\mu\overline{z}_0)(D^\mu z_0) + 2f_0\overline{f}_0\Big] + \lambda\Big[3z_0^2f_0 + 3\overline{z}_0^2\overline{f}_0\Big] + \ldots \end{aligned} \tag{3.42}$$

The covariant derivative is $D_\mu z_0 = \partial_\mu z_0 - \frac{i}{2}qA_\mu z_0$ ($q = 1$) and some gravitino interactions have been omitted. With the dilatation and $U(1)_R$ gauge choice $z_0 = \kappa^{-1}$,

$$\begin{aligned} e^{-1}\mathcal{L} \;=\;& \frac{1}{2\kappa^2}\Big[R\left(\omega(e)\right) + \overline{\psi}_\mu\gamma^{\mu\nu\rho}\widehat{D}_\nu\psi_\rho\Big] - \frac{3}{4\kappa^2}A_\mu A^\mu - 3f_0\overline{f}_0 \\ & +\lambda\Big[3z_0^2f_0 + 3\overline{z}_0^2\overline{f}_0\Big] + \ldots \end{aligned} \tag{3.43}$$

The first line displays the auxiliary fields A_μ and f_0 of old minimal supergravity. Eliminating them leads finally to

$$e^{-1}\mathcal{L} = \frac{1}{\kappa^2}\left[\frac{1}{2}R\left(\omega(e)\right) + \frac{1}{2}\overline{\psi}_\mu\gamma^{\mu\nu\rho}\widehat{D}_\nu\psi_\rho + 3\frac{\lambda^2}{\kappa^2}\right] + \ldots, \tag{3.44}$$

with an Anti-de Sitter cosmological constant $\Lambda = -3\lambda^2\kappa^{-2}$ induced by the F–density. The related gravitino mass-like term required by supersymmetry[18] is actually generated by the omitted gravitino term

$$\frac{\lambda}{4}\, z_0^3\,\overline{\psi}_\mu\gamma^{\mu\nu}\psi_\nu + \text{h.c.}$$

omitted in the F–density $\lambda[S_0^3]_F$.

3.6. Four-dimensional supergravities for all $\mathcal{N}$

Massless supermultiplets of $\mathcal{N}$–extended supersymmetry fall in three categories: matter multiplets with |helicities| $\leq 1/2$, gauge or Yang-Mills multiplets with |helicities| ≤ 1 and supergravity multiplets with |helicities| ≤ 2. The following table indicates, as a function of the number $\mathcal{N}$ of supersymmetries:[19]

- The number of (on-shell) helicity states in supergravity, gauge and matter multiplets.
- Supermultiplets with scalar fields. These theories are potentially able, at ground states with nonzero scalar expectation values, to offer various patterns of symmetry and supersymmetry breakings (indicated by *).
- That while Yang-Mills multiplets can gauge all symmetry groups, chirality of the fermion representation can only be obtained in the matter (chiral) multiplet of $\mathcal{N} = 1$ supersymmetry.

SUSY	Supergravity	$\lvert$Hel.$\rvert \leq 1$	$\lvert$Hel.$\rvert \leq 1/2$	Chirality
$N=1$	$2_B + 2_F$	$2_B + 2_F$	$2_B + 2_F$ *	✓
$N=2$	$4_B + 4_F$	$4_B + 4_F$ *	$4_B + 4_F$ *	-
$N=3$	$8_B + 8_F$	$8_B + 8_F$ *	-	-
$N=4$	$16_B + 16_F$ *	$8_B + 8_F$ *	-	-
$N=5$	$32_B + 32_F$ *	-	-	-
$N=6$	$64_B + 64_F$ *	-	-	-
$N=8$	$128_B + 128_F$ *	-	-	-

Supergravity field theories with $\mathcal{N} > 1$ are much harder to construct. The number of fields of all helicities increases fast with $\mathcal{N}$ and off-shell representations do not exist in general. The flexibility in the choice of gauge group and matter representation decreases fast with increasing $\mathcal{N}$. Arbitrary representations are allowed with $\mathcal{N} = 1$ only, arbitrary non-chiral representations with $\mathcal{N} \leq 2$, and for $\mathcal{N} \geq 3$, only the adjoint representation is admitted. Arbitrary gauge

[18] As in eq. (3.36).
[19] The $\mathcal{N} = 7$ theory does not exist: the eighth supersymmetry arises automatically and cannot be decoupled.

Journal of Physics: Conference Series **631** (2015) 012009 doi:10.1088/1742-6596/631/1/012009

groups are allowed for $\mathcal{N} \leq 4$, while for higher $\mathcal{N}$, gauged supergravities exclusively depend on the vector fields present in the supergravity multiplet.

The number of vector fields in the supergravity multiplet is $\mathcal{N}(\mathcal{N}-1)/2$. The choice of possible gaugings increases then rapidly with $\mathcal{N}$, and also taking advantage of electric-magnetic duality. These gauged algebras cannot be identified with the compact Lie algebras used in the Standard Model or its extensions, but their breaking patterns is a subject of interest due to relations with properties found in superstrings. This is an area where fundamental developments of supergravity theories is a subject of present researches.

Chirality of the $SU(3)_c \times SU(2)_L \times U(1)_Y$ fermion representation in the Standard Model, associated with parity violation by weak interactions, appears to be a fundamental property. This has given a particular importance to $N = 1$ supergravity coupled to a Yang-Mills multiplet gauging any symmetry algebra, and allowing any anomaly-free representation of this gauged symmetry. The most general form of this theory has been derived in 1982 [37, 17].

4. $\mathcal{N} = 1$ supergravity–matter couplings

The most general interaction of chiral, gauge and supergravity $\mathcal{N} = 1$ multiplets is defined by two ingredients. Firstly, the choice of a gauge group G and of the representation R of chiral supermultiplets. The only constraint would be the absence of chiral anomaly, even if supergravity is not a quantum field theory. The representation can be chiral and one can then couple the Standard Model to $\mathcal{N} = 1$ supergravity, adding only a sector in which local supersymmetry is spontaneously broken (the *super-higgs* mechanism [40, 41, 42][20]) which should also mediate breaking contributions into the supersymmetric Standard Model (generation of *soft breaking terms*). Secondly, the choice of three gauge-invariant (or gauge-covariant) functions of the scalar fields in chiral supermultiplets. The first function, the real Kähler potential $\mathcal{K}$, defines the kinetic lagrangian of chiral superfields. The holomorphic superpotential W defines the interactions of chiral supermultiplets and the holomorphic $\mathcal{F}$ defines the gauge kinetic (super-Yang-Mills) lagrangian.

These ingredients are known from Poincaré global supersymmetry: the most general gauged nonlinear supersymmetric sigma model is defined in terms of identical ingredients. There is however a subtlety related to the AdS case. In global supersymmetry in an AdS space-time with cosmological constant $-3M^2$, the theory depends on the combination $M\mathcal{K} + W + \overline{W}$ [43]. Then, since supergravity naturally describes AdS *and* the limiting Minkowski case, a similar phenomenon would not be a surprise: one actually finds that the supergravity theory depends on[21]

$$\mathcal{G} = \mathcal{K} + \ln(W\overline{W}), \tag{4.1}$$

[20] I use the lower case "higgs" for Higgs-Brout-Englert ...
[21] This combination always used in recent literature corresponds to $-\mathcal{G}$ in ref. [37].

and of its derivatives (if the superpotential does not vanish). This fact can be loosely traced to the fact that Kähler (or R) and dilatation symmetries in the superconformal algebra are not compatible (do not commute) with the AdS superalgebra.

The best procedure to derive the lagrangian is probably[22] to start from the observation that all supermultiplets of $\mathcal{N} = 1$ Poincaré or AdS supersymmetry are also representations of the $\mathcal{N} = 1$ superconformal symmetry. The method is described in full detail in the book recently published by Dan Freedman and Toine Van Proeyen [38]. We only give here a symbolic explanation and focus on the gravitino and scalar sectors. Very schematically, it is as follows:

- Consider all supermultiplets denoted as Φ^i (chiral, helicities $\pm 1/2, 0, 0$) and $\mathcal{W}_\alpha$ (gauge, helicities $\pm 1, \pm 1/2$) as representations of the superconforma algebra. A Weyl weight and a $U(1)_R$ charge are then associated with each supermultiplet. Supergravity fields e_μ^a and $\psi_{\alpha\mu}$ are part of superconformal gauge fields.
- Add a compensating supermultiplet which is used to gauge fix the unwanted superconformal symmetries. Here: we symbolically describe old minimal supergravity with a chiral compensating multiplet S_0. It provides the most general coupling (up to two derivatives and up to some generalizations of minor importance) to supergravity [25].
- Use tensor calculus methods, as explained in [38, 39], to generate the locally superconformal lagrangian.
- Gauge fix superconformal symmetries absent in the Poincaré or AdS symmetries and eliminate all auxiliary fields. In this step, a gravity frame (Einstein, Jordan, string) is chosen, see below.
- Identify the ground state(s) of the theory from the analysis of the scalar potential. It defines the background geometry (the cosmological constant) and decides if supersymmetry or symmetries in general are spontaneously broken.

Symbolically, the superconformal lagrangian is represented by

$$\mathcal{L} = -\frac{3}{2}\Big[S_0\overline{S}_0 \exp\Big\{-\frac{1}{3}\mathcal{K}(\Phi^i, \overline{\Phi}_i e^{\mathcal{A}})\Big\}\Big]_D + \Big[S_0^3 W(\Phi^i) + \frac{1}{4}\mathcal{F}(\Phi^i)\mathcal{W}\mathcal{W}\Big]_F \tag{4.2}$$

where $[\ldots]_D$ and $[\ldots]_F$ denote the real and chiral invariant densities expressed in terms of the supermultiplet components and the superconformal gauge fields [38, 39]. The Weyl weights (scale dimensions) of the supermultiplets are $w = 1,\ 0,\ 3/2$ for S_0, Φ^i, $\mathcal{W}$ respectively and the D and F densities apply to supermultiplets with weights 2 and 3: this (with reality and chirality) dictates the occurences of S_0.

[22] *I.e.* in my opinion.

4.1. The scalar sector

The bosonic part of the lagrangian density (4.2) also depends on the bosonic gauge fields of the superconformal algebra, some of them being algebraic (like the spin connection) or gauge-fixed (the dilatation gauge field for instance). The Poincaré theory retains the vierbein e_μ^a or metric tensor $g_{\mu\nu}$, the gravitino ψ_μ and 6_B auxiliary fields: the gauge field A_μ of the (gauge-fixed) $U(1)_R$ superconformal symmetry and the complex scalar f_0 in the chiral compensator S_0.

After the elimination of all auxiliary fields, a convenient expression for the scalar part of this theory is

$$e^{-1}\mathcal{L}_{scalar.} = \frac{1}{2}(z_0\overline{z}_0\mathcal{H})R - \frac{3}{4}(z_0\overline{z}_0\mathcal{H})\left[\partial_\mu \log(z_0\overline{z}_0\mathcal{H})\right]^2 + (z_0\overline{z}_0\mathcal{H})\mathcal{K}^i_j(\partial_\mu z^j)(\partial^\mu \overline{z}_i) - V_0 \qquad (4.3)$$

with $\mathcal{H} = \exp[-\mathcal{K}/3]$ a function of the scalar fields $z^i, \overline{z}_i$ and with

$$\mathcal{K}^i_j = \frac{\partial^2}{\partial z^j \partial \overline{z}_i}\mathcal{K}. \qquad (4.4)$$

We have kept the complex compensating scalar z_0 with scale dimension $w = 1$: its value fixes the dilatation and $U(1)_R$ gauges. As we can see in the first term, $z_0\overline{z}_0\mathcal{H}$ defines the gravity frame, and the Einstein frame is the gauge condition

$$\frac{1}{\kappa^2} = z_0\overline{z}_0\mathcal{H} = z_0\overline{z}_0 \exp[-\mathcal{K}/3]. \qquad (4.5)$$

In the Einstein frame,

$$e^{-1}\mathcal{L}_{scalar.} = \frac{1}{2\kappa^2}R + \frac{1}{\kappa^2}\mathcal{K}^i_j(\partial_\mu z^j)(\partial^\mu \overline{z}_i) - V_0. \qquad (4.6)$$

The scalar fields z^i are then Kähler coordinates: their kinetic metric $\mathcal{K}^i_j$ derives from the *Kähler potential* $\mathcal{K}$. Notice that this is only true in the Einstein frame.

The scalar potential is generated by the elimination of auxiliary fields f^i (chiral), D^A (gauge) and f_0 (in compensator S_0). The auxiliary fields are: [23]

$$\begin{aligned} f^i &= -(z_0\overline{z}_0\mathcal{H})^{-1}\,\overline{z}_0^3\,(\mathcal{K}^{-1})^i_j\left[\overline{W}^j(\overline{z}_i) + \mathcal{K}^j\,\overline{W}(\overline{z}_i)\right], \qquad W_i = \frac{\partial W}{\partial z^i}, \\ D^A &= -[\mathrm{Re}\,\mathcal{F}(z^i)]^{-1}(z_0\overline{z}_0\mathcal{H})\,\overline{z}_i(T^A_R)^i{}_j\mathcal{K}^j\,, \\ \widetilde{f} &= f_0 - \tfrac{1}{3}z_0\,\mathcal{K}_i\,f^i \;=\; e^{\mathcal{K}/3}\,\overline{z}_0^2\,\overline{W}(\overline{z}_i). \end{aligned} \qquad (4.7)$$

And the scalar potential reads

$$V_0 = (z_0\overline{z}_0\mathcal{H})\,K^i_j\overline{f}_i f^j + \frac{1}{2}\,\mathrm{Re}\,\mathcal{F}D^A D^A - 3\mathcal{H}\tilde{f}^*\tilde{f}. \qquad (4.8)$$

As in global supersymmetry, each auxiliary field f^i of a chiral or D^A of a gauge multiplet produces a positive contribution. A nonzero value at the vacuum state of the potential would

[23] T^A_R: generators of the representation R of chiral multiplets. Possible Fayet-Ilopoulos terms are omitted.

spontaneously break supersymmetry. But, in contrary to the case of global supersymmetry, the supergravity auxiliary field f^0 produces, via $\widetilde{f}$, a negative contribution which drives the theory into Anti-de Sitter space. It is then much easier to create a potential with a ground state breaking supersymmetry, and even at zero cosmological constant. Notice once again that *unbroken supersymmetry* is always associated with either Minkowski or AdS geometry.

The final form of the scalar potential, before fixing the gravity frame is

$$\begin{aligned} V_0 &= (z_0\overline{z}_0\mathcal{H})^2\Big\{e^{\mathcal{K}}\mathcal{K}^i_j[W_i+\mathcal{K}_iW][\overline{W}^j+\mathcal{K}^j\overline{W}]+\frac{1}{2\,\mathrm{Re}\,\mathcal{F}}\sum_A[\overline{z}_i(T^A_R)^i{}_j\mathcal{K}^j]^2 \\ &\qquad -3\,e^{\mathcal{K}}W\overline{W}\Big\}, \end{aligned} \tag{4.9}$$

inserting expressions (4.7) into the original form (4.8). In the Einstein frame (4.5), the prefactor $(z_0\overline{z}_0\mathcal{H})^2$ is simply κ^{-4}.

But we may as well choose another gravity (Jordan) frame with

$$\frac{e^{-2\varphi}}{\kappa^2}=z_0\overline{z}_0\mathcal{H}=z_0\overline{z}_0\exp[-\mathcal{K}/3]. \tag{4.10}$$

One should understand φ as one of the scalar fields in the theory. The supergravity lagrangian reads then

$$\begin{aligned} \mathcal{L} &= \frac{e^{-2\varphi}}{\kappa^2}e\Big[\frac{1}{2}R-3(\partial_\mu\varphi)(\partial^\mu\varphi)+\mathcal{K}^i_j(\partial_\mu z^j)(\partial^\mu\overline{z}_i)\Big] \\ &\quad -\frac{e^{-4\varphi}}{\kappa^4}e\Big[e^{\mathcal{K}}\mathcal{K}^i_j[W_i+\mathcal{K}_iW][\overline{W}^j+\mathcal{K}^j\overline{W}]-3e^{\mathcal{K}}W\overline{W} \\ &\qquad +\frac{1}{2\,\mathrm{Re}\,\mathcal{F}}\sum_A[\overline{z}_i(T^A_R)^i{}_j\mathcal{K}^j]^2\Big]+\text{gauge and fermion contributions.} \end{aligned} \tag{4.11}$$

Obviously, different frames are related by rescalings of the vierbein field.

In general, one can show that if the scalar potential has a supersymmetric stationary point, with values $\langle f^i\rangle=\langle D^A\rangle=0$ at this point, it is then stable under small field fluctuations: a supersymmetric vacuum is stable. This does not apply to non supersymmetric stationary points. We will briefly consider some examples with spontaneously broken supersymmetry below.

4.2. The gravitino sector

In the case of pure $\mathcal{N}=1$ supergravity, we found that a deformation of the Poincaré theory leads to a negative cosmological constant term associated with an arbitrary energy scale parameter M. These are eqs. (3.36) and (3.35).

Let us write these gravitino terms as

$$e^{-1}\mathcal{L}_{3/2}=\frac{1}{\kappa^2}\left[\frac{1}{2}\overline{\psi}_\mu\gamma^{\mu\nu\rho}D_\nu\psi_\rho+\frac{1}{2}m_{3/2}\overline{\psi}_\mu\gamma^{\mu\nu}\psi_\nu+3\,m^2_{3/2}\right], \tag{4.12}$$

where $m_{3/2}$ is the quantity appearing in the mass-like term and in the negative or zero cosmological constant

$$\Lambda = -3m_{3/2}^2. \tag{4.13}$$

As explained earlier, the relative coefficient 3 is imposed by supersymmetry and ensures that gravitino waves have "light-like" propagation in the AdS geometry. In simple Anti-de Sitter supergravity, $m_{3/2}$ is the constant M appearing in the variation $\delta\psi_\mu = -\frac{1}{2}M\gamma_\mu\epsilon + \ldots$

The supergravity theory coupled to gauge and matter multiplets considered previously and defined by the superconformal expression (4.2) actually contains a mass-like term for the gravitino with a field-dependent

$$m_{3/2} = \kappa^2|z_0|^3 W = \frac{1}{\kappa}e^{\mathcal{K}/2}W, \tag{4.14}$$

if the theory is formulated in the Einstein frame. For a supersymmetric ground state, the expectation value of the scalar potential and the induced cosmological constants are then

$$\langle V\rangle = -\frac{3}{\kappa^4}e^{\mathcal{K}}W\overline{W} \qquad \Lambda = -\kappa^2\langle e^{-1}\mathcal{L}\rangle = \kappa^2\langle V\rangle = -3\langle |m_{3/2}|^2\rangle, \tag{4.15}$$

as required. A violation at the vacuum state of these relations would indicate spontaneous supersymmetry breaking, and generate a physical mass for the gravitino.

5. A no-scale model, dilaton supergravity

Breaking spontaneously supersymmetry in supergravity is easy. A superpotential is first of all needed. Here is an example taken in the class of "no-scale" models [44, 45]. Consider a theory describing two chiral supermultiplets with scalar fields S and T, defined by

$$\begin{aligned} &\mathcal{K} = -n\ln(T+\overline{T}) + \widehat{\mathcal{K}}(S,\overline{S}) && \text{Kähler potential,} \\ &W = W(S) && \text{Superpotential.} \end{aligned} \tag{5.16}$$

The scalar potential reads (using $\kappa = 1$ in Einstein frame)

$$V = (T+\overline{T})^{-n}\, e^{\widehat{\mathcal{K}}}\left[\widehat{\mathcal{K}}_{S\overline{S}}^{-1}|W_S + \widehat{\mathcal{K}}_S W|^2 + (n-3)W\overline{W}\right]. \tag{5.17}$$

The value $n = 3$ is particular: the scalar potential is positive or zero, a solution of

$$W_S + \widehat{\mathcal{K}}_S W = 0, \tag{5.18}$$

if it exists, is an absolute minimum and a stable ground state in Minkowski geometry. This minimum condition fixes in general the value of $\langle S\rangle$ and cancels the auxiliary field f_S: supersymmetry is not broken by S. But the value of $\langle T\rangle$ remains arbitrary and the auxiliary field

$$f_T = (T+\overline{T})^{-1/2}e^{\widehat{\mathcal{K}}/2}\,\overline{W} \tag{5.19}$$

Journal of Physics: Conference Series **631** (2015) 012009 doi:10.1088/1742-6596/631/1/012009

does not vanish if the superpotential is not zero at the ground state. In this case, supersymmetry is broken by T and, since the value of $\langle T\rangle$ is not fixed by the potential, the scale of supersymmetry breaking is arbitrary and unrelated to any fixed scale of the theory. The gravitino mass

$$m_{3/2} = \langle e^{\mathcal{K}/2}W\rangle = \langle (T+\overline{T})^{-3/2}e^{\widehat{\mathcal{K}}/2}W\rangle \tag{5.20}$$

is the order parameter of supersymmetry breaking. Since supersymmetry is broken with zero cosmological constant, $m_{3/2}$ is the true gravitino mass.

The Kähler potential $-3\ln(T+\overline{T})$ commonly appears in compactifications from ten dimensions, T being the volume modulus of the compact space. The S field in a superstring context could arise from the dilaton scalar (the string coupling field) partner of the metric $g_{\mu\nu}$ and of an antisymmetric tensor $B_{\mu\nu}$. In this case however, the superpotential does not depend on S and eq. (5.18) cannot be solved. The consequence is in general the absence of a ground state. Non-perturbative corrections are necessary to create the dependence on S and a minkowskian ground state.

It is maybe of interest to examine "dilaton supergravity" more precisely. In four space-time dimensions, an antisymmetric tensor $B_{\mu\nu}$ with gauge invariance and (free) wave equation

$$\delta B_{\mu\nu} = \partial_\mu\Lambda_\nu - \partial_\nu\Lambda_\mu, \qquad \partial^\mu H_{\mu\nu\rho} = 0, \qquad H_{\mu\nu\rho} = 3\,\partial_{[\mu}B_{\nu\rho]} \tag{5.21}$$

describes 3_B off-shell (since gauge invariance removes three fields) and 1_B on-shell states. The single massless on-shell state has of course helicity zero. Combined with a real scalar C and a Majorana spinor χ_α, the three fields form an off-shell representation of supersymmetry without any auxiliary field. Actually, at the level of global $\mathcal{N}=1$ supersymmetry, the variations

$$\begin{aligned} &\delta C = i\epsilon\chi - i\overline{\epsilon\chi}, \qquad \delta B_{\mu\nu} = \tfrac{i}{2\sqrt{2}}\Big(\overline{\epsilon}[\overline{\sigma}^\mu,\sigma^\nu]\overline{\chi} - \epsilon[\sigma^\mu,\overline{\sigma}^\nu]\chi\Big), \\ &\delta\chi_\alpha = -i(\sigma^\mu\overline{\epsilon})_\alpha\Big(\tfrac{1}{\sqrt{2}}\epsilon_{\mu\nu\rho\sigma}H^{\nu\rho\sigma} - i\partial_\mu C\Big) \end{aligned} \tag{5.22}$$

close the supersymmetry algebra without using field equations. This *linear* multiplet can be coupled to supergravity and is also a representation of the superconformal algebra with Weyl weight $w=2$ for C. The variations (5.22) indicate that a constant C does not break supersymmetry and that the spinor χ cannot be a Goldstino spinor. Hence a linear multiplet is not a source for supersymmetry breaking and it does not contribute to the scalar potential.

A duality transformation can always, in principle, transform the antisymmetric tensor with gauge symmetry into a real scalar τ with shift symmetry. Schematically,

$$H_{\mu\nu\rho} = 3\,\partial_{[\mu}B_{\nu\rho]} \qquad \longleftrightarrow \qquad \epsilon_{\mu\nu\rho\sigma}\partial^\sigma\tau \tag{5.23}$$

with symmetry $\delta\tau=$ constant. In other words, the dual chiral theory has a Kähler potential $\mathcal{K}(S+\overline{S})$, $\tau = \mathrm{Im}\,S$.

On one side of the duality, the linear L does not have an auxiliary field and cannot break supersymmetry. On the other side, the chiral dual S has an auxiliary f_S in principle able to induce supersymmetry breaking. In global supersymmetry, what happens is that either f_S is identically zero or, if other chiral multiplets are present, f_S is a linear combination of the other auxiliary fields f^i. In any case, f_S does not provide an *independent* source of supersymmetry breaking. Hence, models with supersymmetry breaking induced exclusively by the chiral S dual to the string dilaton and the antisymmetric tensor $B_{\mu\nu}$ do not exist.

In supergravity with a linear multiplet, the situation is different. If the superpotential is constant (this is meaningless in global supersymmetry), there is always a scalar potential generated by the supergravity auxiliary field f_0 (which however cannot break supersymmetry). In the dual version with the chiral S, the auxiliary field f_S is proportional to f_0, but f_S is now in principle able to break supersymmetry, since $\overline{f}_S \sim \mathcal{K}_T W$ is not zero in general. What happens now if that the potential is in general unstable if $W \neq 0$: this is the runaway behaviour naturally expected from the dilaton, which in turn raises the problem of its stabilisation. Notice that supergravity with the Kähler potential $-3\ln(T+\overline{T})$, which has identically zero potential and broken supersymmetry if $W \neq 0$, *cannot* be transformed into a linear superfield: the supersymmetric duality transformation between S and L does not exist for precisely this Kähler potential. If other chiral multiplets are present, f_S is a linear combination of the chiral auxiliary fields f^i and of the supergravity f_0. Again, f_S is not an independent source of breaking. But stability remains a non simple issue.

The simplest Calabi-Yau compactifications of heterotic superstrings with $\mathcal{N} = 1$ four-dimensional supersymmetry provide a concrete realization of the mechanisms described in this section. Retaining the overall (complex) volume modulus T in a chiral multiplet and the dilaton–antisymmetric tensor supermultiplet, we have two dual descriptions,

$$L \ \text{(linear)} \ \text{and} \ T \qquad \Longleftrightarrow \qquad S \ \text{(chiral)} \ \text{and} \ T.$$

In the chiral version, at lowest order of string perturbation theory, the Kähler potential defining the effective supergravity is of no-scale type [46, 47, 48, 49],

$$\mathcal{K} = -\ln(S+\overline{S}) - 3\ln(T+\overline{T}). \tag{5.24}$$

There are two primary sources for a superpotential [46, 48]. Firstly, at the perturbative level, ten-dimensional sixteen-supercharge supergravity has a gauge-invariant three-form H_{MNP}. It generates $H_{\mu\nu\rho} = 3\,\partial_{[\mu}B_{\nu\rho]}$ in the dilaton sector and an order parameter H_{ijk}, leading to a constant superpotential $W = \langle H \rangle$, since the Calabi-Yau space has a holomorphic three-form. At this stage, the minimum equation (5.18) cannot be solved and the potentiel

$$V = (T+\overline{T})^{-3}(S+\overline{S})W\overline{W} \tag{5.25}$$

is unstable ("run-away behaviour"), as expected from dilaton supergravity. The second source of superpotential is nonperturbative gaugino condensation in a hidden gauge sector, and it is described to a good approximation by the addition to the constant $\langle H \rangle$ of a term of the form ae^{bS}:

$$W(S) = \langle H \rangle + ae^{bS}. \tag{5.26}$$

The equation $f_S = 0$ can now be solved and supersymmetry breaks in the T sector, with Minkowski geometry and order parameter (4.14) controlled by the arbitrary value of $T + \overline{T}$.

It should however be observed that heterotic string perturbation theory is organized in powers of the linear supermultiplet L [50], with its scalar C directly related to the string dilaton, and not as an expansion in S. Since the superpotential cannot depend on L, the description of gaugino condensation uses then a different effective lagrangian, with almost identical phenomenology as long as supersymmetry breaking and scales are concerned [51]. In any case, the $S \sim L$ duality is a useful tool in the effective description of the universal string dilaton sector.

6. Final words

After (almost) forty years, supergravity has certainly found its way into the toolbox of theoretical physicists. Its development is far from complete and gauge-gravity dualities, in particular, have recently suggested new directions and research projects. In the context of superstring compactifications, finding methods to classify supergravity gaugings, and the corresponding symmetry and supersymmetry breaking patterns would allow a better control of flux compactifications, with supergravity providing in this case the "bottom-up" approach to select candidate fluxes from specific low-energy properties. In the unification programme, supergravity cannot claim to be the fundamental theory. But it is certainly on the "supersymmetric path" to quantum gravity. And there are open questions concerning the quantum status of the maximal $\mathcal{N} = 8$ supergravity. Supersymmetric theories have exceptional ultraviolet properties. The maximal ($\mathcal{N} = 4$) super-Yang-Mills theory is known to be finite. Brilliant works have shown that divergences plausible in $\mathcal{N} = 8$ arise in perturbation theory at higher orders than expected. This suggests that the $\mathcal{N} = 8$ theory could maybe display ingredients of a consistent quantum gravity, in a much simpler theoretical framework than superstrings.

Acknowledgments

I wish to thank the organizers of Discrete 2014, and especially Nick Mavromatos, for the invitation to present in their conference a subject which fascinates me since many years. This work has been in part supported by the Swiss National Science Foundation.

References

[1] J. Wess and B. Zumino, Nucl. Phys. B **70** (1974) 39; Phys. Lett. B **49** (1974) 52; Nucl. Phys. B **78** (1974) 1.
[2] S. Ferrara and B. Zumino, Nucl. Phys. B **79** (1974) 413.
[3] L. Brink, J. H. Schwarz and J. Scherk, Nucl. Phys. B **121** (1977) 77.
[4] J. Iliopoulos and B. Zumino, Nucl. Phys. B **76** (1974) 310.
[5] P. Fayet and J. Iliopoulos, Phys. Lett. B **51** (1974) 461.
[6] L. O'Raifeartaigh, Nucl. Phys. B **96** (1975) 331.
[7] S. Ferrara and B. Zumino, Nucl. Phys. B **87** (1975) 207.
[8] A. Salam and J. A. Strathdee, Nucl. Phys. B **76** (1974) 477.
[9] S. Ferrara, J. Wess and B. Zumino, Phys. Lett. B **51** (1974) 239.
[10] Y. A. Golfand and E. P. Likhtman, JETP Lett. **13** (1971) 323 [Pisma Zh. Eksp. Teor. Fiz. **13** (1971) 452].
[11] R. Haag, J. T. Lopuszanski and M. Sohnius, Nucl. Phys. B **88** (1975) 257.
[12] P. Fayet, Nucl. Phys. B **90** (1975) 104; Phys. Lett. B **64** (1976) 159.
[13] D. Z. Freedman, P. van Nieuwenhuizen and S. Ferrara, Phys. Rev. D **13** (1976) 3214.
[14] S. Deser and B. Zumino, Phys. Lett. B **62** (1976) 335.
[15] M. T. Grisaru, W. Siegel and M. Rocek, Nucl. Phys. B **159** (1979) 429.
[16] R. Barbieri, S. Ferrara and C. A. Savoy, Phys. Lett. B **119** (1982) 343.
[17] A. H. Chamseddine, R. L. Arnowitt and P. Nath, Phys. Rev. Lett. **49** (1982) 970.
[18] L. Girardello and M. T. Grisaru, Nucl. Phys. B **194** (1982) 65.
[19] L. E. Ibáñez and G. G. Ross, Phys. Lett. B **110** (1982) 215.
[20] L. Alvarez-Gaume, M. Claudson and M. B. Wise, Nucl. Phys. B **207** (1982) 96.
[21] L. Alvarez-Gaume, J. Polchinski and M. B. Wise, Nucl. Phys. B **221** (1983) 495.
[22] L. E. Ibáñez and C. Lopez, Phys. Lett. B **126** (1983) 54.
[23] J. R. Ellis, J. S. Hagelin, D. V. Nanopoulos and K. Tamvakis, Phys. Lett. B **125** (1983) 275.
[24] S. Ferrara and L. Maiani, CERN-TH-4232/85.
[25] S. Ferrara, L. Girardello, T. Kugo and A. Van Proeyen, Nucl. Phys. B **223** (1983) 191.
[26] K. S. Stelle and P. C. West, Phys. Lett. B **74** (1978) 330; **77** (1978) 376.
[27] S. Ferrara and P. van Nieuwenhuizen, Phys. Lett. B **74** (1978) 333; **76** (1978) 404.
[28] M. F. Sohnius and P. C. West, Phys. Lett. B **105** (1981) 353.
[29] G. Girardi, R. Grimm, M. Muller and J. Wess, Phys. Lett. B **147** (1984) 81.
[30] C. S. Aulakh, J. P. Derendinger and S. Ouvry, Phys. Lett. B **169** (1986) 201.
[31] W. Siegel, Class. Quant. Grav. **3** (1986) L47.
[32] W. Lang, J. Louis and B. A. Ovrut, Phys. Lett. B **158** (1985) 40.
[33] Z. Komargodski and N. Seiberg, JHEP **1007** (2010) 017 [arXiv:1002.2228 [hep-th]].
[34] D. Arnold, J. P. Derendinger and J. Hartong, Nucl. Phys. B **867** (2013) 370 [arXiv:1208.1648 [hep-th]].
[35] J. Ellis and N. E. Mavromatos, Phys. Rev. D **88** (2013) 8, 085029 [arXiv:1308.1906 [hep-th]].
[36] N. E. Mavromatos, EPJ Web Conf. **95** (2015) 03023 [arXiv:1412.6437 [gr-qc]].
[37] E. Cremmer, S. Ferrara, L. Girardello and A. Van Proeyen, Nucl. Phys. B **212** (1983) 413.
[38] D. Z. Freedman and A. Van Proeyen, *Supergravity*, Cambridge University Press, Cambridge, (2012) 607 p.
[39] T. Kugo and S. Uehara, Nucl. Phys. B **226** (1983) 49; B **222** (1983) 125; B **226** (1983) 93.
[40] D. V. Volkov and V. A. Soroka, JETP Lett. **18** (1973) 312 [Pisma Zh. Eksp. Teor. Fiz. **18** (1973) 529].
[41] S. Deser and B. Zumino, Phys. Rev. Lett. **38** (1977) 1433.
[42] E. Cremmer, B. Julia, J. Scherk, P. van Nieuwenhuizen, S. Ferrara and L. Girardello, Phys. Lett. B **79** (1978) 231; Nucl. Phys. B **147** (1979) 105.
[43] A. Adams, H. Jockers, V. Kumar and J. M. Lapan, JHEP **1112** (2011) 042 [arXiv:1104.3155 [hep-th]].
[44] E. Cremmer, S. Ferrara, C. Kounnas and D. V. Nanopoulos, Phys. Lett. B **133** (1983) 61.
[45] A. B. Lahanas and D. V. Nanopoulos, Phys. Rept. **145** (1987) 1.
[46] J. P. Derendinger, L. E. Ibáñez and H. P. Nilles, Phys. Lett. B **155** (1985) 65.
[47] E. Witten, Phys. Lett. B **155** (1985) 151.
[48] M. Dine, R. Rohm, N. Seiberg and E. Witten, Phys. Lett. B **156** (1985) 55.
[49] J. P. Derendinger, L. E. Ibáñez and H. P. Nilles, Nucl. Phys. B **267** (1986) 365.
[50] S. Cecotti, S. Ferrara and M. Villasante, Int. J. Mod. Phys. A **2** (1987) 1839.
[51] C. P. Burgess, J.-P. Derendinger, F. Quevedo and M. Quirós, Phys. Lett. B **348** (1995) 428 [hep-th/9501065].

Hamilton–Jacobi meet Möbius

Alon E. Faraggi[1] **and Marco Matone**[2]

[1] Department of Mathematical Sciences, University of Liverpool, Liverpool L69 7ZL, UK
[2] Dipartimento di Fisica, Universit'a di Padova, Via Marzolo 8, I–35131 Padova, Italy

E-mail: `alon.faraggi@liv.ac.uk; marco.matone@pd.infn.it`

Abstract.
Adaptation of the Hamilton–Jacobi formalism to quantum mechanics leads to a cocycle condition, which is invariant under D–dimensional Möbius transformations with Euclidean or Minkowski metrics. In this paper we aim to provide a pedagogical presentation of the proof of the Möbius symmetry underlying the cocycle condition. The Möbius symmetry implies energy quantization and undefinability of quantum trajectories, without assigning any prior interpretation to the wave function. As such, the Hamilton–Jacobi formalism, augmented with the global Möbius symmetry, provides an alternative starting point, to the axiomatic probability interpretation of the wave function, for the formulation of quantum mechanics and the quantum spacetime. The Möbius symmetry can only be implemented consistently if spatial space is compact, and correspondingly if there exist a finite ultraviolet length scale. Evidence for non–trivial space topology may exist in the cosmic microwave background radiation.

1. Introduction

The possibility that the Standard Model of particle physics provides a viable parameterisation of subatomic data up to the Planck scale received substantial support from the observation of a Higgs–like particle at the LHC. While the properties of this Higgs–like particle will be the subject of experimental scrutiny in the decades to come, it is clear that fundamental understanding of the Standard Model parameters can only be gained by incorporating gravity into the picture. Alas, the synthesis of gravity and quantum mechanics remains an enigma. The majority of contemporary efforts entail the quantisation of gravity, *i.e.* the quantisation of the metric in Einstein's theory of general relativity. The most developed approach in this endeavour is string theory [1]. While many alternative approaches exist, which in principle should be regarded on equal footing, the key advantage of string theory is that, while producing a consistent approach to quantum gravity, it gives rise to the matter and gauge structures that arise in the Standard Model. Several examples of quasi–realistic string models have been constructed over the years, and include the free fermionic standard–like models [2]. This class of models produce solely the states of the Minimal Supersymmetric Standard Model in the Standard Model charged sector, and produced a successful prediction of the top quark mass [3]. String theory provides an effective perturbative approach to quantum gravity. Important properties of string theory include the various perturbative [4] and non–perturbative dualities [5]. While string theory, due to its relation to observational data, can be regarded as an important step on the road to the synthesis of gravity and quantum mechanics, it does not provide a satisfactory final answer. What we would like to have is a formulation of quantum gravity, which follows from a basic physical hypothesis, akin to equivalence principle underlying general relativity.

An alternative approach to the quantisation of gravity may be pursued by seeking the geometrisation of quantum mechanics. This is the aim of the equivalence postulate (EP) approach to quantum mechanics [6, 7, 8, 9, 10, 11, 12, 13]. The EP formalism may be regarded as adaptation of the Hamilton–Jacobi (HJ) formalism to quantum mechanics. In the classical HJ theory the solution to the mechanical problem is obtained by performing canonical transformation to a system in which the phase space variables are constants of the motion. The Hamiltonian in the new system therefore vanishes identically and the Hamilton equations of motion are zero. The transformation from the old set of to the new set treat the phase space variables as independent variables. The solution to this problem is given by the Classical Hamilton–Jacobi Equation (CHJE). The functional relation between the phase space variables is then extracted from the solution of the CHJE $S(q)$. We may a pose a similar question, but rather than performing canonical transformations that treat the phase space variables as independent variables, we seek a trivialising coordinate transformation from the system q^a to the system q^b, and the induced transformation $p^a = \partial_a S^a(q^a) \to p^b = \partial_b S^b(q^b)$. Without loss of generality we further impose that $S^a(q^a) = S^b(q^b)$, *i.e.* that $S(q)$ transform as a scalar function under the transformation. By trivialising we mean that in the new coordinate system the kinetic and potential energy vanish. However, it is clear that this proposition would not make sense in classical mechanics. The reason is that in classical mechanics we have the solution $S(q) = constant$ for the state with vanishing kinetic energy and vanishing potential. In this case we have $p \equiv 0$, and this will remain the case after the transformation. This does not make sense because if we insist that all the physical states should be connectable to the free state with vanishing energy and vanishing potential, consistency dictates that the inverse transformation should exist as well. This means that we have to remove the solution $S(q) = constant$ from the space of allowed solution. We therefore seek a modification of the CHJE that will remove the state $S(q) = constant$ from the space of allowed solutions. In the stationary case the CHJE is given by

$$\frac{1}{2m}\left(\frac{\partial S_0}{\partial q}\right)^2 + V(q) - E = 0. \tag{1}$$

We consider the modification of the CHJE by adding the function $Q(q)$ to the CHJE, whose properties are yet to be determined. We further require that the all the physical systems labelled by $W(q) = V(q) - E$ can be connected by coordinate transformations to the trivial state $W(q) \equiv 0$. The condition that $S(q) \neq constant$ and that $S(q) = constant$ can be reached only in the classical limit, entails that the classical limit coincides with the limit $Q(q) \to 0$.

Imposing these conditions consistently leads to a cocyle condition. The modified Hamilton–Jacobi equation corresponds to the Quantum Hamilton–Jacobi Equation (QHJE), and is related to the Schrödinger equation. The cocycle condition is invariant under D–dimensional Möbius transformations. This invariance is the key property of quantum mechanics in the EP approach, and is equivalent to the condition that $S(q) \neq constant$. The Möbius symmetry implies that spatial space is compact and the the decompactification limit coincides with the classical limit, *i.e.* the limit in which $Q(q) \to 0$. Furthermore, it implies the existence of a finite length scale in the ultraviolet and correspondingly a finite length scale in the infrared. Hence, the Möbius symmetry has far reaching implications on the structure of the geometry, and correspondingly on the structure of the quantum space time. In this paper our aim is to provide a pedagogical presentation of the proof of the Möbius invariance of the cocycle condition.

2. The Hamilton–Jacobi Theory

Physics is first and foremost an experimental science, and may be defined as the mathematical modelling of experimental observations. A successful mathematical model is the one that can account for a wider range of experimental data. One may contemplate the possibility that there exist a representation of the physical laws, which is devoid of any experimental input, and follows from rigorous mathematical reasoning. The relation between the circumference of a circle to its diameter may attest to this possibility. Given the finite period allocated to an experiment, its resolution is similarly limited. Infinite resolution requires infinite time and the proposition that physics may have a completely mathematically rigorous representation, without resource to experimental input, is theological.

The methodology of the mathematical modelling of experimental observations is constructed as follows. Start with some initial conditions of a set of physical observables. Build a mathematical model for how this set of variables may evolve in the course of the physical process. Measure the set of variables in some experimental apparatus. Confront the observations by using the experimental apparatus with the prediction of the mathematical model. The key to the experimental methodology is that, given a proper set of instructions for the preparation of the initial conditions and the experimental apparatus, the outcome of the experimental measurements will be identical.

The mathematical modelling therefore rests on identifying a set of variables that are to be measured in the experiment. In the Newtonian system these may be the position and velocities of some objects. Say, a ball is dropped at some height h, with zero initial velocity. Then measure its position and velocity as a function of time. In this type of experiment the position and velocity change with time, which makes the experiment difficult. What we would like to do is to measure quantities that do not change with time. For example, if we can measure the initial energy and we know that the energy does not change with time, then we can measure the energy at the end of the experiment and we should get the same value. We can model the physical process from the start of the experiment and how the total energy is distributed among the constituents of the mathematical model. Ultimately, whatever happens between the start and the end of the experiment the total energy should be the same. What we are after are the constants of the motion. The Hamilton–Jacobi (HJ) formulation of classical mechanics is completely equivalent to Newtonian mechanics, and provides a general method to extract the constants of the motion for any physical system.

To achieve this fiat in classical mechanics we have to make a change of variables from configuration space to phase space. In configuration space the set of variables are the positions and velocities. In phase space the set of variables are the coordinates and the momenta. What we are after is to get rid of the explicit time dependence of the variables, so that the constants of the motion can be more readily extracted.

The Hamiltonian in classical mechanics is a function of the phase space variables. In systems in which the Hamiltonian does not depend explicitly on time the Hamiltonian corresponds to the total energy of the physical system and is a constant of the motion. The Hamilton equations of motion are given by

$$\dot{q} = \frac{\partial H}{\partial p} \;, \; \dot{p} = -\frac{\partial H}{\partial q} \tag{2}$$

The Hamilton–Jacobi procedure then follows by transforming the Hamiltonian to a trivial Hamiltonian, with the result that the new phase space–variables are constants of the motion, *i.e.*

$$H(q,p) \longrightarrow K(Q,P) \equiv 0 \qquad \Longrightarrow \qquad \dot{Q} = \frac{\partial K}{\partial P} \equiv 0 \;, \; \dot{P} = -\frac{\partial K}{\partial Q} \equiv 0$$

The solution is the Classical Hamilton–Jacobi Equation

$$H(q,p) \longrightarrow K(Q,P) \;=\; H(q, p = \frac{\partial S}{\partial q}) + \frac{\partial S}{\partial t} \;=\; 0 \;\Rightarrow\; \text{CHJE},$$

which in the stationary case becomes eq. (1). The transformations from the old to the new phase space variables, $(q,p) \rightarrow (Q,P)$ are canonical, which treat the phase space variables as

Journal of Physics: Conference Series **631** (2015) 012010 doi:10.1088/1742-6596/631/1/012010

independent variables. Their functional relation is then extracted from the solution of the HJ equation as $p = \partial_q S_0(q)$. We may reverse this order. Namely, we seek a trivialising coordinate transformation, such that $V(q) = E = 0$, but keeping the relation $p = \partial_q S_0(q) =$ when performing the transformations. The transformation are reversible coordinate transformations. Generality, then demands that we should be able to perform the reverse of the transformation from the trivial state to the non–trivial state. Furthermore, from the trivial state, we should be able to transform to any non–trivial state. However, as we discussed above this is not consistent in classical mechanics. As seen from eq. (1), while the first term transforms as quadratic differential, the second term, in general, does not. Furthermore, the state $W(q) = V(q) - E = 0$, corresponding to the solution $S_0 = constant$ with $p = 0$, is a fixed point under the transformations. Therefore, consistent implementation of this trivialising procedure requires the modification of the CHJE. As we elaborate in the next section the modification leads to a cocycle condition which is invariant under D–dimensional Möbius transformations. Our aim is to provide a pedagogical presentation of this proof. The Möbius symmetry is the fundamental property underlying quantum mechanics in our approach. It provides an alternative to the axiomatic probability interpretation of the wave function, and by that it provides a framework for rigorous formulation of quantum mechanics. Furthermore, invariance under the Möbius transformations reveals the existence of an inherent fundamental length in the formalism [10]. Proper implementation of the classical limit then shows that this length scale may be identified with the Planck length [10, 11]. In turn, the existence of a minimal length scale and the Möbius symmetry implies the existence of a maximal length scale. Furthermore, the limit in which the maximal length scale goes to infinity, *i.e.* the decompactification limit, correspondingly corresponds to the classical limit. The Möbius symmetry underlying the Quantum Hamilton–Jacobi Theory (QHJT), therefore carries within it an intrinsic ultraviolet regularisation length and correspondingly an infrared finite scale. The QHJT, with its underlying Möbius symmetry, provides the arena for the proper understanding of the quantum spacetime.

3. The cocycle condition

We seek the v–transformations

$$(q \, , \, S_0(q) \, , \, p = \frac{\partial S_0}{\partial q}) \longrightarrow (q^v \, , \, S^v{}_0(q^v) \, , \, p^v = \frac{\partial S^v{}_0}{\partial q^v})$$

Such that $W(q) \longrightarrow W^0(q^0) = 0$ exist for all $W(q)$. In the following we impose the conditions

$$S^v(q^v) = S(q), \tag{3}$$

and in particular

$$S^0(q^0) = S(q). \tag{4}$$

The CSHJE for a particle of mass m and energy E moving in N dimensions under the influence of a static velocity independent potential $V(q)$ is

$$\frac{1}{2m}\sum_{i=1}^{N}\left(\frac{\partial S}{\partial q_i}\right)^2 + W(q) = 0, \tag{5}$$

where $W \equiv V - E$, and where the q_i are Cartesian coordinates. Under a change of coordinates $q \to q^v$ we have by (3)

$$\frac{\partial S^v(q^v)}{\partial q^v_j} = \frac{\partial S(q)}{\partial q^v_j} = \sum_i \frac{\partial S(q)}{\partial q_i}\frac{\partial q_i}{\partial q^v_j}, \tag{6}$$

which we can write as

$$\boldsymbol{p}^v = \mathbf{J}^v \boldsymbol{p} \tag{7}$$

where

$$J^v_{ij} = \frac{\partial q_i}{\partial q^v_j}$$

is the Jacobian matrix connecting the coordinate systems q and q^v and $p_i = \frac{\partial S}{\partial q_i}$. Then

$$\begin{aligned} \sum_j \left(\frac{\partial S^v}{\partial q^v_j}\right)^2 &= |\boldsymbol{p}^v|^2 \\ &= \left(\frac{|\boldsymbol{p}^v|^2}{|\boldsymbol{p}|^2}\right)|\boldsymbol{p}|^2 \\ &= (p^v|p)|\boldsymbol{p}|^2, \end{aligned} \tag{8}$$

where we defined

$$(p^v|p) \equiv \frac{|\boldsymbol{p}^v|^2}{|\boldsymbol{p}|^2} \tag{9}$$

$$\begin{aligned} &= \frac{\boldsymbol{p}^{v\mathsf{T}}\boldsymbol{p}^v}{\boldsymbol{p}^\mathsf{T}\boldsymbol{p}} \\ &= \frac{\boldsymbol{p}^\mathsf{T}\mathbf{J}^{v\mathsf{T}}\mathbf{J}^v\boldsymbol{p}}{\boldsymbol{p}^\mathsf{T}\boldsymbol{p}}. \end{aligned} \tag{10}$$

Note also that by (7)

$$\begin{aligned} \boldsymbol{p} &= \mathbf{J}^{v-1}\boldsymbol{p}^v \\ \Rightarrow \quad p_i &= \sum_j (\mathbf{J}^{v-1})_{ij} p^v_j. \end{aligned} \tag{11}$$

But simply relabelling (6)

$$p_i = \frac{\partial S(q)}{\partial q_i} = \frac{\partial S^v(q^v)}{\partial q_i} = \sum_j \frac{\partial S^v(q^v)}{\partial q^v_j}\frac{\partial q^v_j}{\partial q_i} = \sum_j \frac{\partial q^v_j}{\partial q_i} p^v_j. \tag{12}$$

Thus $(\mathbf{J}^{v-1})_{ij} = \frac{\partial q^v_j}{\partial q_i}$ and we have

$$(\mathbf{J}^{v-1})_{ij} = (J^v_{ij})^{-1}. \tag{13}$$

Then using (12) in (9) we have

$$[(p^v|p)]^{-1} = (p|p^v) = \frac{\boldsymbol{p}^{v\mathsf{T}}\mathbf{J}^{v-1\mathsf{T}}\mathbf{J}^{v-1}\boldsymbol{p}^v}{\boldsymbol{p}^{v\mathsf{T}}\boldsymbol{p}^v}, \tag{14}$$

where by (13)

$$\begin{aligned} [\mathbf{J}^{v-1\mathsf{T}}\mathbf{J}^{v-1}]_{jk} &= \sum_i \mathbf{J}^{v-1}_{ik}\mathbf{J}^{v-1}_{ij} \\ &= \sum_i \left[J^v_{ik}J^v_{ij}\right]^{-1}. \end{aligned} \tag{15}$$

Having determined that the first term in (5) transforms as a quadratic differential under v-maps, we require that for consistency the second term transforms similarly That is:

$$W^v(q^v) = (p^v|p)W(q), \tag{16}$$

and

$$W^0(q^0) = (p^0|p)W(q), \tag{17}$$

where $W^0 \equiv 0$ corresponds to the free particle at rest, with $V = 0$ and $E = 0$. Substituting $W^0 = 0$ in the left–hand side of eq. (17) yields

$$0 = (p^0|p)W(q). \tag{18}$$

Hence, W^0 is a fixed point under the v–maps and cannot be connected to other states. Therefore, classical mechanics in not compatible with the equivalence postulate.

As the CSHJE is not consistent with the equivalence postulate, we consider the modification

$$W \to W + Q,$$

The Modified Stationary Hamilton Jacobi Equation takes the form

$$\frac{1}{2m}\sum_i\left(\frac{\partial S}{\partial q_i}\right)^2 + W(q) + Q(q) = 0. \tag{19}$$

Covariance under v–map imposes the transformation property

$$W^v + Q^v = (p^v|p)\left[W + Q\right]. \tag{20}$$

In order that the v–map may connect W^0 to any other state, W must transform with an inhomogeneous term,

$$W^v = (p^v|p)W + (q; q^v), \tag{21}$$

and by eq, (20)

$$Q^v = (p^v|p)Q - (q; q^v). \tag{22}$$

We note that for an arbitrary state

$$\begin{aligned} W &= (p|p^0)W^0 + (q^0; q) \\ &= 0 + (q^0; q) \\ &= (q^0; q), \end{aligned} \tag{23}$$

so the inhomogeneous term for the trivialising map generates all other W states. Let us consider the properties of the inhomogeneous term. First, we note from (21) that in the case of the identity transformation $q \to q$

$$\begin{aligned} W &= (p|p)W + (q; q) \\ \Rightarrow (q; q) &= 0. \end{aligned} \tag{24}$$

Second, from eq. (21) we can write

$$W^b = (p^b|p^a)W^a + (q^a; q^b), \tag{25}$$

$$W^c = (p^c|p^a)W^a + (q^a; q^c), \tag{26}$$

and

$$W^c = (p^c|p^b)W^b + (q^b; q^c), \tag{27}$$

for three states a, b, c. Substituting (25) into (27) and equating this with (26) we obtain

$$\begin{aligned}(p^c|p^a)W^a + (q^a; q^c) &= (p^c|p^b)\left[(p^b|p^a)W^a + (q^a; q^b)\right] + (q^b; q^c)\\ \Rightarrow (q^a; q^c) - (q^b; q^c) &= (p^c|p^b)(q^a; q^b),\end{aligned} \tag{28}$$

or

$$(q^a; q^b) = (p^b|p^c)\left[(q^a; q^c) - (q^b; q^c)\right]. \tag{29}$$

Eq. 29) is our celebrated *cocycle condition*. It expresses the essence of the quantum mechanics in the equivalence postulate approach. We further note that if we set $q^a = q^c$ in (29) we get

$$(q^a; q^b) = -(p^b|p^a)(q^b; q^a). \tag{30}$$

4. Higher dimensional Möbius group

The form of the cocycle condition has far–reaching implications. Here we reproduce the arguments introduced in [12] which reveal a symmetry of the inhomogeneous term $(q^a; q^b)$, under D–dimensional Möbius transformations. Our aim is to provide a more pedagogical presentation of the proof of these properties.

We denote by $q = (q_1, \cdots, q_D)$ an arbitrary point in R^D. A similarity is a D–dimensional transformation that includes translations, rotations and dilatations. Similarities are naturally extended to the compactified space $\hat{R}^D = R^D \cup \{\infty\}$. A similarity maps ∞ to itself. Setting

$$r^2 = q_1^2 + \cdots + q_D^2, \tag{31}$$

the last generator of the Möbius group is the *inversion* or reflection in the unit sphere S^{D-1}. If $q \neq 0$

$$q \longrightarrow q^* = \frac{q}{r^2}, \tag{32}$$

otherwise

$$0 \longrightarrow \infty, \qquad \infty \longrightarrow 0. \tag{33}$$

Summarising, an arbitrary Möbius transformation of an N-dimensional vector $q = (q_1, \ldots, q_i, \ldots, q_N)$ is made up of any combination of:

- a 'translation' $q \to q^B = q + B$ (B a vector in R^D),
- a 'dilatation' $q \to q^A = Aq$ (A a real number),
- a 'rotation' $q \to q^R = Rq$ (R an orthogonal $D \times D$matrix),
- an 'inversion' $q \to q^* = \frac{q}{r^2}$ ($r^2 \equiv \sum_i q_i^2$).

The Möbius transformations naturally extend to the compactified space $\hat{R}^D = R^D \cup \{\infty\}$. A similarity maps ∞ to itself. The Möbius group $M(\hat{R}^D)$ is defined as the set of transformations generated by all similarities together with the inversion. A general Möbius transformation is the combination of a number of reflections and inversions. A Möbius transformation is conformal with respect to the Euclidean metric. A theorem due to Liouville states that the conformal group. $M(\hat{R}^D)$ actually coincide[1] for $D > 2$.

[1] for review of the D–dimensional Möbius group, see *e.g.* [14].

4.1. Transformation Factors

In the subsections bellow we derive the transformations factors $(p^M|p)$ in the case of the Möbius transformations $q \to q^M$ discussed in section 4.

4.1.1. Translation The Jacobian for a translation is

$$J^B_{ij} = \frac{\partial q_i}{\partial q^B_j} = \frac{\partial q_i}{\partial(q_j + B_j)} = \frac{\partial q_i}{\partial q_j} \times \left(\frac{\partial(q_j + B_j)}{\partial q_j}\right)^{-1} = \delta_{ij}. \tag{34}$$

Hence, $\mathbf{J}^B = I_N$, where I_N is the $N \times N$ identity matrix. The transformation factor is therefore given by

$$(p^B|p) = \frac{\mathbf{p}^\mathsf{T} I_N^\mathsf{T} I_N \mathbf{p}}{\mathbf{p}^\mathsf{T}\mathbf{p}} = 1. \tag{35}$$

4.1.2. Dilatation The Jacobian for a dilatation is

$$J^A_{ij} = \frac{\partial q_i}{\partial q^A_j} = \frac{\partial q_i}{\partial(A q_j)} = A^{-1}\delta_{ij}. \tag{36}$$

Then $\mathbf{J}^A = A^{-1} I_N$ and

$$(p^A|p) = \frac{\mathbf{p}^\mathsf{T} A^{-1} I_N^\mathsf{T} A^{-1} I_N \mathbf{p}}{\mathbf{p}^\mathsf{T}\mathbf{p}} = A^{-2}, \tag{37}$$

or

$$(p|p^A) = A^2. \tag{38}$$

4.1.3. Rotation We have

$$\begin{aligned} q^R &= Rq \\ \Rightarrow \quad q &= R^{-1} q^R \\ \text{and} \quad q_i &= \sum_k \left(R^{-1}\right)_{ik} q^R_k, \\ \text{so that} \quad J^R_{ij} = \frac{\partial q_i}{\partial q^R_j} &= \left(R^{-1}\right)_{ij}. \end{aligned}$$

Furthermore, since R is orthogonal

$$\begin{aligned} R^\mathsf{T} R = R R^\mathsf{T} &= I_N \\ \Rightarrow \quad R^\mathsf{T} &= R^{-1} \end{aligned}$$
$$\text{so that} \quad \mathbf{J}^R = R^\mathsf{T} \quad \text{and} \quad \mathbf{J}^{R^\mathsf{T}} = R.$$

This leads to

$$(p^R|p) = \frac{\mathbf{p}^\mathsf{T} R R^\mathsf{T} \mathbf{p}}{\mathbf{p}^\mathsf{T}\mathbf{p}} = \frac{\mathbf{p}^\mathsf{T} I_N \mathbf{p}}{\mathbf{p}^\mathsf{T}\mathbf{p}} = 1. \tag{39}$$

4.1.4. Inversion This requires a modicum of care. First, observe that

$$\begin{aligned} \frac{\partial r^2}{\partial q_i} &= \frac{\partial}{\partial q_i}\sum_k q_k^2 \\ \Rightarrow \quad 2r\frac{\partial r}{\partial q_i} &= 2q_i \\ \Rightarrow \quad \frac{\partial r}{\partial q_i} &= \frac{q_i}{r}. \end{aligned} \tag{40}$$

For the Jacobian we get

$$\begin{aligned} J^*_{ij} = \frac{\partial q_i}{\partial q^*_j} &= \left(\frac{\partial q^*_j}{\partial q_i}\right)^{-1} \\ \Rightarrow \quad \left(J^*_{ij}\right)^{-1} = \frac{\partial q^*_j}{\partial q_i} &= \frac{\partial}{\partial q_i}\left(\frac{q_j}{r^2}\right) \\ &= \frac{\delta_{ij}}{r^2} - \frac{2q_j}{r^3}\frac{\partial r}{\partial q_i} \\ &= \frac{\delta_{ij}}{r^2} - \frac{2}{r^4}q_i q_j. \end{aligned} \tag{41}$$

Now, using (14)

$$(p|p^*) = \frac{\boldsymbol{p}^{*\mathsf{T}}\mathbf{J}^{*-1\mathsf{T}}\mathbf{J}^{*-1}\boldsymbol{p}^*}{\boldsymbol{p}^{*\mathsf{T}}\boldsymbol{p}^*}, \tag{42}$$

where

$$\begin{aligned} [\mathbf{J}^{*-1\mathsf{T}}\mathbf{J}^{*-1}]_{kj} = \sum_i \left[J^*_{ik}J^*_{ij}\right]^{-1} &= \sum_i \left(\frac{\delta_{ik}}{r^2} - \frac{2}{r^4}q_i q_k\right)\left(\frac{\delta_{ij}}{r^2} - \frac{2}{r^4}q_i q_j\right) \\ &= \sum_i \frac{\delta_{ik}\delta_{ij}}{r^4} - \frac{2}{r^6}\sum_i q_i(q_j\delta_{ik} + q_k\delta_{ij}) + \frac{4}{r^8}\sum_i q_i^2 q_k q_j \\ &= \frac{\delta_{kj}}{r^4} - \frac{2}{r^6}q_j q_k - \frac{2}{r^6}q_k q_j + \frac{4r^2}{r^8}q_k q_j \\ &= \frac{\delta_{kj}}{r^4} \\ \Rightarrow \quad \mathbf{J}^{*-1\mathsf{T}}\mathbf{J}^{*-1} &= r^{-4}I_N. \end{aligned} \tag{43}$$

Substituting this into (42) gives us

$$(p|p^*) = \frac{\boldsymbol{p}^{*\mathsf{T}}r^{-4}I_N\boldsymbol{p}^*}{\boldsymbol{p}^{*\mathsf{T}}\boldsymbol{p}^*} = r^{-4} \quad \Rightarrow \quad (p^*|p) = r^4. \tag{44}$$

5. Inhomogeneous Term

In this section we prove the invariance of the inhomogeneous term, and hence of the cocycle condition under D–dimensional Möbius transformations.

5.1. Translation

The cocycle condition (29) can be written

$$(q^a; q^b) = (p^b|p^c)(q^a; q^c) + (q^c; q^b), \tag{45}$$

Journal of Physics: Conference Series **631** (2015) 012010 doi:10.1088/1742-6596/631/1/012010

which tells us that

$$\begin{aligned}(q+B+C;q) &= (p|p^B)(q+B+C;q+B)+(q+B;q) \\ &= (q+B+C;q+B)+(q+B;q),\end{aligned} \tag{46}$$

or,

$$\begin{aligned}(q+B+C;q) &= (p|p^C)(q+B+C;q+C)+(q+C;q) \\ &= (q+C+B;q+C)+(q+C;q),\end{aligned} \tag{47}$$

where B, C are two arbitrary constant vectors. Now consider restricting B, C so that they each have only a single component along the j-axis:

$$B = (0,0,\ldots,0,B_j,0,\ldots,0) \qquad \text{and} \qquad C = (0,0,\ldots,0,C_j,0,\ldots,0)\ ; \qquad j \in \{1,\ldots,N\}.$$

We can then define a function

$$f(B,q) \equiv (q+B;q), \tag{48}$$

and, equating (46) with (47) we obtain

$$\begin{aligned} f(C,q+B)+f(B,q) &= f(B,q+C)+f(C,q) \\ \Rightarrow \qquad f(C,q+B) &= f(B,q+C)+f(C,q)-f(B,q).\end{aligned} \tag{49}$$

We can differentiate this with respect to B_j. Note first that

$$\frac{\partial(q_j+B_j)}{\partial q_j} = \frac{\partial(q_j+B_j)}{\partial B_j} = 1,$$

so for any function P of the combination $q+B$ we have

$$\begin{aligned}\partial_{B_j}P(q+B) &= \frac{\partial P(q+B)}{\partial q_j} \times \left(\frac{\partial(q_j+B_j)}{\partial q_j}\right)^{-1} \times \frac{\partial(q_j+B_j)}{\partial B_j} \\ &= \partial_{q_j}P(q+B)\ ; \qquad \text{where} \qquad \partial_z \equiv \frac{\partial}{\partial z}.\end{aligned} \tag{50}$$

Differentiating (49) then:

$$\partial_{q_j} f(C,q+B) = \partial_{B_j}\left[f(B,q+C)-f(B,q)\right]. \tag{51}$$

From (24)

$$f(0,q) \equiv (q;q) = 0. \tag{52}$$

We can express f in a general series form which guarantees this. Writing

$$f(B,q) = \sum_{n=1}^{\infty} c_n(q) B_j^n. \tag{53}$$

and substituting (53) into (51) we have

$$\begin{aligned}\partial_{q_j}\sum_{n=1}^{\infty} c_n(q+B)C_j^n &= \partial_{B_j}\sum_{n=1}^{\infty}\left[c_n(q+C)-c_n(q)\right]B_j^n \\ &= \sum_{n=1}^{\infty}\left[c_n(q+C)-c_n(q)\right] n B_j^{n-1}.\end{aligned} \tag{54}$$

Using a Taylor expansion the term on the left hand side becomes

$$\begin{aligned}
\partial_{q_j} \sum_{n=1}^{\infty} c_n(q+B)C_j^n &= \partial_{q_j} \sum_{n=1}^{\infty} \left[\sum_{m=0}^{\infty} \frac{1}{m!} \partial_{q_j}^m c_n(q) B_j^m \right] C_j^n \\
&= \sum_{n=1}^{\infty} \left[\sum_{m=0}^{\infty} \frac{1}{m!} \partial_{q_j}^{m+1} c_n(q) B_j^m \right] C_j^n \\
&= \sum_{n=1}^{\infty} \left[\sum_{l=1}^{\infty} \frac{1}{(l-1)!} \partial_{q_j}^{l} c_n(q) B_j^{l-1} \right] C_j^n , \qquad \text{where } l = m+1, \\
&= \sum_{l,n=1}^{\infty} \frac{1}{(n-1)!} \partial_{q_j}^{n} c_l(q) B_j^{n-1} C_j^l,
\end{aligned} \tag{55}$$

where in the last line we have simply switched indices (n for l) under the double summation. The right hand side of (54) becomes:

$$\begin{aligned}
\sum_{n=1}^{\infty} \left[c_n(q+C) - c_n(q) \right] n B_j^{n-1} &= \sum_{n=1}^{\infty} \left[\sum_{l=0}^{\infty} \frac{1}{l!} \partial_{q_j}^{l} c_n(q) C_j^l - \frac{1}{0!} \partial_{q_j}^{0} c_n(q) C_j^0 \right] n B_j^{n-1} \\
&= \sum_{l,n=1}^{\infty} \frac{1}{l!} \partial_{q_j}^{l} c_n(q) C_j^l n B_j^{n-1}.
\end{aligned} \tag{56}$$

Equating (55) with (56) we find

$$\begin{aligned}
\sum_{l,n=1}^{\infty} \frac{1}{(n-1)!} \partial_{q_j}^{n} c_l(q) B_j^{n-1} C_j^l &= \sum_{l,n=1}^{\infty} \frac{n}{l!} \partial_{q_j}^{l} c_n(q) B_j^{n-1} C_j^l \\
\Rightarrow \qquad \frac{1}{n!} \partial_{q_j}^{n} c_l(q) &= \frac{1}{l!} \partial_{q_j}^{l} c_n(q).
\end{aligned} \tag{57}$$

Setting $l = 1$ gives

$$\partial_{q_j} c_n(q) = \frac{1}{n!} \partial_{q_j}^{n} c_1(q). \tag{58}$$

If we differentiate (53) with respect to q_j and substitute in (58) we obtain

$$\begin{aligned}
\partial_{q_j} f(B,q) = \sum_{n=1}^{\infty} \partial_{q_j} c_n(q) B_j^n &= \sum_{n=1}^{\infty} \frac{1}{n!} \partial_{q_j}^{n} c_1(q) B_j^n \\
&= \sum_{n=0}^{\infty} \frac{1}{n!} \partial_{q_j}^{n} c_1(q) B_j^n - c_1(q) \\
&= c_1(q+B) - c_1(q).
\end{aligned} \tag{59}$$

Integrating this:

$$f(B,q) = c(q+B) - c(q) + g(B,\hat{q}), \tag{60}$$

where we define $c(q)$ such that $\partial_{q_j} c(q) = c_1(q)$ and where $g(B,\hat{q})$ is some function of anything *but* q_j (so $\hat{q}$ denotes all the q_i with $i \neq j$). Recall from (52) that $f(0,q) = 0$. Thus:

$$f(0,q) = c(q) - c(q) + g(0,\hat{q}) = g(0,\hat{q}) = 0. \tag{61}$$

Using the cocycle condition (46) once again:

$$\begin{aligned}
(q+B+C;q) &= (q+B+C;q+B) + (q+B;q) \\
\Rightarrow \qquad f(B+C,q) &= f(C,q+B) + f(B,q),
\end{aligned} \tag{62}$$

Journal of Physics: Conference Series **631** (2015) 012010 doi:10.1088/1742-6596/631/1/012010

that is

$$c(q+B+C)-c(q)+g(B+C,\hat{q}) = c(q+B+C)-c(q+B)+g(C,\hat{q}+\hat{B})+c(q+B)-c(q)+g(B,\hat{q}).$$

Also $\hat{B} = 0$ since by construction $B_i = 0$ for $i \neq j$. So we end up with

$$g(B+C,\hat{q}) = g(C,\hat{q}) + g(B,\hat{q}). \tag{63}$$

We have shown that g is linear in its first argument, and that it vanishes when its first argument is set to zero. We can write g in a general form which guarantees these properties:

$$g(B,\hat{q}) = K(\hat{q})B_j. \tag{64}$$

Now differentiating $f(B,q)$, with respect to B_j this time, we find on the one hand

$$\begin{aligned} \partial_{B_j} f(B,q) &= \partial_{B_j}\left[c(q+B) - c(q) + g(B,\hat{q})\right] \\ &= \partial_{q_j} c(q+B) + \partial_{B_j} K(\hat{q})B_j \\ &= c_1(q+B) + K(\hat{q}), \end{aligned} \tag{65}$$

and on the other hand,

$$\partial_{B_j} f(B,q) = \sum_{n=1}^{\infty} n c_n(q) B_j^{n-1} = c_1(q) + 2c_2(q)B_j + \ldots, \tag{66}$$

so setting $B = 0$ and equating (65) with (66) we have

$$\begin{aligned} c_1(q) + K(\hat{q}) &= c_1(q) \\ \Rightarrow \qquad K(\hat{q}) &= 0. \end{aligned} \tag{67}$$

This means that

$$g(B,\hat{q}) = 0 \times B_j = 0,$$

and so we arrive at the result

$$f(B,q) = c(q+B) - c(q). \tag{68}$$

Next, we consider a general constant vector D (which we need not restrict to having only a single component). We first define the function

$$G(D,q) \equiv (q+D;q). \tag{69}$$

Note that we must have

$$G(B,x) = f(B,x) \tag{70}$$

where as before B has only one component along the j-axis. Again we use the cocycle condition (46,47),

$$\begin{aligned} (q+B+D;q) &= (q+B+D;q+B) + (q+B;q) \\ \textit{and} \qquad (q+B+D;q) &= (q+D+B;q+D) + (q+D;q) \\ \Rightarrow (q+B+D;q+B) - (q+B+D;q+D) &= (q+D;q) - (q+B;q) \\ \text{so} \qquad G(D,q+B) - G(B,q+D) &= G(D,q) - G(B,q), \\ \text{that is} \qquad G(D,q+B) - f(B,q+D) &= G(D,q) - f(B,q) \\ \Rightarrow G(D,q+B) - c(q+D+B) + c(q+D) &= G(D,q) - c(q+B) + c(q). \end{aligned}$$

Differentiating this with respect to B_j we get

$$\partial_{q_j} G(D, q+B) - \partial_{q_j} c(q+D+B) = -\partial_{q_j} c(q+B), \tag{71}$$

which yields, for $B=0$,

$$\partial_{q_j} G(D,q) = \partial_{q_j} \left[c(q+D) - c(q) \right]. \tag{72}$$

Integrating *this* gives

$$G(D,q) = c(q+D) - c(q) + \hat{G}(D,\hat{q}), \tag{73}$$

where $\hat{G}(D,\hat{q})$ is some function of the q_i $(i \neq j)$ which by (70) and (68) we know must vanish if D has only a single component along the j-axis. Putting this back into the cocycle condition again we get

$$\begin{aligned} G(D+B,q) &= G(B,q+D) + G(D,q) \qquad (74)\\ \Rightarrow c(q+D+B) - c(q) + \hat{G}(D+B,\hat{q}) &= c(q+D+B) - c(q+D) + \hat{G}(B,\hat{q}+\hat{D}) + \\ & \quad c(q+D) - c(q) + \hat{G}(D,\hat{q}) \\ \text{so} \quad \hat{G}(D+B,\hat{q}) &= \hat{G}(B,\hat{q}+\hat{D}) + \hat{G}(D,\hat{q}), \end{aligned}$$

so we see that $\hat{G}(D,\hat{q})$ satisfies the same algebra as $G(D,q)$ (only with one fewer variable since $\hat{q}$ excludes q_j). We can therefore apply the same arguments to $\hat{G}(D,\hat{q})$ as we have used for $G(D,q)$: We'll end up with a relation like (73) for $\hat{G}(D,\hat{q})$, with a new function tacked onto the end which is now only a function of $N-2$ of the q_i. Then we put *this* back into the cocycle condition - and so on and so on...

Having worked our way recursively through all N components we obtain

$$G(D,q) \equiv (q+D;q) = F(q+D) - F(q) + H(D), \tag{75}$$

where $H(D)$ is some function which vanishes whenever D has only one component. Since $H(D)$ has no second argument, we find that upon substitution of (75) into (74) we are left with

$$H(D+C) = H(D) + H(C), \tag{76}$$

so H is linear. We can write H in a general form which guarantees this property:

$$H(D) = \sum_{i=1}^{N} a_i D_i \tag{77}$$

(just a linear combination of the components of D). However we require that $H=0$ if $D_i=0$ and $D_j \neq 0$ $(i \neq j)$ for any j, so we must have that $a_j = 0$ for any j, which means that H is identically vanishing. We have determined the form for the inhomogeneous term under arbitrary translations to be

$$(q+D;q) = F(q+D) - F(q). \tag{78}$$

5.2. Dilatation

In a similar way to our previous treatment we define a function

$$h(A,q) \equiv (Aq;q), \tag{79}$$

and use the cocycle condition to examine its structure. From (45), and using (78), we have

$$\begin{aligned} (A[q+B];q) &= (p|p^A)(A[q+B];Aq) + (Aq;q) \\ &= A^2(Aq+AB;Aq) + (Aq;q) \\ &= A^2\left[F(Aq+AB) - F(Aq)\right] + h(A,q), \end{aligned} \tag{80}$$

which may also be written as,

$$\begin{aligned}(A[q+B];q) &= (p|p^B)(A[q+B];q+B)+(q+B;q)\\ &= h(A,q+B)+F(q+B)-F(q).\end{aligned} \tag{81}$$

Equating Eqs. (80) and 81) we obtain,

$$h(A,q+B)-h(A,q)=A^2\left[F(Aq+AB)-F(Aq)\right]-F(q+B)+F(q), \tag{82}$$

and differentiating (82) with respect to B_j we find

$$\partial_{q_j}h(A,q+B)=A^2\partial_{q_j}F(Aq+AB)-\partial_{q_j}F(q+B). \tag{83}$$

Setting $B=0$ gives

$$\partial_{q_j}h(A,q)=\partial_{q_j}\left[A^2F(Aq)-F(q)\right], \tag{84}$$

which upon integration yields

$$h(A,q)=A^2F(Aq)-F(q)+g(A), \tag{85}$$

where $g(A)$ is some function which cannot depend on any of the q_i (as we could have chosen any j in (83)). Now, at the origin ($q=0$) a dilatation can have no effect, so $h(A,0)$ is independent of A and

$$\begin{aligned} h(A,0)=h(1,0)\equiv(q;q) &= 0\\ \Rightarrow h(A,0)=A^2F(A\times 0)-F(0)+g(A) &= 0\\ \Rightarrow \qquad g(A)=F(0)\left[1-A^2\right].\end{aligned}$$

Putting this back into (85) gives

$$h(A,q)=A^2\left[F(Aq)-F(0)\right]-\left[F(q)-F(0)\right], \tag{86}$$

or more concisely

$$h(A,q)\equiv(Aq;q)=A^2F(Aq)-F(q), \tag{87}$$

where we defined $F(0)=0$.

5.3. Rotation

Following similar steps, from the cocycle condition (80,81) we have

$$(R[q+B];q) = (p|p^R)(Rq+RB];Rq)+(Rq;q) \tag{88}$$

$$\text{and} \quad (R[q+B];q) = (p|p^B)(R[q+B];q+B)+(q+B;q) \tag{89}$$

$$\Rightarrow (R[q+B];q+B)-(Rq;q) = (Rq+RB;Rq)-(q+B;q). \tag{90}$$

$$\text{Using (78)}, \quad (R[q+B];q+B)-(Rq;q) = \{F(R[q+B])-F(q+B)\}-\{F(Rq)-F(q)\}.$$

This is satisfied by

$$(Rq;q)=F(Rq)-F(q)+C\,; \qquad C \text{ a constant.} \tag{91}$$

A rotation can have no effect at the origin, so $(Rq;q)|_{q=0}\equiv(q;q)=0$ and

$$F(R\times 0)-F(0)+C=0 \quad \Rightarrow \quad C=0.$$

The form of the inhomogeneous term under rotations is therefore $(Rq;q)=F(Rq)-F(q)$.

5.4. Inversion

To begin with we consider how lengths transform under the generators of the Möbius group.

$$r^{*2} = \sum_{i=1}^{N}(q^*)_i^2 = \sum_{i=1}^{N}\left(\frac{q_i}{r^2}\right)^2 = \frac{1}{r^4}r^2 = \frac{1}{r^2}, \tag{92}$$

$${r^A}^2 = \sum_{i=1}^{N}(Aq)_i^2 = A^2r^2, \tag{93}$$

and trivially, since lengths are preserved by rotations,

$${r^R}^2 = r^2. \tag{94}$$

(92) fixes q as *involutive*:

$$(q^*)_i^* = \left(\frac{q}{r^2}\right)_i^* = \frac{q_i^*}{r^{*2}} = \frac{q_i}{r^2\frac{1}{r^2}} = q_i. \tag{95}$$

From (94) we see that rotation commutes with inversion:

$$(Rq)_i^* = \frac{(Rq)_i}{{r^R}^2} = \frac{\sum_k R_{ik}q_k}{r^2} = \sum_k R_{ik}q_k^* = (Rq^*)_i. \tag{96}$$

Finally, under dilatations we have, using (93),

$$(Aq)_i^* = \frac{(Aq)_i}{{r^A}^2} = \frac{Aq_i}{A^2r^2} = A^{-1}q_i^*. \tag{97}$$

Now, applying (30), we see from (95) that

$$(q^*;q) = -(p|p^*)(q;q^*) = -\frac{1}{r^4}([q^*]^*;q^*), \tag{98}$$

which vanishes when evaluated at any q such that $q = q^*$:

$$(q^*;q)|_{q=q_0} = 0\;; \qquad q_0 = q_0^*. \tag{99}$$

Next, bearing in mind (87), we revisit the cocycle condition (45):

$$\begin{aligned}([Aq]^*;q) &= (p|p^A)([Aq]^*;Aq) + (Aq;q)\\ &= A^2([Aq]^*;Aq) + A^2F(Aq) - F(q),\end{aligned} \tag{100}$$

but using (97), we can equally write

$$([Aq]^*;q) = (p|p^*)([Aq]^*;q^*) + (q^*;q) \tag{101}$$

$$\begin{aligned}&= \frac{1}{r^4}(A^{-1}q^*;q^*) + (q^*;q)\\ &= \frac{1}{r^4}\left[(A^{-1})^2F(A^{-1}q^*) - F(q^*)\right] + (q^*;q).\end{aligned} \tag{102}$$

So, equating (100) with (102) gives:

$$A^2([Aq]^*;Aq) + A^2F(Aq) - F(q) = \frac{1}{r^4}\left[A^{-2}F(A^{-1}q^*) - F(q^*)\right] + (q^*;q). \tag{103}$$

If we choose a point q_0 on the surface of the unit sphere, so that $r_0^2 = \sum_i (q_{0i})^2 = 1$ and $q_0 = q_0^*$, then (by (99)) (103) reduces to

$$\begin{aligned} A^2([Aq_0]^*; Aq_0) &= A^{-2}F(A^{-1}q_0^*) - F(q_0^*) - A^2F(Aq_0) + F(q^0) \\ \Rightarrow \qquad ([Aq_0]^*; Aq_0) &= \frac{1}{A^4}F(A^{-1}q_0^*) - F(Aq_0). \end{aligned} \tag{104}$$

Lastly, we note that any vector q can be expressed in the polar form $r\hat{q}$ where r is the length of q and where $\hat{q}$ is a vector of unit length parallel to q. Clearly $\hat{q}$ automatically has the property required of q_0. The mapping $\hat{q} \to r\hat{q}$ is simply a dilatation of $\hat{q}$ with $A = r$, so we can put $Aq_0 = q$ (and from (97) $A^{-1}q_0^* = [Aq_0]^* = q^*$) in (104) to obtain the final result,

$$(q^*; q) = \frac{1}{r^4}F(q^*) - F(q). \tag{105}$$

6. Summary

To summarise the quantum Hamilton–Jacobi, eq. (19), and the cocycle condition, eq. (29), imply that $(q^a; q^b)$ vanishes if q^a and q^b are related by a Möbius transformation, that is

$$(q + B; q) = 0, \tag{106}$$
$$(Aq; q) = 0, \tag{107}$$
$$(\Lambda q; q) = 0, \tag{108}$$
$$(q^*; q) = 0. \tag{109}$$

These equations are equivalent to $(\gamma(q); q) = 0$, where $\gamma(q)$ is a general Möbius transformation. Moreover, from eq. (29) we have

$$(\gamma(q^a); q^b) = (q^a; q^b), \qquad (q^a; \gamma(q^b)) = (p^{\gamma(b)}|p^b)(q^a; q^b). \tag{110}$$

Considering the Jacobian factor $(p^{\gamma(b)}|p^b)$ we0 observe that the Möbius transformation is conformal with respect to the Euclidean metric, *i.e.*00

$$ds^2 = \sum_{j=1}^{D} d\gamma(q)_j d\gamma(q)_j = \sum_{j,k,l=1}^{D} \frac{\partial\gamma(q)_j}{\partial q_k}\frac{\partial\gamma(q)_j}{\partial q_l} dq_k dq_l = \sum_{j=1}^{D} e^{\phi_\gamma(q)} dq_j dq_j. \tag{111}$$

Hence,

$$(p^{\gamma(b)}|p^b) = e^{-\phi_\gamma(q^b)}. \tag{112}$$

We note that in the case of rotations and translations the conformal re–scaling is the identity. For dilatations $\exp \phi^A = A^2$, whereas for the inversion $\exp \phi^* = r^{-4}$.

We remark that this conformal structure is obtained by setting $S_0^v(q^v) = S_0(q)$. We would like to emphasise that this is not a restriction on the formalism, but merely a convenient choice. Any transformation that we may impose, other than $S_0^v(q^v) = S_0(q)$, would yield the same results. The freedom in setting $S_0^v(q^v) = S_0(q)$ results from the fact that q and q^v represent the spatial coordinates in their own systems. The condition $S_0^v(q^v) = S_0(q)$ can be seen just as the simplest way to set the coordinate transformations from the system with reduced action S_0^v to the one with reduced action S_0. Since the physics is determined by the functional structure of S_0^v, we can denote the coordinate as we like. However, this is not the case in classical mechanics, as for a free particle with vanishing energy we have $S_0(q) = cnst$, and imposing $S_0^v(q^v) = S_0(q)$ does not make sense. Requiring that this is well–defined for any system is synonymous to imposing the equivalence postulate, and the definability of phase space duality for all physical states. The existence of the conformal structure, manifested by the invariance of the inhomogeneous term under Möbius transformations, which in $D \geq 3$ coincides with the conformal group, is at the core of quantum mechanics.

7. Quadratic identities

In the previous sections we discussed the cocycle condition, which is obtained by requiring the QHJE equation retain its form under coordinate transformations. The transformation properties of the kinetic term fix those of the classical and added potential to transform as quadratic differentials. Furthermore, in the one dimensional case the Möbius symmetry uniquely fixes the functional form of the inhomogeneous term to that of the Schwarzian derivative. An identity of Schwarzian derivatives follows from these transformation properties and is given by

$$\left(\frac{\partial S_0}{\partial q}\right)^2 = \frac{\beta^2}{2}\left(\{e^{\frac{i2S_0}{\beta}};q\} - \{S_0;q\}\right)$$

With the identifications

$$W(q) = -\frac{\beta^2}{4m}\{e^{\frac{i2S_0}{\beta}};q\} = V(q) - E \tag{113}$$

$$Q(q) = \frac{\beta^2}{4m}\{S_0;q\}, \tag{114}$$

the modified Hamilton–Jacobi equations becomes

$$\frac{1}{2m}\left(\frac{\partial S_0}{\partial q}\right)^2 + V(q) - E + \frac{\beta^2}{4m}\{S_0;q\} = 0.$$

We therefore note that for the state $W(q) \equiv 0$ the Modified HJ equation admits the solutions

$$S_0 = \pm\frac{\beta}{2}\ln q^0 \neq Aq^0 + B.$$

Hence, quantum mechanics enables consistency of the equivalence postulate by removing the linear solutions from the space of solutions. As we discussed before, since $Q(q)$ is never vanishing quantum mechanics carries within it its own regularisation scheme. Furthermore, from eq. (113) and the properties of the Schwarzian derivative, the function $W(q) = V(q) - E$ is a potential of a second order differential equation given by,

$$\left(-\frac{\beta^2}{2m}\frac{\partial^2}{\partial q^2} + V(q) - E\right)\Psi(q) = 0,$$

which is the Schrödinger equation and we may identify the covariantising parameter β with the Planck constant $\hbar$. The general solutions is given by

$$\Psi(q) = (A\psi_1 + B\psi_2) = \frac{1}{\sqrt{S_0'}}\left(Ae^{+\frac{i}{\hbar}S_0} + Be^{-\frac{i}{\hbar}S_0}\right).$$

The solution of the Schwarzian equation eq. (113) is then given in terms of ratio of the solutions of the Schrödinger equation, *i.e.*

$$e^{+\frac{i2S_0}{\hbar}} = e^{i\alpha}\frac{w + i\bar{\ell}}{w - i\ell} \quad \text{where} \quad w = \frac{\psi_1}{\psi_2},$$

where, due to the symmetries of the Schwarzian derivative, the solution is given up to a Möbius transformation. Furthermore, from the condition that $S_0(q) \neq \textit{constant}$ we have that the constants ℓ and α satisfy $\ell = \ell_1 + i\ell_2$, $\ell_1 \neq 0$ and $\alpha \in R$.

While appearance of the Schwarzian derivative in the one dimensional case may seem a bit esoteric, the multi–dimensional case reveals more clearly the simplicity of the formalism. Consider applying the Laplacian to the function

$$\psi = R(q)\mathrm{e}^{\alpha S_0(q)}, \quad i.e. \quad \Delta\left(R(q)\mathrm{e}^{\alpha S_0(q)}\right).$$

Proper application of the chain rule then leads to a quadratic identity given by

$$\alpha^2(\nabla S_0)\cdot(\nabla S_0) = \frac{\Delta(Re^{\alpha S_0})}{Re^{\alpha S_0}} - \frac{\Delta R}{R} - \frac{\alpha}{R^2}\nabla\cdot\left(R^2\nabla S_0\right). \tag{115}$$

Setting $\alpha = i/\hbar$ the imaginary part of eq. (115) gives a continuity equation. The first term on the right–hand side of eq. (115) is identified with the classical potential, yielding the D–dimensional nonrelativistic Schrödinger equation, with the general solution given by

$$\Psi(q) = (A\psi_1 + B\psi_2) = R(q)\left(A\mathrm{e}^{+\frac{i}{\hbar}S_0} + B\mathrm{e}^{-\frac{i}{\hbar}S_0}\right), \tag{116}$$

where q_i are now the D–dimensional coordinates. These identifications produce the D–dimensional stationary nonrelativistic Quantum Hamilton–Jacobi Equation given by

$$\frac{1}{2m}(\nabla S_0)\cdot(\nabla S_0) + V(q) - E - \frac{\hbar^2}{2m}\frac{\Delta R}{R} = 0.$$

8. Time parameterisation

We note that the QHJE is reminiscent of Bohm's approach to quantum mechanics [15]. However, there is a crucial difference, which is precisely related to the Möbius symmetry underlying quantum mechanics in the equivalence postulate approach. As is well known Bohm's approach argues for the existence of a trajectory representation of quantum mechanics, in which time parameterisation of trajectories is obtained. In Bohm's approach the wave function is identified with $A = 0$ and $B \neq 0$ in eq. (116). In that case one identifies the conjugate momentum as

$$p = \hbar \mathrm{Im}\frac{\nabla\psi}{\psi}, \tag{117}$$

which can be used to define a trajectory representation by identifying the conjugate momentum the mechanical momentum, *i.e.* by setting $p = m\dot{q}$. However, the choice $A = 0$ and $B \neq 0$ is not consistent with the Möbius symmetry underlying quantum mechanics, which necessitates that $A \neq$ **and** $B \neq 0$. Alternatively, the Möbius symmetry implies that space is compact, in which case the boundary conditions are not compatible with the choice $A = 0$ and $B \neq 0$ but impose that both must be included in the solution. Therefore, the Möbius symmetry of the QHJE implies that

$$\nabla S \neq \hbar \mathrm{Im}\frac{\nabla\psi}{\psi}$$

and Bohm's definition of trajectories is not valid [16]. An alternative proposal [17] to define trajectories in quantum mechanics suggests to use Jacobi's theorem that identifies time as the derivative of the S_0 with respect to E, and by replacing the solution of the CHJE, with the solution of the QHJE, *i.e.*

$$t - t_0 = \frac{\partial S_0^{\mathrm{qm}}}{\partial E}. \tag{118}$$

Time parameterisation of the trajectory can then be obtained by inverting $t(q) \to q(t)$. However, the Möbius symmetry that underlies the QHJE dictates that the energy levels are always quantised [16]. Hence, differentiation with respect to the energy is not well defined. Time in quantum mechanics can be thought of as a classical background parameter, but not as a fundamental quantum variable. At the quantum level trajectories may only have a probabilistic interpretation rather than a deterministic representation. It should be stressed, however, that while a deterministic time parameterisation is not consistent with the Möbius symmetry that underlies the QHJE, time parameterisation à la Bohm [15] or via the bi–polar representation [18, 19] provides a useful semi–classical approximation.

The compactness of space, imposed by the Möbius symmetry underlying the QHJE, implies that the energy levels are always quantised. However, in quantum mechanics this does not suffice. The probability interpretation of the wave function implies that the wave function for bound states is square integrable. In general, the differential equations associated with the quantum mechanical problems for bound states admit solutions that are not square integrable. In the one dimensional case it was proven rigoursly that trivialising transformations $q \to q^0 = \psi_1/psi_2$, where ψ_1 and ψ_2 are the two solutions of the Schrödinger equation, has to be continuous on the extended real line, *i.e.* the real line plus the point at infinity. This requirement is synonymous to the requirement that the Möbius symmetry is preserved. It is then shown [9, 11] that this condition is satisfied iff the corresponding Schrödinger equation admits a square integrable solution. Thus, the same physical states that are selected in conventional quantum mechanics by the axiomatic probability interpretation of the wave function, are selected by consistency in the EP approach. The EP approach may therefore be regarded as reproducing the basic properties of conventional quantum mechanics, with the caveat that spatial space is compact.

9. Conclusions

The requirement that the HJ equation retain its form under coordinate transformation led to the cocycle condition and the QHJE. In turn this led to the removal of the classical solution $S_0(q) = const$ from the space of admissible solutions and consequently the requirement that $p \neq 0$. These properties are intimately related to a duality in phase space that is defined in terms of the involutive nature of the Legendre transformations [6, 11]. Now, the Legendre transformations are not defined for linear functions. The solutions admitted by the QHJE in the case of the trivial classical potential correspond to the self–dual states under the phase space duality. Thus, the QHJE enables the consistency of the phase space duality for the entire space of solutions.

The Möbius symmetry underlying the QHJE is the fundamental property of quantum mechanics in the equivalence postulate approach. It provides an alternative to the axiomatic formulation of quantum mechanics. It necessitates the existence of the quantum potential, which is never vanishing. The quantum potential may be interpreted as an internal curvature term of elementary particles [8, 20], which can therefore be seen to be a direct consequence of the Möbius symmetry underlying the QHJE. In turn, the Möbius symmetry, and the duality structure that it enforces, implies the existence of a fundamental length scale in the formalism [10, 11]. compatibility with the classical limit then implies that the fundamental length scale may be identified with the Planck length. In turn, the existence of a fundamental length scale implies the admission of an ultraviolet cutoff. Similarly, the Möbius symmetry dictates that spatial space is compact. Thus, the Möbius symmetry provides an intrinsic quantum mechanical regularisation scheme in the ultraviolet, as well as the infrared. Furthermore, it leads to the phenomenological characteristics of quantum mechanics [9, 11] without assuming any prior interpretation of the wave function. The HJ formalism, augmented with the Möbius symmetry, therefore provides an alternative starting point to the axiomatic formulation of of quantum mechanics, based on the probability interpretation of the wave function. In that respect, while a fundamental appreciation of the geometrical role of the wave function is yet to be developed, a key guide may lie in duality relations between the wave function and the space coordinates [21]. We further note that the universality of the quantum potential implies that it corresponds to a universal force acting on elementary particles [22]. The Möbius symmetry underlying quantum mechanics implies that spatial space is compact, which may have left a remnant in the Cosmic Microwave Background Radiation [23]. Additionally, it leads to modified energy dispersion relations [24], which may affect the propagation of gamma rays over astrophysical distances [25].

Acknowledgements
This work is supported in part by the STFC under contract ST/L000431/1.

References

[1] See *e.g.*: Becker K, Becker M and Schwarz J H 2007, *String theory and M-theory: A modern introduction* (Cambridge University Press).
[2] Faraggi A E, Nanopoulos D V and Yuan K 1990 *Nucl. Phys.* **B335** 347;
Faraggi A E 1992 *Phys. Lett.* **B278** 131; 1992 *Nucl. Phys.* **B387** 239;
Cleaver G B, Faraggi A E and Nanopoulos D V 1999 *Phys. Lett.* **B455** 135;
Faraggi A E, Manno E and Timiraziu C 2007 *Eur. Phys. Jour.* **C50** 701;
Christodoulides K, Faraggi A E and Rizos J 2011 *Phys. Lett.* **B702** 81.
[3] Faraggi A E 1992 *Phys. Lett.* **B274** 47; 1997 *Nucl. Phys.* **B487** 55.
[4] See *e.g.*: Giveon A, Porrati M and Rabinovici E 1994 *Phys. Rep.* **244** 77.
[5] See *e.g.*: Hull C M and Townsend P K 1995 *Nucl. Phys.* **B438** 109;
Witten E 1995 1995 *Nucl. Phys.* **B443** 85.
[6] Faraggi A E and Matone M 1999 *Phys. Lett.* **B450** 34.
[7] Faraggi A E and Matone M 1998 *Phys. Lett.* **B437** 369.
[8] Faraggi A E and Matone M 1998 *Phys. Lett.* **A249** 180.
[9] Faraggi A E and Matone M 1999 *Phys. Lett.* **B445** 357.
[10] Faraggi A E and Matone M 1998 *Phys. Lett.* **B445** 77.
[11] Faraggi A E and Matone M 2000 *Int. J. Mod. Phys.* **A15** 1869.
[12] Bertoldi G, Faraggi A E and Matone M 2000 Class. Quant. Grav. **17** 3965.
[13] Faraggi A E and Matone M, arXiv:0912.1225.
[14] Vuorinen M 1988 *Conformal Geometry and Quasiregular Mappings, Lect. Notes Math.* **1319**, Springer–Verlag
[15] *For review and references see e.g.:* Holland P R 1993 *Phys. Rep.* **224** 95.
[16] Faraggi A E and Matone M 2014 *Eur. Phys. Jour.* **C74** 2694; arXiv:1502.04456
[17] Floyd E R 1982 *Phys. Rev.* **D26** 1339.
[18] Wyatt R E 2005 *Quantum dynamics with trajectories. Introduction to quantum hydrodynamics.* (Springer).
[19] Poirier B 2004 J. Chem. Phys. **121** 4501.
[20] Faraggi A E 2013 Adv. High Energy Phys. **2013**, 957394; arXiv:1305.0044.
[21] Faraggi A E and Matone M 1997 *Phys. Rev. Lett.* **78** 163.
[22] Matone M 2002 *Found. Phys. Lett.* **15** 311.
[23] Aslanyan G and Manohar A V 2012 *JCAP* **1206** 003;
Ben-David A, Rathaus B and Itzhaki N 2012 *JCAP* **1211** 020.
[24] Faraggi A E 2012 *Eur. Phys. Jour.* **C72** 1944.
[25] Amelino-Camelia G, Ellis J, Mavromatos N E, Nanopoulos D V and Sarkar S 1998 *Nature* **393** 763.

Lorentz symmetry breaking: phenomenology and constraints

Stefano Liberati
SISSA and INFN, Via Bonomea 265, 34136 Trieste, Italy

Abstract.
In this talk I shall review several motivations for considering departures from exact Lorentz invariance and the different theoretical frameworks adopted to describe these departures. Among these, I shall focus on an effective field theory approach and discuss the phenomenology and constraints of Lorentz symmetry breaking in the Standard Model as well as in Gravity. In particular I will focus on current constraints on UV breaking inspired by quantum gravity scenarios and briefly discuss the open issues and future perspectives for this field of research.

1. Introduction

Understanding the nature of spacetime at the Planck scale $M_{\rm Pl} = 1.2\times10^{28}$ eV and beyond is still one of the greatest challenges in modern physics. It is in fact at this scale that present theories fails to be predictive and quantum aspects of gravitation are expected to be relevant. While this endeavour has been in the past partially hampered by lack of experimental/observational tests, this situation has changed in the last fifteen years. Indeed, quantum gravity models have shown that there can be several sub-Planckian physics "relic signatures" of quantum gravitational effects able to produce potentially observable deviations from the standard predictions.

Some of these new phenomena, which comprise what is often termed "QG phenomenology", include tests of Quantum decoherence and state collapse [1], QG imprint on initial cosmological perturbations [2], Cosmological variation of couplings [3, 4], TeV Black Holes related to extra-dimensions [5], Planck scale spacetime fuzziness [6], Generalized uncertainty principle [7, 8, 9], Violation of discrete symmetries [10] and Violation of space-time symmetries [11].

Among the above fields a preeminent role was played by physics associated to violations of space-time symmetries in the form of departures from exact local Lorentz invariance (LI). In these proceedings I will provide a succinct description of the Effective Field theory (EFT) approach to this problem and provide a concise summary of its phenomenology and associated constraints. A more extensive recent review on the same subject can be found in [12]. See also [11, 8, 13].

2. Forms of Lorentz violations

The introduction of a set of observers endowed with standard clocks and rods and the agreement on a suitable procedure of synchronisation of such clocks allows to rigorously define operatively spacetime. Within such a setting the usual von Ignatovsky derivation (see *e.g.* [12] and references therein) of Lorentz transformations is based on four basic axioms: i) Pre-causality; ii) Spatial and temporal homogeneity; iii) Spatial isotropy; iv) Relativity principle. In view of such hypotheses, one can prove that the most general change of coordinates between two inertial reference frames reduce to the familiar Lorentz transformations with an invariant limit speed C to be determined by observations. Notice that the speed scale C is in principle arbitrary and can even be infinite — in this latter case one finds back the usual Galilean transformations.

In extending this construction to the case of a spacetime that change its nature beyond some scale it is worth revising the implications of the above mentioned assumptions [14]. Assumption i) guarantees that the temporal order of events along the history of an observer cannot be upturned by a change of reference frame. Hypothesis ii) enforces that the laws of transformation relating coordinates in two inertial frames have to be linear, i.e. independent of the location of space and time where they are performed. As for assumption iii), it mirrors the indifference of physical phenomena with respect to the orientation of reference axes. Assumption iv) embodies the notion of the equivalence of all the inertial frames. It also guarantees that the transformations gain a group structure. Finally, assumptions iii) and iv) yield the so-called Principle of Reciprocity (see *e.g.* [12] and references therein).

While assumption i) and ii) are generally assumed to hold (see however [14]) the hypothesis iii) can be relaxed, and a preferred direction can be introduced, generally reducing the numbers of generators and symmetries for the system. The resulting theory is commonly referred to as Very Special Relativity [15], and the set of transformations associated to this end up leaving invariant the four-dimensional spacetime manifold a Finslerian structure [16], i.e. explicit dependencies on the velocities are present in the symmetric metric tensor accounting for measurements of space and time intervals. In what follows we shall instead relax assumption iv) so we shall consider the case in which quantum gravitational physics could provide a preferred frame.

3. The EFT approach to Lorentz Violation

In the absence of a definitive theory of quantum gravity, several ideas have been present for discussing possible Lorentz breaking scenarios. As said, we shall focus here on the most conservative approach based on an EFT framework within which to describe and study the empirical effects of small deviations from exact LI. Such a framework goes under the name of Standard Model Extension (SME). This consists in supplementing the standard model of particle physics with all the possible Lorentz violating operators that can be written without changing the field content or violating gauge symmetry. The operators appearing in the SME can be conveniently classified according to their mass dimension and behaviour under $\mathcal{CPT}$ given that Lorentz violation does not imply $\mathcal{CPT}$ violation for local EFTs, while $\mathcal{CPT}$ violation does imply Lorentz violation in local EFT [17, 18].

The total SME consists of hundreds of operators, many of which are very tightly constrained, however it is usually simplified by imposing that subgroups of the Lorentz group or discrete symmetries such as $\mathcal{CPT}$ are preserved. As often done, we shall consider here that rotational invariance is a symmetry of nature in one particular frame which is generally identified with the rest frame of the cosmic background radiation (CBR). Given that our motion with respect to this frame is only $\beta \simeq 10^{-3}$ this does not significantly affect our results.

With the assumption of rotational invariance the number of possible operators for the SME drops significantly and it makes sense, from an EFT point of view, to classify the operators for their relative mass dimensions. In this sense one generally considers separately the case of operators of mass dimension 3,4 and that of higher order operators (mass dimension 5 and 6). The Minimal Standard Model Extension or mSME, is the subset of the SME which considers only mass dimension 3 and 4 operators i.e. only those which naively (without considering anisotropic scaling techniques [19]) power counting renormalizable. In what follows we shall discuss in detail the resulting EFT for leptons and photons (QED) and neutrinos given that these elementary particles are the most relevant for the current phenomenological constraints.

3.1. QED sector

The high energy ($M_{\rm Pl} \gg E \gg m$) rotational invariant dispersion relations e.g. for the mSME of QED can be expressed as (see e.g. [11] and references therein)

$$E_{\rm el}^2 = m_e^2 + p^2 + f_e^{(1)} p + f_e^{(2)} p^2 \quad \text{electrons} \tag{1}$$

$$E_\gamma^2 = (1 + f_\gamma^{(2)}) p^2 \quad \text{photons} \tag{2}$$

Noticeably in this limit the mSME reduces to the phenomenological theory for Lorentz breaking proposed by Coleman and Glashow where particles dispersion relation pertain their relativistic form but are characterised by different limit speeds [20, 21, 22].

Going further one can consider dimension 5 operators. In [23] it was found that there are essentially only three operators of dimension five, quadratic in the fields, that can be added to the QED Lagrangian preserving rotational and gauge invariance, but breaking local LI. These are all $\mathcal{CPT}$ odd. The corresponding high energy dispersion relations respectively for photons and electrons are

$$\omega_\pm^2 = k^2 \pm \xi \frac{k^3}{M_{\rm Pl}} \,, \tag{3}$$

where the + and − signs denote right and left circular polarisation, and

$$E_\pm^2 = p^2 + m^2 + \eta_\pm \frac{p^3}{M_{\rm Pl}} \,, \tag{4}$$

where the + and − signs now denoting positive and negative helicity states.

It is quite important to note that for positrons, the $\mathcal{CPT}$ odd character of the relevant operators implies that the same dispersion relation holds, with $\eta_{\bar{q}_\pm} = -\eta_{q_\mp}$ where $\bar{q}$ and q denote respectively anti-fermion and fermion [24, 25].

Finally, one can consider $\mathcal{CPT}$ even, rotationally invariant mass dimension five and six LIV terms has been computed in [26]. From these operators, the dispersion relations of fermions and photons can be easily computed, yielding

$$\begin{aligned} E^2 - p^2 - m^2 = & \frac{\alpha_R^{(6)} E^3}{M_{\rm Pl}^2}(E + sp) + \frac{\alpha_L^{(6)} E^3}{M_{\rm Pl}^2}(E - sp) + \\ & + \frac{m}{M_{\rm Pl}}(\alpha_R^{(5)} + \alpha_L^{(5)}) p^2 + \alpha_R^{(5)} \alpha_L^{(5)} \frac{p^4}{M_{\rm Pl}^2} \end{aligned} \tag{5}$$

$$\omega^2 - k^2 = \beta^{(6)} \frac{k^4}{M_{\rm Pl}^2} \,, \tag{6}$$

where m is the electron mass and where $s = \sigma \cdot \mathbf{p}/|\mathbf{p}|$ is the helicity of the fermions, and we have neglected terms of order $m^2/M_{\rm Pl}^2$ stemming from the $\tilde{\alpha}^{(6)}$ terms as they are highly suppressed.

We can further simplify eq. (5) by noting that high-energy ($E \sim p$) fermions states are almost exactly chiral, and by grouping terms with same powers of the momentum. In this case we get

$$E^2 = p^2 + m^2 + \frac{m}{M_{\rm Pl}} \eta^{(2)} p^2 + \eta_\pm^{(4)} \frac{p^4}{M_{\rm Pl}^2} \tag{7}$$

where $R = +$, $L = -$ and we have labelled $\eta^{(n)}$ is the dispersion coefficient of the LIV p^n term in the dispersion relation for the fermion. We choose $\eta^{(n)}$ as the fermion coefficient symbol as this

Journal of Physics: Conference Series **631** (2015) 012011 doi:10.1088/1742-6596/631/1/012011

nomenclature is common in the literature. Similarly, we shall use $\xi^{(n)}$ for the generic dispersion coefficient for a photon (so in (6) we shall take $\beta^{(6)} = \xi^{(4)}$).

In (7) the quadratic modification generated by the dimension five operator is suppressed by a factor of order $m/M_{\rm Pl}$ and hence it can be safely neglected, provided that $E > \sqrt{mM_{\rm Pl}}$. By $\mathcal{CPT}$, the dispersion relation of the anti-fermion is given by (5), with the replacements $s \to -s$ and $p \to -p$. If $q, \overline{q}$ denote a charge fermion and anti-fermion, then the relevant anti-fermion coefficient $\eta^{(6)}_{\overline{q}}$ is such that $\eta^{(6)}_{\overline{q}_\pm} = \eta^{(6)}_{q_\mp}$, where $\overline{q}_\pm$ indicates an anti-fermion of positive/negative helicity (and similarly for the $q_\pm$). This different behaviour between even and odd powers of "n" type dispersion relations leads to quite distinct phenomenologies.

3.2. Neutrino sector

Neutrino physics can be affected in several ways by Lorentz breaking operators. The most common channels for casting constraints are observations concerning the speed of neutrinos with respect to light, modified oscillations between neutrinos flavors and threshold reactions.

The generic neutrino LIV operators, at any mass dimension, have been categorized in [27]. Also in this case, a significant reduction in the number of terms can be achieved by requiring that the LIV operators are rotationally symmetric. Let us then focus on the Lagrangian for neutrinos with LIV operators of mass dimension up to six involving a vector field u^a coupled to a Dirac neutrino ψ [1] in a mass eigenstate with mass m that are quadratic in matter fields and hence modify the free field equations. We shall further assume that the Lorentz violation is diagonal in the mass basis. This might be justified by the idea that any theory of quantum gravity inducing such LIV must reduce to general relativity in the infrared, hence any Lorentz violation induced by quantum gravity would be primarily controlled by the charges that couple to gravity. Of course, this does not meant that the coefficients for each mass eigenstate are the same, as RG effects would not allow them to be the same at any energy.

With these assumptions, yield a high energy neutrino dispersion relation of the form [12]

$$E_i^2 = p^2 + N_i^2 \tag{8}$$

$$N_i^2 = m_i^2 + 2(a_i + b_i)p - (c_i + d_i)p^2 + 2(\zeta_{i,1} - \zeta_{i,2})\frac{p^3}{M_{\rm Pl}} + 2\frac{\alpha^{(6)}_{i,L}p^4}{M_{\rm Pl}^2}$$

As before, the term proportional to $m/M_{\rm Pl}$ has been ignored. The net result of the LIV terms is to modify the dispersion relation of each mass eigenstate according to

$$E^2 - p^2 - (m_i)^2 = \sum_{n=1}^{4} \xi_i^{(n)} \frac{|p|^n}{M_{\rm Pl}^{n-2}}, \tag{9}$$

where $\xi_i^{(n)}$ is a coefficient that depends on the relevant terms in the Lagrangian (8), so that constraints on the $\xi_i^{(n)}$ can always be translated in constraints on the coefficients of Eq. (8). The corresponding anti-particle dispersion relation is easily derived by considering the behaviour of each term under CPT and given by

$$E^2 - p^2 - (m_i)^2 = \sum_{n=1}^{4} (-1)^n \xi_i^{(n)} \frac{|p|^n}{M_{\rm Pl}^{n-2}}. \tag{10}$$

One can then leave the index n as a free phenomenological parameter and consider the cases $n = 2, 3, 4$ separately (the case $n = 1$ would produce huge effects at low energy and is strongly constrained).

[1] No Majorana, LIV operators can be constructed in the rotational invariant case [27] albeit it is not clear if they could be dynamically generated e.g. by the same mechanism providing mass to such neutrinos.

Note also that many existing neutrino oscillation experiments measure the transition probability between neutrino flavours P_{IJ} (I and J here labelling neutrino flavours) which is affected by the modified dispersion relation (8).

$$P_{IJ} = \delta_{IJ} - \sum_{i,j>i} 4F_{IJij} \sin^2\left(\frac{\delta N_{ij}^2 L}{4E}\right) + 2G_{IJij} \sin^2\left(\frac{\delta N_{ij}^2 L}{2E}\right) , \tag{11}$$

with $\delta N_{ij}^2 = N_i^2 - N_j^2$ and F_{IJij} and G_{IJij} are functions of the mixing matrixes. Many of these experiments also quote results on a deviation of the neutrino speed from that of light, i.e.

$$\left(\frac{\Delta c}{c}\right)_{ij}^{LIV} = E^{-2}(\delta N_{ij}^2 - \delta m_{ij}^2) \tag{12}$$

which can be easily translated into constraints on the coefficients of Eq. (8).

4. Phenomenology

The development of a systematic EFT-based approach as discussed above represented a milestone in searches for deviations from exact Lorentz invariance. The phenomenological toolkit that was developed also thanks to this systematic approach is now quite rich and can be split in two big subsets: terrestrial experiments and astrophysical observations.

4.1. Terrestrial experiments

It is an observational fact that Nature as we probe it well below the Planck scale it is Lorentz invariant to a very high degree. It is hence logic that in seeking for low energy deviations from this symmetry, as those systematically described by the minimal Standard Model extension (mSME), one has to resort to very high precision, and hence Earth based, experiments (some of them testing rotational breaking operators as well). An incomplete list includes (for more details see e.g. [11, 28, 29, 12]).

Clock comparison Experiments Two co-local atomic transition frequencies can be considered as two clocks. As the clocks move in space, they pick out different components of the Lorentz violating tensors in the mSME. This in turn yields a sidereal drift between the two clocks which could be constrained by measuring the difference between the frequencies over long periods. This technique allows to cast very high precision limits on some parameters in the mSME (generally for for protons and neutrons).

Cavity Experiments The technique adopted in these experimenys casts constraints on the variation of the cavity resonance frequency (with respect to a stationary frequency standard) as its orientation changes in space. While this is intrinsically similar to clock comparison experiments, these kind of experiments allows to cast constraints also on the electromagnetic sector of the mSME (as one of the "clocks" in this case involves photons).

Neutral mesons Experiments The mass difference of neutral mesons is one of the most accurately quantities in the SM. The SME operators do affect such a quantity in a Lorentz breaking way, hence one generically expects an orientation dependent change which can be constrained by looking for sidereal variations and other orientation effects. Also lifetime directional dependence has been investigated (see e.g. [30]).

Penning traps In a Penning trap a combination of static magnetic and electric fields confines a charged particle for long times. One can then problem possible deviation from exact Lorentz invariance by monitoring the particle cyclotron motion in the magnetic field and its Larmor precession due to the spin. In fact the relevant frequencies for both these motions are affected by some mSME operators and Penning traps can be set up so to make them very sensitive to differences in these frequencies.

Spin polarized torsion balances Spin-torsion balances are a very effective tools for constraining the electron sector of the mSME. An example of such balances consists in an octagonal pattern of magnets which is constructed so to have an overall spin polarization in the octagons plane. Four of these octagons are suspended from a torsion fibre in a vacuum chamber so to give an estimated net spin polarization equivalent to $\approx 10^{23}$ aligned electron spins. In order to detect Lorentz breaking effects one has again to look for orientation dependent phenomena.

Nuclear Spin Experiments The absence of a preferred direction is also checked with great precision using nuclear spin, which translates into more stringent limits on LIV operators for the light quarks and photons (the photon LIV operator will also contribute because of the electromagnetic interactions inside the nucleon). These methods are currently able to constraints the coefficients of quarks and photons at $O(10^{-7} \div 10^{-8})$.

4.2. Astrophysical observations
Testing the higher mass dimensions operators (5 or 6) of the SME is obviously a task that requires to probe much higher energies than those achieved in terrestrial experiments. Energies up to 10^{20} eV are achieved in high energy astrophysics and as such this field has played an eminent role in QG phenomenology. One comment is in order before we further describe these tests and it concerns our parametrisation. We have introduced dimensionless coefficients $\eta^{(n)}$ and expressed the higher order terms in the dispersion relations via suitable rations of the particle momentum to the Planck scale. Of course nothing guarantees that models of quantum/emergent spacetime will predict the Lorentz breaking scale to be coincident with the Planck scale. So, missing a derivation of our dispersion relation from a specific QG model, this has to be considered just a convenient parametrisation.

A brief list of the most commonly used observational constraints in high energy astrophysics are (for more details see again [11, 12])

Vacuum Birefringence The fact that in the "LIV extended QED" with dimension 5, $\mathcal{CPT}$ odd operators, opposite "helicities" have slightly different group velocities, implies that the polarisation vector of a linearly polarised plane wave with energy k rotates, during the wave propagation over a distance d. The angle of rotation will be different for different photon energies hence this effect could potentially disrupt the amount of polarisation present in a some polarised light travelling over long distances. This methods has been applied to several astrophysical sources like GRBs and Pulsar Wind Nebula (closer but more energetic). Note also that this method cannot be applied to the case of dimension 6, $\mathcal{CPT}$ even operators as in this case the two felicities of the photon have the same LIV coefficient, $\xi^{(4)}$.

Photon time of flight A photon dispersion relation in the form of (3) implies that photons of different colours (wave vectors k_1 and k_2) travel at slightly different speeds. We shall assume here that no birefringent effects are present, so that $\xi_+^{(n)} = \xi_-^{(n)}$ (for the case of dimension 5 $\mathcal{CPT}$ odd operators one can consider one helicity at a time or analyse the effect of a realistic photon beam in a more detailed way, see e.g. [31]). When propagating on a cosmological distance d, the effect of energy dependence of the photon group velocity will generically produces an energy, and travelled distance, dependent time delay which can be constrained using current astrophysical observations mainly from gamma ray bursts (GRBs) and active galactic nuclei (i.e. high redshift, high energy sources).

Threshold reactions LIV corrections are quite important in threshold processes because the LIV term (which as a first approximation can be considered as an additional mass term) only needs to be comparable to the (invariant) mass of the particles produced in the final state for strongly affecting this kind of reactions. A quite rich phenomenology of threshold reactions is introduced by LIV in EFT and threshold theorems can be generalized in this context [32, 33]. The main conclusions of the investigation into threshold reactions are that [34, 12]

- Threshold configurations still corresponds to head-on incoming particles and parallel outgoing ones
- The threshold energy of existing threshold reactions can shift, and upper thresholds (i.e. maximal incoming momenta at which the reaction can happen in any configuration) can appear
- Pair production can occur with unequal outgoing momenta
- New, normally forbidden, reactions can be viable

All of these physical situations can be exploited to cast constraints as we shall see in the next section.

Synchrotron radiation Synchrotron emission is strongly affected by LIV [35, 36, 37, 38, 39, 40, 41] and hence effective constraints can be obtained by confronting observed and expected synchrotron spectra from energetic astrophysical sources (e.g. the Crab Nebula). For Planck scale LIV and observed energies, strong constraints can be casted for mass dimension four and five LIV operators. Dimension six operators can be effectively constrained within theoretical frameworks for which the LIV scale is supposedly lower than the Planck scale (see e.g. [41]).

5. Constraints

We shall now briefly discuss the (mainly astrophysical) constraints on Lorentz breaking mass dimension 5 and 6 for different sectors of the SME.

5.1. QED with rotational invariant CPT odd dimension 5 operators

There is a very rich literature devoted to constraining the Myers-Pospelov version of the SME. We shall not attempt here a detailed discussion but just briefly summarise the available constraints.

Synchrotron radiation constraints from the Crab Nebula: Presently a very important object in this sense is the Crab nebula (CN). This pulsar wind nebula had its origin by a supernova explosion observed in 1054 A.D. Its distance from Earth is approximately 1.9 kpc and it is one of the best studied sources of diffuse radio, optical and X-ray radiation. The Nebula emits an extremely broad-band spectrum (21 decades in frequency, see [40] for a comprehensive list of relevant observations) that is produced by two major radiation mechanisms. The emission from radio to low energy γ-rays ($E < 1$ GeV) is thought to be synchrotron radiation from relativistic electrons, whereas inverse Compton (IC) scattering by these electrons is the favoured explanation for the higher energy γ-rays. From a theoretical point of view, the current understanding of the whole environment is based on the so called Kennel–Coroniti [42], which accounts quite accurately for the general features observed in the CN spectrum.

Developing on the initial idea discussed in [35], a full reconstruction of the synchrotron emission processes in the CN within the SME with dimension 5 CPT operators (see Eqs. (3) and (4)) has been performed in [40]. This procedure requires fixing most of the model parameters using radio to soft X-rays observations (which are basically unaffected by LIV), a careful step by step revisitation of the basic assumptions made in the Kennel–Coroniti model and also to take into account the contribution of novel LIV effects such as vacuum Čerenkov and helicity decay. Finally, a χ^2 analysis can be performed to quantify the agreement between models and data [40]. From this analysis, one can conclude that the LIV parameters for the leptons are both constrained, at 95% CL, to be $|\eta_\pm| < 10^{-5}$.

Birefringence constraints: The best constraints on the photon sector are instead obtained by using birefringence effects associated with the CPT odd nature of the relevant LIV operators. Strong constraints came again from the CN [43], from which a value $|\xi^{(3)}| \lesssim 6 \times 10^{-10}$ at 95% Confidence Level (CL) was obtained by considering the observed polarization of hard-X rays [44] (see also [45]). Polarized light from GRBs has also been detected and given their cosmological distribution they could be ideal sources for improving the above mentioned constraints from birefringence. Attempts in this sense were done in the past [24, 46] but the observed of polarisation was later deemed controversial. Furthermore, GRBs for which the polarization is detected and the spectral redshift is precisely determined are scarce. In [47] this problem was circumvented by using indirect methods (the same used to use GRBs as standard candles) for the estimate of the redshift. This leads to a possibly less robust but striking constraint $|\xi^{(3)}| \lesssim 2.4 \times 10^{-15}$. A similar constraint $|\xi^{(3)}| \lesssim 10^{-15}$ was obtained in [48] by making use of two GRBs observed by the Japanese satellite IKAROS. Also in this case the redshift was determined using indirect methods such as the correlation between peak energy and luminosity of the GRB prompt emission. Remarkably, the above mentioned constraints were recently improved by using the INTEGRAL/IBIS observation of the GRB 061122, for which a redshift $z \approx 1.33$ was derived thanks to the determination of the GRB's host galaxy. In this case a constraint $|\xi^{(3)}| \lesssim 3.4 \times 10^{-16}$ was derived [49] (see also [50] for a similar previous constraint).

5.2. QED with rotational invariant CPT even dimension 5 and 6 operators

As we have shown before, this kind of operators lead to modified dispersion relations characterised by quartic terms in the particle momenta suppressed by the squared Planck mass. Furthermore the CPT even nature of the operators prevent birefringent features in the photon sector. This feature basically implies the impossibility to use any of the above discussed observation for casting effective constraints (e.g. the synchrotron emission from the CN would not provide a constraint better that 10^6 on the electron parameter given the $(E/M_{\rm Pl})^2$ suppression). However, there are available observations exploring much higher energies, i.e. those related to the so called Ultra High Energy Cosmic Rays (UHECR).

Astrophysical constraints from GZK reaction secondaries: The GZK cut off [51, 52], is a suppression of the high-energy tail of the UHECR spectrum arising from interactions with CMB photons: $p\gamma \to \Delta^+ \to p\pi^0(n\pi^+)$. When Lorentz invariance holds, this process has a threshold energy $E_{\rm th} \simeq 5 \times 10^{19}$ $(\omega_0/1.3~{\rm meV})^{-1}$ eV (where ω_0 is the target photon energy). Experimentally, the presence of a suppression of the UHECR flux was claimed by the HIRES collaboration [53, 54] (but was initially claimed to be absent by the AGASA collaboration [55]). Although the cut off could be also due to the finite acceleration power of the UHECR sources, the fact that it occurs at the expected energy favours the GZK explanation. The results presented in [56] seemed to further strengthen this hypothesis (but see further discussion in the conclusions).

Significant limits on $\xi = \xi^{(4)}$ and $\eta = \eta^{(4)}$ for the electron/positron can be derived by considering UHE photons generated as secondary products of the GZK reaction [57, 58]. These UHE photons originate because the GZK process leads to the production of neutral pions that subsequently decay into photon pairs. These photons are mainly absorbed by pair production onto the CMB and radio background. Thus, the fraction of UHE photons in UHECRs is theoretically predicted to be less than 1% at 10^{19} eV [59]. Several experiments imposed limits on the presence of photons in the UHECR spectrum. In particular, the photon fraction is less than 2.0%, 5.1%, 31% and 36% (95% C.L) at $E = 10$, 20, 40, 100 EeV respectively [60, 61] (at lower energies even stricter limits 0.4%, 0.5%, 1.0%, 2.6% and 8.9% (95% C.L) above $E = 1$, 2, 3, 5 and 10 EeV respectively were recently found [62]).

The key idea for casting constraints here, relies on the strong dependence of the photon pair production on on LIV modifications. In particular, the (lower) threshold energy can be shifted and in general an upper threshold can be introduced [34]. If the upper threshold energy is lower than 10^{20} eV, then GZK secondary UHE photons are no longer attenuated by the CMB and can reach the Earth. Hence, they would constitute a significant fraction of the total UHECR flux and thereby they would violate the above mentioned experimental bounds [57, 58, 63].

To complete the analysis one can also notice that the γ-decay process can also imply a significant constraint. Indeed, if some UHE photon ($E_\gamma \simeq 10^{19}$ eV) is detected by experiments (and the Pierre Auger Observatory, PAO, will be able to do so in few years [60]), then γ-decay must be forbidden above 10^{19} eV [58] (note that for $n = 4$ the two photons felicities travel at the same speed, so there is in this case a single population).

The combination of above mentioned "upper threshold constraint" and the expected constraint from the absence of gamma decay would bound the QED LIV parameters at order $n = 4$ to the roughly rectangular region (see Figure 2, of [31] for more details) $-10^{-7} \lesssim \xi^{(4)} \lesssim 10^{-8}$ and $-10^{-7} \lesssim \eta^{(4)} \lesssim 10^{-6}$. Note that the absence of gamma decay can improve on the determination of $\eta = \eta^{(3)}$ as provided by the synchrotron constraint, leading to a double sided bound $|\eta^{(3)}| \lesssim 10^{-16}$ (see Figure 3 [58]). All of this of course relies on the actual observation of the GZK cutoff which has instead been questioned recently, see Section 7 for a dedicated discussion.

5.3. SME: Hadronic sector

The very same GZK photo-pion production is strongly affected by LIV and the constraints implied by the detection of this effect have been extensively considered in the literature [64, 65, 34, 26, 66, 67]. Nonetheless, a detailed LIV study of the GZK feature is hard to perform, because of the many astrophysical uncertainties related to the modelling of the propagation and the interactions of UHECRs. In fact, the Lorentz breaking operators could lead to relevant modifications of the GZK mean free path. Consequently, the propagated UHECR spectrum can display new features, like bumps at specific energies, suppression at low energy, recovery at energies above the cutoff. These are all features which cannot be easily conciliated with the observed spectrum, even taking into account experimental uncertainties. Furthermore, the emission of Cherenkov γ-rays and pions in vacuum would lead to sharp suppression of the spectrum above the relevant threshold energy. After a detailed statistical analysis of the agreement between the observed UHECR spectrum and the theoretically predicted one in the presence of LIV and assuming pure proton composition, the final constraints implied by UHECR physics are (at 99% CL) [68] $-10^{-3} \lesssim \eta_p^{(4)} \lesssim 10^{-6}$ and $-10^{-3} \lesssim \eta_\pi^{(4)} \lesssim 10^{-1}$ for $\eta_p^{(4)} > 0$ or $-10^{-3} \lesssim \eta_\pi^{(4)} \lesssim 10^{-6}$ for $\eta_p^{(4)} < 0$. Obviously the same analysis can be applied also to dimension five, CPT odd, operators leading to much stronger constraints (order $O(10^{-14})$).

5.4. SME: Neutrinos

Neutrinos physics have recently gained a prominent role in modern tests of Lorentz invariance due to the growing experimental effort devoted to their study as well as for their light mass that makes their threshold reaction very sensitive to LIV at relatively accessible energies. here we shall succinctly review the main constraints so far cast on this sector of the Standard Model.

Constraints from neutrino time of flights: In assessing the limit speed of neutrinos with respect to that of light, we have to date only a single event to rely on, the supernova SN1987a. This was a peculiar event which allowed to detect the almost simultaneous (within a few hours) arrival of electronic antineutrinos and photons. Although only few electronic antineutrinos at MeV energies was detected by the experiments KamiokaII, IMB and Baksan, it was enough to establish a constraint $(\Delta c/c)^{TOF} \lesssim 10^{-8}$ [69] or $(\Delta c/c)^{TOF} \lesssim 2 \times 10^{-9}$ [70] by looking at the difference in arrival time between antineutrinos and optical photons over a baseline distance of

Journal of Physics: Conference Series **631** (2015) 012011 doi:10.1088/1742-6596/631/1/012011

1.5×10^5 ly. Further analyses of the time structure of the neutrino signal strengthened this constraint down to $\sim 10^{-10}$ [71, 72]. The scarcity of the detected neutrino did not allow the reconstruction of the full energy spectrum and of its time evolution in this sense one should probably consider constraints purely based on the difference in the arrival time with respect to photons more conservative and robust. Unfortunately adopting $\Delta c/c \lesssim 10^{-8}$, the SN constraint implies very weak constraints, $\xi_\nu^{(3)} \lesssim 10^{13}$ and $\xi_\nu^{(4)} \lesssim 10^{34}$. Note that observational constraints on $\Delta c/c$ translate in constraints on the LIV parameter via the formula [73]

$$\xi_\nu^{(n)} = \frac{2}{n-1}\left(\left.\frac{\Delta c}{c}\right|_{\rm Obs}^{TOF} + \frac{m^2}{2p^2}\right) \times \left(\frac{M_{\rm Pl}}{p}\right)^{(n-2)} . \tag{13}$$

Constraints from neutrino oscillations: At odd with the previous case, we do have a wealth of information only about neutrino oscillations which however constraints only the differences among LIV coefficients of different flavors. The best constraint to date comes from survival of atmospheric muon neutrinos observed by the former IceCube detector AMANDA-II in the energy range 100 GeV to 10 TeV [74], which searched for a generic LIV in the neutrino sector [75] and achieved $(\Delta c/c)_{ij} \leq 2.8 \times 10^{-27}$ at 90% confidence level assuming maximal mixing for some of the combinations i, j. The same constraint applies to the corresponding antiparticles as IceCube does not distinguish neutrinos from antineutrinos. IceCube is also expected to improve this constraint to $(\Delta c/c)_{ij} \leq 9 \times 10^{-28}$ in the next few years [76]. The lack of sidereal variations in the atmospheric neutrino flux also yields comparable constraints on some combinations of SME parameters [77]. Putting all together, it seems that no-flavour dependent SME can be tolerated and many studies just assume flavour independent LIV as they starting base of their analysis.

Constraints from neutrino threshold reactions: Several threshold processes have been considered in the literature, most prominently the neutrino Čerekov emission $\nu \to \gamma\,\nu$, the neutrino splitting $\nu \to \nu\,\nu\bar{\nu}$ and the neutrino electron/positron pair production $\nu \to \nu\, e^- e^+$. They are all very similar, so we shall discuss here only the latter for illustrational purposes and assume no LIV modification in the electron/positron sector as we have already seen that LIV in this sector is strongly constrained. The threshold energy is for arbitrary n is then

$$E_{th,(n)}^2 = \frac{4m_e^2}{\delta_{(n)}} \qquad \text{with} \quad \delta_{(n)} = \xi_\nu^{(n)}\left(\frac{E_{th,(n)}}{M_{\rm Pl}}\right)^{n-2} . \tag{14}$$

The rate of this reaction was firstly computed in [78] for $n = 2$ but can be easily generated to arbitrary n [73] (see also [79] for a more general treatment and detailed considerations). The generic energy loss time-scale then reads (dropping purely numerical factors)

$$\tau_{\nu-{\rm pair}} \simeq \frac{m_Z^4 \cos^4\theta_w}{g^4 E^5 \xi_\nu^3}\left(\frac{M_{\rm Pl}}{E}\right)^{3(n-2)} , \tag{15}$$

where g is the weak coupling and θ_w is Weinberg's angle.

The observation of upward-going atmospheric neutrinos up to 400 TeV by the experiment IceCube implies that the free path of these particles is at least longer than the Earth radius implies a constraint $\xi_\nu^{(3)} \lesssim 40$ (taking a conservative baseline of about 6000 Km). No effective constraint can be obtained for $n = 4$ LIV, however in this case neutrino splitting (which has the further advantage to be purely dependent on neutrinos LIV) could be used on the "cosmogenic" neutrino flux which one expects from the decay of charged pions produced by the aforementioned GZK reaction.

It is easy to see that the neutrino splitting should modify the spectrum of the ultra high energy neutrinos by suppressing the flux at the highest energies and enhancing it at the lowest ones. In [80] it was shown that future experiments like ARIANNA [81] will achieve the required sensitivity to cast a constraint of order $\xi_\nu^{(4)} \lesssim 10^{-4}$. Note however, that the rate for neutrino splitting computed in [80] was recently recognised to be underestimated by a factor $O(E/M)^2$ [82]. An improved analysis can be found in [73] leading to an expected constraint from AUGER $\xi_\nu^{(4)} \lesssim 10^{-7}$ (note that this constraint is cast using the expected flux which won't be able to distinguish different flavours).

However, we have just seen that experimental observations of the depth of the shower maximum of UHECR interactions in the atmosphere hinted at the possible presence of nuclei heavier than protons in UHECRs [83]. In this case, pion production would be suppressed at UHE and hence the UHE neutrino flux could be much smaller than the expectation from pure proton composition.

A final comment is devoted to the so called pion decay channel $\pi^+ \to \nu_\mu \mu^+$ that was extensively explored in dealing with the recent OPERA claim [84, 85]. In this case it was found that for superluminal neutrinos and unmodified pion and muon, the process could be actually forbidden hence suppressing muonic neutrino pair production in the CERN bean. In [84] it was noticed that the detection of up to 40 GeV neutrinos at OPERA would imply a bound $\xi_\nu^{(2)} \lesssim 10^{-7}$. Also in [84, 85] a constraint $\xi_\nu^{(2)} \lesssim 10^{-13}$ was derived from atmospheric neutrinos using neutrino pair production and neutrino Čerenkov emission.

While OPERA detection of superluminal neutrinos [86] was soon recognised as flawed (see revised version of [86]), it is worth reconsider it here as a case study. Indeed, the excitement created by the OPERA's initial report has obviously subsided for the greater physics community. For the quantum gravity community that focuses on possible experimental signatures of quantum gravity, however, technical issues were raised in how to analyse these types of accelerator based experiments properly. In particular, the detailed physics of anomalous reactions, and how they reduce the intensity of a particle beam from source to detector became central to the discussion.

The most convincing theoretical objection to the initial OPERA result was produced by Cohen and Glashow [78] shortly after the OPERA report and involved just such an anomalous reaction. Cohen and Glashow used the fact that superluminal neutrinos should emit electron-positron pairs (see section 3.2) to argue that the OPERA results were not even self-consistent: any neutrino with the speed reported by OPERA should have lost most of its energy to pair production while it propagated from CERN to the detector at Gran Sasso. The maximum energy in the beam would therefore have dropped to be below some termination energy E_T, and Cohen and Glashow showed that E_T for the OPERA beam was less than the average ~ 17 GeV energy reported by OPERA.

More specifically by integrating the energy loss rate from pair production (as deducible from Eq. (15)) over a distance L and by assuming that the typical energy loss length be much smaller than L one obtains

$$E^{-3n+1} - E_0^{-3n+1} = (3n-1)\xi_\nu^3 E_{\rm ref}^{-3(n-2)} k \frac{G_F^2}{192\pi^3} L \equiv E_T^{-3n+1} , \qquad (16)$$

where E is the energy on a neutrino starting with energy E_0 after propagation over the distance L and $E_{\rm ref}$ is the energy at which we normalize the parameter ξ_ν. The factor $k = 25/448$ was computed in [78] for the case $n = 2$, while for the general case it can be found in [79]. The "termination" energy E_T corresponds to the energy that a neutrino would approach after propagation over a distance L if it started with $E_0 \gg E_T$. We remark here that the termination energy E_T is a mildly varying function of n and of the energy scale $E_{\rm ref}$. For a LIV extension of the standard model at order $n = 2$ (see the Coleman-Glashow model discussed briefly above) it was then shown that the Opera claim of $\xi_\nu \sim 5 \times 10^{-5}$ for $E_{\rm ref} \sim 10$ to 30 GeV implied

$E_T \simeq 12.5$ GeV: a value of E_T obviously incompatible with the observation of neutrinos up and above 40 GeV in Opera. Similar considerations lead to an expectation of $E_T \simeq 15.0$ GeV for $n = 3$.

Note that this estimate was done without taking into account the similar mechanism of neutrino splitting. While this reaction is strictly forbidden in the $n = 2$ case, it is not so at higher orders and needs to be taken into account given that the energy loss rate of this process is comparable to the one for pair production loss (see e.g. [73, 82, 79]), and hence is not negligible for $n > 2$. In particular, this extra reaction can lead to energy losses without generating large numbers or electron-positron pairs as those searched for in an early ICARUS analysis of the Opera beam [87].

The physics of the Cohen-Glashow argument was correct, however the authors did not worry about adjusting for the finite size of the baseline. A finite baseline can be of the same order as the energy loss length of neutrinos undergoing pair production. This allows for some neutrinos to undergo only one or a few Cherenkov emissions within their time of flight. Therefore the most energetic neutrinos of the injection beam can still reach the end of the baseline with an energy larger than E_T [73]. It is then necessary, in order to cast a robust constraint on LIV by using long baseline experiments, to run a full MonteCarlo simulation of the propagation of neutrinos aimed at computing the neutrino spectrum on arrival in the presence of this energy loss process. While this was not an issue for the Cohen and Glashow result, as it was one piece of a number of experimental and theoretical concerns about OPERA [88, 87, 69, 70], if one wishes to use time of flight experiments alone to set robust constraints on neutrino LIV, the issue must be addressed.

In [73] a complete analysis for the case of OPERA in the cases $n = 2$ and $n = 3$, taking into account neutrino pair creation and neutrino splitting (for $n = 3$) has been performed. It was there found that the propagated spectrum does indeed show a pronounced bump at the expected E_T, but is also characterised by a high energy tail that extends well above E_T and has an amplitude about 10% of the amplitude of the bump. Hence, the simple calculation of E_T is not per se conclusive for casting constraints, although the reconstruction of the propagated spectra in [73] demonstrated that in the special case of OPERA the detection of neutrinos with $E > 40$ GeV would have still pointed out an incompatibility between the adopted LIV framework and the experimental observation.

6. Overview of the constraints on the Matter Sector

We can now summarize the current status of the constraints for the LIV SME (rotational invariant) in Table 1. See [12] for further discussion.

Order	photon	e^-/e^+	Protrons	Neutrinos[a]
n=2	N.A.	$O(10^{-16})$	$O(10^{-20})$ (CR)	$O(10^{-8} \div 10^{-10})$
n=3	$O(10^{-16})$ (GRB)	$O(10^{-16})$ (CR)	$O(10^{-14})$ (CR)	$O(40)$
n=4	$O(10^{-8})$ (CR)	$O(10^{-8})$ (CR)	$O(10^{-6})$ (CR)	$O(10^{-7})^*$ (CR)

Table 1. Summary of typical strengths of the available constrains on the SME at different n orders for rotational invariant, neutrino flavour independent LIV operators. GRB=gamma rays burst, CR=cosmic rays. [a] From neutrino oscillations we have constraints on the difference of LIV coefficients of different flavors up to $O(10^{-28})$ on dim 4, $O(10^{-8})$ and expected up to $O(10^{-14})$ on dim 5 (ICE3), expected up to $O(10^{-4})$ on dim 6 op. * Just expected constraint from future experiments probing cosmogenic neutrinos up to 10^{20} eV [73].

Of course at first sight this might seem a quite satisfactory state of the art, so much so that

Journal of Physics: Conference Series **631** (2015) 012011 doi:10.1088/1742-6596/631/1/012011

one might ask if we haven't tests Lorentz violations enough and should now move one towards new phenomenology. As usual, the answer is not a sharp one. Let us further elaborate on this point.

7. Open issues

Let's first stick to tests of violation of Lorentz invariance in the SME. Here the main open issue is provided by the lasting uncertainty about the ultra high energy cosmic rays (UHECR) composition and hence the actual observation of the GZK cutoff (see e.g. [89]). Given that all of the best constraints on $n = 4$ (massi dimension six LIV operators) are currently based on UHECR physics it is clear that addressing such uncertainty it should be a crucial task in the next coming years. For the moment, it would be however unfair to play down the present uncertainties on UHECR and place these constraints at the same level of robustness e.g. of those cast at order $n = 3$ by using the synchrotron radiation from the Crab nebula. In this sense any UHECR-independent constraint on the dimension 6 operators of the SME would be more than welcomed.

7.1. The naturalness of Lorentz violations

Another open issue is of course the naturalness problem of Lorentz breaking in the SME: in EFT radiative corrections will generically allow the percolation of higher dimension Lorentz violating terms into the lower dimension terms due to the interactions of particles [90]. Indeed, EFT loop integrals will be naturally cut-off at the EFT breaking scale, if such scale is as well the Lorentz breaking scale present in the dispersion relation, the two will effectively cancel leading to unsuppressed, coupling dependent, contributions to the mass dimension four kinetic terms that generate the usual propagators. Hence radiative corrections will not allow a dispersion relation with only p^3/M or p^4/M^2 Lorentz breaking terms but will automatically induce extra unsuppressed LIV terms in p and p^2 which will be naturally dominant. Several mechanisms have been proposed for protecting the lowest order operators. While no proposal can be deemed as definitive it is worth noticing that in most cases these "protective mechanisms" entail extra energy scales beyond the Planck one and hence imply possible new ways to constraint LIV scenarios (for a summary see [12]).

7.2. A possible LIV explanation of the IceCube UHE neutrinos anomaly?

So far, the IceCube collaboration has identified 87^{+14}_{-10} events from neutrinos of astrophysical origin with energies above 10 TeV, (the error being due to the uncertainty in the number of astrophysical events determined by the modelled subtraction of both conventional and atmospheric neutrinos particularly at energies below 60 TeV [91]). Above 60 TeV, the IceCube data are consistent with a spectrum given by $E_\nu^2(dN_\nu/dE_\nu) \simeq 10^{-8}$ GeVcm^{-2}s^{-1}. Spectra steeper than E_ν^{-2} do not give a good fit to the existing data in the 60 TeV to 2 PeV energy range [92]. However, no neutrino induced events have been seen above $\sim$ 2 PeV, as would be expected from extending an E_ν^{-2} spectrum beyond $\sim$ 2 PeV. In particular, IceCube has not detected any neutrino induced events from the Glashow resonance effect at 6.3 PeV. This enhancement leads to an increased IceCube effective area for detecting the sum of the ν_e's, *i.e.*, ν_e's plus $\bar{\nu}_e$'s by a factor of $\sim$ 10 [93]. It is usually expected that 1/3 of the potential 6.3 PeV neutrinos would be ν_e's plus $\bar{\nu}_e$'s unless new physics is involved. Thus, the enhancement in the overall effective area expected is a factor of $\sim$3. Taking account of the increased effective area between 2 and 6 PeV and a decrease from an assumed neutrino energy spectrum of E_ν^{-2}, we would expect about 3 events at the Glashow resonance provided that the number of $\bar{\nu}_e$'s is equal to the number of ν_e's. Even without considering the Glashow resonance effect, several neutrino events above 2 PeV would be expected if the E_ν^{-2} spectrum extended to higher energies. Thus,

Journal of Physics: Conference Series **631** (2015) 012011 doi:10.1088/1742-6596/631/1/012011

the lack of neutrinos above 2 PeV energy and at the 6.3 PeV resonance may be indications of a cutoff in the neutrino spectrum.

In [94] the effects of for LIV operators of mass dimension greater than four on the propagation and resulting energy spectrum of superluminal neutrinos of extragalactic origin was explored in particular by taking into account the possible effect of electron-positron pair production and neutrino splitting.

Interestingly, a high-energy drop off in a propagated superluminal neutrino spectrum above ~ 2 PeV can result from above mentioned processes in the $\mathcal{CPT}$-conserving cases. The drop off matches the observed neutrino spectra for energy dependent values of $\xi_{\mu}^{(4)}$ of $O(10^3)$. This drop off would be preceded by a non-negligible pile up in the neutrino spectrum which could be used as a signature of this kind of LIV phenomenology. If instead mass dimension five, $\mathcal{CPT}$-violating, operators dominate, it was found [94] an absence of a clear cutoff in the propagated neutrino spectrum. On the other hand, if the cutoff will be in the end explained by a natural break in the neutrino spectra of the astrophysical neutrino sources, the above number would become a constraint.

8. Lorentz breaking in the gravity sector

Of course, given that are general quantum gravity arguments the main motivation for considering departures from exact Lorentz invariance at very high energies one might wonder why the gravitational sector should not be the main one affected by this new physics. Indeed, one can introduce an analogue of the mSME for the gravitational sector, i.e. the so called Einstein–Aether theory of gravity and then try extend it in the UV.

8.1. Einstein–Aether theory

The SME is constructed by coupling matter terms to non-zero LIV tensors in vacuum. However, when dealing with extensions of this framework to gravitational phenomena it is clear that leaving this tensors non-dynamical, would break general covariance. In this respect one can proceed along two alternative approaches, the first being the introduction of a suitable dynamics for the LIV tensor, the other being the the acceptance of an explicit breaking of the four dimensional diffeomorphism invariance of GR. In what follows we shall discuss the most representative models in this respect and they relation.

If one choses to preserve general covariance by promoting the LIV tensors to dynamical fields and if restricts his/her attention to rotational invariance, then it is natural to generate LIV couplings by including in the action either a scalar or a timelike vector field that takes a vacuum expectation value. In the case of a scalar, one can use a shift symmetry ($\phi(x) \to \phi(x) + \phi_0$) to construct actions for which the derivative of the scalar takes a non-zero value [95]. In the vector case, one simply puts a potential for the vector field such that the vector acquires a vev.

We concentrate on the vector case here as it is the simplest model that allows for rotationally invariant Lorentz violation [96] and it is also the natural extension of the framework considered so far for the rotationally invariant LIV matter sector. It is the most general theory for a unit timelike vector field coupled to gravity (but not to matter), which is second order in derivatives.

Let us the express again the aether vector field by u^α and, in analogy with what we did for the SME, write the most general theory for a unit timelike vector field coupled to gravity. If we limit ourself to second order terms in derivatives (or equivalently low energies) this take the form

$$\mathcal{S} = \mathcal{S}_{EH} + \mathcal{S}_u = \frac{1}{16\pi G_{æ}} \int d^4x\sqrt{-g}\ (R + \mathcal{L}_u)\,. \tag{17}$$

where $\mathcal{L}_u$ is given by

$$\mathcal{L}_u = -Z^{\alpha\beta}{}_{\gamma\delta}(\nabla_\alpha u^\gamma)(\nabla_\beta u^\delta) + V(u). \tag{18}$$

The tensor $Z^{\alpha\beta}_{\gamma\delta}$ is defined as [97]

$$Z^{\alpha\beta}{}_{\gamma\delta} = c_1 g^{\alpha\beta} g_{\gamma\delta} + c_2 \delta^{\alpha}{}_{\gamma} \delta^{\beta}{}_{\delta} + c_3 \delta^{\alpha}{}_{\delta} \delta^{\beta}{}_{\gamma} - c_4 u^{\alpha} u^{\beta} g_{\gamma\delta} \,, \tag{19}$$

where c_i, $i = 1, \ldots, 4$ are simple coefficients of the various kinetic terms and $V(u)$ is a potential term such that it generates a non-zero vev for u^{α}. Note the indicial symmetry $Z^{\beta\alpha}{}_{\delta\gamma} = Z^{\alpha\beta}{}_{\gamma\delta}$. An additional term, $R_{ab}u^a u^b$ is a combination of the above terms when integrated by parts, and hence is not explicitly included here.

The potential in (18) is normally fixed by the requirement to remove the ghost excitation associated to the one of the vector components which will necessarily acquire a wrong sign in the kinetic term. The choice $V(u) = \lambda(u^2 + 1)$, where λ is a Lagrange multiplier, fixes the norm of u^{α} and removes the ghost excitation (c.f. the discussions in [98] and [99]). It is this choice of the potential that it is normally associated to the so called "Einstein–Aether theory" [96] which we shall consider here.

8.1.1. Constraints on Einstein–Aether gravity: For what regards the constraints on Einstein–Aether gravity, they can be divided in those on the aether kinetic terms and those on the aether-matter couplings. Given the the latters can be reduced to mSME constraints, we shall here discuss specifically only the first kind (one can find a discussion of the second kind in e.g. [100]).

Before of doing so, we can note however that the couplings between aether and the standard model field content while being the same as those discussed before in this review, they do have new features as the aether field has now dynamics, so there can be position dependent violations of Lorentz symmetry. Interestingly, some of the couplings to matter that are unobservable for a single fermion field can have relevant effects when the aether varies. For example, the $-au_{\mu}\overline{\psi}\gamma^{\mu}\psi$ term in the mSME could be removed by making a phase change for the fermion. However, once u^{α} is dynamical and varies with position, only a single component of the term can actually be removed by a phase change [101]. This leads to a new type of term which requires gravitational/position dependent tests in the matter sector [101].

In order to constrain instead aether kinetic terms one can adopt two different approaches. One consists of course in adopting, as normally done any modified gravity theory, a PPN analysis that allows to compare the theory with observations e.g. solar system constraints. Alternatively, one can use the fact that the theory predicts extra degrees of freedom in the gravity sector (there are naively four, but the unit constraint removes one) and that these excitations strongly couple to the metric via the unit constraint (for a more detailed discussion see [97, 100]).

Constraints from PPN analysis: All the PPN parameters vanish except for α_1, α_2 which describe preferred frame effects. α_1 and α_2 were calculated in [102]

$$\alpha_1 = \frac{-8(c_3^2 + c_1 c_4)}{2c_1 - c_1^2 + c_3^2} \tag{20}$$

$$\alpha_2 = \frac{\alpha_1}{2} - \frac{(c_1 + 2c_3 - c_4)(2c_1 + 3c_2 + c_3 + c_4)}{(c_1 + c_2 + c_3)(2 - c_1 - c_4)} . \tag{21}$$

Current constraints are $\alpha_1 < 10^{-4}$ and $\alpha_2 < 4 \times 10^{-7}$ [103] and so from a PPN analysis alone there is still a large 2-d region of parameter space that remains consistent with available tests of GR. Furthermore, recently the orbital evolution of binary pulsars was studied in this framework and was found to induce a much more rapid decay of the binarys orbital period due to the emission of dipolar radiation and modifies quadruple emission [104, 105]. This can be used to cast strong constraint on $(c_+, c_-) \approx O(10^{-2}, 10^{-3})$ where $c_{\pm} = c_1 \pm c_3$ which, in the limit of small $\alpha_{1,2}$, are the only remain free LIV parameters of the theory.

Constraints from gravity-aether wave modes: The combined aether-metric modes consist of the two usual transverse traceless graviton modes, a vector mode, and a scalar mode [106]. The speeds of each of the modes can differ from the speed of light. Hence if the speeds are less than unity, high energy cosmic rays will emit vacuum gravitational Čerenkov radiation [107]. If we denote the speeds of the spin-2, spin-1 and spin-0 modes by s_2, s_1, s_0 then we have [106]

$$s_2^2 = (1 - c_1 - c_3)^{-1} \tag{22}$$

$$s_1^2 = \frac{2c_1 - c_1^2 + c_3^2}{2(c_1 + c_4)(1 - c_1 - c_3)} \tag{23}$$

$$s_0^2 = \frac{(c_1 + c_2 + c_3)(2 - c_1 - c_4)}{(c_1 + c_4)(1 - c_1 - c_3)(2 + 3c_2 + c_1 + c_3)}. \tag{24}$$

The requirement that all these speeds are greater than unity therefore puts constraints on a combination of the c_i coefficients. However, even after imposing all of the above constraints there is still a large region of parameter space allowed. Indeed, the PPN and gravitational Čerenkov constraints are all satisfied provided quite lose conditions on the model coefficients are satisfied [97, 100]. Hence, the gravitational sector is only minimally constrained compared to aether-matter couplings.

8.2. Hořava–Lifshitz gravity

The underlying idea of the Hořava–Lifshitz (HL) gravity (see *e.g.* [108] for a review) is to achieve power-counting renormalizability by modifying the graviton propagator in the ultraviolet by adding to the action terms containing higher order spatial derivatives of the metric, but not higher order time derivatives, so to preserve unitarity. This procedure naturally leads to a space-time foliation into spacelike surfaces, labeled by the t coordinate and with x^i being the coordinates on each surface. The resulting theory is then invariant only under the reduced set of diffeomorphisms that leave this foliation intact, $t \to \tilde{t}(t)$ and $x^i \to \tilde{x}^i(t, x^i)$.

It was shown that power counting renormalizability requires the action to includes terms with at least 6 spatial derivatives in 4 dimensions [109, 19]. Of course, all lower order operators compatible with the symmetry of the theory are expected to be generated by radiative corrections, so the most general action takes the form [110]

$$S_{HL} = \frac{M_{\rm Pl}^2}{2} \int dt d^3x \, N\sqrt{h} \left(L_2 + \frac{1}{M_\star^2} L_4 + \frac{1}{M_\star^4} L_6 \right), \tag{25}$$

where h is the determinant of the induced metric h_{ij} on the spacelike hypersurfaces,

$$L_2 = K_{ij}K^{ij} - \lambda K^2 + \xi\, {}^{(3)}\!R + \eta a_i a^i , \tag{26}$$

where K is the trace of the extrinsic curvature K_{ij}, ${}^{(3)}\!R$ is the Ricci scalar of h_{ij}, N is the lapse function, and $a_i = \partial_i \ln N$. L_4 and L_6 denote a collection of 4th and 6th order operators respectively and $M_\star$ is the scale that suppresses these operators which does not coincide *a priori* with $M_{\rm Pl}$.

It is perhaps tempting to call $M_\star$ the Lorentz breaking scale, but the theory exhibits Lorentz violations (LIV) at all scales, as L_2 already contains LIV operators. These Infrared (IR) Lorentz violations are controlled by three dimensionless parameters that take the values $\lambda = 1$, $\xi = 1$ and $\eta = 0$ in General Relativity (GR). While, ξ can be set to 1 by a suitable coordinate rescaling, it is presently unclear if the running of the remaining two parameters will converge on the GR values in the IR. Nonetheless, it seems that the the theory could still be viable and consistent for suitable choices of the dimensionless parameters λ, η which admits the GR values as extremal limits of the allowed range [111].

Action (25) does present, however, the unappealing feature to contain (in L_4 and L_6) a very large number, $O(10^2)$, of operators and independent coupling parameters. In remedy of this situation, restrictions to the theory have been proposed which would limit the proliferation of independent couplings. We shall not deal with such restrictions here (but see e.g [108, 112] for a concise review) as they will not be determinant for the phenomenological discussion on HL that we shall present later on in this review.

8.3. Constraints on Hořava gravity

Coming to constraints on HL gravity theory, it was shown in [113] (but see also [114]) that the Einstein–Aether theory is equivalent to the the infrared limit of HL gravity if the aether is assumed to be hypersurface orthogonal before the variation. Then it should be obvious that given the relation in the IR between hypersuface orthogonal Einstein–Aether and Hořava–Lifshiftz gravity, the previously presented constraints on Einstein–Aether gravity can be related to constraints for the latter theory (see e.g. [104, 105]).

Looking then at the UV complete the theory from the point of view of QG phenomenology, it then interesting to know the available constraints on the Lorentz breaking scale of the theory $M_\star$. Remarkably, this scale happens to be bounded both below and above. Indeed, for the theory to preserve power counting renormalizablity and be at the same time compatible with current observations (microgravity experiments and solar system tests) one has to require $O(1)\,\text{meV} < M_\star < 10^{16}$ GeV [111, 108, 41], a quite broad opportunity window for the theory.

We have seen however, that radiative corrections will always allow for LIV operators to percolate from a SM sector to the other, gravity being no exception. In this sense it can be generically expected for the above theory to induce Lorentz breaking operators in the matter sector at all orders. Let us assume here that no LIV is present in the matter sector at tree level and that again some protective mechanism will prevent the percolation of the Lorentz breaking terms to the lowest order (mass dimension 3 and 4) operators of the matter sector. Also, one can assume that the CPT and Parity (P) invariance of the gravitational action is preserved in the matter sector. This assumption forbids helicity dependent terms and allows only even power of the momentum in the matter dispersion relation. Hence matter and anti-matter are expected to share the same dispersion relation which within this framework one can then expect to be

$$E^2 = m^2 + p^2 + \eta \frac{p^4}{M_{\rm LIV}^2} + O\left(\frac{p^6}{M_{\rm LIV}^4}\right) . \tag{27}$$

From a logic point of view, there are now two options: (a) $M_\star = M_{\rm LIV}$, i.e. $M_\star$ is a universal scale; (b) $M_\star \ll M_{\rm LIV}$ (as we have seen in this review current phenomenological constraints already rule out the case $M_{\rm LIV} \ll M_\star < 10^{16}$ GeV). Clearly, option (b) requires some mechanism which suppresses the percolation of gravity LIVs into the matter sector (see e.g. [115]). Here, reporting the work done in [41], we shall conclude the necessity for such mechanisms.

For doing so one can adopt option (a) and demonstrate that, in this case, matter LIV constraints imply $M_\star = M_{\rm LIV} > 10^{16}$ GeV, thus closing the available window for $M_\star$. Of course this could be easily achieved using the constraints on QED for modified dispersion relation of order $n = 4$ using UHECR. However, we have seen that these constrains are somewhat questionable nowadays while further evidence about the nature of the highest energy particles is awaited. However, a more robust observation is that of the synchrotron radiation from the Crab Nebula, which for Lorentz breaking scales of order 10^{16} GeV can be sensitive to modified dispersion relation of order $n = 4$! In [41] the possible LIV induced modifications to the standard Fermi mechanism (which is thought to be responsible for the formation of the spectrum of energetic electrons in the CN) were considered and the synchrotron spectrum of the CN was recomputed taking into account all the new, LIV induced, phenomena. This was done for $n = 4$ and the specific modified dispersion relation Eq. (27).

The free parameters of the model (electron/positron density and spectrum and magnetic field strength) were fixed in order to reproduce the low energy part of the spectrum, which is not affected by LIV. This allows to reconstruct how LIV affects the higher energy part of the spectrum ($E \gtrsim 100$ keV) and to use a χ^2 statistics to measure when deviations from the observed spectrum due to LIV become unacceptably large. By considering the offset from the minimum of the reduced χ^2 exclusion limits at 90%, 95% and 99% Confidence Level (CL), according to [116] it was shown in [41] that mass scales $M_{\rm LIV} \lesssim 2 \times 10^{16}$ GeV are excluded at 95% CL.

9. Perspectives

Perhaps a final comment is due to the present state of the art of the field. Lorentz breaking phenomenology has been a remarkable success, a community effort which has built (in a bit over a decade) a wealth of knowledge, methods and constraints that are now at our disposal for efficiently testing candidate theories of QG once their low energy limit is known. However, this state of affair is not and cannot be the end of the story as EFT phenomenology is not yet a true quantum gravity phenomenology. There are probably many more phenomenological consequences of QG beyond and apart LIV and maybe some of them will go beyond the realm of local EFT. For example, causal sets models seems to entail non-locality as a counterweight to the requirement of a Lorentz invariant discretisation of spacetime [117]. The development of more specific phenomenological studies tailored on specific quantum gravity scenarios is probably what we should be heading to in the years to come. Hopefully, the training provided by Lorentz breaking effective field theories would help us facing this sort of future challenges.

Acknowledgments

I wish to thank my main collaborators on this subject David Mattingly and Luca Maccione, and the DISCRETE2014 organisation for inviting me to participate to the workshop. This project/publication was made possible through the support of a grant from the John Templeton Foundation. The opinions expressed in this publication are those of the author and do not necessarily reflect the views of the John Templeton Foundation.

References

[1] Mavromatos N E 2005 *Lect. Notes Phys.* **669** 245–320 (*Preprint* arXiv:gr-qc/0407005)
[2] Weinberg S 2005 *Phys. Rev.* **D72** 043514 (*Preprint* arXiv:hep-th/0506236)
[3] Damour T and Polyakov A M 1994 *Nucl. Phys.* **B423** 532–558 (*Preprint* arXiv:hep-th/9401069)
[4] Barrow J D 1997 (*Preprint* arXiv:gr-qc/9711084)
[5] Bleicher M, Hofmann S, Hossenfelder S and Stoecker H 2002 *Phys. Lett.* **B548** 73–76 (*Preprint* arXiv:hep-ph/0112186)
[6] Amelino-Camelia G 1999 *Nature* **398** 216–218 (*Preprint* gr-qc/9808029)
[7] Garay L J 1995 *Int.J.Mod.Phys.* **A10** 145–166 (*Preprint* arXiv:gr-qc/9403008)
[8] Hossenfelder S 2012 (*Preprint* arXiv:gr-qc/12036191)
[9] Marin F, Marino F, Bonaldi M, Cerdonio M, Conti L *et al.* 2013 *Nature Phys.* **9** 71–73
[10] Kostelecky V A 2004 *Phys. Rev.* **D69** 105009 (*Preprint* arXiv:hep-th/0312310)
[11] Mattingly D 2005 *Living Rev. Rel.* **8** 5 (*Preprint* arXiv:gr-qc/0502097)
[12] Liberati S 2013 *Class.Quant.Grav.* **30** 133001 (*Preprint* 1304.5795)
[13] Amelino-Camelia G 2008 (*Preprint* arXiv:gr-qc/0806.0339)
[14] Di Casola E, Liberati S and Sonego S 2014 (*Preprint* 1405.5085)
[15] Cohen A G and Glashow S L 2006 *Phys. Rev. Lett.* **97** 021601 (*Preprint* arXiv:hep-ph/0601236)
[16] Gibbons G, Gomis J and Pope C 2007 *Phys.Rev.* **D76** 081701 (*Preprint* 0707.2174)
[17] Greenberg O W 2002 *Phys. Rev. Lett.* **89** 231602 (*Preprint* arXiv:hep-ph/0201258)
[18] Chaichian M, Dolgov A D, Novikov V A and Tureanu A 2011 *Phys.Lett.* **B699** 177–180 (*Preprint* arXiv:hep-th/1103.0168)
[19] Visser M 2009 *Phys.Rev.* **D80** 025011 (*Preprint* arXiv:hep-th/09020590)

[20] Coleman S R and Glashow S L 1997 *Phys.Lett.* **B405** 249–252 (*Preprint* arXiv:hep-ph/9703240)
[21] Coleman S R and Glashow S L 1999 *Phys.Rev.* **D59** 116008 (*Preprint* arXiv:hep-ph/9812418)
[22] Coleman S R and Glashow S L 1998 (*Preprint* arXiv:hep-ph/9808446)
[23] Myers R C and Pospelov M 2003 *Phys. Rev. Lett.* **90** 211601 (*Preprint* arXiv:hep-ph/0301124)
[24] Jacobson T A, Liberati S, Mattingly D and Stecker F W 2004 *Phys. Rev. Lett.* **93** 021101 (*Preprint* arXiv:astro-ph/0309681)
[25] Jacobson T, Liberati S and Mattingly D 2006 *Annals Phys.* **321** 150–196 (*Preprint* arXiv:astro-ph/0505267)
[26] Mattingly D 2008 (*Preprint* arXiv:gr-qc/0802.1561)
[27] Kostelecky A and Mewes M 2012 *Phys.Rev.* **D85** 096005 (*Preprint* arXiv:hep-ph/1112.6395)
[28] Kostelecky V Alan E Prepared for 3rd Meeting on CPT and Lorentz Symmetry (CPT 04), Bloomington, Indiana, 8-11 Aug 2007
[29] Kostelecky V A and Russell N 2008 (*Preprint* arXiv:hep-ph/08010287)
[30] De Angelis A, De Maria M, Antonelli M and Dreucci M 2011 *Nuovo Cim.* **C034N3** 323–326 (*Preprint* 1011.3720)
[31] Liberati S and Maccione L 2009 *Ann.Rev.Nucl.Part.Sci.* **59** 245–267 (*Preprint* arXiv:astro-ph.HE/09060681)
[32] Mattingly D, Jacobson T and Liberati S 2003 *Phys. Rev.* **D67** 124012 (*Preprint* arXiv:hep-ph/0211466)
[33] Baccetti V, Tate K and Visser M 2011 (*Preprint* arXiv:hep-ph/1111.6340)
[34] Jacobson T, Liberati S and Mattingly D 2003 *Phys. Rev.* **D67** 124011 (*Preprint* arXiv:hep-ph/0209264)
[35] Jacobson T, Liberati S and Mattingly D 2003 *Nature* **424** 1019–1021 (*Preprint* arXiv:astro-ph/0212190)
[36] Ellis J R, Mavromatos N E and Sakharov A S 2004 *Astropart. Phys.* **20** 669–682 (*Preprint* arXiv:astro-ph/0308403)
[37] Montemayor R and Urrutia L F 2005 *Phys. Lett.* **B606** 86–94 (*Preprint* arXiv:hep-ph/0410143)
[38] Montemayor R and Urrutia L F 2005 *Phys. Rev.* **D72** 045018 (*Preprint* arXiv:hep-ph/0505135)
[39] Altschul B 2006 *Phys.Rev.* **D74** 083003 (*Preprint* arXiv:hep-ph/0608332)
[40] Maccione L, Liberati S, Celotti A and Kirk J 2007 *Journal of Cosmology and Astroparticle Physics* **2007** 013 (*Preprint* arXiv:astro-ph/07072673)
[41] Liberati S, Maccione L and Sotiriou T P 2012 *Phys.Rev.Lett.* **109** 151602 (*Preprint* arXiv:gr-qc/12070670)
[42] Kennel C F and Coroniti F V 1984 *Astrophys. J.* **283** 694
[43] Maccione L, Liberati S, Celotti A, Kirk J G and Ubertini P 2008 (*Preprint* arXiv:astro-ph/08090220)
[44] Dean A J, Clark D J, Stephen J B, McBride V A, Bassani L, Bazzano A, Bird A J, Hill A B, Shaw S E and Ubertini P 2008 *Science* **321** 1183–1185
[45] Forot M, Laurent P, Grenier I A, Gouiffes C and Lebrun F 2008 (*Preprint* arXiv:astro-ph/08091292)
[46] Mitrofanov I G 2003 *Nature* **426** 139
[47] Stecker F W 2011 *Astropart.Phys.* **35** 95–97 (*Preprint* arXiv:astro-ph.HE/1102.2784)
[48] Toma K, Mukohyama S, Yonetoku D, Murakami T, Gunji S *et al.* 2012 *Phys.Rev.Lett.* **109** 241104 (*Preprint* arXiv:astro-ph.HE/1208.5288)
[49] Gotz D, Covino S, Fernandez-Soto A, Laurent P and Bosnjak Z 2013 (*Preprint* 1303.4186)
[50] Laurent P, Gotz D, Binetruy P, Covino S and Fernandez-Soto A 2011 (*Preprint* arXiv:astro-ph.HE/11061068)
[51] Greisen K 1966 *Phys. Rev. Lett.* **16** 748–750
[52] Zatsepin G T and Kuz'min V A 1969 On the interaction of cosmic rays with photons. *Cosmic rays, Moscow, No. 11, p. 45 - 47* vol 11 pp 45–47
[53] Abbasi R *et al.* (HiRes) 2007 (*Preprint* arXiv:astro-ph/0703099)
[54] Roth M (Pierre Auger) 2007 (*Preprint* arXiv:0706.2096 [astro-ph])
[55] Takeda M, Hayashida N, Honda K, Inoue N, Kadota K *et al.* 1998 *Phys.Rev.Lett.* **81** 1163–1166 (*Preprint* arXiv:astro-ph/9807193)
[56] Abraham J *et al.* (Pierre Auger) 2007 *Science* **318** 938–943 (*Preprint* arXiv:astro-ph/0711.2256)
[57] Galaverni M and Sigl G 2008 *Phys. Rev. Lett.* **100** 021102 (*Preprint* arXiv:astro-ph/07081737)
[58] Maccione L and Liberati S 2008 *JCAP* **0808** 027 (*Preprint* arXiv:astro-ph/08052548)
[59] Gelmini G, Kalashev O and Semikoz D V 2005 (*Preprint* arXiv:astro-ph/0506128)
[60] Abraham J *et al.* (Pierre Auger) 2008 *Astropart. Phys.* **29** 243–256 (*Preprint* arXiv:astro-ph/07121147)
[61] Rubtsov G I *et al.* 2006 *Phys. Rev.* **D73** 063009 (*Preprint* arXiv:astro-ph/0601449)
[62] Kampert K H (Pierre Auger Collaboration) 2012 (*Preprint* arXiv:astro-ph.HE/12074823)
[63] Galaverni M and Sigl G 2008 *Phys.Rev.* **D78** 063003 (*Preprint* arXiv:astro-ph/0807.1210)
[64] Aloisio R, Blasi P, Ghia P L and Grillo A F 2000 *Phys. Rev.* **D62** 053010 (*Preprint* arXiv:astro-ph/0001258)
[65] Alfaro J and Palma G 2003 *Phys. Rev.* **D67** 083003 (*Preprint* arXiv:hep-th/0208193)

[66] Scully S and Stecker F 2009 *Astropart.Phys.* **31** 220–225 (*Preprint* arXiv:astro-ph/0811.2230)
[67] Stecker F W and Scully S T 2009 *New J.Phys.* **11** 085003 (*Preprint* arXiv:astro-ph.HE/0906.1735)
[68] Maccione L, Taylor A M, Mattingly D M and Liberati S 2009 *JCAP* **0904** 022 (*Preprint* arXiv:astro-ph.HE/0902.1756)
[69] Stodolsky L 1988 *Phys.Lett.* **B201** 353
[70] Longo M J 1988 *Phys.Rev.Lett.* **60** 173
[71] Ellis J R, Harries N, Meregaglia A, Rubbia A and Sakharov A 2008 *Phys.Rev.* **D78** 033013 (*Preprint* arXiv:hep-ph/0805.0253)
[72] Sakharov A, Ellis J, Harries N, Meregaglia A and Rubbia A 2009 *J.Phys.Conf.Ser.* **171** 012039 (*Preprint* arXiv:hep-ph/09035048)
[73] Maccione L, Liberati S and Mattingly D M 2011 (*Preprint* arXiv:hep-ph/1110.0783v.2)
[74] Kelley J (IceCube Collaboration) 2009 *Nucl.Phys.* **A827** 507C–509C
[75] Gonzalez-Garcia M and Maltoni M 2004 *Phys.Rev.* **D70** 033010 (*Preprint* arXiv:hep-ph/0404085)
[76] Huelsnitz W and Kelley J (IceCube Collaboration) 2009
[77] Abbasi R *et al.* (IceCube Collaboration) 2010 *Phys.Rev.* **D82** 112003 (*Preprint* arXiv:astro-ph.HE/1010.4096)
[78] Cohen A G and Glashow S L 2011 *Phys.Rev.Lett.* **107** 181803 (*Preprint* arXiv:hep-ph/11096562)
[79] Carmona J, Cortes J and Mazon D 2012 (*Preprint* arXiv:physics.gen-ph/12032585)
[80] Mattingly D M, Maccione L, Galaverni M, Liberati S and Sigl G 2010 *JCAP* **1002** 007 (*Preprint* arXiv:hep-ph/0911.0521)
[81] Barwick S W 2007 *J.Phys.Conf.Ser.* **60** 276–283 (*Preprint* arXiv:astro-ph/0610631)
[82] Ward B 2012 (*Preprint* arXiv:hep-ph/1201.1322)
[83] Abraham J *et al.* (Pierre Auger Observatory Collaboration) 2010 *Phys.Rev.Lett.* **104** 091101 (*Preprint* arXiv:astro-ph.HE/10020699)
[84] Bi X J, Yin P F, Yu Z H and Yuan Q 2011 *Phys.Rev.Lett.* **107** 241802 (*Preprint* arXiv:hep-ph/11096667)
[85] Cowsik R, Madziwa-Nussinov T, Nussinov S and Sarkar U 2012 *Phys.Rev.* **D86** 045024 (*Preprint* arXiv:hep-ph/12060713)
[86] Adam T *et al.* (OPERA Collaboration) 2012 *JHEP* **1210** 093 (*Preprint* arXiv:hep-ex/1109.4897)
[87] Antonello M *et al.* (ICARUS Collaboration) 2012 (*Preprint* arXiv:hep-ex/12033433)
[88] Antonello M, Baibussinov B, Benetti P, Calligarich E, Canci N *et al.* 2012 *JHEP* **1211** 049 (*Preprint* arXiv:hep-ex/1208.2629)
[89] Aloisio R, Boncioli D, di Matteo A, Ghia P L, Grillo A F *et al.* 2014 (*Preprint* 1408.5213)
[90] Collins J, Perez A, Sudarsky D, Urrutia L and Vucetich H 2004 *Phys. Rev. Lett.* **93** 191301 (*Preprint* arXiv:gr-qc/0403053)
[91] Aartsen M *et al.* (IceCube Collaboration) 2015 *Phys.Rev.* **D91** 022001 (*Preprint* 1410.1749)
[92] Chen C Y, Bhupal Dev P and Soni A 2014 *Phys.Rev.* **D89** 033012 (*Preprint* 1309.1764)
[93] Aartsen M *et al.* (IceCube) 2013 *Science* **342** 1242856 (*Preprint* 1311.5238)
[94] Stecker F W, Scully S T, Liberati S and Mattingly D 2015 *Phys.Rev.* **D91** 045009 (*Preprint* 1411.5889)
[95] Arkani-Hamed N, Cheng H C, Luty M A and Mukohyama S 2004 *JHEP* **05** 074 (*Preprint* arXiv:hep-th/0312099)
[96] Jacobson T and Mattingly D 2001 *Phys.Rev.* **D64** 024028 (*Preprint* arXiv:gr-qc/0007031)
[97] Jacobson T 2007 *PoS* **QG-PH** 020 (*Preprint* arXiv:gr-qc/08011547)
[98] Eling C, Jacobson T and Mattingly D 2004 (*Preprint* arXiv:gr-qc/0410001)
[99] Bluhm R, Gagne N L, Potting R and Vrublevskis A 2008 *Phys.Rev.* **D77** 125007 (*Preprint* arXiv:hep-th/0802.4071)
[100] Liberati S and Mattingly D 2012 (*Preprint* arXiv:gr-qc/12081071)
[101] Kostelecky V A and Tasson J 2009 *Phys.Rev.Lett.* **102** 010402 (*Preprint* arXiv:gr-qc/08101459)
[102] Foster B Z and Jacobson T 2006 *Phys.Rev.* **D73** 064015 (*Preprint* arXiv:gr-qc/0509083)
[103] Will C M 2006 *Living Rev.Rel.* **9** 3 (*Preprint* arXiv:gr-qc/0510072)
[104] Yagi K, Blas D, Barausse E and Yunes N 2014 *Phys.Rev.* **D89** 084067 (*Preprint* 1311.7144)
[105] Yagi K, Blas D, Yunes N and Barausse E 2014 *Phys.Rev.Lett.* **112** 161101 (*Preprint* 1307.6219)
[106] Jacobson T and Mattingly D 2004 *Phys.Rev.* **D70** 024003 (*Preprint* arXiv:gr-qc/0402005)
[107] Elliott J W, Moore G D and Stoica H 2005 *JHEP* **08** 066 (*Preprint* arXiv:hep-ph/0505211)
[108] Sotiriou T P 2011 *J.Phys.Conf.Ser.* **283** 012034 (*Preprint* arXiv:hep-th/10103218)
[109] Horava P 2009 *Phys.Rev.* **D79** 084008 (*Preprint* arXiv:hep-th/09013775)
[110] Blas D, Pujolas O and Sibiryakov S 2010 *Phys.Rev.Lett.* **104** 181302 (*Preprint* arXiv:hep-th/09093525)
[111] Blas D, Pujolas O and Sibiryakov S 2011 *JHEP* **1104** 018 (*Preprint* arXiv:hep-th/10073503)
[112] Vernieri D and Sotiriou T P 2012 *Phys.Rev.* **D85** 064003 (*Preprint* 1112.3385)
[113] Jacobson T 2010 *Phys.Rev.* **D81** 101502 (*Preprint* arXiv:hep-th/10014823)

[114] Afshordi N 2009 *Phys.Rev.* **D80** 081502 (*Preprint* 0907.5201)
[115] Pospelov M and Shang Y 2010 (*Preprint* arXiv:hep-th/10105249)
[116] Yao W M *et al.* (Particle Data Group) 2006 *J. Phys.* **G33** 1–1232
[117] Dowker F 2011 *J.Phys.Conf.Ser.* **306** 012016

Journal of Physics: Conference Series **631** (2015) 012012 doi:10.1088/1742-6596/631/1/012012

Aspects of the Bosonic Spectral Action

Mairi Sakellariadou

Physics Department, King's College London, University of London
Strand, London WC2R 2LS, UK

E-mail: `Mairi.Sakellariadou@kcl.ac.uk`

Abstract. A brief description of the elements of noncommutative spectral geometry as an approach to unification is presented. The physical implications of the doubling of the algebra are discussed. Some high energy phenomenological as well as various cosmological consequences are presented. A constraint in one of the three free parameters, namely the one related to the coupling constants at unification, is obtained, and the possible rôle of scalar fields is highlighted. A novel spectral action approach based upon zeta function regularisation, in order to address some of the issues of the traditional bosonic spectral action based on a cutoff function and a cutoff scale, is discussed.

1. Introduction

The Standard Model (SM) of strong and electroweak interactions remains the most successful particle physics model we have at hand and its validity has been recently confirmed by the discovery of the Higgs boson. However, several conceptual questions remain unanswered, while it may be necessary to go beyond the SM, possibly relating it to a theory of Quantum Gravity. Noncommutative Spectral Geometry (NCSG) aims at explaining some of the conceptual issues of the SM, whilst it offers a new geometrical framework to address physics at the Quantum Gravity regime. To construct a Quantum theory of Gravity coupled to matter, one may either neglect matter altogether (as for instance within the framework of Loop Quantum Gravity), or consider instead that the interaction between gravity and matter is the most important ingredient to define the dynamics. Noncommutative Spectral Geometry [1, 2] follows the latter approach, aiming at defining the noncommutative algebra of observables of a Quantum theory of Gravity.

Noncommutative spectral geometry starts from the following remark. At energy scales much below the Planck scale it is reasonable to assume that physics can be described by continuum fields and an effective action (the sum of the Einstein-Hilbert and SM actions), but close to Planck scale this assumption is no longer valid and Quantum Gravity effects may imply that space-time is a heavily noncommutative manifold. Remaining close but below Planck scale, one can however consider that the algebra of coordinates is only a lightly noncommutative algebra of matrix valued functions, and by chosen properly this algebra NCSG leads to a purely geometric explanation of the SM coupled to gravity [3]. In the context of NCSG, gravity and the SM fields are packaged into geometry and matter on a Kaluza-Klein noncommutative space and using well-established experimental results at the electroweak scale, we can guess the small-scale space-time structure avoiding an *ad hoc* proposal. In that sense, noncommutative geometry can be considered as a bottom-up approach, complimentary to the top-down string theory approach.

Noncommutative spectral geometry proposes to consider the SM as a phenomenological model which dictates the space-time geometry. In this way, the geometric space is defined as the product of a 4-dimensional compact Riemannian manifold $\mathcal{M}$, describing the geometry of space-time, with an internal zero-dimensional discrete finite internal Kaluza-Klein space $\mathcal{F}$, describing the internal geometry, attached to each point. Such simple noncommutative spaces $\mathcal{M} \times \mathcal{F}$, where the noncommutative algebra describing space is the algebra of functions over ordinary space-time, are called *almost commutative manifolds*. Note that such spaces are different from the more general noncommutative spaces such as the Moyal plane for which $[x^i, x^j] = i\theta^{ij}$, where θ^{ij} is an anti-symmetric real $d \times d$ matrix (with d the space-time dimensionality) representing the fuzziness of space-time. In NCSG, the description of ordinary Riemannian manifolds in terms of spectral data, is extended for noncommutative manifolds. Hence, one defines the almost commutative manifold $\mathcal{M} \times \mathcal{F}$ by a spectral tripe, and dynamics are given by a spectral action that sums up all frequencies of vibration of space. At last, one aims at answering whether we can hear the shape of such a spectral triple, called a *spinorial drum*.

In the following, we will give a short description of the main elements of NCSG [4]-[7] , discuss its phenomenological particle physics consequences, the physical implications of the precise construction of the almost commutative manifold [8, 9], and examine the cosmological consequences of the gravitational sector of the theory [10]-[18]. We will then address some issues regarding the traditional bosonic spectral action approach and highlight a new proposal [19].

2. Elements of Noncommutative Spectral Geometry

Let us start with the case of spin (in order to be able to describe spinors) manifolds in NCSG. Given a compact 4-dimensional Riemannian spin manifold $\mathcal{M}$, consider the set $C^\infty(\mathcal{M})$ of smooth infinitely differentiable functions, and the Hilbert space $\mathcal{H} = L^2(\mathcal{M}, S)$ of square-integrable spinors S on $\mathcal{M}$. One can show that the set $C^\infty(\mathcal{M})$ is an algebra $\mathcal{A}$ under point-wise multiplication, acting on $\mathcal{H}$ as multiplication operators. Let us also consider $\not{\partial}$, the Dirac operator $-i\gamma^\mu \nabla^s_\mu$, acting as first order differential operator on the spinors. The algebra, Hilbert space and Dirac operator form the canonical triple $(C^\infty(\mathcal{M}), L^2(\mathcal{M}, S), \not{\partial})$. In addition, we consider the γ_5 operator with $\gamma_5^2 = 1, \gamma_5^* = \gamma_5$, which is just a $\mathbb{Z}_2$-grading, that decomposes the Hilbert space $\mathcal{H}$ into a positive and negative eigenspace $L^2(\mathcal{M}, S) = L^2(\mathcal{M}, S)^+ \oplus L^2(\mathcal{M}, S)^-$, hence playing the rôle of a chirality operator. We also consider an antilinear isomorphism $J_\mathcal{M}$ with $J_\mathcal{M}^2 = -1, J_\mathcal{M}\not{\partial} = \not{\partial} J_\mathcal{M}, J_\mathcal{M}\gamma_5 = \gamma_5 J_\mathcal{M}$, as the charge conjugation operator on spinors.

Consider now an almost-commutative manifold $\mathcal{M} \times \mathcal{F}$. The canonical triple defining $\mathcal{M}$ encodes the space-time structure, whereas the triple $(\mathcal{A}_\mathcal{F}, \mathcal{H}_\mathcal{F}, D_\mathcal{F})$ encodes the internal degrees of freedom at each point in space-time, allowing a description of a gauge theory on the spin manifold $\mathcal{M}$. To obtain the SM, the most important ingredient is the choice of the matrix algebra $\mathcal{A}_\mathcal{F}$, acting on the Hilbert space $\mathcal{H}_\mathcal{F}$ via matrix multiplication. The operator $\mathcal{D}_\mathcal{F}$ is a 96×96 matrix expressed in terms of the 3×3 Yukawa mixing matrices and a real constant responsible for the neutrino mass terms. This operator corresponds to the inverse of the Euclidean propagator of fermions. In addition, we consider a $\gamma_\mathcal{F}$ grading such that $\gamma_\mathcal{F} = +1$ for left-handed fermions and $\gamma_\mathcal{F} = -1$ for right-handed ones, and a conjugation operator $J_\mathcal{F}$ for the finite space $\mathcal{F}$. The almost-commutative manifold $\mathcal{M} \times \mathcal{F}$ is expressed by the spectral triple $(\mathcal{A}, \mathcal{H}, D)$:

$$\mathcal{M} \times \mathcal{F} := (C^\infty(\mathcal{M}, \mathcal{A}_\mathcal{F}), L^2(\mathcal{M}, S) \otimes \mathcal{H}_\mathcal{F}, \not{\partial} \otimes \mathbb{I} + \gamma_5 \otimes \mathcal{D}_\mathcal{F}) \ .$$

The choice of the algebra $\mathcal{A}_\mathcal{F}$ is the most important input of the NCSG approach to the SM, and has to be chosen appropriately. For instance, $\mathcal{A}_\mathcal{F}$ cannot be right-handed symmetric. It has been shown [20] that this algebra has to be the product of the algebra of quaternions and the algebra of the complex $k \times k$ matrices with k an even number $k = 2a$:

$$\mathcal{A}_\mathcal{F} = M_a(\mathbb{H}) \oplus M_k(\mathbb{C}) \ .$$

Journal of Physics: Conference Series **631** (2015) 012012 doi:10.1088/1742-6596/631/1/012012

The first value of k that produces the correct number of fermions (namely 16) in each of the three generations, is $k = 4$. Hence, NCSG predicts that the number of fermions is the square of an even integer, while the existence of three generations is just a physical input. Note that the particular choice of Hilbert space is of no importance, since all separable infinite-dimensional Hilbert spaces are isomorphic. Hence, the fermions of the SM provide the Hilbert space of a spectral triple for the algebra, while the boson of the SM, including the Higgs boson, are obtained through inner fluctuations of the Dirac operator of the product $\mathcal{M} \times \mathcal{F}$ geometry. Thus, the Higgs boson becomes just a gauge field corresponding to a finite difference.

To derive a physical Lagrangian one then applies the spectral action principle, stating that the action functional depends only on the spectrum of the fluctuated Dirac operator $\mathcal{D}_A$:

$$\mathcal{D}_A = \mathcal{D} + A + \epsilon' J A J^{-1} , \tag{1}$$

with $A = A^\star$ a self-adjoint operator of the form

$$A = \sum_j a_j [\mathcal{D}, b_j] \ ; \ a_j, b_j \in \mathcal{A} , \tag{2}$$

J an anti-unitary operator such that $J^2 = 1$ and $\epsilon \in \{\pm 1\}$, and is of the form

$$\mathrm{Tr}(f(\mathcal{D}_A^2/\Lambda^2)) , \tag{3}$$

with f a cut-off function and Λ denoting the energy scale at which this Lagrangian is valid. More precisely, f is a positive function that falls to zero at large values of its argument, so that the integrals $\int_0^\infty f(u)udu$ and $\int_0^\infty f(u)du$ are finite. Typical cut-off functions f used in the literature are $f(x) = 1$ for $x \leq \Lambda$, or $f(x) = e^{-x}$. The action given in Eq. (3) above, sums up all eigenvalues of the fluctuated Dirac operator $\mathcal{D}_A$ which are smaller than the cut-off energy scale Λ. This trace can be then evaluated using heat kernel techniques and thus expressed through the Seeley-de Witt coefficients a_n, known for any second order elliptic differential operator, as $\sum_{n=0}^\infty F_{4-n}\Lambda^{4-n}a_n$ where F is defined as $f(\mathcal{D}_A^2)$.

The spectral action can be expanded in powers of the scale Λ in the form [21]

$$\mathrm{Tr}\left(f\left(\frac{\mathcal{D}_A^2}{\Lambda^2}\right)\right) \sim \sum_{k\in \mathrm{DimSp}} f_{2k}\Lambda^{2k} {\int\!\!\!\!\!\!-} |\mathcal{D}_A|^{-2k} + f(0)\zeta_{\mathcal{D}_A^2}(0) + \mathcal{O}(1) , \tag{4}$$

where f_{2k} are the momenta of the function f, defined as

$$f_{2k} \equiv \int_0^\infty f(u)u^{2k-1}\mathrm{d}u \quad , \quad \text{for} \ \ k > 0 ,$$

and $f_0 \equiv f(0)$. The noncommutative integration is defined in terms of residues of zeta functions, $\zeta_{\mathcal{D}_A^2}(s) = \mathrm{Tr}(|\mathcal{D}_A|^{-2s})$ at poles of the zeta function, and the sum is over points in the *dimension spectrum* of the spectral triple.

Since f is a cut-off function, its Taylor expansion vanishes at zero, implying that the asymptotic expansion of the trace, namely

$$\mathrm{Tr}(f(\mathcal{D}_A^2/\Lambda^2)) \sim 2f_4\Lambda^4 a_0(\mathcal{D}_A^2) + 2f_2\Lambda^2 a_2(\mathcal{D}_A^2) + f(0)a_4(\mathcal{D}_A^2) + \mathcal{O}(\Lambda^{-2}) , \tag{5}$$

can be only given from the three first terms of the expansion. The cut-off function plays a rôle through only three of its momenta:

$$f_4 = \int_0^\infty f(u)u^3du \ ; \ f_2 = \int_0^\infty f(u)udu \ ; f_0 = f(0) . \tag{6}$$

related to the cosmological constant, the gravitational constant and the coupling constants at unification, respectively.

The bosonic spectral action, Eq. (3), must be seen *à la* Wilson, hence as the bare action at the mass scale Λ. This action only accounts for the bosonic part. Hence, to account for the terms involving fermions and their coupling to the bosons, one needs to include the fermionic part, which for a KO-dimension 2 almost commutative manifold reads

$$(1/2)\langle J\Psi, \mathcal{D}_A\Psi\rangle \; ; \; \Psi \in \mathcal{H}^+ \; . \tag{7}$$

After a long calculation one eventually obtains that the bosonic spectral action at the cutoff scale Λ and using the cutoff normalisation through the cutoff function f, reads

$$S_\Lambda = \frac{-2af_2\Lambda^2 + ef_0}{\pi^2}\int|\phi|^2\sqrt{g}d^4x + \frac{f_0}{2\pi^2}\int a|D_\mu\phi|^2\sqrt{g}d^4x - \frac{f_0}{12\pi^2}\int aR|\phi|^2\sqrt{g}d^4x$$
$$-\frac{f_0}{2\pi^2}\int\left(g_3^2G^i_{\mu}G^{\mu i} + g_2^2F^a_{\mu}F^{\mu\nu a} + \frac{5}{3}g_1^2B_\mu B^\mu\right)\sqrt{g}d^4x + \frac{f_0}{2\pi^2}\int b|\phi|^4\sqrt{g}d^4x + \mathcal{O}(\Lambda^{-2}) \; , \tag{8}$$

with a, b, c, d, e constants depending on the Yukawa parameters. Adding to the above action the fermionic part, as indicated in Eq. (7), one obtains [3] the full SM Lagrangian. Since this spectral action is characterised by the cutoff function f and the cutoff scale Λ, we will call it the *cutoff bosonic spectral action*, to differentiate it from another regularisation procedure we will highlight later.

To discuss the particle physics phenomenological consequences of NCSG let us briefly discuss the obtained Lagrangian. Its coefficients are given in terms of the three momenta $f(0), f_2, f_4$ of the cut-off function f, of the cut-off scale Λ, of the vacuum expectation value of the Higgs field ϕ, and of the coefficients a, b, c, d, e, which are determined by the mass matrices in the Dirac operator $\mathcal{D}_{\mathcal{F}}$. Given that among the various relations connecting the coefficients a, b, c, d, e, one finds $g_2^2 = g_3^2 = (5/3)g_1^2$, which holds in several Grand Unified Theories (GUTs) (like SU(5)), one may assume that the theory is valid at the GUT scale. One then uses standard renormalisation group flow techniques to obtain predictions for the SM phenomenology. For instance, one finds that the top quark mass is $m_{\rm t} \leq 180$ GeV. The NCSG approach leads to a Higgs doublet with a negative mass term and a positive quartic term, hence implying the existence of a spontaneously symmetry breaking mechanism of the electroweak symmetry. Let us comment on the predicted value of the Higgs mass. The NCSG model involves three scalars, namely a Higgs field, a singlet and a dilaton. The singlet is a real scalar field associated with the Majorana mass of the right-handed neutrino, having a nontrivially mixing with the Higgs field. In the original approach [3], the singlet was integrated out being replaced by its vacuum expectation value, leading to an incorrect prediction of the Higgs mass, namely 167 GeV $\leq m_{\rm h} \leq$ 176 GeV. This conflict was resolved in the subsequent approach [22] where this assumption was relaxed. Hence considering the mixing between the Higgs doublet and singlet, consistency with the experimental result of a 125 GeV Higgs mass and a 170 GeV top quark mass was achieved. Note that the rôle of the singlet field was already mentioned previously [23]. Moreover, the experimentally found Higgs mass can be accommodated by either considering a model based on a larger symmetry, the *grand symmetry*, where the algebra is $A_G = M_4(\mathbb{H}) \oplus M_8(\mathbb{C})$ [24], or by generalising the inner fluctuations to real spectral triples that fail on the first order condition, leading to a Pati-Salam type of model $\mathrm{SU(2)_R \times SU(2)_L \times SU(4)}$ [25].

Assuming the big desert hypothesis, one-loop renormalisation group analysis for the three gauge couplings and the Newton constant, has shown [3] that they do not exactly meet at a point; the error being just a few percent. Hence, the big desert hypothesis is only approximately valid, and one may expect new physics between unification and present energy scales. Finally,

NCSG predicts the existence of a see-saw mechanism for neutrino masses with large right-handed neutrino mass of the order of the cutoff scale Λ.

In conclusion, NCSG offers an elegant geometric interpretation of the SM coupled to gravity. Applying Einstein's theory of General Relativity (GR) within Riemannian geometry, one obtains the familiar gravitational theory. As we have discussed above, applying the spectral action approach within the context of an almost commutative geometry, one gets gravity combined with Yang-Mills and Higgs. What remains to be done in this programme, is to construct the appropriate tools we need to apply within a fully noncommutative geometry and then deduce the theory to which they will lead us.

3. Physical meaning of the doubling of the algebra

Let us highlight the physical implications of choosing an almost commutative manifold. The geometry is specified by the product $\mathcal{M}\times\mathcal{F}$ given from the spectral triple

$$(\mathcal{A},\mathcal{H},\mathcal{D},J,\gamma)=(C^\infty(\mathcal{M}),L^2(\mathcal{M},S),\partial\!\!\!/_\mathcal{M},J_\mathcal{M},\gamma_5)\otimes(\mathcal{A}_\mathcal{F},\mathcal{H}_\mathcal{F},\mathcal{D}_\mathcal{F},J_\mathcal{F},\gamma_\mathcal{F})\ ,$$

defined as

$$(\mathcal{A},\mathcal{H},\mathcal{D},J,\gamma)=(\mathcal{A}_1,\mathcal{H}_1,\mathcal{D}_1,J_1,\gamma_1)\otimes(\mathcal{A}_2,\mathcal{H}_2,\mathcal{D}_2,J_2,\gamma_2)\ ,$$

with

$$\mathcal{A}=\mathcal{A}_1\otimes\mathcal{A}_2\ ,\ \ \mathcal{H}=\mathcal{H}_1\otimes\mathcal{H}_2\ ,\ \ \mathcal{D}=\mathcal{D}_1\otimes 1+\gamma_1\otimes\mathcal{D}_2\ ,\gamma=\gamma_1\otimes\gamma_2\ ,\ \ J=J_1\otimes J_2\ ,$$

where $J^2=-1, [J,\mathcal{D}]=0, [J_1,\gamma_1]=0$ and $\{J,\gamma\}=0$. The doubling of the algebra is intimately related to dissipation, gauge field structure (necessary to address the physics of the SM), as well as neutrino mixing, while it incorporates the seeds of quantisation [8, 9].

Consider the classical Brownian motion of a particle of mass m with equation of motion

$$m\ddot{x}(t)+\gamma\dot{x}(t)=f(t)\ , \tag{9}$$

where $f(t)$ denotes a random Gaussian distributed force. This equation of motion can be derived from a Lagrangian in a canonical procedure, using a delta functional classical constraint representation as a functional integral. It is easy to see [8] that the constraint condition at the classical level introduces a new coordinate, called y, with the y-system being the time-reversed of the x-one, so that the equations of motion read

$$m\ddot{x}+\gamma\dot{x}=f\ ,\ \ m\ddot{y}-\gamma\dot{y}=0\ . \tag{10}$$

The x-system represents an *open* (dissipating) system, while the $\{x,y\}$-system is a *closed* one. This doubling, discussed here in a completely classical context, is necessary in order to build a canonical formalism for dissipative systems [8].

To argue the relation between the doubling of the algebra and the gauge field structure, consider

$$m\ddot{x}+\gamma\dot{x}+kx=0\ , \tag{11}$$

the equation of a classical one-dimensional damped harmonic oscillator, with time independent quantities m,γ,k. Following the previous discussion, we will complement the x-system with its time-reversed image, called y-system, as

$$m\ddot{y}-\gamma\dot{y}+ky=0\ , \tag{12}$$

in order to build a well-defined Lagrangian formalism. Equation (12) above is that of a one-dimensional amplified harmonic oscillator.

The Lagrangian of the closed $\{x, y\}$-system can be then written as

$$L = \frac{1}{2m}(m\dot{x}_1 + \frac{e_1}{c}A_1)^2 - \frac{1}{2m}(m\dot{x}_2 + \frac{e_2}{c}A_2)^2 - \frac{e^2}{2mc^2}(A_1^2 + A_2^2) - e\Phi \; , \tag{13}$$

where we have introduced the coordinates x_1, x_2 through

$$x_1(t) = \frac{x(t) + y(t)}{\sqrt{2}} \; , \; x_2(t) = \frac{x(t) - y(t)}{\sqrt{2}} \; , \tag{14}$$

and the vector potential

$$A_i = \frac{B}{2}\epsilon_{ij}x_j \;\text{ for } i, j = 1, 2 \;\text{ with } B \equiv \frac{\gamma c}{e} \; , \; \epsilon_{ii} = 0 \; , \; \epsilon_{12} = -\epsilon_{21} = 1 \; . \tag{15}$$

The Lagrangian (14) describes two particles having opposite charges $e_1 = -e_2 \equiv e$ in the potential $\Phi = \Phi_1 - \Phi_2$, where $\Phi_i \equiv (k/2/e)x_i^2$ in the constant magnetic field $\mathbf{B} = \nabla \times \mathbf{A}$. Identifying the doubled coordinate with the x_2, we observe that it acts as the gauge field component A_1 to which the original x_1 coordinate is coupled. We hence conclude that energy dissipated by one of the two systems is gained by the other one, so that the gauge field can be seen as the *reservoir* in which the system is embedded.

Dissipation may also lead to a quantum evolution. This can be easily shown by using 't Hooft's conjecture, saying that loss of information (i.e., dissipation) in a regime of deterministic dynamics may lead to a quantum mechanical evolution. We consider again the classical damped harmonic x-oscillator and its time-reversed image, the y-oscillator, discussed above. The Hamiltonian of the $\{x, y\}$-system can be schematically written as

$$H = H_{\rm I} - H_{\rm II} \;\text{ with the constraint }\; H_{\rm II}|\psi\rangle = 0 \; , \tag{16}$$

in order to define physical states ψ and guarantee that the Hamiltonian is bounded from below. The physical consequence of this constraint is *information loss*. Physical states are invariant under time reversal and periodical, implying that

$${}_H\langle\psi(\tau)|\psi(0)\rangle_H = e^{i\alpha\pi} \; , \tag{17}$$

where $\tau = 2\pi/\Omega$ (with Ω expressed in terms of m, k, γ) stands for the period, and α is a real constant. Hence

$$\langle\psi_n(\tau)|H|\psi_n(\tau)\rangle = \hbar\Omega(n + \alpha/2) = \hbar\Omega n + E_0 \; , \tag{18}$$

with $E_0 = (\hbar/2)\Omega\alpha$ the zero point ($n = 0$) energy. Note that the index n above signals the n-dependence of the state and the corresponding energy. In conclusion, the zero point quantum contribution to the spectrum of physical states found above, results from information loss, imposed by the underlying dissipative dynamics [8].

The algebra doubling can also lead to neutrino oscillations. Linking the algebra doubling to the deformed Hopf algebra, one can build Bogogliubov operators as linear combinations of the co-product operators defined in terms of the deformation parameter obtained from the doubled algebra, and show the emergence of neutrino mixing [9]. In particular, one can write the mixing transformations connecting the flavour fields $\psi_{\rm f}$ to the neutrino fields with nonvanishing masses $\psi_{\rm m}$ as

$$\nu_e(x) = G_\theta^{-1}(t)\nu_1(x)G_\theta(t) \;\; ; \;\; \nu_\mu(x) = G_\theta^{-1}(t)\nu_2(x)G_\theta(t) \; , \tag{19}$$

through the generator of field mixing transformations $G_\theta(t)$. Note that for simplicity, and no loss of generality, we have only used two neutrino species. Then writing $\psi_{\rm m}$ in terms of flavour creation/annihilation operators, and similarly writing $\psi_{\rm m}$ in terms of mass creation/annihilation operators, one finds that $G_\theta(t)$ contains rotation operator terms and Bogogliubov transformation operator terms. Since deformed co-products are a basis of Bogogliubov transformations, one concludes that field mixing arises from the algebraic structure of the deformed co-product in the noncommutative Hopf algebra embedded in the algebra doubling of noncommutative spectral geometry. We can hence conclude that the SM derived from NCSG, includes neutrino mixing by construction [9].

4. NCSG leading to an extended gravitational theory

We are currently living in a very exciting time for early universe cosmology, since our models can be now tested with a variety of very precise astrophysical and high energy physics data, and in particular with the Cosmic Microwave Background temperature anisotropies data and the Large Hadron Collider results. However, despite the present golden era of cosmology, a number of questions are still awaiting for a definite answer. For instance, one does not know the origin of dark matter and dark energy, whilst the search for a natural and well-motivated inflationary model (or plausible alternatives to the inflationary paradigm) still remains unsuccessful.

The main approaches to build early universe cosmological models have been based to string/M-theory or some non perturbative approach to Quantum Gravity, with Loop Quantum Cosmology being the leading candidate. Noncommutative spectral geometry can provide another proposal, since the model lives by construction at the GUT scale.

The bosonic action in Euclidean signature, favoured by the formalism of spectral triples, is [3]

$$\begin{aligned}\mathcal{S}^{\rm E} = \int \Big(& \frac{1}{2\kappa_0^2}R + \alpha_0 C_{\mu\nu\rho\sigma}C^{\mu\nu\rho\sigma} + \gamma_0 + \tau_0 R^\star R^\star + \frac{1}{4}G^i_{\mu\nu}G^{\mu\nu i} + \frac{1}{4}F^\alpha_{\mu\nu}F^{\mu\nu\alpha} \\ & + \frac{1}{4}B^{\mu\nu}B_{\mu\nu} + \frac{1}{2}|D_\mu \mathbf{H}|^2 - \mu_0^2|\mathbf{H}|^2 - \xi_0 R|\mathbf{H}|^2 + \lambda_0|\mathbf{H}|^4\Big)\sqrt{g}\ d^4x\ , \end{aligned} \tag{20}$$

where

$$\begin{aligned} \kappa_0^2 &= \frac{12\pi^2}{96 f_2\Lambda^2 - f_0\mathfrak{c}} \quad , \quad \alpha_0 = -\frac{3f_0}{10\pi^2}\ , \\ \gamma_0 &= \frac{1}{\pi^2}\left(48 f_4\Lambda^4 - f_2\Lambda^2\mathfrak{c} + \frac{f_0}{4}\mathfrak{d}\right) \quad , \quad \tau_0 = \frac{11 f_0}{60\pi^2}\ , \\ \mu_0^2 &= 2\Lambda^2\frac{f_2}{f_0} - \frac{\mathfrak{e}}{\mathfrak{a}} \quad , \quad \xi_0 = \frac{1}{12}\ , \\ \lambda_0 &= \frac{\pi^2\mathfrak{b}}{2f_0\mathfrak{a}^2} \quad , \quad \mathbf{H} = (\sqrt{\mathfrak{a}f_0}/\pi)\phi\ ; \end{aligned} \tag{21}$$

$\mathbf{H}$ a rescaling of the Higgs field ϕ to normalize the kinetic energy, and the momentum f_0 is physically related to the coupling constants at unification. The geometric parameters $\mathfrak{a}, \mathfrak{b}, \mathfrak{c}, \mathfrak{d}, \mathfrak{e}$ correspond to the (running) Yukawa parameters of the particle physics model and the Majorana terms for the right-handed neutrinos. The first two terms in Eq. (20) depend only on the Riemann curvature tensor. The first is the Einstein-Hilbert term and the second is the Weyl curvature term; hence they are the Riemannian curvature terms. The third one is the cosmological term, while the fourth term

$$R^\star R^\star = \frac{1}{4}\epsilon^{\mu\nu\rho\sigma}\epsilon_{\alpha\beta\gamma\delta}R^{\alpha\beta}_{\mu\nu}R^{\gamma\delta}_{\rho\sigma}\ ,$$

is the topological term that integrates to the Euler characteristic, hence is nondynamical. The three next terms are the Yang-Mills terms. The eighth term is the scalar minimal coupling term,

Journal of Physics: Conference Series **631** (2015) 012012 doi:10.1088/1742-6596/631/1/012012

the next one is the scalar mass term, and the last one is the scalar quartic potential term. There is one more term, the $-\xi_0 R|\mathbf{H}|^2$, that couples gravity with the SM. For $\xi_0 = 1/12$, this term encodes the conformal coupling between the Higgs field and the Ricci curvature.

Hence, the Lagrangian obtained through the NCSG approach contains, in addition to the full SM Lagrangian, the Einstein-Hilbert action with a cosmological term, a topological term related to the Euler characteristic of the space-time manifold, a conformal Weyl term and a conformal coupling of the Higgs field to gravity. Within the NCSG context the Higgs field appears as the vector boson of the internal noncommutative degrees of freedom.

At this point, let us make a few remarks. The relations given in Eq. (21) above, are tied to the cutoff scale Λ, hence *a priori* there is no reason for those to hold at any other scale. Since the action Eq. (20) includes only the first three terms in the asymptotic expansion, one must be cautious keeping in mind that there are scales for which the neglected nonperturbative effects become important. Since to study physical consequences of the NCSG proposal one must use a Lorentzian signature, we will assume that a Wick rotation to imaginary time can be achieved. Noticing the absence of quadratic terms in the curvature — there is only the term quadratic in the Weyl curvature and the topological term $R^\star R^\star$ — we immediately conclude that for Friedmann-Lemaître-Robertson-Walker geometries, the Weyl term vanishes. Finally, notice the term that couples gravity with the SM, a term which should always be present when one considers gravity coupled to scalar fields.

The gravitational part of the asymptotic formula for the bosonic sector of the NCSG, including the coupling between the Higgs field and the Ricci curvature scalar, in Lorentzian signature, reads

$$\mathcal{S}_{\rm grav}^{\rm L} = \int \left(\frac{1}{2\kappa_0^2} R + \alpha_0 C_{\mu\nu\rho\sigma} C^{\mu\nu\rho\sigma} + \tau_0 R^\star R^\star - \xi_0 R |\mathbf{H}|^2 \right) \sqrt{-g}\; d^4 x\ . \tag{22}$$

It will lead to the following equations of motion [10]:

$$R^{\mu\nu} - \frac{1}{2} g^{\mu\nu} R + \frac{1}{\beta^2} \delta_{\rm cc} \left[2 C^{\mu\lambda\nu\kappa}_{\ \ ;\lambda;\kappa} + C^{\mu\lambda\nu\kappa} R_{\lambda\kappa} \right] = \kappa_0^2 \delta_{\rm cc} T^{\mu\nu}_{\rm matter}\ , \tag{23}$$

where

$$\beta^2 \equiv -\frac{1}{4\kappa_0^2 \alpha_0} \quad \text{and} \quad \delta_{\rm cc} \equiv [1 - 2\kappa_0^2 \xi_0 \mathbf{H}^2]^{-1}\ . \tag{24}$$

The definition $\delta_{\rm cc}$ captures the conformal coupling between the Ricci scalar and the Higgs field.

In the low energy weak curvature regime, one may neglect the nonminimal coupling term between the background geometry and the Higgs field, getting $\delta_{\rm cc} = 1$. Hence, for a cosmological context, namely Friedmann-Lemaître-Robertson-Walker space-time, the Weyl tensor vanishes and the noncommutative spectral geometry corrections to the Einstein equation vanish [10]. Consequently, any modifications to the background equation may appear at leading order only for anisotropic and inhomogeneous models, such as a Bianchi type-V model defined by the space-time metric

$$g_{\mu\nu} = {\rm diag}\left[-1, \{a_1(t)\}^2 e^{-2nz}, \{a_2(t)\}^2 e^{-2nz}, \{a_3(t)\}^2 \right]\ , \tag{25}$$

where $a_i(t)$ with $i = 1, 2, 3$, arbitrary functions, denoting the three scale factors, and n an integer. In this metric, the modified Friedmann equation reads [10]:

$$\begin{aligned}
\kappa_0^2 T_{00} = \\
-\dot{A}_3\left(\dot{A}_1+\dot{A}_2\right) - n^2 e^{-2A_3}\left(\dot{A}_1\dot{A}_2-3\right) \\
+\frac{8\alpha_0\kappa_0^2 n^2}{3}e^{-2A_3}\left[5\left(\dot{A}_1\right)^2+5\left(\dot{A}_2\right)^2-\left(\dot{A}_3\right)^2\right. \\
\left.-\dot{A}_1\dot{A}_2-\dot{A}_2\dot{A}_3-\dot{A}_3\dot{A}_1-\ddot{A}_1-\ddot{A}_2-\ddot{A}_3+3\right] - \frac{4\alpha_0\kappa_0^2}{3}\sum_i\left\{\dot{A}_1\dot{A}_2\dot{A}_3\dot{A}_i\right. \\
+\dot{A}_i\dot{A}_{i+1}\left(\left(\dot{A}_i-\dot{A}_{i+1}\right)^2-\dot{A}_i\dot{A}_{i+1}\right)+\left(\ddot{A}_i+\left(\dot{A}_i\right)^2\right)\left[-\ddot{A}_i-\left(\dot{A}_i\right)^2+\frac{1}{2}\left(\ddot{A}_{i+1}+\ddot{A}_{i+2}\right)\right. \\
\left.+\frac{1}{2}\left(\left(\dot{A}_{i+1}\right)^2+\left(\dot{A}_{i+2}\right)^2\right)\right]+\left[\dddot{A}_i+3\dot{A}_i\ddot{A}_i-\left(\ddot{A}_i+\left(\dot{A}_i\right)^2\right)\left(\dot{A}_i-\dot{A}_{i+1}-\dot{A}_{i+2}\right)\right] \\
\left.\times\left[2\dot{A}_i-\dot{A}_{i+1}-\dot{A}_{i+2}\right]\right\}, \quad (26)
\end{aligned}$$

where we have defined $A_i(t) = \ln a_i(t)$ with $i = 1, 2, 3$. One then immediately observes that the correction terms in Eq. (26) above come in two types. Those which are fourth order in time derivatives, and those that are at the same order as the ones derived from the standard Einstein-Hilbert action. The former can be considered as small corrections since usually in cosmology we have slowly varying functions. The latter are proportional to n^2, hence they vanish for homogeneous versions of Bianchi type-V. In conclusion, the corrections to Einstein's equations are only present in inhomogeneous and anisotropic space-times [10].

As energies are approaching the Higgs scale, the nonminimal coupling of the Higgs field to the curvature can no longer be neglected. Then the equations of motion read [10]

$$R^{\mu\nu} - \frac{1}{2}g^{\mu\nu}R = \kappa_0^2\left[\frac{1}{1-\kappa_0^2|\mathbf{H}|^2/6}\right]T^{\mu\nu}_{\text{matter}}\,, \quad (27)$$

where for simplicity we set $\beta = 0$. We thus conclude that $|\mathbf{H}|$ leads to an effective gravitational constant. A different way to see the rôle of the nonminimal coupling is by examining its effect on the equations of motion for the Higgs field in a constant gravitational field. Hence, since

$$-\mu_0|\mathbf{H}|^2 \to -\left(\mu_0+\frac{R}{12}\right)|\mathbf{H}|^2\,, \quad (28)$$

we conclude that for constant curvature, the self interaction of the Higgs field is increased.

Finally, one may find links to dilatonic gravity and chameleon cosmology [10]. Firstly, the action

$$\mathcal{L}_{|\mathbf{H}|} = -\frac{R}{12}|\mathbf{H}|^2 + \frac{1}{2}|D^\alpha\mathbf{H}||D^\beta\mathbf{H}|g_{\alpha\beta} - \mu_0|\mathbf{H}|^2 + \lambda_0|\mathbf{H}|^4 \quad (29)$$

(where D^α denotes covariant derivative) for the pure Higgs field $\mathbf{H}$, can be written in the form of four-dimensional dilatonic gravity as

$$\mathcal{L}_{\tilde{\phi}} = e^{-2\tilde{\phi}}\left[-R+6D^\alpha\tilde{\phi}D^\beta\tilde{\phi}g_{\alpha\beta}-12\left(\mu_0-12\lambda_0 e^{-2\tilde{\phi}}\right)\right]\,, \quad (30)$$

by a redefinition of the Higgs field as

$$\tilde{\phi} = -\ln\left(|\mathbf{H}|/(2\sqrt{3})\right)\,. \quad (31)$$

Secondly, chameleon models are characterised by the existence of a scalar field having a nonminimal coupling to the standard matter content (thus evading solar system tests of GR). In the context of NCSG, the Higgs field has a nonzero coupling to the background geometry. If the equations of motion can be approximated by Einstein's equations, then the background geometry will be approximately given by standard matter, making the mass and dynamics of the Higgs field explicitly dependent of the local matter content.

Exploring the possible rôle of scalar fields appearing in the NCSG action, one may wonder whether the Higgs field, through its nonminimal coupling to the background geometry, can be the inflaton. The Gravity-Higgs sector of the asymptotic expansion of the spectral action, in Lorentzian signature reads

$$S^{\rm L}_{\rm GH} = \int \Big[\frac{1 - 2\kappa_0^2\xi_0 H^2}{2\kappa_0^2}R - \frac{1}{2}(\nabla H)^2 - V(H)\Big]\sqrt{-g}\ d^4x\ , \tag{32}$$

where

$$V(H) = \lambda_0 H^4 - \mu_0^2 H^2\ , \tag{33}$$

with μ_0 and λ_0 subject to radiative corrections as functions of energy. For each value of the top quark mass, there is a value of the Higgs mass where the Higgs potential is locally flattened [13]. However, since the flat region is narrow, the slow-roll must be very slow, otherwise the quasi-exponential expansion will not last long enough. Moreover, the amplitude of density perturbations in the Cosmic Microwave Background must be in agreement with the measured one.

Calculating the renormalisation of the Higgs self-coupling up to two-loops and constructing an effective potential which fits the renormalisation group improved potential around the flat region, one concludes that while the Higgs potential can lead to the slow-roll conditions being satisfied, the constraints imposed from the CMB data make the predictions of such a scenario incompatible with the measured value of the top quark mass [13].

The gravitational sector of the NCSG action provides a proposal for an extended theory of gravity. Studying the astrophysical consequences of such a theory, one is able to constrain one of the three momenta of the cutoff function, namely f_0 (or equivalently $a_0 = -3f_0/(10\pi^2)$ in Eq. (22)) which specifies the initial conditions on the gauge couplings [14]-[18]. Hence we will get a restriction on the particle physics at unification. Note that one cannot constrain the other two free parameters (the momenta f_2, f_4) without an *ad hoc* assumption on the running of the coefficients in the action functional.

To simplify the analysis and with no loss of generality, let us neglect in the following the conformal coupling between the Ricci curvature and the Higgs field. Hence to find nonzero correction terms we have to go beyond the homogeneous and isotropic case. The equations of motion read [10]

$$G^{\mu\nu} + \frac{1}{\beta^2}[2\nabla_\lambda\nabla_\kappa C^{\mu\nu\lambda\kappa} + C^{\mu\lambda\nu\kappa}R_{\lambda\kappa}] = \kappa^2 T^{\mu\nu}_{(\rm matter)}\ , \tag{34}$$

where $\kappa^2 \equiv 8\pi G$, $G^{\mu\nu}$ is the (zero order) Einstein tensor, $T^{\mu\nu}_{\rm matter}$ the energy-momentum tensor of matter and $\beta^2 = 5\pi^2/(6\kappa^2 f_0)$. Performing a detailed analysis of linear perturbations

$$g_{\mu\nu} = \eta_{\mu\nu} + \gamma_{\mu\nu}\ , \tag{35}$$

around a Minkowski background metric $\eta_{\mu\nu}$, one can show that the linearised equation of motion, derived within the NCSG context, reads [10]

$$\left(1 - \frac{1}{\beta^2}\Box_\eta\right)\Box_\eta \bar{h}^{\mu\nu} = -2\kappa^2 T^{\mu\nu}_{\rm matter}\ , \tag{36}$$

with $T^{\mu\nu}_{\rm matter}$ taken to lowest order in $\gamma^{\mu\nu}$, so that it is independent of $\gamma^{\mu\nu}$ and satisfies the conservation equation $\partial_\mu T^{\mu\nu}_{\rm (matter)} = 0$. The equation of motion above has been written in terms of the tensor [15]

$$\bar{h}_{\mu\nu} = \bar{\gamma}_{\mu\nu} - \frac{1}{3\beta^2}\,\mathcal{Q}^{-1}\left(\eta_{\mu\nu}\Box_\eta - \partial_\mu\partial_\nu\right)\gamma\,, \tag{37}$$

with

$$\mathcal{Q} \equiv 1 - \frac{1}{\beta^2}\Box_\eta\,, \tag{38}$$

and having defined $\bar{\gamma}_{\mu\nu}$ the *trace reverse* of $\gamma_{\mu\nu}$:

$$\bar{\gamma}_{\mu\nu} = \gamma_{\mu\nu} - \frac{1}{2}\eta_{\mu\nu}\gamma\,. \tag{39}$$

Note that we constrain $\alpha_0 < 0$, hence $\beta^2 > 0$, so that Minkowski is a stable vacuum of the theory. The general solution [15]

$$h^{\mu\nu} = 2\beta^2\kappa\int \mathrm{d}S(x')G_{\rm R}(x,x')T^{\mu\nu}(x')\,, \tag{40}$$

to Eq. (36) is given in terms of Green's functions $G_{\rm R}(x,x')$ which satisfy the fourth-order partial differential equation

$$\left(\Box - \beta^2\right)\Box G_{\rm R}(x,x') \quad = \quad 4\pi\delta^{(4)}(x-x')\,; \tag{41}$$

the operators $\Box$ above are acting on x. Consequently, the field is given by [15]

$$h^{\mu\nu}(\mathbf{r},t) = \frac{4G\beta}{c^4}\int \mathrm{d}\mathbf{r}'\mathrm{d}t'\frac{\Theta(T)}{\sqrt{(cT)^2 - |\mathbf{R}|^2}}\mathcal{J}_1\left(\beta\sqrt{(cT)^2 - |\mathbf{R}|^2}\right)T^{\mu\nu}(\mathbf{r}',t')\,\Theta(cT - |\mathbf{R}|)\,, \tag{42}$$

where $J_1(x)$ is the first order Bessel function of the first kind; Θ is the Heavyside step function; $T = t - t'$ is the difference between the time of observation t and time of emission t' of the perturbation; and $\mathbf{R} = \mathbf{r} - \mathbf{r}'$ denotes the difference between the location $\mathbf{r}$ of the observer and the location $\mathbf{r}'$ of the emitter. One can then calculate the propagation of gravitational waves and investigate discrepancies from the results obtained within standard GR.

In the far-field limit $|\mathbf{r}| \approx |\mathbf{r} - \mathbf{r}'|$, the spatial components of the $h^{\mu\nu}$ field are [15]

$$h^{ik}(\mathbf{r},t) \approx \frac{2G\beta}{3c^4}\int_{-\infty}^{t-\frac{1}{c}|\mathbf{r}|}\frac{\mathrm{d}t'}{\sqrt{c^2(t-t')^2 - |\mathbf{r}|^2}}\mathcal{J}_1\left(\beta\sqrt{c^2(t-t')^2 - |\mathbf{r}|^2}\right)\dddot{D}^{ik}(t')\,, \tag{43}$$

where D^{ik} stands for the quadrupole moment,

$$D^{ik}(t) \equiv \frac{3}{c^2}\int \mathrm{d}\mathbf{r}\; x^i x^k T^{00}(\mathbf{r},t)\,. \tag{44}$$

In this limit, the rate of energy loss from a circular binary system, under the assumption that the internal structure of the pair of masses m_1 and m_2 can be neglected, reads

$$-\frac{\mathrm{d}\mathcal{E}}{\mathrm{d}t} \approx \frac{c^2}{20G}|\mathbf{r}|^2\dot{h}_{ij}\dot{h}^{ij}\,, \tag{45}$$

where the time derivatives of the spatial components of the field are

$$\dot{h}^{ij} = \frac{4G\beta A^{ij}\omega^{ij}}{3c^4}\left[\sin\left(\omega^{ij}t+\phi^{ij}\right) f_c\left(\beta|\mathbf{r}|, \frac{\omega^{ij}}{\beta c}\right) + \cos\left(\omega^{ij}t+\phi^{ij}\right) f_s\left(\beta|\mathbf{r}|, \frac{\omega^{ij}}{\beta c}\right)\right] , \quad (46)$$

(note that no summation is implied) and we have defined the functions $f_s(x,z), f_c(x,z)$ as

$$\begin{aligned} f_s(x,z) &\equiv \int_0^\infty \frac{ds}{\sqrt{s^2+x^2}} \mathcal{J}_1(s) \sin\left(z\sqrt{s^2+x^2}\right) , \\ f_c(x,z) &\equiv \int_0^\infty \frac{ds}{\sqrt{s^2+x^2}} \mathcal{J}_1(s) \cos\left(z\sqrt{s^2+x^2}\right) . \end{aligned} \quad (47)$$

The integrals in Eq. (47) show a strong resonance behaviour at $z = 1$ corresponding to a critical frequency

$$2\omega_c = c\beta = c(-\alpha_0 G)^{-1} , \quad (48)$$

close to which strong deviations from the standard results of GR are expected; these integrals are easily evaluated numerically for $z > 1$ and $z < 1$.

In the large $|\mathbf{r}|$ limit, the rate of energy loss to gravitational radiation by a circular (for simplicity) binary system of masses m_1, m_2 (we denote by μ the reduced mass of the system) at a separation vector of magnitude ρ reads [14]

$$-\frac{d\mathcal{E}}{dt} \approx \frac{32G\mu^2\rho^4\omega^6}{5c^5} \times \begin{cases} 1 + \frac{C}{\beta|\mathbf{r}|\left(1-\frac{\omega}{\omega_c}\right)} \mathcal{J}_1\left(\beta|\mathbf{r}| - \frac{\omega}{\omega_c}\right) + \ldots & ; \omega < \omega_c \\ 4\sin^2\left(\beta|\mathbf{r}|\tilde{f}\left(\frac{\omega}{\omega_c}\right)\right) & ; \omega > \omega_c \end{cases} , \quad (49)$$

where in the $\omega < \omega_c$ case the dots refer to higher powers of $1/(\beta|\mathbf{r}|)$.

Hence, for orbital frequencies small compared to the critical one ω_c, any deviation from the standard GR result is suppressed by the distance to the source. In this case, the $\beta \to \infty$ limit leads, as expected, to the GR result. This is not the case however for $\omega > \omega_c$, since then the GR result in only recovered if $\beta|\mathbf{r}|\tilde{f}(\omega/\omega_c) = \pi/3$. We will thus only consider the $\omega < \omega_c$ physically interesting case, and restrict β (equivalently f_0) by requiring that the energy lost to gravitational radiation agrees with the one predicted by GR to within observational uncertainties. Note that the presence of the Bessel function implies that the amplitude of the deviation from the result obtained within standard GR will oscillate with frequencies as well as distances, however the effect will be suppressed by the $|\mathbf{r}|^{-1}$ factor.

Considering binary pulsar systems, for which the rate of change of the orbital frequency is well known, the observational constraint is $\beta \gtrsim 7.55 \times 10^{-13}\text{m}^{-1}$ [14]. This (weak) limit can be improved through future observations of rapidly orbiting binaries relatively close to the Earth.

One can also set a lower bound on the Weyl term appearing in the NCSG action using results from the Gravity Probe B [17] and LAser RElativity Satellite (LARES) [18] experiments. Gravity Probe B satellite contains a set of four gyroscopes and has tested the geodesic and the frame-dragging (Lense-Thirring) effects of GR with extreme precision. The LARES mission is designed to test these two effects to within 1% of the value predicted within the theory of GR.

Let us write the metric in terms of the metric potentials Φ, Ψ and the vector potential $\mathbf{A}$ as

$$ds^2 = -(1+2\Phi)dt^2 + 2\mathbf{A}\cdot d\mathbf{x}dt + (1+2\Psi)d\mathbf{x}^2 . \quad (50)$$

The rate of an orbiting gyroscope precession can be then splitted into a part generated by the metric potentials and one generated by the vector potential. The obtained spin equation of motion for the gyro-spin three-vector is hence expressed as the sum of the instantaneous geodesic and Lense-Thirring precessions. Each of these two precessions can be then written as the sum of two terms, one obtained within GR and one coming from NCSG. Setting the geodesic precession (equivalently for the Lense-Thirring precession) to be the one predicted from standard GR and requiring that the NCSG contribution is within the accuracy of its measured value, Gravity Probe B results imply [17] $\beta \gtrsim 7.1 \times 10^{-5}\mathrm{m}^{-1}$, and LARES experiment sets [18] $\beta \gtrsim 1.2 \times 10^{-6}\mathrm{m}^{-1}$.

A much stronger constraint can be imposed to β using the torsion balance experiments. The modifications induced by the NCSG action to the Newtonian potentials Φ, Φ lead to the following expressions for the components of $\gamma_{\mu\nu}$ [17]:

$$\begin{aligned}
\gamma_{00} &= -2\Phi = \frac{2GM}{r}\left(1 - \frac{4}{3}e^{-\beta r}\right) , \\
\gamma_{0i} &= \gamma_{i0} = A_i = -\frac{4G}{r^3}[1 - (1 + \beta r)e^{-\beta r}](\mathbf{r} \wedge \mathbf{J})_i , \\
\gamma_{ij} &= 2\Psi\delta_{ij} = \frac{2GM}{r}\left[1 + \frac{5}{9}e^{-\beta r}\right]\delta_{ij} . \qquad (51)
\end{aligned}$$

These modifications are similar to those induced by a fifth-force through a potential

$$V(r) = -\frac{GMm}{r}\left(1 + \alpha e^{-r/\lambda}\right) , \qquad (52)$$

where α is a dimensionless strength parameter and λ a length scale. The tightest constraint on λ from the latest torsion balance experiments is $\lambda \lesssim 10^{-4}\mathrm{m}$ leading to [17]

$$\beta \gtrsim 10^4 \mathrm{m}^{-1} , \qquad (53)$$

a much stronger constraint than the ones obtained through pulsar timings, Gravity Probe B or LARES experiments.

5. A new approach: The zeta function regularisation

The cutoff bosonic spectral action is a quite successful and promising scheme, worth to be further investigated. Based upon an elegant mathematical theory, it offers a description of geometry in terms of spectral properties of operators and leads to a model of particle interactions which is very close to the real phenomenology as revealed from high energy physics experiments. Whilst the Standard Model and the Pati-Salam gauge groups fit into the NCSG model, the SU(5) or SO(10) groups do not; and absence of large groups is interesting since it prevents proton decay. Following the NCSG scheme, one is able to infer quantities related to the Higgs boson based only upon the input from the fermionic parameters in the fluctuated Dirac operator $\mathcal{D}_A$, which defines also the fermionic part of the bosonic spectral action.

However, despite its success, the cutoff spectral action faces some issues. It is calculated via the asymptotic heat kernel expansion and is only valid when the fields and their derivatives are small with respect to the cutoff energy scale Λ. Hence, the asymptotic expansion leading to the appearance of only three of the momenta of the cutoff function f is only valid in the weak-field approximation and one may wonder what does it happen in the ultraviolet regime when high momenta are the dominant ones. Issues with super-renormalisability have been addressed in the literature, indicating that high energy bosons do not propagate [26]. In short, within the traditional cutoff bosonic spectral action approach, it is not clear what is the meaning of the cutoff scale Λ nor what does it happen at scales beyond Λ. Moreover, the cutoff bosonic spectral action depends (even though not in a very strong matter) on the particular choice of the cutoff f function. Finally, there is an issue with the magnitude of the dimensionful parameters in the model, namely the cosmological constant, the Higgs vacuum expectation value and the gravitational coupling. The natural value for the cosmological constant obtained through the spectral action approach is $\sim \Lambda^4$, which is clearly much bigger than its observational value. Thus, to render it compatible with the observational value of the cosmological constant, one should add by hand an appropriate term. The heat expansion does not lead to a minimum of the Higgs potential for all natural choices of the cutoff function. Hence one must add by hand to the H^2 term, a quadratic term with a large coefficient, to provide a minimum of the potential with a Higgs vacuum expectation value which is many orders of magnitude smaller than Λ. There is also a problem with the value of the gravitational constant given by the coefficient in front of the scalar curvature. The value obtained through the cutoff bosonic spectral action is at least one order of magnitude smaller than its experimental value. Therefore, one has again to add by hand an appropriate term. In conclusion, the physical values of these dimensionful parameters necessitate an experimental input beyond the NCSG approach. A different way of phrasing this problem is by calling it *the naturalness problem*.

To cure the dependence on the cutoff scale and the cutoff function, it has been recently proposed [19] another way to regularise the infinite sum of the eigenvalues of the (unbounded) fluctuated Dirac operator, based on the ζ function. More precisely, the zeta bosonic spectral action can be defined as [19]

$$S_\zeta \equiv \lim_{s\to 0} \mathrm{Tr}\, \mathcal{D}^{-2s} \equiv \zeta(0, \mathcal{D}^2) \ , \tag{54}$$

with the zeta function given by the a_4 heat kernel coefficient associated with the Laplace type operator $\mathcal{D}^2$:

$$S_\zeta = a_4\left[\mathcal{D}^2\right] = \int d^4x\,\sqrt{g}\,L \ \text{ with } \ L(x) = a_4(\mathcal{D}^2, x) \ . \tag{55}$$

This ζ spectral function leads to the following Lagrangian density [19]:

$$\begin{aligned} L(x) &= \alpha_1 M^4 + \alpha_2 M^2 R + \alpha_3 M^2 H^2 + \alpha_4 B_{\mu\nu}B^{\mu\nu} + \alpha_5 W^{\alpha}_{\mu\nu}W^{\mu\nu\,\alpha} + \alpha_6 G^{a}_{\mu\nu}G^{\mu\nu\,a} \\ &\quad + \alpha_7 H\left(-\nabla^2 - \frac{R}{6}\right)H + \alpha_8 H^4 + \alpha_9 C_{\mu\nu\rho\sigma}C^{\mu\nu\rho\sigma} + \alpha_{10} R^* R^* \ . \end{aligned} \tag{56}$$

Note that $B_{\mu\nu}$, $W_{\mu\nu}$ and $G_{\mu\nu}$ stand for the field strength tensors of the corresponding U(1), SU(2) and SU(3) gauge fields; $\alpha_1, .., \alpha_{10}$ are dimensionless constants determined by the Dirac operator; R^*R^* is the Gauss-Bonnet density and C is the Weyl tensor.

Up to this point, the dimensionful quantity M appearing in the position corresponding to the Majorana mass of the right-handed neutrino in the Dirac operator, is just a constant. All dimensionless constants are normalised to their spectral action values, whilst the three dimensionful parameters are normalised from experiments. In analogy with the cutoff bosonic spectral action we consider the zeta bosonic spectral action as valid at the scale $\Lambda \sim (10^{14} - 10^{17})$ GeV, but emphasising that the action is itself independent of the scale Λ. Since the zeta spectral action Eq. (56) does not contain higher than 4-dimensional operators, it is renormalisable and local. Moreover, there are no issues about asymptotic expansion or convergence, and the zeta spectral action is purely spectral with no dependence on a cutoff function.

Within the cutoff spectral action approach, if the full momentum-dependence of the propagators is considered, then the spectral dimension becomes independent of the spin and vanishes identically [27]. A physical interpretation of this behaviour can be summarised in the statement that *high energy bosons do not propagate* [26]. However, the zeta spectral action leads to nontrivial spectral dimensions. The Higgs scalar part of the bosonic action has the same bahaviour in the ultraviolet as in the infrared limit, so that the spectral dimension coincides with the topological dimension of the manifold. The same holds for the gauge fields. Hence the spectral dimension of the Higgs scalar and the gauge fields is equal to four. To calculate the gravitational spectral dimension one needs to do an analytic continuation [19]; we will highlight the computation below.

The spectral dimension $D_{\rm s}$ is defined as

$$D_{\rm s} \equiv \lim_{T\to 0}\left[-2\frac{\partial \log P(T)}{\partial \log T}\right] , \tag{57}$$

where $P(T)$ stands for the value of the heat kernel $P(T,x,x')$, corresponding to the quadratic part of the gravitational part of the action for transverse and traceless fluctuations $h_{\mu\nu}$ of the metric tensor $g_{\mu\nu}$, at $x = x'$, and T denotes the diffusion time. Such a heat kernel is given by the integral

$$P(T,x,x') = \int \frac{d^4p}{(2\pi^4)} e^{ip(x-x')} e^{-(p^2-ap^4)T} , \tag{58}$$

and one can show that there exists an analytic continuation of the relevant integral in a region of positive a [19]. The obtained gravitational spectral dimension for all nonzero real a is [19]

$$D_{\rm s} = 2 ; \tag{59}$$

in agreement with the fact that here the gravitational propagators decrease faster in infinity due to the presence of fourth derivatives. Finally, there exists a *low energy* limit of the gravitational spectral dimension, valid for all real a, with [19]

$$D_{\rm s}^{\rm low} = 4 , \tag{60}$$

as expected since at very low energies the dynamics does not feel the fourth derivative terms.

6. Conclusions

Noncommutative spectral geometry offers a geometric framework for the description of the Standard Model of strong and electroweak interactions, based upon a purely algebraic description. The constructed spectral action for fermions and bosons based on the spectral properties of the generalised Dirac operator and for an appropriately chosen algebra led to phenomenological results very close to the experimental ones.

The doubling of the algebra, which can be interpreted as considering a geometric space formed by two copies of a four-dimensional manifold, has profound physical implications. In particular, the doubling of the algebra is required in order to incorporate gauge symmetries, a fundamental ingredient of the Standard Model, and it is also the main element to explain neutrino mixing. Following 't Hooft's conjecture, one can also show that the NCSG classical construction carries implicit in its doubling of the algebra the seeds of quantisation.

Considering the gravitational sector of the spectral action, one can constrain one of the momenta of the cutoff function, namely the one related to the coupling constants at unification. The strongest constraint on the coefficient of the curvature squared term is obtained through torsion balance experiments, while using data from binary pulsars, GPB and LARES experiments the corresponding constraint is much weaker.

To address the issues of renormalisability and spectral dimensions, a novel definition of the bosonic spectral action has been proposed, based upon the ζ function regularisation. While the zeta spectral action shares the same predictive power with the traditional cutoff spectral action, only the former leads to a local, unitary and renormalisable theory. In this new proposal, the open aspect that remains to be addressed is the dynamical generation of the three dimensionful fundamental constants, namely the cosmological constant, the Higgs vacuum expectation value, and the gravitational constant.

References

[1] Connes A 1994 *Noncommutative Geometry* (New York: Academic Press)
[2] Connes A and Marcolli M 2008 *Noncommutative Geometry, Quantum Fields and Motives* (India: Hindustan Book Agency)
[3] Chamseddine A H, Connes A and Marcolli M 2007 *Adv. Theor. Math. Phys.* **11** 991
[4] van Suijlekom W D 2014 *Mathematical Physics Studies Springer*; ISBN: 978-94-017-9161-8 (Print) 978-94-017-9162-5 (Online) a geometric space fromed by two copies of a four-dimensional manifold
[5] Sakellariadou M 2010 *PoS CNCFG* **2010** 028
[6] Sakellariadou M 2011 *Int. J. Mod. Phys.* D **20** 785
[7] Sakellariadou M 2011 *PoS CORFU* **2011** 053
[8] Sakellariadou M, Stabile A and Vitiello G 2011 *Phys. Rev.* D **84** 045026
[9] Gargiulo M V, Sakellariadou M and Vitiello G 2014 *Eur. Phys. J.* C **74** 2695
[10] Nelson W and Sakellariadou M 2010 *Phys. Rev.* D **81** 085038
[11] Nelson W and Sakellariadou M 2009 *Phys. Lett.* B **680** 263
[12] Marcolli M and Pierpaoli E 2010 *Adv. Theor. Math. Phys.* **14**
[13] Buck M, Fairbairn M and Sakellariadou M 2010 *Phys. Rev.* D **82** 043509
[14] Nelson W, Ochoa J and Sakellariadou M 2010 *Phys. Rev. Lett.* **105** 101602
[15] Nelson W, Ochoa J and Sakellariadou M 2010 *Phys. Rev.* D **82** 085021
[16] Sakellariadou M 2011 *Int. J. Mod. Phys.* D **20** 785
[17] Lambiase G, Sakellariadou M and Stabile A 2013 *J. Cosm. & Astrop. Phys.* **12** 020
[18] Capozziello S, Lambiase G, Sakellariadou M, Stabile A and Stabile An 2015 *Phys. Rev.* D **91** 044012
[19] Kurkov M A, Lizzi F, Sakellariadou M and Watcharangkool A 2015 *Phys. Rev.* D *in press Preprint* arXiv:1412.4669 [hep-th].
[20] Chamseddine A H and Connes A 2007 *Phys. Rev. Lett.* **99** 191601
[21] Chamseddine A H and Connes A 1997 *Comm. Math. Phys.* **186** 731
[22] Chamseddine A H and Connes A 2012 *JHEP* **1209** 104
[23] Stephan C A 2009 *Phys. Rev.* D **79** 065013
[24] Devastato A, Lizz Fi and Martinetti P 2014 *JHEP* **1401** 042
[25] Chamseddine A H, Connes A and van Suijlekom W D 2013 *J. Geom. Phys.* **73** 222
[26] Kurkov M A, Lizzi F and Vassilevich D 2014 *Phys. Lett.* B **731** 311
[27] Alkofer N, Saueressig F and Zanusso O 2015 *Phys. Rev.* D **91** 025025

Constructing Galileons

Mark Trodden

Center for Particle Cosmology, Department of Physics and Astronomy, University of Pennsylvania, Philadelphia, PA 19104, USA

E-mail: trodden@physics.upenn.edu

Abstract.
In this plenary talk delivered at the DISCRETE 2014 conference in London, I briefly summarize the ideas behind and attractive properties of the Galileon field theories, and describe a broad class of scalar field theories that share these properties. After describing how Galileons arise, and commenting on their fascinating properties, in the latter half of the talk I focus on novel ways of constructing Galileon-like theories, using both the probe brane construction, and the coset construction.

1. Introduction

The problem of cosmic acceleration has prompted the search for both new and well-behaved forms of dark energy and ghost free infrared modifications of General Relativity (GR). At the level of effective field theory (EFT), around a given background, either approach boils down to a field theory of interacting fields, about which we may ask a variety of questions of theoretical consistency (and, of course, observational viability). These include requirements that the theory be ghost-free, that it remain in the weakly-coupled, calculable regime in the environments we are interested in, and that the theory be technically natural. There are also more technical questions that the theory may face, such as whether there are any obstacles to a Lorentz-invariant ultraviolet (UV) completion, arising, for example, from certain analyticity properties of the S-matrix.

These requirements, coupled with those from demanding that any new long-range force of gravitational strength not ruin precision tests of GR within the solar system and local universe, have proven extremely restrictive, with very few new ideas remaining viable. On the other hand, this search has yielded a variety of fascinating new ideas that remain of interest for cosmic acceleration, but also present intriguing new theoretical challenges all on their own. The more intriguing of these ideas have arisen from studies of extra-dimensional brane-localized gravity theories [1], and ghost free massive gravity [2, 3]. In both cases, there exists a limit in which a fascinating scalar field theory with higher-derivative interaction results - the *Galileons*.

In this summary of a plenary lecture, delivered at the DISCRETE 2014 conference, I briefly introduce the idea of Galileons, their attractive properties and their interesting features. I then describe how one may construct theories such as these from two different viewpoints. The first is the probe brane construction, which I show can be generalized to yield new theories, and the second is the coset construction, which can she new light on the topological properties of such theories. As is usual in a proceedings, I will have space only to reference those papers of most direct relevance to my own thinking on the subject, and will not attempt to be comprehensive.

2. Galileons, the DGP Model and Massive Gravity

An easy starting point from which to motivate Galileons is the Dvali-Gabadadze-Porrati (DGP) model [1], in which a $3+1$-dimensional brane is embedded in a 5d bulk, with action

$$S = \frac{M_5^3}{2}\int d^5X\,\sqrt{-G}\ R[G] + \frac{M_4^2}{2}\int d^4x\,\sqrt{-g}\ R[g]\ . \tag{1}$$

In this model gravity is modified on large distances, and there exists a branch of cosmological solutions that exhibit self-acceleration at late times, although this branch contains a ghost. There also exists a ghost-free "normal" branch, for which a 4d effective action can be derived which, in a particular "decoupling" limit, reduces to a theory of a single scalar π, with a cubic self-interaction term $\sim (\partial\pi)^2\Box\pi$. This term yields second order field equations and is invariant (up to a total derivative) under the *Galilean* transformations

$$\pi(x) \to \pi(x) + c + b_\mu x^\mu\ , \tag{2}$$

with c and b_μ constants.

It is natural to ask what other terms are allowed in a four dimensional field theory with this same symmetry [4]. Referring to the field of such a theory as the *Galileon*, it turns out that, in addition to an infinite number of terms in which two derivatives act on every field (trivially invariant under the galilean symmetry) there are a finite number of terms that have fewer numbers of derivatives per field. These possess second order equations of motion, and there can exist regimes in which these special terms are important, and the infinity of other possible terms within the effective field theory are not. Along with a non-renormalization theorem for Galileons [5–7], this means that we are able, in some circumstances, to understand non-linear effects that are quantum mechanically exact.

For $n \geq 1$, the $(n+1)$-th order Galileon Lagrangian (where "order" refers to the number of copies of the scalar field π that appear in the term) is

$$\mathcal{L}_{n+1} = n\eta^{\mu_1\nu_1\mu_2\nu_2\cdots\mu_n\nu_n}\left(\partial_{\mu_1}\pi\partial_{\nu_1}\pi\partial_{\mu_2}\partial_{\nu_2}\pi\cdots\partial_{\mu_n}\partial_{\nu_n}\pi\right)\ , \tag{3}$$

where

$$\eta^{\mu_1\nu_1\mu_2\nu_2\cdots\mu_n\nu_n} \equiv \frac{1}{n!}\sum_p (-1)^p\,\eta^{\mu_1 p(\nu_1)}\eta^{\mu_2 p(\nu_2)}\cdots\eta^{\mu_n p(\nu_n)} \tag{4}$$

and the sum is over all permutations of the ν indices, with $(-1)^p$ the sign of the permutation. The tensor (4) is antisymmetric in the μ indices, antisymmetric in the ν indices, and symmetric under interchange of any μ, ν pair with any other. These Lagrangians are unique up to total derivatives and overall constants. Because of the antisymmetry requirement on η, only the first n of these Galileons are non-trivial in n-dimensions. In addition, the tadpole term, π, is Galilean invariant, and we therefore include it as the first-order Galileon.

These Galileon actions can be generalized to the multi-field case, where there is a multiplet π^I of fields [6, 8–12]. It should be remembered that all these theories are not renormalizable, and should be considered as effective field theories with some cutoff Λ, above which some UV completion is required. However, as I have mentioned, there can exist regimes in which the above terms dominate, since the symmetries forbid any renormalizable terms, and other terms will have more derivatives. One important consequence of these properties is that galileon theories exhibit a particular type of screening mechanism - the Vainshtein mechanism [13] - that allows them to mediate long range forces, competing with gravity at cosmic scales, while having negligible physical effects at the scales of local tests of gravity. This is the key to their use in modified gravity applications (see [14] for a recent review of these mechanisms and their effects and tests).

3. The Probe Brane Construction and its Generalization

One way to construct the Galileons is to generalize the DGP model to a general action for a probe brane floating in a higher dimensional flat bulk, and then to take a limit in which only a scalar field degree of freedom remains [15]. This *probe brane* construction can then be generalized in a variety of ways [16–20] to generate new galileon-like theories possessing many of the attractive features of the original model.[1]

One generalization is to introduce additional co-dimensions into the construction. One can then show that the resulting theory is a 4-d field theory of multiple scalar fields with a single possible interaction term, and thus a single free coupling constant. As for the original galileons, this term is not renormalized to any order in perturbation theory [6].

Another generalization of the probe brane construction can be obtained [16–18] by relaxing the assumption of flat geometry in the original method. To see this, consider a dynamical 3-brane in a fixed (4+1)-dimensional background, but now with a general bulk metric. The dynamical variables are the brane embedding $X^A(x)$, which are five functions of the world-volume coordinates x^μ. The bulk metric $G_{AB}(X)$ allows us to construct the induced metric $\bar{g}_{\mu\nu}(x)$ and the extrinsic curvature $K_{\mu\nu}(x)$ via

$$\bar{g}_{\mu\nu} = e^A_{\ \mu} e^B_{\ \nu} G_{AB}(X) \, , \tag{5}$$

$$K_{\mu\nu} = e^A_{\ \mu} e^B_{\ \nu} \nabla_A n_B \, , \tag{6}$$

where n^A is the unit normal vector, and $e^A_{\ \mu} = \frac{\partial X^A}{\partial x^\mu}$ are the tangent vectors to the brane.

To ensure that the world-volume action is gauge invariant under brane reparametrizations, we write the action as a diffeomorphism scalar, F, of $\bar{g}_{\mu\nu}$, $K_{\mu\nu}$, the covariant derivative $\bar{\nabla}_\mu$ and the

[1] See [21] for a more comprehensive review of this.

curvature $\bar{R}^{\alpha}{}_{\beta\mu\nu}$ constructed from $\bar{g}_{\mu\nu}$.

$$S = \int d^4x \sqrt{-\bar{g}} F\left(\bar{g}_{\mu\nu}, \bar{\nabla}_{\mu}, \bar{R}^{\alpha}{}_{\beta\mu\nu}, K_{\mu\nu}\right) . \tag{7}$$

Requiring that the resulting equations of motion are second order, and hence avoid obvious ghosts, the allowed actions turn out to be the Lovelock [22] and boundary [23–25] terms.

Now, global symmetries of this action exist only if the bulk metric has Killing symmetries. To exploit this, we choose the gauge

$$X^{\mu}(x) = x^{\mu}, \quad X^5(x) \equiv \pi(x) , \tag{8}$$

such that the bulk is foliated by time-like slices given by the surfaces $X^5(x) =$ constant. This completely fixes the gauge symmetry, and the leaves of the foliation are then parametrized by the arbitrary choices of the remaining coordinates X^{μ}. The coordinate $\pi(x)$ measures the transverse position of the brane relative to the foliation in this gauge, and the resulting action for π is

$$S = \int d^4x \sqrt{-\bar{g}} F\left(\bar{g}_{\mu\nu}, \bar{\nabla}_{\mu}, \bar{R}^{\alpha}{}_{\beta\mu\nu}, K_{\mu\nu}\right)\big|_{X^{\mu}=x^{\mu},\ X^5=\pi} . \tag{9}$$

Any global symmetries of (7) are global symmetries of (9), but their form depends on the gauge. If a Killing vector K^A generates a transformation, then we may restore the gauge (8) by making a compensating gauge transformation $\delta_{g,\mathrm{comp}} x^{\mu} = -K^{\mu}$. These two symmetries then combine to yield the following transformation of π

$$(\delta_K + \delta_{g,\mathrm{comp}})\pi = -K^{\mu}(x,\pi)\partial_{\mu}\pi + K^5(x,\pi) , \tag{10}$$

which is a symmetry of the gauge fixed action (9).

Simple, yet illuminating examples can be obtained by making a few assumptions. We assume our foliation is Gaussian normal with respect to G_{AB}, and we further demand that the extrinsic curvature on each slice be proportional to the induced metric. The metric then takes the form

$$G_{AB} dX^A dX^B = d\rho^2 + f(\rho)^2 g_{\mu\nu}(x) dx^{\mu} dx^{\nu} , \tag{11}$$

where $X^5 = \rho$ denotes the transverse coordinate, and $g_{\mu\nu}(x)$ is an arbitrary brane metric. This special case includes all examples in which a maximally symmetric ambient space is foliated by maximally symmetric slices. Killing symmetries which preserve the foliation will be linearly realized, whereas those that do not are realized nonlinearly. Thus, the algebra of all Killing vectors is spontaneously broken to the subalgebra of Killing vectors preserving the foliation.

In the gauge (8), the induced metric is $\bar{g}_{\mu\nu} = f(\pi)^2 g_{\mu\nu} + \nabla_{\mu}\pi\nabla_{\nu}\pi$. Defining the quantity $\gamma = 1/\sqrt{1 + \frac{1}{f^2}(\nabla\pi)^2}$, the extrinsic curvature is then

$$K_{\mu\nu} = \gamma\left(-\nabla_{\mu}\nabla_{\nu}\pi + ff' g_{\mu\nu} + 2\frac{f'}{f}\nabla_{\mu}\pi\nabla_{\nu}\pi\right) . \tag{12}$$

On the 4-dimensional brane, as described, we can add four Lovelock and boundary terms, plus a tadpole term,

$$
\begin{aligned}
\mathcal{L}_1 &= \sqrt{-g}\int^{\pi} d\pi' f(\pi')^4, \\
\mathcal{L}_2 &= -\sqrt{-\bar{g}}\ , \\
\mathcal{L}_3 &= \sqrt{-\bar{g}}K\ , \\
\mathcal{L}_4 &= -\sqrt{-\bar{g}}\bar{R}\ , \\
\mathcal{L}_5 &= \frac{3}{2}\sqrt{-\bar{g}}\mathcal{K}_{\rm GB}\ ,
\end{aligned} \tag{13}
$$

where the explicit form of the Gauss-Bonnet boundary term is $\mathcal{K}_{\rm GB} = -\frac{1}{3}K^3 + K^2_{\mu\nu}K - \frac{2}{3}K^3_{\mu\nu} - 2\left(\bar{R}_{\mu\nu} - \frac{1}{2}\bar{R}\bar{g}_{\mu\nu}\right)K^{\mu\nu}$.

$\mathcal{L}_1$ is the zero derivative tadpole term, which is the proper volume between any ρ = constant surface and the brane position, $\pi(x)$ [16]. While different in origin from the other terms, it too has the symmetry (10). Each of these terms may appear in a general Lagrangian with an arbitrary coefficient.

Evaluating these expressions for the metric (11) involves a lengthy calculation which ultimately yields

$$
\begin{aligned}
\mathcal{L}_1 &= \sqrt{-g}\int^{\pi} d\pi' f(\pi')^4, \\
\mathcal{L}_2 &= -\sqrt{-g}f^4\sqrt{1+\frac{1}{f^2}(\partial\pi)^2}, \\
\mathcal{L}_3 &= \sqrt{-g}\left[f^3f'(5-\gamma^2) - f^2[\Pi] + \gamma^2[\pi^3]\right], \\
\mathcal{L}_4 &= -\sqrt{-g}\Big\{\frac{1}{\gamma}f^2R - 2\gamma R_{\mu\nu}\nabla^\mu\pi\nabla^\nu\pi \\
&\quad +\gamma\left[[\Pi]^2 - [\Pi^2] + 2\frac{\gamma^2}{f^2}\left(-[\Pi][\pi^3] + [\pi^4]\right)\right] \\
&\quad +6\frac{f^3f''}{\gamma}\left(-1+\gamma^2\right) \\
&\quad +2\gamma f f'\left[-4[\Pi] + \frac{\gamma^2}{f^2}\left(f^2[\Pi] + 4[\pi^3]\right)\right] \\
&\quad -6\frac{f^2f'^2}{\gamma}\left(1-2\gamma^2+\gamma^4\right)\Big\}\ ,
\end{aligned} \tag{14}
$$

where the expression for $\mathcal{L}_5$ is too long to write here, but its full form is given in [16]. In these expressions, all curvatures and covariant derivatives are those of the background metric $g_{\mu\nu}$, and Π denotes the matrix of covariant second dervatives $\Pi_{\mu\nu} \equiv \nabla_\mu\nabla_\nu\pi$. We use the notation $[\Pi^n] \equiv Tr(\Pi^n)$, e.g. $[\Pi] = \Box\pi$, $[\Pi^2] = \nabla_\mu\nabla_\nu\pi\nabla^\mu\nabla^\nu\pi$, and $[\pi^n] \equiv \nabla\pi\cdot\Pi^{n-2}\cdot\nabla\pi$, i.e. $[\pi^2] = \nabla_\mu\pi\nabla^\mu\pi$, $[\pi^3] = \nabla_\mu\pi\nabla^\mu\nabla^\nu\pi\nabla_\nu\pi$. The equations of motion contain no more than two derivatives on each field, ensuring that no extra degrees of freedom propagate around any background.

There are interesting cases in which the 5d background metric has 15 global symmetries, the maximal number – the bulk can be either 5d anti-de Sitter space AdS_5 with isometry algebra $so(4,2)$, 5d de-Sitter space dS_5 with isometry algebra $so(5,1)$, or flat 5d Minkowski space M_5 with isometry algebra the five dimensional Poincaré algebra $p(4,1)$. In addition, there are cases in which the brane metric $g_{\mu\nu}$, and hence the extrinsic curvature, are maximally symmetric, so that the unbroken subalgebra has the maximal number of generators, 10. This means that the leaves of the foliation are either 4d anti-de Sitter space AdS_4 with isometry algebra $so(3,2)$, 4d de-Sitter space dS_4 with isometry algebra $so(4,1)$, or flat 4d Minkowski space M_4 with isometry algebra $p(3,1)$. In fact, there are only 6 such possible foliations of 5d maximally symmetric spaces by 4d maximally symmetric time-like slices, such that the metric takes the form (11). Flat M_5 can be foliated by flat M_4 slices or by dS_4 slices; dS_5 can be foliated by flat M_4 slices, dS_4 slices, or AdS_4 slices; and AdS_5 can only be foliated by AdS_4 slices. Each of these 6 foliations, through the construction leading to (9), will generate a class of theories living on an AdS_4, M_4 or dS_4 background and having 15 global symmetries broken to the 10 isometries of the brane, the same numbers as those of the original galileon.

To see analogues of the galileons emerge, we expand the Lagrangians in powers of λ around some constant background, $\pi \to \pi_0 + \lambda\pi$. There exist linear combinations of the Lagrangians, $\bar{\mathcal{L}}_n = c_1\mathcal{L}_1 + \ldots + c_n\mathcal{L}_n$, for which all terms $\mathcal{O}\left(\lambda^{n-1}\right)$ or lower are total derivatives. Carrying this out for the remaining four maximally symmetric cases in which the 4d background is curved, new classes of theories are produced, and canonically normalizing via $\hat{\pi} = \frac{1}{L^2}\pi$, where L is the dS_4 or AdS_4 radius, the Lagrangians become

$$
\begin{aligned}
\hat{\mathcal{L}}_1 &= \sqrt{-g}\hat{\pi}\ , \\
\hat{\mathcal{L}}_2 &= -\frac{1}{2}\sqrt{-g}\left((\partial\hat{\pi})^2 - \frac{R}{3}\hat{\pi}^2\right)\ , \\
\hat{\mathcal{L}}_3 &= \sqrt{-g}\left(-\frac{1}{2}(\partial\hat{\pi})^2[\hat{\Pi}] - \frac{R}{4}(\partial\hat{\pi})^2\hat{\pi} + \frac{R^2}{36}\hat{\pi}^3\right)\ , \\
\hat{\mathcal{L}}_4 &= \sqrt{-g}\Big[-\frac{1}{2}(\partial\hat{\pi})^2\Big([\hat{\Pi}]^2 - [\hat{\Pi}^2] + \frac{R}{24}(\partial\hat{\pi})^2 \\
&\qquad + \frac{R}{2}\hat{\pi}[\hat{\Pi}] + \frac{R^2}{8}\hat{\pi}^2\Big) + \frac{R^3}{288}\hat{\pi}^4\Big]\ ,
\end{aligned}
\tag{15}
$$

with $\mathcal{L}_5$ again found in [16].

In the above $R = \pm\frac{12}{L^2}$ is the Ricci curvature of the dS_4 or AdS_4 background. These Lagrangians describe Galileons that propagate on curved space yet retain the same number of symmetries as the full theory, whose form comes from expanding (10) in appropriate powers of λ. For example, in the case of a dS_4 background in conformal inflationary coordinates (u, y^i), the non-linear symmetries are

$$
\delta_+\hat{\pi} = \frac{1}{u}\left(u^2 - y^2\right), \quad \delta_-\hat{\pi} = -\frac{1}{u}, \quad \delta_i\hat{\pi} = \frac{y_i}{u}\ . \tag{16}
$$

One interesting feature of these models which is not true for the flat space Galileon theories is the existence of potentials with couplings determined by the symmetries (10). Specifically, the scalar field acquires a mass of order the dS_4 or AdS_4 radius, with a value protected by the symmetries (16). This means that small masses can be made technically natural, which can be handy for cosmological applications.

The methods presented in this section can be used to create large numbers of galileon-like models, with potentially extremely complex symmetries, which would be hard to discover in other ways.

4. The Coset Construction and Wess-Zumino Terms

A different, and complementary, way to investigate galileon and galileon-like theories is to take advantage of the fact that the relevant symmetries are not realized linearly on the fields. For example, the basic galilean symmetry, $\pi \to \pi + c + b_\mu x^\mu$, is a combination of two transformations - a shift symmetry and a shift symmetry acting on the derivative, neither of which is linearly realized on the field π. One way to think about nonlinearly realized symmetries is in the language of spontaneous symmetry breaking. If we have a symmetry group, G, broken down to a subgroup H, then generically the broken phase linearly realizes the preserved subgroup and nonlinearly realizes the elements G/H that are no longer preserved. Given the breaking pattern, there exists an algorithmic method - the "coset construction" [26–30] - for constructing the most general Lagrangians linearly realizing H and nonlinearly realizing the remaining symmetries. The low energy physics of a spontaneously broken system is described by the Goldstone modes that result from this construction.

The coset construction can be used even when the dynamics leading to symmetry breaking are unknown or not understood, such as in the classic application to the physics of pions [31]. Furthermore, this construction can be applied to gauge theories by noticing that they nonlinearly realize the local versions of their symmetry groups, and this application can be extended to the physics of such theories in the Higgs phase [32].

Consider a Lie group G, broken down to a subgroup H, and represent the generators of H by $\{V_I\}$, and those of the remaining – broken – generators by $\{Z_a\}$. Assuming that the commutator of an element of $\{V_I\}$ with an element of $\{Z_a\}$ does not contain another generator from the $\{V_I\}$, then we may write a canonical representative element of the coset G/H as $g(\xi) \equiv \exp(\xi^a Z_a)$, where the Goldstone fields are represented by the ξ^a.

Any element $g' \in G$ generates a transformation $g(\xi) \mapsto \tilde{g}(\xi, g')$, where we define $\tilde{g}(\xi, g')$ by

$$g' g(\xi) = \tilde{g}(\xi, g') h(\xi, g') \ , \tag{17}$$

where $h(\xi, g) \in H$. In this manner, as required, the Goldstone fields linearly realize the symmetries of the preserved subgroup H and nonlinearly realize the remaining broken symmetries. This can be seen by writing $\tilde{g}(\xi, g') \equiv \exp(\tilde{\xi}^a Z_a)$, and noting that the relation between $\tilde{\xi}^a$ and ξ^a is linear if $g' \in H$, but nonlinear otherwise.

The first step in constructing actions is to build a Lie algebra-valued 1-form known as the Maurer–Cartan (MC) form $\Omega \equiv g^{-1}(\xi) dg(\xi)$. Decomposing this into its parts along the broken and unbroken generators as

$$\Omega \equiv \Omega^a Z_a + \Omega^I V_I \equiv \Omega_Z + \Omega_V \ , \tag{18}$$

one sees that an arbitrary element $g' \in G$ induces a transformation under which the components of the MC form transform as

$$g' : \begin{cases} \Omega_Z & \longmapsto h(\xi, g')\Omega_Z h^{-1}(\xi, g'), \\ \Omega_V & \longmapsto h(\xi, g')\left(\Omega_V + \mathrm{d}\right) h^{-1}(\xi, g'). \end{cases} \tag{19}$$

These transformation properties under the action of the group G are what make the MC form so useful. To build Lagrangians that are invariant under *all* the symmetries of G we combine factors of Ω_Z together and trace over group indices. The operator that results is invariant under (19), *i.e.*, it is H-invariant. We then couple the Goldstone fields to other matter fields that transform in some representation of H using the Ω_V components, which transform as a connection.

At this point, we must deal with an issue that arises whenever spacetime symmetries are among those that are nonlinearly realized - the removal of fields via inverse Higgs (IH) constraints [33]. Naïvely, for a breaking pattern $G \to H$, we have $\dim(G/H)$ fields $\{\xi^a\}$ in the representative element g, which is the appropriate number of Goldstone modes for the case of internal symmetry breaking. However, this counting does not always work when spacetime symmetries are involved [34]. In practice, if the commutator between a preserved translation generator, P_μ, and a broken generator, Z_1, contains a second broken generator, Z_2, then we can set some part of the MC component along Z_2 to zero, and hence eliminate the Goldstone field corresponding to Z_1. These inverse Higgs constraints allow us to consistently reduce the number of fields while still realizing all the symmetries of G.

Another subtlety concerns how we deal with any preserved translation generators, P_μ. When treating a case of spacetime symmetry breaking, we treat these generators on the same footing as the broken generators, since translations are nonlinearly realized on the spacetime coordinates. Taking this into account, the MC form is thus written as

$$\Omega = \Omega^a Z_a + \Omega^I V_I + \Omega^\mu P_\mu \ . \tag{20}$$

We may construct invariant actions by combining forms using the wedge product, and we may also form a covariant derivative from these objects. The components along P_μ define a vielbein, $e_\nu{}^\mu$, via $\Omega^\mu P_\mu \equiv dx^\nu e_\nu{}^\mu P_\mu$, which may be used to define an invariant measure, and define a covariant derivative of the Goldstone fields via $\Omega^a Z_a = dx^\nu e_\nu{}^\mu \mathcal{D}_\mu \xi^a Z_a$. Terms in the invariant actions are then formed by contracting factors of covariant derivatives into H-invariant combinations, and finally integrating with the invariant measure.

So what happens when we try to use the coset construction for galileons in four dimensions? To start, recall that the galileons nonlinearly realize the following symmetries

$$\delta_C \pi = 1 \ , \qquad \delta_{B^\mu} \pi = x^\mu \ , \tag{21}$$

with non-trivial commutators

$$[P_\mu, B_\nu] = \eta_{\mu\nu} C \ , \qquad [J_{\rho\sigma}, B_\nu] = \eta_{\rho\nu} B_\sigma - \eta_{\sigma\nu} B_\rho \ . \tag{22}$$

Along with Poincaré transformations, these comprise the galileon algebra $\mathfrak{Gal}(3+1,1)$, and so in $4d$ galileons nonlinearly realize the breaking pattern

$$\mathfrak{Gal}(3+1,1) \longrightarrow \mathfrak{iso}(3,1)\ . \tag{23}$$

The coset is therefore parameterized by

$$\tilde{V} = e^{x\cdot P} e^{\pi C+\xi\cdot B}\ . \tag{24}$$

Because the linearly realized generators are the Lorentz transformations, the coset is

$$\mathrm{Gal}(3+1,1)/\mathrm{SO}(3,1)\ . \tag{25}$$

The coefficients of the components of the MC form are

$$\omega_P^\mu = \mathrm{d}x^\mu\ , \qquad \omega_C = \mathrm{d}\pi + \xi_\mu \mathrm{d}x^\mu\ , \qquad \omega_B^\mu = \mathrm{d}\xi^\mu\ , \qquad \omega_J^{\mu\nu} = 0\ . \tag{26}$$

Although there are two broken generators, V_μ and C, we only have a single Goldstone mode π because of an inverse Higgs constraint – the commutator $[P_\mu, B_\nu] = \eta_{\mu\nu}C$ implies that we can eliminate the ξ_μ field by setting $\omega_C = 0$, yielding

$$\xi_\mu = -\partial_\mu \pi\ . \tag{27}$$

The components of the Maurer–Cartan form can then be written as

$$\omega_P^\mu = \mathrm{d}x^\mu\ , \qquad \omega_B^\mu = -\mathrm{d}x^\nu \partial_\nu \partial^\mu \pi\ . \tag{28}$$

Now something interesting happens. Because we can only build Lagrangians by using these ingredients and higher covariant derivatives, the field π must always appear with at least 2 derivatives per field. Thus the galileons can never be obtained from this construction, since the galileon terms all have fewer than two derivatives per field. While this is puzzling at first, it is possible to show that the $4d$ galileons can nevertheless be realized as *Wess-Zumino terms.*

To see this, we work on the coset space, in which π and ξ^μ are considered as coordinates in addition to those of space-time. We obtain the Lagrangian by integrating a *Wess–Zumino* form on the subspace on which $\pi = \pi(x)$ and $\xi^\mu = \xi^\mu(x)$. The symmetries on the coset space are generated by the vector fields

$$C = \partial_\pi\ , \qquad B_\mu = \partial_{\xi^\mu} - x_\mu \partial_\pi\ , \qquad P_\mu = \partial_\mu\ . \tag{29}$$

The components of the MC form (26) are the (left) invariant 1-forms on the coset space parametrized by $\{\pi, \xi^\mu, x^\mu\}$, obeying $\pounds_X \omega = 0$, where X is any of the vector fields (29) and ω is any of the forms (26).

Given this, we construct the Wess–Zumino terms by creating invariant 5-forms by wedging together the 1-forms (26) so that they are invariant under the Lorentz transformations $\mathfrak{so}(3,1)$, and hence are well defined on the coset. This requires that the Lorentz indices in (26) are contracted using the Lorentz invariant tensors $\eta_{\mu\nu}$ and $\epsilon_{\mu\nu\rho\sigma}$.

As an example, consider the invariant 5-form

$$\omega_1^{\rm wz} = \epsilon_{\mu\nu\rho\sigma}\, \omega_C \wedge \omega_P^\mu \wedge \omega_P^\nu \wedge \omega_P^\rho \wedge \omega_P^\sigma = \epsilon_{\mu\nu\rho\sigma} {\rm d}\pi \wedge {\rm d}x^\mu \wedge {\rm d}x^\nu \wedge {\rm d}x^\rho \wedge {\rm d}x^\sigma \ . \tag{30}$$

We may write this as the exterior derivative of a 4-form,

$$\omega_1^{\rm wz} = {\rm d}\beta_1^{\rm wz}, \qquad \beta_1^{\rm wz} = \epsilon_{\mu\nu\rho\sigma}\pi {\rm d}x^\mu \wedge {\rm d}x^\nu \wedge {\rm d}x^\rho \wedge {\rm d}x^\sigma \ , \tag{31}$$

which may be pulled back to the space-time manifold M and then integrated, to yield

$$S_1^{\rm wz} = \int_M \beta_1^{\rm wz} = \int_M \pi\epsilon_{\mu\nu\rho\sigma} {\rm d}x^\mu \wedge {\rm d}x^\nu \wedge {\rm d}x^\rho \wedge {\rm d}x^\sigma \sim \int {\rm d}^4x \ \pi \ . \tag{32}$$

This is the tadpole term - the first galileon.

We continue by considering

$$\omega_2^{\rm wz} = \epsilon_{\mu\nu\rho\sigma}\, \omega_C \wedge \omega_B^\mu \wedge \omega_P^\nu \wedge \omega_P^\rho \wedge \omega_P^\sigma = \epsilon_{\mu\nu\rho\sigma} \left({\rm d}\pi + \xi_\lambda {\rm d}x^\lambda\right) \wedge {\rm d}\xi^\mu \wedge {\rm d}x^\nu \wedge {\rm d}x^\rho \wedge {\rm d}x^\sigma \ , \tag{33}$$

which is also the exterior derivative of a 4-form

$$\omega_2^{\rm wz} = {\rm d}\beta_2^{\rm wz}, \quad \beta_2^{\rm wz} = \epsilon_{\mu\nu\rho\sigma} \left(\pi {\rm d}\xi^\mu - \frac{1}{8}\xi^2 {\rm d}x^\mu\right) \wedge {\rm d}x^\nu \wedge {\rm d}x^\rho \wedge {\rm d}x^\sigma \ . \tag{34}$$

Pulling back and integrating, we now obtain

$$S_2^{\rm wz} = \int_M \beta_2^{\rm wz} = 3! \int {\rm d}^4x \left(\pi\partial_\mu\xi^\mu - \frac{1}{2}\xi^2\right) \ . \tag{35}$$

Using the Higgs constraint $\xi_\mu = -\partial_\mu\pi$, we obtain the kinetic term - the second galileon,

$$S_2^{\rm wz} \sim \int {\rm d}^4x \ (\partial\pi)^2 \ . \tag{36}$$

As a final, and non-trivial, example, let's see how the construction of $\mathcal{L}_3$ works. Consider

$$\omega_3^{\rm wz} = \epsilon_{\mu\nu\rho\sigma}\, \omega_C \wedge \omega_B^\mu \wedge \omega_B^\nu \wedge \omega_P^\rho \wedge \omega_P^\sigma = \epsilon_{\mu\nu\rho\sigma} \left({\rm d}\pi + \xi_\lambda {\rm d}x^\lambda\right) \wedge {\rm d}\xi^\mu \wedge {\rm d}\xi^\nu \wedge {\rm d}x^\rho \wedge {\rm d}x^\sigma \ , \tag{37}$$

which is the exterior derivative of the 4-form

$$\omega_3^{\rm wz} = {\rm d}\beta_3^{\rm wz}, \quad \beta_3^{\rm wz} = \epsilon_{\mu\nu\rho\sigma} \left(\pi {\rm d}\xi^\mu \wedge {\rm d}\xi^\nu \wedge {\rm d}x^\rho \wedge {\rm d}x^\sigma - \frac{1}{3}\xi^2 {\rm d}\xi^\mu \wedge {\rm d}x^\nu \wedge {\rm d}x^\rho \wedge {\rm d}x^\sigma\right) \ . \tag{38}$$

Pulling back and integrating gives

$$S_3^{\rm wz} = \int_M \beta_3^{\rm wz} = \int_M {\rm d}^4x \left[-2\pi \left[(\partial_\mu\xi^\mu)^2 - \partial_\mu\xi^\nu\partial_\nu\xi^\mu\right] + 2\xi_\alpha\xi^\alpha\partial_\mu\xi^\mu \right] \ , \tag{39}$$

and applying the inverse Higgs constraint $\xi_\mu = -\partial_\mu \pi$, and integrating by parts, we obtain the cubic galileon,

$$S_3^{\text{wz}} \sim \int_M \mathrm{d}^4x \, \Box\pi(\partial\pi)^2 \,. \tag{40}$$

The higher 4d galileons can be obtained from the forms

$$\begin{aligned} \omega_4^{\text{wz}} &= \epsilon_{\mu\nu\rho\sigma}\omega_C \wedge \omega_B^\mu \wedge \omega_B^\nu \wedge \omega_B^\rho \wedge \omega_P^\sigma \,, \\ \omega_5^{\text{wz}} &= \epsilon_{\mu\nu\rho\sigma}\omega_C \wedge \omega_B^\mu \wedge \omega_B^\nu \wedge \omega_B^\rho \wedge \omega_B^\sigma \,. \end{aligned} \tag{41}$$

It follows from this construction that the galileon terms are members of the relative Lie algebra cohomology group

$$H^5\left(\mathfrak{Gal}(3+1,1), \mathfrak{so}(3,1)\right) \,, \tag{42}$$

in a similar way to that in which the usual Wess-Zumino terms can be characterized via de Rham cohomology.

The coset methods discussed in this section provide a fresh perspective on the construction of galileons, and some insight into the mathematical structure underlying them. They are a useful counterpart to the probe brane methods of the previous section.

5. Summary

Attempts to modify gravity with an eye to providing explanations for cosmic acceleration have yielded fascinating new theoretical ideas such as the DGP model and massive gravity. Fascinatingly, these theories share similar behaviors in certain limits, described by a scalar field theory – the galileon theory – that is interesting in its own right. In this talk I described the original Galileon idea, and then discussed work I have been involved with in recent years to generalize ways to construct galileon-like theories with more complicated structures but similar attractive properties.

One way is to use the probe brane construction, the embedding of a particular brane in a suitable ambient space to yield an interesting 4-dimensional theory. In this case, the resulting symmetries are inherited from combinations of five-dimensional Poincaré invariance and brane reparametrization invariance. In the case of Galileons on a curved background there is a rich story, with maximally symmetric choices for the bulk and brane metrics yielding complicated four dimensional symmetry groups.

Another way to construct galileon-like theories is to use the coset construction to take advantage of the fact that the relevant symmetries are nonlinearly realized. For the simple galileons, the relevant terms arise as Wess-Zumino terms, which can then be categorized through relative Lie algebra cohomology groups. In even more recent work, we have shown [35] that massive gravity [2, 3] itself arises through the coset construction, although interestingly there the relevant terms are not Wess-Zumino terms, only becoming so in a particular limit.

There remain many open questions for these field theories. In this talk I was only able to cover the basic idea of construction techniques, but it is possible that fascinating new applications are right around the corner. But it is also possible that further investigation of the theoretical constraints on these models may render them inconsistent. Many of us are working hard to try to figure out which of these might be true.

Acknowledgments

I would like to thank the organizers of the DISCRETE 2014 conference, at King's College London, for a very enjoyable meeting and their wonderful hospitality. I am also indebted to my collaborators in the work described here — Garrett Goon, Austin Joyce, Kurt Hinterbichler and Daniel Wesley. I would also like to thank Melinda Andrews, Justin Khoury and James Stokes for helpful discussions and collaboration. This work was supported by the US Department of Energy grant DE-FG02-95ER40893, and by NASA ATP grant NNX11AI95G.

References

[1] G. R. Dvali, G. Gabadadze and M. Porrati, Phys. Lett. B **485**, 208 (2000) [arXiv:hep-th/0005016].

[2] C. de Rham and G. Gabadadze, Phys. Rev. D **82**, 044020 (2010) [arXiv:1007.0443 [hep-th]].

[3] C. de Rham, G. Gabadadze and A. J. Tolley, Phys. Rev. Lett. **106**, 231101 (2011) [arXiv:1011.1232 [hep-th]].

[4] A. Nicolis, R. Rattazzi and E. Trincherini, Phys. Rev. D **79**, 064036 (2009) [arXiv:0811.2197 [hep-th]].

[5] M. A. Luty, M. Porrati and R. Rattazzi, JHEP **0309**, 029 (2003) [arXiv:hep-th/0303116].

[6] K. Hinterbichler, M. Trodden and D. Wesley, Phys. Rev. D **82**, 124018 (2010) [arXiv:1008.1305 [hep-th]].

[7] C. Burrage, C. de Rham, D. Seery and A. J. Tolley, JCAP **1101**, 014 (2011) [arXiv:1009.2497 [hep-th]].

[8] C. Deffayet, S. Deser and G. Esposito-Farese, Phys. Rev. D **82**, 061501 (2010) [arXiv:1007.5278 [gr-qc]].

[9] A. Padilla, P. M. Saffin and S. Y. Zhou, JHEP **1012**, 031 (2010) [arXiv:1007.5424 [hep-th]].

[10] A. Padilla, P. M. Saffin and S. Y. Zhou, Phys. Rev. D **83**, 045009 (2011) [arXiv:1008.0745 [hep-th]].

[11] A. Padilla, P. M. Saffin and S. Y. Zhou, JHEP **1101**, 099 (2011) [arXiv:1008.3312 [hep-th]].

[12] S. Y. Zhou, Phys. Rev. D **83**, 064005 (2011) [arXiv:1011.0863 [hep-th]].

[13] A. I. Vainshtein, Phys. Lett. B **39**, 393 (1972).

[14] A. Joyce, B. Jain, J. Khoury and M. Trodden, Phys. Rept. **568**, 1 (2015) [arXiv:1407.0059 [astro-ph.CO]].

[15] C. de Rham and A. J. Tolley, JCAP **1005**, 015 (2010) [arXiv:1003.5917 [hep-th]].

[16] G. Goon, K. Hinterbichler and M. Trodden, JCAP **1107**, 017 (2011) [arXiv:1103.5745 [hep-th]].

[17] G. Goon, K. Hinterbichler and M. Trodden, Phys. Rev. Lett. **106**, 231102 (2011) [arXiv:1103.6029 [hep-th]].

[18] G. Goon, K. Hinterbichler and M. Trodden, JCAP **1112**, 004 (2011) [arXiv:1109.3450 [hep-th]].

[19] G. Goon, K. Hinterbichler, A. Joyce and M. Trodden, Phys. Lett. B **714**, 115 (2012) [arXiv:1201.0015 [hep-th]].

[20] C. Burrage, C. de Rham and L. Heisenberg, JCAP **1105**, 025 (2011) [arXiv:1104.0155 [hep-th]].

[21] M. Trodden and K. Hinterbichler, Class. Quant. Grav. **28**, 204003 (2011) [arXiv:1104.2088 [hep-th]].

[22] D. Lovelock, J. Math. Phys. **12**, 498 (1971).

[23] G. W. Gibbons and S. W. Hawking, Phys. Rev. D **15**, 2752 (1977).

[24] J. W. York, Phys. Rev. Lett. **28**, 1082 (1972).

[25] R. C. Myers, Phys. Rev. D **36**, 392 (1987).

[26] S. Coleman, J. Wess and B. Zumino, Phys. Rev. **177**, 2239 (1969)

[27] C. Callan, S. Coleman, J. Wess and B. Zumino, Phys. Rev. **177**, 2247 (1969)

[28] D. V. Volkov, Sov. J. Particles and Nuclei **4**, 3 (1973)

[29] V. I. Ogievetsky, Proc. of X–th Winter School of Theoretical Physics in Karpacz, Vol. 1, Wroclaw 227 (1974)

[30] E. A. Ivanov, V. I. Ogievetsky, Teor. Mat. Fiz. **25**, 164-177 (1975).

[31] S. Weinberg, Phys. Rev. **166**, 1568 (1968).

[32] G. Goon, A. Joyce and M. Trodden, Phys. Rev. D **90**, no. 2, 025022 (2014) [arXiv:1405.5532 [hep-th]].

[33] E. A. Ivanov and V. I. Ogievetsky, Teor. Mat. Fiz. **25**, 164 (1975).

[34] I. Low and A. V. Manohar, Phys. Rev. Lett. **88**, 101602 (2002) [hep-th/0110285].

[35] G. Goon, K. Hinterbichler, A. Joyce and M. Trodden, arXiv:1412.6098 [hep-th].

Journal of Physics: Conference Series **631** (2015) 012014 doi:10.1088/1742-6596/631/1/012014

The MoEDAL Experiment at the LHC – a New Light on the Terascale Frontier

J L Pinfold
Physics Department, University of Alberta,
Edmonton, Alberta T6G 2E1, CANADA.

E-mail: jpinfold@ualberta.ca

Abstract. MoEDAL is a pioneering experiment designed to search for highly ionizing avatars of new physics such as magnetic monopoles or massive (pseudo-)stable charged particles. Its groundbreaking physics program defines a number of scenarios that yield potentially revolutionary insights into such foundational questions as: are there extra dimensions or new symmetries; what is the mechanism for the generation of mass; does magnetic charge exist; what is the nature of dark matter; and, how did the big-bang develop. MoEDAL's purpose is to meet such far-reaching challenges at the frontier of the field. The innovative MoEDAL detector employs unconventional methodologies tuned to the prospect of discovery physics. The largely passive MoEDAL detector, deployed at Point 8 on the LHC ring, has a dual nature. First, it acts like a giant camera, comprised of nuclear track detectors - analyzed offline by ultra fast scanning microscopes - sensitive only to new physics. Second, it is uniquely able to trap the particle messengers of physics beyond the Standard Model for further study. MoEDAL's radiation environment is monitored by a state-of-the-art real-time TimePix pixel detector array. A new MoEDAL sub-detector to extend MoEDAL's reach to millicharged, minimally ionizing, particles (MMIPs) is under study.

1. Introduction

In 2010 the MoEDAL experiment [1] at the Large Hadron Collider (LHC) was unanimously approved by CERN's Research Board to start data taking in 2015. MoEDAL is a pioneering experiment designed to search for highly ionizing avatars of new physics such as magnetic monopoles or massive (pseudo-)stable charged particles. Its groundbreaking physics program defines over 30 scenarios that yield potentially revolutionary insights into such foundational questions as: are there extra dimensions or new symmetries; what is the mechanism for the generation of mass; does magnetic charge exist; what is the nature of dark matter; and, how did the big-bang develop.

The LHC machine [2] brings its counter rotating beams nominally consisting of 2808 bunches of protons each with roughly 10^{11} protons per bunch. The proton beams are brought into collision at four points around the 27 km circumference LHC ring. The biggest of the LHC's seven experiments are ATLAS [3] and CMS [4], general-purpose detectors designed to investigate the largest range of physics possible. ALICE [5] and LHCb [6] have detectors specialized for focussing on specific phenomena such as quark–gluon plasma formation and the investigation of the matter-antimatter asymmetry using b-hadron physics, respectively. These four detectors in huge caverns sit roughly 100m underground on the LHC ring.

Prior to the approval of the MoEDAL experiment the smallest experiments on the LHC were TOTEM [7] and LHCf [8], which focus on "forward particles" – protons or heavy ions that brush past each other rather than meeting head on when the beams collide. TOTEM uses detectors positioned on either side of the CMS interaction point, while LHCf is made up of two detectors that sit along the LHC beam line, at 140 metres either side of the ATLAS collision point. MoEDAL the seventh and latest LHC experiment shares an interaction point with the LHCb experiment. It is designed to significantly expand the discovery horizon of the general-purpose LHC detectors, ATLAS and CMS in a complementary manner.

The innovative MoEDAL detector employs unconventional methodologies tuned to the prospect of discovery physics. The largely passive MoEDAL detector, deployed at Point 8 on the LHC ring, as shown in Figure 1, has a dual nature. First, it acts like a giant camera, comprised of Nuclear Track Detectors (NTDs) - analyzed offline by ultra fast scanning microscopes - sensitive only to new physics. Second, it is uniquely able to trap the particle messengers of physics beyond the Standard Model for further study. MoEDAL's radiation environment is monitored by a state-of-the-art real-time TimePix pixel detector array. In addition, we are preparing a proposal for a third active detector element capable of detecting the faintest flicker of light - single photon signals - produced by the passage of Mllicharged Minimally Ionizing Particles (MMIPs) with charge as small as one thousandth that of an electron.

Figure 1. The position of the MoEDAL experiment on the LHC ring

2. Highly Ionizing Particles – Messengers of New Physics

MoEDAL's groundbreaking physics prospectus [9] covers 34 fundamentally important Beyond the Standard Model (BSM) scenarios involving electrically and magnetically charged particles. It is designed to exploit the use of Highly Ionizing Particles (HIPs) as wide ranging avatars of new physics. One of the main objectives for MoEDAL is the state-of-the-art search for the magnetic charge carried by the monopole, just as the search for the Higgs was the prime motivation for the LHC and its general-purpose experiments.

To be detected a particle must interact with the material of the detector losing energy as they travel. Except when they are moving very near to the speed of light charged particles heavier than electrons primarily lose energy by interacting with the atoms of the material by exciting atomic electrons or ionizing atoms. The Bethe formula describes the energy loss per distance travelled of fast charged particles that are more massive than electrons such as muons, protons, pions and ions (atomic nuclei). A simplified form of the Bethe equation is:

$$-\frac{dE}{dx} = z^2 \frac{KZ}{A}\frac{1}{\beta^2}\left[\frac{1}{2}\ln\left(\frac{2m_e\beta^2\gamma^2}{I^2}\right) - \beta^2\right]$$

where, β is the speed of the particles expressed as a fraction of the speed of light ($\beta = v/c$) and $K = 4\pi N_A r_e^2 m^e c^2$. The properties of the medium are the atomic number Z, the atomic mass A and the mean excitation energy I. The constants are the electron mass, m_e, the classical electron radius, r_e and Avogardo's number N_A.

All stable or long-lived Standard Model charged particles produced in proton-proton collisions at the LHC have a single charge ($ze \rightarrow e$), are relatively light i.e. at most the mass of a proton and are moving near to the speed of light ($\beta \sim 1$). An inspection of the Bethe formula and Figure 3 shows us that such particles will be near to or slightly more than a "minimum ionizing" ($\beta\gamma = 3 \rightarrow 4$ or dE/dx ~ 1→2 MeV g^{-1} cm^2) we shall call Minimum Ionizing Particles (MIPs).

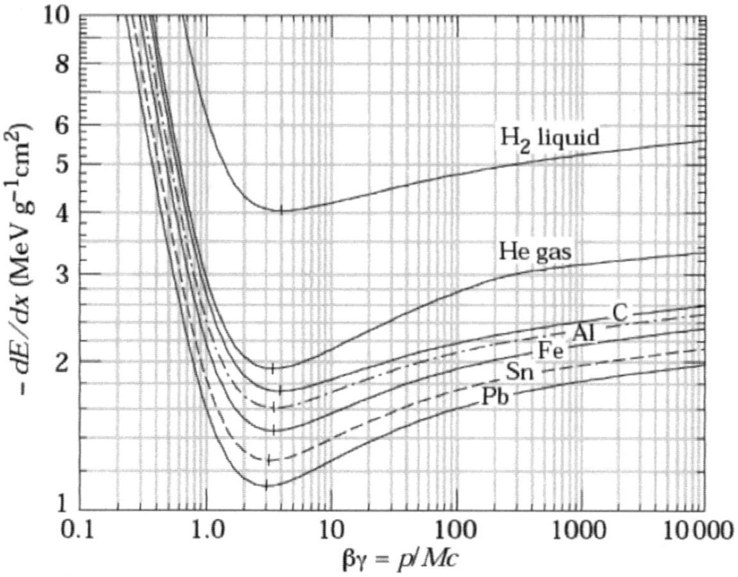

Figure 2. The energy loss curve for a proton travelling through various materials.

In proton-proton collisions as the LHC electrically charged particles can only be highly ionizing if they are massive and thus moving very slowly with $\beta\gamma < 1$, and/or multiply charged. However, there are no known massive stable or long-lived particles that are multiply charged. The only way that a Standard Model particle can be highly ionizing is if it is slowing down to stop. Thus, massive stable or pseudo-stable HIPs traversing the detector are clear messengers of new physics with no obvious Standard Model Backgrounds.

A prime example of a HIP from BSM is the magnetic monopole. The ionization energy loss of a magnetically charged particle can be obtained from the Bethe formula by replacing the electric charge ze by $ng\beta$, where g is the magnetic charge and n is the number of Dirac magnetic charges:

$$-\frac{dE}{dx} = (ng/e)^2 \frac{KZ}{A}\left[\frac{1}{2}\ln\left(\frac{2m_e\beta^2\gamma^2}{I^2}\right) - \beta^2\right]$$

A single Dirac (magnetic) charge is g ≈ 68.5e. As we can see from the Bethe formula for magnetic charge given above, the energy loss of a singly charged (n=1) very relativistic magnetic monopole (β ≈ 1) is amazingly about 4700 times that of a proton!

The MoEDAL experiment detects HIPs using stacks of passive plastic Nuclear Track Detectors (NTDs). The HIP damages the plastic as it traverses the NTD stack. Etching the plastic in a hot sodium hydroxide solution can reveal this damage as two collinear etch-pits each with diameter ~ 5 - 10 μm in each of the six layers of the stack. The NTD plastic employed by MoEDAL has a threshold of around Z/β of 5 (or 5 MIPs) where Z is the charge number of the particle and β is the velocity of the particle divided by the speed of light (v/c). A HIP messenger of new physics would leave a tell tale signature of twelve tiny etch pits (for each stack of ten NTD sheets) forming a precisely defined line pointing back to the collision. There is no known Standard Model particle that can give a trail of collinear etch pits in a MoEDAL stack.

3. Directly Detecting Magnetic Charge

Clearly, a unique property of the magnetic monopole is that it has magnetic charge. Thus a moving magnetic monopole will be ringed by an electric field just as a moving electric charge is ringed by a magnetic field. Imagine that a magnetic monopole traverses the superconducting wire coil of a SQUID. As the monopole approaches the coil it drives an electrical current within the coil that in turn creates a changing magnetic field. This changing magnetic field creates an electric field that counters the original electric field of the moving monopole. As the wire in the coil has no resistance the overall effect is to make the total electric field vanish in the coil. Meanwhile the current continues to flow in the coil because the wire is superconducting and without resistance. One can use Faraday's Law to calculate the magnitude of the current that would be generated:

$$I = -\frac{\mu_0 q_m}{L}.$$

where, L is the inductance of the coil, μ_0 is the permeability of free space and q_m is the magnetic charge. Consequently, the current induced only depends on the magnetic charge and is independent of the speed, mass or direction of the magnetic monopole. A visualization of the generation of a current by the passage of a monopole though a squid is shown in Figure 3.

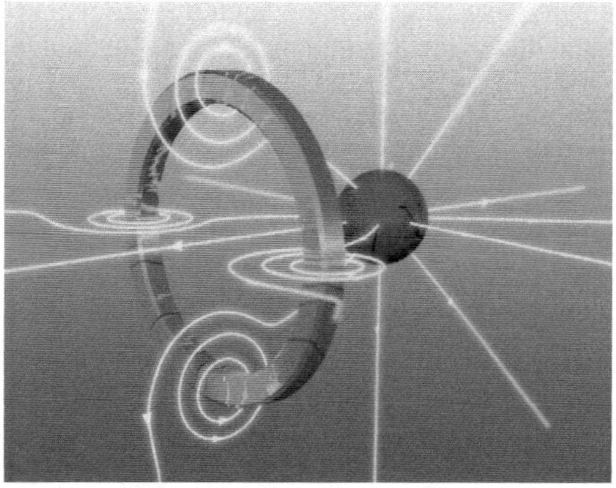

Figure 3. Passage of a magnetic monopole (North pole) through a superconducting coil.

In the early 1980's Blas Cabrera was one of the first experimenters to deploy a SQUID device in order to detect magnetic monopoles from the cosmos. His initial detector employed a four-turn, 5-cm diameter loop, positioned with its axis vertical, connected to the superconducting input coil of a SQUID magnetometer. The whole device was shielded from external magnetic fields by a 20-cm-diameter, 1m long cylindrical superconducting shield mounted inside a single Mumetal cylinder. As we have seen above the passage of a single Dirac charge through the loop would result in a current of a definite amount to start flowing as soon as the Monopole had passed through the device. As the coil is superconducting the current would continue to flow indefinitely.

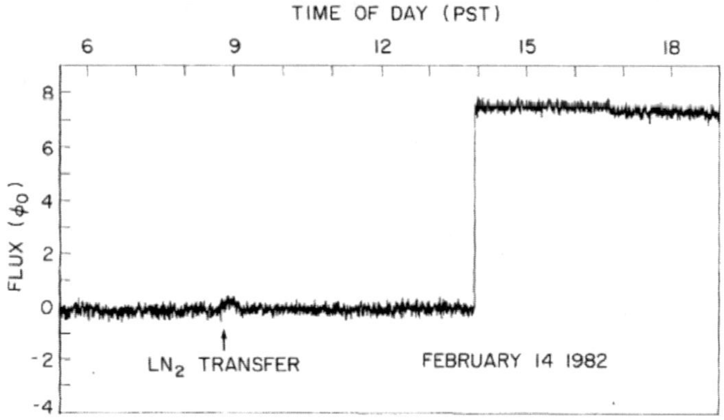

Figure 4. Cabrera's famous St Valentine's eve event that looks exactly as if it is due to the passage of a magnetically charged monopole passing through his detector.

Amazingly, on St Valentine's day 1982, his detector recorded an event, which had exactly the expected signal properties of a magnetic monopole traversing the detector. This event is shown in Figure 4. After he published his discovery [10] Cabrera and a number of other research groups built bigger and better detectors. It is often stated that no other monopole candidate events were found. However, a group at Imperial College London observed another perfect candidate magnetic monopole event on the 11th of August 1985. To this day it is still not clear what background process can explain these signals.

Journal of Physics: Conference Series **631** (2015) 012014 doi:10.1088/1742-6596/631/1/012014

4. A Description of the MoEDAL Detector

The novel MoEDAL detector is deployed at IP8 on the LHC ring, depicted in Figure 5, is comprised of three key subsystems. The main component of MoEDAL's NTD subsystem consists of a low threshold (LT) NTD array - comprised of an array of (400) plastic NTD stacks (25 x 25 cm^2) each stack made up of CR39 (3) and MAKROFOL (3) NTD sheets. This is the largest NTD array every deployed at an accelerator. The CR39 NTDs have an excellent charge resolution (≥ 0.2 of a single electric charge, based on a single measurement) and a low threshold, allowing the detection of HIPs with and ionizing power equal to or more than ~5 times that of a Minimum Ionizing Particle (MIP). An illustration of the LT-NTD stack is given in Figure 6 (Left)

The LT-NTD array has been enhanced by the Very High Charge Catcher (VHCC) subdetector – with threshold around 50 MIPs. The VHCC subdetector – comprised of two flexible low mass stacks of MAKROFOL in an aluminium foil envelope - is deployed in the forward acceptance of the LHCb experiment just after the LHCb RICH detector. The overall acceptance of the NTD system is ~60$. The exposed NTD detectors will be etched to reveal the passage of a HIP as

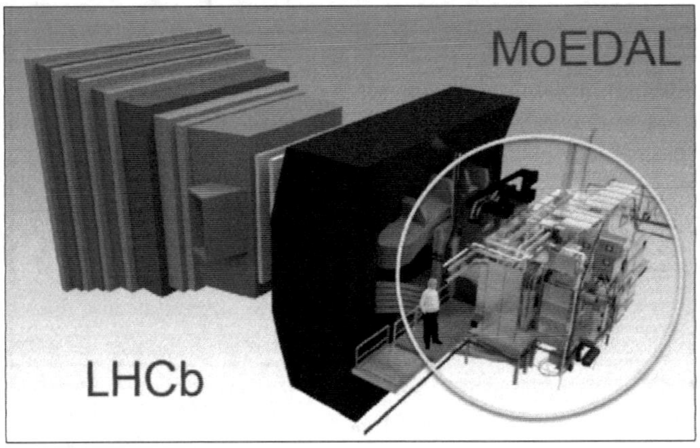

Figure 5. A view of the MoEDAL detector deployed around the LHCb's VELO detector at IP8

illustrated in Figure 6 (right). The etched plastic will then be analysed at Bologna and CERN using state-of-the art ultra-fast automated optical scanning microscopes developed by the Muenster and INFN-Bologna MoEDAL groups, combined with cutting edge image analysis software. This enables low NTD thresholds to be maintained despite large beam related backgrounds. The NTD detectors will be calibrated using the heavy-ion (Fe) beam at NASA's Space Radiation Laboratory Laboratory (NSRL).

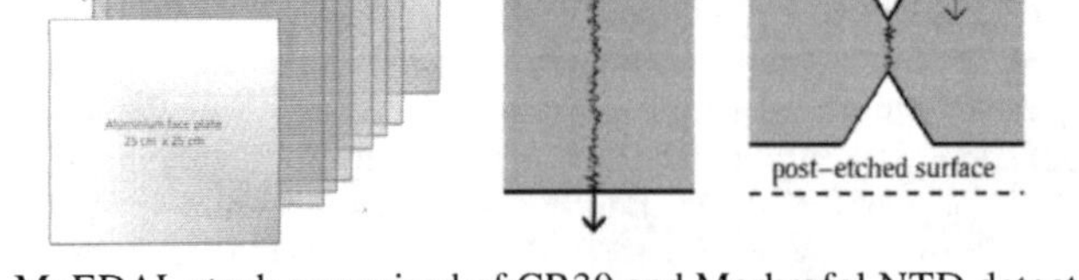

Figure 6. (Left) A MoEDAL stack comprised of CR39 and Markrofol NTD detectors. (Right) The damage zone created by the HIP that is subsequently revealed by etching in a hot NaOH solution

The Magnetic Monopole Trapper (MMT) is the third MoEDAL sub-detector system. Its sensitive volume consists of roughly 1 tonne of pure aluminium trapping volumes deployed around the MoEDAL cavern at IP8. Aluminium is well suited as it has an anomalously large nuclear magnetic moment. A fraction of the massive HIPs created will stop and be captured in the MMT detector as shown in Figure 7. After exposure the MMT trapping volumes will be monitored at the ETH SQUID facility for the presence captured monopoles. A schematic description of this facility is shown in Figure 8 (left). Figure 8 (right) shows the results of an exposure of a MMT test detector at IP8 on the LHC ring. As can be seen the SQUID can detect magnetic charges greater than one tenth of a Dirac charge. Test solenoids can be used to calibrate the response of the SQUID to a trapped monopole. After the SQUID scan has been performed the trapping volumes will be sent to an underground laboratory, for example SNOLAB, to be monitored for the decays of very long lived electrically charged particles.

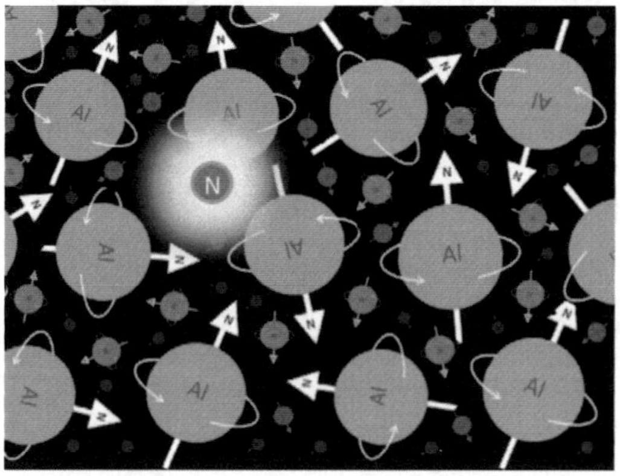

Figure 7. A monopole (a North pole captured within the aluminium volume of the MMT detector

The use of SQUIDS to detect trapped magnetic charge has also been thoroughly tested in particle and astroparticle experiments where the search has been performed on "found" or alternate use (such as beampipes) objects on a "one-off" basis. But, MoEDAL is the first experiment ever to use purpose-made replaceable trapping volumes to capture magnetic charge. After the SQUID scan has been performed the trapping volumes will be sent to SNOLAB – 2km underground – to be monitored for the decays of very long lived electrically charged particles.

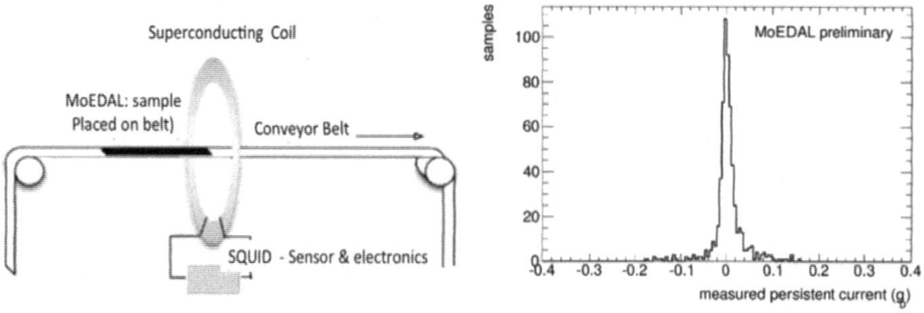

Figure 8. (Left) A schematic depiction of the use of a SQUID magnetometer to detect the presence of a trapped magnetic charge. (Right) The results of a SQUID scan of MMT elements exposed to roughly a year of LHC collisions at IP8 -as part of a MoEDAL test MMT detector deployment.

The fourth and only active sub-detector system is the TimePix pixel device array (TMPX), consisting of ~8 devices distributed throughout the MoEDAL cavern at IP8. TMPX will be used to monitor highly ionizing beam related backgrounds. Each pixel of the innovative TimePix chip contains a preamp, a discriminator with threshold adjustment, synchronization logic and a 14-bit counter. MoEDAL uses TimePix Time-over-Threshold modes where each pixel acts as an ADC for energy measurement. Effectively, the TimePix device is a tiny electronic bubble-chamber providing a real-time "colour" (energy dependent) movie of the background as illustrated in Figure 9.

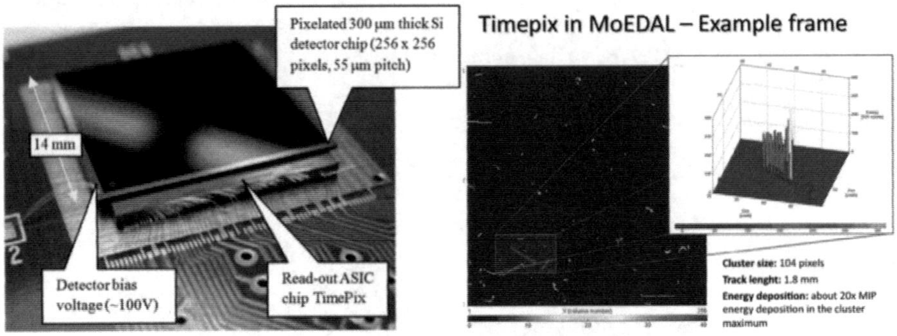

Figure 9. Left) a photograph of a TimePix pixel device. (Right) A view of a low energy highly ionizing event in the TimePix chip's silicon sensitive volume.

4.1. A Proposed Millicharged Particle Detector for MoEDAL

A key aspect of the MoEDAL experiment is that is sensitive to HIP avatars of new physics for which the standard general-purpose LHC detectors are not optimized. MoEDAL is preparing a proposal to add a new subdetector that is sensitive to MMIPs. This addition is consistent with the MoEDAL collaboration's aim to extend the physics reach of the LHC in a way that is complementary to the existing capability provided by the main LHC detectors since they have minimal sensitivity to MMIPs.

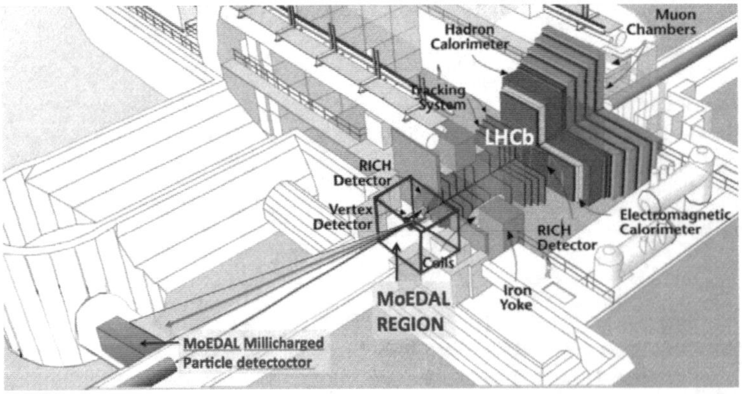

Figure 10. MoEDAL is preparing a proposal to instal a millicharged particle detector positioned in the forward region of IP8 some 20m from the LHCb/MoEDAL intersection point (IP8).

The proposed detector is similar in principle to that employed on an earlier beam dump search for MMIPs at SLAC [11]. It will employ three deep liquid scintillator layers in triple coincidence with an acceptance of 0.01%. The baseline plan has detectors deployed in the position shown in Figure 6. The liquid scintillator and low noise photomultiplier tubes used – manufactured to have very low natural backgrounds - will be the same as those utilized by the SNO+ experiment. A prototype MMIP detector of the sort described here is currently under construction.

5. A Comparison of the MoEDAL Detector with Conventional LHC Detectors

MoEDAL's exposed NTD detectors are sent for analysis to INFN-Bologna and CERN and for calibration at the NASA Space Radiation Laboratory (NSRL) heavy-ion beam facility. The exposed MMT trapping detectors are monitored for the presence of trapped magnetic charge at the ETH Zurich SQUID Facility and at SNOLAB in Canada for captured long-lived massive electrically charged particles. In this way MoEDAL's unparalleled use of passive tracking and trapping detectors has the extraordinary effect of allowing the physics to be brought to the detectors, rather than the usual situation of bringing the detector to the physics. This unique approach has the effect of *adding* the capabilities of the remote facilities to that of the deployed MoEDAL detector and allowing us to significantly expand the physics discovery reach of the LHC in a cost effective way.

Standard collider detectors are not optimized for the detection of HIPs such as magnetic monopoles and massive long-lived or stable charged particles. Rather, the general-purpose detectors, such as ATLAS and CMS, are designed to detect minimum-ionizing particles moving near to the speed of light. Effects arising from the particles low velocity and high-density energy deposition, such as electronics saturation, light quenching in scintillators, and adjacent hits from delta electrons, are extremely challenging to deal with. Indeed, in some cases it may be impossible to make an accurate measurement of the effective charge of the particle. For example, the resulting dead time as a result of electronics saturation may be of the order of the bunch crossing time. In addition, HIPS will be quickly absorbed within the detector mass even before they penetrate far into the inner tracking detectors.

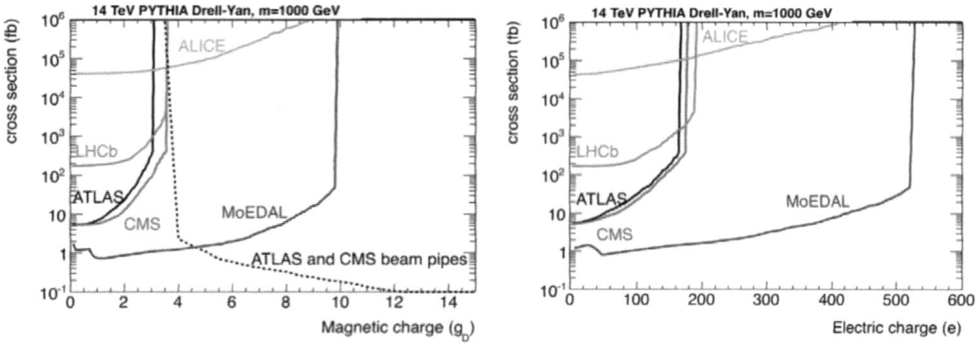

Figure11. The expected reach [12] of the search for direct monopole–anti-monopole Drell-Yan pair production process at the LHC (Ecm =14 TeV). Assuming the luminosity taken by LHCb/MoEDAL is 2 fb^{-1}, by ATLAS & CMS 20 fb^{-1}, and by ALICE 0.004 fb^{-1}.

On the other hand, MoEDAL does not suffer from the problems that would severely hamper detection of slow moving HIPS at the main LHC detectors. The NTD based sub-detectors have the demonstrated ability to measure accurately the ionizing power of a HIP. The MMT sub-detector, read out through a remote SQUID, can directly detect magnetic charge - no other Collider detector has this capability. MoEDAL's largely passive detection systems require no trigger, electronics, high voltage or gas supply. Its light and uniform detection technology requires minimal support structure and provides maximum cost effective coverage close to IP8 with few experimental compromises.

MoEDAL's sensitivity to highly ionizing messengers of new physics, together with its ability to make a permanent "photographic" record of HIP BSM particles in its NTD detectors and capture such particles in its trapping detectors, is unique. This set of capabilities combined with the absence of SM backgrounds allows the definitive identification of HIP avatars of new physics with only a few candidate events detected. In an independent study [12] a comparison was made of the response of the LHC detectors to HIPs, as shown in Fig. 11 MoEDAL's sensitivity clearly surpasses that of the other LHC detectors.

6. MoEDAL's Physics Program – a New Light on the Terascale Frontier

MoEDAL's pioneering physics prospectus [9] covers more than 34 fundamentally important BSM scenarios allowing MoEDAL to significantly expand the LHC's discovery horizon in a complementary way. This program is illustrated in the pie chart shown in Figure 11. One of the main objectives for MoEDAL is the state-of-the-art search for the magnetic charge, just as the search for the Higgs was the prime motivation for the LHC and its general-purpose experiments. For example, the work of our Theory Board has renewed interest in the LHC search for a Terascale EW Monopole [13] arising from the SM. If discovered, this would be the first topological particle to be observed – with the utmost consequences for our understanding of the Universe. But MoEDAL is designed to do much more as indicated in Figure 12 and as discussed below.

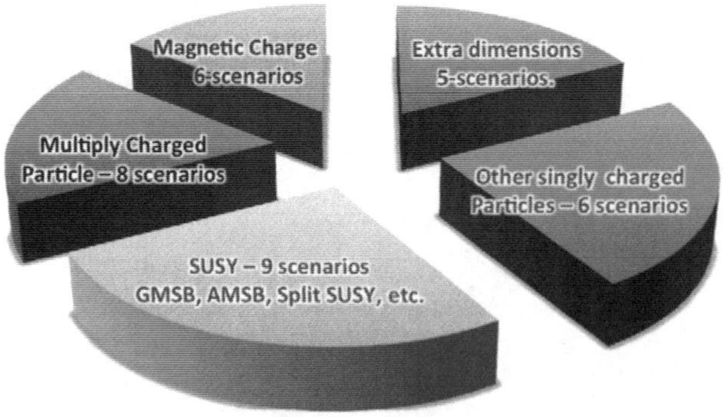

Figure 4. The MoEDAL Physics Program defined in Ref. [4] broken down by topic.

6.1 The Hunt For Magnetic Charge

It was Dirac who first hypothesized that a magnetic monopole could exist within the framework of quantum mechanics [14]. He saw the monopole as the end of an infinitely thin infinitely long solenoid called a Dirac string. The annoying string does not pose a problem as long as it cannot be detected. Dirac surmised that the Dirac string would not detectable if the phase picked up by the wave function

Journal of Physics: Conference Series **631** (2015) 012014 doi:10.1088/1742-6596/631/1/012014

of the electron as it travels completely around the string is equal to one – in other words it has no affect. He calculated that the condition for the string *not* to be detected:

$$e^{iqg} = 1 \Rightarrow qg = 2\pi n \quad (n = 1,2,3.....)$$

Dirac was quick to realize that the above relation between electric charge (q) and magnetic charge (g) explains the quantization of electric charge, as long as at least one monopole exists. The numerical value of the magnetic charge is $g = n/2\alpha = n68.5q$ - using $1/\alpha = 4\pi/q^2$ where α is the fine structure constant (~1/137) giving rise to the extremely highly ionizing nature of the magnetic monopole that we discussed above.

Julian Schwinger [15] generalized the quantization condition to dyons, particles that carry both electric and magnetic charge. If we consider two dyons with electric and magnetic charges (q_1, g_1) and (q_2, g_2) Schwinger found that each dyon is unable to detect the Dirac string of the other iff:

$$q_1 g_2 - q_2 g_1 = n$$

He found that the minimum charge for a magnetic monopole should be twice the Dirac charge.

6.1.1 The 't Hooft Polyakov Monopole

In the Standard Model electroweak unification was achieved using a gauge theory developed by Sheldon Glashow, Steven Weinberg and Abdus Salam (GWS) [16]. This theory had two components called U(1) - that has the same Abelian gauge symmetry structure as Maxwell's electromagnetism and SU(2), that has a non-Abelian gauge symmetry that describes the weak interaction. These symmetries are broken at low energies by the Higgs field. In 1974 Gerard 't Hooft [17] and Alexander Polyakov [18] wrote their famous papers on a new kind of magnetic monopole that didn't have a problematic "Dirac String" with a point-like singularity at the origin. Their monopole had its origin in Georgi's and Glashow's earlier model of EW unification [19] later supplanted by GWS theory.

The Georgi Glashow model has only one gauge symmetry, a non-Abelian symmetry known as SO(3), that is spontaneously broken by the Higgs field down to the U(1) gauge symmetry of Maxwell's equations. The quantization of charge arises naturally in such theories without the need for a magnetic monopole as was the case in Dirac's theory. However, a major implication of 't Hooft's and Polyakov's work was that any unified gauge theory like Georgi's and Glashow's inevitably possesses monopole solutions. So, magnetic monopoles that were only "nice" in Abelian theories like Maxwell's electromagnetism are now *necessary* in some non-Abelian theories.

The 't Hooft and Polyakov monopole arose in the situation where the Higgs field vector points away from the origin everywhere in the "hedgehog" configuration shown in Figure 13. In contrast to the Dirac monopole, which necessitates a singular point-like magnetic source at the origin, the 't Hooft-Polyakov monopole is everywhere smooth. In this case the field configuration is topologically stable, like a knot that cannot unravel. Remarkably, the Higgs-field configuration corresponding to this magnetic monopole is an example of a topological soliton, a localized and stable field excitation that behaves like an elementary particle. In the microscopic heart of the 't Hooft-Polyakov monopole the full unbroken SO(3) gauge symmetry of the unified theory reigns supreme. Far away the 't Hooft-Polyakov monopole would be indistinguishable from a Dirac monopole except that its magnetic charge is twice as much.

The roughly 100 GeV/c^2 mass of these monopoles — determined by the energy scale of the weak interaction – was ruled out by existing experimental results. But 't Hooft and Polyakov observed that the structure of the SU(5) Grand Unified Theory (GUT), which attempts to describe all known elementary particle forces except gravity, was similar in some respects to that of the Georgi-Glashow model. It was not then surprising that the SU(5) GUT model also contains a 't Hooft-Polyakov magnetic monopole solution, but with the much higher mass of 10^{16} GeV. The existence of magnetic monopoles is an unavoidable consequence of any theory that would describe all strong and electroweak forces by one GUT interaction. Unfortunately, GUT monopoles are far too massive be produced at any conceivable man-made accelerator.

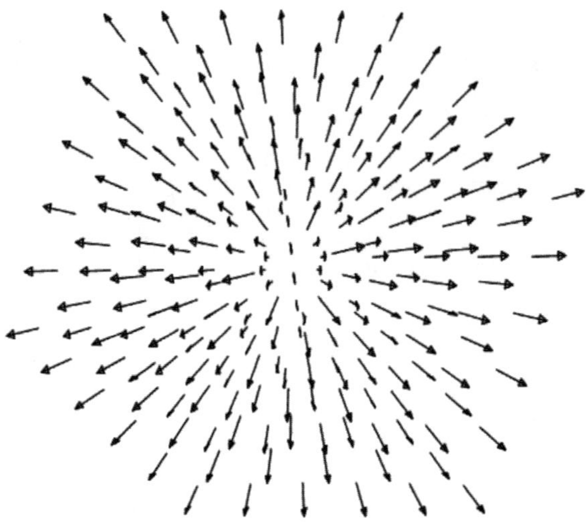

Figure 13. The "hedgehog" Higgs field configuration of the `t Hooft Polyakov magnetic monopole

6.1.1 The Electroweak Monopole

The discovery of what increasingly appears to be the Standard Model Higgs boson in July 2012 by the ATLAS and CMS [20] experiments at the Large Hadron Collider (LHC) at CERN has led to the possibility that the last piece of the Standard Model puzzle may now be in place. This discovery reinforces the electroweak theory of Glashow, Salam and Weinberg, a central pillar of the successful Standard Model. A monopole-like solution within the framework of the Standard Model would explain from first principles the quantization of the electric charge, something that the present Standard Model cannot do. In 1986 Cho and Maison published a paper suggesting that such a magnetic monopole could exist.

Cho and Maison's "Electroweak" monopole is a hybrid of the Dirac Monopole and the 't Hooft Polyakov monopole. Like the 't Hooft-Polyakov monopole its magnetic charge is twice that of the Dirac monopole. According to a recent estimate [9] the mass of the Electroweak Monopole could lie in the range 4-7 GeV/c^2, possibly low enough to be detectable at the LHC by the MoEDAL experiment. Thus, the LHC could be the first collider to produce Electroweak monopoles. Naively, one might expect the production rate would be comparable to a typical electroweak process such as WW production, above the threshold energy. If that is so then MoEDAL experiment should easily be able to detect the EW monopole, if it the threshold is within reach at the LHC. If detected the Electroweak monopole would be the first topological particle to be observed in nature.

7. Electrically Charged Messengers of New Physics

Electrically charged stable or pseudo-stable massive particles (SMPs), with mass greater than the proton mass do not exist within the framework of the Standard Model – they are true and certain messengers of new physics, predicted in a plethora of beyond the Standard Model scenarios. As we have seen such charged massive particles can be highly ionizing, if they have a combination of low velocity and/or multiple charge, and are thus detectable by MoEDAL. The most obvious possibility for an SMP is that one or more new states exist which carry a new conserved, or almost conserved, global quantum number. SUSY with R-parity, extra dimensions with KK-parity, and several other models fall into this category. The lightest of the new states will be stable, due to the conservation of this new "parity", and depending on quantum numbers, mass spectra, and interaction strengths, one or more higher-lying states may also be stable or meta-stable.

In general, electrically charged stable states are on the face of it highly constrained by cosmology. Coloured particles are also strongly constrained. Consequently, most models are usually constructed to provide non-charged and colourless stable dark-matter candidates. For those models where Weakly Interacting Massive Particles (WIMPs) are the dark matter candidate, SMPs must be meta-stable states that are higher in the hierarchy of particles. Models with long-lived or stable SMPs either have a non-WIMP dark matter candidate or else do not address the problem of particulate dark matter at all. Our recent paper [9] describing MoEDAL's physics program describes nearly thirty such new physics scenarios involving electrically charged massive particles that are "detector stable" and can have a Z/β or at least around five.

7.1 Highly Ionizing Particles from Supersymmetric Scenarios

One key problem of the Standard Model is that it does not include gravity, one of the four fundamental forces. The model also fails to explain why gravity is so much weaker than the electromagnetic or nuclear forces. For example, an electrostatically charged hair comb can counteract the gravitational attraction of a whole planet on a small scrap of paper. This difference in the strength of fundamental forces is one aspect of the "Hierarchy Problem".

A prime candidate for a deeper more fundamental theory that would resolve many of the important problems with the Standard Model is Supersymmetry (SUSY), where for every Standard Model particle there is a superpartner (sparticle) whose spin differs by ½ - for excellent reviews see, e.g. [21]. By doubling the number of particles in this way one obtains cancellations and degrees of freedom with which we can tackle such the Hierarchy Problem. Indeed, SUSY appears to be a necessary ingredient in the only available candidate for quantum theory of gravity, string theory.

If SUSY were an exact symmetry, SUSY particles would have the same masses as their SM partners. But as no SUSY particles have been observed to date we can surmise that the SUSY partners must be much heavier than the Standard Model particles and therefore SUSY must be a broken symmetry. The mass gap between Standard Model and SUSY particles cannot be much larger than ~ 1 TeV without destroying SUSY's ability to solve the hierarchy problem. Other important benefits of SUSY are that it allows for the unification of the electromagnetic, weak and strong gauge couplings of the Standard Model – a necessary and predicts particles that are natural dark matter candidates. The Standard Model fails in both these arenas.

At present, little is known for certain about the nature of the SUSY-breaking mechanism. As a result the MoEDAL physics program explores many supersymmetric scenarios [9]. The most common SUSY framework is divided into two sectors. The usual fields of the Minimal Supersymmetric extension to the Standard Model (MSSM) i.e. quarks, leptons, gauge bosons, and two Higgs doublets, together with their supersymmetric partners, belong to the visible sector. While the second, hidden, sector contains additional fields – yet unobserved quantum superfields and the corresponding hypothetical particles, which do not interact via the Standard Model force messengers (gluons, photons, and W and Z bosons) - that lead to the breaking of Supersymmetry. The SUSY breaking, occurring at some higher energy scale in the hidden sector, is mediated to the visible sector by the exchange of weakly interacting ``messenger'' particles.

As an example we consider the search for long lived stau slepton, the superpartner of the tau lepton, just one of many examples of SUSY scenarios that give rise to SMPs detectable by MoEDAL [9]. At the LHC, the signals of the supersymmetric particles depend on the nature of the lightest SUSY particle. It may be the lightest neutralino, which escapes from detection and leads to missing transverse energy in the detector. Another possibility for the LSP is the gravitino ($\tilde{G}$), the superpartner of the graviton. If the LSP is the gravitino the lifetime of the NLSP slepton, decaying via $\tilde{l} \rightarrow \tilde{G}l$, is given by [22]

$$\mathrm{T} = \frac{1}{\Gamma(\tilde{l} \rightarrow \tilde{G}l)} = \frac{1}{48\pi M_{Pl}} \frac{m_{\tilde{l}}^5}{M_{\tilde{G}}^2} \left[1 - \frac{M_{\tilde{G}}^2}{m_{\tilde{l}}^2}\right]^4$$

where, M_{Pl} is the Planck Scale, $M_{\tilde{G}}$ is the mass of the gravitino, $m_{\tilde{l}}$ is the mass of the slepton for example a stau, the superpartner of the tau, and m_l is the mass of the lepton resulting from the slepton decay. Since M_{Pl} is much larger than the electroweak scale, the NLSP lifetime is naturally long, perhaps reaching into years, as can be seen in Figure 14.

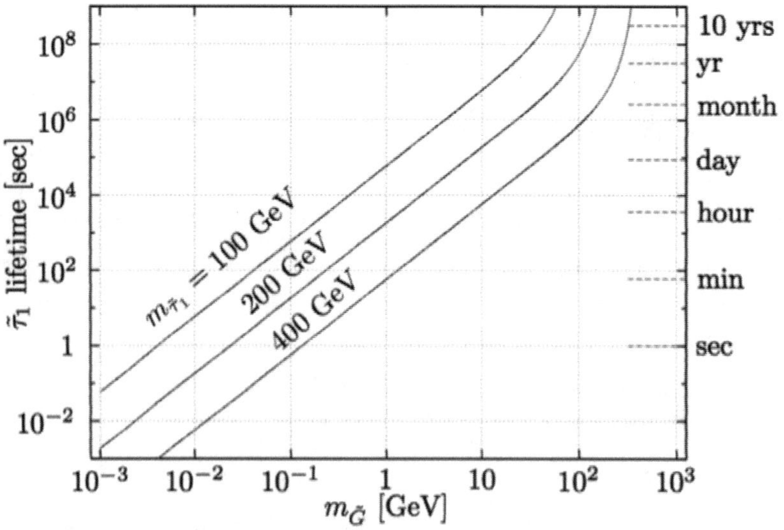

Figure 14. The stau lifetime in a gravitino LSP scenario, for different values of the stau ($\tilde{\tau}$) mass and the gravitino ($\tilde{G}$) mass [22].

MoEDAL can detect massive slowly moving (β < 0.2) very long-lived sleptons in two ways. First, via the high ionization which will leave a signal in MoEDAL's NTD detectors. Secondly, a fraction of the sleptons produced will be captured in its trapping detectors. After exposure at IP8 the trapping volumes will be removed and monitored deep underground, where cosmic ray backgrounds are minimised, for the slow decays of any trapped sleptons. MoEDAL's sensitivity to such events complements that of the standard general-purpose LHC detectors ATLAS and CMS that are optimized for minimum ionizing relativistic particles. On the other hand, MoEDAL is optimized for HIPs.

8. Millicharged Matter

Although Charged Massive Particles (ChaMPs) have been more or less ruled out as dark matter candidates, this is not easily the case for particles with tiny charge. Millicharged particles (mQPs) have been firstly proposed as a solution of the dark matter puzzle long time ago [23]. It is possible to simply add mQPs to the Standard Model for example as Standard Model neutrinos that have tiny electric charges. A small modification of the hypercharge assignments of the Standard Model particles can accommodate tiny electric charges to neutrinos (for a review see e.g. [24]).

Holdom [25] has shown that mQPs will arise naturally, and without violating charge quantization, in certain models involving what are called "shadow universes" or "mirror universes" that are hidden from our own. If there is a shadow-U(1) gauge group i.e a shadow electromagnetism, it follows there is such a thing as a shadow photon, also called a paraphoton. Holdom showed that if the photon and the paraphoton mix a particle charged under the Shadow-U(1) appears to have a very small coupling to the photon – it is a mQP.

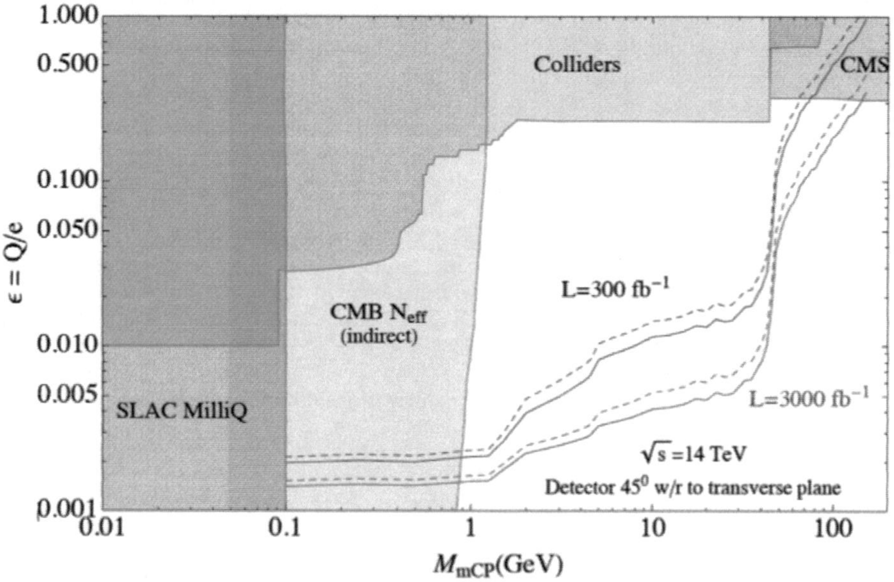

Figure 15. The existing bounds are taken from Ref. [26]. The black line shows the expected 95% C.L. exclusion (solid) and 3σ sensitivity (dashed) of the proposed experiment at = 14 TeV, assuming 300 fb^{-1} of luminosity. In blue the corresponding lines achievable in an LHC run with 3000 fb^{-1} [13].

The mQP parameter space as defined by mass and charge is highly constrained by direct searches by accelerator experiments [27-30] as well as indirect astrophysical [27,31-33] and cosmological observations [26,31,34]. Of course, indirect observations can be evaded by adding extra degrees of freedom or by using different assumptions. In addition, the parameter space for mQPs with masses between 0.1 GeV and 100 GeV is largely unexplored by accelerator based experiments. MoEDAL's proposed mQP subdetector should be able to push into uncovered territory towards the discovery limits shown in Figure 14 [35] for mQPs with charge ≥ 0.001e.

9. Summary and Conclusion

The MoEDAL detector employing unconventional methodologies and tuned to the prospect of discovery physics was installed at Point 8 on the LHC rings in the Winter of 2014. The largely passive MoEDAL detector, has a dual nature. First, it acts like a giant camera, comprised of nuclear track detectors - analyzed offline by ultra fast scanning microscopes - sensitive only to new physics. Second, it is uniquely able to trap the particle messengers of physics beyond the Standard Model for further study. MoEDAL's radiation environment is monitored by a state-of-the-art real-time TimePix pixel detector array. The possibility of adding a new MoEDAL subdetector sensitive to mQPs is being studied.

In 2015 the LHC will restart operations at the unparalleled energy of nearly 7 TeV beam energy and the enormous collision rate of around a billion collisions per second. The LHC has been compared to an immense microscope allowing us to peer deeper into the Terascale quantum universe than ever before, to explore the inner structure of matter at a scale down to $\sim 10^{-19}$ m. The LHC will recreate the conditions that occurred when the Universe was a picosecond old providing an unprecedented laboratory for the study of the cosmology of the nascent Universe at the earliest times. In parallel, non-accelerator experiments are also exploring the Terauniverse via high-energy astrophysics.

At present it looks very much as though the Higgs boson - the last piece of the Standard Model - has been discovered. But, the LHC is only just getting started. The quest for the deeper underlying theory has now begun in earnest that brings many fundamental questions to the fore. Are there new symmetries of nature? Are there extra spatial dimensions? Is there a deeper substructure? Does magnetic charge exist? Why is gravity so weak compared to the other fundamental forces? What is the nature of dark matter? What was the physics of the earliest era of the universe? The list goes on.

The MoEDAL experiment has a huge discovery potential designed to expand the physics horizons of the LHC and provide paradigm-shifting insights into foundational questions in a complementary way to the other LHC experiments due to its enhanced sensitivity to the *highly* ionizing avatars of new physics such as magnetic monopoles or massive (pseudo-)stable charged particles. We are investigating the possibility to further enhance the complementary extension of the LHC's physics reach by incorporating a new sub-detector to search for *minimally* ionizing particles. The unprecedented but complementary nature of the MoEDAL detector and its systematics – exemplified by its ability to retain a permanent record, and even capture new particles for further study - will make MoEDAL an invaluable asset in the elucidation of any Terascale Beyond the Standard Model (BSM) scenario covered by its extensive physics repertoire.

References

[1] J. L. Pinfold, MoEDAL Collaboration, *Technical Design Report of the MoEDAL Experiment* MoEDAL Collaboration Jun 8, 2009 CERN-LHCC-2009-006, MoEDAL-TDR-001.
[2] Large Hadron Collider Official Site: http://home.web.cern.ch/topics/large-hadron-collider
[3] G. Aad *et al.* [ATLAS Collaboration] 2008 JINST **3** S08003.
[4] S. Chatrchyan *et al.* [CMS Collaboration] 2008 JINST **3** S08004.
[5] K. Aamodt *et al.* [ALICE Collaboration] 2008 JINST **3** S08002
[6] A. A. Alves, Jr. *et al.* [LHCb Collaboration] 2008 JINST **3** S08005.
[7] G. Anelli *et al.* [TOTEM Collaboration] 2008 JINST **3** S08007.
[8] O. Adriani *et al.* [LHCf Collaboration] 2008 JINST **3** S08006.
[9] B. Acharya *et al.*, MoEDAL Collaboration 2014 *Int. J. Mod. Phys.* A**29,** 1430050.
[10] B. Cabrera 1982 Phys. Rev. Lett. **48** no. 20 1378.
[11] A.A. Prinz, R. Baggs, J. Ballam, S. Ecklund, C. Fertig, J.A. Jaros, K. Kase, A. Kulikov, Willem G. J. Langeveld, R. Leonard *et al.*, 1998 *Phys. Rev. Lett.* **81** 1175.
[12] A. de Roeck *et al.*, 2012 Eur. Phys. J. **C72**.
[13] K. Kimm, J. H. Yoon and Y. M. Cho 2015 *Eur. Phys. J.* **C75** 67
[14] P. A. M. Dirac, 1931 *Proc. Roy. Soc. Lond.* **A133** 1931 60.

[15] J. S. Schwinger 1969 *Science* **165** 757.
[16] S. L. Glashow 1961 *Nucl. Phys.* **22** no. 4 579; S. Weinberg 1967 *Phys. Rev. Lett.* **19** 1264; A. Salam 1968 *Weak and electromagnetic interactions, in Elementary particle physics: relativistic groups and analyticity*, N. Svartholm, ed., p. 367. Almqvist & Wiskell, Proceedings of the eighth Nobel symposium; G. 't Hooft and M. Veltman 1972 *Nucl. Phys.* **B44** 189.
[17] G. 't Hooft 1974 *Nucl. Phys.* **B79** 276.
[18] A. M. Polyakov 1974 *JETP Lett.* **20** 194.
[19] H. Georgi and S. Glashow 1974 *Phys. Rev. Lett.* **32** 438.
[20] G. Aad ATLAS Collaboration 2012 *Phys. Lett.* **B716** 1; S. Chatrchya *et al.* CMS Collaboration 2012 *Phys. Lett.* **B716** 30.
[21] S.P. Martin 1997 hep-ph/9709356; X. Tata 1997 hep-ph/9706307.
[22] K. Hamaguchi, M. M. Nojiri and A. de Roeck 2007 JHEP 0703 046.
[23] H. Goldberg and L. J. Hall 1986 *Phys. Lett.* **B174** 151; A. De Rujula, S. L. Glashow and U. Sarid 1990 *Nucl. Phys.* **B333**, 173; S. Dimopoulos, D. Eichler, R. Esmailzadeh and G. D. Starkman 1990 *Phys. Rev.* **D41**, 2388.
[24] R. Foot, G. C. Joshi, H. Lew and R. R. Volkas *1990 Mod. Phys. Lett.* **A5**, 95 (1990) [Erratum-ibid. 1990 A **5**, 2085]; R. Foot, H. Lew and R. R. Volkas 1993 *J. Phys.* **G19** 361 [Erratum-ibid. 1993 G **19**, 1067]; S. Davidson, S. Hannestad and G. Raffelt 2000 *JHEP* **0005**, 003.
[25] B. Holdom 1986 *Phys. Lett.* **B166** 196.
[26] H. Vogel and J. Redondo, 2014 JCAP **1402** 029.
[27] S. Davidson, S. Hannestad, and G. Raffelt 2000 *JHEP* **0005** 003.
[28] A. Prinz, R. Baggs, J. Ballam, S. Ecklund, C. Fertig, et al. 1998 Phys. Rev. Lett. **81** 1175.
[29] A. Badertscher *et al.* 2007 *Phys.Rev.* **D75**, 032004.
[30] S. Chatrchyan et al. (CMS Collaboration) 2013 *Phys. Rev.* **D87** 092008.
[31] S. Davidson, B. Campbell, and D. C. Bailey 1991 *Phys. Rev.* **D43**, 2314.
[32] R. Mohapatra and I. Rothstein 1990 *Phys. Lett.* **B247** 593.
[33] S. Davidson and M. E. Peskin 1994 *Phys. Rev.* **D49**, 2114.
[34] A. Dolgov, S. Dubovsky, G. Rubtsov, and I. Tkachev 2013 *Phys. Rev.* **D88** 117701.
[35] A. Haas *et al.* 2014 [hep-ph] arXiv.1410.6816v1.

Discrete Symmetries CP, T, CPT

J Bernabeu

Department of Theoretical Physics, University of Valencia,
IFIC, Joint Centre Univ. Valencia-CSIC, E-46100 Burjassot, Spain

E-mail: jose.bernabeu@uv.es

Abstract. CP-Violation (CPV) was discovered in 1964 in the weak decays of neutral Kaons. This Symmetry Breaking is understood in the Standard Model as a consequence of its Particle Content, codified in the 3×3 Cabibbo-Kobayashi-Maskawa Unitary Mixing Matrix. The measured CPV asymmetries for Kaons and B's are well described by this Mechanism, but the current level of experimental accuracy and theoretical uncertainties leaves room for additional sources of CPV, as demanded by Baryogenesis. A direct Time-Reversal-Violation (TRV) Effect was clearly observed in 2012 in the time evolution of neutral B-mesons. The conceptual basis to bypass the irreversibility of decays, and prepare in the B-Factories both the Reference and the T-reverse transitions, is provided by the quantum properties of Entanglement and the Decay as a Filtering Measurement. Flavour and CP eigenstate decay channels allow a separate independent measurement of CP, T and CPT asymmetries. Prospects for extending such a programme to any pair of B-decays and for Kaons in DAPHNE are discussed. The interest in measuring CPTV asymmetries in transitions is emphasized. This includes the search for CPT-breaking as well as the case of not having a well-defined CPT operator.

1. Introduction

Symmetries have been an essential ingredient in the understanding of the physical laws of Nature. The assumption of the form invariance of the dynamical equations under a symmetry transformation of the physical magnitudes leads to observable consequences, with regularities, conservation laws and invariant observables that act as guiding components for the dynamics. In the case of local gauge symmetries, the requirement of the symmetry leads to the determination of the dynamics. Even more interesting, Symmetry Breaking through a definite mechanism is a source of new phenomena and new physics. In this talk I will concentrate on the Discrete Symmetries CP, T and CPT.

In 2014 we have celebrated 50 years of CP-Violation (CPV). On July 10th 1964 Cronin, Fitch, Christenson and Turlay submitted a paper to Phys. Rev. Lett. "Evidence for the $\pi\pi$ decay of the KL meson" [1], announcing the discovery of CPV in the weak decays of neutral Kaons. For this discovery, the Nobel Prize in Physics 1980 was awarded to James Cronin and Val Fitch. The CP-Symmetry Breaking is understood as a consequence of the particle content in the Standard Model (SM). In 1973 Kobayashi and Maskawa published the paper "CP-Violation in the Renormalized Theory of Weak Interaction" in Progress of Theoretical Physics [2] and they shared one half of the Nobel Prize in Physics 2008 "for the discovery of the origin of the broken symmetry which predicts the existence of at least three families of quarks in nature". The direct evidence of Time-Reversal-Violation (TRV) was clearly established in the time evolution of neutral B-mesons by the BABAR Collaboration and in 2012 the paper "Observation of Time-Reversal Violation in the B^0 Meson System" was published in Phys. Rev. Lett. [3], following the lines of the conceptual basis for entangled neutral mesons in the B-factories discussed in the paper [4] "CP, T and CPT versus temporal asymmetries for entangled states of the B_d-system". Up to now all CPT-Violation searches have given negative results, with the best limits coming from the mass difference between K^0 *and* $\bar{K}^0$ [5].

In Section 2 we discuss the fact that CPV asymmetries are well described by the Cabibbo-Kobayashi-Maskawa Quark Mixing Matrix Mechanism, but the current level of experimental accuracy and theoretical uncertainties leave room for additional sources of CPV, as required by Baryogenesis in the Universe. We identify potential transitions in B-physics, using Flavour-Changing-Neutral. Current (FCNC) processes and CPV asymmetries, able to incorporate virtual contributions of New Physics. Section 3 is devoted to TRV concepts and results: what is "Time Reversal" in classical and quantum mechanics, the NO-GO argument for its search with unstable particles, its By-pass using Entanglement and the Decays as Filtering Measurements for entangled $B^0 - \bar{B}^0$ system in the B-Factories, and $K^0 - \bar{K}^0$ system in the Φ-Factory, the role of time-ordered decay channels to Flavour and CP-eigenstates for disentangling genuine separate independent asymmetries for CP, T and CPT and the 14σ observation of TRV by the BABAR experiment. Section 4 emphasizes the interest in the search of CPTV for transitions and not only in the expectation values of masses and lifetimes of elementary particles. We distinguish explicit CPT symmetry breaking mechanisms from physical scenarios in which the CPT- operator implementing the symmetry is ill-defined. For entangled systems this alternative leads to the ω-effect, a component of the wrong symmetry in the coherent state of neutral mesons. Finally, Section 5 presents our results, conclusions and prospects.

2. CP Violation

It is well known since 1957 that weak interactions have little respect for symmetries. That year space inversion (parity, P) symmetry was discovered to be broken in β decays [6-8]. Then, there was the hope that the combination of P with charge conjugation (CP) was a good symmetry. But just a few years later, in 1964, there was discovered a small but unambiguous violation of the CP symmetry in K meson decays [1,9]. More recently, in 2001, the B factory experiments BABAR and Belle, observed that CP is violated in B mesons [10,11].

The now well established CP violation in the quark sector can be successfully accommodated within the Standard Model (SM) of particles and fields through the three-family Cabibbo-Kobayashi-Maskawa (CKM) quark-mixing mechanism [12,2]. It describes the coupling of the W boson to up and down quarks and conveys the fact that the quarks with definite properties under charged-current weak interactions are linear combinations of the quark mass eigenstates [13]. For three families, the unitarity

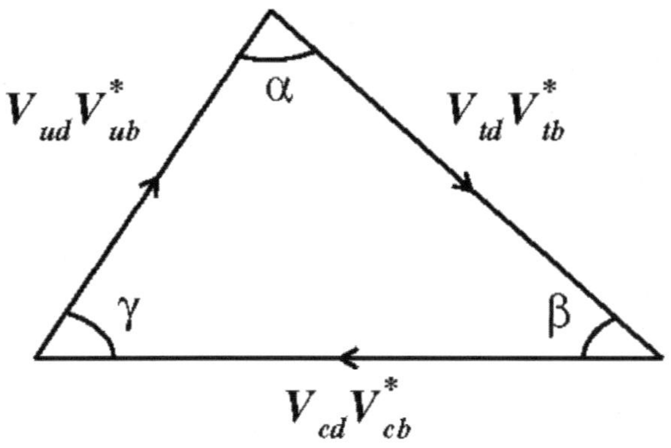

Figure 1. The bd unitarity triangle representing the CKM unitarity conditions. The three sides are determined from semileptonic and non-leptonic B decays, including $B^0 - \bar{B}^0$ oscillations. Since they are of comparable length the angles are sizeable and one expects large CP asymmetries in B decays in the SM. There are other two triangles which almost collapse to a line. This gives an intuitive understanding of why CP violation is small in the leading K decays (ds triangle) and in the leading B_s decays (bs triangle).

conditions of the quark-mixing matrix V are represented by triangles in the complex plane, as illustrated in figure 1, and lead to four fundamental parameters: three magnitudes and one single irreducible phase.

In the Wolfenstein parameterization [14] we can write V to $O(\lambda_c^4)$ as

$$V = \begin{bmatrix} 1-\lambda_c^2/2 & \lambda_c & A\lambda_c^3(\rho - i\eta) \\ -\lambda_c & 1-\lambda_c^2/2 & A\lambda_c^2 \\ A\lambda_c^3(1-\rho-i\eta) & -A\lambda_c^2 & 1 \end{bmatrix} \quad (1)$$

where the matrix line runs over d, s, and b quarks, and $\lambda c \approx 0.226$, $A \approx 0.814$, $\varrho \approx 0.135$ and $\eta \approx 0.34931$.

Extensive tests of the CKM mechanism using all experimental data show a high degree of consistency [15]. Historically, Kobayashi and Maskawa extended in 1973 the 2×2 Cabibbo mixing matrix to 3×3 to explain the CP-violation discovered nine years before, thus anticipating the existence of the third family of quarks, quickly confirmed with the discovery of the τ lepton in 1975 [16] and of the fifth quark, the b, two years later [17].

At the asymmetric B factories, electron and positron beams collide with high luminosity at a c.m. energy of 10.58 GeV corresponding to the mass of the Υ(4S) resonance, a vector particle with $J^{PC}= 1^{--}$. The Υ(4S) is a bound state of a b and a $\bar{b}$ quark, that decays exclusively to a pair of B and $\bar{B}$ mesons. Since the mass of the Υ(4S) is only slightly higher than twice the mass of the B meson, the two B mesons have low momenta (about 330 MeV/c) and are produced almost at rest in the Υ(4S) reference frame with no additional particles besides those associated to the B decays. The energy of the electron beam is adjusted to be between twice and three times larger than that of the positrons, so that the c.m. frame has a Lorentz boost along the collision axis. Two B factory colliders, PEP-II at SLAC in California and KEKB at KEK in Japan, with their corresponding detectors, BABAR [18,19] and Belle [20], have been operating during the last decade, accumulating an integrated luminosity of data exceeding 500 fb^{-1} and 1 ab^{-1}, respectively.

To the B factory programme we add in the last few years the LHCb experiment at CERN [21] with new and complementary information about rare decays, FCNC processes and CP Violation.

For the B_d system CPV in the mixing is very small. The interference of mixing and no-mixing amplitudes in the B^0 decays is observed in decay products which are accessible by both B^0 and $\bar{B}^0$. The corresponding CPV asymmetry can be written as

$$A_f(t) = S_f \sin(\Delta m t) - C_f \cos(\Delta m t) \quad (2)$$

where $\Delta\Gamma = 0$ is assumed. The most precise asymmetries are measured in the tree-dominated b → ccs transitions, such as $B^0 \to \psi K^0$, and are given by [22] (we use K^0 throughout to denote results that combine K_S and K_L modes): $S\psi K^0 = +0.682 \pm 0.019$. The penguin contributions are very small, so that one has the interpretation in the SM $S = \sin(2\beta)$, C=0.

CP violation in the interference of mixing and decay in the $B^0 \to \pi^+ \pi^-$ mode is given by [22] $S\pi^+\pi^- = -0.66 \pm 0.06$. It is interpreted in the SM as $S = \sin(2\alpha)$.

For the B^0 system, the phase of the mixing amplitude is determined from the intermediate top-quark exchange in the box diagram, so that the SM interpretations of these interferences in terms of the β and α CP angles of the unitarity triangle are apparent. On the contrary, the phase γ in the unitarity triangle involves the interference of the sides for decays with charm and up quark constituents, without any relation to the mixing. The CP angle γ is thus a measure of Direct CPV. Its measurement has been undertaken by BABAR, BELLE and LHCb with the decay $B^+ \to D K^+$ and other related transitions. The extraction of γ needs a detailed analysis involving in addition the presence of strong phases associated to final-state hadronic interactions.

An ideal experiment would be one in B factories using Entanglement and detecting the pair of decays $\rightarrow \psi\, K^0$ and $\rightarrow \pi\pi$ at equal times without any effect of the mixing phase. It remains to be seen whether this gedanken experiment can become a real experiment in the upgraded SuperBELLE.

The rare decay $B_s \rightarrow \mu\mu$ has been observed by LHCb and CMS. Based on the presence of a FCNC penguin amplitude induced by Z-exchange as seen in the diagram

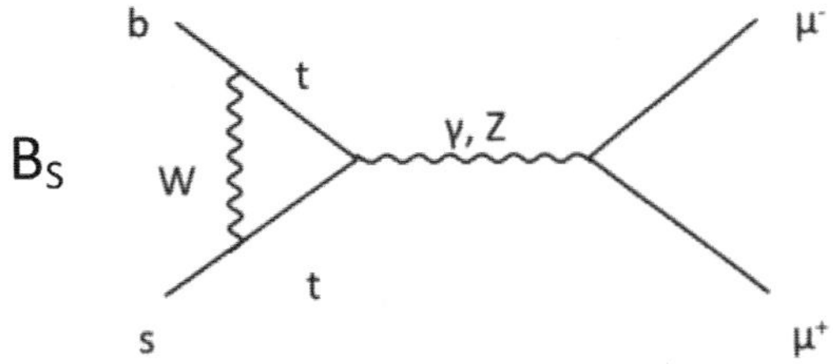

one could expect a priori New Physics virtual effects induced by Non-Decoupling of longitudinal contributions. The result [23] presented in figure 2 has represented the latest disappointment of the scientific community.

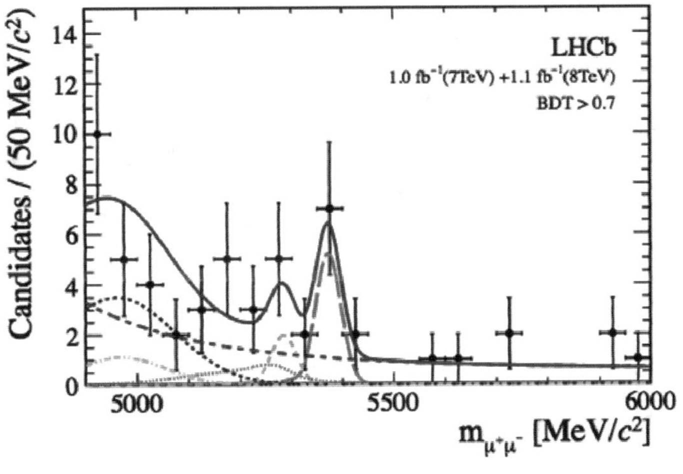

Figure 2. The invariant mass distribution of the $\mu\mu$ system showing the Bs-peak.

The experimental Branching Ratio is given by

$$B(B_s^0 \rightarrow \mu^+ \mu^-) = \left(3.2^{+1.4}_{-1.2}(stat)^{+0.5}_{-0.3}(syst)\right) x\ 10^{-9}$$

in perfect agreement with the SM value $(3.23 \pm 0.27)x10^{-9}$. In some models, like SuperSymmetry, the result could have been orders of magnitude different from the SM value. One has to be aware, however, that in this process the FCNC penguin is projected to a pseudoscalar, so that its contraction with the leptonic vertex leads to helicity suppression and an amplitude proportional to the mass of the lepton. The FCNC b-s penguin vertex can be probed under more general conditions by searching for the processes $B \rightarrow K(K^*)\, l^+ l^-$, which open new effective current operators in scalar-scalar and scalar-vector matrix elements. In fact, the process has been observed and the analysis, using the experimental results of LHCb, making use of the OPE formalism leads to intriguing tensions with the SM expectations in some of the effective operators [24]. If these discrepancies are associated to longitudinal amplitudes of the mediators, one should seriously consider the search of the process

$B \to K(K^*)\nu\nu$ in order to avoid the amplitudes mediated by γ-exchange. As mentioned before, the hadronic vertex for these transitions leads to an amplitude which is not proportional to the neutrino mass.

Non-Decoupling from longitudinal contributions which are not giving an helicity-suppressed amplitude can also be obtained from the process $B_s \to \nu\nu\gamma$, as shown in the diagram

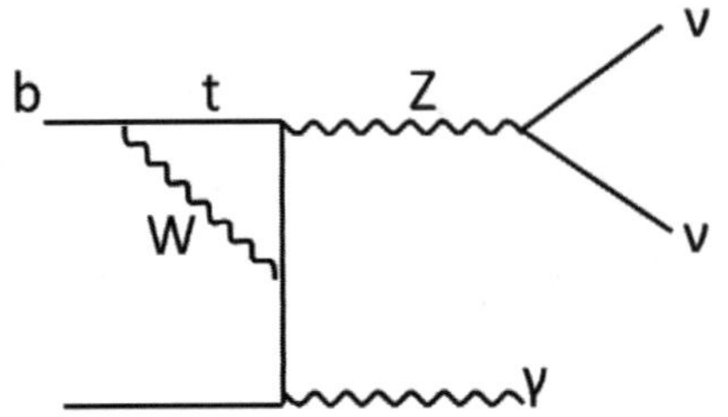

An existing upper limit [25] by BABAR gives $Br < 4.7 \times 10^{-5}$, 90%.

3. Time-Reversal Violation

The symmetry transformation that changes a physical system with a given sense of the time evolution into another with the opposite sense is called Time-Reversal T. It corresponds to changing the sign of the velocity vector **v** or the momentum **p**, without changing the position **r**. In the dynamical equations of motions, or their solutions, such a transformation corresponds formally to replacing t by −t. The T transformation changes the sign of other dynamical variables such as angular momentum. For fields, the magnetic field changes its sign under time reversal, whereas the electric field does not.

The time-reversal transformation in classical mechanics corresponds to substitute for each trajectory **r**(t) the trajectory **r**(−t), i.e. to moving along the given trajectory with the opposite velocity at each point, as illustrated in figure 3. It is not obvious that the dynamics remains invariant under this T transformation. If the original trajectory is dynamically possible, d**p**/dt = **F** with a force **F** depending on the sense (sign) of the velocity leads to a violation of T invariance. This observed violation would be a fake violation if there is an external agent which is not T-invariant, like friction in general or a magnetic field for charged particles.

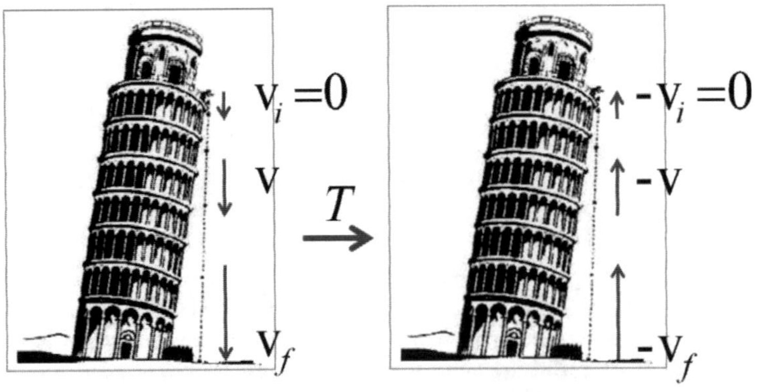

Figure 3. (a) Trajectory of a stone falling from the leaning tower.
(b) Trajectory after time-reversal transformation.

In quantum mechanics, Wigner's time-reversal transformation [26],

$$\psi(t) \rightarrow T\psi(t) \equiv \psi_T(t) = \psi^*(-t) \tag{3}$$

keeps the Schrodinger equation, $i\hbar\partial\psi(t)/\partial t = H\psi(t)$, invariant under a T transformation if the Hamiltonian H is real. This has three fundamental consequences:
First, the T operator is antiunitary. This property can be seen, for example, evaluating the scalar product of two states,

$$\langle\psi_T(t)|\phi_T(t)\rangle = \langle\psi(-t)|\phi(-t)\rangle^* = \langle\phi(-t)|\psi(-t)\rangle \tag{4}$$

Thus, time reversal has to do with interchange of bra and ket states. Second, the complex conjugation implies that time reversal does not have observable and conserved eigenvalues. Third, for a plane wave with momentum **p**, $\psi(\mathbf{r};t) = \exp[i(\mathbf{p}\cdot\mathbf{r} - Et)/\hbar]$, the time-reverse wavefunction is $\psi^*(\mathbf{r},-t) = \exp[i(-\mathbf{p}\cdot\mathbf{r} - Et)/\hbar]$, i.e. the T-transformed function describes a particle with momentum **-p** and energy E, thus it is not necessary to interpret the transformed function as a particle going backwards in time. For this reason the T transformation is often referred to as "motion reversal" rather than "time reversal".

The T transformation is implemented in the space of states by the antiunitary operator U_T in such a way that, for spinless particles,

$$U_T \boldsymbol{r} U_T^\dagger = \boldsymbol{r}, \qquad U_T \boldsymbol{p} U_T^\dagger = -\boldsymbol{p} \tag{5}$$

and $\psi_T(t) = U_T\,\psi(-t)$. Equation (5) guarantees the invariance of the commutation rule between **r** and **p**, thus we might say that T transforms quantum mechanics into quantum mechanics. For a Hamiltonian H invariant under time reversal, $[H;U_T] = 0$, the time-evolution operator $\mathcal{U}(t;t_0)$ transforms as

$$U_T\mathcal{U}(t,t_0)U_T^\dagger = \mathcal{U}^\dagger(t,t_0) \tag{6}$$

The antiunitary character of U_T allows to write $U_T = UK$, where U is unitary $(U^{-1} = U^\dagger)$ and K is an operator which complex conjugates all complex numbers. For the matrix elements of time-dependent transitions we have

$$\langle f|\mathcal{U}(t,t_0)|i\rangle = \langle f|U_T^\dagger U_T\mathcal{U}(t,t_0)U_T^\dagger U_T|i\rangle = \langle U_T f|\mathcal{U}^\dagger(t,t_0)|U_T i\rangle^* = \langle U_T i|\mathcal{U}(t,t_0)|U_T f\rangle, \tag{7}$$

where time-reversal invariance is assumed in equation (7). As a consequence, the comparison between $i \rightarrow f$ and $U_T f \rightarrow U_T i$ transitions is a genuine test of this invariance. It is because of these special properties that the role of time reversal is distinct from that of any other symmetry operation in physics, and makes its experimental investigation significantly more difficult than other symmetries.

A direct consequence of quantum dynamics is the negative-exponential time behavior of the decay of any unstable system into two or more particles, as given by the Fermi golden rule. The reversal of the exponential decay law reveals that the T transformation is not defined for a decaying state [27], thus it appears that the decay prevents proofs of motion reversal. The decay is irreversible, and it behaves like the arrow of time explained by the increase of entropy. The TRV asymmetry should not contain the decay as an essential ingredient.

There are two main types of experiments or observables that can be used to detect directly time-reversal non-invariance [28-30]. First, a non-zero expectation value of a T-odd operator for a non-degenerate stationary state. This is the case for an electric dipole moment (e.d.m.) of a particle with spin, which is also a P-odd, C-even quantity. Thus, if either parity or time reversal are good symmetries, the particle cannot have an e.d.m. A non-zero e.d.m. can be generated by either strong interaction T violation, unless it is annulled by a Peccei-Quinn symmetry leaving the axion as remnant [31], or by T violation in weak interactions. In the SM with the CKM mechanism, a non-vanishing e.d.m. of the neutron only appears at three-loop level. Hence, these experiments probe for physics beyond the SM. To date, no signals for e.d.m. have been found, although there are strong limits, as for the neutron and the electron, $|d_n| = \langle 2.9 \times 10^{-26} e - cm\ and\ |d_e| = \langle 1.05 \times 10^{-27}\ e - cm$[32].

Second: in transitions, with the requirement of exchanging initial and final states to compare the T-reverse transition with the Reference transition. We might consider this exchange in the mixing of the pseudoscalar neutral K, B and D mesons. In this case one compares the probability of a flavor eigenstate (say) K^0 transforming into a $\overline{K}^0$, and viceversa. Since the states K^0 and $\overline{K}^0$ are particle and antiparticle, the two transitions are connected by both T and CP transformations. Even if CPT symmetry would be broken, there exists no difference between CP and T in this case. Thus the two symmetry transformations are experimentally identical and lead to the same asymmetry. This flavor-mixing or Kabir asymmetry [33] is independent of time since the two processes have identical time dependence, and it is induced by the interference between the dispersive, M_{12}, and absorptive, Γ_{12}, contributions in the effective Hamiltonian. Evidence at 4σ level for this asymmetry was found by the CPLEAR experiment [34] at CERN in 1998 for $K^0 \rightarrow \overline{K}^0$ in p p collisions. The strangeness (strange and antistrange flavor content) of the K^0 and $\overline{K}^0$ mesons at production time was determined by the charge of the accompanying charged kaon. Since weak interactions do not conserve strangeness, the K^0 and $\overline{K}^0$ may subsequently transform into each other. The strangeness of the neutral kaon at decay time is determined through the semileptonic decay $K^0 \rightarrow e^+ \pi^- \nu$. The asymmetry, shown in figure 4, is effectively independent of time, i.e., not built in the time evolution, and reveals a net value with respect to zero.

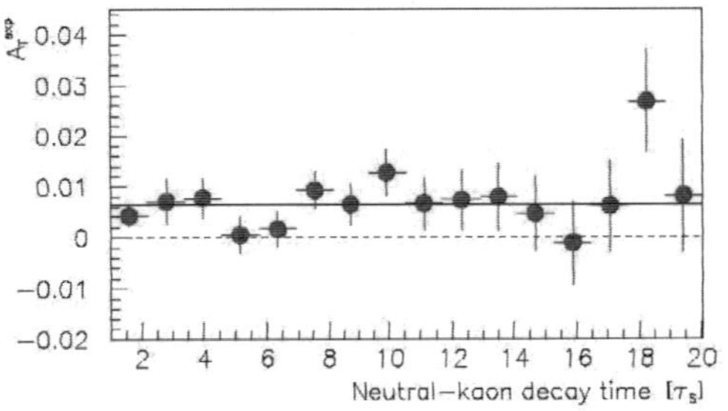

Figure 4. The CPLEAR asymmetry versus the neutral-Kaon decay time (in units of the KS lifetime) [34]. The solid line represents the average.

The interpretation of the asymmetry relies on two main aspects. Firstly, in the framework of the Weisskopf-Wigner approach [35,36], the effect comes from the overlap (non-orthogonality) of the "stationary" K_S and K_L states of definite mass and lifetime. Secondly, the decay plays an essential role; indeed, the dispersive and absorptive contributions to $K - \overline{K}^0$ mixing are at leading order proportional to the mass and decay width differences between K_L and K_S, respectively. The presence of the decay as an initial state interaction, essential to construct a non-vanishing interference for this observable, has been argued by Wolfenstein to claim that this asymmetry "is not as direct a test of time-reversal violation as one might like" [27,28]. In the neutral B-system, where the decay width difference between the B^0 mass eigenstates is negligible, the measurement of the asymmetry has, in fact, brought negative results [37,38]. Other authors, however, have argued that its interpretation as a genuine signal for T-violation does not get affected by these arguments [39-41].

A direct evidence for TRV would mean an experiment which, by itself, is able to establish a non-vanishing genuine TRV asymmetry independent of CPV or CPT invariance. The problem is then the filtering of definite initial and final states of the neutral meson for the Reference and T-reverse transitions.

The solution [4,42-45] arises from the quantum mechanical properties imposed by the EPR entanglement [46,47] between the two neutral B mesons produced in the Υ(4S) resonance decay. As the individual state of each meson in the entangled system is not defined before the first decay, just as one B meson in the entangled pair is prepared in the definite flavour B^0 or $\bar{B}^0$ states at the time when the other B is projected into its orthogonal by the observation of a semileptonic decay, the first decay of one B into the definite CP final states J/ψ K_S or J/ψ K_L prepares the other living B into well defined, orthogonal linear combination B_- or B_+ of B^0 and $\bar{B}^0$ states. In fact, this idea offers the opportunity to explore separately time reversal T, CP and CPT asymmetries, selecting appropriately different transitions defined by different decay channels.

The entangled state of the two mesons produced by the Υ(4S) decay can be written in terms of any pair of orthogonal states of the individual B mesons as

$$|\Upsilon\rangle = \frac{1}{\sqrt{2}}\left[|B^0(t_1)\rangle|\bar{B}^0{}_-(t_2)\rangle - |\bar{B}^0{}_-(t_1)\rangle|B^0(t_2)\rangle\right] \quad (8)$$
$$= \frac{1}{\sqrt{2}}\left[|B_+(t_1)\rangle|B_-(t_2)\rangle - |B_-(t_1)\rangle|B_+(t_2)\rangle\right]$$

Its time evolution, including mixing, keeps only, by virtue of the antisymmetry, terms $B^0, \bar{B}^0$ or B_+, B_-. What is T-transformation experimentally? As illustrated in figure 5, suppose that the Reference transition is defined by the time-ordered decay channels l^+ first, J/ψ K_S later, as shown in the left-hand side of figure 5. The use of Entanglement plus the Decay as a Filtering Measurement tells us that the meson transition corresponds to $\bar{B}^0 \to B_-$. In terms of meson states, the T-reverse transition is then $B_- \to \bar{B}^0$ and the question arises: Which are the time-ordered decay channels which correspond to this T-reverse transition? For definite flavour and CP eigenstates, orthogonality of $B^0, \bar{B}^0$ and B_+, B_- provides the solution: J/ψ K_L first, l^- later.

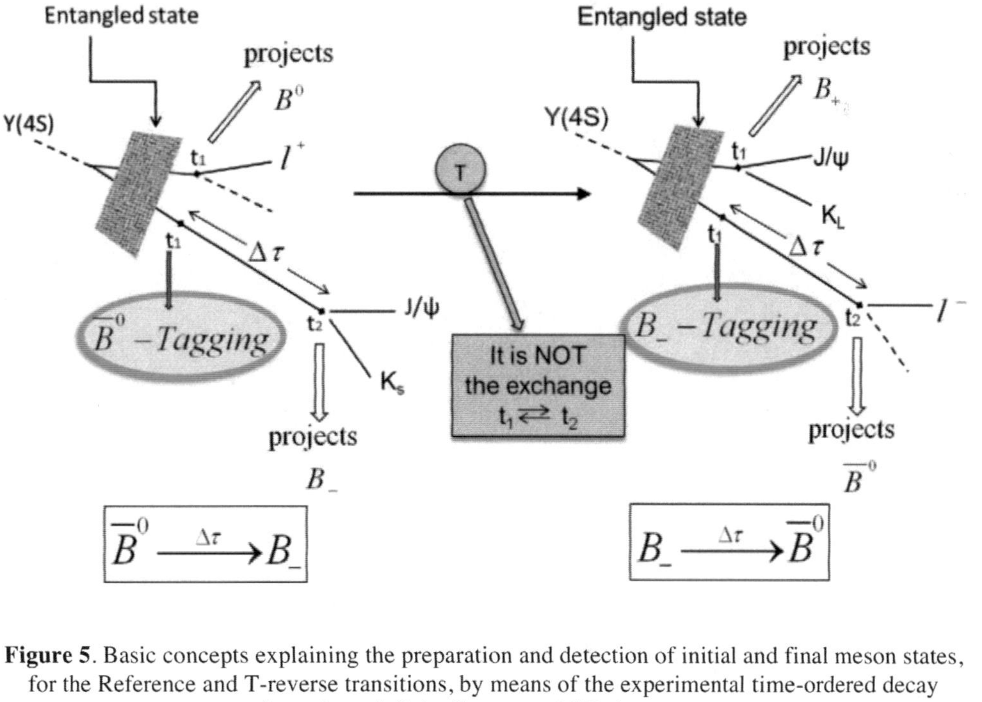

Figure 5. Basic concepts explaining the preparation and detection of initial and final meson states, for the Reference and T-reverse transitions, by means of the experimental time-ordered decay channels to definite flavour and CP eigenstates.

There are eight processes of this kind: 2 for flavour × 2 for CP × 2 time-orderings. Each of them has a time dependent probability distribution of the following form

$$I_i(\Delta t) \sim e^{-\Gamma\Delta t}\{C_i\, cos(\Delta m\Delta t) + S_i sin(\Delta m\Delta t)\} \qquad (9)$$

with the only use of quantum mechanics, and $\Delta\Gamma = 0$ for the neutral B_d mesons.

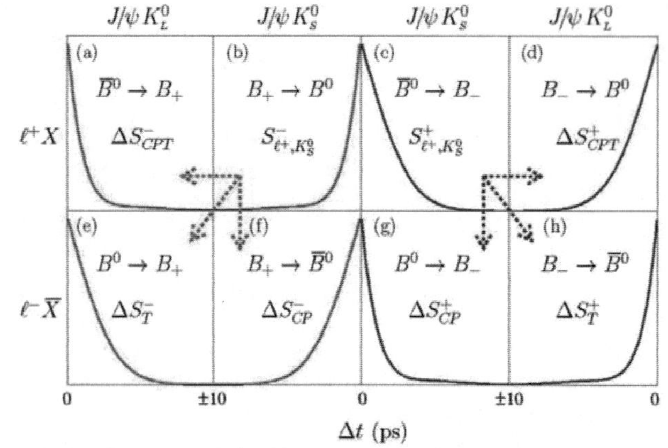

Figure 6. Time-dependent probability distributions for the eight time-ordered event classes associated to flavour-CP eigenstate decay channels, shown as a planar map. The independent CP, T and CPT asymmetries from a given Reference transition are apparent.

BABAR reported in November 2012 the measurement [3] of these eight time-ordered event classes. From the eight pairs (S,C) of signal coefficients, one might construct two sets of three pairs each of independent asymmetry parameters, $(\Delta S_T^{\pm};\ \Delta C_T^{\pm}), (\Delta S_{CP}^{\pm};\ \Delta C_{CP}^{\pm})\ and\ (\Delta S_{CPT}^{\pm};\ \Delta C_{CPT}^{\pm})$. They are given in the left panel of Table 1,

Table 1.

Parameter	Final result	SM expected val.
ΔS_T^+	$-1.37 \pm 0.14 \pm 0.06$	-1.36
ΔS_T^-	$1.17 \pm 0.18 \pm 0.11$	1.36
ΔC_T^+	$0.10 \pm 0.14 \pm 0.08$	0.
ΔC_T^-	$0.04 \pm 0.14 \pm 0.08$	0.
ΔS_{CP}^+	$-1.30 \pm 0.11 \pm 0.07$	-1.36
ΔS_{CP}^-	$1.33 \pm 0.12 \pm 0.06$	1.36
ΔC_{CP}^+	$0.07 \pm 0.09 \pm 0.03$	0.
ΔC_T^-	$0.08 \pm 0.10 \pm 0.04$	0.
ΔS_{CPT}^+	$0.16 \pm 0.21 \pm 0.09$	0.
ΔS_{CPT}^-	$-0.03 \pm 0.13 \pm 0.06$	0.
ΔC_{CPT}^+	$0.14 \pm 0.15 \pm 0.07$	0.
ΔC_{CPT}^-	$0.03 \pm 0.12 \pm 0.08$	0.
$S^+_{\ell^+,K_S^0}$	$0.55 \pm 0.09 \pm 0.06$	0.68
$S^-_{\ell^+,K_S^0}$	$-0.66 \pm 0.06 \pm 0.04$	-0.68
$C^+_{\ell^+,K_S^0}$	$0.01 \pm 0.07 \pm 0.05$	0.
$C^-_{\ell^+,K_S^0}$	$-0.05 \pm 0.06 \pm 0.03$	0.

The Reference transitions are taken by convention $B_+ \rightarrow B^{\,0} and\, B^{\,0} \rightarrow B_-$. This choice has the advantage that the breaking of time-reversal symmetry would directly manifest itself through any nonzero value of $\Delta S_T^{\pm}$ or any difference between $\Delta S_{CP}^{\pm}$ and $\Delta S_{CPT}^{\pm}$. It follows that the experimental results given in Table 1 demonstrate an unambiguous, direct evidence of time-reversal violation in the time evolution of neutral B mesons, obtained through motion reversal in transitions that are not CP conjugate to each other. The reported TRV asymmetries do not need the decay as an essential ingredient and originate in the interference between mixing and no-mixing amplitudes in the time evolution of the system. The asymmetry parameters $\Delta S_T^{\pm}$ are independent of the value of $\Delta\Gamma$.

The experimental results are, on the other hand, compatible with all C-coefficients vanishing. This a proof that the decay amplitudes have the same absolute values for particle and antiparticle, implying a single weak phase and no direct CP-violation in the decay. A recent analysis [48] of the conditions under which the measured motion reversal asymmetry is a true TRV effect reinforces the adequacy of the method. The comparison of the experimental results with the right-handed panel of Table 1 tells us the ability of the SM for reproducing all observations.

The four TRV independent raw symmetries measured by the BABAR experiment [3] are given in figure 7

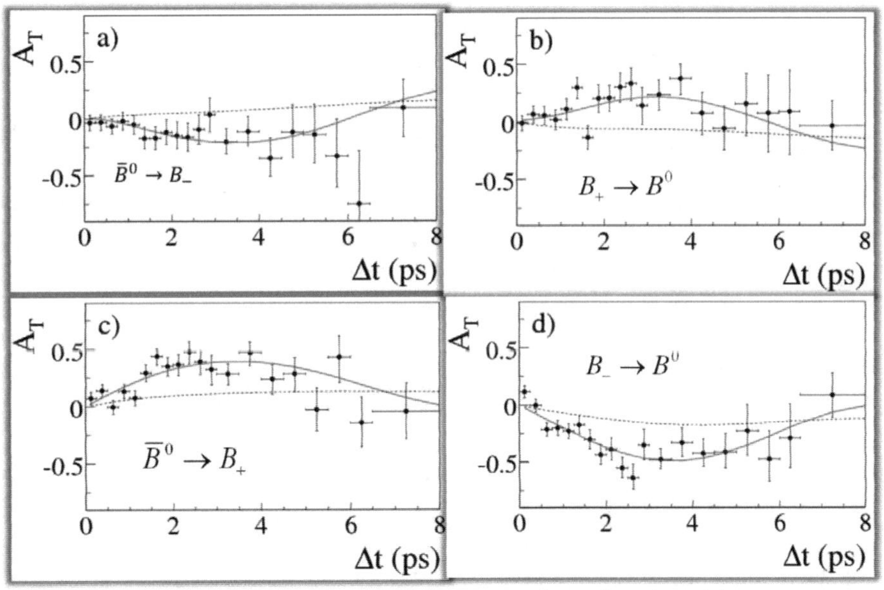

Figure 7. Experimental TRV asymmetries as function of Δt

They correspond to the asymmetries between the transitions $\bar{B}^0 \rightarrow B_-$, $B_+ \rightarrow B^{\,0}$, $\bar{B}^0 \rightarrow B_+$, $B_- \rightarrow B^0$, and their T-reverse transitions. The Δt dependence demonstrates that the TRV effect is built during the time evolution of the neutral B. The combined significance of the four measured non-vanishing asymmetries provides a conclusive 14σ TRV effect.

This discovery was made possible thanks to the spectacular quantum properties of the Einstein-Podolsky-Rosen entangled state. In the epistemological language for the nature of Reality one could say that "The reality of two entangled neutral B's is much more than the sum of the two separate neutral B-meson local realities". The appropriate preparation of initial and final meson estates, as required by a genuine measurement of a TRV effect, is based on: 1) Entanglement to transfer the information from the first decaying meson to its living partner; 2) the Decay as a Filtering Measurement for the meson state.

The conceptual basis which is at the origin of the TRV effect for neutral B mesons has been adapted [49] to the treatment of entangled neutral Kaons at the Φ-factory DAΦNE. The expectations for a measurement of the corresponding asymmetries at the upgraded facility have been calculated. Like for B's, one considers the pair of decay channels to definite Flavour and CP eigestate decay products for the preparation of the appropriate neutral Kaon states. The question arises: Can the method be extended to any pair of decay channels? There is an "orthoganility problem" for identifying which are the decay channels associated to the T-reverse transition when the decay products are no longer either Flavour or CP eigenstates. In order to bypass this problem, there is a recent suggestion [50] of taking a modified Reference meson transition from the detected pair of decay channels.

4. CPT Invariance

The CPT theorem [51] establishes that interactions described by a Local Quantum Field Theory, with Lorentz invariance and Hermiticity, are CPT invariant. The simplest tests of CPT invariance are the equality of masses and lifetimes for a particle and its antiparticle. The best test comes from the mass difference between K^0 *and* $\bar{K}^0$ [5].

It is very important to move the tests of CPT invariance beyond the comparison of diagonal terms for particle and antiparticle by considering these searches for transitions. For entangled systems, the proposed tests of separate CP, T, CPT symmetries in the neutral meson systems are based on the EPR-Entanglement existing in the Meson Factories as a consequence of Particle Identity: $K^0, \bar{K}^0$ are two states of identical particles, connected by CPT.

Besides the permutation operation P for space-time properties, the strangeness charge connection is made by C, so that for bosons the indistinguishibility requirement is C P= +.

In neutral meson factories, $K^0, \bar{K}^0$ are produced by Φ-decay with J=1, S=0. This implies L=1 and C= -, so that P = -, i.e., an antisymmetric wave function. This antisymmetry is responsible for preserving $K^0, \bar{K}^0$ terms only in the time evolution of the two-body system, including the Mixing $K^0 \leftrightarrow \bar{K}^0$. Similarly for K_+, K_- terms only. This correlation is perfect for tagging: Flavour-Tag, CP-Tag [56].

The question is [51]: What if the $K^0, \bar{K}^0$ Identity is lost? In this case, the two particle system would not satisfy the requirement CP = +. In perturbation theory, if still J=1 with C=-, this breaking leads to a mixing of the "forbidden" P = + symmetric state into the "allowed" P = - antisymmetric state:

$$|i\rangle = |antisymmtric\rangle + \omega\ |symmetric\rangle \quad \textbf{(10)}$$

This perturbative mixing is the ω-effect: In the time evolution of the system one finds now ω K^0 K^0, ω K_+ K_+ terms,..., i.e., a Demise of Tagging.

In some Quantum Gravity models, matter propagation in topologically non-trivial space-time vacua sufflers a possible loss of quantum coherence or "decoherence". This effect can be originated by space-time foam backgrounds [52]. The matter quantum system is an open system, interacting with the "environment" of quantum gravitational degrees of freedom: this interaction leads to an apparent loss of unitarity for low-energy observers. As a consequence, there is not a well-defined S-matrix between asymptotic states and the CPT operator is not well defined [53].

This kind of CPT-non definition should be disentangled from the case of effective theories for Lorentz invariance violation [54], in which CPT-breaking means [Heff, CPT] ≠ 0. The CPT-Violation discussed here would be an "intrinsic" microscopic time irreversibility, so that $\bar{K}^0$ is not "well-defined" from K^0. It implies: 1) a modified single $K^0, \bar{K}^0$ evolution, leading to the (α, β, γ) parameterization [55]; 2) for entangled Kaon states in a Φ-factory, the ω-effect.

Looking for observables which are a signal of the ω-effect, consider [51] the Φ-decay amplitude described by Figure 8 and equation (11):

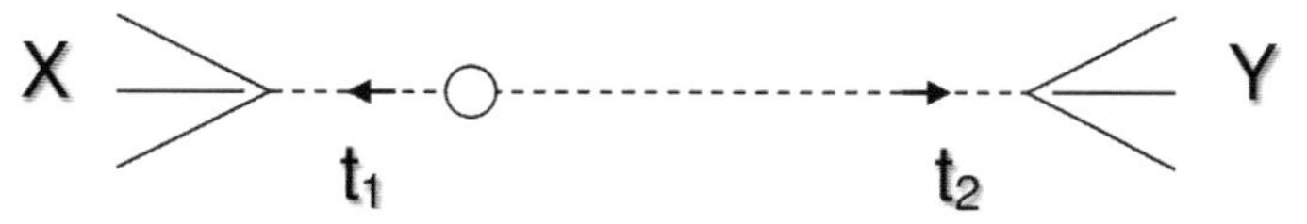

Figure 8. $K^0 - \overline{K}^0$ decay to (X, Y) at times t_1 and t_2, respectively

$$A(X,Y) = \langle X|K_S\rangle\langle Y|K_S\rangle N(A_1A_2)$$
$$A_1 = e^{-\frac{i(\lambda_L+\lambda_S)t}{2}}\left[\eta_X e^{-\frac{i\Delta\lambda\Delta t}{2}} - \eta_Y e^{\frac{i\Delta\lambda\Delta t}{2}}\right] \quad \textbf{(11)}$$
$$A_2 = \omega\left[e^{-i\lambda_S t} - \eta_X\eta_Y e^{-i\lambda_L t}\right]$$

where A_1 is the "allowed" amplitude, whereas A_2 is the "forbidden" amplitude proportional to ω.

An inspection of equation (11) tells us that the good strategy to enhance the relative effect of ω is the selection of a decay channel suppressed by the η's, the ratio between the K_L and K_S decay amplitudes. This requirement is fulfilled by the choice X = Y = $\pi^+\pi^-$, a CP "forbidden" channel in K-physics, with η a small ratio. The relative ω-effect in equation (11) is thus $\omega/|\eta_{+-}|$.

From equation (11) we can calculate the Intensity as function of Δt

$$I(\Delta t) = \frac{1}{2}\int_{\Delta t}^{\infty} dt\,|A(X,Y)|^2 \quad \textbf{(12)}$$

For the ($\pi^+\pi^-$, $\pi^+\pi^-$) channel, and writing $\omega = |\omega|e^{i\Omega}, \eta_{+-} = |\eta_{+-}|e^{i\Phi_{+-}}$, the Intensity is plotted in figure 9 for ω = 0 (continuous line) and an ω-value comparable to η_{+-} (broken line).

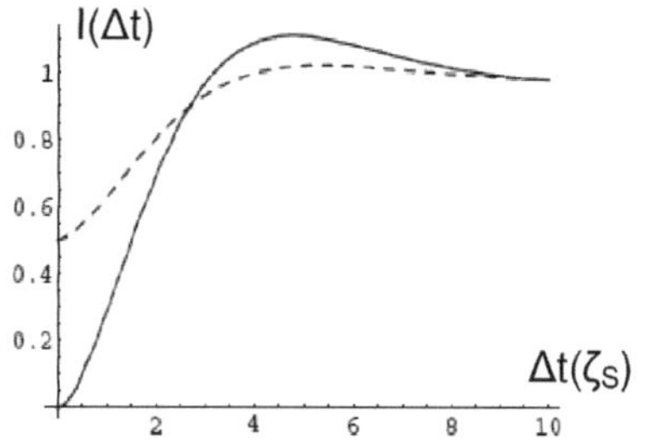

Figure 9. The Intensity I(Δt) for the decay of $K^0 - \overline{K}^0$ to ($\pi^+\pi^-$, $\pi^+\pi^-$), for ω=0 (—) and $|\omega| = |\eta_{+-}|, \Omega = \Phi_{+-} - 0.16\pi$ (---)

The most prominent effect is the breaking of I(Δt) ~ 0 for small values of Δt, a result that was a consequence of the particle identity anti-correlation: no identical states at t_1=t_2.

The KLOE experiment has obtained the first measurement of the ω-parameter [57]:

$$\left.\begin{aligned}\mathrm{Re}(\omega) &= \left(-1.6^{+3.0}_{-2.1\,stat} \pm 0.4_{syst}\right)\times 10^{-4}\\ \mathrm{Im}(\omega) &= \left(-1.7^{+3.3}_{-3.0\,stat} \pm 1.2_{syst}\right)\times 10^{-4}\end{aligned}\right\} \quad |\omega| < 1.0\times 10^{-3}\,at\ 95\%\ CL \quad \textbf{(13)}$$

At least one order of magnitude improvement is expected with KLOE-2 at the upgraded DAΦNE.

All decoherence effects, including the ω-effect, manifest as a deviation from the quantum mechanical prediction of the EPR-correlation $I(\pi^+\pi^-, \pi^+\pi^-; \Delta t = 0) = 0$. Hence the reconstruction of events in the region near $\Delta t \sim 0$ is crucial. Experimentally, one needs vertex resolution. In figure 9 one finds a Monte Carlo simulation of $I(\pi^+\pi^-, \pi^+\pi^-; \Delta t)$, with both the KLOE resolution $\sigma\Delta t \approx \zeta S$ (black histogram) and the expected KLOE-2 resolution $\sigma\Delta t \approx 0.3\ \zeta S$ (red histogram)

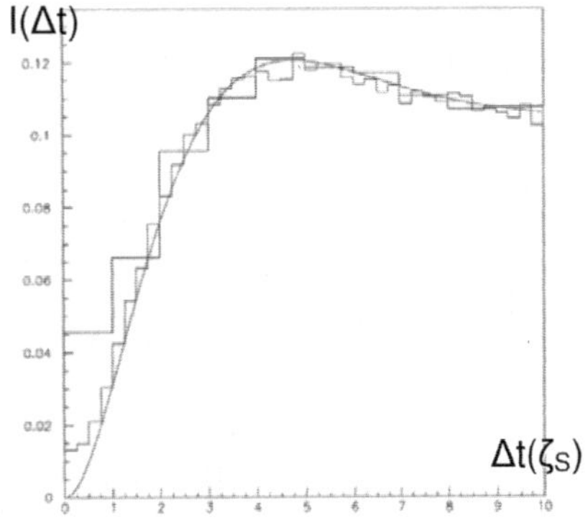

Figure 10. Comparision of the $I(\pi^+\pi^-, \pi^+\pi^-)$ simulation for KLOE Δt resolution and KLOE-2 Δt-resolution

In B-factories, there is no privileged channel to enhance the ω-effect. With currently available data from BABAR and BELLE, the CP-violating semileptonic charge asymmetry, in the equal sign dilepton channels $I(l^\pm, l^\pm; \Delta t)$ gives the bounds [58]

$$-0.0084 \le \mathrm{Re}(\omega) \le 0.0100 \text{ at } 95\%\mathrm{CL} \quad \textbf{(14)}$$

The ω-effect has been described as the result of local distortions of space-time in the neighborhood of defects, interacting with matter [59]. The recoil of the Planck-mass defect leads to a metric deformation

$$g_{0i} \sim \Delta k^i / M_P = \zeta k^i / M_P \quad \textbf{(15)}$$

In this framework, Lorentz invariance still holds macroscopically $< \zeta k^i > = 0$, but one has non-trivial quantum fluctuations $< \zeta^2 k^i k^j > \sim \delta_{ij} \zeta^2 |\vec{k}|^2$. The stochastic effects of the space-time foam lead to an ω-value

$$|\omega|^2 \sim \frac{\zeta^2 |\vec{k}|^4}{M_P^2 \Delta m^2} \quad \textbf{(16)}$$

which is enhanced by the quasi-degeneracy of the mass eigenstates.

At the DAΦNE energy, equation (16) gives $|\omega| \sim 10^{-4}\zeta$, which lies within the sensitivity of KLOE-2 for not much small values of the momentum transfer fraction ζ.

5. Conclusions

CPV violating asymmetries are well described up to now by the SM mechanism of the CKM Quark Mixing Matrix. It is of highest importance the search of virtual effects of New Physics in rare decays that are sensitive to the details of the FCNC transitions and CPV, like in penguin vertices. The current level of experimental accuracy and theoretical uncertainties leaves room for additional sources of CPV. The motivation for new physics contributions includes the fact that the SM is unable to explain the Matter-Antimatter asymmetry of the Universe.

The arguments used to preclude a true search of TRV in unstable particles have been by-passed by the transfer of quantum information, implied by Entanglement of the pair of mesons in meson factories, from the decaying particle to its living partner. In these entangled systems, the preparation of the initial and final meson states required by a genuine test of Time-Reversal is made by using the pair of time-ordered decays as filtering measurements. Combining the eight independent intensities associated to time-ordered flavour-CP eigenstates of the decay products, BABAR has measured separate asymmetries for each of CP, T, CPT transformations. TRV has been observed, at 14σ level, in the time evolution of the neutral B-meson. The measurement of the TRV asymmetries for neutral Kaons in DAΦNE is envisaged.

The search of CPTV effects in transitions, beyond the comparison of masses and lifetimes of particle and antiparticle, has a peculiar implication for entangled meson systems. If CPT is ill-defined, as suggested by certain models of quantum gravity, there is a "ω-effect" which leads to a component of the wrong symmetry in the entangled state. The best known transition to search for such ω-effect is the time-dependent coherent decay of $K^0 - \bar{K}^0$ to ($\pi^+\pi^-$, $\pi^+\pi^-$) in DAΦNE.

In addition to the precision studies of CPV asymmetries, there are well motivated arguments for a full experimental programme on TRV and CPT tests.

Acknowledgments

I would like to thank Professor N. E. Mavromatos, as Chair of the Organizing Committee for DISCRETE 2014, for the opportunity to discuss these problems in the stimulating atmosphere of the Symposium. This research has been supported by Grants of the Spanish Science Plan with the Project FPA-2011-23596 and of Generalitat Valenciana with the Project GVPROMETEOII 2013-017.

References

[1] Christenson, J H, Cronin J W, Fitch V L and Turlay R, *Phys.Rev* **140**, B74 (1965)
[2] Kobayashi M and Maskawa T, *Prog.Theor.Phys.* **49**, 652 (1973).
[3] BABAR Collaboration, Lees J L et al., *Phys. Rev. Lett.* **109**(2012)211801.
[4] Banuls M C and Bernabeu J, *Phys.Lett.*B **464**, 117 (1999).
[5] Carossi R et al., *Phys. Lett.* B **237**(1990)303
Abonzaid E et al., *Phys. Rev.* D **83**(2011)092001.
[6] Lee T D and Yang C-N, *Phys.Rev.* **104**, 254 (1956).
[7] Wu C S, Ambler E, Hayward R W, Hoppes D D, and R.P. Hudson, *Phys.Rev.* **105**, 1413 (1957).
[8] Garwin R L, Lederman L M, and Weinrich M, *Phys.Rev.* **105**, 1415 (1957).
[9] Christenson J H, Cronin J W, Fitch V L, and Turlay R, *Phys.Rev.* **140**, B74 (1965).
[10] Aubert B et al. (BABAR Collaboration), *Phys.Rev.Lett.* **87**, 091801 (2001).
[11] Abe K et al. (Belle Collaboration), *Phys.Rev.Lett.* **87**, 091802 (2001).
[12] Cabibbo N, *Phys.Rev.Lett.* **10**, 531 (1963).
[13] Beringer J et al. (Particle Data Group), *Phys.Rev.*D **86**, 010001 (2012).
[14] Wolfenstein L, *Phys.Rev.Lett.* **51**, 1945 (1983).
[15] Antonelli M, Asner D M, Bauer D A, Becher T G, Beneke M, et al., *Phys.Rept.* **494**, 197 (2010).
[16] Perl M et al., *Phys.Rev.Lett.* **35**, 1489 (1975).
[17] Herb S W et al., *Phys.Rev.Lett.* **39**, 252 (1977).
[18] Aubert B et al. (BABAR Collaboration), *Nucl.Instrum.Meth.*A, 615(2013).
[19] B. Aubert et al. (BABAR Collaboration), *Nucl.Instrum.Meth.*A **479**, 1 (2002).
[20] Abashian A et al., *Nucl.Instrum.Meth.*A **479**, 117 (2002).
[21] The LHCb Collaboration, JINST 3, S008 005(2008).
[22] Gershon T, Nir Y, in RPP, *Chin. Phys.* **38**, Number 9 (2014).
[23] LHCb Collaboration, Aaij R et al., *Phys. Rev. Lett.* **110**, 028101 (2013)
[24] Matias J, Proceedings DISCRETE 2014.
[25] BABAR Collaboration, Lees J P, *Phys. Rev.* D **86** 051105(2012)
[26] Wigner E P, *Nachr. Ges. Wiss.* Götingen **32**, 35(1932).
[27] Wolfenstein L, *Phys.Rev.Lett.* **83**, 911 (1999).

[28] Wolfenstein L, Int.J.*Mod.Phys*.E **8**, 501 (1999).
[29] Henley E M, Int.J.*Mod.Phys*.E **22**, 1330010 (2013).
[30] Sachs R G, *"The Physics of Time Reversal"*, The University of Chicago Press, 1-325(1987).
[31] Peccei R D and Quinn H R, *Phys.Rev.Lett.* **38**, 1440 (1977).
[32] Beringer J et al. (Particle Data Group), *Phys.Rev*.D **86**, 010001(2012).
[33] Kabir P K, *Phys.Rev*.D **2**, 540 (1970).
[34] Angelopoulos A et al. (CPLEAR Collaboration), *Phys.Lett*.B **444**, 43 (1998).
[35] Weisskopf V and Wigner E P , *Z.Phys*. **63**, 54 (1930).
[36] Weisskopf W and Wigner E P, *Z.Phys*. **65**, 18 (1930).
[37] Lees J P et al. (BABAR Collaboration), *Phys.Rev.Lett.* **111**,101802 (2013).
[38] Nakano E et al. (Belle Collaboration), *Phys.Rev*.D **73**, 112002 (2006).
[39] Alvarez-Gaume L, Kounnas C, Lola S, and Pavlopoulos P, *Phys.Lett*.B **458**, 347 (1999).
[40] Ellis J R and Mavromatos N E, *Phys.Rept*. **320**, 341 (1999).
[41] Gerber H J, *Eur.Phys*.J.C **35**, 195 (2004).
[42] Banuls M C and Bernabeu J, *Nucl.Phys*.B **590**, 19 (2000).
[43] Wolfenstein L, *Int.J.Mod.Phys*.E **8**, 501 (1999).
[44] Quinn H R, J.*Phys.Conf.Ser*. **171**, 011001 (2009).
[45] Bernabeu J, Martinez-Vidal F, and Villanueva-Perez P, *JHEP* **1208**, 064 (2012).
[46] Einstein A, Podolsky B, and Rosen N, *Phys.Rev*. **47**, 777(1935).
[47] Reid M D, Drummond P D, Bowen W P, Cavalcanti E G, Lam P K, et al., *Rev.Mod.Phys*. **81**, 1727 (2009)
[48] Applebaum E, Efrati A, Grossman Y, Nir Y, and Soreq Y, *Phys.Rev*.D **89**, 076011 (2014).
[49] Bernabeu J, Di Domenico A, and Villanueva-Perez P, *Nucl.Phys*.B **868**, 102 (2013).
[50] Bernabeu J, Botella F J, and Nebot M, *Phys.Lett*.B **728**, 95 (2014).
Botella F J et al., Proceedings DISCRETE 2014.
[51] Bernabeu J, Mavromatos N E and Papavassiliou J, *Phys. Rev.Lett*. **92**:131601(2004)
[52] For a review, see Wheeler J A and Ford K, Geons, black holes and quantum foam: A life in physics (New York, USA; Norton, 1998);
Ellis J R, Hagelin J S, Nanopoulos D V and Srednicki M, Nucl. Phys. B 241(1984)381
[53] Wald R, *Phys. Rev.D* **21**(1980)2742
[54] Altschul B, *this Proceedings*
[55] Lindblad G, *Comm. Math. Phys*. **48**(1976)119
[56] Bañuls M C and Bernabeu J, *JHEP* 9906:032(1999)
[57] Di Domenico A, 0904.1976 (2009)
[58] Alvarez E, Bernabeu J and Nebot M, *JHEP* 0611(2006)087
[59] Bernabeu J, Mavromatos N E and Sarkar S, *Phys. Rev. D* **74**(2006)045014

Subtleties in the BaBar measurement of time-reversal violation

Aielet Efrati
Department of Particle Physics and Astrophysics
Weizmann Institute of Science, Rehovot 76100, Israel
E-mail: aielet.efrati@weizmann.ac.il

Abstract. A first measurement of time-reversal (T) asymmetries that are not also CP asymmetries has been recently achieved by the *BABAR* collaboration. In this talk, which follows the work done in Ref. [1], I discuss the subtleties of this measurement in the presence of direct CP violation, CPT violation, wrong strangeness decays and wrong sign semi-leptonic decays. In particular, I explain why, in order to identify the measured asymmetries with time-reversal violation, one needs to assume (i) the absence of wrong strangeness decays or of CPT violation in strangeness changing decays, and (ii) the absence of wrong sign decays.

1. Introduction

The *BABAR* collaboration has recently announced the first direct observation of time-reversal violation in the neutral B meson system [2]. The basic idea is to compare the time-dependent rates of two processes that differ by exchange of initial and final states. The measurement makes use of the EPR effect in the entangled B mesons produced in $\Upsilon(4S)$ decays [3, 4, 5, 6, 7]. For example, one rate, $\Gamma_{(\psi K_L)_\perp,\ell^- X}$, involves the decay of one of the neutral B's into a ψK_L state, and, after time t, the decay of the other B into $\ell^- X$. The other rate, $\Gamma_{(\ell^+ X)_\perp,\psi K_S}$, involves the decay of one of the neutral B's into $\ell^+ X$, and, after time t, the decay of the other B into ψK_S. Under certain assumptions, to be spelled out below, this is a comparison between the rates of $B_- \to \overline{B}^0$ and $\overline{B}^0 \to B_-$, where $\overline{B}^0$ has a well defined flavor content $(b\bar{d})$ and B_- is a CP-odd state.

Time reversal violation had been observed earlier in the neutral K meson system by CPLEAR [8]. The measurement involves the processes $p\bar{p} \to K^-\pi^+K^0$ and $p\bar{p} \to K^+\pi^-\overline{K}^0$. Again, one aims to compare rates of processes that are related by exchange of initial and final states. One rate, Γ_{K^-,e^-}, involves a production of K^- and a neutral K that after time t decay into $e^-\pi^+\bar{\nu}$. The other rate, Γ_{K^+,ℓ^+}, involves the production of K^+ and a neutral $\overline{K}$ that after time t decay into $e^+\pi^-\nu$. Under certain assumptions, this is a comparison between the rates of $K^0 \to \overline{K}^0$ and $\overline{K}^0 \to K^0$ [9]. The CPLEAR asymmetry is also a CP asymmetry, since the initial and final states are CP-conjugate. In contrast, the *BABAR* asymmetry is not a CP asymmetry.

When aiming to demonstrate that time-reversal is violated, one needs to allow for CPT violation [10, 11]. (Otherwise, T violation follows from CP violation.) For two processes to be related by time-reversal, the initial state in each of them should be the time-reversal conjugate of the final state in the other. The subtleties related to this requirement, in the context of the *BABAR* measurements, are clarified in this work. To do so, it is helpful to use only parameters

Journal of Physics: Conference Series **631** (2015) 012016 doi:10.1088/1742-6596/631/1/012016

that have well-defined transformation properties under all three relevant symmetries: CP, T and CPT. In the presence of CPT violation, most of the parameters used in the literature do not have well-defined transformation under T and CPT. In particular, ΔS_T^+, which is used to demonstrate time reversal violation by the *BABAR* collaboration, has, apart from a T-odd CPT-even part, also a T-even CPT-odd part. We introduce parameters with well-defined transformation properties under these discrete symmetries and formulate the assumptions one needs to make in order to identify the asymmetries measured by *BABAR* with time reversal violation.

2. Parameter definitions and transformation properties

Our definitions and notations, which use a formalism that allows for CPT violation, are described in detail in Ref. [1]. The following summarizes the necessary and novel ingredients used in our work.

2.1. CPT violation

We define decay amplitudes and inverse-decay amplitudes

$$\begin{aligned} A_f \equiv A(B^0 \to f) &= \langle f|T|B^0\rangle\,, & \bar{A}_f \equiv A(\overline{B}{}^0 \to f) = \langle f|T|\overline{B}{}^0\rangle\,, \\ A_f^{\rm ID} \equiv A(f^T \to B^0) &= \langle B^0|T|f^T\rangle\,, & \bar{A}_f^{\rm ID} \equiv A(f^T \to \overline{B}{}^0) = \langle \overline{B}{}^0|T|f^T\rangle\,, \end{aligned} \quad (1)$$

where f^T is T-conjugate (*i.e.* reversed spins and momenta) to f. Using these amplitudes, we define complex parameters, θ_f, representing CPT violation in the decay:

$$\theta_f = \theta_f^R + i\theta_f^I \equiv \frac{A_f^{\rm ID}/\bar{A}_f^{\rm ID} - \bar{A}_f/A_f}{A_f^{\rm ID}/\bar{A}_f^{\rm ID} + \bar{A}_f/A_f}\,. \quad (2)$$

For final CP eigenstates, $\theta_f \neq 0$ breaks CPT and CP, but not T. The complex parameter $z \equiv z^R + iz^I$ represents CP and CPT violation in mixing, while the real parameter, R_M, represents T and CP violation in mixing. For these two parameters we use the standard definitions which can be found in Ref. [1]. We further define the phase convention independent combination of amplitudes and mixing parameters,

$$\lambda_f \equiv \frac{q}{p}\frac{\bar{A}_f}{A_f}\sqrt{\frac{1+\theta_f}{1-\theta_f}} = \frac{q}{p}\frac{A_f^{\rm ID}}{\bar{A}_f^{\rm ID}}\sqrt{\frac{1-\theta_f}{1+\theta_f}}\,. \quad (3)$$

In the CPT limit $\theta_f = 0$ and the standard definition of λ_f is recovered. It is convenient to introduce the following functions of λ_f:

$$C_f \equiv \frac{1-|\lambda_f|^2}{1+|\lambda_f|^2}\,, \qquad S_f \equiv \frac{2\,\mathcal{I}m(\lambda_f)}{1+|\lambda_f|^2}\,, \qquad G_f \equiv \frac{2\,\mathcal{R}e(\lambda_f)}{1+|\lambda_f|^2}\,, \quad (4)$$

with $C_f^2 + G_f^2 + S_f^2 = 1$. The transformation rules for these parameters under T, CP and CPT, are summarized in Table 1. As explained above, it is very convenient for our purposes to use parameters that are either even or odd under all three transformations. Using superscript $+$ for T-even, and $-$ for T-odd, we define the following combinations:

$$C_f^{\pm} = \frac{1}{2}(C_f \mp C_{\bar{f}})\,, \quad S_f^{\pm} = \frac{1}{2}(S_f \mp S_{\bar{f}})\,, \quad G_f^{\pm} = \frac{1}{2}(G_f \pm G_{\bar{f}})\,, \quad \theta_f^{\pm} = \frac{1}{2}(\theta_f \pm \theta_{\bar{f}})\,. \quad (5)$$

A summary of the transformation properties of these parameters is provided in Table 2.

We emphasize that the definition of λ_f in Eq. (3) and, consequently, the definitions of S_f, C_f and G_f in Eq. (4) differ from the standard definitions of these parameters in the literature.

Table 1. Transformation rules of the various parameters under T, CP and CPT.

Parameter	T	CP	CPT
R_M	$-R_M$	$-R_M$	R_M
z	z	$-z$	$-z$
λ_f	$1/\lambda_{\bar{f}}$	$1/\lambda_{\bar{f}}$	λ_f
S_f	$-S_{\bar{f}}$	$-S_{\bar{f}}$	S_f
C_f	$-C_{\bar{f}}$	$-C_{\bar{f}}$	C_f
G_f	$G_{\bar{f}}$	$G_{\bar{f}}$	G_f
θ_f	$\theta_{\bar{f}}$	$-\theta_{\bar{f}}$	$-\theta_f$

Table 2. Eigenvalues of the various parameters under T, CP and CPT.

Parameter	T	CP	CPT
R_M, S_f^-, C_f^-, G_f^-	$-$	$-$	$+$
z, θ_f^+	$+$	$-$	$-$
θ_f^-	$-$	$+$	$-$
S_f^+ ,C_f^+ ,G_f^+	$+$	$+$	$+$
$S_{f_{CP}}$, $C_{f_{CP}}$	$-$	$-$	$+$
$\theta_{f_{CP}}$	$+$	$-$	$-$
$G_{f_{CP}}$	$+$	$+$	$+$

Our definition lends itself straightforwardly to the theoretical analysis that we are doing. The standard definition lends itself straightforwardly to the description of the experimentally measured rates. In practice, inverse decays are not accessible to the experiments. In particular, experiments are not sensitive to λ_f, as defined in Eq. (3), but to the related observable λ_f^e, defined via

$$\lambda_f^e \equiv \frac{q}{p}\frac{\bar{A}_f}{A_f} = \lambda_f (1-\theta_f) \,, \tag{6}$$

where the second equation holds to first order in θ_f. Accordingly, the experiments are sensitive to the following parameters:

$$\begin{aligned} C_f^e &= C_f + (1-C_f^2)\theta_f^R \,, \\ S_f^e &= S_f(1-C_f\theta_f^R) - G_f\theta_f^I \,, \\ G_f^e &= G_f(1-C_f\theta_f^R) + S_f\theta_f^I \,. \end{aligned} \tag{7}$$

The distinction between the "theoretical" and "experimental" parameters is crucial for understanding the subtleties in the interpretation of the *BABAR* measurements. Of course, in the absence of CPT violation, the two sets of definitions coincide.

2.2. *Wrong-strangeness decays*

Among the final CP eigenstates, we focus on decays into the final $\psi K_{S,L}$ states (neglecting effects of ϵ_K). We distinguish between the right strangeness decays and the wrong strangeness decays, and define

$$\begin{aligned}
\hat{C}_{\psi K} &\equiv \frac{1}{2}(C_{\psi K_S}+C_{\psi K_L}), & \Delta C_{\psi K} &\equiv \frac{1}{2}(C_{\psi K_S}-C_{\psi K_L}), \\
\hat{S}_{\psi K} &\equiv \frac{1}{2}(S_{\psi K_S}-S_{\psi K_L}), & \Delta S_{\psi K} &\equiv \frac{1}{2}(S_{\psi K_S}+S_{\psi K_L}), \\
\hat{G}_{\psi K} &\equiv \frac{1}{2}(G_{\psi K_S}-G_{\psi K_L}), & \Delta G_{\psi K} &\equiv \frac{1}{2}(G_{\psi K_S}+G_{\psi K_L}), \\
\hat{\theta}_{\psi K} &\equiv \frac{1}{2}(\theta_{\psi K_L}-\theta_{\psi K_S}), & \Delta \theta_{\psi K} &\equiv \frac{1}{2}(\theta_{\psi K_L}+\theta_{\psi K_S}).
\end{aligned} \tag{8}$$

In the limit of no wrong strangeness decay, $\lambda_{\psi K_S} = -\lambda_{\psi K_L}$ [12] (Ref. [12] assumes CPT conservation, but this is not a necessary assumption) and, consequently, $\Delta C_{\psi K}, \Delta G_{\psi K}, \Delta S_{\psi K}$ and $\Delta\theta_{\psi K}$ vanish.

2.3. Wrong-sign decays

Among the flavor specific final states, we focus on decays into final $\ell^{\pm}X$ states. Here we distinguish between the right sign decays and the wrong sign decays, and define $C_\ell^{\pm}$, $S_\ell^{\pm}$ and $G_\ell^{\pm}$ according to Eq. (5), with $f=\ell^+$, and a super-index $+$ $(-)$ denoting a T conserving (violating) combination. Taking the wrong sign decays to be much smaller in magnitude than the right sign decays, we have $|\lambda_{\ell^+}| \ll 1$ and $|\lambda_{\ell^-}^{-1}| \ll 1$. We will work to first order in $|\lambda_{\ell^+}|$ and in $|\lambda_{\ell^-}^{-1}|$, which means that we set $C_\ell^+=1$ and $C_\ell^-=0$. On the other hand, the four other relevant parameters are linear in $|\lambda_{\ell^+}|$ and in $|\lambda_{\ell^-}^{-1}|$:

$$S_\ell^{\pm} \simeq \mathcal{I}m(\lambda_{\ell^+} \pm \lambda_{\ell^-}^{-1}), \qquad G_\ell^{\pm} \simeq \mathcal{R}e(\lambda_{\ell^+} \pm \lambda_{\ell^-}^{-1}). \tag{9}$$

Equipped with these definitions, we are now ready to analyze the asymmetries measured by the *BABAR* collaboration.

3. The *BABAR* T-asymmetry

Consider a pair of B-mesons produced in $\Upsilon(4S)$ decay, where one of the B-mesons decays at time t_1 to a final state f_1, and the other B-meson decays at a later time, $t_2 = t_1 + t$, to a final state f_2. The time dependent rate for this process (to zeroth order in y) is given by

$$\Gamma_{(f_1)_\perp, f_2} = N_{(1)_\perp,2}e^{-\Gamma(t_1+t_2)} \times \left[\kappa_{(1)_\perp,2} + C_{(1)_\perp,2}\cos(x\Gamma t) + S_{(1)_\perp,2}\sin(x\Gamma t)\right]. \tag{10}$$

where the coefficients $N_{(1)_\perp,2}$, $\kappa_{(1)_\perp,2}$, $C_{(1)_\perp,2}$ and $S_{(1)_\perp,2}$ are defined in Ref. [1]. We work to linear order in the following small parameters:

$$R_M,\ z^R,\ z^I,\ \theta_f^R,\ \theta_f^I,\ \hat{C}_{\psi K},\ \Delta C_{\psi K},\ \Delta S_{\psi K},\ \Delta G_{\psi K}, S_\ell^{\pm},\ G_\ell^{\pm}. \tag{11}$$

The analysis performed by *BABAR* , as described in Ref. [2], is as follows. The time dependent decay rates are measured and fitted to time-dependence of the form (10), approximating (as we do) $y=0$. The quantities ΔS_T^+ and ΔC_T^+ correspond to the following combinations:

$$\begin{aligned}
\Delta S_T^+ &= \frac{S_{(\psi K_L)_\perp,\ell^- X}}{\kappa_{(\psi K_L)_\perp,\ell^- X}} - \frac{S_{(\ell^+ X)_\perp,\psi K_S}}{\kappa_{(\ell^+ X)_\perp,\psi K_S}} \equiv S_{\ell^-,K_L}^- - S_{\ell^+,K_S}^+, \\
\Delta C_T^+ &= \frac{C_{(\psi K_L)_\perp,\ell^- X}}{\kappa_{(\psi K_L)_\perp,\ell^- X}} - \frac{C_{(\ell^+ X)_\perp,\psi K_S}}{\kappa_{(\ell^+ X)_\perp,\psi K_S}} \equiv C_{\ell^-,K_L}^- - C_{\ell^+,K_S}^+.
\end{aligned} \tag{12}$$

We note that the normalization of Eqs. (12) removes the dependence on the total production rates and effects such as direct CP violation in leptonic decays.

We obtain the following expressions for these observables:

$$\Delta S_T^+ = -2\left[(\hat{S}_{\psi K} - \hat{G}_{\psi K}\Delta\theta_{\psi K}^I) + \hat{S}_{\psi K}\hat{G}_{\psi K}(G_\ell^- - z^R)\right],$$
$$\Delta C_T^+ = 2\left[(\hat{C}_{\psi K} + \Delta\theta_{\psi K}^R) + \hat{S}_{\psi K}(S_\ell^- - z^I)\right]. \tag{13}$$

If we switch off all the T-odd parameters, we are left with the following T conserving (TC) contributions:

$$\Delta S_T^+(\text{T-odd parameters} = 0) = 2\hat{G}_{\psi K}\Delta\theta_{\psi K}^I,$$
$$\Delta C_T^+(\text{T-odd parameters} = 0) = 2\Delta\theta_{\psi K}^R. \tag{14}$$

These contributions are CPT violating, and they vanish in the limit of no wrong strangeness decays. Yet, since $\Delta\theta_{\psi K}$ involves inverse decays, we are not aware of any way to experimentally bound it, and to exclude the possibility that it generates the measured value of ΔS_T^+. We would like to emphasize, however, the following three points.

- The appearance of the T conserving, CPT violating effects should come as no surprise. As explained in the discussion of Eq. (7), experiments can only probe $S_{\psi K}^e$ and $C_{\psi K}^e$, which include these terms.
- While we are not aware of any way to constrain $\Delta\theta_{\psi K}$ from tree level processes, it is quite possible that it affects measurable effects, such as CPT violation in mixing, via loop effects. In the absence of a rigorous framework that incorporates CPT violation, it is impossible to calculate such effects.
- It would of course be extremely exciting if the *BABAR* measurement is affected by CPT violating effects.

An additional interesting feature of Eqs. (13) is the appearance of terms that are quadratic in T-odd parameters,

$$\Delta S_T^+(\text{quadratic in T-odd parameters}) = -2\hat{G}_{\psi K}\hat{S}_{\psi K}G_\ell^-,$$
$$\Delta C_T^+(\text{quadratic in T-odd parameters}) = 2\hat{S}_{\psi K}S_\ell^-. \tag{15}$$

While these terms would vanish if we take all T-odd parameters to zero, they are still T-conserving. Note that since we expand to linear order in all T-odd parameters, except for $\hat{S}_{\psi K_S}$, there are additional T conserving, $\hat{S}_{\psi K}$-independent, contributions that are quadratic in T-odd parameters that are not presented in Eqs. (15). Since $\hat{G}_{\psi K}^2 + \hat{S}_{\psi K}^2 \leq 1$, the maximal absolute value of the term on the right hand side of Eq. (15) for ΔS_T^+ is 1, $|2\hat{G}_{\psi K}\hat{S}_{\psi K}G_\ell^-| \leq 1$. Thus, if experiments establish $|\Delta S_T^+| > 1$, such a result cannot be explained by this term alone.

We are now also able to formulate the conditions under which the *BABAR* measurement would demonstrate T violation unambiguously:

$$\Delta\theta_{\psi K} = G_\ell^- = S_\ell^- = 0. \tag{16}$$

In words, the necessary conditions are the following:

- The absence of wrong strangeness decays and of wrong strangeness inverse decays or, if such wrong strangeness processes occur, the absence of CPT violation in strangeness-changing decays.

- The absence of wrong sign decays or, if wrong sign decays occur, the absence of direct CP violation in semileptonic decays.

We further note that not only the T-asymmetries get T-even contributions in the presence of the aforementioned phenomena, but also the CP asymmetries get CP-even contributions, and the CPT asymmetries get CPT-even contributions. All of these effects vanish if there are neither wrong strangeness nor wrong sign (inverse) decays.

A-priori, one would expect that direct CP violation in right-strangeness decays is enough to allow for $A_T \neq 0$ even in the T-symmetry limit. In $\Gamma_{(\psi K_L)_\perp,\ell^- X}$, the initial B-meson state is orthogonal to the one that decays to ψK_L. In $\Gamma_{(\ell^+ X)_\perp,\psi K_S}$, the final state is the one that decays into ψK_S. Are these two states identical? They would be if the state that does not decay to ψK_L, $|B_{(\to\psi K_L)_\perp}\rangle$, and the state that does not decay into ψK_S, $|B_{(\to\psi K_S)_\perp}\rangle$, were orthogonal to each other. In fact, this is the case if there is no direct CP violation in the $B \to \psi K$ decays. This is presumably the reason why the theoretical paper [5] and the experimental paper [2] explicitly state that they neglect direct CP violation.

However, the correct question is not whether the state that does not decay to ψK_L is the same as the state that decays to ψK_S. Instead, the question is whether it is the same as the state generated in the inverse decay of ψK_S. We find that if wrong strangeness decays can be neglected, then the two processes are related by exchange of the initial and final CP-tagged states, as required for time-reversal conjugate processes.

One can raise an analogous question for the lepton-tagged states. The question to be asked is then whether the state that does not decay to $\ell^+ X$ is orthogonal to the state that is not generated in the inverse decay of $\ell^- X$. We find that if wrong sign decays can be neglected, then the two processes are related by exchange of the initial and final lepton-tagged states, as required for time-reversal conjugate processes.

We conclude that if wrong strangeness and wrong sign decays can be neglected, then the two processes measured by *BABAR* represent two T-conjugate processes, and then there should be no T conserving contributions to ΔS_T^+ and ΔS_T^-, consistent with Eqs. (14) and (15). In particular, one need not assume the absence of direct CP violation.

4. Isolating parameters of interest

Combinations of observables measured by *BABAR* allow us to constrain various parameters of interest. From the results of [2] we get

$$\hat{S}_{\psi K} - \hat{G}_{\psi K}\Delta\theta^I_{\psi K} = 0.69 \pm 0.04 . \tag{17}$$

and the following bounds can be deduced

$$\begin{aligned} &|\hat{C}_{\psi K} + \Delta\theta^R_{\psi K}| < 0.07 , \\ &|G_{\psi K_{S,L}} S_{\psi K_{S,L}} \left(G_\ell^- - z^R\right)| < 0.10 , \\ &|S_{\psi K_{S,L}} \left(S_\ell^- - z^I\right)| < 0.06 , \end{aligned} \tag{18}$$

at 2σ level. In case we assume no CPT violation naive combination of the above will lead to

$$|S_\ell^-| < 0.10 , \qquad |G_\ell^-| < 0.21 , \tag{19}$$

at 2σ level.

Journal of Physics: Conference Series **631** (2015) 012016 doi:10.1088/1742-6596/631/1/012016

5. Conclusions

The *BABAR* collaboration has measured time-reversal asymmetries in B decays. Two main ingredients — the EPR effect between the two B-mesons produced in $\Upsilon(4S)$ decays and the availability of both lepton-tagging and CP-tagging — allow the experimenters to approximately realize the main principle of time-reversal conjugate processes: exchanging initial and final states.

A precise exchange is impossible. The final state is identified by the B-meson decay, and the T-conjugate process requires, therefore, that a B-meson is produced in the corresponding inverse decay. Instead, the experimenters measure a process where the initial B-meson is identified by the decay of the other, entangled B-meson. We found however that the initial B-meson prepared by lepton tagging, and the one that would be produced in the appropriate inverse decay are not identical only if there are wrong-sign decays. The initial B-meson prepared by CP tagging, and the one that would be produced in the appropriate inverse decay, are not identical only if there are wrong-strangeness contributions and, furthermore, if there is direct CP violation or the presence of CPT violation in decays.

The effect of CPT violation in decay has gained very little attention in the literature. One reason is that it can only be probed by measuring both decay rates and inverse decay rates, but the latter are practically inaccessible to experiments. For precisely this reason, there are no bounds on these effects. In principle, they could play a significant role in the asymmetries measured by *BABAR* , in spite of the fact that they are T conserving.

Both wrong-sign and wrong-strangeness effects are expected to be very small. If so, then the asymmetry measured by *BABAR* is indeed a time-reversal asymmetry to a very good approximation. The existing experimental upper bounds on these effects are rather weak. The same set of measurements used for the time-reversal asymmetries can be used (in different combinations) to constrain also the wrong-sign and wrong-strangeness contributions.

While in this work we concentrated on a very specific measurements in B decays, our results are more general. They apply straightforwardly, with minor changes, to other meson systems. The main ideas also apply to neutrino oscillation experiments. Observation of $P(\nu_\alpha \to \nu_\beta) \neq P(\nu_\beta \to \nu_\alpha)$ has been advocated as a way to establish T-violation. Such a result can arise, however, also from non-standard interactions in the production or the detection processes [13, 14, 15].

References

[1] E. Applebaum, A. Efrati, Y. Grossman, Y. Nir and Y. Soreq, Phys. Rev. D **89**, no. 7, 076011 (2014) [arXiv:1312.4164 [hep-ph]].

[2] J. P. Lees *et al.* [BaBar Collaboration], Phys. Rev. Lett. **109**, 211801 (2012) [arXiv:1207.5832 [hep-ex]].

[3] M. C. Banuls and J. Bernabeu, Phys. Lett. B **464**, 117 (1999) [hep-ph/9908353]; M. C. Banuls and J. Bernabeu, Nucl. Phys. B **590**, 19 (2000) [hep-ph/0005323].

[4] L. Wolfenstein, Int. J. Mod. Phys. E **8**, 501 (1999).

[5] J. Bernabeu, F. Martinez-Vidal and P. Villanueva-Perez, JHEP **1208**, 064 (2012) [arXiv:1203.0171 [hep-ph]].

[6] E. Alvarez and A. Szynkman, Mod. Phys. Lett. A **23**, 2085 (2008) [hep-ph/0611370].

[7] J. Bernabeu, F. J. Botella and M. Nebot, [arXiv:1309.0439 [hep-ph]].

[8] A. Angelopoulos *et al.* [CPLEAR Collaboration], Phys. Lett. B **444**, 43 (1998).

[9] P. K. Kabir, Phys. Rev. D **2**, 540 (1970).

[10] K. R. Schubert, B. Wolff, J. C. Chollet, J. M. Gaillard, M. R. Jane, T. J. Ratcliffe and J. P. Repellin, Phys. Lett. B **31**, 662 (1970).

[11] L. Lavoura, Annals Phys. **207**, 428 (1991).

[12] Y. Grossman, A. L. Kagan and Z. Ligeti, Phys. Lett. B **538**, 327 (2002) [hep-ph/0204212].

[13] Y. Grossman, Phys. Lett. B **359**, 141 (1995) [hep-ph/9507344].

[14] M. C. Gonzalez-Garcia, Y. Grossman, A. Gusso and Y. Nir, Phys. Rev. D **64**, 096006 (2001) [hep-ph/0105159].

[15] B. Bellazzini, Y. Grossman, I. Nachshon and P. Paradisi, JHEP **1106**, 104 (2011) [arXiv:1012.3759 [hep-ph]].

Novel T-violation observable open to any decay channel at meson factories

José Bernabeu[1]**, Francisco J Botella**[1] **and Miguel Nebot**[2]
[1]Departament de Física Teòrica and IFIC, Universitat de València - CSIC, E-46100, Burjassot, Spain
[2]Centro de Física Teórica de Partículas – CFTP, Instituto Superior Técnico – IST, Universidade de Lisboa – UL, Avenida Rovisco Pais, 1049-001 Lisboa, Portugal

E-mail: fbotella@uv.es, jose.bernabeu@uv.es, miguel.nebot@uv.es

Abstract. Two genuine Quantum phenomena: Entanglement and Filtering Measurement are at the origin of the first direct observation of Time-Reversal-Violation in the time evolution of the B neutral meson system by BaBar. The used meson transitions are directly connected to semileptonic and CP-eigenstate decay channels. We analyse the possibility of extending the observable asymmetries to more decay channels. We propose an alternative T-Violation Asymmetry in the meson factories which allows its opening to any pair of decay channels. The new asymmetry needs also the measurement of the time dependent total survival probability of the initial tagged states. By combining several asymmetries these total survival probabilities can be avoided. We analyse carefully the channels that can be used and the sufficient condition in order the proposed asymmetry be a genuine T violation measurement.

1. Introduction

The BABAR Collaboration reported, in the $B^0 - \bar{B}^0$system, the first direct observation of T-violation [1], in the time evolution of any system with high statistical significance. The measurement is based in a method described in [2] following the concepts originally proposed in [3, 4]

In the $\{|P^0\rangle, |\bar{P}^0\rangle\}$ system with evolution operator $U(t,0)$, where P^0 stands for a neutral meson K^0, D^0, B_d^0 *or* B_s^0 (and $\bar{P}^0$ for the corresponding antimeson), two arbitrary states are:

$$
\begin{aligned}
|P_1\rangle &= p_1{}^0|P^0\rangle + \bar{p}_1{}^0|\bar{P}^0\rangle \\
|P_2\rangle &= p_2{}^0|P^0\rangle + \bar{p}_2{}^0|\bar{P}^0\rangle
\end{aligned}
\tag{1}
$$

The probability that an initially prepared state P_1, evolving after time t to $P_1(t)$, behave like state P_2 is

$$
\Pr[P_1 \to P_2(t)] = |\langle P_2|P_1(t)\rangle|^2 = |\langle P_2|U(t,0)|P_1\rangle|^2 \tag{2}
$$

The T-violation observable is related to compare the reversed probabilities

$$
\Pr[P_1 \to P_2(t)] - \Pr[P_2 \to P_1(t)] \tag{3}
$$

The BABAR asymmetry was built from $P_1 = B_d^0, \bar{B}_d^0$ and $P_2 = B_\pm$ where $B_\pm$ are the B-states tagged or filtered by the decays to CP-eigenstates with definite flavour content. These asymmetries are experimentally independent of CP violation. The Kabir asymmetry [5], with $P_1, P_2 = K^0, \bar{K}^0$ was measured by the CPLEAR Collaboration [6] with a non-vanishing value near 4 standard deviations.

2. Reference transition at C = - initial Entangled State

We want to understand which is the Reference Transition $P_1 \to P_2(t)$, between meson states, associated to a given pair of decays f_1 at t_1 and f_2 at $t_2 > t_1$.
Two Quantum effects are relevant to answer this question:

- Entanglement between the two neutral mesons produced at a meson factory is essential to prepare the initial sate.
- The filtering measurement induced by the meson decay is used to connect to the final state.

The entangled state of the two mesons is in an antisymmetric combination of individual orthogonal states

$$|\Phi_{(C=-)}\rangle = \frac{1}{\sqrt{2}}\{|P^0(\vec{k})\rangle|\bar{P}^0(-\vec{k})\rangle - |\bar{P}^0(\vec{k})\rangle|P^0(-\vec{k})\rangle\} \tag{4}$$

If at time t_1 we observe at one side the decay product f, the (still living) meson at time t_1 is tagged as *the state that does not decay into* f [7, 8]

$$|P_{\nrightarrow f}\rangle = \frac{1}{\sqrt{\left(|A_f|^2 + |\bar{A}_f|^2\right)}}\{\bar{A}_f|P^0\rangle - A_f|\bar{P}^0\rangle\} \tag{5}$$

Where A_f ($\bar{A}_f$) is the decay amplitude from $P^0(\bar{P}^0)$ to f:

$$A_f = \langle f|W|P^0\rangle; \bar{A}_f = \langle f|W|\bar{P}^0\rangle \tag{6}$$

to first order in the weak Hamiltonian H_W and to all orders in strong interactions $W = U_S(\infty, 0)H_W$. $U_S(\infty, 0)$ is the strong evolution operator and is equal to the identity if we can neglect final sate interactions. (For hadronic decays we will assume transitions with one helicity amplitude: $0 \to 0 + j$).
The corresponding orthogonal state $\langle P_{\nrightarrow f}^{\perp}|P_{\nrightarrow f}\rangle = 0$ is given by

$$|P_{\nrightarrow f}^{\perp}\rangle = \frac{1}{\sqrt{\left(|A_f|^2 + |\bar{A}_f|^2\right)}}\{A_f^*\,|P^0\,\rangle + \bar{A}_f^*\,|\bar{P}^0\rangle\} \tag{7}$$

and it is the one filtered by the decay. What we call the "filtering identity" defines the precise meaning of this statement:

$$|\langle P_{\nrightarrow f_2}^{\perp}|P_1(t)\rangle|^2 = \frac{|\langle f_2|W|P_1(t)\rangle|^2}{\left(|A_{f_2}|^2 + |\bar{A}_{f_2}|^2\right)} \tag{8}$$

Journal of Physics: Conference Series **631** (2015) 012017 doi:10.1088/1742-6596/631/1/012017

Experimentally, the Reference Transition $P_1 \rightarrow P_2(t)$ is therefore directly connected to $P_1 = P_{\nrightarrow f_1}$ and $P_2 = P^{\perp}_{\nrightarrow f_2}$, i.e., $P_{\nrightarrow f_1}(t_1) \rightarrow P^{\perp}_{\nrightarrow f_2}(t_2)$. And the T transformed transition (*reversed*) $P^{\perp}_{\nrightarrow f_2}(t_1) \rightarrow P_{\nrightarrow f_1}(t_2)$ does not correspond to the pair of decays f_2 at t_1 , f_1 at $t_2 > t_1$, neither in the initial nor in the final decays. This is the *orthogonality problem* – previously defined in reference [9] - that prevent taking an arbitrary pair of decay channels. To connect the T transformed transition with experiment, we need to find a pair of decay channels such that, for each of them:

$$Given\ f \rightarrow \exists f'/|P_{\nrightarrow f'}\rangle = |P^{\perp}_{\nrightarrow f}\rangle \tag{9}$$

This orthogonality condition is satisfied by either CP conjugate decay channels $(P^0, \bar{P}^0)$ or CP eigenstates of opposite sign with the same flavour content (P_+, P_-) and no direct CP violation. The precise condition to be fulfilled by f and $f'$with the definitions provided in the next section is $(\lambda_{f'}\lambda_f^* = -|q/p|^2 \sim -1)$. Hence the exceptionality of the transitions between semileptonic and CP eigenstate decays. As a consequence, the orthogonality condition limits the pair of decay channels suitable for T-symmetry tests if we start, as a Reference, from the "experimental" transition $P_{\nrightarrow f_1}(t_1) \rightarrow P^{\perp}_{\nrightarrow f_2}(t_2)$.

Having identified the orthogonality problem, we give a bypass consisting in having an alternative reference transition once we fix the two channels f_1, f_2 . We make the following replacements

	BABAR		NEW
REFERENCE TRANSITION	$P_{\nrightarrow f_1}(t_1) \rightarrow P^{\perp}_{\nrightarrow f_2}(t_2)$	$\Rightarrow$	$P_{\nrightarrow f_1}(t_1) \rightarrow P_{\nrightarrow f_2}(t_2)$
T-TRANSFORMED TRANSITION	$P^{\perp}_{\nrightarrow f_2}(t_1) \rightarrow P_{\nrightarrow f_1}(t_2)$	$\Rightarrow$	$P_{\nrightarrow f_2}(t_1) \rightarrow P_{\nrightarrow f_1}(t_2)$

Whereas now the two initial P-states are directly connected to experiment, the price to be paid is that the two final states are not. One has to work out what is the connection of the novel genuine theoretical T-asymmetry observable to experimental measurements.

3. The New T-violating Asymmetry

The probability associated to the new reference transition is

$$P_{12}(t) \equiv |\langle\langle P_{\nrightarrow f_2}|P_{\nrightarrow f_1}(t)\rangle\rangle|^2 \tag{10}$$

It is not directly connected to the double decay rate as measured at meson factories

$$I_{12}(t) \equiv \frac{|\langle f_2|W|P_{\nrightarrow f_1}(t)\rangle|^2}{\left(|A_{f_2}|^2 + |\bar{A}_{f_2}|^2\right)} = |\langle P^{\perp}_{\nrightarrow f_2}|P_{\nrightarrow f_1}(t)\rangle|^2 \tag{11}$$

but using closure it easy to relate both quantities:

$$\begin{aligned} P_{12}(t) &= \langle P_{\nrightarrow f_1}(t)|P_{\nrightarrow f_2}\rangle\langle P_{\nrightarrow f_2}|P_{\nrightarrow f_1}(t)\rangle \\ &= \langle P_{\nrightarrow f_1}(t)|\left[I - |P^{\perp}_{\nrightarrow f_2}\rangle\langle P^{\perp}_{\nrightarrow f_2}|\right]|P_{\nrightarrow f_1}(t)\rangle \end{aligned} \tag{12}$$

The key result is

$$P_{12}(t) = N_1(t) - I_{12}(t) \tag{13}$$

Where $N_1(t)$ is the total survival probability of the $P_{\nrightarrow f_1}(t)$ state after a time t:

$$N_1(t) = \langle P_{\nrightarrow f_1}(t) | P_{\nrightarrow f_1}(t) \rangle \tag{14}$$

a well-defined and measurable quantity: the ratio between the number of mesons that have not decayed after a time t and the initial number tagged at t_1 as $|P_{\nrightarrow f_1}\rangle$ by the observation of the first decay f_1. We may then call this term the *total survival probability* of the state $|P_{\nrightarrow f_1}\rangle$.
With the probability of the reference transition and its time reversed theoretically and experimentally well-defined we present the Motion Reversal Asymmetry [9]:

$$A_R(f_1, f_2; t) = P_{12}(t) - P_{21}(t) \tag{15}$$

This new observable becomes entirely measurable in a 1^{--} meson factory for any pair of decay channels f_1, f_2. We have not imposed any particular condition to the pair of decay channels f_1, f_2, so one is not forced to use flavour specific or CP eigenstate decay channels. We will discuss later whether the measurable $A_R(f_1, f_2; t)$ becomes a genuine Time-Reversal-Violating Asymmetry for any pair of decay channels. This asymmetry can be written explicitly as

$$A_R(f_1, f_2; t) = P_{12}(t) - P_{21}(t) = [N_1(t) - N_2(t)] + [I_{21}(t) - I_{12}(t)] \tag{16}$$

The second piece in the r.h.s. corresponds to the typical $t \leftrightarrow -t$ asymmetry of the double decay rate in the meson factory: comparing the rate of a process decaying first one side to the final state f_1 and after a time t the other side decays to f_2 with the one with f_i reversed. The first piece is the difference of the total survival probability of the two states $|P_{\nrightarrow f_i}\rangle$: this is the new piece to be measured that allows working with the here proposed new reference transition.

4. The theoretical expressions

Using the time evolution imposed by Quantum Mechanics we know the time dependent structure of the needed measurable quantities in terms of Δm, Γ and $\Delta\Gamma$, determined by the eigenvalues of the entire Hamiltonian for the $(P^0, \bar{P}^0)$. The motion reversal asymmetry is then given by:

$$A_R(f_1, f_2; t) = e^{-\Gamma t} \left\{ \begin{array}{c} C^N[f_1, f_2](cosh\left(\frac{\Delta\Gamma t}{2}\right) - cos(\Delta m t)) + \\ +S_h^N[f_1, f_2] sinh\left(\frac{\Delta\Gamma t}{2}\right) + S_c^N[f_1, f_2] sin(\Delta m t) \end{array} \right\} \tag{17}$$

and depends on three "asymmetry parameters": the non-vanishing value of any of these asymmetry parameters would be a signal of Time-Reversal-Violation.
The theoretical connection of these measurable parameters to the matrix element of the meson evolution Hamiltonian H_{ij} involved in the Weisskoppf-Wigner Approach (WWA) [10, 11, 12] for the $(P^0, \bar{P}^0)$ system is

$$\begin{aligned} C^N[f_1,f_2] &= \delta(C_{f_1}-C_{f_2}) - Im(\theta)(S_{f_1}-S_{f_2}) \\ S_h^N[f_1,f_2] &= \delta(C_{f_1}R_{f_2}-C_{f_2}R_{f_1}) - Im(\theta)(S_{f_1}R_{f_2}-S_{f_2}R_{f_1}) \\ S_c^N[f_1,f_2] &= (C_{f_1}S_{f_2}-C_{f_2}S_{f_1})\{1+\delta(C_{f_1}+C_{f_2})\} + \\ &+\delta(S_{f_1}-S_{f_2}) - Re(\theta)(S_{f_1}R_{f_2}-S_{f_2}R_{f_1}) \end{aligned} \tag{18}$$

We have presented the results at leading order in terms of the usual mixing parameters -see the book by G. Branco et al. [8]- $\delta = (1-|q/p|^2)/(1+|q/p|^2)$, that is real, T and CP violating. As usual $(q/p)^2 = H_{21}/H_{12}$, and the complex parameter θ, that is CPT and CP violating, $\theta = (H_{22}-H_{11})/(\Delta m - i\Delta\Gamma/2)$. We also use the parameter R_i, S_i and C_i defined in terms of the well-known $\lambda_i = q\bar{A}_{f_i}/pA_{f_i}$

$$C_i = \frac{(1-|\lambda_i|^2)}{(1+|\lambda_i|^2)} \quad ; \quad S_i = \frac{2Im(\lambda_i)}{(1+|\lambda_i|^2)} \quad ; \quad R_i = \frac{2Re(\lambda_i)}{(1+|\lambda_i|^2)} \tag{19}$$

Note that these parameters are not all independent $C_i^2 + S_i^2 + R_i^2 = 1$.
It is worthwhile to point out that all these pieces in equation (18) should be small, except the term $(C_{f_1}S_{f_2} - C_{f_2}S_{f_1})$ that could be of order 1. In fact this term is the reminiscent of the BaBar T-violating measurement translated to the present context. If the asymmetry $A_R(f_1,f_2;t)$ were always a time reversal asymmetry this observable would represent the generalization of the T violating program to "any pair of decay channels"

5. Conditions to have a true T violating asymmetry

The principle of microreversibility [13] strictly means that T-invariance implies the cancellation of the asymmetry

$$A_T(f_1,f_2;t) = P_{12}(t) - P_{2_T1_T}(t) \tag{20}$$

Where

$$P_{1_T2_T}(t) = |\langle U_T P_{\nrightarrow f_2}|U_T P_{\nrightarrow f_1}(t)\rangle|^2 \tag{21}$$

This asymmetry is not exactly $A_R(f_1,f_2;t)$. U_T is the antiunitary time reversal operator $U_T = \widehat{U}_T K$, where $\widehat{U}_T$ is unitary and K is the complex conjugation operator. For pseudoscalars, the effect of $\widehat{U}_T$ is safe with the only change $\vec{p} \to -\vec{p}$ and using rotational invariance. The possible difficulty in assigning a genuine character of T-violation to the asymmetry $A_R(f_1,f_2;t)$ is concentrated on K. It is therefore clear that a sufficient condition that must satisfy *both* channels in order to be $A_R(f_1,f_2;t)$ a true T-violating asymmetry is that in the T invariant limit

$$U_T\,|P_{\nrightarrow f_i}\rangle = e^{i\phi}\,|P_{\nrightarrow f_i}\rangle \tag{22}$$

In the $(P^0, \bar{P}^0)$ meson space one can take

$$\widehat{U}_T = \begin{pmatrix} e^{i(\nu-\xi)} & 0 \\ 0 & e^{i(\nu+\xi)} \end{pmatrix} \tag{23}$$

where ν is the phase of the CPT operator $U_{CPT}|P^0\rangle = e^{i\nu}|\bar{P}^0\rangle$ and $U_{CP}|P^0\rangle = e^{-i\xi}|\bar{P}^0\rangle(U_{CP}|\bar{P}^0\rangle = e^{-i\xi}|P^0\rangle)$ defines de phase of the CP operator [8]. Therefore we can write

$$|P_{\nrightarrow f_i}\rangle = N^*_{f_i} A_{f_i} \left(\frac{\bar{A}_{f_i}}{A_{f_i}} |P^0\rangle - |\bar{P}^0\rangle \right)$$
$$U_T |P_{\nrightarrow f_i}\rangle = N^*_{f_i} A^*_{f_i} e^{i(\nu+\xi)} \left(\frac{\bar{A}^*_{f_i}}{A^*_{f_i}} e^{-2i\xi} |P^0\rangle - |\bar{P}^0\rangle \right) \tag{24}$$

And imposing equation (22) we can conclude that in order the motion reversal asymmetry $A_R(f_1, f_2; t)$ be a true T-violating asymmetry both channels f_i should satisfy, in the T invariant limit, the relation:

$$A^*_f \bar{A}_f = \bar{A}^*_f A_f e^{-2i\xi} \tag{25}$$

In the limit of T invariance it is well-known that $(q/p)^2 = e^{2i\xi}$ [8] therefore we have to choose decay channels f_i in $A_R(f_1, f_2; t)$ such that they verify in the T invariant limit

$$\lambda_{f_i} = \lambda^*_{f_i} \tag{26}$$

Or equivalently: channels that in the T invariant limit verifies

$$S_{f_i} = 0 \tag{27}$$

This condition seems quite reasonable but reveals the difficulties associated to T violating observables. Namely in the absence of final state interactions there only remains weak phases that should vanish in the T-invariant limit: therefore $S_{f_i} = 0$. But by the same token we discovered that also in these observables final state interactions can give fake T violation. If we look at the expression of $A_R(f_1, f_2; t)$ in equations (17) and (18) we discovered that $A_R(f_1, f_2; t)$=0 provided the quantity δ - violating T- is equal to zero, together with the condition $S_{f_1} = S_{f_2} = 0$. Therefore if the chosen channels f_i are such that S_{f_i} are T violating quantities, then $A_R(f_1, f_2; t)$ is a T violating quantity.

6. Channels that make the Motion Reversal Asymmetry a True T-Violation Asymmetry

1. Flavour specific channels verify $S_{f_i} = 0$ always. Therefore flavour specific channels can be used in $A_R(f_1, f_2; t)$.
2. If $U_T\,|f\rangle = e^{i\upsilon_f}\,|f'\rangle$, and we take rotationally invariant states $f = f'$ (for example a two body state with one spin zero particle and total spin zero) and *neglecting final state interactions* (FSI) $W = H_w$ one can show that in the T invariant limit

$$\bar{A}_f = \langle f | U_T^\dagger U_T H_w U_T^\dagger U_T |\bar{P}^0\rangle = e^{-i\upsilon_f} \bar{A}^*_f e^{-i(\nu+\xi)}$$
$$A_f = \langle f | U_T^\dagger U_T H_w U_T^\dagger U_T |P^0\rangle = e^{-i\upsilon_f} A^*_f e^{-i(\nu-\xi)} \tag{28}$$

that reproduces equation (25). This confirms our previous comment in a precise way: *any channel f_i where we can neglect FSI can be used in $A_R(f_1, f_2; t)$*.
3. As in the precedent case but including final state interactions one get under T invariance [13]

$$A_{\bar{f}}^{*} = \langle U_T f | S^{\dagger} U(\infty, 0) H_w | U_T P^0 \rangle = \sum_{\beta} \langle U_T f | S^{\dagger} | \beta \rangle \, e^{i(\nu - \xi)} A_{\beta}$$
$$\bar{A}_{\bar{f}}^{*} = \langle U_T f | S^{\dagger} U(\infty, 0) H_w | U_T \bar{P}^0 \rangle = \sum_{\beta} \langle U_T f | S^{\dagger} | \beta \rangle \, e^{i(\nu + \xi)} \bar{A}_{\beta} \quad (29)$$

where we have used the strong scattering matrix $S = U(+\infty, -\infty)$. If f is an eigenstate of the S matrix $< \beta | S | f >= e^{-i2\delta_f} \delta_{f,\beta}$ then it is straightforward for rotationally invariant f states to reproduce equation (25). We conclude that one can also use in $A_R(f_1, f_2; t)$ f_i eigenstates of the strong scattering S matrix. This can be relevant to study T violation at Φ factories [14] where kaons can decay to a two pion state with well-defined isospin: $(\pi\pi)_I$ that is an eigenstate of the strong S matrix.

4. If the CP operator acting on the state f is such that $CP|f\rangle = e^{-i\xi_f}|\bar{f}\rangle$ it is easy to prove that in the CP invariant limit in the decay amplitudes we get

$$\bar{A}_{\bar{f}} = e^{i(\xi_f - \xi)} A_f \quad ; \quad A_{\bar{f}} = e^{i(\xi_f + \xi)} \bar{A}_f \quad (30)$$

therefore in the case of CP eigenstates $f = \bar{f}$, and $e^{-i\xi_f} = \eta_f = \pm 1$ and we reproduce again the sufficient condition equation (25). Therefore we can use in our T violating asymmetry all the CP eigenstates channels f where we can neglect CP violation in the corresponding decay amplitudes. This argument can be transformed in a more general one related to CPT.

5. If we consider just decays to stays $f = \bar{f}$ mediated by CPT invariant decay amplitudes, then in the T invariant limit we also have CP invariance and consequently we get again the desired condition equation (25). An important difference with the previous case is that assuming CP invariance in the decay we just can use the decay $B^0 \to (\pi\pi)_{I=2}$ in the two pion mode. Under CPT invariance in the decay we can use both $B^0 \to \pi^+\pi^-$ and $B^0 \to \pi^0\pi^0$ [15]. Assuming CPT invariance in the decay amplitude we can use any CP eigenstate in our T violating asymmetry $A_R(f_1, f_2; t)$. Even if the asymmetries we are studying here are different from the one measured by BabBar, this result is conceptually in agreement with the conditions obtained in reference [16]
6. It is clear that if f is not a CP eigenstates, strong phases can make $S_f \neq 0$ even in the absence of weak phases (in the T invariant limit). This means that our asymmetry $A_R(f_1, f_2; t)$ cannot be used alone with this non CP eigenstate f. But let us assume CP invariance of the decay amplitudes, then it can be shown [8] that

$$\lambda_{\bar{f}} = \frac{q\bar{A}_{\bar{f}}}{pA_{\bar{f}}} = \frac{q}{p} \frac{e^{i(\xi_f - \xi)} A_f}{e^{i(\xi_f + \xi)} \bar{A}_f} = e^{-2i\xi} \frac{q}{p} \frac{A_f}{\bar{A}_f} = e^{-2i\xi} \left(\frac{q}{p}\right)^2 \frac{1}{\lambda_f} \quad (31)$$

and in the T invariant limit we get

$$\lambda_f \lambda_{\bar{f}} = 1 \quad (32)$$

From here it is straightforward to prove that assuming CP invariance in the decay amplitudes A_f and $\bar{A}_f$ the violation of the following identities will be a clear signal of T violation

$$C_f + C_{\bar{f}} = 0\,;\; S_f + S_{\bar{f}} = 0\,;\; R_f - R_{\bar{f}} = 0 \tag{33}$$

7. As in the case of CP eigenstates we can exchange the role of CP by CPT and still use the violation of equation (33) as a departure of T invariance. The fact that for example $S_f + S_{\bar{f}} \neq 0$ is a signal of T violation for non CP eigenstates suggest us to combine two asymmetries:

$$\langle A_R(f,g;t)\rangle_f = A_R(f,g;t) + A_R(\bar{f},g;t) \tag{34}$$

in such a way that if g fulfill the conditions to participate then the new *averaged* asymmetry is a T violating one provided the decay amplitudes A_f and $\bar{A}_f$ are invariant under CPT.

7. Avoiding the measurement of total survival probabilities

The T violating program here presented relays on the measurement of the asymmetry

$$A_R(f_1,f_2;t) = [N_1(t) - N_2(t)] + [I_{21}(t) - I_{12}(t)] \tag{35}$$

The second piece on the r.h.s. corresponds to the typical $t \leftrightarrow -t$ asymmetry of the double decay rate in the meson factory. The first piece is the difference of the total survival probability of the two states $|P_{\nrightarrow f_i}\rangle$. In order to measured $N_i(t)$ one has to have a good control of all decay channels f_i to be sure that our $|P_{\nrightarrow f_i}\rangle$ has not decay at time t. In several cases a very good approximation is $N_i(t) \sim e^{-\Gamma t}$ and therefore the contribution of the total survival probability almost cancels out. Nevertheless we have presented a method to eliminate these $N_i(t)$ contributions. The idea is to sum three T violating asymmetries to cancel out theses unwanted pieces in the following way:

$$\begin{aligned} A_3(f_1,f_2,f_3;t) &= A_R(f_1,f_2;t) + A_R(f_2,f_3;t) + A_R(f_3,f_1;t) = \\ &= I_{21}(t) - I_{12}(t) + I_{32}(t) - I_{23}(t) + I_{13}(t) - I_{31}(t) \end{aligned} \tag{36}$$

As long as the three asymmetries are T violating, we have constructed a T violating observable by comparing three time-inverted and normalized doubled decay rate processes at a meson factory. The result we obtain do not contain terms whose time dependence goes with $cosh(\Delta\Gamma t/2)$ or $cos(\Delta mt)$, these pieces coming from the total survival probabilities cancel out. We get therefore

$$A_3(f_1,f_2,f_3;t) = e^{-\Gamma t}\left\{S_h^N[f_1,f_2,f_3]sinh\left(\frac{\Delta\Gamma t}{2}\right) + S_c^N[f_1,f_2,f_3]sin(\Delta mt)\right\} \tag{37}$$

To simplify we choose Flavour Specific Channels (FS): $f_3 = FS$ and $f_2 = \overline{FS}$. Then if we chose f a CP eigenstates and assume CPT invariance in the decay to this channel, for any channel $f = f_{CP}$ it turns out that $A_3(f_{CP},\overline{FS},FS;t)$ is a genuine T violating asymmetry. *This procedure allows extending to any CP eigenstate the T violanting program -initiated by Babar- in any meson factory like Belle II, Daphne II etc*... The result in this case is

$$A_3(f,\overline{FS},FS;t) = e^{-\Gamma t}\left\{2\delta R_f \sinh\left(\frac{\Delta\Gamma t}{2}\right) + 2S_f(1+\delta C_f)sin(\Delta mt)\right\} \tag{38}$$

Controlled by δ and S_f as it should. The program can be extended to non CP eigenstates with the combination

$$A_3(f,\overline{FS},FS;t) + A_3(\bar{f},\overline{FS},FS;t) = \\ = e^{-\Gamma t}\left\{2\delta(R_f + R_{\bar{f}})sinh\left(\frac{\Delta\Gamma t}{2}\right) + 2(S_f + S_{\bar{f}})\left(1+\delta(C_f + C_{\bar{f}})\right)\sin(\Delta m t)\right\} \tag{39}$$

where the T violating pieces are proportional to δ or $(S_f + S_{\bar{f}})$ consistently with our previous analysis. This second formula represents *the extension of the T violating program to any non CP eigenstate*.
We have to stress that equations (17) and (34) already represent the generalizations of the BaBar program to any channel, with the appropriate assumptions. The difference with equations (37) and (39) involves the measurement of the total survival probabilities defined in equation (14) instead of combining several motion reversal asymmetries $A_R(f_1, f_2; t)$

8. Conclusions

To summarize, we have extended the T -Violation tests to any decay channel by means of a novel strategy: instead of searching which is the pair of decay channels associated to the time-reversed meson transition, we build a new asymmetry by selecting the decay channels in such a way that automatically tags the initial states of both the Reference and the T -reverse meson transitions. The connection to the experimental quantities requires the determination of the total survival probability of the tagged meson, as well as the filtering probability for the considered decay. We have analyzed the precise condition under which the proposed asymmetry is a genuinely T violating one. By combining appropriate asymmetries we can avoid the measurement of the total survival probabilities. With the proposal discussed here, the way is open to a full experimental program of studies of T -Violation observables at meson factories [17], using the quantum principles of entanglement and the decay as a filtering measurement.

Acknowledgments

We would like to acknowledge Nick Mavromatos for the great organization of Discrete 2014 and the excellent atmosphere we enjoyed at King College. This work has been supported by Spanish MINECO (FEDER) under grant FPA2011-23596 and by Generalitat Valenciana under grants PROMETEOII 2014-049 and PROMETEOII 2013-017. FJ B would like to acknowledge enlightening discussions with Aielet Efrati.

References

[1] Lees J, et al., Observation of Time Reversal Violation in the B0 Meson System, *Phys.Rev.Lett.* 109 (2012) 211801 (arXiv:1207.5832)
[2] Bernabeu J, Martinez-Vidal F, Villanueva-Perez P, Time Reversal Violation from the entangled B0-antiB0 system, *J. High Energy Phys.* JHEP 1208 (2012) **064** (arXiv:1203.0171).
[3] Banuls M, Bernabeu J, CP, T and CPT versus temporal asymmetries for entangled states of the B(d) system, *Phys.Lett.* B464 (1999) **117**-122 (hep-ph/9908353).
[4] Banuls M, Bernabeu J, Studying indirect violation of CP, T and CPT in a B factory, *Nucl.Phys.* B590 (2000) **19**-36 (hep-ph/0005323).
[5] Kabir P, What is not invariant under time reversal?, *Phys.Rev.* D2 (1970) **540**-542.
[6] Angelopoulos A et al., First direct observation of time reversal noninvariance in the neutral kaon system, *Phys.Lett.* B444 (1998) **43**-51.

[7] Lipkin H J, Simple symmetries in EPR correlated decays of kaon and B meson pairs with CP violation, *Phys.Lett.* B219 (1989) **474**

[8] Branco G C, Lavoura L, Silva J P, CP Violation, Inter. Ser. Monogr. Phys vol. 103, Oxford University Press, 1999.

[9] Bernabeu J, Botella F J, Nebot M, Novel T -Violation observable open to any pair of decay channels at meson factories, *Phys.Lett. B728* (2014) **95**-98 (arXiv:1309.0439)

[10] Weisskopf V, Wigner E P, Calculation of the natural brightness of spectral lines on the basis of Dirac's theory, *Z.Phys.* 63 (1930) **54**-73.

[11] Weisskopf V, Wigner, Over the natural line width in the radiation of the harmonius oscillator, *Z.Phys.* 65 (1930) **18**-29

[12] Lee T D, Oehme R, Yang C N, Remarks on Possible Noninvariance Under Time Reversal and Charge Conjugation, *Phys.Rev.* 106 (1957) **340**-345

[13] Sachs R G, The Physics of Time Reversal, The University of Chicago Press, 1-325(1987)

[14] Bernabeu J, Di Domenico A, Villanueva-Perez P, Direct test of time reversal symmetry in the entangled neutral kaon system at a Phi-factory, *Nucl.Phys.* B868 (2013) **102**-119, (arXiv: 1208.0773).

[15] Baek S, Botella F J, London D and Silva J P, Can one detect new physics in I=0 and/or I=2 contributions to the decays B -> pi pi?, *Phys.Rev.* D 72 (2005) 036004 (hep-ph/0506075).

[16] Applebaum E, Efrati A, Grossman Y, Nir Y, and Soreq Y, *Phys.Rev.*D89 (2014) 076011 (arXiv:1312.4164)

[17] Bevan A, Inguglia G, Zoccali M, Neutral meson tests of time-reversal symmetry invariance (arXiv:1302.4191)

A direct test of $\mathcal{T}$ symmetry in the neutral K meson system at KLOE-2

Aleksander Gajos on behalf of the KLOE-2 Collaboration
Marian Smoluchowski Institute of Physics, Jagiellonian University
Łojasiewicza 11, 30-348 Cracow, Poland

E-mail: aleksander.gajos@uj.edu.pl

Abstract. This work presents prospects for conducting a novel direct test of time-reversal symmetry at the KLOE-2 experiment. Quantum entanglement of neutral K meson pairs uniquely available at KLOE-2 allows to probe the T symmetry directly and independently of CP violation. This is achieved by a comparison of probabilities for a transition and its inverse obtained through an exchange of initial and final states. Such transitions between flavor and CP-definite states of the neutral kaons are only connected by the T conjugation which ensures the CP-independence of the test. While a similar measurement was performed by the BaBar experiment with neutral B mesons, the KLOE-2 detector can test T-violation in the neutral kaons system. Such a test requires i.a. reconstruction of the $K_L \to 3\pi^0$ decay accompanied by $K_S \to \pi^{\pm}\ell^{\mp}\nu$ with good timing information. Therefore a new reconstruction method for the $K_L \to 3\pi^0$ decay is also presented which is capable of reconstructing this process with decay time resolution of $\mathcal{O}(1\tau_S)$.

1. Introduction

The fact that $\mathcal{CP}$ symmetry is not conserved in the neutral meson system has been well known for almost fifty years. Conversely, the time-reversal symmetry in this system, although its violation should follow from the well confirmed $\mathcal{CPT}$ invariance, lacked a direct experimental evidence for decades after $\mathcal{CP}$ violation discovery with neutral kaons in 1964 [1]. The reason for this is that a direct measurement of $\mathcal{T}$ non-invariance would require observation of a probability asymmetry between a process and the same process inverted in time and such an experiment is difficult to realize for any unstable system.

Since for neutral mesons, as zero-spin particles, the transitions to compare are conjugated by an exchange of initial and final states, the oscillation phenomenon is one of the few processes that can be used to obtain both transitions. It was first used to test the time-reversal symmetry by the CPLEAR experiment in 1998, yielding a non-zero probability asymmetry in $K^0 \to \bar{K^0}$ and $\bar{K^0} \to K^0$ oscillations [2]. However, the fact that the initial and final states in this case are conjugated both by the $\mathcal{T}$ and $\mathcal{CP}$ operations has lead to a controversy as to whether this result can be attributed solely to $\mathcal{T}$ violation independently of $\mathcal{CP}$ violation. While some authors pointed out the role of decay as initial state interaction in this process [3, 4] and others argued that is not relevant in this case [5, 6], it remained clear that another way to directly test the time reversal symmetry violation independently of $\mathcal{CP}$ would be highly desirable [4].

A suitable measurement comparing selected transitions between $\mathcal{CP}$-definite and flavour-definite states with their time-reversal conjugates was proposed for the neutral B meson system by Bernabeu *et al.* in 2012 [7] and soon followed by a similar proposition realizable with neutral kaons [8]. For the transitions' reversibility, quantum entanglement in the system must be used which is available at B-factories and at the DAΦNE ϕ-factory. Among the former, the BaBar experiment has already measured a $\mathcal{T}$-violating asymmetry at 14σ level [9] constituting the first direct observation of $\mathcal{T}$-violation in transitions that are only connected by the time-reversal transformation. In case of K mesons, since entangled $K_S K_L$ pairs are uniquely available at the DAΦNE collider, KLOE-2 is the only experiment presently able to provide the first time-reversal violation evidence in the neutral kaon system. Prospects for such a measurement at KLOE-2 and first analysis steps are presented in the remainder of this work.

2. A $\mathcal{T}$-symmetry test independent of $\mathcal{CP}$

The principle of the time-reversal symmetry test at KLOE-2 is based on defining transitions between states of neutral kaons being either states with definite strangeness $\{K^0\ (S=+1), \bar{K^0}\ (S=+1)\}$ or eigenstates of the $\mathcal{CP}$ operator which can be expressed using the former as:

$$|K_+\rangle = \frac{1}{\sqrt{2}}\left[|K^0\rangle + |\bar{K^0}\rangle\right] \qquad (\mathcal{CP}=+1), \tag{1}$$

$$|K_-\rangle = \frac{1}{\sqrt{2}}\left[|K^0\rangle - |\bar{K^0}\rangle\right] \qquad (\mathcal{CP}=-1). \tag{2}$$

State of a neutral kaon can be identified in one of these two bases at the moment of its decay by observation of the decay final state. With an assumption of the $\Delta Q = \Delta S$ rule (well tested in semileptonic kaon decays [10]), each charge of secondary lepton can only be created from a strangeness-definite state of the decaying kaon, as shown in the diagrams in Figure 1. In the following, final states with a positively and negatively charged leptons will be indicated as ℓ^+ and ℓ^-, respectively.

Figure 1. Diagrams of semileptonic decays of neutral kaons. The $\Delta S = \Delta Q$ rule guarantees that the state with a positively-charged lepton comes from K^0 decay whereas $\bar{K^0}$ always decays into a final state with a lepton of negative charge.

On the other hand, hadronic final states which are $\mathcal{CP}$-even (e.g. two pions, hereafter named the $\pi\pi$ final state) can only be produced from the K_+ kaon state and $\mathcal{CP}$-odd states like $3\pi^0$ come from decays of K_-. It can be shown that $\mathcal{CP}$ violation may be safely neglected in these considerations [8].

Although the above facts allow for identification of the decaying kaon, observing a transition between two kaon states also requires tagging of the kaon state also a certain time before its decay. This can be achieved using neutral kaon pairs available at KLOE-2 which exhibit quantum entanglement of their states. Once the state of the first decaying kaon is identified by observation of its decay, its entangled and still-living partner is known to be in an orthogonal state at the same time. Consequently, observation of its decay after a time interval Δt can lead to the observation of a kaon transition where both initial and final state are tagged in the strangeness or $\mathcal{CP}$-basis.

3. Time-reversal symmetry violation observables at KLOE-2

There are four possible transitions between flavour and $\mathcal{CP}$-definite states of neutral K mesons, as listed in Table 1. As these transitions are not irreversible, there can be a $\mathcal{T}$-conjugated process defined for each of them. Comparison of rates between each transition and its time-reversal conjugate would constitute a direct $\mathcal{T}$ symmetry test and its independence of $\mathcal{CP}$ violation follows from a comparison of the $\mathcal{T}$-and $\mathcal{CP}$-conjugated processes shown in Table 1, which are not identical.

Table 1. Four possible transitions between flavour and $\mathcal{CP}$-definite states of neutral kaons and their time-reversal conjugates obtained by exchange of initial and final states. The $\mathcal{T}$-conjugated transitions are not identical with the $\mathcal{CP}$ conjugates given in the last column. Each of the transitions is experimentally identified by a time-ordered pair of decays given in parentheses.

	Transition	$\mathcal{T}$-conjugate	$\mathcal{CP}$-conjugate
1	$\mathrm{K}^0 \to \mathrm{K}_+$ $(\ell^-, \pi\pi)$	$\mathrm{K}_+ \to \mathrm{K}^0$ $(3\pi^0, \ell^+)$	$\bar{\mathrm{K}^0} \to \mathrm{K}_+$
2	$\mathrm{K}^0 \to \mathrm{K}_-$ $(\ell^-, 3\pi^0)$	$\mathrm{K}_- \to \mathrm{K}^0$ $(\pi\pi, \ell^+)$	$\bar{\mathrm{K}^0} \to \mathrm{K}_-$
3	$\bar{\mathrm{K}^0} \to \mathrm{K}_+$ $(\ell^+, \pi\pi)$	$\mathrm{K}_+ \to \bar{\mathrm{K}^0}$ $(3\pi^0, \ell^-)$	$\mathrm{K}^0 \to \mathrm{K}_+$
4	$\bar{\mathrm{K}^0} \to \mathrm{K}_-$ $(\ell^+, 3\pi^0)$	$\mathrm{K}_- \to \bar{\mathrm{K}^0}$ $(\pi\pi, \ell^-)$	$\mathrm{K}^0 \to \mathrm{K}_-$

The time-reversal asymmetry can be defined using ratios of probabilities of transitions at time Δt to probabilities of their $\mathcal{T}$-inverses at the same time. Four theoretical ratios may be defined this way:

$$\begin{aligned} R_1(\Delta t) &= \frac{P[\mathrm{K}^0(0) \to \mathrm{K}_+(\Delta t)]}{P[\mathrm{K}_+(0) \to \mathrm{K}^0(\Delta t)]}, & R_2(\Delta t) &= \frac{P[\mathrm{K}^0(0) \to \mathrm{K}_-(\Delta t)]}{P[\mathrm{K}_-(0) \to \mathrm{K}^0(\Delta t)]}, \\ R_3(\Delta t) &= \frac{P[\bar{\mathrm{K}^0}(0) \to \mathrm{K}_+(\Delta t)]}{P[\mathrm{K}_+(0) \to \bar{\mathrm{K}^0}(\Delta t)]}, & R_4(\Delta t) &= \frac{P[\bar{\mathrm{K}^0}(0) \to \mathrm{K}_-(\Delta t)]}{P[\mathrm{K}_-(0) \to \bar{\mathrm{K}^0}(\Delta t)]}. \end{aligned} \tag{3}$$

Each of the transitions used in the above ratios is experimentally identified by a time-ordered pair of kaon decays given in brackets in Table 1. For transitions 1 and 3, the conjugated process would involve the first kaon to decay into $3\pi^0$, for which little statistics is expected as the search for $\mathcal{CP}$-violating decay $\mathrm{K_S} \to 3\pi^0$ at KLOE yielded no candidates [11]. For the remaining transitions 2 and 4, however, large statistics is available at KLOE-2 and their probabilities can be measured through numbers of double kaon decays to states f_1, f_2 separated by time Δt, denoted as $I(f_1, f_2; \Delta t)$. Experimental equivalents of ratios R_2 and R_4 are thus defined as follows:

$$R_2^{exp}(\Delta t) = \frac{\mathrm{I}(l^-, 3\pi^0; \Delta t)}{\mathrm{I}(\pi\pi, l^+; \Delta t)} = R_2(\Delta t) \times \frac{C(l^-, 3\pi^0)}{C(\pi\pi, l^+)}, \tag{4}$$

$$R_4^{exp}(\Delta t) = \frac{\mathrm{I}(l^+, 3\pi^0; \Delta t)}{\mathrm{I}(\pi\pi, l^-; \Delta t)} = R_4(\Delta t) \times \frac{C(l^+, 3\pi^0)}{C(\pi\pi, l^-)}. \tag{5}$$

The theoretical ratios $(R_i(\Delta t))$ are easily extracted from R_2^{exp} and R_4^{exp} as shown in Eq. 4, 5 by using coefficients dependent on neutral kaon branching fractions and widths [8]:

$$\frac{C(l^-, 3\pi^0)}{C(\pi\pi, l^+)} \simeq \frac{C(l^+, 3\pi^0)}{C(\pi\pi, l^-)} \simeq \frac{\mathrm{BR}(\mathrm{K_L} \to 3\pi^0)\,\Gamma_L}{\mathrm{BR}(\mathrm{K_S} \to \pi\pi)\,\Gamma_S}, \tag{6}$$

all of which have been measured by KLOE [12] and whose precision should be further improved by KLOE-2 [13].

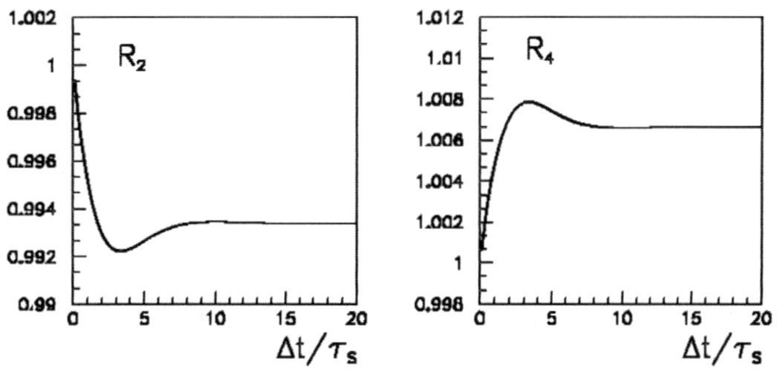

Figure 2. Expected behaviour of the R_i ratios as a function of decay time difference. The figure was adapted from [8].

Finally, determination of an asymptotic behaviour of theoretical ratios $R_2(\Delta t)$ and $R_4(\Delta t)$ would measure violation of time-reversal symmetry in the neutral kaon system. Figure 2 shows the expected dependence of these ratios with assumed time-reversal violation. If $\mathcal{T}$ was conserved, the ratios should tend to unity for large time intervals Δt. Indeed, it can be shown [8] that asymptotic discrepancy of $R_2(\Delta t)$ and $R_4(\Delta t)$ from 1 is related to the $\mathcal{T}$-violating parameter ϵ as:

$$R_2(\Delta t \gg \tau_S) \simeq 1 - 4\Re\epsilon, \quad (7)$$

$$R_4(\Delta t \gg \tau_S) \simeq 1 + 4\Re\epsilon. \quad (8)$$

Therefore, a measurement of R_2^{exp} and R_4^{exp} as functions of Δt for large time differences at KLOE-2 will provide measurement of degree of $\mathcal{T}$ symmetry violation in the neutral K meson system.

4. KLOE-2 and DAΦNE

4.1. The DAΦNE ϕ-factory

DAΦNE is an electron-positron collider located at the accelerator complex of INFN National Laboratory of Frascati (LNF). It is composed of two separate storage rings, each storing a beam of 0.51GeV. The rings intersect at two regions, one of which provides events to the KLOE detector (Figure 3). The center-of-mass energy of the colliding beams ($\sqrt{s}$ ≈1020MeV) is the mass of the phi-meson resonance, which is produced with a cross-section of about 3μb.

The ϕ-mesons are produced almost at rest ($\beta_\phi \approx 0.015$), with only a small momentum component in the direction perpendicular to the beam axis. Their almost immediate decays ($\tau_\phi = 1.55 \pm 0.01 \times 10^{-22}$s) provide pairs of K mesons, either charged (with a branching fraction of 48.9%) or neutral (34.2%). The K^+K^- and K_LK_S pairs are widely used to study kaon properties as the possibility of tagging kaons by their partner's decay allows absolute branching ratios to be measured. Furthermore, the neutral kaon pairs provided by DAΦNE have an additional unique feature. Due to quantum numbers conservation in a strong ϕ meson decay, the K_SK_L pairs are produced in an anti-symmetric zero-strangeness state:

$$|i\rangle = \frac{1}{\sqrt{2}}\left(\left|K^0(+\vec{p})\right\rangle \left|\bar{K^0}(-\vec{p})\right\rangle - \left|\bar{K^0}(+\vec{p})\right\rangle \left|K^0(-\vec{p})\right\rangle\right), \quad (9)$$

or as expressed in the basis of $\mathcal{CP}$ eigenstates:

$$|i\rangle = \frac{1}{\sqrt{2}}\left(|\mathrm{K}_+(+\vec{p})\rangle |\mathrm{K}_-(-\vec{p})\rangle - |\mathrm{K}_-(+\vec{p})\rangle |\mathrm{K}_+(-\vec{p})\rangle\right). \quad (10)$$

This initial state of the neutral kaon system produced by DAΦNE exhibits quantum entanglement of the two kaon states in the genuine EPR sense [14]. What follows is that whereas state of each kaon is undefined until the observation of the first decay, measurement of one of the states (by observation of its decay) guarantees the partner particle to be in the orthogonal state at the same time.

This feature allows for a broad range of kaon interferometry based tests of fundamental symmetries [15] such as CPT and Lorentz Symmetry tests [16]. Moreover, it is the property which opens a possibility of a time-reversal symmetry at KLOE-2.

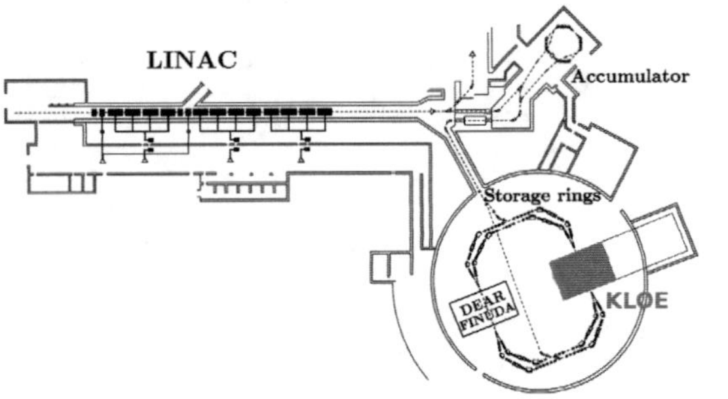

Figure 3. The accelerator complex at LNF. The DAΦNE collider is composed of two storage rings intersecting at two points, at one of which the KLOE detector is located. The figure was adapted from [12].

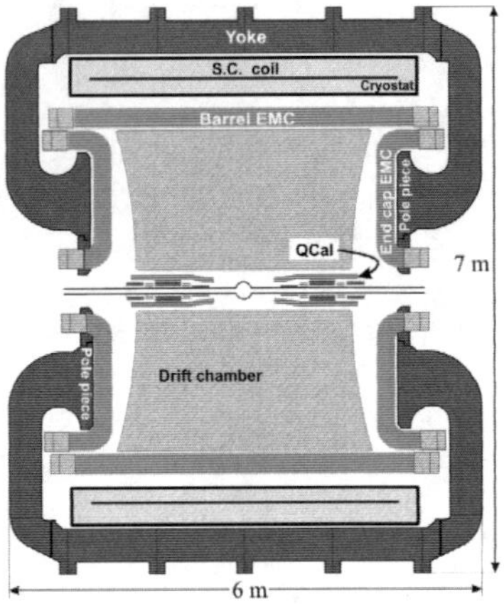

Figure 4. Longitudinal section of the KLOE detector. The figure was adapted from [12].

4.2. The KLOE detector

Tho KLOE (K LOng Experiment) detector is shaped as a barrel surrounding one of the DAΦNE interaction points, with a radius of 2 m and length of almost 3.5 m. Its large size is dictated by the mean path travelled by the long-lived neutral kaons produced there, which is about 3.4 m. The detector is constituted by a cylindrical drift chamber surrounded by an electromagnetic calorimeter (see the longitudinal detector section in Figure 4) which provides good coverage around the interaction point (98% of solid angle).

The drift chamber [17] uses a gas mixture based on helium (90%) and isobutane (10%) which ensures small material budget in order to prevent K_L regeneration and photon conversion in the gas. The resolution of the KLOE Drift Chamber (DC) is 150 μm in the transverse plane and 2 mm in the z direction. Momentum of charged particles can be reconstructed with the relative resolution of 0.4% whereas vertexing resolution is of the order of 1 mm [17]. The whole detector is immersed in a magnetic field of 0.52 T provided by a superconducting coil.

The calorimeter [18] is a sampling electromagnetic detector with scintillating fibers as active material and passive lead layers to enhance electromagnetic shower production. It provides an excellent timing resolution and high detection efficiency for photons in the 20–500 MeV energy range [18], which is crucial for the reconstruction presented in section 5.2. Spatial and time resolution of the KLOE calorimeter is given below:

$$\sigma_t = \frac{54\,ps}{\sqrt{E[GeV]}} \oplus 140ps, \quad \sigma_E = \frac{5.7\%E}{\sqrt{E[GeV]}}, \quad \sigma_x = \sigma_y = 1\,cm, \quad \sigma_z = \frac{1.2\,cm}{\sqrt{E[GeV]}} \qquad (11)$$

4.3. Upgrade of KLOE to KLOE-2

KLOE started its operation at the DAΦNE collider in 1999 and was taking data collecting a total integrated luminosity of 2.5 fb^{-1} which corresponds to about 10^{10} of produced ϕ mesons. During KLOE runs DAΦNE reached a peak value of instantaneous luminosity of $1.4 \cdot 10^{32}\ cm^{-2}s^{-1}$.

In the recent years, the KLOE detector has undergone a thorough upgrade to start new measurements as KLOE-2. The upgrades involved the addition of new calorimeters at small angles around the beam axis to increase acceptance for photons and instrument the final focusing region [19, 20]. Moreover, new sets of detectors have been installed close to the beam line for tagging $\gamma\gamma$ events through detection of scattered high and low energy e^+e^- [21, 22]. Finally, a novel tracking device was constructed and installed in the region between the interaction point and drift chamber inner wall. This KLOE-2 Inner Tracker is a pioneer construction of a cylindrical Gas Electron Multiplier (GEM) tracker [23] composed of 4 layers of a triple-GEM detector barrel-shaped around the KLOE-2 interaction point. Its addition to KLOE-2 improves vertexing capabilities and increases acceptance for tracks with low transverse momentum.

The KLOE-2 detector restarted operation in 2014 in view of collecting a sample of the order of 10 fb^{-1} in the next years. The physics programme of KLOE-2 is rich [13] and includes the possibility to conduct the first direct $\mathcal{T}$ symmetry test with neutral kaons described in this work.

5. Prospects for experimental realization of the $\mathcal{T}$ test at KLOE-2

5.1. Required reconstruction of events

In order to measure the double decay rates used in R_2^{exp} and R_4^{exp} (Eq. 4–5) reconstruction must be performed for the following two classes of events, each composed of a pair of kaon decays separated by a time interval of Δt:

$$\phi \to \mathrm{K_S K_L} \quad \to 2\pi\ \pi^{\mp}\ell^{\pm}\nu, \tag{12}$$

$$\phi \to \mathrm{K_S K_L} \quad \to \pi^{\pm}\ell^{\mp}\nu\ 3\pi^0. \tag{13}$$

As the time-reversal symmetry test observables (Eq. 4–5) are time-dependent, it is important to reconstruct the time interval between decays with good accuracy. In turn, the resolution of kaon decay times in the asymptotic plateau region of $\Delta t \gg \tau_S$ should be $\mathcal{O}(1\tau_S)$ which to translates to decay point spatial resolution $\mathcal{O}(1\ cm)$.

Figure 5 schematically shows the reconstruction of these two classes of events. In case of a two-pion final state, charged pions ($\pi^+\pi^-$) may be chosen rather than $\pi^0\pi^0$ to profit from pion tracks reconstructed by the KLOE drift chamber. Both kaon decay vertices can then be easily reconstructed using charged particle tracks (Figure 5, left). The other class of processes, shown in Figure 5, right, is significantly more challenging to reconstruct as one of the decays, $\mathrm{K_L} \to 3\pi^0$, only involves neutral particles. Moreover, as the kinematics of the accompanying $\mathrm{K_S}$ decay is not closed due to a missing neutrino, reconstruction of $\mathrm{K_L} \to 3\pi^0$ must rely solely on information on clusters created in the EMC by photons originating in π^0 decays. Therefore a special vertex reconstruction method was devised for this decay.

5.2. New $\mathrm{K_L} \to 3\pi^0$ reconstruction method for $\mathrm{K_S K_L} \to \pi^{\pm}e^{\mp}\nu_e\ 3\pi^0$

The new reconstruction method presented in this work aims at providing spatial coordinates of the $\mathrm{K_L}$ decay point and $\mathrm{K_L}$ decay time by using only the photons from $\mathrm{K_L} \to 3\pi^0 \to 6\gamma$ recorded by the electromagnetic calorimeter (Figure 6).

Each EMC cluster contains information on its location as well as recording time (X_i, Y_i, Z_i, T_i). Thus, one might consider a set of possible points at which the corresponding photon originated. Such a set constitutes a sphere centered at cluster position with a radius equal to path length travelled by the γ. The latter, however, is unknown due to its dependence on the unknown decay time of the kaon t, which is also the γ creation time if the π^0 lifetime

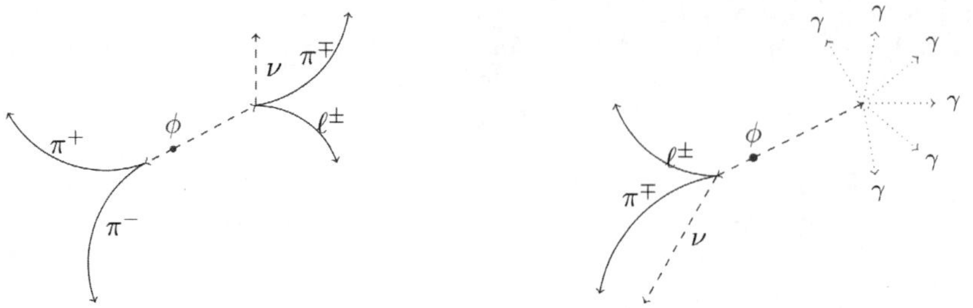

Figure 5. Schematic view of the two classes of processes whose reconstruction is required to measure R_2^{exp} and R_4^{exp} for the $\mathcal{T}$ test. Continuous lines denote DC tracks of charged particles while dashed and dotted lines correspond to neutral particles. Process shown in the right poses a reconstruction challenge due to the $\mathrm{K_L} \to 3\pi^0$ decay involving neutral particles only.

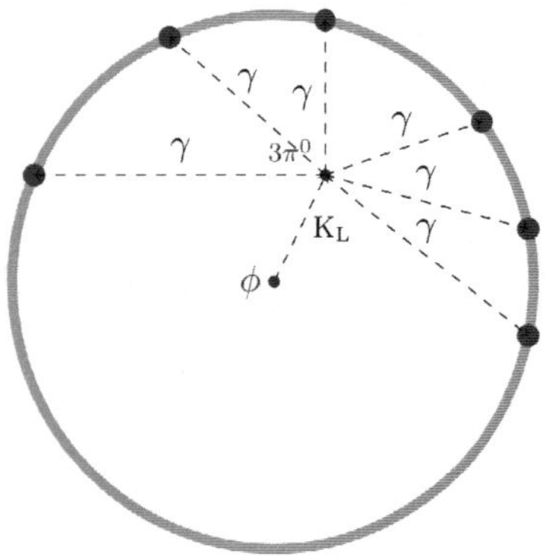

Figure 6. The $\mathrm{K_L} \to 3\pi^0 \to 6\gamma$ process in the cross-section view of the detector. The neutral kaon nor the pions are recorded by the tracking detectors and the only experimental information on this decay are up to six clusters in the KLOE electromagnetic calorimeter (grey band) created by photons to which the neutral pions decay almost immediately.

is neglected (Figure 7). The equation describing the sphere of origin points for i-th calorimeter cluster is:

$$(T_i - t)^2 c^2 = (X_i - x)^2 + (Y_i - y)^2 + (Z_i - z)^2. \tag{14}$$

Such a sphere can be defined for each recorded EMC cluster. The $\mathrm{K_L}$ decay vertex is the common origin of all the photons and therefore can be reconstructed as the intersection of the photon spheres as shown in Figure 6. If at least 4 photons are recorded, their corresponding system of equations (Eq. 14) can be solved analytically. This yields two solutions for the $\mathrm{K_L}$ decay vertex coordinates and decay time, among which the physical solution is identified by a set of criteria and the other one is rejected as a mathematical artifact. Possible presence of up to six recorded photons is used to numerically solve an overdetermined system of equations to obtain improved accuracy.

It should be stressed that this reconstruction method directly yields the kaon decay time in addition to vertex location, which eliminates the need to calculate it with kaon travelled path length and momentum which would introduce additional uncertainty to its evaluation. In conjunction with the excellent timing properties of the KLOE electromagnetic calorimeter, this reconstruction yields a good resolution of long-lived kaon decay time although being based on calorimetric information only.

Journal of Physics: Conference Series **631** (2015) 012018 doi:10.1088/1742-6596/631/1/012018

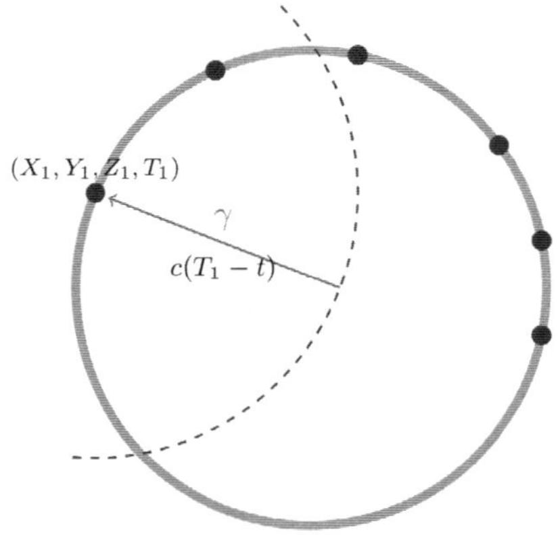

Figure 7. For each of the EMC clusters its position and recording time is reconstructed by the calorimeter with a good resolution (see 11). This allows to define a set of possible origin points of the incident photon as a sphere centered at the cluster whose radius is proportional to the difference of cluster time and K_L decay time.

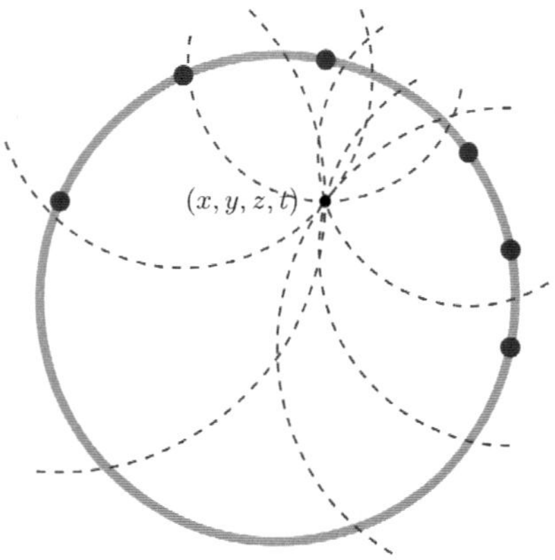

Figure 8. The K_L decay vertex is also a common origin points of the 6 photons which hit the calorimeter. Therefore, its location can be found analytically as the intersection of at least 4 of the spheres defined for each EMC clusters.

5.3. Resolution of K_L *decay time reconstruction*

The reconstruction was tested using a MC simulated sample of $K_L \to 3\pi^0$ events (generated with the official KLOE software [24]) including complete detector response and run-by-run data taking conditions. Time resolution, crucial for the time-reversal symmetry test at KLOE-2, was studied as a function of the decay vertex distance from the ϕ-meson decay vertex, i.e. the path length travelled by the kaon before its decay. The result, shown in Figure 9 proves that this reconstruction algorithm has an almost constant resolution at the level of $2\tau_S$ independently of the decay vertex location. This is a promising result for the future test of time-reversal symmetry at KLOE-2 and studies on the required processes will continue towards this test. This will allow $\mathcal{T}$ to be tested in the asymptotic region of $\Delta t \gg \tau_S$ as stated in Section 5.1. This new reconstruction method for the challenging $K_L \to 3\pi^0$ decay opens the way for KLOE-2 to provide the first direct evidence for time-reversal symmetry violation in the neutral kaon system.

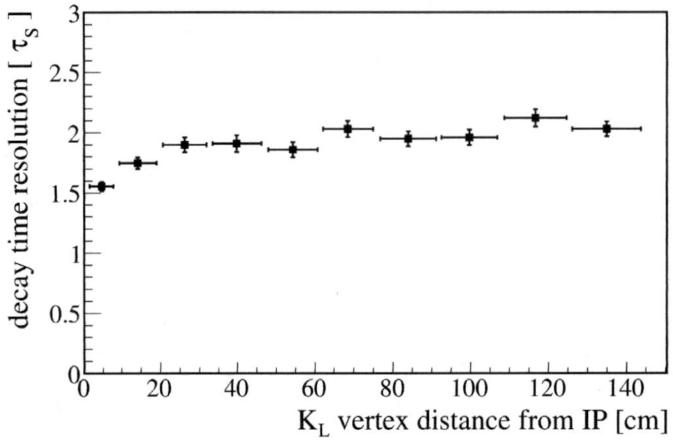

Figure 9. The K_L decay time resolution as a function of the decay distance from ϕ decay point.

Acknowledgments

We warmly thank our former KLOE colleagues for the access to the data collected during the KLOE data taking campaign. We thank the DAΦNE team for their efforts in maintaining low background running conditions and their collaboration during all data taking. We want to thank our technical staff: G.F. Fortugno and F. Sborzacchi for their dedication in ensuring efficient operation of the KLOE computing facilities; M. Anelli for his continuous attention to the gas system and detector safety; A. Balla, M. Gatta, G. Corradi and G. Papalino for electronics maintenance; M. Santoni, G. Paoluzzi and R. Rosellini for general detector support; C. Piscitelli for his help during major maintenance periods.

This work was supported in part by the EU Integrated Infrastructure Initiative Hadron Physics Project under contract number RII3-CT- 2004-506078; by the European Commission under the 7th Framework Programme through the *Research Infrastructures* action of the *Capacities* Programme, Call: FP7-INFRASTRUCTURES-2008-1, Grant Agreement No. 227431; by the Polish National Science Centre through the Grants No. 2011/03/N/ST2/02641, 2011/01/D/ST2/ 00748, 2011/03/N/ST2/02652, 2013/08/M/ST2/00323, DEC-2014/12/S/ST2/00459, 2013/11/B/ST2/04245 and by the Foundation for Polish Science through the MPD programme and the project HOMING PLUS BIS/2011-4/3.

References

[1] Christenson J, Cronin J, Fitch V and Turlay R 1964 *Phys.Rev.Lett.* **13** 138–140
[2] Angelopoulos A *et al.* (CPLEAR Collaboration) 1998 *Phys.Lett.* **B444** 43–51
[3] Wolfenstein L 1999 *Int.J.Mod.Phys.* **E8** 501–511
[4] Bernabeu J and Martinez-Vidal F 2014 (*Preprint* 1410.1742)
[5] Ellis J R and Mavromatos N 1999 *Phys.Rept.* **320** 341–354 (*Preprint* hep-ph/9903386)
[6] Gerber H 2004 *Eur.Phys.J.* **C35** 195–196
[7] Bernabeu J, Martinez-Vidal F and Villanueva-Perez P 2012 *JHEP* **1208** 064 (*Preprint* 1203.0171)
[8] Bernabeu J, Di Domenico A and Villanueva-Perez P 2013 *Nucl.Phys.* **B868** 102–119 (*Preprint* 1208.0773)
[9] Lees J *et al.* (BaBar Collaboration) 2012 *Phys.Rev.Lett.* **109** 211801 (*Preprint* 1207.5832)
[10] Beringer J *et al.* (Particle Data Group) 2012 *Phys.Rev.* **D86** 010001
[11] Babusci D *et al.* (KLOE Collaboration) 2013 *Phys.Lett.* **B723** 54–60 (*Preprint* 1301.7623)

[12] Bossi F, De Lucia E, Lee-Franzini J, Miscetti S and Palutan M (KLOE Collaboration) 2008 *Riv.Nuovo Cim.* **31** 531–623 (*Preprint* 0811.1929)

[13] Amelino-Camelia G, Archilli F, Babusci D, Badoni D, Bencivenni G *et al.* 2010 *Eur.Phys.J.* **C68** 619–681 (*Preprint* 1003.3868)

[14] Einstein A, Podolsky B and Rosen N 1935 *Phys. Rev.* **47** 777–780

[15] Di Domenico A 2007 *Frascati Physics Series* **43** 1–38

[16] Babusci D *et al.* (KLOE-2 Collaboration) 2014 *Phys.Lett.* **B730** 89–94 (*Preprint* 1312.6818)

[17] Adinolfi M, Ambrosino F, Andryakov A, Antonelli A, Antonelli M *et al.* 2002 *Nucl.Instrum.Meth.* **A488** 51–73

[18] Adinolfi M, Ambrosino F, Antonelli A, Antonelli M, Anulli F *et al.* 2002 *Nucl.Instrum.Meth.* **A482** 364–386

[19] Cordelli M, Corradi G, Happacher F, Martini M, Miscetti S *et al.* 2010 *Nucl.Instrum.Meth.* **A617** 105–106 (*Preprint* 0906.1133)

[20] Cordelli M, Dan E, Giovannella S, Gatta M, Happacher F *et al.* 2013 *Nucl.Instrum.Meth.* **A718** 81–82

[21] Babusci D, Gonnella F, Iafolla L, Iannarelli M, Mascolo M *et al.* 2013 *Nucl.Instrum.Meth.* **A718** 577–579 (*Preprint* 1206.0680)

[22] Babusci D, Bini C, Ciambrone P, Corradi G, De Santis A *et al.* 2010 *Nucl.Instrum.Meth.* **A617** 81–84 (*Preprint* 0906.0875)

[23] Balla A, Bencivenni G, Branchini P, Budano A, Capodiferro M *et al.* 2013 *Nucl.Instrum.Meth.* **A732** 221–224

[24] Ambrosino F, Antonelli A, Antonelli M, Bini C, Bloise C *et al.* 2004 *Nucl.Instrum.Meth.* **A534** 403–433 (*Preprint* physics/0404100)

Lepton Mixing Predictions from (Generalised) CP and Discrete Flavour Symmetry

Thomas Neder

School of Physics and Astronomy, University of Southampton, Southampton, SO17 1BJ, U.K.

E-mail: t.neder@soton.ac.uk

Abstract. An important class of flavour groups that are subgroups of $U(3)$ and that predict experimentally viable lepton mixing parameters including Majorana phases, is the $\Delta(6n^2)$ series. The most well-known member is $\Delta(24) = S_4$. I present results of several extensive studies of lepton mixing predictions obtained in models with a $\Delta(6n^2)$ flavour group that preserve either the full Klein symmetry or a Z_2 subgroup for neutrinos and can include a generalised CP symmetry. Predictions include mixing angles and the Dirac CP phase generally; and if invariance under a generalised CP symmetry is included, also Majorana phases. For this, the interplay of flavour group and generalised CP symmetry has to be studied carefully.

1. Introduction

This presentation concerns itself with two concepts that at a first glance seem to be entirely separate, whose interplay however is actually non-trivial and rather interesting.

The first concept is the idea of promoting simultaneous conjugation of charge and parity of fields (CP) to a symmetry of a model: This means that at some high energy one requires the Lagrangian of the model not to change when every field is replaced by its CP-conjugate. In the Standard Model (SM), this is not the case and the question is whether this violation of CP-invariance happens spontaneously. To study this question, one needs to look at models that are explicitly invariant under CP conjugation at some high energy. In the work presented here, CP is added to the symmetries of the Standard Model in the unbroken state. However, it will not be discussed how exactly the breaking of CP invariance happens, because it is possible to determine the possible low-energy predictions independently of the breaking mechanism. Generally, these predictions would be quite general, but in the interplay with discrete flavour symmetries they become discrete (i.e. predictive) and specific. In particular will it be possible to obtain predictions for the values of all mixing parameters, including the Majorana phases.

In the Standard Model, the three generations (or flavours, or families) of fermions are completely independent and are merely copies of each other. Flavour symmetries [1, 2] provide a link between the three generations by assigning them to representations of a flavour group in one way or another and demanding that the Lagrangian remains unchanged when fields are transformed according to their representation under the flavour group. In this article, only left-handed leptons are considered because this is already sufficient to arrive at low energy predictions for their mixing matrix. Furthermore, the three doublets of leptons are assigned to a 3-dimensional irreducible representation of the flavour group G_F. Thus in the unbroken state the theory is to be invariant under all of the following transformations allowed by G_F

$$\begin{pmatrix} (e,\nu_e) \\ (\mu,\nu_\mu) \\ (\tau,\nu_\tau) \end{pmatrix} \mapsto \rho(g) \begin{pmatrix} (e,\nu_e) \\ (\mu,\nu_\mu) \\ (\tau,\nu_\tau) \end{pmatrix} \tag{1}$$

where $\rho(g)$ is a 3×3 matrix that correponds to group elements g of the flavour group G_F in the representation under which the three generations of leptons transform. The invariance under the flavour group will eventually need to be broken and from breaking it differently for charged leptons and neutrinos the mixing matrix will arise. Again independently of the actual mechanism, imagine therefore that G_F is broken to different subgroups, namely G_ν for neutrinos and G_e for charged leptons: The Lagrangian will be invariant under the following transformations in the broken state

$$\begin{pmatrix} e \\ \mu \\ \tau \end{pmatrix} \mapsto \rho(g_e) \begin{pmatrix} e \\ \mu \\ \tau \end{pmatrix} \quad \text{and} \quad \begin{pmatrix} \nu_e \\ \nu_\mu \\ \nu_\tau \end{pmatrix} \mapsto \rho(g_\nu) \begin{pmatrix} \nu_e \\ \nu_\mu \\ \nu_\tau \end{pmatrix} \tag{2}$$

where now $\rho(g_e)$ and $\rho(g_\nu)$ are 3×3-matrices that correspond to elements of the subgroups that remain unbroken but only in their sector, G_e and G_ν respectively, again in the same representation as in the unbroken state.

After having defined flavour symmetries above, the same needs to and will be done for CP transformations in the next section. There, it will also be discussed why one has to be careful how to define CP transformations in the presence of several generations of fermions and in particular if the theory is to be invariant under a flavour symmetry. This leads to a consistency requirement which CP and flavour transformations have to obey if not one is to accidentally violate invariance under the other.

In the section thereafter, it will be explained how predictions for the mixing matrix can be obtained from this symmetry-based approach in the different types of models with flavour symmetries. For the most predictive class of models with flavour symmetries, only $\Delta(6n^2)$ groups remain viable flavour symmetry groups.[1] After that, the predictions for mixing parameters including Majorana phases will be presented for two classes of models employing $\Delta(6n^2)$ groups with no and one free unitary rotation, respectively. It will turn out that despite the free continuous parameters, the predictions are rather constrained.

In the final section, conclusions are drawn and possible directions for future work are indicated.

2. (Generalised) CP Transformations

The kind of CP transformations that are sometimes called canonical CP transformations act on a scalar field $\phi(x)$ in the following way:

$$\phi(x) \mapsto \phi^{CP} = e^{i\varphi}\phi^*(x^P) \tag{3}$$

where φ is a phase that might depend on the kind of field and $x^P = (t, -\mathbf{x})$ is the parity-conjugated spacetime coordinate. But now imagine that several copies of the field ϕ exist in such a way that the Lagrangian allows for unitary rotations transforming these copies into each other. If all of the copies of ϕ would still transform under CP in the canonical way, then one could think that the CP properties of the set of copies would be basis-dependent, which is unphysical. However, this argument is wrong, because the unitary rotations mix the different states in an

[1] The $\Delta(6n^2)$ group are not always the minimal flavour group for a certain mixing pattern, however, all candidate groups are subgroups of a $\Delta(6n^2)$ group and the 3-dimensional irreducible representations are inherited by $\Delta(6n^2)$.

arbitrary way which is incompatible with canonical CP transformations acting diagonally on the copies of fields. So already the existence of several copies of a field that allow changes of basis demands that a CP transformation allows for these changes of basis [3, 4]. This can be achieved by promoting the phase $e^{i\varphi}$ to a unitary matrix, here called X, where now Φ is the multiplet of the copies of the field ϕ:

$$\Phi(x) \mapsto \Phi^{CP} = X\Phi^*(x^P). \tag{4}$$

Indeed it can be shown, that in situations similar to the one above, canonical CP can be violated while observables of CP show no CP violation because generalised CP is conserved correctly [6]. This means that in such a situation the correct definition of CP is the generalised one because it accommodates for changes of basis. The above applies exactly in the same way to several copies of fermions because in CP transformations of fermions only an additional matrix appears that acts on spinor indices and not among copies of the fields and thus can be neglected without loss of generality.

This situation is complicated further if the theory already is invariant under some symmetry group that acts on the same multiplets that allow for generalised CP transformations. In this case flavour and CP symmetry transformations need to fulfill a consistency condition [5, 6, 4]: Assume that the Lagrangian is invariant under flavour transformations like in Eq.(1) of a flavour group G_F. A matrix X can only appear in a generalised CP transformation acting on the multiplet of leptons if G_F contains elements g and g' such that the following equation is fulfilled

$$X^\dagger \rho^*(g) X = \rho(g') \tag{5}$$

where $\rho(g)$ and $\rho(g')$ are the 3×3-matrices that correspond to g and g' in the representation of the leptons. Using this condition, all X matrices allowed in the presence of some group G_F (not necessarily discrete) can be listed. Firstly, for ρ a faithful representation of G_F, i.e. if ρ as a function from the group into the space of unitary[2] matrices of the right dimension is injective, the consistency condition defines a mapping of the group onto itself:

$$u(g) := \rho^{-1}\left(X^\dagger \rho^*(g) X\right) \tag{6}$$

This mapping can be shown to be an automorphism of G_F [5, 6]. Furthermore, for unitary representations ρ of G_F, u maps group elements into the class of the inverse element. To prove this, take the trace of the consistency equation Eq.(5) and use the fact that group elements in the same conjugacy class have the same trace:

$$\mathrm{tr}(\rho(g')) = \mathrm{tr}(X^\dagger \rho(g)^* X) = \mathrm{tr}(\rho(g)^*) = \mathrm{tr}(\rho(g^{-1})). \tag{7}$$

Thus all unitary matrices X for which u maps every group element into the class of its inverse facilitate generalised CP transformations.

For the groups considered in this paper, the X matrices allowed in CP transformations correspond exactly to the representation matrices of the flavour group [6, 7]:

$$\{X\} = e^{i\alpha}\rho(G) \tag{8}$$

with an additional phase α that will cancel in physical predictions but is shown here for completeness.

When the flavour group G_F is broken to its subgroups G_e and G_ν in the different sectors, then invariance under CP can be spontaneously broken as well. This follows from the fact that X matrices that are consistent with G_e or G_ν in one sector respectively are not necessarily consistent in the other sector.

[2] For finite groups, for any representation a basis exists where the representation matrices are unitary.

3. Residual Symmetries and Flavour Symmetries

Consider a multiplet of Majorana or Dirac fields Ψ with projections on left/right-handed components $\Psi_{L/R}$ that have the following mass terms after all symmetry breakings have occured:

$$\mathcal{L}_{\text{Majorana}} = \Psi_L^T M_\nu \Psi_L \text{ or } \mathcal{L}_{\text{Dirac}} = \Psi_L^\dagger M_e \Psi_R + \text{h.c.} \tag{9}$$

Both Majorana and Dirac mass matrices can be diagonalised by appropriate unitary matrices $U_{e,L/R} = (u_1^e, u_2^e, u_3^e)$ or $U_\nu = (u_1^\nu, u_2^\nu, u_3^\nu)$ respectively, where the lower case $u_i^{\nu/e}$ denote the columns of the matrices. Invariance under the transformations in Eq.(2) puts the following constraints on the mass matrices:

$$\rho(g_\nu)^T M_\nu \rho(g_\nu) = M_\nu \text{ and } \rho(g_e)^\dagger M_e M_e^\dagger \rho(g_e) = M_e M_e^\dagger. \tag{10}$$

The convention chosen is that the mixing matrices act actively on the multiplets of fields such that the physical mixing matrix that appears in the vertex of W-boson, neutrino, and antifermion becomes $U_{\text{PMNS}} = U_{e,L}^\dagger U_\nu$.

For Majorana fermions one can define the following matrices:

$$G_1 = +u_1^\nu (u_1^\nu)^T - u_2^\nu (u_2^\nu)^T - u_3^\nu (u_3^\nu)^T, \tag{11}$$

$$G_2 = -u_1^\nu (u_1^\nu)^T + u_2^\nu (u_2^\nu)^T - u_3^\nu (u_3^\nu)^T, \tag{12}$$

$$G_3 = -u_1^\nu (u_1^\nu)^T - u_2^\nu (u_2^\nu)^T + u_3^\nu (u_3^\nu)^T. \tag{13}$$

Together with the identity matrix these form a $Z_2 \times Z_2$ group. For 3 massive and non-degenerate neutrinos this is the maximal residual symmetry in the following sense: The subgroup G_ν of G_F that is conserved by the Majorana mass term needs to be a subgroup of a $Z_2 \times Z_2$ group (or equal to it) or the constraints imposed on M_ν from the invariance under G_ν force neutrinos to be degenerate or even massless.[3]

Depending on how much of this residual symmetry is identified with a subgroup of G_F, models can be classified and will have different properties: Firstly, models where the whole residual $Z_2 \times Z_2$ is identified with G_ν are called direct models. In this case, all 3 G_i are known in a certain basis and the mixing matrix of the neutrinos is fixed in that basis, even including the Dirac phase but not including the Majorana phases. The reason for this is that the ith column of the mixing matrix is an eigenvector of G_i with eigenvalue +1:

$$G_i u_i^\nu = +u_i^\nu. \tag{14}$$

The Majorana phases remain free because they are just overall phases on the columns. In this case, all mixing matrices that are allowed in a G_F-symmetric model can be obtained by listing all $Z_2 \times Z_2$ subgroups of G_F. If on the other hand $G_\nu \simeq Z_2$ and G_ν is identified with one of the Z_2 factors of the residual $Z_2 \times Z_2$, a model would be called semidirect and only one of the columns of the mixing matrix is determined from symmetry as an eigenvector of the corresponding G_i. Again, it is possible to list all Z_2 subgroups of G_F. Every Z_2 subgroup gives rise to a class of mixing matrices that can be parametrized as

$$U = U_{Z_2} U_{2\times 2} \tag{15}$$

where U_{Z_2} is a unitary matrix with one of the three columns fixed as an eigenvector of the G_i coming from G_F and the other columns arbitrary but fixed throughout the analysis. $U_{2\times2}$ is a

[3] Technically one could extend the residual symmetry by the negative identity matrix to $Z_2 \times Z_2 \times Z_2$ but this would not impose any further constraints on the mixing matrix.

unitary matrix that rotates the two free columns and e.g. in case of having the first column of U_{Z_2} determined by $Z_2 \subset G_F$ has the form

$$U_{2\times 2} = \begin{pmatrix} 1 & 0 & 0 \\ 0 & \cos(\theta) & \sin(\theta)e^{i\varphi} \\ 0 & -\sin(\theta)e^{-i\varphi} & \cos(\theta) \end{pmatrix} \tag{16}$$

where θ and ϕ are real parameters and analogous for the other columns.

If the flavour group G_F is broken completely in the sector of neutrinos, $G_\nu \simeq \{1\}$, then no constraints on the mixing matrix follow from the invariance under G_F in this immediate way from symmetry but can be obtained in a different way [2]. In this work, direct and semidirect models are considered.

What is peculiar about the residual symmetry is that it is an accidental symmetry that always exists as defined by Eqs.(11-13) independently of it being incorporated into a bigger group.[4]

Similar to the way described in this section so far, also the mass term of a multiplet of Dirac fermions has a maximal residual symmetry which forms a $U(1) \times U(1)$ group[5]

$$\left\{ U_e \begin{pmatrix} e^{i\alpha} & 0 & 0 \\ 0 & e^{i\beta} & 0 \\ 0 & 0 & e^{-i\alpha-i\beta} \end{pmatrix} U_e^\dagger : \alpha, \beta \in [0, 2\pi) \right\} \simeq U(1) \times U(1). \tag{17}$$

In particular, $U(1) \times U(1)$ contains all groups of the kind $Z_p \times Z_q$ as subgroups with q, p natural numbers. If the form of one of the Z_q factors is known from G_F and $q \geq 3$, then the mixing matrix can be completely determined as the eigenvectors of the generator of this Z_q group. Only if $q = 2$, then similarly to the semidirect case for Majorana fermions, only one column of the matrix U_e is fixed. The latter case will be called charged-lepton-semidirect and in this case the mixing matrix of the charged leptons, U_e, will contain a free unitary rotation as in Eqs.(15) and (16). This is interesting because in contrast to the neutrino-semidirect case, in the physical mixing matrix the free rotation will act on rows instead of columns which opens the field for new mixing predictions as will be seen in the discussion of semidirect models.

The discreteness of the residual symmetry for Majorana neutrinos could be seen as a motivation for considering discrete flavour groups G_F: Generally (and as will be shown for an example later) continuous groups allow for an infinite number of $Z_2 \times Z_2$ subgroups and thus a discrete and specific prediction for the mixing matrix from symmetry would become impossible. On the other hand, Dirac fermions would be perfectly happy with a continuous residual symmetry.

Many discrete groups were considered as candidates for G_F in direct, semidirect and indirect models, often motivated by the conjecture that the reactor neutrino mixing angle is zero, $\theta_{13} = 0$. After the measurement of a non-zero θ_{13} [8, 9, 10], many of these models were ruled out and searches for other groups started, in particular using the group theory program GAP [11]. In two such searches [12, 13], it was discovered that only groups that belong to the class of groups called $\Delta(6n^2)$ [14] could provide columns that would remain viable in direct or semidirect models. Motivated by this finding all $\Delta(6n^2)$ groups were studied for their possible predictions in direct and semidirect models [7, 15, 16, 17, 18]. Later it was proved [19] by listing all possible mixing matrices in direct models for general finite groups, that only $\Delta(6n^2)$ groups and certain

[4] Although also semidirect models are considered in this work, this could lead to the conjecture that if it is impossible to identify the residual symmetry with a subgroup of G_F that one has chosen the wrong G_F or that it is simply not big enough. In a way, direct models thus seem more natural, because all accidental symmetries are accounted for.

[5] Again, one could add another $U(1)$ which corresponds to an overall phase on the mass matrix which will not impact the subsequent discussion.

subgroups thereof can provide mixing matrices that remain experimentally viable. As mentioned earlier, the procedure of deriving the structure of mixing matrices from a partially broken flavour group cannot predict Majorana phases. On the other hand, by combining a flavour symmetry with a CP symmetry consistent with it, all phases of the mixing matrices can be calculated and are predicted to particular discrete values in a direct model or can additionally only depend on up to two parameters that are strongly constrained by measurements in semidirect models. Taking this as an incentive, the possible predictions for mixing matrices from $\Delta(6n^2)$ groups combined with their consistent CP transformations were examined in [7, 17, 18]. It is the results and consequences of these studies, starting from [14], that will be reported in the following.

The groups $\Delta(6n^2)$ are isomorphic to a semidirect product

$$\Delta(6n^2) \simeq (Z_n \times Z_n) \rtimes S_3 \tag{18}$$

and are not as obscure as the name or notation suggests. An easy way to present these groups is in terms of four generators called a, b, c, d, where a and b generate the S_3 subgroup and c and d each generate one Z_n subgroup. c and d commutate, while a and b have to fulfill the rules of cycling three objects and interchanging two:

$$a^3 = b^2 = (ab)^2 = 1 \tag{19}$$

The semidirect product, "$\rtimes$", imposes conditions on products of a, b with c, d [14]. This presentation has the advantage that every group element $g \in \Delta(6n^2)$ can be expressed by the generators in the following way

$$g = a^\alpha b^\beta c^\gamma d^\delta \tag{20}$$

with $\alpha = 0, 1, 2$; $\beta = 0, 1$; $\gamma, \delta = 0, \ldots, n-1$. In a way the groups can be thought of as two cyclic groups with n elements with permutations of three objects acting on them. The $\Delta(6n^2)$ groups for small n are well-known: $\Delta(6) = S_3$, $\Delta(24) = S_4$, $\Delta(54)$ and $\Delta(96)$. Constraining the representation of lepton doublets to a 3-dimensional one, the $\Delta(6n^2)$ groups still have to offer $2n-1$ different representations. However, it can be shown that it is sufficient to only consider one of these representations [15]. In the representation chosen without loss of generality for this analysis, the generators have the following matrix form:

$$a = \begin{pmatrix} 0 & 1 & 0 \\ 0 & 0 & 1 \\ 1 & 0 & 0 \end{pmatrix}, b = -\begin{pmatrix} 0 & 0 & 1 \\ 0 & 1 & 0 \\ 1 & 0 & 0 \end{pmatrix}, c = \begin{pmatrix} \eta & 0 & 0 \\ 0 & \eta^{-1} & 0 \\ 0 & 0 & 1 \end{pmatrix}, d = \begin{pmatrix} 1 & 0 & 0 \\ 0 & \eta & 0 \\ 0 & 0 & \eta^{-1} \end{pmatrix} \tag{21}$$

with $\eta = e^{2\pi i/n}$. To arrive at the mixing matrix predictions, first all $Z_2 \times Z_2$ subgroups are listed as candidates for G_ν. Concerning G_e, only Z_3 groups turn out to produce experimentally viable mixing matrices, although other candidates for G_e had been considered [15]. Knowing G_ν and G_e in a direct model completely determines the mixing matrix including the Dirac CP phase except for the ordering of rows and columns. It is found that one column always is trimaximal, i.e. $\propto (1, 1, 1)^T$ and has to serve as middle column of the mixing matrix. Additionally, one requires that the entry with the smallest absolute value becomes U_{13}^{PMNS}. The only freedom that remains after that is interchanging the second and third row, which corresponds to two different predictions for θ_{23}. As the Majorana phases cannot be predicted using this symmetry approach, the effect of additionally imposing CP transformations consistent with $\Delta(6n^2)$ was examined. It can be shown that the allowed X matrices are identical to the representation matrices under which the leptons transform. Only CP transformations that are consistent with G_ν can be imposed onto the neutrino mass matrix M_ν, while analogously only CP transformations that are consistent with G_e can be used to constrain the charged lepton mass matrix. After this procedure, in the PDG convention [20] all experimentally viable mixing matrices have the form

$$U_{\text{PMNS}} = \begin{pmatrix} \sqrt{\frac{2}{3}}\cos\left(\frac{\pi\gamma}{n}\right) & \frac{1}{\sqrt{3}} & \sqrt{\frac{2}{3}}\sin\left(\frac{\pi\gamma}{n}\right) \\ -\sqrt{\frac{2}{3}}\sin\left(\pi\left(\frac{\gamma}{n}+\frac{1}{6}\right)\right) & \frac{1}{\sqrt{3}} & \sqrt{\frac{2}{3}}\cos\left(\pi\left(\frac{\gamma}{n}+\frac{1}{6}\right)\right) \\ \sqrt{\frac{2}{3}}\sin\left(\pi\left(\frac{1}{6}-\frac{\gamma}{n}\right)\right) & -\frac{1}{\sqrt{3}} & \sqrt{\frac{2}{3}}\cos\left(\pi\left(\frac{1}{6}-\frac{\gamma}{n}\right)\right) \end{pmatrix} \begin{pmatrix} 1 & 0 & 0 \\ 0 & [i]ie^{-i6\pi(\gamma+x)/n} & 0 \\ 0 & 0 & [i]i \end{pmatrix} \tag{22}$$

with $\gamma, x = 0, \ldots, n-1$ and the parts in square brackets correspond to different possible choices of the Majorana phases. Each value of γ correponds to a different $Z_2 \times Z_2$ subgroup. In figure 1 the possible predictions for the value of $|U_{13}^{\text{PMNS}}|$ in direct models with $\Delta(6n^2)$ flavour groups is shown. There, each dot corresponds to a different choice of γ and thus a different $Z_2 \times Z_2$ subgroup. When 3 divides n, points are missing because for such $\Delta(6n^2)$ groups those subgroups are missing which in a way renders such groups more predictive.[6] In the limit $n \to \infty$, a prediction of the value of θ_{13} becomes impossible. As

$$\Delta(6n^2) \simeq (Z_n \times Z_n) \rtimes S_3 \xrightarrow{n\to\infty} (U(1) \times U(1)) \rtimes S_3, \tag{23}$$

this constitutes an example for a continuous group that does not predict discrete and specific values of a parameter in a direct model. For an arbitrary but fix γ, it is currently not possible to determine the ordering of the second and third row from experiment. Thus for each value of θ_{13} two values of θ_{23} are predicted. This can also be expressed by the sum rule

$$\theta_{23} = 45^\circ \mp \theta_{13}/\sqrt{2}. \tag{24}$$

A further important prediction is that the Dirac CP phase is zero: $\delta_{CP} = 0$ whereas global fits are starting to show a slight preference for a non-zero value [22, 23, 24]. To summarize, these predictions are general for all finite groups where the lepton-doublets transform in a 3-dimensional representation and where the flavour group is broken to a $Z_2 \times Z_2 2$ subgroup for neutrinos. If the lepton mixing parameters are experimentally found to differ from the above predictions and corrections through renormalisation cannot account for the difference, then direct flavour models with finite groups are excluded as a paradigm.[7]

The previous discussion exhausts the topic of direct models with discrete flavour groups. In the following we focus on the possible predictions for the lepton mixing matrix that can be obtained in semidirect models with a $\Delta(6n^2)$ flavour group. We consider breaking the flavour group in such a way that only one unitary rotation is free as in Eqs.(15) and (16). This will entail neutrino-semidirect breaking, where $G_\nu \simeq Z_2$ while the charged lepton mixing matrix is completely fixed, i.e. $G_e \simeq Z_2 \times Z_2, Z_p$, as well as charged-lepton-semidirect breaking, where $G_e \simeq Z_2$ and $G_\nu \simeq Z_2 \times Z_2$. Each combination of G_e and G_ν will provide one column of the physical mixing matrix in the neutrino-semidirect case and one row of the mixing matrix in the charged-lepton-semidirect case. In table 3 the fixed columns for each combination are shown for $G_\nu \simeq Z_2$. A tick mark denotes that this column is experimentally viable as one of the columns of the physical mixing matrix, possibly requiring reordering of rows of the matrix. In table 3 the predicted rows of the mixing matrix are shown, again not barring reordering.

[6] The group with n=42 produces no predictions within the three sigma range, contrasting well-regarded hints in the literature [21].

[7] In a way, as continuous groups will not produce specific predictions in a direct model, one could think that the most important motivation for considering discrete groups lies in the Majorana nature of neutrinos. If discrete groups fail to predict the correct mixing matrix one could thus be tempted to consider doubting the Majorana nature of neutrinos.

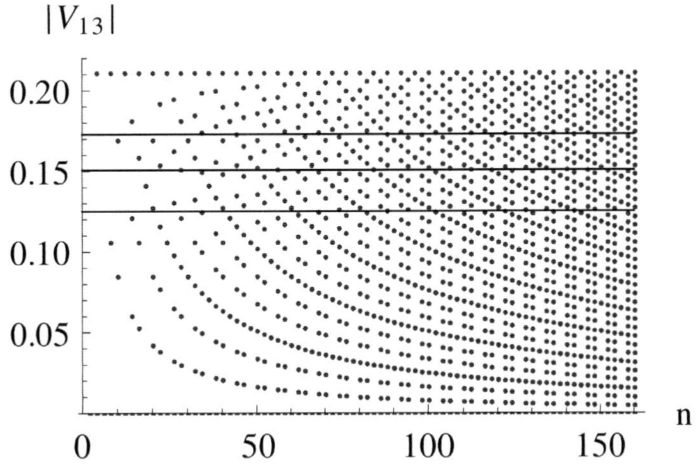

Figure 1. The possible predictions for the value of $|U_{13}^{\rm PMNS}|$ in direct models with $\Delta(6n^2)$ flavour groups. Each dot corresponds to a different choice $Z_2 \times Z_2$ subgroup and the horizontal lines indicate the central value and three sigma range from [23]. When 3 divides n, points are missing because for such $\Delta(6n^2)$ groups those subgroups are missing.

Out of all possible combinations of subgroups shown in tables 1 and 2 only three different columns are found for $G_\nu \simeq Z_2$ that are not experimentally excluded, while only one distinct row is found for $G_e \simeq Z_2$. Even when combining these columns and rows with a free unitary rotation of the remaining two columns or rows respectively and allowing for arbitrary permutations of the mixing matrices, only five classes of mixing matrices are produced that are not immediately excluded by experiment. The full form of all matrices can be found in [17]. There, the mixing matrices originating from $G_l = \langle ac^s d^t \rangle$ and $G_\nu \simeq Z_2^{bc^x d^x}$ allow for two permutations that produce viable mixing parameters. In figures 3-6, these are shown in the left and middle panel in the top row for each observable where only values of mixing parameters are shown for combinations of group parameters for which every measured observable lies within the experimental limits. The predictions from $G_e = \langle abc^s d^t \rangle$ and $G_\nu = Z_2^{bc^x d^x}$ are shown in the top right panel for each observable, the predictions from $G_e = \langle ac^s d^t \rangle$ and $G_\nu = Z_2^{c^{n/2}}$ in the bottom left panel and the predictions from $G_e = \langle ac^s d^t \rangle$ and $G_\nu = Z_2^{c^{n/2}}$ in the bottom right panel.

In figures 3-6, it can be seen that despite the additional free parameters, the predictions for mixing parameters can often only take discrete or small ranges of values. The predictions are specific, i.e. if certain experimental values are measured, first the broad class of mixing matrices can be identified, further the particular n that identifies a group and eventually the unbroken subgroups G_e and G_ν. For very large n, the predictions tend to lie increasingly dense until a certain continuous range is filled in the limit $n \to \infty$. Nevertheless, correlations between parameters that are distinct for the different classes of mixing matrices allow to differentiate them.

4. Conclusions and Future Work

In this presentation, lepton mixing predictions in direct and semidirect models invariant under $\Delta(6n^2)$ groups and consistent CP transformations have been examined. In direct models, all mixing angles and the Dirac CP phase are predicted purely from symmetry principles. The inclusion of broken invariance under consistent CP transformations additionally allows to predict

	$G_\nu = Z_2^{bc^x d^x}$	$G_\nu = Z_2^{c^{n/2}}$
$G_l = \langle c^s d^t \rangle$	$\frac{1}{\sqrt{2}}\begin{pmatrix} 0 \\ -1 \\ 1 \end{pmatrix}$ ✗	$\begin{pmatrix} 0 \\ 0 \\ 1 \end{pmatrix}$ ✗
$G_l = \langle bc^s d^t \rangle$	$\begin{pmatrix} 0 \\ \cos\left(\frac{s+t-2x}{2n}\pi\right) \\ \sin\left(\frac{s+t-2x}{2n}\pi\right) \end{pmatrix}$ ✗	$\frac{1}{\sqrt{2}}\begin{pmatrix} 0 \\ -1 \\ 1 \end{pmatrix}$ ✗
$G_l = \langle ac^s d^t \rangle$	$\sqrt{\frac{2}{3}}\begin{pmatrix} \sin\left(\frac{s-x}{n}\pi\right) \\ \cos\left(\frac{\pi}{6} - \frac{s-x}{n}\pi\right) \\ \cos\left(\frac{\pi}{6} + \frac{s-x}{n}\pi\right) \end{pmatrix}$ ✓	$\frac{1}{\sqrt{3}}\begin{pmatrix} 1 \\ 1 \\ 1 \end{pmatrix}$ ✓
$G_l = \langle abc^s d^t \rangle$	$\frac{1}{2}\begin{pmatrix} 1 \\ 1 \\ -\sqrt{2} \end{pmatrix}$ ✓	$\begin{pmatrix} 0 \\ 0 \\ 1 \end{pmatrix}$ ✗
$G_l = \langle a^2bc^s d^t \rangle$	$\frac{1}{2}\begin{pmatrix} 1 \\ 1 \\ -\sqrt{2} \end{pmatrix}$ ✓	$\frac{1}{\sqrt{2}}\begin{pmatrix} 0 \\ -1 \\ 1 \end{pmatrix}$ ✗

Table 1. Candidate columns predicted in the neutrino-semidirect case, i.e. $G_\nu \simeq Z_2$ for all combinations of subgroups. A tick mark means that this column is viable as one of the columns of the physical mixing matrix, possibly requiring reordering of rows and columns of the mixing matrix. A fail mark indicates that this column is not viable.

	$G_l = Z_2^{bc^{x'} d^{x'}}$	$G_l = Z_2^{c^{n/2}}$
$G_\nu = K_4^{(c^{n/2}, d^{n/2})}$	$\frac{1}{\sqrt{2}}\begin{pmatrix} 1 \\ -1 \\ 0 \end{pmatrix}^T$ ✗	$\begin{pmatrix} 1 \\ 0 \\ 0 \end{pmatrix}^T$ ✗
$G_\nu = K_4^{(c^{n/2}, abc^y)}$	$\frac{1}{2}\begin{pmatrix} 1 \\ 1 \\ -\sqrt{2} \end{pmatrix}^T$ ✓	$\begin{pmatrix} 1 \\ 0 \\ 0 \end{pmatrix}^T$ ✗
$G_\nu = K_4^{(d^{n/2}, a^2bd^z)}$	$\frac{1}{2}\begin{pmatrix} 1 \\ 1 \\ -\sqrt{2} \end{pmatrix}^T$ ✓	$\frac{1}{\sqrt{2}}\begin{pmatrix} 1 \\ -1 \\ 0 \end{pmatrix}^T$ ✗
$G_\nu = K_4^{(c^{n/2}d^{n/2}, bc^x d^x)}$	$\begin{pmatrix} \cos\left(\frac{x-x'}{n}\pi\right) \\ -i\sin\left(\frac{x-x'}{n}\pi\right) \\ 0 \end{pmatrix}^T$ ✗	$\frac{1}{\sqrt{2}}\begin{pmatrix} 1 \\ -1 \\ 0 \end{pmatrix}^T$ ✗

Table 2. Candidate columns predicted in the charged-lepton-semidirect case, i.e. $G_l \simeq Z_2$ for all combinations of subgroups. A tick mark means that this column is viable as one of the columns of the physical mixing matrix, possibly requiring reordering of rows and columns of the mixing matrix. A fail mark indicates that this column is not viable.

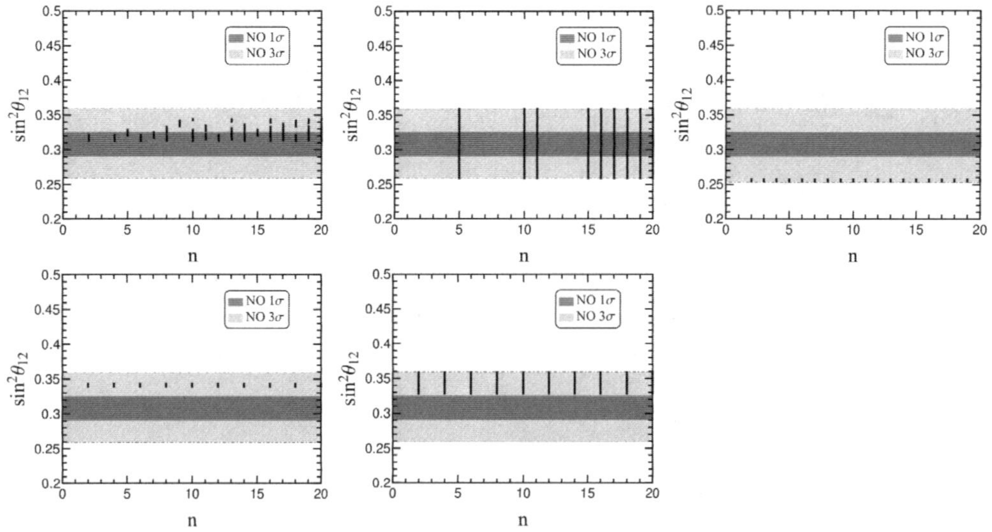

Figure 2. Predictions for $\sin^2\theta_{12}$ for the five classes of experimentally viable mixing matrices for semidirect models with a $\Delta(6n^2)$ flavour group. The coloured bands show the one and three sigma bands respectively.

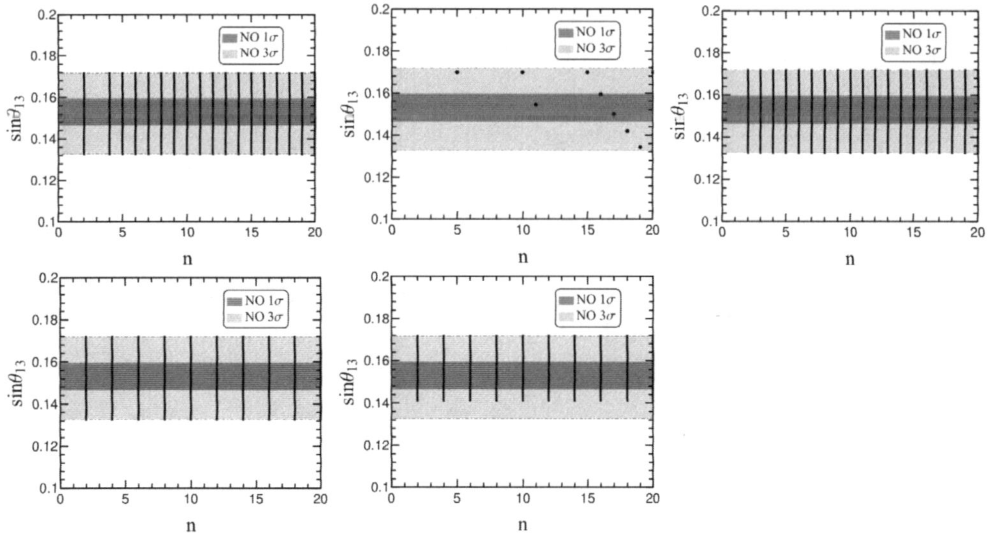

Figure 3. Predictions for $\sin\theta_{13}$ for the five classes of experimentally viable mixing matrices for semidirect models with a $\Delta(6n^2)$ flavour group. The coloured bands show the one and three sigma bands respectively.

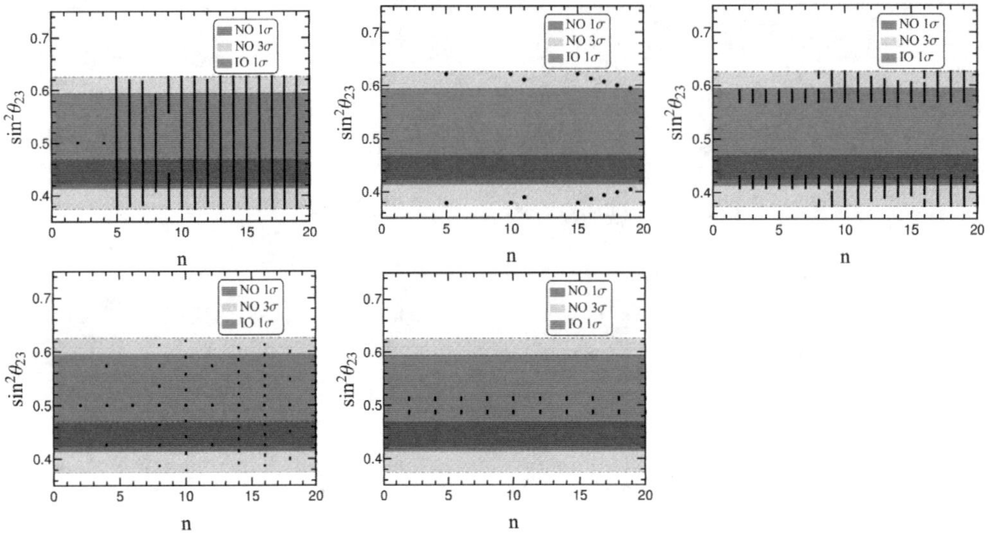

Figure 4. Predictions for $\sin^2\theta_{23}$ for the five classes of experimentally viable mixing matrices for semidirect models with a $\Delta(6n^2)$ flavour group. The coloured bands show the one and three sigma bands respectively.

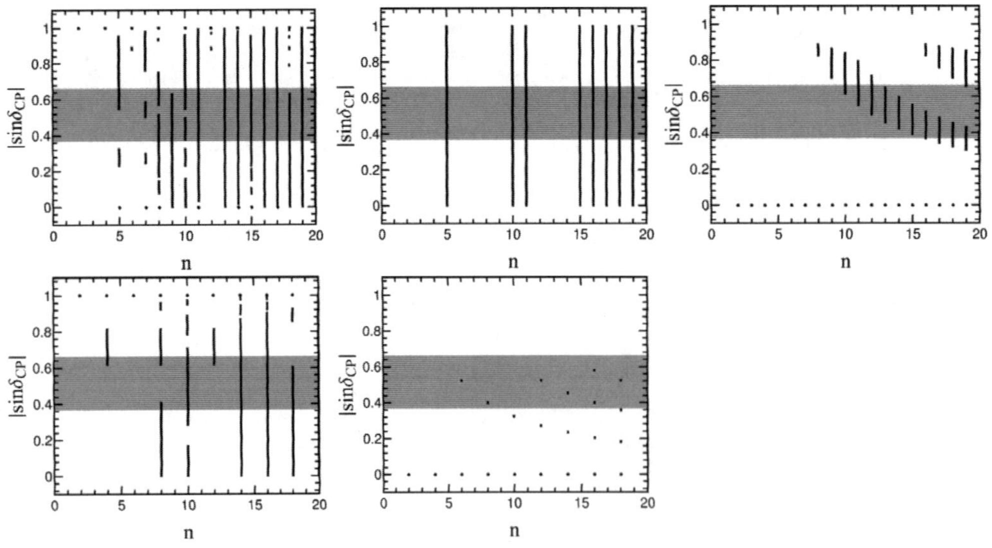

Figure 5. Predictions for $|\sin\delta_{\rm CP}|$ for the five classes of experimentally viable mixing matrices for semidirect models with a $\Delta(6n^2)$ flavour group. The band shows the one sigma experimental preference.

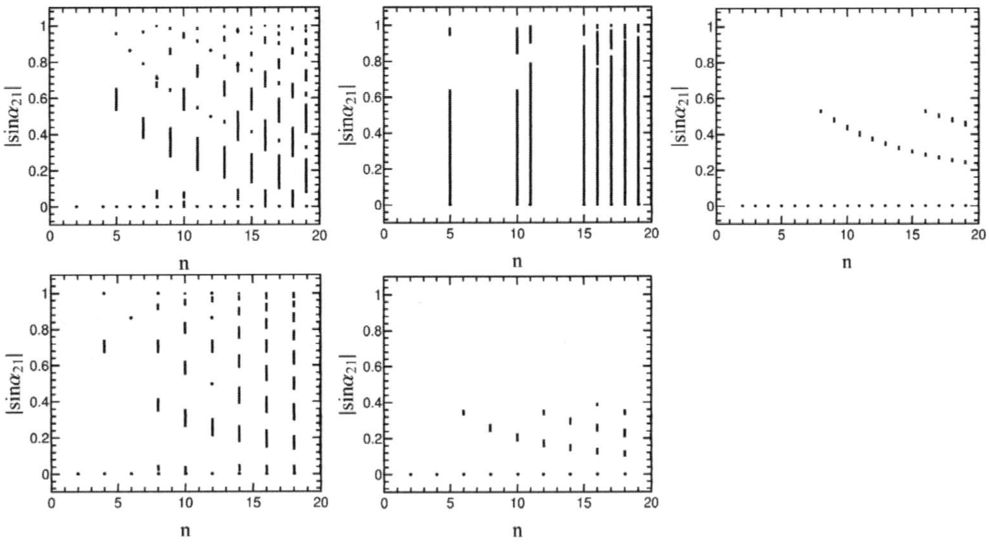

Figure 6. Predictions for $|\sin\alpha_{21}|$ for the five classes of experimentally viable mixing matrices for semidirect models with a $\Delta(6n^2)$ flavour group.

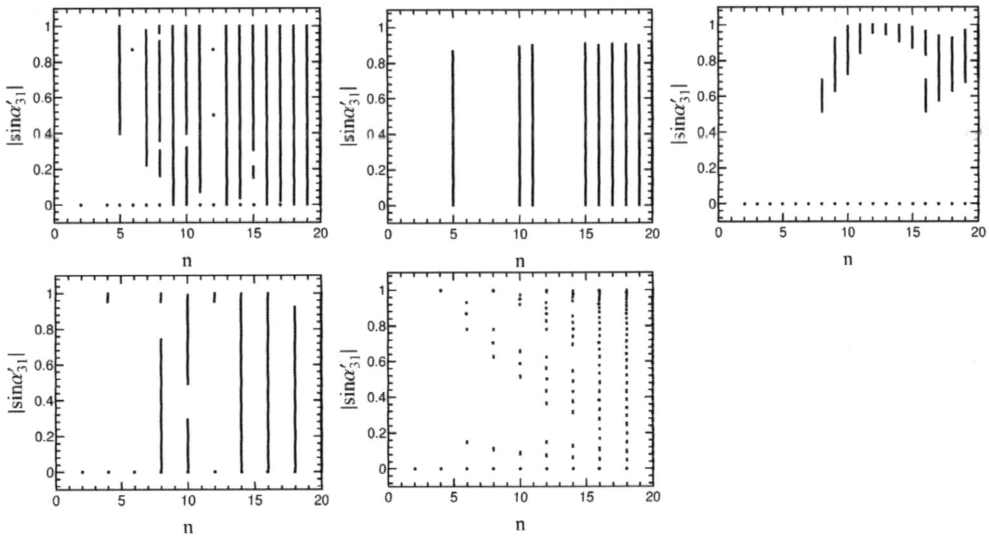

Figure 7. Predictions for $|\sin\alpha'_{31}|$ for the five classes of experimentally viable mixing matrices for semidirect models with a $\Delta(6n^2)$ flavour group. α'_{31} is defined through $\alpha'_{31} = \alpha_{31} - 2\delta_{\rm CP}$.

Majorana phases. Furthermore, only $\Delta(6n^2)$ groups remain viable flavour symmetry groups in direct models. As all members of the $\Delta(6n^2)$ group series can be analysed simultaneously for all n, this makes it possible to test the paradigm of direct flavour models with discrete invariance groups generally. In direct models with $\Delta(6n^2)$ groups, discrete values for θ_{13} are predicted which depend on group parameters. Additionally, predictions that are independent of n can be obtained: The Dirac CP phase is predicted to be zero and the middle column is required to be trimaximal. These predictions are currently compatible with global fits and will be tested experimentally in the near future.

Furthermore, possible predictions of semidirect models, i.e. models in which the flavour group is broken to a Z_2 subgroup either for neutrinos or charged leptons are analysed. In this case, despite a variety of possible combinations of subgroups and additional free parameters, discrete or constrained predictions are obtained for all mixing parameters as well as general predictions that allow to differentiate between classes of mixing matrices or to eventually exlude this approach.

In the analyses presented here, no renormalisation corrections to the mixing parameters are taken into account. With increasing experimental accuracy, it might become necessary to consider these.

If experimental results are not excluding models based on $\Delta(6n^2)$ flavour groups, or are even pointing towards specific values of n, then the phenomenology of these models is going to be studied in detail, in particular concerning possible breaking mechanisms.

Acknowledgments

TN would like to thank the organisers of the Discrete2014 conference. TN acknowledges support from the European Union FP7 ITN-INVISIBLES (Marie Curie Actions, PITN- GA-2011- 289442).

References

[1] G. Altarelli and F. Feruglio, Rev. Mod. Phys. **82**, 2701 (2010) [arXiv:1002.0211 [hep-ph]]; H. Ishimori, T. Kobayashi, H. Ohki, Y. Shimizu, H. Okada and M. Tanimoto, Prog. Theor. Phys. Suppl. **183**, 1 (2010) [arXiv:1003.3552 [hep-th]]; W. Grimus and P. O. Ludl, J. Phys. A **45**, 233001 (2012) [arXiv:1110.6376 [hep-ph]].

[2] S. F. King and C. Luhn, Rept. Prog. Phys. **76** (2013) 056201 [arXiv:1301.1340 [hep-ph]]; S. F. King, A. Merle, S. Morisi, Y. Shimizu and M. Tanimoto, arXiv:1402.4271 [hep-ph].

[3] W. Grimus and M. N. Rebelo, Phys. Rept. **281**, 239 (1997) [arXiv:9506272[hep-ph]].

[4] F. Feruglio, C. Hagedorn and R. Ziegler, JHEP **1307**, 027 (2013) [arXiv:1211.5560 [hep-ph]].

[5] M. Holthausen, M. Lindner and M. A. Schmidt, JHEP **1304**, 122 (2013) [arXiv:1211.6953 [hep-ph]].

[6] M. C. Chen, M. Fallbacher, K. T. Mahanthappa, M. Ratz and A. Trautner, Nucl. Phys. B **883**, 267 (2014) [arXiv:1402.0507 [hep-ph]].

[7] S. F. King and T. Neder, arXiv:1403.1758 [hep-ph].

[8] F. P. An *et al.* [DAYA-BAY Collaboration], Phys. Rev. Lett. **108**, 171803 (2012) [arXiv:1203.1669 [hep-ex]]; Chin. Phys. C **37**, 011001 (2013) [arXiv:1210.6327 [hep-ex]].

[9] J. K. Ahn *et al.* [RENO Collaboration], Phys. Rev. Lett. **108**, 191802 (2012) [arXiv:1204.0626 [hep-ex]].

[10] Y. Abe *et al.* [DOUBLE-CHOOZ Collaboration], Phys. Rev. Lett. **108**, 131801 (2012) [arXiv:1112.6353 [hep-ex]]; Phys. Rev. D **86**, 052008 (2012) [arXiv:1207.6632 [hep-ex]].

[11] The GAP Group, *GAP – Groups, Algorithms, and Programming, Version 4.4.12*, 2008. http://www.gap-system.org.

[12] C. S. Lam, Phys. Rev. D **87** (2013) 013001 [arXiv:1208.5527 [hep-ph]].

[13] M. Holthausen, K. S. Lim and M. Lindner, Phys. Lett. B **721** (2013) 61 [arXiv:1212.2411 [hep-ph]].

[14] J. A. Escobar and C. Luhn, J. Math. Phys. **50**, 013524 (2009) [arXiv:0809.0639 [hep-th]].

[15] S. F. King, T. Neder and A. J. Stuart, Phys. Lett. B **726**, 312 (2013). [arXiv:1305.3200 [hep-ph]].

[16] T. Neder, [arXiv:1405.0019 [hep-ph]].

[17] G. J. Ding, S. F. King and T. Neder, JHEP **1412**, 007 (2014). [arXiv:1409.8005 [hep-ph]].

[18] C. Hagedorn, A. Meroni and E. Molinaro, arXiv:1408.7118 [hep-ph].

[19] R. M. Fonseca and W. Grimus, JHEP **1409**, 033 (2014). [arXiv:1405.3678 [hep-ph]].

[20] K. A. Olive *et al.* [Particle Data Group Collaboration], Chin. Phys. C **38**, 090001 (2014).
[21] D. Adams, "The Hitchhiker's Guide to the Galaxy", Pan Books (1979)
[22] D. V. Forero, M. Tortola and J. W. F. Valle, Phys. Rev. D **86**, 073012 (2012) [arXiv:1205.4018 [hep-ph]].
[23] F. Capozzi, G. L. Fogli, E. Lisi, A. Marrone, D. Montanino and A. Palazzo, Phys. Rev. D **89**, 093018 (2014) [arXiv:1312.2878 [hep-ph]].
[24] M. C. Gonzalez-Garcia, M. Maltoni and T. Schwetz, arXiv:1409.5439 [hep-ph].

Adding CP to flavour symmetries

I. de Medeiros Varzielas

School of Physics and Astronomy, University of Southampton, Southampton, SO17 1BJ, U.K.

E-mail: ivo.de@soton.ac.uk

Abstract. I propose the use of CP-odd invariants, which are independent of basis and valid for any choice of CP transformation, as a powerful approach to study CP in the presence of flavour symmetries. As examples of the approach I focus on Lagrangians invariant under $\Delta(27)$. I comment on the consequences of adding a specific CP symmetry to a Lagrangian and distinguish cases where several $\Delta(27)$ singlets are present depending on how they couple to the triplets. One of the examples included is a very simple toy model with explicit CP violation with calculable phases, which is referred to as explicit geometrical CP violation by comparison with previously known cases of (spontaneous) geometrical CP violation.

1. Introduction

This contribution to the proceedings of DISCRETE 2014 follows closely the layout of seminar I presented in the conference. I include here an expanded discussion of situations with $\Delta(27)$ singlets, including cases with explicit geometrical CP violation (first identified recently, in [1]). Some aspects discussed here are to appear also in a subsequent publication.

1.1. The invariant approach

I refer to the Invariant Approach (IA) to CP [2] as an approach that starts by splitting the Lagrangian into $\mathcal{L}_{CP}$, a part that automatically conserves CP (e.g. kinetic terms, gauge interactions) and the remaining part $\mathcal{L}_{rem.}$:

$$\mathcal{L} = \mathcal{L}_{CP} + \mathcal{L}_{rem.} \,. \tag{1}$$

The next steps are to

- Impose the most general CP transformations (that leave $\mathcal{L}_{CP}$ invariant).
- Apply them and see if it restricts $\mathcal{L}_{rem.}$.

Only if the most general CP transformations restrict the shape of $\mathcal{L}_{rem.}$ can CP be violated. An example of this type of restrictions is if the most general CP transformations force some coefficient to be real.

The IA is powerful because:

- Gets results just from the Lagrangian.
- Independent of basis.
- Shows relevant quantities for physical processes.

1.2. The invariant approach for Standard Model leptons

As a brief review of the IA, I apply it to a study of CP for Standard Model (SM) leptons. The mass Lagrangian is

$$-\mathcal{L}_m = m_l \bar{e}_L e_R + \tfrac{1}{2} m_\nu \bar{\nu}_L \nu_L^c + h.c.\,, \tag{2}$$

where $L = (e_L, \nu_L)$ stand for the left-handed neutrino and charged lepton fields in a weak basis; e_R is the right-handed counterpart.

Due to the $SU(2)_L$ interactions (inside $\mathcal{L}_{\mathcal{CP}}$), the most general CP transformations are:

$$(CP)L(CP)^\dagger = iU\gamma^0 \mathcal{C}\bar{L}^T, \tag{3}$$

$$(CP)e_R(CP)^\dagger = iV\gamma^0 \mathcal{C}\bar{e}_R^T. \tag{4}$$

I adopt a less precise notation that is more convenient to work with:

$$L \rightarrow UL^*, \tag{5}$$

$$e_R \rightarrow Ve_R^*. \tag{6}$$

I use this notation throughout, particularly as for simplicity I will mostly consider scalar fields in future sections, where the shorter notation is precise.

In order for $\mathcal{L}_m$ to be CP invariant, under eq(3), eq(4) the terms shown in eq(2) go into the respective $h.c.$ and vice-versa:

$$U^\dagger m_\nu U^* = m_\nu^*, \qquad U^\dagger m_l V = m_l^*\,. \tag{7}$$

Defining $H_\nu \equiv m_\nu m_\nu^\dagger$ and $H_l \equiv m_l m_l^\dagger$, I have:

$$U^\dagger H_\nu U = H_\nu^*, \qquad U^\dagger H_l U = H_l^*\,. \tag{8}$$

At this stage I follow [2] and build CP-odd invariants (CPI) by constructing combinations where U and V do not appear. First, I note that from eq(8), I can obtain $\mathrm{Tr}(H_\nu H_l) = \mathrm{Tr}(H_\nu H_l)^*$, which does not depend on U, V. As the matrices are Hermitian, $\mathrm{Tr}(H_\nu H_l)^* = \mathrm{Tr}(H_\nu^T H_l^T) = \mathrm{Tr}(H_l H_\nu)^T = \mathrm{Tr}(H_l H_\nu)$, concluding that for any CP transformations U, V, $\mathrm{Tr}((H_\nu H_l) - (H_l H_\nu)) = 0$ is required for CP conservation. Given that this is the trace of a commutator, this particular CPI automatically vanishes, meaning it is not very useful. A more useful alternative is the necessary condition for CP conservation:

$$I_1 \equiv \mathrm{Tr}\,[H_\nu, H_l]^3 = 0\,, \tag{9}$$

valid for any number of fermion generations. For 3 generations (the SM case), it can be shown that eq(9) is a sufficient condition to have no Dirac-type CP violation in the lepton sector.

2. The invariant approach and flavour symmetries

One of the main points of my talk at the conference and of [1] is that the IA is very useful for analysing flavour symmetry models. In order to illustrate this, I present some examples.

2.1. Toy model with 4 couplings

I start by considering a version of the toy model presented in section 3.1.1. of [3]. As the aim here is to show the IA in action, I replace all fermions with scalars to avoid unnecessary complications. The Lagrangian (with fermions) as presented by [3] is shown in figure 1.

I refer to a similar Lagrangian (with scalars) as $\mathcal{L}_4$ (due to its 4 couplings):

$$\mathcal{L}_4 = S\bar{\Psi}F\Sigma + X\bar{\Psi}G\Sigma + Y\bar{\Psi}H_\Psi\Psi + Y\bar{\Sigma}H_\Sigma\Sigma + h.c.\,, \tag{10}$$

$$\begin{aligned}\mathscr{L}_{\text{toy}} &= f\left[S_{\mathbf{1}_0}\otimes\left(\overline{\Psi}\otimes\Sigma\right)_{\mathbf{1}_0}\right]_{\mathbf{1}_0} + g\left[X_{\mathbf{1}_1}\otimes\left(\overline{\Psi}\otimes\Sigma\right)_{\mathbf{1}_2}\right]_{\mathbf{1}_0}\\ &+ h_\Psi\left[Y_{\mathbf{1}_3}\otimes\left(\overline{\Psi}\otimes\Psi\right)_{\mathbf{1}_6}\right]_{\mathbf{1}_0} + h_\Sigma\left[Y_{\mathbf{1}_3}\otimes\left(\overline{\Sigma}\otimes\Sigma\right)_{\mathbf{1}_6}\right]_{\mathbf{1}_0} + \text{h.c.}\\ &= \boxed{F^{ij}\,S\,\overline{\Psi}_i\Sigma_j + G^{ij}\,X\,\overline{\Psi}_i\Sigma_j + H_\Psi^{ij}\,Y\,\overline{\Psi}_i\Psi_j + H_\Sigma^{ij}\,Y\,\overline{\Sigma}_i\Sigma_j + \text{h.c.}}\,.\end{aligned}$$

Figure 1. Toy model Lagrangian from [3].

where scalar fields S, X, Y have just one generation, whereas scalar fields Ψ and Σ have n generations, meaning F, G, H_Ψ, H_Σ are $n \times n$ coupling matrices. In the original toy model, Ψ and Σ are fermions (with $n = 3$ generations) and I use the notation $\bar{\Psi} = \Psi^\dagger$, $\bar{\Sigma} = \Sigma^\dagger$ in eq(10) for easier comparison with the box in figure 1.[1]

$\mathcal{L}_4$ is the $\mathcal{L}_{rem.}$ of this toy model and the most general CP transformations (consistent with the respective $\mathcal{L}_{CP}$) are independent unitary transformations for each of the fields - phases for S, X, Y and $n \times n$ unitary matrices Q and R for Ψ and Σ:

$$\begin{aligned} S &\to e^{is} S^*\,, && (11)\\ X &\to e^{ix} X^*\,, && (12)\\ Y &\to e^{iy} Y^*\,, && (13)\\ \Psi &\to Q\Psi^*\,, && (14)\\ \Sigma &\to R\Sigma^*\,. && (15)\end{aligned}$$

Imposing CP conservation requires $\mathcal{L}_4$ to be invariant under these, which implies that the terms displayed in eq(10) go into their $h.c.$ and vice-versa. Starting with $Y\bar{\Sigma}H_\Sigma\Sigma$, I have

$$\mathcal{L}_4 \supset Y\bar{\Sigma}H_\Sigma\Sigma + Y^*\bar{\Sigma}^*H_\Sigma^*\Sigma^* \tag{16}$$

and the relevant CP transformations

$$\begin{aligned} Y &\to e^{iy} Y^*\,, && (17)\\ \Sigma &\to R\Sigma^*\,, && (18)\end{aligned}$$

act on $Y\bar{\Sigma}H_\Sigma\Sigma$:

$$Y\bar{\Sigma}H_\Sigma\Sigma \to e^{iy}Y^*\bar{\Sigma}^*R^\dagger H_\Sigma R\Sigma^*\,. \tag{19}$$

Comparing with the $h.c.$ I conclude that if $\mathcal{L}_4$ remains invariant under CP, $e^{iy}R^\dagger H_\Sigma R = H_\Sigma^*$. I repeat the procedure for the other 3 couplings and obtain the 4 relations

$$\begin{aligned} e^{is}Q^\dagger F R &= F^*\,, && (20)\\ e^{ix}Q^\dagger G R &= G^*\,, && (21)\\ e^{iy}Q^\dagger H_\Psi Q &= H_\Psi^*\,, && (22)\\ e^{iy}R^\dagger H_\Sigma R &= H_\Sigma^*\,. && (23)\end{aligned}$$

These 4 relations are necessary and sufficient for $\mathcal{L}_4$ to conserve CP. At this stage I build CPIs by combining the 4 relations to obtain equations where the general CP transformations cancel

[1] I reconsider $\mathcal{L}_4$ in section 2.3.2 making it invariant under a flavour symmetry, as done in [3].

out, meaning they are independent of s, x, y, Q and R. A relevant example is obtained by multiplying in order 1. the dagger of eq(20), 2. eq(22), 3. eq(20) and 4. the dagger of eq(23):

$$R^\dagger F^\dagger Q Q^\dagger H_\Psi Q Q^\dagger F R R^\dagger H_\Sigma^\dagger R \, e^{i(-s+y+s-y)} = (F^\dagger H_\Psi F H_\Sigma^\dagger)^* \,, \tag{24}$$

removing the phases and leaving unitary matrices only outside the product of couplings:

$$R^\dagger F^\dagger H_\Psi F H_\Sigma^\dagger R = (F^\dagger H_\Psi F H_\Sigma^\dagger)^* \,. \tag{25}$$

Doing the same with eq(21) in steps 1. and 3. I would obtain similarly:

$$R^\dagger G^\dagger H_\Psi G H_\Sigma^\dagger R = (G^\dagger H_\Psi G H_\Sigma^\dagger)^* \,. \tag{26}$$

The remaining dependence on unitary matrix R can be eliminated by taking the trace:

$$\mathrm{Tr}\left[F^\dagger H_\Psi F H_\Sigma^\dagger\right] = \mathrm{Tr}\left[F^\dagger H_\Psi F H_\Sigma^\dagger\right]^* \quad \rightarrow \quad \mathrm{Im}\,\mathrm{Tr}\left[F^\dagger H_\Psi F H_\Sigma^\dagger\right] = 0 \,, \tag{27}$$

$$\mathrm{Tr}\left[G^\dagger H_\Psi G H_\Sigma^\dagger\right] = \mathrm{Tr}\left[G^\dagger H_\Psi G H_\Sigma^\dagger\right]^* \quad \rightarrow \quad \mathrm{Im}\,\mathrm{Tr}\left[G^\dagger H_\Psi G H_\Sigma^\dagger\right] = 0 \,. \tag{28}$$

These conditions illustrate the power of the IA. Simply from studying the CP properties of $\mathcal{L}_4$ I have found a set of necessary conditions for CP conservation (they are not necessarily sufficient). The conditions are basis independent, valid for *any* choice of CP transformation and for any number of generations n. They also directly constrain quantities that are relevant for physical processes.

In [3], the authors compute the CP asymmetry in the decay $Y \to \bar{\Psi}\Psi$, as shown in the text and equation displayed in figure 2. By computing only a single CP asymmetry one might conclude that CP conservation can be obtained from a cancellation of the two quantities. Instead, by applying the IA to $\mathcal{L}_4$ I conclude that there are at least 2 independent necessary conditions for CP conservation, eq(27), eq(28).

Let us now study the decay $Y \to \bar{\Psi}\Psi$. Interference between tree–level and one–loop diagrams (figures 3(a)–3(c)) leads to a CP asymmetry $\varepsilon_{Y\to\bar{\Psi}\Psi}$, which is proportional to

$$\begin{aligned} \varepsilon_{Y\to\bar{\Psi}\Psi} &\propto \mathrm{Im}\,[I_S]\;\mathrm{Im}\left[\mathrm{tr}\left(F^\dagger\, H_\Psi\, F\, H_\Sigma^\dagger\right)\right] + \mathrm{Im}\,[I_X]\;\mathrm{Im}\left[\mathrm{tr}\left(G^\dagger\, H_\Psi\, G\, H_\Sigma^\dagger\right)\right] \\ &= |f|^2\;\mathrm{Im}\,[I_S]\;\mathrm{Im}\,[h_\Psi\, h_\Sigma^*] + |g|^2\;\mathrm{Im}\,[I_X]\;\mathrm{Im}\,[\omega\, h_\Psi\, h_\Sigma^*] \;. \end{aligned} \tag{3.3}$$

Figure 2. Decay of $Y \to \bar{\Psi}\Psi$ as computed in [3].

2.2. Δ(27) and adding CP

2.2.1. $\Delta(27)$ I discuss now the group theory of $\Delta(27)$ that is required for the remaining sections. I define $\omega \equiv e^{i2\pi/3}$, the cyclic generator c and diagonal generator d of the group ($\omega^3 = 1$, $c^3 = d^3 = 1$). There is an additional generator but it is not directly relevant for the discussion here. The group has irreducible representations that are either 1 or 3 dimensional - referred as singlets and triplets. The action of generators on singlets is simply multiplying them by a phase: $c 1_{ij} = \omega^i 1_{ij}$ and $d 1_{ij} = \omega^j 1_{ij}$, where $i, j = 0, 1, 2$ - there are 9 distinct singlets. In a convenient basis the action of the generators on a 3_{01} triplet $A = (a_1, a_2, a_3)_{01}$ or a 3_{02} triplet $\bar{B} = (\bar{b}_1, \bar{b}_2, \bar{b}_3)_{02}$ is:

$$c_{3_{0j}} = \begin{pmatrix} 0 & 1 & 0 \\ 0 & 0 & 1 \\ 1 & 0 & 0 \end{pmatrix} \,, \quad c_{3_{01}} \begin{pmatrix} a_1 \\ a_2 \\ a_3 \end{pmatrix} = \begin{pmatrix} a_2 \\ a_3 \\ a_1 \end{pmatrix} \,, \tag{29}$$

$$d_{3_{01}} = \begin{pmatrix} 1 & 0 & 0 \\ 0 & \omega & 0 \\ 0 & 0 & \omega^2 \end{pmatrix} , \quad d_{3_{02}} = \begin{pmatrix} 1 & 0 & 0 \\ 0 & \omega^2 & 0 \\ 0 & 0 & \omega \end{pmatrix} . \tag{30}$$

My nomenclature follows from the action of generators on triplets. The generator d distinguishes the triplets 3_{01} and 3_{02} according to their subscripts, which are the powers of ω on the first two diagonal entries of the respective matrix. Hereafter I often refer to 3_{01} as the triplet representation and to 3_{02} as the anti-triplet representation. The cyclic generator acts equally on triplet and anti-triplet by cyclic permutation of the components.

The product of singlet with singlet leads to another singlet transforming as the sum of indices (modulo 3): $1_{ij} \times 1_{kl}$ transforms as $1_{(i+k)(j+l)}$. The product of triplet and anti-triplet gives a sum of all nine singlets. In the following it will be necessary to know how the singlets 1_{i0} and 1_{0j} are built from the product of triplet and anti-triplet. 1_{00} is the trivial singlet transforming trivially under all generators and is formed from the $SU(3)$ contraction:

$$(A\bar{B})_{00} \equiv (a_1\bar{b}_1 + a_2\bar{b}_2 + a_3\bar{b}_3)_{00} \, . \tag{31}$$

The 1_{i0} singlets are built as

$$(A\bar{B})_{10} \equiv (a_1\bar{b}_1 + \omega^2 a_2\bar{b}_2 + \omega a_3\bar{b}_3)_{10} \, , \tag{32}$$

$$(A\bar{B})_{20} \equiv (a_1\bar{b}_1 + \omega a_2\bar{b}_2 + \omega^2 a_3\bar{b}_3)_{20} \, , \tag{33}$$

as acting with c on A and $\bar{B}$ leads to multiplication by ω, ω^2 respectively:

$$(A\bar{B})_{10} \to (a_2\bar{b}_2 + \omega^2 a_3\bar{b}_3 + \omega a_1\bar{b}_1)_{10} \, , \tag{34}$$

$$(A\bar{B})_{20} \to (a_2\bar{b}_2 + \omega a_3\bar{b}_3 + \omega^2 a_1\bar{b}_1)_{20} \, . \tag{35}$$

In turn, the 1_{0j} are built as

$$(A\bar{B})_{01} \equiv (a_2\bar{b}_1 + a_3\bar{b}_2 + a_1\bar{b}_3)_{01} \, , \tag{36}$$

$$(A\bar{B})_{02} \equiv (a_1\bar{b}_2 + a_2\bar{b}_3 + a_3\bar{b}_1)_{02} \, , \tag{37}$$

as acting with d on A and $\bar{B}$ leads to multiplication by ω, ω^2 respectively:

$$(A\bar{B})_{01} \to (\omega a_2\bar{b}_1 + \omega^2 a_3 \omega^2 \bar{b}_2 + a_1 \omega \bar{b}_3)_{01} \, , \tag{38}$$

$$(A\bar{B})_{02} \to (a_1 \omega^2 \bar{b}_2 + \omega a_2 \omega \bar{b}_3 + \omega^2 a_3\bar{b}_1)_{02} \, . \tag{39}$$

2.2.2. Adding CP I consider now a specific $\Delta(27)$ invariant Lagrangian and study its CP properties. The field content is triplet A, anti-triplet $\bar{B}$, and singlets C, D (transforming respectively as 3_{01}, 3_{02}, 1_{10}, 1_{01}). The $\Delta(27)$ invariant Lagrangian for this field content contains one 3-field invariant between triplet, anti-triplet and each singlet:

$$\mathcal{L}_{CD} = y_c(A\bar{B})_{20}C_{10} + y_d(A\bar{B})_{02}D_{01} + h.c. \, . \tag{40}$$

An additional Z_N or $U(1)$ symmetry can be added to guarantee the absence of additional terms coupling C, D to AA^* or $\bar{B}^*\bar{B}$. Focusing on the CP properties of $\mathcal{L}_{CD}$, I start by adding a specific CP transformation. A simple option is the trivial CP transformation CP_1, defined by the action on A, $\bar{B}$, C and D:

$$CP_1 A = A^* = (a_1^*, a_2^*, a_3^*)_{02} \, , \tag{41}$$

$$CP_1 \bar{B} = \bar{B}^* = (\bar{b}_1^*, \bar{b}_2^*, \bar{b}_3^*)_{01} \, , \tag{42}$$

$$CP_1 C_{10} = C_{20}^* \, , \tag{43}$$

$$CP_1 D_{01} = D_{02}^* \, , \tag{44}$$

where A^*, $\bar{B}^*$, C^*, D^* transform respectively as 3_{02}, 3_{01}, 1_{20}, 1_{02} (reflected by the subscripts).

If I impose invariance under CP_1 on $\mathcal{L}_{CD}$, the y_c term which transforms to:

$$\to y_c(a_1^*\bar{b}_1^* + \omega a_2^*\bar{b}_2^* + \omega^2 a_3^*\bar{b}_3^*)_{20} C_{20}^* \,, \tag{45}$$

should become the $h.c.$, which features y_c^*:

$$y_c^*(a_1^*\bar{b}_1^* + \omega^2 a_2^*\bar{b}_2^* + \omega a_3^*\bar{b}_3^*)_{10} C_{20}^* \,. \tag{46}$$

In addition to the conjugated coefficient, the expressions inside the parentheses are different, as denounced by their subscripts. In turn, under CP_1 the y_d term transforms into:

$$\to y_d(a_1^*\bar{b}_2^* + a_2^*\bar{b}_3^* + a_3^*\bar{b}_1^*)_{01} D_{02}^* \,, \tag{47}$$

and comparing to its $h.c.$ with y_d^*

$$y_d^*(a_1^*\bar{b}_2^* + a_2^*\bar{b}_3^* + a_3^*\bar{b}_1^*)_{01} D_{02}^* \,, \tag{48}$$

shows that apart from swapping y_d to y_d^* the expressions are the same.

A closer look at eq(45) shows that the transformed quantity is no longer invariant under $\Delta(27)$ (the subscripts do not add up to make a trivial singlet). One might state that, for this field content, $\Delta(27)$ is inconsistent with CP_1. A more precise statement is that for $\mathcal{L}_{CD}$ to be invariant under both $\Delta(27)$ and CP_1 requires $y_c = 0$ (and y_d to be real) or alternatively, that insisting that $y_c \neq 0$ explicitly violates either $\Delta(27)$ or CP_1. That y_c is forced to vanish by adding a specific CP symmetry may appear drastic, but this is rather an usual consequence of adding symmetries to a Lagrangian. For example, one could also force $y_c = 0$ in $\mathcal{L}_{CD}$ simply by having only the field C transform non-trivially under an additional Z_2 symmetry.

One important point is that although imposing a specific CP transformation can force coefficients to vanish this needs not mean that CP violation occurs if those coefficients do not vanish. Indeed, $\mathcal{L}_{CD}$ with arbitrary y_c and y_d is CP conserving. I prefer to see this using the IA, and rewrite:

$$\mathcal{L}_{CD} = A_i Y_{10}^{ij} \bar{B}_j C + A_i Y_{01}^{ij} \bar{B}_j D + h.c. \,, \tag{49}$$

with

$$Y_{10} = y_c \begin{pmatrix} 1 & 0 & 0 \\ 0 & \omega & 0 \\ 0 & 0 & \omega^2 \end{pmatrix}; \quad Y_{01} = y_d \begin{pmatrix} 0 & 1 & 0 \\ 0 & 0 & 1 \\ 1 & 0 & 0 \end{pmatrix}. \tag{50}$$

Then I take the most general transformations

$$A \to U^* A^*; \quad \bar{B} \to V \bar{B}^*; \quad C \to e^{ip_{10}} C^*; \quad D \to e^{ip_{01}} D^* \,, \tag{51}$$

and obtain the conditions for CP conservation

$$U^\dagger Y_{01} V e^{ip_{01}} = Y_{01}^* \,, \tag{52}$$

$$U^\dagger Y_{10} V e^{ip_{10}} = Y_{10}^* \,. \tag{53}$$

By building CPIs I conclude they are of the form

$$\mathrm{Im}\ \ \mathrm{Tr}\ \ [(Y_{01}^\dagger Y_{01})^{n_1} (Y_{10}^\dagger Y_{10})^{n_2} (Y_{01}^\dagger Y_{01})^{n_3} (...)] \,, \tag{54}$$

$$\mathrm{Im}\ \ \mathrm{Tr}\ \ [(Y_{01} Y_{01}^\dagger)^{n_1} (Y_{10} Y_{10}^\dagger)^{n_2} (Y_{01} Y_{01}^\dagger)^{n_3} (...)] \,, \tag{55}$$

where n_i are positive integers. These CPIs automatically vanish due to $\Delta(27)$, as $(Y_{01}^\dagger Y_{01})$, $(Y_{10}^\dagger Y_{10})$, $(Y_{01}Y_{01}^\dagger)$ and $(Y_{10}Y_{10}^\dagger)$ are proportional to the identity matrix[2] with either $|y_c|^2$ or $|y_d|^2$. The conclusion is that CP is conserved for any y_c, y_d. Therefore, there must be at least one CP symmetry that leaves $\mathcal{L}_{CD}$ invariant regardless of arbitrary couplings (even though CP_1 does not). One explicit example is:

$$U = \begin{pmatrix} 1 & 0 & 0 \\ 0 & \omega^2 & 0 \\ 0 & 0 & 1 \end{pmatrix}; \quad V = \begin{pmatrix} 1 & 0 & 0 \\ 0 & 1 & 0 \\ 0 & 0 & \omega^2 \end{pmatrix}; \quad p_{10} = -2\text{Arg}(y_c); \quad p_{01} = -2\text{Arg}(y_d). \tag{56}$$

This example is within the possibilities listed in [4] for CP transformations consistent with $\Delta(27)$ triplets. While eq(56) applies to singlets 1_{01} and 1_{10}, the reasoning based on the IA can be easily applied to any choice of two $\Delta(27)$ singlets.

2.3. Additional singlets

In the context of $\Delta(27)$ models with *spontaneous geometrical CP violation*, meaning CP that is spontaneously broken with *calculable phases* [5], adding $\Delta(27)$ singlets coupling to triplet and anti-triplet was originally explored in [6]. The goal was to obtain additional Yukawa couplings for SM fermions and the most promising choices considered the SM quark doublets as different singlets of $\Delta(27)$.[3]

Geometrical CP violation is a very interesting topic that continued to be explored after [6]. Non-renormalisable terms in the scalar potential were considered in [7], with focus on their effects on the calculable phases. The first viable model with SM quark doublets transforming as non-trivial $\Delta(27)$ singlets was realised in [8], featuring geometrical CP violation. Subsequently it was shown in [9] how an additional symmetry can prevent the additional singlets from endangering the calculable phases and simultaneously explain the quark mass hierarchies. In [10] an extension of this type of model with the complete SM fermion sector was realised (see also [11] for a different proposal for the leptons).

Geometrical CP violation was further explored in the context of multi-Higgs models with symmetries other than $\Delta(27)$ in [12, 13] (see also [14]). The $\Delta(27)$ scalar potential for one triplet (invariant under $\Delta(54)$ [6]) was also analysed extensively with different approaches [15, 16, 17].

In [6, 7, 8, 9, 10] CP is broken spontaneously, therefore the Lagrangian should conserve CP. As pointed out in [8], one must take care in adding extra singlets that couple to the triplet and anti-triplet as that may be incompatible with CP conservation.

One way to approach the constraints arising from adding singlets is by studying the outer automorphisms of the group, as discussed in [15] and also [3]. Alternatively, the reason why coupling more singlets to triplets may lead to CP violation becomes very clear in the IA: additional couplings enable more CPIs to be built. Eventually, adding an extra coupled singlet leads to a CPI that does not automatically vanish due to $\Delta(27)$, meaning one of 3 possibilities: $\Delta(27)$ is explicitly violated; CP is explicitly violated; or specific relations on the couplings are imposed. The last possibility is relevant in the context of [6, 7, 8, 9, 10], where one wants the symmetries to be broken spontaneously. In analogy to CP_1 forcing $y_c = 0$, one option that allows preserving both symmetries would be to have the couplings involving the triplets and some of the singlets vanish, and this is also understood clearly through the IA: the additional CPIs that do not vanish automatically can vanish due to the couplings.

Indeed, if one considers a set of singlets including two or three 1_{0j} singlets as in [8], the possible CP transformations on the triplets are already so constrained that one can not couple

[2] For the same reason, CPIs like $\text{Im Tr}[(Y_{01}^\dagger Y_{01})^{n_1} Y_{10}^\dagger Y_{01} (Y_{10}^\dagger Y_{10})^{n_2} Y_{01}^\dagger Y_{10}]$ give the same result as those in eq(54).
[3] In [6] it was pointed out that one triplet of $\Delta(54)$ has the same scalar potential as one triplet of $\Delta(27)$, and options with irreducible representations of $\Delta(54)$ were also considered therein.

Journal of Physics: Conference Series **631** (2015) 012020 doi:10.1088/1742-6596/631/1/012020

the triplets to even a single additional singlet 1_{ij} with $i \neq 0$ and preserve CP. Partly, this is why only 1_{0j} singlets were considered therein.

Strictly, the statement in [8] that singlets 1_{ij} with $i \neq 0$ generate no coupling due to CP conservation is not mathematically rigorous. This statement is valid e.g. when CP_1 is imposed but not in general, as was pointed out recently in [17]. Nonetheless, for 3-field couplings between $\Delta(27)$ triplets and singlets (as in $\mathcal{L}_{CD}$), the physics of CP conserving situations - with at most 3 independent couplings - is contained in the choices considered in [8], which effectively corresponds to working in a basis where CP conservation is reflected on the Clebsch-Gordan (CG) coefficients being real. An analysis of this issue also clarifies why the presence of any 2 singlets coupling in the manner of $\mathcal{L}_{CD}$ leads automatically to CP conservation.

2.3.1. Changing basis Starting with just 1 singlet S_{ij} and term $y_{ij}(A\bar{B})_{(-i)(-j)}S_{ij}$, I can always change the basis of A such that the $(A\bar{B})_{(-i)(-j)}$ contraction looks like the corresponding contraction $(A\bar{B})_{(0)(-j)}$ in the original basis, which has real CG coefficients. An explicit example is $y_c(A\bar{B})_{20}C_{10}$, where the change of basis $(a_1, a_2, a_3) \rightarrow (a_1, \omega^2 a_2, \omega a_3)$ does precisely this:

$$(a_1\bar{b}_1 + \omega a_2\bar{b}_2 + \omega^2 a_3\bar{b}_3)_{20} \rightarrow (a_1\bar{b}_1 + a_2\bar{b}_2 + a_3\bar{b}_3)\,. \tag{57}$$

As far as the $y_{ij}(A\bar{B})_{(-i)(-j)}S_{ij}$ coupling is concerned it is equivalent to take a singlet in the set 1_{0j}. Note that other couplings can distinguish the singlets, e.g. if $j = 0$, terms 1_{00} and 1_{00}^2 are $\Delta(27)$ invariants whereas the same does not apply for 1_{0j}. But restricting ourselves to Lagrangian terms of the form of those in $\mathcal{L}_{CD}$ implies there are other symmetries that forbid such terms (such as the SM gauge group, in [8]).

With 2 singlets, changing only the basis for A may simply move the complex CG coefficients from one contraction into the other. $\mathcal{L}_{CD}$ is an explicit example of this as $(a_1, a_2, a_3) \rightarrow (a_1, \omega^2 a_2, \omega a_3)$ takes

$$(a_1\bar{b}_1 + \omega a_2\bar{b}_2 + \omega^2 a_3\bar{b}_3)_{20} \rightarrow (a_1\bar{b}_1 + a_2\bar{b}_2 + a_3\bar{b}_3)\,, \tag{58}$$

$$(a_1\bar{b}_2 + a_2\bar{b}_3 + a_3\bar{b}_1)_{02} \rightarrow (a_1\bar{b}_2 + \omega^2 a_2\bar{b}_3 + \omega a_3\bar{b}_1)\,. \tag{59}$$

But if one uses the change of basis:

$$(a_1, a_2, a_3) \rightarrow (a_1, \omega^2 a_2, a_3)\,, \tag{60}$$

$$(\bar{b}_1, \bar{b}_2, \bar{b}_3) \rightarrow (\bar{b}_1, \bar{b}_2, \omega\bar{b}_3)\,, \tag{61}$$

then both singlets couple to triplets in $\mathcal{L}_{CD}$ with real CG coefficients:

$$(a_1\bar{b}_1 + \omega a_2\bar{b}_2 + \omega^2 a_3\bar{b}_3)_{20} \rightarrow (a_1\bar{b}_1 + a_2\bar{b}_2 + a_3\bar{b}_3)\,, \tag{62}$$

$$(a_1\bar{b}_2 + a_2\bar{b}_3 + a_3\bar{b}_1)_{02} \rightarrow (a_1\bar{b}_2 + a_2\bar{b}_3 + a_3\bar{b}_1)\,. \tag{63}$$

This change of basis takes U and V in eq(56) to the identity matrices of CP_1.

In a situation with 3 singlets coupling in the manner of $\mathcal{L}_{CD}$, the possibility of explicit CP violation depends on whether the freedom to change the basis of A and $\bar{B}$ is enough to eliminate complex CG coefficients or not. Most choices of singlets can explicitly violate CP, but for 12 sets (out of 84 combinations) this is not possible. For these 12 sets there is at least one non-trivial element of $\Delta(27)$ which does not distinguish the 3 singlets. The special 12 sets can be identified in the notation I use here by summing the two generator indices over the 3 singlets - if both sums add up to 0 (modulo 3), an appropriate change of basis makes the CG coefficients real. I demonstrate with the set 1_{00}, 1_{10} and 1_{20}:

$$(A\bar{B})_{00} \equiv (a_1\bar{b}_1 + a_2\bar{b}_2 + a_3\bar{b}_3)_{00}\,, \tag{64}$$

$$(A\bar{B})_{10} \equiv (a_1\bar{b}_1 + \omega^2 a_2\bar{b}_2 + \omega a_3\bar{b}_3)_{10}\,, \tag{65}$$

$$(A\bar{B})_{20} \equiv (a_1\bar{b}_1 + \omega a_2\bar{b}_2 + \omega^2 a_3\bar{b}_3)_{20}\,, \tag{66}$$

The required basis change is not readily seen from the expressions, but noting the 3 singlets in the set are distinguished only by generator c, the basis change to eigenstates of $c_{3_{0j}}$:

$$(a_1, a_2, a_3) \rightarrow (a_1 + a_2 + a_3, a_1 + \omega a_2 + \omega^2 a_3, a_1 + \omega^2 a_2 + \omega a_3)/\sqrt{3}\,, \tag{67}$$
$$(\bar{b}_1, \bar{b}_2, \bar{b}_3) \rightarrow (\bar{b}_1 + \bar{b}_2 + \bar{b}_3, \bar{b}_1 + \omega^2 \bar{b}_2 + \omega \bar{b}_3, \bar{b}_1 + \omega \bar{b}_2 + \omega^2 \bar{b}_3)/\sqrt{3}\,, \tag{68}$$

takes the expressions to those of $(A\bar{B})_{0j}$, with real CG coefficients:

$$(a_1\bar{b}_1 + a_2\bar{b}_2 + a_3\bar{b}_3)_{00} \rightarrow (a_1\bar{b}_1 + a_2\bar{b}_2 + a_3\bar{b}_3)\,, \tag{69}$$
$$(a_1\bar{b}_1 + \omega^2 a_2\bar{b}_2 + \omega a_3\bar{b}_3)_{10} \rightarrow (a_2\bar{b}_1 + a_3\bar{b}_2 + a_1\bar{b}_3)\,, \tag{70}$$
$$(a_1\bar{b}_1 + \omega a_2\bar{b}_2 + \omega^2 a_3\bar{b}_3)_{20} \rightarrow (a_1\bar{b}_2 + a_2\bar{b}_3 + a_3\bar{b}_1)\,. \tag{71}$$

The generalisation of the change of basis for sets of 3 singlets sharing a non-zero index is relatively straightforward for sets 1_{ij} sharing a fixed $i \neq 0$ (distinct only under generator d), where one has a diagonal change of basis (in analogy with eq(60), eq(61)). An explicit example is for singlets with a fixed $i = 1$, with triplets contracting as:

$$(A\bar{B})_{20} \equiv (a_1\bar{b}_1 + \omega a_2\bar{b}_2 + \omega^2 a_3\bar{b}_3)_{20}\,, \tag{72}$$
$$(A\bar{B})_{21} \equiv (a_2\bar{b}_1 + \omega a_3\bar{b}_2 + \omega^2 a_1\bar{b}_3)_{21}\,, \tag{73}$$
$$(A\bar{B})_{22} \equiv (a_3\bar{b}_1 + \omega a_1\bar{b}_2 + \omega^2 a_2\bar{b}_3)_{22}\,, \tag{74}$$

where a change to a basis with real CG is $(\bar{b}_1, \bar{b}_2, \bar{b}_3) \rightarrow (\bar{b}_1, \omega^2\bar{b}_2, \omega\bar{b}_3)$. For sets 1_{ij} sharing a fixed $j \neq 0$ (distinct only under generator c), the generalisation of the change of basis involves a mix of eq(67), eq(68) and the diagonal type similar to eq(60), eq(61), or equivalently, reordering the eigenstates of $c_{3_{0j}}$ in eq(67), eq(68). For sets not sharing an index, but for which the sum over indices both sum up to 0 (modulo 3) the change of basis is possible but requires an additional redefinition of one of the 3 singlets (in addition to the triplet representations). Fortunately, the IA produces results that are basis independent, so for a given Lagrangian one can avoid checking whether basis changes that lead to real CG exist or not.

Either using basis changes or the IA, the conclusion for Lagrangians with singlets coupling to triplet and anti-triplet in the manner of $\mathcal{L}_{CD}$ is the same. There are 12 sets of 3 singlets that conserve CP, starting from 1_{00}, 1_{01}, 1_{02} and ending with 1_{20}, 1_{21}, 1_{22}. The sets can be identified whenever the sum of both indices over the 3 singlets adds up to 0 (modulo 3), meaning that there is one non-trivial element of $\Delta(27)$ that does not distinguish the 3 singlets and it is then possible to choose that element to be the generator c in another basis. As far as the 3-field couplings are concerned these 12 sets are equivalent through a change of basis to the choice with $i = 0$, which is why this was the only set considered in [8]. For the other 72 choices of 3 singlets, or for 4 or more singlets, the complex CG coefficients can only be moved around by the change of basis, but not eliminated. In such situations, the coupling of the additional singlets to triplets is not allowed due to CP invariance of the Lagrangian, cf. [8].

A similar conclusion, based on an analysis of the automorphisms of $\Delta(27)$, was presented later in [3]: that adding more than two non-trivial singlets (to a setting with just triplet representations) no longer allows a consistent CP transformation to be defined.[4]

[4] Strictly, how the singlets couple to triplet representations is very relevant, as discussed above cf. cases like the 8 sets of 3 non-trivial singlets that conserve CP automatically. Furthermore, this type of statement assumes all singlets have non-vanishing couplings to the triplets. This is not a spurious assumption as CP can be conserved even in settings with triplets and more than three non-trivial singlets, where triplet-decoupled singlets are still relevant due to coupling to other singlets. This can be a natural outcome if the vanishing couplings are enforced by a symmetry, which can be a specific CP symmetry as illustrated by CP_1 leading to $y_c = 0$ for $\mathcal{L}_{CD}$.

Journal of Physics: Conference Series **631** (2015) 012020 doi:10.1088/1742-6596/631/1/012020

2.3.2. $\Delta(27)$ and $\mathcal{L}_4$ In [3] a $\Delta(27)$ toy model with the trivial singlet and two non-trivial singlets was considered. This is actually the model illustrated here in figure 1, which was the starting point for the scalar field Lagrangian $\mathcal{L}_4$ I used to exemplify the IA in section 2.1. The authors of [3] employed a $U(1)$ symmetry to restrict the allowed couplings and used the structures imposed by $\Delta(27)$ on the coupling matrices F, G, H_Ψ and H_Σ to compute the bottom row in figure 2.

The field S is associated with coupling matrix F proportional to the identity so in their notation the 1_0 singlet corresponds to the trivial singlet 1_{00} here. The fields X, Y are associated to G proportional to Y_{01} and H_Ψ, H_Σ proportional to $Y_{20} = Y_{10}^\dagger$ (Y_{01} and Y_{10} are shown in eq(50)), based on these couplings I identify that the singlets 1_1 and 1_3 in their notation correspond to singlets 1_{01} and 1_{20} here. With these coupling matrices and the couplings f, g, h_Ψ and h_Σ defined in their Lagrangian as shown in figure 1, calculating the CPIs in eq(27), eq(28) leads to:

$$\mathrm{Im}\,\mathrm{Tr}\left[F^\dagger H_\Psi F H_\Sigma^\dagger\right] = \mathrm{Im}\left(|f|^2 h_\Psi h_\Sigma^*\right), \tag{75}$$

$$\mathrm{Im}\,\mathrm{Tr}\left[G^\dagger H_\Psi G H_\Sigma^\dagger\right] = \mathrm{Im}\left(\omega |g|^2 h_\Psi h_\Sigma^*\right), \tag{76}$$

cf. figure 2. Both CPIs depend on the phase $\mathrm{Arg}(h_\Psi h_\Sigma^*)$ (the relative phase between two arbitrary Lagrangian parameters). It is clear that no value of this phase can make both CPIs vanish so the conclusion is again that $\Delta(27)$ is explicitly violated, or CP is explicitly violated, or at least one coupling vanishes. If I impose $f = 0$ (or $g = 0$) due to CP conservation, then S (or X) decouples from the triplets Ψ, Σ and CP can be conserved for specific values of the phase $\mathrm{Arg}(h_\Psi h_\Sigma^*)$. Note that the respective CP transformations for $f = 0$ will differ from those for $g = 0$. The triplet-decoupled singlet in either CP conserving case is still coupled to the other singlet scalars through quartic couplings that are unconstrained by the $U(1)$ symmetry, $SS^\dagger XX^\dagger$, $SS^\dagger YY^\dagger$, $XX^\dagger YY^\dagger$.

2.4. Explicit geometrical CP violation

I propose now a toy model similar to $\mathcal{L}_{CD}$ but where the field content is reduced to contain only triplet A and singlets C_{10}, D_{01}, and there are no $U(1)$ or Z_N symmetries forbidding singlets from coupling to (AA^*). Then the Lagrangian has terms:

$$\mathcal{L}_A = y_c (AA^*)_{20} C_{10} + y_d (AA^*)_{02} D_{01} + h.c., \tag{77}$$

In contrast with $\mathcal{L}_{CD}$ which has the same singlets, there is no $\bar{B}$. The situation is to some extent similar to adding to $\mathcal{L}_{CD}$ the trivial singlet, in the sense that with couplings to a triplet in the manner of $\mathcal{L}_A$, the only pairs of singlets that automatically conserve CP are the 8 pairs including 1_{00}, and the four pairs with 2 non-trivial singlets 1_{01}, 1_{02}; 1_{10}, 1_{20}; 1_{11}, 1_{22}; 1_{12}, 1_{21} (where the sum over the 2 singlets of both indices adds up to 0 modulo 3).

Instead of dwelling further on basis changes I use the IA to study the CP properties of $\mathcal{L}_A$. The most general transformations are the same from eq(51) (ignoring $\bar{B}$) and the CP invariance conditions coming from L_A are similar to eq(52), eq(53):

$$U^\dagger Y_{01} U e^{ip_{01}} = Y_{01}^*, \tag{78}$$

$$U^\dagger Y_{10} U e^{ip_{10}} = Y_{10}^*. \tag{79}$$

Rather than trying to find unitary matrices that may not exist, it is often better to skip directly to building CPIs that do not depend on them. In this case a relevant CPI is:

$$I_2 \equiv \mathrm{Im}\,\mathrm{Tr}(Y_{01} Y_{10}^\dagger Y_{01}^\dagger Y_{10}), \tag{80}$$

which has to vanish for CP conservation. Using Y_{01}, Y_{10} from eq(50) I find:

$$I_2 = \text{Im}(3\omega^2 |y_c|^2 |y_d|^2) \tag{81}$$

This means that CP can be *explicitly* violated, in a minimal model with only 2 $\Delta(27)$ singlets. But furthermore the IA also shows that in this model, CP is violated by a *calculable phase* that is *entirely determined by the symmetry of the Lagrangian* (and not by arbitrary parameters of the Lagrangian). The phases of the arbitrary $\mathcal{L}_A$ parameters y_c and y_d do not contribute, as shown very clearly in eq(81). This situation is directly comparable to the original definition of *calculable phases* in [5], where special cases with spontaneous CP violation were referred to as *geometrical.* In analogy with the original definition, it is reasonable to refer to cases like this as *explicit* geometrical CP violation.

Explicit geometrical CP violation was first identified in [1]. The model presented therein was not a scalar toy model but a physical multi-Higgs doublet model with scalars h_{00}, h_{01} and h_{10}. It contains fermions L (SM lepton doublets) and ν^c (SM singlet neutrinos) transforming under $\Delta(27)$ as triplet and anti-triplet. The neutrino Lagrangian is

$$\mathcal{L}_3 = y_{00}(L\nu^c)_{00}h_{00} + y_{01}(L\nu^c)_{02}h_{01} + y_{10}(L\nu^c)_{20}h_{10} + h.c.. \tag{82}$$

CP is explicitly violated due to the presence of the 3 coupled singlets and the relevant CPI is naturally sensitive to all 3 couplings:

$$I_3 \equiv \text{Im}\,\text{Tr}(Y_{00}Y_{01}^\dagger Y_{10}Y_{00}^\dagger Y_{01}Y_{10}^\dagger)\,, \tag{83}$$

where $\Delta(27)$ imposes Y_{00} proportional to the identity and I rename in eq(50) $y_c = y_{10}$ and $y_d = y_{01}$ to match the notation from [1] used in eq(82). Then:

$$I_3 = \text{Im}(3\omega^2 |y_{00}|^2 |y_{01}|^2 |y_{10}|^2)\,, \tag{84}$$

showing CP is *explicitly violated by a phase only originating from the group structure, and not from arbitrary couplings* - the arbitrary phases of y_{00}, y_{01} and y_{10} do not affect I_3.

3. Conclusions

The main conclusion to be drawn is that the invariant approach is a powerful method to study the CP properties of specific Lagrangians, particularly in the presence of flavour symmetries. The CP-odd invariants built from a Lagrangian do not require detailed knowledge of group theory and require relations between the couplings for the Lagrangian to conserve CP. One can then insert into the relevant CP-odd invariants the couplings that respect the flavour symmetry, and obtain a basis independent answer if CP is violated by those couplings.

I have illustrated the use of the invariant approach with several examples, mostly based on the $\Delta(27)$ symmetry. For a given Lagrangian I commented on the consequences of adding a specific CP symmetry. I also clarified what are the possible outcomes when adding more $\Delta(27)$ singlets to different models, noting that it is relevant to distinguish how the singlets couple to triplets. For 3 coupled singlets, a model with distinct triplet and anti-triplet (or two distinct triplets) can explicitly violate CP for 72 choices of 3 singlets; the other 12 choices lead to automatic CP conservation, which occurs when the 3 singlets are undistinguished by at least one non-trivial element of $\Delta(27)$ (this case includes 8 choices where the 3 singlets are non-trivial). In contrast, in a model with just one triplet the possibility for explicit CP violation exists already with 2 coupled non-trivial singlets. Finally, I used a simple toy model with a triplet and 2 coupled non-trivial singlets as an example of explicit geometrical CP violation, followed by the more realistic example with 3 singlets proposed in [1].

Acknowledgments
This project is supported by the European Union's Seventh Framework Programme for research, technological development and demonstration under grant agreement no PIEF-GA-2012-327195 SIFT. I thank the organisers of DISCRETE 2014 for hosting a very interesting conference, and also G. C. Branco, S. F. King for helpful discussions.

References
[1] G. C. Branco, I. de Medeiros Varzielas and S. F. King, arXiv:1502.03105 [hep-ph].
[2] J. Bernabeu, G. C. Branco and M. Gronau, Phys. Lett. B **169**, 243 (1986).
[3] M. C. Chen, M. Fallbacher, K. T. Mahanthappa, M. Ratz and A. Trautner, Nucl. Phys. B **883** (2014) 267 [arXiv:1402.0507 [hep-ph]].
[4] C. C. Nishi, Phys. Rev. D **88** (2013) 3, 033010 [arXiv:1306.0877 [hep-ph]].
[5] G. C. Branco, J. M. Gerard and W. Grimus, Phys. Lett. B **136**, 383 (1984).
[6] I. de Medeiros Varzielas and D. Emmanuel-Costa, Phys. Rev. D **84** (2011) 117901 [arXiv:1106.5477 [hep-ph]].
[7] I. de Medeiros Varzielas, D. Emmanuel-Costa and P. Leser, Phys. Lett. B **716** (2012) 193 [arXiv:1204.3633 [hep-ph]].
[8] G. Bhattacharyya, I. de Medeiros Varzielas and P. Leser, Phys. Rev. Lett. **109** (2012) 241603 [arXiv:1210.0545 [hep-ph]].
[9] I. de Medeiros Varzielas and D. Pidt, J. Phys. G **41**, 025004 (2014) [arXiv:1307.0711 [hep-ph]].
[10] I. Medeiros Varzielas and D. Pidt, JHEP **1311**, 206 (2013) [arXiv:1307.6545 [hep-ph], arXiv:1307.6545].
[11] E. Ma, Phys. Lett. B **723** (2013) 161 [arXiv:1304.1603 [hep-ph]].
[12] I. de Medeiros Varzielas, JHEP **1208** (2012) 055 [arXiv:1205.3780 [hep-ph]].
[13] I. P. Ivanov and L. Lavoura, Eur. Phys. J. C **73** (2013) 4, 2416 [arXiv:1302.3656 [hep-ph]].
[14] I. de Medeiros Varzielas, J. Phys. Conf. Ser. **447** (2013) 012038 [arXiv:1302.3991 [hep-ph]].
[15] M. Holthausen, M. Lindner and M. A. Schmidt, JHEP **1304** (2013) 122 [arXiv:1211.6953 [hep-ph]].
[16] I. P. Ivanov and C. C. Nishi, JHEP **1501** (2015) 021 [arXiv:1410.6139 [hep-ph]].
[17] M. Fallbacher and A. Trautner, arXiv:1502.01829 [hep-ph].

Tests of Lorentz and CPT invariance with neutrinos and photons

Jorge S Díaz

Institute for Theoretical Physics, Karlsruhe Institute of Technology, 76128 Karlsruhe, Germany

E-mail: jorge.diaz@kit.edu

Abstract. Lorentz symmetry is a cornerstone of modern physics. As the spacetime symmetry of special relativity, Lorentz invariance is a basic component of the standard model of particle physics and general relativity, which to date constitute our most successful descriptions of nature. Deviations from exact symmetry would radically change our view of the universe and current experiments allow us to test the validity of this assumption. The propagation of neutrinos and photons can be used to search for deviations from exact Lorentz and CPT invariance.

1. Introduction

The nonconservation of parity (P) violation in weak interactions motivates the search for new physics by using different experimental setups that can be sensitive to the potential breakdown of other combinations of discrete symmetries. In fact, our matter-dominated Universe suggests that minute differences matter and antimatter properties can have profound consequences. Today we search for deviations from the invariance under parity and charge conjugation (CP) in different sectors of the Standard Model (SM) because the size of the CP-violating effects observed in the quark sector cannot fully account for the matter-antimatter imbalance. A more drastic possibility is the potential violation of CPT invariance, this is the combination of parity, charge conjugation, and time-reversal (T). In a realistic quantum field theory, in which unitarity and microcausality are preserved, CPT violation implies the violation of Lorentz symmetry [1].

During the last decade, experimental searches and phenomenological consideration for spacetime anisotropies have been considered [2, 3, 4, 5]. A systematic classification of the most general ways to incorporate Lorentz-violating operators in the SM action is an effective field theory called Standard-Model Extension (SME) [6, 7]. In this general framework, Lorentz-violating terms are introduced at the action level by contracting operators of conventional fields in the SM with controlling coefficients to preserve coordinate invariance. A subset of Lorentz-violating operators also break CPT. This framework can also be written in a curved spacetime and incorporate Lorentz violation in gravity [8]. Since the SME allows the identification of the relevant observable effects in different experiments, the search for possible deviations from exact Lorentz invariance has become a worldwide experimental program [9].

The interferometric nature of neutrino oscillations makes this phenomenon an attractive method to search for novel effects. Similarly, the long distance and the high energy of astrophysical neutrinos and photons makes these two particles cosmic messengers of astrophysical phenomena as well as probes of fundamental physics.

In what follows, the key signatures of deviations of Lorentz and CPT invariance will be studied within the context of the SME; nevertheless, different approaches for CPT-violating physics can be implemented with and without Lorentz invariance [10, 11, 12, 13, 14, 15, 16, 17].

2. Lorentz-violating neutrinos

Active neutrinos in the SME are characterized by a 6×6 effective Hamiltonian, in which a 3×3 diagonal block h_{ab} describes three left-handed neutrinos, whereas the other 3×3 diagonal block $h_{\bar{a}\bar{b}}$ describes three right-handed antineutrinos. The remaining 3×3 off-diagonal block $h_{a\bar{b}}$ corresponds to mixing between neutrinos and antineutrinos arising due to Lorentz violation. The indices $a, b = e, \mu, \tau$ and $\bar{a}, \bar{b} = \bar{e}, \bar{\mu}, \bar{\tau}$ denote the flavors of active neutrinos and antineutrinos, respectively. The neutrino block can be explicitly written in the form [18]

$$h_{ab} = |\boldsymbol{p}|\delta_{ab} + \frac{m^2_{ab}}{2|\boldsymbol{p}|} + (a_L)^{\alpha}_{ab}\hat{p}_{\alpha} - (c_L)^{\alpha\beta}_{ab}\hat{p}_{\alpha}\hat{p}_{\beta}|\boldsymbol{p}|, \tag{1}$$

where at leading order the neutrino momentum is given by the energy $|\boldsymbol{p}| \approx E$ [19]. The first two terms correspond to a Lorentz-invariant description, in which the Pontecorvo-Maki-Nakagawa-Sakata (PMNS) matrix [20, 21] is used to parametrize the flavor mixing and mass-squared differences m^2_{ij} lead to conventional neutrino oscillations. The third element in the Hamiltonian (1) is an energy-independent term in which the complex matrix $(a_L)^{\alpha}_{ab}$ controls CPT-odd Lorentz violation, while the last term grows linearly with the neutrino energy and describes CPT-even Lorentz violation through the complex matrix $(c_L)^{\alpha\beta}_{ab}$. The components of these 3×3 matrices are called coefficients for Lorentz violation. The four-vector depends on the neutrino direction of propagation $\hat{p}$ as $\hat{p}^{\alpha} = (1; \hat{p})$. Operators of arbitrary dimension can also be incorporated in the theory, which would lead to higher powers of the neutrino energy in the Hamiltonian [22, 23].

The coefficients for Lorentz violation $(a_L)^{\alpha}_{ab}$ and $(c_L)^{\alpha\beta}_{ab}$ behave like fixed background fields that produce measurable effects when the experiment is boosted or rotated. In order to systematically compare different experiments searching for Lorentz-violation, a common inertial frame must be chosen. Experimental results are conventionally reported in the Sun-centered equatorial frame [9]. In this frame, a neutrino beam rotates with sidereal frequency $\omega_{\oplus} \simeq 2\pi/(23 \text{ h } 56 \text{ min})$ with respect to the background fields for experiments with detector and source on the surface of the Earth. This time dependence can be used to parametrize the relevant observable in terms of harmonics of the sidereal angle $\omega_{\oplus}T_{\oplus}$. In addition to searching for generic effects of Lorentz violation in neutrinos, some studies have considered the possibility of using the Hamiltonian (1) to describe all the neutrino-oscillation data as an alternative to the conventional mass-driven oscillations. These global models based on the SME have interesting features that could accommodate all the established data as well as some anomalous results reported by different experiments [24, 25, 26, 27, 28, 29].

As simple classification of the different experimental signatures of the many SME coefficients, neutrino experiments can be classified into two types: flavor mixing and oscillation-free effects. As the name suggests, mixing effects arise from off-diagonal elements in flavor space and leading to neutrino oscillations. Oscillation-free effects, on the other hand, correspond to modifications that affect the behavior of the three neutrino flavors in the same form, leaving oscillations unaffected because they arise from factors proportional to the identity matrix in flavor space. These oscillation-free effects can be observed in kinematical measurements such as neutrino velocity and weak decays.

3. Neutrino oscillations

Flavor mixing makes neutrinos natural interferometers that can be used as sensitive tools to study the coefficients in the Hamiltonian (1). Perturbative methods can be implemented for the determination of the relevant oscillation probabilities, depending on the details of the experiment of interest.

Experiments with long baselines have a far detector located where the first oscillation maximum is expected according to the conventional massive-neutrino description. Given the propagation distance from the source to the detector, in these experiments the oscillation signal due to neutrino masses is large and Lorentz-violating effects can be incorporated as a small perturbation to the mass-driven oscillations. Using conventional time-dependent perturbation theory, we can write the oscillation probabilities as a power series [30]

$$P_{\nu_b\to\nu_a} = P^{(0)}_{\nu_b\to\nu_a} + P^{(1)}_{\nu_b\to\nu_a} + P^{(2)}_{\nu_b\to\nu_a} + \ldots, \tag{2}$$

where the conventional oscillation probability from the massive-neutrino model is corresponds to the leading-order term $P^{(0)}_{\nu_b\to\nu_a}$, whereas the next term, $P^{(1)}_{\nu_b\to\nu_a}$, arises from the interference between the mass-driven oscillations and Lorentz violation. Each term in the series (2) can be calculated order by order. In the Sun-centered frame mentioned earlier, the orientation of the neutrino beam varies with sidereal phase $\omega_\oplus T_\oplus$ with respect to the fixed background fields $(a_L)^\alpha_{ab}$ and $(c_L)^{\alpha\beta}_{ab}$, which also introduces direction-dependent effects. The sub-leading term in the series (2), for example, can be written as an expansion on harmonics of the sidereal phase in the form

$$\begin{aligned}\frac{P^{(1)}_{\nu_b\to\nu_a}}{2L} &= (P^{(1)}_{\mathcal{C}})_{ab} + (P^{(1)}_{\mathcal{A}_s})_{ab}\sin\omega_\oplus T_\oplus + (P^{(1)}_{\mathcal{A}_c})_{ab}\cos\omega_\oplus T_\oplus \\ &\quad +(P^{(1)}_{\mathcal{B}_s})_{ab}\sin 2\omega_\oplus T_\oplus + (P^{(1)}_{\mathcal{B}_c})_{ab}\cos 2\omega_\oplus T_\oplus .\end{aligned} \tag{3}$$

The amplitude of each term in this expansion has units of energy and depends on the coefficients for Lorentz violation as well as properties of the experiment such as the orientation of the neutrino beam, the location of the detector, and the neutrino energy [30, 31]. Corresponding expressions for antineutrino oscillations can be obtained directly by replacing $(a_L)^\alpha_{ab} \to -(a_L)^{\alpha*}_{ab}$ and $(c_L)^{\alpha\beta}_{ab} \to (c_L)^{\alpha\beta*}_{ab}$.

In high-energy experiments with a short baseline, no oscillations are expected according to the conventional massive-neutrino model. In this case, the first two terms in the series (2) vanish and any oscillation signal would be a consequence of Lorentz violation. Direct calculation shows that for appearance experiments the leading-order driving the oscillation probability for neutrinos takes the explicit form [31]

$$P_{\nu_b\to\nu_a} \simeq L^2\left|(a_L)^\alpha_{ab}\hat{p}_\alpha - (c_L)^{\alpha\beta}_{ab}\hat{p}_\alpha\hat{p}_\beta E\right|^2, \quad a\neq b, \tag{4}$$

where L is the baseline and E the neutrino energy.

The two limits described above have been implemented to perform several independent searches for Lorentz and CPT violation by Double Chooz [32], IceCube [33], LSND [34], MiniBooNE [35, 36], MINOS [37, 38, 39], and Super-Kamiokande [40]. The complementarity of these searches is due to the study of different oscillation channels, which depend on similar coefficients but with different flavor indices. To date no compelling evidence for Lorentz violation has been found and tight constraints on several SME coefficients have been determined in these experimental searches [9].

There is also a type of oscillation that can occur in the SME context that is absent in the conventional Lorentz-invariant case. As mentioned at the beginning of the previous section, in the presence of Lorentz violation neutrinos and antineutrinos can also mix. The oscillation between left-handed neutrinos and right-handed antineutrinos would be caused by an independent set of SME coefficients. These oscillations always exhibit direction dependence and are absent in the first two terms of the series (2). In other words, the SME coefficients that produce neutrino-antineutrino mixing appear quadratically in the oscillation probability. For these oscillations, the probability can be written in the same form as the linear term (3). However, being a second-order effect the sidereal decomposition will include up to fourth harmonics [30]. Experimental searches of these neutrino-antineutrino oscillations have been performed using accelerator [41] and reactor neutrinos [42].

4. Oscillation-free signals

Modifications to the neutrino propagation properties that change the three flavors in the same way produce no effects in oscillations because they would correspond to a uniform shift of the eigenstate energies (oscillations only depend on eigenstate energy differences). These oscillation-free effects can be studied in experiments that can measure the neutrino velocity. In the SME, the neutrino velocity can be written as [22]

$$v_\nu \approx 1 - \frac{|m|^2}{2E^2} + \sum_{djm} (d-3) E^{d-4}\, e^{im\omega_\oplus T_\oplus}\, {}_0\mathcal{N}_{jm} \big((a_{\text{of}}^{(d)})_{jm} - (c_{\text{of}}^{(d)})_{jm} \big), \tag{5}$$

where the real mass parameter $|m|^2$ does not participate in oscillations, E the neutrino energy, and we have used a spherical decomposition for the Lorentz-violating part. This spherical decomposition is very handy to describe transformation properties under rotations of the independent coefficients indicated by the sum The index d denotes the mass dimension of the operator in the theory, whereas the angular momentum indices jm label the rotation properties of the oscillation-free coefficients for CPT-odd $(a_{\text{of}}^{(d)})_{jm}$ and CPT-even $(c_{\text{of}}^{(d)})_{jm}$ for Lorentz violation. The information regarding the orientation of the neutrino beam as well as the location of the laboratory is encoded in the angular factors ${}_0\mathcal{N}_{jm}$ and the harmonics of the sidereal phase $\omega_\oplus T_\oplus$ is controlled by the index m. For this reason, the neutrino velocity becomes time dependent for $m \neq 0$. Similarly, direction-dependent effects appear for $j \neq 0$ due to the loss of invariance under rotations. The isotropic limit corresponds to the particular case $j = 0$.

It should be noticed that the sum (5) only one of the coefficients appear for a given value of the index d. Explicitly, $(a_{\text{of}}^{(d)})_{jm} = 0$ for d even and $(c_{\text{of}}^{(d)})_{jm} = 0$ for d odd. Here we notice that for odd d, CPT violation makes neutrinos and antineutrinos move at different speed (for antineutrinos the coefficient $(a_{\text{of}}^{(d)})_{jm}$ has opposite sign). The expression for the neutrino velocity (5) can be directly applied to this type of studies in beam experiments [43, 44, 45, 46, 47, 48, 49].

For operators of dimension $d > 5$ in the theory, the neutrino velocity (5) becomes energy dependent, which can produce novel effects including the dispersion of a neutrino burst due to the the independent speed of neutrinos of different energies. Dispersive effects and time delays due to modifications of the neutrino speed can be tested to high precision using astrophysical sources of neutrinos because the long distances traveled enhance the small effects of the coefficients for Lorentz violation [22, 50].

Furthermore, the modified neutrino (and antineutrino) energy and momentum relation exhibited by the Hamiltonian (1) indicates that the oscillation-free coefficients can also be studied by exploring the properties of modified dispersion relations in different processes. One of the remarkable effects of Lorentz violation is the possibility of making kinematically forbidden reactions allowed and vice versa. This means that nonzero SME coefficients can open the phase space of conventionally forbidden processes as well as blocking certain processes that would be otherwise allowed. Consider the leptonic decay of a charged meson M^+ of the form $M^+ \to l^+ + \nu_l$, which becomes forbidden above some threshold energy E_{th} [22, 51, 52, 53, 54, 55, 56, 57, 58]. If the decay products are observed at a given energy E_0, then the size of the SME coefficients must lead to a threshold energy such that $E_{\text{th}} > E_0$. This threshold condition can be written directly from energy-momentum conservation relations and the Hamiltonian (1) in the form [22]

$$\sum_{djm} E_0^{d-2} Y_{jm}(\hat{\boldsymbol{p}})\big[\pm(a_{\rm of}^{(d)})_{jm} - (c_{\rm of}^{(d)})_{jm}\big] < \frac{1}{2}(M_M - m_{l^\pm})^2, \tag{6}$$

where M_M denotes the mass of the parent charged meson, $m_{l^\pm}$ indicates the mass of the charged lepton, and the sign $+$ $(-)$ refers to neutrinos (antineutrinos). The spherical harmonics Y_{jm} encode possible direction-dependent effects (for $j \neq 0$). This expression shows that better constraints on the SME coefficients can be obtained by observing energetic neutrinos. The observation of energetic neutrinos in IceCube [59, 60] serves as a very sensitive probe of Lorentz and CPT invariance using this threshold analysis. In particular, the observation of several events can be used to go beyond conventional isotropic studies and to explore direction-dependent effects, which requires a spread of events in the sky [61].

Following a similar approach, forbidden processes can become allowed due to Lorentz and CPT violation. For instance, some coefficients in the neutrino velocity (5) can produce superluminal neutrinos, which could lose energy in the form of Cherenkov radiation [22, 62, 63, 64, 65, 66, 67, 68, 69, 70, 71, 72]. The determination of a characteristic energy-loss distance $D(E) = -E/(dE/dx)$ due to Cherenkov radiation can be used to write the condition $L < D(E)$, where L is the distance traveled by the neutrino [22]. For the electron-positron emission in the process $\nu \to \nu + e^- + e^+$ the rate of energy loss takes the explicit form [22, 61]

$$-\frac{dE}{dx} = \frac{C}{8}\int \frac{\kappa^0\,\boldsymbol{\kappa}'^2}{(\kappa^2 - M_Z^2)^2}\frac{\partial|\boldsymbol{\kappa}'|}{\partial\kappa_0}\frac{q\cdot k\; q'\cdot k'}{q_0 k_0 q_0' k_0'}\, d^3p'\, d\Omega_{\kappa'}, \tag{7}$$

where $\kappa = k + k'$ and $\kappa' = k - k'$ are auxiliary 4-vectors defined in terms of the momentum of the electron (k) and the positron (k'), C is a constant, and $q/q_0 = (1, \hat{\boldsymbol{p}})$, $q'/q_0' = (1, \hat{\boldsymbol{p}}')$ encode the Lorentz-violating effects in the neutrino dispersion relation. This formula can be used to search for Lorentz violation through the observation of high-energy astrophysical neutrinos [61, 73, 74, 75, 76].

There one type of oscillation-free coefficient, however, that cannot be studied in experiments sensitive to modifications of the neutrino velocity (5). Arising from operators of dimension $d = 3$, the coefficients $(a_{\rm of}^{(3)})_{jm}$ leave the neutrino velocity unchanged; however, their experimental signatures would appear weak decays [77]. For single beta decay, the observable signatures of Lorentz and CPT violation appear due to the modified dispersion relation of the antineutrino as well as the altered spinor solutions of the equation of motion [78]. Additionally, a CPT-violating Majorana couplings in the SME can trigger neutrinoless double beta decay even for a negligible Majorana mass [79].

5. Conclusions

Potential violations of Lorentz and CPT invariance have been studied both theoretically and experimentally in a wide range of systems. Neutrinos offer the possibility to perform a great variety of tests of the validity of these spacetime symmetries, from the high precision of low-energy experiments studying weak interactions to the great power of neutrino oscillations thanks to their interferometric nature as well as the high energy of astrophysical neutrinos traveling long distances before reaching our detectors. Lorentz and CPT invariance have survived the several tests performed using the SME as a general framework; nonetheless, there are many other effects that remain unexplored.

Neutrinos have amazed us since they were proposed to save the conservation of energy, today neutrinos offer the opportunity to challenge the cornerstone of modern physics.

Acknowledgments
This work was supported in part by the German Research Foundation (DFG) under Grant no. KL 1103/4-1.

References

[1] O. W. Greenberg, Phys. Rev. Lett. **89**, 231602 (2002).
[2] H. B. Nielsen and I. Picek, Phys. Lett. B **114**, 141 (1982).
[3] H. B. Nielsen and I. Picek, Nucl. Phys. B **211**, 269 (1983) [Addendum-ibid. B **242**, 542 (1984)].
[4] R. Huerta and J. L. Lucio, Phys. Lett. B **131**, 471 (1983).
[5] R. Huerta and J. L. Lucio, Phys. Lett. B **124**, 369 (1983).
[6] D. Colladay and V. A. Kostelecký, Phys. Rev. D **55**, 6760 (1997).
[7] D. Colladay and V. A. Kostelecký, Phys. Rev. D **58**, 116002 (1998).
[8] V. A. Kostelecký, Phys. Rev. D **69**, 105009 (2004).
[9] V. A. Kostelecký and N. Russell, Rev. Mod. Phys. **83**, 11 (2011).
[10] J. R. Ellis and N. E. Mavromatos, Phys. Rept. **320**, 341 (1999).
[11] F. R. Klinkhamer, Nucl. Phys. B **578**, 277 (2000).
[12] C. Adam and F. R. Klinkhamer, Nucl. Phys. B **607**, 247 (2001).
[13] G. Barenboim, L. Borissov, J. D. Lykken and A. Y. Smirnov, JHEP **0210**, 001 (2002).
[14] G. Barenboim, J. F. Beacom, L. Borissov and B. Kayser, Phys. Lett. B **537**, 227 (2002).
[15] F. R. Klinkhamer and C. Rupp, Phys. Rev. D **70**, 045020 (2004).
[16] N. E. Mavromatos, Lect. Notes Phys. **669**, 245 (2005).
[17] M. Chaichian, A. D. Dolgov, V. A. Novikov and A. Tureanu, Phys. Lett. B **699**, 177 (2011).
[18] V. A. Kostelecký and M. Mewes, Phys. Rev. D **69**, 016005 (2004).
[19] J. S. Díaz, Adv. High Energy Phys. **2014**, 962410 (2014).
[20] Z. Maki, M. Nakagawa, and S. Sakata, Prog. Theor. Phys. **28**, 870 (1962).
[21] B. Pontecorvo, Zh. Eksp. Teor. Fiz. **53**, 1717 (1967); Sov. Phys. JETP **26**, 984 (1968).
[22] V. A. Kostelecký and M. Mewes, Phys. Rev. D **85**, 096005 (2012).
[23] V. A. Kostelecký and M. Mewes, Phys. Rev. D **88**, 096006 (2013).
[24] V. A. Kostelecký and M. Mewes, Phys. Rev. D **70**, 031902 (2004).
[25] T. Katori *et al.*, Phys. Rev. D **74**, 105009 (2006).
[26] V. Barger, D. Marfatia and K. Whisnant, Phys. Lett. B **653**, 267 (2007).
[27] V. Barger, J. Liao, D. Marfatia and K. Whisnant, Phys. Rev. D **84**, 056014 (2011).
[28] J. S. Díaz and V. A. Kostelecký, Phys. Lett. B **700**, 25 (2011); Phys. Rev. D **85**, 016013 (2012).
[29] S. -J. Rong and Q. -Y. Liu, Chin. Phys. Lett. **29**, 041402 (2012).
[30] J. S. Díaz *et al.*, Phys. Rev. D **80**, 076007 (2009).
[31] V. A. Kostelecký and M. Mewes, Phys. Rev. D **70**, 076002(R) (2004).
[32] Y. Abe *et al.* [Double Chooz Collaboration], Phys. Rev. D **86**, 112009 (2012).
[33] R. Abbasi *et al.* [IceCube Collaboration], Phys. Rev. D **82**, 112003 (2010).
[34] L. B. Auerbach *et al.* [LSND Collaboration], Phys. Rev. D **72**, 076004 (2005).
[35] A. A. Aguilar-Arevalo *et al.* [MiniBooNE Collaboration], Phys. Lett. B **718**, 1303 (2013).
[36] T. Katori [MiniBooNE Collaboration], Mod. Phys. Lett. A **27**, 1230024 (2012).
[37] P. Adamson *et al.* [MINOS Collaboration], Phys. Rev. Lett. **101**, 151601 (2008).
[38] P. Adamson *et al.* [MINOS Collaboration], Phys. Rev. D **85**, 031101 (2012).
[39] P. Adamson *et al.* [MINOS Collaboration], Phys. Rev. Lett. **105**, 151601 (2010).
[40] K. Abe *et al.* [Super-Kamiokande Collaboration], Phys. Rev. D, in press (2015), arXiv:1410.4267.
[41] B. Rebel and S. Mufson, Astropart. Phys. **48**, 78 (2013).
[42] J. S. Díaz, T. Katori, J. Spitz and J. M. Conrad, Phys. Lett. B **727**, 412 (2013).
[43] J. Alspector *et al.*, Phys. Rev. Lett. **36**, 837 (1976).
[44] G. R. Kalbfleisch, N. Baggett, E. C. Fowler and J. Alspector, Phys. Rev. Lett. **43**, 1361 (1979).
[45] P. Adamson *et al.* [MINOS Collaboration], Phys. Rev. D **76**, 072005 (2007).
[46] T. Adam *et al.* [OPERA Collaboration], JHEP **1210**, 093 (2012).
[47] M. Antonello *et al.* [ICARUS Collaboration], Phys. Lett. B **713**, 17 (2012).
[48] P. Alvarez Sanchez *et al.* [Borexino Collaboration], Phys. Lett. B **716**, 401 (2012).
[49] N. Y. Agafonova *et al.* [LVD Collaboration], Phys. Rev. Lett. **109**, 070801 (2012).
[50] M. J. Longo, Phys. Rev. D **36**, 3276 (1987).
[51] B. Altschul, Phys. Rev. D **84**, 091902 (2011).
[52] A. Chodos, V. A. Kostelecký, R. Potting and E. Gates, Mod. Phys. Lett. A **7**, 467 (1992).
[53] A. Chodos and V. A. Kostelecký, Phys. Lett. B **336**, 295 (1994).

[54] L. Gonzalez-Mestres, arXiv:1109.6630.
[55] X. -J. Bi, P. -F. Yin, Z. -H. Yu and Q. Yuan, Phys. Rev. Lett. **107**, 241802 (2011).
[56] R. Cowsik, S. Nussinov and U. Sarkar, Phys. Rev. Lett. **107**, 251801 (2011).
[57] M. Mannarelli, M. Mitra, F. L. Villante and F. Vissani, JHEP **1201**, 136 (2012).
[58] V. Baccetti, K. Tate and M. Visser, JHEP **1203**, 087 (2012).
[59] M. G. Aartsen *et al.* [IceCube Collaboration], Phys. Rev. Lett. **111**, 021103 (2013).
[60] M. G. Aartsen *et al.* [IceCube Collaboration], Science 342, 1242856 (2013).
[61] J. S. Díaz, V. A. Kostelecký and M. Mewes, Phys. Rev. D **89**, 043005 (2014).
[62] S. R. Coleman and S. L. Glashow, Phys. Rev. D **59**, 116008 (1999).
[63] T. Jacobson, S. Liberati and D. Mattingly, Phys. Rev. D **67**, 124011 (2003).
[64] D. M. Mattingly, L. Maccione, M. Galaverni, S. Liberati and G. Sigl, JCAP **1002**, 007 (2010).
[65] J. Alexandre, J. Ellis and N. E. Mavromatos, Phys. Lett. B **706**, 456 (2012).
[66] A. G. Cohen and S. L. Glashow, Phys. Rev. Lett. **107**, 181803 (2011).
[67] J. M. Carmona and J. L. Cortes, arXiv:1110.0430.
[68] L. Maccione, S. Liberati and D. M. Mattingly, JCAP **1303**, 039 (2013).
[69] S. Mohanty and S. Rao, arXiv:1111.2725.
[70] M. Li, D. Liu, J. Meng, T. Wang and L. Zhou, arXiv:1111.3294.
[71] Y. Huo, T. Li, Y. Liao, D. V. Nanopoulos and Y. Qi, Phys. Rev. D **85**, 034022 (2012).
[72] S. J. Brodsky and S. Gardner, arXiv:1112.1090.
[73] E. Borriello, S. Chakraborty, A. Mirizzi and P. D. Serpico, Phys. Rev. D **87**, 116009 (2013).
[74] F. W. Stecker, Astropart. Phys. **56**, 16 (2014).
[75] F. W. Stecker and S. T. Scully, Phys. Rev. D **90**, 043012 (2014).
[76] D. Mazón, Phys. Rev. D **89**, 056012 (2014).
[77] J. S. Díaz *et al.*, Phys. Rev. D **88**, 071902(R) (2013).
[78] J. S. Díaz, Adv. High Energy Phys. **2014**, 305298 (2014).
[79] J. S. Díaz, Phys. Rev. D **89**, 036002 (2014).

Non-minimally flavour violating dark matter

Monika Blanke

CERN Theory Division, CH-1211 Geneva 23, Switzerland

Institut für Theoretische Teilchenphysik, Karlsruhe Institute of Technology, Engesserstraße 7, D-76128 Karlsruhe, Germany

Institut für Kernphysik, Karlsruhe Institute of Technology, Hermann-von-Helmholtz-Platz 1, D-76344 Eggenstein-Leopoldshafen, Germany

E-mail: monika.blanke@kit.edu

Abstract. Flavour symmetries provide an appealing mechanism to stabilize the dark matter particle. I present a simple model of quark flavoured dark matter that goes beyond the framework of minimal flavour violation. I discuss the phenomenological implications for direct and indirect dark matter detection experiments, high energy collider searches as well as flavour violating precision data.

1. Flavour symmetries in the dark sector

While astrophysical and cosmological observations provide convincing evidence for the existence of dark matter (DM), little is known about its particle properties. Besides its gravitational interactions, appealing arguments exist that the DM particles should interact weakly with the Standard Model (SM). The reason for this belief is referred to as the "WIMP miracle": a DM particle with weak scale mass, $m_{\mathrm{DM}} \sim \mathcal{O}(10^2\,\mathrm{GeV})$ and weak coupling automatically generates the right DM relic abundance. The WIMP hypothesis receives further support from the electroweak (EW) naturalness problem, requiring an extension of the SM at the TeV scale in order to stabilize the (EW) symmetry breaking scale.

Adopting the WIMP hypothesis many properties of the DM particle are still unknown. It has to be neutral under strong and electromagnetic interactions, yet its transformation under EW symmetry can be non-trivial. Indeed the WIMP hypothesis seems to suggest a coupling to the weak gauge bosons. Nevertheless it is equally justified to introduce some new type of interactions through which an EW singlet DM particle couples to the SM.

Non-trivial symmetry transformation properties of DM, both for global and local groups, are possible and worthwhile to be studied. In this article, summarizing the results of [1], we focus on a particular type of global symmetry, namely a flavour group.

The idea of flavoured DM is not new, see e. g. [2] for a review of early studies. However most studies so far assumed that the DM is charged under the SM flavour group and its interactions satisfy the Minimal Flavour Violation (MFV) [3, 4] hypothesis, so that the SM Yukawa couplings $Y_{u,d}$ remain the only source of flavour breaking. This approach has several benefits: apart from avoiding the stringent constraints from flavour changing neutral current (FCNC) processes, it can be shown [5] that the MFV principle implies the stability of DM, provided the latter transforms under certain representations of the flavour group.

The drawback of the MFV assumption is that no interesting effects in flavour violating observables can be expected, thus closing a potential discovery channel of flavoured DM. Therefore in [1] we studied the possibility of non-minimally flavour violating DM. The results of that paper have been presented at the DISCRETE 2014 conference and are summarized in the present article.

2. Dark Minimal Flavour Violation and its implications

2.1. The Dark Minimal Flavour Violation hypothesis

The Dark Minimal Flavour Violation (DMFV) hypothesis [1] is a conceptually straightforward extension of the well-known MFV principle to the DM sector. In complete analogy to the SM matter content we assume that the DM χ transforms under the fundamental representation of a new global flavour symmetry $U(3)_\chi$. The full flavour symmetry group[1] then reads

$$G_F = U(3)_q \times U(3)_u \times U(3)_d \times U(3)_\ell \times U(3)_e \times U(3)_\chi \,. \tag{1}$$

The crucial ingredient for DMFV is then that this global flavour symmetry is broken *only* by the SM Yukawa couplings Y_u, Y_d, Y_e, and the DM coupling λ to the SM fermions. Depending on which type of SM fermion the DM field χ couples to, and depending on the Dirac structure of the coupling, different classes of DMFV models can be obtained.

We are thus left with a huge variety of models with interesting and quite distinct phenomenological implications. It is important to note that while DMFV is conceptually close to the MFV ansatz, the phenomenology of DMFV models differs vastly from the MFV ones. This is due to the new coupling matrix λ whose structure is unrelated to the SM Yukawa couplings, and upon which no hierarchical structure is imposed.

2.2. A minimal model

For concreteness, following [1], let us focus on a concrete class of DMFV models in what follows. We introduce DM as a Dirac fermion χ that transforms as a triplet under the global flavour symmetry $U(3)_\chi$, and that couples to right-handed down-type quarks via a scalar mediator ϕ. The field ϕ is singlet under the flavour group G_F, however it transforms as

$$\phi \sim (\mathbf{3},\mathbf{1})_{-1/3} \tag{2}$$

under the SM gauge group $SU(3)_C \times SU(2)_L \times U(1)_Y$. The DM field χ is a gauge singlet. In order to render the coupling

$$\lambda \bar{d}_R \chi \phi \tag{3}$$

formally invariant under the flavour group G_F, λ is promoted to a spurion field which transforms as triplet under $U(3)_d$ and as anti-triplet under $U(3)_\chi$. Table 1 summarizes the transformation properties of all matter and spurion fields present in this DMFV framework.

The minimal DMFV (mDMFV) model coupling DM to down-type quarks and containing only the matter content summarized in table 1 is fully described by the renormalisable Lagrangian

$$\begin{aligned}\mathcal{L} &= \mathcal{L}_{\rm SM} + i\bar{\chi}\partial\!\!\!/\chi - m_\chi\bar{\chi}\chi - (\lambda_{ij}\bar{d}_{Ri}\chi_j\phi + \text{h.c.}) \\ &\quad + (D_\mu\phi)^\dagger(D^\mu\phi) - m_\phi^2\phi^\dagger\phi + \lambda_{H\phi}\,\phi^\dagger\phi\, H^\dagger H + \lambda_{\phi\phi}\,\phi^\dagger\phi\,\phi^\dagger\phi \,.\end{aligned} \tag{4}$$

Here $i,j = 1,2,3$ are flavour indices. We stress that the 3×3 coupling matrix λ constitutes a new source of flavour violation with respect to MFV models. Before we turn to summarizing the phenomenological implications of this model, we recapitulate a number of direct implications of the DMFV ansatz. We stress that these implications follow from symmetry principles alone and therefore hold beyond the minimal model.

[1] in the absence of right-handed neutrinos

	$SU(3)_c$	$SU(2)_L$	$U(1)_Y$	$U(3)_q$	$U(3)_u$	$U(3)_d$	$U(3)_\chi$
q_L	**3**	**2**	1/6	**3**	**1**	**1**	**1**
u_R	**3**	**1**	2/3	**1**	**3**	**1**	**1**
d_R	**3**	**1**	-1/3	**1**	**1**	**3**	**1**
ℓ_L	**1**	**2**	-1/2	**1**	**1**	**1**	**1**
e_R	**1**	**1**	-1	**1**	**1**	**1**	**1**
H	**1**	**2**	1/2	**1**	**1**	**1**	**1**
ϕ	**3**	**1**	-1/3	**1**	**1**	**1**	**1**
χ_L	**1**	**1**	0	**1**	**1**	**1**	**3**
χ_R	**1**	**1**	0	**1**	**1**	**1**	**3**
Y_u	**1**	**1**	0	**3**	$\bar{\mathbf{3}}$	**1**	**1**
Y_d	**1**	**1**	0	**3**	**1**	$\bar{\mathbf{3}}$	**1**
λ	**1**	**1**	0	**1**	**1**	**3**	$\bar{\mathbf{3}}$

Table 1. Symmetry transformation properties of the minimal DMFV matter content and the Yukawa spurions.

2.3. Implications of DMFV

The attentive reader might have already noticed that the DM mass term m_χ in (4) does not carry flavour indices. Indeed the DMFV principle ensures the universality of masses for the three DM flavours. Yet this statement strictly speaking is true only at the leading order in the DMFV spurion expansion. Including the leading non-trivial contribution the mass matrix becomes

$$m_{ij} = m_\chi(\mathbb{1} + \eta\,\lambda^\dagger\lambda + \cdots)_{ij}\,. \tag{5}$$

The flavour violating spurion λ thus introduces a mass splitting between the various DM flavours, and the size of this splitting is intimately tied to the size of deviations from a universal DM coupling to quarks. Furthermore DMFV ensures that the DM mass matrix is aligned with the coupling λ in the $U(3)_\chi$ flavour space.

The coefficient η in (5) is a free parameter of the low energy theory, and can be predicted only in a UV-complete flavour model. If the correction to m_χ is generated at the tree level, it is expected to be of $\mathcal{O}(1)$. Tree level contributions might be absent, depending on the details of the UV completion. A mass splitting of the form $\lambda^\dagger\lambda$ is however unavoidably generated at the one loop level, leaving us with the constraint $|\eta| \gtrsim \mathcal{O}(10^{-2})$. In our phenomenological analysis we will be agnostic about the size of η. In order to ensure perturbativity of the DMFV expansion we will however impose that the mass splittings generated by $\eta\,\lambda^\dagger\lambda$ are at most 30%.

Another welcome implication of the DMFV assumption is the reduction of free parameters in the flavour sector. The $U(3)_\chi$ flavour symmetry can be used to remove a number of non-physical parameters from the coupling matrix λ. The latter can then be parametrised in terms of six real parameters and three CP violating phases as follows. We split λ into a unitary matrix U_λ and a positive diagonal matrix D_λ,

$$\lambda = U_\lambda D_\lambda\,, \tag{6}$$

and then parametrise U_λ in terms of three mixing angles $s^\lambda_{12,13,23}$ and three phases $\delta^\lambda_{12,13,23}$ as in [6]. For D_λ we use two parametrisations that can easily translated into each other

$$D_\lambda \equiv \mathrm{diag}(D_{\lambda,11}, D_{\lambda,22}, D_{\lambda,33}) = \lambda_0 \cdot \mathbb{1} + \mathrm{diag}(\lambda_1, \lambda_2, -(\lambda_1+\lambda_2))\,. \tag{7}$$

The most interesting consequence of the DMFV ansatz, as far as the DM phenomenology is concerned, is the implied stability of the lightest state in the DM sector. An explicit analysis (see appendix B of [1]) shows that the flavour symmetry G_F broken only by the SM Yukawa couplings and the new coupling λ, together with the unbroken $SU(3)_C$ gauge symmetry, contain an unbroken $\mathbb{Z}_3$ subgroup under which only χ and ϕ transform non-trivially. Consequently neither ϕ nor any of the flavours of χ can decay into SM particles only, and the lightest χ flavour is predicted to be stable. At the same time, none of these states can be singly produced at the LHC, and their decay will give rise to missing energy signatures, similar to the ones of supersymmetric squarks.

Last but not least the DMFV hypothesis also ensures that the results of our flavour analysis, while obtained within the mDMFV model, hold beyond the minimal particle content. We will get back to this point in the following section.

3. Phenomenology of the minimal DMFV model

3.1. Flavour constraints

The coupling λ of DM to SM quarks is a new source of flavour violation and therefore generates new contributions to FCNC observables. This gives rise to an exciting new discovery opportunity, in addition to the usual channels in direct and indirect detection experiments and collider searches. See figure 1 for schematic diagrams. At the same time however we also have to deal with the stringent constraints from well measured flavour violating decays.

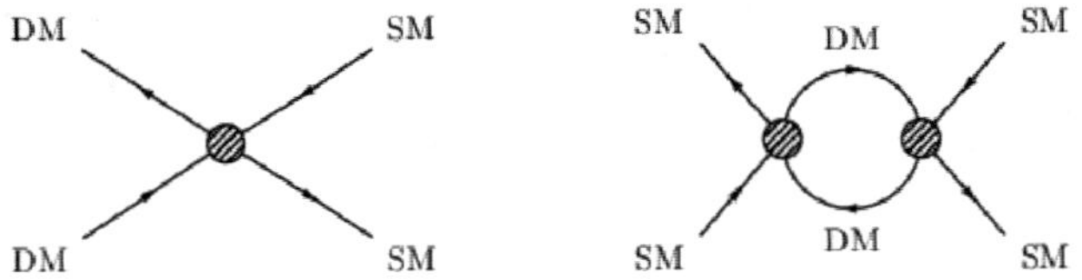

Figure 1. Schematic diagrams contributing to experimental constraints on flavoured DM [1].

Like in many extensions of the SM, the most restrictive constraints on the mDMFV parameter space arise from meson anti-meson mixing observables. New contributions are generated by the exchange of the new particles ϕ and χ at the loop level, see figure 2. The full expression for the mixing amplitude can be found in [1]. While the exact expression for the loop function has to obtained from an explicit calculation, the flavour violating structure can be obtained from the DMFV principle by taking the leading term in the spurion expansion. We find

$$M_{12}^{K,\text{new}} \sim \left((\lambda\lambda^\dagger)_{sd}\right)^2 F(x)\,, \quad M_{12}^{d,\text{new}} \sim \left((\lambda\lambda^\dagger)_{bd}\right)^2 F(x)\,, \quad M_{12}^{s,\text{new}} \sim \left((\lambda\lambda^\dagger)_{bs}\right)^2 F(x)\,, \quad (8)$$

where $F(x)$ is the flavour universal loop function.

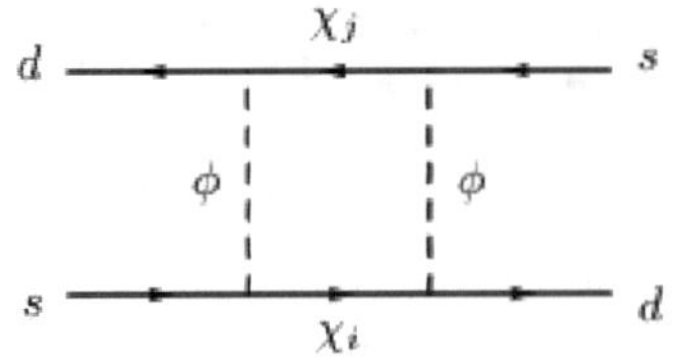

Figure 2. New contribution to $K^0 - \bar{K}^0$ mixing in the mDMFV model [1].

Equation (8) shows that the mDMFV contributions to the $\Delta F = 2$ mixing amplitudes – and in fact to all FCNC transitions – are driven by the off-diagonal elements of $\lambda\lambda^\dagger$. A number of scenarios for the structure of λ can thus be derived, and confirmed in a numerical parameter scan (c. f. figure 3).

(i) **universality scenario** (black): $\lambda_1 \simeq \lambda_2 \simeq 0$
In this case $\lambda \simeq U_\lambda \cdot \lambda_0$ so that $\lambda\lambda^\dagger \simeq \lambda_0^2 \cdot \mathbb{1}$.

(ii) **12-degeneracy** (blue): $\lambda_1 \simeq \lambda_2$
If the first two generations of DM fermions are quasi-degenerate, then the mixing angle s_{12}^λ can be generic while $s_{13,23}^\lambda$ have to be small. This can be understood by taking the limit $\lambda_1 = \lambda_2$, in which case the mixing angle s_{12}^λ becomes non-physical.

(iii) **13-degeneracy** (red): $\lambda_2 \simeq -2\lambda_1$
In the case $\lambda_2 \simeq -2\lambda_1$ the first and third DM flavour are quasi-degenerate, and consequently s_{13}^λ is unconstrained. In order to suppress the remaining flavour violating effects both s_{12}^λ and s_{23}^λ have to be small.

(iv) **23-degeneracy** (green): $\lambda_2 \simeq -1/2\lambda_1$
Finally if $\lambda_2 \simeq -1/2\lambda_1$, the second and third DM flavour are quasi degenerate. Consequently the mixing angle s_{23}^λ is arbitrary, while s_{12}^λ and s_{13}^λ have to be small.

(v) **small mixing scenario** (yellow): arbitrary D_λ
Last but not least if D_λ does not exhibit any degeneracies, then FCNC effects have to be suppressed by the smallness of all three mixing angles $s_{12}^\lambda \simeq s_{13}^\lambda \simeq s_{23}^\lambda \simeq 0$. This scenario, shown by the yellow points in figure 3, corresponds to a diagonal but non-degenerate coupling matrix λ.

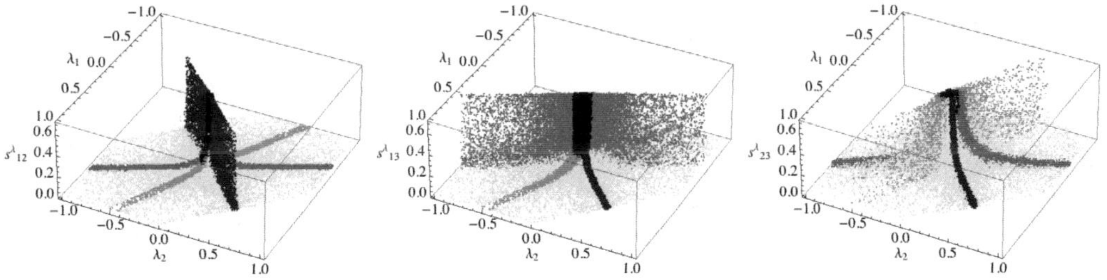

Figure 3. Scenarios for the structure of λ: Universality (black), 12-degeneracy (blue), 13-degeneracy (red), 23-degeneracy (green), small mixing (yellow) [1].

It is important to note that equation (8) is a direct consequence of the DMFV ansatz, without further assumption on the particle content of the model. The above scenarios will therefore also hold in non-minimal models with additional particles, as long as the coupling λ remains the only new source of flavour violation.

While meson anti-meson mixing observables place severe bounds on the mDMFV parameter space, the new contributions to rare K and B decays turn out to be small. One the one hand we do not need to worry about the constraints from well-measured decays like $B \to X_s\gamma$, yet on the other hand at least within the minimal model we can neither explain the anomaly in the $B \to K^+\mu^+\mu^-$ data nor can we hope for a spectacular new physics signature in rare decays that have not yet been measured, like $K \to \pi\nu\bar{\nu}$.

3.2. Dark matter phenomenology

In order to proceed with the DM phenomenology of the mDMFV model we have to make a choice for the mass spectrum in the DM sector, i. e. which of the χ flavours is the lightest. We identify the various flavours by which quark flavour d, s or b they couple to dominantly. From a theoretical point of view, all three choices for the lightest DM flavour are equally well justified. The case of d-flavoured DM is however disfavoured by the lack of signal in the direct detection experiments, at least if we assume DM to be a thermal relic. b- and s-flavoured DM are similar to each other, as far as direct detection experiments are concerned. However b-flavoured DM can explain the recently observed γ ray excess from the galactic centre [7]. It also gives rise to interesting b-jet signatures at the LHC. We therefore restrict our attention to this case, as also done in [1].

While it would be straightforward to circumvent this constraint from a model building point of view, we assume for simplicity that the DM relic abundance is generated by thermal freeze-out. Depending on the mass splitting in the DM sector different scenarios have to be considered. When their masses are nearly degenerate, then all DM flavours are present during freeze-out and can be treated together. Otherwise only the lightest flavour of DM remains in the thermal bath. The relic abundance condition then constrains the couplings of all DM flavours present in the thermal bath, as a function of the DM mass and the mediator mass.

Significant constraints on the parameter space arise from the LUX experiment, searching for DM scattering off nuclei. In the mDMFV model with b-flavoured DM the relevant diagrams contributing to WIMP-nucleon scattering are shown in figure 4.

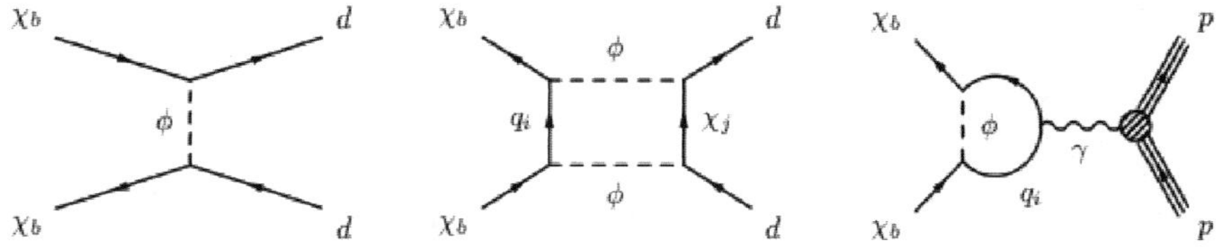

Figure 4. Diagrams contributing to WIMP-nucleon scattering in the mDMFV model [1].

In order to efficiently suppress the tree level contribution, the mixing between the first and third DM generation, s^{λ}_{13}, has to be small. Non-negligible effects arise however from the one-loop box and penguin diagrams. These can not be suppressed by choosing small flavour mixing angles, as they are present even in the flavour conserving case. However they interfere destructively with each other thanks to an overall minus sign of the leading logarithmic term in the penguin diagram. This cancellation becomes effective if $D_{\lambda,11}$, the first generation DM coupling to quarks, lies in a certain range.

This effect can be observed in figure 5. The region covered by yellow points is the one that remains valid after LUX and flavour constraints have been taken into account. The left panel shows the allowed range for the coupling $D_{\lambda,11}$ as a function of the DM mass m_{χ_b} in the case of a single DM flavour freeze-out. For small DM mass where the LUX bound is strongest, we can clearly see that only a small range for the coupling $D_{\lambda,11} \sim 1$ is allowed. Interestingly this constraint is not visible from the red points, where only the LUX bound but not the flavour constraints has been imposed.

The restriction to a certain range of $D_{\lambda,11}$ has a striking consequence in the case where the first and third DM generations are quasi-degenerate. In this case the relic abundance constraint and the LUX bound become incompatible below DM masses below $m_{\chi_b} \sim 100\,\text{GeV}$. Lower DM masses are therefore excluded in the 13-degeneracy scenario, as can be seen from the right panel of figure 5.

Journal of Physics: Conference Series **631** (2015) 012022 doi:10.1088/1742-6596/631/1/012022

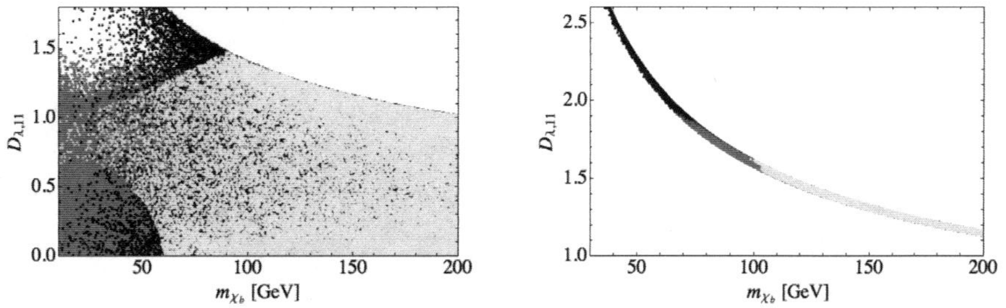

Figure 5. First generation coupling $D_{\lambda,11}$ as functions of the DM mass m_{χ_b} for significant mass splitting $m_{\chi_{d,s}} > 1.1 m_{\chi_b}$ scenario (left), and in the 13-degeneracy scenario $m_{\chi_d} \simeq m_{\chi_b}$ (right). The red points satisfy the bound from LUX, while the blue points satisfy the $\Delta F = 2$ constraints. For the yellow points both LUX and $\Delta F = 2$ constraints are imposed [1].

3.3. Bounds from LHC searches

While the new particles of the DMFV framework have not yet been the subject of experimental analyses, interesting and complementary constraints on the mDMFV parameter space can be obtained from existing searches for new particles at the LHC. Similarly to supersymmetric models with conserved R-parity, also the mDMFV states have to be pair-produced and lead to missing energy signatures.

We can therefore obtain a bound on the masses m_ϕ and m_χ from the experimental searches for squark pair production. Figure 6 shows the limits on the mDMFV model obtained from the CMS 19.5 fb^{-1} sbottom [9] and squark [10] searches. To this end both the modified production cross section and the branching ratios to the various DM flavours have been taken into account. Since the heavier DM flavours' decay leads to additional soft particles in the final state, which are not considered in the experimental analysis, all three flavours have been treated as missing energy. The DM coupling to the b quark, $D_{\lambda,33}$, has been fixed everywhere by the corresponding relic abundance constraint, and all mixing angles are set to zero for simplicity. We can see that mediator masses up to $m_\phi \sim 850$ GeV are excluded for light DM masses m_χ, however the bounds become weaker for heavier χ.

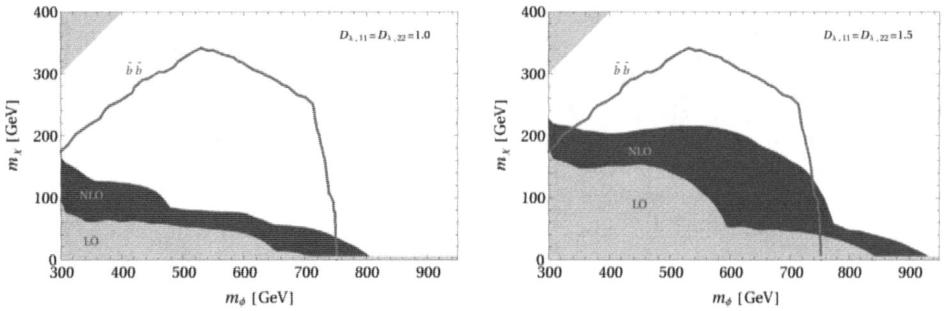

Figure 6. Limits on the mDMFV model from CMS squark searches [1].

The CMS search for light stops decaying into a charm and a neutralino [11] places a constraint on a different region of mDMFV parameter space, see figure 7. An excluded region appears when the mediator ϕ and the DM m_χ are close to degenerate, so that the decay products of ϕ are soft. The constraint reaches up to $m_\phi \sim 250\,\text{GeV}$.

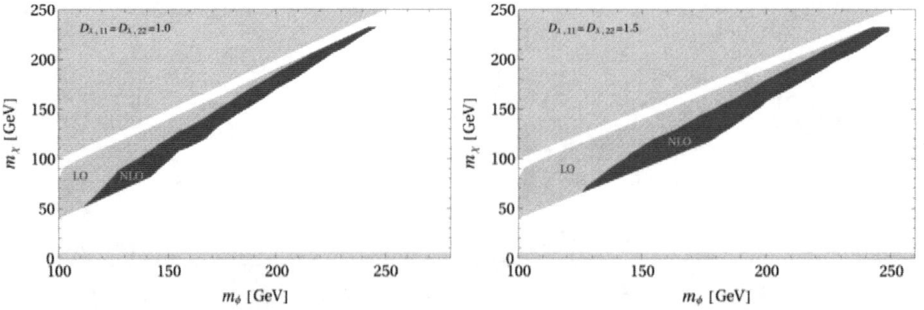

Figure 7. Limits on the mDMFV model from the CMS 19.7 fb^{-1} search [11] for stops decaying to a charm and a neutralino, using the monojet + MET final state [1].

Last but not least monojet searches that place constraints on effective DM interactions are also sensitive to flavoured DM. In figure 8 we show the result of reinterpreting the CMS monojet search [12] in terms of the mDMFV model. Note that in this case the obtained constraints on the mediator mass m_ϕ are rather insensitive to the DM mass m_χ.

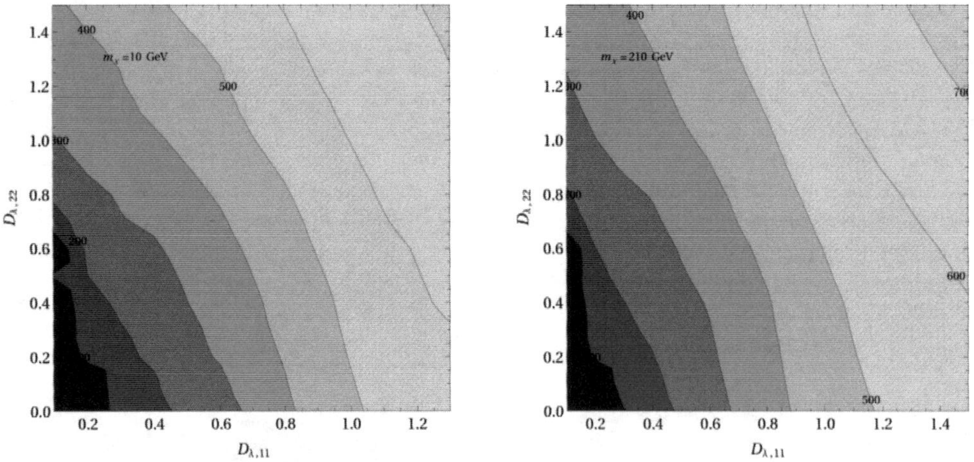

Figure 8. Limits on the mDMFV model in terms of mass of ϕ (in GeV) from the CMS monojet search [12] [1].

4. Summary

Flavoured DM is an exciting scenario on the astroparticle physics playground, that has interesting implications for flavour physics, collider physics, and direct DM searches, with a non-trivial interplay between the various areas. In this article we have reviewed a simplified model of quark flavoured DM beyond the MFV hypothesis, in which the flavour violating DM coupling to quarks is the only new source of flavour violation. The mDMFV model realises only one out of many possibilities to realize the flavoured DM paradigm in a simplified model scenario. Distinct yet equally interesting implications can be expected in other implementations that deserve further study.

References

[1] P. Agrawal, M. Blanke and K. Gemmler, JHEP **1410** (2014) 72 [arXiv:1405.6709 [hep-ph]].
[2] J. Kile, Mod. Phys. Lett. A **28** (2013) 1330031 [arXiv:1308.0584 [hep-ph]].
[3] A. J. Buras, P. Gambino, M. Gorbahn, S. Jager and L. Silvestrini, Phys. Lett. B **500** (2001) 161 [hep-ph/0007085].
[4] G. D'Ambrosio, G. F. Giudice, G. Isidori and A. Strumia, Nucl. Phys. B **645** (2002) 155 [hep-ph/0207036].
[5] B. Batell, J. Pradler and M. Spannowsky, JHEP **1108** (2011) 038 [arXiv:1105.1781 [hep-ph]].
[6] M. Blanke, A. J. Buras, A. Poschenrieder, S. Recksiegel, C. Tarantino, S. Uhlig and A. Weiler, Phys. Lett. B **646** (2007) 253 [hep-ph/0609284].
[7] P. Agrawal, B. Batell, D. Hooper and T. Lin, Phys. Rev. D **90** (2014) 063512 [arXiv:1404.1373 [hep-ph]].
[8] D. S. Akerib *et al.* [LUX Collaboration], Phys. Rev. Lett. **112** (2014) 9, 091303 [arXiv:1310.8214 [astro-ph.CO]].
[9] CMS Collaboration [CMS Collaboration], CMS-PAS-SUS-13-018.
[10] S. Chatrchyan *et al.* [CMS Collaboration], JHEP **1406** (2014) 055 [arXiv:1402.4770 [hep-ex]].
[11] CMS Collaboration [CMS Collaboration], CMS-PAS-SUS-13-009.
[12] [CMS Collaboration], CMS-PAS-EXO-12-048.

Time-reversal and the Bessel equation

Eleonora Alfinito[1] **and Giuseppe Vitiello**[2]

[1]Innovation Energy Department, Lecce University, Campus universitario,Corpo O 73100 Lecce
[2]Dipartimento di Fisica "E.R.Caianiello" Universitá di Salerno and INFN Gruppo collegato di Salerno, Fisciano (SA) - 84084, Italy

E-mail: vitiello@sa.infn.it

Abstract. The system of two damped/amplified oscillator equations is of widespread interest in the study of many physical problems and phenomena, from inflationary models of the Universe to thermal field theories, in condensed matter physics as well in high energy physics, and also in neuroscience. In this report we review the equivalence, in a suitable parametrization, between such a system of equations and the Bessel equations. In this connection, we discuss the breakdown of loop-antiloop symmetry, its relation with time-reversal symmetry and the mechanism of group contraction. Euclidean algebras such as $e(2)$ and $e(3)$ are also discussed in relation with Virasoro-like algebra.

1. Introduction

Eugene Wigner has remarked [1] that "the role which is common to all special functions is to be matrix elements of representations of the simplest Lie groups". Such a statement could constitute by itself a good motivation to study special functions such as, for example, Bessel functions, whose relation with time reversal symmetry and group contraction we are going to review in this paper. Further motivations, however, come from the specific interest to many physical problems of the so-called Bateman system of the couple of damped/amplified oscillators [2], which, as we will see, is equivalent, in a specific parametrization, to Bessel equations. We will also see that the Euclideans groups $E(2)$ and $E(3)$, whose representations are given in terms of planar and spherical Bessel functions, respectively, are obtained by the mechanism of group contraction [3]. Such a mechanism is involved in the breakdown of time-reversal symmetry and of the loop-antiloop symmetry around a preferred axis, which in turn is related to loop-algebras, such as the Virasoro-like algebra.

In this paper we report results whose details can be found in ref. [4]). They are object of renewed interest in these days, since according to Freeman's observations the distribution functions of impulse responses of cortex to electric shocks resemble Bessel functions [5]. The cortex dynamics can be then studied by considering the equivalence of spherical Bessel equation, in a given parametrization, to two oscillator equations, one damped and one amplified oscillator. The study of such a couple of equations, which are at the basis of the formulation of the dissipative many-body model of brain [6, 7, 8], reveals the structure of the root loci of poles and zeros of solutions of Bessel equations [5]. Therefore, it seems to us interesting to discuss in this paper the mathematics underlying the relation between Bessel function and the system of damped/amplified oscillators.

Journal of Physics: Conference Series **631** (2015) 012023 doi:10.1088/1742-6596/631/1/012023

Following thus ref. [4] let me recall that Bessel functions describe solutions with different Pontryagin number in the punctured plane $R^2/(0)$ [9]. The elements of the homotopy group, Π_n, are there represented by differential operators acting on analytic functions:

$$\Pi_n \equiv \frac{\partial^n}{\partial z^n} , \quad n \in N , \tag{1}$$

n is the loop number around the hole and $\Pi_n \cdot \Pi_m = \Pi_{n+m}$. There are two different kinds of behavior, corresponding to the two different functions $\varphi_m(z)$ and $\psi_m(z)$:

$$\frac{\partial^n}{z\partial z^n} \varphi_m(z) = (-)^m \varphi_{m+n}(z) , \quad \varphi_m(z) = \frac{j_m(z)}{z^m} \tag{2}$$

and

$$\frac{\partial^n}{z\partial z^n} \psi_m(z) = \psi_{m-n}(z) , \quad \psi_m(z) = z^m j_m(z) , \tag{3}$$

respectively. On φ, Π_n acts in counter-clockwise way while on ψ it acts in clockwise way. $j_m(z)$ is the planar Bessel function (Bessel function of integer order). Eqs.(2) and (3) are indeed the differential formulae for the planar Bessel functions [10]. Similar formulae are true for the spherical Bessel functions [9] J_m:

$$\varphi_m(z) = J_m(z) z^{-m} , \qquad \psi_m(z) = z^{(m+1)} J_m(z) . \tag{4}$$

For the planar Bessel functions, the "raising" and "lowering" functions coincide for $m = 0$ (no loops). This is not true with the spherical Bessel functions.

In the following we will exploit these topological properties of the φ and ψ functions. In our presentation we will closely follow ref. [4]. We will start considering the spherical Bessel equation in Section 2. Group contraction and loop algebras are discussed in Section 3 and 4, respectively. Section 5 is devoted to conclusions.

2. The Bessel equation and the damped/amplified oscillator system

The spherical Bessel equation of order n is:

$$\eta^2 J_{n;\eta\eta} + 2\eta J_{n;\eta} + [\eta^2 - n(n+1)] J_n = 0 , \tag{5}$$

where n is zero or an integer, $n = 0, \pm 1, \pm 2, ...$, the labels "$;\eta$" and "$;\eta\eta$" denote first and second order derivatives, respectively. The solutions of Eq. (5) constitute a complete set of (parametric) decaying functions [10], the so called spherical Bessel functions, and can be expressed in terms of the first and second kind Bessel functions and their linear combinations (the Hankel functions).

Eq. (5) is invariant under the transformation $n \to -(n+1)$: J_n and $J_{-(n+1)}$ are both solutions of the same equation. In other words, J_n and $J_{-(n+1)}$ are degenerate solutions corresponding to the same eigenvalue $n(n+1)$ of the operator $\eta^2 \frac{d^2}{d\eta^2} + 2\eta \frac{d}{d\eta} + \eta^2$.

Here we show that by means of suitable transformations, Eq. (5) turns out to be equivalent to a set of two equations representing a couple of damped/amplified parametric oscillators. This is true also for the planar Bessel equation and will be discussed in the following.

Perform the change of variables in Eq. (5): $\eta \to \eta \equiv \epsilon x$ with $x \equiv e^{-t/\alpha}$, with arbitrary parameters ϵ and α. The new variable t may be thought to denote the time variable. By using $w_{n,l} \equiv J_n \cdot (x)^{-l}$, Eq.(5) then becomes:

$$\ddot{w}_{n,l} - \frac{2l+1}{\alpha} \dot{w}_{n,l} + \left[\frac{l(l+1) - n(n+1)}{\alpha^2} + (\frac{\epsilon}{\alpha})^2 e^{-2t/\alpha}\right] w_{n,l} = 0. \tag{6}$$

$\dot{w}$ denotes derivative with respect to time t. Chosing $l(l+1) = n(n+1)$ the degeneracy between the solutions J_n and $J_{-(n+1)}$ is removed and thus a partition is induced between the two solution sectors $\{J_n\}$ and $\{J_{-(n+1)}\}$. Correspondingly, two different sets of equations are obtained [4], one for $w_{n,l}$ and the other one for $w_{-(n+1),l}$, respectively. Moreover, each set is composed by two *different* equations, one for $l = -(n+1)$ and the other one for $l = n$. The symmetry under the transformation $n \to -(n+1)$ gets broken.

Chose then the arbitrary parameters α and ϵ to be n-dependent: $\alpha \to \alpha_n$ and $\epsilon \to \epsilon_n$. Thus we use $\eta \to \eta_n \equiv \epsilon_n x_n$ with $x_n \equiv e^{-t/\alpha_n}$. Perform such a choice in such a way that $\frac{2n+1}{\alpha_n} \equiv L$ and $\frac{\epsilon_n}{\alpha_n} \equiv \omega_0$ depend neither on n, nor on time. By setting $u_n \equiv w_{n,-(n+1)}$, and $v_n \equiv w_{n,n}$, the couple of equations for the damped/amplified parametric oscillators is finally obtained:

$$\begin{aligned} \ddot{u}_n + L\,\dot{u}_n + {\omega_n}^2(t) u_n &= 0, \\ \ddot{v}_n - L\,\dot{v}_n + {\omega_n}^2(t) v_n &= 0. \end{aligned} \tag{7}$$

The frequency $\omega_n(t)$ is $\omega_n(t) = \omega_0\, e^{-\frac{Lt}{2n+1}}$. We see that $\omega_n(t)$ approaches to the time-independent value ω_0 for $n \to \infty$: the frequency time-dependence is thus "graded" by the order n of the original Bessel equation. L and ω_0, which may be arbitrarily chosen, are characteristic parameters of the oscillator system.

Note that, since keeping L independent of n implies that $\alpha_{-(n+1)} = -\alpha_n$, the transformation $n \to -(n+1)$ leads to solutions $J_{-(n+1)}$ with frequencies exponentially increasing in time. These solutions can be obtained from the ones of Eqs. (7) by time-reversal $t \to -t$ and exchanging u with v (referred to as "charge conjugation"). In the large n limit, $\omega_n \to \omega_0$, u_n and v_n are reciprocally time-reversed and the two sectors $\{J_i\}$, $i = n, -(n+1)$ are mapped one into the other one in that limit.

Note also that $w_{n,n}$ and $w_{n,-(n+1)}$ are "harmonically conjugate" functions in the sense that they may be represented as $w_{n,-(n+1)}(t) = \frac{1}{\sqrt{2}} r_n(t) e^{\frac{-Lt}{2}}$ and $w_{n,n}(t) = \frac{1}{\sqrt{2}} r_n(t) e^{\frac{Lt}{2}}$, with $r_n(t)$ solution of the single parametric oscillator

$$\ddot{r}_n + \Omega_n^2(t) r_n = 0\ . \tag{8}$$

The common frequency Ω_n is:

$$\Omega_n(t) = \left(\omega_n^2(t) - \frac{L^2}{4}\right)^{\frac{1}{2}}. \tag{9}$$

In ref. [4] it has been shown also that the Bessel-like equation

$$J_{;\eta\eta} + \frac{\alpha}{\eta}\, J_{;\eta} + \left(1 - \frac{\beta^2}{\eta^2}\right) J = 0\ , \tag{10}$$

where $\alpha = 1$ or 2 and β is an arbitrary real constant, may represent parametric oscillators with constant damping and with different functional choices for the frequency. Vice-versa, from the equation of the damped parametric oscillator, by use of convenient transformations, one can always obtain the Bessel-like equation (10) [4]. In this report we do not discuss more on this point.

Summarizing, under convenient parametrization, the spherical Bessel equation is represented by a two-fold hierarchy of couples of parametric oscillators with time dependent frequency graded by n and by $-(n+1)$ and constant damping/amplification given by L. The transition $n \to -(n+1)$ corresponds to time-reversal $t \to -t$ combined with "charge conjugation" $u \to v$. The breakdown of the $n \to -(n+1)$ symmetry of the spherical Bessel equation corresponds to the breakdown of time-reversal symmetry. In the manifold of the the spherical Bessel functions

$\{J_i\}$, $i = n, -(n+1)$ the *arrow of time* emerges, corresponding to the arrow of time in each of the damped or amplified oscillator equations.

In the following Section such a feature will be seen to have its root in the $E(3)$ group, whose representations can be indeed constructed by means of the spherical Bessel functions. The breakdown of the $n \to -(n+1)$ symmetry there appears as the breakdown of the loop-antiloop symmetry around a preferred direction in the 4D-space.

Finally, we remark that in expanding geometry (inflationary) models of the Universe the first of Eqs. (7), with $n = 1$ and L denoting the Hubble constant, is commonly used [11].

3. Group contraction

As already mentioned, some of the features studied in the previous Section are found to be rooted in the Euclidean groups, whose representations can be constructed in terms of Bessel functions [1, 10].

We consider the Euclidean groups in two and in three dimensions, $E(2)$ and $E(3)$, respectively and their relation with the Laplace equations [1]. We start by considering $E(2)$.

3.1. The planar Bessel equations and $E(2)$

$E(2)$ is the group of the translations $T(\vec{v})$ in the plane by the vector $\vec{v}$ ($\vec{v} \equiv (a,b)$) and the rotations $R(\theta)$ of the plane around the origin by the angle θ. The Lie algebra is given in terms of the two translation generators P_a, P_b and of the rotation generator M:

$$[P_a, P_b] = 0, \quad [P_a, M] = -P_b, \quad [P_b, M] = P_a. \tag{11}$$

$P^2 = P_a^2 + P_b^2 = P_+\, P_- = P_-\, P_+$, with $P_\pm \equiv P_a \pm iP_b$, which has non-positive eigenvalue, $-p^2$, is the invariant operator of $E(2)$.

On square integrable functions f, the group transformations are represented by the action of the operator $D(\vec{v},\theta)$ [1]:

$$\begin{aligned} D(\vec{v},0)f(\phi) &= e^{i\vec{p}\cdot\vec{v}}f(\phi), \\ D(0,\theta)f(\phi) &= f(\phi-\theta)\ . \end{aligned} \tag{12}$$

Assume as basis functions the complete set of normalized eigenfunctions of the rotation subgroup $\{f_n(\phi)\}$, $f_n \equiv (2\pi)^{-\frac{1}{2}}\, i^{-n}\, e^{in\phi}$: $D(0,\theta)f_n(\phi) = e^{in\theta}f_n(\phi)$. The representation of $E(2)$ in terms of the planar Bessel functions j_m [1] is expressed in the form:

$$\begin{aligned} D(\vec{v},\theta)f_n = \textstyle\sum_m \Delta(\vec{v},\theta)_{mn} f_m, \\ \Delta(\vec{v},\theta)_{mn} = (-1)^{m-n}e^{-im\beta}\, j_{m-n}(p\,r)e^{in(\beta-\theta)}. \end{aligned} \tag{13}$$

(r,β) are the polar coordinates of $\vec{v}$ and j_m is the planar Bessel function of order m.

In order to study the P^2 eigenvalue equation we consider the 3D-Laplace equation in an isotropic and homogeneous 3D-space. In the cylindrical coordinates we have:

$$\left(\frac{\partial^2}{\partial r^2} + \frac{\partial}{r\partial r} + \frac{\partial^2}{r^2\partial\theta^2} + \frac{\partial^2}{\partial x_3^2}\right)\psi = 0. \tag{14}$$

We are interested in solutions of the type:

$$\psi(r,\theta,x_3) = \varphi(r,\theta)\cdot\sigma(x_3), \qquad \frac{\partial^2}{\partial x_3^2}\sigma(x_3) \equiv p^2\sigma(x_3). \tag{15}$$

For p positive, by selecting, for positive x_3, the solution: $\sigma = e^{-x_3 p}$ and for negative x_3 the solution: $\sigma = e^{x_3 p}$, one may obtain square integrable eigenfunctions.

Journal of Physics: Conference Series **631** (2015) 012023 doi:10.1088/1742-6596/631/1/012023

Clearly, the choice of the cylindrical coordinates, instead of the spherical one, breaks the symmetry of 3D-spatial rotation group $SO(3)$: *when cylindrical coordinates are chosen, x_3 is differently treated with respect to the two remaining coordinates and this singles out a privileged axis for rotations.* The resulting symmetry group is $E(2)$, the group contraction of $SO(3)$. Indeed, the Laplace equation (14) reduces to the eigenvalue equation for P^2 (in polar coordinates), i.e. the 2D-Helmholtz equation:

$$P_+ \, P_- \, \varphi = \left(\frac{\partial^2}{\partial r^2} + \frac{\partial}{r\partial r} + \frac{\partial^2}{r^2 \partial \theta^2} \right) \varphi = -p^2 \varphi. \tag{16}$$

Assuming $\varphi(r,\theta) = f_n(r) \cdot e^{in\theta}$, solving this equation gives $f_n(r) = j_n(pr)$, with $j_n(pr)$ solution of the planar Bessel equation of order n:

$$j_{n;\eta\eta} + \frac{1}{\eta} \, j_{n;\eta} + [1 - \frac{n^2}{\eta^2}] \, j_n \;=\; 0, \tag{17}$$

with $\eta = pr$. Positive/negative n values correspond to positive/negative rotations (loop/antiloop) around the x_3 axis, namely to different orientations of the x_3 axis and thus to different solutions [10]. A similar situation occurs in the case of spherical Bessel functions. Thus, in a natural way we are led to consider topological properties of Bessel functions related with loop operators and loop-algebras, according to our comments in the Introduction.

We remark that although the solutions are not symmetric under the reversal of the x_3 axis (i.e. under time-reversal in our choice $x_3 \equiv t$), Eq. (17) is invariant under the $n \to -n$ exchange. Such an invariance reflects indeed the existence of the two sets, D^n and D^{-n-1}, of the $SO(3)$ independent representations [12].

As in the case of the spherical Bessel equation (5) in Section 2, the planar Bessel equation (17), by setting $w_{n,l}(\eta) = (\eta)^{-l} J_n(\eta)$, $\eta = \epsilon e^{-t/\alpha}$, and by introducing the mirror parameter $l = \pm n$, gives the couple of damped/amplified harmonic oscillators:

$$\begin{aligned} l = -n: \quad & \ddot{w}_{n,-n} + \tfrac{2n}{\alpha} \, \dot{w}_{n,-n} + \left[(\tfrac{\epsilon}{\alpha})^2 \, e^{-2t/\alpha} \right] w_{n,-n} = 0, \\ l = n: \quad & \ddot{w}_{n,n} - \tfrac{2n}{\alpha} \, \dot{w}_{n,n} + \left[(\tfrac{\epsilon}{\alpha})^2 \, e^{-2t/\alpha} \right] w_{n,n} = 0, \end{aligned} \tag{18}$$

where the value of l is connected with the choice of negative/positive eigenvalues of the rotation generator M.

We may thus summarize our discussion and results up to this point in the following way: the breakdown of the rotational symmetry of $SO(3)$, which leads to its group contraction $E(2)$, has been shown to introduce a crucial difference in the double choice of the x_3 axis orientation, namely the loss of loop-antiloop symmetry. This results, in turn, in the difference between the planar Bessel functions or order $+n$ and the ones of order $-n$, in terms of which the $E(2)$ representations can be built. Then, we have shown that the planar Bessel equation (17) can be cast, by a convenient re-parametrization, into the set of Eqs. (18) for the damped/amplified harmonic oscillators: the mirror index $\pm n$ of the Bessel functions is associated to the *couple* of damped/amplified harmonic oscillators: it is a time-mirror index[1].

The physical content of the contraction of $SO(3)$ to $E(2)$ can be geometrically depicted as the projection of the sphere, corresponding to $SO(3)$, on the plane tangent to one of the poles. The radius ρ of the sphere acts as a "scale": the $E(2)$ translations in the tangent plane are "good" approximations of rotations in the limit $\rho \to \infty$, namely for distances much smaller than

[1] Identifying x_3 with the time t, p acquires the dimensions of an energy over an action; see, e.g., the eigenfunctions $\sigma = e^{\pm x_3 p}$.

ρ. The physical meaning of this is that the $SO(3)$ contraction to $E(2)$ manifests itself in local observations. However, in local observations the orientation of the x_3 axis is "locked", which means that the symmetry of the solutions under the $n \to -n$ exchange is lost (breakdown of the loop-antiloop symmetry). Specifying the direction of the x_3 axis, namely, choosing one of the two possible forms for σ in Eq. (15), produces topologically inequivalent configurations [9].

Finally note that the "spinor" representation of $SO(3)$, which involves half-integer n and uses a representation in terms of the Pauli matrices $\sigma_i, i = 1, 2, 3$, maps the planar Bessel equation into the harmonic oscillator equation. In fact, for $n = \pm\frac{1}{2}$ and with $w \equiv \sqrt{\eta}\ j_{\pm\frac{1}{2}}$, Eq.(17) reduces to the harmonic oscillator equation

$$\frac{d^2w(r)}{dr^2} + p^2\ w(r) = 0, \qquad r = \frac{\eta}{p}. \tag{19}$$

with frequency p and r playing the role of time. The oscillator so obtained may be thought as a possible classical analogue of a Fermi oscillator based on a 'rotation system' [13].

3.2. *The spherical Bessel equation and $E(3)$*

$E(3)$ is the group contraction of SO(4). The algebra $e(3)$ has six translation and rotation generators P_i and M_i, $i = 1, 2, 3$, respectively, with commutation relations:

$$[P_i, P_j] = 0, \quad [M_i, M_j] = \epsilon_{ijk}M_k, \quad [P_i, M_j] = \epsilon_{ijk}P_k; \tag{20}$$

The $SO(3)$ subgroup generated by the $M_i's$ is left unchanged in the contraction process. The algebra has two invariants, $P^2 = \Sigma P_i^2$ and $\Sigma P_i \cdot M_i$.

In the $4D$-space the Laplace equation may be solved by using the position $\psi(x_1, x_2, x_3, x_4) = \varphi(r, \theta, \phi) \cdot \sigma(x_4)$, with r, θ, ϕ spherical coordinates, $\sigma = e^{\pm x_4 p}$. The role of time t may be played by x_4. The solution is a functions $\varphi = Y_{n,m}(\theta, \phi) \cdot J_n(pr)$, with $Y_{n,m}$ the spherical harmonics and J_n solution of the spherical Bessel equation. The spherical Bessel functions depend on the continuous eigenvalue p^2 of P^2. Their label n (integer or zero) is related to the discrete eigenvalue $n(n+1)$ of the rotation operator $\mathbf{M}^2$ and classifies the representations D^n of the compact subgroup $SO(3)$ of $E(3)$. The existence of two sets of independent representations [12], D^n and D^{-n-1}, is reflected the invariance of Eq. (5) under the transformation $n \to -(n+1)$. For $n = 0$, or $n = -1$ Eq. (5) reduces to the harmonic oscillator equation with frequency p:

$$\frac{d^2w(r)}{dr^2} + p^2\ w(r) = 0, \qquad r = \frac{\eta}{p}, \tag{21}$$

where $w = \eta\ J_0$ or $\eta\ J_{-1}$. The harmonic oscillator thus appears to be related to the "ground state" (in the D^n or D^{-n-1} spectra) of the Bessel system. Here, the so-called "true" representation of $SO(3)$, i.e. the one with integer values of n [12], has been used.

The breakdown of the symmetry under the transformation $n \to -(n+1)$ is built in in the geometrical structure of the $E(3)$ group: it corresponds to the breakdown of the x_4 axis reversal symmetry, or loop-antiloop symmetry, i.e. of time-reversal symmetry when x_4 is identified with time. Again, the physical meaning is that the $SO(4)$ contraction to $E(3)$ manifests itself in local observations and the x_4 axis orientation then gets "locked", turning in the loss of the symmetry of the solutions under the $n \to -(n+1)$.

4. Loop algebras

It is known that infinite-dimensional algebras and the so-called "loop-algebras" can be constructed on some finite-dimensional group as SU(1,1) [14]. The algebra $e(3)$, which, as we have seen, is related to the topological properties of the Bessel functions, is related to a particular structure of loop-algebras. We focus on the Virasoro algebra, which plays a central role in the conformal field theories.

We will show that a contraction procedure, based on the so called graded contraction method [15, 16], maps the Virasoro algebra into a generalization of the Euclidean algebra $e(3)$.

The commutation relations of the Virasoro algebra $\mathcal{L}$ of central charge c (c commuting with all the T's) are

$$[T_n, T_m] = (n-m)T_{n+m} + \frac{c}{12}(n^3 - n)\delta_{n+m,0} , \quad m, n \in \mathbf{Z} . \tag{22}$$

The $\mathbf{Z}_2$-grading of the algebra consists in dividing the set of the T_n generators into an even set $L_0 \equiv \{A_n, c\}$ and an odd set $L_1 \equiv \{B_n\}$, so that $\mathcal{L} = L_0 \bigoplus L_1$ and

$$A_n = \frac{1}{2}\left(T_{2n} + \frac{c}{8}\delta_{n,0}\right) , \quad B_n = \frac{1}{2}T_{2n+1} , \tag{23}$$

$$[L_0, L_0] \subseteq L_0 , \quad [L_0, L_1] \subseteq L_1 , \quad [L_1, L_1] \subseteq L_0 . \tag{24}$$

The commutation relations of the graded generators are given by [17]

$$[A_n, A_m] = (n-m)A_{n+m} + \frac{2c}{12}(n^3 - n)\delta_{n+m,0} , \tag{25}$$

$$[B_n, B_m] = (n-m)A_{n+m+1} + \frac{2c}{12}(n - \frac{1}{2})(n + \frac{1}{2})(n + \frac{3}{2})\delta_{n+m+1,0} , \tag{26}$$

$$[A_n, B_m] = (n - m - \frac{1}{2})B_{n+m} . \tag{27}$$

Eq. (25) shows that $\{A_n, c\}$ is again a Virasoro algebra but with central charge $2c$.

The $\mathbf{Z}_2$-graded contraction of the algebra (25)-(27) is obtained [17] by putting equal to zero the commutator $[B_n, B_m]$:

$$[B_n, B_m] = 0 , \tag{28}$$

In the centerless case ($c = 0$), the A_0 and $A_{\pm 1}$ generators close the algebra isomorphic to $so(3) \sim su(2)$ and the set of these three generators and the operators $B_{-\frac{1}{2}}$, $B_{\frac{1}{2}}$ and $B_{-\frac{3}{2}}$ close the $e(3)$ isomorphic algebra. This is shown by setting:

$$\begin{aligned} M_+ &\equiv A_1 , \quad M_- \equiv A_{-1} , \quad M_3 \equiv iA_0 , \\ P_+ &\equiv B_{\frac{1}{2}}, \quad P_- \equiv B_{-\frac{3}{2}} , \quad P_3 \equiv iB_{-\frac{1}{2}} , \end{aligned} \tag{29}$$

with the Ms and Ps generators satisfying indeed the $e(3)$ commutation relations (20).

A general extension of this result is the following: the algebra $\mathcal{E}_n \equiv \{A_0, A_{\pm n}\} \bigoplus \{B_{-\frac{1}{2}}, B_{\pm n - \frac{1}{2}}\}$ reproduces the $e(3)$ algebra for each integer value of n, provided the following positions are satisfied:

$$\begin{aligned} M_+ &\equiv \tfrac{1}{n}A_n , \quad M_- \equiv \tfrac{1}{n}A_{-n} , \quad M_3 \equiv \tfrac{i}{n}A_0 , \\ P_+ &\equiv B_{n-\frac{1}{2}} , \quad P_- \equiv B_{-n-\frac{1}{2}} , \quad P_3 \equiv iB_{-\frac{1}{2}} . \end{aligned} \tag{30}$$

We remark that, by choosing $A_{\pm n} = 0$, for non-zero values of n, the $e(2)$ algebra can be obtained as a subalgebra of (30).

Summarizing, the extension of the Virasoro algebra by means of its $\mathbf{Z}_2$-grading with the subsequent step of the $\mathbf{Z}_2$-graded contraction leads to a n-graded hierarchy of Euclidean algebras. Considering the discussion in the previous Sections, an interesting relation then emerges between the couple of damped/amplified parametric oscillators graded by n and the loop algebras.

5. Conclusions

We have shown that a relation exists between spherical and planar Bessel equations and dissipation/amplification processes. Some topological properties of the Bessel functions in connection with Virasoro-like loop-algebras have been also considered. After a convenient parametrization, the breakdown of the $n \to -(n+1)$ $(n \to -n)$ symmetry of the spherical (planar) Bessel equation corresponds to the breakdown of time-reversal symmetry: the *arrow of time* emerges in the manifold of the Bessel functions $\{J_i\}$, $i = n, -(n+1)$ $(i = n, -n)$. Such a breakdown is realized through the mechanism of group contraction leading to the Euclidean groups $E(3)$ and $E(2)$, whose representations are constructed by means of the spherical and the planar Bessel functions, respectively.

We thus find that in our parametrization the symmetry of the Bessel equation is not the symmetry of their solutions. In quantum field theory (QFT) the spontaneous breakdown of symmetry occurs when the *continuous* symmetry of the dynamical equations is not the symmetry of the physical vacuum. In the Bessel equation case the *discrete* time-reversal symmetry is broken. However, since breakdown of time-reversal symmetry implies dissipation/amplification phenomena, also continuous time translational symmetry is broken. Moreover, in QFT the effects of spontaneous breakdown of symmetry are solely observable in the contraction limit, namely at the observation scale [18]. In the Bessel equation case, the breakdown of time-reversal symmetry manifests itself at observational level as dissipation/amplification phenomena.

We have considered only the classical sector. The quantization of the damped harmonic oscillator has been extensively studied in a number of papers [19] - [27]. The couple of damped/amplified oscillators has been used in inflationary models of the Universe [11], thermal field theories [21], Chern-Simons gauge theory [26], Bloch electrons in metals [26], the dissipative quantum model of brain [6, 7, 27], etc.

The canonical quantization of the simple damped harmonic oscillator requires the "doubling" of the system degrees of freedom by introducing the amplified oscillator, representing the environment, as the time-reversed copy of the damped oscillator [20, 21, 28]. Such a doubling is not required at a classical level, however, where it is possible to solve the damped oscillator equation by ignoring the environment. From our discussion it emerges that apparently Bessel equation may represent both the damped and the amplified oscillator as an inseparable "doublet" also at the classical level.

References

[1] Talman J D 1968 *Special Functions. A group theoretical approach* New York: Benjamin
[2] Bateman H 1931 *Phys. Rev.* **38** 815
[3] Inönü E and Wigner E P 1953 *Proc. Nat. Acad. Sci. U.S.A.* **39** 510
[4] Alfinito E and Vitiello G 2003 *Mod. Phys. Lett. B* **17** 1
[5] Freeman W J, Capolupo A, Kozma R, Olivares del Campo A, Vitiello G 2015 Bessel functions in mass action modeling of memories and remembrances Preprint
[6] Vitiello G 1995 *Int. J. Mod. Phys.* **B9** 973
[7] Vitiello G 2001 *My Double Unveiled* Amsterdam: John Benjamins Publ Co
[8] Vitiello G 2014 *Current Opinion in Neurobiology* CONEUR1426 PII: S0959-4388(14)00145-7 Doi: 10.1016/j.conb.2014.07.017
[9] Mekhfi M 2000 *Int.J. Theor. Phys.* **39** No.4 1163
[10] Abramowitz M Stegun I eds 1970 *Handbook of Mathematical Functions* New York: Dover Inc
[11] Alfinito E and Vitiello G 2000 *Class. Quant. Grav.* **17** 93
[12] Wybourne B G 1974 *Classical groups for physicists* New York: J Wiley and Sons
[13] Martin J L 1959 *Proc. Roy. Soc.* **A 251** 536
[14] Fairlie D B, Nuyts J and Zachos C K 1988 *Phys.Lett.* **B202** 320
[15] de Montigny M and Patera J 1991 *J. Phys.* **A 24** 525
[16] Moody R V and Patera J 1991 *J. Phys.* **A 24** 2227
[17] Kostyakov I V, Gromov N A and Kuratov V V 2001 *Nucl.Phys.* **B102** (Proc. Suppl.) 316
[18] De Concini C and Vitiello G 1976 *Nucl. Phys.* **B116** 141

[19] Dekker H 1981 *Phys. Rep.* **80** 1
[20] Feshbach H and Tikochinski Y 1977 *Trans. NY Acad. Sci.* **38** 44
[21] Celeghini E, Rasetti M and Vitiello G 1992 *Ann. Phys.* **215** 156
[22] Tarasov V E 2001 *Phys. Lett.* **A288** 173
[23] Blasone M, Jizba P and Vitiello G 2001 *Phys. Lett.* **A287** 205
[24] Blasone M, Celeghini E, Jizba P and Vitiello G 2003 *Phys. Lett.* **A310** 393
[25] Blasone M and Jizba P 2002 *Canadian J. of Phys.* **80** 645
[26] Blasone M, Graziano E, Pashaev O K and Vitiello G 1996 *Ann. Phys. (N.Y.)* **252** 115
[27] Alfinito E and Vitiello G 2000 *Int. J. Mod. Phys.* **B14** 853
[28] Blasone M, Jizba P, Vitiello G 2011 *Quantum Field theory and its macroscopic manifestations* London: Imperial College Press

Journal of Physics: Conference Series **631** (2015) 012024 doi:10.1088/1742-6596/631/1/012024

CKM and PMNS mixing matrices from discrete subgroups of SU(2)

Franklin Potter

Sciencegems.com, 8642 Marvale Drive, Huntington Beach, CA 92646 USA

E-mail: frank11hb@yahoo.com

Abstract. Remaining within the realm of the Standard Model(SM) local gauge group, this first principles derivation of both the PMNS and CKM matrices utilizes quaternion generators of the three discrete (i.e., finite) binary rotational subgroups of SU(2) called [3,3,2], [4,3,2], and [5,3,2] for three lepton families in R^3 and four related discrete binary rotational subgroups [3,3,3], [4,3,3], [3,4,3], and [5,3,3] represented by four quark families in R^4. The traditional 3x3 CKM matrix is extracted as a submatrix of the 4x4 CKM4 matrix. If these two additional quarks b' and t' of a 4th quark family exist, there is the possibility that the SM lagrangian may apply all the way down to the Planck scale. There are then numerous other important consequences. The Weinberg angle is derived using these same quaternion generators, and the triangle anomaly cancellation is satisfied even though there is an obvious mismatch of three lepton families to four quark families. In a discrete space, one can also use these generators to derive a *unique* connection from the electroweak local gauge group $SU(2)_L$ x $U(1)_Y$ acting in R^4 to the discrete group Weyl E_8 in R^8. By considering Lorentz transformations in discrete (3,1)-D spacetime, one obtains another Weyl E_8 discrete symmetry group in R^8, so that the combined symmetry is Weyl E_8 x Weyl E_8 = "discrete" SO(9,1) in 10-D spacetime. This *unique* connection is in direct contrast to the 10^{500} possible connections for superstring theory!

1. Introduction

One of the greatest challenges in particle physics is to determine the first principles origin of the quark and lepton mixing matrices CKM and PMNS that relate the flavor states to the mass states. Within the realm of the Standard Model (SM) but considering a specific SU(2) discrete (i.e., finite) subgroup for each lepton and quark family, I determine that the true source of these mixings [1,2] is the mismatch of one Pauli generator for SU(2) with the corresponding generator for each proposed family subgroup. By using a linear combination of the generators for these specific discrete binary rotational subgroups of the electroweak (EW) gauge group $SU(2)_L$ x $U(1)_Y$, the mixing angles and matrices are determined directly.

One realizes immediately that the PMNS and CKM matrices are telling us that the very successful SM is an excellent approximation that requires but a small geometrical reconciliation in order to agree with the proposed discrete symmetry properties of the physical lepton and quark families. I repeat that these derivations are done within the realm of the SM and no alternative theoretical framework beyond the SM is required.

Many talks at this conference are devoted to deriving the neutrino PMNS mixing matrix and the quark CKM mixing matrix from various additional discrete *horizontal* symmetries. I suggest that they are missing the true origin of the mixing matrices: these matrices connect the lepton and quark weak isospin $\pm\frac{1}{2}$ states representing *discrete* individual family symmetries acting collectively to the isospin $\pm\frac{1}{2}$ states of SU(2), the traditionally assumed continuous symmetry group having two flavor states per family.

The Standard Model (SM) is based upon symmetry properties of the local gauge group $SU(2)_L$ x $U(1)_Y$ x $SU(3)_C$, which defines an electroweak (EW) interaction part and a color interaction part acting in a continuous internal symmetry space. However, we know that the SM is incomplete, having 28 parameters. So, something important is absent in our traditional understanding. I strongly suspect that the SM gauge group may be the best one possible but requiring only the proposed correction in order to find agreement with the discrete symmetries represented by the leptons and quarks.

Perhaps the major assumption of the SM is that space is continuous. At present, no one knows whether space at the ultimate scale, the Planck scale of about 10^{-35} meters, is continuous or discrete. Although there is no evidence presently, one can suggest that space is actually discrete at the Planck scale and is described by an C^2 (= R^4) lattice of mathematical nodes. The nodes themselves would possess *no measurable physical properties*, so the established physical properties of the fundamental leptons and quarks must "emerge" from the collection of nodes that as a unit exhibit the specific discrete rotational symmetries for each particle, a lepton or quark. In this manner, there is an end to the "Russian doll" hierarchy of particles within particles.

I point out that a discrete space at the Planck scale is not required absolutely for the mathematical and physical relationships and concepts that I discuss, but such a viewpoint does offer a reasonable avenue for trying to understand Nature at its ultimate scale.

2. Brief mathematical review

Each family of leptons and of quarks has two isospin flavor states that are the basis states in C^2 for SU(2). The subgroups of SU(2) for the leptons and quarks also have two orthogonal basis states per family.

In order to calculate the PMNS and CKM mixing matrix values in the following sections, I use unit quaternions instead of the more familiar and equivalent unitary 2x2 complex matrices of SU(2). The unit quaternion q = a + b**i** + c**j** + d**k**, where the coefficients a, b, c, d are real numbers for the one real and three imaginary axes. The unit quaternion generators are equivalent to the SU(2) generators, with the three Pauli generators σ_x, σ_y, σ_z, when multiplied by i, corresponding to the quaternion generators k, j, and i, respectively. These three quaternion generators are the keys to deriving the PMNS and CKM matrices from first principles [1,2].

The unit quaternion spans the space R^4 while the imaginary prime part spans the subspace R^3. With $i^2 = j^2 = k^2 = ijk = -1$, the quaternion q can be expressed as an SU(2) matrix

$$\begin{bmatrix} a+bi & c+di \\ -c+di & a-bi \end{bmatrix}.$$

Both the quaternions and the SU(2) matrices operate in the unitary plane C^2 with its two orthogonal complex axes, so the quaternion can be written also as q = u + v**j**, with u = a + b**i** and v = c + d**i**.

In previous research [3,4] on possible mathematical lepton family origins I have determined that each lepton family represents its own unique discrete rotational subgroup of SU(2), i.e., a discrete quaternion group. That is, one requires three specific discrete (i.e., finite) binary rotational subgroups of the EW gauge group $SU(2)_L$ x $U(1)_Y$, one group for each lepton family, thereby remaining within the realm of the SM lagrangian. The only finite (i.e., discrete) quaternion groups are [5,6]

$$2I, \quad 2O, \quad 2T, \quad 2D_{2n}, \quad 2C_n, \quad 1C_n \; (n \; odd) \tag{1}$$

with the 2 in front meaning binary (double) group, the double cover of the normal 3-D rotation group by SU(2) over SO(3).

The three lepton family groups are specifically the binary rotational groups called [3,3,2], [4,3,2], and [5,3,2], (often labeled 2T, 2O, and 2I, respectively), which have discrete rotational symmetries in R^3. Some properties of these groups are given in Table 1.

Table 1. Lepton Family Group Assignments

Fam.	Group	Order	N_i	Mass (ℓ_i) (MeV)	Mass (ν_i) (MeV)
ν_e,e	[332] 2T	24	1	0.511	~ 0
ν_μ,μ	[432] 2O	48	108	105.7	~ 0
ν_τ,τ	[532] 2I	120	1728	1776.8	~ 0

My assignment of each particular binary rotational group to a specific lepton family in Table 1 is based upon the N_i value for each group. Here the integer N_i arises from the mathematical syzygy for each group relating the j-invariant of elliptic modular functions (and of the Monster Group) to functions of two complex variables that remain invariant under operations of the group. The origin of these different N_i values is derived in a famous book by F. Klein in 1884 titled "The Icosahedron and solutions of equations of the fifth degree", originally in German but available in English in a Dover edition.

One can see a remarkable similarity of the mass values in column 5 of the three charged leptons to the three N_i values associated with discrete groups [332], [432], and [532]. That was my first clue that these discrete binary rotational subgroups of SU(2) could be related to the fundamental leptons of the SM.

Each group has two degenerate basis states which must be taken in linear superposition to form the two actual orthogonal fermion flavor states in each family, i.e., (ν_e, e), (ν_μ, μ), and (ν_τ, τ). That is, from basis states |1> and |2> having the same degenerate energy E_0, one forms the new orthogonal flavor states | I > and | II > of different energies

$$|I>= [|1> -|2>]/\sqrt{2} \qquad E_I = E_0 + A_i \tag{2}$$
$$|II>= [|1> +|2>]/\sqrt{2} \qquad E_{II} = E_0 - A_i \tag{3}$$

where the quantity $-A_i$ is associated with the quantum mechanical probability amplitude for state |1> to end up as state |2> and vice-versa. As a possibility, each of the different A_i values could be a function of the local spacetime environment. For an example, if the "up" flavor state | I > is a neutrino state $|\nu>$, its passage through the material of the Sun or through the Earth may have measurable effects in different amounts for each of the families.

3. The PMNS matrix derivation

This section reviews the mathematical procedure used in my 2013 derivation [1] of the PMNS matrix from first principles. As you know, the three quaternions i, j, and k, can generate all rotations in R^3 about a chosen axis or, equivalently, all rotations in the plane perpendicular to this axis. For example, the quaternion k is a binary rotation by 180° in the i-j plane.

One now must construct the three SU(2) generators, U_1 = j, U_2 = k, and U_3 = i, from the three quaternion generators from each of the discrete subgroups [3,3,2], [4,3,2], and [5,3,2] for the three lepton families. The complete mathematical description [7] for the generators R_s operating on the unit vector x in R^3 extending from the origin to the surface of the unit sphere S^2 is given by R_s = i x U_s where s = 1, 2, 3 and

$$U_1 = j, \; U_2 = -icos\frac{\pi}{q} - jcos\frac{\pi}{p} + ksin\frac{\pi}{h}, \; U_3 = i, \tag{4}$$

with h = 4, 6, 10 for the three lepton family groups [p,q,2], respectively. Their U_2 generators are listed in Table 2, with the quantity $\phi = (\sqrt{5}+1)/2$, the golden ratio.

My three lepton family binary rotational groups, [3,3,2], [4,3,2], and [5,3,2], all have the same SU(2) generators U_1 = j and U_3 = i, but one sees immediately in Table 2 that each U_2 is a different quaternion generator operating in R^3. One obtains the correct neutrino PMNS mixing angles from the linear superposition of their U_2's by making the total U_2 = k, agreeing with the SU(2) generator. This particular combination of three discrete angle rotations is now equivalent to a rotation in the i-j plane by the quaternion k.

Table 2. Lepton Family Quaternion Generators U_2

Fam.	Grp.	Generator	Factor	Angle°
ν_e,e	332	$-\frac{1}{2}i - \frac{1}{2}j + \frac{1}{\sqrt{2}}k$	-0.2645	105.337
ν_μ,μ	432	$-\frac{1}{2}i - \frac{1}{\sqrt{2}}k + \frac{1}{2}j$	0.8012	36.755
ν_τ,τ	532	$-\frac{1}{2}i - \frac{\phi}{2}j + \frac{\phi^{-1}}{2}k$	-0.5367	122.459

The total contribution of all three U_2 generators should be k, so there are three equations for the three unknown factors, which are determined to be: -5.537, 16.773, and -11.236, which become normalized to the values in the table. The resulting angles in Table 2 are the arccosines of these factors, i.e., their projections to the k-axis, but they are twice the rotation angles required in R^3, a property of quaternion rotations.

Using one-half of these angles produces

$$\theta_1 = 52.67°, \; \theta_2 = 18.38°, \; \theta_3 = 61.23°, \tag{5}$$

then taking differences results in predicted mixing angles

$$\theta_{12} = 34.29°, \; \theta_{13} = -8.56°, \; \theta_{23} = -42.85°. \tag{6}$$

The absolute values of these mixing angles are all within the 1σ range of their values for the normal mass hierarchy [8,9] as determined from several experiments:

$$\theta_{12} = \pm 34.47°, \; \theta_{13} = \pm 8.5°, \; \theta_{23} = \pm(38.39° - 45.81°). \tag{7}$$

The $\pm$ signs arise from the squares of the sines of the angles determined by the experiments.

For three lepton families, assuming that the charged lepton matrix is the identity, the neutrino flavor states ν_e, ν_μ, ν_τ, and the mass states ν_1, ν_2, ν_3, are related by the PMNS matrix V_{ij}

$$\begin{bmatrix} \nu_e \\ \nu_\mu \\ \nu_\tau \end{bmatrix} = \begin{bmatrix} V_{e1} & V_{e2} & V_{e3} \\ V_{\mu 1} & V_{\mu 2} & V_{\mu 3} \\ V_{\tau 1} & V_{\tau 2} & V_{\tau 3} \end{bmatrix} \begin{bmatrix} \nu_1 \\ \nu_2 \\ \nu_3 \end{bmatrix}.$$

The PMNS entries are the products of the sines and cosines of the predicted angles using the standard parametrization of the matrix, producing:

$$\begin{bmatrix} 0.817 & 0.557 & -0.149e^{-i\delta} \\ -0.413 - 0.084e^{i\delta} & 0.605 - 0.057e^{i\delta} & -0.673 \\ -0.383 + 0.090e^{i\delta} & 0.562 + 0.061e^{i\delta} & 0.725 \end{bmatrix}.$$

For direct comparison, the empirically estimated PMNS matrix for the *normal hierarchy* of neutrino masses is

$$\begin{bmatrix} 0.822 & 0.547 & -0.150 + 0.038i \\ -0.356 + 0.0198i & 0.704 + 0.0131i & 0.614 \\ 0.442 + 0.0248i & -0.452 + 0.0166i & 0.774 \end{bmatrix}$$

Comparing the V_{e3} elements from each, the phase angle δ is confined to be $0° \leq \delta \leq \pm 14.8°$, an angle in agreement with the T2K collaboration value of $\delta \approx 0$ but quite different from other proposed $\delta \approx \pi$ values.

The derivation of the neutrino mixing angles from first principles means that deviations from these values could indicate that the charged lepton mixing matrix is not the identity as assumed.

4. What has been learned?

At this point I pause to review what has been learned by the successful derivation of the neutrino mixing angles and the PMNS matrix when each lepton family has its own specific discrete binary rotational group in R^3, namely, the discrete quaternion groups 2T, 2O, and 2I.

(i) The three neutrino mixing angles derive from the three U_2 generator projections to the k-axis in order to collectively act like the SU(2) quaternion generator k.

(ii) Therefore, we have confirmation that the 3 lepton families represent the 3 discrete binary rotational subgroups [3,3,2], [4,3,2], and [5,3,2], of SU(2).

(iii) Leptons are 3-D entities existing in R^3 as a subspace of R^4 and C^2.

(iv) There are no more lepton families because there are no more discrete binary rotational subgroups in R^3 that contain a volume, in agreement with the predictions from Z^0 decays.

(v) Their mass ratios are related via syzygies to the j-invariant of elliptic modular functions, which in turn are related to Möbius transformations and the Monster group.

(vi) With θ_{23} = -42.85° and the matrix agreement above, one can predict the normal mass hierarchy for the neutrino mass states.

(vii) Space could be discrete at the Planck scale.

5. More background

In order to have a consistent geometrical approach toward understanding the SM, I have proposed also [3,4] that the quark families represent discrete binary rotational groups. However, one must move up one spatial dimension from R^3 to R^4 and use the related four discrete binary rotational subgroups [3,3,3], [4,3,3], [3,4,3], and [5,3,3], (or 5-cell, 16-cell, 24-cell, and 600-cell), for the quarks, thereby dictating four quark families. Recall that both R^3 and R^4 are subspaces of the unitary space C^2.

Journal of Physics: Conference Series **631** (2015) 012024 doi:10.1088/1742-6596/631/1/012024

Table 3. Lepton and quark families for the discrete binary rotational groups [abc], their j-invariant constant N, and the predicted mass values for the quarks based upon group-to-group N ratios (n.b. The mass values are the original published values from 1992).

Leptons					Quarks				
group	family	N	Pred. Mass (MeV)	Emp. Mass (MeV)	**group**	family	N	Pred. Mass (GeV)	Emp. Mass (GeV)
				.					.
				.	**[333]**	$d^{-1/3}$	1/4	0.011	0.007
				.		$u^{+2/3}$		0.38	0.004
[332]	e^-	1	[1]	0.511	**[433]**	$s^{-1/3}$	1	0.046	0.2
	ν_e		0?	0.0?		$c^{+2/3}$		[1.5]	1.5
[432]	μ^-	108	108	103.5	**[343]**	$b^{-1/3}$	108	[5]	5.0
	ν_μ		0?	0.0?		$t^{+2/3}$		$\sim$ 160	171.4
[532]	τ^-	1728	1728	1771.0	**[533]**	$b'^{-1/3}$	1728	$\sim$ 80	?.?
	ν_τ		0?	0.0?		$t'^{+2/3}$		$\sim$ 2600	?.?

Again, my assignment of a particular binary rotational group to a specific quark family in Table 3 is based upon the N_i value for each group. As for the leptons, the mass ratios of the fundamental quark families are determined by the ratios of the N_i values. For the quarks, the values in brackets, [1.5 GeV] and [5 GeV] in the right-hand mass columns, were used in 1992 to predict the top quark mass of $\sim$ 160 GeV and the mass values for the b' and t' quarks of the proposed 4th quark family.

These seven closely-related groups representing specific discrete rotational symmetries dictate the three known lepton families in R^3 and four geometrically related quark families in R^4. Consequently, being representatives of 3-D and 4-D entities, neither leptons nor quarks are to be considered as point objects. If my geometrical derivation of both the PMNS and CKM mixing matrices is based upon the correct reason for the mixing of flavor states to make the mass states, then one must reconcile the empirical data with the prediction of a fourth quark family. This fourth quark family of the t' quark and the b' quark has yet to be discovered.

My proposal that leptons are 3-D entities and that quarks are 4-D entities has several advantages. There is a clear distinction between leptons and quarks determined by inherent geometrical properties, with the specific consequence that leptons do not experience the color interaction via $SU(3)_C$ because gluons and quarks are the only physical entities involving 4-D rotations associated with the three color charges that can be defined in R^4. Also, one now has a geometrical reason for there being more than one family of leptons and of quarks, as well as there being a reason for an additional quark family to be discovered at the LHC.

However, a new problem arises with regard to there being 4 quark families and only 3 lepton families. The triangle anomalies do not cancel in the normal manner, i.e., across a generation with the lepton contributions being exactly the same value but opposite sign of the quark contributions. Remarkably, this triangle anomaly problem has a successful solution to be discussed in a later section, a solution closely related to the derivation of the PMNS and CKM matrices using the quaternion generators!

6. The CKM4 matrix derivation

The success of the above geometrical procedure for deriving the lepton PMNS matrix by using the quaternion generators from the 3 discrete binary rotation groups demands that the same approach should work for the quark families in R^4 using the proposed 4 discrete binary rotation groups [3,3,3], [4,3,3], [3,4,3], and [5,3,3]. If this procedure succeeds in deriving the CKM matrix elements as a 3x3 submatrix of CKM4, then a fourth sequential quark family, call its quark states b' and t', should exist in Nature.

These 4 discrete binary rotational groups for the quark families each have rotation subgroups of SO(4) = SO(3) x SO(3), and they also have the double covering SU(2) x SU(2). The SO(4) is the rotation group of the unit hypersphere S^3 in R^4, with every 4-D rotation being simultaneous rotations in *two orthogonal planes.*

Mathematically, the 4 discrete binary groups for the quark families each can be identified [5,6] as (L/L_K; R/R_K) with the homomorphism $L/L_K = R/R_K$. Here L and R are the specific discrete groups of quaternions listed above in section 2 and L_K and R_K are their kernels.

P. DuVal [5] established that one only needs the cyclic groups $2C_n$ and $1C_n$ when considering the four discrete rotational symmetry groups, i.e., the ones I am using for the quark families. Essentially, vertices on the 4-D regular polytope can be projected to be a regular polygon on each of the two orthogonal planes in R^4.

There will be 6 quaternion generators for each of the 4 groups, producing simultaneous rotations in two orthogonal planes. The two sets of Pauli matrices for producing continuous rotations can be identified with quaternion generators i, j, k, and another i, j, k, but they act on the two different S^2 spheres, i.e, in the two orthogonal planes. One considers this 4-D rotational transformation as the result of a bi-quaternion operation [10], or equivalently, a bi-spinor or Ivanenko-Landau-Kähler spinor or Dirac-Kähler spinor operation.

For three quark families, one has the "down" flavor states d', s', b', and their mass states d, s, b, related by the CKM matrix. This quark mixing matrix for the left-handed components is defined in the standard way as

$$V = U_L D_L^\dagger, \tag{8}$$

but for four quark families in R^4 the mathematics is a little different, for one must consider the bi-quaternion case in which there will be Bogoliubov mixing [10], producing two subfactors for each component, i.e.,

$$U_L = W^u_{14,23} W^u_{12,34}, \quad D_L = W^d_{14,23} W^d_{12,34} \tag{9}$$

with the W^u and W^d subfactor on the right mixing the 1st and 2nd generations and, separately, mixing the 3rd and 4th generations. The Bogoliubov mixing in the subfactor on the left mixes the 1st and 4th generations and, separately, the 2nd and 3rd generations. Therefore, the CKM4 matrix derives from

$$V_{CKM4} = U_L D_L^\dagger = W^u_{14,23} W^u_{12,34} (W^d_{14,23} W^d_{12,34})^\dagger. \tag{10}$$

The product $W^u_{12,34} W^{d\dagger}_{12,34}$ is given by

$$W^u_{12,34} W^{d\dagger}_{12,34} = \begin{bmatrix} x_1 & y_1 & 0 & 0 \\ z_1 & w_1 & 0 & 0 \\ 0 & 0 & x_2 & y_2 \\ 0 & 0 & z_2 & w_2 \end{bmatrix}.$$

The upper left block is an SU(2) matrix that mixes generations 1 and 2 while the lower right block is an SU(2) matrix that mixes generations 3 and 4. Each 2x2 block relates the rotation angles and the phases via

$$\begin{bmatrix} x & y \\ z & w \end{bmatrix} = \begin{bmatrix} cos\theta\, e^{i\alpha} & -sin\theta\, e^{i\beta} \\ sin\theta\, e^{i\gamma} & cos\theta\, e^{i\delta} \end{bmatrix}.$$

The 4x4 matrix that achieves the Bogoliubov mixing has four possible forms for the four possible isospin cases obeying SU(2) x SU(2): (0, 0), (1/2, 0), (0, 1/2), and (1/2, 1/2). The (1/2, 1/2) is the one for equal, simultaneous, isospin 1/2 rotations in the two orthogonal planes for CKM4:

$$W^{u,d}_{14,23} = \frac{1}{\sqrt{2}} \begin{bmatrix} 1 & 0 & -1 & 0 \\ 0 & 1 & 0 & -1 \\ 1 & 0 & 1 & 0 \\ 0 & 1 & 0 & 1 \end{bmatrix}.$$

Multiplying out these three 4x4 bi-quaternion mixing matrices,

$$V_{CKM4} = \frac{1}{2} \begin{bmatrix} x_1 + x_2 & y_1 + y_2 & x_1 - x_2 & y_1 - y_2 \\ z_1 + z_2 & w_1 + w_2 & z_1 - z_2 & w_1 - w_2 \\ x_1 - x_2 & y_1 - y_2 & x_1 + x_2 & y_1 + y_2 \\ z_1 - z_2 & w_1 - w_2 & z_1 + z_2 & w_1 + w_2 \end{bmatrix},$$

in which the phases $\alpha, \beta, \gamma, \delta$ have been ignored.

One determines the angles θ_1 and θ_2 from the quaternion generators of the 4 discrete binary rotation groups for the quark families. Projections of each of the four discrete symmetry 4-D entities onto the two orthogonal planes produces a regular polygon [5,11] with the U_2 generator iexp[2πj/h], as given in Table 4, where the h values are 5, 8, 12, 30, for the [3,3,3], [4,3,3], [3,4,3], and [5,3,3], respectively. Again, U_1 = j and U_3 = i, as for the leptons.

I need to determine the contribution from each U_2 group generator that will make the sum add to 180°, i.e., make their collective action produce the rotation U_2 = k. Expanding out the exponentials in terms of sines and cosines reveals four unknowns but only two equations. Alternately, because the four rotation angles sum to only 159°, we can use the same factor for each group, i.e., the ratio 180°/159° = 1.132.

Table 4. Quark Family Discrete Group Assignments for U_2

Fam.	Grp.	Generator	Angle°	Factor	Angle°
u,d	333	exp[2πi/5]	72	1.132	81.504
c,s	433	exp[2πi/8]	45	1.132	50.940
t,b	343	exp[2πi/12]	30	1.132	33.960
t',b'	533	exp[2πi/30]	12	1.132	13.584

In the last column of Table 4 are the normalized angles which are twice the angle required. Therefore, taking the appropriate half-angle differences, i.e., between groups 1 and 2 and then between groups 3 and 4, produces the two required mixing angles

$$\theta_1 = 15.282°, \quad \theta_2 = 10.188°. \tag{11}$$

Substituting the cosines and sines of these two derived angles into the CKM4 matrix form above produces a mixing matrix symmetrical about the diagonal. Remember that I have ignored up to eight possible phases in the 2x2 blocks.

$$V_{CKM4} = \begin{bmatrix} 0.9744 & 0.2203 & 0.0098 & 0.0433 \\ 0.2203 & 0.9744 & 0.0433 & 0.0098 \\ 0.0098 & 0.0433 & 0.9744 & 0.2203 \\ 0.0433 & 0.0098 & 0.2203 & 0.9744 \end{bmatrix}$$

One can compare the upper left 3x3 submatrix to the recent *estimated absolute values* [12]

$$V_{CKM} = \begin{bmatrix} 0.9745 & 0.2246 & 0.0036 \\ 0.2244 & 0.9736 & 0.0415 \\ 0.0088 & 0.0407 & 0.9991 \end{bmatrix}.$$

Note that most of these estimated V_{CKM} values are probably good to within a few percent but some could have uncertainties as large as 10% or more.

Of some concern are my low values of 0.2203 for V_{us} and V_{cd}. However, according to the Particle Data Group (2013) there are two possible values [12]: 0.2253 and 2204, the latter from tau decays. A recent report [13] further confirms the 0.2204 ± 0.0014 value from tau decays. Also, my derived symmetric CKM4 matrix V_{ub} value is high while the V_{td} value is reasonable, i.e., V_{td} at 0.0098 compares well with the estimated value of 0.0088.

The main diagonal V_{tb} element of CKM4 is 0.9744, quite a bit smaller than the suggested 0.9991 V_{tb} value for the 3x3 CKM matrix. However, if one imposes the unitarity condition on the rows and columns of the extracted 3x3 CKM submatrix, the new value for this V_{tb} matrix element would be 0.999, in agreement.

My final comment is that when I calculate CKM using only the first three quark groups [3,3,3], [4,3,3], and [3,4,3], the resulting 3x3 CKM matrix will disagree significantly with the known CKM matrix. Therefore, one cannot eliminate a fourth quark family when discrete binary rotational groups are considered for the lepton and quark families.

7. What has been learned?

At this point I pause to review what has been learned by the successful derivation of the quark CKM4 mixing matrix and its CKM submatrix when each quark family has its own specific discrete binary rotational group in R^4.

- (i) The quark mixing angles derive from the four U_2 generator projections to the k-axis in order to collectively act like the SU(2) quaternion generator k. CKM is a submatrix of CKM4.
- (ii) Therefore, we have confirmation that 4 quark families represent the 4 discrete binary rotational subgroups [333], [433], [343], and [532].
- (iii) Quarks are 4-D entities defined in R^4 and C^2. (Unlike the 3-D leptons, they must combine to make a 3-D entity, as in QCD, as briefly explained in section 11.)
- (iv) There are 4 quark families predicted, but where is the 4th quark family?
- (v) The quark mass ratios are related to the j-invariant of elliptic modular functions, Möbius transformations, and the Monster group.
- (vi) Space could be discrete at the Planck scale.
- (vii) There would be an end to the "Russian doll" hierarchy at the lattice nodes, which would possess no measurable physical parameters individually.

Journal of Physics: Conference Series **631** (2015) 012024 doi:10.1088/1742-6596/631/1/012024

8. Discussion about the predicted b' quark

The EW symmetry group $SU(2)_L$ x $U(1)_Y$ works extremely well, meaning that all its predictions agree with experiments so far. However, in this context there is no reason for Nature to have more than one fermion family, and certainly no reason for having 3 lepton families and at least 3 quark families. As far as I know, the traditional interpretation of the SM provides no answer that dictates the actual number of families, although the upper limit of 3 lepton families with low mass neutrinos is well established via Z^0 decays and via analysis of the CMB background.

My geometrical approach with *discrete* symmetries alters the default reliance upon SU(2) and its continuous symmetry transformations, for I utilize discrete binary rotational subgroups of SU(2) for the fundamental fermion states, a different subgroup for each lepton family and for each quark family. In this scenario one can surmise that the enormous success of the SM occurs because $SU(2)_L$ x $U(1)_Y$ is acting like a mathematical "cover group" for the actual underlying discrete rotations operating on the lepton states and quark states.

Assuming that my above matrix derivations are correct, the important question is: Where is the b' quark of the predicted 4th quark family? In 1992 I predicted a top quark mass of about 160 GeV, a b' quark mass of 65–80 GeV, and a t' quark at a whopping 2600 GeV. These mass predictions were based upon the mass ratios being determined by the j-invariant function of elliptic modular functions and of fractional linear transformations, i.e., Möbius transformations. Note that all seven discrete groups I have for the fermions are related to the j-invariant and Möbius transformations, which have direct connections to practically all areas of fundamental mathematics.

With a predicted b' mass that is much smaller than the top quark mass of 173.3 GeV and even smaller than the W mass at 80.4 GeV, one would have expected some production of the b' at LEP, Fermilab, and the LHC. Yet, no clear indication of a b' quark production and decay has appeared.

Perhaps the b' quark has escaped detection at the LHC and lies hidden in the stored data from the runs at 7 TeV and 8 TeV. With a mass value below the W and Z masses, the b' quark must decay via flavor changing neutral current (FCNC) decay channels such as b' $\rightarrow$ b + γ and b' $\rightarrow$ b + gluon. The b' could have an average lifetime too long for the colliders to have detected a reasonable number of its decays within the detector volumes and/or the energy cuts and angle cuts. However, the b' quark and t' quark would affect certain other decays that depend upon the heaviest "top" quark in a box or penguin diagram.

Another possibility is that a long lifetime might allow the formation of the quarkonium bound state b'-anti-b', which has its own specific decay channels, to bb-bar, gg, $\gamma\gamma$, and WW* $\rightarrow \nu\nu\ell\ell$. Depending upon the actual quarkonium bound state, the spin and parity $J^{PC} = 0^{++}$ or 0^{-+}. The relevant decay rates need to be calculated for comparison to the data.

And finally, as brought up in the question session of my presentation, there are two important theoretical considerations that need attention. First, associated with the mismatch of three lepton families to four quark families, e.g., the famous triangle anomalies do not cancel their infinities in the traditional manner. Second, EWSB involves the Weinberg angle θ_W, i.e., the weak mixing angle. What is the origin of θ_W in my geometrical approach? The solutions to these problems are discussed in the next sections, to be followed by the 'bigger picture'.

9. Triangle anomaly cancellation solution

The traditional cancellation procedure of matching each lepton family with a quark family "generation by generation" is known to produce the required triangle anomaly cancellation by summing the appropriate $U(1)_Y$, $SU(2)_L$, and $SU(3)_C$ generators, producing the "generation" cancellation. But such an approach is not expected to work when there is the numerical mismatch of 3 lepton families to 4 quark families.

However, I now suspect that this "generation" idea is incomplete and actually incorrect.

Why? Because the derivation of the lepton and quark mixing matrices from the U_2 generators of the discrete binary subgroups of SU(2) as shown above dictates that the 3 lepton families act as *one collective lepton family* for $SU(2)_L$ x $U(1)_Y$ and that the 4 quark families act as *one collective quark family.*

I have now created an effective single "generation" with one collective quark family matching one collective lepton family, so there is now the previously heralded "generation cancellation" of the triangle anomalies with the traditional summation of generator eigenvalues [14]. Furthermore, in the SU(3) representations one can show also that the quark and antiquark contributions cancel. Therefore, there are no SU(3)xSU(3)xU(1), SU(2)xSU(2)xU(1), U(1)xU(1)xU(1), or mixed U(1)-gravitational anomalies remaining.

There was always the suspicion that the traditional "generation" labeling was fortuitous because there was no specific reason for dictating the particular pairings of the lepton families to the quark families within the SM. That is, one could have paired the 1st lepton family with the 3rd quark family and still obtained the same cancellation result. Now, with the leptons and quarks representing the specific discrete binary rotational groups and acting collectively, a better understanding of how the families are related within the SM can be achieved.

10. Derivation of the Weinberg angle

Also considered in the question session was the statement that the Weinberg angle is determined by experiment only and not derivable from first principles. However, by using my quaternion generator approach that successfully derived the CKM4 and the PMNS mixing matrices, the Weinberg angle θ_W, i.e., the weak mixing angle, is derivable from first principles [14].

The four electroweak generators of the SM local gauge group $SU(2)_L$ x $U(1)_Y$ are typically labeled W^+, W^0, W^-, and B^0, but they can be defined equivalently as the quaternion generators i, j, k and b. But we do not require the full SU(2) to act upon the flavor states $\pm\frac{1}{2}$ for discrete rotations in the unitary plane C^2 because the lepton and quark families represent specific discrete binary rotational symmetry subgroups of SU(2). That is, we require just a discrete subgroup of $SU(2)_L$ x $U(1)_Y$.

One might suspect that the large 2I subgroup would be able to perform all the discrete symmetry rotations, but 2I omits some of the rotations in 2O. Instead, one finds that 2I x 2I' works for the leptons and the quarks, where 2I' provides the "reciprocal" rotations, i.e., the third generator U_2 of 2I becomes the third generator U'_2 for 2I' by interchanging ϕ and ϕ^{-1}:

$$U_2 = -\frac{1}{2}i - \frac{\phi}{2}j + \frac{\phi^{-1}}{2}k, \quad U_2' = -\frac{1}{2}i - \frac{\phi^{-1}}{2}j + \frac{\phi}{2}k. \tag{12}$$

Consider the three SU(2) generators i, j, k and their three simplest products: i x i = -1, j x j = -1, and k x k = -1. Now compare the three corresponding 2I x 2I' discrete generator products: i x i = -1, j x j = -1, and

$$U_2\, U_2' = -0.75 + 0.559i - 0.25j + 0.25k, \tag{13}$$

definitely not equal to -1. The reverse product U'_2U_2 just interchanges signs on the i, j, k, terms.

One needs to multiply this product quaternion $U_2U'_2$ by

$$P = 0.75 + 0.559i - 0.25j + 0.25k \tag{14}$$

to make the result -1. Again, P' has opposite signs for the i, j, k, terms only.

Given any unit quaternion q = Cos θ + $\hat{n}$ Sin θ, its power can be written as q^α = Cos $\alpha\theta$ + $\hat{n}$ Sin $\alpha\theta$. Consider P to be a squared quaternion P = Cos 2θ + $\hat{n}$ Sin 2θ because we have the product of two quaternions U_2 and U'_2 (since we need the first term in P only). Therefore, the *quaternion* square root of P has Cos $\theta = \sqrt{0.75} = 0.866$, rotating the U_2 (and U'_2) in the unitary plane C^2 by the quaternion angle of 30° so that each third generator becomes k. Thus the Weinberg angle, i.e., the weak mixing angle,

$$\theta_W = 30^\circ. \tag{15}$$

Therefore, the Weinberg angle derives from the mismatch of the third generator of 2I x 2I' to the SU(2) third generator k.

The empirical value of θ_W ranges from 28.1° to 28.8°, values less than the predicted 30°. The reason for the discrepancy is unknown (but see [15]), although one can surmise either (1) that in determining the Weinberg angle from the empirical data perhaps some contributions have been left out, or (2) the calculated θ_W is its value at the Planck scale at which the internal symmetry space and spacetime could be discrete instead of continuous.

11. Origin of quark color

With quark states being defined in R^4, there is a geometrical origin to color charge and gluon interactions. One must consider 4-D rotations in two orthogonal planes as the source of the color interaction and of color charge. Indeed, there are exactly 3 pairs of rotation planes: [wx,yz], [xy,zw], and [yw,xz], which represent the *red, green*, and *blue* color charges. One sees immediately that they are an *exact symmetry*, in agreement with QCD. One can use the appropriate 4x4 real matrices to represent the gluon rotations from one color charge to another. In fact, eight specific 4x4 real matrices are equivalent to the SU(3) matrices traditionally used for the gluon interactions.

Therefore, the 3-D leptons do not experience the color interaction because only 4-D entities can have the 4-D color properties. In addition, particular geometrical combinations of the 4-D quarks can result in the 4x4 identity matrix that represents no 4-D rotation, i.e., one now has combined quarks (and anti-quarks) into hadrons which are 3-D entities in the R^3 subspace. Thereby, hadrons exist in our (3,1)-D spacetime.

We have therefore explained quark confinement as the inability of a 4-D quark state to exist in a 3-D space as a single entity. Instead, quarks must form the appropriate combinations called hadrons, again in agreement with QCD.

12. Why are there leptons?

At the DISCRETE '08 conference in Valencia, Spain, Cecilia Jarlskog asked me to explain why there are leptons. She said that QCD is self-contained and could be a world by itself without the need for additional particles and interactions.

My answer is still the same six years later, that the mathematics called graph theory dictates my response. That is, the discrete groups that I have for the leptons and quarks are related to specific mathematical graphs of nodes and their connections. Kuratowski's theorem states that a graph is planar (i.e., cannot remain more than 2-dimensional) if and only if the graph does not contain the K_5 or $K_{3,3}$ subgraph.

Fortunately for us all, my discrete binary group for the first quark family, the up and down quark states, is [3,3,3] with graph {3,3,3}, which is the K_5 graph! All other quark families must decay down to the first family representing [3,3,3] and K_5. But the color interaction changes color charge only and does not produce quark decays from one family to another. Therefore, one needs the weak interaction and additional particles called leptons.

The first lepton family, the electron and its neutrino state, represent the discrete binary group [3,3,2] with graph {3,3,2}, a subgraph of K_5 and also protected by Kuratowski's theorem.

13. Some properties of the EW interaction

The above derivation of the Weinberg angle required the use of 2I x 2I' in place of the traditional $SU(2)_L$ x $U(1)_Y$ local gauge group. We needed the substitute group because discrete rotations are required. Because the EW operators are quaternions and the lepton and quark states are represented also by quaternions, one has quaternions operating on quaternions, an action that mathematically produces left-handed doublets (LH) and right-handed (RH) singlets. Therefore, one has the origin of the maximal parity violation of the weak interaction.

In addition, because my geometrical approach to the SM occurs in R^4, there exist two equivalent 4-D real spaces: the particle space R^4 and the anti-particle space R'^4. The general concept from Clifford algebra is that all real spaces of dimension 4n, where n is an integer, actually have two equivalent spaces while all others are a single space of that dimension. Bott periodicity tells us that their mathematical properties repeat every 8 dimensions.

Therefore, the anti-particles are revealing to us that our space is 4-D or a multiple of 4D. I outline in the next section how our 4-D space is mathematically related to much larger spaces.

14. A brief look at the bigger picture!

Here I present a *very brief outline* of the bigger picture, suggesting how several mathematical connections to higher dimensional spaces and larger symmetry groups could dictate the SM behavior in R^3, R^4, and C^2.

We know from experiment and from the ability to derive the PMNS and CKM4 mixing matrices that the SM is an excellent approximation for understanding the behavior of leptons, quarks, and the interaction bosons in the lower energy region when the spatial resolution is not better than 10^{-24} meters. At smaller distance scales, perhaps one needs to consider a discrete space-time, for which the discrete binary rotation groups that I have suggested for the fundamental particles would reveal their true identity even more clearly. Quite possibly, with this emphasis on discrete subgroups of the local gauge group, the SM lagrangian will hold all the way down to the Planck scale.

If indeed the SM applies at the Planck scale, then one can show [15] that the Monster group dictates all of physics! One then must accept two surprising consequences: (1) The Universe *is* mathematics and is *unique*. (2) We humans are mathematics!

The second consequence is not easily appreciated but the argument would develop as follows: the fundamental leptons and quarks represent specific mathematical group properties → atoms are made of leptons and quarks → molecules are made of atoms → we humans are made of molecules → therefore we are mathematics!

The connection to the Monster Group is present already in determining the lepton and quark mass ratios, which are proportional to the j-invariant of elliptic modular functions, the same j-invariant that is the partition function for the Monster Group in a quantum field theory [17].

The mathematics of these discrete groups does even more for us, however, for there is a direct connection [16] from the lepton groups [3,3,2], [4,3,2], [5,3,2], and the quark groups [3,3,3], [4,3,3], [3,4,3], [5,3,3], in R^3 and R^4, respectively, and from the EW operations of 2I x 2I' via special quaternions called icosians to the discrete space R^8, thereby forming the E_8 lattice. Its discrete symmetry group is Weyl E_8, not the continuous symmetry group E_8 of M-theory.

One then separately considers the *discrete* (3,1)-D space-time of our existence and its discrete Lorentz transformations, which introduces another discrete R^8 lattice for relativistic space-time transformations via the same telescoping up from R^4 to R^8 via iscosians.

Mathematically, the two discrete R^8 spaces combine into a 10-D discrete space-time obeying the discrete symmetry transformations of "Weyl" SO(9,1) = Weyl E_8 x Weyl E_8. This proposed *unique* connection to "Weyl" SO(9,1) was a surprise to me because one has two 8-D spaces combining to make a 10-D space-time! Its direct and unique relationship back to the SM and

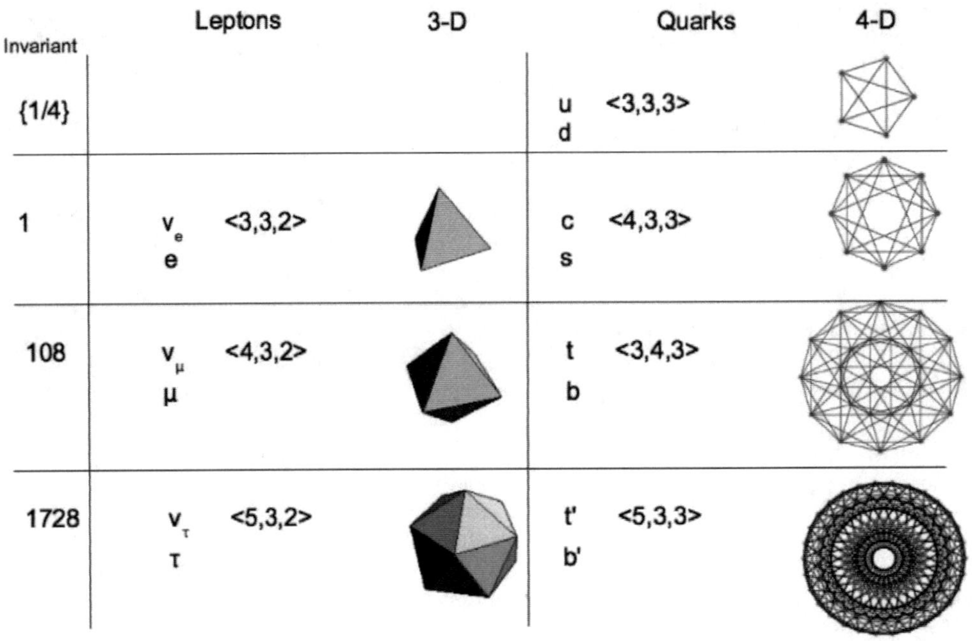

Figure 1. Discrete symmetries for the lepton and quark families

Lorentz transformations certainly is a welcome replacement to the 10^{500} ways possible for M-theory. One has eliminated the multiverse conjecture also.

Finally, one must not forget that having a fourth family of quarks provides a possible explanation for the baryon asymmetry of the Universe (BAU). From both the CKM and the PMNS matrices, one learns that their predicted CP violation (CPV) is at least 10 orders of magnitude too small to explain the BAU. That is, the important quantity called the Jarlskog value is much too small. But a 4th quark family resolves this issue [18] because substituting the fourth quark family mass values into the Jarslkog expression increases the CPV value by more than 10^{13}! Voilà. One now has penguin diagrams distinguishing the particle and antiparticle decays with sufficient difference to have the particle dominated Universe we experience.

15. The geometrical representations

Figure 1 shows the table of the lepton and quark families representing their discrete geometrical entities. I displayed this same table at the DISCRETE '08 conference and these geometrical figures remain applicable for my presentation here at DISCRETE 2014. One can surmise that these discrete entities strongly suggest that space is discrete at the Planck scale; however, their symmetry properties can remain true in a continuous space also.

16. Summary

The neutrino mixing angles and the PMNS mixing matrix have been derived successfully from the mismatch of the SU(2) generators to the corresponding generators for the discrete groups [3,3,2], [4,3,2], and [5,3,2], one for each of the lepton families but now acting in combination. The quark mixing matrix CKM4 has been derived via the same procedure using the appropriate generators from four specific related discrete binary rotational groups [3,3,3], [4,3,3], [3,4,3], and [5,3,3], corresponding to four quark families. The traditional CKM matrix is a submatrix of CKM4. However, neither quark of the 4th quark family has been detected yet. Their appearance could mean that the Standard Model lagrangian might be a good approximation to the ultimate lagrangian all the way down to the Planck scale, particularly if space-time is discrete.

In addition, now that the first principles origin of the lepton and quark mixings is understood in my geometrical approach to the SM, the Weinberg angle θ_W was derived and found to be 30°. Also, there was serious concern raised in the question session about the cancellation of the triangle anomaly because the traditional lepton-quark generational canceling no longer works with 3 lepton families and 4 quark families. However, the mixing angle derivation reveals that the 3 lepton families are acting in combination as one effective lepton family to cancel exactly the triangle anomaly contributions of 4 quark families acting as one effective quark family.

Finally, the important mathematical relationships originating from the SM acting in R^4 can be telescoped up to R^8 and higher dimensional spaces via icosians. I briefly introduced the fact that there exists a *unique* mathematical connection from the SM and Lorentz transformations to the discrete symmetry group in 10-D space-time called Weyl E_8 x Weyl E_8 = "Weyl" SO(9,1), in contrast to the 10^{500} ways possible for M-theory.

Acknowledgments
I wish to thank the DISCRETE 2014 selection committee for an opportunity to present these ideas and for their hospitality and generosity. Many participants considered this conference to be the best one they have attended in the past ten years because there was a great variety of interesting topics in several diverse research areas that were presented at the appropriate level for all of us to enjoy, discuss, and think about.

References
[1] Potter F 2013 Geometrical Derivation of the Lepton PMNS Matrix Values *Prog. in Phys.* **9(3)** 29–30
[2] Potter F 2014 CKM and PMNS Mixing Matrices from Discrete Subgroups of SU(2) *Prog. in Phys.* **10(1)** 1–5
[3] Potter F 1994 Geometrical Basis for the Standard Model *Int. J. of Theor. Phys.* **33** 279–305
[4] Potter F 2006 Unification of Interactions in Discrete Spacetime *Prog. in Phys.* **2** 3–9
[5] Du Val P 1964 *Homographies, Quaternions, and Rotations* (Oxford: Oxford University Press) pp 49–54
[6] Conway J H and Smith D A 2003 *On Quaternions and Octonions: Their Geometry, Arithmetic, and Symmetry* (Wellesly, Massachusetts: A.K. Peters) pp 33, 41–47
[7] Coxeter H S M 1974 *Regular Complex Polytopes* (Cambridge: Cambridge University Press) p 92
[8] Naumov D V (on behalf of the Daya Bay collaboration) 2014 Recent results from Daya Bay experiment *arXiv* 1412.7806v1
[9] Haigh M D 2015 Results from T2K *arXiv* 1501.04283v1
[10] Jourjine A 2014 Scalar Spin of Elementary Fermions *Phys. Lett.* B **728** 347–357 (*arXiv* 1307.2694)
[11] Coxeter H S M 1973 *Regular Polytopes* (New York, N.Y.: Dover Publications, Inc.) p 221
[12] Beringer J et al. (Particle Data Group) 2012 and 2013 partial update The Review of Particle Physics: V_{ud}, V_{us}, the Cabbibo angle, and CKM Unitarity *Phys. Rev.* D **86** 010001
[13] Lusiani A 2014 Determination of $|V_{us}|$ from τ branching fractions *arXiv* 1411.4526v1
[14] Potter F 2015 Weinberg Angle Derivation from Discrete Subgroups of SU(2) and All That *Prog. in Phys.* **11(1)** 76–77
[15] Faessler M A 2013 Weinberg Angle and Integer Electric Charges of Quarks *arXiv* 1308.5900
[16] Potter F 2011 Our Mathematical Universe: I. How the Monster Group Dictates All of Physics *Prog. in Phys.* **7(4)** 47–54
[17] Witten E 2007 Three-Dimensional Gravity Reconsidered *arXiv* 0706.3359
[18] Hou W S 2011 Source of CP Violation for the Baryon Asymmetry of the Universe *Int. J. of Mod. Phys.* D **20** 1521–1532 (*arXiv* 1101.2161v1)

Journal of Physics: Conference Series **631** (2015) 012025 doi:10.1088/1742-6596/631/1/012025

Hunting for New Physics with Up Vector-like Quarks

Miguel Nebot
Centro de Física Teórica de Partículas,
Instituto Superior Técnico, Universidade Técnica de Lisboa, Av. Rovisco Pais, P-1049-001 Lisboa, Portugal

E-mail: `nebot@cftp.ist.utl.pt`

Abstract. An interesting class of scenarios beyond the Standard Model extends the fermionic content of the theory through the addition of vector-like isosinglet quarks. Through a detailed analysis of available experimental constraints, potential deviations from the Standard Model expectations are addressed in observables such as the time-dependent CP asymmetry in $B_s \to J/\Psi\Phi$ decays, the D0 same charge dimuon asymmetry in B meson systems A^b_{SL}, rare kaon and B decays and deviations from a 3×3 unitary mixing matrix.

1. Introduction

We consider an extension of the Standard Model (SM) where one isosinglet vector-like quark T with charge $Q = 2/3$ is added to the spectrum [1, 2, 3]. In addition to the usual Yukawa terms[1],

$$\mathscr{L}_{Y_{SM}} = -\bar{q}_{0L\mathbf{i}}\, \tilde{\Phi}\, {Y_u}^{\mathbf{i}}_{\ \mathbf{j}}\, u^{\mathbf{j}}_{0R} - \bar{q}_{0L\mathbf{i}}\, \Phi\, {Y_d}^{\mathbf{i}}_{\ \mathbf{j}}\, d^{\mathbf{j}}_{0R} + \text{h.c.} \tag{1}$$

for an up vector-like quark, we have additional terms:

$$\mathscr{L}_T = -\bar{q}_{0L\mathbf{i}}\, \tilde{\Phi}\, Y^{\mathbf{i}}_T\, T_{0R} - \bar{T}_{0L}\, \mu_{T\mathbf{i}}\, u^{\mathbf{i}}_{0R} - M_{0T}\, \bar{T}_{0L}\, T_{0R} + \text{h.c.}\,. \tag{2}$$

After diagonalization of the up and down mass matrices, the 3×3 mixing matrix connecting u, c, t and d, s, b quarks is no longer unitary, but a submatrix of a larger 4×4 unitary matrix U. The charged and neutral current interactions have the form

$$\mathscr{L}_W = -\frac{g}{\sqrt{2}} \bar{\mathbf{u}}_L\, \gamma^\mu\, V\, \mathbf{d}_L\, W_\mu + \text{h.c.}\,, \tag{3}$$

$$\mathscr{L}_Z = -\frac{g}{2\cos\theta_W} \left[\bar{\mathbf{u}}_L\, \gamma^\mu \left(VV^\dagger\right) \mathbf{u}_L - \bar{\mathbf{d}}_L\, \gamma^\mu\, \mathbf{d}_L - 2\sin^2\theta_W\, J^\mu_{em}\right] Z_\mu\,, \tag{4}$$

where $\mathbf{d} \equiv (d, s, b)$, $\mathbf{u} \equiv (u, c, t, T)$ and V is a 4×3 submatrix of the matrix U:

$$U = \begin{pmatrix} V_{ud} & V_{us} & V_{ub} & U_{u4} \\ V_{cd} & V_{cs} & V_{cb} & U_{c4} \\ V_{td} & V_{ts} & V_{tb} & U_{t4} \\ V_{Td} & V_{Ts} & V_{Tb} & U_{T4} \end{pmatrix}. \tag{5}$$

[1] Subindex 0 labels *weak* eigenstates, indices **i** and **j** run over the three usual generations.

Journal of Physics: Conference Series **631** (2015) 012025 doi:10.1088/1742-6596/631/1/012025

The submatrix $V_{(3\times3)}$, i.e. the upper left 3×3 block within U, is not a unitary matrix, since $V_{(3\times3)}V^{\dagger}_{(3\times3)} \neq \mathbf{1}_{(3\times3)}$. These deviations of unitarity of the "would-be standard" mixing matrix lead to flavour changing neutral currents (FCNC) which are only present in the up sector and controlled by[2]

$$\left(VV^{\dagger}\right)_{ij} = \delta_{ij} - U_{i4}U^{*}_{j4}\ . \tag{6}$$

The addition of one isosinglet vector-like up quark provides:

- A new mass eigenstate in the up sector which can give new contributions to amplitudes involving virtual up quarks, as in neutral meson mixings and decays.
- A mixing matrix V which is not 3×3 unitary anymore, allowing for deviations of the mixing elements V_{ij} from SM values.
- Modified couplings to the Z boson in the up sector, including tree level flavour changing couplings, and reduced flavour conserving ones.

2. Experimental constraints

To reflect the abundant experimental information that constrains modifications of the flavour sector such as the ones introduced in the present scenario, we have considered the following observables.

- Tree level observables, whose extraction from experiment is presumably unaffected by New Physics (NP) effects. These observables include moduli of the CKM elements in the first and second rows. For the third row the only relevant measurement, from single top production, concerns V_{tb}. The physical phase γ is also obtained from tree level information. The input values for the analysis are collected in table 1.

Table 1. Tree level observables [4,6].

$\lvert V_{ud}\rvert$	0.97425 ± 0.00022	$\lvert V_{us}\rvert$	0.2252 ± 0.0009
$\lvert V_{cd}\rvert$	0.230 ± 0.011	$\lvert V_{cs}\rvert$	1.023 ± 0.036
$\lvert V_{ub}\rvert$	0.00375 ± 0.00046	$\lvert V_{cb}\rvert$	0.0406 ± 0.0013
$\sum \lvert V_{q_u q_d}\rvert^2$	2.002 ± 0.027	$\lvert V_{tb}\rvert$	1.00 ± 0.07
γ	$(73.2 \pm 7.0)^{\circ}$		

- Concerning B_d^0–$\bar{B}_d^0$ and B_s^0–$\bar{B}_s^0$ mixings: we consider time-dependent CP asymmetries $A_{J/\Psi K_S}$ and $A_{J/\Psi\Phi}$ (the "golden" channel in each system), mass and width differences ΔM_{B_d}, $\Delta\Gamma_d$, and ΔM_{B_s}, $\Delta\Gamma_s$, additional CP asymmetries involving *different* combinations of invariant phases, $\sin(2\bar{\alpha})$, $\sin(2\bar{\beta}+\gamma)$, the individual semileptonic asymmetries A^d_{SL}, A^s_{SL} and the same charge dimuon asymmetry A^b_{SL}. Input values are collected in table 2.
- Representative rare decays of B mesons (table 3).
- Observables from the kaon sector: rare decays and CP violation in the mixing of neutral kaons (table 4).
- Electroweak precision observables: the oblique parameters S and T (the role of U is negligible):

$$\Delta S = 0.02 \pm 0.11\,, \qquad \Delta T = 0.05 \pm 0.12\ , \text{ with a correlation } 0.879.$$

[2] They are naturally suppressed by ratios m^2/m_T^2, where m denotes generically the standard quark masses [1]. This natural suppression of FCNC is crucial in order to make the model plausible.

Table 2. B_d and B_s mixing-related observables [4, 5, 7].

$A_{J/\Psi K_S}$	0.682 ± 0.017	ΔM_{B_d}	(0.510 ± 0.003) ps^{-1}
$A_{J/\Psi\Phi}$	0.015 ± 0.035	ΔM_{B_s}	(17.757 ± 0.021) ps^{-1}
$\sin(2\bar{\alpha})$	0.00 ± 0.12	$\sin(2\beta + \gamma)$	1.0 ± 0.16
$\cos(2\beta)$	0.87 ± 0.13		
A^d_{SL}	-0.0015 ± 0.0017	$\Delta\Gamma_d/\Gamma_d$	0.0007 ± 0.0012
A^s_{SL}	-0.0075 ± 0.0041	$\Delta\Gamma_s$	(0.077 ± 0.007) ps^{-1}
A^b_{SL}	-0.00496 ± 0.00169		

Table 3. B_d and B_s rare decays [4, 5, 8].

$\mathrm{Br}(B \to X_s\gamma)$	$(3.56 \pm 0.25) \times 10^{-4}$
$\mathrm{Br}(B_s \to \mu^+\mu^-)$	$(2.8 \pm 0.7) \times 10^{-9}$
$\mathrm{Br}(B_d \to \mu^+\mu^-)$	$(3.6 \pm 1.5) \times 10^{-10}$

Table 4. Kaon mixing and rare decays [10, 11].

ϵ_K	$(2.228 \pm 0.011) \times 10^{-3}$
ε'/ϵ_K	$(1.67 \pm 0.16) \times 10^{-3}$
$\mathrm{Br}(K^+ \to \pi^+\nu\bar{\nu})$	$(1.73 \pm 1.05) \times 10^{-10}$
$\mathrm{Br}(K_L \to \pi^0\nu\bar{\nu})$	$< \mathcal{O}(10^{-8})$
$\mathrm{Br}(K_L \to \mu^+\mu^-)_{SD}$	$(6.84 \pm 0.11) \times 10^{-9}$

- D^0–$\bar{D}^0$ mixing: the Z-mediated, short distance contribution to the D^0–$\bar{D}^0$ mixing amplitude does not give a larger than observed mixing parameter x_D. Notice that, as long distance contributions to D^0–$\bar{D}^0$ are also present, we are not requiring that the short distance ones fully account for the observed $x_D = (0.8 \pm 0.2) \cdot 10^{-2}$.

3. Results

In the following I present a number of selected results from a full numerical exploration of the available parameter space of the model with an up vector-like quark. In figure 1, the physical phases β and β_s of the CKM matrix are shown together with selected moduli of mixing elements. Deviations from the SM values can be significant in β_s and $|V_{tb}|$. Figure 1 shows that despite the tight and abundant experimental information, there is still room for deviations from the SM picture with a 3×3 unitary CKM matrix, and thus New Physics effects could be expected in some observables.

In figure 2 additional information on the deviation with respect to the 3×3 unitary mixing is presented: mixing elements controlling the couplings of the new quark T are shown in subfigures 2(a), 2(b) and 2(c). Since $|V_{Tb}|$ is typically constrained to be 10-20 times smaller than $|V_{tb}|$, B physics flavour constraints dominated in the SM by top quark contributions can be satisfied even for values of $|V_{Td}|$ and $|V_{Ts}|$ as large as the ones shown, especially when compared to the values allowed for $|V_{td}|$ and $|V_{ts}|$. Subfigures 2(d) and 2(e) show the deviations from unitarity of the u and c rows,

$$\Delta_u = 1 - |V_{ud}|^2 - |V_{us}|^2 - |V_{ub}|^2 , \qquad \Delta_c = 1 - |V_{cd}|^2 - |V_{cs}|^2 - |V_{cb}|^2 .$$

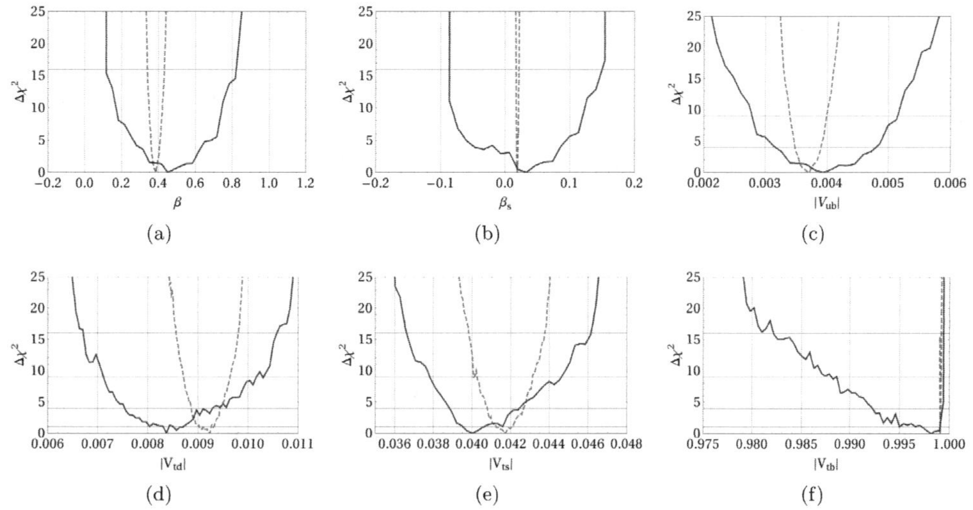

Figure 1. $\Delta\chi^2$ profiles of selected phases and moduli in the CKM matrix: solid line (blue) for the VL scenario, dashed line (red) for the SM.

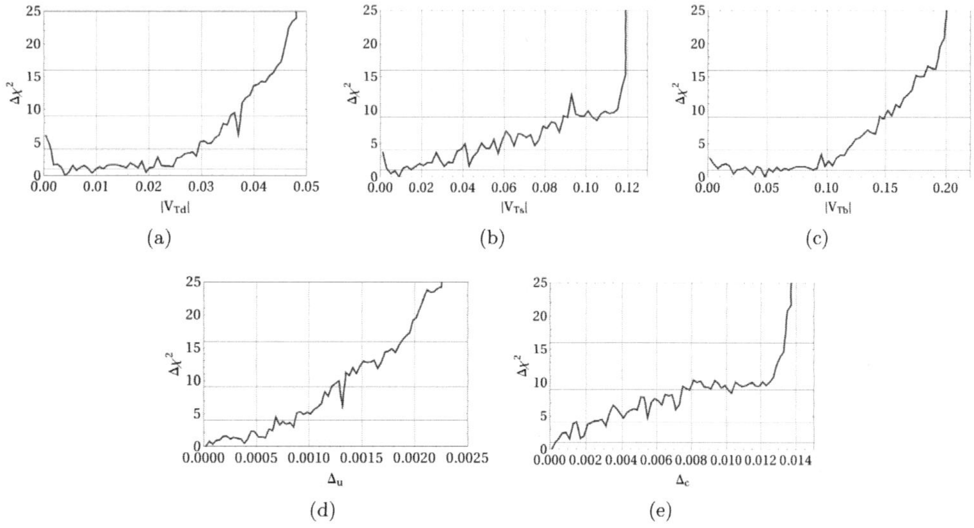

Figure 2. $\Delta\chi^2$ profiles of mixings beyond 3×3 unitarity and deviations in the u and c rows.

While figures 1 and 2 show the possibility to accommodate deviations from the SM in the mixing matrix, figure 3 focuses then on what can be obtained for different observables.

- Figure 3(a) illustrates how a significant enhancement of the same charge dimuon asymmetry A^b_{SL} can be obtained in this scenario; nevertheless, values at the -5×10^{-3} level, as required by the D0 measurements, cannot be obtained just from the 3×3 unitarity deviations [9].

- Figure 3(b) shows that deviations from the SM expectation $A_{J/\Psi\Phi} \sim 0.04$ can be introduced at the level of uncertainty of the current LHCb measurement.
- Figures 3(c) and 3(d) illustrate how loop induced rare decays can be either enhanced or supressed, with respect to SM expectations, at levels which could be probed in future experiments.
- Tree level flavour changing couplings of up-type quarks and the Z, shown in eq. (4) and (6), allow decays highly suppressed in the SM such as $t \to cZ$ or $t \to uZ$. According to figure 3(e), the addition of an up vector-like quark allows branching ratios that could reach the 10^{-5}–10^{-4} level, that could be explored at the LHC.

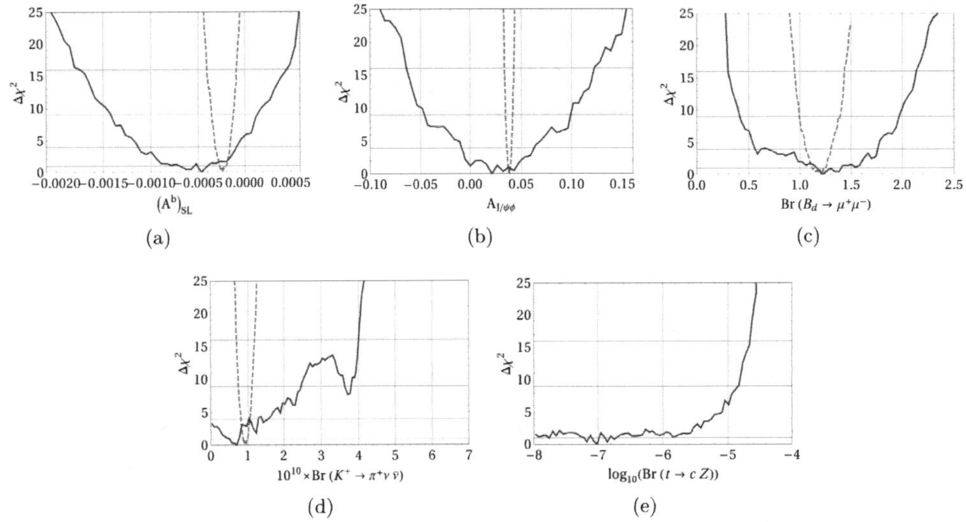

Figure 3. $\Delta\chi^2$ profiles of selected observables: solid line (blue) for the VL scenario, dashed line (red) for the SM.

Figures 1, 2 and 3 display different $\Delta\chi^2$ profiles of individual quantities. Beyond individual quantities, one can exploit correlations among different quantities to characterize the pattern of potential SM deviations associated to this New Physics scenario. Figures 4 to 7 show different joint $\Delta\chi^2$ regions.

In terms of deviations from a 3×3 unitary matrix, $|V_{ts}|$ and $|V_{td}|$ in figure 4(a) are typically expected to be reduced with respect to SM values. Their counterparts for the new quark T couplings, $|V_{Ts}|$ and $|V_{Td}|$, are shown in fig. 4(b): although, as anticipated in figs. 2(a) and 2(b), they can reach sizable values, it is now clear that such values cannot be obtained for both simultaneously.

Concerning the same charge dimuon asymmetry A^b_{SL}, fig. 3(a) establishes that the scenario under study cannot produce values in agreement with the D0 measurement. Figure 5 shows in addition that the partial enhancement in A^b_{SL} can only be obtained if $|V_{ub}|$ does also deviate to larger-than-standard values.

For loop induced rare decays, figure 6 displays to contrasting cases. According to 6(a), although both $\text{Br}(B_s \to \mu^+\mu^-)$ and $\text{Br}(B_d \to \mu^+\mu^-)$ can deviate from SM expectations, it may be hard to disentangle those deviations from a SM value. For rare kaon decays, figure 6(b) shows

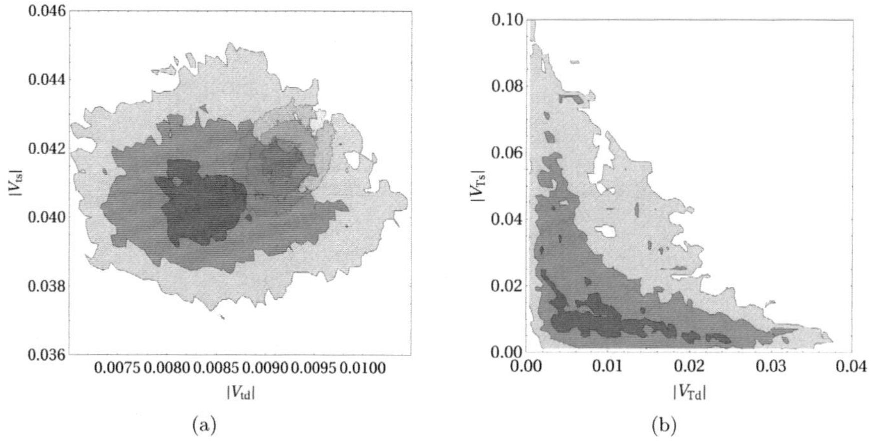

Figure 4. $\Delta\chi^2$ profiles, 68%, 95% and 99% CL regions are shown: blue regions for the VL scenario, red regions for the SM.

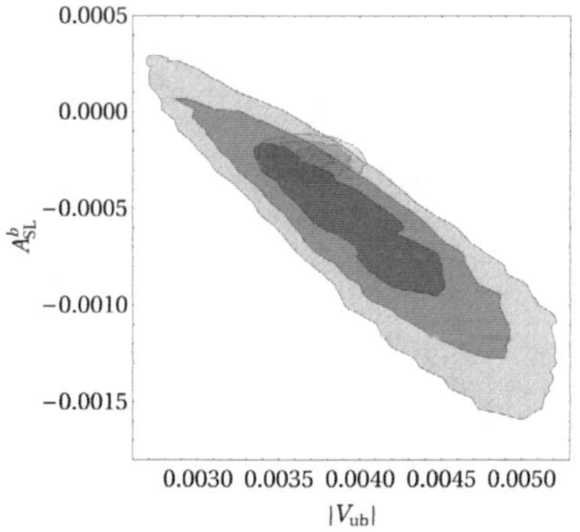

Figure 5. $\Delta\chi^2$ profiles, 68%, 95% and 99% CL regions are shown: blue regions for the VL scenario, red regions for the SM.

two connected branches: $K_L \to \pi^0\nu\bar{\nu}$ can only be enhanced if the charged mode $K^+ \to \pi^+\nu\bar{\nu}$ is also enhanced. There is also room for the model to keep at SM level or below the neutral mode while the charged mode can be increased up to 2-3 times the SM value.

For decays induced by the Z flavour changing tree couplings, $t \to cZ$ and $t \to uZ$, figure 7 shows them in association with the mass m_T of the new quark T. Although the branching ratios of those rare decays can reach values within reach of the LHC, it is clear that this also requires light values of the mass m_T. This aspect illustrates the power of detailed analyses: even though the $t \to qZ$ branching ratios do not depend on the value of m_T, full use of the available experimental information can establish such correlations.

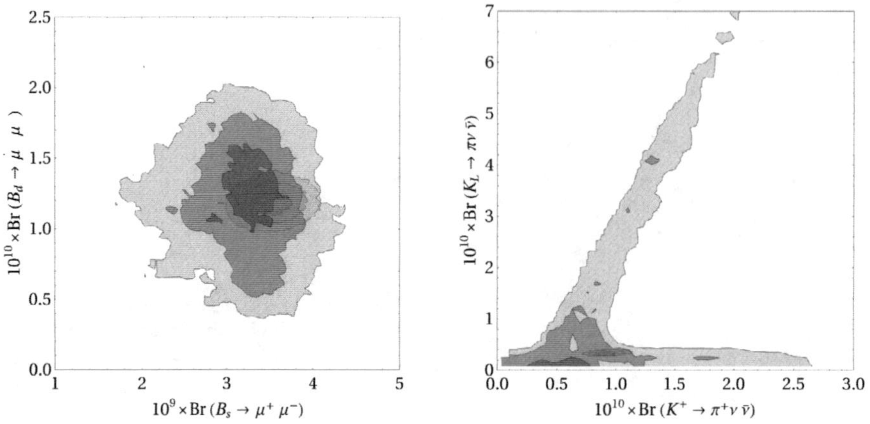

(a) (b)

Figure 6. $\Delta\chi^2$ profiles, 68%, 95% and 99% CL regions are shown: blue regions for the VL scenario, red regions for the SM.

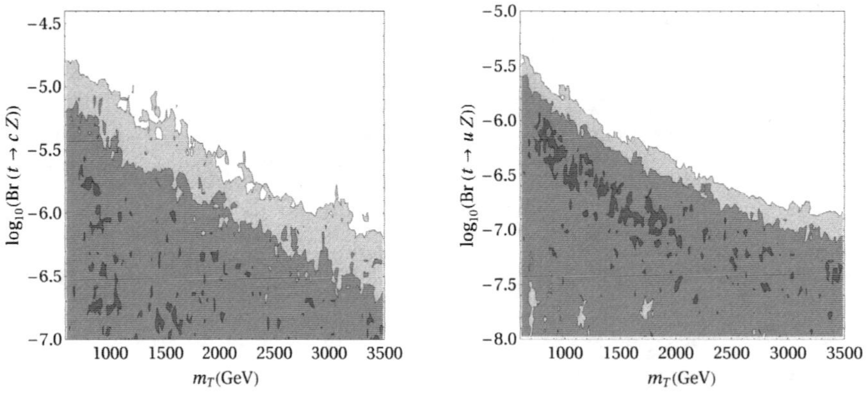

(a) (b)

Figure 7. $\Delta\chi^2$ profiles, 68%, 95% and 99% CL regions are shown: blue regions for the VL scenario, red regions for the SM.

4. Conclusions

An overview of a detailed analysis of flavour data in the context of a simple extension of the Standard Model, that includes an additional $Q = 2/3$ vector-like isosinglet quark is presented. Experimental constraints from all the relevant quark flavour sectors are imposed and yet deviations from Standard Model expectations can be accommodated. This is illustrated through different individual observables and the power of correlations to characterize the New Physics scenario under study.

Acknowledgments
This work was partially supported by *Fundação para a Ciência e a Tecnologia* (FCT, Portugal) through the projects CFTP-FCT Unit 777 and CERN/FP/123580/2011. The author is very grateful for the hospitality and organization of *Discrete 2014*.

References
[1] Langacker P and London D 1988 *Phys.Rev.* **D38** 886; del Aguila F and Bowick M 1983 *Nucl.Phys.* **B224** 107; del Aguila F, Laermann E and Zerwas P 1988 *Nucl.Phys.* **B297** 1; Cheng T and Li L F 1992 *Phys.Rev.* **D45** 1708–1710.

[2] del Aguila F and Cortes J 1985 *Phys.Lett.* **B156** 243; del Aguila F, Chase M and Cortes J 1986 *Nucl.Phys.* **B271** 61; Branco G and Lavoura L 1986 *Nucl.Phys.* **B278** 738; Buchmuller W and Gronau M 1989 *Phys.Lett.* **B220** 641; Nir Y and Silverman D J 1990 *Phys.Rev.* **D42** 1477–1484; Nardi E, Roulet E and Tommasini D 1992 *Nucl.Phys.* **B386** 239–266; Silverman D 1992 *Phys.Rev.* **D45** 1800–1803; Branco G, Parada P, Morozumi T and Rebelo M 1993 *Phys.Lett.* **B306** 398–402; Branco G, Morozumi T, Parada P and Rebelo M 1993 *Phys.Rev.* **D48** 1167–1175; Branco G, Parada P and Rebelo M 1995 *Phys.Rev.* **D52** 4217–4222 (*Preprint* `hep-ph/9501347`); Barger V, Berger M and Phillips R 1995 *Phys.Rev.* **D52** 1663–1683 (*Preprint* `hep-ph/9503204`); del Aguila F, Aguilar-Saavedra J and Branco G 1998 *Nucl.Phys.* **B510** 39–60 (*Preprint* `hep-ph/9703410`); Barenboim G and Botella F 1998 *Phys.Lett.* **B433** 385–395 (*Preprint* `hep-ph/9708209`); Barenboim G, Botella F, Branco G and Vives O 1998 *Phys.Lett.* **B422** 277–286 (*Preprint* `hep-ph/9709369`); Kakebe I and Yamamoto K 1998 *Phys.Lett.* **B416** 184–191 (*Preprint* `hep-ph/9705203`); Barenboim G, Botella F and Vives O 2001 *Phys.Rev.* **D64** 015007 (*Preprint* `hep-ph/0012197`); Higuchi K and Yamamoto K 2000 *Phys.Rev.* **D62** 073005 (*Preprint* `hep-ph/0004065`); Barenboim G, Botella F and Vives O 2001 *Nucl.Phys.* **B613** 285–305 (*Preprint* `hep-ph/0105306`); Aguilar-Saavedra J 2003 *Phys.Rev.* **D67** 035003 (*Preprint* `hep-ph/0210112`); Aguilar-Saavedra J, Botella F, Branco G and Nebot M 2005 *Nucl.Phys.* **B706** 204–220 (*Preprint* `hep-ph/0406151`); Botella F J, Branco G C and Nebot M 2009 *Phys.Rev.* **D79** 096009 (*Preprint* `0805.3995`); Higuchi K and Yamamoto K 2010 *Phys.Rev.* **D81** 015009 (*Preprint* `0911.1175`); Frampton P H, Hung P and Sher M 2000 *Phys.Rept.* **330** 263 (*Preprint* `hep-ph/9903387`).

[3] Botella F, Branco G and Nebot M 2012 *JHEP* **1212** 040 (*Preprint* `1207.4440`)

[4] Beringer J and others (Particle Data Group) (Particle Data Group) 2012 *Phys.Rev.* **D86** 010001

[5] Amhis Y and others (Heavy Flavor Averaging Group) (Heavy Flavor Averaging Group) 2012 (*Preprint* `1207.1158`)

[6] Aubert B and others (BABAR Collaboration) (BaBar Collaboration) 2007 *Phys.Rev.Lett.* **99** 251801 (*Preprint* `hep-ex/0703037`); Poluektov A and others (Belle Collaboration) (Belle Collaboration) 2010 *Phys.Rev.* **D81** 112002 (*Preprint* `1003.3360`); Ikado K and others (Belle Collaboration) (Belle Collaboration) 2006 *Phys.Rev.Lett.* **97** 251802 (*Preprint* `hep-ex/0604018`); Hara K and others (Belle Collaboration) (Belle collaboration) 2010 *Phys.Rev.* **D82** 071101 (*Preprint* `1006.4201`); Aubert B and others (BABAR Collaboration) (BABAR Collaboration) 2008 *Phys.Rev.* **D77** 011107 (*Preprint* `0708.2260`); Lees J and others (BABAR Collaboration) (BABAR Collaboration) 2012 (*Preprint* `1207.0698`).

[7] Aaij R and others (LHCb Collaboration) (LHCb Collaboration) 2012 *Phys.Rev.Lett.* **108** 101803 (*Preprint* `1112.3183`); Aaij R and others (LHCb Collaboration) (LHCb Collaboration) 2012 *Phys.Lett.* **B707** 497–505 (*Preprint* `1112.3056`); Aubert B and others (BABAR Collaboration) (BABAR Collaboration) 2009 *Phys.Rev.* **D79** 072009 (*Preprint* `0902.1708`); Adachi I, Aihara H, Asner D, Aulchenko V, Aushev T and others (Belle Collaboration) (Belle Collaboration) 2012 *Phys.Rev.Lett.* **108** 171802 (*Preprint* `1201.4643`); Aad G and others (ATLAS Collaboration) (ATLAS Collaboration) 2012 *Phys. Lett.* **B713** 387 (*Preprint* `1204.0735`).

[8] Chatrchyan S and others (CMS Collaboration) (CMS Collaboration) 2011 *Phys.Rev.Lett.* **107** 191802 (*Preprint* `1107.5834`); Chatrchyan S and others (CMS Collaboration) (CMS Collaboration) 2012 *JHEP* **1204** 033 (*Preprint* `1203.3976`); Aaij R and others (LHCb Collaboration) (LHCb collaboration) 2012 *Phys. Rev. Lett. 108,* **231801** (*Preprint* `1203.4493`); Aaij R and others (LHCb Collaboration) (LHCb Collaboration) 2012 *Phys.Lett.* **B708** 55–67 (*Preprint* `1112.1600`).

[9] F. J. Botella, G. C. Branco, M. Nebot and A. Sánchez, *Phys. Rev.* **D91** (2015) 3, 035013 (Preprint `1402.1181`).

[10] Alavi-Harati A and others (KTeV Collaboration) (KTeV Collaboration) 1999 *Phys.Rev.Lett.* **83** 22–27 (*Preprint* `hep-ex/9905060`); Ambrose D and others (E871 Collaboration) (E871 Collaboration) 2000 *Phys.Rev.Lett.* **84** 1389–1392; Ahn J and others (E391a Collaboration) (E391a Collaboration) 2008 *Phys.Rev.Lett.* **100** 201802 (*Preprint* `0712.4164`); Artamonov A and others (E949 Collaboration)

(E949 Collaboration) 2008 *Phys.Rev.Lett.* **101** 191802 (*Preprint* 0808.2459); Abouzaid E and others (KTeV Collaboration) (KTeV Collaboration) 2011 *Phys.Rev.* **D83** 092001 (*Preprint* 1011.0127).

[11] Cirigliano V, Ecker G, Neufeld H, Pich A and Portoles J 2012 *Rev.Mod.Phys.* **84** 399 (*Preprint* 1107.6001)

Order and anarchy hand in hand in 5D SO(10)

D Vicino[1 2]

[1] Dipartimento di Fisica e Astronomia 'G. Galilei', Università di Padova, Via Marzolo 8, I-35131 Padua, Italy

[2] Istituto Nazionale di Fisica Nucleare, Sezione di Padova, I-35131 Padua, Italy

E-mail: denise.vicino@pd.infn.it

Abstract. A mechanism to generate flavour hierarchy via 5D wave-function localization is revisited in the context of SO(10) grand unified theory. In an extra-dimension compactified on an orbifold, fermions (living in the same 16 representation of SO(10)) result having exponential zero-modes profiles, localized around one of the brane. The breaking of SO(10) down to SU(5) $\times$ U(1)$_X$ provides the key parameter that distinguishes the profiles of the different SU(5) components inside the same 16 representation. Utilizing a suitable set of scalar fields, a predictive model for fermion masses and mixing is constructed and shown to be viable with the current data through a detailed numerical analysis. The scalar field content of the model is also suitable to solve the doublet-triplet splitting problem through the missing partner mechanism. All the Yukawa couplings in the model are anarchical and of order unity, while the hierarchies among different fermions result only from zero-mode profiles. The naturalness of Anarchical Yukawa couplings is studied, showing a preference for a normal ordered neutrino spectrum; predictions for various observables in the lepton sector are also derived.

1. Introduction

It is well known that the Standard Model (SM), despite being so far the most complete and successful theory of particles interactions, still doesn't account for some aspects of the observed phenomenology or doesn't provide explanation to how particular structures of interaction originate. Among these, the flavour sector of SM remains one of the most obscure aspects: besides the necessity of introducing 3 generations of fermions, constituting the matter fields, lots of parameters are needed to describe their mass terms and flavour mixing interactions. These parameters, given in the so-called Yukawa couplings, are simply fixed by experiments, from which a peculiar pattern and a spanning of several orders of magnitude emerge, without any theoretical explanation: a huge hierarchy among the generations characterizes the masses of the quarks, especially the up-type ones ($\sim$ 1 MeV-170 GeV), while a milder hierarchy rules the down-type quarks ($\sim$ 5 MeV-5 GeV) and the charged leptons ($\sim$ 0.5 MeV-2 GeV); besides this, neutrinos are known to have a very small mass ($\sim$ eV) that is not taken into account in the SM; then, strong hierarchy appears in the flavour mixing of the quarks, while, on the other side, the flavour mixing of the leptons results mostly anarchical. This setup is summarized by the following relations, expressed in terms of the parameter $\lambda \approx 0.27$, identified as the Cabibbo mixing angle:

$$\begin{aligned} m_u : m_c : m_t &\approx \lambda^8 : \lambda^4 : 1; \\ m_d : m_s : m_b &\approx \lambda^5 : \lambda^3 : 1; \\ m_e : m_\mu : m_\tau &\approx \lambda^6 : \lambda^2 : 1; \end{aligned} \tag{1}$$

$$\frac{m_{\nu 2}^2 - m_{\nu 1}^2}{| m_{\nu 3}^2 - m_{\nu 2}^2 |} \approx \lambda^2; \tag{2}$$

$$V_{CKM} \approx \begin{pmatrix} 1 & \lambda & \lambda^3 \\ \lambda & 1 & \lambda^2 \\ \lambda^3 & \lambda^2 & 1 \end{pmatrix}; \tag{3}$$

$$U_{PMNS} \approx \begin{pmatrix} 0.8 & 0.5 & 0.2 \\ 0.5 & 0.6 & 0.6 \\ 0.3 & 0.6 & 0.7 \end{pmatrix}. \tag{4}$$

All these unexplained features in the flavour sector characterize the so called “flavour problem” and, together with other open issues, necessarily require to extend the SM within new scenarios. This work [1] analyzes the theoretical advantages and the phenomenological viability of realizing this extension to a supersymmetric SO(10) grand unified theory (GUT) in extra dimensions, by updating a primordial model originally proposed by Kitano and Li [2].

The first motivation of this scenario is the possibility of perfectly accommodating the particle content of the SM and its gauge group into an SO(10) GUT, whose supersymmetric (SUSY) version allows for gauge coupling unification without need of intermediate scales. All fermions of each generation result unified in the 16 spinorial representation of SO(10), nicely explaining the origin of the fermions quantum numbers under the SM gauge group. The description of fermion masses and mixing angles, anyway, remains as complicated as in the SM, requiring still lots of parameters spanning a vast range of magnitude, but some advantages arise from the structure of SO(10) GUT:

(A) an extra component present in the 16 would be a right-handed (RH) neutrino, naturally allowing a description of the light neutrino masses in terms of the seesaw mechanism;

(B) the possible embedding of intermediate symmetry breaking down to SU(5) would allow for unified representation of charged leptons and down-type quarks, a picture compatible with the similar hierarchy manifested between their masses and with the known unification of tau lepton and bottom quark masses approximately realized at high energy in the minimal supersymmetric standard model (MSSM).

The Yukawa interactions in SO(10) couple two 16 fermions with a Higgs field in the possible representations: 10, 120 or $\overline{126}$ (any combination of them). Viable SO(10) models requires to combine at least two of these representations (10-$\overline{126}$ or 120-$\overline{126}$), in order to fit all the fermion masses and mixing angles (see *e.g.* [3, 4]). The use of $\overline{126}$ allows also a Majorana mass term for RH neutrinos, implementing the light neutrino masses through the seesaw mechanism.

Overall, there is no qualitative difference with respect to the SM, since the number of free parameters in the flavour sector remains very large (so that no predictions are available) and

Journal of Physics: Conference Series **631** (2015) 012026 doi:10.1088/1742-6596/631/1/012026

the best fit parameters span several orders of magnitude, as much as in the SM.
It would be desirable to have a natural theory of flavour, in which all the couplings are of the same order magnitude.
Extra dimensions are introduced in this context exactly with this purpose, by accounting for the hierarchies of the charged fermion masses and of the quark mixing angles in terms of an irreducible set of order-one parameters that don't require to be fine-tuned. The constraints given by SUSY, moreover, let some advantages in the theoretical structure, permitting to reduce the number of free parameters with respect to the non-supersymmetric case.
The idea is to build up the observed hierarchical structure of the Yukawas starting from fundamental Yukawas with anarchical entries of order one, combined with a mechanism that can "reorder" these entries introducing the wanted hierarchy: this mechanism exploits the possibility to associate a hierarchical structure in the generation space to the matter fields, in such a way to "sandwich" the fundamental anarchical Yukawa between two diagonal hierarchical matrices. To give roughly the idea, the fundamental Yukawa would be:

$$\mathcal{Y} \approx \begin{pmatrix} \varepsilon^2 & 0 & 0 \\ 0 & \varepsilon & 0 \\ 0 & 0 & 1 \end{pmatrix} \begin{pmatrix} \mathcal{O}(1) & \mathcal{O}(1) & \mathcal{O}(1) \\ \mathcal{O}(1) & \mathcal{O}(1) & \mathcal{O}(1) \\ \mathcal{O}(1) & \mathcal{O}(1) & \mathcal{O}(1) \end{pmatrix} \begin{pmatrix} \hat{\varepsilon}^2 & 0 & 0 \\ 0 & \hat{\varepsilon} & 0 \\ 0 & 0 & 1 \end{pmatrix} \tag{5}$$

where ε and $\hat{\varepsilon}$ are just example parameters to express the hierarchy associated to the fields by some mechanism, and the $\mathcal{O}(1)$ matrix is the natural, fundamental Yukawa.
This kind of scenario[1] is emerging introducing a compactified extra dimension, provided the matter fields are propagating in the bulk with a hierarchical profile: putting the Higgs field (and so the Yukawa coupling) on a brane, the coupling results hierarchically modulated by the value assumed by the fermions profiles on that brane, playing the role of the ε parameters above. Exponential profiles along the extra dimension naturally emerge by introducing a bulk-mass parameter (a 5-dimensional mass term) and hierarchical Yukawa couplings are generated by small ($< \mathcal{O}(1)$) difference of these parameters.

2. Flavour hierarchy from extra dimension: an abelian model

Before introducing the specific features of our model in SO(10), let's first consider the basic setup of a 5-dimensional (5D) U(1) gauge theory with N=1 SUSY. The extra dimension is considered compactified on an orbifold S^1/Z_2 of radius R, $y \in [0, \pi R]$. As shown in [6], 5D N=1 SUSY can conveniently be written in terms of 4-dimensional N=2 SUSY formalism. Gauge fields are described by a 5D vector supermultiplet, containing an $N=1$ chiral multiplet Φ and a vector multiplet V; matter fields are described by a 5D hypermultiplet, containing a pair of $N=1$ chiral multiplets, H and H^c. The U(1) gauge invariant action of interacting hypermultiplet and

[1] A similar scenario is realized, for example, in Froggatt-Nielsen models [5] by adding a $U(1)_{FN}$ flavour group, introducing a coupling with flavon field and proper assignment of charges to the fields. Hierarchy is created by different powers of the flavon VEV breaking the symmetry.

vector multiplet is written as: [6, 7, 8]

$$\begin{aligned} S_5 &= \int dy\, d^4x \left[\int d^4\theta \left(\partial_y V - \frac{1}{\sqrt{2}}(\Phi + \bar{\Phi}) \right)^2 + \frac{1}{4} \int (d^2\theta\, W^\alpha W_\alpha + \text{h.c.}) \right. \\ &+ \int d^4\theta \, (\bar{H}_i e^{2g_5 QV} H_i + \bar{H}_i^c e^{-2g_5 QV} H_i^c) \\ &+ \left. \left(\int d^2\theta \, H_i^c \left(\hat{m}_i + \partial_y - \sqrt{2} g_5 Q \Phi \right) H_i + \text{h.c.} \right) \right], \end{aligned} \tag{6}$$

where W^α is a field strength, g_5 is the 5D gauge coupling constant, $\hat{m}$ is the bulk mass and Q is the U(1) charge of the chiral multiplet H; i is the index in the generation space. The first line describes the 5D kinetic term for the gauge fields; the second line describes the 4D part of the kinetic term for matter fields and the gauge interactions; the last line complete the previous one with the derivative with respect to the extra dimension, the gauge interaction with the field Φ, related to the fifth component of the 5D gauge multiplet, and a 5D mass term for each generation.
Z_2 parity is assigned by boundary conditions of the fields, as: $(H; H^c) = (+; -)$ and $(V, \Phi) = (+; -)$, in such a way to break explicitly N=2 down to N=1 SUSY on the branes. To have invariance, $\hat{m}$ must be odd under Z_2 and the simplest choice is $\hat{m}(y) = m\ \text{sgn}(y)$, m being a real constant.
Eq.(6) describes the action in the bulk and, besides this, there can be contributions strictly localized on the branes, which should only respect $N = 1$ SUSY. Here we discuss the theory in the ideal limit of exact $N = 1$ SUSY.

By performing the Kaluza-Klein (KK) expansion of 5D bulk fields (for the chiral multiplet, $H(x, y) = \sum_n H_n(x) f_n(y)$) and by reduction to the 4D theory, we can obtain the profiles $f_n(y)$ of the various modes using the equations of motion (e.o.m.) and imposing the Z_2-boundary conditions. For the chiral superfields, e.o.m. come only from the last line of the action (6), leading to the 0-mode profiles:

$$f_0^i(y) = \sqrt{\frac{2m_i}{1 - e^{-2m_i \pi R}}} e^{-m_i y} \text{ and } f_0^c(y) = 0. \tag{7}$$

Only the Z_2-even fields have a 0-mode, that results exponentially modulated by the bulk mass parameter m_i, localized mostly on the $y = 0$ brane or the $y = \pi R$ brane, according to the sign of m_i. This feature is an essential ingredient: localizing the Higgs sector on one of the branes (*e.g.* $y = 0$) with $\mathcal{O}(1)$ Yukawa couplings, the effective 4D Yukawa results being suppressed or enhanced depending on the profile value at that brane. The bulk masses, for a field of a given representation of the gauge group (in this case, for assigned U(1) charge) permitt to distinguish the profiles with respect to the 3 generations. In this way the hierarchical pattern observed in fermion masses and mixing angles can be explained without appealing to small ad hoc parameters [9, 10, 11].
At this point it is worth to note that the VEV of the scalar field contained in the chiral superfield Φ of the gauge sector, can correct the bulk mass term by:

$$m_i \to m_i - \sqrt{2} g_5 Q \langle \phi \rangle \tag{8}$$

in a way proportional on the U(1) charge Q. The only interaction with Φ allowed by $N = 1$ 5D SUSY is of gauge type, so that this field couples to H and H^c through g_5, universally with respect to the generations of matter fields. The correction of Eq.(8) is then universal as well.

3. The SO(10) model

The previous model can be implemented in the SO(10) framework, following the original work of [2], that we improved in [1] by making the theory more predictive and testing its phenomenological viability. In this context the compactification radius of the extra dimension is taken above the GUT scale ($\frac{1}{R} > M_{\rm GUT} \approx 10^{16}$ GeV).

In this framework H and H^c are replaced by three copies of $\mathbf{16}$ and $\mathbf{16}^c$, accomodating the three generations of SM fermions and RH neutrinos, transforming as $\mathbf{16}$ and $\overline{\mathbf{16}}$ under SO(10) respectively. The vector supermultiplet, comprising $\mathbf{45}_V$ and $\mathbf{45}_\Phi$, transforms in the adjoint of SO(10).

The relevant part of the action for getting the 0-modes of matter fields is given, as in the previous model, by the superpotential in the bulk, that is written in the SO(10) theory as:

$$\mathcal{W}_{\rm bulk} = \mathbf{16}^c_i \left[\hat{m}_i + \partial_y - \sqrt{2} g_5\, \mathbf{45}_\Phi\right] \mathbf{16}_i \tag{9}$$

This superpotential leads to the same 0-mode profile in Eq.(7), identical for all the components of the $\mathbf{16_i}$ multiplet of a given generation. To exploit completely the mechanism of addressing the flavour hierarchy through extra dimension, it is not sufficient to distiguish the profiles only with respect to the generations, but it would be desirable, indeed, to split the profiles with respect to the different components of the $\mathbf{16}$, implementing a breaking of the SO(10) symmetry. With this purpose, we exploit the mechanism described for the abelian model in Eq.(8), using the fact that the $\mathbf{45}_\Phi$ field contains a SU(5) singlet which, taking a VEV, is responsible of breaking SO(10) down to SU(5)×U(1)$_{\rm X}$. Such a VEV generates different contributions to the bulk masses of the SU(5) components of each $\mathbf{16}$ bulk multiplet, proportional to their U(1)$_{\rm X}$ charges, causing the splitting of the profiles. Under SU(5)× U(1)$_{\rm X}$ the $\mathbf{16}$ decomposes as

$$\mathbf{16} = 10_{-1} + \bar{5}_3 + 1_{-5} \ , \tag{10}$$

with the U(1)$_{\rm X}$ charges Q^r_X given in subscript. Each SU(5) multiplet gets an effective bulk mass m^r_i ($r = 10, \bar{5}, 1$) given by

$$m^r_i = m_i - \sqrt{2} g_5 Q^r_X v^{3/2}_\Phi \ , \tag{11}$$

Schematically, from the initial unified picture of the $\mathbf{16}$, we end up with the splitting of bulk masses and relative profiles, for each generation, as:

$$m^{16}_i \stackrel{\langle \mathbf{45}_\Phi \rangle}{\longrightarrow} m^{10}_i,\ m^{\bar{5}}_i,\ m^1_i$$

$$f^{16}_i \stackrel{\langle \mathbf{45}_\Phi \rangle}{\longrightarrow} f^{10}_i,\ f^{\bar{5}}_i,\ f^1_i$$

The effective Yukawas for the charged fermions are then, resembling the scheme in Eq.(5):

$$\mathcal{Y}_u = F_{10} Y_u F_{10} \ , \ \mathcal{Y}_d = F_{10} Y_d F_{\bar{5}} \ , \ \mathcal{Y}_e = F_{\bar{5}} Y_e F_{10} \tag{12}$$

where the entries of diagonal matrices F_r are the zero-mode profiles evaluated at the $y = 0$ brane:

$$F_r = \mathtt{diag}(f_1^r, f_2^r, f_3^r)\ ,\ f_i^r = \sqrt{\frac{2m_i^r}{1 - e^{-2m_i^r \pi R}}}. \tag{13}$$

The mass matrix of light neutrinos is obtained through the type I seesaw mechanism and is proportional to

$$m_\nu \propto F_{\bar{5}}\ Y_\nu Y_R^{-1} Y_\nu^T\ F_{\bar{5}}\ . \tag{14}$$

where Y_ν is the Yukawa of the Dirac mass term for neutrinos and Y_R the one of Majorana mass term for RH neutrinos.

It's quite remarkable that in this mechanism, designed by the gauge group structure and by the SUSY contraints[2], only one parameter, v_Φ, is responsible of the profiles splitting into the three SU(5) submultiplets, so that, overall, only 4 parameters (m_i, v_Φ) describe 9 different profiles. Implementing a compactified extra dimension at the level of the SM gauge group would require, instead, to introduce 15 different bulk masses (one for each representation and generation) without even describing RH neutrinos. Thus, the present SO(10) model would be much more predictive, but we have to verify its viability by fitting the fermion masses and mixing angles: this fit is not so trivial, since the profiles are still unified in the SU(5) multiplets and the splitting is not so flexible, but dictated by the fixed $\mathrm{U}(1)_\mathrm{X}$ charges.

Let's now discuss the particular Higgs sector of our model to see how this scenario is concretely implemented.

The following expression describes the superpotential at the branes as chosen in our framework [1] and characterizing the main difference with respect to the original model [2]:

$$\begin{aligned}\mathcal{W}_{\text{branes}} = & \frac{\delta(y)}{\Lambda}\left[Y_{10}^{ij}\mathbf{16}_i\mathbf{16}_j\mathbf{10}_H + Y_{120}^{ij}\mathbf{16}_i\mathbf{16}_j\mathbf{120}_H + Y_{126}^{ij}\mathbf{16}_i\mathbf{16}_j\overline{\mathbf{126}}_H + ...\right] \\ & + \ \delta(y)w_0(\mathbf{45}_H, \mathbf{10}_H, \mathbf{126}_H, \overline{\mathbf{126}}_H, \mathbf{120}_H) \\ & + \ \delta(y-\pi R)w_\pi(\mathbf{126}'_H, \overline{\mathbf{126}}'_H)\ .\end{aligned} \tag{15}$$

The first line describes the Yukawa interactions localized at the $y = 0$ brane. $\Lambda > M_{GUT}$ is a cut-off scale and here all the Yukawas arise from operators of the same dimensionality [3]. The $\overline{\mathbf{126}}_H$, which contains an SU(5) singlet, has the following roles: 1) it provides a Majorana mass term to the RH neutrinos; 2) its VEV is responsible for breaking the residual $\mathrm{U}(1)_X$ symmetry; 3) together with a $\mathbf{126}_H$ on the same brane and another pair of $\overline{\mathbf{126}}'_H$ and $\mathbf{126}'_H$ located on the other brane, it is responsible for restoring the N=1 SUSY on the branes which otherwise is violated by the non-vanishing D-terms arising from the VEV $\langle \mathbf{45}_\phi \rangle$, because of its non trivial profile (for more details, see [1]); 4) it is a heavy field, with a mass $\sim \mathcal{O}(M_{GUT})$, an important feature for solving the doublet-triplet splitting problem, as we are going to discuss.

The field $\mathbf{45}_H$ does not appear in the Yukawas, but it is responsible of breaking SU(5) down to the SM gauge group; it is an heavy field with mass $\sim \mathcal{O}(M_{GUT})$, also entering in the mechanism

[2] 5D N=1 SUSY, equivalent to 4D N=2 SUSY in the bulk, forbids to couple to **16** and $\mathbf{16}^c$ another chiral multiplet in the **45** representation, whose coupling would not be universal, but a 3×3 matrix in the flavour space, introducing much more parameters.

[3] In the original model [2], instead, higher dimension operators are necessary to correct the Yukawas, with Λ required to be $\mathcal{O}(M_{GUT})$, spoiling the consistency of the effective theory approach underlying the whole model.

solving the doublet-triplet splitting problem.
The $\mathbf{10}_H$ and $\mathbf{120}_H$ contains pairs of weak doublets that play the role of the MSSM Higgses, being responsible of thw EW symmetry breaking and giving mass to the charged SM fermions; the Yukawa couplings of charged fermions are then linear combinations of Y_{10} and Y_{120}. The combination of these two fields permits to distinguish the Yukawa couplings of down-type quarks from those of the charged leptons[4] [12], that otherwise would result equal because of the unified representation under SU(5): this choice permits to maintain viable the splitting of the profiles down to the SU(5) multiplets, without need to go further and distinguishing with respect to the SM representations (for which a similar mechanism of splitting through spontaneous symmetry breaking cannot be implemented).
The above Higgs content on the brane, with the distinction made between heavy and light fields, can address the doublet-triplet splitting problem through the missing partner mechanism [13]. As can be seen from the decomposition under the SM gauge group, the set of light fields ($\mathbf{10}_H$ and $\mathbf{120}_H$) contains three pairs of weak doublets and three pairs of color triplets, while the set of heavy fields ($\mathbf{126}_H$, $\overline{\mathbf{126}}_H$, $\mathbf{45}_H$) contains three pairs of triplets but only two pairs of doublets[5]. All the doublets and triplets of the different sectors get mixed with each other when $\mathbf{45}_H$ takes a VEV (from operators coupling these fields in the superpotential w_0, see [1] and [13]): the light triplets get mixed with the same number of heavy ones and all of them get a mass at the GUT scale, while in the case of doublets, all pairs get a mass at GUT scale but one linear combination of doublets remains massless because of the "missing partner" in the heavy sector: it is this combination to be identified as the MSSM Higgs doublets.

Explicitely, the Yukawa couplings of Dirac type fermions, entering Eq.(12), are given by:

$$\begin{aligned}
Y_u &= 2\sqrt{2}\alpha_1 Y_{10} + (-2\sqrt{2}\alpha_2 + \frac{2}{3}i\sqrt{6}\alpha_3)Y_{120} \, , \\
Y_d &= 2\sqrt{2}\bar{\alpha}_1 Y_{10} + (-2\sqrt{2}\bar{\alpha}_2 + \frac{2}{3}i\sqrt{6}\bar{\alpha}_3)Y_{120} \, , \\
Y_\nu &= 2\sqrt{2}\alpha_1 Y_{10} + (-2\sqrt{2}\alpha_2 - 2i\sqrt{6}\alpha_3)Y_{120} \, , \\
Y_e &= 2\sqrt{2}\bar{\alpha}_1 Y_{10} + (-2\sqrt{2}\bar{\alpha}_2 - 2i\sqrt{6}\bar{\alpha}_3)Y_{120} \, ,
\end{aligned} \tag{16}$$

where α_i and $\bar{\alpha}_i$ parameters are coefficients of the linear combinations of doublets describing the H_u and H_d Higgs fields of MSSM (up and down type), satisfying $\sum_{i=1}^{3}|\alpha_i|^2 = \sum_{i=1}^{3}|\bar{\alpha}_i|^2 = 1$. The Yukawa for RH neutrino, from the Majorana mass term, is simply $Y_R = Y_{126}$.
All these couplings, considering also the symmetry properties of some representations, introduce 35 real parameters (for more details, see [1]), to be added to the 4 real parameters describing the profiles. With this set of parameters we must fit 17 observables (fermion masses and mixing angles). Despite the large number of parameters, the fit is not trivial at all, since all the 35 parameters from the Yukawa sector are required to be anarchical and $\mathcal{O}(1)$ for naturalness: the

[4] This happens because $\mathbf{120}_H$ contains the Higgs doublets pair in different representations of SU(5), in particular the 45 and $\overline{45}$ of SU(5), that is known to couple to down-type quarks and charged leptons with different Clebsch-Gordan coefficients.
[5] The unequal content of doublets and triplets in the heavy sector arises from the 50, $\overline{50}$ of SU(5) residing in $\mathbf{126}_H$, $\overline{\mathbf{126}}_H$ which contain only triplets.

parameters with the leading role remain only the 4 describing the profiles (m_i, v_Φ), responsible of creating the hiararchies.

4. Viability test, naturalness test and predictions

With the above setup, we have performed the numerical fit for testing the viability of the model. As input data we used the experimental values of the fermion masses and mixing parameters extrapolated at the GUT scale in the MSSM. All the parameters of the Yukawas are restricted within the narrow range $0.5 \leq| Y_{ij} |\leq 1.5$. In a first approach, we have optimized all the free parameters, including the Yukawa entries, by the tecnique of χ^2 minimization (for technical details we refer the reader to [1]). A good fit results possible only for large value of $\tan\beta$, in agreement with the unification of bottom quark and tau lepton masses, resulting from the SU(5) representations. Then, fixing $\tan\beta = 50$, the fit was done for normal ordering (NO) and inverted ordering (IO) of the neutrino masses, obtaining in both cases an acceptable solution, with $\chi^2 \approx 0$ for NO and $\chi^2 \approx 5.8$ for IO. Anyway these solutions could still be fine-tuned in the Yukawa entries so, in order to test the naturalness of the $\mathcal{O}(1)$ Yukawa couplings, we proceed by taking a flat random distribution for all the Yukawa parameters in the range extablished above, and optimizing by χ^2 minimization only the remaining parameters, *i.e.* the Higgs mixing parameters and the set of bulk masses and VEV parametrizing the profiles. In fig. 1 we plotted the distribution of the minimal χ^2/ν, where ν is the number of independent degrees of freedom, obtained in the case of NO and IO. From this plot it results very clearly how the NO is favoured,

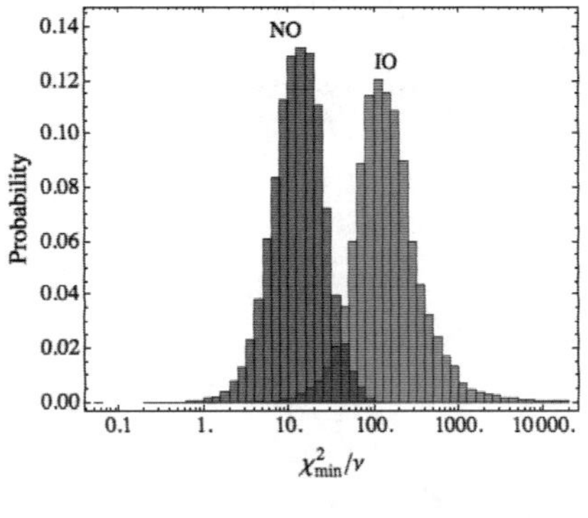

Figure 1. The distribution of minimized χ^2/ν for NO and IO neutrino masses, with $\tan\beta = 50$

having a distribution peaked on lower χ^2 values. In terms of goodness of fit characterized by the p-value, we find $p > 0.05$ for 2.2% of cases in the NO and only for 10^{-3}% in the IO. From this picture we conclude that the proposed model has an acceptable naturalness only for the NO case.

For the selected cases of NO with $p > 0.05$, which fit well the experimental data and respect naturalness, we find the distribution for some physical observables that can be predicted after the fit:

- the lightest neutrino mass, shown in fig. 2, results restricted to be $\lesssim 5$ meV, corresponding to a hierarchical neutrino mass spectrum. Unfortunately this prediction is far from the sensitivity of the current experiments.

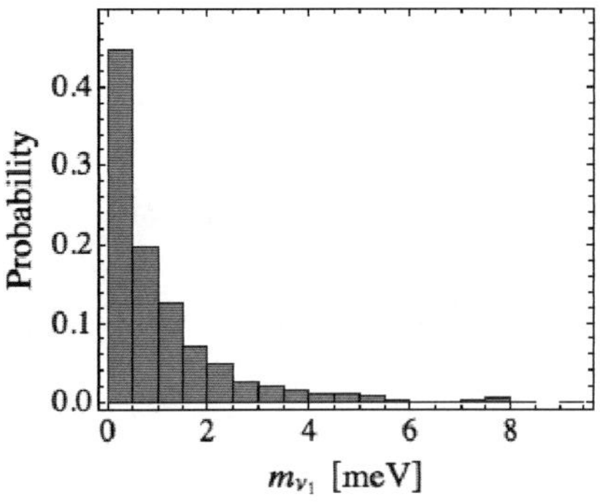

Figure 2. The distribution for the lightest neutrino mass (NO, $\tan\beta = 50$, selected cases with p-value> 0.05)

- the effective mass $m_{\beta\beta}$, shown in fig. 3, is predicted in the range 0.1-5 meV. Also this prediction is far from the sensitivity of the current experiments.

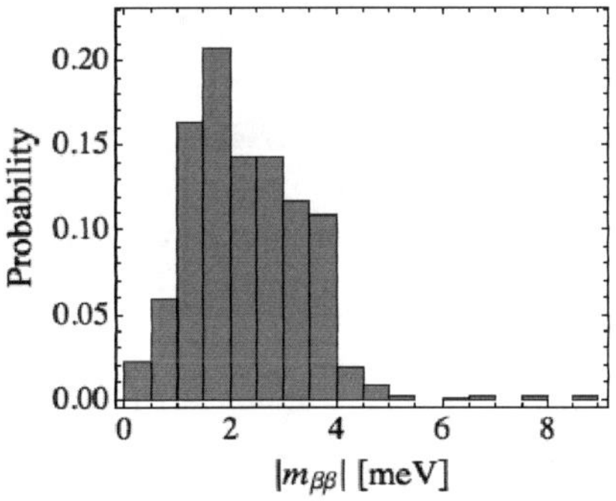

Figure 3. The distribution for the effective mass $m_{\beta\beta}$ (NO, $\tan\beta = 50$, selected cases with p-value> 0.05)

- the Dirac CP violating phase, shown in fig. 4, has an almost uniform distribution, allowing the entire range in CP phase.
- the RH neutrino masses, shown in fig. 5, result in an extremely hierarchical spectrum, $M_{N_1} \ll M_{N_2} < 10^9$ GeV. This kind of spectrum turns out to be incompatible with the models of standard thermal leptogenesis, both in a flavour independent or dependent scheme, requiring $M_{N_1} \gg 10^9$ or 10^{12} GeV $\gtrsim M_{N_2} \gtrsim 10^9$ GeV and $M_{N_1} \ll 10^9$ GeV.

5. Conclusion

In conclusion, we have demonstrated that a model realized on 5D SO(10) SUSY GUT can nicely address the flavour problem taking fundamental anarchical $\mathcal{O}(1)$ Yukawas and creating the observed hierarchies by introduction of only four parameters: three bulk masses describing the profiles along the extra dimension for the three generations of **16** (unified representation of all the

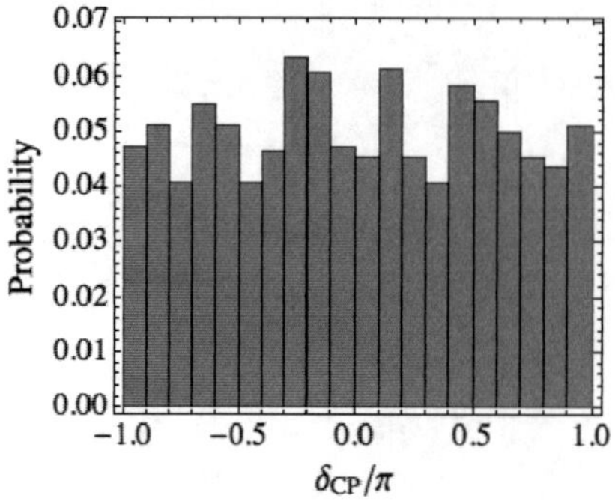

Figure 4. The distribution for Dirac CP violating phase (NO, $\tan\beta = 50$, selected cases with p-value> 0.05)

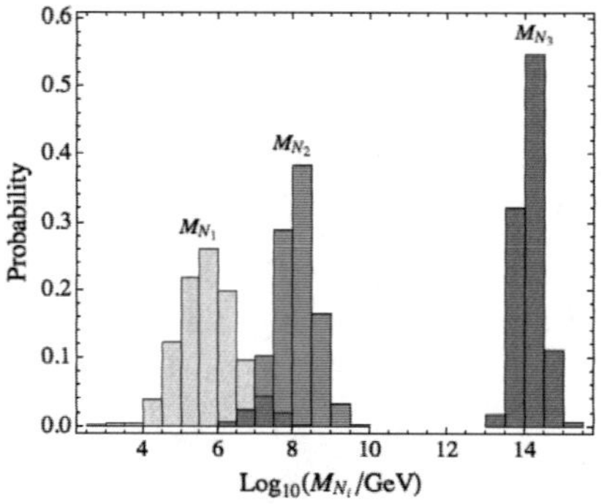

Figure 5. The distribution for the RH neutrino masses (NO, $\tan\beta = 50$, selected cases with p-value> 0.05)

SM fermions), plus a VEV responsible of splitting the profiles, within a mechanism of symmetry breaking, with respect to the SU(5) submultiplets. The scalar sector of the theory is chosen also to address the doublet-triplet splitting problem through the missing partner mechanism. It's significative that in this 5D context the MSSM Higgs fields can be chosen in the 10 and the 120 representations of SO(10), a combination that in 4D doesn't allow to fit the data well.

The viability of the model has been checked by fitting the the fermion masses and mixing angles and a test of naturalness has be performed, showing a clear preference for the case of NO in neutrino masses.

The model has some predictions for unobserved quantities, that result unfortunately not measurable in the current or next-future experiments. This makes the model not possible to be tested directly, but still falsifiable.

Let's finally remark that this model has been analyzed in the ideal limit of exact N=1 SUSY, neglecting soft SUSY breaking contributions with a characteristic scale in the range $1 \div 10$ TeV. It would be desirible, in a further project, to make the model more realistic by proposing a viable mechanism of N=1 SUSY breaking (possibly exploiting the extra dimension) and specifying the sparticle spectrum.

Aknowledgments
D. V. would like to thank the organizers of the DISCRETE 2014 conference for having offered the opportunity of giving this talk and for the excellent organization.

References
[1] Feruglio F, Patel K M and Vicino D 2014 *JHEP* **1409** 095 (*Preprint* 1407.2913)
[2] Kitano R and Li T j 2003 *Phys.Rev.* **D67** 116004 (*Preprint* hep-ph/0302073)
[3] Altarelli G and Blankenburg G 2011 *JHEP* **1103** 133 (*Preprint* 1012.2697)
[4] Joshipura A S and Patel K M 2011 *Phys.Rev.* **D83** 095002 (*Preprint* 1102.5148)
[5] C Froggatt H B N 1979 *Nucl.Phys.* **B147** 277
[6] Arkani-Hamed N, Gregoire T and Wacker J G 2002 *JHEP* **0203** 055 (*Preprint* hep-th/0101233)
[7] Pomarol A and Quiros M 1998 *Phys.Lett.* **B438** 255–260 (*Preprint* hep-ph/9806263)
[8] Barbieri R, Contino R, Creminelli P, Rattazzi R and Scrucca C 2002 *Phys.Rev.* **D66** 024025 (*Preprint* hep-th/0203039)
[9] Arkani-Hamed N and Schmaltz M 2000 *Phys.Rev.* **D61** 033005 (*Preprint* hep-ph/9903417)
[10] Mirabelli E A and Schmaltz M 2000 *Phys.Rev.* **D61** 113011 (*Preprint* hep-ph/9912265)
[11] Kaplan D E and Tait T M 2001 *JHEP* **0111** 051 (*Preprint* hep-ph/0110126)
[12] R Barbieri D V Nanopoulos G M 1980 *Phys.Lett.* **B90** 91
[13] Babu K, Gogoladze I, Nath P and Syed R M 2012 *Phys.Rev.* **D85** 075002 (*Preprint* 1112.5387)

Theoretical status of $B \to K^* \mu^+ \mu^-$: The path towards New Physics

Sébastien Descotes-Genon[1], **Lars Hofer**[2], **Joaquim Matias**[2] **and Javier Virto**[3]

[1] Laboratoire de Physique Théorique, CNRS/Univ. Paris-Sud 11 (UMR 8627) 91405 Orsay Cedex, France
[2] Universitat Autònoma de Barcelona, 08193 Bellaterra, Barcelona
[3] Theoretische Physik 1, Naturwissenschaftlich-Technische Fakultät, Universität Siegen, 57068 Siegen, Germany

E-mail: matias@ifae.es

Abstract. We summarize the status of the analysis of the 4-body angular distribution $B \to K^*(\to K\pi)\mu^+\mu^-$ using a basis of optimized observables. We provide a New Physics interpretation of the pattern of deviations observed in the 1fb^{-1} dataset with the coefficient of the semileptonic operator O_9 being the main responsible for this pattern. Also other scenarios involving more Wilson coefficients are briefly discussed. Further, we present a detailed description of each of the hadronic uncertainties entering our predictions and suggest possible tests for alternative hadronic explanations. Our most accurate SM predictions for the optimized observables in various bins including all uncertainties are also provided.

1. Introduction

After the long awaited discovery of the Higgs particle, it was assumed that this new scalar particle would be instrumental as a handle that could open a portal for New Physics (NP). Unfortunately, it was found that this scalar resembles very much the SM Higgs particle with SM-like couplings up to the present precision. In this sense it will possibly be a long-term task to find some clear evidence of NP in this direction. In the meanwhile, other portals for NP, like those constructed upon rare B decays can open new directions in these searches. Recently, it was found that one of those rare B decays, $B \to K^*(\to K\pi)\mu^+\mu^-$, seems particularly promising. Its full 4-body angular distribution provides sensitivity to many different Wilson coefficients, of electromagnetic, semileptonic and scalar operators (including their chiral counterparts). It thus will help to draw a picture of the flavour sector of the fundamental theory that lies beyond the SM. A large variety of angular observables [1–10] will allow to test some of the above-mentioned Wilson coefficients with an unprecedented precision.

The keypoint in this type of searches for NP is to find the cleanest possible procedure to extract the interesting short distance electroweak information disentangling it from any possible QCD polluting source. At leading order (LO) one can classify observables in form factor independent (FFI) and form factor dependent ones (FFD). Some time ago, a procedure [8] was proposed to factorize the information contained in the 4-body angular distribution into products of a set of FFI observables times FFD ones like the longitudinal polarization fraction

Journal of Physics: Conference Series **631** (2015) 012027 doi:10.1088/1742-6596/631/1/012027

(F_L). The short distance information can then be extracted from the less QCD-polluted FFI observables, which were named clean or optimized observables $P_i^{(\prime)}$ [9, 10]. The first of them, named P_1 or A_T^2 [2], was designed to detect the presence of right-handed currents (RHC). Soon after, other observables followed, like A_T^{re} [7] (or P_2 [8]) that correspond to the clean version of A_{FB}, and P_3 sensitive to the presence of NP weak phases. The set was completed by the P_i' observables with $i = 4, 5, 6, 8$. Particularly interesting are the observables P_4' and P_5' that represent the interference between the longitudinal amplitude of the K^* with the parallel and the perpendicular one, respectively. In these last observables at low q^2 (q^2 is the square of the invariant mass of the dilepton pair) but in the region above 5-6 GeV^2, the semileptonic operators dominate opening an excellent window to test them. In parallel, the observable P_2 testing the interference between perpendicular and parallel amplitudes, exhibits a window of sensitivity to these semileptonic coefficients in a different region, namely at the position of its maximum around 2 GeV^2 [35] and at the position of its zero near 4 GeV^2.

In 2013 in a ground-breaking effort going beyond traditional analyses (studying only F_L, A_{FB} and the branching ratio) and borrowing techniques from different fields, LHCb presented the first analysis of this set of optimized observables [12]. A first interpretation of this measurement presented in [13] pointed out a *coherent pattern* of small and large tensions in different observables, mainly in P_2 (and A_{FB}) and P_5' that could be explained under the hypothesis that the coefficient C_9 of the semileptonic operator and possibly also the coefficient C_7 of the electromagnetic operator receive a NP contribution.

The main outcome of the analysis in Ref. [13] based on the data from the observables

i) $P_{1,2}$, $P'_{4,5,6,8}$, A_{FB} of $B \to K^*\mu^+\mu^-$ in the q^2 bins [0.1,2], [2,4.3], [4.3,8.68], [14.18,16], [16,19],

ii) radiative and dileptonic B decays: $\mathcal{B}_{B\to X_s\gamma}$, $\mathcal{B}_{B\to X_s\mu^+\mu^-}$, $\mathcal{B}_{B_s\to\mu^+\mu^-}$, $A_I(B \to K^*\gamma)$,

was to find that the deviations from the SM values of the Wilson coefficients, $C_i^{\rm NP} = C_i - C_i^{\rm SM}$, at 68.3% C.L. lie within the following ranges:

$$\begin{aligned} C_9^{\rm NP} &\in [-1.6, -0.9], & C_7^{\rm NP} &\in [-0.05, -0.01], & C_{10}^{\rm NP} &\in [-0.4, 1.0], \\ C_9' &\in [-0.2, 0.8], & C_7'^{\rm NP} &\in [-0.04, 0.02], & C_{10}' &\in [-0.4, 0.4]. \end{aligned} \tag{1}$$

Here, $C_9^{\rm NP}$ is consistent with zero (SM) only above 3σ, $C_7^{\rm NP}$ is consistent with the SM at 2σ, while the rest are consistent with the SM already at 1σ. In this solution one observes that $C_9^{\rm NP} + C_9' < 0$, $C_7^{\rm NP} < 0$ and $C_{10}^{\rm NP}$ seems to show a mild preference for positive values due to $\mathcal{B}_{Bs\to\mu^+\mu^-}$. Remarkably, different groups [14–16] using different observables, techniques and statistical approaches also confirmed the large contribution to $C_9^{\rm NP}$ of order 25% with respect to the SM value. It was also claimed by some groups [14, 18] using, for instance, the low-recoil 1fb^{-1} data on $B^+ \to K^+\mu^+\mu^-$ [17] that at the same time also the chirally flipped semileptonic coefficient was getting a large NP contribution such that $C_9' + C_9^{\rm NP} \simeq 0$ in order to have a SM-like BR for the $B^+ \to K^+\mu^+\mu^-$ mode. However, this scenario was in conflict [13, 19] with the anomaly in P_5' (4σ deviation in the third bin of P_5') that required C_9' to be zero or small such that $C_9^{\rm NP} + C_9' < 0$ in order not to increase this anomaly. Fortunately an updated result from LHCb with 3fb^{-1} data on $B^+ \to K^+\mu^+\mu^-$ and $B^0 \to K^0\mu^+\mu^-$ [20] has solved this tension since both modes point in the low-q^2 and large-q^2 region to the solution $C_9^{\rm NP} + C_9' < 0$ in agreement with the anomaly in P_5'.

In summary, the present situation exhibits a coherent pattern of deviations in $B \to K^*\mu^+\mu^-$, $B^+ \to K^+\mu^+\mu^-$, $B^0 \to K^0\mu^+\mu^-$ but also $B_s \to \phi\mu^+\mu^-$ and the radiative and dileptonic observables mentioned above, pointing to the semileptonic coefficient C_9 as the main responsible, with all other coefficients not deviating significantly from zero and playing a less important role. It is important to emphasize already at this point that this conclusion obviously does not preclude the possibility that the new data can switch on with enough significance other Wilson coefficients.

Our main goal here will be to discuss all new corrections that we have included in our predictions [21] and in parallel analyse alternative explanations that has been raised to explain some of the deviations suggesting possible tests or cross checks to discriminate them from a NP solution.

The outline of the proceedings is the following: in Section 2 we discuss the theoretical framework at low q^2 where the strongest deviations from the SM have been found by LHCb. In Section 3 we detail our method to include factorizable and non-factorizable (including charm-loop) corrections. We also suggest some tests to check the presence of huge charm-loop contributions and comment on some of the possible consistency tests that can be done on data. Section 4 is devoted to S-wave pollution. In Section 5 we comment on and explore other scenarios. Finally we conclude in Section 6. We provide in an appendix our most accurate predictions for the optimized basis of observables in various bins and in two different parametrizations including all corrections discussed before based on the method developed in [21].

2. Theoretical Framework at low-q^2

In this section we will focus on the theoretical framework used to describe the large-recoil region where the strongest deviations have been observed. Still for completeness we will also comment at the end of this section on some aspects concerning the size of possible uncertainties that might affect either the large or the low recoil region.

A keypoint in the evaluation of the angular observables is to keep track of the correlations among form factors. In the large-recoil region there are basically two approaches:

- "Improved QCDF approach": In this framework [10] the large-recoil symmetries between form factors are used to implement the dominant correlations among them. This is a transparent and general approach that is easy to cross-check and valid for any form factor parametrization (e.g. for the light-cone sum rules parametrizations [22] or [23]). The symmetries allow the 7 form factors to be written in terms of only two, so called soft form factors $\xi_{\perp,\parallel}$ [24]:

$$\frac{m_B}{m_B+m_{K^*}}V(q^2) = \frac{m_B+m_{K^*}}{2E}A_1(q^2) = T_1(q^2) = \frac{m_B}{2E}T_2(q^2) = \xi_\perp(E),$$

$$\frac{m_{K^*}}{E}A_0(q^2) = \frac{m_B+m_{K^*}}{2E}A_1(q^2) - \frac{m_B-m_{K^*}}{m_B}A_2(q^2) = \frac{m_B}{2E}T_2(q^2) - T_3(q^2) = \xi_\parallel(E). \quad (2)$$

  Then the soft form factors can be computed in a specific parametrization. In [21] we employed this approach considering all symmetry breaking corrections to the relations Eq.(2). Our predictions take into account factorizable and non-factorizable α_s corrections computed within QCDF [25–27], as well as factorizable and non-factorizable power corrections, discussed in Sec.3, including charm-loop effects. The analytic structure of the large-recoil correlations led to the construction of the FFI observables P_i and P_i^{CP} [2,8–10] that exhibit a sensitivity to the soft form factor suppressed in α_s or Λ/m_b. All bins are traditionally considered in our analysis.
- "Ball-Zwicky Form Factor approach": Here a specific set of full form factors (Ball-Zwicky [23]) is used. Factorizable α_s and factorizable power corrections are automatically included with correlations associated to this particular parametrization. All other corrections have to be included and/or estimated exactly as in the previous approach, non-factorizable α_s correction taken from QCDF (here also soft form factors are necessary), non-factorizable power corrections (see [29]) and charm-loop effects. This approach has been employed in [14] where the basis S_i and A_i [28] is used and only bins below 6 GeV^2 are considered.

Both approaches are useful and complementary, should converge and give comparable results and error sizes. Of course, both methods can be used to compute both type of observables P_i and S_i. For a comparison one can apply the general first approach (QCDF) to the particular form factor parametrization (BZ) used in the second one and the observables studied in the second approach [29]. We will show an example of this comparison in Sec. 3.1.

In the low-q^2 region, where we focused the previous discussion, the large set of observables measured provide a nice cross-check of the observed deviations. For this reason we think it is important to consider all available bins. Sometimes it is stated that in bins beyond 6 GeV^2 soft-gluon corrections to the virtual charm-loop could have a large impact. However, there are two important remarks: first, the partial soft-gluon charm correction computed in Ref. [22] is found to be positive implying that its effect is to enlarge and not reduce the anomaly, requiring an even larger NP contribution to compensate for it. Second, it is manifest in the plots of Fig. 5 and 7 of Ref. [22] that the impact is moderate up to 8 GeV^2. For some amplitudes (M_3) it is even smaller than the impact at very low q^2. In our predictions we take this calculation as an estimate of the effect and allow both signs for this correction as explained later on. In any case it is advisably to cut the experimental data at 8 GeV^2 where the impact of the charm-loop contribution calculated in [22] is still moderate.

In the large-q^2 region, one of the main sources of uncertainty and model dependence is related to the treatment of the observed set of resonances by LHCb in the data of the partner channel $B^+ \to K^+\mu^+\mu^-$. This observation prevents one from taking small bins afflicted by the resonance structures. In [30] a quantitative estimate of duality violation is given. Unavoidably, one needs to use a model for this estimate, still the result is that the integrated single low recoil bin gets a duality-violation impact of a few percent (5 % in [31] or 2% in [30]). On top of that, there is certainly an impact from the choice of the starting point for the integrated bin that might coincidence with the position of a resonance. It is thus under debate how large this violation should be taken and where to start the long bin. In [30] the estimate for $B \to K\mu^+\mu^-$ is given for $q^2 \geq 15$ GeV^2 while first data on $B \to K^*\mu^+\mu^-$, where a similar problem will arise, starts at 14.18 GeV^2. An interesting exercise to show how sensitive and dependent on the starting point the result is, was done in [16]. There lattice form factors were used to analyze the low-recoil region for $B \to K^*\mu^+\mu^-$ and $B_s \to \phi\mu^+\mu^-$. It was found that if the two bins were taken the best fit value was given by a solution with C_9^{NP} negative and C_9' positive (both with a significance slightly above 1 σ), while if only the last bin was used (starting at 16 GeV^2) the result on C_9^{NP} was practically unaffected while C_9' became consistent with zero. In this sense again it would be interesting that the new data on $B \to K^*\mu^+\mu^-$ starts at 15 GeV^2 or above as it has been done for the 3 fb^{-1} data of $B \to K\mu^+\mu^-$.

3. Methodology beyond QCDF and alternative explanations

In this section we discuss in detail the method to include factorizable and non-factorizable power corrections (including charm-loops) that we used for our predictions [21]. We also provide some tests to show the robustness of the predictions and point out possible ways to check, using the new data, some alternative explanations (huge charm loops) proposed to explain one of the deviations. We also discuss model independent relations between observables that allow to perform clear tests on the data, assuming the absence of new CP violating phases beyond the SM.

3.1. Factorizable power corrections

The decomposition of a full form factor contains in general three pieces,

$$F(q^2) = F^{\rm soft}(\xi_{\perp,\|}(q^2)) + \Delta F^{\alpha_s}(q^2) + \Delta F^{\Lambda}(q^2), \tag{3}$$

Journal of Physics: Conference Series **631** (2015) 012027 doi:10.1088/1742-6596/631/1/012027

where the first piece includes the soft form factors that implement automatically the dominant correlations among form factor given in Eq.(2), $\Delta F^{\alpha_s}(q^2)$ represents the known QCDF correction induced by hard gluons, and $\Delta F^{\Lambda}(q^2)$ are the factorizable power corrections of $\mathcal{O}(\Lambda/m_b)$. Following [32] we parametrize the latter correction, which is expected to be small compared to hadronic uncertainties, as an expansion in q^2:

$$\Delta F^{\Lambda}(q^2) = a_F + b_F \frac{q^2}{m_B^2} + c_F \frac{q^4}{m_B^4} + ... \tag{4}$$

The decomposition (3) is not unique, one can always reshuffle the contributions among the different terms. It is necessary to fix a renormalization scheme, namely define $\xi_{\perp,\|}$ in terms of full form factors. Our implementation of this correction [21] makes particularly emphasis on two aspects. The first aspect is the importance of respecting the correlations among *all* the parameters of Eq.(4) of different form factors. There are exact kinematical relations at $q^2 = 0$, like $T_1(0) = T_2(0)$ implying $a_{T1} = a_{T2}$, but also correlations among the a_{F_i}, b_{F_i} and c_{F_i} coefficients that originate from the definition of the soft form factors in terms of the full form factors. For this reason it is very important to always work consistently within a form factor parametrization at a time not mixing them. The second aspect is the relevance of choosing the most appropriate scheme. The choice of scheme determines which part of the $\mathcal{O}(\alpha_s, \Lambda/m_b)$ correction will remain in the function ΔF^{α_s} or ΔF^{Λ} and which part will be absorbed in to the $F^{\rm soft}(\xi_{\perp,\|}(q^2))$. While the perturbative ΔF^{α_s} correction can be computed explicitly in each scheme, the lack of a full control of the correlations among the errors of ΔF^{Λ} induces a scheme dependence at order $\mathcal{O}(\Lambda/m_b)$ on the otherwise scheme independent observables. In this situation a proper choice of scheme is mandatory in such a way that the effects of unknown power corrections get absorbed as much as possible into the soft form factors in order not to artificially inflate uncertainties . A simple example illustrates this point: assume an observable that only depends on a single form factor F_1. A good scheme would be the one that takes this form factor directly as input, so that all power corrections are absorbed and would not appear from the beginning. A bad scheme would fix as input another form factor F_2, and in the process of relating it to F_1 unknown power corrections would be introduced increasing the uncertainty of the result artificially. Our choice to determine the soft form factor $\xi_\perp(q^2)$ is given by

$$\xi_\perp(q^2) \equiv \frac{m_B}{m_B + m_{K^*}} V(q^2). \tag{5}$$

This definition eliminates all corrections to the form factor V leading to $\Delta V^{\alpha_s}(q^2) = \Delta V^{\Lambda}(q^2) = 0$. A different choice consists in defining $\xi_\perp(q^2)$ from T_1 like in [32] leads to larger uncertainties in P_5'.

The soft form factor $\xi_\|$ can be defined as

$$\xi_\|(q^2) \equiv \frac{m_B + m_{K^*}}{2E} A_1(q^2) - \frac{m_B - m_{K^*}}{m_B} A_2(q^2), \tag{6}$$

as done for example in Ref. [25]. This definition minimizes power corrections in the form factors $A_{1,2}$ by correlating $\Delta A_1^{\alpha_s}(q^2), \Delta A_1^{\Lambda}(q^2)$ with $\Delta A_2^{\alpha_s}(q^2)$ and $\Delta A_2^{\Lambda}(q^2)$. Another possible determination of this form factor uses the form factor $A_0(q^2)$ [32], albeit perfectly acceptable as a choice this FF does not enter by construction any of the optimized observables.

In our computation we first determine the parameters a_F, b_F and c_F such that the central value of the full form factor is exactly reproduced. We perform a fit to these parameters and keep the correlated (non-zero) central values obtained: $\hat{a}_F, \hat{b}_F, \hat{c}_F$. We have applied this procedure to two different form factor determinations KMPW [22] and BZ [23], taking always one at each time to respect correlations. This induces a shift in the central values as compared to the previous predictions in [13]. For the errors associated to the factorizable power corrections we take $\Delta F^\Lambda \sim F \times \mathcal{O}(\Lambda/m_b) \sim 0.1F$ which amounts to assign an error of order 100% with respect to the fitted central values $\hat{a}_F, \hat{b}_F$ and $\hat{c}_F$. Finally, we vary the parameters within the range $\hat{a}_F - \Delta\hat{a}_F \leq a_F \leq \hat{a}_F + \Delta\hat{a}_F$ and similarly for b_F and c_F. Note that our error estimate for the factorizable power corrections is solely based on dimensional arguments and exhibits a marginal dependence on the form factor parametrization used.

Finally to close this section on factorizable power corrections we compare the results obtained using the two different approaches (QCDF versus Ball-Zwicky FF)[1] . Let us take, for instance, the bin [2,4.3] for the observable S_5 using the improved QCDF approach for the particular case of the BZ parametrization. Our prediction from [21] for this observable is $S_{5\,BZ}^{[2,4.3]} = -0.17 \pm 0.04$. Now we compare this prediction with the one in the recent paper [29] obtained using the different method based on the correlated BZ form factors. The result there is $S_5^{[2,4.3]} = -0.17 \pm 0.04$ exactly the same.[2] This perfect agreement is remarkable given the different treatment of the error associated to factorizable power corrections. However, if one does the same exercise using the value of P_5' in the same bin given in [32] and transforming it into a prediction for S_5 *assuming zero error* from F_L, one gets $S_{5\,JC}^{[2,4.3]} = -0.11^{+0.13}_{-0.11}$ with the error being a factor of 3 larger as compared to the predictions from [29] and [21]. We have also done the same exercise for the [1-6] bin of S_5 with identical outcome (see [33] for further details).

3.2. Non-Factorizable power corrections

Concerning non-factorizable contributions, we will separate the non-perturbative contributions appearing at subleading order in the $1/m_b$ expansion into two types. First we will single out the hadronic contribution not related to Wilson coefficients steaming from the piece of the three hadronic $\mathcal{T}_i(q^2)$ that parametrize the matrix element $\langle K^*\gamma^*|H_{eff}|B\rangle$. We obtain it taking the limit $\mathcal{T}_i^{\rm had} = \mathcal{T}_i|_{C_7^{(\prime)}\to 0}$. Finally, we multiply each of these amplitudes, that will serve as a normalization, with a complex q^2-dependent factor

$$\mathcal{T}_i^{\rm had} \to \big(1 + r_i(q^2)\big)\mathcal{T}_i^{\rm had}, \tag{7}$$

where

$$r_i(s) = r_i^a e^{i\phi_i^a} + r_i^b e^{i\phi_i^b}(s/m_B^2) + r_i^c e^{i\phi_i^c}(s/m_B^2)^2. \tag{8}$$

We define our central values as the ones with $r_i(q^2) \equiv 0$, and estimate the uncertainties from non-factorizable power corrections by varying $r_i^{a,b,c} \in [0, 0.1]$ and $\phi_i^{a,b,c} \in [-\pi, \pi]$ independently, corresponding to a $\sim$ 10% correction with an arbitrary phase. The uncertainties for each observable are then obtained by performing a random scan and taking the maximum deviation from the central values to each side, to obtain upward and downward error bars. A separated discussion of the treatment of non-factorizable contributions coming from charm loops is given in next section.

3.3. Charm-loop pollution

Part of the $c\bar{c}$-loop contributions have been already included in the non-factorizable contributions (hard-gluon exchange). The remaining long-distance contributions from $c\bar{c}$ loops are still under

[1] We will not include charm-loop effects in this comparison in our prediction because we only computed them for KMPW and not for BZ. In the same way we will not take it into account in [32]

[2] Notice that if we use the KMPW parametrization our prediction becomes $S_{5\,KMPW}^{[2,4.3]} = -0.17^{+0.08}_{-0.06}$ where the difference comes as it is well known from the larger hadronic form factor errors in KMPW, while the error associated with factorizable power corrections is similar in size as with BZ form factors. This is the parametrization we take for our predictions.

debate. We will rely for these corrections on the partial computation [22]. It is important to remark that the soft-gluon contribution of [22] coming from 4-quark and penguin operators induces a *positive* contribution to $C_9^{\rm eff}$ whose effect is to enhance the anomaly. We are interested only in the long-distance contribution $\delta C_9^{\rm LD}(q^2)$, so we subtract the perturbative LO part and include the shift due to a different reference value for m_c (see [21] for more details on the procedure used). In order to be conservative, and in particular given the discussion on the sign of this contribution, we use the result of [22] as an order of magnitude estimate, including it as

$$\mathcal{C}_9 \to \mathcal{C}_9 + s_i \delta C_9^{\rm LD}(q^2) \ , \tag{9}$$

with s_i scanned in the range $[-1, 1]$ individually for each amplitude M_i, with $i = 1, 2, 3$ (remember that the result of the calculation is $s_i = 1$).

In an attempt of explaining mainly P_5', the possibility of a huge charm contribution has been proposed [34]. As will be explained below the fundamental problem of this explanation is to focus on a single observable, while it is important to consider the global picture and the combined input from all observables that can put in serious difficulty this explanation. At a practical level this explanation transfers the responsibility of the deviation from $C_9^{\rm NP}$ to a modified $Y(q^2)$ (with a prefactor) and adds a similar function (and prefactor) also to C_9'. Here $Y(q^2)$ is the 4-quark contribution to the effective Wilson coefficient $C_9^{\rm eff}$. While $C_9^{\rm NP}$ is a global and constant effect entering all bins, the modified $Y(q^2)$ is a q^2 dependent function that is expected to give a small contribution around 2 GeV^2. Several examples of the impact of an hypothetical parametrization for this function were presented in [34]. All the illustrative examples shown there are disfavoured for some observable, either P_2, P_5' or P_1, this shows the importance of performing a global analysis.

It is precisely the question of a global versus a local q^2 dependent effect which discriminates between a NP scenario and a charm-loop effect. For instance, as explained in [35] the position of the maximum of the observable P_2 is practically unaffected by a hypothetical large charm contribution if one takes into account the experimental constraint on the zero of P_2 (the same as A_{FB}). This implies that while the shift in the position of the zero of P_2 could be explained with this hypothetical huge contribution, this is not the case for the maximum. On the contrary, a C_9^{NP} contribution would shift in a precise way *both* the position of the maximum and the position of the zero of P_2. Present data seems to prefer a shift of both maximum and zero in contrast to the hypothetical large charm contribution. If new data confirms the double shift, the claim about a huge charm-loop contribution will be challenged.

A second fundamental problem of a large charm explanation is the recently measured R_K [36]. It is clear that, given its universal character that does not distinguish muons from electrons, the charm loop cannot explain R_K, while the same NP scenario that is favoured by the anomaly in $B \to K^*\mu^+\mu^-$ can explain it if NP couples only to muons [37]. Finally, the strong q^2 dependence of the charm loop proposed in [34] tends to generate more than one zero in the region of low q^2 for P_2 and P_5' (see plots of Fig.12 in [34]). If no second zero is found, many of those solutions will be severely constrained. If $P'_{5[6,8]}$ is equal or above $P'_{5[4,6]}$, it would provide a positive test in favour of this alternative explanation, and an important negative test otherwise.

3.4. Large statistical fluctuation and consistency tests on data

Another possible explanation of the strong deviation in the third bin of P_5' is the possibility of a purely statistical fluctuation or an isolated experimental problem in that bin. This is possible and very plausible if one decouples the third bin of P_5' from tensions in the rest of observables. The aim behind the paper [38] was to explore if it is possible to establish a connection between different observables under reasonable and well defined assumptions, but independently of any effective Hamiltonian computation, just relying on symmetry arguments of the distribution. If such a relation exists and is confirmed by experimental data, the anomaly in P_5' cannot be considered as an isolated problem.

A simple but powerful relation obtained in [38] under the assumption of real Wilson coefficients shows that

$$[P_4'^2 + \beta^2 P_5'^2]_{q^2=q_0^2} = 1 + \eta(q_0^2), \tag{10}$$

where q_0^2 is the position of the zero of P_2 and $\eta(q_0^2) \sim 10^{-3}$ if no RHC are present. This equation establishes a non-trivial link between the zero of P_2 and the anomaly in P_5'. If the position of the zero of P_2 is at a higher q^2 value than predicted by the SM ($q_{0SM}^2 = 4$ GeV^2), then taking into account that the measured P_4' has a strong positive slope towards 1, the value of $P_5'(q_0^2)$ should be close to 0. This implies that the small value of P_5' in that region and the shift of the zero of A_{FB} are two sides of the same coin. This establishes a strong link between two observables based on symmetries [8, 39].

Also the more general relation [38]

$$P_2 = \frac{1}{2}\left[P_4'P_5' + \frac{1}{\beta}\sqrt{(-1 + P_1 + P_4'^2)(-1 - P_1 + \beta^2 P_5'^2)}\right], \tag{11}$$

together with exact bounds like [38]

$$P_5'^2 - 1 \leq P_1 \leq 1 - P_4'^2, \tag{12}$$

allows to perform several tests on the data. Let us briefly mention some of them:

i) A comparison between the measured value of P_2 and the value obtained from Eq.(11) using the measured values of P_1 and $P_{4,5}'$ show an excellent agreement for the second bin (see Fig.1). The third bin shows a tension of 2.4σ between the measured value of P_2 and the one obtained from Eq.(11). This implies that data on the third bin should change either in P_2 or in P_1, $P_{4,5}'$ to get a better consistency between the measured and the obtained value from the other measurements.

ii) The exact bound Eq.(12) implies that, if one writes $P_4' = 1 + \delta$ in the third large recoil bin, then $P_1 \leq -2\delta + \mathcal{O}(\delta^2)$. If P_1 is positive (as data suggest) in that third bin, δ should be negative and P_4' be slightly below 1. This is consistent with data given the large error bars, although the higher positive values of P_1 would be in tension with that bound.

iii) Finally, the first low-recoil bin [14.18-16.00] GeV^2 shows a large discrepancy of 3.7σ between the measured value for P_2 and the value obtained from the data on P_1, P_4' and P_5'. In this case, it is very plausible that the responsible of this disagreement can be traced to the value of P_4' in that bin.

A simple consequence for data in the third bin, pointed out in [21] (see Appendix A3), can be easily found by combining i) and ii). P_2 has a zero at q_0^2,implying that $P_2 = -\epsilon$ (with $\epsilon > 0$) for $q^2 > q_0^2$, and from Eq.(11) it follows that

$$P_5' < -2\frac{\epsilon}{1+\delta}. \tag{13}$$

This implies that one should naturally expects P_5' to be below P_2 in that region, and this is the reason for the tension of 2.4σ found in Fig.1. Consequently we expect that new data should rearrange the third bins of P_2 and P_5' making more manifest the ordering of these two observables.

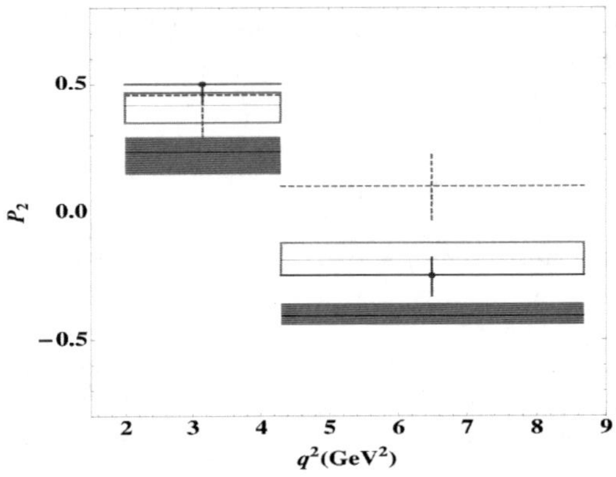

Figure 1. Comparison between the measured value of P_2 (blue cross) and the value obtained from the measurements of P_1, $P'_{4,5}$ using Eq.(11) (green cross). Gray band is SM.

4. S-wave pollution

Another source of pollution is due to events from the S-wave decay $B \to K_0^*(\to K\pi)\mu^+\mu^-$ where K_0^* is a scalar resonance and its interference with the P-wave decay $B \to K^*(\to K\pi)\mu^+\mu^-$. In Ref. [40] a detailed and complete calculation of the S-wave background was performed and it was concluded that any observable will unavoidably suffer from its pollution. While this conclusion is correct in the case of uniangular distributions, it does not apply to full or folded distributions where the P- and the S-wave parts can be separated according to their different angular dependence. As shown in Ref. [41], S-wave pollution can easily be avoided for the $P_i^{(\prime)}$ observables if folded distributions are used instead of uniangular ones. A discussion of the experimental implications of the S-wave contribution was presented in Ref. [42] (see also Ref. [43]). A parametrization of these terms were given in [10] together with a set of bounds on their size. Recently, it has been shown in [35] using the symmetries of the distribution that only 4 (out of 6) of such parameters are independent and the explicit correlations among the over-complete six observables have been provided.

5. Z' particle, other scenarios and the importance of considering all bins

In [13] the existence of a Z' boson was proposed as a possible explanation of the large negative contribution to C_9, with couplings $\Delta_R^{sb} \sim 0$ (in the notation of [44,45]) not to give contributions to C'_9, with Δ_L^{sb} having the same phase as $V_{tb}V_{ts}^*$ to avoid large contributions to the B_s mixing phase ϕ_s, and a left-right symmetric coupling to muons $\Delta^\mu = \Delta_L^{\mu\mu} = \Delta_R^{\mu\mu}$ not to contribute to C_{10}. Finally, the constraint coming from Δm_s fixes the flavour-changing coupling Δ_L^{sb}. A particle with couplings to muons Δ^μ of order 0.1-0.2 and with a mass around 1-2 TeV would give the right contribution to C_9. Of course, this is an ad-hoc model and moreover, some of the coefficients that were taken to be zero in the first dataset may be switched on with more data. Later on there have been several attempts to embed a Z' particle inside specific models (see [46–49]).

Recently, using a large set of observables it was confirmed in [29] that a solution with $C_9^{\rm NP} < 0$ gives the best fit. In their analysis for the $B \to K^*\mu^+\mu^-$ mode the authors of [29] do not use the long [1-6] bin which has a lower sensitivity to some Wilson coefficients but smaller bins (two out of the three possible bins). Another interesting outcome of this work [29], is the proposal of a second possible solution. There it is found that $C_9^{\rm NP} = -C_{10}^{\rm NP}$ could also provide an explanation (with only slightly worse χ^2 as compared to a fit where only $C_9^{\rm NP}$ is switched on, see Table 2 in [29]) with all other coefficients switched off. This solution is an interesting possibility from a model-building point of view, but also because it would introduce NP in a cleaner coefficient C_{10} (notice that this coefficient cannot receive any charm contribution). More in general, it has been found in [50], under the assumption that the scale of NP is well above the electroweak scale, that if one imposes that the operators of the effective lagrangian originate from manifestly $SU(3)_c \times SU(2)_L \times U(1)_Y$ invariant high-scale operators, in the process of integrating the heavy degrees of freedom a set of relations between the coefficients of the effective operators arises at low scale. In particular, no tensor contribution are generated and scalar and pseudoscalar coefficients are correlated. The remaining semileptonic coefficients are still independent in general, but functions of the coefficients of the high scale operators. If some of those coefficients of the dimension 6 operators vanish, the coefficients at the low-scale would become correlated.

We have performed an exploratory scan analysis of the scenario $C_9^{\rm NP} = -C_{10}^{\rm NP}$, considering all three bins of the relevant observables P_1, P_2, $P'_{4,5}$ together with $\mathcal{B}_{Bs\to\mu^+\mu^-}$, and taking 2σ for the theory errors and enlarging the experimental errors of the third bins to 2σ to cover possible changes in the data (see Section 3.4 and Fig. 1). The result of this very preliminary analysis is:

- If only the first two bins of $B \to K^*\mu^+\mu^-$ for the relevant observables (and $B_s \to \mu^+\mu^-$) are included we find points that fulfill the constraints

$$C_9^{\rm NP} = -C_{10}^{\rm NP} < 0, \quad C'_i = 0, \tag{14}$$

  confirming the result in [29] using our method and observables. Notice that this only proves that points exist that fulfill the constraints, but we did not test the quality of the fit for this exploratory scan. Of course, one should check if those points survive constraints from other observables like $B \to K\mu^+\mu^-$ as it was done in [29].
- On the other hand, if also the third bins for the relevant observables are included in the scan (not included in [29]), all the previous points disappear. The reason is easy to understand: in the same way as a large $C'_9 > 0$ substantially worsens the anomaly (3rd bin) in P'_5, also a positive $C_{10}^{\rm NP}$ tends to enhance the anomaly. This suggests the need to switch one other coefficients like C'_9, C'_{10}.

The new 3 fb^{-1} data on $B \to K^*\mu^+\mu^-$ could easily confirm or erase and draw completely new scenarios.

6. Conclusions and Outlook

Using a basis of optimized observables LHCb released in 2013 the 1 fb^{-1} data of $B \to K^*\mu^+\mu^-$ which exhibited a set of coherent tensions focused mainly in the observables P_2 and P'_5. The analysis in [13] using all large-recoil and low-recoil bins pointed clearly to a large negative NP contribution to the Wilson coefficient C_9, and with less significance also to a negative contribution to C_7, while all other coefficients were consistent with the SM. Later on, analyses done by other groups using different observables and methods [14–16, 29] and including the new 3 fb^{-1} data of $B \to K\mu^+\mu^-$ and $B_s \to \phi\mu^+\mu^-$ confirmed this general pattern.

Here we have discussed additional subleading corrections that we included in the theoretical predictions given in [21] to confront the new data. Particular emphasis has been put on factorizable and non-factorizable (specially coming from charm loops) power corrections. It is important to control these factorizable power corrections because they affect all P_i and all S_i observables if computed from QCDF approach as done in [21, 32]. We have given details on our method to include the factorizable power corrections based on model-independent correlations and using an adequate scheme to define soft form factors. Moreover, we have explicitly shown using the example of the observable S_5 that the predictions obtained using two different methods to deal with those corrections [21] and [29] (including correlations in a specific form factor parametrization or using symmetries to include the dominant ones together with dimensional arguments) are in excellent agreement.

We suggest simple tests to be done on the new data to test an alternative explanation that was suggested to explain the P_5' anomaly by the presence of huge charm-loop contributions. The simplest one consists in a comparison between the bins [4,6] and [6,8] of P_5'. Moreover, the confirmation of the R_K tension that could be explained simultaneously with the anomaly in $B \to K^*\mu^+\mu^-$ within the same NP scenario [37] (with couplings only to muons) but not via a large charm-loop contribution, would further challenge this alternative explanation.

We have also presented tests to perform on data, like comparing the measured value of P_2 with the one obtained from the measured values of P_1, $P_{4,5}'$ that indicates that future data should evolve such that the order of the bins between P_5' and P_2 in the region above 4 GeV^2 should flip, with P_2 moving above P_5'.

Finally, we have done an exploratory scan using only a subset of relevant observables to see if we could find solutions consistent with the scenario $C_9^{\rm NP} = -C_{10}^{\rm NP}$ and $C_9' = C_{10}' = 0$ recently proposed in [29]. While we confirmed that this scenario is perfectly viable if only the first two bins in $B \to K^*\mu^+\mu^-$ are taken into account (in agreement with [29]), we could not find solutions once the third bins are included.

We expect/hope that the new data can help to find the next step in the path towards NP and start switching on with sufficient significance other Wilson coefficients besides $C_9^{\rm NP}$.

Acknowledgements

J.V. is funded by the Deutsche Forschungsgemeinschaft (DFG) within research unit FOR 1873 (QFET). J.M. and L.H. acknowledge support from FPA2011-25948, 2014 SGR 1450.

References

[1] F. Kruger, L. M. Sehgal, N. Sinha and R. Sinha, Phys. Rev. D **61**, 114028 (2000) [Erratum-ibid. D **63**, 019901 (2001)] [hep-ph/9907386].
[2] F. Kruger and J. Matias, Phys. Rev. D **71**, 094009 (2005) [hep-ph/0502060].
[3] C. Bobeth, G. Hiller and G. Piranishvili, JHEP **0807** (2008) 106, arXiv:0805.2525 [hep-ph].
[4] W. Altmannshofer, P. Ball, A. Bharucha, A. J. Buras, D. M. Straub and M. Wick, JHEP **0901** (2009) 019, arXiv:0811.1214 [hep-ph].
[5] C. Bobeth, G. Hiller and D. van Dyk, JHEP **1007**, 098 (2010), arXiv:1006.5013 [hep-ph].
[6] C. Bobeth, G. Hiller and D. van Dyk, JHEP **1107**, 067 (2011), arXiv:1105.0376 [hep-ph].
[7] D. Becirevic and E. Schneider, Nucl. Phys. B **854**, 321 (2012), arXiv:1106.3283 [hep-ph].
[8] J. Matias, F. Mescia, M. Ramon, J. Virto, JHEP **1204** (2012) 104, arXiv:1202.4266 [hep-ph].
[9] S. Descotes-Genon, J. Matias, M. Ramon and J. Virto, JHEP **1301**, 048 (2013), arXiv:1207.2753 [hep-ph].
[10] S. Descotes-Genon, T. Hurth, J. Matias and J. Virto, JHEP **1305**, 137 (2013), arXiv:1303.5794 [hep-ph].
[11] L. Hofer and J. Matias, arXiv:1502.00920 [hep-ph].
[12] R. Aaij *et al.* [LHCb Collaboration], Phys. Rev. Lett. **111** (2013) 19, 191801 [arXiv:1308.1707 [hep-ex]].
[13] S. Descotes-Genon, J. Matias and J. Virto, Phys. Rev. D **88**, no. 7, 074002 (2013), arXiv:1307.5683 [hep-ph].
[14] W. Altmannshofer and D. M. Straub, Eur. Phys. J. C **73**, no. 12, 2646 (2013), arXiv:1308.1501 [hep-ph].
[15] F. Beaujean, C. Bobeth and D. van Dyk, Eur. Phys. J. C **74**, no. 6, 2897 (2014) [Erratum-ibid. C **74**, no. 12, 3179 (2014)]
[16] R. R. Horgan, Z. Liu, S. Meinel and M. Wingate, Phys. Rev. Lett. **112**, 212003 (2014), arXiv:1310.3887 [hep-ph].
[17] R. Aaij *et al.* [LHCb Collaboration], JHEP **1302** (2013) 105 [arXiv:1209.4284 [hep-ex]].
[18] C. Hambrock, G. Hiller, S. Schacht and R. Zwicky, Phys. Rev. D **89** (2014) 7, 074014 [arXiv:1308.4379 [hep-ph]].
[19] S. Descotes-Genon, J. Matias and J. Virto, PoS EPS **-HEP2013**, 361 (2013), arXiv:1311.3876 [hep-ph].
[20] R. Aaij *et al.* [LHCb Collaboration], JHEP **1406** (2014) 133 [arXiv:1403.8044 [hep-ex]].

[21] S. Descotes-Genon, L. Hofer, J. Matias and J. Virto, JHEP **1412**, 125 (2014), arXiv:1407.8526 [hep-ph].
[22] A. Khodjamirian, T. Mannel, A. A. Pivovarov and Y.-M. Wang, JHEP **1009** (2010) 089 [arXiv:1006.4945 [hep-ph]].
[23] P. Ball and R. Zwicky, Phys. Rev. D **71**, 014029 (2005) [hep-ph/0412079].
[24] J. Charles, A. Le Yaouanc, L. Oliver, O. Pene and J. C. Raynal, Phys. Rev. D **60**, 014001 (1999) [hep-ph/9812358].
[25] M. Beneke, T. Feldmann and D. Seidel, Eur. Phys. J. C **41**, 173 (2005) [hep-ph/0412400].
[26] M. Beneke and T. Feldmann, Nucl. Phys. B **592**, 3 (2001) [hep-ph/0008255].
[27] M. Beneke, T. Feldmann and D. Seidel, Nucl. Phys. B **612**, 25 (2001) [hep-ph/0106067].
[28] W. Altmannshofer, P. Ball, A. Bharucha, A. J. Buras, D. M. Straub and M. Wick, JHEP **0901** (2009) 019, arXiv:0811.1214 [hep-ph].
[29] W. Altmannshofer and D. M. Straub, arXiv:1411.3161 [hep-ph].
[30] M. Beylich, G. Buchalla and T. Feldmann, Eur. Phys. J. C **71**, 1635 (2011), arXiv:1101.5118 [hep-ph].
[31] B. Grinstein and D. Pirjol, Phys. Rev. D **70**, 114005 (2004) [hep-ph/0404250].
[32] S. Jäger and J. Martin Camalich, JHEP **1305** (2013) 043 [arXiv:1212.2263 [hep-ph]].
[33] See talk from J. Matias at ZPW2015 at Zurich University.
[34] J. Lyon and R. Zwicky, arXiv:1406.0566 [hep-ph].
[35] L. Hofer and J. Matias, arXiv:1502.00920 [hep-ph].
[36] **LHCb** Collaboration, Phys. Rev. Lett. **113**, no. 15, 151601 (2014), arXiv:1406.6482 [hep-ex].
[37] D. Ghosh, M. Nardecchia and S. A. Renner, JHEP **1412** (2014) 131 [arXiv:1408.4097 [hep-ph]].
[38] J. Matias and N. Serra, Phys. Rev. D **90**, no. 3, 034002 (2014), arXiv:1402.6855 [hep-ph].
[39] U. Egede, T. Hurth, J. Matias, M. Ramon and W. Reece, JHEP **1010**, 056 (2010), arXiv:1005.0571 [hep-ph].
[40] D. Becirevic and A. Tayduganov, Nucl. Phys. B **868** (2013) 368 [arXiv:1207.4004 [hep-ph]].
[41] J. Matias, Phys. Rev. D **86** (2012) 094024 [arXiv:1209.1525 [hep-ph]].
[42] T. Blake, U. Egede and A. Shires, JHEP **1303** (2013) 027 [arXiv:1210.5279 [hep-ph]].
[43] M. Döring, U. G. Meissner and W. Wang, JHEP **1310** (2013) 011 [arXiv:1307.0947 [hep-ph]].
[44] A. J. Buras, F. De Fazio and J. Girrbach, JHEP **1302**, 116 (2013), arXiv:1211.1896 [hep-ph].
[45] A. J. Buras and J. Girrbach, JHEP **1312**, 009 (2013), arXiv:1309.2466 [hep-ph].
[46] R. Gauld, F. Goertz and U. Haisch, JHEP **1401**, 069 (2014), arXiv:1310.1082 [hep-ph].
[47] A. J. Buras, F. De Fazio and J. Girrbach, JHEP **1402**, 112 (2014), arXiv:1311.6729 [hep-ph].
[48] A. J. Buras, F. De Fazio and J. Girrbach-Noe, JHEP **1408**, 039 (2014), arXiv:1405.3850 [hep-ph].
[49] W. Altmannshofer, S. Gori, M. Pospelov and I. Yavin, Phys. Rev. D **89**, no. 9, 095033 (2014), arXiv:1403.1269 [hep-ph].
[50] R. Alonso, B. Grinstein and J. Martin Camalich, Phys. Rev. Lett. **113** (2014) 24, 241802 [arXiv:1407.7044 [hep-ph]].

Appendix: *SM predictions for $B \to K^* \mu^+ \mu^-$ Observables*

We provide here our SM predictions obtained using the method described in [21] for two different form factor parametrizations [22] and [23] including all theoretical errors: parametric errors, errors from form factors, factorizable power corrections, non-factorizable power corrections and charm-loop (only for KMPW) respectively. Notice that the error for the P_i observables coming from form factors in BZ parametrization is below the permille level.

Observable $\langle P_1 \rangle$	KMPW - scheme 1	BZ - scheme 1
$[0.1, 0.98]$	$0.027^{+0.005+0.006+0.009+0.035+0.061}_{-0.003-0.007-0.009-0.044-0.083}$	$0.036^{+0.006+0.000+0.009+0.036+---}_{-0.004-0.000-0.009-0.046----}$
$[1.1, 2]$	$0.001^{+0.003+0.016+0.020+0.033+0.091}_{-0.003-0.021-0.024-0.043-0.122}$	$0.030^{+0.005+0.000+0.015+0.033+---}_{-0.004-0.000-0.020-0.043----}$
$[2, 3]$	$-0.008^{+0.003+0.006+0.024+0.019+0.074}_{-0.002-0.013-0.025-0.025-0.096}$	$0.001^{+0.003+0.000+0.031+0.016+---}_{-0.004-0.000-0.039-0.021----}$
$[3, 4]$	$0.005^{+0.005+0.012+0.050+0.005+0.025}_{-0.004-0.015-0.051-0.006-0.028}$	$-0.034^{+0.003+0.000+0.060+0.003+---}_{-0.003-0.000-0.067-0.003----}$
$[4, 5]$	$0.019^{+0.005+0.033+0.062+0.002+0.021}_{-0.003-0.029-0.065-0.002-0.013}$	$-0.058^{+0.001+0.000+0.075+0.003+---}_{-0.002-0.000-0.075-0.003----}$
$[5, 6]$	$0.023^{+0.005+0.044+0.071+0.003+0.050}_{-0.002-0.036-0.071-0.003-0.035}$	$-0.077^{+0.001+0.000+0.079+0.004+---}_{-0.002-0.000-0.078-0.004----}$
$[6, 7]$	$0.020^{+0.005+0.050+0.074+0.004+0.072}_{-0.001-0.038-0.072-0.004-0.051}$	$-0.094^{+0.002+0.000+0.078+0.004+---}_{-0.003-0.001-0.076-0.004----}$
$[7, 8]$	$0.012^{+0.003+0.052+0.074+0.005+0.100}_{-0.001-0.039-0.070-0.006-0.070}$	$-0.113^{+0.002+0.001+0.076+0.006+---}_{-0.004-0.001-0.074-0.006----}$
$[1.1, 2.5]$	$-0.002^{+0.003+0.015+0.016+0.030+0.089}_{-0.002-0.020-0.021-0.039-0.119}$	$0.025^{+0.004+0.000+0.010+0.030+---}_{-0.004-0.000-0.014-0.039----}$
$[2.5, 4]$	$0.001^{+0.005+0.006+0.045+0.008+0.037}_{-0.003-0.010-0.045-0.010-0.045}$	$-0.026^{+0.003+0.000+0.054+0.006+---}_{-0.004-0.000-0.062-0.006----}$
$[4, 6]$	$0.021^{+0.005+0.039+0.066+0.003+0.037}_{-0.002-0.033-0.068-0.003-0.025}$	$-0.069^{+0.001+0.000+0.077+0.004+---}_{-0.002-0.000-0.077-0.003----}$
$[6, 8]$	$0.016^{+0.004+0.051+0.074+0.004+0.087}_{-0.001-0.039-0.071-0.005-0.062}$	$-0.104^{+0.002+0.001+0.077+0.005+---}_{-0.004-0.001-0.075-0.005----}$

Observable $\langle P_2 \rangle$	KMPW - scheme 1	BZ - scheme 1
$[0.1, 0.98]$	$0.118^{+0.007+0.004+0.015+0.002+0.004}_{-0.006-0.005-0.012-0.002-0.003}$	$0.125^{+0.007+0.000+0.014+0.002+---}_{-0.006-0.000-0.013-0.002----}$
$[1.1, 2]$	$0.428^{+0.018+0.009+0.029+0.005+0.012}_{-0.017-0.013-0.032-0.005-0.011}$	$0.447^{+0.015+0.000+0.020+0.004+---}_{-0.016-0.000-0.031-0.005----}$
$[2, 3]$	$0.438^{+0.012+0.019+0.037+0.006+0.023}_{-0.029-0.018-0.067-0.008-0.034}$	$0.386^{+0.022+0.000+0.063+0.009+---}_{-0.040-0.000-0.083-0.012----}$
$[3, 4]$	$0.153^{+0.038+0.049+0.103+0.012+0.066}_{-0.061-0.038-0.107-0.015-0.077}$	$0.051^{+0.039+0.000+0.115+0.012+---}_{-0.059-0.000-0.099-0.015----}$
$[4, 5]$	$-0.102^{+0.036+0.047+0.092+0.009+0.070}_{-0.066-0.033-0.078-0.011-0.073}$	$-0.189^{+0.030+0.000+0.086+0.007+---}_{-0.057-0.000-0.064-0.009----}$
$[5, 6]$	$-0.255^{+0.027+0.036+0.064+0.006+0.061}_{-0.065-0.024-0.049-0.006-0.058}$	$-0.318^{+0.021+0.000+0.054+0.004+---}_{-0.053-0.000-0.037-0.005----}$
$[6, 7]$	$-0.345^{+0.025+0.026+0.043+0.004+0.053}_{-0.073-0.017-0.031-0.003-0.046}$	$-0.389^{+0.019+0.000+0.032+0.003+---}_{-0.056-0.000-0.022-0.003----}$
$[7, 8]$	$-0.402^{+0.027+0.019+0.028+0.004+0.049}_{-0.054-0.012-0.019-0.003-0.040}$	$-0.432^{+0.020+0.000+0.019+0.003+---}_{-0.040-0.000-0.013-0.002----}$
$[1.1, 2.5]$	$0.443^{+0.009+0.004+0.011+0.003+0.004}_{-0.013-0.007-0.021-0.003-0.006}$	$0.450^{+0.004+0.000+0.002+0.002+---}_{-0.009-0.000-0.014-0.003----}$
$[2.5, 4]$	$0.227^{+0.035+0.045+0.093+0.011+0.058}_{-0.058-0.036-0.105-0.015-0.071}$	$0.130^{+0.039+0.000+0.111+0.012+---}_{-0.059-0.000-0.103-0.016----}$
$[4, 6]$	$-0.189^{+0.031+0.041+0.077+0.007+0.066}_{-0.067-0.028-0.062-0.008-0.065}$	$-0.262^{+0.025+0.000+0.068+0.005+---}_{-0.055-0.000-0.049-0.007----}$
$[6, 8]$	$-0.377^{+0.026+0.022+0.034+0.004+0.051}_{-0.057-0.014-0.024-0.003-0.042}$	$-0.412^{+0.020+0.000+0.025+0.003+---}_{-0.043-0.000-0.017-0.002----}$

Observable $\langle P_3 \rangle$	KMPW - scheme 1	BZ - scheme 1
$[0.1, 0.98]$	$-0.001^{+0.000+0.000+0.000+0.017+0.004}_{-0.001-0.000-0.001-0.019-0.003}$	$-0.001^{+0.001+0.000+0.000+0.018+---}_{-0.001-0.000-0.001-0.020----}$
$[1.1, 2]$	$0.000^{+0.000+0.001+0.002+0.016+0.008}_{-0.000-0.001-0.002-0.018-0.005}$	$-0.002^{+0.001+0.000+0.002+0.017+---}_{-0.001-0.000-0.003-0.018----}$
$[2, 3]$	$0.002^{+0.001+0.003+0.004+0.010+0.010}_{-0.001-0.002-0.004-0.010-0.007}$	$-0.002^{+0.001+0.000+0.004+0.008+---}_{-0.002-0.000-0.004-0.009----}$
$[3, 4]$	$0.002^{+0.002+0.003+0.004+0.003+0.009}_{-0.001-0.002-0.004-0.003-0.007}$	$-0.002^{+0.001+0.000+0.004+0.001+---}_{-0.001-0.000-0.003-0.001----}$
$[4, 5]$	$0.002^{+0.002+0.003+0.004+0.001+0.007}_{-0.001-0.001-0.003-0.001-0.005}$	$-0.001^{+0.001+0.000+0.003+0.002+---}_{-0.001-0.000-0.003-0.002----}$
$[5, 6]$	$0.002^{+0.003+0.002+0.003+0.002+0.006}_{-0.001-0.001-0.003-0.002-0.005}$	$-0.001^{+0.001+0.000+0.003+0.002+---}_{-0.001-0.000-0.002-0.002----}$
$[6, 7]$	$0.002^{+0.004+0.002+0.003+0.002+0.006}_{-0.002-0.001-0.002-0.002-0.005}$	$-0.001^{+0.001+0.000+0.002+0.002+---}_{-0.002-0.000-0.002-0.002----}$
$[7, 8]$	$0.001^{+0.003+0.001+0.003+0.004+0.007}_{-0.007-0.001-0.002-0.003-0.006}$	$-0.001^{+0.003+0.000+0.002+0.004+---}_{-0.002-0.000-0.002-0.003----}$
$[1.1, 2.5]$	$0.000^{+0.000+0.002+0.003+0.015+0.008}_{-0.000-0.001-0.003-0.016-0.006}$	$-0.002^{+0.001+0.000+0.003+0.015+---}_{-0.001-0.000-0.003-0.016----}$
$[2.5, 4]$	$0.002^{+0.002+0.003+0.004+0.004+0.009}_{-0.001-0.002-0.004-0.004-0.007}$	$-0.002^{+0.001+0.000+0.004+0.003+---}_{-0.001-0.000-0.003-0.003----}$
$[4, 6]$	$0.002^{+0.003+0.002+0.003+0.002+0.007}_{-0.001-0.001-0.003-0.001-0.005}$	$-0.001^{+0.001+0.000+0.003+0.002+---}_{-0.001-0.000-0.002-0.002----}$
$[6, 8]$	$0.001^{+0.002+0.002+0.003+0.003+0.007}_{-0.005-0.001-0.002-0.002-0.005}$	$-0.001^{+0.002+0.000+0.002+0.003+---}_{-0.001-0.000-0.002-0.002----}$

Observable $\langle P_4' \rangle$	KMPW - scheme 1	BZ - scheme 1
$[0.1, 0.98]$	$-0.495^{+0.013+0.029+0.022+0.010+0.335}_{-0.016-0.017-0.023-0.010-0.211}$	$-0.476^{+0.016+0.001+0.020+0.010+---}_{-0.017-0.001-0.022-0.011----}$
$[1.1, 2]$	$-0.162^{+0.036+0.055+0.096+0.011+0.117}_{-0.025-0.050-0.073-0.011-0.107}$	$-0.076^{+0.046+0.002+0.101+0.012+---}_{-0.030-0.002-0.086-0.012----}$
$[2, 3]$	$0.271^{+0.052+0.075+0.144+0.012+0.110}_{-0.043-0.082-0.126-0.011-0.097}$	$0.417^{+0.051+0.002+0.135+0.014+---}_{-0.045-0.002-0.143-0.012----}$
$[3, 4]$	$0.611^{+0.042+0.068+0.117+0.009+0.089}_{-0.036-0.083-0.120-0.008-0.087}$	$0.753^{+0.034+0.001+0.096+0.008+---}_{-0.030-0.001-0.119-0.008----}$
$[4, 5]$	$0.785^{+0.026+0.054+0.082+0.005+0.072}_{-0.022-0.071-0.092-0.005-0.078}$	$0.905^{+0.018+0.001+0.065+0.005+---}_{-0.016-0.001-0.083-0.004----}$
$[5, 6]$	$0.871^{+0.016+0.044+0.061+0.004+0.061}_{-0.013-0.060-0.070-0.004-0.077}$	$0.974^{+0.011+0.000+0.051+0.004+---}_{-0.009-0.000-0.061-0.003----}$
$[6, 7]$	$0.918^{+0.013+0.037+0.050+0.004+0.055}_{-0.017-0.052-0.056-0.004-0.079}$	$1.011^{+0.009+0.000+0.043+0.004+---}_{-0.014-0.000-0.048-0.003----}$
$[7, 8]$	$0.949^{+0.012+0.032+0.043+0.006+0.057}_{-0.017-0.046-0.049-0.005-0.091}$	$1.035^{+0.009+0.000+0.038+0.005+---}_{-0.016-0.000-0.042-0.004----}$
$[1.1, 2.5]$	$-0.058^{+0.039+0.064+0.113+0.011+0.117}_{-0.030-0.059-0.088-0.011-0.102}$	$0.043^{+0.050+0.002+0.116+0.013+---}_{-0.036-0.002-0.104-0.012----}$
$[2.5, 4]$	$0.537^{+0.047+0.073+0.127+0.010+0.094}_{-0.039-0.085-0.125-0.009-0.092}$	$0.683^{+0.040+0.001+0.107+0.010+---}_{-0.035-0.002-0.129-0.009----}$
$[4, 6]$	$0.830^{+0.020+0.050+0.071+0.005+0.066}_{-0.017-0.065-0.080-0.004-0.077}$	$0.940^{+0.013+0.001+0.057+0.004+---}_{-0.012-0.001-0.071-0.004----}$
$[6, 8]$	$0.933^{+0.013+0.035+0.046+0.005+0.055}_{-0.011-0.048-0.052-0.004-0.086}$	$1.022^{+0.009+0.000+0.040+0.005+---}_{-0.010-0.000-0.045-0.003----}$

Observable $\langle P_5' \rangle$	KMPW - scheme 1	BZ - scheme 1
$[0.1, 0.98]$	$0.675^{+0.011+0.014+0.035+0.012+0.152}_{-0.016-0.013-0.037-0.013-0.187}$	$0.679^{+0.011+0.001+0.030+0.013+---}_{-0.017-0.001-0.034-0.015----}$
$[1.1, 2]$	$0.308^{+0.032+0.013+0.078+0.016+0.092}_{-0.048-0.006-0.084-0.018-0.126}$	$0.289^{+0.034+0.002+0.076+0.017+---}_{-0.051-0.002-0.088-0.019----}$
$[2, 3]$	$-0.164^{+0.053+0.011+0.114+0.019+0.103}_{-0.076-0.012-0.113-0.023-0.126}$	$-0.204^{+0.054+0.002+0.107+0.019+---}_{-0.075-0.002-0.112-0.024----}$
$[3, 4]$	$-0.553^{+0.047+0.011+0.103+0.014+0.095}_{-0.072-0.005-0.094-0.018-0.109}$	$-0.570^{+0.043+0.001+0.085+0.014+---}_{-0.066-0.002-0.088-0.016----}$
$[4, 5]$	$-0.761^{+0.035+0.016+0.073+0.009+0.077}_{-0.064-0.012-0.069-0.011-0.089}$	$-0.746^{+0.030+0.001+0.068+0.008+---}_{-0.056-0.001-0.068-0.009----}$
$[5, 6]$	$-0.865^{+0.025+0.019+0.054+0.006+0.063}_{-0.058-0.020-0.052-0.007-0.077}$	$-0.828^{+0.022+0.001+0.057+0.006+---}_{-0.050-0.001-0.056-0.007----}$
$[6, 7]$	$-0.919^{+0.023+0.021+0.043+0.005+0.056}_{-0.062-0.024-0.041-0.006-0.076}$	$-0.870^{+0.020+0.000+0.049+0.005+---}_{-0.053-0.000-0.048-0.005----}$
$[7, 8]$	$-0.951^{+0.025+0.021+0.037+0.006+0.062}_{-0.044-0.026-0.037-0.006-0.082}$	$-0.893^{+0.022+0.000+0.043+0.006+---}_{-0.037-0.000-0.042-0.005----}$
$[1.1, 2.5]$	$0.196^{+0.038+0.010+0.088+0.017+0.094}_{-0.057-0.010-0.094-0.019-0.125}$	$0.171^{+0.041+0.002+0.087+0.018+---}_{-0.060-0.002-0.098-0.021----}$
$[2.5, 4]$	$-0.467^{+0.050+0.009+0.109+0.016+0.099}_{-0.073-0.008-0.100-0.020-0.114}$	$-0.491^{+0.047+0.002+0.093+0.015+---}_{-0.068-0.002-0.095-0.019----}$
$[4, 6]$	$-0.816^{+0.029+0.017+0.061+0.007+0.069}_{-0.061-0.017-0.060-0.008-0.082}$	$-0.788^{+0.025+0.001+0.062+0.007+---}_{-0.053-0.001-0.061-0.008----}$
$[6, 8]$	$-0.935^{+0.024+0.021+0.039+0.006+0.059}_{-0.048-0.025-0.039-0.006-0.079}$	$-0.881^{+0.021+0.000+0.046+0.006+---}_{-0.040-0.000-0.045-0.005----}$

Observable $\langle P_6' \rangle$	KMPW - scheme 1	BZ - scheme 1
$[0.1, 0.98]$	$-0.058^{+0.022+0.007+0.005+0.016+0.015}_{-0.031-0.008-0.006-0.017-0.015}$	$-0.060^{+0.024+0.001+0.005+0.017+---}_{-0.031-0.001-0.006-0.018----}$
$[1.1, 2]$	$-0.072^{+0.027+0.009+0.006+0.016+0.009}_{-0.037-0.010-0.008-0.016-0.012}$	$-0.076^{+0.030+0.001+0.007+0.017+---}_{-0.038-0.001-0.008-0.017----}$
$[2, 3]$	$-0.070^{+0.029+0.009+0.004+0.013+0.007}_{-0.038-0.011-0.004-0.012-0.009}$	$-0.073^{+0.031+0.001+0.004+0.013+---}_{-0.039-0.001-0.004-0.013----}$
$[3, 4]$	$-0.056^{+0.027+0.009+0.003+0.008+0.005}_{-0.035-0.011-0.002-0.008-0.005}$	$-0.056^{+0.028+0.001+0.003+0.008+---}_{-0.034-0.001-0.003-0.007----}$
$[4, 5]$	$-0.041^{+0.024+0.007+0.003+0.007+0.004}_{-0.034-0.010-0.003-0.005-0.003}$	$-0.040^{+0.024+0.001+0.004+0.007+---}_{-0.031-0.001-0.004-0.006----}$
$[5, 6]$	$-0.030^{+0.022+0.007+0.003+0.008+0.003}_{-0.042-0.009-0.004-0.009-0.003}$	$-0.029^{+0.022+0.001+0.003+0.008+---}_{-0.040-0.001-0.004-0.009----}$
$[6, 7]$	$-0.022^{+0.041+0.006+0.003+0.011+0.004}_{-0.075-0.008-0.003-0.012-0.004}$	$-0.020^{+0.039+0.000+0.003+0.011+---}_{-0.067-0.001-0.004-0.013----}$
$[7, 8]$	$-0.015^{+0.095+0.005+0.003+0.014+0.005}_{-0.070-0.007-0.004-0.017-0.005}$	$-0.014^{+0.090+0.000+0.003+0.015+---}_{-0.069-0.000-0.004-0.017----}$
$[1.1, 2.5]$	$-0.072^{+0.027+0.009+0.006+0.015+0.008}_{-0.037-0.010-0.007-0.015-0.011}$	$-0.076^{+0.030+0.001+0.007+0.016+---}_{-0.038-0.001-0.007-0.016----}$
$[2.5, 4]$	$-0.060^{+0.028+0.009+0.002+0.009+0.005}_{-0.036-0.011-0.002-0.009-0.006}$	$-0.060^{+0.029+0.001+0.003+0.009+---}_{-0.035-0.001-0.002-0.008----}$
$[4, 6]$	$-0.035^{+0.023+0.007+0.003+0.007+0.003}_{-0.039-0.009-0.004-0.007-0.003}$	$-0.034^{+0.022+0.001+0.004+0.007+---}_{-0.036-0.001-0.004-0.008----}$
$[6, 8]$	$-0.018^{+0.056+0.006+0.003+0.013+0.004}_{-0.046-0.007-0.003-0.014-0.004}$	$-0.017^{+0.053+0.000+0.003+0.013+---}_{-0.045-0.000-0.004-0.015----}$

Observable $\langle P_8' \rangle$	KMPW - scheme 1	BZ - scheme 1
$[0.1, 0.98]$	$0.020^{+0.027+0.009+0.008+0.016+0.021}_{-0.017-0.008-0.007-0.017-0.030}$	$0.023^{+0.026+0.001+0.007+0.016+---}_{-0.017-0.001-0.006-0.017----}$
$[1.1, 2]$	$0.041^{+0.035+0.009+0.010+0.016+0.010}_{-0.022-0.007-0.009-0.018-0.009}$	$0.044^{+0.035+0.001+0.009+0.016+---}_{-0.024-0.001-0.008-0.018----}$
$[2, 3]$	$0.048^{+0.038+0.009+0.007+0.014+0.007}_{-0.024-0.008-0.007-0.015-0.007}$	$0.049^{+0.037+0.001+0.007+0.014+---}_{-0.025-0.001-0.006-0.015----}$
$[3, 4]$	$0.042^{+0.036+0.009+0.004+0.010+0.005}_{-0.023-0.007-0.004-0.011-0.005}$	$0.041^{+0.033+0.001+0.004+0.010+---}_{-0.023-0.001-0.004-0.010----}$
$[4, 5]$	$0.033^{+0.034+0.008+0.003+0.007+0.003}_{-0.020-0.006-0.003-0.008-0.003}$	$0.032^{+0.031+0.001+0.003+0.007+---}_{-0.019-0.000-0.002-0.008----}$
$[5, 6]$	$0.026^{+0.039+0.007+0.002+0.006+0.003}_{-0.019-0.005-0.002-0.007-0.003}$	$0.024^{+0.037+0.000+0.002+0.006+---}_{-0.018-0.000-0.002-0.007----}$
$[6, 7]$	$0.020^{+0.063+0.006+0.002+0.006+0.003}_{-0.036-0.004-0.002-0.008-0.003}$	$0.019^{+0.056+0.000+0.002+0.005+---}_{-0.033-0.000-0.002-0.007----}$
$[7, 8]$	$0.015^{+0.058+0.005+0.002+0.009+0.003}_{-0.090-0.004-0.002-0.009-0.004}$	$0.014^{+0.056+0.000+0.002+0.008+---}_{-0.082-0.000-0.002-0.009----}$
$[1.1, 2.5]$	$0.043^{+0.036+0.009+0.009+0.015+0.009}_{-0.023-0.007-0.008-0.017-0.009}$	$0.045^{+0.036+0.001+0.009+0.015+---}_{-0.024-0.001-0.008-0.018----}$
$[2.5, 4]$	$0.044^{+0.036+0.009+0.005+0.011+0.005}_{-0.023-0.008-0.005-0.012-0.005}$	$0.043^{+0.034+0.001+0.005+0.011+---}_{-0.023-0.001-0.004-0.012----}$
$[4, 6]$	$0.029^{+0.037+0.007+0.002+0.006+0.003}_{-0.019-0.005-0.002-0.008-0.003}$	$0.028^{+0.034+0.000+0.002+0.006+---}_{-0.019-0.000-0.002-0.007----}$
$[6, 8]$	$0.017^{+0.039+0.006+0.002+0.007+0.003}_{-0.059-0.004-0.002-0.009-0.004}$	$0.016^{+0.037+0.000+0.002+0.007+---}_{-0.054-0.000-0.002-0.008----}$

Observable $\langle A_{\mathrm{FB}} \rangle$	KMPW - scheme 1	BZ - scheme 1
$[0.1, 0.98]$	$-0.099^{+0.002+0.036+0.006+0.001+0.014}_{-0.002-0.026-0.006-0.001-0.009}$	$0.097^{+0.003+0.004+0.005+0.001+---}_{-0.002-0.004-0.004-0.000----}$
$[1.1, 2]$	$-0.199^{+0.012+0.123+0.022+0.004+0.023}_{-0.007-0.182-0.020-0.003-0.021}$	$-0.172^{+0.013+0.016+0.021+0.005+---}_{-0.010-0.015-0.019-0.003----}$
$[2, 3]$	$-0.140^{+0.019+0.090+0.031+0.006+0.018}_{-0.013-0.152-0.030-0.005-0.022}$	$-0.106^{+0.017+0.010+0.029+0.006+---}_{-0.014-0.010-0.030-0.004----}$
$[3, 4]$	$-0.048^{+0.021+0.027+0.034+0.005+0.024}_{-0.014-0.032-0.035-0.004-0.023}$	$-0.015^{+0.018+0.001+0.030+0.005+---}_{-0.012-0.001-0.035-0.004----}$
$[4, 5]$	$0.038^{+0.023+0.066+0.031+0.004+0.030}_{-0.013-0.034-0.034-0.003-0.027}$	$0.070^{+0.019+0.006+0.027+0.003+---}_{-0.011-0.006-0.033-0.003----}$
$[5, 6]$	$0.111^{+0.026+0.126+0.027+0.002+0.037}_{-0.011-0.086-0.030-0.002-0.032}$	$0.142^{+0.021+0.010+0.023+0.002+---}_{-0.010-0.011-0.030-0.002----}$
$[6, 7]$	$0.174^{+0.034+0.159+0.023+0.002+0.044}_{-0.012-0.130-0.026-0.002-0.038}$	$0.204^{+0.026+0.013+0.019+0.002+---}_{-0.010-0.014-0.025-0.002----}$
$[7, 8]$	$0.228^{+0.028+0.176+0.019+0.004+0.057}_{-0.014-0.166-0.022-0.003-0.046}$	$0.256^{+0.023+0.015+0.015+0.004+---}_{-0.012-0.016-0.021-0.003----}$
$[1.1, 2.5]$	$-0.187^{+0.014+0.117+0.024+0.005+0.022}_{-0.009-0.182-0.023-0.004-0.021}$	$-0.157^{+0.015+0.015+0.022+0.005+---}_{-0.012-0.014-0.021-0.004----}$
$[2.5, 4]$	$-0.071^{+0.020+0.043+0.033+0.006+0.022}_{-0.014-0.061-0.035-0.004-0.021}$	$-0.038^{+0.018+0.003+0.030+0.005+---}_{-0.013-0.003-0.034-0.004----}$
$[4, 6]$	$0.076^{+0.025+0.100+0.029+0.003+0.034}_{-0.012-0.061-0.032-0.002-0.030}$	$0.107^{+0.020+0.008+0.025+0.002+---}_{-0.010-0.009-0.032-0.002----}$
$[6, 8]$	$0.202^{+0.028+0.169+0.021+0.003+0.051}_{-0.013-0.149-0.024-0.003-0.042}$	$0.231^{+0.023+0.014+0.017+0.003+---}_{-0.012-0.016-0.023-0.003----}$

Observable $\langle F_L \rangle$	KMPW - scheme 1	BZ - scheme 1
$[0.1, 0.98]$	$0.213^{+0.021+0.214+0.039+0.008+0.078}_{-0.016-0.137-0.033-0.006-0.043}$	$0.261^{+0.024+0.025+0.040+0.009+---}_{-0.021-0.021-0.036-0.007----}$
$[1.1, 2]$	$0.646^{+0.030+0.184+0.047+0.010+0.038}_{-0.025-0.267-0.049-0.008-0.039}$	$0.704^{+0.027+0.023+0.038+0.009+---}_{-0.025-0.021-0.044-0.008----}$
$[2, 3]$	$0.761^{+0.016+0.140+0.025+0.005+0.022}_{-0.015-0.244-0.029-0.004-0.025}$	$0.794^{+0.015+0.017+0.020+0.004+---}_{-0.014-0.016-0.021-0.004----}$
$[3, 4]$	$0.770^{+0.010+0.147+0.015+0.003+0.017}_{-0.007-0.249-0.016-0.003-0.017}$	$0.786^{+0.010+0.018+0.014+0.002+---}_{-0.009-0.017-0.014-0.002----}$
$[4, 5]$	$0.739^{+0.007+0.175+0.012+0.003+0.021}_{-0.004-0.265-0.013-0.002-0.021}$	$0.744^{+0.009+0.021+0.013+0.002+---}_{-0.008-0.020-0.014-0.002----}$
$[5, 6]$	$0.697^{+0.006+0.208+0.012+0.003+0.026}_{-0.005-0.275-0.014-0.003-0.029}$	$0.693^{+0.009+0.023+0.015+0.003+---}_{-0.008-0.021-0.016-0.003----}$
$[6, 7]$	$0.654^{+0.007+0.240+0.013+0.004+0.032}_{-0.005-0.276-0.014-0.004-0.037}$	$0.644^{+0.009+0.025+0.015+0.004+---}_{-0.007-0.022-0.016-0.003----}$
$[7, 8]$	$0.613^{+0.008+0.267+0.013+0.005+0.043}_{-0.005-0.271-0.013-0.005-0.051}$	$0.598^{+0.010+0.025+0.015+0.005+---}_{-0.007-0.023-0.016-0.005----}$
$[1.1, 2.5]$	$0.681^{+0.027+0.172+0.041+0.009+0.033}_{-0.023-0.263-0.044-0.007-0.035}$	$0.732^{+0.024+0.021+0.033+0.008+---}_{-0.023-0.020-0.038-0.007----}$
$[2.5, 4]$	$0.771^{+0.011+0.143+0.017+0.003+0.017}_{-0.009-0.247-0.019-0.003-0.019}$	$0.791^{+0.010+0.018+0.015+0.002+---}_{-0.010-0.017-0.015-0.002----}$
$[4, 6]$	$0.717^{+0.006+0.193+0.012+0.003+0.024}_{-0.005-0.271-0.014-0.003-0.025}$	$0.717^{+0.009+0.022+0.014+0.003+---}_{-0.008-0.021-0.015-0.002----}$
$[6, 8]$	$0.632^{+0.007+0.254+0.013+0.004+0.038}_{-0.005-0.274-0.013-0.004-0.045}$	$0.620^{+0.010+0.025+0.015+0.004+---}_{-0.007-0.023-0.016-0.004----}$

Higgs physics: Review of recent results and prospects from ATLAS and CMS

Martin Flechl on behalf of the ATLAS and CMS collaborations
Institute of High Energy Physics, Austrian Academy of Sciences, Wohllebengasse 12-14, 1040 Vienna, Austria
E-mail: martin.flechl@cern.ch

Abstract. An overview of recent results in Higgs boson physics obtained with the ATLAS and CMS experiments at the Large Hadron Collider located at CERN, Geneva, is presented. The focus is on measurements of the properties of the recently discovered Higgs boson with a mass of about 125 GeV. A brief selection of results in searches for Higgs bosons beyond the Standard Model is given, and prospects of future Higgs boson measurements and searches at the LHC are discussed.

1. Introduction

In July 2012, the ATLAS [1] and CMS [2] collaborations announced the discovery of a new particle compatible with being the Higgs boson of the Standard Model (SM) [3, 4]. Subsequent measurements of the properties of this particle are all consistent with the SM Higgs boson interpretation. In the following sections, the most recent results from studies of this Higgs boson using the data collected with the ATLAS and CMS experiments are presented. First, the results in various Higgs boson production and decay channels are given, followed by the measurement of Higgs boson properties using a combination of these results. Searches for Higgs bosons predicted by theories beyond the Standard Model (BSM) are briefly summarized and prospects for future Higgs boson property measurements are discussed. Most of the results are based on the full data set recorded in 2011 and 2012 at center-of-mass energies of 7 TeV and 8 TeV, respectively, corresponding to about 25 fb^{-1} per experiment.

2. Higgs boson search results

The dominant Higgs boson production modes are gluon-gluon fusion (ggF), vector boson fusion (VBF), production in association with a vector boson V (VH, $V = W$ or $V = Z$), and in association with top quarks (ttH). The five most sensitive Higgs boson decay channels at the LHC are the modes $\gamma\gamma$, 4-leptons, WW, $\tau\tau$ and $b\bar{b}$ [5, 6, 7]. Results for these channels are summarized in the following.

2.1. $H \to \gamma\gamma$

The $H \to \gamma\gamma$ channel is characterized by relatively high total event counts but a low signal-over-background ratio. Furthermore, it offers a high mass resolution ($m_{\gamma\gamma}$) and thus contributes strongly to the Higgs boson mass measurement. The analyses [8, 9] proceed by selecting events with two photons in various categories, aimed at different production modes. This is done

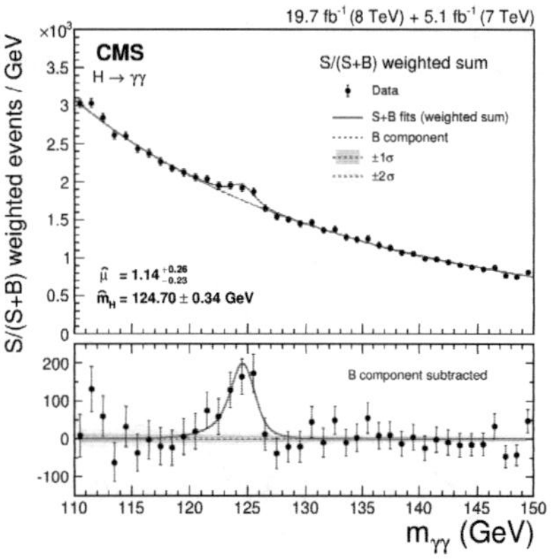

Figure 1. The CMS $m_{\gamma\gamma}$ distribution as weighted sum of all categories [8]. S and B are the number of signal and background events in a small mass window around m_H for each event category, respectively.

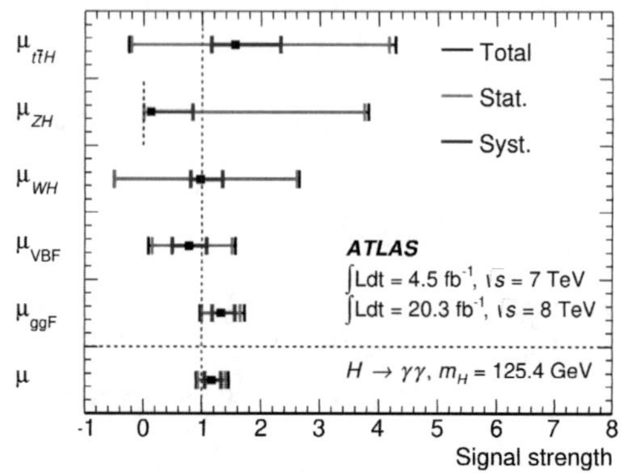

Figure 2. Signal strength in the $H \to \gamma\gamma$ channel. Shown is the measurement for the individual Higgs production modes as well as the combined ATLAS value [9].

by additionally requiring two jets with a high rapidity gap (VBF), additional leptons and in some cases missing energy (VH) or an event topology consistent with an additional top quark pair event (ttH). The final discriminating variable is the $m_{\gamma\gamma}$ estimator. The background, dominantly continuum $\gamma\gamma$, γ+jet and di-jet events, is estimated by fitting the sidebands of the $m_{\gamma\gamma}$ distribution, as illustrated in Fig. 1. The signal is visible on top of the estimated background at $m_{\gamma\gamma} \approx 125$ GeV.

Both ATLAS and CMS observe slightly more $H \to \gamma\gamma$ candidate events than expected; however, the measurement is within one Gaussian standard deviation (σ) of the SM expectation. Both experiments measure the signal strength μ (observed cross section times branching ratio divided by the SM expectation) also individually for the various Higgs boson production modes, see Fig. 2. All values are in agreement with the SM expectation of $\mu = 1$. The observed signal significance is 5.7σ (expected: 5.2σ) for CMS and 5.2σ (expected: 4.6σ) for ATLAS.

2.2. $H \to 4l$

The expected rate of $H \to 4$ lepton events is low compared to the other channels presented here; however, this is compensated by the highest signal-over-background ratio. This channel is further characterized by a high mass resolution (m_{4l}) and dominates the Higgs boson mass measurement together with $H \to \gamma\gamma$. The analyses [10, 11] first require four light leptons and are then pursued in subchannels depending on the flavor of the two lepton pairs. Like the $H \to \gamma\gamma$ analysis, it is split in categories targeting different production modes. However, the main sensitivity rests in the 0/1 jet category aimed at gluon-gluon fusion production. The main background is continuum ZZ^* production which is estimated from simulation and validated in control regions. The combined m_{4l} distribution for ATLAS is shown in Fig. 3. The mass peak at about 125 GeV is clearly visible.

Both the ATLAS and CMS results are in very good agreement with the SM expectation. This is illustrated in Fig. 4 where the signal strength for the ggF and ttH production mode is measured with respect to the VBF and VH production modes. The observed signal significance

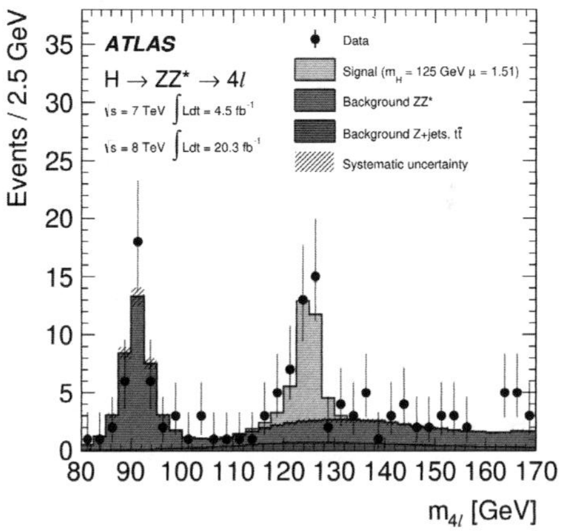

Figure 3. The ATLAS combined m_{4l} distribution [10].

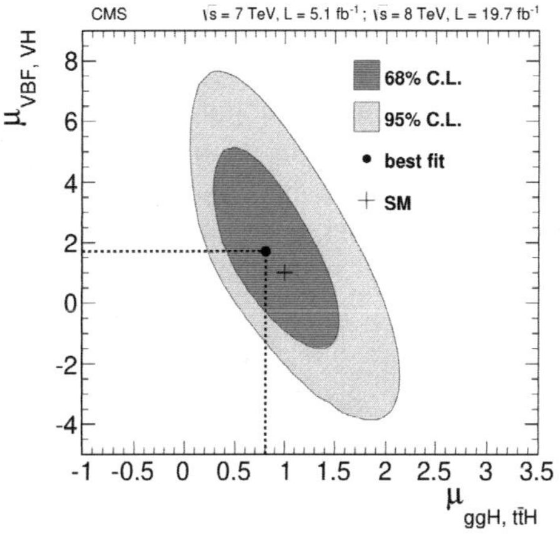

Figure 4. The CMS measurement of the signal strength associated to ggF and ttH production versus VBF and VH production [11].

is 6.8σ (expected: 6.7σ) for CMS and 8.1σ (expected: 6.2σ) for ATLAS.

2.3. $H \to WW$

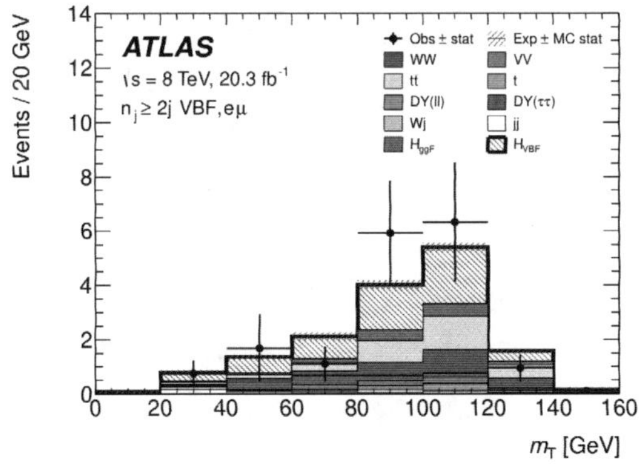

Figure 5. The ATLAS m_T distribution in the category "$e\nu\mu\nu$ + 2 (or more) jets" [12]. The events are weighted by the value $\ln(1+S/B)$ of the corresponding bin of the final analysis discriminant, where S and B are the number of expected signal and background events, respectively.

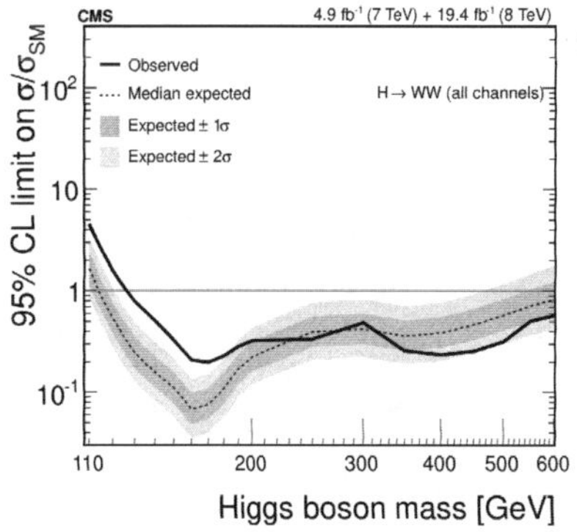

Figure 6. The CMS exclusion versus Higgs boson mass in the WW channel [13], with no assumptions on the presence of a Higgs boson.

Higgs boson decays to a W boson pair offer a relatively high expected rate but suffer from a high irreducible WW continuum background which is hard to suppress due to the low m_{WW} resolution, caused by the presence of neutrinos in the most sensitive W decay modes. The

W+jets and top quark pair background is also sizable. The most sensitive subchannel features a $e\nu\mu\nu$ final state with low jet multiplicity [12, 13]. Requirements of additional jets or leptons aim to add extra sensitivity by exploiting production modes other than ggF. The m_T distribution in the category "$e\nu\mu\nu$ + 2 (or more) jets" is shown in Fig. 5. This is the WW category with the highest expected signal purity.

Both the observed ATLAS and CMS rates are within about 1σ of the SM expectation. In Fig. 6, the Higgs boson exclusion versus m_H is shown. Visible are both the excess at about 125 GeV and the exclusion of other, heavy SM-like Higgs bosons.

2.4. $H \to \tau\tau$

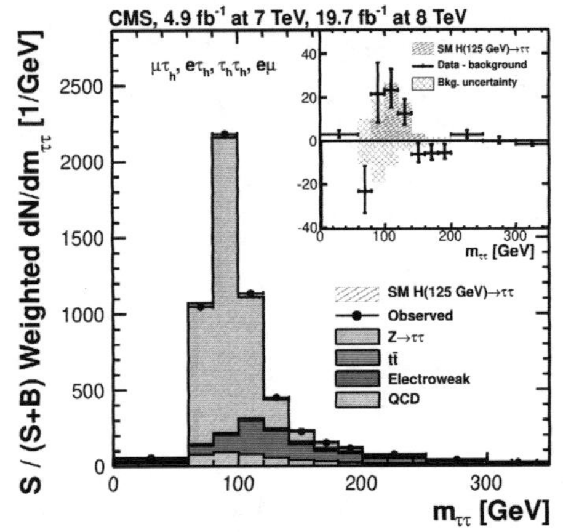

Figure 7. The $m_{\tau\tau}$ distribution estimated by CMS [14]. Events from several channels are weighted by signal purity and then combined. The insert at the upper right shows the background-subtracted distribution.

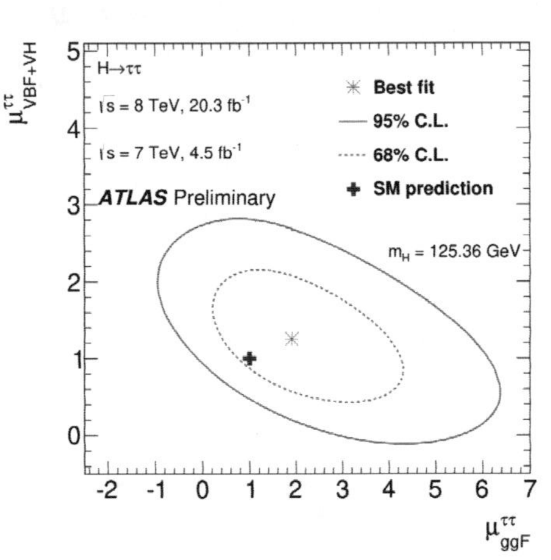

Figure 8. The ATLAS measurement of the signal strength in ggF versus VBF/VH [15].

The $H \to \tau\tau$ analyses are among the most complex analyses of LHC data: Six different channels, depending on the combination of τ lepton decays, are investigated: ee, $e\mu$, $\mu\mu$, $e\tau$, $\mu\tau$ and $\tau\tau$ (here, τ signifies a hadronic τ lepton decay, and neutrinos are omitted). For all these channels, categories motivated by the production mode can be established: 0/1-jet, boosted (boson candidate with high p_T), VBF and VH. CMS has also investigated ttH with $H \to \tau\tau$ [16]. The CMS analysis [14] which is based on orthogonal selection requirements implements almost one hundred categories. ATLAS [15] uses only six; however, here, the analysis is based on a Boosted Decision Tree (BDT). Both ATLAS and CMS use sophisticated $m_{\tau\tau}$ estimators using kinematic information of the whole decay chain to reconstruct a mass value in spite of the presence of a number of undetected neutrinos. The combined $m_{\tau\tau}$ distribution is shown in Fig. 7. The main background is due to $Z/\gamma^* \to \tau\tau$ events which are estimated by replacing muons in $Z \to \mu\mu$ collision data events with simulated τ leptons.

Both ATLAS and CMS report evidence for $H \to \tau\tau$ decays. ATLAS observes 4.5σ (expected: 3.5σ), CMS observes 3.2σ (expected: 3.7σ). This is the first evidence for Higgs boson decays to τ leptons. The sensitivity is mostly driven by the VBF category, as can be seen in Fig. 8.

While the signal strength in ggF is only weakly constrained, VBF/VH decays are established at a 2σ-level.

2.5. $H \to bb$

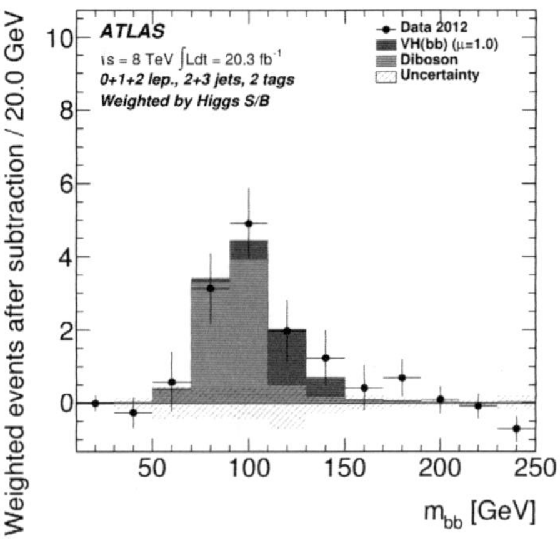

Figure 9. The m_{bb} distribution of the ATLAS analysis after subtracting all backgrounds except diboson events [17].

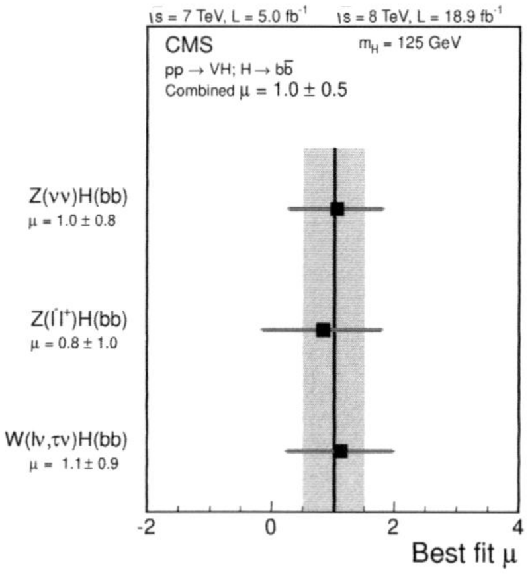

Figure 10. The signal strength in the $H \to bb$ channel as measured by CMS, split by production modes [18].

The $H \to bb$ analysis is extremely challenging at the LHC, despite the high expected branching ratio of 0.58 at $m_H = 125$ GeV. The ggF production is not accessible as the signal cannot be separated from the overwhelming non-resonant $b\bar{b}$ background. The situation for VBF production is only marginally better. The main workhorse are thus VH events; in addition, ttH is also investigated. Both experiments use a BDT-based analysis [17, 18] to isolate the signal from the overwhelming background of Z, W, $t\bar{t}$, VV and multi-jet events. The m_{bb} distribution after background subtraction is shown in Fig. 9.

Both observed and expected significance by CMS are 2.1σ. ATLAS observes 1.4σ while the expectation is 2.6σ. The non-significant ATLAS deficit is driven by the $\sqrt{s} = 7$ TeV data. The CMS measurement of the signal strength is shown in Fig. 10.

2.6. ttH

The ttH channel poses very particular challenges and is thus in most cases not treated in the context of the usual decay mode analyses by ATLAS and CMS. Instead, dedicated ttH analyses and combinations are produced. CMS investigates six different final states [16], see Fig. 11. The measured combined signal strength is $\mu = 2.8 \pm 1.0$ which is about 2σ high compared to the SM expectation. The small excess is almost entirely driven by the same-sign two-lepton category. The BDT output for this category is shown in Fig. 12 and illustrates this point. ATLAS investigates and combines [19] two final states in ttH production, namely bb and $\gamma\gamma$. The combined ATLAS limit on ttH production is in good agreement with the SM.

3. Higgs boson property measurements

The Higgs boson property measurements use the results of the analyses aiming at the various production and decay modes presented in the previous section. For each measurement, the

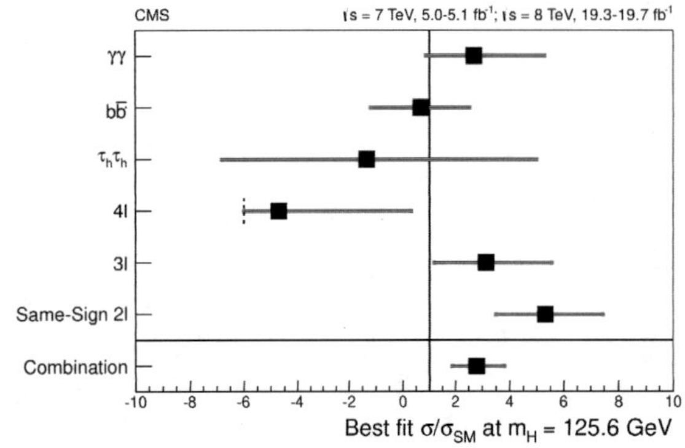

Figure 11. The measured signal strength in the CMS ttH analyses [16].

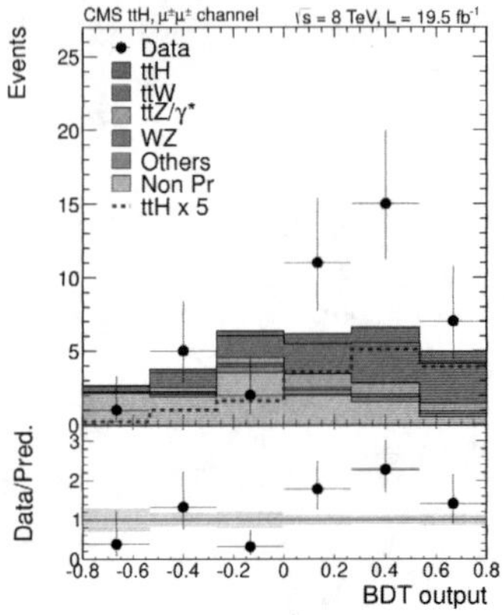

Figure 12. The BDT output for the CMS ttH analysis in the same-sign two-lepton channel [16].

channels sensitive to the property in question are combined. Among the studied properties are mass, width, signal strength (i.e., normalized cross section times branching ratio), coupling strength and tensor coupling structure (CP properties) of the Higgs boson in relation to other SM particles. In addition, differential cross sections allow further insight and additional tests of BSM scenarios.

3.1. Mass

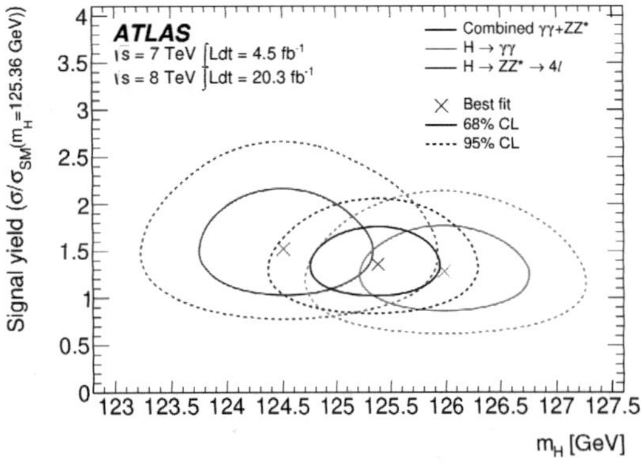

Figure 13. ATLAS m_H measurement versus the signal strength evaluated at the best-fit mass [20].

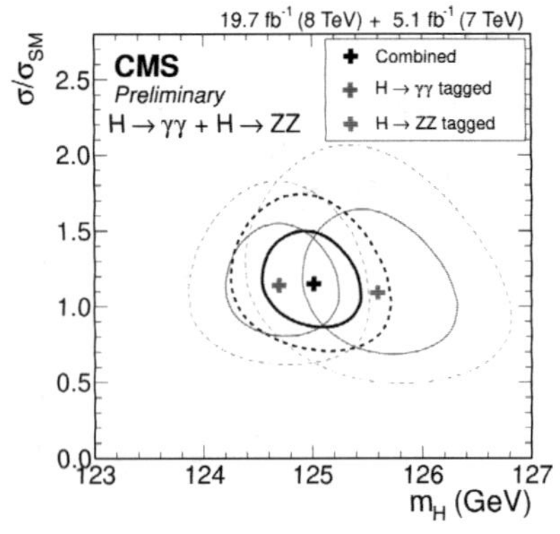

Figure 14. CMS m_H measurement versus the signal strength [21].

The Higgs boson mass measurement uses input from the $\gamma\gamma$ and 4-lepton decay modes. In both channels, the Higgs boson candidate mass can be reconstructed with high precision directly from its visible decay products. The ATLAS measurement [20], illustrated in Fig. 13, yields $m_H = 125.36 \pm 0.37(\text{stat}) \pm 0.18(\text{syst})$. The CMS measurement [21], see Fig. 14, is consistent with this result: $m_H = 125.03^{+0.26}_{-0.27}(\text{stat})^{+0.13}_{-0.15}(\text{syst})$. Historically, there has been a tension between the two ATLAS measurements. However, this tension is gradually decreasing and currently the $\gamma\gamma$ mass is only about 2σ high with respect to the four-lepton mass. The two CMS mass measurements have a 1.4σ tension in the opposite direction, i.e. $m_{4l} < m_{\gamma\gamma}$.

3.2. Width

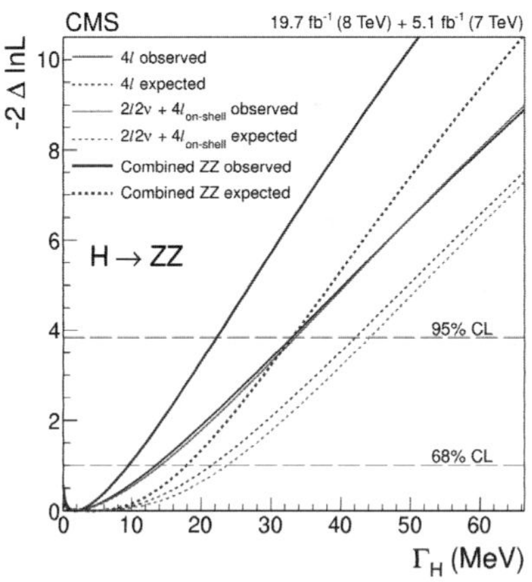

Figure 15. The CMS Higgs boson width constraints from off-shell $H \rightarrow 4$ leptons signal strength measurements [22].

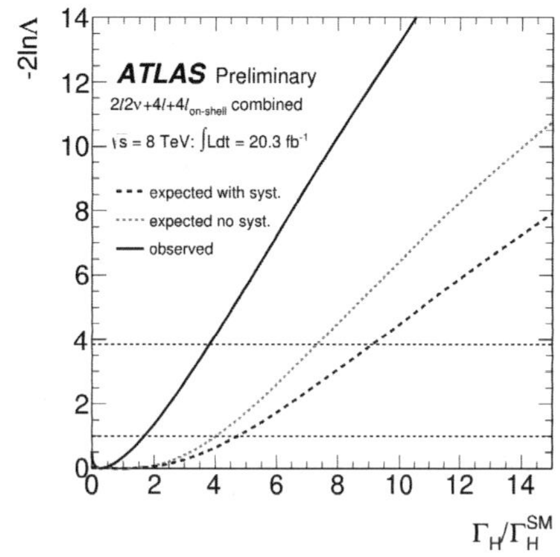

Figure 16. The ATLAS Higgs boson width constraints from off-shell $H \rightarrow 4$ leptons signal strength measurements [23].

The SM expectation for the Higgs boson width at $m_H = 125$ GeV is about 4 MeV. There are several ways to access the Higgs boson width experimentally. Directly, the width can be measured by analyzing the width of the m_{4l} and $m_{\gamma\gamma}$ distributions. This method is limited by the experimental resolution which is about three orders of magnitude higher than the width predicted by the SM. The direct ATLAS width limits are 5.0 GeV ($H \rightarrow \gamma\gamma$), 2.6 GeV ($H \rightarrow 4$ leptons) [20]; the corresponding CMS limits are 2.4 GeV ($H \rightarrow \gamma\gamma$) [8] and 3.4 GeV ($H \rightarrow 4$ leptons) [11].

Indirect limits can be obtained by Higgs boson coupling fits (by leaving the invisible width as free fit parameter). However, the most precise indirect measurement is based on the comparison of the on- and off-shell $H \rightarrow 4$ leptons signal strength. The CMS limit on the Higgs width using this technique is 22 MeV (expected: 33 MeV), which corresponds to 5.4 (expected: 8.0) times the SM expectation [22]. ATLAS uses a more flexible model and excludes 5-8 (expected: 7-12) times the SM expectation, depending on the ratio of the leading-order to next-to-leading order cross section for on- and off-shell production [23].

3.3. Signal strength

The signal strength μ, defined as measured cross section times branching ratio of a given process divided by the SM expectation, is an important test for the validity of the SM. With the data accumulated in the LHC Run 1, no significant deviation is observed. The combined CMS result is $\mu = 1.00 \pm 0.13$, and the results for each decay channel are consistent with the SM within 1σ [21], see Fig. 17. The ATLAS result is slightly high, $\mu = 1.30 \pm 0.18$. Note that this combined value has not been updated after the release of the most recent results which are summarized

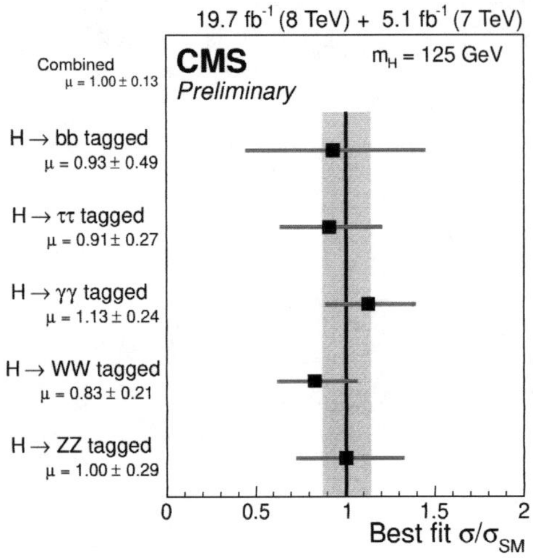

Figure 17. CMS measurement of the signal strength μ [21].

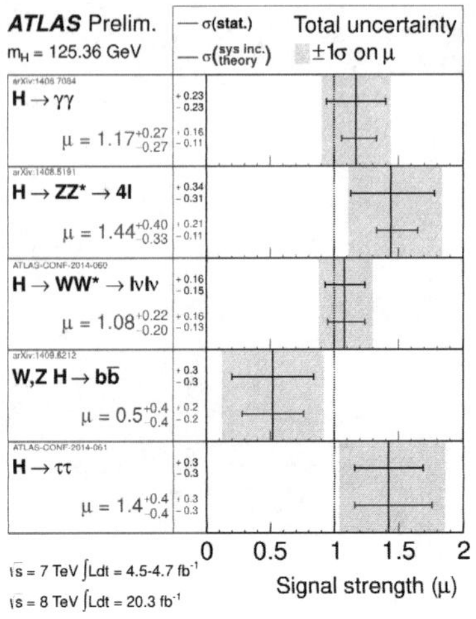

Figure 18. ATLAS measurement of the signal strength μ [24].

in Fig. 18. The ATLAS results for all channels agree with the SM expectation within about 1σ [24].

3.4. Coupling strength

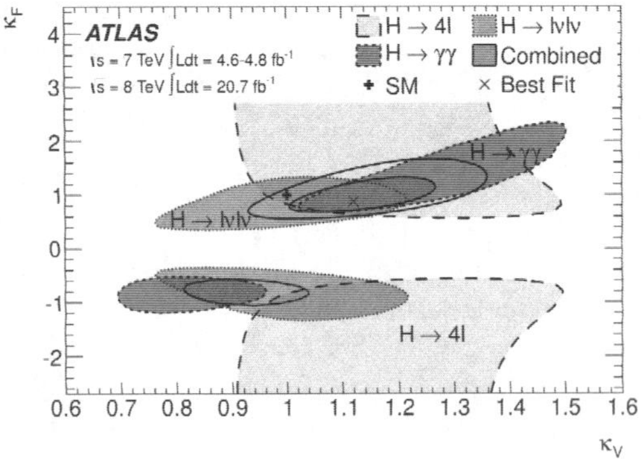

Figure 19. ATLAS measurement of κ_V versus κ_F [25].

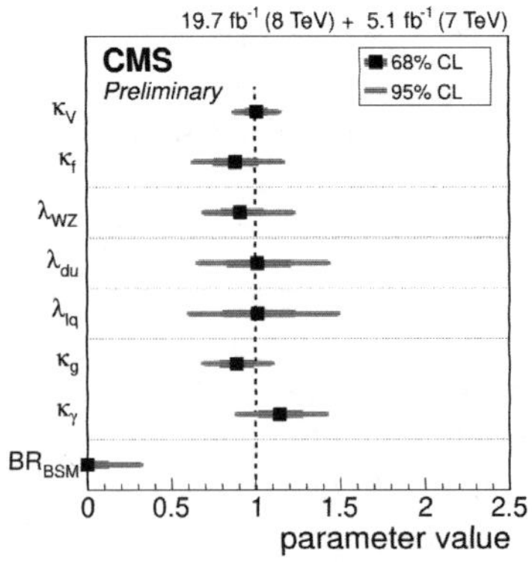

Figure 20. CMS measurement of coupling strength parameters [21].

The coupling strength factors κ_i are a leading-order-inspired parametrisation of the Higgs boson coupling to a particle or particle class i with respect to the SM expectation. λ_{jk} is used to denote the ratio of two values κ_j and κ_k. The measurements are typically a result of a global fit of a large subset of the Higgs boson analyses and due to the large number of free parameters, some assumptions have to be made to obtain sensible results with the data of the LHC run 1.

Usual assumptions are those on the Higgs width (to be SM-like) or of a universal coupling strength scaling, e.g. of all fermions or all up-type quarks. An important SM test is the scaling of fermion couplings versus boson couplings, shown in Fig. 19. The result is consistent with the SM expectation [25]. Further tests involve the custodial symmetry (W versus Z coupling), up- versus down-type quark couplings or lepton versus quark couplings. As shown in Fig. 20, all results agree with the SM prediction [21].

3.5. Tensor coupling structure

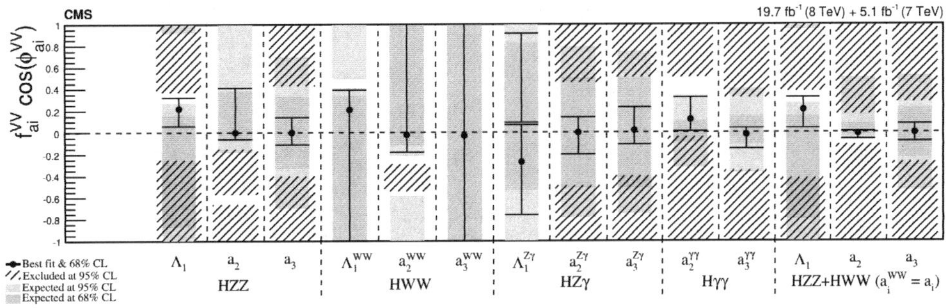

Figure 21. CMS test for anomalous components of the tensor coupling structure of Higgs bosons to various other particles [26].

The study of the spin and CP properties of the discovered boson are essential for the claim of the discovery of the first fundamental scalar. To this date, all measurements are in agreement with the SM but since the parameter space of alternative models is continuous in several dimensions, no generic exclusion of e.g. spin-2 models has been possible. However, a large set of the best-motivated models has been tested with the result that the SM hypothesis is favored [26, 27]. This is illustrated in Fig. 21 where measurements of several parameters for the spin-0 case are made and all are found to be consistent with the SM expectation. Today, all alternatives to a spin-0 boson are disfavored and large anomalous contributions to the CP structure of the Higgs boson couplings are excluded. However, small or medium-sized CP-odd or anomalous CP-even admixtures are still feasible.

3.6. Differential cross sections

By the end of LHC Run 1, sufficient data have been collected to allow for the first publications of differential distributions in Higgs boson events, namely in ATLAS studies of the $H \to \gamma\gamma$ [28] and $H \to 4$ leptons [29] channels. The resulting contributions are compared to predictions of the SM, but model builders can compare them to predictions of any competing models. Some of the aspects which are probed are the jet multiplicity in Higgs boson events (to test the QCD structure of SM calculations and models), see Fig. 22, the transverse momentum of the Higgs boson candidate (to test for BSM effects hidden in loops, e.g. involving top quarks), see Fig. 23, or the difference of the azimuthal angle of two of the jets in the event (to test for effects of an anomalous tensor coupling structure). The results, within large statistical uncertainties, are broadly in line with the SM predictions.

4. BSM Higgs boson searches

The programme of BSM Higgs boson searches at the LHC is very extensive and cannot be presented exhaustively here. In the following, highlights of searches for Higgs bosons predicted

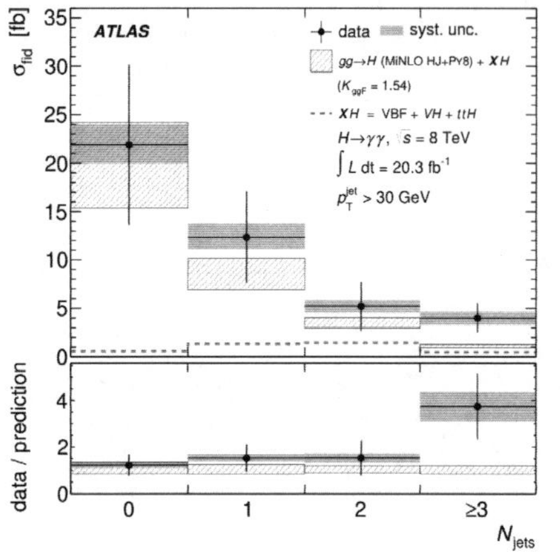

Figure 22. The jet multiplicity in ATLAS $H \to \gamma\gamma$ events [28].

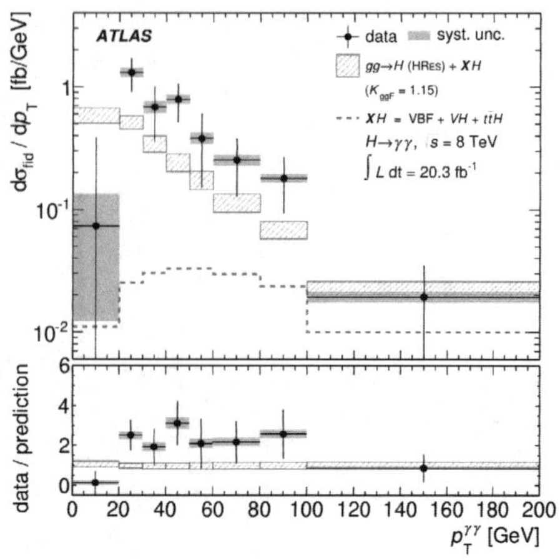

Figure 23. The p_T distribution of the Higgs boson candidate in ATLAS $H \to \gamma\gamma$ events [28].

by the Minimal Supersymmetric extension of the SM (MSSM) are presented, and other BSM Higgs boson searches are listed and their conclusions are summarized.

4.1. MSSM Higgs boson searches

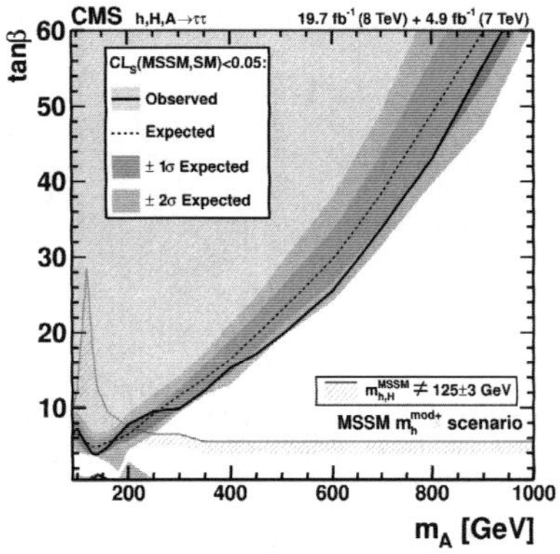

Figure 24. The CMS m_A-$\tan\beta$ exclusion contour from neutral MSSM Higgs boson searches [30], for the $m_h^{\mathrm{mod+}}$ scenario.

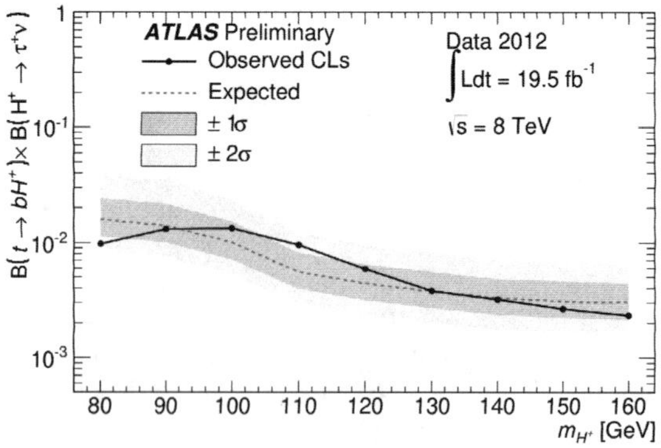

Figure 25. ATLAS upper limit on the branching ratio BR($t \to bH^+$) [31].

The MSSM predicts the existence of five Higgs bosons, three of them neutral and a charged pair. The most sensitive search for neutral MSSM Higgs bosons for most of the parameter space is via their $\tau\tau$ decays. The current CMS limit [30] in the $m_h^{\mathrm{mod+}}$ scenario [32] is shown in Fig. 24.

Due to the charged Higgs boson search result, the option to identify the discovered Higgs boson (at $m_H = 125$) GeV with the heavy CP-even neutral MSSM Higgs boson has been strongly disfavored. Presently, BR($t \to bH^+$) above about 1% are excluded for $m_{H^+} < 160$ GeV. For charged Higgs bosons heavier than the top quark, a sizable region at moderate and high $\tan\beta$ has been excluded [31].

4.2. Other BSM Higgs boson searches

There is a vast number of BSM Higgs boson searches with ATLAS and CMS in addition to the $A/H \to \tau\tau$ and $H+ \to \tau\nu$ searches introduced in the previous section. A non-exhaustive list of these searches is given here:

- Other (N)MSSM-inspired searches
  - $H^+ \to c\bar{s}$ [33, 34]
  - $H^+ \to t\bar{b}$ [35]
  - $H/A \to \mu\mu$ [36, 37]
  - $H/A \to b\bar{b}$ [38]
  - $H/A \to WW$ [39]
  - $a_1 \to \gamma\gamma$ [40]
  - $a_1 \to \mu\mu$ [41, 42]
- Generic Higgs boson searches
  - Heavy Higgs, $H \to WW/ZZ$ [43, 44]
  - Invisible Higgs, ZH [45, 46]
  - Doubly charged Higgs, H^{++} [47, 48]
  - Lepton flavor violation, $H \to \tau\mu$ [49]
  - Flavor-changing neutral current, $t \to cH$ [50, 51]
  - Fermiophobic Higgs [52, 53]
  - Higgs in 4th-generation models [54, 53]
- Indirect search via Higgs boson property measurements [55, 21]
- Higgs-to-Higgs decays, Higgs pair production
  - HH or $X \to HH$ [56, 57, 58, 59]
  - $H \to aa$ [60]
  - $A \to ZH$ [61]
  - $H \to H^+W$ with $H^+ \to Wh$ [62]
- Exotic Higgs boson searches
  - Higgs boson decays to long-lived particles [63]
  - Higgs boson decays to electron jets [64]
  - Higgs boson decays to displaced muon jets [65]

All searches have in common that no significant deviation from the SM expectation has been found and that the phase space for BSM Higgs bosons has been constrained. However, the higher kinematic reach at $\sqrt{s} = 13 - 14$ TeV and the extension of the current data set by 1-2 orders of magnitude will allow to explore a magnitude scenarios over the next few years.

5. Prospects for Higgs boson searches

The present LHC programme projects the delivery of about 300 fb^{-1} of data at $\sqrt{s} = 13 - 14$ TeV per experiment by the year 2022. One of the options is that this will be followed by a high-luminosity LHC (HL-LHC), producing 3000 fb^{-1} of data at $\sqrt{s} = 14$ TeV until 2035 [66]. In the following, projections of Higgs boson property measurements assuming these conditions are presented. Typically, it is assumed that the detector and reconstruction performance is comparable to that of LHC Run 1 due to improved reconstruction algorithms and detector upgrades countering the effect of an increase in concurrent proton-proton interactions ("pile-up"). Systematic uncertainties are assumed to be similar to the current status (CMS Scenario 1 and ATLAS) or to go down (CMS Scenario 2; by a factor 0.5 for theoretical uncertainties, and by the inverse of the square root of the integrated luminosity for other systematic uncertainties) [66, 67].

5.1. Signal strength and coupling strength projections

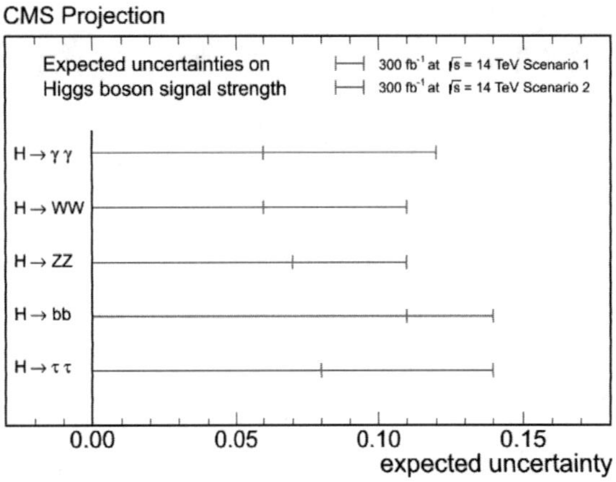

Figure 26. CMS projection for the precision on Higgs boson signal strength measurements with 300 fb^{-1} [67]. The two scenarios are explained in the text.

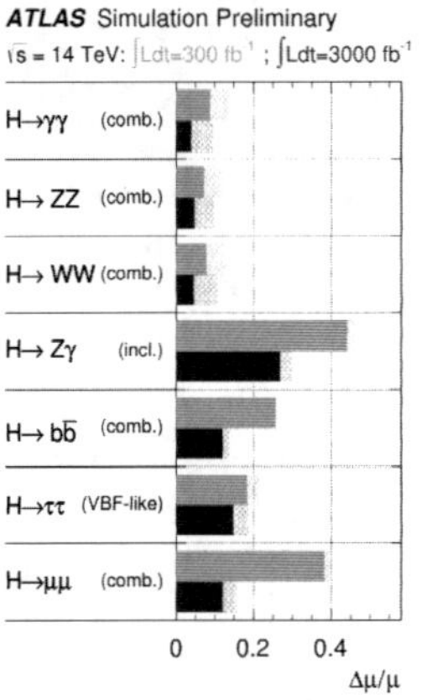

Figure 27. ATLAS projection for the precision on Higgs boson signal strength measurements with 300 fb^{-1} and 3000 fb^{-1} [66]. The hatched area is the effect of theoretical uncertainties.

The CMS estimate for the first 300 fb^{-1} is a precision of roughly 10% on the signal strength for the five most sensitive channels, see Fig. 26. The predicted uncertainties are between 6%–14%, with lower values for the bosonic decay modes and higher ones for the fermionic modes, and depending on the scenario [67]. The ATLAS projection for 300 fb^{-1} is slightly more conservative, expecting a precision of about 10% for the bosonic modes, and about 20% for the fermionic modes. In addition, a precision of about 40% on the rare decay channels $Z\gamma$ and $\mu\mu$ is estimated, see Fig. 27. Sizable improvements with 3000 fb^{-1} are in particular expected for the bb, $Z\gamma$ and $\mu\mu$ measurements but the precision for all signal strength measurements is expected to improve significantly [66].

CMS estimates a precision of 4%–8% on the Higgs boson coupling strength to elementary bosons and the τ lepton, and of 10%–15% to the bottom and the top quark with 300 fb^{-1} [67], as illustrated in Fig. 28. The ATLAS estimates are comparable but in all cases slightly more pessimistic [66]. In Fig. 29, the projected uncertainty for a fit with two free parameters is shown: κ_V and κ_F. Assuming only these two universal scaling factors, a precision of about 2%–4% on κ_V (when κ_F is profiled) and of 3%–9% on κ_F (when κ_V is profiled) can be achieved, depending on the integrated luminosity and the assumptions on theory uncertainties. The potential improvement on the precision of κ_F with the HL-LHC is almost a factor of two.

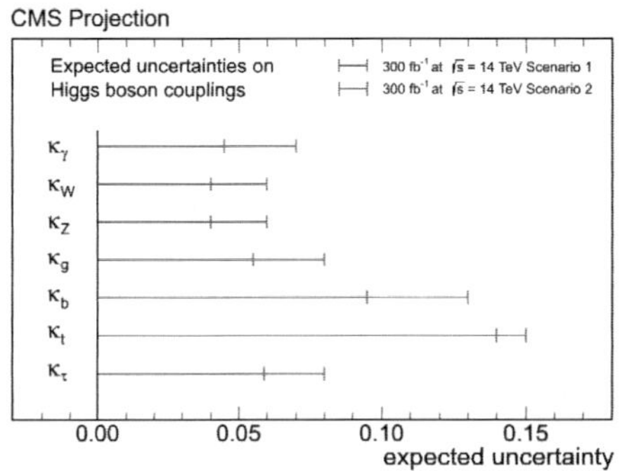

Figure 28. CMS projection for the precision on the Higgs boson coupling strength measurements with 300 fb^{-1} [67].

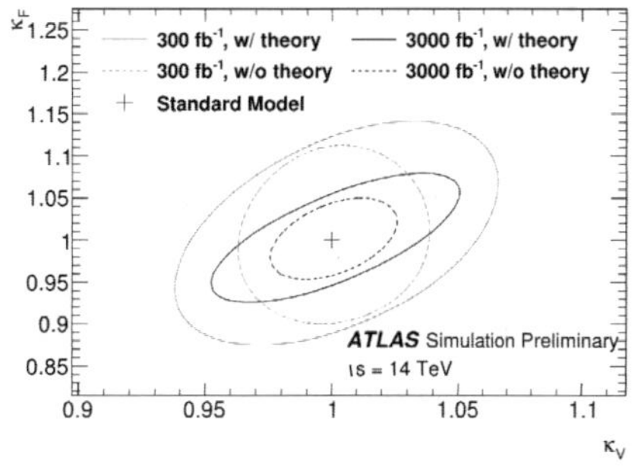

Figure 29. ATLAS projection for the precision on the measurement of κ_V versus κ_F with 300 fb^{-1} and 3000 fb^{-1}. The precision corresponds to a confidence level of 68% [66].

5.2. Double-Higgs production and Higgs self-coupling projections

There are two kinds of contributions to double-Higgs production: With, and without a triple-Higgs vertex. A measurement of the cross section of this process thus allows to draw conclusions about the Higgs self-coupling. Due to the negative interference of these two contributions in the SM, only about 10 signal events are expected in the final state $HH \to bb\gamma\gamma$ in 3000 fb^{-1} of data. According to ATLAS projections, the expected significance for HH production in this channel for the whole HL-LHC data set is about 1.3σ [68]. CMS estimates a relative uncertainty on the HH cross section measurement of about 60% [69] for the same final state. As of today it is thus unclear if any meaningful measurement of the Higgs self-coupling at the LHC is possible.

6. Conclusions

Both ATLAS and CMS offer a rich Higgs physics programme. Following the Higgs boson discovery in 2012, recent years have seen a focus on measuring the properties of this Higgs boson. All measurements so far are consistent with the SM expectation but only the sizable improvements on their precision expected in the near future will allow to distinguish between the SM and many of its alternatives. In searches for BSM Higgs bosons, no evidence for a new state has been found, severely constraining a wide range of BSM models.

References

[1] ATLAS Collaboration *JINST* **3** S08003
[2] CMS Collaboration *JINST* **3** S08004
[3] ATLAS Collaboration *Phys.Lett.* **B716** 1–29 (*Preprint* 1207.7214)
[4] CMS Collaboration *Phys.Lett.* **B716** 30–61 (*Preprint* 1207.7235)
[5] LHC Higgs Cross Section Working Group, Dittmaier S, Mariotti C, Passarino G and Tanaka R E *CERN-2011-002* (*Preprint* 1101.0593)
[6] LHC Higgs Cross Section Working Group, Dittmaier S, Mariotti C, Passarino G and Tanaka R E *CERN-2012-002* (*Preprint* 1201.3084)
[7] LHC Higgs Cross Section Working Group, Heinemeyer S, Mariotti C, Passarino G and Tanaka R E *CERN-2013-004* (*Preprint* 1307.1347)
[8] CMS Collaboration *Eur.Phys.J.* **C74** 3076 (*Preprint* 1407.0558)
[9] ATLAS Collaboration *Phys.Rev.* **D90** 112015 (*Preprint* 1408.7084)
[10] ATLAS Collaboration *Phys.Rev.* **D91** 012006 (*Preprint* 1408.5191)
[11] CMS Collaboration *Phys.Rev.* **D89** 092007 (*Preprint* 1312.5353)
[12] ATLAS Collaboration (*Preprint* 1412.2641)

[13] CMS Collaboration *JHEP* **1401** 096 (*Preprint* 1312.1129)
[14] CMS Collaboration *JHEP* **1405** 104 (*Preprint* 1401.5041)
[15] ATLAS Collaboration *ATLAS-CONF-2014-061*
[16] CMS Collaboration *JHEP* **1409** 087 (*Preprint* 1408.1682)
[17] ATLAS Collaboration (*Preprint* 1409.6212)
[18] CMS Collaboration *Phys.Rev.* **D89** 012003 (*Preprint* 1310.3687)
[19] ATLAS Collaboration *ATLAS-CONF-2014-043*
[20] ATLAS Collaboration *Phys.Rev.* **D90** 052004 (*Preprint* 1406.3827)
[21] CMS Collaboration *CMS-PAS-HIG-14-009*
[22] CMS Collaboration *Phys.Lett.* **B736** 64 (*Preprint* 1405.3455)
[23] ATLAS Collaboration *ATLAS-CONF-2014-042*
[24] ATLAS Collaboration Summary plots from the ATLAS Higgs physics group 10/2014 URL https://atlas.web.cern.ch/Atlas/GROUPS/PHYSICS/CombinedSummaryPlots/HIGGS/
[25] ATLAS Collaboration *Phys.Lett.* **B726** 88–119 (*Preprint* 1307.1427)
[26] CMS Collaboration (*Preprint* 1411.3441)
[27] ATLAS Collaboration *Phys.Lett.* **B726** 120–144 (*Preprint* 1307.1432)
[28] ATLAS Collaboration *JHEP* **1409** 112 (*Preprint* 1407.4222)
[29] ATLAS Collaboration *Phys.Lett.* **B738** 234–253 (*Preprint* 1408.3226)
[30] CMS Collaboration *JHEP* **1410** 160 (*Preprint* 1408.3316)
[31] ATLAS Collaboration *ATLAS-CONF-2014-050*
[32] Carena M, Heinemeyer S, Stål O, Wagner C and Weiglein G *Eur.Phys.J.* **C73** 2552 (*Preprint* 1302.7033)
[33] ATLAS Collaboration *Eur.Phys.J.* **C73** 2465 (*Preprint* 1302.3694)
[34] CMS Collaboration *CMS-PAS-HIG-13-035*
[35] CMS Collaboration *CMS-PAS-HIG-13-026*
[36] ATLAS Collaboration *ATLAS-CONF-2012-094*
[37] CMS Collaboration *CMS-PAS-HIG-12-011*
[38] CMS Collaboration *Phys.Lett.* **B722** 207–232 (*Preprint* 1302.2892)
[39] ATLAS Collaboration *ATLAS-CONF-2013-027*
[40] ATLAS Collaboration *ATLAS-CONF-2012-079*
[41] ATLAS Collaboration *ATLAS-CONF-2011-020*
[42] CMS Collaboration *Phys.Rev.Lett.* **109** 121801 (*Preprint* 1206.6326)
[43] ATLAS Collaboration *ATLAS-CONF-2013-067*
[44] CMS Collaboration *Phys.Rev.* **D90** 112013 (*Preprint* 1410.2751)
[45] ATLAS Collaboration *Phys.Rev.Lett.* **112** 201802 (*Preprint* 1402.3244)
[46] CMS Collaboration *Eur.Phys.J.* **C74** 2980 (*Preprint* 1404.1344)
[47] ATLAS Collaboration (*Preprint* 1412.0237)
[48] CMS Collaboration *Eur.Phys.J.* **C72** 2189 (*Preprint* 1207.2666)
[49] CMS Collaboration *CMS-PAS-HIG-14-005*
[50] ATLAS Collaboration *JHEP* **1406** 008 (*Preprint* 1403.6293)
[51] CMS Collaboration *CMS-PAS-HIG-13-034*
[52] ATLAS Collaboration *Eur.Phys.J.* **C72** 2157 (*Preprint* 1205.0701)
[53] CMS Collaboration *Phys.Lett.* **B725** 36–59 (*Preprint* 1302.1764)
[54] ATLAS Collaboration *ATLAS-CONF-2011-135*
[55] ATLAS Collaboration *ATLAS-CONF-2014-010*
[56] ATLAS Collaboration (*Preprint* 1406.5053)
[57] ATLAS Collaboration *ATLAS-CONF-2014-005*
[58] CMS Collaboration *CMS-PAS-HIG-14-013*
[59] CMS Collaboration *CMS-PAS-HIG-13-032*
[60] CMS Collaboration *CMS-PAS-HIG-13-010*
[61] CMS Collaboration *CMS-PAS-HIG-14-011*
[62] ATLAS Collaboration *Phys.Rev.* **D89** 032002 (*Preprint* 1312.1956)
[63] ATLAS Collaboration *Phys.Rev.Lett.* **108** 251801 (*Preprint* 1203.1303)
[64] ATLAS Collaboration *New J.Phys.* **15** 043009 (*Preprint* 1302.4403)
[65] ATLAS Collaboration *Phys.Lett.* **B721** 32–50 (*Preprint* 1210.0435)
[66] ATLAS Collaboration *ATL-PHYS-PUB-2014-016*
[67] CMS Collaboration *CMS-NOTE-13-002* (*Preprint* 1307.7135)
[68] ATLAS Collaboration *ATL-PHYS-PUB-2014-019*
[69] CMS Collaboration Physics Performance for 2nd ECFA workshop 10/2014 URL https://twiki.cern.ch/twiki/pub/CMSPublic/PhysicsResultsFP/ECFA-CMSPublicResults.pdf

Implication of Higgs mediated Flavour Changing Neutral Currents with Minimal Flavour Violation

M N Rebelo

Centro de Física Teórica de Partículas (CFTP), Instituto Superior Técnico (IST), U. de Lisboa (UL), Av. Rovisco Pais, P-1049-001 Lisboa, Portugal.

E-mail: rebelo@tecnico.ulisboa.pt

Abstract. We analise phenomenological implications of two Higgs doublet models with Higgs flavour changing neutral currents suppressed in the quark sector by small entries of the Cabibbo-Kokayashi-Maskawa matrix. This suppression occurs in a natural way since it is the result of a symmetry applied to the Lagrangian. These type of models were proposed some time ago by Branco Grimus and Lavoura. Our results clearly show that these class of models allow for new physical scalars, with masses which are reachable at the LHC. The imposed symmetry severely reduces the number of free parameters and allows for predictions. Therefore these models can eventually be proved right or eliminated experimentally.

1. Introduction

There are several good motivations to consider models with two Higgs doublets. Such models allow for new sources of CP violation and for the possibility of having spontaneous CP violation. It is by now established that the SM cannot account for the observed baryon asymmetry of the universe and that new sources of CP violation are required. Spontaneous CP violation was first proposed by Lee [1] in the context of two Higgs doublet models (2HDM) and puts CP violation on the same footing as the electroweak symmetry breaking. Models with two Higgs doublets may also provide a solution to the strong CP problem of the type proposed by Peccei and Quinn [2]. Furthermore, supersymmetric extensions of the Standard Model (SM) also require the existence of two Higgs doublets.

The discovery of a scalar boson in the run 1 of LHC by Atlas [3] and CMS [4] immediately raises the question of whether this is the SM Higgs boson or part of a multi-Higgs theory. The properties of the Higgs boson already discovered are up till now in good agreement with those predicted by the SM but more precise determinations may show evidence for New Physics. There is also the possibility that the LHC will soon discover a charged Higgs or additional neutral scalars thus confirming the need to extend the scalar sector of the SM.

There are important experimental constraints on 2HDM. Such models have potentially large Higgs mediated flavour changing neutral currents (FCNC) [5], [6] [7]. Effects due to these FCNC are severely constrained and therefore some mechanism to suppress these effects is required. It is possible to eliminate tree level FCNC by imposing for instance natural flavour conservation [8] or alignment [9]. Another possibility, which was proposed some time ago by Branco Grimus and Lavoura (BGL) [10], consists on imposing a symmetry on the Lagrangian allowing for tree level FCNC in the quark sector suppressed by small entries of the Cabibbo– Kobayashi – Maskawa matrix, V_{CKM}. Later-on, BGL models were extended to the leptonic sector [11] and their relation to Minimal Flavour Violation models has been studied [12]. Phenomenological implications of these models have been analysed recently [13], [14], [15]. This talk is largely based on work done in Ref. [14].

2. Theoretical Framework

In Ref. [14] we analysed extensions of the SM with two Higgs doublets together with three right-handed neutrinos. We did not add Majorana mass terms to the Lagrangian and as a result neutrinos are Dirac type. However our analysis of phenomenological implications is not sensitive to the character of the neutrinos. The Yukawa interactions can be explicitly written:

$$\begin{aligned}\mathcal{L}_Y &= -\overline{Q_L^0}\,\Gamma_1\Phi_1 d_R^0 - \overline{Q_L^0}\,\Gamma_2\Phi_2 d_R^0 - \overline{Q_L^0}\,\Delta_1\tilde{\Phi}_1 u_R^0 - \overline{Q_L^0}\,\Delta_2\tilde{\Phi}_2 u_R^0 \\ &\quad -\overline{L_L^0}\,\Pi_1\Phi_1\ell_R^0 - \overline{L_L^0}\,\Pi_2\Phi_2\ell_R^0 - \overline{L_L^0}\,\Sigma_1\tilde{\Phi}_1\nu_R^0 - \overline{L_L^0}\,\Sigma_2\tilde{\Phi}_2\nu_R^0 + \text{h.c.},\end{aligned} \tag{1}$$

where Γ_i, Δ_i Π_i and Σ_i denote the Yukawa couplings to the right-handed quarks d_R^0, u_R^0, rigth-handed leptons ℓ_R^0, ν_R^0, respectively, and the Higgs doublets Φ_j. The quark mass matrices generated after spontaneous gauge symmetry breaking are given by:

$$M_d = \frac{1}{\sqrt{2}}(v_1\Gamma_1 + v_2 e^{i\alpha}\Gamma_2), \quad M_u = \frac{1}{\sqrt{2}}(v_1\Delta_1 + v_2 e^{-i\alpha}\Delta_2), \tag{2}$$

where $v_i \equiv |<0|\phi_i^0|0>|$ and α denotes the relative phase of the vacuum expectation values (vevs) of the neutral components of Φ_i. The matrices M_d, M_u are diagonalized by the usual bi-unitary transformations:

$$U_{dL}^\dagger M_d U_{dR} = D_d \equiv diag\ (m_d, m_s, m_b) \tag{3}$$
$$U_{uL}^\dagger M_u U_{uR} = D_u \equiv diag\ (m_u, m_c, m_t) \tag{4}$$

The neutral and the charged Higgs interactions obtained from Eq. (1) for the quark sector are of the form

$$\begin{aligned}\mathcal{L}_Y(\text{quark, Higgs}) &= -\overline{d_L^0}\frac{1}{v}\left[M_d H^0 + N_d^0 R + iN_d^0 I\right] d_R^0 - \\ &- \overline{u_L^0}\frac{1}{v}\left[M_u H^0 + N_u^0 R + iN_u^0 I\right] u_R^0 - \\ &- \frac{\sqrt{2}H^+}{v}(\overline{u_L^0} N_d^0\, d_R^0 - \overline{u_R^0} N_u^{0\dagger}\, d_L^0) + \text{h.c.}\end{aligned} \tag{5}$$

where $v \equiv \sqrt{v_1^2 + v_2^2} \approx 246$ GeV, and H^0, R are orthogonal combinations of the fields ρ_j, arising when one expands [1] the neutral scalar fields around their vacuum expectation values, $\phi_j^0 = \frac{e^{i\alpha_j}}{\sqrt{2}}(v_j + \rho_j + i\eta_j)$, choosing H^0 in such a way that it has couplings to the quarks which are proportional to the mass matrices, as can be seen from Eq. (5). Similarly, I denotes the linear combination of η_j orthogonal to the neutral Goldstone boson. The matrices N_d^0, N_u^0 are given by:

$$N_d^0 = \frac{1}{\sqrt{2}}(v_2\Gamma_1 - v_1 e^{i\alpha}\Gamma_2), \quad N_u^0 = \frac{1}{\sqrt{2}}(v_2\Delta_1 - v_1 e^{-i\alpha}\Delta_2) \tag{6}$$

It is clear from these expressions that the flavour structure of the quark sector of two Higgs doublet models is much richer than that of the SM requiring the four matrices M_d, M_u, N_d^0, N_u^0 in order to be fully specified. For the leptonic sector one can derive similar expressions and the corresponding matrices can be denoted by M_ℓ, M_ν, N_ℓ^0, N_ν^0. In the leptonic sector with Dirac neutrinos the analogy with the quark sector is perfect.

Flavour changing neutral currents in the quark sector are controlled by N_d^0 and N_u^0 while in the leptonic sector they are controlled by N_ℓ^0, N_ν^0.

In terms of physical quarks the neutral and the charged Higgs interactions can be written:

$$\begin{aligned}\mathcal{L}_Y(\text{quark, Higgs}) &= -\frac{\sqrt{2}H^+}{v}\bar{u}\left(VN_d\gamma_R - N_u^\dagger\, V\gamma_L\right)d + \text{h.c.} - \\ &- \frac{H^0}{v}\left(\bar{u}D_u u + \bar{d}D_d\, d\right) - \frac{R}{v}\left[\bar{u}(N_u\gamma_R + N_u^\dagger\gamma_L)u + \bar{d}(N_d\gamma_R + N_d^\dagger\gamma_L)\, d\right] + \\ &+ i\frac{I}{v}\left[\bar{u}(N_u\gamma_R - N_u^\dagger\gamma_L)u - \bar{d}(N_d\gamma_R - N_d^\dagger\gamma_L)\, d\right] \end{aligned} \tag{7}$$

where γ_L and γ_R are the left-handed and right-handed chirality projectors, respectively, and $N_d \equiv U_{dL}^\dagger N_d^0 U_{dR}$, $N_u \equiv U_{uL}^\dagger N_u^0 U_{uR}$, $V \equiv U_{uL}^\dagger U_{dL}$. The matrix V is a simplified notation for the V_{CKM} matrix. There are analogous expressions for the leptonic sector with V_{CKM} replaced by the Pontcorvo-Maki-Nakagawa-Sakata matrix, U_{PMNS}. The physical neutral Higgs fields are combinations of H^0, R and I.

Up till this point the discussion applies to the general two Higgs doublet model with Dirac fermions. The matrices N_d, N_u, N_ℓ and N_ν are entirely arbitrary and the scalar potential is the most general one for two Higgs doublets.

In order suppress the tree level FCNC in the quark sector by means of small entries of V_{CKM} Branco, Grimus and Lavoura imposed the following symmetry on the quark and scalar sector of the Lagrangian [10]:

$$Q_{Lj}^0 \to \exp(i\tau)\, Q_{Lj}^0\,, \qquad u_{Rj}^0 \to \exp(i2\tau)u_{Rj}^0\,, \qquad \Phi_2 \to \exp(i\tau)\Phi_2\,, \tag{8}$$

where $\tau \neq 0, \pi$, with all other quark fields transforming trivially under the symmetry. The index j can be fixed as either 1, 2 or 3. Alternatively the symmetry may be chosen as:

$$Q_{Lj}^0 \to \exp(i\tau)\, Q_{Lj}^0\,, \qquad d_{Rj}^0 \to \exp(i2\tau)d_{Rj}^0\,, \quad \Phi_2 \to \exp(-i\tau)\Phi_2\,. \tag{9}$$

The symmetry given by Eq. (8) leads to Higgs FCNC in the down sector only, whereas the symmetry specified by Eq. (9) leads to Higgs FCNC only in the up sector. These two alternative choices of symmetry combined with the three possible ways of fixing the index j give rise to six different realisations of 2HDM with the flavour structure, in the quark sector, controlled by the V_{CKM} matrix. The models obtained from the symmetry defined by Eq. (8) are called up-type models. In these models N_d and N_u have the simple form:

$$(N_d)_{rs} = \frac{v_2}{v_1}(D_d)_{rs} - \left(\frac{v_2}{v_1} + \frac{v_1}{v_2}\right)(V_{CKM}^\dagger)_{rj}(V_{CKM})_{js}(D_d)_{ss} \tag{10}$$

no sum in j implied, whereas, particularising the index j to be 3, we have:

$$N_u = -\frac{v_1}{v_2}\text{diag}\,(0,0,m_t) + \frac{v_2}{v_1}\text{diag}\,(m_u, m_c, 0) \tag{11}$$

the index j fixes the row of V_{CKM} which suppresses the flavour changing neutral currents. For down-type models, which are those obtained from imposing the symmetry defined by Eq. (9),

Journal of Physics: Conference Series **631** (2015) 012029 doi:10.1088/1742-6596/631/1/012029

the two matrices N_d and N_u exchange rôle and the FCNC are now suppressed by one of the columns of V_{CKM} depending on the index j.

As a result of imposing such a symmetry the matrices N_d and N_u are entirely determined by fermion masses, the V_{CKM} matrix and the angle β defined by $\tan\beta = v_2/v_1$, with no other free parameters. The flavour structure of BGL models depends on parameters already present in the SM apart from the new parameter $\tan\beta$. This characteristic is a defining feature of models that have been later-on denoted as models of Minimal Flavour Violation type [16], [17], [18], [19].

The leptonic sector with Dirac neutrinos is analogous to the quark sector and again there are six possible different realisations. Combining the two sectors one obtains thirty six different models which can be identified by a set of two indices. For example, the model $(\text{up}_3, \ell_2) \equiv (t, \mu)$ will have no tree level neutral flavour changing couplings in the up quark and the charged lepton sectors while the neutral flavour changing couplings in the down quark and neutrino sectors will be controlled, respectively, by $V_{td_i}\ V^*_{td_j}$ and $U_{\mu\nu_a}\ U^*_{\mu\nu_b}$.

The scalar potential is also constrained by the imposed symmetry. With the introduction of a soft symmetry breaking term it will have seven independent parameters which will determine the four scalar masses, the combination $v \equiv \sqrt{v_1^2 + v_2^2}$, $\tan\beta \equiv v_2/v_1$, and α. The angle α is the mixing angle relating the physical neutral CP-even scalars to the fields ρ_1 and ρ_2. One mixing angle is sufficient since this constrained scalar potential does not violate CP neither explicitly nor spontaneously [10] and therefore the field I is already physical. The soft symmetry breaking term prevents the appearance of an would-be Goldstone boson due to an accidental continuous global symmetry of the potential.

In the present work we assumed that H^0 coincides with the observed Higgs boson. Deviations from this assumption are still allowed by the experimental data but are constrained to be small. This assumption corresponds to imposing $\beta - \alpha = \pi/2$. The masses of the extra Higgs bosons must obey constraints coming from electroweak precision tests, in particular the T and S parameters. Bounds on T and S together with direct mass limits, significantly constrain the masses of the new scalar fields in terms of the mass of the charged Higgs $H^{\pm}$ [20] so that once this mass is fixed there is not much freedom left for the masses of the extra neutral scalars. As a result we can approximately scan the whole region of parameter space by varying $\tan\beta$ and and the mass, $m_{H^{\pm}}$, of the charged Higgs boson.

3. Confrontation with the experimental results

We performed an analysis of the thirty six BGL models in order to determine where could the masses of the new scalars lie and how these depend on $\tan\beta$. The masses of all new three scalar fields were treated independently and on an equal footing even though for simplicity we only presented results in terms of $m_{H^{\pm}}$. As stated in the previous section we assumed that the discovered Higgs at the LHC coincides with our H^0 scalar, the one without FCNC.

We imposed present constraints from several relevant flavour observables. In Table 1 we summarise the different types of relevant observables that we took into consideration, indicating where the contributions come from. In some cases the new contributions are only present in some but all of the models. In Ref. [14] we give a detailed description of the analysis that we performed and we list the set of experimental data and bounds that we used.

Figues 1 and 2 present the allowed regions we obtained for each one of the thirty six models.

Some of the BGL models allow for masses of the charged Higgs below 380 GeV which is the constraint from $b \to s\gamma$ on type II 2HDMs [21]. This is due to the different dependence that these models have on $\tan\beta$. However, in general, loop level processes, such as $b \to s\gamma$ as well as $\ell_j \to \ell_i\gamma$ provide important constraints. The same is true for the process $Z \to b\bar{b}$ and the oblique parameters S and T, unlike U.

Journal of Physics: Conference Series **631** (2015) 012029 doi:10.1088/1742-6596/631/1/012029

	BGL - 2HDM				SM	
	Charged $H^{\pm}$		Neutral R, I			
	Tree	Loop	Tree	Loop	Tree	Loop
$M \to \ell\bar{\nu}, M'\ell\bar{\nu}$	X	x		x	X	x
Universality	X	x		x	X	x
$M^0 \to \ell_1^+\ell_2^-$		x	X	x		X
$M^0 \leftrightarrow \bar{M}^0$		x	X	x		X
$\ell_1^- \to \ell_2^-\ell_3^+\ell_4^-$		x	X	x		x
$B \to X_s\gamma$		X		X		X
$\ell_j \to \ell_i\gamma$		X		X		x
EW Precision		X		X		X

Table 1. Summary table of the different types of relevant observables; leading contributions are tagged X while subleading or negligible ones are tagged x.

Concerning electric dipole moments of leptons and quarks [22], [23] BGL models do not give new physics one loop contributions. In [24] it has been shown that the weak basis invariant relevant for the quark EDMs does not acquire an imaginary part. In fact the same applies to two loop contributions within BGL models..

4. Conclusions

BGL models are very constrained since they have a very small number of free parameters and therefore they are highly predictive allowing to establish correlations among different observables. This also means that in principle it will be possible to rule out several of these different scenarios based on the future LHC results. Our results show that there are some very promising BGL implementations, which deserve more attention.

Several of the models allow for scalar masses within the reach of direct searches at the LHC.

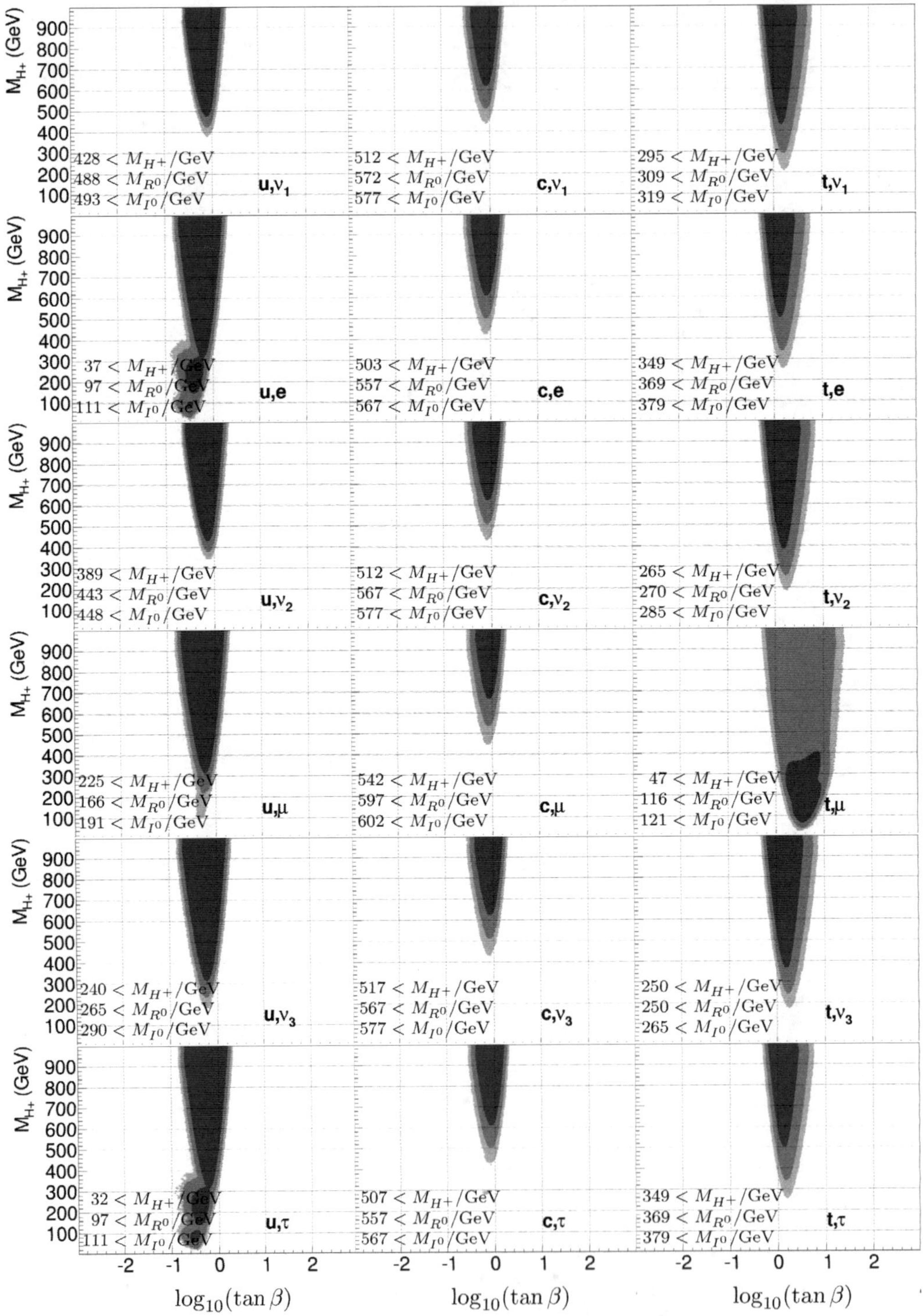

Figure 1. Allowed 68% (black), 95% (gray) and 99% (light gray) CL regions in $m_{H^\pm}$ vs. $\tan\beta$ for BGL models of types (u_i, ν_j) and (u_i, ℓ_j), i.e. for models with FCNC in the down quark sector and in the charged lepton or neutrino sector (respectively). Lower mass values corresponding to 95% CL regions are shown in each case.

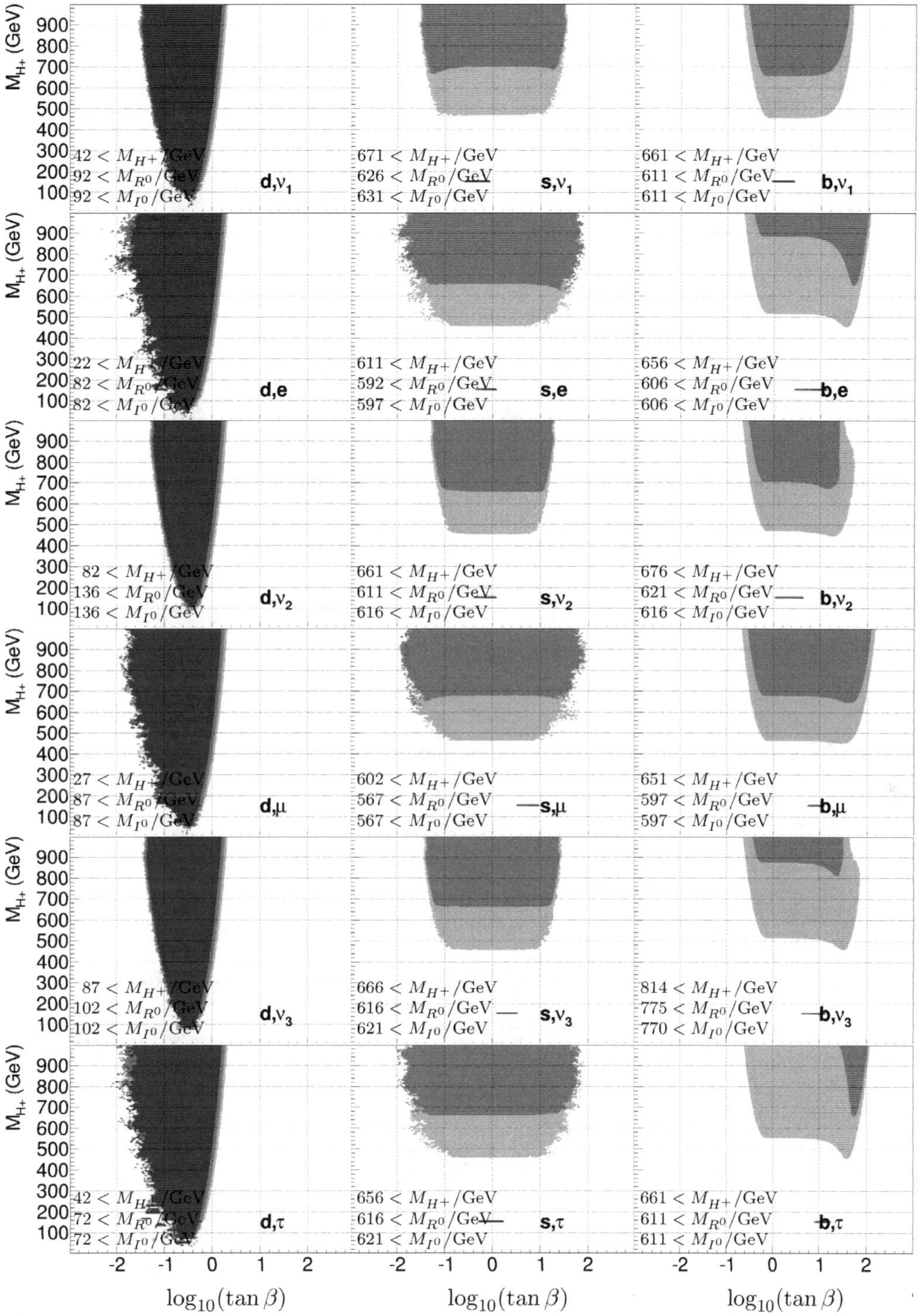

Figure 2. Allowed 68% (black), 95% (gray) and 99% (light gray) CL regions in $m_{H^\pm}$ vs. $\tan\beta$ for BGL models of types (d_i, ν_j) and (u_i, ℓ_j), i.e. for models with FCNC in the up quark sector and in the charged lepton or neutrino sector (respectively). Lower mass values corresponding to 95% CL regions are shown in each case.

Acknowledgments
The author thanks the Organisers of Discrete 2014 for the invitation to present this work for hospitality and for the very stimulating Conference. This work is partially supported by Fundação para a Ciência e a Tecnologia (FCT, Portugal) through the projects CERN/FP/123580/2011, PTDC/FIS-NUC/0548/2012, and CFTP-FCT Unit 777 (UID/FIS/00777/2013) which are partially funded through POCTI (FEDER), COMPETE, QREN and EU.

References

[1] Lee T D 1973 "A Theory of Spontaneous T Violation," *Phys. Rev.* D **8** 1226
[2] Peccei R D and Quinn H R 1977 "CP Conservation in the Presence of Instantons," *Phys. Rev. Lett.* **38** 1440
[3] Aad G *et al.* [ATLAS Collaboration] 2012 "Observation of a new particle in the search for the Standard Model Higgs boson with the ATLAS detector at the LHC," *Phys. Lett.* B **716** 1 [arXiv:1207.7214 [hep-ex]]
[4] Chatrchyan S *et al.* [CMS Collaboration] 2012 "Observation of a new boson at a mass of 125 GeV with the CMS experiment at the LHC," *Phys. Lett.* B **716** 30 [arXiv:1207.7235 [hep-ex]]
[5] Gunion J F, Haber, H E Kane, G L and Dawson S 2000 "The Higgs Hunter's Guide," *Front. Phys.* **80** 1
[6] Branco G C, Ferreira P M , Lavoura L, Rebelo M N , Sher M and Silva J P 2012 "Theory and phenomenology of two-Higgs-doublet models," *Phys. Rept.* **516** 1 [arXiv:1106.0034 [hep-ph]]
[7] Crivellin A Kokulu A and Greub C 2013 "Flavor-phenomenology of two-Higgs-doublet models with generic Yukawa structure," *Phys. Rev.* D **87** 9, 094031 [arXiv:1303.5877 [hep-ph]]
[8] Glashow S L and Weinberg S 1977 "Natural Conservation Laws for Neutral Currents," *Phys. Rev.* D **15** 1958
[9] Pich A and Tuzon P 2009 "Yukawa Alignment in the Two-Higgs-Doublet Model," *Phys. Rev.* D **80** 091702 [arXiv:0908.1554 [hep-ph]]
[10] Branco G C, Grimus W and Lavoura L 1996 "Relating the scalar flavor changing neutral couplings to the CKM matrix," *Phys. Lett.* B **380** 119 [hep-ph/9601383]
[11] Botella F J, Branco G C, Nebot M and Rebelo M N 2011 "Two-Higgs Leptonic Minimal Flavour Violation," it JHEP **1110** 037 [arXiv:1102.0520 [hep-ph]]
[12] Botella F J, Branco G C and Rebelo M N 2010 "Minimal Flavour Violation and Multi-Higgs Models," *Phys. Lett.* B **687** 194 [arXiv:0911.1753 [hep-ph]]
[13] Bhattacharyya G, Das, D, Pal P B and Rebelo M N 2013 "Scalar sector properties of two-Higgs-doublet models with a global U(1) symmetry," *JHEP* **1310** 081 [arXiv:1308.4297 [hep-ph]]
[14] Botella F J, Branco G C, Carmona A, Nebot M, Pedro L and Rebelo M N 2014 "Physical Constraints on a Class of Two-Higgs Doublet Models with FCNC at tree level," *JHEP* **1407** 078 [arXiv:1401.6147 [hep-ph]].
[15] Bhattacharyya G, Das D and Kundu A 2014 "Feasibility of light scalars in a class of two-Higgs-doublet models and their decay signatures," *Phys. Rev.* D **89** 095029 [arXiv:1402.0364 [hep-ph]]
[16] Buras A J, Gambino P, Gorbahn M, Jager S and Silvestrini L 2001 "Universal unitarity triangle and physics beyond the standard model," *Phys. Lett.* B **500** 161 [hep-ph/0007085]
[17] D'Ambrosio G, Giudice G F, Isidori G and Strumia A 2002 "Minimal flavor violation: An Effective field theory approach," *Nucl. Phys.* B **645** 155 [hep-ph/0207036]
[18] Bobeth C, Bona M, Buras A J, Ewerth T, Pierini M, Silvestrini L and Weile, A 2005 "Upper bounds on rare K and B decays from minimal flavor violation," *Nucl. Phys.* B **726** 252 [hep-ph/0505110]
[19] Dery A, Efrati A, Hiller G, Hochberg Y and Nir Y 2013 "Higgs couplings to fermions: 2HDM with MFV," *JHEP* **1308** (2013) 006 [arXiv:1304.6727 [hep-ph]]
[20] Grimus W, Lavoura L Ogreid, O M and Osland P 2008 "The Oblique parameters in multi-Higgs-doublet models," *Nucl. Phys.* B **801** 81 [arXiv:0802.4353 [hep-ph]]
[21] Hermann T, Misiak M and Steinhauser M 2012 "$\bar{B} \to X_s \gamma$ in the Two Higgs Doublet Model up to Next-to-Next-to-Leading Order in QCD," *JHEP* **1211** 036 [arXiv:1208.2788 [hep-ph]]
[22] Raidal M, van der Schaaf A, Bigi I, Mangano M L, Semertzidis Y K, Abel S, Albino S and Antusch S, *et al.* 2008 "Flavour physics of leptons and dipole moments," *Eur. Phys. J.* C **57** 13 [arXiv:0801.1826 [hep-ph]].
[23] Jung M and Pich A 2014 "Electric Dipole Moments in Two-Higgs-Doublet Models," *JHEP* **1404** 076 [arXiv:1308.6283 [hep-ph]]
[24] Botella F J, Branco G C and Rebelo M N 2013 "Invariants and Flavour in the General Two-Higgs Doublet Model," *Phys. Lett.* B **722** 76 [arXiv:1210.8163 [hep-ph]]

Journal of Physics: Conference Series **631** (2015) 012030 doi:10.1088/1742-6596/631/1/012030

Natural Standard Model Alignment in the Two Higgs Doublet Model

P. S. Bhupal Dev and Apostolos Pilaftsis

Consortium for Fundamental Physics, School of Physics and Astronomy, University of Manchester, Manchester M13 9PL, United Kingdom.

Abstract. In order to satisfy the current LHC Higgs data, which require the couplings of the observed 125 GeV Higgs boson to be close to the Standard Model (SM) expectations, any extended Higgs sector must lead to the so-called SM *alignment limit*. In the context of the Two Higgs Doublet Model (2HDM), this alignment is often associated with either decoupling of the heavy Higgs sector or accidental cancellations in the 2HDM potential. We present a symmetry justification for 'natural' alignment without decoupling or fine-tuning. We show that in Type II 2HDM, there exist only *three* different symmetry realizations which could lead to a natural alignment. We discuss some phenomenological implications of the Maximally-Symmetric 2HDM based on SO(5) symmetry group and analyze new collider signals for the heavy Higgs sector, involving third-generation quarks, which can be a useful observational tool during the Run-II phase of the LHC.

1. Introduction

The discovery of a Higgs boson with mass around 125 GeV [1] is the main highlight of the Run-I phase of the LHC, as it provides the first experimental evidence for the Higgs mechanism [2]. Although the measured couplings of the discovered Higgs boson show remarkable compatibility with those predicted by the Standard Model (SM) [3], the current experimental data still leave open the possibility of an extended Higgs sector. In fact, several well-motivated new-physics scenarios necessarily come with an enlarged Higgs sector, such as supersymmetry [4] and axion models [5], in order to address a number of theoretical and cosmological issues, including the gauge hierarchy problem, the origin of the Dark Matter (DM) and matter-antimatter asymmetry in our Universe. Here we consider the simplest extension of the standard Higgs mechanism, namely the Two Higgs Doublet Model (2HDM) [6], where the SM Higgs doublet is supplemented by another isodoublet with hypercharge $Y = 1$. This model can provide new sources of spontaneous [7] or explicit [8] CP violation, viable DM candidates [9] and a strong first order phase transition for electroweak baryogenesis [10].

In the doublet field space $\Phi_{1,2}$, where $\Phi_i = (\phi_i^+, \phi_i^0)^{\mathsf{T}}$, the general 2HDM potential reads

$$\begin{aligned} V \ = \ & -\mu_1^2(\Phi_1^\dagger\Phi_1) - \mu_2^2(\Phi_2^\dagger\Phi_2) - \Big[m_{12}^2(\Phi_1^\dagger\Phi_2) + \text{H.c.}\Big] \\ & + \lambda_1(\Phi_1^\dagger\Phi_1)^2 + \lambda_2(\Phi_2^\dagger\Phi_2)^2 + \lambda_3(\Phi_1^\dagger\Phi_1)(\Phi_2^\dagger\Phi_2) + \lambda_4(\Phi_1^\dagger\Phi_2)(\Phi_2^\dagger\Phi_1) \\ & + \left[\frac{\lambda_5}{2}(\Phi_1^\dagger\Phi_2)^2 + \lambda_6(\Phi_1^\dagger\Phi_1)(\Phi_1^\dagger\Phi_2) + \lambda_7(\Phi_1^\dagger\Phi_2)(\Phi_2^\dagger\Phi_2) + \text{H.c.}\right], \qquad (1) \end{aligned}$$

which contains four real mass parameters $\mu_{1,2}^2$, $\mathrm{Re}(m_{12}^2)$, $\mathrm{Im}(m_{12}^2)$, and ten real quartic couplings $\lambda_{1,2,3,4}$, $\mathrm{Re}(\lambda_{5,6,7})$, and $\mathrm{Im}(\lambda_{5,6,7})$. Thus, the vacuum structure of the general 2HDM can be quite rich [11], as compared to the SM.

The quark-sector Yukawa Lagrangian in the general 2HDM is given by

$$-\mathcal{L}_Y^q \;=\; \bar{Q}_L(h_1^u\Phi_1 + h_2^u\Phi_2)u_R \;+\; \bar{Q}_L(h_1^d\widetilde{\Phi}_1 + h_2^d\widetilde{\Phi}_2)d_R\,, \tag{2}$$

where $\widetilde{\Phi}_i = \mathrm{i}\sigma_2\Phi_i^*$ are the isospin conjugates of Φ_i, $Q_L = (u_L, d_L)^\mathsf{T}$ is the $SU(2)_L$ quark doublet and u_R, d_R are right-handed quark singlets. Due to the Yukawa interactions in (2), the neutral scalar bosons often induce unacceptably large flavor-changing neutral current (FCNC) processes at the tree level. This is usually avoided by imposing a discrete Z_2 symmetry [12] under which

$$\Phi_1 \to -\Phi_1, \quad \Phi_2 \to \Phi_2, \quad u_{Ra} \to u_{Ra}, \quad d_{Ra} \to d_{Ra} \;\text{ or }\; d_{Ra} \to -d_{Ra}\,, \tag{3}$$

($a = 1, 2, 3$ being the quark family index) so that only Φ_2 gives mass to up-quarks, and only Φ_1 or only Φ_2 gives mass to down-quarks. In this case, the scalar boson couplings to quarks are proportional to the quark mass matrix, as in the SM, and therefore, there is no tree-level FCNC process. The Z_2 symmetry (3) is satisfied by four discrete choices of tree-level Yukawa couplings between the Higgs doublets and SM fermions, which are known as Type I, II, X (lepton-specific) and Y (flipped) 2HDMs [6]. Global fits to the current LHC Higgs data [13, 14] suggest that all four types of discrete 2HDM are constrained to lie close to the so-called SM *alignment limit*, where the mass eigenbasis of the CP-even scalar sector aligns with the SM gauge eigenbasis. Specifically, in the Type-II (MSSM-type) 2HDM, the coupling of the SM-like Higgs to vector bosons is constrained to lie within 10% of the SM value at 95% CL [15].

Naively, the SM alignment is often associated with the decoupling limit, in which all the non-standard Higgs bosons are assumed to be much heavier than the electroweak scale so that the lightest CP-even scalar behaves just like the SM Higgs boson. This SM alignment limit can also be achieved, without decoupling [16–21]. However, for small $\tan\beta$ values, this is usually attributed to accidental cancellations in the 2HDM potential [19]. We present a symmetry argument to naturally justify the alignment limit [22], independently of the kinematic parameters of the theory, such as the heavy Higgs mass and $\tan\beta$. In the context of Type-II 2HDM, we show that there exist only *three* possible symmetry realizations of the scalar potential having natural alignment. We explicitly analyze the simplest case, namely the maximally symmetric 2HDM (MS-2HDM) with SO(5) symmetry. We show that the renormalization group (RG) effects due to the hypercharge gauge coupling g' and third-generation Yukawa couplings, as well as soft-breaking mass parameters, induce relevant deviations from the SO(5) limit, which lead to distinct predictions for the Higgs spectrum of the MS-2HDM. In particular, the heavy Higgs sector is predicted to be *quasi-degenerate*, which is a distinct feature of the SO(5) limit, apart from being *gaugephobic*, which is a generic feature in the alignment limit. Moreover, the current experimental constraints force the heavy Higgs sector to lie above the top-quark threshold in the MS-2HDM. Thus, the dominant collider signal for this sector involves final states with third-generation quarks. We study some of these collider signals for the upcoming run of the LHC.

The plan of this proceedings is as follows: In Section 2, we present the natural alignment condition for a generic 2HDM scalar potential. In Section 3, we list the symmetry classifications of the 2HDM potential and identify the symmetries leading to a natural alignment. In Section 4, we analyze the MS-2HDM in presence of custodial symmetry and soft breaking effects. In Section 5, we discuss some collider phenomenology of the heavy Higgs sector in the alignment limit, with particular emphasis on the heavy Higgs sector beyond the top-threshold. Our conclusions are given in Section 6.

2. Natural Alignment Condition

For simplicity, we will consider the 2HDM potential 1 with CP-conserving vacua; the results derived in this section can be easily generalized to the CP-violating 2HDM potential. We start with the linear decomposition of the two Higgs doublets in terms of eight real scalar fields:

$$\Phi_j = \begin{pmatrix} \phi_j^+ \\ \frac{1}{\sqrt{2}}(v_j + \phi_j + i a_j) \end{pmatrix}, \tag{4}$$

where $v_{1,2}$ are the vacuum expectation values (VEVs) and $v = \sqrt{v_1^2 + v_2^2} = 246.2$ GeV is the SM electroweak VEV. After spontaneous symmetry breaking, there are three Goldstone modes $(G^\pm, G^0)$, which become the longitudinal components of the $W^\pm$ and Z bosons. Thus, there are five remaining physical scalar mass eigenstates: two CP-even (h, H), one CP-odd (a) and two charged $(h^\pm)$ scalars. The corresponding physical mass eigenvalues are given by [23, 24]

$$M_{h^\pm}^2 = \frac{m_{12}^2}{s_\beta c_\beta} - \frac{v^2}{2}(\lambda_4 + \lambda_5) + \frac{v^2}{2 s_\beta c_\beta}\left(\lambda_6 c_\beta^2 + \lambda_7 s_\beta^2\right), \tag{5a}$$

$$M_a^2 = M_{h^\pm}^2 + \frac{v^2}{2}(\lambda_4 - \lambda_5), \tag{5b}$$

$$M_H^2 = \frac{1}{2}\left[(A+B) - \sqrt{(A-B)^2 + 4C^2}\right], \tag{5c}$$

$$M_h^2 = \frac{1}{2}\left[(A+B) + \sqrt{(A-B)^2 + 4C^2}\right], \tag{5d}$$

where we have used the short-hand notations $c_\beta \equiv \cos\beta$ and $s_\beta \equiv \sin\beta$ with $\tan\beta = v_2/v_1$ and

$$A = M_a^2 s_\beta^2 + v^2\left(2\lambda_1 c_\beta^2 + \lambda_5 s_\beta^2 + 2\lambda_6 s_\beta c_\beta\right), \tag{6a}$$

$$B = M_a^2 c_\beta^2 + v^2\left(2\lambda_2 s_\beta^2 + \lambda_5 c_\beta^2 + 2\lambda_7 s_\beta c_\beta\right), \tag{6b}$$

$$C = -M_a^2 s_\beta c_\beta + v^2\left(\lambda_{34} s_\beta c_\beta + \lambda_6 c_\beta^2 + \lambda_7 s_\beta^2\right). \tag{6c}$$

with $\lambda_{34} = \lambda_3 + \lambda_4$. The mixing between the mass eigenstates in the CP-odd and charged sectors is governed by the angle β, whereas in the CP-even sector, it is governed by an a priori different angle α, where $\tan 2\alpha = 2C/(A-B)$.

The SM Higgs field can be identified as the linear combination

$$H_{\rm SM} = \phi_1 \cos\beta + \phi_2 \sin\beta = H\cos(\beta - \alpha) + h \sin(\beta - \alpha). \tag{7}$$

From (7), we see that the couplings of h and H to the gauge bosons $(V = W^\pm, Z)$ with respect to the SM Higgs couplings $g_{H_{\rm SM}VV}$ will be

$$g_{hVV} = \sin(\beta - \alpha), \qquad g_{HVV} = \cos(\beta - \alpha). \tag{8}$$

Thus, the *SM alignment limit* is defined as the limit $\alpha \to \beta$ (or $\alpha \to \beta - \pi/2$) when H (h) couples to the vector bosons exactly like in the SM, whereas h (H) becomes gaugephobic. For notational clarity, we will take the alignment limit to be $\alpha \to \beta$ in the following.

To derive the alignment condition, we rewrite the CP-even scalar mass matrix as

$$M_S^2 = \begin{pmatrix} A & C \\ C & B \end{pmatrix} = \begin{pmatrix} c_\beta & -s_\beta \\ s_\beta & c_\beta \end{pmatrix} \begin{pmatrix} \widehat{A} & \widehat{C} \\ \widehat{C} & \widehat{B} \end{pmatrix} \begin{pmatrix} c_\beta & s_\beta \\ -s_\beta & c_\beta \end{pmatrix}, \tag{9}$$

where

$$\widehat{A} = 2v^2\Big[c_\beta^4\lambda_1 + s_\beta^2 c_\beta^2\lambda_{345} + s_\beta^4\lambda_2 + 2s_\beta c_\beta\Big(c_\beta^2\lambda_6 + s_\beta^2\lambda_7\Big)\Big]\,, \tag{10a}$$

$$\widehat{B} = M_a^2 + \lambda_5 v^2 + 2v^2\Big[s_\beta^2 c_\beta^2\Big(\lambda_1 + \lambda_2 - \lambda_{345}\Big) - s_\beta c_\beta\Big(c_\beta^2 - s_\beta^2\Big)\Big(\lambda_6 - \lambda_7\Big)\Big]\,, \tag{10b}$$

$$\widehat{C} = v^2\Big[s_\beta^3 c_\beta\Big(2\lambda_2 - \lambda_{345}\Big) - c_\beta^3 s_\beta\Big(2\lambda_1 - \lambda_{345}\Big) + c_\beta^2\Big(1 - 4s_\beta^2\Big)\lambda_6 + s_\beta^2\Big(4c_\beta^2 - 1\Big)\lambda_7\Big]. \tag{10c}$$

Here we have used the short-hand notation: $\lambda_{345} \equiv \lambda_3 + \lambda_4 + \lambda_5$. Evidently, the SM alignment limit $\alpha \to \beta$ is obtained when $\widehat{C} = 0$ in (9) [17]. From (10c), this yields the quartic equation

$$\lambda_7 t_\beta^4 - (2\lambda_2 - \lambda_{345})t_\beta^3 + 3(\lambda_6 - \lambda_7)t_\beta^2 + (2\lambda_1 - \lambda_{345})t_\beta - \lambda_6 \;=\; 0\,. \tag{11}$$

For *natural* alignment, (11) should be satisfied for *any* value of $\tan\beta$, which requires the coefficients of the polynomial in $\tan\beta$ to vanish identically. Imposing this restriction, we arrive at the *natural alignment condition* [22]

$$2\lambda_1 \;=\; 2\lambda_2 \;=\; \lambda_{345}\,, \qquad \lambda_6 \;=\; \lambda_7 \;=\; 0\,. \tag{12}$$

In particular, for $\lambda_6 = \lambda_7 = 0$ as in the Z_2-symmetric 2HDMs, (11) has a solution

$$\tan^2\beta \;=\; \frac{2\lambda_1 - \lambda_{345}}{2\lambda_2 - \lambda_{345}} \;>\; 0\,, \tag{13}$$

independent of M_a. After some algebra, the simple solution (13) to our general alignment condition (11) can be shown to be equivalent to that derived in [19, 25].

3. Symmetry Classifications of the 2HDM Potential

The general 2HDM potential (1) may exhibit three different classes of accidental symmetries. The first class of symmetries pertains to transformations of the Higgs doublets $\Phi_{1,2}$ only, but not their complex conjugates $\Phi^*_{1,2}$, and are known as the Higgs family (HF) symmetries [18, 26]. The second class of symmetry transformations relates the fields $\Phi_{1,2}$ to their complex conjugates $\Phi^*_{1,2}$ and are generically termed as CP symmetries [26]. The third class of symmetries utilize mixed HF and CP transformations that leave the $SU(2)_L$ gauge kinetic terms of $\Phi_{1,2}$ canonical [11].

In order to identify all accidental symmetries of the 2HDM potential, it is convenient to work in the bilinear scalar field formalism [27] by introducing an 8-dimensional complex multiplet [11, 28, 29]:

$$\mathbf{\Phi} \;\equiv\; \begin{pmatrix} \Phi_1 \\ \Phi_2 \\ \widetilde{\Phi}_1 \\ \widetilde{\Phi}_2 \end{pmatrix}\,, \tag{14}$$

where $\widetilde{\Phi}_i = \mathrm{i}\sigma^2\Phi_i^*$ (with $i = 1, 2$) and σ^2 is the second Pauli matrix. In terms of the $\mathbf{\Phi}$-multiplet, the following *null* 6-dimensional Lorentz vector can be defined [11, 29]:

$$R^A \;\equiv\; \mathbf{\Phi}^\dagger\Sigma^A\mathbf{\Phi}\,, \tag{15}$$

where $A = 0, 1, ..., 5$ and the six 8×8-dimensional matrices Σ^A may be expressed in terms of the three Pauli matrices $\sigma^{1,2,3}$ and the identity matrix $\mathbf{1}_{2\times 2} \equiv \sigma^0$, as follows:

$$\Sigma^{0,1,3} \;=\; \frac{1}{2}\sigma^0\otimes\sigma^{0,1,3}\otimes\sigma^0, \quad \Sigma^2 \;=\; \frac{1}{2}\sigma^3\otimes\sigma^2\otimes\sigma^0,$$
$$\Sigma^4 \;=\; -\frac{1}{2}\sigma^2\otimes\sigma^2\otimes\sigma^0, \quad \Sigma^5 \;=\; -\frac{1}{2}\sigma^1\otimes\sigma^2\otimes\sigma^0. \tag{16}$$

symmetry	μ_1^2	μ_2^2	m_{12}^2	λ_1	λ_2	λ_3	λ_4	$\mathrm{Re}(\lambda_5)$	$\lambda_6 = \lambda_7$
$Z_2\times$ O(2)	-	-	Real	-	-	-	-	-	Real
$(Z_2)^2\times$SO(2)	-	-	0	-	-	-	-	-	0
$(Z_2)^3\times$O(2)	-	μ_1^2	0	-	λ_1	-	-	-	0
O(2) $\times$O(2)	-	-	0	-	-	-	-	0	0
$Z_2\times$ $[\mathrm{O}(2)]^2$	-	μ_1^2	0	-	λ_1	-	-	$2\lambda_1 - \lambda_{34}$	0
O(3)$\times$O(2)	-	μ_1^2	0	-	λ_1	-	$2\lambda_1 - \lambda_3$	0	0
SO(3)	-	-	Real	-	-	-	-	λ_4	Real
$Z_2\times$O(3)	-	μ_1^2	Real	-	λ_1	-	-	λ_4	Real
$(Z_2)^2\times$SO(3)	-	μ_1^2	0	-	λ_1	-	-	$\pm\lambda_4$	0
O(2)$\times$O(3)	-	μ_1^2	0	-	λ_1	$2\lambda_1$	-	0	0
SO(4)	-	-	0	-	-	-	0	0	0
$Z_2\times$O(4)	-	μ_1^2	0	-	λ_1	-	0	0	0
SO(5)	-	μ_1^2	0	-	λ_1	$2\lambda_1$	0	0	0

Table 1: Relations between the parameters of the $U(1)_Y$-invariant 2HDM potential (1) for the 13 accidental symmetries [29] in a diagonally reduced basis, where $\mathrm{Im}(\lambda_5) = 0$ and $\lambda_6 = \lambda_7$. A dash signifies the absence of a constraint for that parameter.

Note that the bilinear field space spanned by the 6-vector R^A realizes an *orthochronous* $\mathrm{SO}(1,5)$ symmetry group.

In terms of the null-vector R^A defined in (15), the 2HDM potential (1) takes on a simple quadratic form:

$$V \;=\; -\frac{1}{2}\, M_A\, R^A \;+\; \frac{1}{4}\, L_{AB}\, R^A R^B\,, \tag{17}$$

where M_A and L_{AB} are $\mathrm{SO}(1,5)$ constant 'tensors' that depend on the mass parameters and quartic couplings given in (1) and their explicit forms may be found in [29–32]. Requiring that the $\mathrm{SU}(2)_L$ gauge-kinetic term of the $\mathbf{\Phi}$-multiplet remains canonical restricts the allowed set of rotations from SO(1,5) to SO(5), where only the spatial components R^I (with $I = 1, ..., 5$) transform and the zeroth component R^0 remains invariant. Consequently, in the absence of the hypercharge gauge coupling and fermion Yukawa couplings, the maximal symmetry group of the 2HDM is $G^R_{\rm 2HDM} = \mathrm{SO}(5)$. Including all its proper, improper and semi-simple subgroups of SO(5), all accidental symmetries for the 2HDM potential were classified in [11, 29], as shown in Table 1. Here we have used a diagonally reduced basis [33, 34], where $\mathrm{Im}(\lambda_5) = 0$ and $\lambda_6 = \lambda_7$, thus reducing the number of independent quartic couplings to seven. Each of the symmetries listed in Table 1 leads to certain constraints on the mass and/or coupling parameters.

From Table 1, we observe that in the context of Type-II 2HDM, there are *only* three symmetries, namely (i) $Z_2 \times [\mathrm{O}(2)]^2$, (ii) O(3)$\times$ O(2) and (iii) SO(5), which satisfy the natural alignment condition given by (12).[1] Note that in all the three naturally aligned scenarios, $\tan\beta$ as given in (13) 'consistently' gives an *indefinite* answer 0/0. In what follows, we focus on the simplest realization of the SM alignment, namely, the MS-2HDM based on the SO(5) group. A detailed study of the other two cases will be presented elsewhere.

[1] In Type-I 2HDM, there exists an additional possibility of realizing an exact Z_2 symmetry [35] which leads to an exact alignment, i.e. in the context of the so-called inert 2HDM [36].

Journal of Physics: Conference Series **631** (2015) 012030 doi:10.1088/1742-6596/631/1/012030

4. Maximally Symmetric 2HDM

From Table 1, we see that the maximal symmetry group in the bilinear field space is SO(5), in which case the parameters of the 2HDM potential (1) satisfy the following relations:

$$\begin{aligned} \mu_1^2 &= \mu_2^2\,, \quad m_{12}^2 = 0\,, \\ \lambda_2 &= \lambda_1\,, \quad \lambda_3 = 2\lambda_1\,, \quad \lambda_4 = \mathrm{Re}(\lambda_5) = \lambda_6 = \lambda_7 = 0\,, \end{aligned} \tag{18}$$

Thus, in this case, the 2HDM potential (1) is parametrized by just a *single* mass parameter $\mu_1^2 = \mu_2^2 \equiv \mu^2$ and a *single* quartic coupling $\lambda_1 = \lambda_2 = \lambda_3/2 \equiv \lambda$, as in the SM:

$$V = -\mu^2\left(|\Phi_1|^2 + |\Phi_2|^2\right) + \lambda\left(|\Phi_1|^2 + |\Phi_2|^2\right)^2 = -\frac{\mu^2}{2}\,\mathbf{\Phi}^\dagger\mathbf{\Phi} + \frac{\lambda}{4}\left(\mathbf{\Phi}^\dagger\mathbf{\Phi}\right)^2. \tag{19}$$

Note that the MS-2HDM scalar potential in (19) is more minimal than the respective potential of the MSSM at the tree level. Even in the custodial symmetric limit $g' \to 0$, the latter possesses a smaller symmetry: $\mathrm{O}(2) \times \mathrm{O}(3) \subset \mathrm{SO}(5)$, in the 5-dimensional bilinear R^I space.

Given the group isomorphy $\mathrm{SO}(5) \sim \mathrm{Sp}(4)/Z_2$, the maximal symmetry group of the 2HDM in the original $\mathbf{\Phi}$-field space is $\mathrm{G}^{\mathbf{\Phi}}_{\mathrm{2HDM}} = [\mathrm{Sp}(4)/Z_2] \times \mathrm{SU}(2)_L$ [22,29] in the custodial symmetry limit of vanishing g' and fermion Yukawa couplings. We can generalize this result to deduce that in the custodial symmetry limit, the maximal symmetry group for an n Higgs Doublet Model (nHDM) will be $\mathrm{G}^{\mathbf{\Phi}}_{n\mathrm{HDM}} = [\mathrm{Sp}(2n)/Z_2] \times \mathrm{SU}(2)_L$.

4.1. Scalar Spectrum in the MS-2HDM

Using the parameter relations given by (18), we find from (5a)-(5d) that in the MS-2HDM, the CP-even Higgs H has mass $M_H^2 = 2\lambda_2 v^2$, whilst the remaining four scalar fields, denoted hereafter as h, a and $h^\pm$, are massless. This is a consequence of the Goldstone theorem [37], since after electroweak symmetry breaking, $\mathrm{SO}(5) \xrightarrow{\langle\Phi_{1,2}\rangle \neq 0} \mathrm{SO}(4)$. Thus, we identify H as the SM-like Higgs boson with the mixing angle $\alpha = \beta$ [cf. (7)], i.e. the SM alignment limit can be naturally attributed to the SO(5) symmetry of the theory.

In the exact SO(5)-symmetric limit, the scalar spectrum of the MS-2HDM is experimentally unacceptable. This is because the four massless pseudo-Goldstone particles, viz. h, a and $h^\pm$, have sizable couplings to the SM Z and $W^\pm$ bosons, and could induce additional decay channels, such as $Z \to ha$ and $W^\pm \to h^\pm h$, which are experimentally excluded [38]. As we will see in the next subsection, the SO(5) symmetry may be violated predominantly by renormalization group (RG) effects due to g' and third-generation Yukawa couplings, as well as by soft SO(5)-breaking mass parameters, thereby lifting the masses of these pseudo-Goldstone particles.

4.2. RG and Soft Breaking Effects

To calculate the RG and soft-breaking effects in a technically natural manner, we assume that the SO(5) symmetry is realized at some high scale $\mu_X \gg v$. The physical mass spectrum at the electroweak scale is then obtained by the RG evolution of the 2HDM parameters given by (1). Using state-of-the-art two-loop RG equations given in [22], we first examine the deviation of the Higgs spectrum from the SO(5)-symmetric limit due to g' and Yukawa coupling effects, in the absence of the soft-breaking term. This is illustrated in Figure 1 for a typical choice of parameters in a Type-II realization of the 2HDM. We find that the RG-induced g' effects only lift the charged Higgs-boson mass $M_{h^\pm}$, while the corresponding Yukawa coupling effects also lift slightly the mass of the non-SM CP-even pseudo-Goldstone boson h. However, they still leave the CP-odd scalar a massless, which can be identified as a $\mathrm{U}(1)_{\mathrm{PQ}}$ axion [39].

Therefore, g' and Yukawa coupling effects are *not* sufficient to yield a viable Higgs spectrum at the weak scale, starting from a SO(5)-invariant boundary condition at some high scale μ_X. To

Journal of Physics: Conference Series **631** (2015) 012030 doi:10.1088/1742-6596/631/1/012030

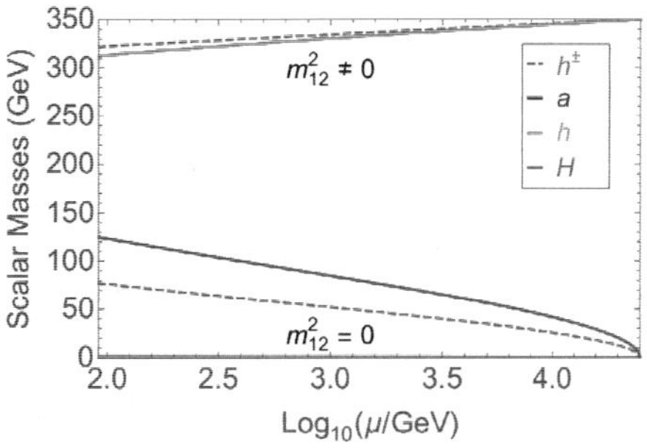

Figure 1: The Higgs mass spectrum in the MS-2HDM without and with soft breaking effects induced by m_{12}^2. For $m_{12}^2 = 0$, the CP-odd scalar a remains massless at tree-level, whereas h and $h^\pm$ receive small masses due to the g' and Yukawa coupling effects. For $m_{12}^2 \neq 0$, one obtains a quasi-degenerate heavy Higgs spectrum, cf. (20). Here we have chosen $\mu_X = 2.5 \times 10^4$ GeV, $\lambda(\mu_X) = 0$ and $\tan\beta = 50$ for illustration.

minimally circumvent this problem, we include soft SO(5)-breaking effects, by assuming a non-zero soft-breaking term $\mathrm{Re}(m_{12}^2)$. In the SO(5)-symmetric limit for the scalar quartic couplings, but with $\mathrm{Re}(m_{12}^2) \neq 0$, we obtain the following mass spectrum [cf. (5a)-(5d)]:

$$M_H^2 \;=\; 2\lambda_2 v^2 \,, \qquad M_h^2 \;=\; M_a^2 \;=\; M_{h^\pm}^2 \;=\; \frac{\mathrm{Re}(m_{12}^2)}{s_\beta c_\beta} \,, \tag{20}$$

as well as an equality between the CP-even and CP-odd mixing angles: $\alpha = \beta$, thus predicting an *exact* alignment for the SM-like Higgs boson H, simultaneously with an experimentally allowed heavy Higgs spectra (see Figure 1 for $m_{12}^2 \neq 0$ case). Note that in the alignment limit, the heavy Higgs sector is exactly degenerate [cf. (20)] at the SO(5) symmetry-breaking scale, and at the low-energy scale, this degeneracy is mildly broken by the RG effects. Thus, we obtain a quasi-degenerate heavy Higgs spectrum, which is a unique prediction of the MS-2HDM, valid even in the non-decoupling limit, and can be used to distinguish this model from other 2HDM scenarios.

4.3. Misalignment Predictions

As discussed in Section 4.2, there will be some deviation from the alignment limit in the low-energy Higgs spectrum of the MS-2HDM due to RG and soft-breaking effects. By requiring that the mass and couplings of the SM-like Higgs boson H are consistent with the latest Higgs data from the LHC [3], we derive predictions for the remaining scalar spectrum and compare them with the existing (in)direct limits on the heavy Higgs sector. For the SM-like Higgs boson mass, we use the 3σ allowed range from the recent CMS and ATLAS Higgs mass measurements [3,40]: $M_H \in \big[124.1,\ 126.6\big]$ GeV. For the Higgs couplings to the SM vector bosons and fermions, we use the constraints in the $(\tan\beta,\ \beta - \alpha)$ plane derived from a recent global fit for the Type-II 2HDM [15]. For a given set of SO(5) boundary conditions $\big\{\mu_X, \tan\beta(\mu_X), \lambda(\mu_X)\big\}$, we thus require that the RG-evolved 2HDM parameters at the weak scale must satisfy the above constraints on the lightest CP-even Higgs boson sector. This requirement of *alignment* with the SM Higgs sector puts stringent constraints on the MS-2HDM parameter space, as shown in Figure 2 by the blue shaded region. In the red shaded region, there is no viable solution to the RG equations. We ensure that the remaining allowed (white) region satisfies the necessary theoretical constraints, i.e. positivity and vacuum stability of the Higgs potential, and perturbativity of the Higgs self-couplings [6]. From Figure 2, we find that there exists an upper

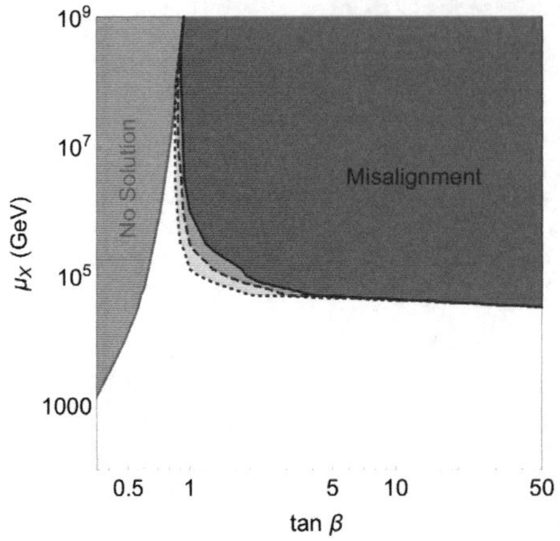

Figure 2: The 1σ (dotted), 2σ (dashed) and 3σ (solid) exclusion contours (blue shaded region) from the alignment constraints in MS-2HDM. The red shaded region is theoretically excluded, as there is no solution to the RG equations up to two-loop order in this region.

limit of $\mu_X \lesssim 10^9$ GeV on the SO(5)-breaking scale of the 2HDM potential, beyond which an ultraviolet completion of the theory must be invoked. Moreover, for 10^5 GeV $\lesssim \mu_X \lesssim 10^9$ GeV, only a narrow range of $\tan\beta$ values are allowed.

For the allowed parameter space of our MS-2HDM as shown in Figure 2, we obtain concrete predictions for the remaining Higgs spectrum. In particular, the alignment condition imposes a *lower* bound on the soft breaking parameter $\text{Re}(m_{12}^2)$, and hence, on the heavy Higgs spectrum. The comparison of the existing global fit limit on the charged Higgs-boson mass as a function of $\tan\beta$ [15] with our predicted limits from the alignment condition in the MS-2HDM for a typical value of the boundary scale $\mu_X = 3 \times 10^4$ GeV is shown in Figure 3 (left panel). It is clear that the alignment limits are stronger than the global fit limits, except in the very small and very large $\tan\beta$ regimes. For $\tan\beta \lesssim 1$ region, the indirect limit obtained from the $Z \to b\bar{b}$ precision observable becomes the strictest [15,41]. Similarly, for the large $\tan\beta \gtrsim 30$ case, the alignment limit can be easily obtained [cf. (10c)] without requiring a large soft-breaking parameter m_{12}^2, and therefore, the lower limit on the charged Higgs mass derived from the misalignment condition becomes somewhat weaker in this regime.

From Figure 2, it should be noted that for $\mu_X \gtrsim 10^5$ GeV, phenomenologically acceptable alignment is not possible in the MS-2HDM for large $\tan\beta$ *and* large m_{12}^2. Therefore, we also get an *upper* bound on the charged Higgs-boson mass $M_{h^\pm}$ from the misalignment condition, depending on $\tan\beta$. This is illustrated in Figure 3 (right panel) for $\mu_X = 10^5$ GeV.

Similar alignment constraints are obtained for the heavy neutral pseudo-Goldstone bosons h and a, which are predicted to be quasi-degenerate with the charged Higgs boson $h^\pm$ in the MS-2HDM [cf. (20)]. The current experimental lower limits on the heavy neutral Higgs sector [38] are much weaker than the alignment constraints in this case. Thus, the MS-2HDM scenario provides a natural reason for the absence of a heavy Higgs signal below the top-quark threshold, and this has important consequences for the heavy Higgs searches in the run-II phase of the LHC, as discussed in the following section.

5. Collider Signatures in the Alignment Limit

In the alignment limit, the couplings of the lightest CP-even Higgs boson are exactly similar to the SM Higgs couplings, while the heavy CP-even Higgs boson is gaugephobic [cf. (8)].

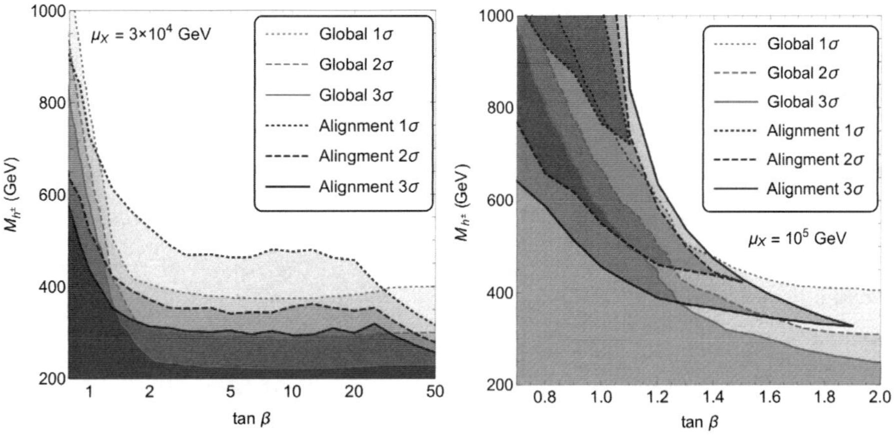

Figure 3: *Left:* The 1σ (dotted), 2σ (dashed) and 3σ (solid) *lower* limits on the charged Higgs mass obtained from the alignment condition (blue lines) in the MS-2HDM with $\mu_X = 3 \times 10^4$ GeV. *Right:* The 1σ (dotted), 2σ (dashed) and 3σ (solid) *allowed* regions from the alignment condition (blue lines) for $\mu_X = 10^5$ GeV. For comparison, the corresponding lower limits from a global fit are also shown (red lines).

Therefore, two of the relevant Higgs production mechanisms at the LHC, namely, the vector boson fusion and Higgsstrahlung processes are suppressed for the heavy neutral Higgs sector. As a consequence, the only relevant production channels to probe the neutral Higgs sector of the MS-2HDM are the gluon-gluon fusion and $t\bar{t}h$ ($b\bar{b}h$) associated production mechanisms at low (high) $\tan\beta$. For the charged Higgs sector of the MS-2HDM, the dominant production mode is the associated production process: $gg \to \bar{t}bh^+ + t\bar{b}h^-$, irrespective of $\tan\beta$.

Similarly, for the decay modes of the heavy neutral Higgs bosons in the MS-2HDM, the $t\bar{t}$ ($b\bar{b}$) channel is the dominant one for low (high) $\tan\beta$ values, whereas for the charged Higgs boson $h^{+(-)}$, the $t\bar{b}(\bar{t}b)$ mode is the dominant one for any $\tan\beta$. Thus, the heavy Higgs sector of the MS-2HDM can be effectively probed at the LHC through the final states involving third-generation quarks.

5.1. Charged Higgs Signal

The most promising channel at the LHC for the charged Higgs boson in the MS-2HDM is

$$gg \;\to\; \bar{t}bh^+ + t\bar{b}h^- \;\to\; t\bar{t}b\bar{b}\,. \tag{21}$$

Experimentally, this is a challenging mode due to large QCD backgrounds and the non-trivial event topology, involving at least four b-jets [42]. Nevertheless, a recent CMS study [43] has presented for the first time a realistic analysis of this process, in the leptonic decay mode of the W's coming from top decays:

$$gg \;\to\; h^\pm tb \;\to\; (\ell\nu_\ell bb)(\ell'\nu_{\ell'}b)b \tag{22}$$

(ℓ, ℓ' beings electrons or muons). Using the $\sqrt{s} = 8$ TeV LHC data, they have derived 95% CL upper limits on the production cross section $\sigma(gg \to h^\pm tb)$ times the branching ratio $\mathrm{BR}(h^\pm \to tb)$ as a function of the charged Higgs mass, as shown in Figure 4. In the same Figure, we show the corresponding predictions at $\sqrt{s} = 14$ TeV LHC in the Type-II MS-2HDM

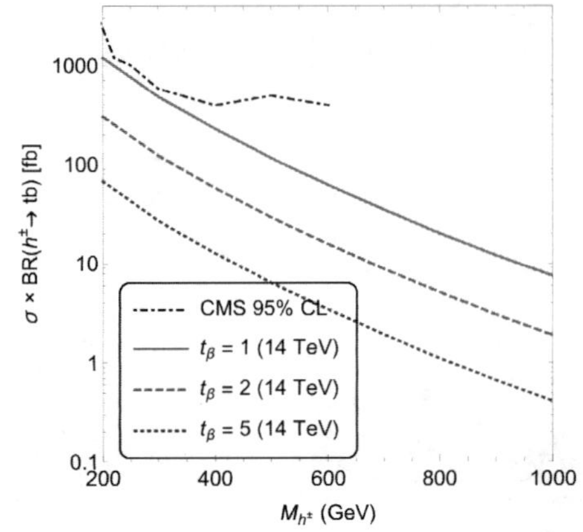

Figure 4: Predictions for the cross section of the process (21) in the Type-II MS-2HDM at $\sqrt{s} = 14$ TeV LHC for various values of $\tan\beta$. For comparison, we have also shown the current 95% CL CMS upper limit from the $\sqrt{s} = 8$ TeV data [43].

for some representative values of $\tan\beta$. The cross section predictions were obtained at leading order (LO) by implementing the 2HDM in `MadGraph5_aMC@NLO` [44] and using the `NNPDF2.3` PDF sets [45]. A comparison of these cross sections with the CMS limit suggests that the run-II phase of the LHC might be able to probe the low $\tan\beta$ region of the MS-2HDM parameter space using the process (21). Note that the production cross section $\sigma(gg \to \bar{t}bh^+)$ decreases rapidly with increasing $\tan\beta$ due to the Yukawa coupling suppression, even though $\mathrm{BR}(h^+ \to \bar{t}b)$ remains close to 100%. Therefore, this channel is only effective for low $\tan\beta$ values.

In order to make a rough estimate of the $\sqrt{s} = 14$ TeV LHC sensitivity to the charged Higgs signal (21) in the MS-2HDM, we perform a parton level simulation of the signal and background events using `MadGraph5` [44]. For the event reconstruction, we use some basic selection cuts on the transverse momentum, pseudo-rapidity and dilepton invariant mass, following the CMS analysis [43]:

$$\begin{aligned} p_T^\ell &> 20 \text{ GeV}, \quad |\eta^\ell| < 2.5, \quad p_T^j > 30 \text{ GeV}, \quad |\eta^j| < 2.4, \quad \not{E}_T > 40 \text{ GeV} \\ \Delta R^{\ell\ell} &> 0.4, \quad \Delta R^{\ell j} > 0.4, \quad M_{\ell\ell} > 12 \text{ GeV}, \quad |M_{\ell\ell} - M_Z| > 10 \text{ GeV}. \end{aligned} \tag{23}$$

Jets are reconstructed using the anti-k_T clustering algorithm [46] with a distance parameter of 0.5. Since four b-jets are expected in the final state, at least two b-tagged jets are required in the signal events, and we assume the b-tagging efficiency for each of them to be 70%.

The inclusive SM cross section for $pp \to t\bar{t}b\bar{b} + X$ is $\sim$ 18 pb at NLO, with roughly 30% uncertainty due to higher order QCD corrections [47]. Most of the QCD background for the $4b + 2\ell + \not{E}_T$ final state given by (22) can be reduced significantly by reconstructing at least one top-quark. The remaining irreducible background due to SM $t\bar{t}b\bar{b}$ production can be suppressed with respect to the signal by reconstructing the charged Higgs boson mass, once a valid signal region is defined, e.g. in terms of an observed excess of events at the LHC in future. For the semi-leptonic decay mode of top-quarks as in (22), one cannot directly use an invariant mass observable to infer $M_{h^\pm}$, as both the neutrinos in the final state give rise to missing momentum. A useful quantity in this case is the M_{T2} variable [48], defined as

$$M_{T2} \;=\; \min_{\left\{\not{p}_{T_a} + \not{p}_{T_b} = \not{p}_T\right\}} \left[\max\left\{m_{T_a}, m_{T_b}\right\}\right], \tag{24}$$

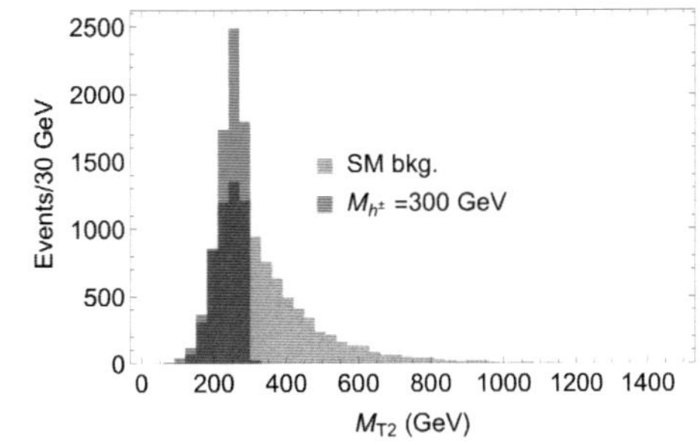

Figure 5: An illustration of the charged Higgs boson mass reconstruction using the M_{T2} variable. The irreducible SM background distribution is also shown for comparison.

where $\{\mathrm{a}\}, \{\mathrm{b}\}$ stand for the two sets of particles in the final state, each containing a neutrino with part of the missing transverse momentum ($\not{\mathbf{p}}_{T_{\mathrm{a,b}}}$). Minimization over all possible sums of these two momenta gives the observed missing transverse momentum $\not{\mathbf{p}}_T$, whose magnitude is the same as $\not{E}_T$ in our specific case. In (24), m_{T_i} (with i =a,b) is the usual transverse mass variable for the system $\{i\}$, defined as

$$m_{T_i}^2 = \left(\sum_{\text{visible}} E_{T_i} + \not{E}_{T_i}\right)^2 - \left(\sum_{\text{visible}} \mathbf{p}_{T_i} + \not{\mathbf{p}}_{T_i}\right)^2 . \tag{25}$$

For the correct combination of the final state particles in (22), i.e. for $\{\mathrm{a}\} = (\ell\nu_\ell bb)$ and $\{\mathrm{b}\} = (\ell'\nu_{\ell'} bb)$ in (24), the maximum value of M_{T2} represents the charged Higgs boson mass, with the M_{T2} distribution smoothly dropping to zero at this point. This is illustrated in Figure 5 for a typical choice of $M_{h^\pm} = 300$ GeV. For comparison, we also show the M_{T2} distribution for the SM background, which obviously does not have a sharp endpoint. Thus, for a given hypothesized signal region defined in terms of an excess due to $M_{h^\pm}$, we may impose an additional cut on $M_{T2} \leq M_{h^\pm}$ to enhance the signal (22) over the irreducible SM background.

Assuming that the charged Higgs boson mass can be reconstructed efficiently, we present an estimate of the signal to background ratio for the charged Higgs signal given by (21) at $\sqrt{s} = 14$ TeV LHC with 300 fb^{-1} for some typical values of $\tan\beta$ in Figure 6. Since the mass of the charged Higgs boson is a priori unknown, we vary the charged Higgs mass, and for each value of $M_{h^\pm}$, we assume that it can be reconstructed around its actual value within 30 GeV uncertainty.

5.2. Heavy Neutral Higgs Signal

So far there have been no direct searches for heavy neutral Higgs bosons involving $t\bar{t}$ and/or $b\bar{b}$ final states, mainly due to the challenges associated with uncertainties in the jet energy scales and the combinatorics arising from complicated multiparticle final states in a busy QCD environment. Nevertheless, these channels become pronounced in the MS-2HDM scenario, and hence, we have made a preliminary attempt to study them in [22]. In particular, we focus on the search channel

$$gg \to t\bar{t}h \to t\bar{t}t\bar{t} . \tag{26}$$

Such four top final states have been proposed before in the context of other exotic searches at the LHC (see e.g. [49]). However, their relevance for heavy Higgs searches have not been explored so far. We note here that the existing 95% CL experimental upper limit on the four top production cross section is 59 fb from ATLAS [50] and 32 fb from CMS [51], whereas the SM prediction for the inclusive cross section of the process $pp \to t\bar{t}t\bar{t} + X$ is about 10-15 fb [52].

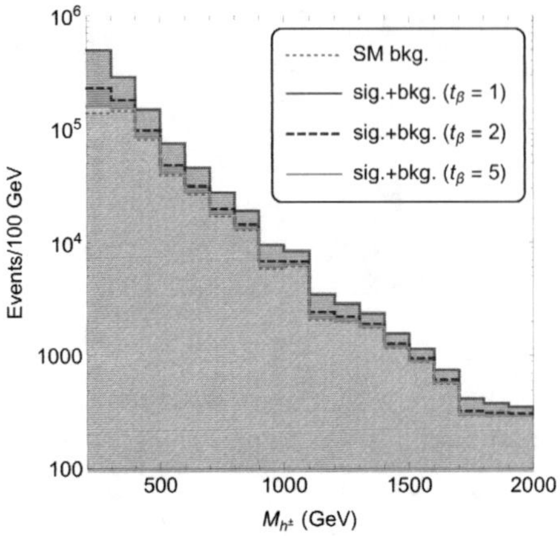

Figure 6: Predicted number of events for the $t\bar{t}b\bar{b}$ signal from the charged pseudo-Goldstone boson in the MS-2HDM at $\sqrt{s}$ = 14 TeV LHC with 300 fb^{-1} integrated luminosity. The irreducible SM background (red shaded) is controlled by assuming an efficient mass reconstruction technique [22].

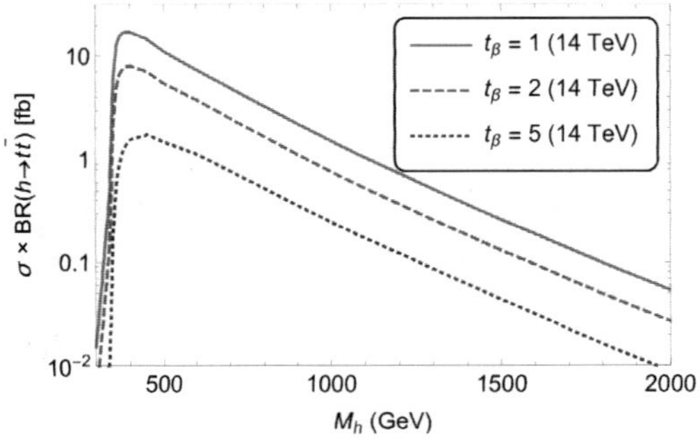

Figure 7: Predictions for the cross section of the process (26) in the Type-II MS-2HDM at $\sqrt{s} = 14$ TeV LHC for various values of $\tan\beta$.

To get a rough estimate of the signal to background ratio for our four-top signal (26), we perform a parton-level simulation of the signal and background events at LO in QCD using `MadGraph5_aMC@NLO` [44] with `NNPDF2.3` PDF sets [45]. For the inclusive SM cross section for the four-top final state at $\sqrt{s} = 14$ TeV LHC, we obtain 11.85 fb, whereas our proposed four-top signal cross sections are found to be comparable or smaller depending on M_h and $\tan\beta$, as shown in Figure 7. However, since we expect one of the $t\bar{t}$ pairs coming from an on-shell h decay to have an invariant mass around M_h, we can use this information to significantly boost the signal over the irreducible SM background. Note that all the predicted cross sections shown in Figure 7 are well below the current experimental upper bound [51].

Depending on the W decay mode from $t \to Wb$, there are 35 final states for four top decays. According to a recent ATLAS analysis [53], the experimentally favored channel is the semi-leptonic/hadronic final state with two same-sign isolated leptons. Although the branching fraction for this topology (4.19%) is smaller than most of the other channels, the presence of two same-sign leptons in the final state allows us to reduce the large QCD background substantially,

Journal of Physics: Conference Series **631** (2015) 012030 doi:10.1088/1742-6596/631/1/012030

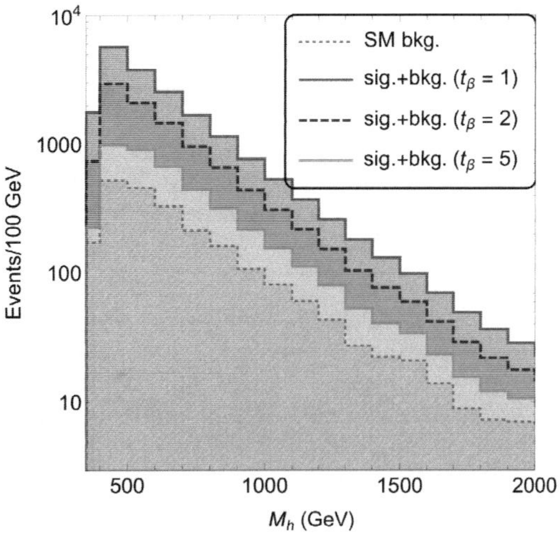

Figure 8: Predicted number of events for the $t\bar{t}t\bar{t}$ signal from the neutral pseudo-Goldstone boson in the MS-2HDM at $\sqrt{s}$ = 14 TeV LHC with 300 fb^{-1} integrated luminosity. The results are shown for three different values of $\tan\beta$ =1 (green solid), 2 (blue dashed) and 5 (orange solid). The SM background (red dotted) is controlled by assuming an efficient mass reconstruction technique, as outlined in the text.

including that due to the SM production of $t\bar{t}b\bar{b}$+jets [53]. Therefore, we will only consider the following decay chain in our preliminary analysis:

$$gg \to t\bar{t}h \to (t\bar{t})(t\bar{t}) \to \Big((\ell^{\pm}\nu_\ell b)(jjb)\Big)\Big((\ell'^{\pm}\nu_{\ell'} b)(jjb)\Big) . \tag{27}$$

For event reconstruction, we will use the same selection cuts as in (23), and in addition, following [53], we require the scalar sum of the p_T of all leptons and jets (defined as H_T) to exceed 350 GeV.

As in the charged Higgs boson case (cf. Figure 5), the heavy Higgs mass can be reconstructed from the signal given by (27) using the M_{T2} endpoint technique, and therefore, an additional selection cut on $M_{T2} \leq M_h$ can be used to enhance the signal over the irreducible background. Our simulation results for the predicted number of signal and background events for the process (27) at $\sqrt{s}$ = 14 TeV LHC with 300 fb^{-1} luminosity are shown in Figure 8. The signal events are shown for three representative values of $\tan\beta$. Here we vary the a priori unknown heavy Higgs mass, and for each value of M_h, we assume that it can be reconstructed around its actual value within 30 GeV uncertainty. From this preliminary analysis, we find that the $t\bar{t}t\bar{t}$ channel provides the most promising collider signal to probe the heavy Higgs sector in the MS-2HDM at low values of $\tan\beta \lesssim 5$.

The above analysis is also applicable for the CP-odd Higgs boson a, which has similar production cross sections and $t\bar{t}$ branching fractions as the CP-even Higgs h. However, the $t\bar{t}h(a)$ production cross section as well as the $h(a) \to t\bar{t}$ branching ratio decreases with increasing $\tan\beta$. This is due to the fact that the $ht\bar{t}$ coupling in the alignment limit is $\cos\alpha/\sin\beta \sim \cot\beta$, which is same as the $at\bar{t}$ coupling. Thus, the high $\tan\beta$ region of the MS-2HDM cannot be searched via the $t\bar{t}t\bar{t}$ channel proposed above, and one needs to consider the channels involving down-sector Yukawa couplings, e.g. $b\bar{b}b\bar{b}$ and $b\bar{b}\tau^+\tau^-$ [42]. It is also worth commenting here that the simpler process $pp \to h/a \to t\bar{t}$ ($b\bar{b}$) at low (high) $\tan\beta$ suffers from a huge SM $t\bar{t}$ ($b\bar{b}$) QCD background, even after imposing an $M_{t\bar{t}\ (b\bar{b})}$ cut. Some parton-level studies of this signal in the context of MSSM have been performed in [54].

We should clarify that the results obtained in this section are valid only at the parton level. In a realistic detector environment, the sharp features of the signal [see e.g., Figure 5] used to derive the sensitivity reach in Figures 6 and 8 may not survive, and therefore, the signal-to-background ratio might get somewhat reduced than that shown here. A detailed detector-level analysis of these signals, including realistic top reconstruction efficiencies and smearing effects, is currently being pursued in a separate dedicated study.

6. Conclusions

We provide a symmetry justification of the so-called SM alignment limit, independently of the heavy Higgs spectrum and the value of $\tan\beta$ in the 2HDM. We show that in the Type-II 2HDM, there exist *only* three different symmetry realizations, which could lead to the SM alignment by satisfying the natural alignment condition (12) for *any* value of $\tan\beta$. In the context of the Maximally Symmetric 2HDM based on the SO(5) group, we demonstrate how small deviations from this alignment limit are naturally induced by RG effects due to the hypercharge gauge coupling g' and third generation Yukawa couplings, which also break the custodial symmetry of the theory. In addition, a non-zero soft SO(5)-breaking mass parameter is required to yield a viable Higgs spectrum consistent with the existing experimental constraints. Employing the current Higgs signal strength data from the LHC, which disfavor large deviations from the alignment limit, we derive important constraints on the 2HDM parameter space. In particular, we predict lower limits on the heavy Higgs spectrum, which prevail the present global fit limits in a wide range of parameter space. Depending on the scale where the maximal symmetry could be realized in nature, we also obtain an upper limit on the heavy Higgs masses in certain cases, which could be probed during the run-II phase of the LHC. Finally, we have studied the collider signatures of the heavy Higgs sector in the alignment limit beyond the top-quark threshold. We find that the final states involving third-generation quark final states can become a valuable observational tool to directly probe the heavy Higgs sector of the 2HDM in the alignment limit for low values of $\tan\beta$.

Acknowledgments

This work was supported by the Lancaster-Manchester-Sheffield Consortium for Fundamental Physics under STFC grant ST/L000520/1. P.S.B.D. would like to acknowledge the local hospitality provided by the CFTP and IST, Lisbon where part of this proceedings was written.

References

[1] G. Aad *et al.* [ATLAS Collaboration], Phys. Lett. B **716**, 1 (2012); S. Chatrchyan *et al.* [CMS Collaboration], Phys. Lett. B **716**, 30 (2012).
[2] P. W. Higgs, Phys. Rev. Lett. **13**, 508 (1964); F. Englert and R. Brout, Phys. Rev. Lett. **13**, 321 (1964); G. S. Guralnik, C. R. Hagen and T. W. B. Kibble, Phys. Rev. Lett. **13**, 585 (1964).
[3] ATLAS collaboration, ATLAS-CONF-2014-009; V. Khachatryan *et al.* [CMS Collaboration], arXiv:1412.8662 [hep-ex].
[4] H. E. Haber and G. L. Kane, Phys. Rept. **117**, 75 (1985).
[5] J. E. Kim, Phys. Rept. **150**, 1 (1987).
[6] G. C. Branco *et al.*, Phys. Rept. **516**, 1 (2012).
[7] T. D. Lee, Phys. Rev. D **8**, 1226 (1973); Phys. Rept. **9**, 143 (1974).
[8] H. Georgi, Hadronic J. **1**, 155 (1978); Y. L. Wu and L. Wolfenstein, Phys. Rev. Lett. **73**, 1762 (1994).
[9] V. Silveira and A. Zee, Phys. Lett. B **161**, 136 (1985).
[10] V. A. Kuzmin, V. A. Rubakov and M. E. Shaposhnikov, Phys. Lett. B **155**, 36 (1985).
[11] R. A. Battye, G. D. Brawn and A. Pilaftsis, JHEP **1108**, 020 (2011).
[12] S. L. Glashow and S. Weinberg, Phys. Rev. D **15**, 1958 (1977); E. A. Paschos, Phys. Rev. D **15**, 1966 (1977).

[13] ATLAS collaboration, ATLAS-CONF-2014-010; V. Khachatryan *et al.* [CMS Collaboration], Phys. Rev. D **90**, 112013 (2014).
[14] A. Celis, V. Ilisie and A. Pich, JHEP **1307**, 053 (2013); C. W. Chiang and K. Yagyu, JHEP **1307**, 160 (2013); C.-Y. Chen, S. Dawson and M. Sher, Phys. Rev. D **88**, 015018 (2013); N. Craig, J. Galloway and S. Thomas, arXiv:1305.2424 [hep-ph]; K. Cheung, J. S. Lee and P.-Y. Tseng, JHEP **1401**, 085 (2014); L. Wang and X.-F. Han, JHEP **1411**, 085 (2014); B. Dumont, J. F. Gunion, Y. Jiang and S. Kraml, Phys. Rev. D **90**, 035021 (2014); arXiv:1409.4088 [hep-ph]; S. Kanemura, K. Tsumura, K. Yagyu and H. Yokoya, Phys. Rev. D **90**, 075001 (2014).
[15] O. Eberhardt, U. Nierste and M. Wiebusch, JHEP **1307**, 118 (2013); J. Baglio, O. Eberhardt, U. Nierste and M. Wiebusch, Phys. Rev. D **90**, 015008 (2014).
[16] P. H. Chankowski, T. Farris, B. Grzadkowski, J. F. Gunion, J. Kalinowski and M. Krawczyk, Phys. Lett. B **496**, 195 (2000).
[17] J. F. Gunion and H. E. Haber, Phys. Rev. D **67**, 075019 (2003).
[18] I. F. Ginzburg and M. Krawczyk, Phys. Rev. D **72**, 115013 (2005);
[19] M. Carena, I. Low, N. R. Shah and C. E. M. Wagner, JHEP **1404**, 015 (2014).
[20] G. Bhattacharyya and D. Das, Phys. Rev. D **91**, 015005 (2015).
[21] A. Delgado, G. Nardini and M. Quiros, JHEP **1307**, 054 (2013).
[22] P. S. B. Dev and A. Pilaftsis, JHEP **1412**, 024 (2014).
[23] H. E. Haber and R. Hempfling, Phys. Rev. D **48**, 4280 (1993).
[24] A. Pilaftsis and C. E. M. Wagner, Nucl. Phys. B **553**, 3 (1999).
[25] M. Carena, H. E. Haber, I. Low, N. R. Shah and C. E. M. Wagner, Phys. Rev. D **91**, 035003 (2015).
[26] P. M. Ferreira, H. E. Haber and J. P. Silva, Phys. Rev. D **79**, 116004 (2009); P. M. Ferreira, H. E. Haber, M. Maniatis, O. Nachtmann and J. P. Silva, Int. J. Mod. Phys. A **26**, 769 (2011).
[27] M. Maniatis, A. von Manteuffel, O. Nachtmann and F. Nagel, Eur. Phys. J. C **48**, 805 (2006).
[28] C. C. Nishi, Phys. Rev. D **83**, 095005 (2011).
[29] A. Pilaftsis, Phys. Lett. B **706**, 465 (2012).
[30] M. Maniatis, A. von Manteuffel and O. Nachtmann, Eur. Phys. J. C **57** (2008) 719.
[31] I. P. Ivanov, Phys. Rev. D **77** (2008) 015017.
[32] C. C. Nishi, Phys. Rev. D **77** (2008) 055009.
[33] J. F. Gunion and H. E. Haber, Phys. Rev. D **72**, 095002 (2005).
[34] M. Maniatis and O. Nachtmann, JHEP **1111**, 151 (2011).
[35] N. G. Deshpande and E. Ma, Phys. Rev. D **18**, 2574 (1978).
[36] R. Barbieri, L. J. Hall and V. S. Rychkov, Phys. Rev. D **74**, 015007 (2006).
[37] J. Goldstone, Nuovo Cim. **19**, 154 (1961).
[38] K. A. Olive *et al.* (Particle Data Group), Chin. Phys. C **38**, 090001 (2014) [`http://pdg.lbl.gov/`].
[39] R. D. Peccei and H. R. Quinn, Phys. Rev. Lett. **38**, 1440 (1977).
[40] G. Aad *et al.* [ATLAS Collaboration], Phys. Rev. D **90**, 052004 (2014).
[41] O. Deschamps, S. Descotes-Genon, S. Monteil, V. Niess, S. T'Jampens and V. Tisserand, Phys. Rev. D **82**, 073012 (2010).
[42] S. Heinemeyer *et al.* [LHC Higgs Cross Section Working Group Collaboration], arXiv:1307.1347 [hep-ph].
[43] CMS Collaboration, CMS-PAS-HIG-13-026.
[44] J. Alwall *et al.*, JHEP **1407**, 079 (2014); `https://launchpad.net/mg5amcnlo`.
[45] R. D. Ball *et al.*, Nucl. Phys. B **867**, 244 (2013); `https://nnpdf.hepforge.org/`
[46] M. Cacciari, G. P. Salam and G. Soyez, JHEP **0804**, 063 (2008).
[47] G. Bevilacqua, M. Czakon, C. G. Papadopoulos, R. Pittau and M. Worek, JHEP **0909**, 109 (2009).
[48] C. G. Lester and D. J. Summers, Phys. Lett. B **463**, 99 (1999).
[49] M. Spira and J. D. Wells, Nucl. Phys. B **523**, 3 (1998); A. Pomarol and J. Serra, Phys. Rev. D **78**, 074026 (2008); K. Kumar, T. M. P. Tait and R. Vega-Morales, JHEP **0905**, 022 (2009); S. Jung and J. D. Wells, JHEP **1011**, 001 (2010); G. L. Kane, E. Kuflik, R. Lu and L. T. Wang, Phys. Rev. D **84**, 095004 (2011); G. Cacciapaglia, R. Chierici, A. Deandrea, L. Panizzi, S. Perries and S. Tosi, JHEP **1110**, 042 (2011).
[50] ATLAS Collaboration, ATLAS-CONF-2013-051.
[51] V. Khachatryan *et al.* [CMS Collaboration], JHEP **1411**, 154 (2014).
[52] G. Bevilacqua and M. Worek, JHEP **1207**, 111 (2012).
[53] D. Paredes [ATLAS Collaboration], PhD Thesis, Université Blaise Pascal (2013), CERN-THESIS-2013-202.
[54] S. Moretti and D. A. Ross, Phys. Lett. B **712**, 245 (2012); A. Djouadi and J. Quevillon, JHEP **1310**, 028 (2013); A. Djouadi, L. Maiani, A. Polosa, J. Quevillon and V. Riquer, arXiv:1502.05653 [hep-ph].

Degeneracy on a two-Higgs CP non-invariant Higgs sector

M Gómez-Bock[1], E Hernández[2], A Mondragón[2], M Mondragón[2].

[1] Universidad de las Américas Puebla. Ex-hda. Sta. Catarina Mártir S/N. C.P. 72810. San Andrés Cholula, Puebla. México.

[2] Instituto de Física, Universidad Nacional Autónoma de México, Apdo. Postal 20-364, 01000 México D.F., México.

E-mail: melina.gomez@udlap.mx, queta@fisica.unam.mx, mondra@fisica.unam.mx, myriam@fisica.unam.mx

Abstract. Many models with extended Higgs sectors have states where masses are quasi-degenerate at tree-level. Higher order corrections may separate this masses, but they may also bring them closer, opening the possibility of total or near degeneracy of the physical masses. If, in addition, the Higgs sector does not conserve CP symmetry, then the neutral Higgs physical states are no longer defined with a CP charge, they would be a mixture of CP defined states given at tree level. Near degeneracy in the heavy neutral Higgs system with CP violation could imply mixing of coherent overlapping states, which at exact degeneracy will show up as a double pole in the propagator matrix of the neutral Higgs sector. This possibility has consequences on the phenomenology and in the type of expected experimental signatures for the nearly and exact Higgs degeneracy, for example, the lineshape will be different than expected for the fermion scattering amplitude via $\not{CP}$ Higgs, $\sigma(\mu\bar{\mu} \to H_i^{\not{CP}} \to f'\bar{f}')$.

1. Introduction

It is well known that CP (charge-parity) is not an exact symmetry of the hadronic sector and, at higher energies, the non-conservation of CP could be present in other sectors as well, for instance the scalar Higgs sector. Besides this *natural* reason, new sources of CP violation are a required condition for baryogenesis [1]. Models that consider direct CP non-invariance in the Higgs sector imply mixing of $CP-even$ and $CP-odd$ neutral states.

In the case of the Minimal Supersymmetric Standard Model (MSSM), CP conservation at first order could be broken by higher order terms when the soft-SUSY couplings are complex numbers. For the Two Higgs Doublet Model (THDM) the couplings are not as restricted as in the MSSM and some can be considered complex even at tree level.

In the MSSM the tree level Higgs states H^0 and A^0 are nearly degenerate in mass and have been widely studied [2]. In the context of the radiative corrected CP non-invariant self-energies, the exact degeneracy of the non-physical states H^0 and A^0 has been considered [3]. The Argand diagram of a complex degeneracy is calculated in a THDM with a top dominating diagram and also for a supersymmetric model considering stops contributions. This $\not{CP}$ system could affect $\gamma\gamma$ Higgs production. In [4] stop and sbottom as dominant radiative corrections using Renormalization Group Equations (RGE) to $\not{CP}$ neutral Higgs bosons masses have been performed,

Journal of Physics: Conference Series **631** (2015) 012031 doi:10.1088/1742-6596/631/1/012031

they considering values for the parameters which move the physical states away from degeneracy.

A renormalized procedure using the Feynman Diagrammatic Approach and dominant m_t^4 diagrams $\not{CP}$ in the Higgs mass matrix was performed in [5]. The most complete work done so far considering all the one-loop diagrams and dominant two-loop, to obtain renormalized Higgs masses in this $\not{CP}$ MSSM (or CMSSM) context is presented in [6], where the authors found that the real part of the propagator has an important contribution from the off-diagonal neutral Higgs masses.

The $\not{CP}$ Higgs neutral system might show coherence of Higgs states as a manifestation of a geometrical degeneration. A geometrical degeneracy is characterized by coalescence of eigenvalues and of the corresponding eigenstates [7, 8, 9]. Associated to this degeneracy a branch point or exceptional point appears in the non-Hermitian Hamiltonian spectra as function of the parameters of the model [10]. In this specific point of non-Hermitian degeneracy the basis of the space states is not complete, the system is no longer diagonalizable by a similarity transformation with a unitary matrix, and implies that the states are self-orthogonal.

One of the first studies which shed light on the presence of this branch point singularities was done by C. Bender and T. Wu [12] in a $\lambda\phi^4$ field theory. In this model the perturbation series for the ground state energy diverges. The authors carried out a study for the structure of energy levels extending the analyticity to the complex $\lambda-$plane based on a WKB method and the property of self-orthogonality of the functions, and found that the energy levels $E(\lambda)$ have an infinite number of branch points. For each branch point a crossing energy level occurs (two-fold degeneracy) associated to a self-orthogonal state.

This property of branch points has been studied in the THDM model in ref. [11]. The effects of interference due to widths of two or more near neutral Higgs resonances in a general multi-Higgs doublet models with CP non invariance has been studied in [13].

The presence of branch points have been studied in nuclear physics [21] and meson physics, [22], as well as in string theory [23]. Their physical properties have been observed in non-relativistic physical systems, as for example in microwave cavities [14, 15, 16, 17, 18], in electronic systems [19], and in molecular photo-dissociation [20].

The goal of this work is to study the manifestation of accidental degeneracy in the Higgs heavy neutral system. In section 3 we calculate the transition matrix for the s-channel via Higgs $\not{CP}$ considering a top dominant loop showing the analytical form at exact degeneracy. The coherence of the Higgs boson states might show up as non-hermitian resonances in the Higgs lineshape via fermionic scattering. Considering a complete calculation with $s-$dependence will be relevant when analyzing possible $\not{CP}$ experimental signals at colliders.

2. Masses as poles of the propagator

The first order squared mass matrix $M^{(0)2}$ of neutral Higgs fields, which is Hermitian and symmetric by CPT invariance, is readily derived from the second derivative of the Higgs potential. In order to renormalize the masses it has to be supplemented by the anti-hermitian decay matrix $-iM\Gamma(s)$ [26, 27], which includes the widths of the Higgs states in the diagonal elements, as well as the transition matrix elements for any combination of pairs of states in the off diagonal elements, these last elements are non zero in the $\not{CP}$ case.

$$\mathcal{M}^2(s) = \mathcal{M}^2(s) - iM\ \Gamma(s) = M^{(0)2} + \hat{\Pi}(s), \tag{1}$$

where $M^{(0)2}$ is the bare mass and $\hat{\Pi}(s)$ contains the quantum corrections. The mass matrix elements $\hat{\Pi}(s)$ are constructed explicitly from the contribution of the self-energy diagrams, *i.e*

Journal of Physics: Conference Series **631** (2015) 012031 doi:10.1088/1742-6596/631/1/012031

radiative mass corrections. It is important to notice that the physical masses, defined as the poles of the propagator of the transition matrix, where the complete renormalized mass matrix should be considered, will depend on the model as new loop corrections coming from extra particles may contribute to Higgs radiative corrections; it will also depend on the way non-Hermitian couplings appear in this mass matrix.

To obtain renormalized masses there are different approaches [28]: **a)** The effective potential method, **b)** The renormalization group method, **c)** Complete one-loop calculation.
Specifically for the case of CP violation of neutral Higgs sector this issue has been treated with a variety of approaches in different models:

(i) Only Standard Model loop contributions resumed in the *Pinch Technique* framework [29, 30].
(ii) In the MSSM considering that the $\not{CP}$ contributions are the diagrams dominated by complex trilineal A-terms couplings to the stop and sbottom [31].
(iii) Considering domain m_t^4 contributions [5]
(iv) All complex-MSSM contributions using Feynman Diagrammatic approach to the self-energies considering complex couplings were calculated numerically by Frank *et. al.* [6].

Whichever way it is calculated, the mass matrix should be gauge independent, display only physical thresholds, have the correct unitarity, analyticity and renormalization group properties and satisfy the equivalence principle.
For the purpose of this paper we will consider the Higgs masses on a simplified toy model given in [3]. A complete analysis of degeneracy for the renormalized Higgs masses in the CMSSM will be presented in a forthcoming paper.

2.1. Neutral Higgs bosons as s-channel resonances
The detailed study of the $H_2 - H_3$ system is particularly interesting because in most beyond the Standard Model scenarios, such as SUSY, the masses of these two particles are almost degenerate. Therefore, their resonant production is expected to be detected as two relatively closely spaced or even overlapping resonances in processes such as the production of fermion and anti-fermion pairs in collisions of muons and anti-muons. In the case of non-coherence, the lineshape of these resonances (fig. (1)) in the muon cross section via Higgs boson exchange, would indicate the presence or absence of CP mixing in the $H_2 - H_3$ system [32]. The lineshape of non-complex couplings on the MSSM is given in ref. [34].

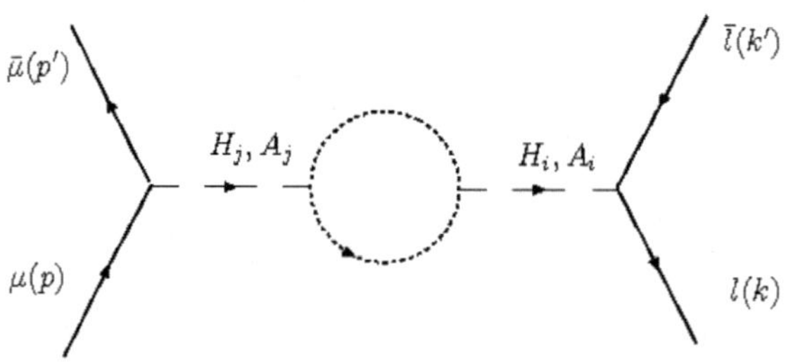

Figure 1. s-channel one-loop muon cross section through neutral $\not{CP}$ Higgs

The transition amplitude for the production of a fermion-antifermion pair f and $\bar{f}$ in a $\mu\mu$ collision is a two channel contribution and may be written as

$$\mathcal{T} = \mathcal{T}^s + \mathcal{T}^t = V_i^P \hat{\Delta}_{ij} V_j^D + \mathcal{T}^t, \tag{2}$$

where $i, j = A, H$ for the known CP invariant basis and $i, j = H_2, H_3$ for the physical basis. V_i^P and V_j^D are the production and decay amplitudes of the processes. At one-loop level, the production and decay vertices, $V_i^P(s)$ and $V_j^D(s)$ coincide with the corresponding tree level vertices.

In eq.(2), $\hat{\Delta}_{ij}$ is the propagator matrix in the presence of mixing in the $s-$channel, and $\mathcal{T}^t$ represents all the $t-$channel and box diagrams. In the resonant region, on which we concentrate in this section, the $t-$channel and box diagrams are sub-dominant and can be safely ignored [30].

Then, in the electroweak basis, the transition amplitude matrix via resonant excitation of CP-non conserving Higgs exchange, with CP violation at one loop is given by

$$\mathcal{T}^s(s) = V^P \frac{1}{s - \mathcal{M}_H^2} V^D. \tag{3}$$

where we have consider the decoupling limit approximation for the neutral Higgs system, in which we obtain the light Higgs boson h^0 as a light SM-like Higgs, leaving two neutral heavy Higgs states which at tree level are almost degenerate. Then the mixing of CP is within the two heavier states and the mass matrix in eq. (3) is given by

$$\mathcal{M}_H^2 = \mathcal{M}_{H_2H_3}^2(s) = \begin{pmatrix} m_H^2 + \hat{\Pi}_{HH}(s) & \hat{\Pi}_{HA}(s) \\ \hat{\Pi}_{HA}(s) & m_A^2 + \hat{\Pi}_{AA}(s) \end{pmatrix}, \tag{4}$$

This decoupling limit given in terms of the parameters of the Higgs potential, is defined by the inequality [33]

$$m_A^2 >> |\lambda_i| v^2, \tag{5}$$

where λ_i are the couplings of the Lagrangian [3]. This means that the mixing between the light state, $H_1 \approx h^0$, and the heavy states, H_2 and H_3, is small, compared with the mixing of the nearly degenerate heavy Higgs states H_2 and H_3. Now, the physical masses of the neutral heavy Higgs bosons are identified with the poles of the propagator matrix $\hat{\Delta}_{H_2H_3}(s)$. Hence, the masses of the neutral, heavy Higgs bosons are defined as the solutions of the implicit equation

$$det\left[\hat{\Delta}_{H_2H_3}^{-1}(s)\right] = det\left[(s)\mathbf{1}_{2\times 2} - \mathcal{M}_{H_2H_3}^2(s)\right] = 0, \tag{6}$$

is the mass matrix of the neutral and heavy Higgs bosons in the standard CP defined states basis, where the notation of lower case m_{H_i} implies tree level mass. In the physical basis, $\mathcal{M}_{H_2H_3}^2(s)$ is diagonal with mass eigenvalues given by

$$m_{H_j}^2(s^\star) + \hat{\Pi}_{H_j}(s^\star) := \mu_{H_j}^2(s_j^\star), \qquad j = 2, 3. \tag{7}$$

We identify $s_i^\star$ with the solution of eq. (6), i.e. the value of the pole mass of the neutral heavy Higgs bosons, where $s_i^\star(x_1, x_2)$ is a function of two or more free parameters of the Lagrangian of the particular model. Thus, the two heavy neutral Higgs bosons are mass degenerate if there exists an $s^\star$ such that

$$\mu_{H_2}^2(s^\star) = \mu_{H_3}^2(s^\star). \tag{8}$$

As the determinant is an invariant quantity, we may write eq.(6) in terms of the eigenvalues of this squared mass matrix which, at the pole position $s^\star$, gives

$$(s_2^\star - \mu_{H_2}^2(s_2^\star))(s_3^\star - \mu_{H_3}^2(s_3^\star)) = 0\,, \quad s_2^\star \neq s_3^\star; \tag{9}$$

where $s_i^\star(x_1, x_2)$ is a function of two or more free parameters in $\mathcal{L}$. In our case we use the simplified model where the parameters would be λ as the magnitude of all complex couplings and ϕ as the complex phase.

Finally, in the resonant region, the dispersive (real) part of the self-energy (occurring in the Breit-Wigner denominator of the propagator) can be re-absorbed in the real part of the mass matrix, as is expressed in eq. (1), and the imaginary part will correspond to

$$i\Im m\Pi_{ii}(s) = iM_i\Gamma_i(s), \;\; i = H, A. \tag{10}$$

Then the masses of the neutral heavy Higgs bosons can be written as

$$\mu^2_{H_j}(s) \;\; = \;\; M^2_{H_j} - iM_j\Gamma_j = \frac{1}{2}T \pm \sqrt{(\vec{R} - i\vec{\Gamma})^2} \tag{11}$$

with $j = 2, 3$ and

$$\vec{R} = \left(\frac{1}{2}\left[M_H^2 - M_A^2\right], 0, \Re e\left[\hat{\Pi}_{HA}(s)\right]\right), \;\; \vec{\Gamma} = \left(\tfrac{1}{2}\left[M_H\Gamma_H(s) - M_A\Gamma_A(s)\right], 0, \Im m\left[\hat{\Pi}_{HA}(s)\right]\right)$$

which, in the case of the $\not{C}\not{P}$ Higgs heavy sector may change the structure of the physical states near degeneracy and should be taken into account carefully. In next section we will obtain the mass matrix elements explicitly using a simplified THDM to account for the 1 loop corrections.

3. Branch point in simplified THDM

The branch point or exceptional point is associated to the degeneracy of the masses in the physical basis. At exact degeneracy, the eigenstates and eigenvalues of the mass matrix are equal, and we can find the values of the parameter model at which this occurs.

We will be using the mass matrix parametrization following ref. [3] where one-loop top quark is taken to be dominant. for quantum radiative corrections The two parameters λ and ϕ are the THDM model CP violation parameters coming from the Lagrangian, $\lambda = |\lambda_i|$ as a generalized magnitude of complex couplings and ϕ the generalized phase in the couplings.

The degeneracy can be analyzed through the difference of the physical masses, which may vanish for specific values of the parameters. Other analysis of this kind of singularity can be found in [8, 9, 10]. The complex masses are unfolded generating hypersurfaces which show the branch point behavior of the singularity. The real and imaginary parts of the masses have branch cuts which begin in the same exceptional point but they extend in opposite directions, as is shown in Figure 1 of [11].

The difference of the mass eigenstates (11) can be given in terms of the non-physical basis as

$$\begin{aligned}
\mu_2^2 - \mu_3^2 &= (M^2_{H_2} - M^2_{H_3}) - i(M_{H_2}\Gamma_{H_2} - M_{H_3}\Gamma_{H_3}) \\
&= \Big\{\big[(M_H^2 - M_A^2)^2 - (M_H\Gamma_H(s^\star) - M_A\Gamma_A(s^\star))^2\big] \\
&+ 4\big[(\Re e\Pi_{HA})^2 - (\Im m\Pi_{HA})^2\big] - 2i\big[(M_H^2 - M_A^2)(M_H\Gamma_H(s^\star) \\
&- M_A\Gamma_A(s^\star)) - 4\,(\Re e\Pi_{HA})\,(\Im m\Pi_{HA})\big]\Big\}^{1/2}.
\end{aligned} \tag{12}$$

Considering no s-dependence, the non-physical mass matrix elements differences are given in terms of the Lagrangian parameters by

$$M_H^2 - M_A^2 = \lambda v^2 \cos\phi,$$

$$|M_H\Gamma_H - M_A\Gamma_A| = \frac{1}{32\pi}(\Delta_t + 9\lambda^2 v^2 \cos 2\phi). \tag{13}$$

Here the top quark contribution is contained in $\Delta_t = -12M_{H/A}^2(m_t/v)^2(1-\beta_t^2)\beta_t$, with β_t the rapidity of the top quark. Using the same values of the mases and decay width involved, proposed in the same reference: $M_H = 520GeV$, $M_A = 500GeV$, $\Gamma_H = 2.58GeV$ and $\Gamma_H = 2.58GeV$. The off-diagonal terms are parametrized as

$$\Re e\ \Pi_{HA} = -\frac{1}{2}\lambda v^2 \sin\phi, \quad \text{and} \quad \Im m\ \Pi_{HA} = -\frac{9}{64\pi}\lambda^2 v^2 \sin 2\phi, \tag{14}$$

In terms of this model parametrization, the difference of the squared masses of the heavy neutral system $H_2 - H_3$ becomes

$$\begin{aligned} \mu_2^2 - \mu_3^2 &= \frac{1}{2}\Big\{\Big[\lambda^2 v^4 - \left(\frac{9}{32\pi}\right)^2 \lambda^4 v^2 \\ &- \left(\frac{1}{32\pi}\right)^2 (\Delta_t^2 + 18\Delta_t\lambda^2 v^2 \cos 2\phi)\Big] \\ &- i\frac{1}{16\pi}\Big[\Delta_t \lambda v^2 \cos\phi + 9\lambda^3 v^4 \cos\phi(1 - 4\sin^2\phi)\Big]\Big\}^{1/2} \end{aligned} \tag{15}$$

Solving for $\mu_2^2 - \mu_3^2 = 0$ we find the numerical values of λ and ϕ which accomplish degeneracy conditions

$$\lambda_1^\star = 0.1075 \quad \text{and} \quad \phi = \frac{\pi}{2} \tag{16}$$

which are physically acceptable.

3.1. *Analytic structure of the singularity*

To determine the nature of the mass singularity in the resonance crossing, it is convenient to unfold the exceptional point in surfaces around it. The term under the square root in eq. (15) is a regular function of its arguments and may be expanded in a Taylor series around the exceptional point as

$$\frac{1}{2}\sqrt{c_1^{(1)}(\lambda - \lambda^\star) + c_2^{(1)}(\phi - \phi^\star)}, \tag{17}$$

where the $c_k^{(1)}$'s are the derivatives of the term in the square root in eq. (15). In order to examine the precise nature of the singularity of functions $\mu_{2,3}^2(\lambda,\phi)$ we introduce the following notation in the parameter space

$$\vec{\mathcal{R}} = \begin{pmatrix} \Re e\, c_1^{(1)} & \Re e\, c_2^{(1)} \end{pmatrix}, \quad \vec{\mathcal{I}} = \begin{pmatrix} \Im m\, c_1^{(1)} & \Im m\, c_2^{(1)} \end{pmatrix}, \quad \vec{\zeta} = \begin{pmatrix} \zeta_1 \\ \zeta_2 \end{pmatrix} = \begin{pmatrix} \lambda - \lambda^\star \\ \phi - \phi^\star \end{pmatrix}, \tag{18}$$

and

$$sgn\Big(\Re e\,\mu_{2,3}^2\Big)sgn\Big(\Im m\,\mu_{2,3}^2\Big) = sgn\Big(\vec{\mathcal{I}}\cdot\vec{\zeta}\Big), \tag{19}$$

then we obtain the real and imaginary parts for $\mu_{2,3}^2(\lambda,\phi)$ as

$$\Re e\,\mu_{2,3}^2(\lambda,\phi) = \pm\frac{1}{2\sqrt{2}}\left[\sqrt{(\vec{\mathcal{R}}\cdot\vec{\zeta})^2 + (\vec{\mathcal{I}}\cdot\vec{\zeta})^2} + (\vec{\mathcal{R}}\cdot\vec{\zeta})\right]^{1/2}, \tag{20}$$

$$\Im m\, \mu^2_{2,3}(\lambda,\phi) = \pm\frac{1}{2\sqrt{2}}\left[\sqrt{(\vec{\mathcal{R}}\cdot\vec{\zeta})^2+(\vec{\mathcal{I}}\cdot\vec{\zeta})^2}-(\vec{\mathcal{R}}\cdot\vec{\zeta})\right]^{1/2}, \tag{21}$$

From equation (20) it can be seen that near the exceptional point $\Re e\, \mu^2_{2,3}(\lambda,\phi)$, is a two branch function of (λ,ϕ), which can be represented by a two fold surface Σ_R in a tridimensional euclidean space with coordinates $(\Re e\, \mu^2_{2,3},\lambda,\phi)$. The two branches of $\Re e\, \mu^2_{2,3}(\lambda,\phi)$ are represented by two sheets, which are copies of the (λ,ϕ) plane branches represent twofold sheets as copies of the (λ,ϕ) having a cut a long the line where the two branches join smoothly The cut is defined as the locus of the point where the argument of the square root function vanishes.

Close to the origin of coordinates, this locus is defined by a unit vector $\vec{\zeta}_c$ in the (λ,ϕ) plane, such

$$\vec{\mathcal{I}}\cdot\vec{\zeta}_c = 0 \quad \text{y} \quad \vec{\mathcal{R}}\cdot\vec{\zeta}_c = -|\vec{\mathcal{R}}\cdot\vec{\zeta}_c| \tag{22}$$

Therefore, the real part of the function $\mu^2_{2,3}(\lambda,\phi)$, as a function of the real parameters (λ,ϕ), has an algebraic branch point of square root type at exceptional point, and a branch cut along a line that start at the exceptional point and extend in a positive direction defined by a unit vector $\vec{\zeta}_c$ satisfying eq.(22).

A similar analysis shows that $\Im m\ \mu^2_{2,3}(\lambda,\phi)$, also has an algebraic point of square root type at the exceptional point with coordinates (λ_d,ϕ_d), and also has a branch cut along a line that start at exceptional point and extends in the negative direction defined by the unit vector $\vec{\zeta}_c$, see Fig. 1 of ref. [11]

Discussion about this analysis can be seen in [11] where we show the complex surfaces for the real and imaginary part of the neutral heavy Higgs masses near degeneracy and where the branch point is illustrated.

4. Lineshape of $\not{C}\not{P}$ neutral Higgs bosons

The s-channel resonance cross section, $\sigma(\mu\bar{\mu}\to H_i^{\not{C}\not{P}}\to l\bar{l})$ is obtained from the squared of the transition matrix eq. (3). The differential cross section will be, up to the phase space constants, proportional to the square of the transition amplitude

$$d\sigma \propto |\mathcal{T}^d(s)|^2; \tag{23}$$

or by the Optical Theorem

$$d\sigma(s) \propto -i\left|\mathcal{T}^d(s)-\mathcal{T}^d(s)^\dagger\right|. \tag{24}$$

We will now discuss the form of the transition matrix both off degeneracy and at degeneracy to show its continuity and the consequences on Higgs lineshape of going through the singular point of exact degeneracy.

4.1. Transition matrix with non-coherence of states

Once we are able to write the neutral heavy $\not{C}\not{P}$ Higgs bosons mass matrix (4) in the physical states basis, the transition matrix in eq. (3) as the general transition matrix off-degeneracy for

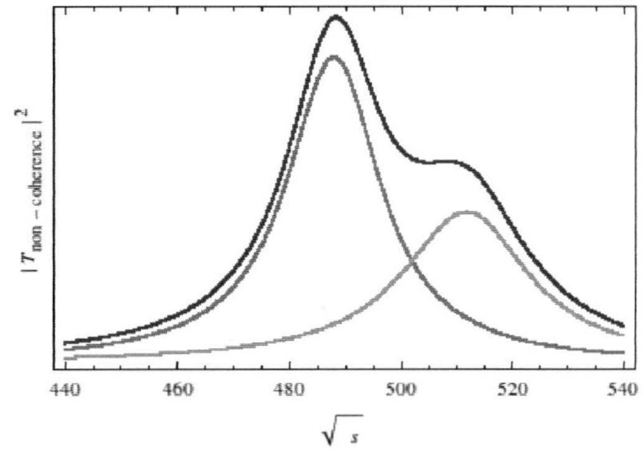

Figure 2. The behavior of non-coherent states, the interference of two resonances associated to Breit-Wigner propagators add constructively.

the two states in the resonance region is given by

$$\begin{aligned}\mathcal{T}_{res}(s) = & \frac{1}{\left(M_{H_3}^2 - M_{H_2}^2\right) - i\left(M_{H_3}\Gamma_{H_3} - M_{H_2}\Gamma_{H_2}\right)} \\ & \times \left(\frac{\left[1-(\mu_{H_2}^2(s_2^\star))'\right]^{-1/2}\left[1-(\mu_{H_3}^2(s_3^\star))'\right]^{-1/2}}{s - M_{H_2}^2 + iM_{H_2}\Gamma_{H_2}(s_2^\star)} \right. \\ & - \left. \frac{\left[1-(\mu_{H_3}^2(s_3^\star))'\right]^{-1/2}\left[1-(\mu_{H_3}^2(s_3^\star))'\right]^{-1/2}}{s - M_{H_3}^2 + iM_{H_3}\Gamma_{H_3}(s_3^\star)} \right) \\ & \times \left\{ V_H^P(s - M_A^2 + iM_A\Gamma_A)V_H^D + V_A^P(s - M_H^2 + iM_H\Gamma_H)V_A^D \right. \\ & + \left. V_H^P\Delta_{HA}^2 V_A^D - V_A^P\Delta_{HA}^2 V_H^D \right\}. \end{aligned} \tag{25}$$

The transition matrix off-degeneracy has the expected behavior in the resonance region, two simple poles in the $s-$plane located at the physical masses of the heavy neutral Higgs bosons H_2 and H_3. It shows no coherence of the states and the interference is constructive, as can be seen as in figure 2.

Within the THDM with only top loop [3] there is no s-dependence on the mass matrix, we thus obtain a simplified expression for the transition matrix (25), as

$$(\mu_{H_i}^2(s_i^\star))' = \left(\frac{d\mu_{H_i}^2(s)}{ds}\right)_{s_i^\star} = 0$$

$$\begin{aligned}\mathcal{T}_{res} = & \frac{1}{\mu_{H_3}^2 - \mu_{H_2}^2}\left[\frac{1}{s-\mu_{H_2}^2} - \frac{1}{s-\mu_{H_3}^2}\right] \times \left\{ V_H^P(s - M_A^2 + iM_A\Gamma_A)V_H^D \right. \\ & \left. + V_A^P(s - M_H^2 + iM_H\Gamma_H)V_A^D + V_H^P\Delta_{HA}^2 V_A^D - V_A^P\Delta_{HA}^2 V_H^D \right\}\end{aligned}$$

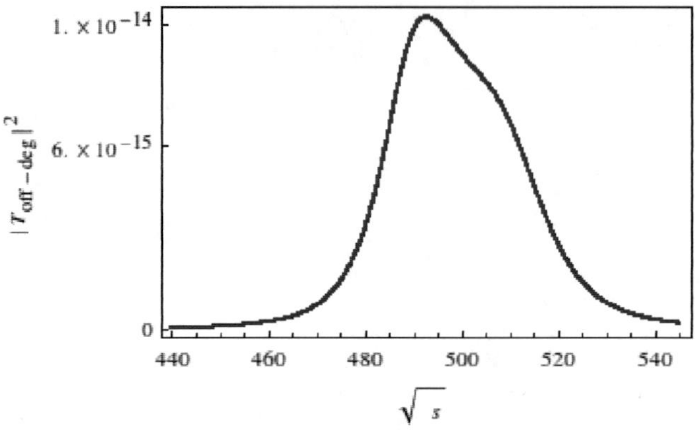

Figure 3. Off degeneracy and non-coherence of states, the heavy and neutral $\not{CP}$ Higgs system will behave as constructive interference of resonances. We are taking $\alpha = \pi/6$ and $\tan\beta = 15$.

$$= -\frac{1}{(s-\mu_{H_3}^2)(s-\mu_{H_2}^2)} \times \left\{ V_H^P(s - M_A^2 + iM_A\Gamma_A)V_H^D + V_A^P(s - M_H^2 + iM_H\Gamma_H)V_A^D \right.$$
$$\left. + V_H^P\Delta_{HA}^2 V_A^D - V_A^P\Delta_{HA}^2 V_H^D \right\}. \quad (26)$$

For the numerator, the vertices correspond to the Yukawa couplings.

$$\begin{aligned} V_H^P = Y_\mu^H &= \frac{gm_\mu \cos\alpha}{2M_W \cos\beta} = V_H^{P\dagger}, \\ V_H^D = Y_l^H &= \frac{gm_l \cos\alpha}{2M_W \cos\beta} = V_H^{D\dagger}; \\ V_A^P = Y_\mu^A &= i\gamma^5 \frac{gm_\mu \tan\beta}{2M_W} = -V_A^{P\dagger}, \\ V_A^D = Y_l^A &= i\gamma^5 \frac{gm_l \tan\beta}{2M_W} = -V_A^{D\dagger}. \end{aligned} \quad (27)$$

Then we find the squared transition matrix will have the form

$$\begin{aligned} |\mathcal{T}_{res}|^2 &= \frac{1}{|s-\mu_{H_3}^2|^2|s-\mu_{H_2}^2|^2} \frac{g^2 m_\mu m_l}{4M_W^2\cos^2\beta} \left\{ \left[(s-M_A^2)^2 + (M_A\Gamma_A)^2\right]\cos^4\alpha \right. \\ &\left. + \left[(s-M_H^2)^2 + (M_H\Gamma_H)^2\right]\sin^4\beta - 2\cos^2\alpha\sin^2\beta\, \Re e\left[\hat{M}_A^2\hat{M}_H^{2*}\right] \right\} \\ &= \frac{1}{|s-\mu_{H_3}^2|^2|s-\mu_{H_2}^2|^2} \frac{g^2 m_\mu m_l}{4M_W^2\cos^2\beta} |\hat{M}_A^2\cos^2\alpha - \hat{M}_H^2\sin^\beta|^2, \end{aligned} \quad (28)$$

where we have defined $\hat{M}_A^2 = (s - M_A^2 + iM_A\Gamma_A)$ and $\hat{M}_H^2 = (s - M_H^2 + iM_H\Gamma_H)$, and $\Re e\left[\hat{M}_A^2\hat{M}_H^{2*}\right] = \left[(s-M_A^2)(s-M_H^2) + M_A\Gamma_A M_H\Gamma_H\right]$.
We can see that off-degeneracy the interference of two particles behave the usual way.

We now run the energy $\sqrt{s}$ for $|\mathcal{T}_{res}|^2$ to find the heavy neutral $\not{CP}$ Higgs state near the accidental degeneracy.

In the case of CP conserving neutral Higgs bosons there are not a coalescence of the states and the states simply overlap as for example in figure 22 of ref. [34]

Journal of Physics: Conference Series **631** (2015) 012031 doi:10.1088/1742-6596/631/1/012031

The next step is to find the continuity of the transition matrix at exact degeneracy and take this into account for the calculation of the neutral heavy Higgs lineshape.

4.2. Transition matrix at degeneracy

Because of the coalescence of the corresponding eigenstates the transition matrix is no longer diagonalizable by a similarity transformation, so we write it as a Jordan block, and we find a generalized eigenstate to complete the basis. As we mentioned the masses can be written as functions of the complex parameters of the Lagrangian and at exact degeneracy will show up as a rank one algebraic branch point. In this section we write down the structure of this singularity which comes from the mixing in the propagator of the neutral and heavy Higgs states H and A. The propagator has a double pole in the non-physical sheet of the complex plane of the squared energy s. We show the form of the propagator at exact degeneracy.

Now, at exact degeneracy $M_{H_2} = M_{H_3} = M_{H_d}$, the propagator matrix in the $H - A$ representation takes the form

$$\hat{\Delta}^{(d)}_{H_2H_3}(s) = \frac{1}{s - M_d^2 + iM_d\Gamma_d}\mathbf{1}_{2\times 2} + \frac{1}{\left(s - M_d^2 + iM_d\Gamma_d\right)^2} M_d\Gamma_d(s^\star)\begin{pmatrix} 1 & i \\ i & -1 \end{pmatrix}. \tag{29}$$

The matrix on the right hand side of this equation may be brought to a Jordan block form by means of a similarity transformation,

$$\hat{\Delta}^{(d)}_{H_2H_3}(s) = \begin{pmatrix} \frac{1}{s-M_d^2+iM_d\Gamma_d} & \frac{iM_d\Gamma_d(s^\star)}{\left(s-M_d^2+iM_d\Gamma_d\right)^2} \\ 0 & \frac{1}{s-M_d^2+iM_d\Gamma_d} \end{pmatrix}, \tag{30}$$

then, the resonant transition matrix at degeneracy, in the mass representation takes the form

$$\mathcal{T}^d(s) = \left(\tilde{V}^P_H, \tilde{V}^P_A\right)\Delta^{(d)}_{H_2H_3}(s)\begin{pmatrix} \tilde{V}^D_H \\ \tilde{V}^D_A \end{pmatrix}. \tag{31}$$

Calculating the squared of this matrix considering parameter values λ and ϕ off-degeneracy, but close to it, and taking also into account fermions in and out of the process we have a quite different behavior of the heavy neutral $\not{CP}$ Higgs lineshape, figure 4.2. Even at not exact degeneracy but close enough, the interference on the coalescent states will cause a lineshape reduction down to zero.

This behavior was analyzed previously by [13] in simplified multi-Higgs models, but it was not considered in previous CP non invariant analysis of neutral heavy Higgs, as for example in figure 3 of ref. [32].

5. Summary and conclusions

In a CP conserving multi-Higgs models, the neutral scalar states will have CP-even charge and pseudoscalar will have CP-odd charge. In general, for a CP non-conserving multi-Higgs models the CP identification of neutral Higgs states in no longer possible. Specifically for models with Higgs complex doublets they mix so having even and odd components in their wave function. In the 2HDM, the neutral Higgs bosons spectrum includes a light state h_1, the SM-like Higgs boson, and two heavy neutral states, H, A. In a $\not{CP}$ context the two heavy neutral Higgs states are no longer CP defined and can strongly mix $H_2 - H_3$. They may be degenerate not only in its mass

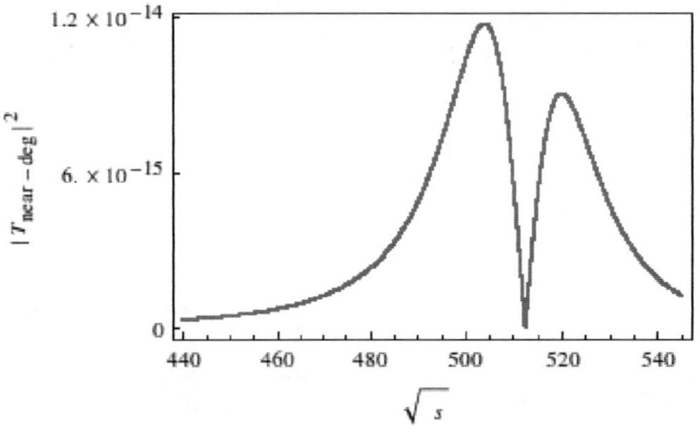

Figure 4. Near degeneracy the coalescence of the states may interfere destructively in the resonance. Considering the values of the parameters not at exact degeneracy but close, the lineshape reduces drastically to zero at the pole for the $\not{C}\not{P}$ Higgs heavy neutral bosons. We are taking $\alpha = \pi/6$ and $\tan\beta = 15$.

but in states. These properties manifest themselves as mixing of two coherent resonances in the transition amplitude for the production of a fermion-antifermion pair in a $\mu^+\mu^-$ collision, for example. From the analysis of the effective mass matrix in the propagator of the heavy neutral system $H_2 - H_3$, we found that, at the degeneracy point, the hypersurfaces that represent the physical masses (pole masses) as functions of the parameters of the system have a rank one algebraic branch point, and the real and imaginary parts of the mass have branch cuts both starting at the same exceptional point but extending in opposite directions. Associated with this singularity, the propagator of the mixed heavy neutral Higgs system, $H_2 - H_3$, has a double pole in the complex $s-$plane. We also examined the continuity of the transition amplitude at the point of exact degeneracy of the two masses. We analyzed the physical masses and possible degeneracy of the heavy neutral Higgs boson system in a THDM with complex couplings. At exact coalescence of two neutral Higgs states, the propagator matrix has two single poles and one double pole. More details on this calculation can be found in ref. [11]. A more detailed CMSSM analysis is in progress.

We analyzed and displayed the behavior of a degeneracy of a non-hermitian matrix which is radically different from a hermitian system. We showed that the transition matrix is continuous at the exact coalescence, allowing thus to calculate the CPV in a specific cross section at the exact degeneracy.

The heavy neutral Higgs and $\not{C}\not{P}$ lineshape would have a radically different behavior than the ones showed in the literature. As an example the process $\mu\mu \to H^{\not{C}\not{P}} \to ff$ in the s-channel resonance goes down to zero at exact degeneracy. This analysis suggests that a detailed experimental study of the lineshape of the resonances associated with the excitation of this Higgs system may provide valuable information on the CP nature of the underlying theory.

Acknowledgments
This work was partially supported by CONACyT México under contracts No.132059 and by DGAPA-UNAM under contracts PAPIIT IN113712 and IN111115.

References

[1] Buchmuller W 2007 Baryogenesis – 40 years later, arXiv:0710.5857 [hep-ph].

[2] For a review see: Gunion J F, Haber H E, Kane G L, and Dawson S 1990 The Higgs hunter's guide *Addison-Wesley, Reading, MA.* Djouadi A 2008 *Phys. Rept.* **459** 1 arXiv:hep-ph/0503173.

[3] Choi S Y, Kalinowski J, Liao Y and Zerwas P M 2005 *Eur. Phys. J.* C **40**, 555 (2005) arXiv:hep-ph/0407347.

[4] Carena M S, Ellis J R, Pilaftsis A and Wagner C E M 2000 *Nucl. Phys.* B **586** 92 arXiv:hep-ph/0003180. Ellis J R, Lee J S and Pilaftsis A 2004 *Phys. Rev.* D **70**, 075010 arXiv:hep-ph/0404167.

[5] Heinemeyer S 2001 *Eur. Phys. J.* C **22** 521 arXiv:hep-ph/0108059.

[6] Frank M, Hahn T, Heinemeyer T, Hollik W, Rzehak H and Weiglein G 2007 *JHEP* **0702** 047

[7] Hernández E and Mondragón A 1993 *J.Phys A: Math.Gen.* **26** 5595

[8] Hernández E, Jáuregui A, Mondragón A and Nellen L 2007 *Int. Jour. Theor. Phys.* **46**, No.6, 1666.

[9] Hernández E, Jáuregui A and Mondragón A 2011 *Phys.Rev E.* **84** 046209

[10] Hernández E, Jáuregui A and Mondragón A 2006 *J. Phys. A: Math. Theor.* **39** 10087

[11] Felix-Beltran O, Gomez-Bock M, Hernandez E, Mondragon A and Mondragon M 2011 *Int. J Theor. Phys.* **50** 2291. Felix-Beltran O, Gomez-Bock M, Hernandez E, Mondragon A and Mondragon M 2009 *Rev.Mex.Fis.Supl.* **55** No.S2, 88-97.

[12] Bender C M and Wu T T 1968 *Phys. Rev. Lett.* **21** 406.

[13] Cacciapaglia G, Deandrea A and De Curtis S 2009 *Phys. Lett. B* **682** 43 arXiv:0906.3417 [hep-ph].

[14] Philipp M, von Brentano P, Pascovici G and Richter A 2000 *Phys. Rev. E* **62** 1922

[15] Dembowski C, Gräf H D, Harney H L, Heine A, Heiss W D, Rehfeld H and Richter A 2001 *Phys. Rev. Lett.* **86** 787

[16] Dembowski C, Dietz B, Gäf H D, Harney H L, Heine A, Heiss W D and Richter A 2004 *Phys. Rev. E* **69** 056216

[17] Dietz B, Friedrich T, Mets J, Miski-Oglu M, Richter A, Schäfer F and Stafford C A 2007 *Phys. Rev. E* **75** 027201

[18] Sang-Buni L, Juhee Y, Songky M, et al. 2009 *Phys. Rev. Lett.* **103** 134101

[19] Stehmann T, Heiss W D and Scholtz F G 2004 *J. Phys. A: Math. Theor.* **37** 7813 arXiv:0312182 [quant-ph]

[20] Lefebvre R, Atabek O, Sindelka M and Moiseyev N 2007 *Phys. Rev.* D **76** 025011 arXiv:hep-th/0703104.

[21] Michel N, Nazarewicz W, Ploszajczak M and Vertse T, 2009 *J.Phys.* G: Nucl. Part. Phys. **36** 013101.

[22] Baupre J V, Coleman T P, Lassila K E and Ruuskanen P V 1968 *Phys. Rev. Lett.* **21** (27) pp.1849.

[23] Dorey N, Hofman D M and Maldacena J M 2007 *Phys. Rev.* D **76** 025011 arXiv:hep-th/0703104.

[24] A. Mostafazadeh 2002 *J. Math. Phys.* **43**, 205-214 math-ph/0107001. Mostafazadeh A 2002 *J. Math. Phys.* **43**, 2814-2816 math-ph/0110016. Mostafazadeh A 2002 *J. Math. Phys.* **43** 3944-3951 math-ph/0203005. Mostafazadeh A 2002 *Nucl. Phys.* B **640** 419 math-ph/0203041.

[25] Bender C M and Mannheim P D 2008 *Phys. Rev.* D **78** 025022 arXiv:0804.4190 [hep-th].

[26] Kaldass H, Bohm A and Wickramasekara S 2002 *Int. J. Mod. Phys.* A **17**, 3749.

[27] Stuart R G 1997 *Phys. Rev. D* **56** 1515

[28] Dabelstein A 1995 *Z. Phys.* C **67** 495 arXiv:hep-ph/9409375.

[29] Binosi D 2004 *J. Phys.* G: Nucl. Part. Phys. **30**, 1021.

[30] Pilaftsis 1997 *Nucl. Phys.* B **504** 61 arXiv:hep-ph/9702393.
Pilaftsis A 1998 *Phys. Lett.* B **435**, 88 arXiv:hep-ph/9805373.

[31] Carena M S, Ellis J R, Pilaftsis A and Wagner C E M 2002 *Nucl. Phys.* B **625** 345 arXiv:hep-ph/0111245.

[32] Bernabeu J, Binosi D and Papavassiliou J 2006 *JHEP* **0609** 023; arXiv:hep-ph/0604046.

[33] Gunion J F and Haber H E 2003 *Phys. Rev.* D **67** 075019 arXiv:hep-ph/0207010.

[34] Barger V D, Berger M S, Gunion J F and Han T 1997 *Phys. Rept.* **286** arXiv:hep-ph/9602415.

Higgs inflation and vacuum stability

Javier Rubio
Institut de Théorie des Phénomènes Physiques, École Polytechnique Fédérale de Lausanne, CH-1015 Lausanne, Switzerland

E-mail: javier.rubio@epfl.ch

Abstract. Inflation is nowadays a well-established paradigm consistent with all the observations. The precise nature of the inflaton is however unknown and its role could be played by any candidate able to imitate a scalar condensate in the slow-roll regime. The discovery of a fundamental scalar in the LHC provides the less speculative candidate. Could the Higgs field itself be responsible for inflation? Do we really need to advocate new physics to explain the properties of the Universe at large scales? Which is the relation between the Standard Model parameters and the inflationary observables? What happens if our vacuum becomes unstable below the scale of inflation? We present an overview of Higgs inflation trying to provide answers to the previous questions with special emphasis on the vacuum stability issue.

1. The Higgs boson in the sky

Nature seems to be extremely simple. The only outcome of LHC experiments till date is a scalar boson with a mass of $125 - 126$ GeV and properties that closely resemble those of the Standard Model Higgs [1, 2]. No sign of new physics beyond the Standard Mode has appeared whatsoever. On top of that, the Planck results favor the simplest realization of the inflationary scenario: a single field model of inflation in which no significant amount of non-gaussianities or isocurvature perturbations are produced [3].

Given the minimalistic scenario presented above, it is certainly tempting to try to identify the scalar field found in the LHC with that responsible for the flatness, homogeneity and isotropy of the Universe at large scales. Unfortunately, the self-coupling of the Higgs field highly exceeds the value needed to generate the right amount of primordial density perturbations [4]. A possible way out is the inclusion of a non-minimal coupling[1] to gravity with no further modifications of the Standard Model Lagrangian. This is the starting point of Higgs-inflation [6]. The relevant part of the Lagrangian density reads

$$\frac{\mathcal{L}}{\sqrt{-g}} = \frac{M_P^2 + \xi h^2}{2} R - \frac{1}{2}(\partial h)^2 - \frac{\lambda}{4}(h^2 - v^2)^2 , \qquad (1)$$

with M_P the Planck mass, h the Higgs field in the unitary gauge and v its vacuum expectation value. The non-minimal coupling ξ is a free parameter that cannot be fixed from the model itself and must be determined by observations. Note that, for sufficiently large values of the Higgs field (namely $h \gg M_P/\sqrt{\xi}$), the theory becomes approximately scale invariant. This asymptotic symmetry will play a central role in the further developments[2].

[1] The existence of this coupling is indeed required by renormalization [5].

[2] The promotion to an exact symmetry of the theory and its phenomenological consequences were considered in

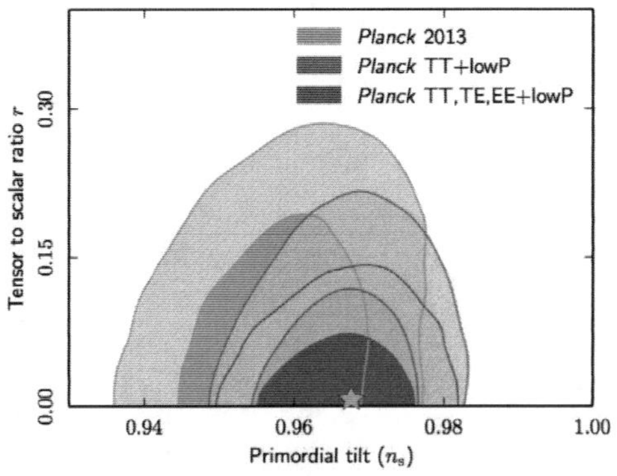

Figure 1. Comparison of the tree level predictions of Higgs inflation (yellow star) and the Planck data [3].

The analysis of inflation is more easily performed in the so-called Einstein frame, in which the gravitational part of the action takes the usual Einstein-Hilbert form. This is obtained by performing a conformal transformation

$$g_{\mu\nu} \to \Omega^2 g_{\mu\nu}\,, \qquad \text{with} \qquad \Omega^2 = 1 + \frac{\xi h^2}{M_P^2}\,, \tag{2}$$

together with a field redefinition

$$\frac{d\chi}{dh} = \sqrt{\frac{\Omega^2 + 6\xi^2 h^2/M_P^2}{\Omega^4}} \qquad \longrightarrow \qquad \chi \simeq \begin{cases} h\,, & h \ll \frac{M_P}{\xi}\,, \\ \sqrt{\frac{3}{2}} M_P \log \Omega^2(h)\,, & h \gg \frac{M_P}{\xi}\,, \end{cases} \tag{3}$$

to make the kinetic term of the Higgs field canonical. All the non-linearities in the original frame are moved to the Higgs potential, which after the field redefinition, becomes non-polynomial (and therefore non-renormalizable)[3]

$$V(\chi) = \begin{cases} \frac{\lambda}{4}(\chi^2 - v^2)^2\,, & \chi \ll \frac{M_P}{\xi}\,, \\ \frac{\lambda M_P^4}{4\xi^2}\left(1 - e^{-\sqrt{2/3}\chi/M_P}\right)^2\,, & \chi \gg \frac{M_P}{\xi}\,. \end{cases} \tag{4}$$

The asymptotic shift-symmetry of the previous expression at large field values is the Einstein-frame manifestation of the approximate scale invariance we started with. With the standard initial chaotic conditions, the flattening potential (4) allows for inflation.

The analysis of inflation follows the usual slow-roll approach. The COBE normalization fixes the amplitude of the potential, $\lambda/\xi^2 \simeq 4 \times 10^{-11}$. For $\lambda \sim \mathcal{O}(1)$, the non-minimal coupling ξ is required to be significantly large, $\xi \sim \mathcal{O}(10^4)$, but still much smaller than the value able to produce noticeable effects at low energies [6]. The model predicts a spectral tilt for the primordial scalar perturbations around $n_s \simeq 0.97$ and a small tensor-to-scalar ratio $r \simeq 0.003$. As shown in Fig. 1, both numbers are fully compatible with the bounds obtained by the Planck

Refs. [7, 8, 9, 10, 11].

[3] We omit the exponentially small correction proportional to the vacuum expectation value of the Higgs field in the expression for the potential at large field values.

collaboration [3] and very close to the predictions of the Starobinsky model of inflation[4] [13, 14]. Since we are dealing with a single field model of inflation with canonical kinetic terms, neither isocurvature perturbations nor non-Gaussianities are produced.

2. A potential problem: vacuum stability

The previous predictions of Higgs inflation are based on the classical action and do not take into account possible changes in the shape of the inflationary potential due to quantum effects. One of the main points to consider is the running of the coupling constants and, in particular, the logarithmic running of the Higgs self-coupling λ.

One should be tempted to think that the precise value of the Higgs coupling at the inflationary scale should not have any effect on the inflationary observables since the non-minimal coupling ξ can be tuned at will in order to satisfy the normalization condition $\lambda/\xi^2 \simeq 4 \times 10^{-11}$. However, the situation turns out to be much more complicated due to the particular value of the Higgs mass.

Among the many possible values that the Higgs mass could have taken, Nature has chosen one that allows to extend the Standard Model all the way up till the Planck scale while staying in the perturbative regime. The scalar self-coupling decreases with energy to reach a minimum at energies close to M_P and starts increasing thereafter. Whether it remains positive till the Planck scale or becomes negative at a smaller scale μ_0 depends on the precise values of the top Yukawa coupling and the Higgs mass (cf. Refs. [15, 16, 17, 18] for details). Although the latest CMS data [19] are perfectly consistent with the absolute stability of the Standard Model within the experimental and theoretical uncertainties (cf. Fig. 2), one should not exclude the possibility that other experiments may be able to establish the metastability of the electroweak vacuum in the future. Should the successful Higgs inflation idea be abandoned in that case? Answering this question is the purpose of the next sections.

3. Higgs inflation as an effective field theory

The detailed analyses about vacuum stability performed in Refs. [15, 16, 17] are based on flat space considerations and should not be extrapolated to the case in which gravitational effects are included since issues such as the determination of the lifetime of the Universe strongly depend on the details of the ultraviolet completion [20, 21, 22].

The presence of gravity makes Higgs inflation non-renormalizable. Quantum corrections should be introduced by interpreting the theory as an effective field theory in which a particular set of higher dimensional operators are included. But which set of operators?

It seems clear that we cannot add arbitrary operators in the Einstein frame, since this would automatically spoil the flatness of the potential. In the absence of an ultraviolet completion for the Standard Model non-minimally coupled gravity, the choice of higher dimensional operators can be only based on the self-consistency of the procedure [23]. To have a (partially) controllable link between the low and high energy parameters of the model, we will only add those counterterms that are required to make the theory finite at every order in perturbation theory, i.e. those with the structure generated by the original Lagrangian via radiative corrections. The finite parts associated to these counter terms are assumed to be small and to have the hierarchy of the loops producing them (i.e. the coefficients in front of the operators coming from two-loop diagrams are much smaller than those coming from one-loop diagrams, etc...). Finally, the procedure will be required to maintain the symmetries of the original theory, and, in particular the approximate scale invariance of Eq. (1) at large field values. This can be accomplished by

[4] The small differences are associated to the different preheating mechanisms in the two models [12].

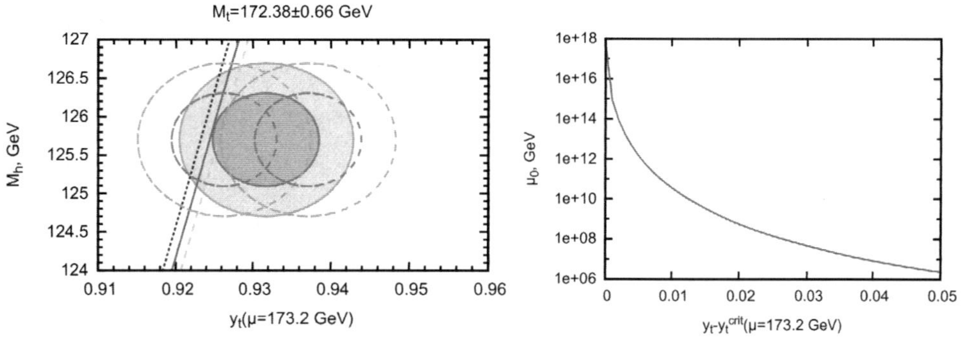

Figure 2. *Left*: Stability/Metastability regions of the Standard Model in the $\overline{\mathrm{MS}}$ scheme at $\mu = 173.2$ GeV [18]. The red diagonal line stands for the critical value of the top Yukawa coupling $y_t^{\rm crit}(M_h)$ leading to a negative self-coupling λ at a scale μ_0 below the Planck scale (dashed lines accounts for the uncertainty in the strong coupling constant α_s). The Standard Model is absolutely stable to the left of these lines and metastable to the right. The filled ellipses correspond to the experimental values of y_t extracted from the latest CMS determination of the Monte-Carlo top quark mass $M_t = 172.38 \pm 0.10\,({\rm stat}) \pm 0.65\,({\rm syst})$ GeV, if this is identified with the pole mass [19]. The dashed ellipses encode the shifts associated to the ambiguous relation between pole and Monte Carlo masses. *Right*: Energy scale μ_0 at which the Higgs self-coupling λ becomes negative as a function of the deviation of the top Yukawa coupling y_t from the critical value $y_t^{\rm crit}$ [18].

using dimensional regularization[5] with an asymptotically scale-free t'Hooft-Veltman parameter[6]

$$\mu^2 \propto M_P^2/2 + \xi H^\dagger H\,. \tag{5}$$

4. From the Standard Model to the Chiral Standard Model

Given the self-consistent framework presented in the previous section, which is the relation between the Standard Model parameters and the inflationary observables? To answer this question consider the propagation of a given fermion, for instance the top quark[7], in the background of the Higgs field. When written in the Einstein frame, the linear Yukawa interaction between these two species becomes modified by a conformal factor

$$\mathcal{L}_F = i\bar\psi_t \partial\!\!\!/ \psi_t + \frac{y_t}{\sqrt{2}} F(\chi)\bar\psi_t\psi_t\,, \qquad \text{with} \qquad F(\chi) \equiv \frac{h(\chi)}{\Omega(\chi)}\,. \tag{6}$$

The divergences associated to the 1-loop diagrams

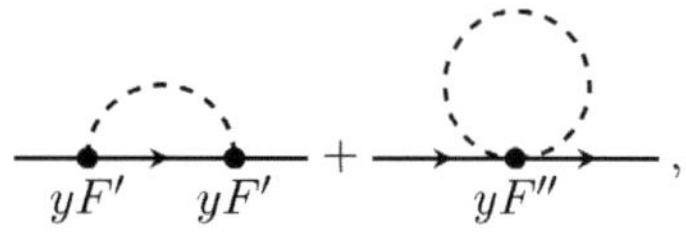

, (7)

[5] This regularization automatically discards all power- law divergences.
[6] This is a Jordan-frame definition.
[7] This will give the most relevant contribution.

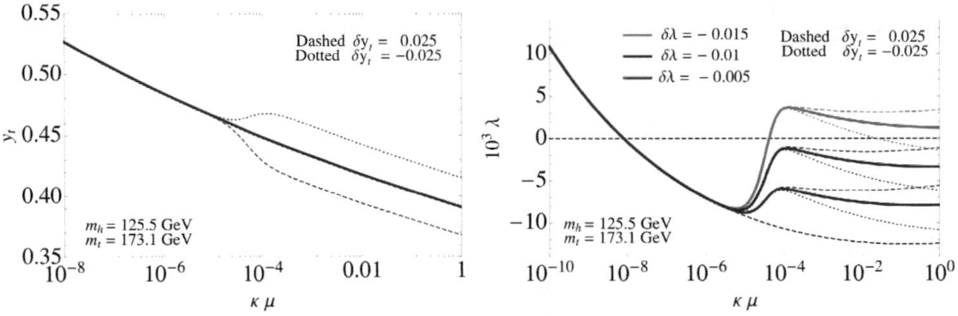

Figure 3. Renormalization group evolution of the top Yukawa coupling y_t and the Higgs self-coupling λ as a function of the field-dependent renormalization scale $\mu(\chi)$ in Planck units ($\kappa = M_P^{-1}$). The effect of the jumps is localized at a given energy scale and does not significantly modify the running of the couplings in the asymptotic low and high energy regions.

are eliminated, as usual, by adding counterterms with the proper coefficients a_1 and a_2 in $1/\bar{\epsilon}$ and arbitrary finite parts δy_{t1} and δy_{t2}

$$\delta\mathcal{L}_{\rm ct} \sim \left(a_1\frac{y_t^3}{\bar{\epsilon}}+\delta y_{t1}\right)F'^2F\bar{\psi}\psi+\left(a_2\frac{y_t\lambda}{\bar{\epsilon}}+\delta y_{t2}\right)F''(F^4)''\bar{\psi}\psi\,. \tag{8}$$

Here $1/\bar{\epsilon}$ stands for $1/\epsilon-\gamma+\ln 4\pi$, with $\gamma=0.5772$ the Euler-Mascheroni constant, and the primes denote derivatives with respect to χ. The structure of the new terms differs from that in Eq. (6), as expected, since we are dealing with a non-renormalizable theory.

At small field values ($\chi \ll M_P/\xi$), the conformal factor $\Omega(h)$ equals one, and the function $F(\chi)$ in Eqs. (6) and (8) becomes linear in the field ($F=\chi$ and $F'=1$). The Higgs-fermion interaction in the Einstein frame reduces then to the standard (renormalizable) Yukawa interaction and the finite part y_{t1} can be reabsorbed in the definition of the top Yukawa coupling, as in any renormalizable theory. This allows us to use the usual Standard Model renormalization group equations to relate the values of the couplings at the electroweak scale with those at $\chi = M_P/\xi$. What happens at higher energies? When approaching $\chi \sim M_P/\xi$, the function $F(\chi)$ becomes

$$F(\chi)=\frac{M_P}{\sqrt{\xi}}\left(1-{\rm e}^{-\sqrt{2/3}\chi/M_P}\right)^{1/2}, \tag{9}$$

and the previously existing counterterms become exponentially suppressed. This gives rise to an effective jump of the top Yukawa coupling at that scale[8] [24]

$$y_t(\mu)\to y_t(\mu)+\delta y_t\left[F'^2-1\right]\,. \tag{10}$$

Beyond that point (in particular for $\chi > M_P$), the function $F(\chi)$ becomes rapidly a constant, $F(\chi)\approx\frac{M_P}{\sqrt{\xi}}$, and the Higgs field decouples from the Standard Model particles. Higgs inflation becomes a chiral Standard Model [25] in which the radial component of the Higgs field has been integrated out.

Threshold effects as those giving rise to (10) appear also in the Higgs sector. The evaluation of the bubble diagrams

$$\text{(dashed loop)} = \frac{1}{2}{\rm Tr}\ln\left[\Box-\left(\frac{\lambda}{4}(F^4)''\right)^2\right], \qquad \text{(solid loop)} = -{\rm Tr}\ln\left[i\partial\!\!\!/+y_tF\right], \tag{11}$$

[8] We neglect the running of δy_{t1}.

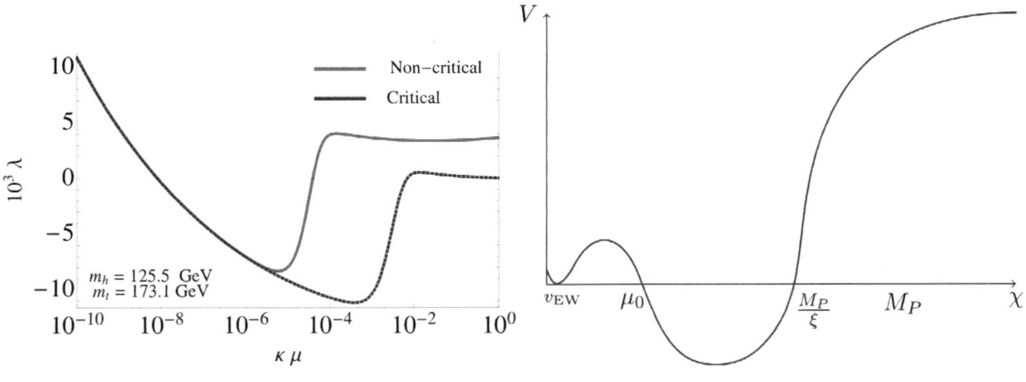

Figure 4. *Left*: Jump-based restoration of a positive self-coupling λ at the inflationary scale in the case of a metastable vacuum. The red and blue lines correspond respectively to the non-critical ($\xi = 1500$, $\delta y_t = 0.025$, $\delta\lambda = -0.015$) and critical ($\xi = 15$, $\delta y_t = 0$, $\delta\lambda = -0.0133$) cases discussed in the text. *Right*: Sketch of the associated effective inflationary potential.

involved in the computation of the effective potential, and the addition of the associated counterterms, produces a rapid change in the self-coupling of the Higgs field at[9] $\chi \sim M_P/\xi$ [24]

$$\lambda(\mu) \to \lambda(\mu) + \delta\lambda \left[\left(F'^2 + \frac{1}{3}F''F\right)^2 - 1\right] . \tag{12}$$

As shown in Fig. 3, the effect of the jumps (10) and (12) is localized at $\chi \sim M_P/\xi$ and does not significantly modify the running of the couplings in the asymptotic low and high energy regions.

5. Restoring the asymptotics

The precise value of the finite parts δy_t and $\delta\lambda$ in Eqs. (10) and (12) is unknown. They depend on the particular ultraviolet completion of Higgs inflation and cannot be determined from the effective field theory itself.

If the jumps $\delta\lambda$ and δy_t are much smaller than the corresponding coupling constants at the inflationary scale –i.e. if $\delta\lambda \ll \lambda(M_P/\xi)$ and $\delta y_t \ll y_t(M_P/\xi)$ – then Higgs inflation requires the absolute stability of the vacuum and provides a clear connection between the Higgs and top masses and the inflationary observables [26].

Note however that the smallness of the self-coupling λ at the inflationary scale appears as the consequence of a non-trivial cancellation between the bosonic and fermionic degrees of freedom in the Standard Model. A situation in which $\delta\lambda \sim \lambda$, $\delta y_t \sim y_t$ should not be, *a priori*, excluded. If that turns out to be the case, the previously discussed connection between low and high energy observables is lost but, at the same time, the model allows for inflation even if our vacuum is not completely stable [24].

As shown in Fig. 4, a small jump $\delta\lambda \sim 10^{-2}$ is enough to convert a negative self-coupling constant below M_P/ξ into a positive one above that scale. The potential in this case displays two minima. The first one (the shallowest and narrowest one) corresponds to the usual electroweak vacuum. The second one (the deepest and widest one) lays between the scale μ_0 at which the self-coupling becomes negative and the transition scale M_P/ξ in which the jumps in the coupling

[9] Again, we neglect the running of the finite part $\delta\lambda$.

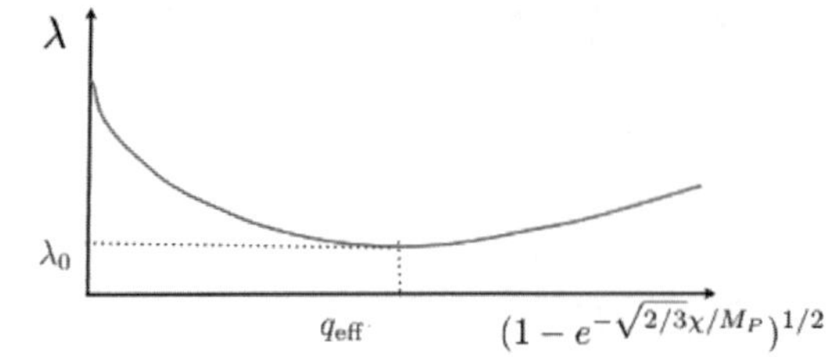

Figure 5. Asymptotic behavior of the Higgs self-coupling around the inflationary scale as a function of the scale $\mu(\chi)$.

constants appear. Beyond $\chi \sim M_P$, the potential becomes asymptotically flat and allows for inflation with the standard chaotic initial conditions.

6. Non-critical vs. critical Higgs inflation

The fate of Universe strongly depends on the interplay between the energy density at the end of inflation, the depth of the deeper minimum and the efficiency of the reheating process. If the backreaction of the particles created during preheating is large enough to modify the shape of the effective potential and to make the wrong vacuum disappear, the Higgs field will be able to roll down to the electroweak minimum. If this is not the case, the Higgs field will end in the minimum at Planckian values and the Universe will eventually collapse.

The relation between the energy density available at the end of inflation and the depth of the wrong minimum depends on the asymptotic behavior of the Higgs self-coupling at the inflationary scale. The last one can be parametrized as (cf. Fig. 5)

$$\lambda(\chi) = \lambda_0 + b \log^2\left(\frac{\sqrt{1-e^{-\alpha\kappa\chi}}}{q_{\rm eff}}\right), \tag{13}$$

with $b \simeq 2.3 \times 10^{-5}$ $\alpha = \sqrt{2/3}$, $\kappa = M_P^{-1}$ and λ_0 and $q_{\rm eff}$ some functions of the top quark pole mass, the Higgs mass and the strong coupling constant *at the inflationary scale* (i.e. after taking into account the rapid changes in the coupling constants at $\chi \sim M_P/\xi$), cf. Ref. [27] for details. The expression (13) reveals the existence of three different regimes (cf. Fig. 6):

- *Non-critical regime*: For $\lambda_0 \gg b/16$, the form of the potential is almost independent of the precise values of the parameters appearing within the logarithmic correction. The potential effectively depends on the combination λ/ξ^2, which, as in the tree-level case, can be fixed by the COBE normalization. The non-minimal coupling ξ associated to a situation like the one presented in Fig. 4 ($\lambda \sim 10^{-3}$) turns out to be of order $\xi \sim 10^3$ [26].
- *Critical regime*: If $\lambda_0 \simeq b/16$, the second derivative of the potential is equal to zero at some intermediate field value between the beginning and the end of inflation and the first derivative becomes very small (but non-zero). The dependence of the running function $\lambda(\chi)$ on the parameters λ_0, ξ and $q_{\rm eff}$ is now essential. The shape of the inflationary potential strongly differs from the tree-level case: it contains a very flat region around the inflection point in which most of the e-folds of inflation take place and a temporally increasing slope at larger field values. The associated slow-roll parameter is non-monotonic and allows to obtain a sizable tensor-to-scalar ratio [27, 28], whose precise value strongly depends on the non-minimal coupling ξ, which is generically required to be small ($\xi \sim 10$).

Journal of Physics: Conference Series **631** (2015) 012032 doi:10.1088/1742-6596/631/1/012032

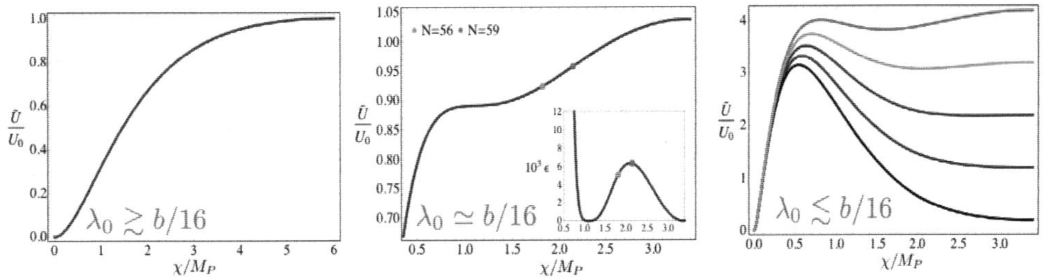

Figure 6. Shape of the inflationary potential in the non-critical (left), critical (center) and forbidden (right) regimes. The behavior of the non-monotonic slow-roll parameter ϵ in the critical case is also shown.

- *Forbidden regime*: If $\lambda_0 \lesssim b/16$, the potential develops a wiggle and inflation is not longer possible.

7. Preheating and symmetry restoration

One of the main differences between non-critical and critical Higgs inflation is the value of the non-minimal coupling ξ. This parameter controls not only the energy scale of inflation, but also the energy scale M_P/ξ at which the jumps in the coupling constants take place and the depth of the vacuum at large field values (cf. Fig. 4) .

7.1. Non-critical Higgs inflation with metastable vacuum

As shown in the left panel of Fig. 7, the energy available at the end of inflation in the non-critical case ($\xi \sim 10^3$) is much larger (by a factor $\sim 10^7$) than the depth of the wrong vacuum. A detailed analysis of the particle production after the end of inflation [24] via the so-called Combined Preheating formalism [29, 30] reveals that the reheating temperature T_{RH}^{NC} exceeds the restoration temperature T_+^{NC} at which the extra minimum at large field values disappears (cf. Fig. 8)

$$T_{RH}^{NC} > T_+^{NC}\,, \qquad \text{with} \qquad T_{RH}^{NC} \simeq 1.8 \times 10^{14}\ \text{GeV}\,, \qquad \text{and} \qquad T_+^{NC} \simeq 7 \times 10^{13}\ \text{GeV}\,. \quad (14)$$

The Higgs field will then be able to relax to the Standard Model vacuum. In the subsequent evolution of the Universe the temperature will decrease and the second minimum will eventually reappear. The probability of decaying via tunneling through the barrier separating the two vacua turns out to be rather small and the lifetime of the electroweak vacuum highly exceeds the life of the Universe [31, 32, 33, 34]. *Non-critical Higgs inflation can take place with a graceful exit even if the Standard Model vacuum is not completely stable.*

7.2. Critical Higgs inflation with metastable vacuum

The situation in the critical case ($\xi \sim 10$) is completely different. The energy stored in the Higgs field after inflation is *comparable* to the height of the barrier separating the two vacua (cf. the right panel of Fig. 7). An upper bound on the reheating temperature $T_{\rm RH}^{C}$ can be obtained by simply converting the energy density of the inflaton field at the end of inflation ($V^{1/4} \simeq 6 \times 10^{16}$ GeV) into an instantaneous radiation temperature. For the number of degrees

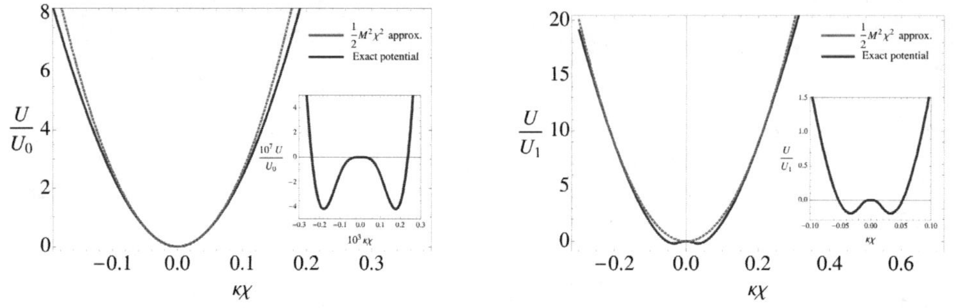

Figure 7. *Left*: Renormalization group enhanced potential in the non-critical case compared to a quadratic approximation $\frac{1}{2}M^2\chi^2$ with $M^2 = \frac{\lambda M_P^2}{3\xi^2}$. The normalization scale U_0 is taken to be $U_0 = (10^{-3}M_P)^4$. *Right*: Similar plot for the critical case. The normalization scale U_1 is given by $U_1 = 10^{-9}M_P^4$.

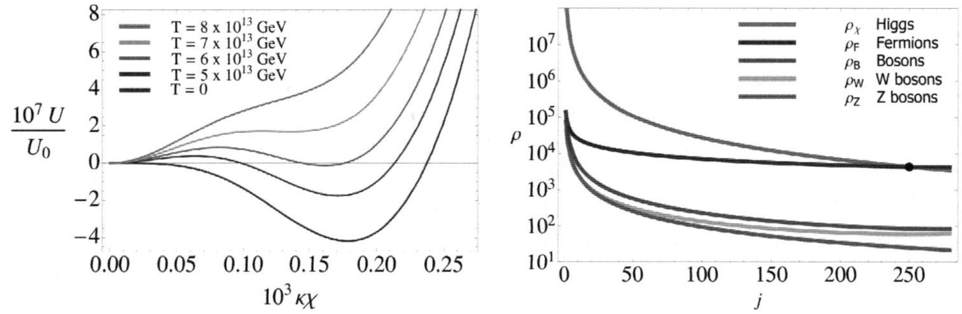

Figure 8. *Left*: High temperature effective potential for non-critical Higgs inflation ($m_H = 125.5\,\mathrm{GeV}$, $m_T = 173.1\,\mathrm{GeV}$, $\xi = 1500$, $\delta y_t = 0.025$, $\delta\lambda = -0.0153$). *Right*: Evolution of the different energy densities (in M^4 units) for the non-critical case ($\lambda = 3.4\times10^{-3}$, $\xi = 1500$) as a function of the number of semioscillations j of the Higgs field around $\chi = 0$.

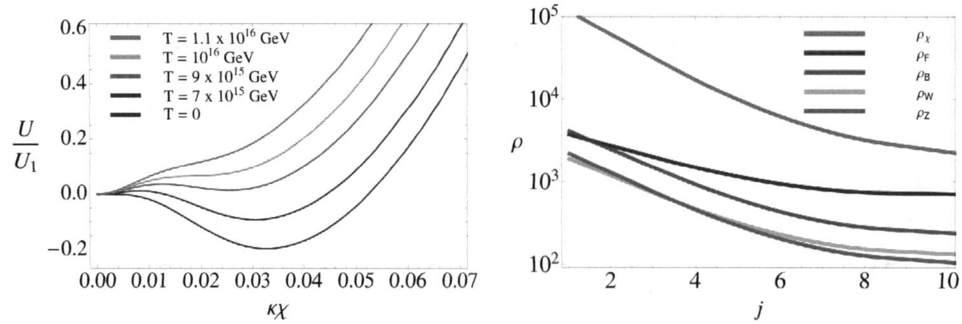

Figure 9. *Left*: High temperature effective potential for critical Higgs inflation ($m_H = 125.5\,\mathrm{GeV}$, $m_T = 173.1\,\mathrm{GeV}$, $\xi = 15$, $\delta y_t = 0$, $\delta\lambda = -0.01325$). *Right*: Evolution of the different energy densities (in M^4 units) for the critical case ($\lambda = 3\times10^{-4}$, $\xi = 15$) as a function of the number of semioscillations j of the Higgs field around $\chi = 0$.

of freedom produced during Combined Preheating, this temperature turns to be smaller than the restoration temperature T_+^C

$$T_{RH}^C < T_+^C\,, \qquad \text{with} \qquad T_{\rm RH}^C < 7.85 g_*^{-1/4} \times 10^{16}\,\text{GeV}\,, \quad \text{and} \quad T_+^C \simeq 10^{16}\,\text{GeV}\,. \tag{15}$$

The shape of the potential in the non-critical case remains unchanged after preheating and the system inevitably relaxes towards the deep minimum at Planckian values. When the energy on the field becomes equal to the amplitude of the barrier, the expansion of the Universe stops and the scale factor begins to shrink till the Universe collapses. *Critical Higgs inflation does necessarily require the absolute stability of the Standard Model vacuum.*

8. Conclusions

The value of the Higgs mass together with the apparent absence of new physics at the LHC allows to speculate with the possibility of consistently extending the Standard Model all the way up till the Planck scale. The Higgs field itself can be responsible for inflation if a minimalistic, and at the same time compelling, non-minimal coupling to gravity is allowed. No new particles beyond the electroweak scale are required. The non-minimal coupling makes the Standard Model non-renormalizable, which requires the addition of an infinite number of counterterms. The absence of an ultraviolet completion introduces some uncertainties associated to the unknown finite parts of the counterterms. If these are small compared to the values of the couplings at the inflationary scale, Higgs inflation requires the absolute stability of the Standard Model vacuum. On the other hand, if the finite parts are comparable to the couplings at the inflationary scale, (non-critical) Higgs inflation can take place even for the case of a metastable vacuum. The associated predictions ($n_s = 0.97$, $r = 0.003$) are universal and in excellent agreement with the latest CMB results.

Acknowledgments

It is a pleasure to thank F. Bezrukov, M. Shaposhnikov and G. Karananas for our collaborations and discussions. This work was partially supported by the Swiss National Science Foundation.

References

[1] G. Aad *et al.* [ATLAS Collaboration], Phys. Lett. B **716** (2012) 1 [arXiv:1207.7214 [hep-ex]].
[2] S. Chatrchyan *et al.* [CMS Collaboration], Phys. Lett. B **716** (2012) 30 [arXiv:1207.72 35 [hep-ex]].
[3] P. A. R. Ade *et al.* [Planck Collaboration], arXiv:1502.02114 [astro-ph.CO].
[4] A. D. Linde, Phys. Lett. B **129** (1983) 177.
[5] N. D. Birrell and P. C. W. Davies,
[6] F. L. Bezrukov and M. Shaposhnikov, Phys. Lett. B **659** (2008) 703 [arXiv:0710.3755 [hep-th]].
[7] M. Shaposhnikov and D. Zenhausern, Phys. Lett. B **671** (2009) 187 [arXiv:0809.3395 [hep-th]].
[8] J. Garcia-Bellido, J. Rubio, M. Shaposhnikov and D. Zenhausern, Phys. Rev. D **84** (2011) 123504 [arXiv:1107.2163 [hep-ph]].
[9] F. Bezrukov, G. K. Karananas, J. Rubio and M. Shaposhnikov, Phys. Rev. D **87** (2013) 9, 096001 [arXiv:1212.4148 [hep-ph]].
[10] J. Rubio and M. Shaposhnikov, Phys. Rev. D **90** (2014) 2, 027307 [arXiv:1406.5182 [hep-ph]].
[11] J. Garcia-Bellido, J. Rubio and M. Shaposhnikov, Phys. Lett. B **718** (2012) 507 [arXiv:1209.2119 [hep-ph]].
[12] F. L. Bezrukov and D. S. Gorbunov, Phys. Lett. B **713** (2012) 365 [arXiv:1111.4397 [hep-ph]].
[13] A. A. Starobinsky, Phys. Lett. B **91** (1980) 99.
[14] V. F. Mukhanov and G. V. Chibisov, JETP Lett. **33** (1981) 532 [Pisma Zh. Eksp. Teor. Fiz. **33** (1981) 549].
[15] F. Bezrukov, M. Y. Kalmykov, B. A. Kniehl and M. Shaposhnikov, JHEP **1210** (2012) 140 [arXiv:1205.2893 [hep-ph]].
[16] G. Degrassi, S. Di Vita, J. Elias-Miro, J. R. Espinosa, G. F. Giudice, G. Isidori and A. Strumia, JHEP **1208** (2012) 098 [arXiv:1205.6497 [hep-ph]].

[17] D. Buttazzo, G. Degrassi, P. P. Giardino, G. F. Giudice, F. Sala, A. Salvio and A. Strumia, JHEP **1312** (2013) 089 [arXiv:1307.3536 [hep-ph]].
[18] F. Bezrukov and M. Shaposhnikov, arXiv:1411.1923 [hep-ph].
[19] https://cds.cern.ch/record/1951019?ln=en
[20] V. Branchina and E. Messina, Phys. Rev. Lett. **111** (2013) 241801 [arXiv:1307.5193 [hep-ph]].
[21] V. Branchina, E. Messina and A. Platania, JHEP **1409** (2014) 182 [arXiv:1407.4112 [hep-ph]].
[22] V. Branchina, E. Messina and M. Sher, arXiv:1408.5302 [hep-ph].
[23] F. Bezrukov, A. Magnin, M. Shaposhnikov and S. Sibiryakov, JHEP **1101** (2011) 016 [arXiv:1008.5157 [hep-ph]].
[24] F. Bezrukov, J. Rubio and M. Shaposhnikov, arXiv:1412.3811 [hep-ph].
[25] A. C. Longhitano, Phys. Rev. D **22** (1980) 1166.
[26] F. Bezrukov and M. Shaposhnikov, JHEP **0907** (2009) 089 [arXiv:0904.1537 [hep-ph]].
[27] F. Bezrukov and M. Shaposhnikov, Phys. Lett. B **734** (2014) 249 [arXiv:1403.6078 [hep-ph]].
[28] Y. Hamada, H. Kawai, K. y. Oda and S. C. Park, Phys. Rev. Lett. **112** (2014) 241301 [arXiv:1403.5043 [hep-ph]].
[29] J. Garcia-Bellido, D. G. Figueroa and J. Rubio, Phys. Rev. D **79** (2009) 063531 [arXiv:0812.4624 [hep-ph]].
[30] F. Bezrukov, D. Gorbunov and M. Shaposhnikov, JCAP **0906** (2009) 029 [arXiv:0812.3622 [hep-ph]].
[31] G. W. Anderson, Phys. Lett. B **243** (1990) 265.
[32] P. B. Arnold and S. Vokos, Phys. Rev. D **44** (1991) 3620.
[33] J. R. Espinosa and M. Quiros, Phys. Lett. B **353** (1995) 257 [hep-ph/9504241].
[34] J. R. Espinosa, G. F. Giudice and A. Riotto, JCAP **0805** (2008) 002 [arXiv:0710.2484 [hep-ph]].

Recent *BABAR* Results

Gerald Eigen on behalf of the *BABAR* Collaboration
Dept. of Physics, University of Bergen, Allegation 55, Bergen, Norway
E-mail: gerald.eigen@ift.uib.no

Abstract. We present herein the most recent *BABAR* results on direct *CP* asymmetry measurements in $B \to X_s\gamma$, on partial branching fraction and *CP* asymmetry measurements in $B \to X_s\ell^+\ell^-$, on a search for $B \to \pi/\eta\ell^+\ell^-$ decays, on a search for lepton number violation in $B^+ \to X^-\ell^+\ell'^+$ modes and a study of $B^0 \to \omega\omega$ and $B^0 \to \omega\phi$ decays.

1. Introduction

The decays $B \to X_{s,d}\gamma$ and $B \to X_{s,d}\ell^+\ell^-$ (with $\ell^+\ell^- = e^+e^-$, $\mu^+\mu^-$) are flavor-changing neutral-current (FCNC) processes that are forbidden in the Standard Model (SM) at tree level. They occur in higher-order processes and are described by an effective Hamiltonian that factorizes short-distance contributions in terms of scale-dependent Wilson coefficients $C_i(\mu)$ [1] from long-distance contributions expressed by local four-fermion operators $\mathcal{O}_i$ that define hadronic matrix elements,

$$H_{\rm eff} = \frac{4G_F}{\sqrt{2}} \sum_i C_i(\mu)\mathcal{O}_i. \quad (1)$$

While Wilson coefficients are calculable perturbatively, the calculation of the hadronic matrix elements requires non-perturbative methods such as the heavy quark expansion [2–4].

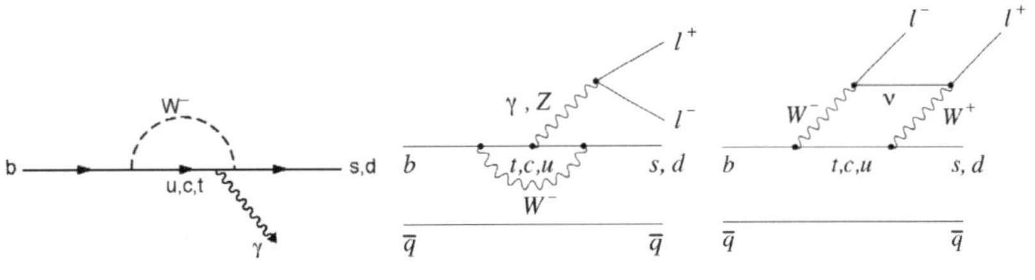

Figure 1. Lowest-order diagrams for $B \to X_{s,d}\gamma$ (left) and $B \to X_{s,d}\ell^+\ell^-$ (middle, right).

Figure 1 shows the lowest order diagrams for these FCNC decays. In $B \to X_{s,d}\gamma$, the electromagnetic penguin loop dominates. The short-distance part is expressed by the effective Wilson coefficient $C_7^{\rm eff}$. Through operator mixing at higher orders, the chromomagnetic penguin enters whose short distance part is parameterized by $C_8^{\rm eff}$. In $B \to X_{s,d}\ell^+\ell^-$modes, the Z penguin and the WW box diagram contribute in addition whose short-distance parts are parametrized in terms of $C_9^{\rm eff}$ and $C_{10}^{\rm eff}$, the vector and axial-vector current contributions of these diagrams. Physics beyond the SM introduces new loops and box diagrams with new particles (e.g. a charged Higgs boson or supersymmetric particles) as shown in Fig. 2 (left, middle). Such contributions modify the Wilson coefficients and may introduce new diagrams with scalar and pseudoscalar current interactions and in turn new Wilson coefficients, C_S and C_P [5]. To determine $C_7^{\rm eff}$, $C_8^{\rm eff}$, $C_9^{\rm eff}$ and $C_{10}^{\rm eff}$ precisely, we need to measure many observables in several radiative and rare semileptonic decays, which potentially can probe new physics at a scale of a few TeV.

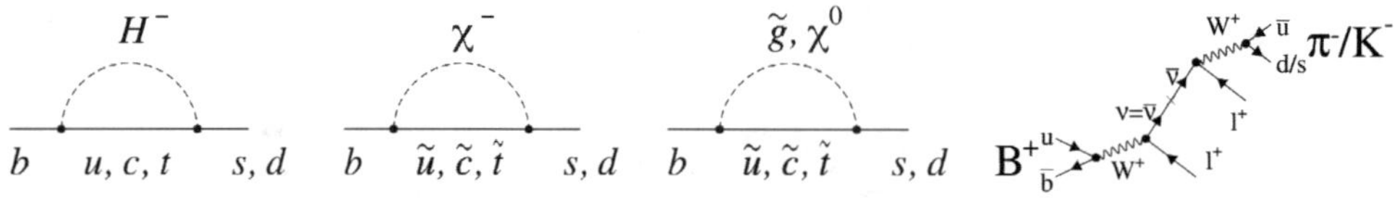

Figure 2. Examples of new physics processes via a charged Higgs boson (left), charginos (middle left), neutralinos (middle right) and via Majorana-type neutrino interactions (right).

Lepton-number-violating decays are highly suppressed in the SM and may need new physics processes. Figure 2 (right) shows a W annihilation diagram into $\ell^+\nu_\ell$ in which the neutrino mixes into an antineutrino producing like-sign leptons that are forbidden in SM interactions. Such processes require Majorana-type neutrinos that are absent in the SM [6].

The decays $B^0 \to \omega\omega$ and $B^0 \to \omega\phi$ also involve FCNC processes that are mediated by gluonic penguin loops included in $H_{\rm eff}$ (see Eqn. (1)). Here, the short-distance contributions are parameterized by the Wilson coefficients C_3, C_4, C_5 and C_6, while the long-distance contributions involve the operators $\mathcal{O}_3$, $\mathcal{O}_4$, $\mathcal{O}_5$ and $\mathcal{O}_6$. Figure 3 shows the lowest-order diagrams for these decays. New physics loops depicted in Fig. 2 may also contribute here. These charmless vector vector decays involve three amplitudes. In the transversity frame, these are the longitudinal amplitude A_0 (S-wave), the transverse amplitude A_T (P-wave) and the parallel amplitude $A_{||}$ (D-wave). For measuring CP violation, they need to be known.

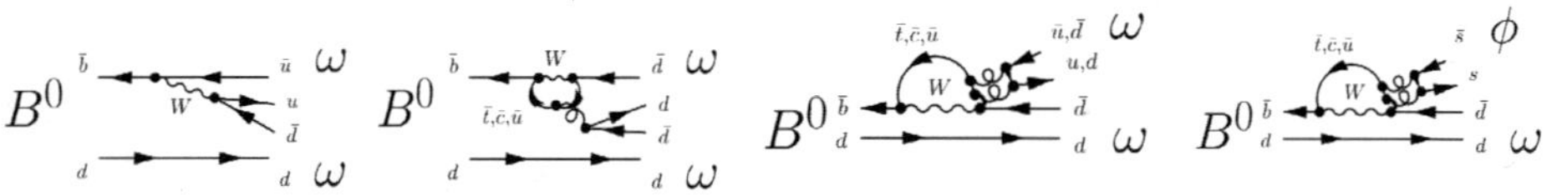

Figure 3. Lowest-order diagrams for the $B^0 \to \omega\omega$ color-suppressed tree (left), singlet penguin (middle left) and gluonic penguin (middle right) and the gluonic penguin for $B^0 \to \omega\phi$ (right).

In chapter 2, we present new *BABAR* measurements of the direct CP asymmetry in $B \to X_s\gamma$ using a semi-inclusive analysis. We extract the ratio of Wilson coefficients $\mathcal{I}m(C_8^{\rm eff}/C_7^{\rm eff})$ from a measurement of the difference in CP asymmetries between charged and neutral B decays. We also show CP asymmetry measurements for $B \to X_{s,d}\gamma$ decays. In chapter 3, we present our branching fraction and CP asymmetry measurements of $B \to X_s\ell^+\ell^-$ decays using a semi-inclusive analysis. In chapter 4, we summarize our branching fraction upper limits on $B \to \pi\ell^+\ell^-$ and $B \to \eta\ell^+\ell^-$. In chapter 5, we summarize our results on searches for lepton number violation in exclusive $B^+ \to X^-\ell^+\ell'^+$ modes. In chapter 6, we present our results on the charmless vector vector decays $B^0 \to \omega\omega$ and $B^0 \to \omega\phi$ and in chapter 7 we end with concluding remarks. Note that *BABAR* performs all analyses blinded meaning that results are sealed until selection criteria and fitting procedures are finalized.

2. Measurement of CP Violation in $B \to X_s\gamma$

In the SM, the $B \to X_s\gamma$ branching fraction is calculated at next-to-next-to-leading order (up to four loops) yielding $\mathcal{B}(B \to X_s\gamma) = (3.14 \pm 0.22) \times 10^{-4}$ for photon energies $E_\gamma^* > 1.6$ GeV in the center-of-mass (CM) frame [7, 8]. For larger values of E^*_γ, the prediction depends the shape of the of the E^*_γ spectrum, which is modeled in terms of a shape function [9] that depends on the Fermi motion of the b quark inside the B meson and thus on the b quark mass. Since the shape function is expected to be similar to that determining the lepton-energy spectrum in $B \to X_u \ell \nu$, precision measurements of the E^*_γ spectrum help to determine $|V_{ub}|$ more precisely [10–12]. The measurement of $\mathcal{B}(B \to X_s\gamma)$ also provides constraints on the charged Higgs mass [13, 14].

Experimentally, the challenge consists of extracting $B \to X_s\gamma$ signal photons from those of π^0 and η decays, copiously produced in $q\bar{q}$ continuum (with $q = u,\ d,\ s,\ c$) and $B\bar{B}$ processes that increase exponentially at smaller photon energies. One strategy consists of summing $b \to s\gamma$ exclusive final states. In a sample of 471×10^{-6} $B\bar{B}$ events collected with the *BABAR* detector [15, 16] at the PEP-II asymmetric storage ring at the SLAC National Laboratory, we reconstruct 38 exclusive final states containing one or three kaons with at most one K^0_S, up to four pions with at most two π^0s and up to one η. We require photon energies in the CM frame of $1.6 < E^*_\gamma < 3.0$ GeV. Previously, we published total and partial branching fractions [17]. Here, we focus on the measurement of direct CP asymmetry, which is defined by

$$\mathcal{A}_{CP}(B \to X_s\gamma) \equiv \frac{\mathcal{B}(\bar{B} \to \bar{X}_s\gamma) - \mathcal{B}(B \to X_s\gamma)}{\mathcal{B}(\bar{B} \to \bar{X}_s\gamma) + \mathcal{B}(B \to X_s\gamma)}. \tag{2}$$

For this analysis [18], we select 16 self-tagging modes, ten $B^+ \to h^+\gamma$[1] and six $B^0 \to h^0\gamma$ final states[2]. We maximize the signal extraction using a bagged decision tree with six input variables. This improves the efficiency considerably with respect to the standard $\Delta E = E^*_B - E^*_{beam}$ selection where E^*_{beam} and E^*_B are the beam energy and B meson energy in the CM frame, respectively. To remove continuum background, we train a separate bagged decision tree using event shape variables. For each X_s mass bin, we optimize the sensitivity $S/\sqrt{(S+B)}$ where $S(B)$ is the signal (background) yield using loosely identified pions and kaons. To extract $\mathcal{A}_{CP}$, we fit the beam-energy-constrained mass $m_{ES} = \sqrt{E^{*2}_{beam} - p^{*2}_B}$ simultaneously for $\bar{B}$-tagged and B-tagged events where p^*_B is the B momentum in the CM frame. After correcting the raw $\mathcal{A}_{CP}$ for detector bias determined from the m_{ES} sideband below the signal region, we measure $\mathcal{A}_{CP}(B \to X_s\gamma) = (1.73 \pm 1.93_{stat} \pm 1.02_{sys})\%$ [18], which agrees well with the SM prediction of $-0.6\% < \mathcal{A}_{CP} < 2.8\%$ at 95% confidence level (CL) [19] and which supersedes the old *BABAR* measurement [20]. Though this result is the most precise single direct $\mathcal{A}_{CP}$ measurement, the uncertainty is sufficiently large to allow for new physics contributions in $C^{\rm eff}_7$. Figure 4 (bottom part) shows our result [18] in comparison to the Belle measurement [21]. The CP asymmetry difference between B^+ and B^0 decays, $\Delta\mathcal{A}_{CP}(B \to X_s\gamma) = \mathcal{A}_{CP}(B^+ \to X^+_s\gamma) - \mathcal{A}_{CP}(B^0 \to X^0_s\gamma)$, is very sensitive to new physics since it originates from the interference between the electromagnetic and the chromomagnetic penguin diagrams in which the latter enters through higher-order corrections. Calculations yield [19]

$$\Delta\mathcal{A}_{CP}(B \to X_s\gamma) \simeq 4\pi^2\alpha_s \frac{\bar{\Lambda}_{78}}{m_b} \mathcal{I}m \frac{C^{eff}_8}{C^{eff}_7} \simeq 0.12 \frac{\bar{\Lambda}_{78}}{100\ \mathrm{MeV}} \mathcal{I}m \frac{C^{eff}_8}{C^{eff}_7}, \tag{3}$$

where $\bar{\Lambda}_{78}$ is the hadronic matrix element of the $\mathcal{O}_7 - \mathcal{O}_8$ interference, predicted to lie in the range 17 MeV $< \bar{\Lambda}_{78} <$ 190 MeV. In the SM, $\Delta\mathcal{A}_{CP}(B \to X_s\gamma)$ vanishes since $C^{\rm eff}_7$ and $C^{\rm eff}_8$ are real. However in new physics models, these Wilson coefficients may have imaginary parts yielding non-vanishing $\Delta\mathcal{A}_{CP}(B \to X_s\gamma)$ [22–24]. From a simultaneous fit to B^+ and B^0 modes, we measure $\Delta\mathcal{A}_{CP}(B \to X_s\gamma) = (5.0 \pm 3.9_{stat} \pm 1.5_{sys})\%$ from which we obtain the constraint $-1.64 < \mathcal{I}m(C^{\rm eff}_8/C^{\rm eff}_7) < 6.52$ at 90% CL. This is the first $\Delta\mathcal{A}_{CP}(B \to X_s\gamma)$ measurement and first constraint on $\mathcal{I}m(C^{\rm eff}_8/C^{\rm eff}_7)$.

[1] $h^+ = K^0_S\pi^+,\ K^+\pi^0,\ K^+\pi^+\pi^-,\ K^0_S\pi^+\pi^0,\ K^+\pi^0\pi^0,\ K^0_S\pi^+\pi^-\pi^+,\ K^+\pi^+\pi^-\pi^0,\ K^0_S\pi^+\pi^0\pi^0,\ K^+\eta,\ K^+K^+K^-$.
[2] $h^0 = K^+\pi^-,\ K^+\pi^-\pi^0,\ K^+\pi^+\pi^-\pi^-,\ K^+\pi^-\pi^0\pi^0,\ K^+\pi^-\eta,\ K^+K^+K^-\pi^-$.

Journal of Physics: Conference Series **631** (2015) 012033 doi:10.1088/1742-6596/631/1/012033

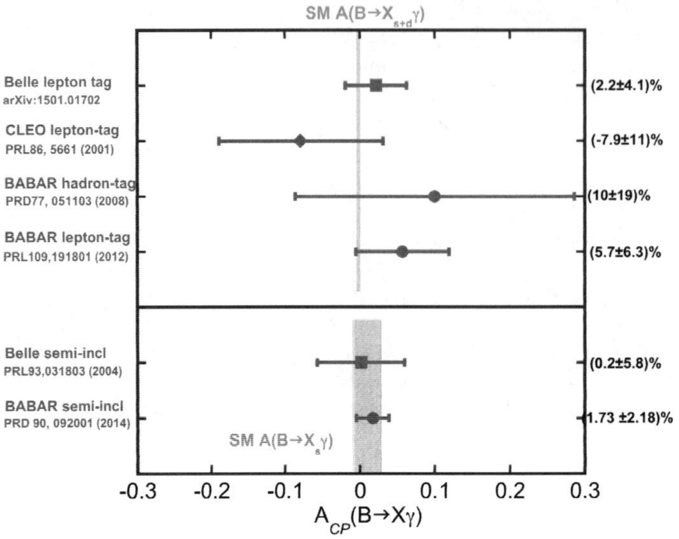

Figure 4. Summary of $\mathcal{A}_{CP}$ measurements for $B \to X_s\gamma$ from semi-inclusive analyses (bottom part) from *BABAR* [18] and Belle [21] and for $B \to X_{s+d}\gamma$ from fully inclusive analyses (top part) from *BABAR* [25–27], Belle [28] and CLEO [29] in comparison to the SM prediction for $B \to X_s\gamma$ [19] and for $B \to X_{s,d}\gamma$ [22, 30], respectively.

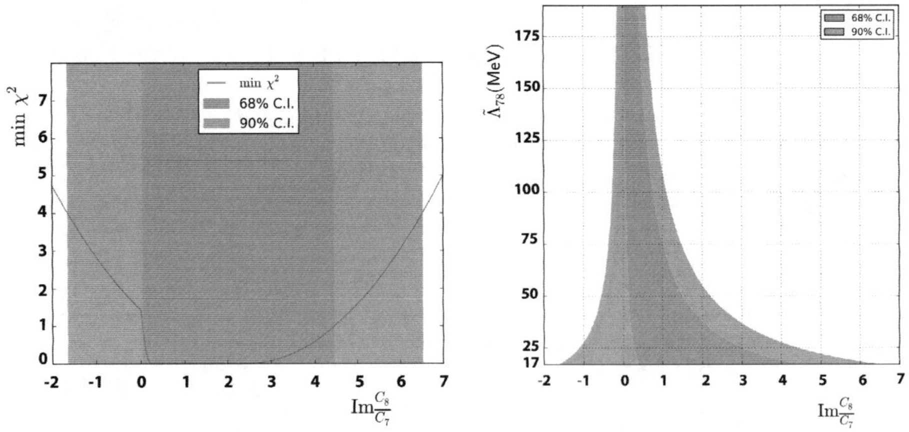

Figure 5. The $\Delta\chi^2$ (left) and $\bar{\Lambda}_{78}$ (right) dependence on $\mathcal{I}m(C_8^{eff}/C_7^{eff})$. The blue dark-shaded (orange light-shaded) region shows the 68% (90%) CL interval.

Figure 5 (left) shows the $\Delta\chi^2$ of the fit as a function of $\mathcal{I}m(C_8^{\rm eff}/C_7^{\rm eff})$. The $\Delta\chi^2$ dependence on $\mathcal{I}m(C_8^{\rm eff}/C_7^{\rm eff})$ is not parabolic indicating that the likelihood has a non-Gaussian shape. The reason is that $\Delta\chi^2$ is determined from all possible values of $\bar{\Lambda}_{78}$. In the region $0.2 < \mathcal{I}m(C_8^{\rm eff}/C_7^{\rm eff}) < 2.6$, a change in $\mathcal{I}m(C_8^{\rm eff}/C_7^{\rm eff})$ can be compensated by a change in $\bar{\Lambda}_{78}$ leaving $\Delta\chi^2$ unchanged. For positive values larger (smaller) than 2.6 (0.2), $\Delta\chi^2$ increases slowly (rapidly) since $\bar{\Lambda}_{78}$ remains nearly constant at the minimum value (increases rapidly). For negative $\mathcal{I}m(C_8^{\rm eff}/C_7^{\rm eff})$ values, $\bar{\Lambda}_{78}$ starts to decrease again, which leads to a change in the $\Delta\chi^2$ shape. Figure 5 (right) shows $\bar{\Lambda}_{78}$ as a function of $\mathcal{I}m(C_8^{\rm eff}/C_7^{\rm eff})$.

In the fully inclusive analysis, $\mathcal{A}_{CP}$ involves contributions from $B \to X_s\gamma$ and $B \to X_d\gamma$ that cannot be separated on an event-by-event basis. Therefore, we define $\mathcal{A}_{CP}$ here as

$$\mathcal{A}_{CP}(B \to X_{s+d}\gamma) \equiv \frac{\mathcal{B}(\bar{B} \to X_{s+d}\gamma) - \mathcal{B}(B \to X_{s+d}\gamma)}{\mathcal{B}(\bar{B} \to X_{s+d}\gamma) + \mathcal{B}(B \to X_{s+d}\gamma)}. \quad (4)$$

We tag the flavor of the non-signal $\bar{B}$ flavor by the lepton charge in semileptonic decays. Using a sample 384×10^6 $B\bar{B}$ events, we measure $\mathcal{A}_{CP}(B \to X_{s+d}\gamma) = 0.057 \pm 0.06_{stat} \pm 0.018_{sys}$ after correcting for charge bias and mistagging [17]. Figure 4 (top part) shows all $\mathcal{A}_{CP}(B \to X_{s+d}\gamma)$ measurements from *BABAR* [25–27], Belle [28] and CLEO [29], which all agree well with the SM prediction [22, 30].

3. Study of $B \to X_s\ell^+\ell^-$ Decays

Using a semi-inclusive approach, we have updated the partial and total branching fraction measurements of $B \to X_s\ell^+\ell^-$ modes ($\ell = e$ or μ) with the full *BABAR* data sample of 471×10^6 $B\bar{B}$ events. We reconstruct 20 exclusive final states: K^+, K^0_S, $K^{*+}(K^+\pi^0, K^0_S\pi^+)$, $K^{*0}(K^+\pi^-, K^0_S\pi^0)$, $K^+\pi^-\pi^0$, $K^0_S\pi^+\pi^-$ with $K^0_S \to \pi^+\pi^-$ recoiling against e^+e^- or $\mu^+\mu^-$ [31]. After accounting for K^0_L modes, $K^0_S \to \pi^0\pi^0$ and π^0 Dalitz decays, the selected decay modes represent 70% of the inclusive rate for hadronic masses $m_{X_s} < 1.8$ GeV/c^2. Using JETSET fragmentation [32] and theory predictions [33–38], we extrapolate for the missing modes and those with $m_{X_s} > 1.8$ GeV/c^2. We impose the requirements $m_{ES} > 5.225$ GeV/c^2 and 0.1 (0.05) $< \Delta E <$ 0.05 (0.05) GeV for $X_se^+e^-$ ($X_s\mu^+\mu^-$) modes. We define six bins of the momentum-squared transferred to the dilepton system $q^2 = m^2_{\ell\ell}$ and four bins in hadronic mass m_{X_s}. Table 1 shows the defined ranges of these bins.

Table 1. Definition of the q^2, $m_{\ell\ell}$ and m_{X_s} bins.

q^2 bin	q^2 range [GeV^2/c^4]	$m_{\ell\ell}$ range [GeV/c^2]	m_{X_s} bin	m_{X_s} range [GeV/c^2]
0	1.0 – 6.0	1.00 – 2.45		
1	0.1–2.0	0.32– 1.41	1	0.4 – 0.6
2	2.0–4.3	1.41–2.07	2	0.6 –1.0
3	4.3–8.1	2.07 –2.6	3	1.0 –1.4
4	10.1 –12.9	3.18–3.59	4	1.4 – 1.8
5	14.2 – $(m_B - m_K^*)^2$	3.77 – $(m_B - m_K^*)$		

To suppress $e^+e^- \to q\bar{q}$ and $B\bar{B}$ combinatorial background, we define boosted decision trees (BDT) for each q^2 bin separately for e^+e^- and $\mu^+\mu^-$ modes. From these BDTs, we determine a likelihood ratio (L_R) to separate signal from $q\bar{q}$ and $B\bar{B}$ backgrounds. We veto J/ψ and $\psi(2S)$ mass regions and use them as control samples. We measure $d\mathcal{B}(B \to X_s\ell^+\ell^-/dq^2$ in six bins of q^2 and four bins of m_{X_s} . We extract the signal in each bin from a two-dimensional fit to m_{ES} and L_R. As examples, Figs. 6 and 7 show the m_{ES} and L_R distributions for e^+e^-modes in bin q_5 and for $\mu^+\mu^-$ modes in bin q_1, respectively. Clear signals are visible both in the m_{ES} and L_R distributions. Figure 8 shows the differential branching faction as a function of q^2 (left) and m_{X_s} (right) [31]. Table 2 summarizes the differential branching fractions in the low and high q^2 regions in comparison to the SM predictions [34, 38–47]. In both q^2 regions, the differential branching fractions are in good agreement with the SM predictions. These results supersede the previous *BABAR* measurements [48] and agree well with the measurements from Belle [49].

The direct *CP* asymmetry is defined by

$$\mathcal{A}_{CP} = \frac{\mathcal{B}(\bar{B} \to \bar{X}_s \ell^+ \ell^-) - \mathcal{B}(B \to X_s \ell^+ \ell^-)}{\mathcal{B}(\bar{B} \to \bar{X}_s \ell^+ \ell^-) + \mathcal{B}(B \to X_s \ell^+ \ell^-)}. \tag{5}$$

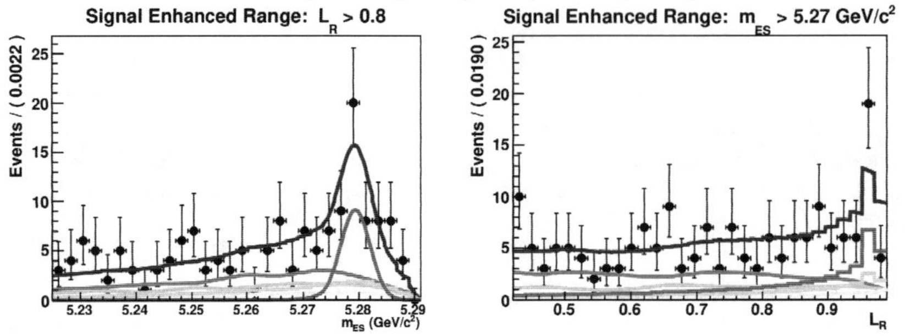

Figure 6. Distributions of m_{ES} (left) and likelihood ratio (right) for $B \to X_s e^+ e^-$ in q^2 bin q_5 showing data (points with error bars), the total fit (thick solid blue curves), signal component (red peaking curves), signal cross feed (cyan curves), $B\bar{B}$ background (magenta curve), $e^+e^- \to q\bar{q}$ background (green curves) and charmonium background (yellow curves).

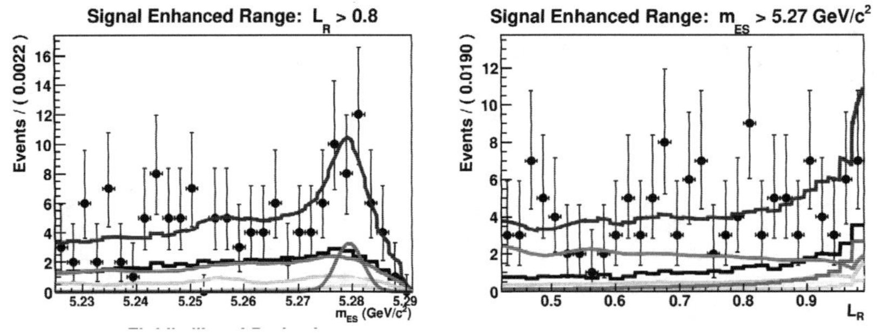

Figure 7. Distributions of m_{ES} (left) and likelihood ratio (right) for $B \to X_s \mu^+ \mu^-$ in q^2 bin q_1 showing data (points with error bars), the total fit (thick solid blue curves), signal component (red peaking curves), signal cross feed (cyan curves), $B\bar{B}$ background (magenta curve), $e^+e^- \to q\bar{q}$ background (green curves) and charmonium background (yellow curves).

We use 14 self-tagging modes consisting of all B^+ modes and B^0 modes with decays to a K^+ to measure $\mathcal{A}_{CP}(B \to X_s \ell^+ \ell^-)$ in five q^2 bins. Due to low statistics, we have combined bins q_4 and q_5. Figure 9 shows the *CP* asymmetry as a function of q^2. The SM prediction of the *CP* asymmetry in the entire q^2 region is close to zero [50–53]. In new physics models, however, $\mathcal{A}_{CP}$ may be significantly enhanced [54, 55]. In the full range of q^2, we measure $\mathcal{A}_{CP}(B \to X_s \ell^+ \ell^-) = 0.04 \pm 0.11_{stat} \pm 0.01_{sys}$ [31], which is in good agreement with the SM prediction. The *CP* asymmetries in the five q^2 bins are also consistent with zero.

4. Search for $B \to \pi/\eta \ell^+ \ell^-$ Decays

In the SM in lowest order, $B \to X_d \ell^+ \ell^-$ modes are also mediated by the electromagnetic penguin, Z penguin and WW box diagrams. However, they are suppressed by $|V_{td}/V_{ts}|^2 \simeq 0.04$ with

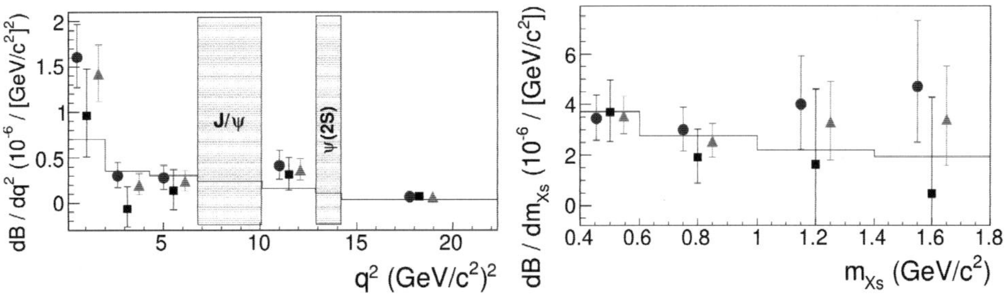

Figure 8. Differential branching fraction of $B \to X_s e^+e^-$ (blue points), $B \to X_s\mu^+\mu^-$ (black squares) and $B \to X_s\ell^+\ell^-$ (red triangles) versus q^2 (top) and versus m_{X_s} (bottom) in comparison to the SM prediction (histogram). Grey-shaded bands show J/ψ and $\psi(2S)$ vetoed regions.

Table 2. Measured $B \to X_s\ell^+\ell^-$ branching fractions in the low and high q^2 regions from *BABAR* [31] and the SM predictions. Uncertainties are statistical, systematic and from model dependence, respectively.

Mode	*BABAR* $[10^{-6}]$	SM $[10^{-6}]$	*BABAR* $[10^{-6}]$	SM $[10^{-6}]$
$q^2[\mathrm{GeV}^2/\mathrm{c}^4]$	$1-6$	$1-6$	> 14.2	> 14.2
$B \to X_s\mu^+\mu^-$	$0.66^{+0.82+0.30}_{-0.76-0.24} \pm 0.07$	1.59 ± 0.11	$0.60^{+0.31+0.05}_{-0.29-0.04} \pm 0.00$	$0.25^{+0.07}_{-0.06}$
$B \to X_s e^+e^-$	$1.93^{+0.47+0.21}_{-0.45-0.16} \pm 0.18$	1.64 ± 0.11	$0.56^{+0.19+0.03}_{-0.18-0.03} \pm 0.00$	$0.25^{+0.07}_{-0.06}$
$B \to X_s\ell^+\ell^-$	$1.60^{+0.41+0.17}_{-0.39-0.13} \pm 0.07$		$0.57^{+0.16+0.03}_{-0.15-0.02} \pm 0.00$	$0.25^{+0.07}_{-0.06}$

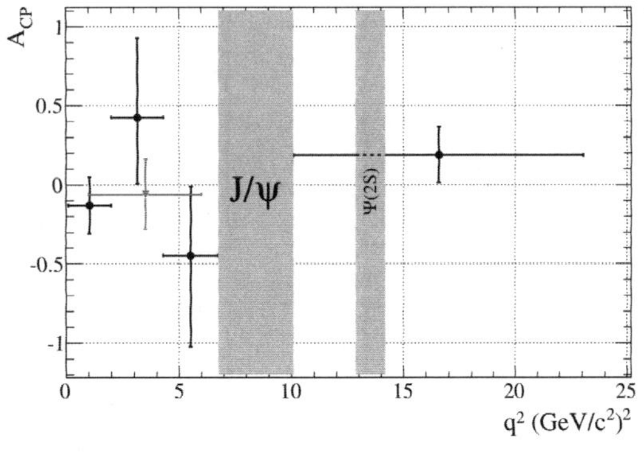

Figure 9. The *CP* asymmetry versus q^2. Grey bands show the J/ψ and $\psi(2S)$ vetoed regions.

respect to the corresponding $B \to X_s\ell^+\ell^-$ decays. In extensions of the SM, rates may increase significantly [56]. Using 471×10^6 $B\bar{B}$ events, we recently updated the search for $B \to \pi\ell^+\ell^-$ modes and performed the first search for $B \to \eta\ell^+\ell^-$ modes [57]. The SM predictions lie in the range $\mathcal{B}(B \to \pi\ell^+\ell^-) = (1.96\text{–}3.30) \times 10^{-8}$ and $\mathcal{B}(B \to \eta\ell^+\ell^-) = (2.5\text{–}3.7) \times 10^{-8}$ where the large ranges result from uncertainties in the $B \to \pi$ form factor calculations [56,58,59] and from a lack of knowledge of $B \to \eta$ form factors [60], respectively.

We fully reconstruct four $B \to \pi\ell^+\ell^-$ and four $B \to \eta\ell^+\ell^-$ final states by selecting $\pi^\pm$, $\pi^0 \to \gamma\gamma$ and $\eta \to \gamma\gamma$, $\pi^+\pi^-\pi^0$ recoiling against e^+e^- or $\mu^+\mu^-$. We select leptons with $p_\ell > 0.3$ GeV/c, recover $e^\pm$ bremsstrahlung losses, remove γ conversions and require good particle identification for $e^\pm$, $\mu^\pm$ and $\pi^\pm$. We select photons with $E_\gamma > 50$ MeV and impose π^0 and η mass constraints of $115 < m_{\gamma\gamma} < 150$ MeV/c^2 and $500\ (535) < m_{\gamma\gamma}\ (m_{3\pi}) < 575\ (565)$ MeV/c^2, respectively. For the $\eta \to \gamma\gamma$ final state, we require $(E_{1,\gamma} - E_{2,\gamma})/(E_{1,\gamma} + E_{2,\gamma}) < 0.8$ to remove asymmetric $q\bar{q}$ background that peaks near one. We veto J/ψ and $\psi(2S)$ mass regions and use four neural networks (NN) to suppress combinatorial $B\bar{B}$ and $q\bar{q}$ continuum backgrounds, separately for e^+e^- and for $\mu^+\mu^-$ modes. The NNs for suppressing $B\bar{B}$ background use 15 (14) input distributions for e^+e^- ($\mu^+\mu^-$) modes, while those for suppressing $q\bar{q}$ continuum use 16 input distributions for both modes. For validations of the fitting procedure and peaking backgrounds, we use pseudo-experiments and the vetoed J/ψ and $\psi(2S)$ samples.

For $B^+ \to \pi^+\ell^+\ell^-$ and $B^0 \to \pi^0\ell^+\ell^-$, we perform simultaneous unbinned maximum likelihood (ML) fits to m_{ES} and ΔE distributions for e^+e^- and $\mu^+\mu^-$ modes separately. We include the $B^+ \to K^+\ell^+\ell^-$ mode in the fit to extract the peaking background contribution in the $B^+ \to \pi^+\ell^+\ell^-$ modes by reconstructing the K^+ as a π^+. For $B^0 \to \eta\ell^+\ell^-$, we perform simultaneous unbinned ML fits to m_{ES} and ΔE distributions, again for e^+e^- and $\mu^+\mu^-$ modes separately. In addition, we perform fits for the isospin-averaged modes $B \to \pi e^+e^-$ and $B \to \pi\mu^+\mu^-$, lepton-flavor-averaged modes $B^+ \to \pi^+\ell^+\ell^-$, $B^0 \to \pi^0\ell^+\ell^-$ and $B^0 \to \eta\ell^+\ell^-$ and both isospin- and lepton-flavor-averaged modes $B \to \pi\ell^+\ell^-$.

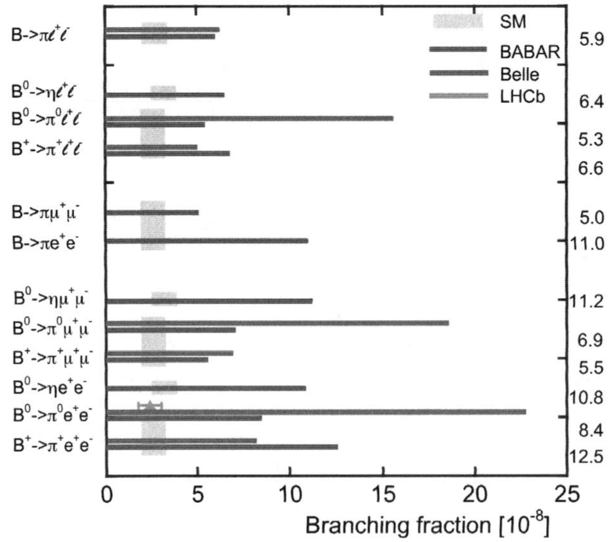

Figure 10. Branching fraction upper limits at 90% CL for $B \to \pi\ell^+\ell^-$ and $B \to \eta\ell^+\ell^-$ modes from *BABAR* [57] and Belle [61] and the measurement of $B^+ \to \pi^+\ell^+\ell^-$ from LHCb [62].

We see no signals in any of these modes and set branching fraction upper limits at 90% CL. Figure 10 shows them in comparison to results from Belle [61] and a measurement of $\mathcal{B}(B^+ \to \pi^+\mu^+\mu^-)$ from LHCb [62]. For $B^0 \to \pi^0\ell^+\ell^-$, our branching fraction upper limit is the lowest and so far only *BABAR* has searched for $B^0 \to \eta\ell^+\ell^-$ modes. The present branching fraction upper limits lie within a factor of two to three of the SM predictions.

5. Search for Lepton Number Violation in $B^+ \to X^- \ell^+ \ell'^+$ Decays

In the SM, lepton number is conserved in low-energy collisions. However, in high-energy and high-density interactions, lepton number may be violated [63]. Many models beyond the SM predict lepton number violation (LNV) with rates [64] that may be accessible already in present data samples. These models also predict Majorana-type neutrinos [6] for which particles and antiparticles are identical. Via oscillation of a neutrino into an antineutrino, lepton-number violating decays become possible such as $B^+ \to K^- \ell^+ \ell'^-$ depicted in Fig. 2 (right). The observation of atmospheric neutrino oscillations confirms hat neutrinos carry mass [65] but we do not know if any Majorana-type neutrinos exist. However, lepton number violation is also a necessary condition to explain the observed baryon asymmetry in the universe [66].

Using the full *BABAR* data set of 471×10^6 $B\bar{B}$ events collected at the $\Upsilon(4S)$ peak, we have searched for lepton number violation in 11 B^+ decays[3] [67]. We select events with more than three charged tracks of which two are identified as like-sign leptons having a combined momentum less than 2.5 GeV/c in the laboratory frame. We remove e^+ and e^- from photon conversions. We define sufficiently wide mass regions around the ρ^- ($0.47 < m_{\pi\pi^0} < 1.07$ GeV/c^2), K^{*-} ($0.770 < m_{K\pi} < 1.01$ GeV/c^2) and D^- ($1.835 < m_{K\pi\pi} < 1.895$ GeV/c^2) mesons to allow reasonable modeling of backgrounds. We combine the π^-, K^-, $\pi^-\pi^0$, $K^-\pi^0$, $K^0_S\pi^-$ and $K^-\pi^+\pi^-$ candidates with the two leptons to form a B candidate. We remove $\ell^\pm h^\mp$ combinations with an invariant mass close to that of the J/ψ as the $h^\mp$ may be a misidentified $\mu^\mp$ from $J/\psi \to \mu^+\mu^-$. The misidentification rate is about 2%.

Table 3. Summary of signal yield after fit bias correction with statistical uncertainty, reconstruction efficiency ϵ, significance S including systematic error, measured branching fraction $\mathcal{B} \times 10^7$ and the 90% CL upper limit $\mathcal{B}_{UL} \times 10^7$. Yields and efficiencies for the $B^+ \to K^{*-} \ell^+ \ell'^+$ modes are for $K^-\pi^0$ and $K^0_S\pi^-$ final states, respectively.

Mode	Yield	ϵ	S	$\mathcal{B}$	$\mathcal{B}_{UL}$
	[events]	[%]	[σ]	[10^{-7}]	[10^{-7}]
$K^{*-}e^+e^+$	$3.8 \pm 3.3,\ 0.8 \pm 3.9$	$11.5 \pm 0.1,\ 12.1 \pm 0.1$	1.2	$1.7 \pm 1.4 \pm 0.1$	4.0
$K^{*-}e^+\mu^+$	$-1.9 \pm 4.7,\ -5.1 \pm 2.6$	$7.9 \pm 0.1,\ 8.5 \pm 0.1$	0.0	$-4.5 \pm 2.6 \pm 0.4$	3.0
$K^{*-}\mu^+\mu^+$	$2.3 \pm 1.8,\ 2.0 \pm 1.8$	$6.1 \pm 0.1,\ 5.8 \pm 0.1$	1.3	$2.4 \pm 1.8 \pm 0.4$	5.9
$\rho^-e^+e^+$	-2.1 ± 5.7	12.1 ± 0.1	0.0	$-0.4 \pm 1.0 \pm 0.1$	1.7
$\rho^-e^+\mu^+$	4.6 ± 11.4	10.3 ± 0.1	0.4	$1.0 \pm 2.4 \pm 0.2$	4.7
$\rho^-\mu^+\mu^+$	2.9 ± 6.8	7.3 ± 0.1	0.5	$0.9 \pm 2.0 \pm 0.3$	4.2
$D^-e^+e^+$	3.9 ± 4.8	10.2 ± 0.1	1.0	$8.8 \pm 8.6 \pm 1.5$	26
$D^-e^+\mu^+$	1.1 ± 3.2	7.7 ± 0.1	0.5	$3.4 \pm 9.4 \pm 1.1$	21
$D^-\mu^+\mu^+$	-1.7 ± 2.5	5.7 ± 0.1	0.0	$-6.5 \pm 9.9 \pm 0.9$	17
$K^-e^+\mu^+$	5.5 ± 3.5	15.2 ± 0.1	1.8	$0.6 \pm 0.5 \pm 0.1$	1.6
$\pi^-e^+\mu^+$	3.8 ± 3.5	16.4 ± 0.2	1.2	$0.5 \pm 0.5 \pm 0.1$	1.5

For each signal mode, we construct BDTs to discriminate signal from $B\bar{B}$ and $q\bar{q}$ backgrounds

[3] $B^+ \to \rho^-(\pi^-\pi^0)\ell^+\ell'^+$, $B^+ \to K^{*-}(K^0_S\pi^-$ and $K^-\pi^0),)\ell^+\ell'^+$, $B^+ \to D^-(K^-\pi^+\pi^-)\ell^+\ell'^+$, $B^+ \to \pi^-e^+\mu^+$ and $B^+ \to \pi^-e^+\mu^+$ where $\ell^+, \ell'^+ = e^+$ or μ^+.

using nine inputs consisting of event shape variables, kinematic observables, flavor tagging and the proper decay time. If more than one candidate is found, we choose the one with the smallest χ^2 in the fit to the B decay vertex. We perform a simultaneous unbinned ML fit to m_{ES}, ΔE and the BDT output distributions. For the $B^+ \to \rho^- \ell^+ \ell'^+$, $B^+ \to K^{*-} \ell^+ \ell'^+$ and $B^+ \to D^- \ell^+ \ell'^+$ final states, we include the $m_{\pi\pi}$, $m_{K\pi}$ and $m_{K\pi\pi}$ mass distributions, respectively. The background PDFs consist of an Argus function [68] for m_{ES}, first- or second-order polynomials (for K^-, π^-, ρ^-, K^{*-} modes) or a Cruijff[4] function (for D^- modes) for ΔE, a non-parametric kernel estimation KEYS algorithm [69] for the BDT output and a first-order polynomial plus a Gaussian function for the resonance masses. The corresponding signal PDFs consist of a Crystal Ball function [70] for m_{ES}, a Crystal Ball function plus a first-order polynomial (for modes with π^0s) for ΔE, the simulated distribution in form of a histogram for the BDT output and for the D, K^*, ρ masses two Gaussians, a relativistic Breit-Wigner function and a Gounaris-Sakurai function [71], respectively. We checked the fit procedure with a simulated background sample having the same size as the on-resonance data sample. We further performed a blinded fit to the on-resonance data sample confirming that the background distributions agreed with the background PDFs. Selection efficiencies vary between 6% and 16% depending on the final state.

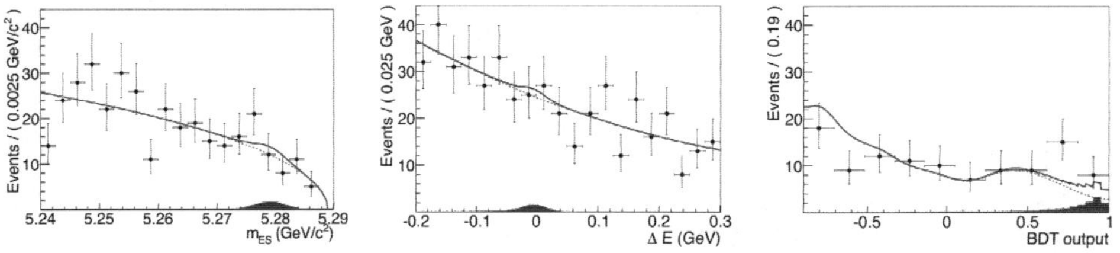

Figure 11. Projections of the multidimensional fit onto m_{ES} (left), ΔE (middle) and the BDT output (right) for $B^+ \to \pi^- e^+ \mu^+$ showing data (points with error bars), total fit (solid blue line), signal PDF (green histogram) and background (dashed magenta line).

Figures 11 and 12 show projections of the fit on the discriminating variables for $B^+ \to \pi^- e^+ \mu^+$ and $B^+ \to K^{*-}(K^0_S \pi^-) \mu^+ \mu^+$, respectively. Table 3 summarizes our results. In all 11 modes, the data are consistent with combinatorial background. We see the highest significance of 1.8σ in $B^+ \to K^- e^+ \mu^+$. We set Bayesian upper limits on the branching fraction at 90% CL using a flat prior (see Tab. 3). The additive systematic uncertainty that includes contributions from the PDF parameterization, fit biases, background yields and efficiencies is mode dependent between 0.2 and 0.7 events. The total multiplicative uncertainty on the branching fraction is 5% or less. The branching fraction upper limits at 90% CL lie in the range 1.5 – 26 $\times 10^{-7}$ where the lowest limit is set in the $B^+ \to \pi^- e^+ \mu^+$ mode. Figure 13 summarizes all results of lepton-number-violating B decays from *BABAR* [67], Belle [72], LHCb [73] and CLEO [74] including results for $B^0 \to \Lambda_c^+ \ell^-$, $B^- \to \Lambda \ell^-$, $B^- \to \bar{\Lambda} \ell^-$ [75] and $B^- \to \pi^-/K^- \ell^+ \ell^+$ [76]. All limits are set at 90% CL except for LHCb whose limits are set at 95% CL.

6. Study of $B^0 \to \omega\omega$ and $B^0 \to \omega\phi$ Decays

The longitudinal polarization fraction $\mathcal{F}_L$ in charmless vector vector decays poses a puzzle. In tree-dominated decays like $B^0 \to \rho^+ \rho^-$ and $B^+ \to \rho^+ \rho^0$, $\mathcal{F}_L$ is nearly 100% while in decays with dominant penguin contributions like $B \to K^* \rho$, $\mathcal{F}_L$ is around 50% [77]. Are there large

[4] The Cruijff function is a centered Gaussian with different left-right resolutions and non-Gaussian tails: $f(x) = \exp((x-m)^2/(2\sigma^2_{L,R} + \alpha_{L,R}(x-m)^2)$.

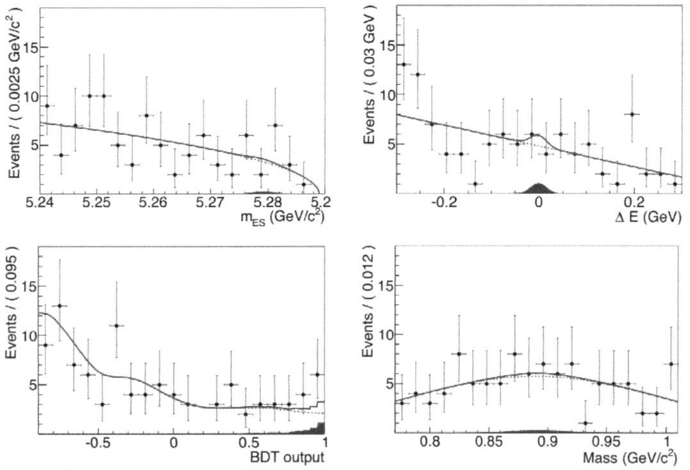

Figure 12. Projections of the multidimensional fit onto m_{ES} (top left), ΔE (top right), the BDT output (bottom left) and K^{*-} mass (bottom right) for $B^+ \to K^{*-}$ $(K_S^0\pi^-)e^+\mu^+$ showing data (points with error bars), total fit (solid blue line), signal PDF (green histogram) and background (dashed magenta line).

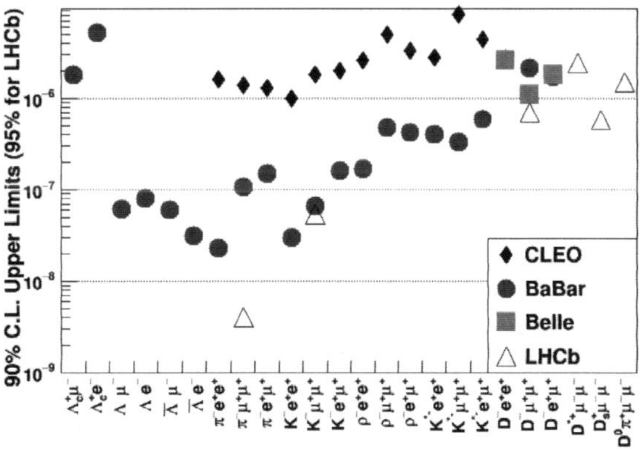

Figure 13. Branching fraction upper limits at 90% CL for LNV decays from *BABAR* [67] (solid blue points), Belle [72] (solid red squares) and CLEO [74] (black diamonds). In addition, LHCb upper limits at 95% CL [73] (open triangles) are shown.

transverse SM contributions that reduce $\mathcal{F}_L$ [78–83] or is this caused by new physics [82,84–94]? Thus, it is interesting to investigate other charmless vector vector decays such as the so far not-observed modes $B^0 \to \omega\omega$ and $B^0 \to \omega\phi$ [95]. In the SM, the branching fractions are expected to be of the order of $\mathcal{O}(10^{-6})$ for $B^0 \to \omega\omega$ and $\mathcal{O}(10^{-7})$ for $B^0 \to \omega\phi$. The SM predicts longitudinal polarization fractions of $\mathcal{F}_L > 0.8$ for both modes [93, 94]. Charmless vector vector modes are also well suited to measure the Unitarity Triangle angle α [96, 97]. The Scan Method group has determined $\alpha - \beta$ contours from a χ^2 fit to measured branching fractions, longitudinal polarizations and *CP* asymmetries using all observed charmless vector vector decays [98]. The decay amplitudes of each mode are expressed in terms of tree, color-suppressed tree, gluonic penguin, singlet penguin, electroweak penguin and W-annihilation/W- exchange amplitudes. For decays involving K^*s, SU(3) breaking is taken into account. All contributions up to order $\mathcal{O}(\lambda^5)$ are considered where $\lambda = \sin\theta_c$ (Cabibbo angle), since the leading amplitude is already at order $\mathcal{O}(\lambda^3)$. Figure 14 shows the 90% CL α–β contour determined from the fit.

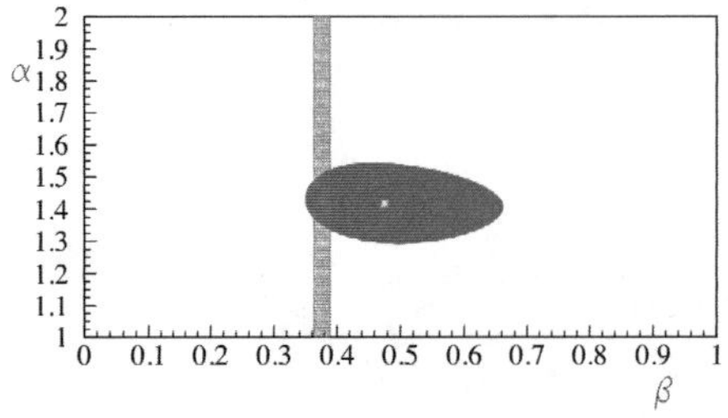

Figure 14. The α–β contour obtained from a ML fit to branching fractions, longitudinal polarizations and CP asymmetries measured in all observed $B \to VV$ modes. The magenta band shows the result of the $\sin 2\beta$ measurements in $B \to c\bar{c}K^0_{S,L}$ decays [77].

Using the full *BABAR* data sample of 471×10^6 $B\bar{B}$ events, we reconstruct the B-daughter candidates via their decays $\omega \to \pi^+\pi^-\pi^0$ with $\pi^0 \to \gamma\gamma$ and $\phi \to K^+K^-$ [99]. If multiple candidates exist, we select the one for which the χ^2 probability of a fit to the two vector meson masses is smallest. The combinatorial background from $e^+e^- \to q\bar{q}$ collisions dominates. We use a tight selection on the angle θ_T between the thrust of the B signal candidate in the $\Upsilon(4S)$ rest frame and that of the rest of the event, requiring $|\cos\theta_T| < 0.8$ (0.9) for $B^0 \to \omega\omega$ and $B^0 \to \omega\phi$ decay modes. Furthermore, we define a Fisher discriminant $\mathcal{F}$ based on four shape and kinetic variables as inputs. We perform an extended unbinned ML fit to extract the signal and background yields from the data. We define the PDF as a product of six individual PDFs including m_{ES}, ΔE, $\mathcal{F}$, masses and helicity angles of the two vector mesons and the decay angle ψ between the π^0 in the dipion rest frame and the ω flight direction:

$$\mathcal{P}^i_j = \mathcal{P}_j(m^i_{ES}) \times \mathcal{P}_j(\Delta E^i) \times \mathcal{P}_j(\mathcal{F}^i) \times \mathcal{P}_j(m^i_{V1}, m^i_{V2}, \theta^i_1, \theta^i_2) \times \mathcal{P}_j(\psi^i_{\omega_1}) \times \mathcal{P}_j(\psi^i_{\omega_2}) \quad (6)$$

where the last term is not present in $B^0 \to \omega\phi$. For signal, we use a sum of two Gaussians for m_{ES} and ΔE, a two-piece normal distribution for $\mathcal{F}$, relativistic Breit-Wigner functions for the m_V distributions, each convolved with two Gaussians to account for detector resolution. We parameterize the helicity angles by the angular distribution:

$$W(\theta_{V_1}, \theta_{V_2}) \propto (1/4)(1 - \mathcal{F}_L)\sin^2\theta_{V_1}\sin^2\theta_{V_2} + \mathcal{F}_L\cos^2\theta_{V_1}\cos^2\theta_{V_2} \quad (7)$$

convolved with a resolution function for each angle. Thus for signal, the PDF is factorized as $\mathcal{P}_j(m^i_{V1}, m^i_{V2}, \theta^i_1, \theta^i_2) = \mathcal{P}_j(m^i_{V_1}) \times \mathcal{P}_j(m^i_{V_2}) \times W(\theta_{V_1}, \theta_{V_2})$. For the angles $\psi_{1,2}$, the PDFs are $\sin^2\psi_{1,2}$ distributions. For combinatorial background, we use an Argus function [68] for m_{ES}, a second-order polynomial for ΔE and a two-piece normal distribution for $\mathcal{F}$. The masses and helicity angles of the two vector mesons are considered to be independent and thus are factorized. We use a third-order polynomial for the m_V distributions. The PDFs of the helicity angles are third-order polynomials for combinatorial background where the parameterization is obtained from on-peak sideband data ($m_{ES} < 5.27$ GeV/c^2). Similarly, we use a third-order polynomial in $\cos\psi$ for the ω decay angle. We also consider $B\bar{B}$ peaking background determined from simulation. We parameterize m_{ES}, ΔE and $\mathcal{F}$ with similar functions as those for the signal. The PDFs for m_V and ψ angle are similar to those of the combinatorial background. For the helicity angles, we use a fourth-order polynomial. For the signal, $B\bar{B}$ and $q\bar{q}$ background components, we determine the PDF parameters from simulation. We study large control samples of $B \to D\pi$ decays of similar topology to verify the simulated resolutions in ΔE and m_{ES}, adjusting the PDFs to account for any differences found.

Figures 15 (left, middle) show the m_{ES} and ΔE projections of the multidimensional ML fit for $B^0 \to \omega\omega$ (top) and $B^0 \to \omega\phi$ (bottom. Figure 15 (right) shows the corresponding distributions of $-2\ln(\mathcal{L}(B)/\mathcal{L}_0)$ for zero signal normalized to the likelihood at the minimum value for the two modes. We fix $\mathcal{F}$ to 0.88 but vary it from 0.58–1.0 in the systematic error determination. Signal and background yields are extracted from the fit. For $B^0 \to \omega\omega$, we observe $N_{\omega\omega} = 69^{+16.4}_{-15.2}$ signal events. Including systematic uncertainties, this yields a 4.4σ effect. For $B^0 \to \omega\phi$, we see no signal. The fit yields $N_{\omega\phi} = -2.8^{+5.7}_{-4.0}$ signal events. Besides the uncertainty from the $\mathcal{F}_L$ variation, systematic uncertainties include contributions from the PDF parameterization, selection efficiency, use of control samples, number of B mesons and B-daughter branching fractions. The largest systematic error comes from yield bias estimation in the fit yielding order $\mathcal{O}$(5 events). We measure a branching fraction of $\mathcal{B}(B \to \omega\omega) = (1.2 \pm 0.3^{+0.3}_{-0.2}) \times 10^{-6}$, which agrees well with the SM prediction. We do not have enough data to measure $\mathcal{F}_L$ in this mode. For the $B^0 \to \omega\phi$ mode, we set a Bayesian upper limit of $\mathcal{B}(B \to \omega\phi) = 0.7 \times 10^{-6}$ at 90% CL assuming a flat prior.

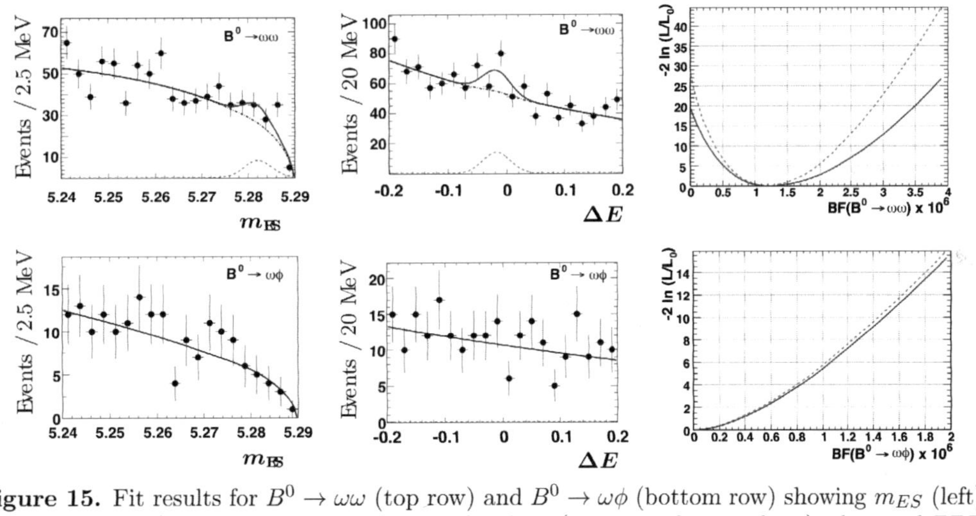

Figure 15. Fit results for $B^0 \to \omega\omega$ (top row) and $B^0 \to \omega\phi$ (bottom row) showing m_{ES} (left) and ΔE (middle) projections of the ML fit for data (points with error bars), the total PDF (solid blue line), background PDF (black dashed line) and signal contribution (red dashed line). The distributions of $-2\ln(\mathcal{L}(B)/\mathcal{L}_0)$ versus branching fraction (right) for statistical errors only (dashed red curve) and statistical and systematic errors combined (solid blue curve).

7. Conclusions

We have performed the most precise direct *CP* asymmetry measurement in the semi-inclusive $B \to X_s\gamma$ decay. The measurement error is still suffiiciently large allowing to accommodate new physics contributions in the Wilson coefficient $C_7^{\rm eff}$. We performed the first determination of $\mathcal{I}m(C_8^{\rm eff}/C_7^{\rm eff})$ by measuring the difference in *CP* asymmetries between charged and neutral B decays. We measured partial branching fractions and *CP* asymmetries in $B \to X_s\ell^+\ell^-$ that agree well with the SM predictions. We find no evidence for $B \to \pi/\eta\ell^+\ell^-$ decays and set branching fraction upper limits at 90% CL that are a factor of two to three above the SM predictions. We also find no evidence for lepton number violation in $B^+ \to X^- \ell^+ \ell'^+$ decays and set stringent branching fraction upper limits at 90% CL. We find first evidence for $B^0 \to \omega\omega$ decays and measure a branching fraction that is consistent with the SM prediction. However, the data sample is too small to extract $\mathcal{F}_L$ from an angular analysis. We set an improved branching fraction upper limit at 90% CL for $B^0 \to \omega\phi$. Significant improvement on these measurements are expected to come from Belle II and for some decays also from LHCb.

Acknowledgments
This work has been supported by the Norwegian Research Council. I would like to thank the *BABAR* collaboration for the opportunity to give this talk. In particular, I would like to thank Justin Albert, Frank Porter, Fergus Wilson for useful comments.

References

[1] K. Wilson, Phys. Rev. **179**, 1499 (1969); K. Wilson and J. Kogut, Phys. Rep. 12, 75 (1974).
[2] N. Isgur et al. Phys.Rev. **D 39**, 799 (1989); N. Isgur and M. Wise, Phys. Lett. B 232, 113 (1989).
[3] H. Georgi, Phys. Lett. B **240**, 447 (1990).
[4] B. Grinstein and D. Pirjol, Phys. Rev. **D 70**, 114005 (2004).
[5] G. Hiller and F. Krüger, Phys.Rev. **D 69**, 074020 (2004).
[6] E. Majorana, Nuo. Cim. 14, **171** (1937).
[7] M. Misiak et al., Phys. Rev. Lett. **98**, 022002 (2007).
[8] M. Misiak and M. Steinhauser, Nucl. Phys. **B 764**, 62 (2007).
[9] M. Neubert, Phys. Rev. **D 72**, 074025 (2005).
[10] B. O. Lange, M. Neubert, and G. Paz, Phys. Rev. **D 72**, 073006 (2005).
[11] C. W. Bauer, et al., Phys. Rev. **D 67**, 054012 (2003).
[12] P. Gambino et al., J. High Energy Phys. **10**, 058 (2007).
[13] K.A. Olive et al., (Particle Data Group), Chin.Phys. **C 38**, 090001 (2014).
[14] G. Eigen, arXiv:0907.4330 [hep-ex] (2009).
[15] B. Aubert et al. (BABAR Collaboration), Nucl. Instrum. Meth. **A 479**, 1 (2002).
[16] B. Aubert et al. (BABAR Collaboration), Nucl. Instrum. Meth. **A 729**, 615 (2013).
[17] J.P Lees et al. (BABAR Collaboration), Phys.Rev. **D 86**, 052012 (2012).
[18] J.P Lees et al. (BABAR Collaboration), Phys.Rev. **D 90**, 092001 (2014).
[19] M. Benzke et al., Phys. Rev. Lett **106**, 141801 (2011).
[20] B. Aubert et al. (BABAR Collaboration), Phys. Rev. Lett. **101**, 171804 (2008).
[21] S. Nishida et al. (Belle Collaboration), Phys. Rev. Lett. **93**, 031803 (2004).
[22] A. L. Kagan and M. Neubert, Phys. Rev. **D 58**, 094012 (1998).
[23] M. Jung, X. -Q. Li and A. Pich, JHEP **1210**, 063 (2012).
[24] A. Hayakawa, et al., Phys. Lett. **B 710**, 446 (2012).
[25] J. P. Lees et al. (BABAR Collaboration), Phys. Rev. Lett. **109**, 191801 (2012).
[26] J. P. Lees et al. (BABAR Collaboration), Phys. Rev. **D 86**, 112008 (2012).
[27] B. Aubert et al. (BABAR Collaboration), Phys. Rev. **D 77**, 051103 (2008).
[28] L. Pesantez et al. (Belle Collaboration), arXiv:1501.01702 [hep-ex] (2014).
[29] T. E. Coan et al. (CLEO Collaboration), Phys. Rev. Lett. **86**, 5661 (2001).
[30] T. Hurth, E. Lunghi and W. Porod, Nucl. Phys. **B 704**, 56 (2005).
[31] J. P. Lees et al. (BABAR Collaboration), Phys. Rev. Lett. **112**, 211802 (2014).
[32] T. Sjostrand, Comput. Phys. Commun. **82**, 74 (1994).
[33] A. Ali, et al., Phys. Rev. **D 66**, 034002 (2002).
[34] T. Huber, et al., Nucl. Phys. **B 740**, 105 (2006).
[35] F. Krüger and L. M. Sehgal, Phys. Lett. **B 380**, 199 (1996).
[36] A. Ali, et al., Phys. Rev. **D 55**, 4105 (1997).
[37] C. Bobeth, M. Misiak, and J. Urban, Nucl. Phys. **B 574**, 291 (2000).
[38] H. H. Asatryan et al., Phys. Rev. **D 65**, 074004 (2002).
[39] H. H. Asatryan et al., Phys. Rev. **D 66**, 034009 (2002).
[40] A. Ghinculov et al., Nucl. Phys. **B 648**, 254 (2003).
[41] P. Gambino, M. Gorbahn and U. Haisch, Nucl. Phys. **B 673**, 238 (2003).
[42] A. Ghinculov et al., Eur. Phys. J. **C 33**, S288 (2004).

[43] C. Bobeth et al., JHEP **0404**, 071 (2004).
[44] A. Ghinculov et al., Nucl. Phys. **B 685**, 351 (2004).
[45] C. Greub, V. Pilipp and C. Schupbach, JHEP **0812**, 040 (2008).
[46] T. Huber, T. Hurth and E. Lunghi, Nucl. Phys. **B 802**, 40 (2008).
[47] M. Beneke, et al., Eur. Phys. J. **C 61**, 439 (2009).
[48] B. Aubert et al. (BABAR Collaboration), Phys. Rev. Lett. **93**, 081802 (2004).
[49] M. Iwasaki et al. (Belle Collaboration), Phys. Rev. **D 72**, 092005 (2005).
[50] D. S. Du and M. Z. Yang, Phys. Rev. **D 54**, 882 (1996).
[51] A. Ali and G. Hiller, Eur. Phys. J. **C 8**, 619 (1999).
[52] C. Bobeth, G. Hiller and G. Piranishvili, JHEP **0807**, 106 (2008).
[53] W. Altmannshofer, et al., JHEP **0901**, 019 (2009).
[54] A. Soni et al., Phys. Rev. **D 82**, 033009 (2010).
[55] A. K. Alok, A. Dighe and S. Ray, Phys. Rev. **D 79**, 034017 (2009).
[56] T. M. Aliev and M. Savci, Phys. Rev. **D 60**, 014005 (1999).
[57] J.P Lees et al. (BABAR collaboration), Phys.Rev. **D 88**, 032012 (2013).
[58] J. J. Wang, et al., Phys. Rev. **D 77**, 014017 (2008).
[59] H. Z. Song, L. X. Lu, and G. R. Lu, Commun. Theor. Phys. **50**, 696 (2008).
[60] G. Erkol and G. Turan, Eur. Phys. Jour. **C 28**, 243 (2003).
[61] J.T. Wei et al., Phys. Rev. **D 78**, 011101 (2008).
[62] R. Aaij et al. (LHCb collaboration), JHEP **1212**, 125 (1212).
[63] F. R. Klinkhamer and N.S. Manton, Phys. Rev. **D 30**, 2212 (1984).
[64] A. Atre, T. Han, S. Pascali, and B. Zhang, JHEP **0905**, 030 (2009).
[65] Y. Fukuda et al. (Super-Kamiokande Collaboration), Phys. Rev. Lett. **81**, 1562 (1998).
[66] S. Davidson, E. Nardi and Y. Nir, Phys. Rep. **466**, 105 (2008).
[67] J.P Lees et al. (BABAR collaboration), Phys.Rev. **D 89**, 011102 (2014).
[68] H. Albrecht et al. (ARGUS Collaboration), Z. Phys. **C 48**, 543 (1990).
[69] K. S. Cranmer, Comput. Phys. Commun. **136**, 198 (2001).
[70] T. Skwarnicki,DESY F31-86-02 (thesis, unpublished) (1986).
[71] G. J. Gounaris and J.J. Sakurai, Phys. Rev. Lett. **21**, 244 (1968).
[72] O. Seon et al. (Belle Collaboration), Phys. Rev. **D 84**, 071106 (2011).
[73] R. Aaij et al. (LHCb Collaboration), Phys. Rev. Lett. **112**, 131802 (2014).
[74] K. W. Edwards et al. (CLEO Collaboration), Phys. Rev. **D 65**, 111102 (2002).
[75] P. del Amo Sanchez et al. (BABAR Collaboration), Phys. Rev. **D 83**, 091101 (2011).
[76] J.P. Lees et al. (BABAR Collaboration), Phys. Rev. **D 85**, 071103 (2012).
[77] D. Asner et al. (Heavy Flavor Averaging Group Collaboration), arXiv:1010.1589 [hep-ex] (2010).
[78] C.W. Bauer et al., Phys. Rev. **D 70**, 054015 (2004).
[79] P. Colangelo, F. De Fazio, and T.N. Pham, Phys. Lett. **B 597**, 291 (2004).
[80] A.L. Kagan, Phys. Lett. **B 601**, 151 (2004).
[81] M. Ladisa et al., Phys. Rev. **D 70**, 114025 (2004).
[82] H. Y. Cheng, C. K. Chua, and A. Soni, Phys. Rev. **D 71**, 014030 (2005).
[83] H.-N. Li and S. Mishima, Phys. Rev. **D 71**, 054025 (2005); Phys. Lett. B 622, 63 (2005).
[84] W. Bensalem and D. London, Phys. Rev. **D 64**, 116003 (2001).
[85] A. K. Giri and R. Mohanta, Phys. Rev. **D 69**, 014008 (2004).
[86] E. Alvarez et al., Phys. Rev. **D 70**, 115014 (2004).
[87] C.-H. Chen and C.-Q. Geng, Phys. Rev. **D 71**, 115004 (2005).
[88] Y.-D. Yang, R. M. Wang and G. R. Lu, Phys. Rev. **D 72**, 094002 (2005).
[89] P. K. Das and K. C. Yang, Phys. Rev. **D 71**, 015009 (2005).
[90] A. K. Giri and R. Mohanta, Eur. Phys. Jour. **C 44**, 249 (2005).
[91] S. Baek et al., Phys. Rev. **D 72**, 094008 (2005).
[92] S.-S. Bou et al., Phys. Rev. **D 77**, 095004 (2008).
[93] Y. Li and C.-D. Lu, Phys. Rev. **D 73**, 014024 (2006).
[94] H.-Y. Cheng and C.-K. Chua, Phys. Rev. **D 80**, 114008 (2009).
[95] S. Oh, Phys. Rev. **D 60**, 034006 (1999).
[96] D. Atwood and A. Soni, Phys. Rev. **D 59**, 013007 (1999); Phys. Rev. D 65, 073018 (2002).
[97] H.-W. Huang et al., Phys. Rev. **D 73**, 014011 (2006).
[98] G. Eigen et al., Phys.Rev. **D 89**, 033004 (2014).
[99] J.P Lees et al. (BABAR collaboration), Phys.Rev. **D 89**, 051101 (2014).

Journal of Physics: Conference Series **631** (2015) 012034 doi:10.1088/1742-6596/631/1/012034

Direct Searches for New Physics Particles at *BABAR*

Gerald Eigen on behalf of the *BABAR* Collaboration
Dept. of Physics, University of Bergen, Allegation 55, Bergen, Norway
E-mail: gerald.eigen@ift.uib.no

Abstract. We present recent *BABAR* results on searches for dark photons, long-lived scalar particles and new π^0-like particles.

1. Introduction

In 2010, the Pamela experiment reported a positron excess for energies greater 10 GeV that is increasing with energy [1]. The excess was confirmed by the Fermi Large Area Telescope in 2012 [2]. Two years later, the AMS experiment published higher-precision data up to 300 GeV confirming an excess of the positron fraction in the 10 – 250 GeV energy range [3]. Figure 1 shows the measurements from Pamela, Fermi and AMS. In the same energy range, Pamela and AMS did not observe any antiproton excess. Though an astrophysical explanation is very likely, theorists have come up with new dark-matter scenarios favoring light particles. In the past, dark-matter particles have been associated with new heavy particles predicted in extensions of the Standard Model (SM), such as neutralinos in supersymmetric models. Physicists have been searching for weakly-interacting massive particles (WIMPs) in dedicated experiments for many years [4–20]. The observation of the positron excess has triggered searches for light dark matter in different accelerator experiments.

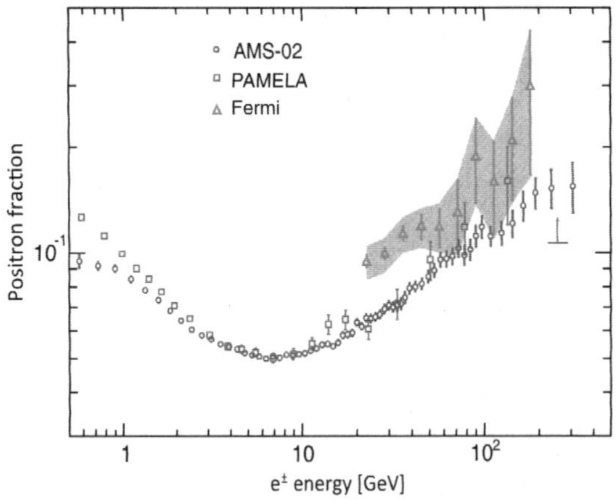

Figure 1. Positron excess observed by Pamela, Fermi and AMS.

The SM may be connected to the dark sector through so-called portals [21]. These are lowest-dimensional operators that may provide coupling of the dark sector to the SM[1]. Table 1 lists these operators for couplings to new vector particles, pseudoscalars, scalars, and neutrinos. At low-energy e^+e^- collisions, the light vector portal is the most accessible portal, but the scalar portal can be probed as well [25]. In the simplest realizations of the dark sector, a new $U(1)'$ symmetry is introduced [26]. The associated gauge boson, called dark photon (A'), couples to the weak hyper charge with a mixing strength ϵ. An effective interaction between the dark photon and the electromagnetic current arises after electroweak symmetry breaking. For an Abelian interaction, this mechanism would explain WIMP annihilation into SM fermions [25, 27, 28]. To accommodate the recent anomalies in cosmic rays, the dark photon mass is constrained to lie in the MeV/c^2 to GeV/c^2 range.

In recent years, we have searched for new physics in e^+e^- collisions collected with the *BABAR* experiment at center-of-mass (CM) energies around 10 GeV [29, 30]. We searched for a dark Higgs h' plus a dark photon A' in the process $e^+e^- \to h'A', h' \to A'A'$. The cross section is proportional to the product of fine structure constant α_D and ϵ^2. Assuming a flat prior, we set 90% confidence level (CL) Bayesian upper limits on $\alpha_D \cdot \epsilon^2$ of 10^{-10} to 10^{-8} for h' and A' masses in the 0.8 – 10 GeV/c^2 and 0.25 – 3 GeV/c^2 regions, respectively [31]. We also searched for dark non-Abelian gauge bosons in $e^+e^- \to W'W' \to e^+e^-e^+e^-, e^+e^-\mu^+\mu^-, \mu^+\mu^-\mu^+\mu^-$ [32].

Furthermore, we searched for a new pseudoscalar A^0 in $\Upsilon(2S, 3S)$ radiative decays: $A^0 \to \mu^+\mu^-$ [33, 34], $\tau^+\tau^-$ [35], hadrons [36], and invisible particles [37]. For the $\mu^+\mu^-$ decay mode, we set 90% CL Bayesian upper limits on the branching fraction of 0.26×10^{-6} to 8.3×10^{-6} for A^0 masses between 0.21 GeV/c^2 and 10.1 GeV/c^2 [33]. For the other modes, we obtained similar results. We also studied the decay $\Upsilon(1S) \to \gamma A^0$, $A^0 \to gg$, $s\bar{s}$ setting 90% CL upper limits on the branching fraction at a level of 10^{-6} to 10^{-2} for A^0 masses in the 1.5 – 9.0 GeV/c^2 range [38].

Table 1. Portals for connecting the dark sector to the Standard Model where $F^Y_{\mu\nu}$ is the strength of the Yang-Mills field, $F'^{\mu\nu}$ is the strength of the dark photon field, ϵ is the mixing parameter of a dark photon to a real photon, a is the axion field, f_a is the scale at which the Peccei-Quinn global $U(1)$ symmetry is spontaneously broken [22–24], $G^{\mu\nu}$ is the gluon field strength, S is a dark scalar field with coupling strengths (λ and μ) to the Higgs field H, and N is the sterile neutrino field with coupling y_N to the Higgs field.

Portal	Particle	Operator
"Vector"	Dark photons	$\epsilon F^Y_{\mu\nu} F'^{\mu\nu}$
"Axion"	Pseudoscalars	$\frac{a}{f_a} F_{\mu\nu}\tilde{F}^{\mu\nu}, \frac{a}{f_a} G_{i\mu\nu}\tilde{G}^{\mu\nu}_i, \frac{\partial_\mu a}{f_a}\bar{\psi}\gamma^\mu\gamma^5\psi$
"Higgs"	Dark scalars	$(\mu S + \lambda S^2)H^\dagger H$
"Neutrino"	Sterile neutrino	$y_N LHN$

2. Search for Dark Photons

In the mass region below 10 GeV/c^2, e^+e^- collisions provide an excellent laboratory for producing dark photons A' [39, 40]. The processes $e^+e^- \to \gamma A' \to \gamma\ell^+\ell^-$ ($\ell^\pm = e^\pm$ or $\mu^\pm$) depicted in Fig. 2 are well suited since they have high sensitivity for A' production because of

[1] higher-dimensional operators are mass suppressed.

high selection efficiencies and large A' decay rates as Fig. 3 shows [39]. The cross section with respect to $e^+e^- \to \gamma\gamma$ is reduced by ϵ^2, where the mixing parameter ϵ is expected to lie in the range of 10^{-5} to 10^{-2}.

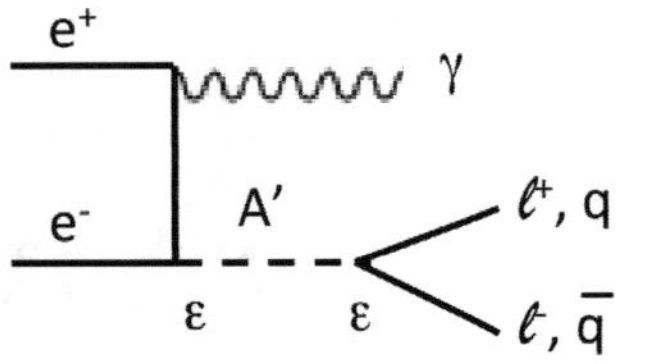

Figure 2. Lowest order diagram for $e^+e^- \to \gamma A' \to \gamma\ell^+\ell^-, \gamma q\bar{q}$.

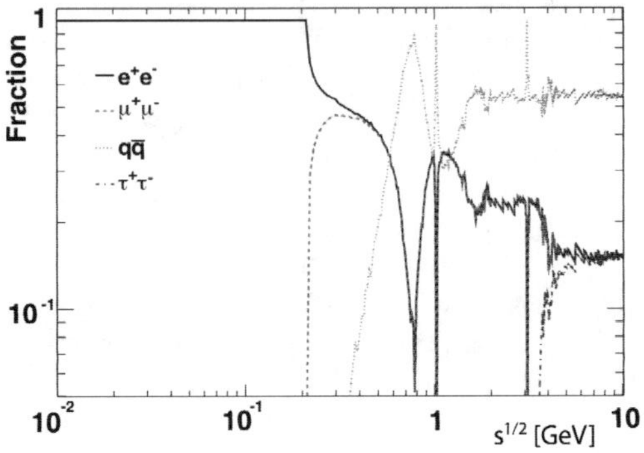

Figure 3. Branching fraction predictions for $A' \to \ell^+\ell^-$ and $A' \to q\bar{q}$ as a function of $m_{A'}$.

We have studied $e^+e^-\gamma$ and $\mu^+\mu^-\gamma$ final states at CM energies larger than 200 MeV using an integrated luminosity of 514 $\mathrm{fb}^{-1}$of e^+e^- collisions [41]. The data were recorded at the $\Upsilon(4S), \Upsilon(3S), \Upsilon(2S)$ peaks and 40 MeV/c^2 below the $\Upsilon(4S)$ peak with the *BABAR* detector [29, 30] at the PEP-II asymmetric storage ring at the SLAC National Laboratory. We select two oppositely-charged leptons plus a photon and apply a constrained fit to the beam energy and interaction point (IP). We use additional kinematic constraints to improve purity, require good quality on the photon, electron and muon and remove electron conversions. We simulate the background processes $e^+e^- \to e^+e^-(\gamma)$ and $e^+e^- \to \gamma\gamma(\gamma)$ with the generator BHWIDE [42], $e^+e^- \to \mu^+\mu^-(\gamma)$ with the generator KK2F [43] and resonance production in initial state radiation $e^+e^- \to \gamma X$ ($X = J/\psi$, $\psi(2S)$, $\Upsilon(1S)$, $\Upsilon(2S)$) using structure function techniques [44, 45]. For the simulation of the detector response, we use GEANT4 [46].

Figure 4 (left) shows the dielectron invariant-mass spectrum in data and in simulation with the generator BHWIDE. Generally, the agreement between data and Monte Carlo (MC) simulation is good except for the low-mass region where the BHWIDE generator fails to reproduce the m_{ee} invariant-mass spectrum. On the other hand, the MADGRAPH generator [47] reproduces the low m_{ee} region in a consistent way but the limited sample size introduces too large uncertainties. Therefore, our signal extraction does not rely on the MC simulation.

For the dimuon sample, we define the reduced mass $m_R = \sqrt{(m^2_{\mu\mu} - 4m^2_\mu)}$ since the turn-on near threshold is smoother. Figure 4 (right) shows the m_R distribution in data and simulations. The KK2F generator [43] reproduces m_R spectrum reasonably well. For both modes, we determine the selection efficiency from simulation yielding $\epsilon_{ee\gamma} = 15\%$ and $\epsilon_{\mu\mu\gamma} = 35\%$. The efficiency reduction in the $ee\gamma$ mode is due to mainly pre-scaling of radiative Bhabha events in the trigger. However, radiative Bhabhas are still the dominant background. Thus, the sensitivity for the dark photon search is dominated by the $\mu^+\mu^-$ mode. We extract the signal yield as a function of $m_{A'}$ by performing independent fits to the m_{ee} and m_R mass distributions for each beam energy covering the range $0.02 < m_{A'} < 10.2$ GeV/c^2 and $0.212 < m_{A'} < 10.2$ GeV/c^2 for the $ee\gamma$ and $\mu\mu\gamma$ mode, respectively. In these fits, we vary the step size to about half the value of the expected mass resolution $\sigma_{m_{A'}}$ and cover a mass range that is at least $20\sigma_{m_{A'}}$. The signal resolution is estimated by fits to a Gaussian function for various A' mass values and is

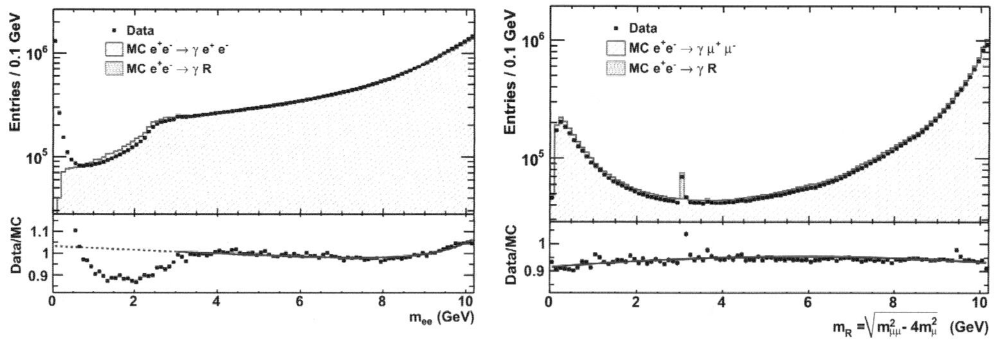

Figure 4. Top: measured m_{ee} (left) and m_R (right) invariant-mass spectra in comparison to simulations; bottom left and right: the data/MC ratio.

interpolated to all other masses. The typical steps sizes vary between 1.5 MeV/c^2 and 8 MeV/c^2 yielding 5704 (5370) mass hypotheses for the $ee\gamma$ ($\mu\mu\gamma$) channel.

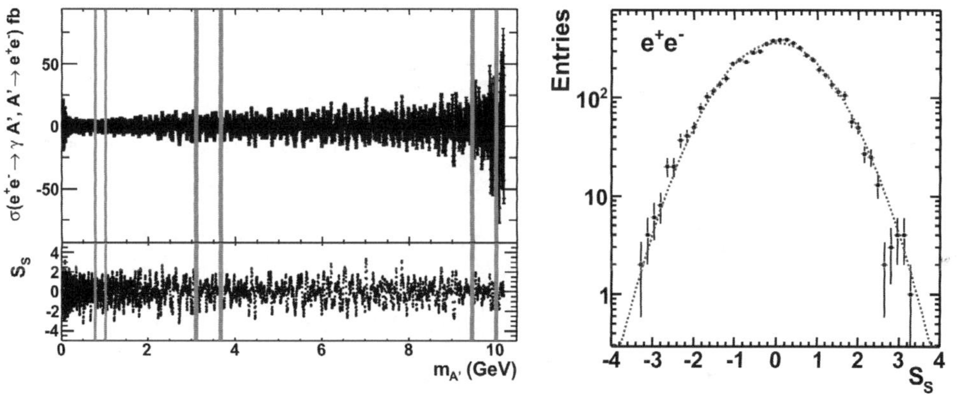

Figure 5. Measurement of $\sigma(e^+e^- \to \gamma A' \to \gamma e^+e^-)$ (left) and significance distribution (right). Grey vertical lines show the excluded regions around the vector mesons.

The likelihood function used to fit the observed spectra contains a signal component, radiative dilepton background and peaking background from vector mesons (ω, ϕ, J/ψ, $\psi(2S)$, $\Upsilon(1S)$, $\Upsilon(2S)$). The signal probability density function (PDF) is modeled directly from the simulated signal mass distribution by a non-parametric kernel pdf and is interpolated between the known simulated masses using an algorithm based on the cumulative mass distributions [48]. We estimate a systematic uncertainty of 5% – 10% for this procedure. The PDFs for the radiative dilepton backgrounds consist of higher-order polynomials. Though we parameterize the vector mesons $\omega, \phi, J/\psi, \psi(2S), \Upsilon(1S)$ and $\Upsilon(2S)$ with Gaussian and Crystal Ball functions, we exclude the mass regions around them in the dark photon search. We determine the $e^+e^- \to \gamma A' \to \gamma e^+e^-$ and $e^+e^- \to \gamma A' \to \gamma\mu^+\mu^-$ cross sections by dividing the extracted signal yield by efficiency, luminosity and dark photon decay branching fractions [39]. Figures 5 (left) and 6 (left) show the measured cross sections as a function of $m_{A'}$ for the $e^+e^-\gamma$ and $\mu^+\mu^-\gamma$ modes, respectively. The results include all data recorded at the $\Upsilon(2S)$, $\Upsilon(3S)$ and $\Upsilon(4S)$ except for 5% data used to optimize the selection criteria. We estimate the significance

Journal of Physics: Conference Series **631** (2015) 012034 doi:10.1088/1742-6596/631/1/012034

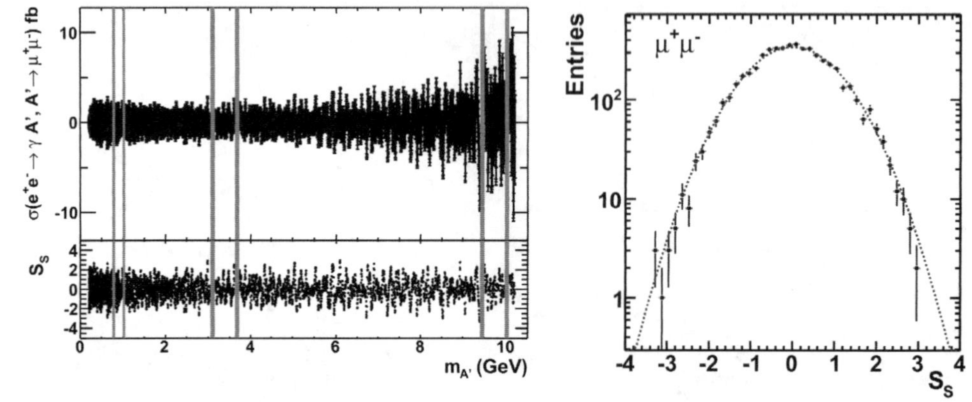

Figure 6. Measurement of $\sigma(e^+e^- \to \gamma A' \to \gamma\mu^+\mu^-)$ (left) and significance distribution (right). Grey vertical lines show the excluded regions around the vector mesons.

for each fit by $\mathcal{S}_S = \sqrt{2\mathcal{L}/\mathcal{L}_0}$ where $\mathcal{L}$ is the likelihood for signal plus background while $\mathcal{L}_0$ is that for pure background. Figures 5 (right) and 6 (right) show the resulting distributions for both modes based on 5704 and 5370 mass hypotheses, respectively. The largest deviation in the e^+e^- invariant-mass spectrum occurs at $m_{ee} = 7.2$ GeV/c^2 with a significance of $\mathcal{S}_S = 3.4$. Including a trial factor determined from the MC, the significance is reduced to 0.6σ. The largest deviation in the $\mu^+\mu^-$ invariant-mass spectrum occurs at $m_{\mu\mu} = 6.09$ GeV/c^2 with a significance of $\mathcal{S}_S = 2.9$. The application of a trial factor reduces this to 0.1σ.

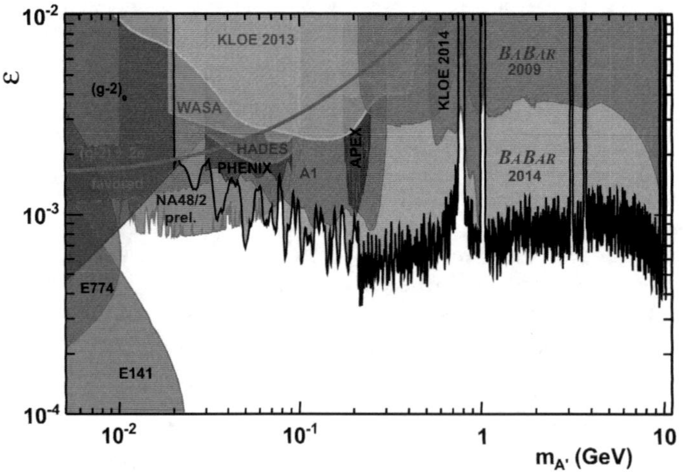

Figure 7. Exclusion region of the mixing parameter ϵ as a function of the dark photon mass for present measurements.

We combine the cross section measurements into a single result. Since we observe no signal, we determine Bayesian upper limits on the cross section at 90% CL assuming a flat prior. We include systematic uncertainties from the uncertainties of the dark photon branching fractions, luminosity and limited MC statistics. The resulting cross section upper limits are typically of

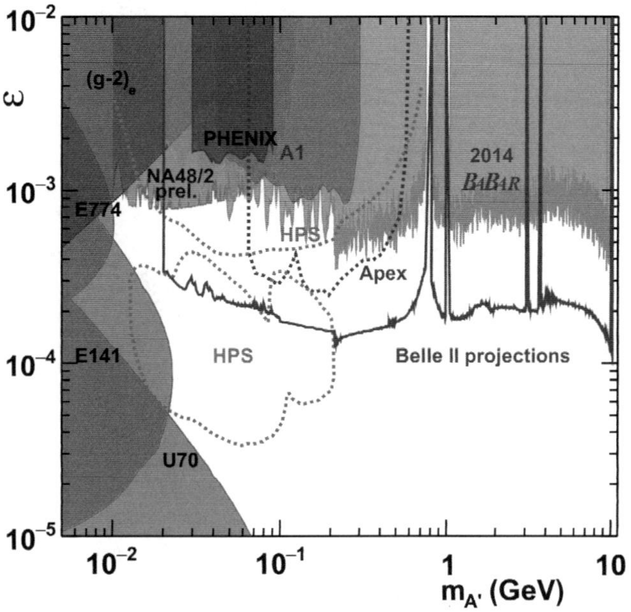

Figure 8. Expected exclusion region on ϵ from results of new dedicated experiments APEX, DarkLight and HPS, MESA, VEPP3 and Belle II.

the order of $\mathcal{O}(1-10$ fb). We then translate these results into 90% CL Bayesian upper limits on the mixing parameter as a function of the dark photon mass. Figure 7 shows our results in comparison to dark photon searches of other experiments: KLOE [49, 50] in $e^+e^- \to \mu^+\mu^-\gamma$ and $\phi(1020) \to \eta e^+e^-$, electron-nucleon fixed target experiments APEX [51] and A1 [52] in $A' \to e^+e^-$, the proton fixed-target experiment HADES [53] in $A' \to e^+e^-$, WASA [54] and NA48 [55] in $\pi^0 \to e^+e^-\gamma$ and electron beam dump experiments [56–58]. In addition, we show the constraints from the measurement of the anomalous electron magnetic moment [59]. The band of favored ϵ versus $m_{A'}$ is obtained if the discrepancy between the measurement and the SM calculation of the muon magnetic moment [60] is attributed to dark photon production. In the $0.03-10$ GeV/c^2 mass region, our results push the exclusion region substantially lower and supersede the results of our previous analysis [61]. Together with the results from NA48, we exclude the entire region favored by the dark-photon scenario for the g-2 measurements[2]. By searching for $A' \to \pi^+\pi^-$, we can further probe the region near the ρ mass. Figure 8 shows the exclusion region of ϵ as a function of $m_{A'}$ that is expected from future measurements such as Belle II [62] for 50 ab^{-1} and several new dedicated experiments including APEX, DarkLight and HPS at Jefferson Laboratory [63, 64], MESA at MAMI in Mainz [65] and VEPP3 at Novosibirsk [66]. Above 200 MeV/c^2, the Belle II data will reduce the exclusion limit on ϵ to $\sim 2\times10^{-4}$ while at lower masses HPS will push the limit below 10^{-4}.

3. Search for Long-Lived Particles

The recent anomalous astrophysical observations have generated also interest in low-mass, long-lived hidden-sector states. Various models discuss this scenario, including those with dark photons [39, 40, 67, 68], an inflaton [69], supersymmetry [70, 71], a dark Higgs boson [72] and

[2] This assumes that decays to invisible particles are small, otherwise the bounds are weakened.

with other dark-sector states [73]. Several multipurpose experiments have conducted searches for long-lived particles in the sub GeV/c^2 [56,74,75] and the multi-GeV/c^2 [76–81] mass regions. In addition, dedicated experiments have been proposed [82] and some are under construction [83]. However, the $\mathcal{O}$(GeV/c^2) mass region has remained mostly unexplored.

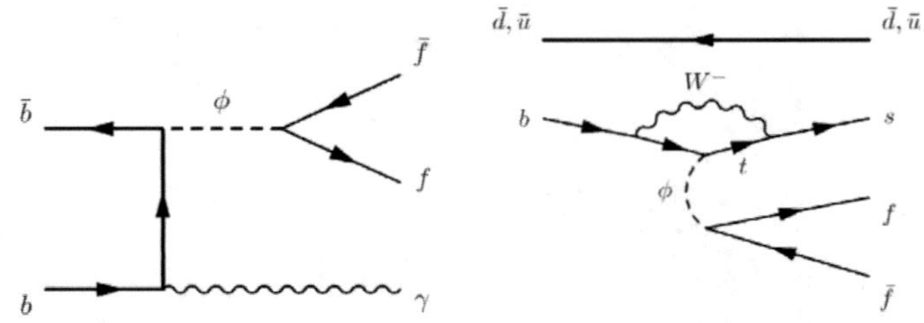

Figure 9. Lowest-order diagrams for the production of a long-lived dark scalar particle in radiative Υ decays (left) and in penguin decays of B mesons (right).

At the B factories, a hidden-sector scalar particle may be produced in an Υ radiative decay or in a B penguin decay and, in turn, may decay into a pair of fermions as shown in Fig. 9. So we performed the first search for a long-lived particle L in the process $e^+e^- \to LX$ where X is any set of particles and L decays to six different final states. Except for 20 fb^{-1}taken at the $\Upsilon(4S)$, we use the entire *BABAR* data collected at the $\Upsilon(4S)$, 40 MeV below the $\Upsilon(4S)$ peak, at the $\Upsilon(2S)$, and $\Upsilon(3S)$ corresponding to a luminosity of 489.1 fb^{-1}. We present the results in two ways. First without any assumptions on the production mechanism, we use the complete data sample and present model-independent results on the product of cross section, branching fraction and reconstruction efficiency of each two-body final state f, $\sigma_f = \sigma(e^+e^- \to LX) \cdot \mathcal{B}(L \to f) \cdot \epsilon_f$. We have produced tables of the reconstruction efficiency as a function the L mass, transverse momentum p_T in the CM frame and the proper decay distance $c\tau$ assuming L to be a scalar particle. Second, we present results for B decays via $B \to LX_s$ where X_s is a hadronic state with strangeness S=-1. This production mechanism is motivated by the Higgs portal [69–72] and axion-portal [84] models of dark matter. In this case, we report model-dependent results on the product branching fraction $\mathcal{B}_{Lf} = \mathcal{B}(B \to LX_s) \cdot \mathcal{B}(L \to f)$.

We use MC simulations of signal and background events to determine the signal efficiency and L mass resolution σ_m. We generate signal events with EvtGen [85] and produce two types of MC samples. In the model-independent approach, we generate $e^+e^- \to Ln\pi$ and $B \to Ln\pi$ where $n \leq 3$. We generate 11 L mass values (0.5, 1, 2, 3, 4, 5, 6, 7, 8, 9, 9.5 GeV/c^2) assuming spin zero. This yields broad L-momentum spectra requiring efficiency tables to test specific models [86]. For the model-dependent approach, we produce $e^+e^- \to B\bar{B}$, $B \to X_sL$ events where X_s is composed of 10% K, 25% K^* and 65% $K^*(1680)$ [87]. The high-mass tail of the X_s mass spectrum is suppressed by phase space limitations of heavy L-states. This choice yields an L-momentum spectrum as a function of m that reproduces the dimuon invariant-mass spectrum for $B \to X_s\mu^+\mu^-$ [85]. In this approach, we generate seven mass values (0.5, 1, 2, 3, 3.5, 4, 4.5 GeV/c^2). In addition, we generate the background processes $e^+e^- \to B\bar{B}$ (EvtGen [85]), $\tau^+\tau^-, \mu^+\mu^-$ (KKF2 [43]), e^+e^- (BHWIDE [42]) and $q\bar{q}$ (JETSET [88]) where ($q = u, d, s, c$). The detector response is simulated with GEANT4 [46].

We reconstruct L from two oppositely-charged particles ($e^+e^-, \mu^+\mu^-, e^\pm\mu^\mp, \pi^+\pi^-, K^+K^-$ and $K^\pm\pi^\mp$) that originate from a common vertex separated from the IP by more than $3\sigma_{PV}$ where σ_{PV} is primary vertex resolution. We reject background from K^0_S and Λ decays as well as low-mass peaking structures in the background by imposing mass thresholds:

$m_{ee} > 0.44$ GeV/c^2, $m_{e\mu} > 0.48$ GeV/c^2, $m_{\mu\mu} > 0.50$ GeV/c^2 and $m_{\mu\mu} < 0.37$ GeV/c^2, $m_{\pi\pi} > 0.86$ GeV/c^2, $m_{K\pi} > 1.05$ GeV/c^2 and $m_{KK} > 1.35$ GeV/c^2. The efficiency varies from 47% for $m = 1$ GeV/c^2, $2 < p_T < 3$ GeV/c^2 and $c\tau = 3$ cm to nearly zero for large m and large $c\tau$ values.

We extract the signal yield for each final state as a function of m using unbinned extended maximum likelihood fits of the m distributions. The strategy is based on the fact that a signal would produce a peak in the m distribution whereas background varies smoothly. Thus, the signal PDF is the pull histogram $P^i_S(m) = H^i_S((m_t - m)/\sigma_m)$ produced from masses closest to the true mass m_t, measured mass m and its uncertainty σ_m. The background PDF $P^i_B(m)$ is a second-order polynomial spline with knots located at the bin boundaries and is determined from the data. The bin width is variable and is defined as 15 times the RMS width of the signal distribution in that region. The factor 15 is chosen since it is sufficiently large to prevent $P^i_B(m)$ conforming to signal peaks but is sufficiently small to produce signal peaks from fluctuations. We scan the data in steps of 2 MeV/c^2 using the PDF $n_S \cdot P_S(m) + n_B \cdot P_B(m)$ where n_S (n_B) is the signal (background) yield determined from the fit. Figure 10 shows the observed number of events in comparison to the expected background for each final state. Two mass points in the dimuon final state show yields with significance $\mathcal{S}_S > 3$. At $m_t = 0.212$ MeV/c^2, we see 13 events with $\mathcal{S}_S = 4.7$. However, most of the vertices lie inside the detector material and the muon tracks have low momenta that are poorly separated from e and π. At $m_t = 1.24$ GeV/c^2, we see ten events with $\mathcal{S}_S = 4.2$. We determine a probability of $P = 0.04\%(0.8\%)$ of seeing yields with $\mathcal{S}_S \geq 4.7(\mathcal{S}_S \geq 4.2)$ in the entire dimuon mass region. Including the other final states, the probabilities are reduced by a factor of six. Thus, the data are consistent with background expectations.

For each scan fit, we determine the systematic error on the signal yield. The dominant systematic error comes from the background PDF. Other contributions are due to the mass resolution, luminosity, particle identification and the number of MC signal events. After adding all contributions in quadrature, we include the total systematic error (σ_{sys}) in the fit by convolving the likelihood function with a Gaussian function with width σ_{sys}. We determine uniform-prior Bayesian upper limits at 90% CL on σ_f as a function of m for each final state. Figure 11 (left) shows the results on σ_f separately for the $\Upsilon(4S)$ and $\Upsilon(3S)+\Upsilon(2S)$ data. Using the efficiency tables [86], these limits can be applied to any production model. For the model-dependent interpretation, we set Bayesian upper limits at 90% CL on the product of branching fractions $\mathcal{B}_{Lf}$. Figure 11 (right) shows the results on $\mathcal{B}_{Lf}$ as a function of m for different values of $c\tau$. These limits exclude a significant region of the parameter space of the inflaton model [69].

4. Search for New π^0-like Particles

The measurement of the π^0 form factor in two-photon collisions by the *BABAR* Collaboration [89] has raised various discussions [90–93]. At sufficiently high squared momentum transfer Q^2, the pion form factor should approach the Brodsky-Lepage limit of $\sqrt{2}f_\pi/Q^2 \simeq 185$ MeV/Q^2 [94]. At $Q^2 > 15$ GeV^2/c^2, the data is expected to be described well by perturbative QCD. Our data, however, show no sign of convergence towards the Brodsky-Lepage limit as Fig. 12 indicates. Though the Belle data [95] show a better agreement with the perturbative prediction, they are consistent with our results. So, maybe a new particle ϕ exists with mass close to that of the π^0 decaying to $\gamma\gamma$ and thus causing the non-convergence of the π^0 form factor. The new particle ϕ may be a scalar (ϕ_S), pseudoscalar (ϕ_P) or a so-called hard-core pion (π^0_{HC}) [96].

We focus on a search for π^0-like particles that are produced in association with a $\tau^+\tau^-$ pair [97]. Figure 13 shows lowest-order Feynman diagrams for the production of π^0-like particles in association with $\tau^+\tau^-$. For $Q^2 > 8$ GeV^2/c^2, predicted cross sections for $e^+e^- \to \tau^+\tau^-\phi$ are large: $\sigma_{\pi^0_{HC}} = 0.25$ pb, $\sigma_{\phi_P} = 2.5$ pb, and $\sigma_{\phi_S} = 68$ pb. Thus, we expect large event yields in our data set: $N_{\pi^0_{HC}} = 1.2 \times 10^5$, $N_{\phi_P} = 1.2 \times 10^6$ and $N_{\phi_S} = 3.2 \times 10^7$, respectively.

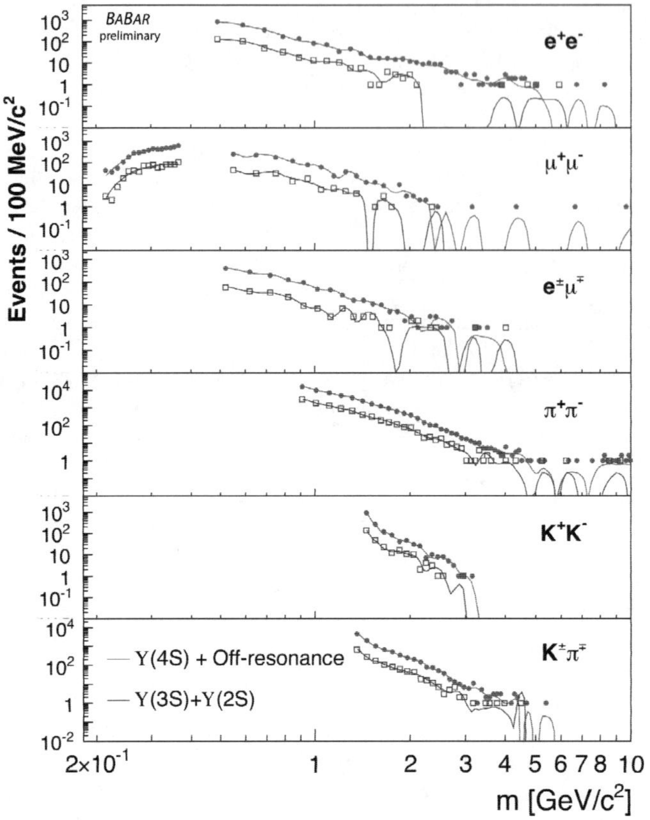

Figure 10. Observed mass distribution for $\Upsilon(4S)+$ off-resonance data (red solid points), for $\Upsilon(3S)+\Upsilon(2S)$ data (blue open squares) as a function of mass m for each final state with the background PDF P_B overlaid (red, blue solid lines).

Using the full *BABAR* $\Upsilon(4S)$ data set of 468 fb^{-1} ($4.3 \times 10^8 \tau$ pairs), we select $\tau^+\tau^-$ events in the decays $\tau^+ \to e^+\bar{\nu}_e\nu_\tau$ and $\tau^- \to \mu^-\nu_\mu\bar{\nu}_\tau$[3] decays requiring $p_T > 0.3$ GeV/c for each lepton. In addition, we require exactly one $\pi^0 \to \gamma\gamma$ with energy $2.2 < E_{\pi^0} < 4.7$ GeV in the laboratory frame. After excluding the energy of the photons from the π^0 decay, we require the remaining energy E_{extra} in the electromagnetic calorimeter to be less than 0.3 GeV. We reduce background from radiative Bhabhas by imposing a minimum photon energy of $E_\gamma > 0.25$ GeV and an opening angle between $e^\pm$ and γ of $30° < \theta(e,\gamma) < 150°$. We further reduce background from semi-leptonic and hadronic $\pi^\pm\pi^0$ decays by requiring $E_{small} + E_\phi > 0.5E_{CM}$ and $m_{\pi^\pm\pi^0} > m_\tau$ where E_{small} is the energy of the lower-energy track and E_{CM} is the CM energy. Figure 14 shows the $E_{small} + E_\phi$ distribution before the requirement of $E_{small} > 0.5E_{CM}$, which is consistent with the expected $e^+e^- \to \tau^+\tau^-$ energy spectrum. We simulate background modes $e^+e^- \to B\bar{B}$ with EvtGen [85], $e^+e^- \to q\bar{q}$ continuum with JETSET [88], $e^+e^- \to e^+e^-$ with BHWIDE [42], $e^+e^- \to \mu^+\mu^-, \tau^+\tau^-$ with KK2F [43] where we use the TAUOLA library [98] to generate τ decays. We model radiative corrections with PHOTOS [99] and the simulation of the detector response with GEANT4 [46].

[3] charge conjugation is implied unless stated otherwise.

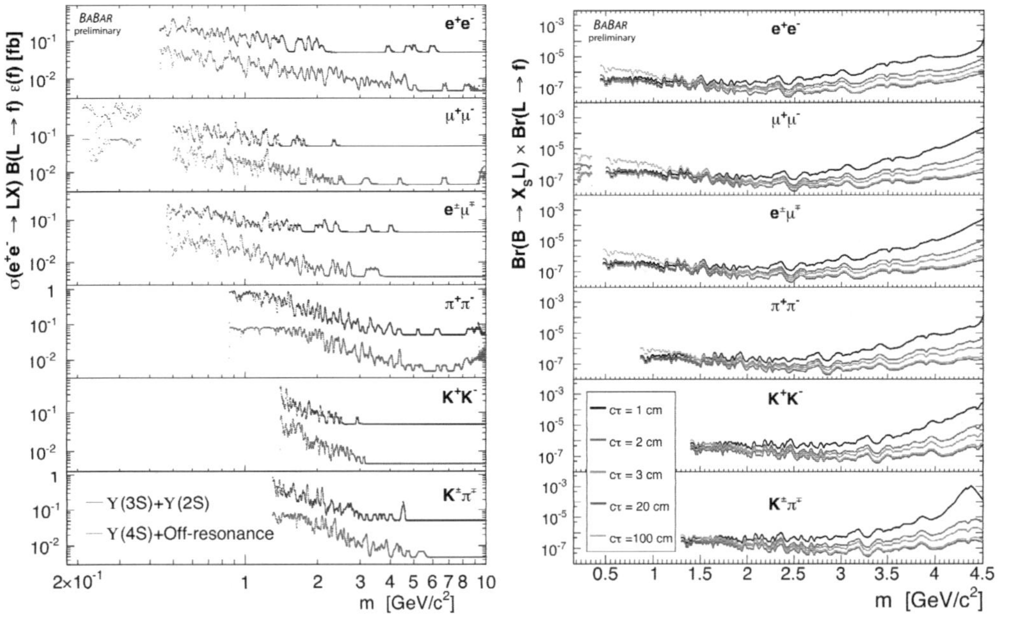

Figure 11. Upper limits on the cross section σ_f as a function of L mass for each final state (left) for $\Upsilon(4S)$ data (lower red curves) and for $\Upsilon(3S + \Upsilon(2S)$ data (upper blue curves) and on the product branching fraction $\mathcal{B}_{Lf}$ (right) for different decay lengths.

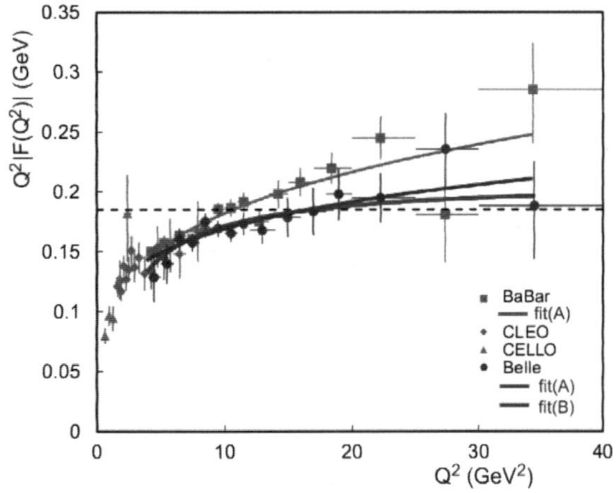

Figure 12. Measurements of the π^0 form factor as a function of Q^2.

Figure 15 shows the entire $m_{\gamma\gamma}$ invariant-mass spectrum after applying all selection criteria. For comparison, we also show a simulated signal and the simulated spectra for $e^+e^- \to \tau^+\tau^-$, $\mu^+\mu^-$, $b\bar{b}$ and $c\bar{c}$ backgrounds. We fit the $m_{\gamma\gamma}$ invariant-mass distribution in the region $50 < m_{\gamma\gamma} < 300$ MeV/c^2 with a Gaussian signal and a linear background: $N(m_{\gamma\gamma}) = N_g\ G(\mu_g, \sigma_g) + N_{lin}\ (1 + a_{lin}\ m_{\gamma\gamma})$. We use extended unbinned maximum log-likelihood fits to extract N_{lin}, N_g and a_{lin}. The width σ_g of the Gaussian function is fixed to 11.1 MeV/c^2

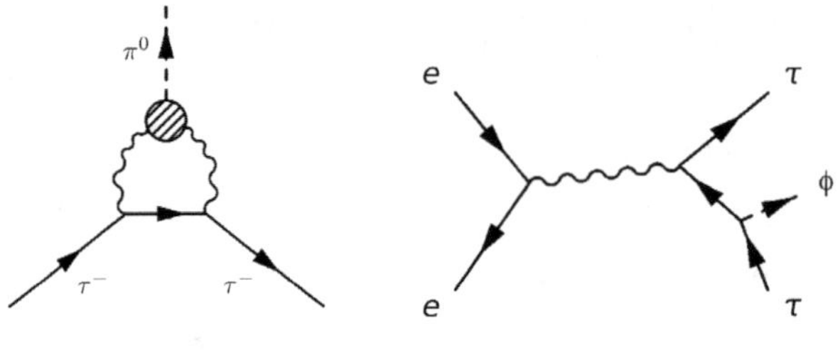

Figure 13. Leading-order SM diagram for $\tau^+\tau^-$ annihilation into a π^0-like particle (left) and for the radiation of a ϕ from a τ lepton in $e^+e^- \to \tau^+\tau^-$ (right).

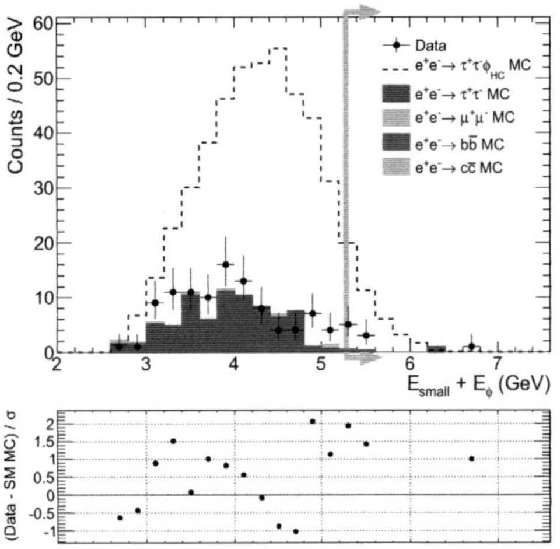

Figure 14. Top: the $E_{small} + E_\phi$ distribution for events with $100 < m_{\gamma\gamma} < 160$ MeV/c^2 in data (points with error bars) and simulations for $e^+e^- \to \tau^+\tau^-$, $\mu^+\mu^-$, $b\bar{b}$, $c\bar{c}$ (colored histograms). The dashed histogram shows the expectation for π^0_{HC} production at $\sigma = 0.254$ pb. Bottom: pull distribution.

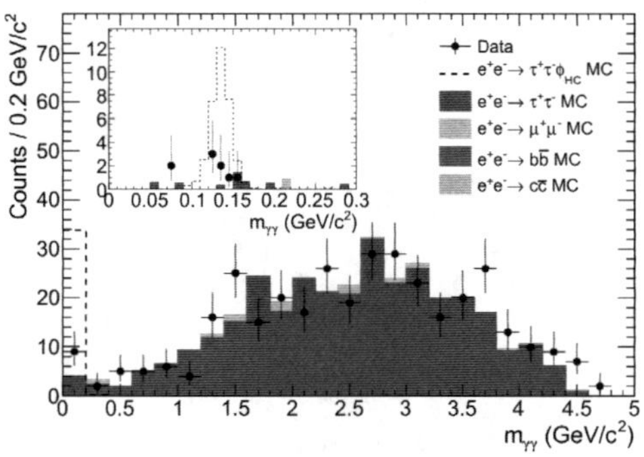

Figure 15. The $m_{\gamma\gamma}$ invariant-mass distribution (right) for data (solid points with error bars), expected signal (dashed histogram) and simulated $e^+e^- \to \tau^+\tau^-$, $\mu^+\mu^-$, $b\bar{b}$, $c\bar{c}$ background spectra (colored histograms). The inset shows the low-mass region in bins of 10 MeV/c^2.

determined from control samples. The mean μ_g represents the mass of the ϕ particle and is fixed in each fit. We scan μ_g values between 110 MeV/c^2 and 160 MeV/c^2 in steps of 0.5 MeV/c^2 and extract N_g from each fit. Figure 16 shows the resulting N_g distribution. Figure 17 shows the fit for the highest yield. We observe a raw yield of $N_g = 6.2 \pm 2.7 \pm 0.06$ events at a mass of 137 MeV/c^2. After subtracting 1.24 ± 0.37 events of peaking background and correcting for a fit bias of -0.06 ± 0.02 events, we find $5.0 \pm 2.7 \pm 0.4$ signal events. We determine identical efficiencies of $(0.455 \pm 0.019)\%$ for π^0_{HC} and ϕ_P. For ϕ_S, the efficiency is reduced to $(0.09 \pm 0.004)\%$. Systematic uncertainties include contributions from the MC sample size, energy scale, π^0 efficiency, energy resolution, particle identification, momentum scale and momentum resolution and amount to 4.2% for π^0_{HC}, ϕ_P and 4.4% for ϕ_S. This leads to measured cross sections of $\sigma_{\pi^0_{HC}, \phi_P} = (38 \pm 21_{stat} \pm 3_{sys})$ fb and $\sigma_{\phi_S} = (190 \pm 100_{stat} \pm 20_{sys})$ fb, respectively.

Since the result is consistent with background, we set Bayesian upper limits on the cross sections assuming a flat prior. We obtain $\sigma_{\pi^0_{HC},\phi_P} < 73$ fb and $\sigma_{\phi_S} < 370$ fb at 90% CL. This should be compared to predictions of 270 – 820 fb and 68 – 1850 fb, respectively.

To check the compatibility of the measured production cross sections with the contribution of particle ϕ to the π^0 form factor, we include this measurement as an additional term in the χ^2 of the fit to the Q^2 dependence of the π^0 in our data. The increase in χ^2 follows a χ^2 distribution with one degree of freedom. Thus, we can determine $p-$values for the process $e^+e^- \to \tau^+\tau^-\phi$ to contribute to the π^0 form factor. We obtain $p_{\pi^0_{HC}} = 5.9 \times 10^{-4}$ for π^0_{HC}, $p_{\phi_P} = 8.8 \times 10^{-10}$ for ϕ_P and $p_{\phi_S} = 2.2 \times 10^{-9}$ for ϕ_S. We conclude that a new particle ϕ with mass close to the π^0 is not a likely explanation of the excess of the π^0 form factor seen in our data at larger Q^2.

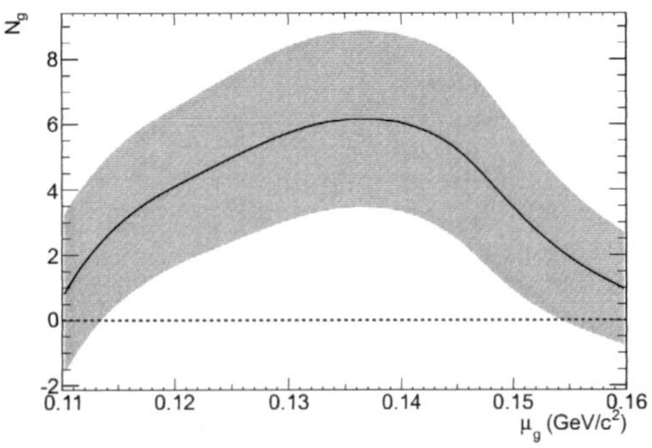

Figure 16. Number of events N_g as a function of the ϕ mass hypothesis μ_g

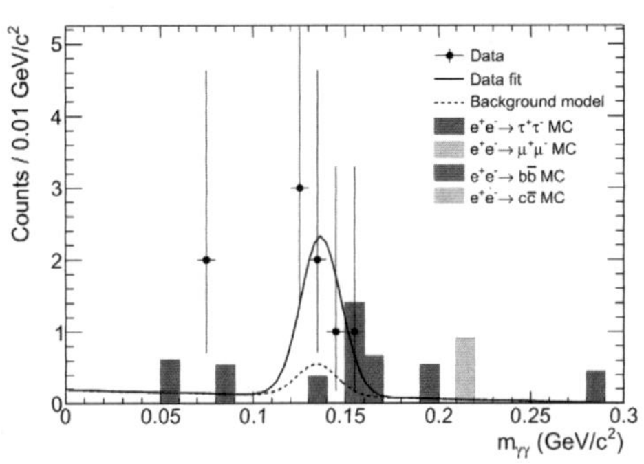

Figure 17. Measured $m_{\gamma\gamma}$ spectrum (points with error bars) with fit overlaid (solid line), peaking background contribution (dotted line) and expected backgrounds from $e^+e^- \to \tau^+\tau^-$, $\mu^+\mu^-$, $b\bar{b}$, $c\bar{c}$ (colored histograms).

5. Conclusion and Outlook

The B factories are an excellent laboratory to search for light dark matter. We see no dark photons in the $0.02 - 10.2$ GeV/c^2 mass region pushing the limit on the mixing parameter ϵ at a level of 10^{-4} to 10^{-3} depending on the dark photon mass. We performed the first search for long-lived particles at a high-luminosity e^+e^- collider and the first search in a heavy-flavor environment. We observe no long-lived particle in the $0.2 < m < 10$ GeV/c^2 mass range and proper decay lengths of $0.5 < c\tau < 100$ cm. In a model-independent approach, we set 90% CL upper limits on the product $\sigma(e^+e^- \to LX) \cdot \mathcal{B}(L \to f) \cdot \epsilon_f$ for different masses, p_T values and proper decay lengths. We also give model-dependent 90% CL upper limits on the product branching fraction $\mathcal{B}(B \to LX_s) \cdot \mathcal{B}(L \to f)$. Finally, we searched for the production of π^0-like particles in association with a $\tau^+\tau^-$ pair and set upper limits on $\sigma(e^+e^- \to \tau^+\tau^-\phi) < 73$ fb at 90% CL. The hypothesis that a π^0-like particle causes the non convergence of the pion form factor has a p-value of 5.9×10^{-4} for π^0_{HC} and thus is rather unlikely. For ϕ_P and ϕ_S production, the p-value is even lower. *BABAR* will continue with searches for new physics. Belle just released first limits on $e^+e^- \to A'h', h' \to A'A'$ [100]. Since Belle's luminosity is about a factor of two larger, the limits are a factor of two lower than the *BABAR* results. However, substantial improvements are expected from Belle II. At 50 ab^{-1}, present *BABAR* limits will be improved by up to two orders of magnitude depending on the final state.

Acknowledgment
This work has been supported by the Norwegian Research Council. I would like to thank the *BABAR* collaboration for the opportunity to give this talk. In particular, I would like to thank Bertrand Echenard, Frank Porter Mike Roney and Abi Soffer for useful comments.

References

[1] O. Adriani *et al.* (Pamela Collaboration), Nature **458**, 607 92009).
[2] M. Ackermann *et al.* (FERMI LAT Collaboration), Phys.Rev.Lett. **108**, 111103 (2012).
[3] M. Aguilar *et al.* (AMS Collaboration), Phys.Rev.Lett. **110**, 141102 (2013).
[4] H.B. Li *et al.* (TEXONO Collaboration), Phys. Rev. Lett. **110**, 261301 (2013).
[5] C. Kelso *et al.*, Phys. Rev. **D 85**, 043515 (2012).
[6] C.E. Aalseth *et al.*, Phys. Rev. Lett. **107**, 141301 (2011).
[7] Z. Ahmed *et al.* (CDMS Collaboration), arXiv:1203.1309 (2012).
[8] R. Bernabei *et al.* (DAMA Collaboration), Eur. Phys. J. **C 67**, 39 (2010).
[9] S.C. Kim *et al.*, Phys. Rev. Lett. **108**, 181301 (2012).
[10] Z. Ahmed *et al.* (CDMS Collaboration), Science **327**, 1619 (2010).
[11] Z. Ahmed *et al.* (CDMS Collaboration), Phys.Rev.Lett. **111**, 251301 (2013).
[12] R. Agnese *et al.* (CDMS Collaboration,), Phys. Rev. **D 88**, 031104 (2013).
[13] R. Agnese *et al.* (CDMS Collaboration), Phys.Rev.Lett. **112**, 041302 (2014).
[14] E. Armengaud *et al.* (EDELWEISS Collaboration), Phys. Rev. **D 86**, 051701 (2012).
[15] Z. Ahmed *et al.* (CDMS Collaboration), Phys. Rev. Lett. **106**, 131302 (2011).
[16] Z. Ahmed *et al.* (EDELWEISS and CDMS Collaboration), Phys. Rev. **D84**, 011102 (2011).
[17] G. Angloher *et al.* (CRESST Collaboration), Eur. Phys. J. **C 72**, 197 (2012).
[18] E. Aprile *et al.* (XENON100 Collaboration), Phys. Rev. Lett. **111**, 021301 (2013).
[19] D.S. Akerib *et al.* (LUX Collaboration), Phys.Rev.Lett. **112**, 091303 (2014).
[20] See also "Dark Matter Portal", http://lpsc.in2p3.fr/mayet/dm.php; http://www.taup-conference.to.infn.it/, http://kicp-workshops.uchicago.edu/IDM2012/overview.php.
[21] R. Essig *et al.*, arXiv:1311.0029 [hep-ph] (2013).
[22] R. D. Peccei and H. R. Quinn, Phys. Rev. Lett. **38**, 1440 (1977).
[23] S. Weinberg, Phys. Rev. Lett. **40**, 223 (1978).
[24] F. Wilczek, Phys. Rev. Lett. **40**, 279 (1978).
[25] M. Pospelov, A. Ritz, and M. B. Voloshin, Phys. Lett. **B 662**, 53 (2008).
[26] B. Holdom, Phys. Lett. **B 166**, 196 (1986).
[27] D. P. Finkbeiner and N.Weiner, Phys. Rev. **D 76**, 083519 (2007).
[28] N. Arkani-Hamed *et al.*, Phys. Rev. **D 79**, 015014 (2009).
[29] B. Aubert *et al.* (*BABAR* Collaboration), Nucl. Instrum. Meth. **A 479**, 1 (2002).
[30] B. Aubert *et al.* (*BABAR* Collaboration), Nucl. Instrum. Meth. **A 729**, 615 (2013).
[31] J.P Lees *et al.* (*BABAR* Collaboration), Phys.Rev.Lett. **108**, 211801 (2012).
[32] B. Aubert et al. (*BABAR* Collaboration), arXiv: 0908.2821 [hep-ex](2009).
[33] J.P Lees *et al.* (*BABAR* Collaboration), Phys. Rev. **D 87**, 031102 (2013).
[34] B. Aubert *et al.* (*BABAR* Collaboration), Phys. Rev. Lett **103**, 181801 (2009).
[35] J.P Lees *et al.* (*BABAR* Collaboration), Phys. Rev. **D 88**, 071102 (2013).
[36] J.P Lees *et al.* (*BABAR* Collaboration), Phys. Rev. Lett. **107**, 221803 (2011).
[37] P. del Amo-Sanchez *et al.* (*BABAR* Collaboration), Phys. Rev. Lett. **107**, 021804 (2011).
[38] J.P Lees *et al.* (*BABAR* Collaboration), Phys. Rev. **D 88**, 031701 (2013).
[39] B. Batell, M. Pospelov, and A. Ritz, Phys. Rev. **D 79**, 115008 (2009).
[40] R. Essig, P. Schuster, and N. Toro, Phys. Rev. **D 80**, 015003 (2009).
[41] J.P Lees *et al.* (*BABAR* Collaboration), Phys.Rev.Lett. **113**, 201801 (2014).
[42] S. Jadach, W. Placzek and B. F. L. Ward, Phys. Lett. **B 390**, 298 (1997).
[43] S. Jadach, B. F. L. Ward and Z. Was, Phys. Rev. **D 63**, 113009 (2001).
[44] A. B. Arbuzov et al., J. High Energy Phys. **9710**, 001 (1997).
[45] M. Caffo, H. Czyz, and E. Remiddi, Nuovo Cim. **A 110**, 515 (1997); Phys. Lett. **B 327**, 369 (1994).
[46] S. Agostinelli et al. (GEANT4 Collab.), Nucl. Instrum. Meth., Phys. Res., Sect. **A 506**, 250 (2003).
[47] J. Alwall *et al.*, JHEP **0709**, 028 (2007).
[48] A. L. Read, Nucl. Instrum. Meth. **A 425**, 357 (1999).
[49] D. Babusci et al. (KLOE-2 Collaboration), Phys. Lett. **B 720**, 111 (2013).
[50] D. Babusci et al. (KLOE-2 Collaboration), Phys.Lett. **B 736**, 459 (2014).
[51] S. Abrahamyan et al. (APEX Collaboration), Phys. Rev. Lett. **107**, 191804 (2011).

[52] H. Merkel *et al.*, Phys.Rev.Lett. **112**, 221802 (2014).
[53] G. Agakishiev et al. (HADES Collaboration), Phys. Lett. **B 731**, 265 (2014).
[54] P. Adlarson et al. (WASA-at-COSY Collaboration), Phys. Lett. **B 726**, 187 (2013).
[55] E. Goudzovski *et a.* (NA48 Collaboration), arXiv:1412.8053 [hep-ex] (2014).
[56] S. Andreas, C. Niebuhr and A. Ringwald, Phys. Rev. **D 86**, 095019 (2012).
[57] A. Bross *etal.*, Phys. Rev. Lett. **67**, 2942 (1991).
[58] E. M. Riordan *et al.*, Phys. Rev. Lett. **59**, 755 (1987).
[59] M. Endo, K. Hamaguchi and G. Mishima, Phys. Rev. **D 86**, 095029 (2012).
[60] M. Pospelov, Phys. Rev. **D 80**, 095002 (2009).
[61] B. Aubert et al. (*BABAR* Collaboration), Phys. Rev. Lett. **103**, 081803 (2009).
[62] A. Rostoman, Task Force Report, BELLE2-Note-34 (2015).
[63] J. R. Boyce (LIPSS, DarkLight, HPS and APEX Collaborations), J. Phys. Conf. Ser. **384**, 012008 (2012).
[64] J. Balewski *et al.*, arXiv:1412.4717 [physics.ins-det] (2014).
[65] M. Molitor, AIP Conf.Proc. **1563**, 140 (2013).
[66] B. Wojtsekhowski, D. Nikolenko and I. Rachek, arXiv:1207.5089 [hep-ex] (2012).
[67] F. Bossi, Adv. High Energy Phys. 2014, 891820 (2014).
[68] P. Schuster, N. Toro and I. Yavin, Phys. Rev. **D 81**, 016002 (2010).
[69] F. Bezrukov and D. Gorbunov, JHEP **1307**, 140 (2013).
[70] C. Cheung and Y. Nomura, JHEP **1011**, 103 (2010).
[71] K. Schmidt-Hoberg, F. Staub and M. W. Winkler, Phys. Lett. **B 727**, 506 (2013).
[72] J. D. Clarke, R. Foot and R. R. Volkas, JHEP **1402**, 123 (2014).
[73] A. E. Nelson and J. Scholtz, arXiv:1311.0040 [hep-ph].
[74] S. N. Gninenko, Phys. Rev. **D 85**, 055027 (2012).
[75] T. Adams et al. [NuTeV Collaboration], Phys. Rev. Lett. **87**, 041801 (2001).
[76] V. M. Abazov et al. [D0 Collaboration], Phys. Rev. Lett. **97**, 161802 (2006).
[77] V. M. Abazov et al. [D0 Collaboration], Phys. Rev. Lett. **103**, 071801 (2009).
[78] F. Abe et al. [CDF Collaboration], Phys. Rev. **D 58**, 051102 (1998).
[79] G. Aad et al. [ATLAS Collaboration], Phys. Lett. **B 720**, 277 (2013).
[80] G. Aad et al. [ATLAS Collaboration], Phys. Rev. Lett. **108**, 251801 (2012).
[81] G. Aad et al. [ATLAS Collaboration], Phys. Lett. **B 719**, 280 (2013).
[82] R. Essig, P. Schuster, N. Toro and B. Wojtsekhowski, JHEP **1102**, 009 (2011).
[83] O. Moreno, arXiv:1310.2060 [physics.ins-det].
[84] M. Freytsis, Z. Ligeti and J. Thaler, Phys. Rev. **D 81**, 034001 (2010).
[85] D. J. Lange, Nucl. Instrum. Methods **A 462**, 152 (2001).
[86] J.P Lees *et al.* (*BABAR* Collaboration), arXiv:1502.02580-[hep-ex], submitted to Phys. Rev. Lett. (2015).
[87] J. Beringer *et al.* (Particle Data Group) Phys. Rev. **D 86**, 010001 (2012) and K.A. Olive *et al.*, Chin.Phys. C38, 090001 (2014).
[88] T. Sjostrand, Comput. Phys. Commun. **82**, 74 (1994).
[89] B. Aubert et al. (BABAR Collaboration), Phys. Rev. **D 80**, 052002 (2009).
[90] A. P. Bakulev *et al.*, Phys. Rev. **D 86**, 031501 (2012).
[91] A. Dorokhov, JETP Letters **91**, 163 (2010).
[92] W. Lucha and D. Melikhov, J. Phys. **G 39**, 045003 (2012).
[93] S. Noguera and V. Vento, EPJ **A 46**, 197 (2010).
[94] S. J. Brodsky, G. P. Lepage, and P. B. MacKenzie, Phys. Rev. **D 28**, 228 (1983).
[95] S. Uehara et al. (Belle Collaboration), Phys. Rev. **D 86**, 092007 (2012).
[96] D. McKeen, M. Pospelov, and J. M. Roney, Phys. Rev. **D 85**, 053002 (2012).
[97] J.P Lees *et al.* (*BABAR* Collaboration), Phys. Rev. **D 90**, 112011 (2014).
[98] S. Jadach *et al.*,Comput. Phys. Commun. **76**, 361 (1993).
[99] E. Barberio and Z.Was, Comput. Phys. Commun. **79**, 291 (1994).
[100] I. Jaegle *et al.* (Belle collaboration), arXiv:1502.00084 [hep-ex] (2015).

Journal of Physics: Conference Series **631** (2015) 012035 doi:10.1088/1742-6596/631/1/012035

Study of CP asymmetry in $B^0\overline{B}^0$ mixing using inclusive dilepton samples

Isabella Garzia on behalf of the *BABAR* Collaboration
INFN-Sezione di Ferrara, via Saragat 1, 44122 Ferrara, Italy
E-mail: garzia@fe.infn.it

Abstract. The asymmetry between same sign inclusive dilepton samples $\ell^+\ell^+$ and $\ell^-\ell^-$ from semileptonic B decays in $\Upsilon(4S) \to B^0\overline{B}^0$ events allows to compare B mixing probabilities $\mathcal{P}(\overline{B}^0 \to B^0)$ and $\mathcal{P}(B^0 \to \overline{B}^0)$, and therefore to test the T and CP invariance. We present the measurement of CP asymmetry in inclusive dilepton samples with the full *BABAR* data set near the $\Upsilon(4S)$ resonance, corresponding to 471 million of $B\overline{B}$ pairs.

1. Introduction

Within the Standard Model (SM), the CP and T symmetries can be violated through weak interaction. For example, a neutral B meson can transform to its antiparticle with a probability $\mathcal{P}(B^0 \to \overline{B}^0)$ which differs from the probability of oscillation of a $\overline{B}^0$ into a B^0: $\mathcal{P}(\overline{B}^0 \to B^0) \neq \mathcal{P}(B^0 \to \overline{B}^0)$. This type of CP violation is called CP violation in the mixing, and was observed first in the kaon system [1, 2], but is not been observed in the neutral B system, for which the SM predicts an asymmetry of the order of 10^{-4} [3]. Therefore, a large measured value could be an indication of new physics. The current experimental average of CP asymmetry for the B^0 system is obtained by averaging *BABAR* [4, 5], DØ [6] and Belle [7] results: $\mathcal{A}_{CP} = (+2.3 \pm 2.6) \times 10^{-3}$. The most recent LHCb result [8], $\mathcal{A}_{CP} = (-0.2 \pm 1.91 \pm 3) \times 10^{-3}$, had not been included in this average. A recent measurement in a mixture of B^0 and B^0_s meson by the DØ collaboration [9] deviates from SM prediction by more than three standard deviations, and so the improving of the experimental precision is needed to understand this discrepancy.

The time evolution of the neutral B mesons can be described by an effective Hamiltonian $\mathbf{H} = \mathbf{M} - i\mathbf{\Gamma}/2$, with $\mathbf{M}$ and $\mathbf{\Gamma}$ two Hermitian matrices describing, respectively, the mass and decay-rate components. Assuming CPT symmetry, the light and heavy mass eigenstates can be written as

$$|B_{L/H}\rangle = p|B^0\rangle \pm q|\overline{B}^0\rangle, \tag{1}$$

where p and q are complex mixing parameters normalized to $|p^2| + |q^2| = 1$. If $|q/p| \neq 1$, both CP and T asymmetries are violated. The B-factories operate at the center-of-mass (c.m.) of the $\Upsilon(4S)$ resonance, which decays into a $B^0\overline{B}^0$ pair about 50% of times. This pair evolves coherently until one B meson decays. In particular, in this analysis, we identify the flavor of one of the B meson at the time of its decay by looking the charge of the lepton (electron or muon) produced from its semileptonic decay. If the second B meson has oscillated, it will produce a lepton with the same charge as the lepton from the first B decay. The CP asymmetry between

$\mathcal{P}(\overline{B}^0 \to B^0)$ and $\mathcal{P}(B^0 \to \overline{B}^0)$ is obtained by measuring the charge asymmetry of the same-sign dilepton event rates $\mathcal{P}_{\ell\ell}^{\pm\pm}$

$$\mathcal{A}_{CP} = \frac{\mathcal{P}_{\ell\ell}^{++} - \mathcal{P}_{\ell\ell}^{--}}{\mathcal{P}_{\ell\ell}^{++} + \mathcal{P}_{\ell\ell}^{--}} = \frac{1-|q/p|^4}{1+|q/p|^4}, \tag{2}$$

which is a quantity independent of the B decay time.

In this analysis, we present an update of $\mathcal{A}_{CP}$ using inclusive dilepton events [10] collected by the *BABAR* detector. In contrast to the previous *BABAR* dilepton analysis [4], we use a new strategy based on a counting procedure, which does not require the proper decay time difference distribution Δt of the two B mesons.

2. The *BABAR* detector and data sample

The *BABAR* detector is described in detail in Ref. [11], so here we give only a brief description of the apparatus. Surrounding the beam-pipe, there is a five-layer silicon vertex detector (SVT), which gives very precise measurements of the vertex decays. Outside the SVT there is a 40-layer drift chamber (DCH) filled with a helium-isobutane (80:20) gas mixture. The DCH and SVT measurements of dE/dx energy loss also contributes to the charged particle identification (PID). After the DCH, there is the detector of internally reflected Cerenkov radiation (DIRC), which gives charged-particle identification. Outside the DIRC, an highly segmented electromagnetic calorimeter (EMC), composed of CsI(Tl) crystals, is used to detect photons and electrons. Finally, the flux return of the superconducting coil surrounding the EMC is instrumented with resistive plate chambers or limited streamer tubes interleaved with iron for the identification of muons and neutral hadrons.

The data set used for this analysis consists of 471×10^6 $B\overline{B}$ pairs produced at the c.m. energy of the $\Upsilon(4S)$ resonance (on-peak data sample), collected by the *BABAR* detector at the PEP-II asymmetric-energy e^+e^- storage ring at the SLAC National Accelerator Laboratory. A smaller sample, about 44 fb^{-1}, of data is collected 40 MeV below the $\Upsilon(4S)$ (off-peak sample), and is used to study background from $e^+e^- \to q\bar{q}$ events (continuum background). The analysis procedure is tested on a Monte Carlo (MC) simulated $B\overline{B}$ event sample.

3. Analysis strategy

We select an event if the two particles with highest momentum are consistent with electron or muon hypothesis. In not otherwise specified, all quantities are evaluated in the c.m frame. We construct the following four leptons combinations: $\ell_1 l_2 = ee, e\mu, \mu e, \mu\mu$, where ℓ_1 (l_2) refers to the highest- (lower-) momentum lepton candidate. Taking into account the four charge combination, we have a total of 16 subsamples. For events originating from two semileptonic B decays, the time-integrated signal yields for same-sign $N_{\ell_1\ell_2}^{\pm\pm}$ (mixed) and opposite-sign $N_{\ell_1\ell_2}^{\pm\mp}$ (unmixed) lepton pairs can be written as

$$N_{\ell_1\ell_2}^{\pm\pm} = \frac{1}{2} N_{\ell_1\ell_2}^0 (1 \pm a_{\ell_1})(1 \pm a_{\ell_2})(1 \pm \mathcal{A}_{CP})\chi_d^{\ell_1\ell_2} \tag{3}$$

$$N_{\ell_1\ell_2}^{\pm\mp} = \frac{1}{2} N_{\ell_1\ell_2}^0 (1 \pm a_{\ell_1})(1 \mp a_{\ell_2})(1 - \chi_d^{\ell_1\ell_2} + r_B), \tag{4}$$

where $N_{\ell_1\ell_2}^0$ is the total time-integrated $B^0\overline{B}^0$ dilepton yield for the $\ell_1\ell_2$ lepton flavor combination, $a_{\ell_j} = (\epsilon_{\ell_j}^+ - \epsilon_{\ell_j}^-)/(\epsilon_{\ell_j}^+ + \epsilon_{\ell_j}^-)$ is the charge asymmetry of the detection efficiency for lepton j, $\chi_d^{\ell_1\ell_2}$ is the effective mixing probability of neutral B mesons including efficiency corrections, and r_B represents the ratio of total B^+B^- yield to the total $B^0\overline{B}^0$ yield.

The number of events observed in the data can be obtained by dividing the total time-integrated yields by the signal fractions $f_{\ell_1\ell_2}$. The signal fraction for same-sign leptons can be written as

$$f^{\pm\pm}_{\ell_1\ell_2} = \frac{S^{\pm\pm}_{\ell_1\ell_2}(1 \pm \mathcal{A}_{CP})}{S^{\pm\pm}_{\ell_1\ell_2}(1 \pm \mathcal{A}_{CP}) + B^{\pm\pm}_{\ell_1\ell_2}} = \frac{1 \pm \mathcal{A}_{CP}}{1 \pm \mathcal{A}_{CP} + B^{\pm\pm}_{\ell_1\ell_2}/S^{\pm\pm}_{\ell_1\ell_2}}, \tag{5}$$

where S and B are the signal and background. Note that, since the signal is generated without CP asymmetry, we need to correct the simulated same-sign signal by a factor $1 \pm \mathcal{A}_{CP}$.

The same-sign background is dominated by unmixed events with a cascade lepton from $B \to X \to \ell Y$ decay, which does not have CP asymmetry. However, a small fraction of same-sign background events originate from mixed events. Some of this mixed events have the B flavor consistent with the measured charge of the lepton pairs, and other events have the opposite B flavor with respect to the measured charge of the lepton pairs. This two types of events could have an opposite effect due to a non-zero $\mathcal{A}_{CP}$. We calculate the dilution factor $\delta_{\ell_1\ell_2}$, defined as the probability of a same-sign background event being consistent with the flavor of the neutral B pairs at the time of their decay after the mixing, i.e. $\ell^+\ell^+$ for B^0B^0 or $\ell^-\ell^-$ for $\overline{B}^0\overline{B}^0$, minus the probability of the opposite case, i.e. $\ell^+\ell^+$ for $\overline{B}^0\overline{B}^0$ or $\ell^-\ell^-$ for B^0B^0. Using this dilution factor, the background events $B^{\pm\pm}_{\ell_1\ell_2}$ in Eq. (5) became $(1 \pm \delta_{\ell_1\ell_2}\mathcal{A}_{CP})B^{\pm\pm}_{\ell_1\ell_2}$.

In the limit of $\mathcal{A}_{CP} \ll 1$ and $a_{\ell_1(\ell_2)} \ll 1$, the number of events observed in the data are

$$M^{\pm\pm}_{\ell_1\ell_2} = \frac{N^{\pm\pm}_{\ell_1\ell_2}}{f^{\pm\pm}_{\ell_1\ell_2}} = \frac{1}{2}N^0_{\ell_1\ell_2}(1 + R^{\pm\pm}_{\ell_1\ell_2})\left[1 \pm a_{\ell_1} \pm a_{\ell_2} \pm \frac{1 + \delta_{\ell_1\ell_2}R^{\pm\pm}_{\ell_1\ell_2}}{1 + R^{\pm\pm}_{\ell_1\ell_2}}\mathcal{A}_{CP}\right]\chi_d^{\ell_1\ell_2}, \tag{6}$$

where $R^{\pm\pm}_{\ell_1\ell_2} = B^{\pm\pm}_{\ell_1\ell_2}/S^{\pm\pm}_{\ell_1\ell_2}$.

For opposite-sing events, the signal comes from unmixed events and, therefore, it is CP symmetric. A small fraction of background comes from mixed events, but its contribution can be neglected in the final fit. In the limit of $\mathcal{A}_{CP} \ll 1$ and $a_{\ell_1(\ell_2)} \ll 1$, the number of opposite-sign events can be written as:

$$M^{\pm\mp}_{\ell_1\ell_2} = \frac{N^{\pm\mp}_{\ell_1\ell_2}}{f^{\pm\mp}_{\ell_1\ell_2}} = \frac{1}{2}N^0_{\ell_1\ell_2}(1 + R^{\pm\mp}_{\ell_1\ell_2})\,(1 \pm a_{\ell_1} \mp a_{\ell_2})\left(1 - \chi_d^{\ell_1\ell_2} + r_B\right). \tag{7}$$

We build a χ^2 fit using the 16 observables so measured. There are thirteen unknowns parameters, $\mathcal{A}_{CP}$, $\chi_d^{\ell_1\ell_2}$, $a_{\ell_1(\ell_2)}$, $N^0_{\ell_1\ell_2}$, and in principle it is possible to fit them at once. However, the strong correlation between the CP asymmetry and the charge asymmetry of lepton detection efficiency produces a very large uncertainties.

We overcome this problem adding another term in the χ^2 fit in order to constrain the detector efficiency $a_\ell \equiv (a_{\ell_1} + a_{\ell_2})/2$. We use single-lepton events (number of lepton ≥ 1) and we measure the asymmetry a_{on} in the on-peak data sample, which is determined mostly by the detection efficiency with a small dependence on $\mathcal{A}_{CP}$. It can be expressed as

$$a_{on} = \alpha + \beta\chi_d\mathcal{A}_{CP} + \gamma a_l, \tag{8}$$

where the α, β, and γ parameters are expressed as a combination of several variables that can be extracted from single lepton data sample, data control sample, and MC sample.
We include Eq. (8) in the χ^2 fit to extract $\mathcal{A}_{CP}$.

4. Event selection

We select events with at least 4 charged tracks and with the normalized second-order Fox-Wolfram momentum [12] $R_2 < 0.6$. We use leptons with momentum 0.6 GeV $< p_\ell <$ 2.2 GeV and with the polar angle θ in the laboratory frame $-0.788 < \cos\theta < 0.961$ for electron and $-0.755 < \cos\theta < 0.956$ for muon candidates. The lepton is rejected if, when combined with

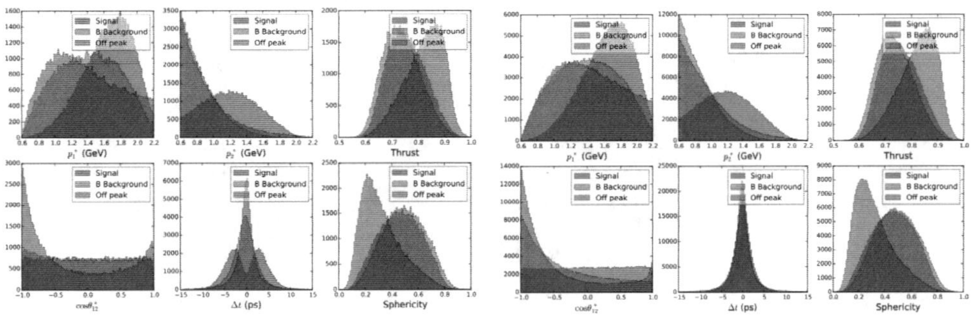

Figure 1. Kinematical variable distributions in same-sign (left) and opposite-sign sample (right).

another lepton of opposite charge, the invariant mass is consistent with the J/ψ or a $\psi(2S)$ meson, or when it arises from a photon conversion.

For dilepton events, the invariant mass of the lepton pairs must be greater that 150 MeV. We require $\Delta t < 15$ ps and its uncertainties $\sigma_{\Delta t} < 3$ ps, where Δt is the proper decay time difference of the two B mesons.

Electrons and muons are identified by two separate multivariate algorithms. The algorithm for electrons is based mainly on the shower shape and the energy deposition in the EMC, while that one for muons uses the track path length and cluster shape in the instrumented flux return. The electron identification efficiency is about 93%, while it ranges from 40% to 80% for muons as a function of the momentum. The probability of a hadron being identified as an electron or muon is less than 0.1% and around 1%, respectively.

5. Background suppression

The main sources of background are continuum events and cascade leptons from $B\overline{B}$ events. The continuum contribution is studied using off-peak data, while a Random Forest multivariate classifier [13] is used to suppress $B\overline{B}$ background.

In the dilepton sample, we use six variables: the c.m. momenta of the two tracks ($p_{1,2}^*$), the thrust and sphericity of the rest of the event, the opening angle of the two tracks (θ_{12}^*) in the c.m. frame, and Δt. The distributions for the same-sign and opposite-sign samples are shown in Fig. 1. The four lepton sample combinations are trained separately, as well as the same sign and opposite sign samples, and Fig. 2 shows the signal probability distributions of the classifiers. We select events with a probability > 0.7 in order to minimize the statistical uncertainty. After this selection, about 2.5% of events come from continuum background, and 35% (8%) from $B\overline{B}$ background in same-sign (opposite-sign) sample.

For the single-lepton sample, we train the Random Forest classifier in order to suppress the continuum background. We use the number of tracks, the event thrust, the R_2 distribution, the difference between the observed energy in the event and the sum of beam energy, the cosine of the angles between the lepton and the axes of the thrust and the sphericity of the rest of the event, and the zeroth-order and second-order polynomial moments L_0 and L_2, where $L_n = \sum_i p_i (\cos\theta_i)^n$, p_i is the momentum of the track in the rest of the event and θ_i is the angle between this track and the single-lepton candidate. We optimize the selection requirement by minimizing the uncertainty of the charge asymmetry after the continuum background subtraction from the on-peak data sample. We found that about 63% of the selected single electrons come from direct semileptonic B decays.

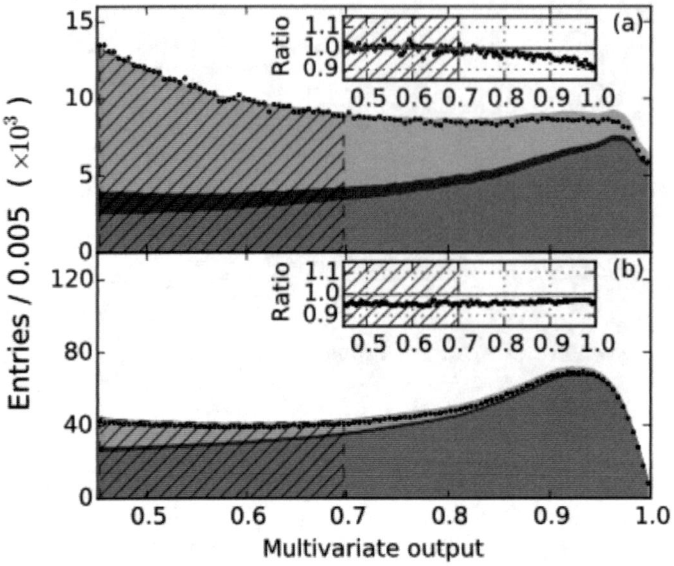

Figure 2. Signal probability distributions from the dilepton multivariate algorithm for (a) same-sign sample and (b) opposite-sign sample. Points represent the continuum-subtracted data, shaded regions from bottom to top indicate the signal, $B\overline{B}$ background with at least one misidentified lepton, and $B\overline{B}$ background with both real leptons. In the inset plots are shown the data/MC ratio. Hatched region is rejected.

5.1. Fake lepton component
In the $B\overline{B}$ background we can distinguish two different components. The first is when both leptons are correctly identified. In this case, the background-to-signal ratio does not depend on the charge asymmetry of the particle identification (PID), because it is cancelled out. The second component originates from the misidentification of at least one lepton, and is called fake lepton background. In this case, the background-to-signal ratio fully depends on the misidentification efficiency.

Approximatively, 3% (0.1%) of the selected muons (electrons) in the dilepton sample are misidentified. From MC simulation, we found that about 98% of mis-identified electrons come from pions, while 87% (12%) of misidentified muons come from pions (kaons). To correct for the difference in the muon misidentification rates between data and MC, we study the muon misidentification efficiency using a clean kaon and muon control samples ($D^* \to D^0\pi^+$, $D^0 \to K^-\pi^+$). The ratio of the efficiencies between data and MC are used to scale the misidentification component in the MC sample. Due to the very low misidentification rate for electrons, a much larger pions control sample from $K_s \to \pi^+\pi^-$ decays is used.

6. Asymmetry results and χ^2 fit
The measurement of the raw asymmetries of the single electrons is $a_{on} = (4.12 \pm 0.14) \times 10^{-3}$ in the on-peak sample, and $a_{off} = (11.1 \pm 1.4) \times 10^{-3}$ in the off-peak data sample. The larger asymmetries observed in the off-peak data is primarily due to the radiative BhaBha events and to the large forward/backward asymmetry of the BABAR detector. We determine the numerical

value of the coefficients in Eq. (8): $a_{on} - \alpha = (2.60 \pm 0.20) \times 10^{-3}$, $\beta \chi_d = 0.057 \pm 0.001$, and $\gamma = 0.8951 \pm 0.0002$.

$\mathcal{A}_{CP}$ is obtained from a χ^2 fit to the 16 continuum-subtracted yields (Eqs. (6), (7)), reported in Table. 1, plus the constraint of Eq. (8).

Table 1. Continuum-subtracted event yields.

	$\ell^+\ell^+$	$\ell^+\ell^-$	$\ell^-\ell^+$	$\ell^-\ell^-$
ee	82303 ± 320	426296 ± 783	425309 ± 782	81586 ± 323
$e\mu$	55277 ± 263	384552 ± 684	378261 ± 660	55878 ± 264
μe	67399 ± 290	467591 ± 737	475363 ± 744	67152 ± 290
$\mu\mu$	47384 ± 243	277936 ± 691	278691 ± 618	48145 ± 247

The fitting procedure is tested on the $B\overline{B}$ MC sample. We measure $\mathcal{A}_{CP}^{MC} = (-1.00 \pm 1.04) \times 10^{-3}$, in agreement with the expectation of no CP asymmetry. In addition, we simulate a non-zero value of asymmetry by reweighing mixed events in the MC sample. We found that the fitting procedure is able to reproduce the asymmetry without bias.

The $\mathcal{A}_{CP}$ measure, after the correction for the small MC bias, is reported in Table 2, where are also summarized all the other parameters extracted from the fit.

Table 2. Parameters of the nominal χ^2 fit. The value of $\mathcal{A}_{CP}$ is corrected for the small MC bias of -1×10^{-3}. Only statistical errors are reported.

$$\mathcal{A}_{CP} = (-3.9 \pm 3.5) \times 10^{-3}$$

	ee	$e\mu$	μe	$\mu\mu$
N^0	430875 ± 515	365343 ± 428	458200 ± 480	268077 ± 391
χ_d	0.2248 ± 0.0006	0.1769 ± 0.0006	0.1754 ± 0.0005	0.2032 ± 0.0007
	a_{e1}	a_{e2}	$a_{\mu1}$	$a_{\mu2}$
a	0.0034 ± 0.0006	0.0030 ± 0.0006	-0.0056 ± 0.0011	-0.0065 ± 0.0011

7. Systematic uncertainties

Several sources of systematic uncertainties are taken into account, and in this section we report those that are significant. The final systematic uncertainty on $\mathcal{A}_{CP}$ is the sum in quadrature of these effects.

Generic MC test. The generic MC is generated with $\mathcal{A}_{CP} = 0$. The fit result is $\mathcal{A}_{CP}^{MC} = (-1.00 \pm 1.04) \times 10^{-3}$, which is consistent with zero. We correct $\mathcal{A}_{CP}$ for this bias and we assign the statistical error as systematic contribution.

MC branching fractions. The branching fractions in B decay chains influence the determination of the background-to-signal ratio, and thus affect $\mathcal{A}_{CP}$. We vary the inclusive charm and semileptonic branching fractions by taking into account the difference between the values reported in PDG [14] and those used by EvtGen in the simulation. The result of $\mathcal{A}_{CP}$ is shifted by $+0.43 \times 10^{-3}$. We take this value as systematic contributions.

Misidentified lepton corrections. We vary the efficiency corrections of leptons within their uncertainties. The systematic contribution is determined by the change in $\mathcal{A}_{CP}$, and is equal to 0.77×10^{-3}. The fake electron asymmetry and fraction also affect $\mathcal{A}_{CP}$ through Eq. (8). The fake fraction is estimated using MC sample and then it is corrected by the data to MC efficiency ratio. By varying this ratio in the single lepton analysis, we found a variation of 0.53×10^{-3} in the measure of $\mathcal{A}_{CP}$, which is taken as systematic uncertainty.

Discrepancy between neutral and charged B mesons. In the single-electron MC sample, the difference between the charge asymmetry of the direct electron in $B^0\overline{B}^0$ and B^+B^- sample is $(0.46 \pm 0.18) \times 10^{-3}$. Since in real data we cannot distinguish neutral and charged B mesons, the single-electron asymmetry measurement is the average of the two asymmetries. We calculate $\mathcal{A}_{CP}$ after shifting the asymmetry in the signal component of the single-electron sample by half of the charge asymmetry difference. The change in $\mathcal{A}_{CP}$ is taken as systematic uncertainty.

Direct/cascade e and μ asymmetry difference. The asymmetry difference between cascade and direct electron (muon) in MC is $\delta_e^{casc} = (-1.16 \pm 0.25) \times 10^{-3}$ ($\delta_\mu^{casc} = (-0.47 \pm 0.28) \times 10^{-3}$). The difference is due to the fact that the charge asymmetry depends on the lepton kinematics. The same trend is observed in the fit results of the dilepton data sample: the asymmetry difference between the lower-momentum and the higher-momentum lepton is negative. We change the asymmetry in the cascade electron (muon) component by the δ_e^{casc} (δ_μ^{casc}) amount, and we take the difference in $\mathcal{A}_{CP}$, $-0.44 \times 10^{-3}(-0.34 \times 10^{-3})$, as systematic uncertainty.

Background-to-signal ratios. The background-to-signal ratio $R_{\ell_1\ell_2}^{\pm\pm(\pm\mp)}$ are obtained from simulation under the condition of $\mathcal{A}_{CP} = 0$. If we vary simultaneously and in the same direction $R_{\ell_1\ell_2}^{++}$ and $R_{\ell_1\ell_2}^{--}$, as well as $R_{\ell_1\ell_2}^{+-}$ and $R_{\ell_1\ell_2}^{-+}$, we found a negligible change in the $\mathcal{A}_{CP}$ value. If they are varied independently, the quadratic sum of the variation in $\mathcal{A}_{CP}$ is 0.68×10^{-3}, which is taken as systematic uncertainty.

Efficiency in random forest cut. The random forest selection efficiency is about 2.6% larger in MC than in data sample for the same-sign events. We vary the random forest selection for MC in order to reduce the selected MC events. The average change in $\mathcal{A}_{CP}$ (0.08×10^{-3}) is taken as systematic uncertainty.

8. Summary and conclusions

We study the CP asymmetry in $B^0\overline{B}^0$ mixing for inclusive dileton decays using the full *BABAR* data set [10]. We measure $\mathcal{A}_{CP} = (-3.9 \pm 3.5 \pm 1.9) \times 10^{-3}$, where the error are statistical and systematic, respectively. The comparison with the current experimental results and averages are shown in Fig. 3. This result is consistent with the SM prediction, the world average [15], and with the recent LHCb measure [8], and represent a significant improvement with respect to our previous results [4], as well as is one of the most precise measurements.

References

[1] Christenson J H, Cronin J W, Fitch V L and Turlay R 1964 *Phys. Rev. Lett.* **13**(4) 138–140
[2] Dorfan D, Enstrom J, Raymond D, Schwartz M, Wojcicki S, Miller D H and Paciotti M 1967 *Phys. Rev. Lett.* **19**(17) 987–993
[3] Lenz A, Nierste U, Charles J, Descotes-Genon S, Jantsch A, Kaufhold C, Lacker H, Monteil S, Niess V and T'Jampens S 2011 *Phys. Rev. D* **83**(3) 036004
[4] Aubert B *et al.* 2006 *Phys. Rev. Lett.* **96**(25) 251802
[5] Lees J P *et al.* 2013 *Phys. Rev. Lett.* **111**(10) 101802
[6] Abazov V M *et al.* 2012 *Phys. Rev. D* **86**(7) 072009

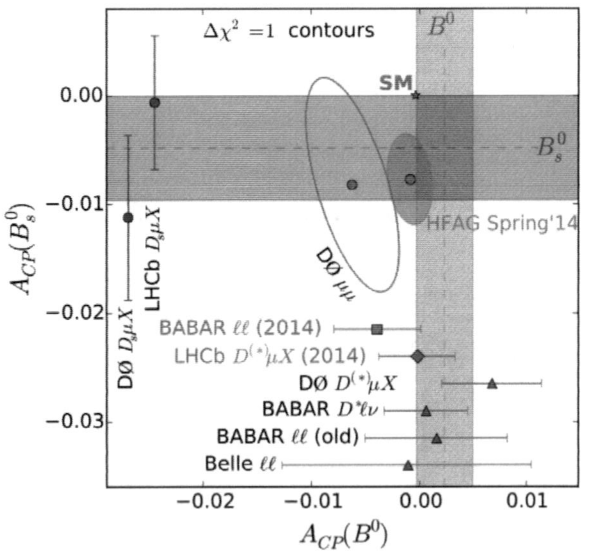

Figure 3. Measurement of CP asymmetry in neutral B mixing. The red square represent this measurement, teal rhombus the recent LHCb result [8], green triangles the B^0 results from Refs. [4, 5, 6, 7] with the vertical band their average, blue circles the B_s^0 results from Refs. [16, 17] with the horizontal band their average, and the magenta point and contour indicate the B^0-B_s^0 mixture from Ref. [9]. The word average "HFAG Spring '14" [15] is also shown by the orange region.

[7] Nakano E *et al.* 2006 *Phys. Rev. D* **73**(11) 112002
[8] Aaij R *et al.* ((LHCb Collaboration)) 2015 *Phys. Rev. Lett.* **114**(4) 041601
[9] Abazov V M and Others ((D0 Collaboration)) 2014 *Phys. Rev. D* **89**(1) 012002
[10] Lees J P *et al.* ((BABAR Collaboration)) 2015 *Phys. Rev. Lett.* **114**(8) 081801
[11] Aubert B *et al.* 2013 *Nuclear Instruments and Methods in Physics Research Section A: Accelerators, Spectrometers, Detectors and Associated Equipment* **729** 615 – 701 ISSN 0168-9002
[12] Fox G C and Wolfram S 1978 *Phys. Rev. Lett.* **41**(23) 1581–1585
[13] Pedregosa F, Varoquaux G, Gramfort A, Michel V, Thirion B, Grisel O, Blondel M, Prettenhofer P, Weiss R, Dubourg V, Vanderplas J, Passos A, Cournapeau D, Brucher M, Perrot M and Duchesnay E 2011 *Journal of Machine Learning Research* **12** 2825–2830
[14] Olive K *et al.* (Particle Data Group) 2014 *Chin.Phys.* **C38** 090001
[15] Y. Amhis *et al.* (Heavy Flavor Averaging Group) (2012), arXiv:1207.1158 and online update `http://www.slac.stanford.edu/xorg/hfag/osc/spring_2014/` accessed: 2010-09-30
[16] Abazov V M *et al.* ((D0 Collaboration)) 2013 *Phys. Rev. Lett.* **110**(1) 011801
[17] 2014 *Physics Letters B* **728** 607 – 615 ISSN 0370-2693

Study of $B \to K\pi\pi\gamma$ decays

Alessandro Pilloni on behalf of the *BABAR* Collaboration

"Sapienza" Università di Roma, Dipartimento di Fisica and INFN, P.le Aldo Moro 5, I-00181 Roma, Italy

E-mail: alessandro.pilloni@roma1.infn.it

Abstract. In $b \to s\gamma$ transitions, the Standard Model predicts that B^0 ($\overline{B}^0$) decays are related predominantly to the presence of right- (left-) handed photons in the final state. Therefore, the mixing-induced *CP* asymmetry in $B \to f_{CP}\gamma$ decays is expected to be small, thus being any sizeable observed asymmetries due to New Physics. In this analysis, we extract information about the $K\pi\pi$ resonant structures by means of an amplitude analysis of the $m_{K\pi\pi}$ and $m_{K\pi}$ spectra in $B^+ \to K^+\pi^-\pi^+\gamma$. The results are used, assuming isospin symmetry, to extract the mixing-induced *CP* parameters of the process $B^0 \to K^0_S\rho^0\gamma$ from the time-dependent analysis of $B^0 \to K^0_S\pi^+\pi^-\gamma$ decays without an explicit amplitude analysis of this mode.

1. Introduction

In $b \to s\gamma$ transitions, the Standard Model predicts that B^0 ($\overline{B}^0$) decays are related predominantly to the presence of right- (left-) handed photons in the final state. Therefore, the mixing-induced *CP* asymmetry in $B \to f_{CP}\gamma$ decays, where f_{CP} is a *CP* eigenstate, is expected to be small (~ 0.03 [1]). This prediction may be altered by New Physics processes, which might emit a photon with both helicity states. Moreover, decays to $K\pi\pi\gamma$ can display an interesting hadronic structure: they have contributions from several kaon resonances decaying to $K\pi\pi$. In this analysis, we extract information about the $K\pi\pi$ resonant structures by means of an amplitude analysis of the $m_{K\pi\pi}$ and $m_{K\pi}$ spectra in the charged decay channel[1] $B^+ \to K^+\pi^-\pi^+\gamma$. The results are used, assuming isospin symmetry, to extract the mixing-induced *CP* parameters of the process $B^0 \to K^0_S\rho^0\gamma$ from the time-dependent analysis of $B^0 \to K^0_S\pi^+\pi^-\gamma$ decays without an explicit amplitude analysis of this mode.

Belle has previously reported a time-dependent *CP* asymmetry measurement of $B^0 \to K^0_S\pi^+\pi^-\gamma$ decays [2]. Similar measurements with $B^0 \to K^0_S\pi^0\gamma$ decays have been reported by *BABAR* [3] and Belle [4]. At this time the measured *CP* asymmetry parameters are compatible with SM predictions, and no evidence of NP has been observed. LHCb has recently performed a study about the photon polarization in $B^+ \to K^+\pi^-\pi^-\gamma$ decay, via the distribution of the angle of the photon with respect to plane defined by the final state hadrons [5]. Studies of the processes $B^+ \to K^+\pi^+\pi^-\gamma$ and $B^0 \to K^0_S\pi^-\pi^+\gamma$ including measurements of the branching fractions have been performed by both *BABAR* [6] and Belle [7]. The latter also finds the branching fraction of the resonant decay $B^+ \to K_1(1270)^+\gamma$.

Journal of Physics: Conference Series **631** (2015) 012036 doi:10.1088/1742-6596/631/1/012036

2. Analysis strategy

The present study aims to perform a time-dependent analysis of $B^0 \to K_S^0\pi^+\pi^-\gamma$ decays to extract the direct and mixing-induced *CP* asymmetry parameters, $\mathcal{C}_{K_S^0\rho\gamma}$ and $\mathcal{S}_{K_S^0\rho\gamma}$, in the $B^0 \to K_S^0\rho^0\gamma$ mode. However, due to the large natural width of the $\rho^0(770)$, a non negligible amount of $B^0 \to K^{*\pm}(K_S^0\pi^\pm)\pi^\mp\gamma$ events, which are not *CP* eigenstates and do not contribute to $\mathcal{S}_{K_S^0\rho\gamma}$, are expected to lie under the $\rho^0(770)$ resonance and dilute $\mathcal{S}_{K_S^0\rho\gamma}$. We can define a dilution factor $\mathcal{D}_{K_S^0\rho\gamma}$ such as

$$\mathcal{D}_{K_S^0\rho\gamma} = \frac{\mathcal{S}_{K_S^0\pi^+\pi^-\gamma}}{\mathcal{S}_{K_S^0\rho\gamma}}, \quad (1)$$

where $\mathcal{S}_{K_S^0\pi^+\pi^-\gamma}$ is the effective value of the mixing-induced *CP* asymmetry measured for the whole $B^0 \to K_S^0\pi^+\pi^-\gamma$. This sample is expected to produce a small number of signal events, so that a proper amplitude analysis to discriminate $B^0 \to K^{*\pm}(K_S^0\pi^\pm)\pi^\mp\gamma$ from $B^0 \to K_S^0\rho^0(\pi^\mp\pi^\pm)\gamma$ is not feasible. Hence the dilution factor needs to be obtained in some other way. We decided to perform the amplitude analysis in the charged decay channel $B^+ \to K^+\pi^-\pi^+\gamma$, which has a richer statistics, and is related to $B^0 \to K_S^0\pi^+\pi^-\gamma$ by isospin symmetry. Assuming that the resonant amplitudes are the same in both modes, the dilution factor is calculated from those of $B^+ \to K^+\pi^-\pi^+\gamma$. Moreover, the amplitude analysis of

[1] Charge conjugation is implicit throughout the document.

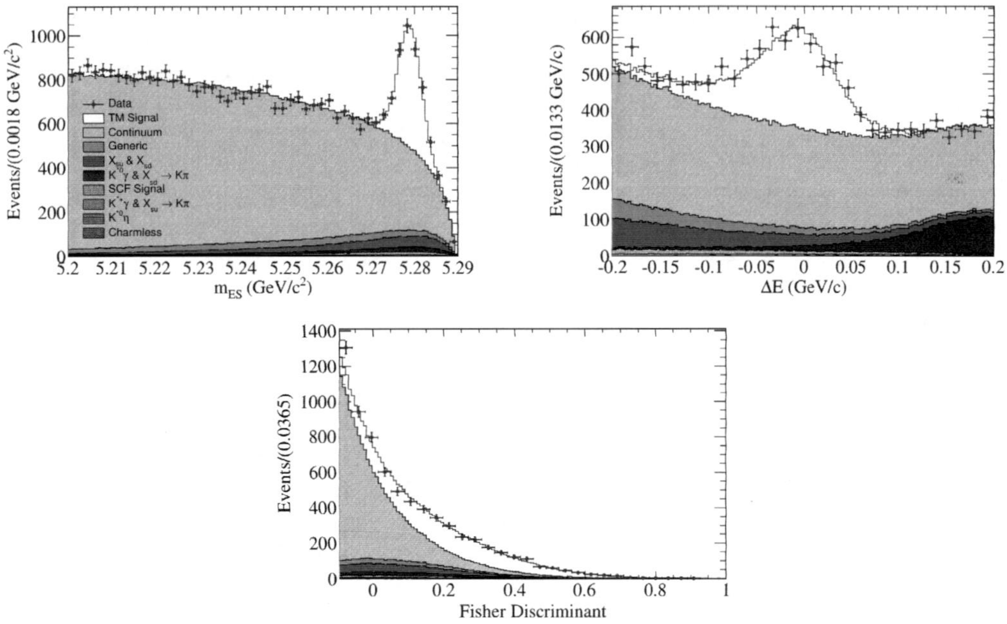

Figure 1. Distributions of $m_{\rm ES}$ (top left), ΔE (top right) and the Fisher-discriminant output (bottom) showing the fit results on the $B^+ \to K^+\pi^-\pi^+\gamma$ data sample. The distributions have their signal/background ratio enhanced by means of the following requirements: $-0.100 < \Delta E < 0.075$ GeV ($m_{\rm ES}$), $m_{\rm ES} > 5.27\,{\rm GeV}/c^2$ (ΔE) and $m_{\rm ES} > 5.27\,{\rm GeV}/c^2$; $-0.100 < \Delta E < 0.075$ GeV (Fisher). Points with error bars give the data. The solid histogram shows the projection of the fit result, and the shaded areas represent the contributions from different backgrounds, as described in the legend. Some of the contributions are hardly visible due to their small fractions.

$B^+ \to K^+\pi^-\pi^+\gamma$ provides also the branching fractions of $B \to K_{\rm res}\gamma$ (where $K_{\rm res}$ designates a kaonic resonance decaying to $K\pi\pi$), which are in general not well known.

We use the full dataset collected at the $\Upsilon(4S)$ resonance with the *BABAR* detector [8] at the PEP-II asymmetric-energy e^+e^- collider at the SLAC National Accelerator Laboratory. This corresponds to a total integrated luminosity of $424\,\mathrm{fb}^{-1}$, and to 471×10^6 $B\overline{B}$ pairs produced.

3. The charged channel $B^+ \to K^+\pi^-\pi^+\gamma$

To select the charged mode candidates, we require the center-of-mass energy of the photon to be between 1.5 and 3.5 GeV. We also require $5.200 < m_{\rm ES} < 5.292\,\mathrm{GeV}/c^2$ and $|\Delta E| < 200\,\mathrm{MeV}$. ΔE is defined as the difference between the expected and reconstructed B meson energy, $\Delta E \equiv E_B^* - \sqrt{s}/2$, where E_B^* is the reconstructed energy of the B in the e^+e^- CM frame, while $m_{\rm ES} = \sqrt{(s/2 - \mathbf{p}_B\mathbf{p}_0)^2/E_0^2 - \mathbf{p}_B^2}$, where $(E_B, \mathbf{p}_B)$ and $(E_0, \mathbf{p}_0)$ are the four-vectors of the B-candidate and of the CM frame, respectively. Charged particle identication is performed to assign the pion and kaon hypotheses. We also reject events where the γ comes from a π^0 or a η.

We first perform an unbinned extended maximum-likelihood fit to data to extract the yield of $B^+ \to K^+\pi^-\pi^+\gamma$ signal events. The fit uses the variables ΔE, $m_{\rm ES}$, and the Fisher-discriminant output to discriminate signal events from background. The Fisher-discriminant combine six discriminating variables related to the shape of the event. Using information from the maximum likelihood fit, the $m_{K\pi\pi}$, $m_{K\pi}$ and $m_{\pi\pi}$ spectra in signal events (where m_X is the invariant mass of the X particle system) are extracted using the ${}_s\mathcal{P}lot$ technique [9]. The results are shown in Figure 1. A signal yield of $2441 \pm 91^{+34}_{-57}$ is observed.

We perform a fit to the signal ${}_s\mathcal{P}lot$ of $m_{K\pi\pi}$ in order to extract from data the branching fractions of kaonic resonances decaying to $K^+\pi^-\pi^+$. We model the $m_{K\pi\pi}$ distribution as a coherent sum of five Relativistic Breit-Wigner (RBW) line shapes, with constant widths. Since helicity angles are not explicitly taken into account, the interference occurs between resonances with same J^P only. The result of the fit is shown in Figure 2, and the branching fractions obtained are given in Table 1.

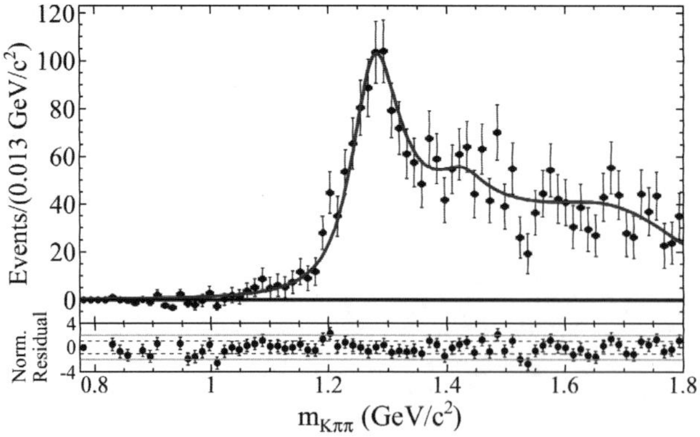

Figure 2. Distribution of $m_{K\pi\pi}$ for $B^+ \to K^+\pi^-\pi^+\gamma$ signal events (${}_s\mathcal{P}lot$), extracted from the maximum likelihood fit to $m_{\rm ES}$, ΔE and the Fisher discriminant. Points with error bars give the sum of ${}_s\mathcal{W}eights$. The blue solid curve is the result of the fit performed directly to this $m_{K\pi\pi}$ distribution to extract the different contributions from kaonic resonances decaying to $K^+\pi^-\pi^+$. Below each bin are shown the residuals, normalized in error units. The parallel dotted and full lines mark the 1σ and 2σ levels, respectively.

Table 1. Branching fractions of the different $K_{\rm res}$ decaying into $K^+\pi^-\pi^+$, extracted from the fit to the $m_{K\pi\pi}$ spectrum. The quoted numbers are obtained using the fit fraction of each component and the corresponding efficiency. To correct for the secondary branching fractions we use the values from PDG [10]. The upper limit reported are at 90% C.L.. The first uncertainty is statistical, the second is systematic, and the third, when present, is due to the uncertainties on the secondary branching fractions.

Mode	$\mathcal{B}(B^+ \to \text{Mode}) \times \mathcal{B}(K_{\rm res} \to K^+\pi^-\pi^+) \times 10^6$	$\mathcal{B}(B^+ \to \text{Mode}) \times 10^6$	PDG values $(\times 10^6)$
Inclusive $B^+ \to K^+\pi^-\pi^+\gamma$	...	$27.2 \pm 1.0^{+1.1}_{-1.3}$	27.6 ± 2.2
$K_1(1270)^+\gamma$	$14.5^{+2.0+1.1}_{-1.3-1.2}$	$44.0^{+6.0+3.5}_{-4.0-3.7} \pm 4.6$	43 ± 13
$K_1(1400)^+\gamma$	$4.1^{+1.9+1.3}_{-1.2-0.8}$	$9.7^{+4.6+3.1}_{-2.9-1.8} \pm 0.6$	< 15
$K^*(1410)^+\gamma$	$9.7^{+2.1+2.4}_{-1.9-0.7}$	$23.8^{+5.2+5.9}_{-4.6-1.4} \pm 2.4$	$\varnothing$
$K_2^*(1430)^+\gamma$	$1.5^{+1.2+0.9}_{-1.0-1.4}$	$10.4^{+8.7+6.3}_{-7.0-9.9} \pm 0.5$	$< 14 \pm 4$
$K^*(1680)^+\gamma$	$17.0^{+1.7+3.5}_{-1.4-3.0}$	$71.7^{+7.2+15}_{-5.7-13} \pm 5.8$	< 1900

We then extract the intermediate resonances decaying into $K^+\pi^-$ and $\pi^+\pi^-$. The limited statistics does not allow to perform a full two-dimensional amplitude analysis, so we perform a binned maximum-likelihood fit to the signal $_s\mathcal{P}lot$ of the projected $m_{K\pi}$ variable. The spectrum is modeled as the projection of two 1^- and one 0^+ components. The vector components are the $\rho^0(770)$ (described with a Gounaris-Sakurai lineshape), and the $K^{*0}(892)$ (described with a Relativistic Breit-Wigner). The scalar component is modeled by the LASS parametrization [11], which consists of the $K_0(1430)^0$ resonance together with an effective range non-resonant component. Both in the fit to the $m_{K\pi\pi}$ and the $m_{K\pi}$ spectra, we measure directly the magnitudes and phases of the different components of the corresponding signal model. The amplitudes of the different kaonic resonances extracted from the fit to $m_{K\pi\pi}$ are used to compute their fit fractions. These, in turn, are used to extract branching fractions of the different $K_{\rm res}$ contributions, that are used as inputs to the $m_{K\pi}$ fit model. The result of the fit is given in Figure 3, and gives the amplitudes of the three $K^+\pi^-$ and $\pi^+\pi^-$ intermediate states. In order to optimize the sensitivity on $\mathcal{S}_{K^0_S\rho\gamma}$, to calculate the dilution factor, we apply the additional cuts $600 < m_{\pi\pi} < 900\,\mathrm{MeV}/c^2$, in order to enhance the proportion of the $\rho^0(770)$, and we apply a veto on the $K^{*0}(892)$ requiring $m_{K\pi} < 845 \wedge m_{K\pi} > 945\,\mathrm{MeV}/c^2$. We obtain the dilution factor:

$$\mathcal{D}_{K^0_S\rho\gamma} = 0.549^{+0.096}_{-0.094}, \tag{2}$$

where the quoted uncertainties are sums in quadrature of statistical and systematic uncertainties. The branching fractions obtained are reported in Table 2.

4. The neutral channel $B^0 \to K^0_S\pi^+\pi^-\gamma$

In the neutral decay mode, the event are selected with the same criteria as in the charged mode analysis. The K^+ is replaced with a $K^0_S \to \pi^+\pi^-$ candidate, which is required to have a mass within $11\,\mathrm{MeV}/c^2$ from the nominal value [10], and a lifetime significance $> 5\sigma$, to ensure the decay vertices of the B^0 and K^0_S to be well separated. In addition, we require the cosine of the angle between the K^0_S flight direction and the vector connecting the B-daughter pions and the K^0_S vertex to be > 0.995, to suppress combinatiorial background. The time difference Δt is obtained from the measured distance between the decay vertices positions of the fully reconstructed B^0 candidate ($B^0_{\rm rec}$) and of the other one ($B^0_{\rm tag}$). The distance is transformed to

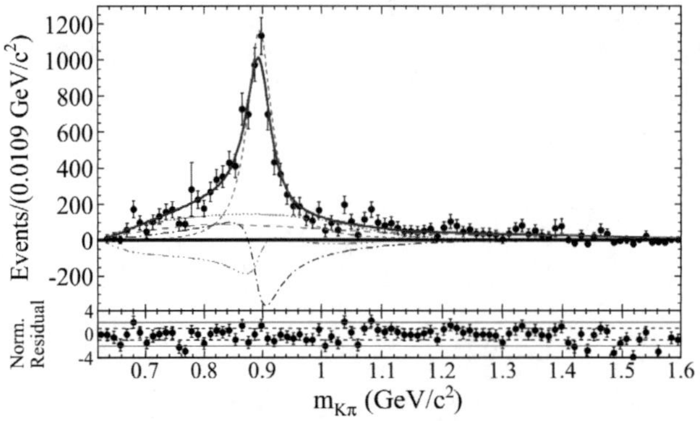

Figure 3. Distribution of $m_{K\pi}$ for $B^+ \to K^+\pi^-\pi^+\gamma$ signal events (${}_s\mathcal{P}lot$), extracted from the maximum likelihood fit to $m_{\rm ES}$, ΔE, and the Fisher discriminant. Points with error bars give the sum of ${}_s\mathcal{W}eights$. The blue solid curve is the result of the fit performed directly to this $m_{K\pi}$ distribution to extract the different contributions from $K\pi$ and $\pi\pi$ intermediate states. The small-dashed red, medium-dashed green and dotted magenta curves correspond to the $K^{*0}(892)$, $\rho^0(770)$ and $(K\pi)$ S-wave contributions, respectively. The dashed-dotted gray curve corresponds to the interferences between the two P-wave components, *i.e.* the $K^{*0}(892)$ and the $\rho^0(770)$, while the dashed-triple-dotted light blue curve corresponds to the interferences between the $(K\pi)$ S-wave and the $\rho^0(770)$.

Table 2. Branching fractions of the different resonances decaying to $K\pi$ and $\pi\pi$, extracted from the fit to the $K^+\pi^-$ spectrum. The quoted numbers are obtained using the fit fraction of each component and the corresponding efficiency. R denotes an intermediate resonant state and h stands for the final state K, π.. To correct for the secondary branching fractions we used the values from PDG [10] and $\mathcal{B}\left(K^{*0}(892) \to K^+\pi^-\right) = \frac{2}{3}$. The first uncertainty is statistical, the second is systematic, and the third, when applicable, is due to the uncertainties on the secondary branching fractions. The last two rows of the table are obtained by separating the contributions from the resonant and the non-resonant part of the LASS parametrization.

Mode	$\mathcal{B}(B^+ \to \text{Mode}) \times \mathcal{B}(R \to hh) \times 10^6$	$\mathcal{B}(B^+ \to \text{Mode}) \times 10^6$	PDG values ($\times 10^6$)
Inclusive $B^+ \to K^+\pi^-\pi^+\gamma$	...	$27.2 \pm 1.0^{+1.1}_{-1.3}$	27.6 ± 2.2
$K^{*0}(892)\pi^+\gamma$	$17.3 \pm 0.9^{+1.2}_{-1.1}$	$26.0^{+1.4}_{-1.3} \pm 1.8$	20^{+7}_{-6}
$K^+\rho^0(770)\gamma$	$9.1^{+0.8}_{-0.7} \pm 1.3$	$9.2^{+0.8}_{-0.7} \pm 1.3 \pm 0.02$	< 20
$(K\pi)^{*0}_0\pi^+\gamma$	$11.3 \pm 1.5^{+2.0}_{-2.6}$	...	$\varnothing$
$(K\pi)^0_0\pi^+\gamma$ (NR)	...	$10.8^{+1.4+1.9}_{-1.5-2.5}$	< 9.2
$K^*_0(1430)^0\pi^+\gamma$	$0.51 \pm 0.07^{+0.09}_{-0.12}$	$0.82 \pm 0.11^{+0.15}_{-0.19} \pm 0.08$	$\varnothing$

Δt using the known boost $\beta\gamma = 0.56$ of the e^+e^- system. The candidates are rejected when $|\Delta t| > 20\,$ps, or if the error on Δt is higher than 2.5 ps. To determine the flavor of $B^0_{\rm tag}$, we use the B flavor-tagging algorithm described in Ref. [12].

We perform an unbinned extended maximum-likelihood fit to data to extract the yield of $B^0 \to K^0_S\pi^+\pi^-\gamma$ signal events. We consider also the additional selection cuts on $m_{\pi\pi}$ and $m_{K\pi}$ considered when computed $\mathcal{D}_{K^0_S\rho\gamma}$. The fit uses the variables ΔE, $m_{\rm ES}$, and the Fisher-discriminant. The including of time-dependence in the fit allows the determination of mixing-induced *CP* violation and provides additional continuum background rejection. The signal time-dependent decay rate is given by

$$\mathcal{P}(\Delta t, \sigma_{\Delta t}) = \frac{e^{-|\Delta t|/\tau_{B^0}}}{4\tau_{B^0}} \times \left[1 + q_{\rm tag}\frac{\Delta D^c}{2} - q_{\rm tag}\langle D\rangle^c \mathcal{C}\cos(\Delta m_d \Delta t) + q_{\rm tag}\langle D\rangle^c \mathcal{S}\sin(\Delta m_d \Delta t)\right] \otimes \mathcal{R}^c(\Delta t, \sigma_{\Delta t}), \quad (3)$$

where $q_{\rm tag}$ is the flavor-tag of the event: $q_{\rm tag} = -1$ for $B_{\rm rec} = B^0$, and $+1$ for $B_{\rm rec} = \overline{B}^0$, τ_{B^0} is the B^0 lifetime and Δm_d is the $B^0\overline{B}^0$ oscillation frequency, fixed to the measurement from Ref. [12]. The parameters $\langle D\rangle^c$ and ΔD^c are related to the mistag probability, and to the performance difference for the two tags, respectively. The resolution function $\mathcal{R}$, takes into account the imperfect measurement of Δt. The coefficients $\mathcal{S}$ and $\mathcal{C}$ are the parameters associated with mixing-induced *CP* violation and direct *CP* violation, respectively.

The maximum-likelihood fit to data yields $245 \pm 24^{+13}_{-16}$ signal events, which results in an inclusive branching fraction of

$$\mathcal{B}(B^0 \to K^0_S\pi^+\pi^-\gamma) = (23.9 \pm 2.4^{+1.6}_{-1.9}) \times 10^{-6}. \quad (4)$$

In both cases, the first quoted uncertainty is statistical and the second is systematic. Figure 4 shows signal-enhanced distributions of the four discriminating variables in the fit. The result of the fit to data for the time-dependent *CP* violation parameters in signal events is

$$\mathcal{S}_{K^0_S\pi^+\pi^-\gamma} = +0.14 \pm 0.25^{+0.04}_{-0.03}, \quad (5)$$

$$\mathcal{C}_{K^0_S\pi^+\pi^-\gamma} = -0.39 \pm 0.20 \pm 0.05. \quad (6)$$

We finally obtain $\mathcal{S}_{K^0_S\rho\gamma}$ dividing $\mathcal{S}_{K^0_S\pi^+\pi^-\gamma}$ by the $\mathcal{D}_{K^0_S\rho\gamma}$ given in Eq. (2):

$$\mathcal{S}_{K^0_S\rho\gamma} = 0.249 \pm 0.455^{+0.076}_{-0.060}, \quad (7)$$

which is compatible with the SM expectation of ~ 0.03 [1].

5. Systematic uncertaintiess

In the analysis of $B^+ \to K^+\pi^-\pi^+\gamma$ decays, we considered different sources of systematic uncertainties affecting the amplitudes in the $m_{K\pi\pi}$ and $m_{K\pi}$ spectra. These have been propagated to the dilution factor $\mathcal{D}_{K^0_S\rho\gamma}$ and to the branching fractions calculated from the amplitudes. As for the $K_{\rm res}$ amplitudes, we account for effects from the fixed parameters in the maximum-likelihood fit, the fixed line-shape parameters of the kaonic resonances in the fit model, the effects of adding other resonances at high masses to the fit model, the procedure of the $m_{K\pi\pi}$ ${}_s\mathcal{P}lot$ extraction, as well as the binning in the fitted distribution. The dominant systematic uncertainty originates from the fixed line-shape parameters of the resonances. The amplitudes extracted from the $m_{K\pi}$ spectrum are affected mostly by the weights of the kaonic resonances obtained in the $m_{K\pi\pi}$ fit. Specific sources of systematic uncertainties are also taken into account for the extraction of branching fractions, like the uncertainties on the number of $B\overline{B}$ pairs recorded in the full *BABAR* dataset, the input branching fractions, and other detector-related systematics. Similarly, for the neutral $B^0 \to K^0_S\pi^+\pi^-\gamma$ channel, systematics due to the fixed parameters in the maximum likelihood fit have been considered.

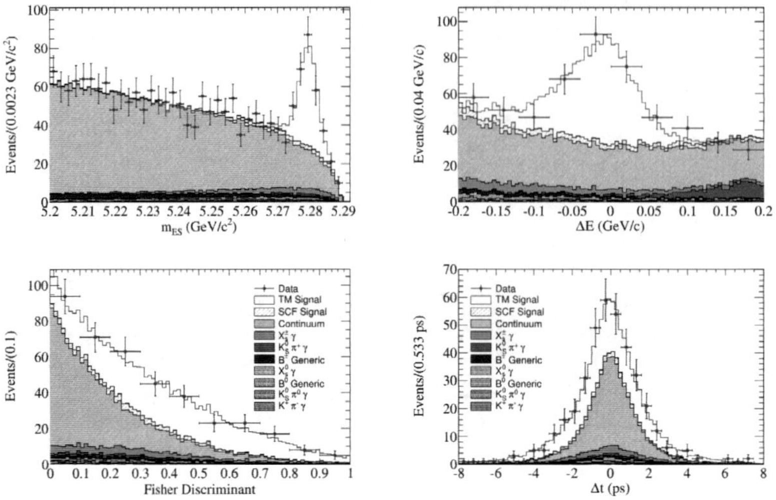

Figure 4. Distributions of $m_{\rm ES}$ (top left), ΔE (top right), the Fisher-discriminant output (bottom left), and Δt (bottom right), showing the fit results on the $B^0 \to K^0_S\pi^+\pi^-\gamma$ data sample. The distributions have their signal/background ratio enhanced by means of the following requirements: $-0.15 \leq \Delta E \leq 0.10\,\rm GeV$ ($m_{\rm ES}$), $m_{\rm ES} > 5.27\,{\rm GeV}/c^2$ (ΔE) and $m_{\rm ES} > 5.27\,{\rm GeV}/c^2$; $-0.15 \leq \Delta E \leq 0.10\,\rm GeV$ (Fisher and Δt).

6. Conclusions

The *BABAR* Collaboration, seven years after the shutdown of the experiment, is still producing competitive results. We have presented a measurement of the time-dependent *CP* asymmetry in the radiative-penguin decay $B^0 \to K^0_S\rho^0\gamma$, using the full *BABAR* dataset recorded at the $\Upsilon(4S)$ peak. We obtain the *CP*-violating parameters $\mathcal{S}_{K^0_S\pi^+\pi^-\gamma} = +0.14 \pm 0.25^{+0.04}_{-0.03}$, and $\mathcal{C}_{K^0_S\pi^+\pi^-\gamma} = -0.39 \pm 0.20 \pm 0.05$, where the first quoted errors are statistical and the second are systematic. We extract from this measurement the time-dependent *CP* asymmetry related to the hadronic *CP* eigenstate $\rho^0 K^0_S$ and obtain $\mathcal{S}_{K^0_S\rho\gamma} = 0.25 \pm 0.46^{+0.08}_{-0.06}$. This observables is related to the photon polarization of the underlying $b \to s\gamma$ transition. The dilution factor has been extracted from the higher statistics charged mode $B^+ \to K^+\pi^-\pi^+\gamma$, assuming isospin symmetry, via the study of the resonant structures. Moreover, the branching fractions of the different $K_{\rm res} \to K\pi\pi$ states and the overall branching fractions of the ρ^0K^+, $K^{*0}\pi^+$, and $(K\pi)_{S\text{-wave}}\pi^+$ are also measured.

References

[1] Atwood D, Gershon T, Hazumi M and Soni A 2005 *Phys.Rev.* **D71** 076003 (*Preprint* hep-ph/0410036)
[2] Li J *et al.* (Belle Collaboration) 2008 *Phys.Rev.Lett.* **101** 251601 (*Preprint* 0806.1980)
[3] Aubert B *et al.* (*BABAR* Collaboration) 2008 *Phys.Rev.* **D78** 071102 (*Preprint* 0807.3103)
[4] Ushiroda Y *et al.* (Belle Collaboration) 2006 *Phys.Rev.* **D74** 111104 (*Preprint* hep-ex/0608017)
[5] Aaij R *et al.* (LHCb Collaboration) 2014 *Phys.Rev.Lett.* **112** 161801 (*Preprint* 1402.6852)
[6] Aubert B *et al.* (*BABAR* Collaboration) 2007 *Phys.Rev.Lett.* **98** 211804 (*Preprint* hep-ex/0507031)
[7] Yang H *et al.* (Belle Collaboration) 2005 *Phys.Rev.Lett.* **94** 111802 (*Preprint* hep-ex/0412039)
[8] Aubert B *et al.* (*BABAR* Collaboration) 2002 *Nucl.Instrum.Meth.* **A479** 1–116 (*Preprint* hep-ex/0105044)
[9] Pivk M and Le Diberder F R 2005 *Nucl.Instrum.Meth.* **A555** 356–369 (*Preprint* physics/0402083)
[10] Beringer J *et al.* (Particle Data Group) 2012 *Phys.Rev.* **D86** 010001
[11] Aston D, Awaji N, Bienz T, Bird F, D'Amore J *et al.* 1988 *Nucl.Phys.* **B296** 493
[12] Aubert B *et al.* (*BABAR* Collaboration) 2007 *Phys.Rev.Lett.* **99** 171803 (*Preprint* hep-ex/0703021)

CP violation in B decays at Belle

L Santelj (for the Belle collaboration)

High Energy Accelerator Research Organization (KEK), Tsukuba, Japan

E-mail: luka.santelj@kek.jp

Abstract. Using the full data sample collected with the Belle detector at the KEKB asymmetric-energy e^+e^- collider, we present three recent measurements of time-dependent CP violation in B decays that are sensitive to potential new physics contributions and a measurement of branching fraction of the $B^0 \to \pi^0\pi^0$ decay. We studied $B \to \omega K$ decays and measured the values of CP violation parameters in $B^0 \to \omega K^0_S$ to be $A_{\omega K^0_S} = 0.36 \pm 0.19 \pm 0.05$ and $S_{\omega K^0_S} = +0.91 \pm 0.32 \pm 0.05$, which gives the first evidence of CP violation in this decay. From the measurement of CP violation parameters in the $B^0 \to \eta' K^0$ decay we obtain $S_{\eta' K^0} = 0.68 \pm 0.07 \pm 0.03$ and $A_{\eta' K^0} = +0.03 \pm 0.05 \pm 0.04$, which are the world's most precise values to date. We present the Belle first measurement of CP violation parameters in $B^0 \to K^0_S\eta\gamma$ decay, where we measure $S_{K^0_S\eta\gamma} = -1.32 \pm 0.77 \pm 0.36$ and $A_{K^0_S\eta\gamma} = -0.48 \pm 0.41 \pm 0.07$ (preliminary). Finally we present our preliminary measurement of the $B^0 \to \pi^0\pi^0$ branching fraction. We measure $\mathcal{B}(B^0 \to \pi^0\pi^0) = (0.89 \pm 0.12 \pm 0.10) \times 10^{-6}$, with significantly improved sensitivity compared to previously available results.

1. Introduction

Measuring the parameters of the unitarity triangle (UT) provides a major test of the Standard Model (SM), in particular of the Cabibbo-Kobayashi-Maskawa (CKM) description of flavor changing currents and CP violation. Angles of the UT related to $B_{u,d}$ decays can be determined by measuring CP asymmetries in various B meson decays, and this was the main motivation for construction of two so-called B factory experiments, Belle and BaBar. In the previous decade both experiments have confirmed the complex phase of the CKM matrix as the main source of CP violation.

In these proceedings we present two recent measurements related to the ϕ_1 angle of the UT, mainly motivated by their sensitivity to possible New Physics contributions ($B^0 \to \eta' K^0$ and $B \to \omega K$), a measurement sensitive to right-handed currents contributions in $b \to s\gamma$ transition ($B^0 \to K^0_S\eta\gamma$), and a measurement of branching fraction of $B^0 \to \pi^0\pi^0$, which is related to determination of the ϕ_2 angle of the UT. All measurements are based on the data sample containing 772 millions $B\bar{B}$ pairs collected by the Belle experiment [1], during its full data taking period (1999-2010).

The angles ϕ_1 and ϕ_2 can be determined by measuring time-dependent asymmetry in decays of B^0 and $\bar{B}^0$ mesons into a common CP eigenstate f_{CP} [2]. At the Belle experiment pairs of B mesons are produced in asymmetric energy collisions of electrons and positrons, through the $e^+e^- \to \Upsilon(4S) \to B\bar{B}$ process. In the measurements of time-dependent asymmetry we demand one of the B mesons (B_{CP}) to decay into a CP eigenstate, and the other one ($B_{\rm tag}$) to decay into a flavor specific final state (i.e. state into which only B^0 or only $\bar{B}^0$ can decay). Since a B

meson pair is in a quantum coherent state, a decay of $B_{\rm tag}$ into a flavor specific final state $f_{\rm tag}$ at $t_{\rm tag}$, determines the flavor of B_{CP} at $t_{\rm tag}$. In this case the CP asymmetry is given by[1]

$$a_{CP}(\Delta t) = \frac{\Gamma(B^0(\Delta t) \to f_{CP}) - \Gamma(\bar{B}^0(\Delta t) \to f_{CP})}{\Gamma(B^0(\Delta t) \to f_{CP}) + \Gamma(\bar{B}^0(\Delta t) \to f_{CP})} = \mathcal{A}_f \cos \Delta M \Delta t + \mathcal{S}_f \sin \Delta M \Delta t, \quad (1)$$

where Δt is the time difference between decays of $B_{\rm tag}$ and B_{CP}, ΔM is the mass difference between the two B^0 mass eigenstates (B_L and B_H), and $\mathcal{A}_f$ and $\mathcal{S}_f$ are the so-called CP violation parameters, which can be within the SM related with the UT angles.

2. The Belle detector, flavor tagging and vertexing

In this section we briefly describe the Belle detector and two basic procedures that are common to all measurements of time-dependent asymmetries.

2.1. The Belle detector

The Belle detector operated at the KEKB asymmetric-energy e^+e^- (3.5 GeV on 8.0 GeV) collider [3]. At KEKB, $B\bar{B}$ pairs are produced with a Lorentz boost of $\beta\gamma = 0.425$ nearly along the $+z$ direction, which is opposite the positron beam direction. The Belle detector is a large-solid-angle magnetic spectrometer consisting of a silicon vertex detector (SVD), a 50-layer central drift chamber (CDC), an array of aerogel threshold Cherenkov counters (ACC), a barrel-like arrangement of time-of-flight scintillation counters (TOF), and an electromagnetic calorimeter (ECL) with CsI(Tl) crystals located inside a superconducting solenoid coil that provides a 1.5 T magnetic field. An iron flux-return yoke (KLM) located outside of the coil is instrumented to detect K_L^0 mesons and to identify muons. The detector is described in detail elsewhere [1, 4]. The data sample used in presented measurements was collected with two different inner detector configurations. The first 152×10^6 of $B\bar{B}$ pairs were collected with a 2.0-cm-radius beampipe and a three-layer silicon vertex detector (SVD1), while the remaining 620×10^6 $B\bar{B}$ pairs were collected with a 1.5-cm-radius beampipe, a four-layer silicon vertex detector (SVD2), and an additional small-cell inner drift chamber.

2.2. Flavor tagging

The b-flavor of the $B_{\rm tag}$ meson is determined from inclusive properties of the particles in the event that are not associated with the B_{CP} reconstruction. The algorithm used is described in detail in Ref. [5]. It is based on the correlation between the charge of decay products and the flavor of decaying B meson. The most relevant are high momentum lepton tracks from $B^0 \to Xl^+\nu$ (or $\bar{B}^0 \to Xl^-\nu$) and K^+ (K^-) tracks from $\bar{b} \to \bar{c} \to \bar{s}$ ($b \to c \to s$) cascade decays. The data are separated into seven bins according to the flavor tagging quality information returned by the algorithm. The fraction of wrongly tagged candidates in each bin is determined from self-tagged semileptonic and hadronic $b \to c$ decays. The presence of wrongly tagged candidates dilutes the expected decay time asymmetry (1) to $a_{CP}^{\rm wt}(\Delta t) = (1 - 2w)a_{CP}(\Delta t)$, where w is the fraction of wrongly tagged candidates. In all measurements presented here the fit to Δt distribution (to extract the CP violation parameters) is done simultaneously to candidates in all flavor tagging quality bins, taking into account different values of w, and also signal-to-background fractions used in the fit are determined for each bin separately (as these fractions differ, this increases sensitivity).

[1] Here $B^0(t)$ ($\bar{B}^0(t)$) denote states that were at $t = 0$ pure B^0 ($\bar{B}^0$) states, but later get mixed due to $B^0 - \bar{B}^0$ mixing.

2.3. Vertexing

Since the B^0 and $\bar{B}^0$ mesons are approximately at rest in the $\Upsilon(4S)$ cms, their decay time difference (Δt) can be inferred from the displacement between z position of the B_{CP} and $B_{\rm tag}$ decay vertices (z_{CP} and $z_{\rm tag}$, respectively):

$$\Delta t \simeq \frac{z_{\rm CP} - z_{\rm tag}}{\beta\gamma c}. \tag{2}$$

In order to reconstruct the decay vertex positions, we use the algorithm that is in more detail described in [6]. In the measurements presented here the vertex of the B_{CP} meson is reconstructed by using the associated charged pion tracks. Pion tracks from K^0_S decay are not used, except in the measurement of $B^0 \to K^0_S\eta\gamma$. The z coordinate resolution for the B_{CP} meson vertex ranges from 70 to 100 μm and the reconstruction efficiency is about 95%, both depending on the final state. To reconstruct the decay vertex of the $B_{\rm tag}$ meson, the tracks not associated with B_{CP} are used. The z-coordinate resolution for these vertices is about 120 μm, and the reconstruction efficiency is $\sim$ 93%.

3. Measurement of CP violation parameters in $B^0 \to \eta' K^0$ decay

The $B^0 \to \eta' K^0$ decays are sensitive to the $\phi_1 = \arg(-V_{cd}V^*_{cb})/(V_{td}V^*_{tb})$ interior angle of the UT. The decay proceeds dominantly by the $b \to s\bar{q}q$ penguin diagram, and within the SM we expect $\mathcal{A}_{\eta' K^0} = 0$ and $\mathcal{S}_{\eta' K^0} = -\xi_f \sin 2\phi$ (where $\xi_f = -1$ (+1) for $B \to \eta' K^0_S$ ($B \to \eta' K^0_L$)), neglecting other contributing CKM-suppressed amplitudes with a different weak phase. However, the contribution of these CKM-suppressed amplitudes may not be negligible, resulting in a non-zero $\mathcal{A}_{\eta' K^0}$ and in a deviation of $\mathcal{S}_{\eta' K^0}$ from $\sin 2\phi_1$. Several theoretical methods were used to estimate the possible size of the deviation $\Delta\mathcal{S}_{\eta' K^0} = \mathcal{S}_{\eta' K^0} - \sin 2\phi_1$ within the SM. For example, the $SU(3)_F$ approach limits $\Delta\mathcal{S}_{\eta' K^0}$ to the range [0.05, 0.09] [7], while QCD factorization constrains it to [0.03, 0.03] [8]. Observing values of $\Delta\mathcal{S}_{\eta' K^0}$ significantly larger than these predictions would be a sign of new physics contributions in the loops of penguin diagram [9].

The measurement presented here was published and in much grater detail described in [10]. The measurement of CP violation parameters is done in two steps. After event reconstruction we first determine the fraction of signal events in the reconstructed sample. In the second step we use the obtained signal probability in the fit to the measured Δt distributions of B^0 and $\bar{B}^0$ tagged events to extract the values of CP violation parameters.

To reconstruct the decay we use $K^0_S \to \pi^+\pi^-, \pi^0\pi^0$ decay modes and also K^0_L candidates reconstructed from KLM clusters, and the following decay modes of η': $\eta' \to \rho^0(\to \pi^+\pi^-)\ \gamma$, $\eta' \to \pi^+\pi^-\eta(\to \gamma\gamma,\ \pi^+\pi^-\pi^0)$. Decay vertices of reconstructed B candidates are obtained from the vertex fit of associated charged pion tracks. Several selection criteria on the masses of intermediate states are applied in order to reduce the number of background candidates which mainly originate from the so-called continuum events, where in e^+e^- collision a pair of lighter quarks (u, d, s, c) is produced. After the event selection we obtain the fraction of signal events by studying the distribution of events in beam-energy-constrained mass $M_{\rm bc} = \sqrt{(E^*_{\rm beam})^2 - (p^*_B)^2}$, energy difference $\Delta E = E^*_B - E^*_{\rm beam}$ and event topology variable $R_{s/b}$ for K^0_S events, and the distribution in p^*_B and $R_{s/b}$ for K^0_L events[2]. Here $E^*_{\rm beam}$ is the beam energy in the center-of-mass system and E^*_B and p^*_B are the measured cms energy and momentum, respectively, of the reconstructed B candidate. The event topology variable $R_{s/b}$ is used in all measurements presented here. It serves to distinguish $B\bar{B}$ from continuum events and is based on the fact that $B\bar{B}$ events have spherical topology, while continuum events tend

[2] Since the energy of the K^0_L cannot be measured, M_{bc} and ΔE cannot be calculated as for the $B^0 \to \eta' K^0_S$ decays.

to be jet-like[3]. In Figure 1 the comparison of the data distribution and the fitted PDF in M_{bc} and ΔE are shown, for the $B^0 \to \eta' K_S^0$ sample. Altogether we reconstruct 2503 ± 63 $B^0 \to \eta' K_S^0$ signal events, and 1041 ± 41 $B^0 \to \eta' K_L^0$ signal events, where the uncertainties are statistical only. Following this, we perform an unbinned maximum likelihood fit to extract the values of CP violation parameters from the measured Δt, q distribution of all reconstructed events ($K_S^0 + K_L^0$). Fit results are

$$\begin{aligned} S_{\eta' K^0} &= +0.68 \pm 0.07(stat) \pm 0.03(syst), \\ A_{\eta' K^0} &= +0.03 \pm 0.05(stat) \pm 0.04(syst), \end{aligned} \tag{3}$$

where the first uncertainty is statistical and the second is systematic. The main contribution to the latter comes from the uncertainties in the Δt resolution function parameters for $\mathcal{S}_{\eta' K^0}$ and from the tag-side interference effect for $\mathcal{A}_{\eta' K^0}$. The comparison of data distribution and the fitted PDF is shown in Figure 1. The measured values of $\mathcal{S}_{\eta' K^0}$ and $\mathcal{A}_{\eta' K^0}$ are the world's most precise values of CP violation parameters in this particular decay, as well as among all $b \to s\bar{q}q$ transition dominated decays. In Figure 2 a comparison of our result with the BaBar measurement and the SM predicted value is shown, and consistency with both can be seen.

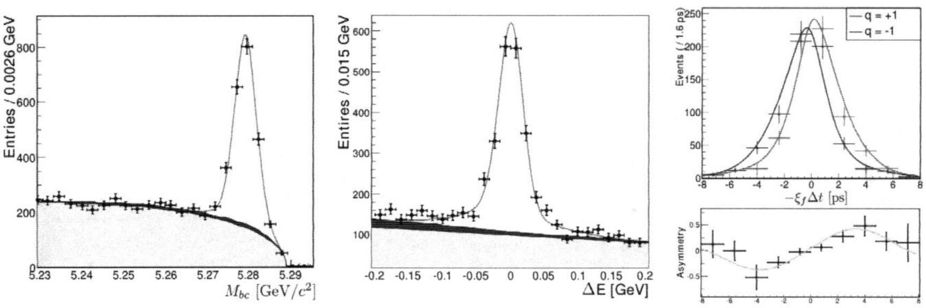

Figure 1. Left two: Distribution of reconstructed $B^0 \to \eta' K_S^0$ events in M_{bc} and ΔE (black points) along with the fitted PDF (the red). The yellow area shows contribution of the continuum background, and the blue area of the background from $B\bar{B}$ events. **Right:** Distribution of the reconstructed events in Δt (events with B^0 (red points) and $\bar{B}^0$ (blue points)) along with the corresponding parts of the fitted PDF; bottom plot shows the asymmetry in data distributions and in the fitted PDF.

4. Measurement of branching fractions and CP violation parameters in $B \to \omega K$ decays

The $B^0 \to \omega K_S^0$ decay also proceeds dominantly by the $b \to s\bar{q}q$ penguin diagram, and within the SM we expect $\mathcal{A}_{\omega K_S^0} \simeq 0$ and $\mathcal{S}_{\omega K_S^0} \simeq \sin 2\phi_1$, neglecting other contributing CKM-suppressed amplitudes. Possible deviations from these values within the SM were studied using several theoretical methods, indicating the expected value of $\mathcal{S}_{\omega K_S^0}$ slightly higher than $\sin 2\phi_1$ [12]. However, current experimental measurements indicate the opposite [13, 14], which might be a consequence of a contribution of new heavy particles in the loop of the penguin diagram [9].

[3] We combine a set of variables that characterize the event topology into a signal (background) likelihood variable $\mathcal{L}_{\rm sig}$ ($\mathcal{L}_{\rm bkg}$), and form the likelihood ratio $\mathcal{R}_{\rm s/b} = \mathcal{L}_{\rm sig}/(\mathcal{L}_{\rm sig} + \mathcal{L}_{\rm bkg})$. The likelihood $\mathcal{L}_{\rm sig\ (bkg)}$ includes a Fisher discriminant $\mathcal{F}$ constructed from the transverse sphericity $S_\perp$ [6], the angle between the thrust axis of the B candidate and the other particles in the cms, and a set of modified Fox-Wolfram moments [11].

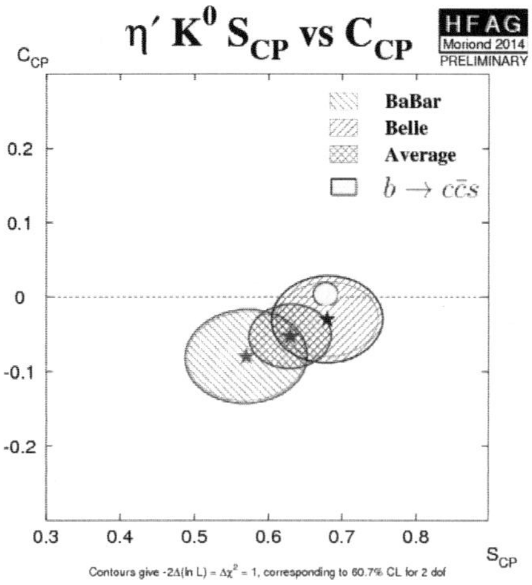

Figure 2. Comparison of our result (blue area) with the Babar result (green area) and the SM model prediction (yellow area) in the $\mathcal{S}_{CP} - (\mathcal{C}_{CP} \equiv -\mathcal{A}_{CP})$ plane.

Here we only give a short summary of the measurement method and results, while a more detailed description is given in [15]. In this measurement we measured two branching fractions (of $B^0 \to \omega K^0_S$ and $B^+ \to \omega K^+$), time-dependent CP violation parameters in $B^0 \to \omega K^0_S$, and the direct CP violation parameter A_{CP} in the charged $B^+ \to \omega K^+$ decay, defined as

$$A_{CP} = \frac{\Gamma(B^- \to \omega K^-) - \Gamma(B^+ \to \omega K^+)}{\Gamma(B^- \to \omega K^-) + \Gamma(B^+ \to \omega K^+)}, \tag{4}$$

where again a deviation from the expected asymmetry could be an indication of New Physics. Furthermore, the measurement of the branching fractions provides an important test of the QCD factorization (QCDF) and perturbative QCD (pQCD) approaches.

To obtain the two branching fractions and CP violation parameters we perform a seven-dimensional unbinned extended maximum likelihood fit to M_{bc}, ΔE, $R_{s/b}$, $m_{3\pi}$ (invariant mass of the reconstructed ω), $H_{3\pi}$ (helicity angle), Δt and q (where $q = +1$ ($q = -1$) for $B_{tag} = B^0$ ($\bar{B}^0$)). The fit is performed simultaneously to $B^0 \to \omega K^0_S$ and $B^+ \to \omega K^+$ data samples, sharing common calibration factors. Following this, the model shape is fixed and the $A_{CP}(B^+ \to \omega K^+)$ parameter is obtained from two further fits to extract the number of B^+ and B^- events. The fit results are

$$\begin{aligned} B(B^0 \to \omega K^0) &= (4.5 \pm 0.4(stat) \pm 0.3(syst)) \times 10^{-6}, \\ B(B^+ \to \omega K^+) &= (6.8 \pm 0.4(stat) \pm 0.4(syst)) \times 10^{-6}, \\ \mathcal{A}_{\omega K^0_S} &= -0.36 \pm 0.19(stat) \pm 0.05(syst), \\ \mathcal{S}_{\omega K^0_S} &= +0.91 \pm 0.32(stat) \pm 0.05(syst), \\ A_{CP}(B^+ \to \omega K^+) &= -0.03 \pm 0.04(stat) \pm 0.01(syst), \end{aligned} \tag{5}$$

where the first uncertainty is statistical and the second is systematic. The latter is dominated by uncertainties of the Δt resolution function parameters for $\mathcal{A}_{\omega K^0_S}$ and $\mathcal{S}_{K^0_S}$, and by parameters of the background PDF shape for the branching fractions. The comparison of data distributions and the fitted PDF is shown in Figure 3. The results given in (5) are the world's most precise measurements of the branching fractions and CP violation parameters in $B \to \omega K$ decays. The observed values of $\mathcal{A}_{\omega K_S^0}$ and $\mathcal{S}_{\omega K_S^0}$ differ from zero with a significance of 3.1 standard deviations, which gives the first evidence of CP violation in the $B^0 \to \omega K_S^0$ decay.

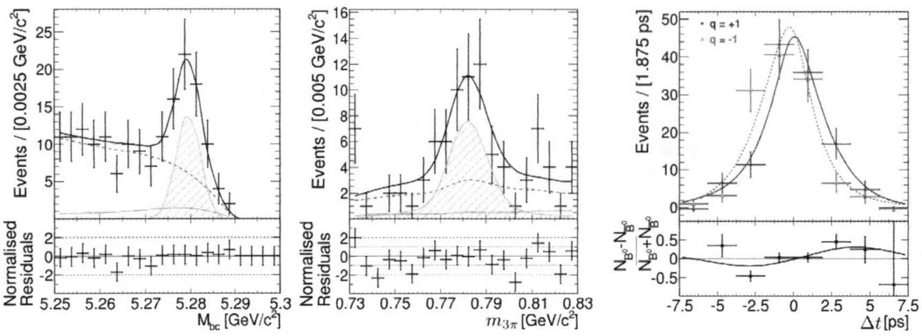

Figure 3. Left two: Distribution of reconstructed events in M_{bc} and $m_{3\pi}$ (black points) along with the fitted PDF (the full line). The dashed line shows contribution of the $q\bar{q}$ background (from $e^+e^- \to q\bar{q}$ ($q = u, d, s, c$) events), the dotted line of the $B\bar{B}$ background. **Right:** Distributions of reconstructed events in Δt (events with $\bar{B}^0$ (blue points) and B^0 (red points)) along with the corresponding parts of the fitted PDF; bottom plot shows the asymmetry in data distributions and in the fitted PDF.

5. Time-dependent CP violation in $B^0 \to K_S^0 \eta\gamma$ decay

The $B^0 \to K_S^0 \eta\gamma$ decay proceeds through the $b \to s\gamma$ transition at the quark level. In the SM the photon emitted in this process is predominantly left-handed. Due to the V-A structure of charged current in the loop of $b \to s\gamma$ penguin diagram, the emission of right-handed photon is suppressed by a factor of m_s/m_b compared to the left-handed photon. However, in many beyond SM theories, for instance Left-Right symmetric models, the emission of right-handed photon can be strongly enhanced [16, 17]. Measuring time-dependent CP violation in $b \to s\gamma$ dominated decays provides a method to detect presence of RH photon component in the final state and therefore examine these beyond the SM theories.

As mentioned in the introduction the time dependent CP violation is induced by the interference of B^0 and $\bar{B}^0$ decays into a common final state (CP eigenstate). In the case of $B^0 \to K_S^0 \eta\gamma$ however, B^0 decays predominantly to $K_S^0 \eta\gamma_R$ and $\bar{B}^0$ decays to $K_S^0 \eta\gamma_L$, which are different final states due to different photon polarizations. As $\bar{B}^0 \to K_S^0 \eta\gamma_R$ is suppressed within the SM, so is the mixing induced CP violation [18]:

$$\mathcal{S}_{K_S^0 \eta\gamma} \simeq 2\frac{m_s}{m_b} \sin 2\phi_1 \ . \tag{6}$$

In the case of enhanced right-handed photon emission in $b \to s\gamma$, as predicted by some NP theories, the enhanced $\bar{B}^0 \to K_S^0 \eta\gamma_R$ rate potentially induces large $\mathcal{S}_{K_S^0 \eta\gamma}$.

In the measurement we reconstruct signal events using the following decay modes: $K_S^0 \to \pi^+\pi^-, \eta \to \gamma\gamma$ and $\eta \to \pi^+\pi^-\pi^0$. Here also the main background contribution arises from continuum events. We first determine the fraction of signal events in the reconstructed sample, by performing a fit to the data distribution in $M_{bc}, \Delta E$ and NB. The latter is the output variable of neural network which mainly utilizes event shape variables. The comparison of the data distribution and the fit model is shown in Figure 4.

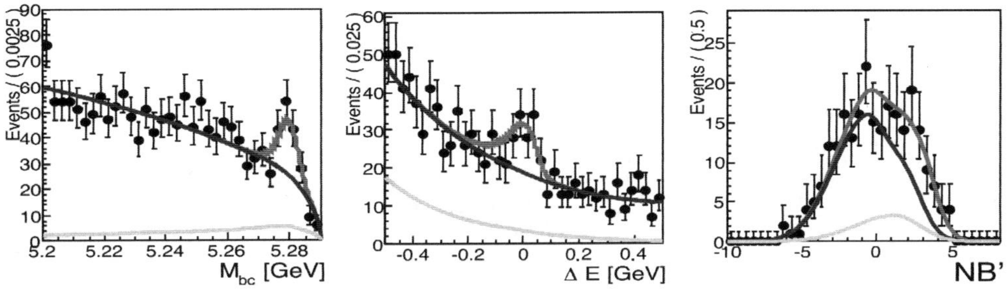

Figure 4. Distribution of reconstructed $B^0 \to K_S^0 \eta\gamma$ events in M_{bc}, ΔE and NB (black points) along with the fitted PDF (the red line). The blue line shows contribution from background events (continuum + $B\bar{B}$), and the green line shows contribution of $B\bar{B}$ background events. The amount and shape of the latter contribution is fixed to the values obtained from study of simulated events.

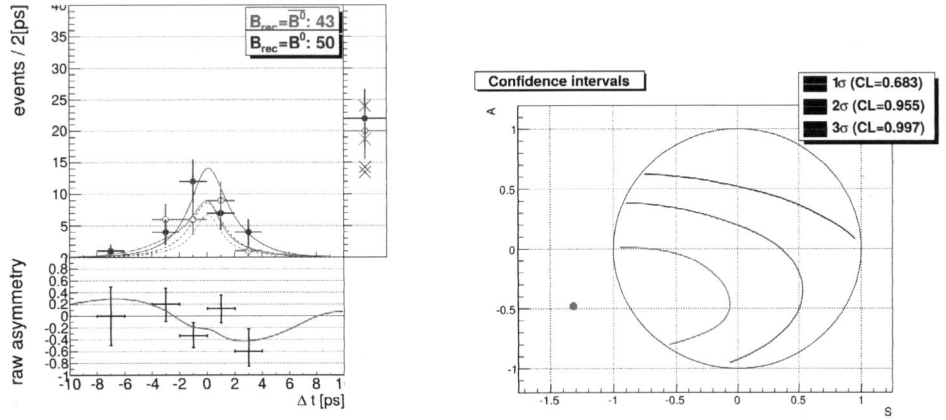

Figure 5. Left: Distribution of the reconstructed events in Δt (events with B^0 (red points) and $\bar{B}^0$ (blue points)) along with the corresponding parts of the fitted PDF; bottom plot shows the asymmetry in data distributions and in the fitted PDF. **Left:** Likelihood scan in the $\mathcal{S} - \mathcal{A}$ plane including systematic uncertainties. The black circle represents the physical boundary of CP violation. The red marker shows the fit result and the concentric curves represent the contours from 1 to 3 standard deviations from the fit result.

Following the signal-to-background fraction fit we perform an unbinned maximum likelihood fit to the Δ distributions of reconstructed candidates, tagged as B^0 and $\bar{B}^0$, to extract the values of CP violation parameters. Our preliminary results are

$$\mathcal{A}_{CP} = -0.48 \pm 0.41(stat) \pm 0.07(syst)$$
$$\mathcal{S}_{CP} = -1.32 \pm 0.77(stat) \pm 0.36(syst), \quad (7)$$

where the first error given is statistical and the second systematic. The main contribution to the systematic error comes from the uncertainties in the Δt resolution function and the reconstructed vertex quality selection criteria. In Figure 5 (left) the distribution of reconstructed candidates in Δt is shown along with the measured asymmetry. We use the Feldman-Cousins approach to determine the significance of the result. In Figure 5 (right) the result along with obtained confidence intervals is shown. The measured values are consistent with no CP violation.

6. Measurement of the branching fraction of $B \to \pi^0\pi^0$ decay

Measuring the branching fraction of $B \to \pi^0\pi^0$ decay is interesting for two main reasons. First, previous measurements indicate somewhat larger value than theoretically predicted by the perturbative QCD and the QCD factorization approaches. Values measured by the BaBar collaboration [20] and the latest published result by Belle collaboration (2005, [21]) are

$$\mathcal{B}(B \to \pi^0\pi^0) = (1.83 \pm 0.21 \pm 0.13) \times 10^{-6} \quad \text{(BaBar)}$$
$$\mathcal{B}(B \to \pi^0\pi^0) = (2.32 \pm 0.5 \pm 0.3) \times 10^{-6} \quad \text{(Belle)},$$

while the theory predictions range up to $< 1 \times 10^{-6}$ [22, 23]. Second motivation for this measurement is related to the measurement of the ϕ_2 angle of the UT. The time-dependent asymmetry in $B^0 \to \pi^+\pi^-$ is proportional to $\sin 2\phi_2$, assuming that only $b \to u$ tree level diagram contributes to the decay. However, the non-negligible $b \to d$ penguin amplitude potentially modifies this asymmetry by introducing a phase difference $\Delta\phi_{\pi\pi}$ in $\sin(2\phi_2 - 2\Delta\phi_{\pi\pi})$ and inducing a non-zero $\mathcal{A}_{\pi\pi}$. Therefore, to extract the value of ϕ_2 penguin amplitude contribution and its relative phase have to be determined. This can be done using isospin relations between several $B \to \pi\pi$ amplitudes, among which is also $B^0 \to \pi^0\pi^0$ [19].

In our measurement we reconstruct π^0 by combining two photon candidate clusters from the ECL. We impose selection criteria on invariant mass of $\gamma\gamma$ pair (115 MeV/c^2 $< M_{\gamma\gamma} <$ 152 MeV/c^2) and on the mass constrained vertex fit quality ($\chi^2 < 50$). The main background contribution arises from continuum events and an event shape variable (T_C) is used to suppress this contribution. In our previous analysis there was also a substantial contribution of background from $B\bar{B}$ events. It was later realized that this background arises from the pile-up of e^+e^- scattering events and $B\bar{B}$ events in the ECL, and we were able to remove 99% of it by applying a cut on the ECL trigger time. The remaining $B\bar{B}$ background comes from other rare charmless B decays (mainly $B^+ \to \rho^+\pi^0$).

We extract the signal yield by performing a fit to the reconstructed candidates distribution in the $M_{bc} - \Delta E - T_C$ space. The signal and $B\bar{B}$ backgrounds are modeled by fitting their distributions from large MC samples, and for continuum background we fit the distribution of candidates in the $M_{bc}, \Delta E$ sidebands of the measured data. The comparison of the fitted PDF projections and the data distributions in the $M_{bc}, \Delta E$, and T_C are shown in Figure 6. Our preliminary result for the branching fraction is

$$\mathcal{B}(B^0 \to \pi^0\pi^0) = (0.89 \pm 0.12 \pm 0.10) \times 10^{-6},$$

where the first uncertainty given is statistic and the second is systematic. The main contribution to the latter comes from the signal region definition, continuum background parametrization, π^0 detection efficiency, and uncertainties in the assumed branching fractions of the contributing rare B decays. The measured branching fraction is the most accurate to date, and significantly more consistent with theoretical predictions than previous measurements.

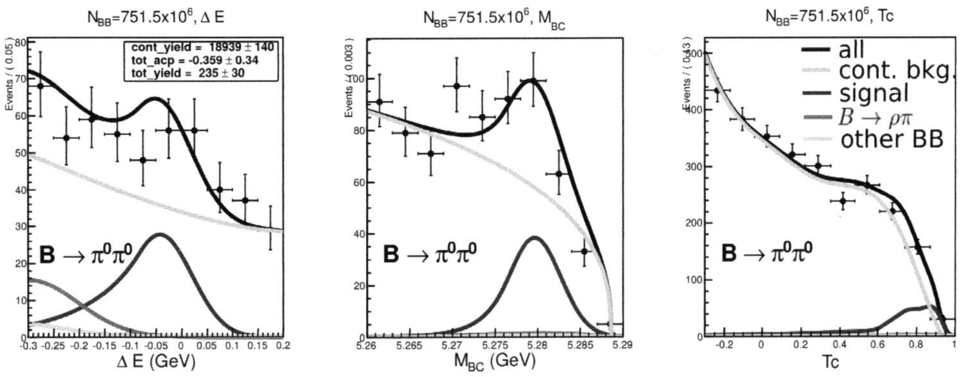

Figure 6. $\Delta E, M_{bc}$ and T_c distributions of $B^0 \to \pi^0\pi^0$ candidates (dots with errors) with the projections of fitted PDF superimposed (black line). Contributions from the signal, continuum background, $B \to \rho\pi$ and other $B\bar{B}$ backgrounds are shown in the blue, green, red and cyan, respectively.

References

[1] Abashian A *et al.* (Belle Collaboration) 2002 *Nucl. Instrum. Methods Phys. Res.* A **479** 117
[2] Bigi I I and Sanda A I 1981 *Nucl. Phys.* B **193** 85
[3] Kurokawa S and Kikutani E 2003 *Nucl. Instrum. Methods Phys. Res.* A **499** 1
[4] Brodzicka J *et al.* 2012 *Prog. Theor. Exp. Phys.* **2012** 04D001 *Preprint* arXiv:1212.5342
[5] Kakuno H *et al.* (Belle Collaboration) 2004 *Nucl. Instrum. Methods Phys. Res.* A **533** 516
[6] Chen K F *et al.* (Belle Collaboration) 2005 *Phys. Rev.* D **72** 012004, *Preprint* hep-ex/0504023
[7] Gronau M, Rosner J L and Zupan J 2006 *Phys. Rev.* D **74** (2006) 093003, *Preprint* hep-ph/0608085
[8] Beneke M 2005 *Phys. Lett.* B **620** 143 *Preprint* hep-ph/0505075
[9] Grossman Y and Worah M P 1997 *Phys. Lett.* B **395** 241
[10] Santelj L and Yusa Y *et al.* (Belle Collaboration) 2014 *J. High Energy Phys.* JHEP1410 (2014) 165 *Preprint* arXiv:1408.5991
[11] Abe K *et al.* (Belle Collaboration) 2001 *Phys. Rev. Lett.* **87** 101801
[12] Chua C -K 2008 *Preprint* arXiv:0807.3596
[13] Chao Y *et al.* (Belle Collaboration) 2007 *Phys. Rev.* D **76** 091103(R)
[14] Aubert B *et al.* (BaBar Collaboration) 2009 *Phys. Rev.* D **79** 052003
[15] Chobanova V *et al.* (Belle Collaboration) 2014 *Physical Review* D **90** 012002 *Preprint* arXiv:1311.6666
[16] Fujikawa K, Babu K S and Yamada A 1994 *Phys. Let.* B **333** 196
[17] Gabbiani F, Gabrielli E, Masiero A and Silvestrini L 1996 *Nucl.Phys.* B **477** 321
[18] Atwood D, Gronau M and Soni A 1997 *Phys.Rev.Lett.* **79** 185 *Preprint* hep-ph/9704272 (1997).
[19] Gronau M and London D 1990 *Phys. Rev. Lett.* **65** 3381
[20] Lees J P *et al.* (BaBar Collaboration) 2013 *Phys. Rev.* D **87** 052009 *Preprint* arXiv:1206.3525
[21] Abe K *et al.* (Belle Collaboration) 2005 *Phys. Rev. Lett.* **94** 181803 *Preprint* hep-ex/0408101
[22] Lu C D, Ukai K and Yang M Z 2001 *Phys. Rev.* D **63** 074009
[23] Beneke M and Yang D 2006 *Nucl. Phys.* B **736** 34

Exclusive $b \to s\gamma$ decays at Belle

D Dutta and B Bhuyan on behalf of the Belle Collaboration
Indian Institute of Technology Guwahati, Assam, India
E-mail: d.dutta@iitg.ac.in; bhuyan@iitg.ac.in

Abstract. We report results on two exclusive $b \to s\gamma$ decays, namely $B_s^0 \to \phi\gamma$ and $B_s^0 \to \gamma\gamma$ at Belle. For this study, we have used 121.4 fb^{-1} of $\Upsilon(5S)$ data collected with the Belle detector at the KEKB asymmetric-energy e^+e^- collider. Calculations within the Standard Model framework predict the $B_s^0 \to \phi\gamma$ branching fraction to be in the range $(3.9 - 4.3) \times 10^{-5}$ and the $B_s^0 \to \gamma\gamma$ branching fraction to be in the range $(0.18 - 2.45) \times 10^{-6}$. The $B_s^0 \to \phi\gamma$ branching fraction is measured as $(3.6 \pm 0.5 \pm 0.3 \pm 0.6) \times 10^{-5}$, where the first uncertainty is statistical, the second is systematic and the last one is due to the uncertainty in the fraction of $B_s^{(*)}\bar{B}_s^{(*)}$ in $b\bar{b}$ events. This result agrees with the theoretical predictions and other experimental results. We observe no statistically significant signal for $B_s^0 \to \gamma\gamma$ and have thus set a 90% confidence-level upper limit on its branching fraction at 3.1×10^{-6}. This improves on the previous result by a factor of about 3 and is the most stringent limit till date.

1. Introduction

$B_s^0 \to \phi\gamma$ and $B_s^0 \to \gamma\gamma$ are exclusive $b \to s\gamma$ and $b \to s\gamma\gamma$ decay processes that are sensitive to physics beyond the Standard Model (SM). The leading order Feynman diagrams for these processes are shown in Figure 1. Precise measurements of the branching fraction of these decays along with other $b \to s\gamma$ modes may help to put stringent limits on the permitted values of Wilson coefficients $\mathcal{C}_7$, which in turn may constrain various new-physics model parameters. The present constraints on $\mathcal{C}_7$ obtained using various decay modes are summarized in Refs. [1, 2].

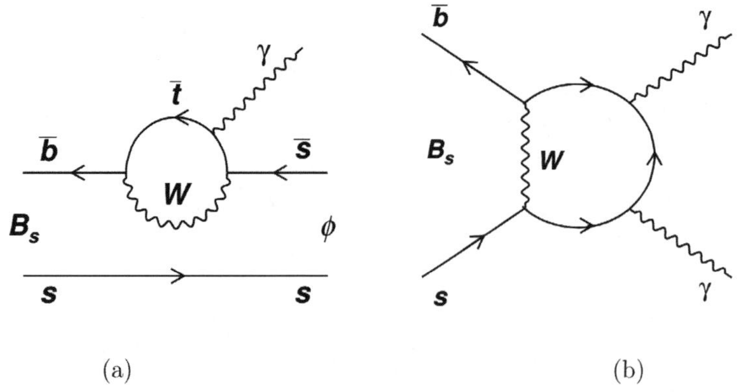

Figure 1. Leading order Feynman diagrams for the decays (a) $B_s^0 \to \phi\gamma$ and (b) $B_s^0 \to \gamma\gamma$.

Calculations of the $B_s^0 \to \phi\gamma$ branching fraction within the SM framework yields a value between (3.9 - 4.3) $\times 10^{-5}$ with about 30% uncertainty [3, 4]. This decay was first observed by the Belle experiment using 23.6 fb^{-1} of data collected at the $\Upsilon(5S)$ resonance with the branching fraction (BF) measured to be $(5.7^{+2.2}_{-1.9}) \times 10^{-5}$ [5]. Recently, the LHCb Collaboration has measured the ratio $\mathcal{B}(B^0 \to K^{*0}\gamma)/\mathcal{B}(B_s^0 \to \phi\gamma)$ and using the world-average value of $\mathcal{B}(B^0 \to K^{*0}\gamma)$ [6], the BF for $B_s^0 \to \phi\gamma$ is estimated to be (3.5 $\pm$ 0.4) $\times 10^{-5}$ [7]. These experimental results are in good agreement with the theoretical ones. The $B_s^0 \to \phi\gamma$ BF is further constrained by a good agreement between theory and experimental results on exclusive $b \to s\gamma$ modes [3, 4, 6, 8] and on inclusive $B^0 \to X_s\gamma$ rates [6, 10, 11, 12]. Small contributions from various new physics effects may still remain hidden within the large uncertainties of this mode [13, 14]. The decay $B_s^0 \to \gamma\gamma$ is yet to be observed and its upper limit at 90% confidence level (CL) is estimated to be 8.7 $\times 10^{-6}$ [5]. This result is an order of magnitude larger than the SM predictions, which lie in the range of $(0.18 - 2.45) \times 10^{-6}$ [15, 16, 17]. Results from $B^0 \to X_s\gamma$ decays constrain the $B_s^0 \to \gamma\gamma$ BF in the R-parity conserving scenario [13]. However, presence of processes mediated by λ-irreducible diagrams [18] may increase its BF by more than an order of magnitude in R-parity violating (RPV) SUSY scenarios [13].

2. The Belle Detector and the Dataset

The results reported here are based on 121.4 fb^{-1} of $\Upsilon(5S)$ data collected with the Belle detector [19, 20] at the KEKB [21] asymmetric-energy B-factory at KEK in Japan. The Belle detector consists of a 4-layer silicon vertex detector (SVD) primarily for an efficient vertex reconstruction, a central drift chamber (CDC) for measuring the momenta of charged particles, Aerogel Cherenkov counters (ACC) to distinguish between charged hadrons, mainly kaons and pions, plastic scintillation counters (TOF) for measuring the time of flight of a particle from the interaction point, and an electromagnetic calorimeter (ECL) for efficient detection of photons. All these detector components are located inside a superconducting solenoid coil that provides a 1.5 T magnetic field. An iron flux return located outside of the coil is instrumented to detect K_L^0 mesons and muons. The detector is described in details elsewhere [19, 20].

The Υ(5S) mass is substantially larger than the $B_s^0\bar{B}_s^0$ production threshold. So, several hadronic processes are possible at the Υ(5S) resonance. The $b\bar{b}$ production cross-section $[\sigma_{b\bar{b}}^{\Upsilon(5S)} = \sigma(e^+e^- \to b\bar{b})]$ at the Υ(5S) resonance is measured to be (0.340 $\pm$ 0.016) nb, by subtracting $e^+e^- \to q\bar{q}$ $(q = u, d, s, c)$ component obtained just above the Υ(4S) resonance [22]. The $b\bar{b}$ pairs can hadronize to produce three categories of events:

- events containing 2 strange B i.e., B_s mesons $(e^+e^- \to B_s^{(*)}\bar{B}_s^{(*)})$
- events containing 2 non-strange B mesons $(e^+e^- \to B^{(*)}\bar{B}^{(*)}(X))$
- events containing no $B_{(s)}$ mesons $(e^+e^- \to non - B_{(s)}\bar{B}_{(s)})$

Charge conjugate modes are implied throughout the proceedings. The fraction of $b\bar{b}$ events that hadronize to $B_s^{(*)}\bar{B}_s^{(*)}$ $\left(f_s = \dfrac{\sigma(e^+e^- \to B_s^{(*)}\bar{B}_s^{(*)})}{\sigma(e^+e^- \to b\bar{b})}\right)$ is determined by the following relations:

$$\frac{\mathcal{B}(\Upsilon(5S) \to D_sX)}{2} = f_s\mathcal{B}(B_s \to D_sX) + (1 - f_s)\mathcal{B}(B \to D_sX) \quad (1)$$

$$\frac{\mathcal{B}(\Upsilon(5S) \to D^0X)}{2} = f_s\mathcal{B}(B_s \to D^0X) + (1 - f_s)\mathcal{B}(B \to D^0X) \quad (2)$$

via a measurement of the inclusive rates $\mathcal{B}(\Upsilon(5S) \to D_s(D^0)X)$ and information on the inclusive BF of the $B_s(B) \to D_s(D^0)X$ decays [23]. The value of f_s is measured to be $(17.2 \pm 3.0)\%$ [22].

The $B_s^{(*)}\bar{B}_s^{(*)}$ pairs include $B_s^{*0}\bar{B}_s^{*0}$, $B_s^{*0}\bar{B}_s^0$ and $B_s^0\bar{B}_s^0$. Their relative abundances, measured using $B_s \to D_s\pi$ [22], are:

$$f_{B_s^{*0}B_s^{*0}} = \frac{\sigma(e^+e^- \to B_s^{*0}\bar{B}_s^{*0})}{\sigma(e^+e^- \to B_s^{(*)}\bar{B}_s^{(*)})} = (87.0 \pm 1.7)\%, \quad (3)$$

$$f_{B_s^{*0}\bar{B}_s^0} = \frac{\sigma(e^+e^- \to B_s^{*0}\bar{B}_s^0)}{\sigma(e^+e^- \to B_s^{(*)}\bar{B}_s^{(*)})} = (7.3 \pm 1.4)\% and \quad (4)$$

$$f_{B_s^0\bar{B}_s^0} = 1 - f_{B_s^{*0}\bar{B}_s^{*0}} - f_{B_s^{*0}\bar{B}_s^0} \quad (5)$$

The B_s can thus be produced directly from $B_s^0\bar{B}_s^0$, or from the excited states $B_s^{*0}\bar{B}_s^{*0}$ and $B_s^{*0}\bar{B}_s^0$, where the B_s^{*0} mesons decay to the ground state B_s^0 mesons through the emission of a photon.

3. Event Generation

MC simulated events for both signal and background processes are generated to optimize the signal selection criteria, estimate the signal reconstruction efficiency, study background sources and to parameterize the signal and background shapes. This is done in two steps: event generation using the EvtGen package [24] and modeling of the detector response using GEANT3 [25]. Beam related backgrounds are added to the simulated samples during detector simulation. Events are generated according to various physics processes using a decay table, which specifies the decay models, modes, BFs, etc for all possible particles involved in the decay chain. MC simulated events for the decay $B^0 \to K^{*0}\gamma$ are also generated for the calibration purposes.

4. Selection Criteria

The background for these decay modes are light quark-antiquark pairs ($u\bar{u}$, $d\bar{d}$, $s\bar{s}$ and $c\bar{c}$), generic B_s ($B_s^{(*)}\bar{B}_s^{(*)}$) and non-B_s ($B^*\bar{B}^*\pi, \Upsilon(4S)\gamma$) decays of $\Upsilon(5S)$ and QED processes (Bhabha scattering or $e^+e^- \to \gamma\gamma$ events). Two-photon processes resulting from the π^0 and η decays can also contribute to the backgrounds. Loose selection criteria based on track multiplicity, photon energies, numbers of clusters and average cluster energy are applied to filter out the non-B events. Charged tracks are required to originate from the interaction point (IP) with impact parameters $|dz| < 3$ cm and $dr < 0.5$ cm, where $|dz|$ and dr are the distances of closest approach to the IP along the z axis (collinear with the e^+ beam) and in the transverse r-ϕ plane, respectively. Charged kaons are identified with about 85% efficiency by requiring:

$$\mathcal{L}(K/\pi) \equiv \frac{\mathcal{L}_K}{\mathcal{L}_K + \mathcal{L}_\pi} > 0.6$$

where $\mathcal{L}_K$ and $\mathcal{L}_\pi$ are the likelihoods of the track being due to a kaon and a pion, respectively. Tracks failing this requirement are assumed to be due to pions. The likelihoods are calculated based on the dE/dx measurement by the CDC, the Cherenkov light yield in the ACC and the time of flight information from TOF by taking a product of the likelihood functions for three discriminants:

$$\mathcal{L}_i = \mathcal{L}_i^{\mathrm{dE/dx}}.\mathcal{L}_i^{\mathrm{ACC}}.\mathcal{L}_i^{\mathrm{TOF}}(i = K, \pi)$$

The ϕ meson candidates are reconstructed by combining a pair of oppositely charged kaons having an invariant mass $M(K^+K^-)$ within ± 12 MeV/c^2 ($\pm$ 2.5 σ) of the nominal ϕ mass. No events with multiple ϕ candidates are seen. Similarly, the K^{*0} candidates for the $B^0 \to K^{*0}\gamma$ control sample are reconstructed with oppositely charged kaon and pion candidates by requiring $|M_{K\pi} - m_{K^{*0}}| < 75$ MeV/c^2, where $M_{K\pi}$ and $m_{K^{*0}}$ are the invariant mass of the kaon-pion pair and the nominal K^{*0} mass, respectively. To remove the multiple K^{*0} candidates, K/π

vertex fit is performed and events with the least χ^2 value is chosen as the best K^{*0} candidate. Photons are reconstructed by identifying energy deposits in the ECL that are not matched to any charged track. Candidate photons are required to have a minimum energy (E_γ) of 100 MeV. Electromagnetic and hadronic showers have different shapes in both the transverse and longitudinal directions. Showers caused by hadrons grow wider than the electromagnetic ones due to multiple showers caused by hadrons. Thus, to reject neutral hadrons like K_L^0 mesons and neutrons and merged photons from π^0 we require the ratio of the energy deposited by a photon candidate in a (3×3) to that in (5×5) ECL crystal array (E_9/E_{25}) to be greater than 0.95. Daughter photons from π^0 and η decays are rejected on the basis of likelihood information [$P_{\pi^0}(\gamma)$ and $P_\eta(\gamma)$] obtained using energy and polar angle of the photons and the diphoton invariant mass, calculated by combining the candidate photon with any other photon in the event. Owing to the long decay time of the CsI crystals, the offtime QED events can leave two perfectly back-to-back clusters in the ECL and can very well mimic the $B_s^0 \to \gamma\gamma$ events. To remove these events, we require the timing characteristics of the clusters used for photon reconstruction to be consistent with the beam collision time, determined at the trigger level for the candidate event. As the timing information is not simulated, a selection criterion based on MC truth information is used to remove such background from the simulated samples with a matching efficiency. Figures 2 and 3 show the $M_{\rm bc}$ distributions before and after applying the MC truth information and timing criteria to background MC and real data, respectively, for the $B_s^0 \to \gamma\gamma$ analysis. For $B_s^0 \to \gamma\gamma$, the candidate photons are selected only from the barrel region ($33° < \theta < 128°$, θ being the photon polar angle), as there is a high level of beam background in the endcaps and it is very unlikely that one photon will be in the barrel and the other in the endcaps for a B_s^0 meson decaying at rest. The B_s^0 candidates are formed by combining a ϕ with a photon candidate for $B_s^0 \to \phi\gamma$ mode or by combining two photon candidates for the $B_s^0 \to \gamma\gamma$ mode. B_s^0 mesons are selected using the beam constrained mass ($M_{\rm bc}$) and energy difference (ΔE), defined as:

$$M_{\rm bc} = \sqrt{(E_{\rm beam}^{\rm CM})^2 - (p_{B_s}^{\rm CM})^2} \tag{6}$$

$$\Delta E = E_{B_s}^{\rm CM} - E_{\rm beam}^{\rm CM} \tag{7}$$

where, $E_{\rm beam}^{\rm CM}$ is the beam energy, $E_{B_s^0}^{\rm CM}$ and $p_{B_s^0}^{\rm CM}$ are the energy and momentum of the B_s^0 candidate, with all quantities evaluated at the e^+e^- center-of-mass (CM) frame. Signal candidates are required to satisfy $M_{\rm bc} > 5.3$ GeV/c^2 and $-0.4 < \Delta E < 0.1$ GeV for $B_s^0 \to \phi\gamma$

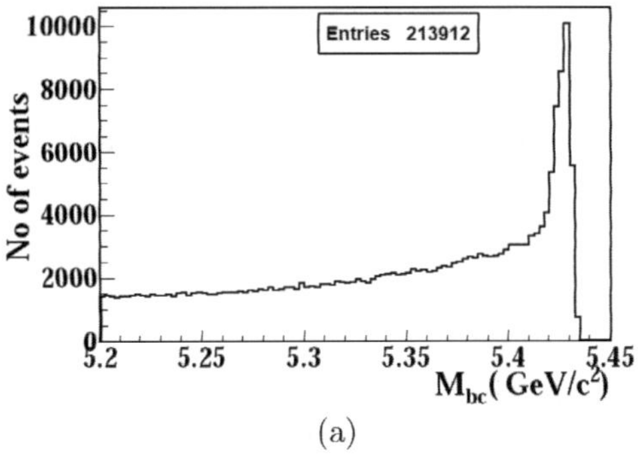

(a)

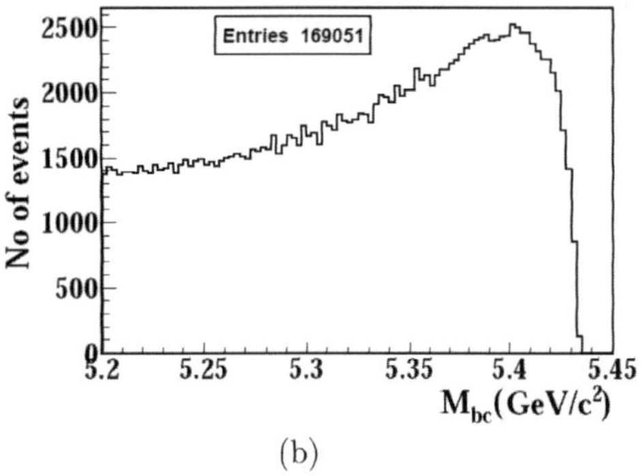

(b)

Figure 2. $M_{\rm bc}$ distributions for background MC (a) before and (b) after the photonID criteria for $B_s^0 \to \gamma\gamma$.

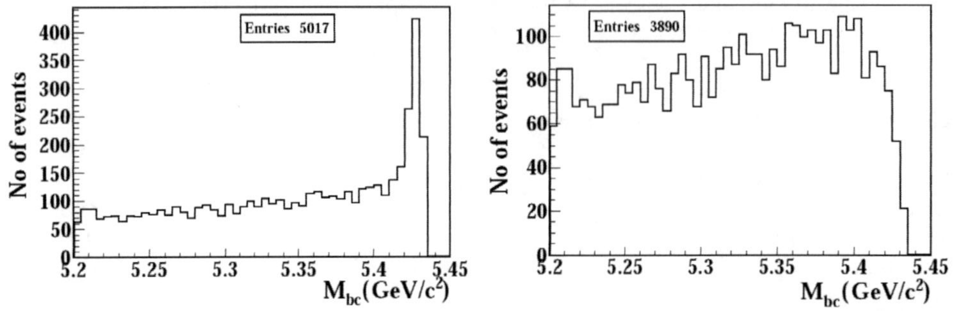

(a) (b)

Figure 3. $M_{\rm bc}$ distributions for 1.86 fb^{-1} real data (a) before and (b) after the timing criteria for $B_s^0 \to \gamma\gamma$.

and $\mathrm{M_{bc}} > 5.3$ GeV/c^2 and $-0.7 < \Delta\mathrm{E} < 0.2$ GeV for the $B_s^0 \to \gamma\gamma$ mode. No events with multiple B_s^0 candidates are seen in signal MC samples, and negligible cases ($< 1.0\%$) of multiple B_s^0 are observed for background MC and real data for both $B_s^0 \to \phi\gamma$ and $B_s^0 \to \gamma\gamma$ analyses. The K^{*0} candidates are combined with photon candidates to form B^0 candidates by applying $M_{\rm bc} > 5.0$ GeV/c^2 and $-0.4 < \Delta\mathrm{E} < 0.2$ GeV for the $B^0 \to K^{*0}\gamma$ control sample. Multiple B_s^0 and B^0 candidates coming from a particular event ($< 1.0\%$) are removed by retaining the one with the more energetic photon.

5. Continuum Suppression

The dominant background for both the decay modes are from $e^+e^- \to q\bar{q}$ continuum events where, $q = u,\ d,\ s$ or c. The mass of a $B_s^{(*)}\bar{B}_s^{(*)}$ pair is less than that of the $\Upsilon(5\mathrm{S})$ resonance by only a few MeVs. So the B_s^0 produced at the $\Upsilon(5\mathrm{S})$ resonance has a very low momentum and its decay products have a random direction, i.e., they have a spherical topology. Since for continuum events, a smaller proportion of e^+e^- energy goes towards particle production, these events have a higher momentum and hence are topologically quite distinct from $\Upsilon(5\mathrm{S}) \to B_s^{(*)}\bar{B}_s^{(*)}$ events. Their decay products are distributed mostly along the beam direction. To reduce these backgrounds, event shape variables such as the modified Fox-Wolfram moments [26] and $|\cos(\theta_{\rm thrust})|$ are used as inputs to a Neural Network (NN) [27] based classifier, where the thrust axis is defined as the direction which maximises the sum of the longitudinal momenta of the particles in the decay. The classifier uses the information of all the discriminating variables and provides a single quantity (NN output) which can then be used as a discriminant instead of using all input variables separately. The NN output ($\mathcal{C}_{\rm NB}$) tends to peak at 1 for signal-like events and at -1 for background-like events. This quantity is used as an additional dimension in the unbinned extended maximum likelihood fit to extract the signal yield for $B_s^0 \to \phi\gamma$ mode. As $\mathcal{C}_{\rm NB}$ peaks sharply at 1 and -1, it is very difficult to model it with a simple analytic function. Thus, to improve the modeling, a modified NeuroBayes output is calculated as:

$$\mathcal{C}'_{\rm NB} = \log\left(\frac{\mathcal{C}_{\rm NB} - \mathcal{C}_{\rm NB_{min}}}{\mathcal{C}_{\rm NB_{max}} - \mathcal{C}_{\rm NB}}\right), \qquad (8)$$

after rejecting events with $\mathcal{C}_{\rm NB} < \mathcal{C}_{\rm NB_{min}}$, where $\mathcal{C}_{\rm NB_{min}} = -0.6$ and $\mathcal{C}_{\rm NB_{max}} \sim 1$ are the lower and upper limits of $\mathcal{C}_{\rm NB}$ for the events used in the fit. For the $B_s^0 \to \gamma\gamma$ mode, an optimized criterion of $\mathcal{C}_{\rm NB} > 0.77$ is applied, since considerable correlations are observed between $\mathcal{C}_{\rm NB}$ with $M_{\rm bc}$ and ΔE distributions. The optimal criterion is taken to be the value for which the signal significance is maximum, where the signal significance defined as:

$$\text{Signal significance} = \frac{\text{S}}{\sqrt{\text{S}+\text{B}}} \tag{9}$$

after applying the cut at different $\mathcal{C}_{\text{NB}}$ values over the range -1 to $+1$. $\text{S} = w_s \times s$ and $\text{B} = w_b \times b$ are the expected signal and background events, w_s and w_b are the weight factors, determined using the signal and background expectations in data, s and b being the number of signal and background events remaining in the MC sample, after applying a particular selection criteria on $\mathcal{C}_{\text{NB}}$.

6. Extended Maximum Likelihood Fit

We perform a four-dimensional (two-dimensional) unbinned extended maximum likelihood fit involving M_{bc}, ΔE, $\cos\theta_{\text{hel}}$ and $\mathcal{C}'_{\text{NB}}$ (M_{bc} and ΔE) to extract the $B_s^0 \to \phi\gamma$ ($B_s^0 \to \gamma\gamma$) signal yields. The ϕ helicity angle (θ_{hel}) is the angle between B_s^0 and any of the ϕ daughters in the ϕ rest frame. The total fit probability distribution function (PDF) consists of two components: signal and $q\bar{q}$ background. The signal further consists of three components: signal coming from $B_s^{*0}\bar{B}_s^{*0}$, $B_s^{*0}\bar{B}_s^0$ and $B_s^0\bar{B}_s^0$ decays, their relative fractions being fixed to the values reported in Ref. [22]. We combine the backgrounds arising from B_s and non-B_s decays with the $q\bar{q}$ continuum as they have a small contribution and have no peaking structure in the signal region. For the $B_s^0 \to \phi\gamma$ mode, the fit regions chosen for different variables are listed below:

- M_{bc} $\longrightarrow$ (5.3 - 5.434) GeV/c^2
- ΔE $\longrightarrow$ (-0.4 - 0.1) GeV
- $\cos(\theta_{\text{hel}})$ $\longrightarrow$ (-1.0 - 1.0)
- $\mathcal{C}'_{\text{NB}}$ $\longrightarrow$ (-10.0 - 10.0)

For $B_s^0 \to \gamma\gamma$ analysis, the fit regions chosen for M_{bc} and ΔE are listed below:

- M_{bc} $\longrightarrow$ (5.3 - 5.434) GeV/c^2
- ΔE $\longrightarrow$ (-0.7 - 0.4) GeV

The signal and background PDFs are determined using MC samples. The PDF for each component is represented by the product of one-dimensional functions since the correlations among the variables are negligible. For the $B_s^0 \to \phi\gamma$ mode, the M_{bc}, ΔE, $\cos\theta_{\text{hel}}$ and $\mathcal{C}'_{\text{NB}}$ distributions of all three signal components are modeled with the sum of a Crystal Ball (CB) [28] and Gaussian function having a common mean, a CB function, a $\sin^2\theta_{\text{hel}}$ distribution and the sum of two Gaussian functions, respectively. The background PDFs are described by an ARGUS function [29] for M_{bc} having its endpoint fixed at 5.434 GeV/c^2, a first-order Chebychev polynomial for ΔE, a parabola for $\cos\theta_{\text{hel}}$ and a Gaussian function for $\mathcal{C}'_{\text{NB}}$. For the $B_s^0 \to \gamma\gamma$ mode, the M_{bc} distributions of all three signal components are parameterized with a combination of CB and Gaussian functions having a common mean and the ΔE distributions are modeled with CB functions. The background is described by an ARGUS function having its endpoint fixed for M_{bc} and a first-order Chebychev polynomial for ΔE. For both analyses, the signal parameters are determined from MC events except for the means and widths of the M_{bc} and ΔE distributions describing the $B_s^*\bar{B}_s^*$ component. For both the decay modes, the widths of M_{bc} and ΔE for the $B_s^{*0}\bar{B}_s^{*0}$ component are calibrated using corrections obtained from the $B^0 \to K^{*0}\gamma$ control sample. As a cross-check, the branching fraction for $B^0 \to K^{*0}\gamma$ is measured and found to be in good agreement with its world average [6]. The M_{bc} mean of the $B_s^{*0}\bar{B}_s^{*0}$ component is similarly adjusted using M_{bc} mean corrections taken from the $B_s \to D_s\pi$ analysis [22]. No calibration is done for the other two components as their contribution is small. The ΔE mean of the $B_s^{*0}\bar{B}_s^{*0}$ component is allowed to float in the $B_s^0 \to \phi\gamma$ analysis. For the $B_s^0 \to \gamma\gamma$ mode, we have fixed the ΔE mean to the signal MC value, as the correction to the ΔE mean obtained from the $B_s^0 \to \phi\gamma$ analysis is found to be within the statistical uncertainties. The uncertainty associated with this procedure is included as a systematic uncertainty for this mode. All the background parameters except the ARGUS endpoint is kept free. In total, we have nine free parameters for the $B_s^0 \to \phi\gamma$ fit: the ARGUS shape parameter, Chebychev polynomial coefficient, both parameters of $\cos(\theta_{\rm hel})$ background, mean and sigma of $\mathcal{C}'_{\rm NB}$ background, ΔE mean, signal and background yields. We have four free parameters for the $B_s^0 \to \gamma\gamma$ fit: the ARGUS shape parameter, Chebychev background parameter, signal and background yields.

7. Signal Significance, Branching Fraction and Upper Limit

In total, we have observed 91^{+14}_{-13} $B_s^0 \to \phi\gamma$ signal events coming from all 3 signal components with a significance of 10.7 σ including systematics. The signal significance is computed assuming a null or background only hypothesis as $\mathrm{sign}(N_{\rm signal})\sqrt{-2\ln(\mathcal{L}_0/\mathcal{L}_{\rm max})}$, where $N_{\rm signal}$ is the signal yield obtained from the fit, $\mathcal{L}_0$ is the likelihood value of the fit when the signal yield is fixed to 0 and $\mathcal{L}_{\rm max}$ is the maximum likelihood value of the likelihood fit. Additive systematic uncertainties are incorporated by convolving the statistical likelihood curve with a Gaussian function of width equal to the total additive systematics. The branching fraction for the decay $B_s^0 \to \phi\gamma$ is calculated as:

$$\mathcal{B}(B_s^0 \to \phi\gamma) = \frac{N(B_s^0 \to \phi\gamma)}{\mathcal{B}(\phi \to K^+K^-) \times \epsilon \times N(B_s^0)}, \text{ with } N(B_s^0) = 2 \times f_s \times \sigma_{b\bar{b}}^{\Upsilon(5S)} \times \mathcal{L}_{\rm int} \quad (10)$$

where, $N(B_s^0 \to \phi\gamma)$ is the signal yield of $B_s^0 \to \phi\gamma$; $\mathcal{B}(\phi \to K^+K^-)$ is the branching fraction of $\phi \to K^+K^-$, ϵ is the signal selection efficiency of $B_s^0 \to \phi\gamma$ signal, f_s is the fraction of $B_s^{(*)}B_s^{(*)}$ events in the $b\bar{b}$ sample, $\sigma_{b\bar{b}}^{\Upsilon(5S)}$ and $\mathcal{L}_{\rm int}$ are the $b\bar{b}$ production cross-section and integrated luminosity at the $\Upsilon(5S)$ energy. The $B_s^0 \to \phi\gamma$ BF is measured as $(3.6 \pm 0.5 \pm 0.3 \pm 0.6) \times 10^{-5}$, where the first uncertainty is statistical, the second is systematic and the third is due to the uncertainty in f_s. We observe no statistically significant signal for the decay $B_s^0 \to \gamma\gamma$ and have rather measured the single-event sensitivity for $B_s^0 \to \gamma\gamma$ to be 0.5×10^{-6}. We use a Bayesian approach and integrate the likelihood curve from 0 to 90% of the total integral under the curve to obtain the 90% CL upper limit of 3.1×10^{-6} on the $B_s^0 \to \gamma\gamma$ branching fraction. The fit results are listed in Table 1, while the fit projections are shown in Figs. 4 and 5. The fits are projected onto the $B_s^{*0}\bar{B}_s^{*0}$ signal region defined as $M_{\rm bc} > 5.4$ GeV/c^2, $-0.2 < \Delta E < 0.02$ GeV, $|\cos\theta_{\rm hel}| < 0.8$ and $0.0 < \mathcal{C}'_{\rm NB} < 10.0$ for $B_s^0 \to \phi\gamma$ and $M_{\rm bc} > 5.4$ GeV/c^2 and $-0.3 < \Delta E < 0.05$ GeV for $B_s^0 \to \gamma\gamma$.

8. Systematic Uncertainties

The systematic uncertainties are classified into two types: additive and multiplicative. The multiplicative uncertainties do not affect the signal yield or signal significance of the decay channel but affects its BF. For both the modes, these uncertainties arise due to the uncertainty in signal reconstruction efficiency and on the number of B_s mesons. The former one comprises

Table 1. Results of the $B_s^0 \to \phi\gamma$ and $B_s^0 \to \gamma\gamma$ analyses.

	$B_s^0 \to \phi\gamma$	$B_s^0 \to \gamma\gamma$
ϵ (%)	36.1 ± 0.1	14.0 ± 0.1
N	91^{+14}_{-13}	$-3.9^{+3.7}_{-2.6}$
$\mathcal{B}(10^{-6})$	$36 \pm 5(\text{stat.}) \pm 3(\text{syst.}) \pm 6(f_s)$	< 3.1 (90% CL)

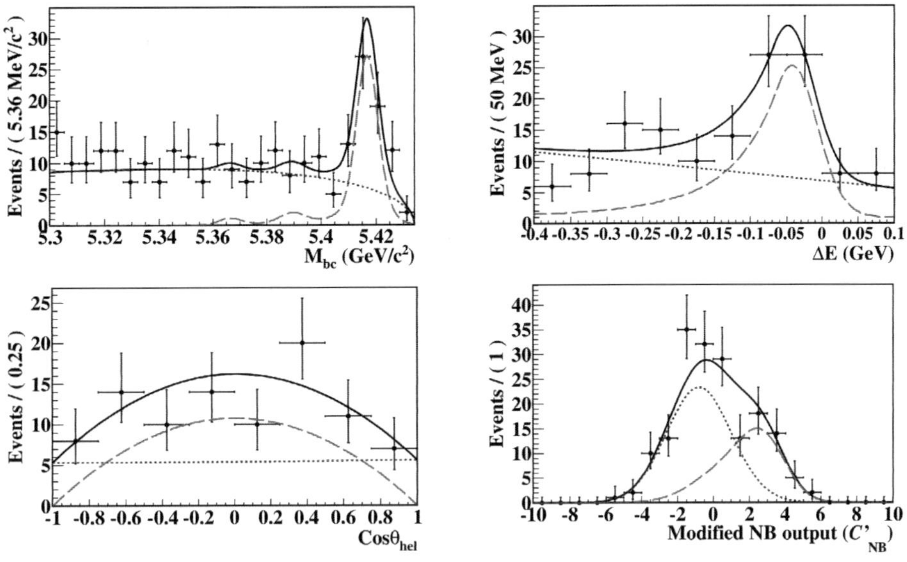

Figure 4. Data fits for the $B_s^0 \to \phi\gamma$ analysis. The projections are shown only for events inside the $B_s^{*0}\bar{B}_s^{*0}$ signal region except for the plotted variable. The points with error bars represent the data, the solid black curve represents the total fit function, the red dashed (blue dotted) curve represents the signal (continuum background) contribution.

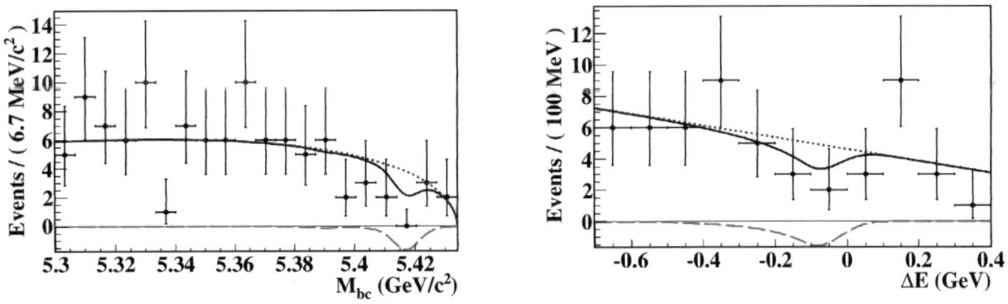

Figure 5. Data fits for the $B_s^0 \to \gamma\gamma$ analysis. The projections are shown only for events inside the $B_s^{*0}\bar{B}_s^{*0}$ signal region except for the plotted variable. The points with error bars represent the data, the solid black curve represents the total fit function, the red dashed (blue dotted) curve represents the signal (continuum background) contribution.

of the uncertainty associated with the photon reconstruction efficiency, kaon identification efficiency, tracking efficiency, the requirement on $\mathcal{C}_{\mathrm{NB}}$ that is estimated by comparing the efficiencies in data and MC simulations with the $B^0 \to K^{*0}\gamma$ control sample and limited MC statistics. The uncertainties due to kaon identification and tracking efficiency are 1.3% and 0.3%, respectively. These are obtained using control samples of $D^{*+} \to D^0\pi^+_{\rm slow} \to K^-\pi^+\pi^+_{\rm slow}$ and $D^{*+} \to D^0\pi^+_{\rm slow}, D^0 \to K^0_S\pi^+\pi^-, K^0_S \to \pi^+\pi^-$ decays, respectively. For $B_s^0 \to \phi\gamma$, the uncertainty due to the $\phi \to K^+K^-$ branching fraction also adds to the multiplicative systematics. This is taken from Ref. [9]. The uncertainty on the number of B_s mesons is the quadratic sum of the uncertainties in the integrated luminosity, $\sigma_{b\bar{b}}^{\Upsilon(5S)}$ and f_s. The additive

Journal of Physics: Conference Series **631** (2015) 012038 doi:10.1088/1742-6596/631/1/012038

uncertainties reduce the signal significance of the observed peak and changes the BF of the decay. That arise due to the fit procedure, which includes the uncertainty due to PDF parametrization and fit bias, respectively. The uncertainty due to PDF parametrization is estimated by varying each fixed parameter of the fit PDF by $\pm 1\sigma$ and estimating the variation in the signal yield and then adding them in quadrature. The calibration uncertainties are also included in the calculation. To estimate the uncertainty due to fit bias, pseudo-experiments are generated according to the PDFs obtained from the fits to data i.e., by fixing all parameters including the background shape parameters and signal and background yields. Events generated from the pseudo-experiments are then fitted to obtain the signal yield and residual distributions. The observed biases of -0.28 ± 0.08 and -0.10 ± 0.07 for $B_s^0 \to \phi\gamma$ and $B_s^0 \to \gamma\gamma$ are corrected and their uncertainties are assigned as systematic uncertainties. The systematic uncertainties are listed in Table 2.

Table 2. Summary of systematic uncertainties.

Additive systematic uncertainties (events)		
Source	$B_s^0 \to \phi\gamma$	$B_s^0 \to \gamma\gamma$
PDF parameterization	$^{+1.6}_{-1.7}$	± 0.4
Fit bias	± 0.1	± 0.1
Total (quadratic sum)	$^{+1.6}_{-1.7}$	± 0.4
Multiplicative systematic uncertainties (%)		
Source	$B_s^0 \to \phi\gamma$	$B_s^0 \to \gamma\gamma$
Photon reconstruction efficiency	2.2	2×2.2
Kaon identification efficiency	2.6	-
Tracking efficiency	0.7	-
$\mathcal{C}_{\rm NB}$ requirement	4.8	8.7
MC statistics	0.2	0.4
$\mathcal{B}(\phi \to K^+K^-)$	1.0	-
$\mathcal{L}_{\rm int}$	1.3	
$\sigma_{b\bar{b}}^{\Upsilon(5S)}$	4.7	
f_s	17.4	
Total (quadratic sum)	19.1	20.6

9. Results and Conclusions

To summarize, we have used the entire Belle $\Upsilon(5S)$ dataset in our study and have measured the $B_s^0 \to \phi\gamma$ branching fraction to be $\mathcal{B}(B_s^0 \to \phi\gamma) = [3.6 \pm 0.5(\text{stat.}) \pm 0.3(\text{syst.}) \pm 0.6(f_s)] \times 10^{-5}$. This measurement supersedes the earlier Belle result [5], and is consistent with theoretical predictions [3, 4] and a recent LHCb measurement [7]. For the $B_s^0 \to \gamma\gamma$ mode, in the absence of any statistically significant signal, we have set the 90% CL upper limit on its BF at 3.1×10^{-6}. This result improves on the previously published result by a factor of about three and is consistent with the expected sensitivity for our data sample. This also rules out large contributions from RPV SUSY in this channel. This decay will possibly be observed at upcoming flavor physics experiments like Belle II with a dedicated run at Υ(5S) resonance.

Acknowledgements
The authors acknowledge the funding received from Department of Science and Technology, Government of India to support this work.

References
[1] W. Altmannshofer and D. M. Straub, JHEP **8**, 1 (2012).
[2] W. Altmannshofer, P. Paradisi and D. M. Straub, JHEP **4**, 1 (2012).
[3] A. Ali, B. D. Pecjak, and C. Greub, Eur. Phys. J. C **55**, 577 (2008).
[4] P. Ball, G. W. Jones, and R. Zwicky, Phys. Rev. D **75**, 054004 (2007).
[5] J. Wicht *et al.* (Belle Collaboration), Phys. Rev. Lett. **100**, 121801 (2008).
[6] Y. Amhis *et al.* (Heavy Flavor Averaging Group), arXiv:1207.1158 (2012).
[7] R. Aaij *et al.* (LHCb Collaboration), Nucl. Phys. B **867**, 1 (2013).
[8] A. Ali and A. Y. Parkhomenko, Eur. Phys. J. C **23**, 89 (2002).
[9] K. A. Olive *et al.* (Particle Data Group), Chin. Phys. C**38**, 090001 (2014).
[10] M. S. Alam *et al.* (CLEO Collaboration), Phys. Rev. Lett. **74**, 2885 (1995).
[11] M. Misiak *et al.*, Phys. Rev. Lett. **98**, 022002 (2007).
[12] M. Misiak and M. Steinhauser, Nucl. Phys. B **764**, 62 (2007).
[13] A. Gemintern, S. Bar-Shalom, and G. Eilam, Phys. Rev. D **70**, 035008 (2004).
[14] P. Colangelo, F. De Fazio, R. Ferrandes, and T. Pham, Phys. Rev. D **77**, 055019 (2008).
[15] L. Reina, G. Ricciardi, and A. Soni, Phys. Rev. D **56**, 5805 (1997).
[16] S. W. Bosch and G. Buchalla, JHEP **2002**, 054 (2002).
[17] R. Mohanta and A. Giri, Phys. Rev. D **85**, 014008 (2012).
[18] A. Gemintern, S. Bar-Shalom, G. Eilam, and F. Krauss, Phys. Rev. D **67**, 115012 (2003).
[19] A. Abashian *et al.* (Belle Collaboration), Nucl. Instrum. Methods Phys. Res., Sect. A **479**, 117 (2002).
[20] J. Brodzicka *et al.* (Belle Collaboration), PTEP **04D001**, (2012).
[21] S. Kurokawa and E. Kikutani, Nucl. Instrum. Methods Phys. Res., Sect. A **499**, 1 (2003).
[22] S. Esen *et al.* (Belle Collaboration), Phys. Rev. D **87**, 031101 (R) (2013).
[23] A. Drutskoy *et al.* (Belle Collaboration), Phys. Rev. Lett. **98**, 052001 (2007).
[24] D. J. Lange, Nucl. Instrum. Methods Phys. Res., Sect. A **462**, 152 (2001).
[25] R. Brun *et al.*, GEANT 3.21. Report No. CERN DD/EE/84-1 (1984).
[26] G. C. Fox and S. Wolfram, Phys. Rev. Lett. **41**, 1581 (1978); S. H. Lee *et al.* (Belle Collaboration), Phys. Rev. Lett. **91**, 261801 (2003).
[27] M. Feindt and U. Kerzel, Nucl. Instrum. Methods Phys. Res., Sect. A **559**, 190 (2006).
[28] T. Skwarnicki, Ph.D. thesis, Institute for Nuclear Physics, Krakow 1986; DESY Internal Report, DESY F31-86-02 (1986).
[29] H. Albrecht *et al.* (ARGUS Collaboration), Phys. Lett. B **241**, 278 (1990).

Chiral Perturbation Theory tests at NA48/2 and NA62 experiments at CERN

Massimo Lenti [1,2]
INFN Sezione di Firenze, Via G. Sansone 1, 50019 Sesto Fiorentino (FI), IT
E-mail: massimo.lenti@fi.infn.it

Abstract. Final results from an analysis of about 400 $K^{\pm} \to \pi^{\pm}\gamma\gamma$ rare decay candidates collected by the NA48/2 and NA62 experiments at CERN during low intensity runs with minimum bias trigger configurations are presented. The results include a model-independent decay rate measurement and fits to Chiral Perturbation Theory (ChPT) descriptions. The data support the ChPT prediction for a cusp in the di-photon invariant mass spectrum at the two pions threshold.

1. Introduction

Experimental studies of radiative non-leptonic decays allow crucial tests of Chiral Perturbation Theory (ChPT) describing weak low energy processes. The rare decay $K^{\pm} \to \pi^{\pm}\gamma\gamma$ ($K_{\pi\gamma\gamma}$) was first observed by the BNL E787 experiment in 1997 [1]: 31 K^+ decay candidates were reported in the kinematic region 100 MeV/c $< p^*_\pi <$ 180 MeV/c, where p^*_π is the π^+ momentum in the K^+ rest frame. A related decay mode $K^{\pm} \to \pi^{\pm}\gamma e^+e^-$ ($K_{\pi\gamma ee}$) has been measured from 120 candidates in the kinematic region $m_{\gamma ee} >$ 260 MeV/c^2 at the NA48/2 experiment [2].

A measurement of the $K_{\pi\gamma\gamma}$ decay based on the combination of data collected by the NA48/experiment in 2003 and 2004 and from the NA62 experiment in 2007 is reported here.

2. The CERN Kaon facility

The NA48/2 experiment at CERN, which took data in 2003 and 2004, used simultaneous K^+ and K^- beams produced by 400 GeV/c primary SPS protons impinging on a beryllium target. An achromatic system of four dipole magnets selected charged particles with (60 ± 3) GeV/c momenta, splitting the two beams in the vertical plane and recombining them on a common axis. The beams were directed through collimators and quadrupole magnets and finally entered a 114 m long cylindrical vacuum tank with increasing diameter (from 1.92 to 2.4 m) which contained the decay region. The K^+/K^- flux ratio was 1.79 and the fraction of kaons decaying in the vacuum tank was 22%.

The beam line and setup of the NA48/2 experiment were used for the NA62 data taking in 2007, with different beam parameters. Secondary beams of positive and negative hadrons with

[1] On behalf of the NA48/2 Collaboration: Cambridge, CERN, Dubna, Chicago, Edinburgh, Ferrara, Firenze, Mainz, Northwestern, Perugia, Pisa, Saclay, Siegen, Torino, Wien

[2] On behalf of the NA62-RK Collaboration: Birmingham, CERN, Dubna, Fairfax, Ferrara, Firenze, Frascati, Mainz, Merced, Moscow, Napoli, Perugia, Pisa, Protvino, Roma I, Roma II, Saclay, San Luis Potosi, Stanford, Sofia, Torino, TRIUMF.

a (74 ± 1.4) GeV/c momentum were selected. The fraction of $K^{\pm}$ in these beams was about 6% and they were delivered, either alternately or simultaneously, into the same vacuum tank. The fraction of beam kaons decaying in the vacuum tank was 18%.

In both experiments the vacuum tank was followed by a magnetic spectrometer contained in a vessel filled with helium at nearly atmospheric pressure, separated from the vacuum by a thin $(0.3\%X_0)$ Kevlar® composite window. An aluminium beam pipe of 158 mm outer diameter and 1.1 mm thickness crossed the centre of the spectrometer and all the following detectors, allowing the undecayed beam particles to continue their path in vacuum. The spectrometer was made by four drift chambers (DCH) with a transverse width of 2.9 m, two of them (DCH1 and DCH2) located upstream and two (DCH3 and DCH4) downstream of a dipole magnet providing a horizontal transverse momentum kick of 120 MeV/c (NA48/2) or 265 MeV/c (NA62) for charged particle. The spectrometer momentum resolution was $\sigma_p/p = (1.02 \oplus 0.044 \cdot p)\%$ (NA48/2) or $\sigma_p/p = (0.48 \oplus 0.009 \cdot p)\%$ (NA62), where p is expressed in GeV/c. A hodoscope (HOD) consisting of two planes of 64 plastic scintillator strips, each plane divided in four quadrants, was placed downstream of the spectrometer and provided trigger signals and time measurements of charged particles with a resolution of 150 ps. The hodoscope was followed by a quasi-homogeneous liquid krypton electromagnetic calorimeter (LKr) with an active volume of 7 m^3, 27 X_0 deep and segmented transversally into 13248 projective 2×2 cm^2 cells with no longitudinal segmentation. The LKr energy resolution was $\sigma_E/E = (3.2/\sqrt{E} \oplus 9/E \oplus 0.42)\%$, its spatial resolution for the transverse coordinates x, y was $\sigma_x = \sigma_y = (4.2/\sqrt{E} \oplus 0.6)$ mm, with E in GeV. A plane of scintillating fibres located inside the LKr calorimeter volume at a depth of about $9.5X_0$, close to the maxima of showers initiated by 10 GeV photons, formed the "neutral hodoscope" (NHOD) which also provided trigger signals. The LKr was followed by a hadronic calorimeter and a muon detector. A detailed description of the detector can be found in [3].

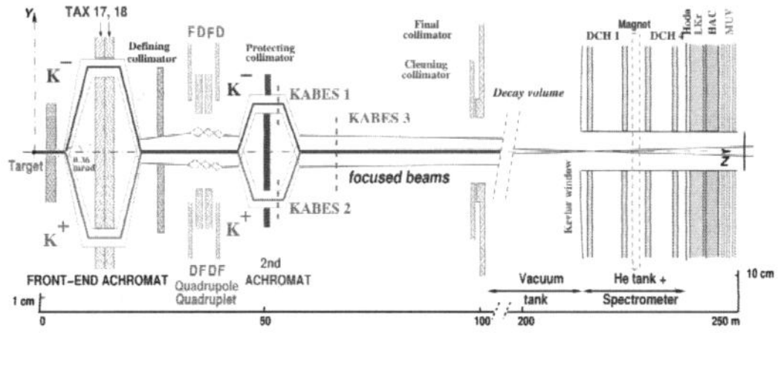

Figure 1. Beam setup of the NA48/2 and NA62 experiments. The NA48/2 experiment used simultaneous K^+ and K^- beams produced by 400 GeV/c primary SPS protons impinging on a beryllium target, the NA62 experiment used K^+ and K^- beams either alternately or simultaneously. The achromat system selected charged particles with (60 ± 3) GeV/c momenta (NA48/2) or (74 ± 1.4) momenta (NA62).

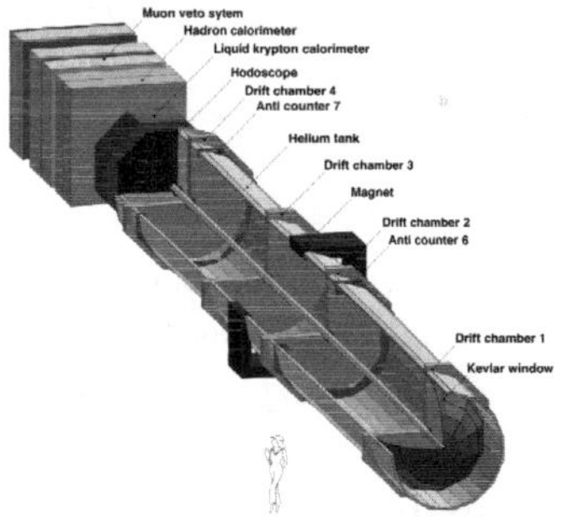

Figure 2. Detector setup for both the NA48/2 and NA62 experiments. Kaons are supposed to come from the bottom right part of the picture.

3. ChPT description of $K^{\pm} \to \pi^{\pm}\gamma\gamma$

The $K^{\pm} \to \pi^{\pm}\gamma\gamma$ decay ($K_{\pi\gamma\gamma}$) can be described by two kinematic variables:

$$z = \frac{(q_1 + q_2)^2}{m_K^2} = \left(\frac{m_{\gamma\gamma}}{m_K}\right)^2, \tag{1}$$

$$y = \frac{p\,(q_1 - q_2)}{m_K^2} \tag{2}$$

where p and $q_{1,2}$ are the four-momenta of the kaon and the two photons, respectively, $m_{\gamma\gamma}$ is the di-photon invariant mass and m_K is the charged kaon mass. The allowed region of the kinematic variables is:

$$0 \le z \le z_{max} = (1 - r_\pi)^2 = 0.515, \tag{3}$$

$$0 \le y \le y_{max} = \frac{1}{2}\sqrt{\lambda\,(1, r_\pi^2, z)} \tag{4}$$

where $r_\pi = m_\pi/m_K$, m_π is the charged pion mass and $\lambda\,(a, b, c) = a^2 + b^2 + c^2 - 2\,(ab + ac + bc)$. In the framework of ChPT the first non-trivial contribution to the $K_{\pi\gamma\gamma}$ decay rate comes

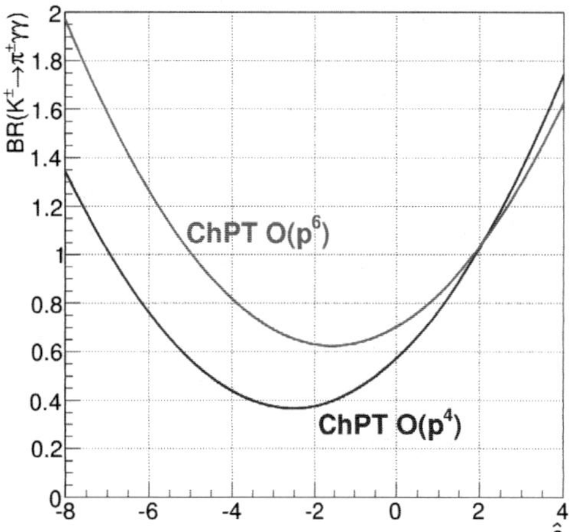

Figure 3. Dependence of the $K_{\pi\gamma\gamma}$ branching ratio on the $\hat{c}$ parameter predicted within ChPT [6].

from next-to-leading order ChPT, i.e. it has no tree-level $\mathcal{O}(p^2)$ contribution [4, 5, 6, 7]. The differential decay rate at leading order $\mathcal{O}(p^4)$ and including $\mathcal{O}(p^6)$ contributions can be written as:

$$\frac{\partial\Gamma}{\partial y \partial z}(\hat{c}, y, z) = \frac{m_K}{2^9\pi^3}\left[z^2\left(|A(\hat{c}, z, y^2) + B(z)|^2 + |C(z)|^2\right) + \left(y^2 - \frac{1}{4}\lambda(1, r_\pi^2, z)\right)^2 |B(z)|^2\right] \tag{5}$$

The decay rate and spectrum depend on a single parameter $\hat{c}$ to be determined experimentally; $A(\hat{c}, z, y^2)$ is an $\mathcal{O}(p^4)$ loop amplitude, $B(z)$ is a loop amplitude which appears only at $\mathcal{O}(p^6)$ order (and dominates the differential rate at low z) and $C(z)$ is a pole amplitude. A cusp is expected in the differential decay rate at the di-pion threshold $z_{th} = 4r_\pi^2 = 0.320$ generated by the pion loop amplitude. The branching ratio in the full kinematic range is expected to be $\mathcal{B}\,(K_{\pi\gamma\gamma}) \simeq 10^{-6}$ and its dependence on $\hat{c}$ is shown in Fig. 3. The differential decay rate for different values of the $\hat{c}$ parameter is shown in Fig. 4 at the $\mathcal{O}(p^4)$ order and in Fig. 5 at the $\mathcal{O}(p^6)$ order.

The ChPT description involves a number of external inputs. The G_8 parameter entering both $\mathcal{O}(p^4)$ and $\mathcal{O}(p^6)$ descriptions is fixed according to [8]. The $\mathcal{O}(p^6)$ framework additionally involves 7 parameters of the $K_{3\pi}$ decay amplitude fixed to those fitted to the experimental data

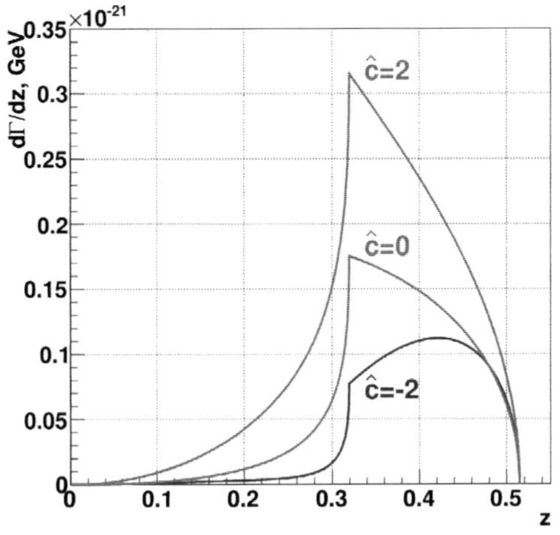

Figure 4. $K_{\pi\gamma\gamma}$ Differential decay rate for different values of the $\hat{c}$ parameter in the $\mathcal{O}(p^4)$ approximation of ChPT.

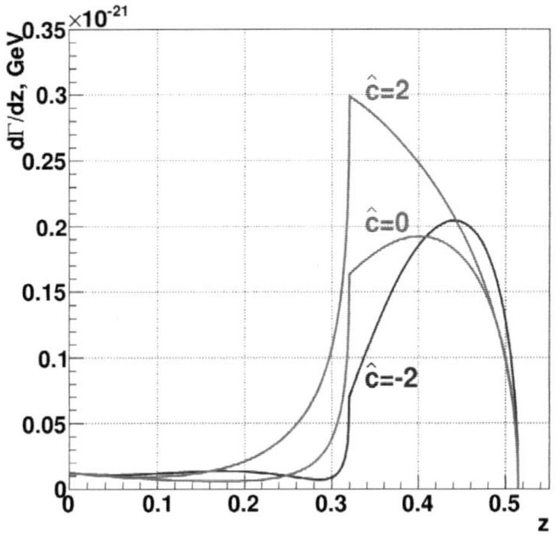

Figure 5. Differential decay rate for different values of the $\hat{c}$ parameter in the $\mathcal{O}(p^6)$ approximation of ChPT.

[9], and 3 polynomial contributions η_i $(i = 1, 2, 3)$ fixed to $\eta_i = 0$. The parameter $\hat{c}$ enters the $\mathcal{O}(p^6)$ differential decay rate via a linear combination:

$$\hat{c}^* \quad = \quad \hat{c} - 2\,(m_\pi/m_K)^2\,\eta_1 - 2\eta_2 - 2\eta_3. \tag{6}$$

Therefore setting $\eta_i = 0$ is equivalent to measuring $\hat{c}^*$ and $\hat{c}$ can be computed for any assumed value of η_i. The values used for the external parameters are listed in table 1.

Table 1. Values of the external parameters used in this analysis. The notation is introduced in [6, 8, 9].

Parameter	Value	Parameter	Value	Parameter	Value
$G_8 m_K^2 \times 10^6$	2.202	$\beta_1 \times 10^8$	−27.06	$\zeta_1 \times 10^8$	−0.40
$\alpha_1 \times 10^8$	93.16	$\beta_3 \times 10^8$	−2.22	$\xi_1 \times 10^8$	−1.83
$\alpha_3 \times 10^8$	−6.72	$\gamma_3 \times 10^8$	2.95	η_i	0

4. Data sample and triggers

The NA48/2 experiment collected data during two high intensity runs in 2003 and 2004 (with about 3×10^6 $K^\pm$ entering the decay volume per SPS spill of 4.8 s duration), in about 100 days of efficient data taking in total dedicated mostly to collect three-pions decays. The $K_{\pi\gamma\gamma}$ measurement is based on two special $K^\pm$ decay samples collected at about 10% the nominal beam intensity during 12 hours in 2003 and 54 hours in 2004 with a minimum bias trigger condition: a time coincidence of signals in both HOD planes within the same quadrant and an energy deposit of at least 10 GeV in the LKr calorimeter.

The NA62 experiment collected data from about 3.5×10^5 SPS spills in 4 months of operation in 2007 with low intensity beams at an instantaneous kaon decay rate in the vacuum tank of about 10^5 Hz; the total number of $K^\pm$ decays in the vacuum tank was about 2×10^{10}. About 27% of them were collected with simultaneous K^+ and K^- beams, 65% (8%) in single-beam K^+ (K^-) mode. A downscaled minimum bias trigger was used to collect $K_{\pi\gamma\gamma}$ decays requiring at least one charged track and/or a minimum calorimetric energy deposit. At least one of the following trigger conditions was required:

- a time coincidence of signals in the two HOD planes within the same quadrant combined with a DCH hit multiplicity signal compatible with a single charged track (~20% of the data sample);
- the above condition in coincidence with a LKr energy deposit of at least 10 GeV (~60% of the data sample);
- a signal from the NHOD detector compatible with an electromagnetic shower energy release (~20% of the data sample).

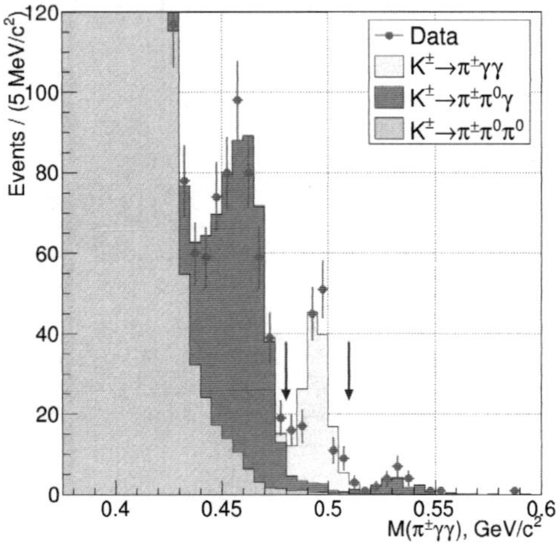

Figure 6. NA48/2 data: invariant mass distributions of $\pi^\pm\gamma\gamma$ compared with the sums of estimated signal and background components. The estimated $K_{\pi\gamma\gamma}$ signal corresponds to the result of a ChPT $\mathcal{O}(p^6)$ fit. The limits of the signal region are indicated with vertical arrows.

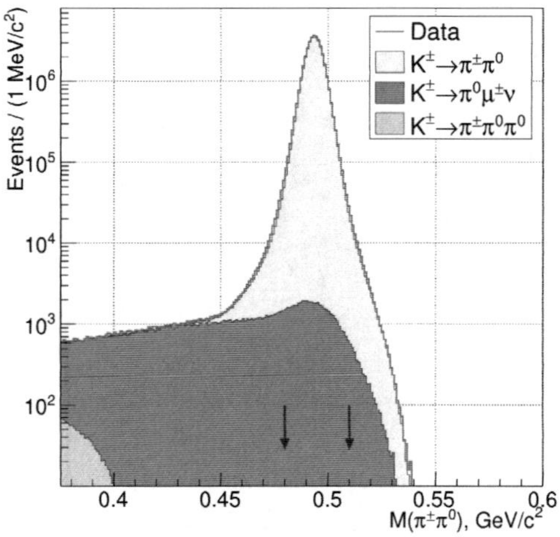

Figure 7. NA48/2 data: invariant mass distributions of $\pi^\pm\pi^0$ compared with the sums of estimated signal and background components. The limits of the signal region are indicated with vertical arrows.

5. Measurement method

The $K_{\pi\gamma\gamma}$ decay rate is measured with respect to the normalization decay chain collected simultaneously with the same trigger logic: $K^\pm \to \pi^\pm\pi^0$ decay ($K_{2\pi}$) followed by $\pi^0 \to \gamma\gamma$ decay ($\pi^0_{\gamma\gamma}$). The kaon beam flux and downscaling factors cancel in the ratio as well as several systematic effects at first order.

The normalization mode branching ratio $\mathcal{B}(K_{2\pi})\mathcal{B}(\pi^0_{\gamma\gamma}) = 0.204 \pm 0.001$ is large and known to a good precision [10].

Journal of Physics: Conference Series **631** (2015) 012039 doi:10.1088/1742-6596/631/1/012039

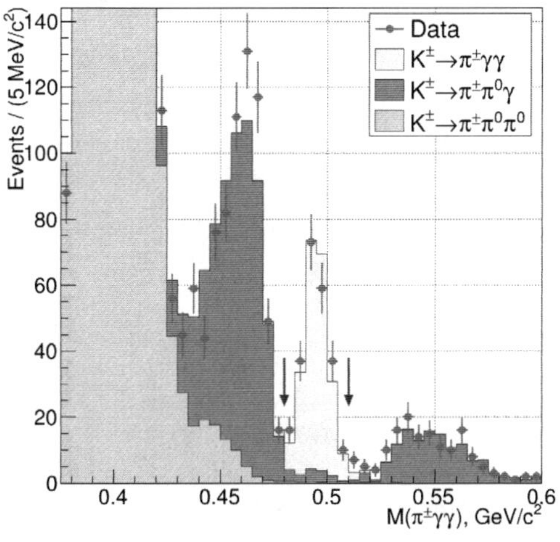

Figure 8. NA62 data: invariant mass distributions of $\pi^{\pm}\gamma\gamma$ compared with the sums of estimated signal and background components. The estimated $K_{\pi\gamma\gamma}$ signal corresponds to the result of a ChPT $\mathcal{O}(p^6)$ fit. The limits of the signal region are indicated with vertical arrows.

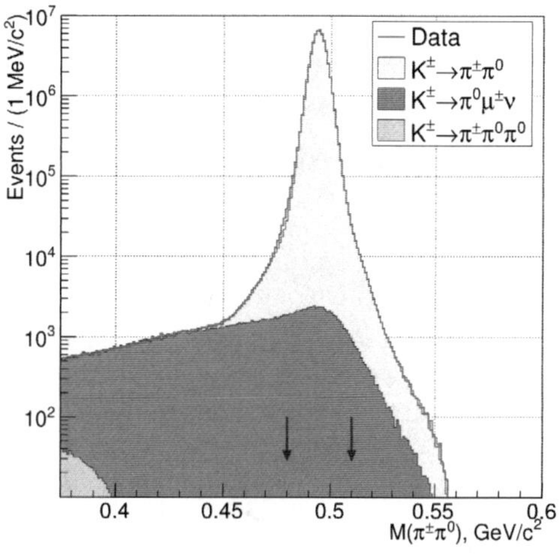

Figure 9. NA62 data: invariant mass distributions of $\pi^{\pm}\pi^0$ compared with the sums of estimated signal and background components. The limits of the signal region are indicated with vertical arrows.

6. Data selection

The signal ($K_{\pi\gamma\gamma}$) and normalization ($K_{2\pi}, \pi^0_{\gamma\gamma}$) decay modes are characterised by the same set of particles in the final states. Therefore the following selection criteria are common for the two modes leading to cancellation of systematic effects.

- A $\pi^{\pm}$ candidate track consistent with originating from a beam $K^{\pm}$ is required. The decay vertex, reconstructed at the point of closest approach of the track and the beam axis, should be located within a 98 m long fiducial volume contained in the vacuum tank.
- Track impact points in the DCH, HOD and LKr calorimeter should be within their fiducial acceptances.
- The reconstructed track momentum is required to be between 10 and 40 GeV/c (NA48/2) or between 8 and 50 GeV/c (NA62). The upper momentum cut is equivalent to a lower limit on the total energy of the two photons ensuring the high efficiency of the LKr trigger conditions.
- The $\pi^{\pm}$ is identified by the ratio of the energy release in the LKr calorimeter to momentum measured by the spectrometer: $E/p < 0.85$.
- Clusters of energy deposition in the LKr calorimeter in time with the track are considered as photon candidates. Exactly two photon candidates are required, with a minimum energy of 3 GeV, a minimum separation of 20 cm and a distance from the track impact point of at least 25 cm.
- The reconstructed $\pi^{\pm}\gamma\gamma$ momentum is required to be between 55 and 65 GeV/c (NA48/2) or between 70 and 78 GeV/c (NA62), while its transverse component relative to the beam axis should be $p_T^2 < 0.5 \times 10^{-3}$ (GeV/c)^2.
- The reconstructed $\pi^{\pm}\gamma\gamma$ ($\pi^{\pm}\pi^0$) invariant mass should be between 480 and 510 MeV/c^2.

The corresponding mass resolutions in NA48/2 are 5.9 (3.9) MeV/c^2 and in NA62 are 5.4 (3.3) MeV/c^2 for the $K_{\pi\gamma\gamma}$ ($K_{2\pi}$) decays.

The $K_{\pi\gamma\gamma}$ and $K_{2\pi}$ selections differ only in the di-photon invariant mass requirement.

- For $K_{\pi\gamma\gamma}$ the signal kinematic region is defined as $z > 0.2$. The low z region is dominated by the π^0 monochromatic line at $z = (m_{\pi^0}/m_K)^2 = 0.075$. The resolution is the z variable increases from $\delta z = 0.005$ at $z = 0.2$ to $\delta z = 0.003$ at $z_{max} = 0.515$.
- For $K_{2\pi}$ it is required that $|m_{\gamma\gamma} - m_{\pi^0}| < 10$ MeV/c^2 ($0.064 < z < 0.086$). The mass resolution is $\delta m_{\gamma\gamma} = 1.6$ MeV/c^2 (corresponding to $\delta z = 0.002$).

The $\pi^\pm\gamma\gamma$ and $\pi^\pm\pi^0$ invariant mass spectra of the selected signal and normalization candidates are shown in Fig. 6 and Fig. 7 for NA48/2 and in Fig. 8 and Fig. 9 for NA62 together with the expectations for the signal and background contributions evaluated with MC simulations.

In the NA48/2 data the number of reconstructed $K_{\pi\gamma\gamma}$ candidates is $N_{\pi\gamma\gamma} = 149$, of which 97 (52) are K^+ (K^-) decay candidates. The number of reconstructed $K_{2\pi}$ candidates is $N_{2\pi} = 3.628 \times 10^7$, of which 2.321 (1.307) $\times 10^7$ are K^+ (K^-) decay candidates.

In the NA62 data the number of reconstructed $K_{\pi\gamma\gamma}$ candidates is $N_{\pi\gamma\gamma} = 232$, of which 179 (53) are K^+ (K^-) decay candidates. The number of reconstructed $K_{2\pi}$ candidates is $N_{2\pi} = 5.488 \times 10^7$, of which 4.431 (1.057) $\times 10^7$ are K^+ (K^-) decay candidates.

The reconstructed z spectrum of the $K_{\pi\gamma\gamma}$ candidates is displayed in Fig. 10 for NA48/2 and in Fig. 11 for NA62.

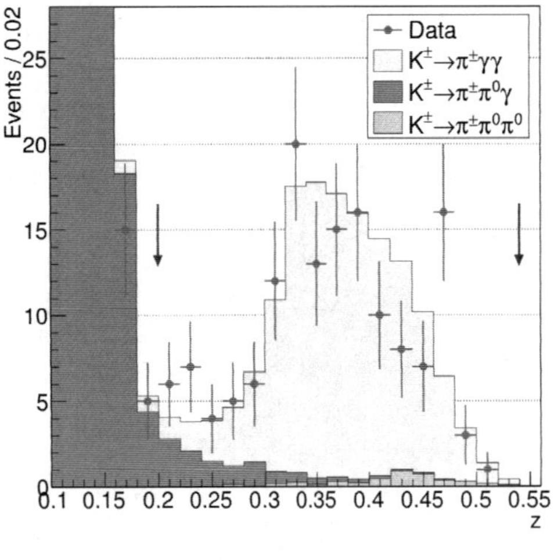

Figure 10. NA48/2 data: reconstructed $z = (m_{\gamma\gamma}/m_K)^2$ distribution for the $K_{\pi\gamma\gamma}$ candidates and estimated signal and background contributions. The estimated signal corresponds to the result of a ChPT $\mathcal{O}(p^6)$ fit. The limits of the signal region are indicated with vertical arrows.

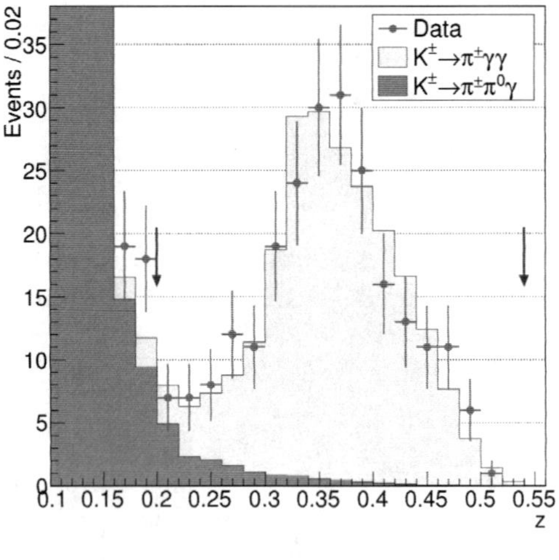

Figure 11. NA62 data: reconstructed $z = (m_{\gamma\gamma}/m_K)^2$ distribution for the $K_{\pi\gamma\gamma}$ candidates and estimated signal and background contributions. The estimated signal corresponds to the result of a ChPT $\mathcal{O}(p^6)$ fit. The limits of the signal region are indicated with vertical arrows.

7. Backgrounds

The main background to the normalization mode ($K_{2\pi}, \pi^0_{\gamma\gamma}$) comes from $K^\pm \to \pi^0\mu^\pm\nu$ decays ($K_{\mu3}$) followed by $\pi^0 \to \gamma\gamma$ decays and it corresponds to 0.13% (NA48/2) or 0.115% (NA62), where the branching ratios $\mathcal{B}(K_{2\pi})$ and $\mathcal{B}(K_{\mu 3})$ are taken from [10]. The acceptances have been evaluated with MC simulations and are 19.18% (NA48/2) or 16.88% (NA62) for $K_{2\pi}$ decays, 0.15% (NA48/2) or 0.12% (NA62) for $K_{\mu 3}$ decays.

The product N_K of the number of $K^{\pm}$ decays in the fiducial volume in the analysed data set and the trigger efficiency for $K_{2\pi}$ sample is computed as:

$$N_K = (0.925 \pm 0.004) \times 10^9 \quad \text{[NA48/2]} \tag{7}$$
$$N_K = (1.592 \pm 0.006) \times 10^9 \quad \text{[NA62]} \tag{8}$$

where the uncertainty is dominated by the limited precision on the external input $\mathcal{B}(K_{2\pi})$.

The principal background in the $K_{\pi\gamma\gamma}$ sample comes from $K^{\pm} \to \pi^{\pm}\pi^0\gamma$ decays followed by $\pi^0_{\gamma\gamma}$ decays. It is due to the merging of LKr energy deposition clusters produced by a photon from the π^0 decay and a photon from the parent $K^{\pm}$ decay. The total background from $K^{\pm} \to \pi^{\pm}\pi^0\gamma$ decays is estimated to be 11.4 ± 0.6 events (NA48/2) or 15.3 ± 1.1 events (NA62), where the uncertainty comes from MC simulation statistics.

Another source of background in the $K_{\pi\gamma\gamma}$ sample is due to $K^{\pm} \to \pi^{\pm}\pi^0\pi^0$ decays followed by $\pi^0_{\gamma\gamma}$ decays. They contribute via photon missing the LKr acceptance as well as LKr cluster merging. This background is estimated to be 4.1 ± 0.4 events (NA48/2) or 2.1 ± 0.3 events (NA62), where the uncertainty comes from MC simulation statistics.

8. Model-independent branching ratio measurement

Partial $K_{\pi\gamma\gamma}$ branching fractions $\mathcal{B}_j$ in 8 bins of the z variable are evaluated as:

$$\mathcal{B}_j = \left(N_j - N_j^B\right) / (N_K A_j) \tag{9}$$

where N_j is the number of reconstructed $K_{\pi\gamma\gamma}$ candidates, N_j^B is the estimated number of background events, A_j is the signal acceptance in bin j and N_K is the number of $K^{\pm}$ decays in the fiducial volume defined in the previous section. The first 7 bins of z are defined in the range $0.20 < z < 0.48$ with $\Delta z = 0.04$, the last bin is for $0.48 < z < z_{max}$ due to the lack of statistics induced by the small acceptance for $\pi^{\pm}$ very close to the beam pipe.

The y-dependence of the differential decay rate expected within the ChPT framework [6, 7] is weak and the y-dependence of acceptance is also weak. The model-independent branching fraction in the kinematic region $z > 0.2$ is computed by summing over the z bins:

$$\mathcal{B}_{MI}\,(z > 0.2) = \sum_{j=1}^{8} \mathcal{B}_j \tag{10}$$

and it is:

$$\mathcal{B}_{MI}\,(z > 0.2) = (0.877 \pm 0.087_{stat} \pm 0.017_{syst}) \times 10^{-6} \quad \text{[NA48/2]} \tag{11}$$
$$\mathcal{B}_{MI}\,(z > 0.2) = (1.088 \pm 0.093_{stat} \pm 0.027_{syst}) \times 10^{-6} \quad \text{[NA62]} \tag{12}$$

where the first error is statistical and the second systematic, dominated by uncertainties on the background subtraction. The $\mathcal{B}_j$ measurement of NA48/2 and NA62 have been combined and then summed, providing the combined result:

$$\mathcal{B}_{MI}\,(z > 0.2) = (0.965 \pm 0.061_{stat} \pm 0.014_{syst}) \times 10^{-6} \quad \text{[NA48/2 and NA62]} \tag{13}$$

The measurements performed separately for K^+ and K^- decays are consistent in the NA48/2 data sample:

$$\mathcal{B}^{+}_{MI}\,(z > 0.2) = (0.881 \pm 0.107_{stat}) \times 10^{-6} \quad \text{[NA48/2]} \tag{14}$$
$$\mathcal{B}^{-}_{MI}\,(z > 0.2) = (0.868 \pm 0.147_{stat}) \times 10^{-6} \quad \text{[NA48/2]} \tag{15}$$

and in the NA62 data sample:

$$\mathcal{B}^{+}_{MI}\,(z > 0.2) \;=\; (1.010 \pm 0.098_{stat}) \times 10^{-6} \quad \text{[NA62]} \tag{16}$$

$$\mathcal{B}^{-}_{MI}\,(z > 0.2) \;=\; (1.417 \pm 0.256_{stat}) \times 10^{-6} \quad \text{[NA62]} \tag{17}$$

The combined $\mathcal{B}^{\pm}_{MI}\,(z > 0.2)$ measurements obtained by averaging the NA48/2 and NA62 results are:

$$\mathcal{B}^{+}_{MI}\,(z > 0.2) \;=\; (0.951 \pm 0.072_{stat}) \times 10^{-6} \quad \text{[NA48/2 and NA62]} \tag{18}$$

$$\mathcal{B}^{-}_{MI}\,(z > 0.2) \;=\; (1.004 \pm 0.127_{stat}) \times 10^{-6} \quad \text{[NA48/2 and NA62]} \tag{19}$$

These values are consistent and the charge asymmetry of the decay rate is:

$$\Delta\left(K_{\pi\gamma\gamma}\right) \;=\; \frac{B^{+}_{MI} - B^{-}_{MI}}{B^{+}_{MI} + B^{-}_{MI}} = -0.03 \pm 0.07 \tag{20}$$

where the sub-dominant systematic uncertainties are neglected.

9. Measurement of the ChPT parameters

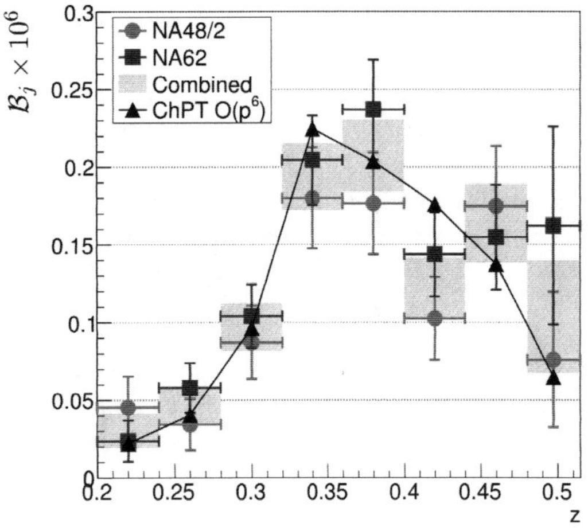

Figure 12. Measurement of the model-independent $K_{\pi\gamma\gamma}$ branching ratio $\mathcal{B}_j$ in z bins: NA48/2, NA62 and the combination of the two. The horizontal bars indicate the bin widths, the vertical bars indicate the statistical errors. The values of $\mathcal{B}_j$ computed within ChPT $\mathcal{O}(p^6)$ formulation [6], obtained by integration of the ChPT differential decay rate for the central value of the combined measurement $\hat{c}_6 = 1.86$ over the bin width, are also shown. The line connecting the markers are drawn to guide the eye.

The $\hat{c}$ parameter described in section 3 is measured in the ChPT $\mathcal{O}(p^4)$ and $\mathcal{O}(p^6)$ framework [6] by fitting the reconstructed z spectrum shown in Fig. 10 for NA48/2 and in Fig. 11 for NA62 with a binned log-likelihood. The fit is performed for $0.2 < z < 0.54$ with a bin width of $\Delta z = 0.02$, with the $\mathcal{O}(p^4)$ and $\mathcal{O}(p^6)$ description separately:

$$\hat{c}_4 \;=\; 1.37 \pm 0.33_{stat} \pm 0.14_{syst}, \qquad \hat{c}_6 = 1.41 \pm 0.38_{stat} \pm 0.11_{syst} \quad \text{[NA48/2]}, \tag{21}$$

$$\hat{c}_4 \;=\; 1.93 \pm 0.26_{stat} \pm 0.08_{syst}, \qquad \hat{c}_6 = 2.10 \pm 0.28_{stat} \pm 0.18_{syst} \quad \text{[NA62]}. \tag{22}$$

The data are consistent with both ChPT descriptions. The z spectrum corresponding to the $\mathcal{O}(p^6)$ fit result is shown in Fig. 10 for NA48/2 and in Fig. 11 for NA62. The measured values of $\hat{c}_6$ translates into the following model-dependent branching ratio in the full kinematic range, obtained by integration of the ChPT $\mathcal{O}(p^6)$ differential decay rate:

$$\mathcal{B}_{ChPT} \;=\; (0.910 \pm 0.072_{stat} \pm 0.022_{syst}) \times 10^{-6} \quad \text{[NA48/2]}, \tag{23}$$

$$\mathcal{B}_{ChPT} \;=\; (1.058 \pm 0.066_{stat} \pm 0.044_{syst}) \times 10^{-6} \quad \text{[NA62]}. \tag{24}$$

The combination of the previous results from NA48/2 and NA62 is:

$$\hat{c}_4 = 1.72 \pm 0.20_{stat} \pm 0.06_{syst}, \quad \text{[NA48/2 and NA62]}, \tag{25}$$
$$\hat{c}_6 = 1.86 \pm 0.23_{stat} \pm 0.11_{syst}, \quad \text{[NA48/2 and NA62]}, \tag{26}$$
$$\mathcal{B}_{ChPT} = (1.003 \pm 0.051_{stat} \pm 0.024_{syst}) \times 10^{-6} \quad \text{[NA48/2 and NA62]}. \tag{27}$$

The corresponding model dependent values of $\mathcal{B}_{ChPT}$ in z bins are displayed in Fig. 12.

10. Conclusions

A model-independent measurement of the $K_{\pi\gamma\gamma}$ decay rate and fits to the ChPT description have been performed by the NA48 experiment and the NA62 experiment and the results are published in [11] and [12], respectively, and have been combined. The model independent branching ratio in a limited kinematic range is $\mathcal{B}_{MI}(z > 0.2) = (0.965 \pm 0.063) \times 10^{-6}$. The observed decay spectrum agrees with the ChPT description and ChPT parameters measured within the considered formulation are $\hat{c}_4 = 1.72 \pm 0.21$ and $\hat{c}_6 = 1.86 \pm 0.25$. The branching ratio in the full kinematic range assuming the $\mathcal{O}(p^6)$ description is $\mathcal{B}_{ChPT} = (1.003 \pm 0.056) \times 10^{-6}$. The uncertainties are dominated by the statistical errors while including a small experimental systematic contribution.

References

[1] Kitching P *et al.* 1997 *Phys. Rev. Lett.* **79** 4079
[2] Batley J R *et al.* 2008 *Phys. Lett.* B **659** 493
[3] Fanti V *et al.* 2007 *Nucl. Instr. Methods* A **574** 433
[4] Sehgal L M 1972 *Phys. Rev.* D **6** 367
[5] Ecker G, Pich A and de Rafael E 1988 *Nucl. Phys.* B **303** 665
[6] D'Ambrosio G and Portolés 1996 *Phys. Lett.* B **386** 403
[7] Gérard J M, Smith C and Trine S 2005 *Nucl. Phys.* B **730** 1
[8] Cirigliano V *et al.* 2012 *Rev. Mod. Phys.* **84** 399
[9] Bijnens J, Dhonte P and Persson F 2003 *Nucl. Phys.* B **648** 317
[10] Beringer J *et al.* (Particle Data Group) 2012 *Phys. Rev.* D **86** 010001
[11] Batley J R *et al.* 2014 *Phys. Lett.* B **730** 141
[12] Lazzeroni C *et al.* 2014 *Phys. Lett.* B **732** 65

Precision tests of the Standard Model with Kaon decays at CERN

Karim Massri [1,2]

University of Birmingham, Edgbaston, Birmingham, United Kingdom

E-mail: `karim.massri@cern.ch`

[1] On behalf of the NA48/2 Collaboration: G. Anzivino, R. Arcidiacono, W. Baldini, S. Balev, J.R. Batley, M. Behler, S. Bifani, C. Biino, A. Bizzeti, B. Bloch-Devaux, G. Bocquet, N. Cabibbo, M. Calvetti, N. Cartiglia, A. Ceccucci, P. Cenci, C. Cerri, C. Cheshkov, J.B. Chèze, M. Clemencic, G. Collazuol, F. Costantini, A. Cotta Ramusino, D. Coward, D. Cundy, A. Dabrowski, P. Dalpiaz, C. Damiani, M. De Beer, J. Derré, H. Dibon, L. DiLella, N. Doble, K. Eppard, V. Falaleev, R. Fantechi, M. Fidecaro, L. Fiorini, M. Fiorini, T. Fonseca Martin, P.L. Frabetti, L. Gatignon, E. Gersabeck, A. Gianoli, S. Giudici, A. Gonidec, E. Goudzovski, S. Goy Lopez, M. Holder, P. Hristov, E. Iacopini, E. Imbergamo, M. Jeitler, G. Kalmus, V. Kekelidze, K. Kleinknecht, V. Kozhuharov, W. Kubischta, G. Lamanna, C. Lazzeroni, M. Lenti, L. Litov, D. Madigozhin, A. Maier, I. Mannelli, F. Marchetto, G. Marel, M. Markytan, P. Marouelli, M. Martini, L. Masetti, E. Mazzucato, A. Michetti, I. Mikulec, N. Molokanova, E. Monnier, U. Moosbrugger, C. Morales Morales, D.J. Munday, A. Nappi, G. Neuhofer, A. Norton, M. Patel, M. Pepe, A. Peters, F. Petrucci, M.C. Petrucci, B. Peyaud, M. Piccini, G. Pierazzini, I. Polenkevich, Yu. Potrebenikov, M. Raggi, B. Renk, P. Rubin, G. Ruggiero, M. Savrié, M. Scarpa, M. Shieh, M.W. Slater, M. Sozzi, S. Stoynev, E. Swallow, M. Szleper, M. Valdata-Nappi, B. Vallage, M. Velasco, M. Veltri, S. Venditti, M. Wache, H. Wahl, A. Walker, R. Wanke, L. Widhalm, A. Winhart, R. Winston, M.D. Wood, S.A. Wotton, A. Zinchenko, M. Ziolkowski.

[2] On behalf of the NA62 Collaboration: G. Aglieri Rinella, R. Aliberti, F. Ambrosino, B. Angelucci, A. Antonelli, G. Anzivino, R. Arcidiacono, I. Azhinenko, S. Balev, J. Bendotti, A. Biagioni, C. Biino, A. Bizzeti, T. Blazek, A. Blik, B. Bloch-Devaux, V. Bolotov, V. Bonaiuto, M. Bragadireanu, D. Britton, G. Britvich, N. Brook, F. Bucci, V. Buescher, F. Butin, E. Capitolo, C. Capoccia, T. Capussela, V. Carassiti, N. Cartiglia, A. Cassese, A. Catinaccio, A. Cecchetti, A. Ceccucci, P. Cenci, V. Cerny, C. Cerri, O. Chikilev, R. Ciaranfi, G. Collazuol, P. Cooke, P. Cooper, G. Corradi, E. Cortina Gil, F. Costantini, A. Cotta Ramusino, D. Coward, G. D'Agostini, J. Dainton, P. Dalpiaz, H. Danielsson, J. Degrange, N. De Simone, D. Di Filippo, L. Di Lella, N. Dixon, N. Doble, V. Duk, V. Elsha, J. Engelfried, T. Enik, V. Falaleev, R. Fantechi, L. Federici, M. Fiorini, J. Fry, A. Fucci, L. Fulton, S. Gallorini, L. Gatignon, A. Gianoli, S. Giudici, L. Glonti, A. Goncalves Martins, F. Gonnella, E. Goudzovski, R. Guida, E. Gushchin, F. Hahn, B. Hallgren, H. Heath, F. Herman, D. Hutchcroft, E. Iacopini, O. Jamet, P. Jarron, K. Kampf, J. Kaplon, V. Karjavin, V. Kekelidze, S. Kholodenko, G. Khoriauli, A. Khudyakov, Yu. Kiryushin, K. Kleinknecht, A. Kluge, M. Koval, V. Kozhuharov, M. Krivda, Y. Kudenko, J. Kunze, G. Lamanna, C. Lazzeroni, R. Leitner, R. Lenci, M. Lenti, E. Leonardi, P. Lichard, R. Lietava, L. Litov, D. Lomidze, A. Lonardo, N. Lurkin, D. Madigozhin, G. Maire, A. Makarov, I. Mannelli, G. Mannocchi, A. Mapelli, F. Marchetto, P. Massarotti, K. Massri, P. Matak, G. Mazza, E. Menichetti, M. Mirra, M. Misheva, N. Molokanova, J. Morant, M. Morel, M. Moulson, S. Movchan, D. Munday, M. Napolitano, F. Newson, A. Norton, M. Noy, G. Nuessle, V. Obraztsov, S. Padolski, R. Page, V. Palladino, A. Pardons, E. Pedreschi, M. Pepe, F. Perez Gomez, M. Perrin-Terrin, P. Petrov, F. Petrucci, R. Piandani, M. Piccini, D. Pietreanu, J. Pinzino, M. Pivanti, I. Polenkevich, I. Popov, Yu. Potrebenikov, D. Protopopescu, F. Raffaelli, M. Raggi, P. Riedler, A. Romano, P. Rubin, G. Ruggiero, V. Russo, V. Ryjov, A. Salamon, G. Salina, V. Samsonov, E. Santovetti, G. Saracino, F. Sargeni, S. Schifano, V. Semenov, A. Sergi, M. Serra, S. Shkarovskiy, A. Sotnikov, V. Sougonyaev, M. Sozzi, T. Spadaro, F. Spinella, R. Staley, M. Statera, P. Sutcliffe, N. Szilasi, D. Tagnani, M. Valdata-Nappi, P. Valente, M. Vasile, V. Vassilieva, B. Velghe, M. Veltri, S. Venditti, M. Vormstein, H. Wahl, R. Wanke, P. Wertelaers, A. Winhart, R. Winston, B. Wrona, O. Yushchenko, M. Zamkovsky, A. Zinchenko.

Journal of Physics: Conference Series **631** (2015) 012040 doi:10.1088/1742-6596/631/1/012040

Abstract. Recent results and prospects for precision tests of the Standard Model in kaon decay-in-flight experiments at CERN are presented. A measurement of the ratio of leptonic decay rates of the charged kaon at the level of 0.4% precision constrains the parameter space of new physics models with extended Higgs sector, a fourth generation of quarks and leptons or sterile neutrinos. Searches for heavy neutrino mass states and the dark photon in the ~ 100 MeV/c^2 mass range based on samples collected in 2003-2007 are in progress and prospects will be discussed. The NA62 experiment, starting in 2014, will search for a range of lepton number and lepton flavour violating decays of the charged kaon and the neutral pion at improved sensitivities down to $\sim 10^{-12}$, which will probe new physics scenarios involving heavy Majorana neutrinos or R-parity violating SUSY.

1. Introduction

Since their discovery in late '40s, kaons have been instrumental in understanding the fundamental interactions of nature. The tremendous impact of kaon physics in the development of the Standard Model includes the observation of kaon decay modes with different parity (the so-called "$\vartheta - \tau$ paradox"), which led to the discovery of the parity symmetry violation in weak interactions, and the evidence for a second generation of quarks. In fact, the quantum number "strangeness" was introduced to explain the anomalously large lifetime of kaons, which successively led to the postulation of the "strange" quark and the quark model, while the existence of the "charm" quark was hypothesised to explain the observed suppression of the $K_L \to \mu^+\mu^-$ with respect to the charged analogous $K^\pm \to \mu^\pm\nu$ (GIM mechanism). Another milestone of the kaon physics was to provide the first evidence of indirect and direct CP violation, by the precision measurement of neutral kaon decay rates.

Nowadays, kaon physics is still a unique probe of the flavour sector of the Standard Model (SM). The clean environment allows precise theoretical predictions, which leads to high experimental sensitivity to new physics. An exemplar quantity with such characteristics is the ratio R_K of the leptonic decay rates of the charged kaons, which probes the lepton universality in the SM and is sensitive to a vast number of beyond the SM (BSM) scenarios. In addition, the large statistics of the collected samples allows the study of rare $K^\pm$ decay modes, as well as the search for dark matter candidates or forbidden Lepton Number/Flavour Violating (LNFV) decays. The recent results obtained from the NA48/2 and NA62 experiments, the two most recent in a long tradition of fixed-target kaon experiments in the CERN North Area (NA), are presented, and the prospects for the forthcoming three-year long data taking, aimed to precisely measure the branching ratio of the $K^+ \to \pi^+\nu\bar{\nu}$ decay, are discussed.

2. Experimental Apparatus and Data Taking conditions

The NA48/2 experiment at CERN SPS was a multi-purpose $K^\pm$ experiment which collected data in 2003-2004, whose main goal was to search for direct CP violation in the $K^\pm \to \pi^\pm\pi^+\pi^-$ and $K^\pm \to \pi^\pm\pi^0\pi^0$ decays [1]. In order to minimise the systematic bias in the comparison between K^+ and K^- decays, simultaneous and collinear K^+ and K^- beams of the same momentum (60 ± 3.7) GeV/c were produced by the 400 GeV/c SPS primary proton beam, which impinged on a Beryllium target, and were steered into a 114 m long decay region, contained in a vacuum (at pressure $< 10^{-4}$ mbar) cylindrical tank, to avoid interactions of kaon decay products before detection. Fig. 1 shows the NA48/2 beam line. The downstream part of the vacuum tank was sealed by a convex Kevlar window, that separated the vacuum from the helium at atmospheric pressure in which a magnetic spectrometer, formed of 4 drift chambers (DCHs) and a dipole magnet providing a horizontal momentum kick $p_t = 120$ MeV/c, was installed. The magnet polarity was inverted on a daily basis, in order to minimise the systematic bias due to different acceptances for the K^+ and K^- decay products. The

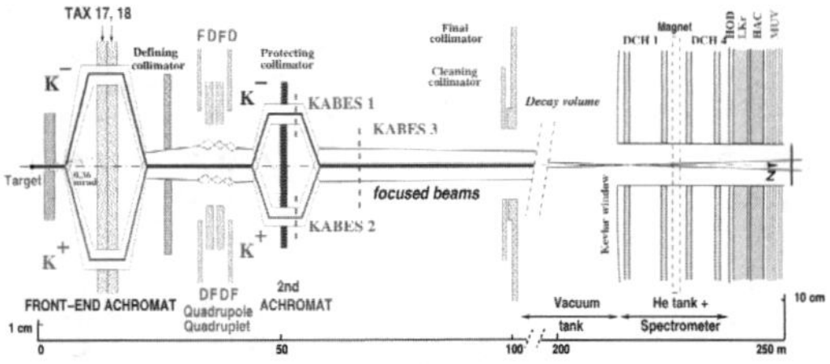

Figure 1. The beam line and detector of both NA48/2 and NA62-R_K experiments.

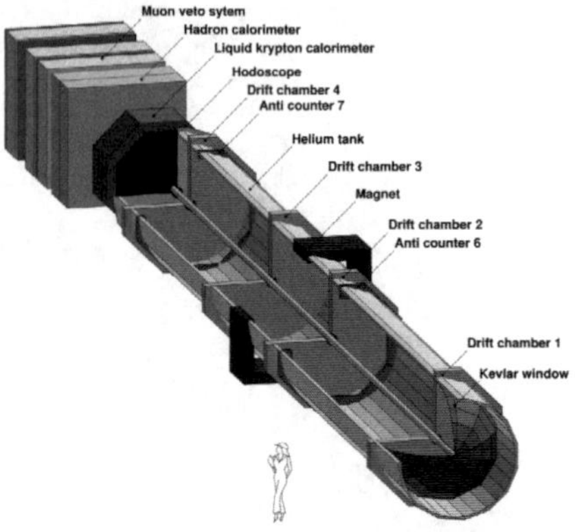

Figure 2. Detector layout of both NA48/2 and NA62-R_K experiments.

spatial resolution of each DCH was $\sigma_x = \sigma_y = 90$ μm, while the momentum resolution of the spectrometer was $\sigma(p)/p = (1.02 \oplus 0.044 \cdot p)\%$, where the momentum p is measured in GeV/c. A hodoscope (HOD), consisting of two planes of 64 scintillator counters each, arranged vertically and horizontally respectively, was placed downstream of the spectrometer and provided fast signals for trigger purposes, as well as time measurements for charged particles with a resolution of ~ 300 ps. The HOD was followed by a quasi-homogeneous ionisation chamber, built for the earlier NA48 experiment as electromagnetic calorimeter, which used ~ 7 m^3 of liquid krypton (LKr) as active material, with a depth of 127 cm, corresponding to 27 radiation lengths. The front plane had an octagonal shape and was segmented in 13248 cells with size 2×2 cm^2. This choice corresponded to a high calorimeter granularity, considering that the LKr Molière radius is 4.7 cm. The LKr calorimeter energy resolution was measured to be $\sigma_E/E = (3.2/\sqrt{E} \oplus 9.0/E \oplus 0.42)\%$, where E is the energy expressed in GeV. The space resolution $\sigma_{x,y}$ of the LKr was $\sigma_{x,y} = (4.2/\sqrt{E} \oplus 0.6)$ mm, and the time resolution on the single shower was $\sigma_t = 2.5$ ns$/\sqrt{E}$. Downstream the LKr calorimeter, a hadronic calorimeter, which was made of iron plates alternated with scintillator planes, and three muon veto planes were used for particle identification. The detector layout is sketched in Fig. 2. A detailed description of the detector can be found in Ref. [2].

The NA62 experiment (R_K phase, denoted as NA62-R_K), whose main goal was to measure the ratio R_K of the rates of the $K^\pm \to \ell^\pm \nu$ decays ($\ell = e, \mu$), collected a large minimum bias data sample in 2007-2008 [3]. It was essentially based on the NA48/2 detector with minor changes. The beam momentum was changed to (74 ± 1.4) GeV/c and the spectrometer momentum kick was increased to $p_t = 270$ MeV/c, which led to an improved momentum resolution $\sigma(p)/p = (0.48 \oplus 0.009 \cdot p)\%$. In addition, the trigger system was modified to collect single track leptonic modes.

3. R_K measurement with the NA62-R_K experiment

Leptonic decays of pseudoscalar mesons ($P^\pm \to \ell^\pm \nu$, denoted $P_{\ell 2}$ below) have rates proportional to the charged lepton mass squared m_ℓ^2, due to helicity considerations. Therefore, the electronic mode P_{e2} is strongly suppressed with respect to the muonic one $P_{\mu 2}$. Moreover, the ratio R_P of leptonic decay rates of the same meson P can be computed very precisely, since it does not dependent on non-perturbative terms, usually the dominant source of errors in the theoretical evaluation of meson decay rates. In particular, the SM prediction for the ratio $R_K = \Gamma(K_{e2})/\Gamma(K_{\mu 2})$ is [4]

$$R_K^{\text{SM}} = \left(\frac{m_e}{m_\mu}\right)^2 \left(\frac{m_K^2 - m_e^2}{m_K^2 - m_\mu^2}\right)^2 (1 + \delta R_{\text{QED}}) = (2.477 \pm 0.001) \times 10^{-5}, \tag{1}$$

where $\delta R_{\text{QED}} = (-3.79 \pm 0.04)\%$ is an electromagnetic correction. Both the helicity suppression and the preciseness of the theoretical prediction make the ratio R_K very sensitive to possible lepton flavour violating terms of many BSM scenarios, probing the SM lepton universality. Within extensions of the SM involving two Higgs doublets, R_K is sensitive to lepton flavour violating effects induced by loop processes with the charged Higgs boson ($H^\pm$) exchange [5, 6]. A recent study [7] has concluded that R_K can be enhanced by $\mathcal{O}(1\%)$ within the Minimal Supersymmetric Standard Model. However, the potential new physics effects are constrained to $\mathcal{O}(1‰)$ by other observables such as $B_s \to \mu^+\mu^-$ and $B_u \to \tau\nu$ decay rates [8]. On the other hand, R_K is sensitive to the neutrino mixing parameters within SM extensions involving a fourth generation of quarks and leptons [9].

3.1. Strategy of the measurement

The analysis strategy is based on counting the numbers of reconstructed K_{e2} and $K_{\mu 2}$ candidates collected concurrently. The simultaneous collection of both samples has the advantage of being independent of the absolute beam flux measurement and less dependent on potential systematic effects: trigger efficiencies and time dependent biases cancel at first order. The study is performed independently for 40 data samples (10 bins of reconstructed lepton momentum and 4 samples with different data taking conditions) by computing the ratio R_K as

$$R_K = \frac{1}{D} \cdot \frac{N(K_{e2}) - N_{bkg}(K_{e2})}{N(K_{\mu 2}) - N_{bkg}(K_{\mu 2})} \cdot \frac{A(K_{\mu 2})}{A(K_{e2})} \cdot \frac{f_\mu \times \epsilon(K_{\mu 2})}{f_e \times \epsilon(K_{e2})} \cdot \frac{1}{f_{\text{LKr}}}, \tag{2}$$

where $N(K_{\ell 2})$ are the numbers of selected $K_{\ell 2}$ candidates ($\ell = e, \mu$), $N_{bkg}(K_{\ell 2})$ are the numbers of background events, $A(K_{\mu 2})/A(K_{e2})$ is the geometric acceptance correction, f_ℓ are the efficiencies of e/μ identification, $\epsilon(K_{\ell 2})$ are the trigger efficiencies, f_{LKr} is the global efficiency of the LKr calorimeter readout (which provides the information used for electron identification), and $D = 150$ is the downscaling factor of the $K_{\mu 2}$ trigger. A Monte Carlo (MC) simulation is used to evaluate the acceptance correction and the geometric part of the acceptances for most background processes entering the computation of $N_{bkg}(K_{\ell 2})$. Particle identification, trigger and readout efficiencies and the beam halo background are measured directly from control data samples. Both f_ℓ and $\epsilon(K_{\ell 2})$ are well above 99%.

3.2. $K^\pm \to \ell^\pm \nu$ selection

Most of the selection criteria are common to both K_{e2} and $K_{\mu 2}$ decay modes. These criteria identify events with a single-track topology and without clusters in the LKr not associated to the track via direct energy deposition or bremsstrahlung. K_{e2} and $K_{\mu 2}$ decays are distinguished by applying kinematic and particle identification criteria. Kinematic identification is based on the reconstructed squared missing mass $M_{miss}^2(\ell) = (P_K - P_\ell)^2$, where P_K and P_ℓ ($\ell = e, \mu$) are the kaon and the lepton 4-momenta, respectively. The mean P_K is monitored on spill basis using fully reconstructed $K^+ \to \pi^+\pi^+\pi^-$ decays; P_ℓ is computed from the measured track momentum, assuming the lepton mass m_ℓ. A selection condition $M_1^2 < M_{miss}^2(\ell) < M_2^2$ is applied; the $M_{1,2}^2$ values vary across the lepton momentum bins according to resolution. Lepton type identification is based on the ratio E/p of the energy E released in the LKr calorimeter by the lepton track to its momentum p measured by the spectrometer. A cut on E/p between 0.95 (0.9 at $p < 25$ GeV/c) and 1.1 identifies the electrons; tracks with $E/p < 0.8$ are classified as muons.

Journal of Physics: Conference Series **631** (2015) 012040 doi:10.1088/1742-6596/631/1/012040

3.3. Results

The distributions of the reconstructed squared missing mass $M^2_{miss}(\ell)$ of $K_{\ell 2}$ candidates ($\ell = e, \mu$) compared with the sums of normalised estimated signal and background components are shown in Figs. 3 and 4. The numbers of selected K_{e2} and $K_{\mu 2}$ candidates are 145,958 and 4.2817×10^7, respectively. The background contamination in the K_{e2} sample has been estimated by MC simulations and, where possible, by direct measurements to be $(10.95 \pm 0.27)\%$. The largest background contribution is the $K_{\mu 2}$ decay with a mis-identified muon via the catastrophic bremsstrahlung process in the LKr calorimeter. To reduce the uncertainty due to background subtraction, the muon mis-identification probability $P_{\mu e}$ has been measured as a function of momentum by reconstructing $K_{\mu 2}$ decays and selecting those muons traversing a lead bar, which was temporary installed in front of the LKr calorimeter to reduce electron contamination and covered about 10% of the calorimeter acceptance. The measured probability has been also corrected for the ionization energy loss and bremsstrahlung of muons passing through the lead. The contributions to the systematic uncertainty of the result include the uncertainties on the backgrounds, amount of material upstream of the spectrometer magnet (which influences the detection efficiency via bremsstrahlung and scattering), beam simulation, spectrometer alignment, particle identification and trigger efficiency. The result of the measurement, combined over the 40 independent samples taking into account correlations between the systematic errors, is

$$R_K = (2.488 \pm 0.007_{stat} \pm 0.007_{syst}) \times 10^{-5} = (2.488 \pm 0.010) \times 10^{-5}. \quad (3)$$

The stability of R_K measurements in lepton momentum bins and for the separate data samples is shown in Figs. 5 and 6. The result is consistent with the SM expectation, and the achieved precision dominates the world average, while its error is still one order of magnitude higher than the theoretical prediction uncertainty.

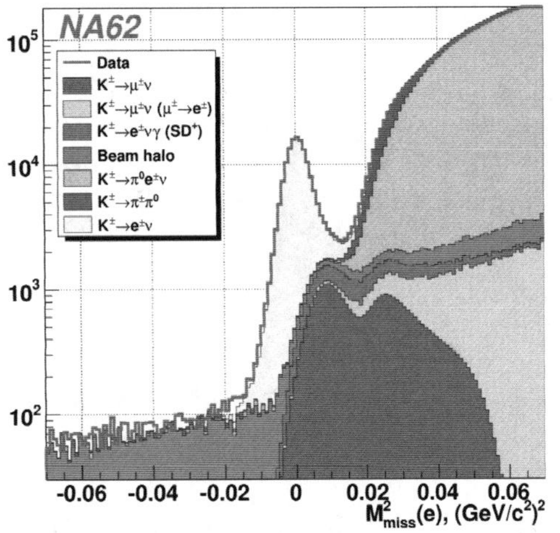

Figure 3. Distribution of the reconstructed squared missing mass $M^2_{miss}(e)$ of K_{e2} candidates compared with the sums of normalised estimated signal and background components.

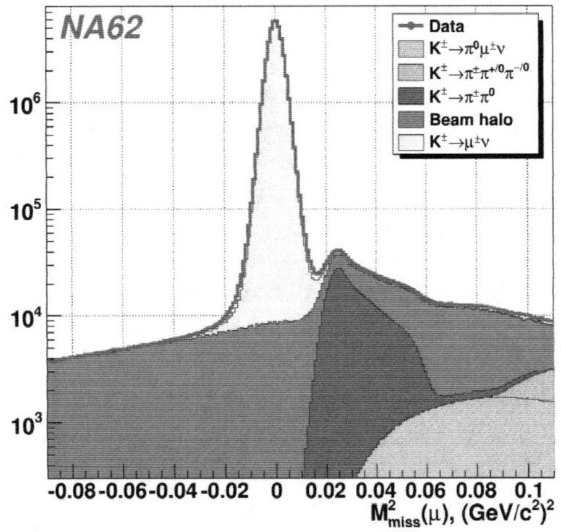

Figure 4. Distribution of the reconstructed squared missing mass $M^2_{miss}(\mu)$ of $K_{\mu 2}$ candidates compared with the sums of normalised estimated signal and background components.

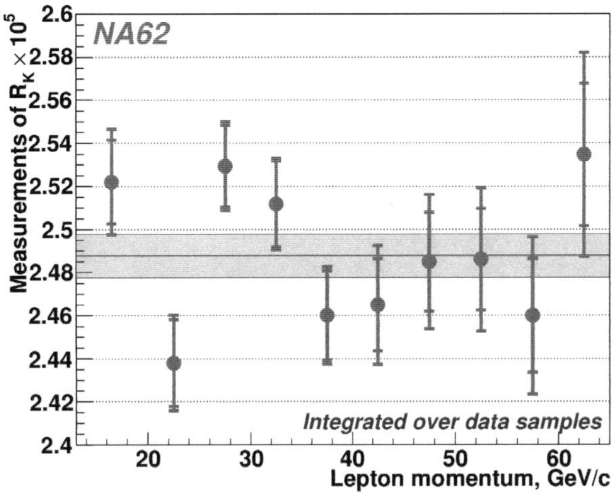

Figure 5. Stability of the R_K measurement versus lepton momentum. The result of the combined fit over the 10 data bins is shown by the horizontal band.

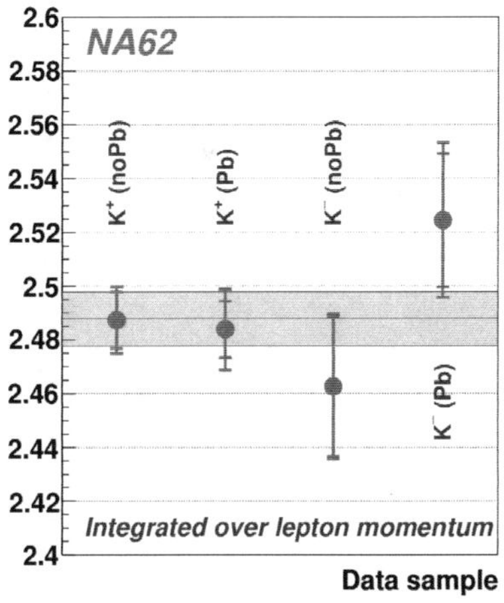

Figure 6. Stability of the R_K measurement for independent data samples. The "Pb/noPb" labels indicate samples with and without the lead bar. The result of the combined fit over the 4 data bins is shown by the horizontal band.

4. Search for Heavy Neutrinos with the NA62-R_K experiment

Neutrinos are strictly massless within the SM, due to the absence of right-handed neutrino states. However, since the observation of neutrino oscillations [10] has unambiguously demonstrated the massive nature of neutrinos, right-handed neutrino states must be included. A natural extension of the SM involves the inclusion of sterile neutrinos which mix with ordinary neutrinos to explain several open questions. An example of such a theory is the Neutrino Minimal Standard Model (νMSM) [11, 12]. In this model, three massive right-handed neutrinos are introduced to explain simultaneously neutrino oscillations, dark matter and baryon asymmetry of the Universe: the lightest (N_1) has mass $\mathcal{O}(1\,\mathrm{keV})$ and is a dark matter candidate; the others (N_2, N_3), with masses ranging from 100 MeV/c^2 to few GeV/c^2, are responsible for the masses of the SM neutrinos (via see-saw mechanism) and introduce extra CP violating phases to account for baryon asymmetry.

Experimentally, searches for heavy sterile neutrinos are performed either by looking for indirect evidence of their production, e.g. by exploring the positive missing mass spectrum of reconstructed meson decays, or directly, by aiming to detect their decay in a possible final state. Decay searches are usually more sensitive than the production ones, but they have the disadvantage of being model-dependent. The possibility of performing a production search for heavy neutrinos in $K^\pm \to \mu^\pm \nu_H$ decays, using a subset of the NA62-R_K data sample collected for the R_K measurement, is currently under investigation. The limits of such a measurement are related to the missing mass resolution and to the background, mainly from beam pions and muon halo. Such analysis is expected to be competitive with the most stringent limits from production searches in kaon decays [13, 14] for heavy neutrino masses above $300\,\mathrm{MeV}/c^2$.

5. Search for Dark Photons with the NA48/2 experiment

A new, light vector gauge boson (called dark photon, A'), with weak couplings to charged SM fermions, has been recently introduced to explain the observed rise in the cosmic-ray positron fraction with energy and the muon gyromagnetic ratio $(g-2)$ measurement [15]. In a rather general set of models, the interaction of the dark photon with the visible sector is through kinetic mixing with the Standard Model hypercharge $U(1)$ [16]. In these models, the new coupling constant ε is proportional to the electric charge and the dark photon couples in exactly the same way to quarks and leptons.

5.1. Strategy of the search

The full NA48/2 data sample is used to search for dark photons produced in π^0 decays and subsequently decayed, using the following chain: $K^\pm \to \pi^\pm\pi^0$, $\pi^0 \to \gamma A'$, $A' \to e^+e^-$, which has three charged particles and a photon in the final state [17]. The expected branching ratio of the π^0 decay is [18]

$$\mathcal{B}(\pi^0 \to \gamma A') = 2\varepsilon^2 \left(1 - \frac{m_{A'}^2}{m_{\pi^0}^2}\right)^3 \mathcal{B}(\pi^0 \to \gamma\gamma), \tag{4}$$

with a kinematic suppression of the decay rate for high dark photon masses $m_{A'}$ approaching m_{π^0}. In the mass range $2m_e \ll m_{A'} < m_{\pi^0}$ accessible in this analysis, the dark photon is below threshold for all decays into SM charged fermions except $A' \to e^+e^-$, while the allowed loop-induced decays ($A' \to 3\gamma$, $A' \to \nu\bar{\nu}$) are highly suppressed. Therefore, assuming that the dark photon decays only into SM particles, $B(A' \to e^+e^-) \approx 1$. The expected total dark photon decay width is then [19]

$$\Gamma(A' \to e^+e^-) = \frac{1}{3}\alpha\varepsilon^2 m_{A'}\sqrt{1 - \frac{4m_e^2}{m_{A'}^2}}\left(1 + \frac{2m_e^2}{m_{A'}^2}\right). \tag{5}$$

From eq. 5 follows that the maximum dark photon mean path in the NA48/2 reference frame for a fully reconstructed event does not exceed 10 cm and can be neglected for $m_{A'} > 10$ MeV/c^2 and $\varepsilon^2 > 5 \times 10^{-7}$, conditions which are satisfied for the present analysis. The dark photon signature is thus characterised by a decay in e^+e^- at the production point (prompt decay), and the 3-track vertex topology can be used without significant acceptance loss. Such a signature, however, is the same of the Dalitz decay $\pi_D^0 \to e^+e^-\gamma$, which therefore represents an irreducible background and determines the sensitivity. The largest π_D^0 sample, and therefore the largest sensitivity, is obtained from the $K^\pm \to \pi^\pm\pi_D^0$ decays (denoted $K_{2\pi D}$ below). The selected candidates mainly originate from $K_{2\pi D}$ decays, with 0.15% coming from the semileptonic $K^\pm \to \pi_D^0\mu^\pm\nu$ decays (denoted $K_{\mu 3D}$ below). Correcting the observed number of candidates for acceptance and trigger efficiency, the total number of $K^\pm$ decays in the 98 m long fiducial decay region for the analysed data sample is found to be $N_K = (1.55 \pm 0.05) \times 10^{11}$, where the quoted error is dominated by the external uncertainty on the π_D^0 decay branching ratio $\mathcal{B}(\pi_D^0)$. A dark photon mass scan (i.e. a search for the dark photon assuming different mass hypotheses with a variable mass step) is performed over the distribution of the e^+e^- invariant mass m_{ee}. A dark photon produced in the π_D^0 decay and decaying promptly to e^+e^- would produce a narrow spike in the spectrum. The mass step of the scan and the width of the dark photon signal mass window around the assumed mass are determined by the resolution δm_{ee} on the e^+e^- invariant mass m_{ee}. The mass window width has been optimised with MC simulations to obtain the highest sensitivity to the dark photon signal, determined by a trade-off between π_D^0 background fluctuation and signal acceptance. In total, 398 dark photon mass hypotheses are tested in the range 10 MeV/$c^2 \le m_{ee} < 125$ MeV/c^2, in which the lower limit is determined by the limited precision of MC simulation of background at low mass, while the upper limit is dictated by the signal acceptance dropping to zero. Upper limits at 90% CL on $\mathcal{B}(\pi^0 \to \gamma A')$ in each dark photon mass hypothesis under the assumption $\mathcal{B}(A' \to e^+e^-) \approx 1$ (which holds for $m_{A'} < 2m_\mu$ if A' decays to SM particles only) are computed using the relation

$$\mathcal{B}(\pi^0 \to \gamma A') = \frac{N_{\rm DP}}{N_K} \left[\mathcal{B}(K_{2\pi})A(K_{2\pi D}) + \mathcal{B}(K_{\mu 3})A(K_{\mu 3D})\right]^{-1}, \qquad (6)$$

where $N_{\rm DP}$ is the number of $A' \to e^+e^-$ decay candidates, while $A(K_{2\pi D})$ and $A(K_{\mu 3D})$ are the acceptances of the $K_{2\pi D}$ and $K_{\mu 3D}$ decays, evaluated with MC simulation for each considered dark photon mass hypothesis. The largest uncertainty ($\approx 3\%$) on the computed $\mathcal{B}(\pi^0 \to \gamma A')$ is the external one due to $\mathcal{B}(\pi^0_D)$ entering via N_K, and is neglected. Upper limits at 90% CL on the mixing parameter ε^2 in each considered dark photon mass hypothesis are calculated from those on $\mathcal{B}(\pi^0 \to \gamma A')$ using eq. 4.

5.2. Event selection

The $K_{2\pi D}$ event selection requires a three-track vertex reconstructed in the fiducial decay region formed of a pion ($\pi^\pm$) candidate track and two opposite-sign electron ($e^\pm$) candidate tracks. Charged particle identification is based on the ratio E/p of the energy E released in the LKr calorimeter by the track to its momentum p measured by the spectrometer. A cut on E/p between 0.85 and 1.1 identifies the electrons; tracks with $E/p < 0.85$ are classified as pions. Furthermore, a single isolated LKr energy deposition cluster is required and considered as the photon candidate. A set of tight selection criteria is applied to the energies of the final state particles, their timing and spatial separations. The total reconstructed $\pi^\pm\pi^0$ momentum is required to be consistent with the beam momentum spectrum, and its transverse component with respect to the nominal beam axis is required to be consistent with no missing momentum. The reconstructed invariant mass of the $\pi^\pm\pi^0$ system (Fig. 7) is required to be consistent with the $K^\pm$ mass. A sample of 4.687×10^6 fully reconstructed π^0_D decay candidates in the e^+e^- invariant mass range $m_{ee} > 10$ MeV$/c^2$ with a negligible background is selected (Fig. 8).

5.3. Preliminary results

Confidence intervals at 90% CL for the number of $A' \to e^+e^-$ decay candidates in each mass hypothesis of the dark photon scan are computed from the numbers of observed data events in the signal region (N_{obs}), the numbers of π^0_D background events expected from MC simulation (N_{exp}) and the uncertainty on the number of expected background events, by using the Rolke-Lopez method [20, 21] assuming Poisson-distributed signal and Gaussian-distributed background. The results obtained in the neighbouring mass hypotheses are highly correlated, as the signal mass window is about 6 times wider than the mass step of the dark photon scan. For the preliminary results, it is assumed conservatively that $N_{obs} = N_{exp}$ in cases when $N_{obs} < N_{exp}$, as the employed implementation of the method (from the ROOT package) has been found to underestimate the upper limits in that case.

The resulting preliminary dark photon exclusion limits, along with constraints from other experiments [22], the band of parameter space where the discrepancy between the measured and calculated muon $g-2$ values falls into the $\pm 2\sigma$ range [15, 23] due to the dark photon contribution, and the region excluded by the electron $g-2$ measurement, are presented in Fig. 9. The obtained upper limits on ε^2 represent an improvement over the existing results in the dark photon mass range 10-60 MeV$/c^2$, and exclude the muon $g-2$ band in the range 10-100 MeV$/c^2$. The most stringent limits (6×10^{-7}) are achieved at $m_{A'} \approx 20$ MeV$/c^2$ where the acceptance for the full decay chain is the highest (reaching 2.5%). The limits weaken at higher $m_{A'}$ due to both the kinematic suppression of the $\pi^0 \to \gamma A'$ decay and the decreasing acceptance.

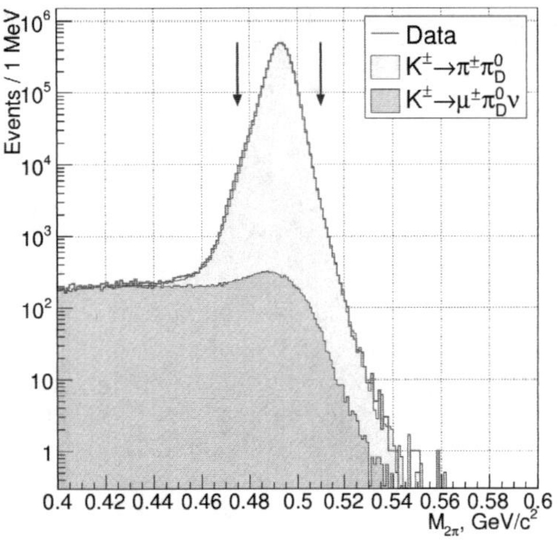

Figure 7. Reconstructed $\pi^{\pm}\pi^0_D$ invariant mass ($m_{2\pi}$) distributions of the data and simulated $K_{2\pi D}$ and $K_{\mu 3D}$ samples. The selection condition 475 MeV/c^2 < $m_{2\pi}$ < 510 MeV/c^2 is illustrated with arrows.

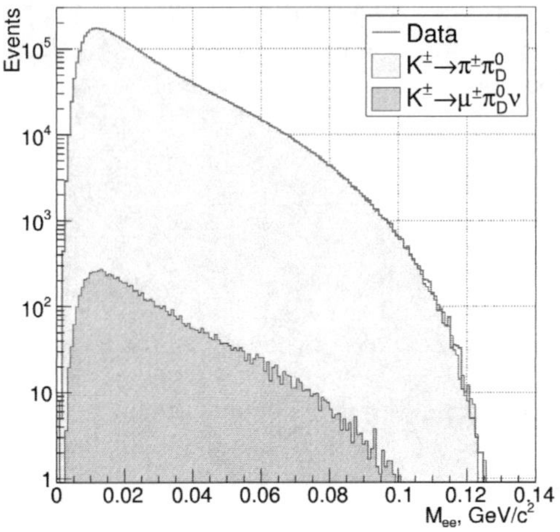

Figure 8. Reconstructed e^+e^- invariant mass distributions of the data and simulated $K_{2\pi D}$ and $K_{\mu 3D}$ samples.

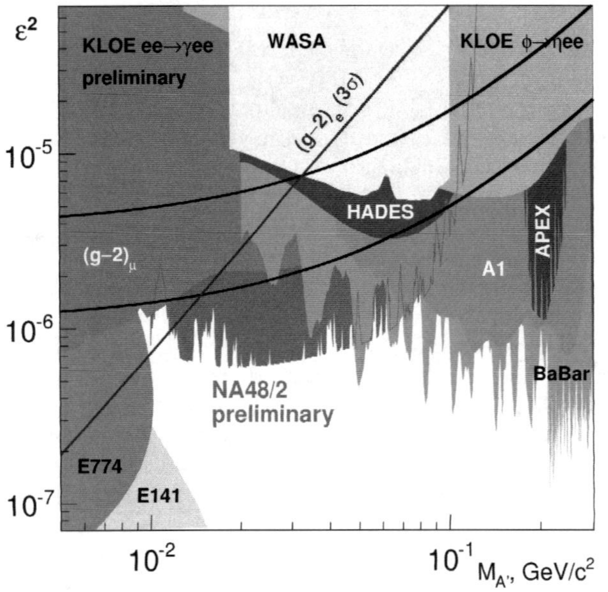

Figure 9. The NA48/2 preliminary upper limits at 90% CL on the mixing parameter ε^2 versus the dark photon mass $m_{A'}$, compared to the other published exclusion limits from meson decay, beam dump and e^+e^- collider experiments [22]. Also shown are the band where the consistency of theoretical and experimental values of muon $g-2$ improves to $\pm 2\sigma$ or less, and the region excluded by the electron $g-2$ measurement [15, 23].

6. Prospects for the new NA62 experiment

A major beam and detector upgrade is currently being completed, aimed to achieve the required physics specifications to perform the precise ($\sim 10\%$) measurement of the $K^+ \to \pi^+\nu\bar{\nu}$ branching ratio [24]. The key features of the new NA62 detector are

- the positive identification, precise time and momentum measurements of kaon candidates;

Journal of Physics: Conference Series **631** (2015) 012040 doi:10.1088/1742-6596/631/1/012040

- a new low-material spectrometer, contained in vacuum;
- the introduction of hermetic photon vetoes up to 50 mrad;
- redundant particle identification for the kaon decay products.

A detailed description of the status of the detector commissioning can be found in Ref. [25]. The NA62 experiment is expected to collect $\sim 1.5 \times 10^{13}$ K^+ decays in three years of data taking, starting from 2015. The enhanced performances of the new experimental apparatus and the unprecedented size of the kaon sample to be collected for the $K^+ \to \pi^+ \nu\bar{\nu}$ analysis allow to substantially improve previous measurements, as the R_K, and search unexplored regions for many forbidden kaon decays.

6.1. The R_K measurement

The NA62 experiment is expected to collect a K_{e2} sample at least one order of magnitude larger than the NA62-R_K one, therefore reducing the statistical uncertainty by at least a factor three. In addition, the background contamination to the K_{e2} sample, which contributes to $\sim 50\%$ of the total systematic error of the current R_K measurement, will be reduced at negligible level, thanks to the combined effect of photon vetoes (on kaon decays with π^0s or γs), particle identification (on $K_{\mu 2}$ decays) and kaon tagging (on the muon halo). These two effects alone already contribute to an improvement in the R_K precision of a factor 2.5, corresponding to an absolute expected precision of 0.16%, to be compared with the current theoretical precision of 0.04% [4].

6.2. Search for Heavy Neutrinos

In NA62 the missing mass resolution will improve by at least a factor of two, thanks to the new spectrometer operated in vacuum, and the background from kaon decays will be better controlled thanks to the full coverage veto system. In addition, kaon indentification will reduce significantly the contribution of the muon halo. Therefore, a production search for heavy neutrinos in $K^\pm \to \mu^\pm \nu_H$ decays with the NA62 detector will improve significantly the current limits [13, 14], at least for neutrino masses above 300 MeV/c^2.

6.3. Search for Dark Photons

The search for the prompt $A' \to e^+e^-$ decay, as the one performed with the NA48/2 experiment, is limited by the irreducible π^0_D background. The sensitivity to ε^2 achievable with the employed method scales as the inverse square root of the integrated beam flux, and therefore this technique is unlikely to advance much below $\varepsilon^2 = 10^{-7}$ in the near future, by exploiting the larger π^0 sample to be collected by the NA62 experiment. On the other hand, a search with the NA62 experiment for a long-lived (i.e. low $m_{A'}$ and low ε^2) dark photon produced in π^0 decays would be less affected by the π^0_D background, due to the different signature involving a displaced vertex, and its sensitivity is worth investigating.

6.4. Search for LNFV K^+ and π^0 decays

The NA62 data sample will consist of $\sim 1.5 \times 10^{13}$ K^+ decays, which will be accompanied by $\sim 3 \times 10^{12}$ π^0 decays from $K^+ \to \pi^+\pi^0$ (BR $\approx 21\%$). Studies of the prospects for searches for lepton-flavour (LF) or -number (LN) violating and other forbidden decays with NA62 are underway. Preliminary estimates of the single-event sensitivities (defined as the inverse of the number of accepted decays) give results at the level of 10^{-12} for K^+ decays to states such as $\pi^+\mu^\pm e^\mp$ (LFV), $\pi^-\mu^+e^+$ (LFNV), and $\pi^-e^+e^+$ or $\pi^-\mu^+\mu^+$ (LNV); and at the level of 10^{-11} for π^0 decays to $\mu^\pm e^\mp$ [26].

7. Conclusions

Our understanding of the fundamental interactions of nature has been deeply enhanced by kaon physics. Nowadays, after more than 60 years since their discovery, kaons still play a unique role in the experimental challenge of probing the SM flavour sector with ever-increasing precision. Both the NA48/2 and NA62 experiments have and still contribute to this effort: their recent results have been presented here, together with the prospects for the forthcoming three-year long data taking. The most precise measurement of R_K to date has been performed from a sample of about $\sim 150 \times 10^3$ candidates collected by the NA62-R_K experiment. The result is $R_K = (2.488 \pm 0.010) \times 10^{-5}$, consistent with the earlier measurements and with the SM expectation. The experimental uncertainty on R_K is still an order of magnitude larger than the uncertainty on the SM prediction. The possibility of performing a production search for heavy neutrinos in $K^\pm \to \mu^\pm \nu_H$ decays, using a subset of the NA62-R_K data sample collected for the R_K measurement, is being investigated. Such analysis is expected to be competitive for heavy neutrino masses above $300\,\mathrm{MeV}/c^2$. Preliminary results on dark photon search in π^0 decays, with the NA48/2 experiment, are reported: no signal is observed, and the obtained upper limits of the order of 10^{-6} (90% CL) on the mixing parameter ε^2 improve over the world data in the mass range 10-60 MeV/c^2. Combined with the other available data, this result rules out the dark photon as an explanation for the muon $(g-2)$ measurement, assuming it couples to SM particles only. The forthcoming three-year long data taking period of the NA62 experiment, after a major beam and detector upgrade, will allow to substantially improve previous measurements, as the R_K, and search unexplored regions for heavy sterile neutrinos, dark photons and many forbidden kaon and π^0 decays, as the LNFV modes.

References

[1] Batley J *et al.* (NA48/2 Collaboration) 2007 *Eur.Phys.J.* **C52** 875–891 (*Preprint* 0707.0697)
[2] Fanti V *et al.* (NA48 Collaboration) 2007 *Nucl.Instrum.Meth.* **A574** 433–471
[3] Lazzeroni C *et al.* (NA62 Collaboration) 2013 *Phys.Lett.* **B719** 326–336 (*Preprint* 1212.4012)
[4] Cirigliano V and Rosell I 2007 *Phys.Rev.Lett.* **99** 231801 (*Preprint* 0707.3439)
[5] Masiero A, Paradisi P and Petronzio R 2006 *Phys.Rev.* **D74** 011701 (*Preprint* hep-ph/0511289)
[6] Masiero A, Paradisi P and Petronzio R 2008 *JHEP* **0811** 042 (*Preprint* 0807.4721)
[7] Girrbach J and Nierste U 2012 (*Preprint* 1202.4906)
[8] Fonseca R, Romao J and Teixeira A 2012 *Eur.Phys.J.* **C72** 2228 (*Preprint* 1205.1411)
[9] Lacker H and Menzel A 2010 *JHEP* **1007** 006 (*Preprint* 1003.4532)
[10] Olive K *et al.* (Particle Data Group) 2014 *Chin.Phys.* **C38** 090001
[11] Asaka T, Blanchet S and Shaposhnikov M 2005 *Phys.Lett.* **B631** 151–156 (*Preprint* hep-ph/0503065)
[12] Asaka T and Shaposhnikov M 2005 *Phys.Lett.* **B620** 17–26 (*Preprint* hep-ph/0505013)
[13] Yamazaki T, Ishikawa T, Akiba Y, Iwasaki M, Tanaka K *et al.* 1984 *Conf.Proc.* **C840719** I.262
[14] Artamonov A *et al.* (E949 Collaboration) 2015 *Phys.Rev.* **D91** 052001 (*Preprint* 1411.3963)
[15] Pospelov M 2009 *Phys.Rev.* **D80** 095002 (*Preprint* 0811.1030)
[16] Holdom B 1986 *Phys.Lett.* **B166** 196
[17] Goudzovski E (NA48/2 Collaboration) 2014 (*Preprint* 1412.8053)
[18] Batell B, Pospelov M and Ritz A 2009 *Phys.Rev.* **D80** 095024 (*Preprint* 0906.5614)
[19] Batell B, Pospelov M and Ritz A 2009 *Phys.Rev.* **D79** 115019 (*Preprint* 0903.3396)
[20] Rolke W A and Lopez A M 2001 *Nucl.Instrum.Meth.* **A458** 745–758 (*Preprint* hep-ph/0005187)
[21] Rolke W A, Lopez A M and Conrad J 2005 *Nucl.Instrum.Meth.* **A551** 493–503 (*Preprint* physics/0403059)
[22] Lees J *et al.* (BaBar Collaboration) 2014 *Phys.Rev.Lett.* **113** 201801 (*Preprint* 1406.2980)
[23] Davoudiasl H, Lee H S and Marciano W J 2014 *Phys.Rev.* **D89** 095006 (*Preprint* 1402.3620)
[24] Ruggiero G (NA62) 2013 *PoS* **KAON13** 032
[25] Hahn F (NA62) 2014 *this conference proceedings*
[26] Moulson M (NA62) 2013 *PoS* **KAON13** 013 (*Preprint* 1306.3361)

Prospects for $K^+ \rightarrow \pi^+ \nu \bar{\nu}$ observation at CERN in NA62

F Hahn on behalf of the the **NA62 Collaboration:**

G Aglieri Rinella, R Aliberti, F Ambrosino, B Angelucci, A Antonelli, G Anzivino, R Arcidiacono, I Azhinenko, S Balev, J Bendotti, A Biagioni, C Biino, A Bizzeti, T Blazek, A Blik, B Bloch-Devaux, V Bolotov, V Bonaiuto, M Bragadireanu, D Britton, G Britvich, N Brook, F Bucci, F Butin, E Capitolo, C Capoccia, T Capussela, V Carassiti, N Cartiglia, A Cassese, A Catinaccio, A Cecchetti, A Ceccucci, P Cenci, V Cerny, C Cerri, O Chikilev, R Ciaranfi, G Collazuol, P Cooke, P Cooper, G Corradi, E Cortina Gil, F Costantini, A Cotta Ramusino, D Coward, G D'Agostini, J Dainton, P Dalpiaz, H Danielsson, J Degrange, N De Simone, D Di Filippo, L Di Lella, N Dixon, N Doble, V Duk, V Elsha, J Engelfried, T Enik, V Falaleev, R Fantechi, L Federici, M Fiorini, J Fry, A Fucci, L Fulton, S Gallorini, L Gatignon, A Gianoli, S Giudici, L Glonti, A Goncalves Martins, F Gonnella, E Goudzovski, R Guida, E Gushchin, F Hahn, B Hallgren, H Heath, F Herman, D Hutchcroft, E Iacopini, O Jamet, P Jarron, K Kampf, J Kaplon, V Karjavin, V Kekelidze, S Kholodenko, G Khoriauli, A Khudyakov, Yu Kiryushin, K Kleinknecht, A Kluge, M Koval, V Kozhuharov, M Krivda, Y Kudenko, J Kunze, G Lamanna, C Lazzeroni, R Leitner, R Lenci, M Lenti, E Leonardi, P Lichard, R Lietava, L Litov, D Lomidze, A Lonardo, N Lurkin, D Madigozhin, G Maire, A Makarov, I Mannelli, G Mannocchi, A Mapelli, F Marchetto, P Massarotti, K Massri, P Matak, G Mazza, E Menichetti, M Mirra, M Misheva, N Molokanova, J Morant, M Morel, M Moulson, S Movchan, D Munday, M Napolitano, F Newson, A Norton, M Noy, G Nuessle, V Obraztsov, S Padolski, R Page, V Palladino, A Pardons, E Pedreschi, M Pepe, F Perez Gomez, M Perrin-Terrin, P Petrov, F Petrucci, R Piandani, M Piccini, D Pietreanu, J Pinzino, M Pivanti, I Polenkevich, I Popov, Yu Potrebenikov, D Protopopescu, F Raffaelli, M Raggi, P Riedler, A Romano, P Rubin, G Ruggiero, V Russo, V Ryjov, A Salamon, G Salina, V Samsonov, E Santovetti, G Saracino, F Sargeni, S Schifano, V Semenov, A Sergi, M Serra, S Shkarovskiy, A Sotnikov, V Sougonyaev, M Sozzi, T Spadaro, F Spinella, R Staley, M Statera, P Sutcliffe, N Szilasi, D Tagnani, M Valdata-Nappi, P Valente, M Vasile, V Vassilieva, B Velghe, M Veltri, S Venditti, M Vormstein, H Wahl, R Wanke, P Wertelaers, A Winhart, R Winston, B Wrona, O Yushchenko, M Zamkovsky, A Zinchenko.

Abstract. The rare decays $K \rightarrow \pi \nu\bar{\nu}$ are excellent processes to probe the Standard Model and indirectly search for new physics complementary to the direct LHC searches. The NA62 experiment at CERN SPS aims to collect and analyse $\mathcal{O}$ (10^{13}) kaon decays before the CERN long-shutdown 2 (in 2018). This will allow to measure the $K^+ \rightarrow \pi^+ \nu\bar{\nu}$ branching ratio to a level of 10% accuracy. The experimental apparatus has been commissioned during a first run in autumn 2014.

1. Introduction

The rare decay $K^+ \rightarrow \pi^+ \nu\bar{\nu}$ is an excellent process to study the physics of flavour because of its very clean nature. The strong suppression to the Standard Model (SM) contributions and the remarkable theoretical precision of the SM rate makes this decay sensitive to possible new degrees of freedom. The NA62 experiment has been designed and built to study this decay mode for which the SM calculations [1] predict a branching ratio of: $BR(K^+ \rightarrow \pi^+ \nu\bar{\nu})(SM) = (7.8 \pm 0.7)\ 10^{-11}$.

The presence of two undetectable neutrinos in the final state reduces the signal signature to one high momentum charged track with nothing else, which has to be discriminated against background coming from all other kaon decays.

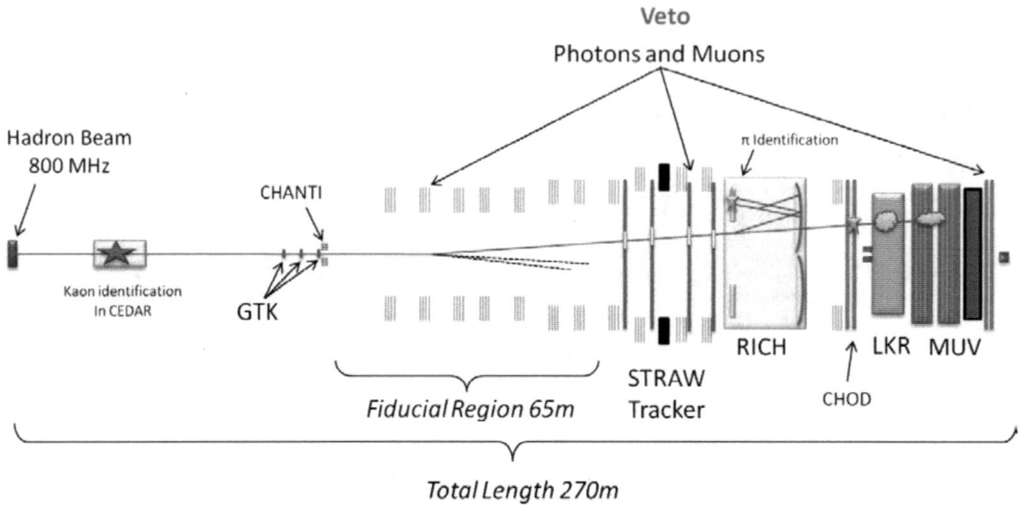

Figure 1 Schematic view of the NA62 experiment showing the main sub-detectors (not to scale). The main elements for the detection of the K^+ decay products are spread along a 170 m long region starting about ~100 m downstream of the beryllium target. Useful K^+ decays will be detected from a 65 m long fiducial region.

On the experimental side NA62 has chosen to use high momentum kaons (75 GeV/c) from a secondary beam produced by the CERN SPS accelerator with a technique of in-flight-decay. The high momentum of the incoming beam improves the background rejection and sets the longitudinal scale of the experiment. (see Figure 1). To achieve the required background suppression different principles have to be combined and the resulting requirements are outlined here:

- **High intensity** and **good timing:** a high intensity kaon beam is essential in order to reach sensitivity to a branching ratio of $\mathcal{O}(10^{-10})$. The incoming secondary beam from the SPS provides a particle rate of 750 MHz, containing about 6% of kaons delivering roughly 45 $*10^6$ decays in the fiducial region per spill. Precise timing (in the range 100 - 150ps) of the K^+ and the π^+ allows precise matching of the particles in the decay. The time resolution is essential in order to keep wrong associations at level below 1%.
- **Low mass Tracking:** The most discriminating variable to distinguish the $K^+ \to \pi^+ \nu\bar{\nu}$ signal from background is the squared missing mass: $m^2_{miss} \equiv (P_K - P_\pi)^2$ where P_K denotes the 4-momentum of the parent particle assumed to be a kaon and P_π is the 4-momentum of the decay particle assumed to be a π^+. In order to exploit this variable, track position and momenta have to be measured with high accuracy in low mass detectors. Low mass detectors are essential because inelastic scattering of beam particles in the detector material can mimic an isolated π^+ appearing like a signal event and hence contribute to the background.
- **Hermetic vetoing for photons and muons:** the kinematic rejection must be accompanied by direct vetoing for photons (in particular for the $K^+ \to \pi^+\pi^0$ background) requiring a typical

inefficiency of 10^{-4} for high energetic photons. The largest fraction of K^+ decays contain muons in the final state, therefore a muon veto system is mandatory both in the trigger and off-line. NA62 uses a plane of fast scintillators (MUV3) for direct muon vetoing, this system is complemented by two sampling hadron calorimeters (MUV1+2), which measure the deposited hadron energy in the event distinguishing hadron showers from muons.

- **Particle ID:** Several detectors (LKR, KTAG, RICH,...) complement the event information with direct evidence on the particle species. The RICH, for example, identifies positively the π^+ in the event providing another factor 100 in the suppression of muons. The LKR measures precisely the electro-magnetic energy with the possibility to identify positrons and pions by distinguished shower profiles. The KTAG tags the incoming kaon with 100ps time resolution and can reject pion or protons from beam gas interactions.

In order to reduce passive material in the acceptance region all the detectors upstream of the RICH are installed inside the beam vacuum. A side effect of the high intensity beam is that most detectors have to be able to cope with high particle rates. In particular the GTK is exposed to 750 MHz of beam particles and the KTAG/CEDAR has to deal with about 50 MHz of kaon rate. The downstream detectors are affected by muon flux of up to 11 MHz which are originating from particle decays or interactions in the target.

The main detectors are described in more detail in the following sections, for completeness, a short summary of all detectors and their features are outlined here (see Figure 1)[2]:

i) **The KTAG** identifies the K^+ component in the beam with respect to the other beam particles by employing an upgraded differential Cherenkov (CEDAR) counter;

ii) The GigaTracker **(GTK)** comprises three Si micro-pixel stations measuring, time, direction and momentum of the beam particles before entering the decay region.

iii) **The STRAW Tracker:** measures the coordinates and momentum of secondary charged particles originating from the decay region. To minimise multiple scattering the chambers are built of ultra-light material and are installed inside the vacuum tank. The four straw chambers are intercepted in the middle by a large aperture dipole magnet (MNP33), providing a vertical B-field of 0.36T.

iv) **The RICH:** is situated downstream of the last straw chamber. It consists of a 17 m long radiator filled with Neon Gas at 1 atm. allowing the separation of pions and muons between 15 and 35 GeV/c.

v) A system of Photon-Veto detectors ensured by:
   - the high-resolution Liquid Krypton electro-magnetic calorimeter (**LKR**),
   - supplemented, at small and forward angles, by an Intermediate Ring Calorimeter (**IRC**) and a Small-Angle Calorimeter, as well as
   - at large angles, by a series of 12 annular photon-veto detectors **(LAV)**.

vi) The Muon-Veto Detectors (**MUV**) are composed of a two-part hadron calorimeter -from which only one module was installed in 2014- followed by additional iron and a scintillator tile array. This system supplements and provides redundancy with respect to the RICH in the detection and rejection of muons.

vii) **The CHANTI:** these detectors are 'guard-ring' scintillating counters surrounding the last GTK station to veto beam particles, which scatter inelastically in the material of GTK3.

viii) **The CHOD:** is a charged-particle hodoscope used in the trigger, covering the acceptance and located between the RICH and the LKR calorimeter.

ix) These detectors are operated and inter-connected with a high-performance trigger and data-acquisition system (**TDAQ**).

2. KTAG

The choice of using a high momentum beam (75 GeV/c) implies a secondary beam with mixed composition because at this energy kaons cannot be efficiently separated from pions and protons. In NA62 the beam intensity is about 17 times higher than the effective kaon rate and, in this context, accidental beam gas events in the vacuum tank can become a critical background for the signal. The KTAG is needed to measure precisely the timing of the mother particle in the decay in order to supress quasi inelastic scattered beam particles, which enter the acceptance of the experiment.

The KTAG consists of a CERN SPS differential Cherenkov counter (called CEDAR) [3], which is upgraded with an external optics and new photo detectors. The CEDAR detectors are optimized to detect Cherenkov light at a specific angle and, at 75 GeV/c, kaons can be selected by choosing the N2 or H2 as radiator gas and the right pressure.
The front extension (called KTAG) is replacing the 8 original photomultipliers (PM's) by light boxes containing each 48 PM's (Figure 2). With this layout the detector can handle the high beam rate and achieves a time resolution of better than 100 ps.

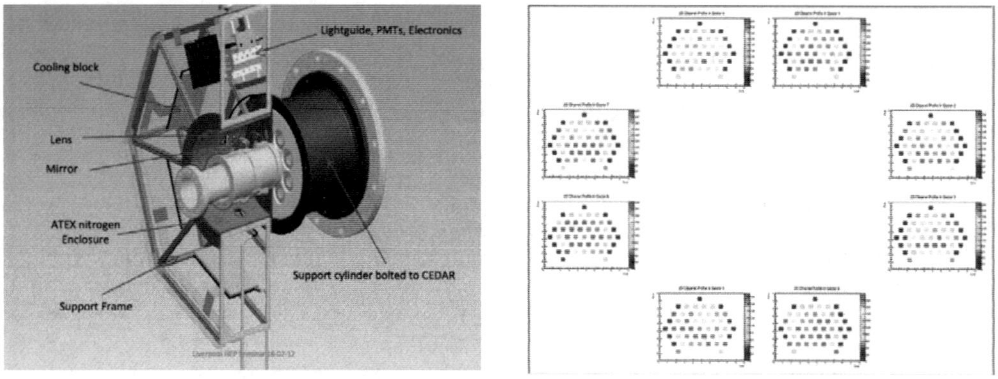

Figure 2 Left: Schematic of the KTAG extension with the new optic and photo-detectors. Right: scatterplots of the 8 light spots (from the 2014 run).

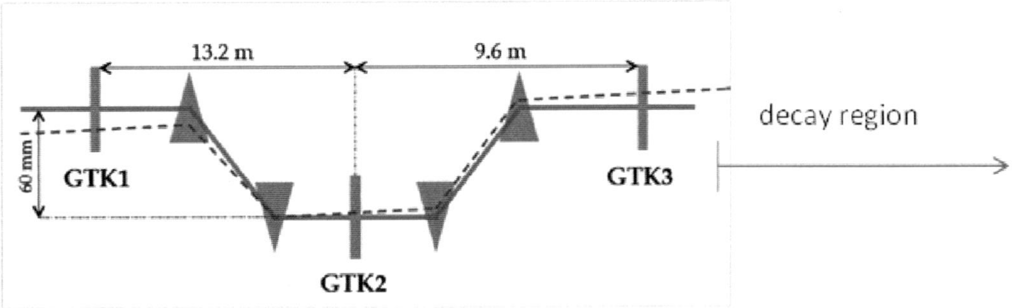

Figure 3 Layout of the three GigaTracker stations with the accompanying achromat magnets.

3. GigaTracker (GTK)

The kaon spectrometer consists of three silicon pixel detectors installed perpendicular to the beam axis. The detectors are traversed by the full beam intensity of 750 MHz and mounted around four achromat magnets as shown in Figure 3 providing measurements of momentum, time and angle of the incoming kaon beam.

The GigaTracker is composed of hybrid silicon pixel detectors (GTK1,2 and 3) and is installed inside the beam vacuum. Each detector has an active area of about 60 x 27 mm^2 containing 18'000 pixels of 300x300 μm^2 size. The full-area sensor is bump bonded to 10 read-out ASIC chips (TDCpix) as shown in Figure 4. The expected fluence for a typical 100 day run time year is 2×10^{14} 1 MeV n equivalent per cm^2 in the central region of the sensor [2].
The pixel dimensions and the distances between stations are adapted to deliver the required momentum and direction resolutions. But most important, and novel for a silicon pixel detector, is the excellent timing performance (σ_t < 200ps per station) allowing to separate sharply the incoming kaon from any other beam particle. To minimize hadronic interactions of the beam the total detector thickness is $\leq$ 0.5mm (see Figure 5) [4].

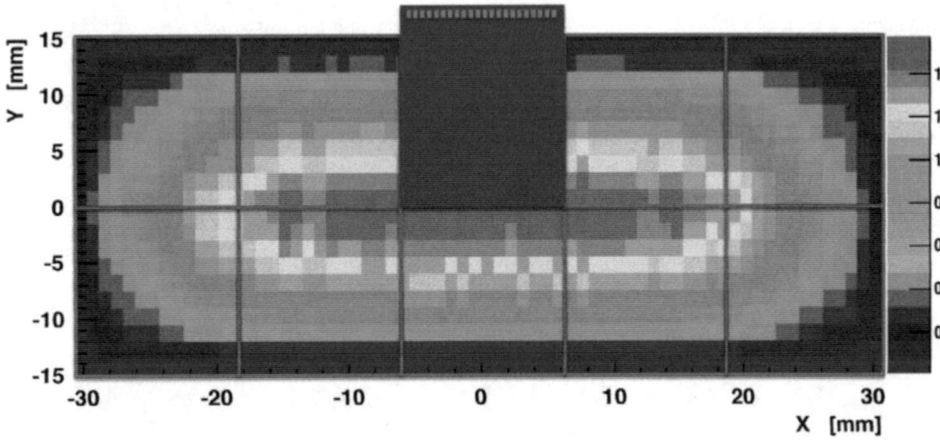

Figure 4 Beam intensity distribution over GTK station 3 (units are MHz/mm^2). One of the ten read-out chip is overlayed.

Component	Material	Thickness [μm]	X_0 [%]
Sensor	Si	200	0.22
Bump Bonds	Pb-Sn	~25	0.001
Readout Chip	Si	100	0.11
Micro Channel plate and cooling fluid	Si + C_6F_{14}	≈150	0.16
Total		< 500	<0.5

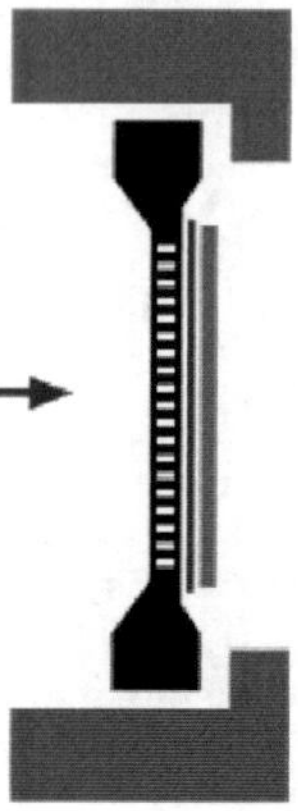

Figure 5 Left side: Material Budget for one station. Right side: sketch of detector cross-section (red = sensor, blue ASIC, black = micro channel plate.

4. Straw Tracker

The purpose of the STRAW Tracker is to measure with good accuracy the direction and the momentum of secondary charged particles originating from the decay region. The spectrometer (see Figure 6) consists of four chambers interleaved in the middle by a large dipole magnet providing a transversal momentum kick of 270 MeV/c vertical B-field of 0.36T. Each chamber is equipped with 1'792 straw tubes, which are positioned in four "Views" providing measurements of four coordinates (see Figure 6 and Figure 7).

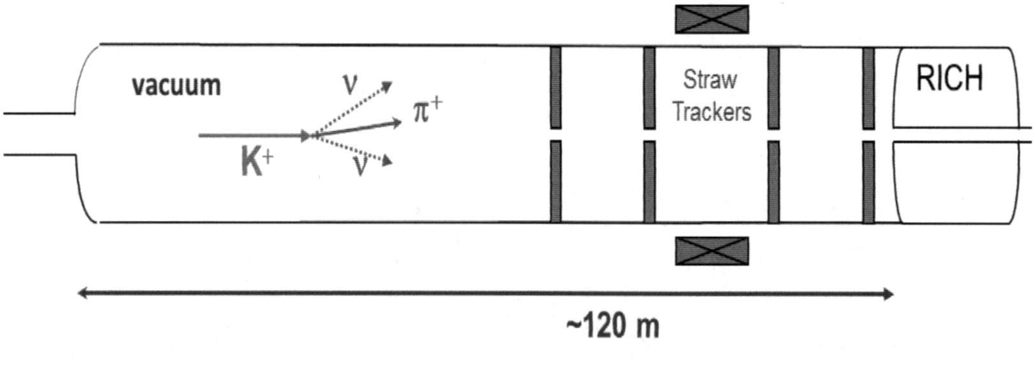

Figure 6 Longitudinal cross-section of the decay vacuum tank, the Straw Spectrometer and the RICH.

In order to minimise multiple scattering, the spectrometer is operated in vacuum without physical separation from the upstream decay volume. The straw tracker technology has been chosen notably because it can be designed without frames and flanges close to the beam, which limits interactions from accompanying beam particles.

The straws are made out of 36 μm thick polyethylene terephthalate film coated with thin layers of copper (50 nm) and gold (20 nm). The straws are fabricated from longitudinal foils through ultrasonic welding in the axial direction [2].

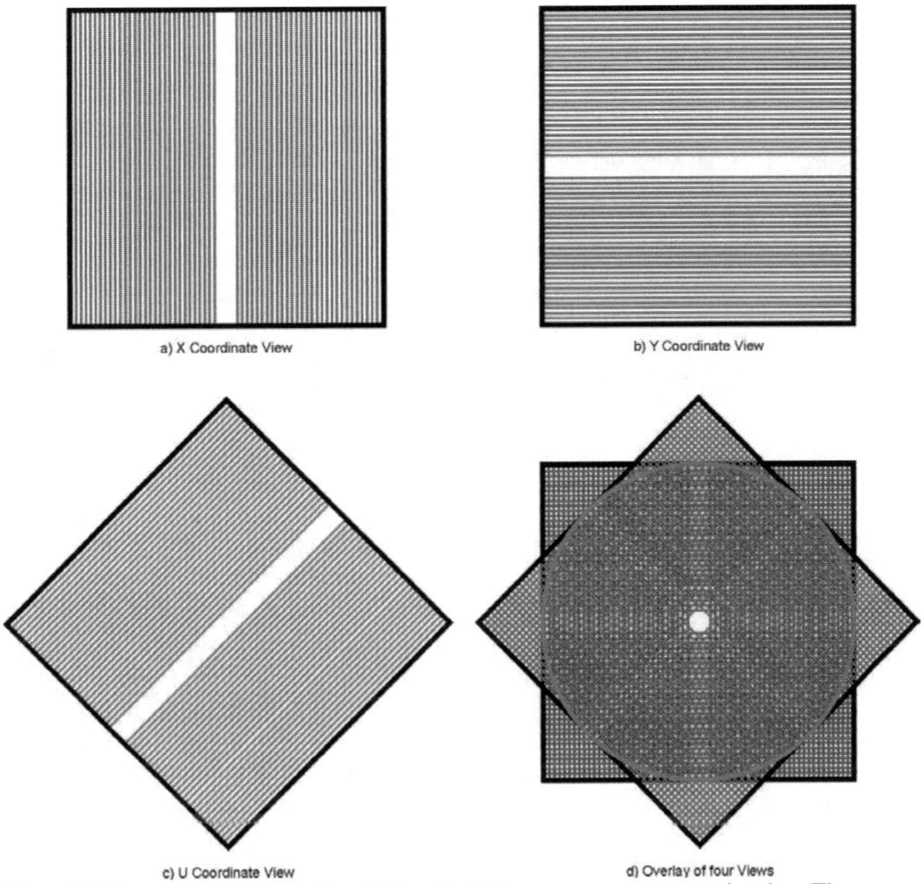

Figure 7 Schematic view of the four "Views" which compose one straw chamber. The gap near the middle is left free forming a hole for the beam passage.

The straw readout is done with an FPGA based TDC imbedded in the front end and a tailored back-end read-out board (SRB) for data extraction and triggering [5].

5. The RICH

A Ring Imaging Cherenkov Counter (RICH) is used to distinguish pions from muons in the momentum range from 15 GeV/*c* to 35 GeV/*c* and to measure the arrival time of final state tracks to better than 100ps.

The RICH (see Figure 8) consists of a steel cylindrical radiator vessel, 17.5 m long and about 4 m in diameter. In the centre it is protruded by a thin aluminium pipe separating the beam vacuum from the radiator gas (neon at 1 atm.). Cherenkov photons are reflected on a mirror mosaic (made of 20 spherical mirrors) and, then, projected on to 2x1'000 photo-multipliers located at 3 and 9 o'clock on the upstream end of the vessel.

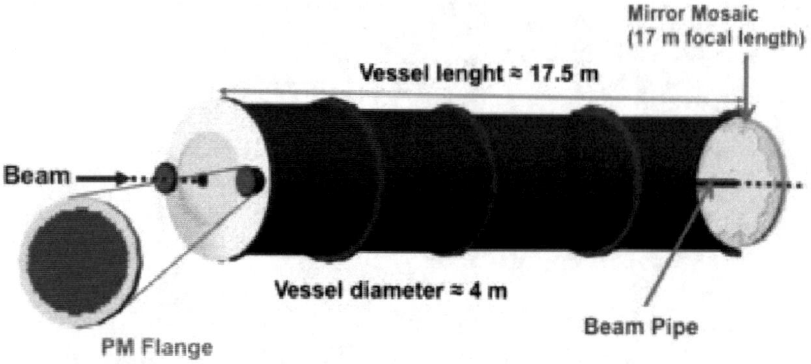

Figure 8 Schematic drawing of the RICH detector. The vessel is filled with neon gas at 1 atm.

6. The Photon-Veto Detectors

Photon vetoes are required to suppress the dominant background originating from the decay $K^+ \rightarrow \pi^+\pi^0$ (BR=20.7%) to the specified level. The average inefficiency for the rejection of the π^0 should be smaller than 10^{-8}. The photon vetoes need to have hermetic geometrical coverage up to 50 mrad for the photons originating from the kaon decays occurring in the decay region (from 5 to 70m after the final collimator). With such a configuration, only about 0.2 % of the $K^+ \rightarrow \pi^+\pi^0$ events have one photon from the π^0 left undetected [2].

The geometry of the experiment is partitioned into three different angular regions, each instrumented by three different detector technologies (see Figure 9):

- Large Angle Vetoes (LAV), covering the angular region between 8.5 mrad and 50 mrad, distributed along the decay volume and spaced by ≈6m in the upstream region and by ≈12m downstream. The LAV is composed of 12 stations (11 inside the vacuum tank), each station consists of ring shaped assemblies of lead glass calorimeter elements with photomultipliers at the end (see Figure 10) [7].
- The NA48 Liquid krypton calorimeter (LKR), covering angles between 1 and 8.5 mrad (see Figure 9) The LKr is a quasi-homogeneous electromagnetic calorimeter, which ensures a very good intrinsic energy and time resolution. Its key functions in the NA62 experiment are to veto photons from K decays and to enhance the hadron identification in conjunction with the muon veto system. For NA62 the LKR calorimeter has been equipped with new back-end electronics (Calorimeter REAdout Module named CREAM). The CREAMs (see Figure 11) provides 40 MHz sampling of 13248 calorimeter channels, data buffering during the SPS spill, zero suppression, and programmable trigger sums for the experiment trigger processor [8].
- Small angle vetoes covering the region down to zero degrees (SAC) and the zone around the inner radius of the LKR (IRC) calorimeter. These will have suitable overlap in the angular acceptance to cover the beam pipe and an inner radius smaller than that of the beam pipe (see Figure 9). Both detectors are sampling calorimeters with scintillating/iron layers and shashlik-type read-out [2].

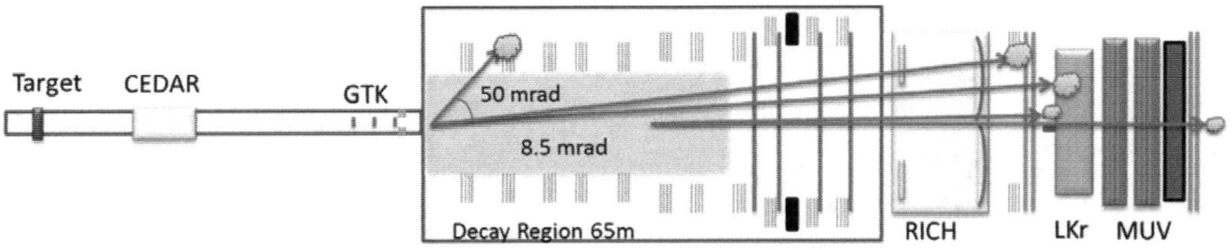

Figure 9 Longitudinal view of the experiment indicating the different region of the photon vetos.

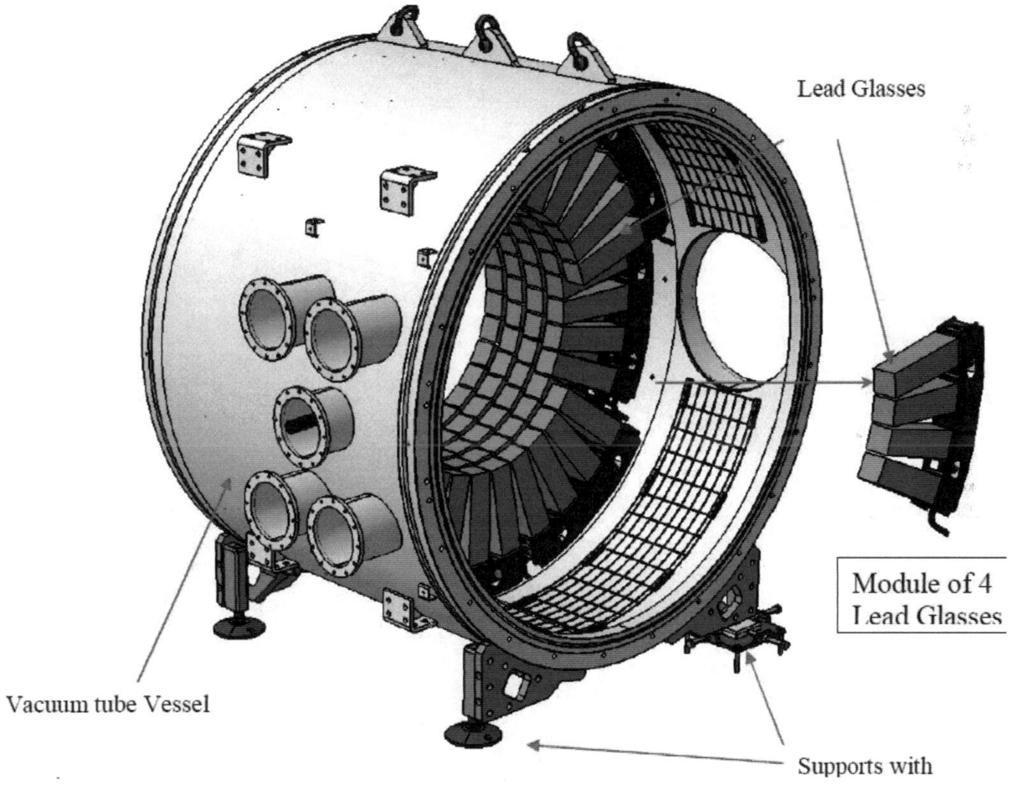

Figure 10 Schematic drawing of one LAV station, the external cylinder is part of the vacuum tank housing inside the ring-shaped lead glass elements and the photomultipliers.

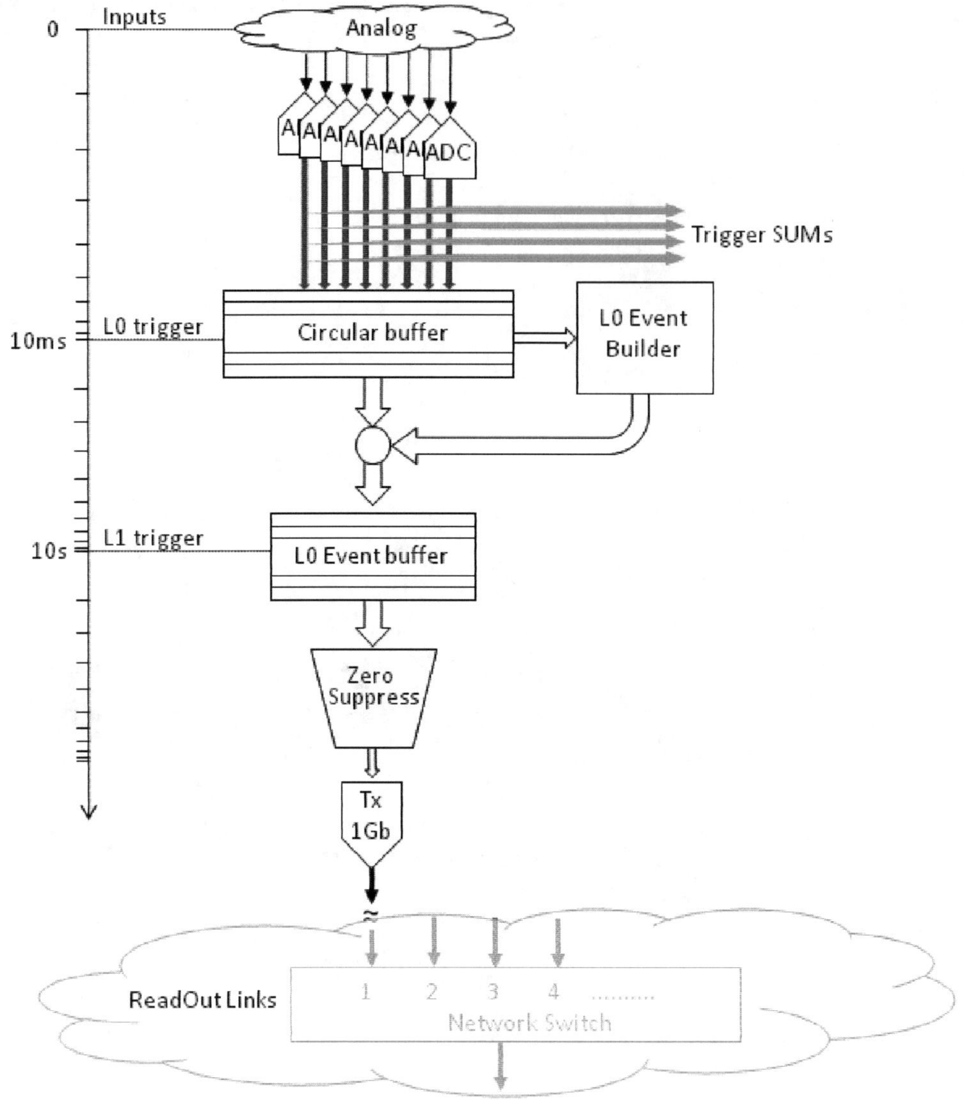

Figure 11 Data flow in the new CREAM readout of the LKR.

7. Experience from the commissioning run

NA62 has had a first commissioning run in autumn 2014 exploiting practically the complete detector at low beam intensity which was an important accomplishment for the first run. Practically all sub-system have been operated and could deliver data of the expected quality. The yet ongoing data analysis shows that the detector performances of the main new systems (e.g. photon and muon vetos, the RICH and the Straw Tracker) meet fully the expectations. Important learnings were obtained concerning the data acquisition system, which will undergo further optimization to increase the acquisition rate, so that the beam intensity can be increased to its nominal rate in in the future.

The first GigaTracker detectors were installed during the run and up to the end three GTK's could be operated together. However, due to technical difficulties in the carrier board, these detectors could be read out only partially (10%), therefore, these data cannot be used in the physics analysis, but they are exploited for detector performance studies. Preliminary results show that the timing properties of the detectors are as expected.

References

[1] Brod J, Gorbahn M, and Stamou E. 2011 *Phys. Rev.* **D83**, 034030.
[2] NA62 Collaboration. 2010 *NA62 Technical Design Document* **NA62-10-07** https://cds.cern.ch/record/1404985
[3] Bovet C, Maleyran R, Placci A, Placidi M. 1978 *IEEE Trans. Nucl. Sci.* **25** 572-576.
[4] Aglieri Rinella G, et al. 2013. *Nucl. Instrum. Methods Phys. Res.*, **A718** (2013) 270-273.
[5] P. Lichard *J. Instrum.* **5** (2010) C12053.
[6] Anzivino G et al. 2013 *Nucl. Instrum. Methods Phys. Res.*, **A732** 342-345
[7] Ambrosini F et al. 2012. *Journal of Physics: Conference Series* **404** 012022
[8] Ceccucci A , Fantechi R, Farthouat, P, Lamanna, G, Ryjov V. 2011 *J. Instrum.* **6** C12017.

Study of K_S semileptonic decays and $\mathcal{CPT}$ test with the KLOE detector

Daria Kamińska on behalf of the KLOE-2 Collaboration
The Marian Smoluchowski Institute of Physics, Jagiellonian University
Łojasiewicza 11, 30-348 Kraków, Poland
E-mail: daria.kaminska@doctoral.uj.edu.pl

Abstract. Study of semileptonic decays of neutral kaons allows to perform a test of discrete symmetries, as well as basic principles of the Standard Model. In this paper a general review on dependency between charge asymmetry constructed for semileptonic decays of short- and long-lived kaons and $\mathcal{CPT}$ symmetry is given. The current status of determination of charge asymmetry for short-lived kaon, obtained by reconstruction of about 10^5 $K_S \to \pi e \nu$ decays collected at DAΦNE with the KLOE detector is also reviewed.

1. Introduction

Discrete symmetries of nature such as charge conjugation ($\mathcal{C}$), parity ($\mathcal{P}$) or time reversal ($\mathcal{T}$) do not lead to new conserved quantities. Nevertheless, all of the mentioned symmetries plays an important role in particle physics, especially in calculations of the cross sections and decay rates. Weak interaction does not conserve the $\mathcal{C}$, $\mathcal{P}$, $\mathcal{T}$ or combined $\mathcal{CP}$ symmetry. However, up to now, there is no indication of $\mathcal{CPT}$ symmetry violation [1], which would also imply the break of Lorentz symmetry [2]. A special role in $\mathcal{CPT}$ violation searches plays a neutral kaon system which, due to a sensitivity to a variety of symmetry violation effects, is one of the best candidates for such kind of studies. One of the possible tests is based on comparison between semileptonic asymmetry in K_S decays (A_S) and the analogous asymmetry in K_L decays (A_L).

2. Test of $\mathcal{CPT}$ symmetry violation through semileptonic decays in neutral kaon system

2.1. Semileptonic decays in neutral kaon system

Neutral kaons are the lightest particles which contain a strange quark. Observed short-lived K_S and long-lived K_L are linear combinations of strange eigenstates (K^0 and $\bar{K^0}$):

$$
\begin{aligned}
|K_L\rangle &= \frac{1}{\sqrt{2(1+|\epsilon_L|^2)}}\left((1+\epsilon_L)\,|K^0\rangle - (1-\epsilon_L)\,|\bar{K^0}\rangle\right), \\
|K_S\rangle &= \frac{1}{\sqrt{2(1+|\epsilon_S|^2)}}\left((1+\epsilon_S)\,|K^0\rangle + (1-\epsilon_S)\,|\bar{K^0}\rangle\right),
\end{aligned}
\tag{1}
$$

where introduced small parameters ϵ_L and ϵ_S can be rewritten to separate $\mathcal{CP}$ and $\mathcal{CPT}$ violation parameters ϵ_K and δ_K, respectively:

$$\begin{aligned} \epsilon_S &= \epsilon_K + \delta_K, \\ \epsilon_L &= \epsilon_K - \delta_K. \end{aligned} \tag{2}$$

In this paper a particular emphasis will be given to semileptonic decays of K-short and K-long states ($K_{S/L} \to \pi e \nu$). According to Eq. 1, only the four possible decays of strange eigenstates should be considered:

$$\begin{aligned} K^0 &\to \pi^- e^+ \nu, & K^0 &\to \pi^+ e^- \bar{\nu}, \\ \bar{K^0} &\to \pi^+ e^- \bar{\nu}, & \bar{K^0} &\to \pi^- e^+ \nu. \end{aligned} \tag{3}$$

However, in the Standard Model the decay of K^0 (or $\bar{K^0}$) state is associated with the transition of the $\bar{s}$ quark into $\bar{u}$ quark (or s into u) and emission of the charged boson. Change of strangeness (ΔS) implies the corresponding change of electric charge (ΔQ) (see Figure 1). This is so called $\Delta S = \Delta Q$ rule. Therefore, decays of $K^0 \to \pi^- e^+ \nu$ and $\bar{K^0} \to \pi^+ e^- \bar{\nu}$ are present but $K^0 \to \pi^+ e^- \bar{\nu}$ and $\bar{K^0} \to \pi^- e^+ \nu$ are forbidden by if $\Delta Q = \Delta S$.

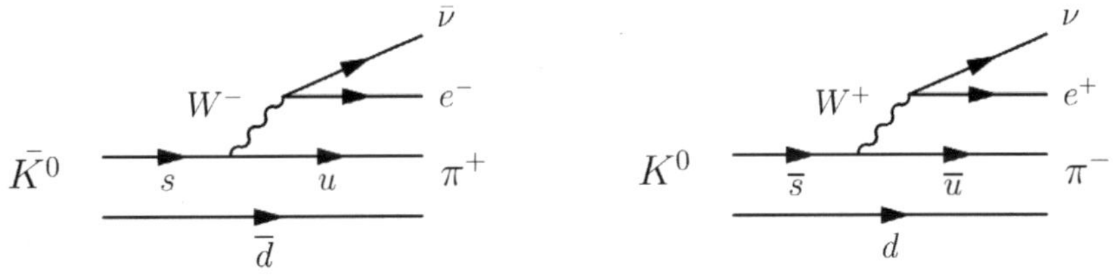

Figure 1. Feynman diagrams for K^0 and $\bar{K^0}$ semileptonic decay.

Decay amplitudes for semileptonic decays of states $|K^0\rangle$ and $|\bar{K^0}\rangle$ can be written as follows [3]:

$$\begin{aligned} \langle \pi^- e^+ \nu | H_{weak} | K^0 \rangle &= \mathcal{A}_+, & \langle \pi^+ e^- \bar{\nu} | H_{weak} | \bar{K^0} \rangle &= \bar{\mathcal{A}}_-, \\ \langle \pi^+ e^- \bar{\nu} | H_{weak} | K^0 \rangle &= \mathcal{A}_-, & \langle \pi^- e^+ \nu | H_{weak} | \bar{K^0} \rangle &= \bar{\mathcal{A}}_+, \end{aligned} \tag{4}$$

where the H_{weak} is the term of Hamiltonian corresponding to the weak interaction and $\mathcal{A}_+, \bar{\mathcal{A}}_-, \mathcal{A}_-, \bar{\mathcal{A}}_+$ parametrize the semileptonic decay amplitudes.

It is useful to introduce the following notation:

$$\begin{aligned} x &= \frac{\bar{\mathcal{A}}_+}{\mathcal{A}_+}, \qquad \bar{x} = \left(\frac{\mathcal{A}_-}{\bar{\mathcal{A}}_-}\right)^*, \qquad y = \frac{\bar{\mathcal{A}}_-^* - \mathcal{A}_+}{\bar{\mathcal{A}}_-^* + \mathcal{A}_+}, \\ x_\pm &= \frac{x \pm \bar{x}^*}{2} = \frac{1}{2}\left[\frac{\bar{\mathcal{A}}_+}{\mathcal{A}_+} \pm \left(\frac{\mathcal{A}_-}{\bar{\mathcal{A}}_-}\right)^*\right]. \end{aligned} \tag{5}$$

For further considerations, rules for applying symmetry operators to amplitudes of two spin zero systems A and B (and corresponding anti-systems $\bar{A}$ and $\bar{B}$) could be summarized as:

$$\begin{aligned} \langle \mathcal{T}\mathbf{B} | \mathcal{T} H_{wk} \mathcal{T}^{-1} | \mathcal{T}\mathbf{A} \rangle &= (\langle \mathbf{B} | \mathcal{T} H_{wk} \mathcal{T}^{-1} | \mathbf{A} \rangle)^* \\ \langle \mathcal{CP}\mathbf{B} | \mathcal{CP} H_{wk} \mathcal{CP}^{-1} | \mathcal{CP}\mathbf{A} \rangle &= \langle \bar{\mathbf{B}} | \mathcal{CP} H_{wk} \mathcal{CP}^{-1} | \bar{\mathbf{A}} \rangle \\ \langle \mathcal{CPT}\mathbf{B} | \mathcal{CPT} H_{wk} \mathcal{CPT}^{-1} | \mathcal{CPT}\mathbf{A} \rangle &= (\langle \bar{\mathbf{B}} | \mathcal{CPT} H_{wk} \mathcal{CPT}^{-1} | \bar{\mathbf{A}} \rangle)^* \end{aligned} \tag{6}$$

One obtains the relation between the semileptonic amplitudes and conservation of a particular symmetry by applying the presented above rules to the states presented in Eq. 4. These considerations are summarized in Table 1.

Journal of Physics: Conference Series **631** (2015) 012042 doi:10.1088/1742-6596/631/1/012042

Table 1. Relations between discrete symmetries and semiletponic amplitudes

Conserved quantity	Required relation
$\Delta S = \Delta Q$ rule	$x = \bar{x} = 0$
$\mathcal{CPT}$ symmetry	$x = \bar{x}^*$, $y = 0$
$\mathcal{CP}$ symmetry	$x = \bar{x}$, $y = Im(y)$
$\mathcal{T}$ symmetry	$y = Re(y)$

2.2. *Charge asymmetry*

Charge asymmetry can be defined for semileptonic decays of K_S and K_L mesons in the following way:

$$A_{S,L} = \frac{\Gamma(K_{S,L} \to \pi^- e^+ \nu) - \Gamma(K_{S,L} \to \pi^+ e^- \bar{\nu})}{\Gamma(K_{S,L} \to \pi^- e^+ \nu) + \Gamma(K_{S,L} \to \pi^+ e^- \bar{\nu})} \tag{7}$$

and can be rewritten in terms of parameters introduced in Eq. 5:

$$A_{S,L} = 2\left[Re\,(\epsilon_K) \pm Re\,(\delta_K) - Re(y) \pm Re(x_-)\right]. \tag{8}$$

In that case, sum and difference of the A_S and A_L allow to search for the $\mathcal{CPT}$ symmetry violation, either in the decay amplitudes through the parameter y or in the mass matrix through the parameter δ_K:

$$\begin{aligned} A_S + A_L &= 4Re(\epsilon) - 4Re\,(y)\,, \\ A_S - A_L &= 4Re(\delta_K) + 4Re\,(x_-)\,. \end{aligned} \tag{9}$$

2.3. *Experimental verification*

The measurement based on 1.9 millions $K_L \to \pi e \nu$ decays produced in collisions of proton beam with a BeO target performed by KTeV Collaboration allowed to determine A_L values [4]:

$$A_L = (3.322 \pm 0.058_{stat} \pm 0.047_{syst}) \times 10^{-3}. \tag{10}$$

At present most accurate measurement of K_S charge asymmetry was conducted by KLOE experiment. Measurement of A_S was performed with 0.41 fb^{-1} total luminosity data sample and the result is [5]:

$$A_S = (1.5 \pm 9.6_{stat} \pm 2.9_{syst}) \times 10^{-3}. \tag{11}$$

Obtained charge asymmetry for K_S decays is consistent in error limits with charge asymmetry for K_L decays. However, the inaccuracy of K_S determination is more than two orders of magnitude bigger than this of the A_L and the error of A_S is dominated by a statistical uncertainty, which is three times larger then systematical one. Therefore, in this work a new measurement of A_S based on around four times bigger data sample collected by means of the KLOE detector in 2004 and 2005, is presented.

3. The K LOng Experiment at DAΦNE

KLOE was mounted at DAΦNE collider in 1999 and collected data during two campaigns in 2001-2002 and 2004-2005. The gathered data sample corresponds to the total luminosity of 2.5 fb^{-1}. Energy of colliding beams (e^- and e^+) was set to the mass of ϕ meson. DAΦNE collider produce ~ 1300 kaon pairs per second which corresponds to the luminosity 5×10^{32} cm^{-2}s^{-1}. In order to obtain efficient detection of the K_L mesons the KLOE detector was constructed with a view to properties of neutral kaon system. A schematic cross-section side view of the KLOE detector is shown in Figure 2. The main components of KLOE are the cylindrical drift chamber and the

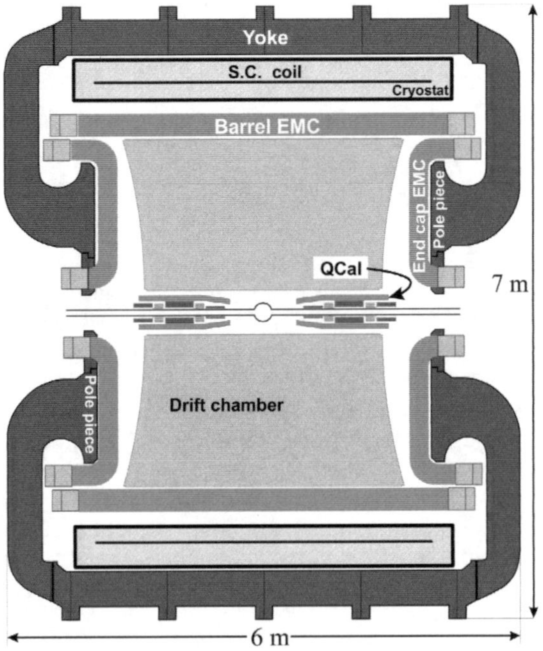

Figure 2. Scheme of the KLOE detector system. Drift chamber in the central part of the KLOE detector is surrounded by the electromagnetic calorimeter. Both detectors are inserted in a magnetic field. Figure adapted from [6].

calorimeter, both surrounding the beam pipe. All elements are immersed in a 0.52 T magnetic field created by superconducting coils that are placed along the beam axis.

The KLOE drift chamber (DC) is a 3.3 m long cylinder with an internal and external radii of 25 cm and 2 m, respectively, which allows to register about 40% of K_L decays inside the chamber while the rest reach the electromagnetic calorimeter. Between the endplates around 12500 sense wires are streched and allows to obtain: a spatial resolution in the r, φ plane better than 200 μm, a resolution along the z axis of $\sim$ 2 mm and on the resolution of the decay vertex determination of $\sim$ 1 mm. Moreover, the curvature of the reconstructed tracks allows to determine the particle momentum with a relative accuracy of 0.4 %. The drift chamber allows to reconstruct information about charged particles while the calorimeter enables recording of both charged and neutral particles. The KLOE calorimeter has been designed to have an excellent accuracy of energy determination and time resolution:

$$\begin{aligned} \frac{\sigma(E)}{E} &= \frac{5.4\%}{\sqrt{E[\mathrm{GeV}]}} \\ \sigma_t &= \frac{54\ \mathrm{ps}}{\sqrt{E[\mathrm{GeV}]}} \oplus 140\ \mathrm{ps} \end{aligned} \quad (12)$$

in order to register the hits of neutral particles and provide a possibility of the Time of Flight (TOF) measurement.

4. Registration of the $\phi \to K_L K_S \to K_L \pi e \nu$ processes at KLOE

4.1. Preselection

Due to the conservation of quantum numbers during decay of ϕ meson the neutral kaons are produced in pairs. In general case it is an entangled anti-symmetric state, which properties allows to perform a fundamental tests of Quantum Mechanics and Lorentz symmetry [7]:

$$|\phi\rangle = \frac{\sqrt{(1+|\epsilon_S|^2)(1+|\epsilon_L|^2)}}{\sqrt{2}(1-\epsilon_S\epsilon_L)} (|K_L(\vec{p})\rangle |K_S(-\vec{p})\rangle - |K_S(\vec{p})\rangle |K_L(-\vec{p})\rangle) \quad (13)$$

where $\vec{p}$ is a momentum in the ϕ meson rest frame. However, when one of the kaon is detected at time $t >> \tau_s$, the state in Eq. 13 factorizes and the system behaves as if the initial state was a mixture of $|K_L(\vec{p})\rangle |K_S(-\vec{p})\rangle$ and $|K_S(\vec{p})\rangle |K_L(-\vec{p})\rangle$. Hence the detection of K_L at large distance from Interaction Point tags K_S state in the opposite direction. Measurement described in this paper is based on identification of K_S through the detection of K_L interaction in the calorimeter. A sketch of typical signal event is shown in Figure 3(left). Selection of K_L candidates takes into account only clusters with high energy deposition and requires that the cluster is not close to any track reconstructed in the drift chamber. Also, velocity of neutral particle that deposits energy in this cluster must be close to the theoretical value of K_L meson velocity in the ϕ meson rest frame ($\beta \sim 0.22$).

In the next step, candidates for K_S meson decays are selected by selection of two oppositely charged particles with tracks forming a vertex close to the interaction point (IP):

$$\begin{aligned} \rho_{vtx} &< 15 \text{ cm}, \\ |z_{vtx}| &< 10 \text{ cm}, \end{aligned} \quad (14)$$

where $\rho_{vtx} = \sqrt{x_{vtx}^2 + y_{vtx}^2}$. Obtained distribution is shown in Figure 3(right). Then, the rejection of main source of background ($K_S \to \pi^+\pi^-$) is conducted by applying the following cuts:

- $70° < \alpha < 175°$
  where α is an angle between charged secondaries in K_S rest frame. Obtained α distribution is shown at the left side of Figure 4. In case of three body decay ($K_S \to \pi e \nu$) it is spanned over a large angle range.
- $300 < M_{inv} < 490$ MeV
  an invariant mass M_{inv} is calculated using momenta of the particles which tracks form a vertex assuming that both particle were pions. Obtained M_{inv} distribution is shown at the right side of Figure 4.

Both tracks reconstructed in the drift chamber must be associated with neighbouring clusters in calorimeter in order to use Time-of-Flight technique.

4.2. $K_S \to \pi e \nu$ events identification

The Time of Flight technique aims at rejection of the background, which at this stage of analysis is due to $K_S \to \pi^+\pi^-$ events, and at identification of the final charge states ($\pi^\pm e^\mp$). For each particle, the difference δ_t between the measured time of associated cluster (t_{cl}) and expected time of flight is calculated assuming a given mass hypothesis, m_x:

$$\begin{aligned} \delta_t(m_x) &= t_{cl} - \frac{L}{c \cdot \beta(m_x)}, \\ \beta(m_x) &= \frac{P}{P^2 + m_x^2}, \end{aligned} \quad (15)$$

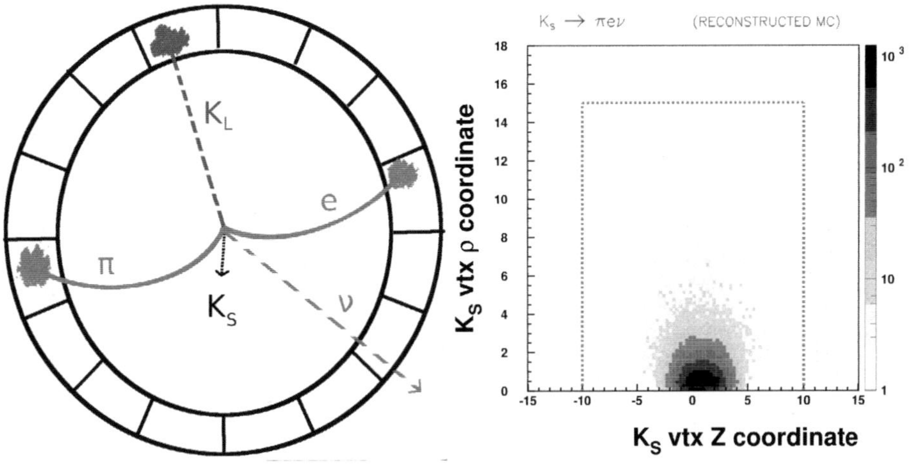

Figure 3. Left: Transverse view of exemplary signal event. The K_S is identified using the K_L interaction in the calorimeter. In the next step two oppositely charged particle tracks are selected which form a vertex near the interaction point. Both tracks must be associated with the colorimeter clusters. Right: Monte Carlo simulation of transversal and longitudinal coordinates of vertex position for $K_S \to \pi e \nu$ events. Dashed line represents applied cuts which preserves $\sim 95\%$ of the signal.

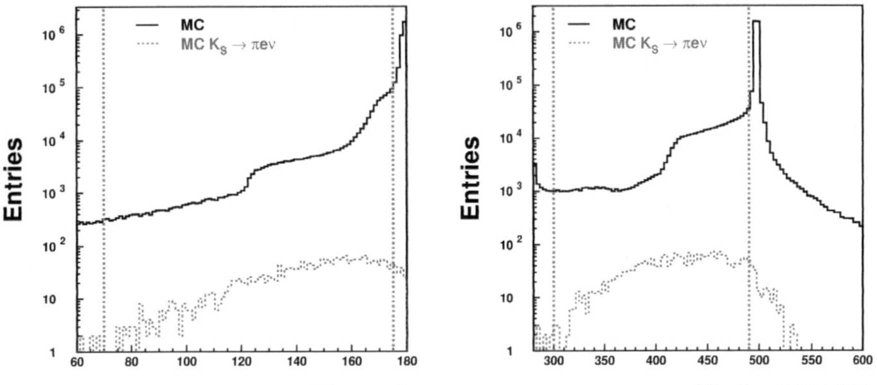

Figure 4. Left: Simulated distribution of angle between charged secondaries in K_S rest frame. Right: Simulated distribution of invariant mass calculated under assumption that both registered particles were pions. In both figures solid and dashed histograms represents all events and semileptonic decays, respectively. Vertical dashed lines represent cuts described in text.

where L is a total length of particle trajectory and P is particle momentum. For further consideration it is useful to introduce the difference of $\delta_t(m_a)$ and $\delta_t(m_b)$:

$$d\delta_{t,ab} = \delta_t(m_a)_1 - \delta_t(m_b)_2. \tag{16}$$

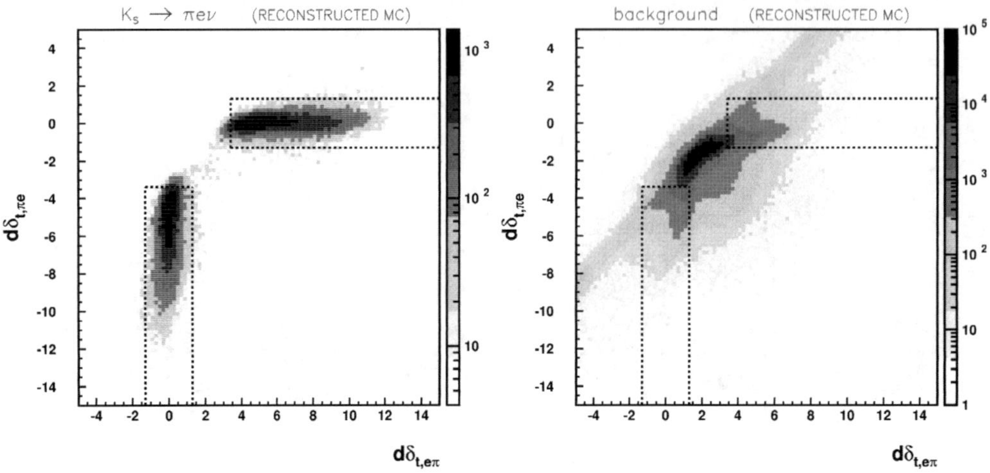

Figure 5. Simulated distributions of time differences $d\delta_{t,\pi e}$ vs $d\delta_{t,e\pi}$ defined in Eq. 18 for $K_S \to \pi e \nu$ events (left) and background events (right), once the $\delta_{t,\pi\pi}$ cut has been applied. The regions delimits by the dashed lines are selected. In case of $e\pi$ (πe) the $d\delta_{t,e\pi}$ ($d\delta_{t,\pi e}$) variable acquires value close to zero.

where suffix 1 and 2 refers to the first and second particle according to the ordering determined by the reconstruction procedure.

Two cuts are further applied. For the first cut, both particles are assumed to be pions and $d\delta_{t,\pi\pi}$ is calculated. For $K_S \to \pi\pi$ events this value is around zero and this fraction of events could be rejected by requiring:

$$|d\delta_{t,\pi\pi}| > 1.5 \text{ ns}. \tag{17}$$

Then, for surviving events, the pion-electron hypothesis is tested:

$$\begin{aligned} d\delta_{t,\pi e} &= \delta_t(m_\pi)_1 - \delta_t(m_e)_2, \\ d\delta_{t,e\pi} &= \delta_t(m_e)_1 - \delta_t(m_\pi)_2. \end{aligned} \tag{18}$$

If a mass assumption is correct then one of the variables above should be close to zero. The obtained Monte Carlo distributions for $K_S \to \pi^{\pm} e^{\mp} \nu$ and background events are presented in Figure 5. Hence, the following cut is applied:

$$\begin{gathered} |d\delta_{t,e\pi}| < 1.3 \text{ ns} \ \wedge \ d\delta_{t,\pi e} < -3.4 \text{ ns} \\ \text{or} \\ d\delta_{t,e\pi} > 3.4 \text{ ns} \ \wedge \ |d\delta_{t,\pi e}| < 1.3 \text{ ns} \end{gathered} \tag{19}$$

The above requirement ensures that the possibility of misidentification of charged particles from signal events is equal to 10^{-4} only.

Obtained distributions for simulated data and experimental KLOE data are shown in Figure 6 (top). The Monte Carlo simulation indicates the signal position around the point (0,0) while the background location is spread at the corners of the obtained distribution. Due to that an additional TOF cut is applied by selecting events within the circle in the $\delta_t(e)$ vs $\delta_t(\pi)$

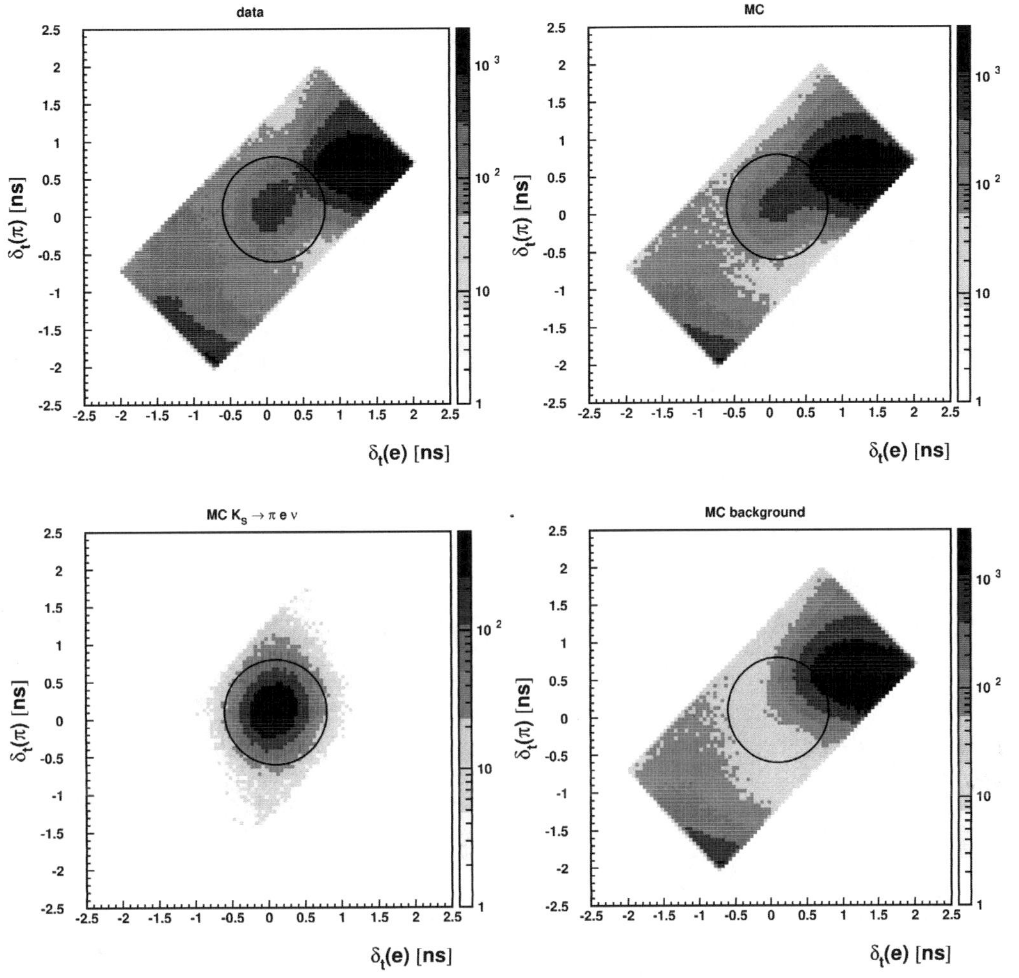

Figure 6. Distributions of the time difference for pion mass hypothesis ($\delta_t(\pi)$) versus the time difference for electron mass hypothesis ($\delta_t(e)$) for experimental data (top-left), total MC events (top-right), MC $K_S \to \pi e \nu$ events (bottom-left) and MC background events (bottom-right). Events within the circle are retained for further analysis.

plane, as shown in Figure 6. This cut allows to preserve 94% of the remaining signal and control the number of selected background events for normalization. The distribution of the difference between the missing energy and momentum ($\Delta E(\pi, e)$) shows the remaining background components (see Figure 7). Based on an integrated luminosity of 1.7fb^{-1} around 10^5 of $K_S \to \pi e \nu$ decays were reconstructed and will be used to determine the charge asymmetry and branching ratio for K_S semileptonic decays. A preliminary analysis shows a potential of reaching a two times better statistical error determination with a sample four times bigger than the previous KLOE analysis. The analysis is still in progress and preliminary results will be available soon.

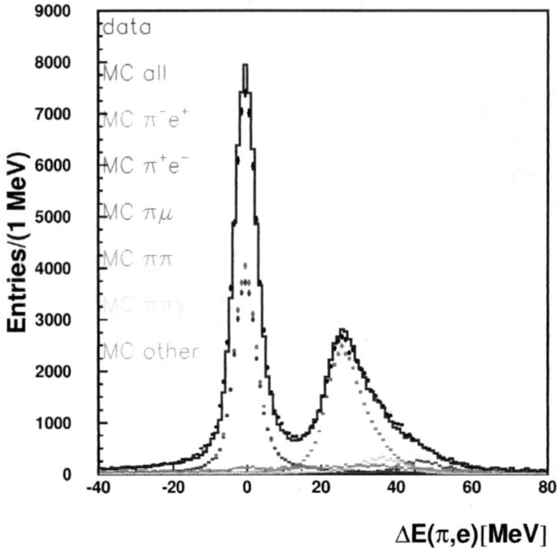

Figure 7. Distribution of $\Delta E(\pi, e) = E_{miss} - p_{miss}$ for all selected events after normalization procedure.

5. KLOE-2 Project

DAΦNE collider in previous years was adapted to increased delivered luminosity by installation of the new interaction region based on the Crabbed Waist compensation together with large Piwinski angle [10]. Those changes should increase by factor of three the amount of the delivered luminosity with respect to the performance reached before. Together with the upgrade of the KLOE detector this will allow to collect by KLOE-2 project the order of 10 fb^{-1} of integrated luminosity. The detector itself was equipped with crystal (CCALT) [11] and tile (QCALT) [12] calorimeters which covers the low polar angles and improve the detection of the photons that are coming from K_L decays in the drift chamber, respectively. Also, there were mounted the two pairs of small angle tagging devices that allows to detect the low (Low Energy Tagger [13]) and high (High Energy Tagger [14]) energy e^+e^- originated from $e^+e^- \to e^+e^-X$ reactions. However, the especially important for studies of $\mathcal{CPT}$ symmetry violation through semileptonic decays in neutral kaons is precise reconstruction of tracks momenta and vertex reconstruction close to the Interaction Point. This will be provided by Inner Tracker detector made out in a novel GEM technology [15]. Those upgrades will also improve the sensitivity of $\mathcal{CPT}$ and Lorentz invariance tests, which were presented by the KLOE experiment and became the most precise measurement of the $\mathcal{CPT}$ violating parameters Δa_μ for neutral kaons in the Standard Model Extension (SME) [16]. Measured values of Δa_μ currently have a precision of 10^{-18} GeV and are proportional to the parameter δ_K [17, 18, 19].

It should be emphasised that KLOE-2 aims to significantly improve the sensitivity of tests of discrete symmetries, through studies of K_S charge asymmetry or quantum interferometry effects in the kaon decays, beyond the presently achieved results.

Acknowledgments
We warmly thank our former KLOE colleagues for the access to the data collected during the KLOE data taking campaign. We thank the DAΦNE team for their efforts in maintaining low background running conditions and their collaboration during all data taking. We want to thank our technical staff: G.F. Fortugno and F. Sborzacchi for their dedication in ensuring efficient operation of the KLOE computing facilities; M. Anelli for his continuous attention to the gas system and detector safety; A. Balla, M. Gatta, G. Corradi and G. Papalino for electronics maintenance; M. Santoni, G. Paoluzzi and R. Rosellini for general detector support; C. Piscitelli for his help during major maintenance periods. This work was supported in part by the EU Integrated Infrastructure Initiative Hadron Physics Project under contract number RII3-CT- 2004-506078; by the European Commission under the 7th Framework Programme through the 'Research Infrastructures' action of the 'Capacities' Programme, Call: FP7-INFRASTRUCTURES-2008-1, Grant Agreement No. 227431; by the Polish National Science Centre through the Grants No. DEC-2011/03/N/ST2/02641, 2011/01/D/ST2/00748, 2011/03/N/ST2/02652, 2013/08/M/ST2/0 0323, DEC-2014/12/S/ST2/00459, 2013/11/B/ST2/04245 and by the Foundation for Polish Science through the MPD programme and the project HOMING PLUS BIS/2011-4/3.

References

[1] V. A. Kostelecky, N. Russell, Rev. Mod. Phys. **83** (2011), 11
[2] O. W. Greenberg, Phys. Rev. Lett. **89** (2002), 231602
[3] L. Maiani, G. Pancheri, N. Paver, INFN-LNF (1995)
[4] A. Alavi-Harati et al. (KTeV Collaboration), Phys. Rev. Lett. **88** (2002), 181601
[5] F. Ambrosino et al. (KLOE Collaboration), Phys. Lett. **B636** (2006), 173
[6] G. Vignola, M. Bassetti, M. E. Biagini, C. Biscari, R. Boni, Conf. Proc. **C960610** (1996), 22
[7] G. Amelino-Camelia et al., Eur. Phys. J. **C68** (2010), 619
[8] Adinolfi, M. et al., (KLOE Collaboration), Nucl. Instrum. Meth. **A461** (2001), 25
[9] Adinolfi, M. et al., (KLOE Collaboration), Nucl. Instrum. Meth. **A482** (2002), 364
[10] Milardi, C. er al., (DAΦNE Collaboration), CERN-2008-006 (2008), 1
[11] M. Cordelli et al., Nucl. Instrum. Meth. **A718** (2013), 81
[12] A. Balla et al., Nucl. Instrum. Meth. **A718** (2013), 95
[13] M. Adinolfi et al., (KLOE Collaboration), Nucl. Instrum. Meth. **A617** (2010), 81
[14] M. Adinolfi et al., (KLOE Collaboration), Nucl. Instrum. Meth. **A617** (2010), 266
[15] G. Morello et al., JINST **9** (2013), C01014
[16] D. Babusci et al., (KLOE Collaboration), Phys. Lett **B730** (2014), 89
[17] D. Colladay and V. A. Kostelecky, Phys. Rev. **D55**, (1997) 6760
[18] D. Colladay and V. A. Kostelecky, Phys. Rev. **D58**, (1998) 116002
[19] V. A. Kostelecky and R. Potting, Phys. Rev. **D51**, (1995) 3923

Journal of Physics: Conference Series **631** (2015) 012043 doi:10.1088/1742-6596/631/1/012043

Recent results, status and prospects for the BESIII experiment

Isabella Garzia for the BESIII Collaboration

INFN-Sezione di Ferrara, via Saragat 1, 44122 Ferrara, Italy

E-mail: garzia@fe.infn.it

Abstract. We report the measurement of the asymmetry $\mathcal{A}_{CP}$ of the branching fractions of $D^0 \to K^-\pi^+$ in the CP-odd and CP-even eigenstates using a data sample of 2.92 fb^{-1} collected with the BESIII detector at the center-of-mass energy $\sqrt{s} = 3.773$ GeV. With the measured $\mathcal{A}_{CP}$, the strong phase difference $\delta_{K\pi}$ between the doubly Cabibbo-suppressed process $\overline{D}^0 \to K^-\pi^+$ and the Cabibbo-favored process $D^0 \to K^-\pi^+$ is extracted. Using world-average values of external parameters, we obtain the most precise measurement of $\delta_{K\pi}$ to date: $\cos\delta_{K\pi} = 1.02 \pm 0.11 \pm 0.06 \pm 0.01$. The first and second uncertainties are statistical and systematic, respectively, while the third arises from external input. Based on the same data sample a preliminary results of the parameter y_{CP} in $D^0\overline{D}^0$ oscillation is obtained. Finally, a summary of the recent results from charmonium spectroscopy is reported. The high statistics accumulated at the $Y(4260)$ and $Y(4360)$ energies help us to understand the nature and the proprieties of the XYZ states.

1. Introduction

Mixing between K^0 and $\overline{K}^0$, B^0 and $\overline{B}^0$, and B_s and $\overline{B}_s$ is well established, and is well described by the Standard Model (SM) box diagrams containing up-type (u, c, t) quarks. In contrast, the $D^0-\overline{D}^0$ mixing amplitude at short distances involves loop containing down-type (d, s, b) quarks. The d and s box amplitudes are suppressed by the GIM mechanism [1], while the contribution from loops involving b quark is suppressed by the CKM factor [2, 3]. Therefore, the short-distance SM prediction is very small, of the order of 10^{-6}, and the charm mixing is expected to be dominated by long-distance processes, consisting of non-perturbative terms, and thus difficult to estimate. The measure of the charm mixing parameters helps to obtain information on the size of the long-distance effects and to the possible presence of new physics. Many experiments in the past decades indicate the that D^0 and $\overline{D}^0$ do mix, and more recently, charm mixing is established by LHCb [4] and CDF [5].

Mixing of neutral mesons occurs when the flavor eigenstates differ from the mass eigenstates. The two mass eigenstates of the neutral D meson system, $|D_{1,2}\rangle$, with masses $m_{1,2}$ and decay widths $\Gamma_{1,2}$, can be expresses as a linear combinations of the flavor eigenstates $|D^0\rangle$ and $|\overline{D}^0\rangle$: $|D_{1,2}\rangle = p|D^0\rangle + q|\overline{D}^0\rangle$, with the p and q complex parameters satisfying the condition $|p|^2 + |q|^2 = 1$. Conventionally, the charm mixing is described by two dimensionless parameters:

$$x = 2\frac{m_1 - m_2}{\Gamma_1 + \Gamma_2}, \quad y = \frac{\Gamma_1 - \Gamma_2}{\Gamma_1 + \Gamma_2}. \tag{1}$$

Using the conventions $CP|D^0\rangle = |\overline{D}^0\rangle$ and $CP|\overline{D}^0\rangle = -|D^0\rangle$, we construct the CP eigenstates as $|D_{CP\pm}\rangle \equiv \frac{|D^0\rangle \pm |\overline{D}^0\rangle}{\sqrt{2}}$, so that we can define the parameter y_{CP} as the difference between the effective lifetime of D decays to CP eigenstates and flavor eigenstates. In the absence of direct CP violation, but allowing small indirect CP violation [6], we have

$$y_{CP} = \frac{1}{2}\left[y\cos\phi\left(\left|\frac{q}{p}\right| + \left|\frac{p}{q}\right|\right) - x\sin\phi\left(\left|\frac{q}{p}\right| - \left|\frac{p}{q}\right|\right)\right], \tag{2}$$

where ϕ is the weak phase characterizing the CP violation in the interference between mixing and decay. Note that if CP is conserved, $|p/q| = 1$ and $\phi = 0$, and hence $y_{CP} = y$.

Most of our knowledge about $D^0 - \overline{D}^0$ mixing comes from time-dependent analysis. In particular, a very precise determination of the size of the mixing is obtained from the studies of the time dependent decay rate of the wrong-size process $D^0 \to K^+\pi^-$, which are sensitive to the quantities $y' = y\cos\delta_{K\pi} - x\sin\delta_{K\pi}$ and $x' = x\cos\delta_{K\pi} + y\sin\delta_{K\pi}$ [4, 5, 7, 8]. The y' and x' "rotated" parameters depend on the the relative strong phase $-\delta_{K\pi}$ between the doubly Cabibbo-suppressed (DCS) decay $D^0 \to K^+\pi^-$ and the corresponding Cabibbo-favored (CF) $\overline{D}^0 \to K^-\pi^+$, and is defined as:

$$\frac{\langle K^-\pi^+|\overline{D}^0\rangle}{\langle K^-\pi^+|D^0\rangle} = -re^{-i\delta_{K\pi}}, \tag{3}$$

where $r = |\langle K^-\pi^+|\overline{D}^0\rangle|/\langle K^-\pi^+|D^0\rangle|$, and $\delta_{K^-\pi^+} \equiv \delta_{K^+\pi^-}$ in the limit of CP conservation. A precise measurement of $\delta_{K\pi}$ is, therefore, necessary to extract x and y from x' and y', and helps to improve accuracy of the γ/ϕ_3 angle measurement in the CKM matrix.

In the first part of this report, we present the measurements of the strong phase $\delta_{K\pi}$ and y_{CP} obtained from a time-independent analysis. We use a sample of 2.92 fb^{-1} of data collected at the center-of-mass energy $\sqrt{s} = 3.773$ GeV in e^+e^- collisions with the BESIII detector [9], and we exploit the quantum-correlation in the mass-threshold production of the process $e^+e^- \to D^0\overline{D}^0$. In this process, the initial system has $J^{PC} = 1^{--}$, and so the D^0 and $\overline{D}^0$ mesons are in a CP-odd quantum-coherent state. Therefore, at any time, the two D mesons have opposite CP-eigenstates until one of them decays.

The second part of the report is dedicated to one of the main physics topics of the BESIII experiment: the charmonium spectroscopy. From 2009, a lot of data were taken in the energy region of about 4 GeV in order to study the X, Y, and Z states recently observed in the charm mass region. These new states have properties that are not in agreement with the theoretical expectations for charmonium states, and they are so called charmonium-like state. There are a lot of interpretations in literature: $c\bar{c}$ hybrids, glueballs, hadronic molecules, tetraquarks, etc., but more experimental informations are needed to reach a conclusion. In particular, in this report we summarize the latest results on the Z_c states.

2. The BESIII detector

The Beijing Spectrometer (BESIII) is a general purpose detector which covers 93% of the solid angle, and operates at the e^+e^- collider BEPCII. From the interaction point to the outside, BESIII is equipped with a main drift chamber (MDC) consisting of 43 layers of drift cells, a time-of-flight counter (TOF) with double-layer scintillators in the barrel and single-layer scintillator in the end-caps, an electromagnetic calorimeter (EMC) composed of 6240 CsI(Tl) crystals, a superconducting solenoid magnet providing a magnetic field of 1 T along the beam direction, and a muon counter (MUC) constituted by multi-layer resistive plate chambers installed in the steel flux-return yoke of the magnet. The spatial resolution achieved by the MDC is about 135 μm, while the momentum resolution is 0.5% for charged particles with transverse momentum of

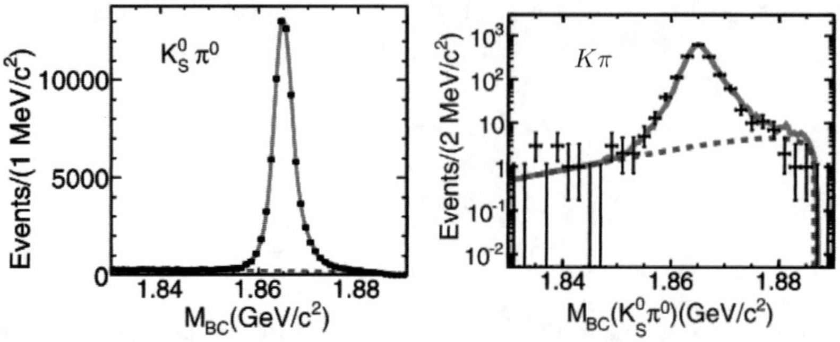

Figure 1. ST M_{BC} distribution of the $D \to K^0_S\pi^0$ final state (left), and DT M_{BC} distribution of the $D \to K^0_S\pi^0$ and $D \to K^-\pi^+$ (right) final states. The red lines represent the fit to the distributions, while blue dashed lines the background contribution.

1 GeV/c. The EMC energy resolution for 1 GeV photon is 2.5% (5%) in the barrel (end-caps) region; the TOF time resolution is about 80 ps (110 ps) for the barrel (end-caps), and provides a K/π separation of 2σ for momenta up to 1GeV/c. More details can be found in Ref. [9].

3. Measurement of the relative strong phase $\delta_{K\pi}$

The relative strong phase can be accessed from [10, 11]

$$2r\cos\delta_{K\pi} + y = (1 + R_{WS}) \cdot \mathcal{A}_{CP}, \tag{4}$$

where R_{WS} is the decay ratio of the wrong-sign process $\overline{D}^0 \to K^-\pi^+$ to the right-sign process $D^0 \to K^-\pi^+$, and

$$\mathcal{A}_{CP} \equiv \frac{\mathcal{B}_{D_{CP-}\to K^-\pi^+} - \mathcal{B}_{D_{CP+}\to K^-\pi^+}}{\mathcal{B}_{D_{CP-}\to K^-\pi^+} + \mathcal{B}_{D_{CP+}\to K^-\pi^+}} \tag{5}$$

is the asymmetry of $CP-$tagged D decay rates to $K^-\pi^+$. Using external parameters for r, R_{WS}, and y, we can extract $\delta_{K\pi}$ after the measurement of $\mathcal{A}_{CP}$. In particular, exploiting the quantum coherence, we use the D-tagging method to obtain the breaching fractions $\mathcal{B}_{D_{CP\pm}\to K\pi}$:

$$\mathcal{B}_{D_{CP\pm}\to K\pi} = \frac{n_{K\pi,CP\pm}}{n_{CP\pm}} \cdot \frac{\epsilon_{CP\pm}}{\epsilon_{K\pi,CP\pm}}, \tag{6}$$

where $n_{CP\pm}$ and $\epsilon_{CP\pm}$ are the yields and detection efficiencies of single tags (ST) of the $CP^\pm$ final state, while $n_{K\pi,CP\pm}$ and $\epsilon_{K\pi,CP\pm}$ are the yields and detection efficiencies of double tag (DT) of the $(CP\pm, K\pi)$ final state, respectively.

In this analysis, we use 5 CP-even D^0 decay modes (K^+K^-, $\pi^+\pi^-$, $K^0_S\pi^0\pi^0$, $\pi^0\pi^0$, $\rho^0\pi^0$), and 3 CP-odd modes ($K^0_S\pi^0$, $K^0_S\eta$, $K^0_S\omega$), with $\pi^0(\eta) \to \gamma\gamma$, $K^0_S \to \pi^+\pi^-$ and $\omega \to \pi^+\pi^-\pi^0$. The beam constrained mass M_{BC} and energy difference ΔE are the two kinematic variables used to identify the signal:

$$M_{BC} \equiv \sqrt{E_0^2/c^2 - |\vec{p}_D|^2/c^2}, \quad \text{and} \quad \Delta\mathrm{E} \equiv \mathrm{E_D} - \mathrm{E_0}, \tag{7}$$

where $\vec{p}_D$ and E_D are the total momentum and energy of the D meson candidate, and E_0 is the beam energy. Figure 1 shows the ST and DT M_{BC} distributions for one of the CP-odd final

state ($D \to K_S^0\pi^0$), for example. For each of the ST mode ($CP+$ or $CP-$), we construct the $D \to K\pi$ among the unused charged tracks. The signal peaks at the nominal D^0 mass, and we perform a maximum likelihood fit to the M_{BC} distributions in order to extract the yields. The signal shape is determined from MC simulation convoluted with a smearing Gaussian function, while the background is described by the ARGUS function.The fit results are shown in Fig. 1. Using Eq. (5), we measure $\mathcal{A}_{CP} = (12.7 \pm 1.3 \pm 0.7)\%$, where the uncertainties are statistical and systematic, respectively.
To extract the strong phase $\delta_{K\pi}$ from Eq. 4, we quote the external input of $r \equiv R_D = (3.47 \pm 0.06)$‰, $y = (6.6 \pm 0.9)\%$ from HFAG [12], and $R_{WS} = (3.80 \pm 0.05)$‰ from PDG [6], and we obtain [13]

$$\cos\delta_{K\pi} = 1.02 \pm 0.11 \pm 0.06 \pm 0.01. \tag{8}$$

The first uncertainty is statistical, the second is systematic, and the third is due to the external input parameters. This result is consistent with CLEO measurement[10] and provides the word best constrain on $\delta_{K\pi}$.

4. Measurement of y_{CP}

The semileptonic branching fraction of the CP-eigenstates $D_{CP\pm}$ can be writes as $\mathcal{B}_{D_{CP\pm}\to l} \sim \mathcal{B}_{D\to l}(1\mp y_{CP})$, where $\mathcal{B}_{D\to l}$ denotes the branching fraction of the semileptonic $D^0 \to hl\nu_l$ decay, which is independent of the CP eigenvalue of the parent D meson. Following Ref. [11], y_{CP} can be obtained as:

$$y_{CP} = \frac{1}{4}\left(\frac{\mathcal{B}_{D_{CP-}\to l}}{\mathcal{B}_{D_{CP+}\to l}} - \frac{\mathcal{B}_{D_{CP+}\to l}}{\mathcal{B}_{D_{CP-}\to l}}\right). \tag{9}$$

As for the $\delta_{K\pi}$ analysis, the branching fractions can be measured as:

$$\mathcal{B}_{D_{CP+}\to l} = \frac{n_{l;CP\pm}}{n_{CP\pm}} \cdot \frac{\epsilon_{l;CP\pm}}{\epsilon_{CP\pm}}, \tag{10}$$

where $n_{CP\pm}$ ($n_{l;CP\pm}$) and $\epsilon_{CP\pm}$ ($\epsilon_{l;CP\pm}$) denote the signal yields and detection efficiencies of ST mode (DT of $D\overline{D} \to l; CP\pm$).

In this analysis [14] , we use 3 CP-even (K^+K^-, $\pi^+\pi^-$, $K_S^0\pi^0\pi^0$) and 3 CP-odd ($K_S^0\pi^0$, $K_S^0\omega$, $K_S\eta$) tag modes, and the semileptonic decays used for the DT are $K^\mp e^\pm\nu_e$ plus $K^\mp\mu^\pm\nu_\mu$. We use the M_{BC} and ΔE to select a fully reconstructed D candidate in the ST sample. In each ST event, we search for semileptonic $D \to Ke(\mu)\nu$ candidates using the unpaired tracks. We define the variable U_{miss} to distinguish signals from backgrounds:

$$\begin{aligned} U_{miss} &\equiv E_{miss} - c|\vec{p}_{miss}| \\ E_{miss} \equiv E_0 - E_K - E_l, \quad |\vec{p}_{miss}| &\equiv -\left[\vec{p}_K + \vec{p}_l + \hat{p}_{ST}\sqrt{E_0^2 + m_D^2}\right], \end{aligned} \tag{11}$$

where $E_{K,l}$ ($\vec{p}_{K,l}$) is the energy (three-momentum) of K (l), $\hat{p}_{ST}$ is the unit vector in the direction of the reconstructed CP-tagged D, and m_D is the nominal D^0 mass [6]. The U_{miss} distributions for correctly-reconstructed signal peak at zero, and in Fig. 2 are reported, for example, the distributions for the $D\overline{D} \to Ke\nu_e; K^+K^-$ and $D\overline{D} \to K\mu\nu_\mu; K^+K^-$ modes.

In the fits of the DT $Ke\nu_e$ modes, the signal shape is modeled using MC shape convoluted with a bifurcated Gaussian, while the background is described with a 1st-order polynomial function. For the $K\mu\nu_\mu$ modes, the signal shape is modeled using the same functions as for the $Ke\nu_e$ modes. However, the background is more complex and consist of three parts. Primary background comes from $D \to K\pi\pi^0$ events. The background shape is taken from a $K\pi\pi^0$ MC control sample and corrected for the data/MC efficiency resolution. The size of the $K\pi\pi^0$

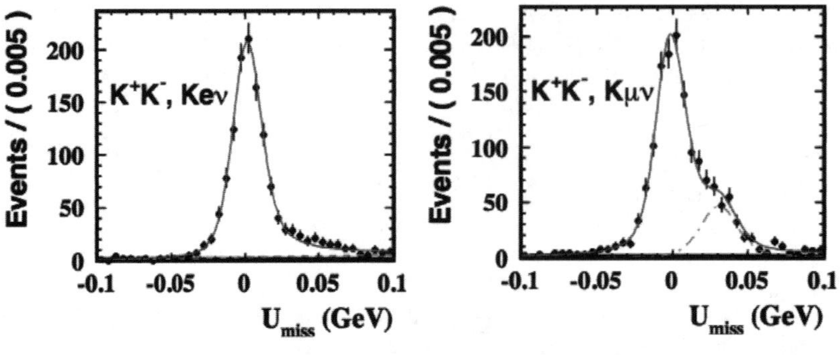

Figure 2. U_{miss} distributions and fits to data for the CP-tagged $Ke\nu_e$ (left) and $K\mu\nu_\mu$ (right) DT samples. In each plot, the solid line is the total fit, the dashed line in $Ke\nu$ is the polynomial background contribution, and the dotted dashed line in $K\mu\nu$ represents the contribution from the $K\pi\pi^0$ background.

background is then fixed by scaling the number of $K\pi\pi^0$ events in the control sample to the number in the signal region according to the simulated relative efficiencies of the selection criteria. The second background component from $Ke\nu$ events in $K\mu\nu$ mode is well modeled using MC events. All the other backgrounds are described with a first-order polynomial function.

To combine the results from different CP modes, we use the standard weighted-last square method [6]. The weighted semileptonic branching fraction $\tilde{\mathcal{B}}_{D_{CP\pm}\to l}$ is determined by minimizing the quantity : $\chi^2 = \sum_\alpha(\tilde{\mathcal{B}}_{D_{CP\pm}\to l} - \mathcal{B}^\alpha_{D_{CP\pm}\to l})/(\sigma^\alpha_{CP\pm})^2$, where α denotes the different CP-tag modes, and $\sigma^\alpha_{CP\pm}$ is the statistical error of $\mathcal{B}^\alpha_{D_{CP\pm}\to l}$ for a given α. Using this $\tilde{\mathcal{B}}$ in Eq. (9), we obtain [14]

$$y_{CP} = (-2.0 \pm 1.3 \pm 0.7)\%,$$

where the quoted uncertainties are statistical and systematic, respectively.

This result is compatible with previous measurements [10, 12, 15] within 2σ, and is the most precise time-integrated measurement of y_{CP} at charm threshold.

5. Observation of Z_c states at BESIII

One of the physics goal of the BESIII experiment is the study of the charmonium-like states: X, Y, and Z states. In particular, in this second part we report a summary on the four Z_c states observed in the last two years: $Z_c(3900)$, $Z_c(4020)$, $Z_c(3885)$, and $Z_c(4025)$. These states have at least four quarks, carry electric charge, and are close to the threshold.

5.1. *Discovery of $Z_c(3900)^+$*

In the study of the process $e^+e^- \to J/\psi\pi^+\pi^-$, using a data sample of 525 pb^{-1} collect at $\sqrt{s} = 4.260$ GeV with the BESIII detector, we observe a peaking structure in the invariant $J/\psi\pi^\pm$ mass distribution [16], with a statistical significance larger that 8σ. This structure, called $Z_c(3900)$ state, is parameterized with a S-wave Breit-Wigner function convoluted with a function accounting for the detector resolution. The fit result is shown in Fig. 3, and the mass and width obtained from the fit are $(3899.0 \pm 3.6(\text{stat.}) \pm 4.9(\text{syst.}))$ MeV/c^2 and $(46 \pm 10(\text{stat.}) \pm 20(\text{syst.}))$ MeV, respectively. The $Z_c(3900)^+$ state was almost simultaneously observed by the Belle collaboration [17] and then confirmed by CLEO-c [18].

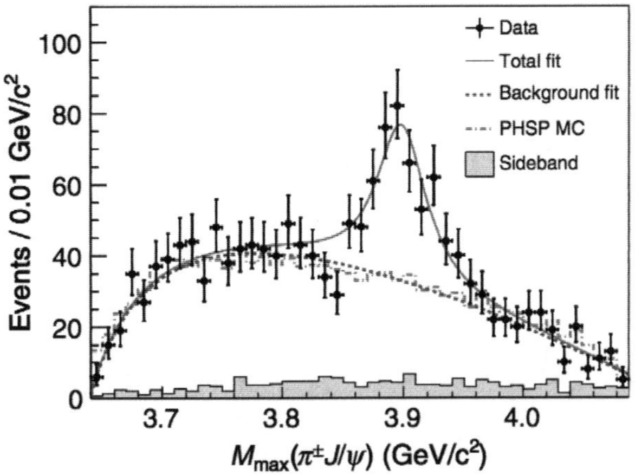

Figure 3. Fit to the $M_{max}(J/\psi\pi)$ distribution, where M_{max} stands for the larger of the two mass combinations $M(J/\psi\pi^+)$ and $M(J/\psi\pi^-)$ in each event. Dots with error bars are data, red solid line shows the fit result, and the blue dotted line the background. In the figure are also reported the results of a phase space MC simulation (dotted-dashed histogram), and the normalized J/ψ sideband events.

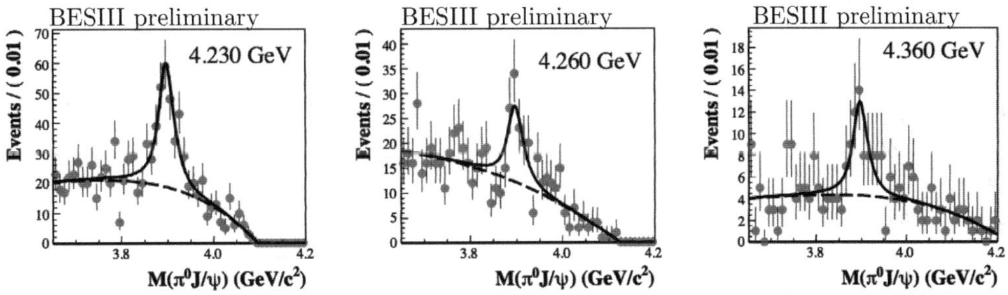

Figure 4. Invariant J/ψ mass spectra for $e^+e^- \to J/\psi\pi^0\pi^0$ at $\sqrt{s} = 4.23$ GeV (left), 4.26 GeV (middle), and 4.36 GeV (right). Data are represented by the red points with error bars, while solid black line shows the fit fit results and the dashed line the background contribution.

Since the $Z_c(3900)$ couples to a $c\bar{c}$ state and carry electric charge, it should be a tetraquark state, as well as a hadronic molecule, or other configuration. In addition, it should be noted that it is very close to the $D\bar{D}^*$ threshold, and this might be related to its nature.

A very recent BESIII analysis search for the neutral isospin partner of $Z_c(3900)^{\pm}$. Evidence for a structure in the invariant $J/\psi\pi^0$ mass distribution is observed in CLEO-c data [18] in the process $e^+e^+ \to J/\psi\pi^0\pi^0$. We study the same channel at three different c.m. energies, $\sqrt{s} = 4.23$, 4.26, and 4.36 GeV, and we observe a clear structure in the invariant $J/\psi\pi^0$ mass, as shown in Fig. 4, where the preliminary results are reported. We fit the signal with a Breit-Wigner function and we found a mass and width values of (3894.8 ± 2.3) MeV/c^2 and (29.6 ± 8.2) MeV, respectively, where the quoted error is statistical only. This result is consistent with the value obtained for the charged $Z_c(3900)^+$ and with the value reported in Ref. [18]. However, the

Journal of Physics: Conference Series **631** (2015) 012043 doi:10.1088/1742-6596/631/1/012043

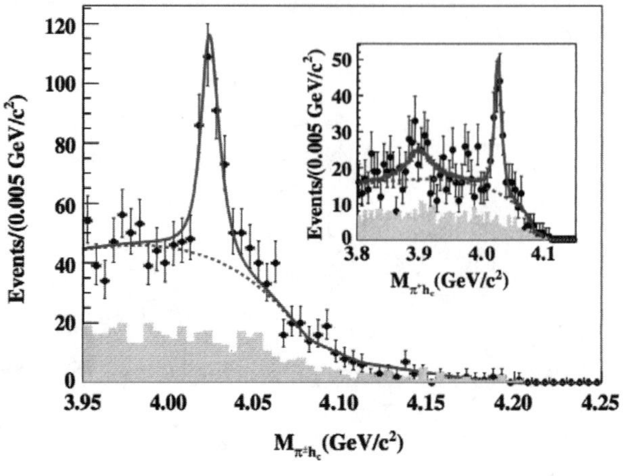

Figure 5. Sum of simultaneous fit to the $M_{h_c\pi^\pm}$ distribution at $\sqrt{s} = 4.23$, 4.26, and 4.36 GeV. In the inset is reported the distribution at 4.23 and 4.26 GeV including the $Z_c(3900)$. Solid line represents the total fit, dashed line the background contribution. Data are represented by points with error bars, and shaded histograms are the normalized sideband background.

determination of the spin-parity of the $Z_c(3900)^+$ and $Z_c(3900)^0$ states is necessary to confirm if the two states establish an isospin triplet.

5.2. Observation of the $Z_c(4020)$

The process $e^+e^- \to h_c\pi^+\pi^-$ is studied at 13 different c.m. energies from 3.90 to 4.42 GeV. A charged structure is observed in the $h_c\pi^\pm$ invariant mass spectrum at 4.02 GeV/c^2 [19], very close to the $D^*\overline{D}^*$ threshold. This structure is called $Z_c(4020)$ hereafter. In this study, we reconstruct the h_c via its electric-dipole transition $h_c \to \gamma\eta_c$, with the η_c reconstructed in 16 exclusive hadronic final states.

In Fig. 7 is shown the fit result to the $h_c\pi^\pm$ mass spectrum obtained by fitting the data at 4.23, 4.26 and 4.36 GeV simultaneously with the same signal function with common mass and width. The signal shape is parameterized as a constant width relativistic Breit-Wigner function convoluted with a Gaussian with a mass resolution determined from data, while the background shape is parameterized as an ARGUS function. The mass and width from the fit are (4022.9 ± 0.8 MeV/c^2) and (7.9 ± 2.7) MeV, respectively, and the statistical significance of the $Z_c(4020)$ signal is calculated to be greater than 8.9σ. There is no significant evidence of the $Z_c(3900)$ structure in the $M_{h_c\pi^\pm}$ mass spectrum: adding the $Z_c(3900)$ mass and width fixed to the BESIII measurement [16], we find a significance of 2.1σ.

A neutral structure, $Z_c(4020)^0$, is expected to couple to the $h_c\pi^0$ final state and to be produced in the process $e^+e^- \to h_c\pi^0\pi^0$. In the study of this process [20] a significant signal was observed in the $M_{h_c\pi^0}$ invariant mass distribution at about the same mass value found for the charged $Z_c(4020)$. Parameterizing the signal with a Breit-Wigner function and fixing the width to the value extracted from Ref [16], the fit yields a mass of ($4023.9 \pm 2.2 \pm 3.8$) MeV/c^2, which is consistent with the mass value of the charged $Z_c(4020)$. However, also in this case, further investigation are needed to understand the nature of $Z_c(4020)$.

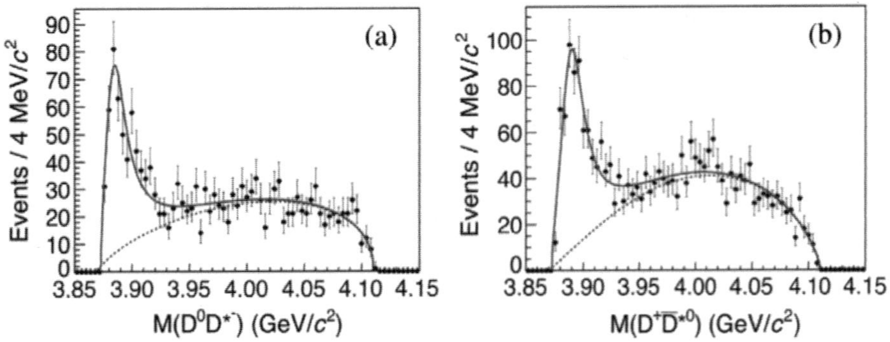

Figure 6. Fit to the invariant mass distributions for (a) D^0D^{*-} and (b) $D^+\bar{D}^{*0}$ events. The dots with error bars are the data, the solid line is the total fit, and dashed line describes the background distribution.

5.3. Observation of the $Z_c(3885)$ in the $D\bar{D}^*$ final state, and $Z_c(4025)$ in the $D^*\bar{D}^*$ final state. Due to the proximity of the $Z_c(3900)$ and $Z_c(4020)$ states to the $D\bar{D}^*$ and $D^*\bar{D}^*$ threshold, it is also interesting to search for similar structures in the corresponding open charm channels. The comparison between the Z_c partial decay widths to a $c\bar{c}$ final state, with the decay widths to a pair of open charm mesons may help to understand their nature. At BESIII, we investigate the processes $e^+e^- \to (D\bar{D}^*)^{\pm}\pi^{\mp}$ [21], and $e^+e^- \to (D^*\bar{D}^*)^{\pm}\pi^{\mp}$ [22] at the $\sqrt{s} = 4.26$ GeV, using a sample of 525 fb^{-1} and 827 fb^{-1}, respectively. The reactions are partially reconstructed: the bachelor $\pi^{\pm}$ is detected, and only one final state D meson are fully reconstructed. The other D meson is inferred from energy momentum conservation.

For the $(D\bar{D}^*)^{\pm}$ system, we reconstruct the D meson from $K\pi$ and $K\pi\pi$ decays. The invariant mass distribution of the $(D\bar{D}^*)$ system shows a clear enhancement at the $(D\bar{D}^*)$ threshold, labeled as $Z_c(3885)^{\pm}$ (see Fig. 7). The fit of the enhancement with a Breit-Wigner function yields a mass and width of $(3883.9 \pm 1.5 \pm 4.2)$ MeV$/c^2$ and $(24.8 \pm 3.3 \pm 11.0)$ MeV, respectively. The bachelor $\pi^{\pm}$ angle distribution is analyzed in order to determine the $Z_c(3885)$ quantum number. The resulting distribution is consistent with a spin-parity assignment of $J^P = 1^+$, and rules out 0^- as well as 1^-. A new preliminary analysis, which uses more statistics and a double-tag technique, improves and confirms the results obtained in Ref. [21]. An important question is whether or not the $Z_c(3900)$ and $Z_c(3885)$ have the same origin. The determination of the J^{PC} quantum number of $Z_c(3900)$ would be necessary to unswear this question. However, if we assume that the $Z_c(3885)$ structure is due to the $Z_c(3900)$, the ratio of partial decay width is determined to be $\Gamma(Z_c(3885) \to D\bar{D}^*)/\Gamma(Z_c(3900) \to J/\psi\pi) = 6.2 \pm 1.1 \pm 2.7$, which is much smaller than typical values for decays of conventional charmonium states above the open charm threshold. The comparison between mass, width, and cross section is reported in Table ?? for the $Z_c(3900)^{\pm}$ and $Z_c(3885)^{\pm}$ states.

For the $e^+e^- \to (D^*\bar{D}^*)^{\pm}\pi^{\mp}$ process, the charged D meson from $D^{*\pm} \to D^{\pm}\pi^0$ is reconstructed by its decay into $K\pi\pi$, and at least one π^0 in the final state is required in order to suppress background events. In the recoil mass spectrum of the bachelor $\pi^{\mp}$, a structure near the $(D^*\bar{D}^*)^{\pm}$ threshold is observed. We assume that this enhancement is due to a particle, labeled as $Z_c(4025)$, and we fit the distribution with a Breit-Wigner function, as shown in Fig. 7. The fit yields a mass of $(4026.3 \pm 2.6 \pm 3.7)$ MeV$/c^2$ and a width of $(24.8 \pm 5.6 \pm 7.7)$ MeV. Unfortunately, the statistics do not allow to perform a spin-parity analysis of $Z_c(4025)$, and

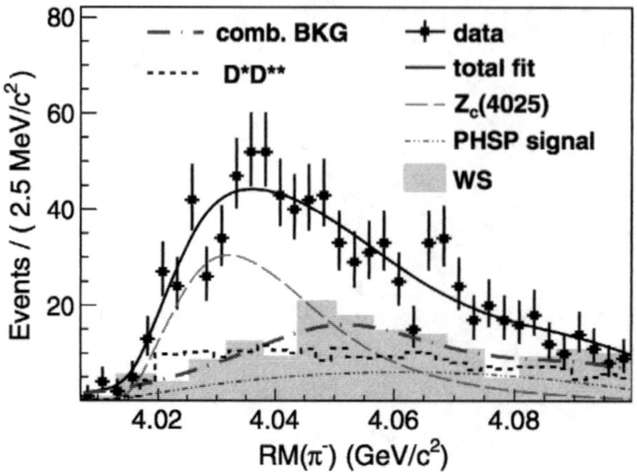

Figure 7. Unbinned maximum likelihood fit to the π^- recoil mass spectrum. Points with error bars are the data, solid line is the best fit result. Backgrounds, $Z_c(4025)^+$ signal, and phase space signal component (PHSP) are also shown.

	$Z_c(3885) \to D\overline{D}^*$	$Z_c(3900) \to J/\psi\pi$
Mass (MeV/c^2)	$3883.9 \pm 1.5 \pm 4.2$	$3899.0 \pm 3.6 \pm 4.9$
Width (MeV)	$24.8 \pm 3.3 \pm 11.0$	$46 \pm 10 \pm 20$
$\sigma \times \mathcal{B}$ (pb)	$83.5 \pm 6.6 \pm 22.0$	$13.5 \pm 2.1 \pm 4.8$

Table 1. Masses, widths and product of the Born cross sections times branching fractions for $Z_c(3885)$ [21] and $Z_c(3900)$ [16].

	$Z_c(4025) \to D^*\overline{D}^*$	$Z_c(4020) \to h_c\pi$
Mass (MeV/c^2)	$4026.3 \pm 2.6 \pm 3.7$	$4022.9 \pm 0.8 \pm 2.7$
Width (MeV)	$24.8 \pm 5.6 \pm 7.7$	$7.9 \pm 2.7 \pm 2.6$
σ (pb)	$139 \pm 9 \pm 15$	$8.7 \pm 1.9 \pm 2.8 \pm 1.4$ at $\sqrt{s} = 4.23$ GeV $7.5 \pm 1.7 \pm 2.1 \pm 1.2$ at $\sqrt{s} = 4.26$ GeV $10.3 \pm 2.3 \pm 3.1 \pm 1.6$ at $\sqrt{s} = 4.36$ GeV

Table 2. Masses, widths and Born cross sections for $Z_c(4025)$ [22] and $Z_c(4020)$ [19].

also in this case further investigation of the $Z_c(4020)$ and $Z_c(4025)$ are mandatory in order to understand if whether or not both are due to the same source. Table 2 summarizes the mass, width and cross section for the $Z_c(4020)^\pm$ and $Z_c(4025)^\pm$ states.

6. Conclusions

Using a data sample of 2.92 fb^{-1} collected with the BESIII detector at $\sqrt{s} = 3.773$ GeV, we measure the strong phase $\cos\delta_{K\pi}$ and the mixing parameter y_{CP} exploiting a quantum-correlated technique. We found $\cos\delta_{K\pi} = 1.02 \pm 0.11 \pm 0.06 \pm 0.01$ and $y_{CP} = (-2.0 \pm 1.3 \pm 0.7)\%$, where the first uncertainty is statistical, the second systematic, and the third is due to the external input parameters. Despite the current precision is statistically limited, these are the most precise time-integrated measurements at charm threshold. Future efforts, both in the global fit procedure than in the data taking, will be implemented in order to reduce the statistical uncertainty by a factor two.

Recent studies focalize the efforts in the charmonium mass region. In particular, four charged and two neutral charmonium-like states have been observed, $Z_c(3885)^{\pm}$, $Z_c(3900)^{\pm}$, $Z_c(3900)^0$, $Z_c(4020)^{\pm}$, $Z_c(4020)^0$, $Z_c(4025)^{\pm}$: these states may be an indication of a new type of hadrons. In the near future, BESIII experiment will take more data in the XYZ region, which will be helpful to shed light on the nature of these new states.

References

[1] Glashow S L, Iliopoulos J and Maiani L 1970 *Phys. Rev. D* **2**(7) 1285–1292
[2] Cabibbo N 1963 *Phys. Rev. Lett.* **10**(12) 531–533
[3] Kobayashi M and Maskawa T 1973 *Prog.Theor.Phys.* **49** 652–657
[4] Aaij R *et al.* ((LHCb Collaboration)) 2013 *Phys. Rev. Lett.* **110**(10) 101802
[5] Aaltonen T *et al.* ((CDF Collaboration)) 2013 *Phys. Rev. Lett.* **111**(23) 231802
[6] Olive K *et al.* (Particle Data Group) 2014 *Chin.Phys.* **C38** 090001
[7] del Amo Sanchez P *et al.* (BABAR Collaboration) 2010 *Phys. Rev. Lett.* **105**(8) 081803
[8] Staric M (Belle Collaboration) 2012 (*Preprint* 1212.3478)
[9] Ablikim M *et al.* ((BESIII Collaboration)) 2010 *Nucl. Instrum. Meth. A* **614** 345 – 399 ISSN 0168-9002
[10] Asner D M *et al.* (CLEO Collaboration) 2012 *Phys. Rev. D* **86**(11) 112001
[11] Xing Z Z 1997 *Phys. Rev. D* **55**(1) 196–218
[12] Y. Amhis *et al.* (Heavy Flavor Averaging Group) (2012), arXiv:1207.1158 and online update http://www.slac.stanford.edu/xorg/hfag/osc/spring_2014/ accessed: 2010-09-30
[13] Ablikim M *et al.* (BESIII Collaboration) 2014 *Physics Letters B* **734** 227 – 233 ISSN 0370-2693
[14] Ablikim M *et al.* (BESIII Collaboration) 2015 (*Preprint* 1501.01378)
[15] Lees J P *et al.* (BaBar Collaboration) 2013 *Phys. Rev. D* **87**(1) 012004
[16] Ablikim M *et al.* (BESIII Collaboration) 2013 *Phys. Rev. Lett.* **110**(25) 252001
[17] Liu Z Q *et al.* (Belle Collaboration) 2013 *Phys. Rev. Lett.* **110**(25) 252002
[18] Xiao T, Dobbs S, Tomaradze A and Seth K K 2013 *Physics Letters B* **727** 366 – 370 ISSN 0370-2693
[19] Ablikim M *et al.* 2013 *Phys. Rev. Lett.* **111**(24) 242001
[20] Ablikim M *et al.* ((BESIII Collaboration)) 2014 *Phys. Rev. Lett.* **113**(21) 212002
[21] Ablikim M *et al.* ((BESIII Collaboration)) 2014 *Phys. Rev. Lett.* **112**(2) 022001
[22] Ablikim M *et al.* ((BESIII Collaboration)) 2014 *Phys. Rev. Lett.* **112**(13) 132001

Journal of Physics: Conference Series **631** (2015) 012044 doi:10.1088/1742-6596/631/1/012044

Status of the PIENU experiment at TRIUMF

S.Ito[1], A.Aguilar-Arevalo[2], M.Aoki[1], M.Blecher[3], D.I.Britton[4], D.A. Bryman[5], D. vom Bruch[5], S.Chen[6], J.Comfort[7], S.Cuen-Rochin[5], L.Doria[8], P.Gumplinger[8], A.Hussein[9], Y.Igarashi[a], S.Kettell[b], L.Kurchaninov[8], L.Littenberg[b], C.Malbrunot[c,5], R.E.Mischke[8], T.Numao[8], D.Protopopescu[4], A.Sher[8], T.Sullivan[5], and D.Vavilov[8]

[1]Osaka University, Toyonaka, Osaka, 560-0043, Japan
[2]Instituto de Ciencias Nucleares, Universidad Nacional Autónoma de Mexico, México
[3]Virginia Tech., Blacksburg, VA 24061, USA
[4]University of Glasgow, Glasgow, UK
[5]University of British Columbia, Vancouver, B.C. V6T 1Z1, Canada
[6]Tsinghua University, Beijing, 100084, China
[7]Arizona State University, Tempe, AZ 85287, USA
[8]TRIUMF, 4004 Wesbrook Mall, Vancouver, B.C. V6T 2A3, Canada
[9]University of Northern British Columbia, Prince George, B.C. V2N 4Z9, Canada
[a]KEK, 1-1 Oho, Tsukuba-shi, Ibaraki, 305-0801, Japan
[b]Brookhaven National Laboratory, Upton, NY 11973-5000, USA
[c]Present address: CERN, Genéve 23 CH-1211, Switzerland

E-mail: s-ito@kuno-g.phys.sci.osaka-u.ac.jp

Abstract. The PIENU experiment at TRIUMF aims to measure the branching ratio of pion decays $R = \Gamma(\pi^+ \to e^+\nu_e + \pi^+ \to e^+\nu_e\gamma)/\Gamma(\pi^+ \to \mu^+\nu_\mu + \pi^+ \to \mu^+\nu_\mu\gamma)$ with precision <0.1%, providing a stringent test of the Standard Model hypothesis of electron-muon universality and a search for new physics.

1. Introduction

The branching ratio of charged pion decays $R = \Gamma(\pi^+ \to e^+\nu_e + \pi^+ \to e^+\nu_e\gamma)/\Gamma(\pi^+ \to \mu^+\nu_\mu + \pi^+ \to \mu^+\nu_\mu\gamma)$ is one of the most precisely calculated observables in the Standard Model (SM) involving quarks[1]. The most recent theoretical evaluation[2] gives

$$\begin{aligned} R_{SM} &= \frac{g_e^2}{g_\mu^2}\frac{m_e^2}{m_\mu^2}\frac{(m_\pi^2 - m_e^2)^2}{(m_\pi^2 - m_\mu^2)^2}(1+\delta)(1+\epsilon) \\ &= (1.2352 \pm 0.0001) \times 10^{-4}. \end{aligned} \tag{1}$$

where g_e and g_μ are the coupling constants between the W boson and the electron or muon, m_e, m_μ, and m_π are the electron, muon, and pion masses, and the terms $(1+\delta)(1+\epsilon)$ come from radiative corrections; inner bremsstrahlung and emission and reabsorption of virtual photons result in $\delta = -(3\alpha/\pi)\ln(m_\mu/m_e)$ and $\epsilon = -0.92(\alpha/\pi)$. Precise measurement of R provides one of the most stringent tests of the hypothesis of electron-muon universality ($g_e = g_\mu$) in weak interactions. The current experimental values of the branching ratio are

$$R_{EXP1} = (1.2265 \pm 0.0034(stat) \pm 0.0044(syst)) \times 10^{-4} (\text{TRIUMF, 1992})[3] \tag{2}$$

and

$$R_{EXP2} = (1.2346 \pm 0.0035(stat) \pm 0.0036(syst)) \times 10^{-4} (\mathrm{PSI,\ 1993})[4] \tag{3}$$

indicating that there is room for improvement by two orders of magnitude in precision.

Table 1 shows the current experimental results of electron-muon universality tests. The goal of the PIENU experiment at TRIUMF is to improve the accuracy of the branching ratio measurement by a factor of 5, to <0.1% resulting in 0.05% precision in the universality test. This precision also allows potential access to new physics up to the mass scale of 1000 TeV for helicity unsuppressed pseudoscalar interactions[1]. Examples of the new physics probed include R-parity violating SUSY[5], heavy neutrino mixing[6], excited gauge bosons, leptoquarks[7], compositeness, and the effects of charged Higgs bosons.

The PIENU experiment collected 6×10^6 $\pi^+ \to e^+ \nu_e$ events during run periods from 2009 to 2012 corresponding to more than 30 times higher statistics than in the previous TRIUMF experiment[3]. The initial analysis discussed here represents 5% of the statistics of the full data.

Table 1. Universality tests.

Decay Mode	g_μ/g_e
$\tau \to \mu / \tau \to e$	1.0018±0.0014[8]
$\pi \to \mu / \pi \to e$	1.0021±0.0016[3][4]
$K \to \mu / K \to e$	0.996±0.005[9]
$K \to \pi\mu / K \to \pi e$	1.002±0.002[10]
$W \to \mu / W \to e$	0.997±0.010[10]

2. The PIENU Experiment

2.1. Strategy

The decays $\pi^+ \to e^+ \nu_e$, and $\pi^+ \to \mu^+ \nu_\mu$ followed by $\mu^+ \to e^+ \nu_e \overline{\nu}_\mu$ decays (the $\pi^+ \to \mu^+ \to e^+$ decay chain) have different energy and time characteristics that can be precisely measured and used to extract the branching ratio. Figure 1 shows a schematic of pion decays and the measurement method. Positive pions stop in a plastic scintillator target and decay to muons or positrons. The muon in $\pi^+ \to \mu^+ \to e^+$ has kinetic energy $T_\mu = 4.1$ MeV and range in plastic scintillator of about 1 mm; the total energy of decay positrons has a continuous distribution from 0.5 to 52.8 MeV. The decay $\pi^+ \to e^+ \nu_e$ produces a mono-energetic positron at 69.8 MeV. In the PIENU experiment, the branching ratio is obtained from the ratio of the positron yields from $\pi^+ \to e^+ \nu_e$ decays and $\pi^+ \to \mu^+ \to e^+$ decays by taking advantage of these differences. Figure 2(a) shows the energy spectra of decay positrons in the calorimeter produced by a Monte Carlo (MC) simulation. Two energy regions, below and above the $\pi^+ \to \mu^+ \to e^+$ distribution shown as the dashed black line in Figure 2(a) are used to determine the time spectra of the decay positron. Figure 2(b) shows the time spectra in the low and high energy regions, respectively. These time spectra are distributed as

$$\pi^+ \to e^+ \nu_e : \varepsilon_{\pi \to e\nu}(t) = \frac{1}{\tau_\pi} \exp(-\frac{t}{\tau_\pi}) \tag{4}$$

$$\pi^+ \to \mu^+ \to e^+ : \varepsilon_{\pi \to \mu \to e}(t) = \frac{1}{\tau_\mu - \tau_\pi} \left\{ \exp(-\frac{t}{\tau_\mu}) - \exp(-\frac{t}{\tau_\pi}) \right\} \tag{5}$$

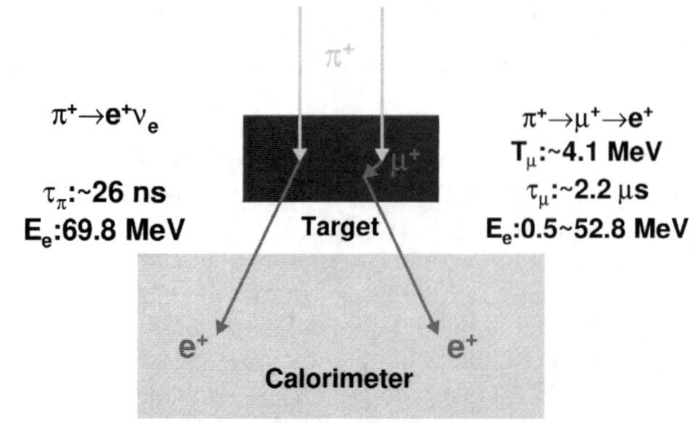

Figure 1. Schematic of the PIENU experiment technique (see text).

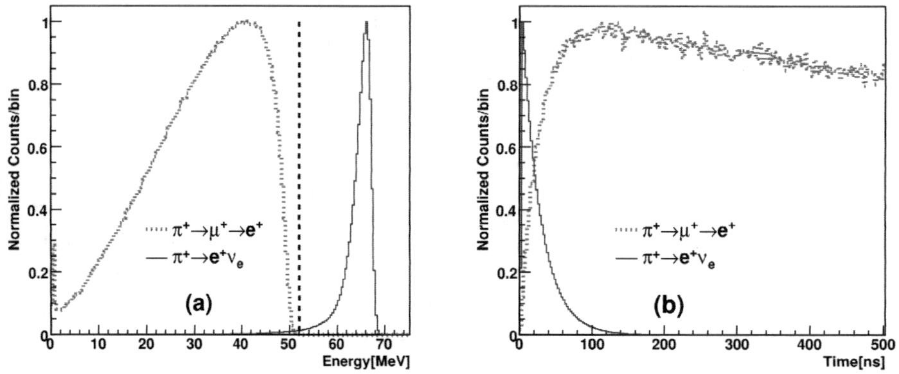

Figure 2. (a) Simulated energy spectra of decay positrons in the calorimeter. The horizontal axis is energy and the vertical axis is the amplitude normalized separately for the two modes. (b) Simulated decay positron time spectra. The horizontal axis is the decay time and the vertical axis is the amplitude normalized separately for the two modes.

where τ_π and τ_μ are the lifetimes of the pion and muon. The time spectra in the two regions are fitted to Eqs. (4) and (5) simultaneously in order to extract the "raw" branching ratio to which corrections are applied. The most important corrections resulting in systemmatic uncertainties include the $\pi^+ \to e^+ \nu_e$ low energy distribution in the calorimeter due to shower leakage and the difference of positron acceptances between $\pi^+ \to e^+ \nu_e$ and $\pi^+ \to \mu^+ \to e^+$ events. The raw branching ratio extraction and these systematic corrections will be discussed in Section 3.

2.2. Beam Line

The PIENU experiment was located at the TRIUMF M13 beam line. The TRIUMF cyclotron delivered a 500 MeV proton beam with an intensity of 120 μA to a 12 mm thick beryllium pion production target every 43 ns with a 4 ns wide pulse. In order to reduce the positron

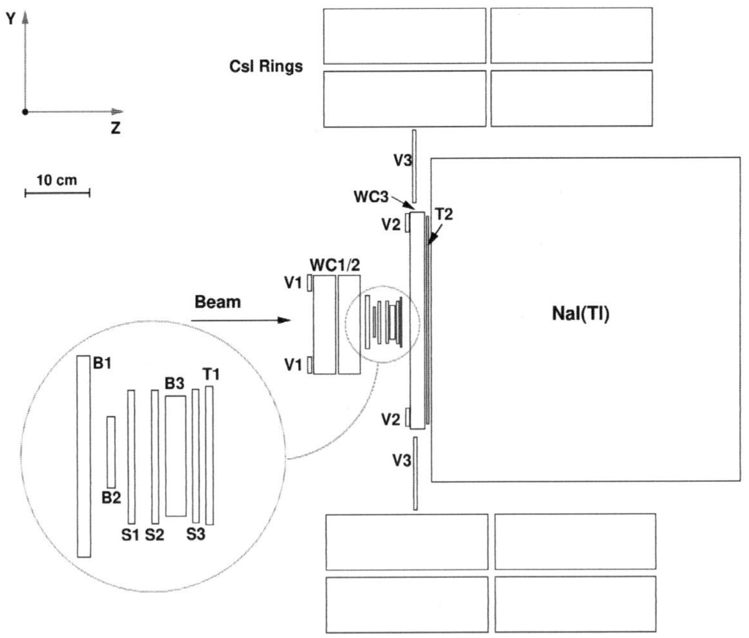

Figure 3. The schematic of the PIENU detector.

contamination in the beam, the M13 channel was modified[11]. A positive charged beam of momentum P=75±1 MeV/c was collected at 135° and transported in vacuum to the PIENU detector. The pion beam rate was about 70 kHz and the beam was composed of 84% π^+, 14% μ^+, and 2% e^+.

2.3. The PIENU Detector

Figure 3 shows the schematic of the PIENU detector. Pion beam tracking was provided by two circular multiwire proportional chambers (WC1 and WC2) each with three planes oriented at 0° and ±120° at the exit of the beam line. Following WC2, the beam was degraded by two thin plastic scintillators B1 and B2 (beam counters) used for time and energy loss measurements in order to identify the beam pions. The beam counters were followed by two sets of single-sided 0.3-mm thick 61×61 mm^2 Si strip detectors (S1 and S2) with strips oriented along the X and Y axes. The pion beams stopped in the center of an 8 mm plastic scintillator target (B3) and decayed at rest.

In order to reconstruct the tracks and define the acceptance of decay positrons, another pair of X-Y Si strips (S3) and three layers of circular multiwire proportional chamber (WC3) were employed. Two thin plastic scintillators (telescope counters T1 and T2) were used to measure decay times and to define the on-line trigger. Triggered decay positrons entered a large single crystal NaI(Tl) calorimeter (48 cm diameter × 48 cm width). Two layers of 97 pure CsI pentagonal crystals of 25 cm in length surrounded the NaI(Tl) crystal for shower leakage detection. Three veto scintillators were installed to cover inactive material. V1 covered the front frame of WC1, V2 covered the frame of WC3, and V3 covered the front flange of the NaI(Tl) crystal. The solid angle acceptance was 20%, about 10 times larger than in the previous TRIUMF experiment[3].

2.4. The Trigger Logic

In the trigger logic, a coincidence of beam counters and B3 defined the pion signal, and a coincidence of positron signals in the telescope counters defined the decay positron signal. A coincidence of pion and decay positron signals within a time window of -300 ns to 540 ns with respect to the pion signal was the basis of the main trigger condition. Since $\pi^+ \to \mu^+ \to e^+$ decays happen much more often than $\pi^+ \to e^+ \nu_e$ decays, only 1/16 of these events were selected by the unbiased trigger (Prescale trigger). On the other hand, $\pi^+ \to e^+ \nu_e$ events were enhanced for the Early trigger and the HE (High Energy) trigger. The Early trigger selected the events in the early time window (4 ns to 40 ns) after pion decay and the HE trigger selected events with $E_e > 45$ MeV. The typical trigger rate was about 600 Hz, of which 240 Hz was for Prescale, 160 Hz for Early, 170 Hz for HE, and 20 Hz for the sum of calibration triggers. The waveforms from all scintillators, S1-3, NaI(Tl) and CsI crystals were digitized and time information from all sense wires of WC1-3 and signals used for the trigger logic were recorded.

3. Analysis -Raw Branching Ratio Extraction-

3.1. Event Selection

The charge and time information of each pulse were extracted from the waveforms. The charge was obtained by integrating the pulse for 40 ns for plastic scintillators, for 160 ns for S1-3 and CsI, and for 1 μs for the NaI. The waveforms in B1, T1, and B3 were fitted to the templates of each waveform for the best timing and particle identification purposes. B1 defined the pion timing (t_{B1}) and T1 defined decay positron timing (t_{T1}). $\pi^+ \to e^+ \nu_e$ decay has two pulses in B3, while $\pi^+ \to \mu^+ \to e^+$ decay has three pulses in B3 because of the additional decay muon.

The beam pions were selected by their energy loss in B1 and B2. The beam positrons and muons were negligible after the energy loss cut. The beam profile in WC1 and WC2 was used to remove particles which had unusual trajectories. Any events with extra hits in the beam counters and telescope counters in the time regions of -6.39 to 1.35 μs with respect to t_{B1} were removed. A radial acceptance cut in WC3 using tracks reconstructed with S3 and WC3 was used to reduce shower leakage from the NaI. This acceptance cut required that the radius at WC3 was < 60 mm.

Figure 4 shows examples of event selection cuts, the energy loss in B1 and the radius distribution of events at WC3. In the PIENU analysis, in order to avoid bias, the raw branching ratio was shifted by a hidden random value within 1% ("blinding"). Before unblinding the analysis, all systematic corrections and cuts are being studied and the stability of the results is being examined.

3.2. Time Spectra Fitting

Figure 5 is the spectrum of the NaI and CsI energies after event selection. It shows the $\pi^+ \to e^+ \nu_e$ peak and the continuous spectrum of $\pi^+ \to \mu^+ \to e^+$ decays. The raw branching ratio was extracted by performing a simultaneous fit of the time spectra using an energy threshold (E_{Cut}) at 52 MeV. Figure 6 shows the time spectra in the low energy region (a) and the high energy region (b). The events in the low energy region were triggered by the Prescale trigger and the Early trigger, while the HE trigger was used for the events in the high energy region.

In the low energy time spectrum, the components included were $\pi^+ \to \mu^+ \to e^+$ decays and muon ($\mu^+ \to e^+ \nu_e \overline{\nu}_\mu$) decays. $\pi^+ \to \mu^+ \to e^+$ decays are described by Eq. (5), and muon decays arising from pion-decay-in-flight (πDIF) and previous pion stops in B3 ("old-muons") are distributed as

$$\mu^+ \to e^+ \nu_e \overline{\nu}_\mu : \varepsilon(t)_{\mu \to e \nu_e \overline{\nu}_\mu} = \frac{1}{\tau_\mu} \exp(-\frac{t}{\tau_\mu}). \tag{6}$$

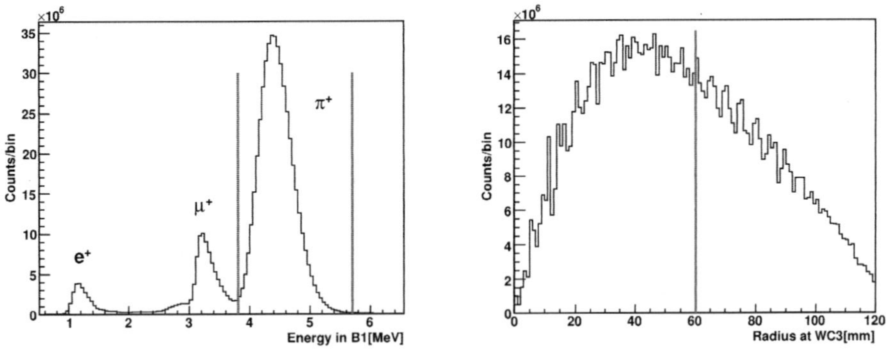

Figure 4. Event selection cuts. Left: Energy loss in B1. Right: Acceptance cut in WC3. Vertical red lines indicate cut values.

The $\pi^+ \to \mu^+ \to e^+$ and πDIF distributions begin at time $t = 0$. The solid red line, dashed dark blue line, and dashed pink line in Figure 6 (a) indicate $\pi^+ \to \mu^+ \to e^+$ decays, πDIF events, and previously arrived muons ("old-muons"). The distributions coming from multiple muon decays and the low energy tail of $\pi^+ \to e^+ \nu_e$ decays were estimated to contribute $< 0.01\%$ to the branching ratio, and were neglected in the time fit.

The main component in the high energy time spectrum was due to the $\pi^+ \to e^+ \nu_e$ decays distributed as Eq. (4). The major backgrounds were muon decays ($\pi^+ \to \mu^+ \to e^+$, πDIF, and old-muons) which were promoted to the high energy time spectrum by the energy resolutions of the NaI and CsI crystals, radiative muon decays in which the γ-ray increases the apparent positron energy, and pileup events in the calorimeter with a flat time distribution (e.g. due to neutrons coming from the production target). These components have the same time distribution in the low energy region. $\pi^+ \to \mu^+ \to e^+$, πDIF, and old-muon decays in the high energy region are shown as the solid blue line, the dashed dark blue line, and the dashed pink line in Figure 6(b) (colors on-line).

Another background in the high energy region came from radiative pion decays $\pi^+ \to \mu^+ \nu_\mu \gamma$ (branching ratio, 2×10^{-4}[12]) followed by $\mu^+ \to e^+ \nu_e \overline{\nu}_\mu$ decays. The time of the γ-ray is different from that of the decay positron. These contributions were estimated by MC simulation by using a waveform template for the NaI and CsI detectors. The amplitude of the $\pi^+ \to \mu^+ \nu_\mu \gamma$ component due to the E_{Cut} was 0.4% of the $\pi^+ \to e^+ \nu_e$ decays. The amplitude was fixed in the fitting function. This distribution ($\mathcal{G}$) is shown as the light blue line in Figure 6(b).

The distribution shown by the dashed violet line in the high energy region in Figure 6 (b) which arises from the $t < 0$ region and decreases in the $t > 0$ region is the pileup component of $\pi^+ \to \mu^+ \to e^+$ decays plus old-muon decays. The shape of this background was obtained by MC simulation using the pulse shapes of the NaI and CsI detectors. This function is represented by $\mathcal{F}_1$.

The pileup cut was based on the pulse shape in T1. If extra hits were observed in T1, the event was rejected. However, two events occurring within the double pulse resolution time of T1 were accepted. The shape of this distribution was the same as that for the pileup components of $\pi^+ \to \mu^+ \to e^+$ decays plus old-muon decays but with a short effective time constant ($\Delta T = 15$ ns). The effect of pileup coming within the double pulse reloluton time in T1 was estimated by artificially increasing the double pulse resolution time up to 150 ns. The amplitude of this component was estimated to be 4% of $\pi^+ \to e^+ \nu_e$ decays. This distribution is shown by the

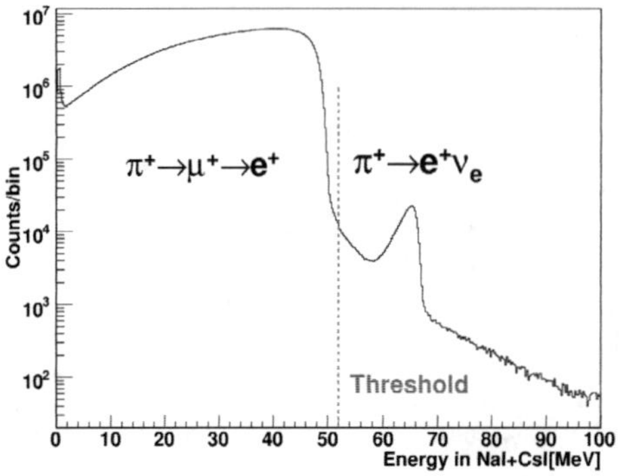

Figure 5. Combined energy spectrum of NaI plus CsI detectors. The vertical red line indicates the E_{Cut}=52 MeV.

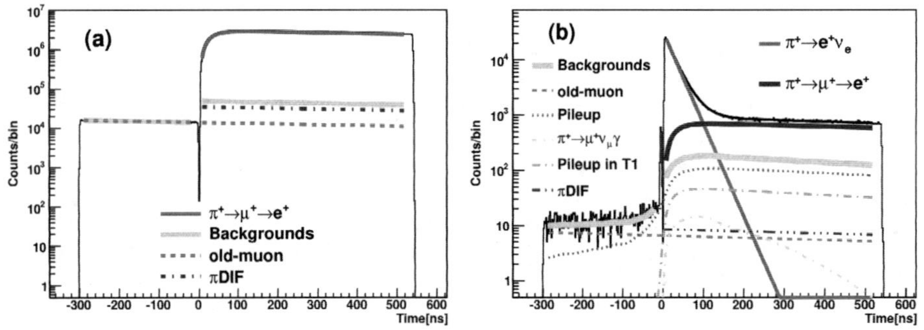

Figure 6. Time spectra in the low (a) and high (b) energy regions. Horizontal axes are decay times ($t_{T1} - t_{B1}$). The solid red lines in the low and high energy regions are $\pi^+ \to \mu^+ \to e^+$ and $\pi^+ \to e^+ \nu_e$, respectively. Other lines indicate background components and the solid green lines in both regions are the sums of the background components. See text for details.

dashed green line in Figure 6(b) and its function is represented by $\mathcal{F}_2$.

The following equations give the fitting functions in the low (PIMUE(t)) and high (PIENU(t)) energy regions as discussed above:

$$\text{PIMUE}(t) = \mathcal{H}(t)\Big[a(1-r)\varepsilon_{\pi\to\mu\to e}(t) + b\varepsilon(t)_{\mu\to e\nu_e\overline{\nu}_\mu}\Big] + c\varepsilon(t)_{\mu\to e\nu_e\overline{\nu}_\mu} \tag{7}$$

$$\begin{aligned}\text{PIENU}(t) = {} & \mathcal{H}(t)\Big[a\{(R+C_{\mu DIF})\varepsilon(t)_{\pi\to e\nu} + d\mathcal{G}(t) + r\varepsilon_{\pi\to\mu\to e}(t)\} + b'\varepsilon(t)_{\mu\to e\nu_e\overline{\nu}_\mu}\Big] \\ & + c'\varepsilon(t)_{\mu\to e\nu_e\overline{\nu}_\mu} + e\mathcal{F}_1 + f\mathcal{F}_2\end{aligned} \tag{8}$$

where R is the raw branching ratio, $\mathcal{H}(t)$ is the Heaviside step function ($\mathcal{H}(t>0)=1$, $\mathcal{H}(t<0)=0$), $\varepsilon(t)_{\pi\to e\nu}$, $\varepsilon_{\pi\to\mu\to e}(t)$, and $\varepsilon(t)_{\mu\to e\nu_e\overline{\nu}_\mu}$ are the same functions as in Eq. (4),(5) and (6), and $t = t' - t_0$ where t' is the measured time and t_0 is the pion stop time which was determined by prompt events and is fixed in the fit. The choice of t_0 doesn't affect the branching ratio. In Eqs.

(7) and (8), the parameter a represents the number of $\pi^+\to\mu^+\to e^+$ decays, and ar indicates the number of $\pi^+\to\mu^+\to e^+$ decays in the high energy region. The other parameters b, b', c, c', d, e, and f are the amplitudes of each function. The free parameters are R, a, r, b, c, c' and e. Parameters d, e, f and $C_{\mu DIF}$ are fixed to the estimated amplitudes. However, the correlation between πDIF decays and old-muon decays or $\mathcal{F}_1$ was significant. Therefore, parameter b' is scaled to the amplitude of πDIF in the low energy region; namely, $b' = rb$. The parameter $C_{\mu DIF}$ was the corrected amplitude for muon-decay-in-flight (μDIF) events, which will be discussed in the next section. The fitting range was from -290 to 520 ns omitting the prompt region of -20 to 5 ns.

The raw branching ratio remains blinded at the time of writing. The statistical uncertainty of the raw branching ratio was 0.2%.

4. Analysis -Systematic Sources-

4.1. Tail Correction

The largest correction to the branching ratio came from knowledge of the low energy $\pi^+\to e^+\nu_e$ tail events below E_{Cut} which were buried under $\pi^+\to\mu^+\to e^+$ distribution. In order to evaluate this correction it was necessary to suppress the dominant $\pi^+\to\mu^+\to e^+$ component. This was done using four suppression factors. First, by exploiting the short pion lifetime compared to muon lifetime, the early time region $5 - 35$ ns after the pion stop enriched $\pi^+\to e^+\nu_e$ decays relative to $\pi^+\to\mu^+\to e^+$ events by a factor of 100. The second factor was obtained using the energy loss and pulse shape information in B3. For $\pi^+\to e^+\nu_e$ decays, the energy loss in B3 was the sum of the kinetic energy of the stopping pion plus a small contribution from the exiting decay positron. On the other hand, $\pi^+\to\mu^+\to e^+$ decays had an additional 4.1 MeV deposited from the kinetic energy of the decay muon which stopped in B3. As shown in Figure 7(a), there are two peaks in the spectrum of total energy deposited by beam pions in B1, B2, S1, S2, and B3. The cut region indicated by two vertical red lines in Figure 7(a) provided a factor of > 100 suppression.

The third $\pi^+\to\mu^+\to e^+$ suppression factor employed beam pion tracking by S1 and S2 in order to reject πDIF events before B3. πDIF events may result in a mismatch of the particle trajectory as determined by WC1 and WC2 compared to that obtained using S1 and S2, as illustrated in Figure 7(b). These events left energy in B3 similar to $\pi^+\to e^+\nu_e$ events but had a larger mismatched ("kink") angle than decay-at-rest events. The fourth factor was found using the energy loss in S3. About 30% of muons from πDIF events traversed B3 and deposited larger energy than the decay positron in S3. The fifth factor was based on the pulse shape fitting in B3. Multiplying the five suppression factors gave a total suppression of 10^5.

Figure 8 shows the suppressed energy spectra. In order to evaluate the correction for the $\pi^+\to e^+\nu_e$ low energy tail, the remaining $\pi^+\to\mu^+\to e^+$ events of the suppressed spectrum were subtracted and the $\pi^+\to e^+\nu_e$ tail was obtained. Additionaly, since the total energy cut removed $\pi^+\to e^+\nu_e$ events with Bhabha scattering and larger energy loss in B3, MC simulation was used to take account of those events. These procedures were based on the method of the previous TRIUMF experiment[3]. The tail fraction found was 3% which provides a lower bound on this correction.

In order to determine an upper bound of the $\pi^+\to e^+\nu_e$ low energy tail empirically, special data sets were taken using a 70 MeV/c positron beam. The left diagram in Figure 9 shows a schematic of the lineshape measurement. The crystal calorimeter was rotated to obtain different angles of entry. The right plot in the figure is the energy spectrum at 0°. It shows the peak of the beam positrons as well as three lower energy bumps due to photoabsorption followed by neutron escape from the NaI crystal[14]. The $\pi^+\to e^+\nu_e$ lineshape distribution using these beam measuements as input was then obtained by simulation. The fraction of the events below E_{Cut} was estimated to be 3% in agreement with the lower bound discussed above. The tail fraction

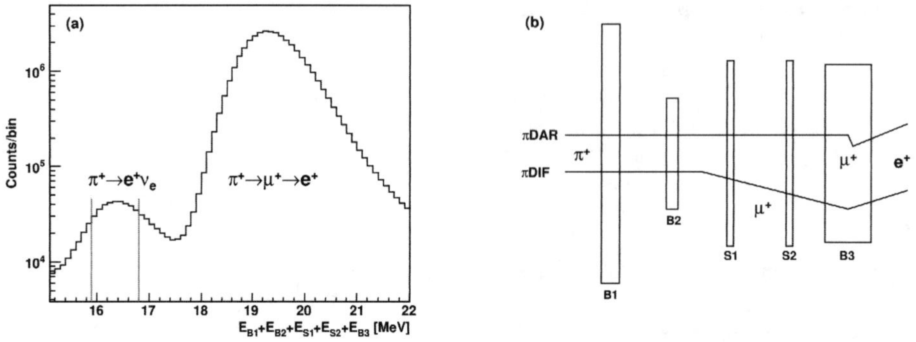

Figure 7. (a) Total energy deposited in B1, B2, S1, S2, and B3. (b) Schematic of π-decay-at-rest (πDAR, upper trajectory) and πDIF (lower trajectory) events.

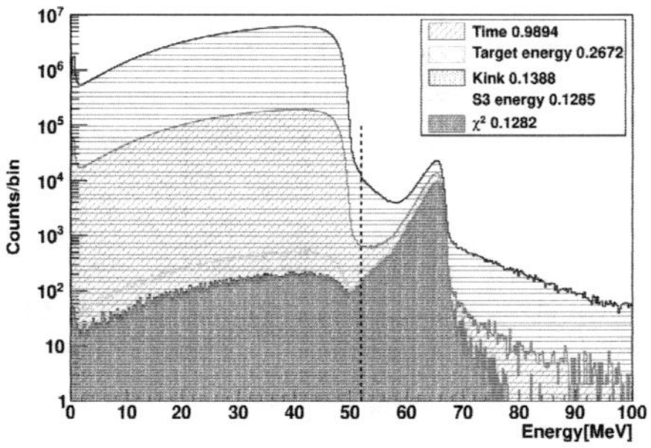

Figure 8. Summed energy spectrum of NaI and CsI after each suppression cut. The black histogram (top line) is before suppression and the pink histogram (filled) is after all suppression cuts. The legend box shows the low energy fraction, which represents the integral of events below E_{Cut}(=52 MeV) divided by the integral of the full energy spectrum.

obtained by this method was used as an upper bound.

Assuming a normal distribution for the probability distribution function, the combined tail fraction was estimated to be 3%.

4.2. Acceptance Correction

Energy dependent effects altered the relative acceptance of $\pi^+ \to e^+ \nu_e$ and $\pi^+ \to \mu^+ \to e^+$ decays. The acceptance correction relied on MC calculations including multiple Coulomb scattering, Bhabha scattering, positron annihilation-in-flight, and trigger losses. The ratio of acceptance of $\pi^+ \to e^+ \nu_e$ and $\pi^+ \to \mu^+ \to e^+$ decays was estimated to be 0.9994 for the radius cut < 60 mm at WC3. The uncertainties of the detector geometry and the pion beam stopping position were included in the error.

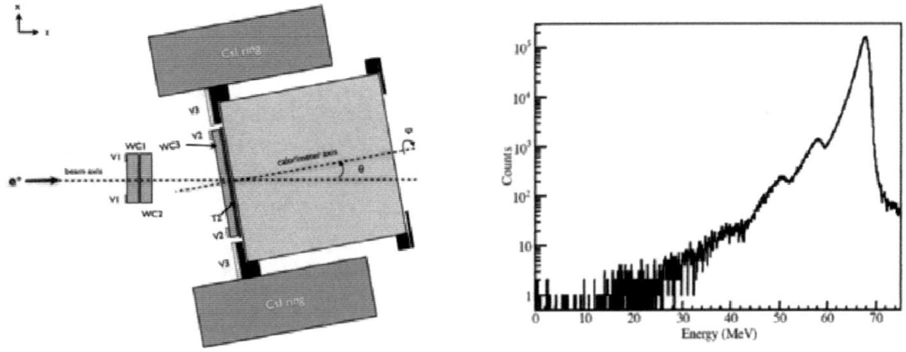

Figure 9. Left: Schematic of the calorimeter lineshape measurement.[13] Right: Energy spectrum in NaI and CsI at 0°[14].

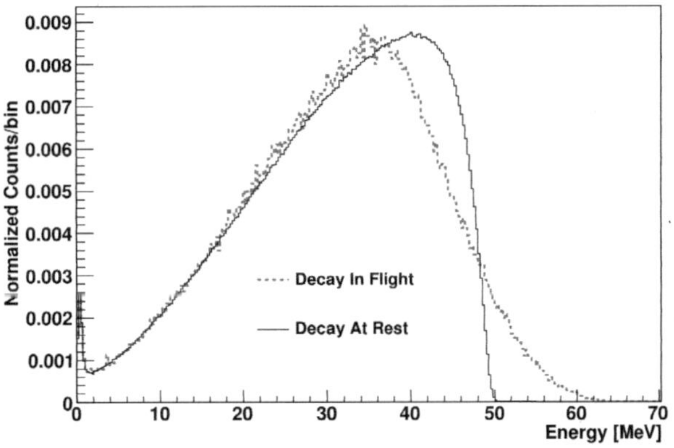

Figure 10. $\pi^+ \to \mu^+ \to e^+$ energy spectra. Red (dashed line): Simulated μDIF events. Black (solid line): Decay-at-rest events.

4.3. Uncertainty of the μDIF Contribution

For the decays-in-flight of muons from $\pi^+ \to \mu^+ \nu_\mu$ in B3, Lorentz boosting raises the positron energy. Figure 10 shows the energy spectra of μDIF and μ-decay-at-rest events determined by MC. Since μDIF had the same time distribution as $\pi^+ \to e^+ \nu_e$ decays, μDIF events inflated the apparent number of $\pi^+ \to e^+ \nu_e$ decays. The contribution of μDIF events $C_{\mu DIF}$ was estimated by MC study and scaled to the number of $\pi^+ \to \mu^+ \to e^+$ decays (parameter A) in Eq.7. The correction value $C_{\mu DIF}$ was found to be 0.2% of $\pi^+ \to e^+ \nu_e$, resulting in an uncertainty on the branching ratio $< 0.01\%$.

4.4. Other Corrections

The timing of the decay positron was based on the signal from energy loss in T1. Possible energy dependence with the pion stop time t_0 was studied using decay positrons from muons stopped at the center of B3. It was also found that the effect of the uncertainty of the muon lifetime[15] on the branching ratio was $< 0.01\%$ while the pion lifetime[15] uncertainty contributed 0.01%.

5. Result and Conclusion

At present, the result for the $\pi^+ \to e^+ \nu_e$ branching ratio for the initial data set analyzed remains blinded. The expected statistical and systematic uncertainties with the current partial data set are both about 0.2%. The full data set is currently under analysis and corresponds to 20 times higher statistics. It is anticipated that the systematic uncertainties will also be substantially improved.

References

[1] Rare Pion and Kaon Decays, D. Bryman, W. Marciano, R. Tschirhart, T. Yamanaka, Ann. Rev. Nucl. Part. Sci. 61, 331 (2011).
[2] V. Cirigliano and I. Rosell,*JHEP* **0710**, 005 (2007).
[3] D.I. Britton *et al.*, *Phys. Rev. Lett.* **68**, 3000 (1992). and D.I. Britton *et al.*, *Phys. Rev.* D**49**, 28 (1994).
[4] G. Czapek *et al.*, *Phys. Rev. Lett.* **70**, 17 (1993).
[5] M.J. Ramsey-Musolf, S. Su and S. Tulin, *Phys. Rev.* D**76**, 095017 (2007).
[6] H. Lacker and A. Menzel, *JHEP* **07**, 006 (2010).
[7] S.Davidson, D. Bailey and B. Campbell, *Z. Phys.* C**61**, 613 (1994).
[8] Alberto Lusiani. Measurements of $|V_{us}|$ and Search for Violation of Lepton Universality and CPT in Tau Decay at BABAR. *PoS*, ICHEP2010:251, 2010. arXiv:1012.3734.
[9] C. Lazzeroni *et al.* Test of Lepton Flavour Universality in $K^+ \to l\nu$ Decays, 2011. arXiv:1101.4805.
[10] A. Pich. Tau Physics: Theory Overview. *Nucl. Phys. Proc. Suppl.*, 181-182:300-305 (2008).
[11] A. Aguilar-Arevalo *et al.*, *Nucl. Instrum. Method* A**609**, (2009) 102.
[12] G. Bressi, *et al.*, *Nucl. Phys.* **B513** (1998) 555.
[13] C.Malbrunot. *Study of* $\pi^+ \to e^+ \nu_e$ *decay*. PhD thesis, University of British Columbia. (2012).
[14] A. Aguilar-Arevalo, *et al.*, Study of a Large NaI(Tl) Crystal, *Nucl. Instrum. Meth.* A**621** (2010) 188191.
[15] K.A. Olive *et al.* (Particle Data Group), *Chin. Phys.* **C38** 090001 (2014).

Journal of Physics: Conference Series **631** (2015) 012045 doi:10.1088/1742-6596/631/1/012045

Non-SUSY Beyond Standard Model Searches: Recent Results from ATLAS and CMS

Fairouz Malek on behalf of the ATLAS and CMS Collaborations
LPSC, CNRS/IN2P3 and Université Grenoble-Alpes, Grenoble, FR
E-mail: fmalek@lpsc.in2p3.fr

Abstract. The Standard Model of particle physics is a sensational success, especially since the discovery of the 125 GeV Higgs boson. However, there are still numerous unanswered questions. Why is the Higgs so light? Do the interactions couplings unify and how can gravity be included? Why three fermion generations? What is dark matter? Theories Beyond the Standard Model (BSM), such as Grand Unified Theories, Extra Dimensions or Technicolour are trying to answer these questions. In these proceedings, we will focus on the most recent results obtained by the ATLAS and CMS experiments at the LHC for BSM searches, excluding Higgs and supersymmetry searches. New results on Dark Matter, heavy narrow-width resonances, new heavy quarks and third generation leptoquarks are presented. A summary of the prospects at 14 TeV and at the High Luminosity LHC period is given.

1. Introduction

The LHC is a proton-proton collider, designed to operate at a center-of-mass energy of $\sqrt{s}$ = 14 TeV and to collect of the order of 100-300 fb^{-1} of data [1]. At the end of 2012, the two multi-purpose LHC experiments, ATLAS [2] and CMS [3], concluded what has come to be known as "Run I" in which the LHC operated at $\sqrt{s}$ = 7 and 8 TeV, collecting integrated luminosities of approximately 5 fb^{-1} and 20 fb^{-1}, respectively at the two energies. Even at these reduced energies, the LHC has well explored the TeV scale and the discovery of a particle in July 2012, which appears to be compatible with the Standard Model (SM) Higgs Boson [4,5], was confirmed.

The search for BSM physics has already been a strong activity at the LHC. So far, these searches show no evidence for new physics and have set stringent limits on many BSM models. The searches are based on distinct experimental signatures, and are sensitive to a wide range of models.

In this paper, we summarize the current status of the most recent searches, excluding BSM Higgs and supersymmetry (SUSY) searches. The review is limited to search results at 8 TeV which have been published or submitted for publication between July and December 2014. The full list of results, both preliminary and published can be found in Refs [6,7]. Also, an exhaustive review of so many signatures in so many models is not possible. The selection made here includes a limited number of signatures for Dark Matter, heavy narrow-width resonances and other new particles such as leptoquarks or Vector-Like quarks.

2. Dark Matter

The existence of Dark Matter (DM) is firmly established by a large number of cosmological and astrophysical observations, but little is known about what it is made of. The most popular candidates are the Weakly Interacting Massive Particles (WIMP). At a collider, one searches for pair production of WIMPs, which escape the detector without interacting. Therefore, an additional object is needed, typically radiated in the initial state and recoiling against the escaping WIMP pair, leading to missing transverse energy (MET) [9–13]. Fig. 1 shows Feynman diagrams of the DM pair production.

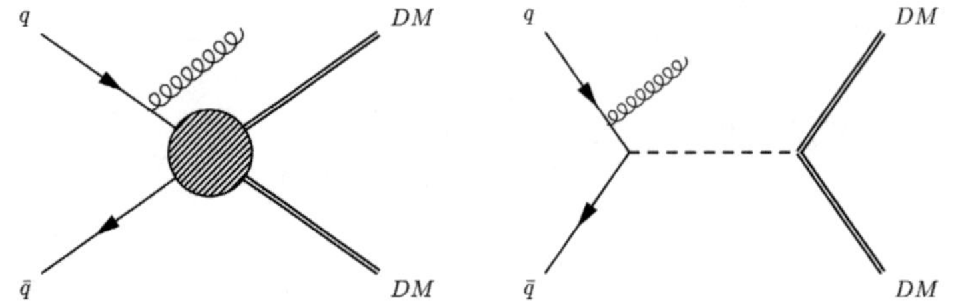

Figure 1. Feynman diagrams for the pair production of DM particles for the case of a contact interaction or heavy mediator exchange (left) and the exchange of a light mediator (right).

These signatures are referred to as mono-X signatures. The ATLAS and CMS experiments have performed searches for DM in the mono-jet, mono-photon, mono-top, mono-W and mono-Z channels. The interpretation of these searches use an effective field theory (EFT) to describe the interaction between SM and DM particles. This makes the assumption that the interaction is mediated by a particle which is itself too heavy to be directly produced at the LHC and can thus be integrated out, leading to a contact interaction. Different operators describe different initial states and different types of interactions [8]. A subset of all possible operators is used in the different mono-X analyses. The cross section for DM pair production via such an operator depends on the mass of the WIMP (m_χ) and the contact interaction scale of the effective theory, M_*. Limits on the M_* can be translated into limits on the χ-nucleon scattering cross section, $\sigma_{\chi-N}$. The virtue of the effective theory is that it allows for a comparison to the results of direct searches, which look for the recoil of a target nuclei due to a WIMP scattering off it.

This section discusses the results of DM searches through three signatures: mono-jet, mono-photon and heavy quark (HQ) plus MET.

2.1. The mono-jet signature

Initial state radiation of a jet has a large cross section at the LHC and therefore the mono-jet analysis is a promising channel for the DM search. It is moreover the only one that is sensitive to the gluon-gluon operator. The example choosen to illustrate that type of search is taken from the CMS experiment [9]. Here, events are required to have one highly energetic jet with a p_T greater than 110 GeV. Events with more than two jets with $p_T > 30$ GeV and pseudorapidity $|\eta|$ less than 4.5 are discarded, thereby significantly reducing background from top-quark pair and QCD multijet events. Processes producing leptons, such as W and Z, dibosons, and top quarks are suppressed by rejecting events with well reconstructed and isolated electrons with $p_T > 10$ GeV, reconstructed muons with $p_T > 10$ GeV and well-identified hadronically decaying tau leptons with $p_T > 20$ GeV. The analysis is performed in seven inclusive regions of MET > 250, 300, 350, 400, 450, 500 and 550 GeV.

The MET spectrum obtained is presented in Fig. 2(left) in comparison to the SM prediction and possible signal distributions. The data are in agreement with the background-only hypothesis. Within the framework of the effective field theory, limits on the contact interaction scale and on $\sigma_{\chi-N}$ are extracted. The confidence level (CL) chosen for these limits is 90%, to enable a direct comparison with the results from the direct detection experiments. Shown in Fig. 2(right) are upper limits on $\sigma_{\chi-N}$ as a function of m_χ compared with previously published results together with results from direct detection experiments, for the vector and scalar operators (spin independent). For a m_χ of 10 GeV, the upper limit on $\sigma_{\chi-N}$ was found to be 7.06×10^{-40} cm^2 for the vector operator (spin independent). Limits on $\sigma_{\chi-N}$ were also derived for the case of the axial-vector operator (spin dependent) and the upper one was found to be 2.70×10^{-41} cm^2 for the same m_χ value.

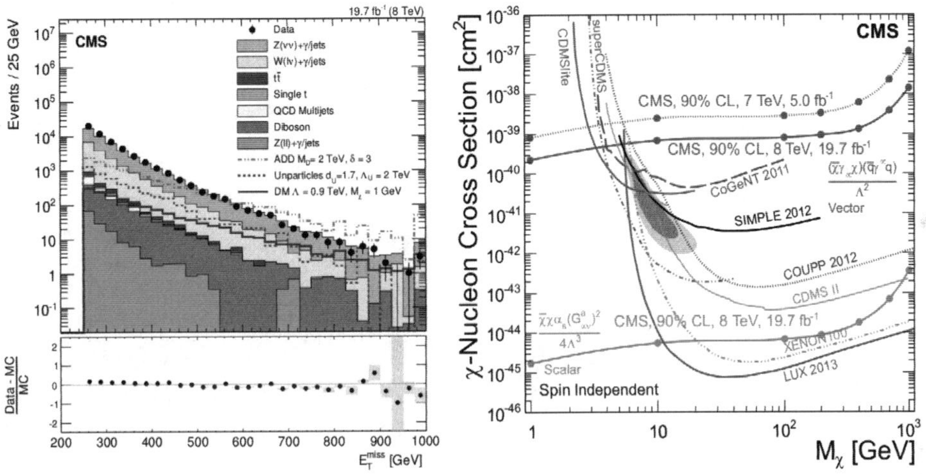

Figure 2. DM searches, the mono-jet signature [9]: (left) Missing transverse energy after all selections for data and SM backgrounds. The processes contributing to the SM background are from simulation, normalized to the estimation from data using a MET threshold of 500 GeV. The shaded bands in the lower panel represent the statistical uncertainty. (right) Upper limits on $\sigma_{\chi-N}$ at 90% CL, plotted against m_χ and compared with previously published CMS results together with results from the CoGeNT, SIMPLE, COUPP, CDMS, SuperCDMS, XENON100, and LUX collaborations. The solid and hatched yellow contours show the 68% and 90% CL contours respectively for a possible signal from CDMS.

2.2. The mono-photon signature

Events with a photon radiated in the initial state have been used to search for DM candidates as well. An analysis by CMS [10] and another by ATLAS [11] are described here. In the CMS analysis, events are selected with a photon with an E_{T} larger than 145 GeV. In order to reduce QCD multijet backgrounds, events are rejected if there is more than one jet with $p_{\mathrm{T}} > 30$ GeV and $\Delta R > 0.5$ (ΔR being the radius of a cone centered on the shower axis) relative to the photon. Events with isolated leptons (electron or muon) with $p_{\mathrm{T}} > 10$ GeV and $\Delta R > 0.5$ relative to the photon, are also rejected to suppress $W\gamma \to l\nu\gamma$ and $W \to l\nu$ backgrounds. The candidate events are required to have a MET larger than 140 GeV. The MET spectrum obtained from that analysis is presented in Fig. 3(left) in comparison to the SM prediction and possible signal distributions. No excess at large MET is observed and limits on the interaction scale and $\sigma_{\chi-N}$ are derived. Fig. 3(right) shows the 90% CL upper limits on $\sigma_{\chi-N}$ as a function of m_χ for spin independent scattering. Also shown are the limits from selected direct detection experiments results. For a m_χ of 10 GeV, $\sigma_{\chi-N}$ is constrained to be less than 9.6×10^{-41} cm^2 (2.6×10^{-39} cm^2) for a spin dependent (spin independent) interaction at 90% CL.

In the same channel, an analysis was performed by ATLAS [11]. The main result gives values of M_* excluded below 760, 760 and 1010 GeV for the spin independent (D5 operator) and spin dependent (D8 and D9 operators) interactions respectively. These limits on M_* can be translated into upper limits on $\sigma_{\chi-N}$. They are shown in Fig. 4 as a function of m_χ for spin independent (left) and spin dependent (right) interactions. The results obtained by ATLAS with 7 TeV data for the same channel are shown for comparison. Also shown are results from various DM direct detection experiments.

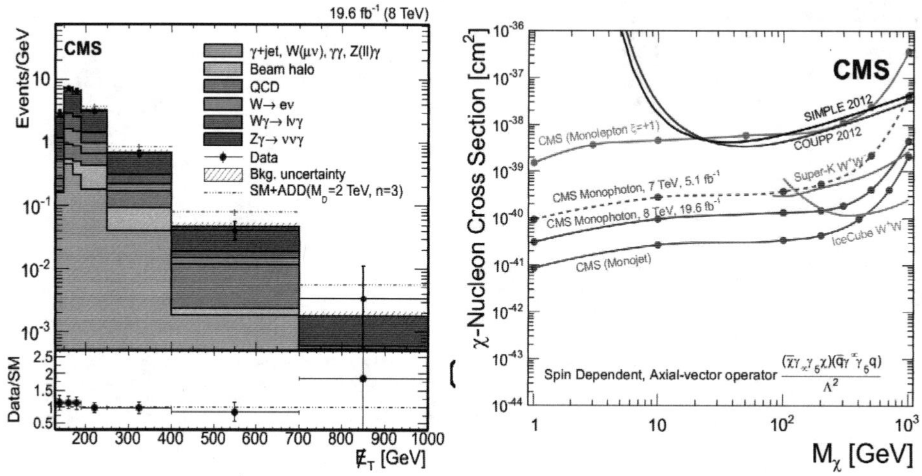

Figure 3. DM searches in CMS, the mono-photon signature [10]: (left) The MET distribution for the candidate sample, compared with estimated contributions from SM. The background uncertainty includes statistical and systematic components. The bottom panel shows the ratio of data and SM predictions. (right) The 90% CL upper limits on $\sigma_{\chi-N}$ as a function of the m_χ for spin independent scattering. Also shown are the limits from selected direct detection experiments results along with 7 and 8 TeV results.

2.3. The HQ plus MET signature

Other signatures such as MET and a top quark or a b quark are also used to search for DM in ATLAS [12] and CMS [13]. The analysis requires a large MET and a central high p_T jet.

In the case of the ATLAS experiment analysis [12], four orthogonal signal regions are defined. To optimize the sensitivity of the analysis, a requirement on the output of the b-tagging algorithm which provides a 60% (70%) b jet efficiency operating point is used in signal regions (SR) 1 and 2 (3 and 4). The first two signal regions focus on events with DM produced in conjunction with one (SR1) or two (SR2) b quarks in the final state. SR3 and SR4 target events in which DM is produced in conjunction with a $t\bar{t}$ pair, where either both top quarks decay hadronically (SR3) or one top quark decays hadronically and the other semileptonically (SR4). Events assigned to SR1 and SR2 are required to have a low jet multiplicity (less than five). The results are interpreted in the framework of an effective field theory to set stringent limits on scalar and tensor interactions between SM and DM particles. The data are found to be consistent with

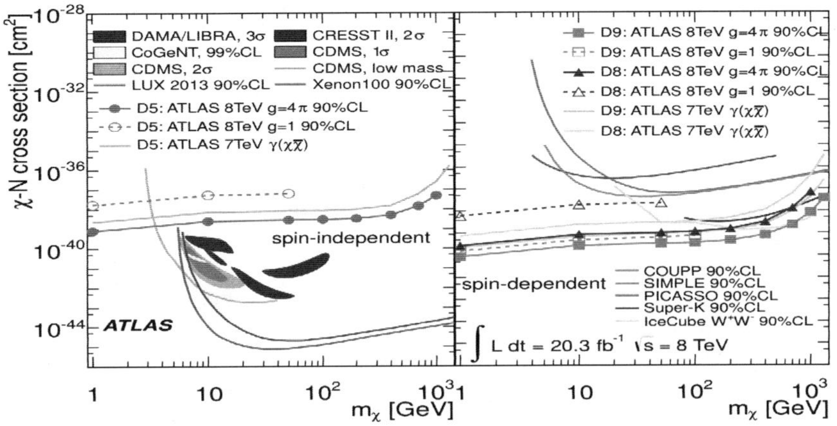

Figure 4. DM searches, the mono-photon signature [11]: Upper limits at 90% CL on $\sigma_{\chi-N}$ as a function of m_χ for spin-independent (left) and spin-dependent (right) interactions, for a coupling strength $g = \sqrt{(g_f * g_\chi)} = 1$. The results obtained from ATLAS with 7 TeV data for the same channel are shown for comparison. Also shown are results from various dark matter direct detection experiments.

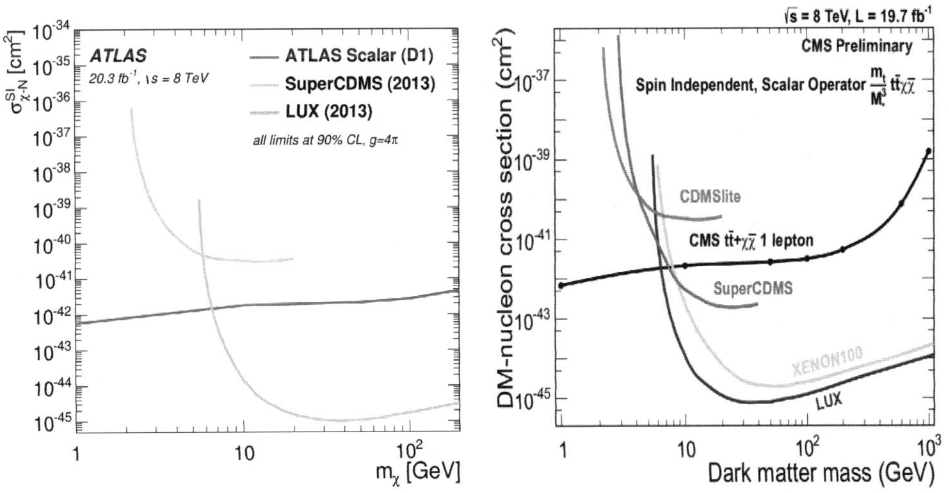

Figure 5. DM searches, HQ+MET signature [12, 13]: (left) ATLAS: Upper limits at 90% CL on the spin independent $\sigma_{\chi-N}$ for the scalar operator D1 (red) as a function of m_χ. The yellow and green curves represent the exclusion limits recently set by the LUX and Super-CDMS collaborations. (right) CMS: The 90% CL upper limits on the spin independent $\sigma_{\chi-N}$ as a function of m_χ for the scalar operator D1. Also shown are 90% CL limits from selected direct detection experiments.

the Standard Model expectations. Limits are also set on the mass scale of the effective field theories.

Fig. 5(left) shows the corresponding 90% CL exclusion curves on $\sigma_{\chi-N}$ for the spin independent interaction for the scalar operator D1 as a function of m_χ. The most stringent limits set by direct detection experiments are also shown. The limits shown are especially strong in the low-mass region where several collaborations (DAMA,COGENT, CDMS) have recently claimed possible observations of DM [14]. The results reported in this analysis improve significantly the sensitivity to WIMP nucleon interactions mediated by the scalar operator D1 (approximately 1×10^{-42} cm^2 for a m_χ of 10 GeV) compared to previous ATLAS results. Limits are also set on the D9 vector operator.

In the case of CMS analysis [13], the search focused on the production of DM particles with a pair of top quarks in the single-lepton channel. Therefore, the events are selected with exactly one identified isolated lepton, at least three jets and at least one b-tagged jet. Signal events are expected to be produced with large MET due to the production of two DM particles which escape from the detector, events are thus preselected with MET > 160 GeV. No excess of events was found above the Standard Model expectation. Interpreting the findings in the context of a scalar interaction between DM particles and top quarks and assuming a DM particle with a mass of 100 GeV, the interaction scale, M_* is excluded at 90% CL below 118 GeV, representing the most stringent limit to date. The limits on the interaction scale are interpreted as limits on $\sigma_{\chi-N}$ as shown in Fig. 5(right) for the spin independent scalar operator D1. More stringent limits are obtained from this analysis compared to direct DM detection in the low χ mass region of less than about 6 GeV. These masses are excluded for $\sigma_{\chi-N}$ higher than $1-2 \times 10^{-42}$ cm^2.

3. New heavy gauge bosons

The existence of heavy resonances is predicted by a variety of BSM models which are alternatives to SUSY in the explanation of the electroweak symmetry breaking. They are heavy new states with integer spin. Left-right symmetric and Extended Gauge models predict spin-1 states [15] while Randall-Sundrum models predict heavy spin-2 states [16].

In general, the analysis basically consists of searching for a bump in the two-body invariant mass spectra. The background is either small or could be modelled from the data control samples. ATLAS and CMS have searched for dilepton (including lepton+ν), diphoton, dijet and jet+γ resonances. Relatively simple topologies allow for a robust analysis to be done rapidly. The most recent searches focused on fully hadronic or semi-hadronic signatures such as $W' \to tb \to q\bar{q}b\bar{b}$ [17] and $W' \to tb \to l\nu b\bar{b}$ [18].

For the fully hadronic final state, the analysis performed by the ATLAS experiment is done in the $W' \to tb \to q\bar{q}b\bar{b}$ mode for W' masses above 1.5 TeV, where the W' decay products are highly boosted [17]. Copious multijets background is present thus necessitating efficient jet substructure techniques, described in the published paper for this analysis, to identify jets from high-momentum top quarks to ensure high sensitivity, independent of W' mass, up to 3 TeV; b-tagging is also used to identify jets originating from b quarks.

The data are consistent with SM expectations and upper limits at 95% CL are set on the $W' \to tb$ cross section times branching ratio ranging from 0.16 pb to 0.33 pb for left-handed W' bosons, and ranging from 0.10 pb to 0.21 pb for W' bosons with purely right-handed couplings. As shown in Fig. 6, the observed limits agree with the expected limits roughly within 1 σ over the whole range for both W' models. For a coupling $g' = g_{SM}$, the limits on the cross section times branching ratio translate to observed (expected) lower limits on the mass to 1.68 TeV (1.63 TeV) and 1.76 TeV (1.85 TeV) in the left- and right-handed models, respectively.

The $W' \to tb \to l\nu b\bar{b}$ channel, analysed by CMS experiment [18], necessitates in addition to the techniques used above for the fully hadronic channel, the selection of a high MET. The search covers masses between 0.5 and 3.0 TeV, for right-handed or left-handed W' bosons. No

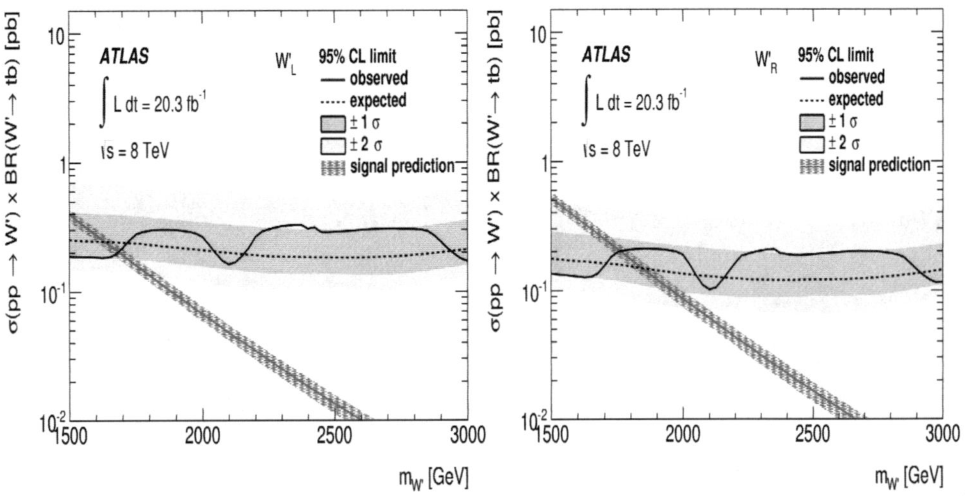

Figure 6. New heavy gauge bosons search, fully hadronic channel [17]: Upper limits at 95% CL on the cross section times branching ratio to tb for the left-handed (left) and for the right-handed (right) W'. The expected cross section for W' production with $g' = g_{SM}$ is also shown (in red).

significant deviation from the Standard Model expectation is observed and limits are set on the $W' \to t\bar{b}$ cross-section times branching ratio and on the W' effective couplings as a function of the W' mass. For a left-handed (right-handed) W', masses below 1.70 (1.92) TeV are excluded at 95% CL. These results are compatible with those from the ATLAS analysis, above.

4. Other new particles

4.1. Leptoquarks

Encountered in various extensions of the Standard Model, such as technicolor theories or GUT, new scalar or vector bosons called leptoquarks (LQ) are colour-triplet bosons that carry both lepton and baryon numbers, and fractional electric charge. Searches for pair-produced LQs assumed to couple only to quarks and leptons of the third SM generation have been issued by ATLAS and CMS recently. The most recent search [19], performed by CMS, looks for a signature of a LQ to a final state involving $b\bar{b}\tau\tau$. A particle-flow (PF) technique [20] is used for the reconstruction of hadronically decaying τ lepton candidates. In the PF approach, information from all subdetectors is combined to reconstruct and identify all final-state particles produced in the collision. The particles are classified as either charged hadrons, neutral hadrons, electrons, muons, or photons. Jets are reconstructed using the anti-k$_t$ algorithm with a size parameter 0.5 using particle candidates reconstructed with the PF technique. The b-tagged jet with the highest p_{T} is selected, and then the remaining jet with the highest p_{T} is selected whether or not it is b-tagged. To discriminate between signal and background in the leptoquark search, the mass of the τ and a jet, denoted M(τ, jet), is required to be greater than 250 GeV. The S_{T} distribution after the final selection is used to extract the limits on the leptoquark signal, where S_{T} is defined as the scalar sum of the p_{T} of the light lepton, the τ, and the two jets.

The S_{T} distribution of the selected events from the data and from the background predictions, combining e and μ channels is shown in Fig. 7(left). The distribution from a 500 GeV signal hypothesis is added to the background to illustrate how a hypothetical signal would appear above the background prediction. The data agree well with the SM prediction.

Fig. 7(right) shows the expected and observed combined upper limits on the third-generation LQ pair production cross section σ times the square of the branching fraction, B^2, at 95% CL, as a function of the LQ mass. Third-generation scalar leptoquarks with masses below 740 GeV are excluded at 95% CL, assuming a 100% branching fraction for the leptoquark decay to a τ lepton and a bottom quark.

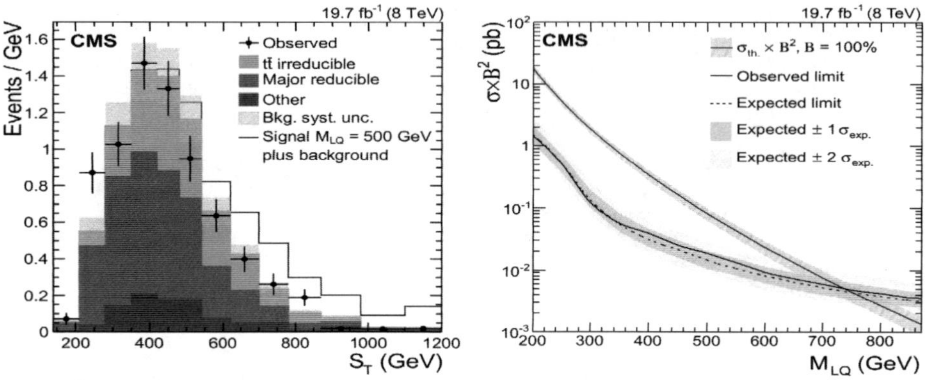

Figure 7. LQ search [19]: (left) The final S_{T} distribution for the leptoquark search with the e and μ channels combined. A signal sample for leptoquarks with a mass of 500 GeV is added on top of the background prediction. The last bin contains the overflow events. The horizontal bar on each observed data point indicates the width of the bin in S_{T}. (right) The expected and observed combined upper limits on the third-generation LQ pair production cross section σ times the square of the branching fraction, B^2, at 95% CL, as a function of the LQ mass. The dark blue curve and the hatched light blue band represent the theoretical LQ pair production cross section, assuming $B = 100\%$, and the uncertainties due to the choice of PDF and renormalization/factorization scales.

4.2. New heavy quarks

New heavy quarks, like the Vector-like Quarks(VLQ), are among the simplest extensions to the SM. They contribute to the regulation of quadratic divergence of one-loop contributions to the Higgs mass, so could provide a solution to the hierarchy problem. They are produced singly or in pairs as shown in Fig. 8(right) for the latter case. The production cross section of the VLQ particles is also shown on that figure, in the right part.

Searches have been performed by ATLAS [24–26] and CMS [27] for both single and pair-produced exotic top partners T or bottom partners B, decaying respectively into Zt, Wb, Ht and Zb, Wt, Hb. The results obtained from the three analyses performed by the ATLAS experiment [24–26] are reported here.

In the first analysis [24], the search looks for the production of new heavy quarks that decay to a Z boson and a third-generation SM quark. In the case of a new charge +2/3 quark T, the decay targeted is $T \to Zt$, while the decay targeted for a new charge -1/3 quark B is $B \to Zb$. The selected events for the analysis should contain a high transverse momentum Z boson candidate reconstructed from a pair of oppositely charged same-flavor leptons (electrons or muons). They are analyzed in two channels defined by the absence or presence of a third lepton. Hadronic jets, in particular those with properties consistent with the decay of a b hadron, are also required to be present in selected events.

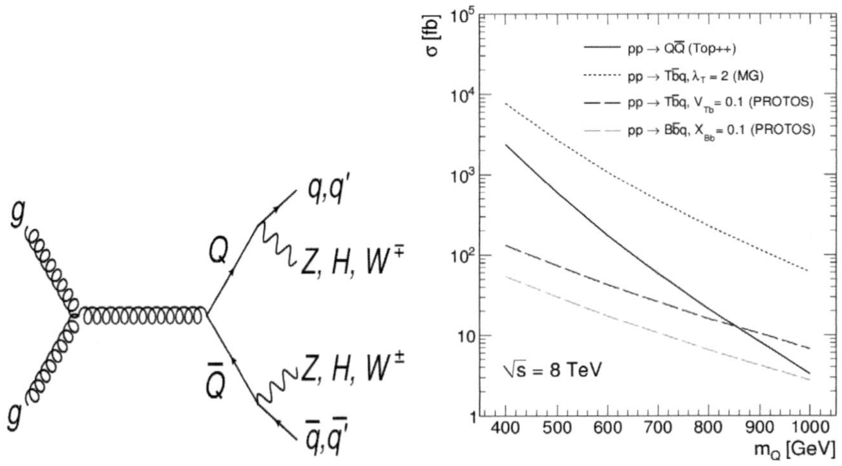

Figure 8. (left) A representative diagram illustrating the pair production and decay modes of a VLQ (Q=T,B). (right) The $\sqrt{s}$=8 TeV LHC cross section versus quark mass for pair production, denoted by the solid line, as well as for the $T\bar{b}q$ and $B\bar{b}q$ single production processes, denoted by dashed lines. The pair production cross section was calculated with TOP++ [21]. The single production cross sections were calculated with PROTOS [22] and MADGRAPH [23] using different electroweak coupling parameters.

The second analysis [25] is a search for pair production of heavy top-like quarks decaying to a high p_{T} W boson and a b quark in the lepton plus jets final state. The selected events for the analysis are characterized by a high p_{T} isolated electron or muon, large missing transverse energy and at least four jets, with at least one of them b tagged. The analysis strategy relies on the substantial boost of the W bosons in the T pair signal when the transverse mass m_T >400 GeV.

The third search [26] is for anomalous production of events with same-sign dileptons and b jets. The final state considered requires the presence of exactly two leptons in the event, both with the same electric charge. In addition, two or more jets, at least one of which is consistent with being a b jet, and sizable MET are required.

No significant excess of events above the Standard Model expectation is observed in the three analyses. Lower limits are derived on the mass of vector-like T and B quarks under various branching ratio hypotheses, as well as upper limits on the magnitude of electroweak coupling parameters.

Using the results from the three analyses and for a given bin in the branching ratio (BR) plane, the strongest of all limits considered is plotted. Fig. 9(left) shows all available limits on the vector-like B quark assuming the pair production hypothesis and presented in the (Wt,Hb) branching ratio plane. No combination is made of the different analyses. Similarly, Fig. 9(right) shows all available limits on the vector-like T quark assuming the pair production hypothesis and presented in the (Wb,Ht) branching ratio plane. No combination is made of the different analyses, except for the $Ht + X$ and $Wb + X$ analyses which are combined.

In the left part of the figure, we see that the sensitivity is higher for a decay of the B pair to $ZbZb$. Therefore, if we assume that the corresponding branching fraction is 100%, one can put the limit on the B mass at $\sim$ 800 GeV. Similarly, on the right part of the figure, we observe a high sensitivity of the T quark pair to decay to $ZtZt$ or to $HtHt$. Therefore, assuming a

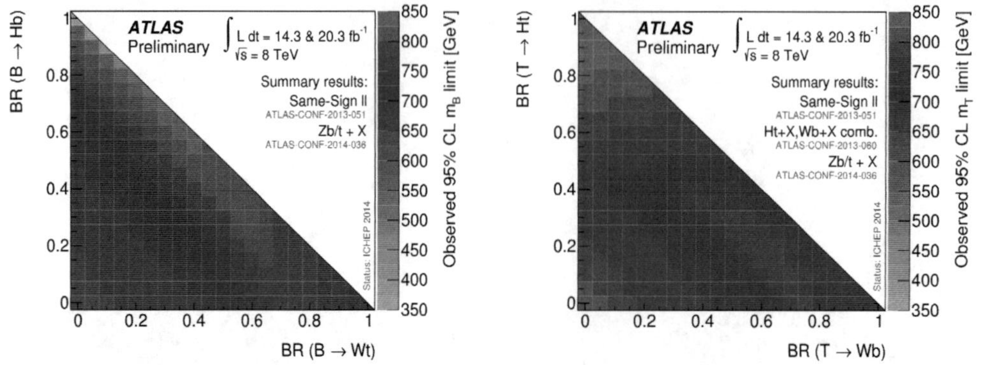

Figure 9. VLQ search [28]: (left) Observed lower limits at 95% CL on the mass of vector-like B quarks for ATLAS searches with 14 fb^{-1} and 20 fb^{-1} of 8 TeV data. For a given bin in the selected BR plane, the strongest of all limits considered is shown (i.e. no combination is made of the different analyses). This figure shows all available limits on the vector-like B quark. (right) Observed lower limits at 95% CL on the mass of vector-like T quarks for ATLAS searches with 14 fb^{-1} and 20 fb^{-1} of 8 TeV data. For a given bin in the selected BR plane, the strongest of all limits considered is shown (i.e. no combination is made of the different analyses, except for the $Ht+X$ and $Wb+X$ analyses which are combined). This figure considers all available limits on the vector-like T quark.

branching ratio of 100% for the decay $T \to Ht$, the limit of the T quark mass can be set to $\sim$ 830 GeV.

5. Prospects

From 2015 to 2017 (Run-II), the ATLAS and CMS experiments will collect about 100 fb^{-1} of data each, at a center-of-mass energy between 13 and 14 TeV. Following this, the accelerator will be upgraded to deliver two to three times more instantaneous luminosity. A total of $\sim$ 300 fb^{-1} of data is expected to be collected by the end of Run-III in 2022. An upgrade to the LHC is proposed to provide an instantaneous luminosity of 5×10^{34} cm^{-2} s^{-1} in 2023, named HL-LHC (High Luminosity LHC).

The HL-LHC will provide a significant increase in discovery reach for new Higgs particles and sensitivity to non-Standard Model effects or direct searches such as new heavy gauge bosons (Z' and W'). Some examples of sensitivity studies to such signals are given here below.

5.1. Sensitivity to Z'

Studies of the sensitivity to a Z' boson in ATLAS were performed to illustrate the reach at the different LHC phases [29]. Only the Drell-Yan background is considered in this study. The Sequential Standard Model (SSM) Z'_{SSM} boson, which has the same fermionic couplings as the Standard Model Z boson, is used as the signal template. The dilepton invariant mass distribution, m_{ll}, for events above 200 GeV, and the resulting limits as a function of Z'_{SSM} pole mass are shown in Fig. 10 for the ee channel. The 95% CL expected limits in the absence of signal, using statistical errors only, are shown in Table 1 for ee and $\mu\mu$ decays. The increase of a factor of ten, from 300 to 3000 fb^{-1} in integrated luminosity, raises the sensitivity to high-mass dilepton resonances by up to 1.3 TeV.

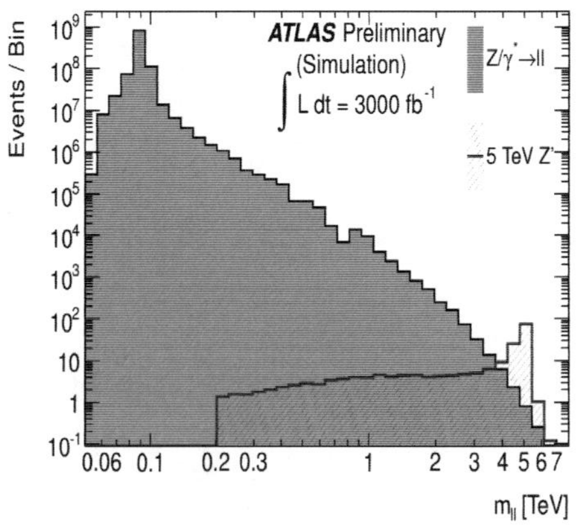

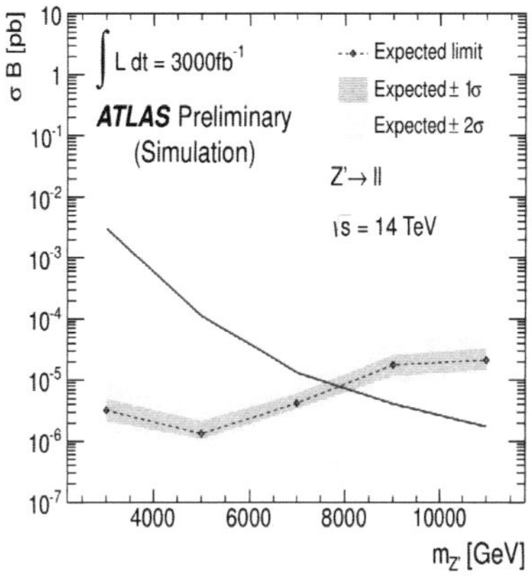

Figure 10. Z' prospects [29]: (left) The reconstructed dielectron mass spectrum for the Z' search with 3000 fb^{-1} of pp collisions at $\sqrt{s} = 14$ TeV. (right) The 95% CL upper limit on the cross section times branching ratio. Also shown is the theoretical expectation for the Z'_{SSM}.

Table 1. Summary of the expected mass limits for $Z'_{SSM} \to ee$ and $Z'_{SSM} \to \mu\mu$ searches in the Sequential Standard Model for pp collisions at $\sqrt{s} = 14$ TeV. All values are quoted in TeV [29].

model	300 fb^{-1}	1000 fb^{-1}	3000 fb^{-1}
$Z'_{SSM} \to ee$	6.5	7.2	7.8
$Z'_{SSM} \to \mu\mu$	6.4	7.1	7.6

5.2. Sensitivity to T quarks

A sensitivity study of the prospects for searching for heavy vector-like charge +2/3 quarks with the upgraded CMS detector has been performed [30]. With 3000 fb^{-1} of *pp* collision data collected at $\sqrt{s} = 14$ TeV with an average of 140 pileup interactions per bunch crossing, the mass reach for the discovery of a heavy T quark at 3σ and 5σ level is expected to be 1.65 TeV, and 1.48 TeV, respectively as shown in Fig. 11. The 95% CL exclusion reach is expected to be around 1.85 TeV, a factor of 2.6 higher than the current lower limit of 700 GeV. This result would have a significant impact on our understanding of the implications of a light Higgs at 125 GeV on composite Higgs models.

5.3. Sensitivity to Dark Matter

A study of the sensitivity of ATLAS to pair production of WIMP Dark Matter with *pp* collision data at $\sqrt{s}$=14 TeV in the mono-jet topology is presented. Different run periods and different integrated luminosities have been considered. As for the previous studies at 8 TeV, two classes of models have been investigated: the Effective Field Theory and the simplified models with an explicit light mediator.

According to the preliminary estimates obtained assuming EFT is a valid approach, the first data delivered early in 2015 may already be sufficient to test for the presence of new physics. Assuming a 5% systematic uncertainty on the background estimation, ATLAS could detect a

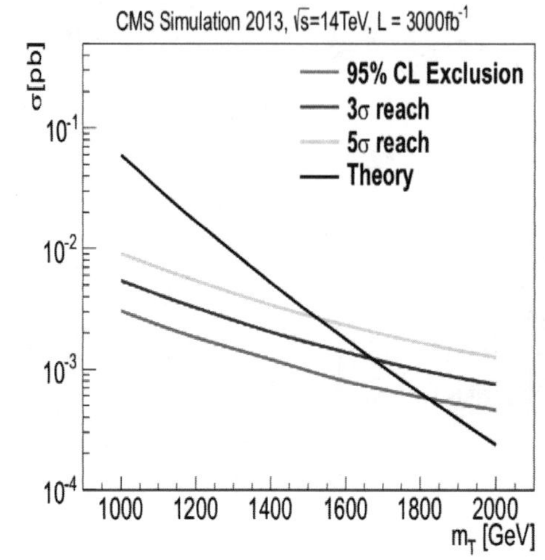

Figure 11. T quark prospects [30]: Expected sensitivity for a T quark pair production signal in the combined multilepton and single-lepton+jets sample. The 95% CL limits, 5σ discovery reach and the 3σ discovery reach are shown as a function of the T quark mass.

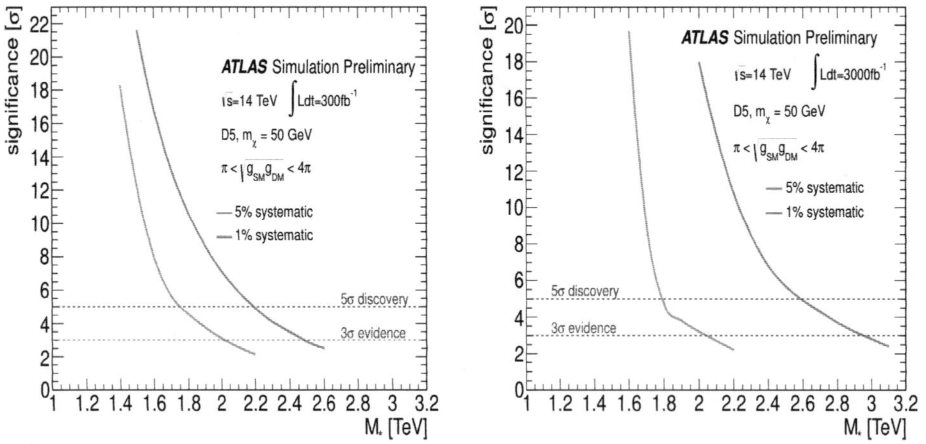

Figure 12. DM search prospects [31]: Discovery potential for DM signal with D5 operator and m_χ=50 GeV with 300 fb^{-1} (left) and 3000 fb^{-1} (right). These results assume that the EFT is a valid approach.

signal within the EFT approach with M_* up to 1500 GeV at 5σ after one year of data-taking. The reach of ATLAS with a full luminosity of 300 fb^{-1} could go up to M_* values of 1800 GeV at 5σ with the same assumed background systematic uncertainty, see Fig. 12(left). Taking into account the expected ultimate precision on the Standard Model background determination (1% total systematic error), the reach extends to M_* ~2200 GeV with an integrated luminosity of

Journal of Physics: Conference Series **631** (2015) 012045 doi:10.1088/1742-6596/631/1/012045

300 fb^{-1} and M_* ~2600 GeV for HL-LHC for a discovery at 5σ, see Fig. 12(right).

6. Conclusion

Using data collected in Run-I, ATLAS and CMS looked all over possible signatures for BSM hints or discoveries. These searches put limits on the mass of new particles, singly produced resonances up to ~ 5 TeV, pair produced new particles up to ~ 0.8 TeV. Limits were also set on the interaction scale in the EFT approach for different operators describing the type of interaction in the Dark Matter searches. No new physics was found.

From the sensitivity studies for 14 TeV runs, one can conclude that the first data delivered early in 2015 may already be sufficient to test for the presence of new physics. However, very important gains in the physics reach are possible with the HL-LHC with 3000 fb^{-1}. The luminosity upgrade would become even more interesting if new phenomena are seen during the 300 fb^{-1} phase of the LHC (Run-II + Run-III), as the ten-fold increase in luminosity would give access to improved measurements of the new physics.

References

[1] L. Evans, The Large Hadron Collider, *Ann.Rev.Nucl.Part.Sci.* **61** (2011) 435-466.

[2] ATLAS Collaboration, The ATLAS Experiment at the CERN Large Hadron Collider, *JINST* **3** (2008) S08003.

[3] CMS Collaboration, The CMS experiment at the CERN LHC, *JINST* **3** (2008) S08004.

[4] ATLAS Collaboration, Observation of a new particle in the search for the Standard Model Higgs boson with the ATLAS detector at the LHC, *Phys.Lett.* B **716** (2012) 1-29.

[5] CMS Collaboration, Observation of a new boson at a mass of 125 GeV with the CMS experiment at the LHC, *Phys.Lett.* B **716** (2012) 30-61.

[6] ATLAS exotics public Wiki page: http://twiki.cern.ch/twiki/bin/view/AtlasPublic/ExoticsPublicResults.

[7] CMS exotica public Wiki page: http://twiki.cern.ch/twiki/bin/view/CMSPublic/PhysicsResultsEXO; B2G list of Publications http://twiki.cern.ch/twiki/bin/view/CMSPublic/PhysicsResultsB2G.

[8] J. Goodman, M. Ibe, A. Rajaraman, W. Shepherd, T. M.P. Tait, H-B. Yu, Constraints on Dark Matter from Colliders, *Phys.Rev.* D **82**:116010, (2010).

[9] CMS Collaboration, Search for dark matter, extra dimensions, and unparticles in monojet events in proton-proton collisions at $\sqrt{s}$ = 8 TeV, arXiv:1408.3583 [hep-ex] (*preprint* CERN-PH-EP-2014-164).

[10] CMS Collaboration, Search for new phenomena in monophoton final states in proton-proton collisions at $\sqrt{s}$ = 8 TeV, arXiv:1410.8812 [hep-ex] (*preprint* CERN-PH-EP-2014-253).

[11] ATLAS Collaboration, Search for new phenomena in events with a photon and missing transverse momentum in pp collisions at $\sqrt{s}$ = 8 TeV with the ATLAS detector, *Phys. Rev.* D **91**, 012008 (2015)

[12] ATLAS Collaboration, Search for dark matter in events with heavy quarks and missing transverse momentum in pp collisions with the ATLAS detector, arXiv:1410.4031 [hep-ex] (*preprint* CERN-PH-EP-2014-229).

[13] CMS Collaboration, Search for the Production of Dark Matter in Association with Top Quark Pairs in the Single-lepton Final State in pp collisions at $\sqrt{s}$ = 8 TeV, CMS-PAS-B2G-14-004: http://cds.cern.ch/record/1749153.

[14] R. Bernabei, et al. (DAMA/LIBRA Collaboration), *Eur. Phys. J. C* **67**(2010) 39;
C. Aalseth, et al. (COGENT Collaboration), *Phys. Rev. Lett.* **107**(2011) 141301;
R. Agnese, et al. (CDMS Collaboration), *Phys. Rev. Lett.* **111** (2013) 251301.

[15] J. C. Pati and A. Salam, *Phys. Rev.* D **10** (1974) 275; R. N. Mohapatra and J. C. Pati, *Phys. Rev.* D **11** , 566 (1975); R. N. Mohapatra and J. C. Pati, *Phys. Rev.* D **11** , 2558 (1975); G. Senjanovic and R. N. Mohapatra, *Phys. Rev.* D **12**, 1502 (1975); G. Senjanovic, *Nucl. Phys.* B **153** (1979) 334.

[16] L. Randall and R. Sundrum, *Phys. Rev. Lett.* **83** (1999) 3370.

[17] ATLAS Collaboration, Search for $W' \to tb \to qqbb$ Decays in pp Collisions at $\sqrt{s}$ = 8 TeV with the ATLAS Detector, arXiv:1408.0886 [hep-ex] (*preprint* CERN-PH-EP-2014-152).

[18] CMS Collaboration, Search for W' to tb decays in the lepton + jets final state in pp collisions at $\sqrt{s}$ = 8 TeV, *J. High Energy Phys.* JHEP05(2014)108.

[19] CMS Collaboration, Search for pair production of third-generation scalar leptoquarks and top squarks in proton-proton collisions at $\sqrt{s}$ = 8 TeV,*Phys. Lett.* B **739** (2014) 229.

[20] CMS Collaboration, Particle-Flow Event Reconstruction in CMS and Performance for Jets, Taus, and MET, CMS-PAS-PFT-09-001: http://cds.cern.ch/record/1194487

[21] M. Czakon and A. Mitov, Top++: a program for the calculation of the top-pair cross-section at hadron colliders, arXiv:1112.5675 [hep-ph]; *Computer Physics Communications* **185**, 11 (2014) 29302938.

[22] J. Aguilar-Saavedra, Protos - PROgram for TOp Simulations, http://jaguilar.web.cern.ch/jaguilar/protos.

[23] J. Alwall, M. Herquet, F. Maltoni, O. Mattelaer, and T. Stelzer, MadGraph 5: going beyond, *J. High Energy Phys.* JHEP1106(2011)128

[24] ATLAS Collaboration, Search for pair and single production of new heavy quarks that decay to a Z boson and a third-generation quark in pp collisions at $\sqrt{s} = 8$ TeV with the ATLAS detector, *J. High Energy Phys.* JHEP11(2014)104.

[25] ATLAS Collaboration, Search for pair production of heavy top-like quarks decaying to a high-p_T W boson and a b quark in the lepton plus jets final state in pp collisions at $\sqrt{s} = 8$ TeV with the ATLAS detector, ATLAS-CONF-2013-060: http://cds.cern.ch/record/1557777.

[26] ATLAS Collaboration, Search for anomalous production of events with same-sign dileptons and b jets in 14.3 fb^{-1} of pp collisions at $\sqrt{s} = 8$ TeV with the ATLAS detector, ATLAS-CONF-2013-051: http://cds.cern.ch/record/1547567.

[27] CMS Collaboration, Inclusive search for a vector-like T quark with charge 2/3 in pp collisions at $\sqrt{s} = 8$ TeV, *Phys. Lett.* B **729** (2014) 149;
CMS Collaboration, Search for pair-produced vector-like quarks of charge -1/3 decaying to bH using boosted Higgs jet-tagging in pp collisions at $\sqrt{s} = 8$ TeV, CMS-PAS-B2G-14-001: http://cds.cern.ch/record/1752557.

[28] ATLAS Collaboration, Summary plots from the ATLAS Exotic physics group, public Twiki page: http://atlas.web.cern.ch/Atlas/GROUPS/PHYSICS/CombinedSummaryPlots/EXOTICS/

[29] ATLAS Collaboration, Studies of Sensitivity to New Dilepton and Ditop Resonances with an upgraded ATLAS detector at a High-Luminosity LHC, ATLAS-PHYS-PUB-2013-003: http://cds.cern.ch/record/1516108.

[30] CMS Collaboration, Sensitivity study for ECFA: heavy vector-like charge 2/3 quarks, CMS-PAS-FTR-13-026: http://cds.cern.ch/record/1605864.

[31] ATLAS Colaboration, Sensitivity to WIMP Dark Matter in the Final States Containing Jets and Missing Transverse Momentum with the ATLAS Detector at 14 TeV LHC, ATL-PHYS-PUB-2014-007: http://cds.cern.ch/record/1708859

AUTHOR INDEX

AUTHOR INDEX

AUTHOR INDEX